ROUTLEDGE HANDBOOK ON THE SCIENCES IN ISLAMICATE SOCIETIES

The *Routledge Handbook on the Sciences in Islamicate Societies* provides a comprehensive survey on science in the Islamic world from the 8th to the 19th century.

Across six sections, a group of subject experts discuss and analyze scientific practices across a wide range of Islamicate societies. The authors take into consideration several contexts in which science was practiced, ranging from intellectual traditions and persuasions to institutions, such as courts, schools, hospitals, and observatories, to the materiality of scientific practices, including the arts and craftsmanship. Chapters also devote attention to scientific practices of minority communities in Muslim majority societies, and Muslim minority groups in societies outside the Islamicate world, thereby allowing readers to better understand the opportunities and constraints of scientific practices under varying local conditions.

Through replacing Islam with Islamicate societies, the book opens up ways to explain similarities and differences between diverse societies ruled by Muslim dynasties. This handbook will be an invaluable resource for both established academics and students looking for an introduction to the field. It will appeal to those involved in the study of the history of science, the history of ideas, intellectual history, social or cultural history, Islamic studies, Middle East and African studies including history, and studies of Muslim communities in Europe and South and East Asia.

Sonja Brentjes is a historian of science in Islamicate societies and Christian Europe; she is an affiliated scholar at the Max Planck Institute for the History of Science, Berlin. Her research includes the history of the mathematical sciences, mapmaking, institutions, cross-cultural exchange of knowledge and the involvement of the arts in the sciences. Among her recent publications are *Teaching and Learning the Sciences in Islamicate Societies, 800–1700* (2018); Brentjes, S., Edis, T. and Richter-Bernburg, L. *1001 Distortions: How (Not) to Narrate the History of Science, Medicine and Technology in Non-Western Cultures* (2016) and Brentjes, S. "MS Paris, Bibliothèque des Missions Étrangères 1069: The French-Arabic Dictionary of François Pétis de la Croix (1653–1713)?" Mediterranea. *International journal on the transfer of knowledge*, 6 (2021), 57–84.

ROUTLEDGE HANDBOOK ON THE SCIENCES IN ISLAMICATE SOCIETIES

Practices from the 2nd/8th to the 13th/19th Centuries

First edition

Edited by Sonja Brentjes

ASSOCIATE EDITOR: PETER BARKER

ASSISTANT EDITOR: RANA BRENTJES

EDITORIAL BOARD:
NAHYAN FANCY, GOTTFRIED HAGEN, MATTHEW MELVIN-KOUSHKI,
LUTZ RICHTER-BERNBURG, ULRICH RUDOLPH

Routledge
Taylor & Francis Group
LONDON AND NEW YORK

Cover image: Spherical astrolabe by Mūsā (Moshe Galeano?), dated 885/1480–81. Inv.49687. © History of Science Museum, University of Oxford.

First published 2022
by Routledge
4 Park Square, Milton Park, Abingdon, Oxon OX14 4RN

and by Routledge
605 Third Avenue, New York, NY 10158

Routledge is an imprint of the Taylor & Francis Group, an informa business

British Library Cataloguing-in-Publication Data
A catalogue record for this book is available from the British Library

Library of Congress Cataloging-in-Publication Data
A catalog record for this book has been requested

ISBN: 978-1-138-04759-4 (hbk)
ISBN: 978-1-032-27162-0 (pbk)
ISBN: 978-1-315-17071-8 (ebk)

DOI: 10.4324/9781315170718

Typeset in Bembo
by Apex CoVantage, LLC

CONTENTS

CONTRIBUTORS

Naomi Aradi received her PhD on medieval Hebrew arithmetic in 2015 from the Hebrew University of Jerusalem. Currently, she is the content manager of the *Mispar* database of medieval Hebrew arithmetic. She published "An Unknown Medieval Hebrew Anonymous Treatise on Arithmetic," *Aleph* 13 (2013) and "The Origins of the *Kalām* Model of Discussion on the Concept of *Tawḥīd*," *Arabic Sciences and Philosophy* 23 (2013).

Taha Yasin Arslan is an assistant professor in the Department of History of Science in Medeniyet University, Istanbul. He studies the transmission of knowledge between Mamluk and Ottoman scholars during the 13th and 16th centuries, focusing on astronomical instruments and related texts. Among other methodological approaches, he reconstructs medieval and early modern practices by reproducing scientific instruments in digital and physical forms. His first major publication is the paper "A 15th-Century Mamluk Astronomer in the Ottoman Realm: ʿUmar al-Dimashqī and his *ʿilm al-mīqāt* corpus the Hamidiye 1453," *Nazariyat: Journal for the History of Islamic Philosophy and Sciences* 4 (2018). Together with Anthony J. Turner and Silke Ackermann, he prepared the new catalogue *Mathematical Instruments in the collections of the Bibliothèque nationale de France.*

Peter Barker is Professor in the Department of the History of Science, Technology and Medicine at the University of Oklahoma, USA. His research includes applying insights from cognitive psychology to conceptual change, and historical studies of the positive role of religion in early modern science and the cultural settings of major figures from the Scientific Revolution. Since learning Persian, he has begun work on knowledge exchanges between Safavid Persia and Mughal India. His recent publications include: "The Social Structure of Islamicate Science," *Journal of World Philosophies* (2017); "The Copernican Revolution since Kuhn," in Wray K. B. (ed.) *Interpreting Kuhn* (2021), and "East-West Passages: European Interest in Islamicate Astronomy during the Scientific Revolution," in Mehl, É. et Pantin, I. (eds.), *De mundi recentioribus phænomenis: Cosmologie et science dans l'Europe des Temps modernes,* XVe-XVIIe siècles.

Alessandro Bausi, Professor of Ethiopian Studies at Universität Hamburg, is a philologist and linguist working on ancient, late antique, and medieval texts and manuscripts. Journal and series editor, he heads several projects in Ethiopian-Eritrean philology, manuscript studies, linguistics, and corpus linguistics at the Hiob Ludolf Centre for Ethiopian and Eritrean Studies, and is a member of the Cluster of Excellence Understanding Written Artefacts. He has especially contributed on the earliest phase of ancient Ethiopic scribal and literary history, on epigraphy, on canonical and hagiographical-liturgical collections, and generally on the textual criticism of Ethiopic texts and on the manuscript culture of Ethiopia and Eritrea.

Michael David Bonner (1952–2019) taught medieval history of the Islamic Middle East at the University of Michigan. He explored the moral and monetary economies of the early Islamic frontier in *Aristocratic Violence and Holy War* (1996) leading to further work on jihad: *Jihad in Islamic History: Doctrines and Practise.* (2008), poverty: Bonner, M., Ener, M. and Singer, A. *Poverty and Charity in Middle Eastern Contexts* (2003), and the markets of late antique Arabia: "'Time Has Come Full Circle': Markets, Fairs and the Calendar in Arabia Before Islam", in Ahmed, A. Q., Bonner, M. and Sadeghi, B. eds. *The Islamic Scholarly Tradition: Studies in History, Law, and Thought in Honor of Professor Michael Allan Cook* (2011). His study of the frontier also resulted in a lifelong interest in geographical literature. His translation of Ibn Ḥawqal's *Book on the Face of the Earth* (*Kitāb Ṣūrat al-arḍ*) is forthcoming posthumously.

Abdelmalek Bouzari received his doctorate in the history of mathematics in 2008 from the University of Lille 1 and his habilitation in 2012 from the University of Constantine. He is the director of the Laboratoire d'Épistémologie et histoire des mathématiques at the École Supérieure Normale in Algiers. He has published "*Les sections coniques d'Apollonius dans la tradition mathématique arabe*," in Barbin, E. and Maltrel J. L., eds. *Les mathématiques méditerranéennes d'une rive et de l'autre* (2015) and "Les coniques en Occident Musulman entre le XIe et le XIVe siècle," *Llull* 32 (2009). Together with Y. Guergour he edited *Actes du 9ème colloque maghrébin sur l'histoire des mathématiques arabes* (2011).

Christopher Braun studied Arabic, French and the history of the Middle East in Berlin and Paris. In 2016, he earned a PhD on Arabic "treasure hunter manuals" at the Warburg Institute in London. He an indpendent scholar. His research interests include the Arabic occult sciences, the cultural history of medieval Islamicate societies, and Arabic lexicography.

Rana Brentjes is a photo designer and curator with a MA in contemporary art history and has submitted her PhD thesis in contemporary German history at Goldsmith College, London, UK. She has curated art exhibitions in Berlin and Brandenburg, written on Palestinian cinematography and co-edits Imagining the Heavens across Eurasia from Antiquity to Early Modernity (2023). Currently, she is digital content curator of the research group "Visualization and Material Cultures of the Heavens" in Department III at the Max Planck Institute for the History of Science, Berlin, Germany.

Paola Buzi is professor of Egyptology and Coptic Studies at Sapienza University of Rome and honorary professor of the same disciplines at Hamburg University. She is Principal Investigator of the European Research Council Advanced project: "PAThs – Tracking papyrus and parchment paths: An Archaeological Atlas of Coptic Literature". She combines historical, literary and codicological interests with archaeological work. Among her publications are *Titoli e autori nella letteratura copta: Studio storico e tipologico* (2005); "Egypt, crossroad of translations and literary interweavings (3rd–6th centuries). A reconsideration of earlier Coptic literature," in Crevatin, F. ed. *Egitto, crocevia di traduzioni* (2018); and with with E. Giorgi, *The Urban Landscape of Bakchias: A Town of the Fayyūm from the Ptolemaic-Roman Period to Late Antiquity* (2020).

Constantin Canavas holds a diploma in chemical engineering and a Dr.-Ing. in system dynamics and control. Since 1993 he has served as professor at the Hamburg University of Applied Sciences in the fields of control and technology assessment, as well as history and philosophy of technology. He has taught history of technology at the University of Athens and the National Technical University of Athens, Greece, as well Arab history at the University of Crete. Among his publications are articles in the *Encyclopedia of Islam, Three* on automata, the compass and Hero of Alexandria.

Anna Caiozzo earned her PhD in history, history of art, and Arabic studies and Islamic civilization in 1998 at the Université IV Sorbonne, Paris and her habilitation in medieval history at the Université d'Orléans. She is a professor of medieval history at the Université d'Orléans. She is a specialist in visual studies of the Islamicate world, in particular in heavenly representations, magic and imagination. She won the Prix Bordin in 2019 (Académie de Inscriptions et Belles Lettres) for the book *Le roi glorieux, les imaginaires de la royauté dans le Shāh Nāma de Firdawsi de Tus* (2018). She has also directed several international projects (in Anthropology of Emotions and Landscape Studies).

Leigh Chipman received her PhD in 2006 from The Hebrew University of Jerusalem. She is an independent scholar, editor and translator based in Jerusalem. She studies aspects of the social and intellectual history of medicine and the sciences in the medieval Islamicate world and is currently working on the afterlife of medieval pharmacy in the late Ottoman and colonial Middle East. She has published two books – *The World of Pharmacy and Pharmacists in Mamlūk Cairo* (2010) and, with Efraim Lev, *Medical Prescriptions in the Cambridge Genizah Collections. Practical Medicine and Pharmacology in Medieval Egypt* (2012) – and some 20 articles, most recently "The reception of Galenic pharmacology in the Arabic tradition," in Zipser, B. and Bouras-Vallianatos, P., eds. *Brill's Companion to the Reception of Galen* (2019).

Ahmed Djebbar is an emeritus professor at the University of Science and Technology, Lille. His research focuses on the history of mathematics in medieval North Africa, al-Andalus and sub-Saharan Africa. He has written or edited about 20 books either alone or in cooperation with colleagues and more than 200 papers. He has also organized major exhibitions, notably "The Golden Age of Arabic Science" in Paris in 2005. He has actively engaged in popularizing research results and in building research and teaching communities for the history of medieval mathematics in Arabic in North African countries. He also was advisor and then a minister of governments of the People's Democratic Republic of Algeria.

Nahyan Fancy is professor of Middle East/comparative history at DePauw University, Indiana, USA. He works on the history of medicine in premodern Islamicate societies, especially the intersections between medicine, philosophy and religion. His first book is *Science and Religion in Mamluk Egypt: Ibn al-Nafīs, Pulmonary Transit and Bodily Resurrection* (Routledge, 2013). He is currently working on a monograph on Arabic medical commentaries on the *Canon of Medicine* produced between 1180 and 1520.

Alexander Fidora studied philosophy at the University of Frankfurt, obtaining his PhD in 2003. He is an ICREA research professor in the Department of Ancient and Medieval Studies of the Universitat Autònoma de Barcelona. His research focuses on medieval philosophy as well as the intercultural and interreligious dimensions of medieval thought, on which he has published more than 50 books and over 150 articles. He has directed the European Research

Council Starting grant "Latin into Hebrew" (2008–2012) and the European Research Council Consolidator project "The Latin Talmud" (2014–2019).

Miquel Forcada is professor of Arabic and Islamic studies at the University of Barcelona. He studies several aspects of the history of the sciences in al-Andalus, and particularly the intellectual and historical contexts of the practice of rational sciences. On these topics he has written "Saphaeae and Hay'āt: the Debate between Instrumentalism and Realism in Al-Andalus," in Rodríguez-Arribas, J., Burnett, C. and Ackermann, S. and Ryan Szpiech, S. eds. *Astrolabes in Medieval Cultures* (2019); "Books from Abroad: The Evolution of Science and Philosophy in Umayyad Al-Andalus," *Intellectual History of the Islamicate World* 5 (2017) and *Ética e ideología de la ciencia: el médico filósofo in al-Andalus* (2011).

Regula Forster is professor of Islamic history and culture at the University of Tübingen. She works on medieval Arabic literature, Quranic exegesis, and the history of science (especially alchemy). Her books include *Das Geheimnis der Geheimnisse. Die arabischen und deutschen Fassungen des pseudo-aristotelischen Sirr al-asrār/Secretum secretorum* (2006) and *Wissensvermittlung im Gespräch. Eine Studie zu klassisch-arabischen Dialogen* (2017).

Amir Mohammad Gamini is an assistant professor in the Institute for the History of Science at the University of Tehran. Educated in history and philosophy of science at the Iranian Institute of Philosophy, he won the 2017 DHST young scholar prize for his PhD thesis titled "Quṭb al-Dīn Shīrāzī and his Role in the Science of Hay'a". He is about to publish an edition of Shīrāzī's *Ikhtīyārāt Muẓaffari*, a Persian comprehensive book in hay'a. As a member of the Commission on the History of Science and Technology in Islamic Societies, his research extends to the reception of early modern science, especially biology and astronomy, in Iran and other Islamic countries.

Nathan P. Gibson received his PhD from the Catholic University of America in 2015. He is currently a research associate in Judaic studies at Ludwig Maximilian University in Munich, and Principal Investigator for the "Communities of Knowledge" project (https://usaybia.net), examining interreligious networks of scholars in Ibn Abī Uṣaybiʿa's *History of Physicians*. Together with Vince Bantu, he is editing *Global Christian Texts*, which will make available in English readings from pre-modern Christian communities throughout Africa, the Middle East, and Asia. His previous work includes a study of Christian-Muslim relations in 9th-century Iraq based on al-Jāḥiẓ's *Refutation of Christians* and co-editing digital reference works for Syriac studies (http://syriaca.org) and the Bibliography of the Arabic Bible (biblia-arabica.com/bibl).

Frank Griffel is the Louis M. Rabinowitz Professor of Religious Studies at Yale University. He has published widely in the fields of Islamic philosophy and theology as well as Muslim intellectual history. In his book *Apostasie und Tolerance im Islam* (2000) he analyzes the development of the judgment of apostasy up to the 6th/12th century. In many of his publications, among them his 2009 book *Al-Ghazali's Philosophical Theology*, he deals with the influential Muslim thinker al-Ghazālī (d. 505/1111) and his impact on Islam and the history of philosophy. His latest work *The Formation of the Post-Classical Philosophy in Islam* appeared in 2021.

S. Irfan Habib was formerly Maulana Azad Chair, National University of Educational Planning and Administration, New Delhi. His publications include *Jihad or Ijtihad: Religious Orthodoxy and Modern Science in Contemporary Islam* (2015), and with Dhruv Raina *Domesticating Modern Science in Colonial India* (2004).

Gottfried Hagen is a professor of history and culture of the Middle East at the University of Michigan, with a focus on Ottoman history and Islam. Besides his research on the veneration of the Prophet Muḥammad in the Ottoman Empire, he has published widely on Ottoman historiography, geography, cartography and travel, including *Ein osmanischer Geograph bei der Arbeit. Entstehung und Gedankenwelt von Kātib Čelebis Ǧihānnümā.* (2003; Turkish tr. 2017); "The Order of Knowledge, the Knowledge of Order: Intellectual Life," in Faroqhi, S. and Fleet, K. eds. *The Cambridge History of Turkey. Volume 2: The Ottoman Empire as a World Power, 1453–1603* (2013); "Ottoman Empire, Geographical Mapping and the Visualization of Space," in Edney, M. H. and Pedley, M. S. eds. *Cartography in the European Enlightenment, History of Cartography*, vol. 4, (2020).

Konrad Hirschler is a professor of Middle Eastern History at Universität Hamburg (Centre for the Study of Manuscript Cultures) and was previously a professor of Middle Eastern History at SOAS (London) and Freie Universität Berlin. Over the last years he has primarily worked on the history of reading, the book and libraries with an emphasis on material culture. He is author of books such as *Owning Books and Preserving Documents in Medieval Jerusalem* (2023), *A Monument to Medieval Syrian Book Culture* (2020), *Plurality and Diversity in an Arabic Library* (2016), *The Written Word in the Medieval Arabic Lands* and *Medieval Arabic Historiography* (2012), co-author of *Owning Books and Preserving Documents in Medieval Jerusalem* (2022) and *Mu'allafat Yūsuf b. Ḥasan Ibn ʿAbd al-Hādī* (2021), as well as co-editor of *The Damascus Fragments* (2020) and *Manuscript Notes as Documentary Sources* (2011).

Yoichi Isahaya received his PhD in Eurasian studies in 2015 from Tokyo University. Currently, he is assistant professor of Eurasian Studies at the Slavic-Eurasian Research Center, Hokkaido University. His recent publications include "Fu Mengzhi: 'The Sage of Cathay,' in Mongol Iran and Astral Sciences along the Silk Roads," in Biran, M. *et al.*, eds. *Along the Silk Roads in Mongol Eurasia* (2020) and "Geometrizing Chinese Astronomy? The View from a Diagram in the *Kashf al-ḥaqā'iq* by al-Nīsābūrī (d. ca. 1330)," in Mak, B. and Huntington, E. eds. *Overlapping Cosmologies in Asia* (2022).

Hossein Kamaly received his PhD in history from Columbia University, New York in 2004. Currently, he holds the Imam Ali Chair in Shia Studies and Dialog Among Islamic Schools of Thought at the Hartford Seminary. He is the author of *God & Man in Tehran: Contending Visions of the Divine from the Qajars to the Islamic Republic* (2018) and *A History of Islam in 21 Women* (2019 in the UK and 2020 in the USA). He has contributed to the *Encyclopedia Iranica*, the *Journal of Iranian Studies*, *International Journal of Middle East Studies*, and the *Journal for the Study of Persianate Societies* and the *Encyclopedia of Islam, Three.* He also writes and publishes research in Persian.

Fabian Käs received his PhD in Semitic studies at the Ludwig Maximilians University, Munich in 2008. Currently, he is a research fellow at the University of Cologne. His publications include *Die Mineralien in der arabischen Pharmakognosie* (2010) and *Al-Maqrīzīs Traktat über die Mineralien* (2015).

Osman Süreyya Kocabaş is a PhD student in history at Hacettepe University in Ankara, Turkey. In 2017, he completed his master's thesis on "The Definitions of Natural Phenomena in Classical Ottoman Thought". His current research focuses on Ottoman perspectives on natural phenomena, which were embraced by Ottoman society in the pre-modern era. He is also interested in the social networks linking Ottoman *ʿulamāʾ* and scholars in the Eurasian cultural and scientific world.

Götz König is a German Iranist whose work focusses on Old and Middle Iranian texts related to Zoroastrianism. Currently he is preparing a new edition of the texts belonging to the *Xorde Avesta* within the project *Corpus Avesticum Berolinense* at the Freie Universität Berlin. He has published books and articles on the intellectual and religious history of ancient and late antique Iran, including *Studien zur Rationalitätsgeschichte im Älteren Iran. Ein Beitrag zur Achsenzeitproblematik* (*Studies on the History of Rationality in Ancient Iran. A Contribution to the Theory of Axial Age*; 2018), and on Zoroastrian texts belonging to the Avesta, the translation of the Avesta, and the Pahlavi literature of the 9th/10th century.

Andreas Lammer is an assistant professor of history of philosophy at Radboud University Nijmegen in the Netherlands. His primary research interests are in Greek, Arabic and Latin natural philosophy in both the Aristotelian and the Avicennian tradition, and more broadly in the transmission of philosophical and scientific literature from Greek into Arabic and from Arabic into Latin. Before his appointment in Nijmegen, he worked as a research associate at the Ludwig Maximilians University, Munich and the Thomas Institute at the University of Cologne, and was junior professor for Arabic philosophy, culture and history at Trier University. He published various papers on the notions of time, creation and nature in ancient, late ancient and Islamic philosophy, and is the author of *The Elements of Avicenna's Physics: Greek Sources and Arabic Innovation* (2018), as well as the co-editor of *Received Opinions: Doxography in Antiquity and the Islamic World* (2022).

Emin Lelić received his PhD from the University of Chicago. He is an assistant professor of history at Salisbury University. His publications include "Physiognomy (*'ilm-i firāsat*) and Ottoman Statecraft. Discerning Morality and Justice," *Arabica* 64 (2017), and "'The Greatest of Tribulations': Constructions of Femininity in Sixteenth-Century Ottoman Physiognomy," in Karateka, H. T., Çipa, H. E. and Anetshofer, H. eds. *Disliking Others: Loathing, Hostility and Distrust in Premodern Ottoman Lands* (2018). At present, he is working on a monograph, tentatively titled *Ottoman Physiognomy: A Window into the Soul of an Empire.*

Boris Liebrenz earned his PhD at the University of Leipzig in 2013. Currently, he is a researcher with the *Bibliotheca Arabica* project at the Saxon Academy of the Sciences and Humanities in Leipzig. His main interest, the history of libraries, books and readers, is the topic of his *Die Rifāʿīya aus Damaskus* (2016), which won the Annemarie Schimmel Research Prize in 2017. Other research areas include papyrology and the history of Oriental Studies in early modern Europe. Recent publications are *The Waqf of a Physician in Late Mamluk Damascus* (2019) and "An Archive in a Book: Documents and Letters from the Early-Mamluk Period," *Der Islam* 97 (2020).

Matteo Martelli has a PhD in Greek philology from the University of Bologna (2007) and in history of science from the University of Pisa (2012). Currently, a professor in history of science at the University of Bologna, he edited and translated the Greek and Syriac sections of Pseudo-Democritus's alchemical books *The Four Books of Pseudo-Democritus* (2014). In Lehmhaus, L. and Martelli, M. eds. *Collecting Recipes: Byzantine and Jewish Pharmacology in Dialogue* (2017), he explored various aspects of Byzantine and Syriac medicine.

Jon McGinnis is a professor of philosophy at the University of Missouri, St. Louis. His research focuses on the history of physics in the Islamicate world. In addition to numerous articles, he is the author of *Avicenna* (2010); translator and editor of *The Physics of* The Healing: *A parallel English-Arabic text* (2009), and co-translator with David C. Reisman of *Classical Arabic Philosophy, An Anthology of Sources* (2007). He has received three National Endowment for the Humanities awards, a Mellon grant, a John Templeton Foundation grant and was a member of the Institute for Advanced Study, Princeton.

Matthew Melvin-Koushki received his from PhD from Yale University. He is currently an assistant professor and McCausland Fellow of History at the University of South Carolina. He specializes in early modern Islamicate intellectual and imperial history, with a focus on the theory and practice of the occult sciences in Timurid-Safavid Iran and the broader Persianate world to the 19th century. He is the author of *The Occult Science of Empire in Aqquyunlu- Safavid Iran: Two Shirazi Lettrists and Their Manuals of Magic* (forthcoming) and editor, with Noah Gardiner, of *Islamicate Occultism: New Perspectives, Arabica*, 64/3–4 (2017) and *Islamicate Occult Sciences in Theory and Practice* (2020).

Simon Mills is a lecturer in early modern history at Newcastle University. He is the author of *A Commerce of Knowledge: Trade, Religion, and Scholarship between England and the Ottoman Empire, c.1600–1760* (2020). His broader interests lie in the religious, cultural and intellectual history of early modern Britain and Europe, with a particular focus on the relationship between Europe and the Ottoman Empire, the histories of biblical and oriental studies and the history of philosophy.

Sébastien Moureau is research associate at the FNRS, and an assistant professor at the Catholic University of Louvain (Belgium). His research is mainly concerned with history of the so-called occult sciences in the medieval Arabic and Latin worlds, with a special focus on alchemy and metallurgy. He is currently involved in the European Research Council project "The Origin and Early Development of Philosophy in Tenth-Century al-Andalus: The Impact of Ill-Defined Materials and Channels of Transmission" (2016) at the Université catholique de Louvain, from 2017 to 2022.

Marc Moyon is a French historian of medieval mathematics at the University of Limoges in France, where he is an associate professor. In 2019, he obtained his habilitation that enables him to direct research. In 2011, he received the Prize for Young Historians awarded by the International Academy of the History of Science. His research focuses on Arabic–Latin mathematical translations. His main recent publications are "L'appropriation des sciences géométriques arabes en occident médiéval," in Agostino, P., ed. *The Diffusion of the Islamic Sciences in the Western World, Micrologus* 28 (2020): 45–67; "The *Liber Restauracionis*: A Newly Discovered Copy of a Mediæval Algebra in Florence," *Historia Mathematica* 46 (2018) and *La géométrie de la mesure dans les traductions arabo-latines médiévales* (2017).

Jeffrey Oaks received his PhD in mathematics at the University of Rochester in 1991. He is a professor of mathematics at the University of Indianapolis, where he has taught since 1992. His research is focused mainly on the history of Arabic algebra, his most recent paper being "Proofs and Algebra in al-Fārisī's Commentary," *Historia Mathematica* 47 (2019).

Eva Orthmann received her MA in Islamic and Iranian studies at the University of Tübingen and did her PhD at Halle University. From 2007 to 2018, she was a professor of Islamic studies at Bonn University and has since become a professor of Iranian studies in Göttingen. Her work focuses mainly on the history of astrology in the Islamic world as well as on Indo-Iranian studies. Among her publications are "Lettrism and magic in an early Mughal text: Muḥammad Ghawth's *k. al-Jawāhir al-khams*," in El-Bizri, N. and Orthmann, E., eds. *The Occult Sciences in Pre-modern Islamic Cultures* (2017); "Court Culture and Cosmology in the Mughal Empire: Humāyūn and the foundation of the dīn-i ilāhī," in Fuess, A. and Hartung, J.-P., eds. *Politics and Patronage. Court Culture in the Muslim World, 7th–19th Centuries* (Routledge, 2011); and "Tarjuma-yi kitāb-i Bārāhī," in Speziale, F. and Ernst, C.W., eds. *Perso-India: An Analytical Survey of Persian Works on Indian Learned Traditions*, available at www.perso-indica.net/work/tarjuma-yi_kitab-i_barahi.

Peter E. Pormann received a D.Phil. and D.Litt. from the University of Oxford. He is a professor of classics and Greco-Arabic studies at the University of Manchester and led two large-scale projects: "Arabic commentaries on the Hippocratic 'Aphorisms'" (European Research Council) and "The Syriac Galen Palimpsest" (Arts and Humanities Research Council, UK). Recent edited collections include special double issues *The Arabic Commentaries on the Hippocratic "Aphorisms"* co-edited with Kamran I. Karimullah, in Oriens 45/1–2 (2017), and *Medical Traditions*, with Leigh Chipman and Miri Shefer- Mossensohn, in *Intellectual History of the Islamicate World*, Part 1: 5/3–4 (2017), Part 2: 6/1–2 (2018). Further recent pbulications are *La construction de la médecine arabe médiévale*, with Pauline Koetschet (Ifao du Caire, 2016); *Philosophy and Medicine in the Formative Period of Islam*, with Peter Adamson (2017); *1001 Cures: Contributions in Medicine and Healthcare from Muslim Civilisation* (2017); the *Cambridge Companion to Hippocrates* (2018); and *Hippocratic Commentaries in the Greek, Latin, Syriac and Arabic Traditions* (2021).

Yves Porter studied oriental languages (Persian, Turkish, Arabic, Pashtu, Hindi) before obtaining his PhD in 1988 and his habilitation à diriger des recherches in 2000. Since 1993, he has taught Islamic Art at Aix Marseille Université. He is a member of the Laboratoire d'Archéologie Méditerranéenne Médiévale et Moderne (LA3M). He joined the Institut Universitaire de France in 2018. He is a specialist in arts and techniques in the Iranian world and Islamized India. His publications (most of them translated into English) include *Painters, Paintings and Books* (1994); *The Glory of the Sultans: Islamic Architecture in India*, with Gérard Degeorge (2009) and *Le prince, l'artiste et l'alchimiste. La céramique dans le monde iranien, Xe-XVIIe siècle*, with Richard Castinel (2011).

Roser Puig Aguilar received her PhD in Arabic language. She is a senior lecturer and researcher at the University of Barcelona, Faculty of Philology and Communication, Department of Classical, Romanic and Semitic Languages (Arabic Section). Her research focuses on Arabic astronomy, astronomical instruments and the Arab linguistic tradition. Two of her books are *Al-šakkāziyya: Ibn al-Naqqāš al-Zarqālluh. Edición, traducción y estudio* (1986) and *Los tratados de construcción y uso de la azafea de Azarquiel* (1987).

Dhruv Raina is a professor at the Jawaharlal Nehru University, New Delhi. He studied physics at the Indian Institute of Technology, Mumbai and received his PhD in philosophy of science from Göteborg University. His research has focused on the politics and cultures of scientific knowledge in South Asia, as well as the history and historiography of mathematics. He has co-edited, with S. Irfan Habib, *Situating the History of Science: Dialogues with Joseph Needham* (1999) and *Social History of Sciences in Colonial India* (2007), and with Feza Günergun, *Science between Europe and Asia* (2010). He is the author of *Images and Contexts* (2003) and co-author, again with S. Irfan Habib, of *Domesticating Modern Science* (2004). *Needham's Indian Network* (2015), focused on the historiography of science in South Asia and the attempts toward institutionalizing the history of science in the region. His more recent work addresses the emergence of inter- and transdisciplinary fields across the natural and social sciences. A volume co-edited with Hans Harder, *Disciplines and Movements: Scientific Exchanges between India and the German Speaking World*, is forthcoming. Among his many positions, he has been a Fellow of the Institute of Advanced Study, Berlin; the first incumbent of the Heinrich Zimmer Professorship of Intellectual History and Indian Philosophy at Heidelberg University and Visiting Faculty at the Universities of Paris and Cambridge. In 2018, he was elected Fellow of the Indian National Science Academy.

Lutz Richter-Bernburg is Professor Emeritus, Islamic Studies, University of Tübingen. He has published on a broad range of topics, including history of medicine and science. Among his most important publications are *Der Syrische Blitz – Saladins Sekretär zwischen Selbstdarstellung und Geschichtsschreibung* (1998); *Eine arabische Version der pseudogalenischen Schrift De theriaca ad Pisonem.* Diss. phil., Göttingen 1969 and "'God created Adam in His likeness' in the Muslim Tradition," in Berthelot, C. and Morgenstern, M. eds. *The Quest for a Common Humanity: Human Dignity and Otherness in the Religious Traditions of the Mediterranean.* (2011).

Mònica Rius-Piniés is a professor of Arabic and Islamic studies at the University of Barcelona. Her research interests center on contemporary Arabic literature, gender studies and social history of science and medicine. Among her publications are *La alquibla en al-Andalus y al-Magrib al-Aqsà* (2000); "Qibla in the Mediterranean" in Ruggles, C.L.N. ed. *Handbook of Archaeoastronomy and Ethnoastronomy* (2015) and "*Ciencia, religión y cultura en al-Ándalus*," *Awraq*, No. 21 (2020).

J. L. Alexis Rivera Luque is a PhD candidate at the Freie Universität Berlin in Byzantine studies. His current research deals with the emergence of *aljamiado* literature in the Iberian Peninsula and the Balkans. Recent publications include a review of Fidora, A. and Polloni, N. eds. *Appropriation, Interpretation and Criticism. Philosophical and Theological Exchanges Between the Arabic, Hebrew and Latin Intellectual Traditions* in *Journal of Transcultural Medieval Studies*, 5.2 (2017) and an article "*Omnes menstruatae sunt*: *tafsīr* at the Service of Polemics in the Translation of the Qurʾān by Robert of Ketton" (in Spanish), to be published in the memoirs of the workshop *Preliminary Considerations on the Corpus coranicum Christianum* held at the Freie Universität Berlin in December 2018.

Ulrich Rudolph is a professor of Islamic studies at the University of Zurich and has published extensively on Islamic philosophy and theology. His recent works include *Al-Māturīdī and the Development of Sunnī Theology in Samarqand* (2015) and *Islamische Philosophie* (4th ed., 2018) a short introduction to the topic which was translated into several languages. He is the editor of *Grundriss der Geschichte der Philosophie (Ueberweg). Philosophie in der islamischen Welt.* planned in four volumes, the first of which appeared in German (2012) and in English (2017).

Liana Saif is an assistant professor at the Centre for the Study of Hermetic Philosophy and Related Currents at the University of Amsterdam. Her work focuses on medieval Islamic esotericism and the occult sciences, with a special interest in the exchange of esoteric and occult knowledge between the Islamicate and Latin worlds in the medieval and early modern periods. Her book *The Arabic Influences on Early Modern Occult Philosophy* was published by Palgrave Macmillan in 2015. Saif is especially interested in the tenth-century secret brotherhood Ikhwān al-Ṣafāʾ (The Brethren of Purity), the pseudo-Aristotelian *Hermetica* and Jābir ibn Ḥayyān (Geber in Latin). She is currently preparing a long-awaited critical translation from Arabic into English of Maslama b. Qāsim al-Qurṭubī's (d. 964) *Ghāyat al-ḥakīm*, known in its Latin translation as the *Picatrix*.

Julio Samsó is professor emeritus of Arabic and Islamic studies at the University of Barcelona. Most of his research deals with the history of medieval astronomy in the Iberian Peninsula and the Maghrib. His most important publications are *Islamic Astronomy and Medieval Spain* (1994), *Astronomy and Astrology in al-Andalus and the Maghrib* (2007), *Las Ciencias de los Antiguos en al-Andalus* (2011) and *On Both Sides of the Straits of Gibraltar: Studies in the History of Medieval Astronomy in the Iberian Peninsula and the Maghrib* (2020).

Deborah Schlein is the Near Eastern Studies librarian at Princeton University, where she obtained her PhD from the department of Near Eastern studies in 2019. She writes on the history of medicine in Mughal and colonial India utilizing the Arabic and Persian manuscripts of these periods to study the social, intellectual, and environmental practices within and surrounding the medical communities of early modern South Asia. Her publications include "In the Ḥakīm's Own Hand: A Study of Colophons and Ownership Notes of the *al-Asbāb wa al-ʿAlāmāt* Commentary Tradition in India," *Journal of Islamic Manuscripts* 9.2–3 (2018).

Charlotte Schriwer is currently an independent scholar who is a historian and art historian of the Middle East. Her research focuses on the history of the eastern Mediterranean. Her publications include "Graffiti Arts and the Arab Spring," in Sadiki, L. ed. *The Routledge Handbook of the Arab Spring: Rethinking Democratization* (Routledge, 2014) and *Water and Technology in Levantine Society 1300–1900: An Historical, Archaeological and Architectural Analysis* (2015). With Nele Lenze, she edited *Participation Culture in the Gulf: Networks, Politics and Identity* (Routledge, 2018).

Ahmet Tunç Şen received his PhD in Ottoman studies in 2017 at the University of Chicago and is now assistant professor of history at Columbia University. His research concerns intellectual life in the Ottoman Empire with a focus on the astral sciences. Among his publications are "Reading the Stars at the Ottoman Court: Bāyezīd II (r. 886/1481–918/1512) and his Celestial Interests," *Arabica* 64/3–4 (2017) and "Practicing Astral Magic in Sixteenth-Century Ottoman Istanbul: A Treatise on Talismans attributed to Ibn Kemāl (d. 1534)," *Journal of Magic, Ritual, and Witchcraft* 12 (2017).

Miri Shefer-Mossensohn is the head of the Zvi Yavetz School of Historical Studies and an associate professor in the Department of Middle Eastern & African History at Tel Aviv University. She is a social historian of medicine, health and well-being in the early modern Ottoman world in its Arabic- and Turkish-speaking regions. Her publications include *Ottoman Medicine: Healing and Medical Institutions 1500–1700* (State University of New York Press, 2009) and *Science among the Ottomans: The Cultural Creation and Exchange of Knowledge* (University of Texas Press, 2015).

Nathan Sidoli received his PhD from the University of Toronto, in the history and philosophy of science and technology and is currently an professor of the history and philosophy of science at Waseda University, Tokyo. His research focuses on the Greek mathematical sciences and their development in Arabic sources. He is author of *Thābit ibn Qurra's Restoration of Euclid's Data* (2018), with Yoichi Isahaya, and *The Spherics of Theodosios* (forthcoming), with Robert Thomas.

Fabrizio Speziale is a professor at the School of Advanced Studies in the Social Sciences, Center for South Asian Studies, Paris. His research interests focus on the history of sciences and the Persianate culture of South Asia. His last book examines the interactions between Ayurvedic and Persianate medical cultures, *Culture persane et médecine ayurvédique en Asie du Sud* (2018). He is the chief editor of *Perso-Indica: An Analytical Survey of Persian Works on Indian Learned Traditions* (www.perso-indica.net).

Kenan Tekin earned his PhD in Middle Eastern, South Asian and African studies in 2016 at Columbia University, New York with a dissertation titled "Reforming Categories of Science and Religion in the Late Ottoman Empire". Currently, he is an assistant professor of Islamic philosophy at Yalova University, Turkey. In 2019–2020, Tekin was a visiting scholar in the Department of Near Eastern Languages and Civilizations at Harvard University, researching the transmission of the classical theory of science to Ottoman scholars. He published a chapter, "A Journal of Science without Boundaries: Ali Suavi's Ulûm Gazetesi," in Forbes, M. ed., *International Perspectives on Publishing Platforms* (Routledge, 2019) and has articles forthcoming in the *British Journal of History of Science* and the *Journal of Ottoman Studies.*

Johannes Thomann earned his PhD in the history of art in 1992 at the University Zürich. He does research at the Institute of Asian and Oriental Studies at the University of Zürich. His main research interest lies in the history of science in the Islamic world. Among his publications are *Studien zum "Speculum physionomie" des Michele Savonarola*, Diss. phil. I (1997), and, with Mattias Vogel, *Schattenspur: Sonnenfinsternisse in Wissenschaft, Kunst und Mythos* (1999). In recent years, his main occupation has been the edition of Arabic astronomical documents, but he has also published on Arabic and Turkish folk literature.

Glen Van Brummelen received his PhD in 1993 at the Simon Fraser University in Vancouver, British Columbia, Canada. Since 2020, he is the dean of science at Trinity Western University in Langley, British Columbia, Canada. He has served twice as president of the Canadian Society for History and Philosophy of Mathematics. A historian of science and mathematics in ancient and medieval cultures, his publications include *The Mathematics of the Heavens and the Earth* (2009); *Heavenly Mathematics* (2013) and *Trigonometry: A Very Short Introduction* (2020). His new book, *The Doctrine of Triangles: A History of Modern Trigonometry*, was published in 2021.

Ronny Vollandt received his PhD in 2011 from the University of Cambridge. Currently, he is a professor of Jewish studies at the Ludwig Maximilians University, Munich. His research focuses on the Arabic versions of the Bible and biblical exegesis in the Arabic language, and, more broadly, the Jewish intellectual heritage in the Near East. His recent publications include *Arabic Versions of the Pentateuch. A comparative study of Jewish, Christian, and Muslim Sources* (2015) and, with Arianna D'Ottone Rambach and Konrad Hirschler, *The Damascus Fragments: Towards a History of the Qubbat al-khazna Corpus of Manuscripts and Documents* forthcoming.

Roy Wagner is a professor of history and philosophy of mathematical sciences at the ETH Zürich. His main interests are the semiotic analysis of mathematical texts and the application of historical case studies to the philosophy of mathematical practice. Among his publications are the Hebrew section of Katz, V. J. *et al.* eds. *Sourcebook in the Mathematics of Medieval Europe and North Africa* (2015) and the monograph *Making and Breaking Mathematical Sense: Histories and Philosophies of Mathematical Practice* (2017).

Dror Weil received his PhD in 2016 from Princeton University. He is currently an assistant professor at the Faculty of History, University of Cambridge. His research interests focus on scientific and other textual exchanges between the Islamicate world and China during the late medieval and early modern periods. Weil's publications include: *Premodern Experience of the Natural World in Translation* (Routledge, 2022); "Islamicated China – China's Participation in the Islamicate Book Culture During the Seventeenth and Eighteenth Centuries," *Intellectual History of the Islamicate World* 4/1–2 (2016); "The Fourteenth-Century Transformation in China's Reception of Arabo-Persian Astronomy," in P. Manning and A. Owen (eds). *Knowledge in Translation: Global Patterns of*

Scientific Exchange, 1000–1800 CE (Pittsburgh University Press, 2018); "Chinese-Muslims as Agents of Astral Knowledge in Late Imperial China," in B. M. Mak and E. Huntington (eds.) *Overlapping Cosmologies in Asia: Transcultural and Interdisciplinary Approaches* (Brill, 2022); and "Unveiling Nature: Liu Zhi's Translation of Arabo-Persian Physiology in Early Modern China" in *Osiris* 38 (2022).

Ronit Yoeli-Tlalim is a reader in the History Department at Goldsmiths, University of London. She is the author of *ReOrienting Histories of Medicine: Encounters along the Silk Roads* (in press); "Galen in Asia?" in Bouras-Vallianatos, P. and Zipser, B., eds., *Brill's Companion to the Reception of Galen* (2019) and "The Silk-Roads as a Model for Exploring Eurasian Transmissions of Medical Knowledge: Views from the Tibetan Medical Manuscripts of Dunhuang," in Smith, P., ed. *Entangled Itineraries of Materials, Practices, and Knowledge: Eurasian Nodes of Convergence and Transformation* (2019).

Arash Zeini obtained his PhD at the School of Oriental and African Studies, University of London in study of religions. He is currently a postdoctoral fellow at the Invisible East project, University of Oxford. His main research interests include the study of ancient Iran; Middle Persian epistolary traditions (Pahlavi Documents); Zoroastrianism, particularly the late antique exegesis of the Avesta and digital humanities. He has published "A Unique Pahlavi Papyrus from Vienna (P. Pehl. 562), With an Introductory Note by Dieter Weber," *Studia Iranica* 45.1(2016); "The King in the Mirror of the Zand: Secrecy in *Sasanian Iran," in Daryaee, T. ed. Sasanian Iran in the Contextof Late Antiquity: The Bahari Lecture Series at the University of Oxford* (Jordan Center for Asian Studies, 2018) and *Zoroastrian Scholasticism in Late Antiquity: The Pahlavi Version ofthe Yasna Haptaŋhāiti* (2020), which received the "AIS Book Prize for Ancient Iranian Studies" in 2022.

ABBREVIATIONS

ann.	annotated
BAV	Biblioteca Apostolica Vaticana
BGA	*Bibliotheca geographorum arabicorum*, de Goeje, M. J., ed. Leiden, 1870–1894 (Reprint: 1967).
BL	British Library, London
BN	Bibliothèque nationale, Algiers
BnF	Bibliothèque nationale de France
BNE	Biblioteca Nacional de España
BNN	Biblioteca Nazionale di Napoli
BSB	Bayerische Staatsbibliothek
CB	Consolidated Bibliography
CNRS	Centre nationale de la recherche scientifique
com.	commented
CSIC	Consejo Superior de Investigaciones Científicas
ed.	edited
EI-1	*Encyclopedia of Islam*. 1st Edition. 1913–1938. Leiden: Brill. 5 vols.
EI-2	*Encyclopaedia of Islam*. 2nd Edition. 1960–2005. eds. Bearman, P. J. *et al.* Leiden: Brill. 12 vols.
EI-3	*Encyclopaedia of Islam*, THREE. 2007 ongoing. eds. Fleet, K. *et al.* Leiden: Brill.
EIr	*Encyclopaedia Iranica*, ongoing. iranicaonline.org
GTISAS	*Geography in the Traditional Islamic and South Asian Societies. History of Cartography* 2:1, Harley J. G. and Woodward, D., eds. Chicago: University of Chicago Press, 1992.
IFRI	Institute Français de Recherche en Iran
IRCICA	Research Centre for Islamic History, Art and Culture
JHA	Journal for the History of Astronomy
MIT	Massachusetts Institute for Technology
ÖNB	Österreichische Nationalbibliothek
tr.	translated
ZB	Zentralbibliothek, University Zurich, Zurich
ZDMG	*Zeitschrift der Deutschen Morgenländischen Gesellschaft*
ZGAIW	*Zeitschrift für die Geschichte der arabisch-islamischen Wissenschaften*

FIGURES

Sources

Adilnor Collection, Sweden
  MS Adilnor Collection
Bayerische Staatsbibliothek, Munich
  Cyrugia Parua Guidones [. . .], Venice 1500
  Johannes Channing, Albucasis De Chirurgia, Oxford 1778
Benaki Museum, Athens
  Astrolabe, inv. no. ΓΕ 13178
Bibliothéque nationale de France Astrolabe, inv. no. Ge A 327 MS arabe 2221
  MS arabe 2346
  MS arabe 2782
  MS arabe 4947
  MS arabe 5036
  MS arabe 5311
  MS arabe 5902
  MS arabe 5968
  MS arabe 6805
  MS persan 174
  MS supplèment turc 242
Bibliothéque nationale de Tunisie
  MS A-MSS-11925
Bodleian Library, University of Oxford, Oxford MS Ar. 90 c.
  MS Greaves 14
  MS Greaves 42
  MS Pococke 226
  MS Pococke 257
  Pierre Belon, Observations de plusieurs singularitez, Giraffe
British Library, London
  MS Add. 25724
  MS Or. 2984
  MS Or. 12988

British Museum, London
fragment of a limestone relief, museum no. EA 5610
incense burner, limestone, museum no. 125111
Cambridge University Library, Cambridge
MS Nn. 3.74
Genizah Collection, T-S K6.181r
Central Library, University of Tehran, Tehran
MS 7070
commons.wikimedia
https://commons.wikimedia.org/wiki/File:Joseph_von_Hammer-Purgstall._Benedetti_(um_1857).jpg
The David Collection, Copenhagen
Box with combination lock, inv. no. 1/1984
Pulley, inv. no. 42/1981
Astrolabe quadrant, inv. no. 16/1988
Balance, inv. no. 12/1994
Ethnologisches Museum der Staatlichen Museen zu Berlin – Preußischer Kulturbesitz, Berlin
Pen box with writing tools, inv. no. I B 100a-i
General Libraries, Texas University, Austin
Map, Ge6530s102518
German National Museum, Nuremberg
Portrait of Athanasius Kircher, inv. no. MP 12693
Getty Open Content Program,
Francesco di Giorgio Martini, Trattato di Architettura (1475–80)
Hasan Paşa İl Halk Library, Corum
MS Arşiv 19, Hk 2989/2
MS Arşiv 19, Hk 3237/2
Istanbul University, Rare Works Library, Istanbul
MS F 1404
Khizāna Ḥasaniyya, Rabat
MS 1393
Lawrence J. Schoenberg Collection of Manuscripts, Kislak Centre for Special Collections, Rare books and manuscripts, University of Pennsylvania, Philadelphia
MS LJS 435
Library, University of Michigan, Ann Arbor
MS Arabic 185
Library of Congress, Washington, D.C.
MS Turkish manuscript 185
Library Mammā Haydarah, Timbuktu
Commentary on al-Rasmūkī's *Epistle on Inheritance and Reckoning* by Abū Sālim al- Samlālī
Malik Library, Tehran
MS 6037
Majlis Library, Tehran
MS 3835
MS 6435
Mathematisch-Physikalischer Salon, Staatliche Kunstsammlungen, Dresden
Celestial Globe, inv. no. E II 1
The Metropolitan Museum of Art, New York

acc. no. 13.152.6 Rogers Fund, 1913
acc. no. 55.121.11. Rogers Fund, 1955
Millet Library, Istanbul
MS Ali Emiri, tip 79/353
Museum for Fine Arts, Boston
acc. no. 14.533
Museum für Islamische Kunst – Staatliche Museen zu Berlin, Berlin
inv. no. 1890, 431
inv. no. I. 1986.229
inv. no. I. 4350
Raza Library, Rampur
MS Kīmiyā' 12, fol. 133b
Research Library, Gotha
MS orient. A 1261
MS orient. A 1521
Science Museum, London
obj 1978–219
Staatsbibliothek zu Berlin – Preußischer Kulturbesitz, Orient Abteilung, Berlin
MS or. quart. 1209
MS Landberg 954
MS Diez A duodez 9a
Sotheby's, London
Sotheby's 2018, Lot. 111
Süleymaniye Library, Istanbul
MS Aya Sofya 2576
MS Haci Beşir Ağa 674/1
MS Esad Efendi 551
Topkapı Palace Library, Istanbul
inv. no. TSMA. E. 5539–1
Victoria and Albert Museum, London
Museum no. M.828PART/1–1928
The Walters Art Museum, Baltimore
MS W.659
Wellcome Collection, London
Geschichte der Chirurgie und ihrer Ausübung. Berlin 1889
Colored lithograph by F. Cazanave, *c.* 1830
Zentralbibliothek, University Zurich
MS Or. 120, fol. 178b
Private
Alexis, generic license, photography sextant, observatory, Samarqand
Ali Kadri, photography, grain mill
Amir Mohammed Gamini, Teheran, planetary models
Behrad, international license, photography, mural quadrant, observatory, Maragha
Charlotte Schriwer, London, photography, water distributing architecture
Dirk Grupe, Munich, planetary model
Emilia Calvo and Roser Puig Aguilar, Barcelona, reconstructed universal plate of Ibn Khalaf
Eva Orthmann, Göttingen, photography, ceiling in Akbar's palace, Dawlatkhāna-yi khāṣṣ
Glen Van Brummelen, Langley, BC, mathematical diagrams

Graham Beards photography, Noria complex in Guadalquivir, Spain
Johannes Thomann, Zurich, diagrams on logic and drawings of Ibn al-Haytham's optical instrument
Marc Moyon, Limoges, photography, Madrasa ʿAlī bin Yūsuf, Marrakesh
Sébastien Moureau and N. Thomas, aludel
Naeinsun, photography, subterranean *qanat*, Iran
Roxana Ashtari, Madrid, photography, Rām Yantra, observatory, New Delhi
Yves Porter, Aix-en-Provence,
three sherds, Egypt, Iran
Üljaytü's Dom, Sultaniyya, Iran
Ulugh Beg's madrasa, Registan square, Samarqand, Uzbekistan
Taha Yasin Arslan, Istanbul, computer models of three instruments
N. Thomas, alembic
Julio Samsó, Barcelona, diagram of the horoscope of the crosses
Fabrizio Speziale, Paris
Niẓāmiyya Ṣadr Shifāʾ-Khāna (1926–1936, Hyderabad)
cover page, Mujarrabāt-i ṭibb-i Sikandarī (1902, Kanpur)
Liana Saif, London
The Book of the Secrets of the Palm of the Hand (*Kitāb Asrār al-Kaff*) by some ʿAbd al-Ḥasan, published in 1303/1886
News Moon, 26 April 2020

TABLES

BOXES

PREFACE

This *handbook* offers an overview on current studies of an important part of the sciences in Islamicate societies by senior, midcareer and junior researchers active in several academic disciplines. They present a kaleidoscope of approaches to the study of the knowledge about the cosmos, nature, the human body, numbers, plane and solid figures, instruments, observations, epistemology, scientific methods, agriculture, occult sciences, balances and automata, as well as institutions and politics in Islamicate societies from the 2nd/8th to the 14th/19th centuries. It brings together surveys of geometrical, arithmetical, medical and other scientific methods and techniques with studies on the societal impact of these varieties of this kind of knowledge and the people who promoted such knowledge or practiced it as experts. Various kinds of practices are discussed, from elementary calculations to designing complicated planetary models, from reading and commenting practices to (al)chemical procedures, from optical experiments to debates on scientific methods and theories of atomism or causation. The editors, the members of the editorial board and the authors have done their best to shed light on how the sciences were practiced in different periods and places: regions, empires, sultanates, provinces or cities going far beyond the traditional focus on the Middle East, North Africa and al-Andalus in the so-called Golden Age, with its unspecific, ever-changing boundaries. Extending the purview of this book into the 13th/19th century serves to highlight the continuities of scientific life in the Islamicate world as a whole, as well as to emphasize the shifts and changes in its geographies, epistemologies and the interests of its participants. The contributors' broadened attention to what have been called the normal sciences has allowed them to include cultural activities such as teaching, political counseling, courtly debates, cross-communal interactions and instrument production, as well as visual representations of knowledge and the artistic enhancement of scientific manuscripts and instruments. Adding to the traditional interest in interactions between Islamicate and Christianate societies in some parts of Europe, several chapters in the handbook examine cultural contacts and their results in West Africa, as well as South and East Asia. These studies re-center and multiply the placement of Islamicate societies in global histories and open new avenues for exploring the activities, interests and practices of their scholars, patrons and artisans. The decision to use the concept of Islamicate societies allowed us to include chapters on minority communities both within and outside Islamicate states, and to address their relations to scholars of majority communities and their works. The price to be paid in a handbook for this enrichment of historical narratives about the sciences in Islamicate societies is the renunciation of chronological continuity, as well as completeness in the presentation of all disciplines that existed during the millennium covered by the book. Nevertheless, it is a price worth paying, as what is gained is a more balanced, simultaneously local, regional and global sense of how these sciences were actually practiced and experienced over time and space, which has so far been lacking in the field for most disciplines.

A second main set of themes encompasses contested as well as widely accepted historiographical interpretations. The extended temporal scope of the handbook effectively contests the long-term belief in the steep decline or even disappearance of Islamicate scientific activities as early as the 5/11th or as late as the 10th/16th century. Likewise, its much broader geographical scope contests the belief in the dominance of scientific activities in the Middle East and to a lesser degree in North Africa and al-Andalus to the detriment of other regions in eastern Europe and Asia East of Iran, as well as sub-Saharan Africa. The use of the term *Islamicate* rather than *Islamic* and the insistence on society as primary analytical category serves to historicize the study of scientific activities and their practitioners in highly specific yet multicultural and plurilingual environments. The Islamization of individual societies and their communities itself followed many different trajectories in different periods and regions. It began and evolved in a variety of contextual conditions. Those differences led to diverse patterns in the evolution of scientific activities, institutions and communities as documented in the chapters on West Africa, al-Andalus, cross-communal and inter-confessional interactions, East Asia, Tibetan medicine and 13th/19th-century India.

Although intellectual, political or social conflicts do not stand at the center of any of the chosen historiographical approaches, they are discussed in the chapters on philosophy, the astral sciences, the determination of the prayer direction (*qibla*), scientific components (for instance, atomism or causation) in *kalām* (dialectical theology) and in the context of accusations of unbelief or heresy. Those chapters show that such conflicts were not simply fights between Muslim believers and scholars dedicated to the ancient (mostly Greek) sciences but struggles over social cohesion, scholarly reputation, political impact and intellectual coherence and consistency. In this sense, those chapters challenge, each in its own way, previous interpretations of the relationship between religious beliefs and scientific theorizing that have seldom gone beyond dichotomies. They also document the impressive progress achieved during the last two or three decades in studies on the history of philosophy and *kalām*. As in any other society on the globe, scholarly life and its practices in Islamicate societies were the result of complex and multifaceted sociocultural processes in which scholars with strong interests in the mathematical sciences could act as heads of courtly offices or as ambassadors of rulers; religious scholars could defend the study of works written by leading Muslim, Christian, Jewish or polytheistic philosophers of the past or present, or promote the calculation of religious acts with scientific tools. But whatever discipline a scholar privileged, he also could denounce an opponent for improper social, including religious, behavior and cheer his imprisonment, social downfall or occasionally even execution.

Another major historiographical position, considered valid over almost half a century and with enormous impact on the history of science and cultural history outside Islamicate societies, is challenged in the first chapters of this handbook. This is the concept of the translation movement in the Abbasid caliphate from the middle of the 2nd/8th to the early 5th/11th centuries. Individual aspects of this narrative and the interpretations of the translations in that period have become controversial since the late 20th century. That is why the first four chapters are devoted to its discussion from different angles, by authors with different opinions, highlighting the difficulties that historians face when trying to reconstruct the past based on incomplete source material. One of the main points in this discussion concerns how the notion *movement* should be defined for societies in the distant past. What are the options for determining criteria that need to be met in order to differentiate between terms provided by the social sciences for adequately labeling social processes? A second point is the lack of actors' categories with respect to the sociocultural meaning of the translations and the reduction of cultural memory to a small number of actors, perhaps too small to constitute a movement. A third concern is the exclusion of many other moments of intense translation activities in other periods and regions, none of which have been labeled movements to date. Additional difficulties are considered in the corresponding chapters.

The inclusion of senior, midcareer and junior colleagues from around the world, not merely a handful of Western countries, and their often divergent views on how to study the sciences in past Islamicate societies troubles too-neat historiographical narratives by showcasing a wide range of possible different attitudes as to how to speak about the past and which aspects to favor. It also offers a small number of graduate students the chance for presenting their first results on so far under-researched themes. A majority of colleagues – but not all – agreed to use the expression Islamicate societies when speaking in general terms about theories, methods, institutions or trends. Most but not all colleagues agreed to abstain from using the notion of translation movement. Numerous colleagues have integrated technical debates into contextual perspectives. Some have focused on content explanations, while others have privileged political and cultural conditions. The majority of the contributors stress regional and even local differences and particularities, while a minority highlights the sharing of practices and their contents over centuries and regions.

Although the handbook is broadly conceived, incorporating many new themes and ways of treating them, it is not all-inclusive. Routledge was extremely generous in allotting an unusual number of chapters to it, with varying lengths appropriate to their task. That some chapters are longer is not an expression of academic verbosity but of the didactic goal to make the many difficult technical and historical aspects of the sciences in Islamicate societies accessible to the intended readership of the book – graduate students and colleagues as well as anybody else interested in surveys and thematic chapters presenting different parts of this vast field of research.

All those points of orientation, change and novelty demonstrate how fruitful and productive the work by historians of science, historians, regional specialists and experts of further academic fields has grown since 1996, when the first multivolume book on the sciences in Islamicate societies was published by Roshdi Rashed and Regis Morelon as a joint venture with thirty colleagues all working in countries of Europe and North America, except for Egypt. The focus then was above all on content covering a broad range of disciplines, themes and methods. Although many new results have been achieved in those domains since the 1990s, Rashed's and Morelon's *Encyclopaedia of the History of Arabic Science* (1996) continues to offer important information about and insights into texts, instruments, maps or technologies written or described in Arabic and the then dominant historiographical positions and interests.

The dates in this handbook follow the academic tradition of giving both Hijra dates and CE dates in the form (Hijra date/CE date). The dates of authors or rulers who lived before the common era will be marked BCE. For those living before or outside of Islamicate societies, common era dates are only marked by CE if misunderstandings could arise from not using the label. Due to the multiplicity of eras used in Islamicate societies for historical and other dates occasionally other eras such as that of Yazdigerd III (r. 632–651), the last Sasanian ruler, are employed and marked by their respective labels. The compilation of the index of authors was – as always in books of this kind – was a difficult enterprise. With the help of several colleagues, I tried to achieve consistency as much as possible. But since the name structures and the use of partial names differ across cultures and period, full consistency remains beyond our grasp.

The transliterations of Greek, Syriac, Middle Persian, Arabic, New Persian, Hebrew, Ottoman Turkish, Urdu, Sanskrit, Hindi, Tibetan or Chinese terms follow standardized academic usage in English. In some cases, such as Middle Persian, Hebrew or Ottoman Turkish, adaptations to transliteration norms of related languages or simplifications are used either for unifying names or expressions shared across several languages or to make it easier for readers not familiar with any such language to read the respective chapters. Particularly difficult was the decision in the case of Ottoman Turkish. I followed here phonetic recommendations of an Ottomanist colleague. Dynastic and geographical names are given in simplified English forms ignoring academic norms of transliteration. Sources used in at least three chapters are presented in the consolidated bibliography (CB) at the end of the book, while specific literature is given at the end of each chapter. Arabic, Persian, Turkish or Hebrew authors mentioned by at least three contributors are included in the consolidated bibliography with all used editions or translations. Entries from encyclopedias are listed in the chapter bibliographies while the titles of the encyclopedias are abbreviated and listed at the beginning of the book in the list of abbreviations. A similar procedure is followed for repeatedly mentioned edited volumes. In the part "Sources" of all bibliographies we translated titles of texts in Oriental languages into English. To translate in the bibliographies all titles of all works from languages other than English was impossible because of their sheer number.

I thank all colleagues who agreed to invest their time and knowledge into the production of this handbook. The various historiographical and didactic challenges caused repeated serious problems for writers and editors and some authors left the project, while others declined to join it. That is why my honest gratitude goes to all who had the patience to stay until the book was finished. I also thank the members of the editorial board, Nahyan Fancy, Gottfried Hagen, Matthew Melvin-Koushki, Lutz Richter-Bernburg and Ulrich Rudolph, who supported me in evaluating the quality of each contribution and offered advice on how to overcome weaknesses and correct mistakes. Special thanks go to Taha Yasin Arslan, Nathan Sidoli, Charles Burnett, Johannes Thomann, Jon McGinnis, Julio Samsó, Glen Van Brummelen, Miri Shefer-Mossensohn, Josep Casulleras, Richard Kremer, Glen Cooper and Katharine Park for their collegial support with regard to individual chapters that proved particularly difficult. My most profound gratitude goes to Peter Barker, the associate editor, and Rana Brentjes, the assistant editor. They not only spent innumerable hours proofreading all chapters (several chapters even more than once; Peter) and ordering, labeling and sorting images, diagrams and tables or listening to my own chapters (Rana), but they also provided me with the emotional support without which I would have given up the project already in its first year. Finally, I thank all authors, all members of the editorial board, my two co-editors and Katharine Park for the tremendous work in proofreading the finished manuscript.

Sonja Brentjes
June 2022

INTRODUCTION

Sonja Brentjes

The key to understanding this multifaceted book is methodological pluralism. That pluralism is both the principled academic stance – taken by myself as editor and my colleagues who agreed to serve as co-editors and on the editorial board – and a pragmatic decision. The many changes in what is regarded as belonging to the history of science for any society or culture, and in theoretical approaches to the study of the past in its many instantiations and their interpretations, necessitate the handbook's emphasis on plurality and inclusiveness rather than on singularity and exclusion. This methodological plurality distinguishes our collection of 64 chapters from previous surveys that name the historical topic they deal with Arabic science or, more broadly, medieval science (Rashed and Morelon 1996[CB];[1] Lindberg and Shank 2013[CB]).

The handbook offers insights into current research on the fields of knowledge subsumed – for the sake of simplicity – under the term *science* or *sciences* and practiced in different forms and under diverse conditions in Islamicate societies between the 2nd/8th and the 13th/19th centuries. It first presents surveys of long-established themes, together with scholarly evaluations that include the latest research. These are followed by thematic chapters dealing with particular, often very recent, research questions, including reflections on how to approach and answer those questions and how to proceed in the future. The book aims not merely to represent a broad range of past scientific achievements but to also explore the historiographical positions behind their interpretations and some of their central problems.

The handbook's contributors tell many stories, covering a greater range of fields of knowledge and local or regional conditions than its predecessors have done. They highlight the multiplicity, complexity and differences that historians encounter in their chosen sources and that demand a growing arsenal of methods in order to be analyzed effectively. Methodological pluralism thus underpins not only matters of historiography and interpretive principles but also the kinds of sources included in the study of past sciences and the methods appropriated from modern sciences and humanities.

The pragmatic aspect of our preference for pluralism arises from the very uneven distribution of research undertaken in the last two decades. Research agendas have, for example, shifted to the previously neglected later centuries, often called the postclassical period. Due to this change in perspective, the handbook's coverage does not stop in some long-ago century but continues up to the late 13th/19th century, when crucial changes in the scholarly world of Islamicate societies were set in motion by colonialism, political reforms, seditions and revolutions. The inclusion of

DOI: 10.4324/9781315170718-1

research on scholarly activities in the occult sciences and natural philosophy between 1300 and 1900 CE is the handbook's second major conceptual novelty, reflecting the profound change in historiographical positions, especially among younger scholars.

Methodological pluralism as practiced in this book allows us to add to the traditional focus on disciplinary content an engagement with a broad spectrum of contextual aspects, ranging from communal interactions to the arts, the materiality of scholarship and many points in between. Our main claim is that a history of science worthy of the name cannot begin and end with the edition, translation and disciplinary analysis of texts or instruments. In addition to those traditional goals, it needs to investigate the changing social, economic, cultural and material conditions for the production and reproduction of disciplinary knowledge. In short, it needs to engage with the contexts that formed, enabled, oriented and reoriented the disciplinary contents studied.

As a result, this handbook's contributors were invited to provide studies of numerous types of context, sometimes small and limited but often surveying one disciplinary aspect over several centuries and across a larger territory. To contextualize the sciences in this way, we must turn away from concepts such as *Arabic science* or *medieval science* or *science in Islam* and toward scholarly activities in concrete circumstances. That might mean working on the level of microhistory – such as a group of scholars at some court, school, hospital or observatory in a specific city or town – or on the medium level of history, such as the entire territory ruled by one dynasty (abbreviated in this book with the term *society*), or on a larger scale, looking at a whole region such as al-Andalus, North Africa or the Indian subcontinent. The shift in territorial and thereby conceptual perspective is the third step forward in the handbook's definition of what the history of science in Islamicate societies should cover and what should be represented in a survey of recent research on past sciences in those societies.

In the effort to incorporate further new aspects into the handbook, the fourth – and, it turned out, most difficult – step is a shift away from representing only results, methods and techniques toward paying attention to the lived practices that shaped the various fields of knowledge. This means asking how past scholars arrived at their questions and answers, how they exercised and shared their knowledge, how they interacted with each other and other members of their societies, and how the things they did in order to produce and reproduce knowledge differed among the disciplines and across space and time.

In the following section, I briefly discuss some of the difficulties with which every historian of science in Islamicate societies has to grapple with. This discussion is not intended to be comprehensive but to prepare the subsequent outline of the conceptual structure of the handbook.

0.2.1 Actors' categories or modern labels?

The terms *science/sciences*, as used in the title of this book and numerous, although not all of its chapters, capture the meaning of the Arabic words *ʿilm/ʿulūm* as given to them by the new language academies in Syria and Egypt in the process of early twentieth-century language reforms. The same signification was chosen in the analogous process in Iran. Urdu speakers and readers in Pakistan and India still use them today in this sense. The exception in the Islamicate world is Turkey, where *fen* (the Turkish version of the Arabic word *fann*) and the Turkish word *bilim* were identified with the modern concept *science*.

In his wonderful book *Knowledge Triumphant: The Concept of Knowledge in Medieval Islam* (Rosenthal 1970), Franz Rosenthal argued that *ʿilm* was of fundamental significance to many (or, as he claimed, most) parts of Islamic culture. Translating the word as *knowledge*, he differentiated between *ʿilm* as an abstract concept, embracing all knowledge, and *ʿilm* as the designation of a concrete discipline of learning, or even only as a concrete collection of known things (*maʿlūmāt*). In historical terms, he located the emergence of the notion of a specific discipline of learning within religious law (*fiqh*) or theology (*kalām*). This new meaning of *ʿilm* as a specific discipline (or field of knowledge) was the starting point for the word's later use as a label for "science or scholarly discipline" (Rosenthal 1970, 43).

The historical process of change in the meaning of the singular form does not, however, explain the emergence and usage of its plural *ʿulūm*, one of the core terms dealt with in this handbook. Rosenthal's chapter "The Plurality of Knowledge" points out the limitations of our knowledge of its genesis and spread. He highlights the multiplicity of meaning and usage in both the singular and the plural (Rosenthal 1970, 43–5). Hence, although his approach to the problem and his views on particular facets differ from those we present, Rosenthal's study of *ʿilm*/*ʿulūm* constitutes the foundation for the stance taken in this handbook.

The position subscribed to here is that there is no single, unproblematic solution to the question of how to translate the *ʿilm*/*ʿulūm* pair or other terms, such as *fann* or *ṣināʿa* and their plurals. The usage of *fann* in modern times, for instance, ranges from *science*, as in Turkish, to *art*, as in Arabic. In Persian, *fann* can mean both *art* and *technology*. The Turkish *sanat* for *art* is derived from the Arabic *ṣanʿa*, a cognate of *ṣināʿa*, while the Turkish terms *teknik* and *teknoloji* are phonetic transliterations of the corresponding French words. In the sources discussed in this handbook, *fann* was – as far as can be said on the basis of the very limited knowledge of the word's history – primarily used to label a concrete discipline or a group of known things; *ṣināʿa* rendered either the Greek terms *technē* and *epistemē* or the meanings *craft*, *handicraft*, or *the skill to manufacture things*, as derived from the verbal root of the nouns *ṣanʿa* and *ṣināʿa*.

The following section explains, mainly using the example of the mathematical sciences, why actors' categories do not necessarily provide a better, clearer or easier solution than the modern decisions I have just sketched.

0.2.2 Historical fields of knowledge, their naming, grouping and changes

Historical actors categorized and labeled the sciences differently in the course of the more than 1,000 years covered in this book. During the first three centuries, sciences were often classed as a special group of disciplines, separated from the religious and philological fields of knowledge. A name preferred for this special group was *sciences of the ancients* or *of the first ones* (*ʿulūm al-mutaqadimīn* or *al-awāʾil*). The religious and philological disciplines as a single group were called *sciences of the moderns* (or *newcomers*) or *of the Arabs* (*ʿulūm al-mutaʾakhirīn* or *al-ʿarab*). Neither the names nor the taxonomy disappeared completely from usage during the subsequent 700 years, but other categories rose to prominence and pushed them to the margins. The most prevalent of these were the terms – and corresponding divisions – *transmitted sciences* (*al-ʿulūm al-naqliyya*), *rational sciences* (*al-ʿulūm al-ʿaqliyya*) and *mathematical sciences*. The mathematical sciences were called either *the sciences to be exercised* (*al-ʿulūm al-riyāḍiyya*) or *the sciences to be taught* (*ʿulūm al-taʿālīm* or *al-ʿulūm al-taʿlimiyya*). The first designation continues to be used today for modern mathematics, while the second has receded into the background. Further groups labeled and classified in those early centuries were the *occult sciences* (*al-ʿulūm al-gharība* plus other terms), the *philosophical sciences* (*falsafa*; later also called *al-ʿulūm al-ḥikmiyya*) and the *professional sciences* (*al-ʿulūm al-miḥniyya*). Some of these names and their groupings were widely recognized; others appealed only to small communities of scholars or appeared only rarely.

The mathematical disciplines began their scholarly life in Islamicate societies in the fields of astral knowledge and surveying, as testified by historical and geographical texts from the 3rd/9th and 4th/10th centuries. Both these fields included arithmetic, that is, different systems of oral and written numeration and calculation. This is reflected in the labels attached to the professional practitioners of the two fields of knowledge: people with knowledge about the stars were called *munajjimūn*; surveyors were called *massāḥūn* and calculators, active in both fields, were called *ḥussāb*. Very soon, a fourth term was coined: *muhandisūn*. According to the early sources, *muhandisūn* participated in building pontoons for troops crossing a river, subterranean water canals or war machines for a siege.

Influenced by the classifications of mathematical knowledge to be found in Greek texts translated into Arabic, the Neoplatonic quadrivium was accepted as the dominant system of classification. It set out four main theoretical disciplines of mathematics: number theory, geometry, astronomy and music (meaning a set of proportions and the rules for calculating with them). Also acquired through the study of translated Greek texts were a number of "branches" within this system, among them optics, mechanics and spherical geometry. At least until the late 4th/10th century, the four main mathematical disciplines could be called by their ancient Greek names transliterated into Arabic letters – *arithmāṭīqī*, *jumaṭriyā*, *asṭrunūmiyā* (vocalized in different forms) and *musīqā*. Only music kept its Greek name over the centuries. The other disciplines soon received Arabic names, mostly derived from translated Greek terms or from appropriated and modified Persian expressions. *Arithmāṭīqī* was rendered as the knowledge of the number (*ʿilm al-ʿadad*), *jumaṭriyā* as the knowledge of *handasa* (this word, believed to come from Middle Persian, seems to have meant the measurement of subterranean canals; it later came to designate engineering and architecture as well), and *asṭrunūmiyā* as the knowledge of the stars (*ʿilm al-nujūm*).

As the main groupings and their names changed over time, so did the names and content of the individual fields of knowledge. Astronomy diversified into *knowledge of the orb* (*ʿilm al-falak*); *knowledge of the zīj* (*ʿilm al-azyāj*), that is, a compendium of short texts and numerous tables on different magnitudes needed in chronology, planetary theory, astrology, geography and religion; *knowledge of the configuration* [of the universe] (*ʿilm al-hayʾa*); *knowledge of times* (*ʿilm al-mīqāt*); and other branches. Number theory was integrated into the *knowledge of calculation* (*ʿilm al-ḥisāb*), often ranked first. Algebra became a separate branch in this class, although its problems and methods were in the oldest surviving texts integral components of *ḥisāb*. From about the late 6th/12th century, some of the occult sciences began to be reclassified under the category *mathematical sciences*, among them *knowledge of letters* (*ʿilm al-ḥurūf*) and *knowledge of sand divination* (*ʿilm al-raml*). Changes similar to those I have described for the mathematical sciences also took place in the other fields of knowledge treated in this book.

These few examples of constant shifts in classifying and naming fields of knowledge point to the difficulties involved in defining the kind of knowledge that the contributors to this handbook call *science* or *sciences*. The examples show that a simple return to historical names or actors' categories is only partially helpful, as is the use of the modern equivalences established in the early 20th century. Accordingly, the contributors differ in their understanding of the fields of knowledge that the book addresses: natural philosophy, *kalām*, the mathematical sciences, medicine, geography, mapmaking, the occult sciences and technical knowledge.

0.2.3 Editorial choices

The decisions as to which fields of knowledge should be included in the handbook entailed a number of boundaries. One important boundary concerns the *rational sciences*. Only those parts of philosophy and *kalām* that address concepts of the universe, the study of nature or certain types of methods are included. A second boundary was set regarding the depth and detail to be provided on the technicalities of the science in question. Although the presentation of content is undoubtedly an important goal of research and thus also an important part of this handbook, this is neither a textbook of past scientific theories, concepts, problems or methods nor a dictionary of oriental languages. Each contributor made her or his own choice on how much information she or he wished to presuppose and how much she or he felt had to be explained in greater detail and given full designations in Arabic, Persian, Turkish, Sanskrit, Chinese or other languages. A third boundary concerns the decision on how much context should be supplied. The handbook presents research on past sciences primarily from the perspective of their practices and their content and context but is not a survey of the many political, cultural or economic aspects of an entire society or even of a particular place. While it would be a fascinating prospect to write a history of the various fields of knowledge as practiced in one city or town during one period of time, this handbook is not the place for such a complex enterprise.

This survey of current research on the history of science in Islamicate societies is not intended as the construction of a new grand narrative, as the one and only authoritative view of what that history was. Three points, at least, speak against such an endeavor. Insights from other parts of the humanities teach us that the main disadvantage of any grand narrative consists in the manifold simplifications that are produced through exclusion, centralization and an interpretive inclination toward cultural hegemons. The second objection to even trying to invent a new grand narrative on the sciences in past Islamicate societies is the incomplete and contradictory state of research, as clearly documented by the numerous gaps this book was not able to fill. Third, and most broadly, it is misleading to teach graduate students and other potential readers that the objective of research is to produce a single story that excludes both profound distinctions and simpler variations between historical sources themselves and among our modern interpretations.

It is not only our academic work that cannot be pressed into a single form or guise. The same constraint applies to the more than 400 different Islamicate societies that formed, disintegrated or survived during the long period from the 2nd/8th to the 13th/19th century. According to our current knowledge, scholarly communities working in the sciences did not emerge in all of them. But thanks to the spread of institutions that fostered sciences as one part of their investment in knowledge in many spatial directions and social groups, there is abundant material to study such activities in more than just a few Islamicate societies and thus narrate new and diverse histories.

In addition to multiplicity, complexity and difference, this book aims to encourage curiosity and flexibility when choosing topics for analysis from the intellectual activities of past actors. It emphasizes the obligation to respect what past actors were interested in and what they achieved in their enterprises, without segregating them into those who deserve our attention and those who do not. "Normal science" and its participants, to use Thomas Kuhn's distinction, have the same right to historical analysis as do paradigm shifts and their instigators. It is our responsibility to discover them and their richness. The goal of this book is to provide a multitude of stories as told in current research and thus to invite and encourage new research.

0.2.4 Practices and their pluralities

This handbook engages in telling the history of science in all its pluralities. Its contributors wish to overcome the decontextualized and largely de-historicized macro-level of study and narrative ("Islam," "Arabic," "medieval") by telling medium- or micro-level histories about a region, a single society or an individual place and the respective actors. The contexts we need to discover in order to understand the roles and courses of the sciences can only be accessed on the lowest historical level, that is, the extant sources. They need to be cautiously and carefully generalized for meaningful temporal and spatial units.

An important conceptual decision structuring the handbook and its contributions was the move toward the notion of *practices*, as noted earlier. In the sciences, some practices are shared across many disciplines; others are specific to certain ones. Practices found in all the sciences are related to literary forms and genres, modes of production, teaching and other social forms of exchanging knowledge and evaluating colleagues and works. The main shared literary forms are prose and didactic poetry; shared genres are textbooks, summaries, dialogues, commentaries and super-commentaries. Teaching took place most often by reading aloud short passages of texts one after the other and explaining their meaning to the student. This could be done in a one-to-one meeting between a teacher and a student in a private setting or in a school. In schools, the number of students could vary between fewer than ten and more than one hundred, although the larger number was found mostly in fields of knowledge not treated in this book, such as the *transmitted sciences* mentioned in Section 0.2.2. Other places of teaching included hospitals, observatories and Sufi convents. There, as in the case of schools, the number of participants could vary considerably. Asking questions of a teacher was permitted, but biographical dictionaries, travel literature and pedagogical writings indicate that challenging him was often regarded as a violation of social codes.

Challenges among colleagues, in contrast, were allowed. They could take the form of discussions in meetings among scholars or other respected men. At courts, in the presence of rulers and their families or in public challenges, they followed more formalized protocols. Opponents were attacked by pointing to shortcomings or lacunae in their works or views. Scholars quoted or paraphrased each other, but they also copied from others' works without mentioning their source. Such practices of conflict, rebuttal or denigration took more combative forms in some disciplines than in others. In some historical and cultural contexts, they inspired intense methodological debates that led to reforms of earlier practices. Another important social practice across all the disciplines presented in this handbook was building hierarchies of scholars and texts by attributing honorary titles to some but not to others or by mentioning them in quotations and other forms of referencing.

Among the occult sciences, alchemy and magic had their own modes of building reputation and spreading ideas. Most of these were modes of behavior inherited from ancient scholarly and artisanal practices. They routinely included secrecy, cover names, enigmatic alphabets and the appropriation of famous historical names for works actually compiled by other people. An impression of isolation, marginalization and the need for protection against exclusion or even persecution arises from such modes of representation, yet this is countered by the impressive number of extant copies of texts, the public recognition of historical authors as authorities in their fields and the widespread use of amulets, talismans, magic shirts and incantation bowls across all social strata.

The modes of technical production shared by the disciplines depended on the material objects that were used by scholars. In their most basic forms, these modes include writing, copying, or formatting and manually working with different kinds of material and tools. At the material level, texts were formatted by means of framing, underlining, rubricating or decorating as elements of layout. At the intellectual level, formatting took place through the inclusion of titles, headings, tables of content or formulas of beginning and ending. All these textual and artistic practices have their own histories and underwent far-reaching changes over time.

Alloys (and, to a lesser extent, pure metals), ceramics, clay, wood, papier-mâché, glass and paper were the materials most often used for instruments and other important material objects such as inkwells and pen boxes. They were also employed for making the bottles, bowls and spoons used in the astral sciences, medicine and alchemy. Temporally and regionally less common materials were papyrus, parchment, palm leaves, bamboo, cotton, silk and stone.

Production processes for shaping these materials into the desired products included casting, molding, hammering, filing and firing. Subsequent steps in production were drawing, engraving, pouncing, tracing, stenciling, coloring or inscribing until the instrument, the book, the map or the image of a constellation was finished. Prescriptions derived from Aristotle's *Analytics*, Euclid's *Elements* and Ptolemy's *Almagest* informed in different ways practices specific for the mathematical sciences, natural philosophy and medicine. They included epistemological rules, forms of modeling, conventions of abbreviation and formalized representations. Other practices employed only in certain disciplines are individual modes of visualization and acknowledged types of experiences, experiments and observations. Examples of discipline-specific epistemological practices are particular modes of argumentation, demonstration and verification such as syllogisms or indirect proofs, or the admission or exclusion and ranking of results achieved by observation. Such practices also encompass hierarchical demarcations from other disciplines and further methods of boundary-drawing and closure. In a few cases, they extend to the creation or elimination of paradigmatic models.

Among the modes of visualization, we may distinguish tabulation, diagrams and treelike or other forms of schematic representation. Conventions for visualizing spatial objects, figurative drawing and other artistic modes reflect not only changing epistemological principles but also cultural preferences and manual skills. Diagrams, although originally found mainly in the mathematical sciences, soon spread into other fields of knowledge, including those not treated in this handbook such as the philological sciences. For a long time, diagrams were not investigated as independent objects of research, but work on geometrical diagrams in the last two decades has enriched our understanding of their typology and purposes. This research has recognized cognitive aspects of diagram production, such as the preference for symmetry and simplicity, a multiplicity of functions and a multitude of relations between the written text and the visualized depiction. Diagrams were used to illustrate an example, explain a claim or summarize a procedure. In other cases, they supplement a textual passage that could also have been understood without a diagram, or they deliver additional, nonverbal information. They might derive from the verbal expression of a piece of knowledge, but they could also be independent of the surrounding text and replace an entire proof. Almost all of these functions and meanings can be found in the vast corpus of Arabic, Persian and Turkish works on Euclid's *Elements*. The possibility of substituting a diagram for a verbal proof or demonstration led to the production of commentaries and editions of the *Elements* that minimized verbal content and primarily consisted of diagrams.

Tables appeared in a variety of forms – rectangular, square, triangular, circular. Their structures ranged from the simplest list format to highly complex formats with multiple sublevels. They served for memorizing operations such as elementary multiplication or items such as the names of stars, asterisms or constellations, mathematical relations or the names of remedies and their ingredients. Other tables were employed for storing information that did not have to be learned by heart but could be checked when needed. A further kind resulted from complicated processes of observation, calculation and conceptual interpretation and served to simplify daily routine work for practitioners of the astral sciences. These examples by no means exhaust the whole spectrum. So far, the study of tables as a type of scientific practice has mainly focused on their content, not on their forms of organization and later functions.

Verbal, numerical or symbolic formulas and forms of designation, along with their developments over time, constitute another body of discipline-specific practices. As well as in the mathematical sciences, such practices can be found in astrology, alchemy, lettrism (the science of letters and their numerical meaning, designated in a variety of ways in the modern literature), sand divination (prognostication through sixteen combinatorial schemes of dots and lines) and magic. They include the conventionalization of signs for fractions, components of equations, the extraction of roots and the arrangement of horoscopes. Furthermore, magical alphabets and alchemical symbolism fall into this category. These practices also appear in geography and mapmaking, where scholars compared coastal lines with parts of their apparel or identified terrestrial and human objects with geometrical figures or stereotyped botanical and cultural symbols.

In addition to the social practices employed in all disciplines, certain activities were performed in only some. For instance, the admission of amateurs or non-scholarly forms of knowledge was a practice in the astral sciences from the 2nd/8th to the 4th/10th century. Such amateurs were asked to witness and comment on scientifically obtained results, in particular observations and measurements, as a way of ensuring their social acceptance. Judges, religious leaders and surveyors could be included in the verification and status formation of new astral knowledge.

Social issues also played a role in teaching the astral and other mathematical sciences, for example, when scientists explained technical terms or translated them into expressions familiar from everyday life. Appropriate social behavior by students, teachers and scholars at courts informed the moral economy of all the sciences. Individuals were appreciated and admired if they showed knowledge of detail, introduced exceptional or surprising solutions to problems, memorized parts of books or entire texts or were acquainted with scholars who themselves had exemplary reputations. Independent thinking and the zeal for solving problems without first having read the authoritative texts were either mentioned as exceptions or condemned as deviations from scholarly norms.

These scholarly practices varied across the time and space covered by this handbook, and not all of them will be discussed in its chapters. Thinking in terms of the entire set of such practices relevant to a particular discipline, or for the solution of a particular group of problems, proved to be challenging – and, indeed, comprehensiveness is not the goal of this volume. Despite intense discussions, we could not always reconcile our views on what the practices were or which of them matter; in such cases, the focus on practices was abandoned in favor of the chapter's contributions on other points.

0.2.5 Constraints

This handbook shows that for more than a thousand years, people in Islamicate societies engaged with intellectual and professional work on the universe, its individual bodies, the atmosphere and the earth, its parts and its inhabitants. They created theories, developed new methods for problem-solving and built new instruments or institutions. We attempted to draw all those features in one form or the other into the book but could not cover them to the same depth everywhere and for the entire millennium. Thanks to the broad scope of scientific activities from which to choose, the goals and content of the 64 chapters differ significantly.

Another reason for the variation between the chapters is that there are remarkable differences between the debates and the state of research in the various fields covered in the volume. Their historical objects, moreover, have been incorporated into the traditional mother discipline "history of science" to varying degrees or even completely excluded from it. These differences in the extent of our knowledge about the various themes come to the fore in the chapters on later periods and on regions far away from the territories around the Mediterranean and in western and central Asia. Thus, the Indian subcontinent features only in chapters on chronologically and thematically limited topics. Also, the Islamicate societies whose scholars wrote their science in Arabic or Persian are so diverse that no general summary of their work can be given – even for a single century, let alone for more than a millennium.

The same applies to Islamicate societies south of the Sahara, along the East African coast, on the Indian Ocean, in Southeast Asia or in eastern Europe. The specialized contributions on some of these regions take this book beyond earlier studies that considered only southern Europe and Byzantium when investigating relations with neighbors and processes of cross-cultural knowledge exchange. They indicate that research in some of these cases is at a very early stage. In other cases, for example in some Asian societies, exchange with Islamicate neighbors seems to have played only a minor role in the cross-cultural construction of knowledge. Such chapters are nevertheless important, since they open up new perspectives and new questions that will help historicize the history of science in Islamicate societies. They point to new plural histories rather than one that is highly aggregated.

The asymmetrical relationship between survey chapters, specialized chapters and chapters focusing on historiographical problems reflects another deficiency of previous, content-focused history of science. Several of the sciences discussed in this handbook found a new institutional home in madrasas after roughly the 6th/12th century. The study of the content of the many thousands of school texts compiled for students in these institutions is only beginning. More research has been done for parts of North Africa and Iran than for other regions, but even there, no systematic surveys of textbook contents are so far available, except for a kind of literary and biographical catalog with brief remarks compiled for North Africa. For Iran, surveys on philosophy are more plentiful for the period from the 10th/15th to the 11th/17th century, but they focus on themes, such as metaphysics, that are not part of this handbook. For the Ottoman Empire, parts of the literature on astral and chronological matters – almanacs – have been recently surveyed, as have texts on agriculture. Because of this partial coverage, survey chapters on entire disciplines were intentionally limited to the first part of the book, which deals with knowledge as practiced between the 2nd/8th and 7th/13th centuries.

0.2.6 The book's structures

The book has more than one structure. On one hand, it is arranged into research fields. On the other, it follows chronological considerations. With regard to the research themes that structure the book we first ask how the sciences came into being in the early Islamicate societies, and survey the narratives we now tell about those processes and the narratives some of the actors told about themselves. This includes not only the results that were achieved in different disciplines or individual communities or texts but also the questions that we have overlooked so far. It is therefore in Part I that most of the disciplinary survey chapters are placed. In addition to the serious historiographical problems associated with the fundamental question of origin and emergence, the two goals of this part are to prepare the later thematic discussions and to connect the various contextual approaches and historiographical challenges with the important, if more traditional, investigation of contents.

In contrast to some scholars in neighboring fields of Islamic studies, I am convinced that the history of ideas has an important place in a broader history of science or knowledge. The study of the content of texts, instruments, maps or visual specimens is a necessary precondition for the study of their authors' roles, positions and achievements in their own lifetimes and locations. Not all kinds of knowledge and beliefs were of equal standing in their historical contexts, nor was their certainty the same. Studying contexts without delving deeply into content is unacceptable for a serious history of science. That is why there are very few chapters in this handbook that focus exclusively on either context or content.

The second set of research themes defining the book's structure concerns questions of how scientific practices and their outcomes became organized, acquired stability and thus survived (Parts II and III). At the core of these questions is the issue of institutionalization. Following the sociological definition of an institution as an organization or other formal social structure that governs a field of action, older history of science often identified institutions with forms of social organization. These were thought to coalesce into buildings, collections, statutes, memberships, certificates and professional titles; prominent examples were schools, libraries, universities, academies, museums, courts, societies, conferences and journals. In the case of Islamicate societies before 1700, this list was much more limited. It included only schools, libraries and courts.

Here, we take a more general view of institutions, as stable patterns of behavior that define, govern or constrain action. The chapters on the emergence of new scholarly languages in the case of New Persian and Ottoman Turkish exemplify this changed understanding, because a new language regulates the participation of new groups of people in established fields of knowledge. Yet we do not treat the entire range of scholarly works in these two new languages, because the point at stake is the formation of new patterns of behavior, not the measurement of a new field of knowledge or the enumeration of its representatives. The same approach applies to the other chapters in Parts II and III, even if the breadth of questions discussed and their focal points vary.

The materiality of scholarship (Part IV) is often discussed in fields outside the history of science, in particular in codicology, art history and the analysis of individual objects using present-day science and technology. In the astral sciences, where material objects occupy an important place in research, it is not the actual materiality of the instruments but their content that has been the focus of investigation. The contributors to the handbook aimed to combine these different aspects. They explain which materials were relevant for which scholarly professions, and how the instruments specific to an individual science were made. They describe different types of instruments, known either from excavations and long-term collecting or from descriptions and depictions in texts. And they discuss what else can be known about the makers or users of such material objects. Some of the authors stress how limited our knowledge about these matters is today and how constrained it is likely to remain in the future. Even so, the inclusion of those chapters is necessary, because intellectual activities and their specific practices are always situated within social and material environments that need to be investigated by whatever means available.

Geographies of knowledge (Part V) raise some of the central historiographical questions mentioned earlier, in particular those of locality and context. The issue of regionality versus universality within the Islamicate world is answered by one contributor to the handbook for some of the mathematical sciences, privileging universality. While acknowledging the importance of specific practices, the claim is made that results and methods were shared all over the Islamicate world within a universally valid frame of disciplines and problems.

Other authors privilege regionality, both in the case of institutions and for themes and methods. Regionality is a concept that has been established for several decades in modern research practices, especially in Spain, North Africa and Turkey. More recently, this orientation was also fostered in Iran and some of the successor states of the Soviet Union. It helped translate current political and academic interests into territorial boundaries for the study of the past. The advantage of such a combination of contemporary and historical interests lies in enhancing the focus on regional and interregional patterns of intellectual and personal mobility and the formation of scholarly communities. Other trends toward regionalization, and thus toward greater historical specification, resulted in the second half of the 20th century from the discovery of thematic and professional specializations in Egypt, Syria, Iran or Yemen, such as the development of non-Ptolemaic planetary models or the emergence of the profession of the timekeeper (*muwaqqit*).

In the handbook, two of these regionalized focal points in modern research activities, al-Andalus and the Maghrib, were chosen due to the breadth and depth of the work carried out on them for more than half a century. A complementary line of research is the study of boundary zones and neighbors South of North Africa, north of al-Andalus, east of the Iberian Peninsula, and east and northeast of the Indian subcontinent. This approach reflects the multiplicity of communities and their cultures, languages and knowledge systems in Islamicate societies and the need for more attention to minority groups within particular societies or during processes of change. A region where those research desiderata are particularly glaring is the Indian subcontinent – in the history of science, an as yet understudied complex of different communities, languages and scholarly practices in diverse environments.

The handbook's concluding part (Part VI) takes up traditional questions and value statements about the impact of conflicts on the sciences in Islamicate societies, combining them with more recent questions about cross-cultural encounters after the 7th/13th century. This section reveals how profoundly academic views of the sciences and their fates in Islamicate societies have changed since the last survey books were published (Rashed and Morelon 1996; Lindberg and Shank 2013). At the same time, it shows how much my own views about what should be included in a full-fledged history of the sciences in those societies and communities differ from the views of some of my senior colleagues and indeed colleagues of my own generation. Whether or not we currently agree, I am all the more truly grateful to everyone who directly or indirectly worked with me and helped bring this book to fruition.

Note

1 Consolidated bibliography.

PART I

Late antiquity, translating, and the formation of the sciences in Islamicate Polities (1st BH–7th/5th–13th centuries)

I.1

TRANSLATION AS AN ENDURING AND WIDESPREAD CULTURAL PRACTICE

Sonja Brentjes

In this chapter, as in most other parts of this handbook, we take translation, not science, as the primary category, when discussing linguistic translations of scientific texts. This means we do not limit our discussion to the translation of scientific texts. We include all groups of people who participated in translation in Islamicate societies. We consider their themes, goals, practices and contexts and examine their linkages as well as their differences (Chapters I.2.–I.4). We accept all translation activities undertaken in societies under Muslim rule as legitimate parts of a history of translation in Islamicate societies. In order to gain a more nuanced understanding of such activities, we also pay attention to previous as well as parallel translation efforts in other than Islamicate societies (Chapters I.2., I.16., V.12.–14., VI.5).

I.1.1 The multiplicity and scope of translation activities in Islamicate societies

The story of translations of scientific, medical, philosophical and occult texts is usually divided in four groups: (1) translations from Greek into Syriac and/or Arabic; (2) translations from Syriac into Arabic; (3) translations from Middle Persian into Arabic; and (4) translations from Sanskrit into Arabic. The early Abbasid caliphate, from the 1st/7th to the first half of the 4th/10th centuries (although other translations of this sort were also made later), is seen as the time frame for these translations. The bulk of Greek mathematical, astronomical, astrological, medical and philosophical texts was translated in this time into Arabic and to a lesser degree into Syriac, though not all texts known today were rendered in the two Semitic languages. For instance, the lack of major works by Archimedes (d. 212 BCE) and many works of Plato (d. 348–7 BCE) may reflect their inaccessibility, due to their absence from teaching in late antiquity (Archimedes) and the shift among Syriac scholars from Plato's texts to those of Pseudo-Dionysius, the Areopagite (early 6th century) (Watt 2013). Translations of Middle Persian texts on history, literature, astrology and medicine into Arabic seem to have primarily occurred in the second half of the 2nd/8th century. But the situation here is less clear than in the case of translations from Greek (Chapters I. 2. and I.16), because most of them appear to be lost. Moreover, the textual transmission of extant translations from Middle Persian is complicated, contradictory and often fragmentary. Contemporaneously, some astronomical, astrological and medical works were translated from Sanskrit into Arabic. Almost all those translations were undertaken in Baghdad (Chapters I.3.–I.4).

DOI: 10.4324/9781315170718-3

Historical accounts of translations from Greek into Syriac in Islamicate societies often consider only the 3rd/9th and 4th/10th centuries. However, translations of this sort also occurred in the 1st/7th century (Watt 2013, 37–8; Chapter I.2), although, because they took place within Syriac monastic circles, these activities are often ignored in discussions about translations in societies under Muslim rule. Equally, because of their subject matter, translations of religious and historical literature from Greek into Syriac or Arabic, Syriac into Arabic or Arabic into Greek, undertaken since the 1st/7th century are usually excluded from reflections on translation activities in science, medicine or philosophy. These two approaches overlook the facts that all those translations happened under Muslim rule and thus belong rightfully to the history of translation activities in Islamicate societies and that some translators of religious and historical literature also translated philosophical or medical works (Griffith 2013, 127).

Translations of scientific, philosophical, medical or occult texts from Arabic into New Persian and vice versa probably began in the early 5th/11th century (Chapter II.1) and became more numerous from the 7th/13th century onward. They were of particular importance in Iran, Central Asia and Anatolia (Chapter V.6., V.8., V.10). Parallel to this development, texts also began to be written in Persian and then translated into Arabic. Translations of Mongol diplomatic and legal material as well as Chinese medical, agricultural and other texts took place in Iran, mostly in Tabriz, during the Ilkhanid dynasty (r. 654–736/1256–1336). Those currently known to us were organized by the grand vizier Rashīd al-Dīn Hamadānī (d. 718/1318) in cooperation with the Mongol court official Bolad ([d. 712/1313]; Allsen 1996).

The Mediterranean was also a zone in constant flux, with a never-ending need for translations, because of the frequent contacts among numerous different states and communities, including commercial, military and diplomatic interactions, as well as the repeatedly shifting subjugation of regions and islands to Christian and Muslim rule, together with pilgrimage, piracy and enforced or voluntary migration. For example, maps and atlases were compiled from sources in Arabic, Latin, Greek, Ottoman Turkish, Catalan, several Italian dialects, Castilian, Dutch, French and possibly English between the 8th/14th and the 13th/19th centuries in Venice, Genoa, Istanbul, Tangier (today Morocco), Tunis, Sfax and Qayrawan (today Tunisia), Gallipoli (today Turkey), Cairo, on the island of Crete and elsewhere (Goodrich 1990; Herrera Casais 2008; Brentjes 2013; Chapter II.8). These maps were the results of many different acts of translation, including the transfer of words and sentences from one language into another, shifts between measurement systems and their representations, reformulations of cultural symbols like banners, flags or religious markers, and the integration and adaptation of nautical signs. Different types of decoration were exchanged, transformed and restructured. The makers of these sources worked with multiple kinds of geographical and political classifications expressing hostilities or alliances between different rulers or communities of faith. This broad set of cross-cultural interactions repeatedly led to shifting boundaries between types of geographical and nautical representations and the inclusion of new fields of knowledge into mapmaking (Brentjes 2009; Brentjes *et al.* 2014).

Translations among Arabic, Persian, Sanskrit, Hindi, Tamil and other Indian languages on the one hand and in interaction with Portuguese, French, Latin and English, on the other, are activities that characterize Islamicate societies and their neighbors and European colonial conquerors in South Asia as late as the 13th/19th century. They include translations of not only medical, mathematical, astronomical, astrological and philosophical texts, but also religious books and works on instruments and technical devices, including weaponry. Books on practical accounting, history, and administrative, political, diplomatic and institutional decision-making also crossed linguistic boundaries (Chapters V.6., V.8., V.10., VI.7).

The Ottoman realm was an expansive multiethnic, multi-confessional and multilingual state. In contrast to many other settings, the Ottoman state itself supported translation activities, in particular after the conquest of Constantinople in 857/1453 (Mavroudi 2013; Pinto 2011) and then in a second wave from the 10th/16th century onward (Hagen 2003[CB];[1] Brentjes 2014; Günergun 2011; Chapter II.8). Other translations took place in circles linked to the court but were not the outcome of planned state activities. A third kind of translation was carried out in scholarly circles. But as in other periods and regions, there are also translations that cannot be linked to any specific context and group of people. As a whole, Ottoman translations included books on the New World, Paracelsian medicine, Ptolemy's *Geography*, books on logarithms, astrology, astronomical tables and instruments, philosophy, fortress construction and, time and again, atlases (Goodrich 1990; Hagen 2003; Günergun 2011; İhsanoğlu *et al.* 1997[CB], 2000[CB]; Özervarlı 2011; Chapter II.8). Translation activities contributed significantly to the emergence of Ottoman Turkish as a scientific language (Chapter II.2), providing audiences outside Arabic-speaking provinces with additional opportunities for accessing classical scholarly knowledge on nature, the heavens, the earth and the human body (Chapter V.9). Slightly later, early modern and modern scientific, medical, technical and philosophical works from various Christian countries in Europe began to be imported into the Ottoman Empire. While the main translators were Jewish, they also included some Muslim emigrants from Christian Spain and Portugal, who were persecuted and expelled in several waves between 1492 and 1613, as well as local Christians and converts from Catholic or Protestant countries in Europe (García-Arenal 2009; İhsanoğlu *et al.* 1997; Rothman 2011). Scholarly travelers, missionaries and ambassadors from Italy, France, England, Hungary and German lands brought books on the new European sciences and instruments to the Ottoman capital, as well as to trading centers such as Aleppo and to cities, like Cairo, famous for their many different cultures (Chapter VI.5).

Many translation activities occurred in connection with patronage by rulers, viziers or courtiers and in cooperation with speakers of different languages. During the formative centuries of Islamicate societies, such cooperative activities connected Christians across three churches, two Syriac and one Greek, despite their Christological differences, as well as Sabians, Zoroastrians, converts to Islam and Arabs, together with people of other linguistic and religious backgrounds. From the 160s/780s to the 200s/820s, a few Buddhists and Hindus also participated (Ibn al-Nadīm 1970[CB], 2014[CB]). In other times and regions, Muslims cooperated with Hindus, Jews interacted with Muslims and Christians or Muslims worked with Christians or Christian converts from both Catholic or Protestant countries in Europe (Pingree 1975; Hagen 2003; Sobers-Khan 2013). While this cross-community interaction among translators, scholars and patrons is usually highlighted for the early Abbasid period during the 2nd/8th and 3rd/9th centuries, later periods of translating and their cross-community interactions have rarely been discussed by historians of science and philosophy, although important work has recently appeared on Mughal India and South Asian medicine (Truschke 2016; Speziale 2018, 2019; Chapter V.8).

Looking at the wider picture we have sketched, translation activities shaped intellectual life in Islamicate societies in a multitude of forms and over the entire period treated in this book. They were undertaken in various communities and, depending on time and place, in at least the 40 languages listed here: Anatolian Turkish, Arabic, the Berber languages, Bugis, Castilian, Catalan, Chagatay, Chinese, Dutch, English, French, Fulani, German, Greek, Hausa, Hebrew, Hindi, Italian, Javanese, Kanembu, Kanuri, Kipčak, Kiswaheli, Latin, Malay, Middle Persian, Mongol, New Persian, Ottoman Turkish, Polish, Portuguese, Russian, Sanskrit, Serbian, Sundanese, Syriac, Tamil, Tibetan, Urdu and Yoruba (Bondarev 2014; Morgan 2012; Ricci 2010, 10–11; Ricci and van der Putten 2014; Haw 2018; Chapters I.2–I.4, II.1–II.2, II.8, II.10, III.2, V.6, V.8–V.14, VI.1, VI.5, VI.7). These translation activities covered religious literature, diplomatic letters, historical chronicles, literature, atlases and maps, military and technical handbooks, dictionaries, grammars, poetry and, time and again, scientific, medical, philosophical and occult texts.

As the languages listed earlier show, these activities took place across the whole Islamicate world from the Atlantic to the Pacific Ocean.

I.1.2 The notion 'translation movement'

Despite the richness of translation activities in Islamicate societies over more than a millennium, translations in the first two Abbasid centuries have attracted by far the greatest attention (see in particular Gutas 1998[CB]; Saliba 2007[CB]). The activities of this time have received their own label, the *translation movement*, and their own focal point, translations from Greek into Arabic, either directly or via Syriac. But many considerations speak against a continued use of this label, in addition to its almost exclusive focus on Greek to Arabic translations (via Syriac or not) of scientific, medical and philosophical works. In fact, only a few Abbasid caliphs and Muslim courtiers appear as major patrons of these activities. Additionally, this approach neglects occult subjects and anonymous as well as fragmentary translations, and it also downplays the importance of translations from Middle Persian for the mathematical sciences and from Syriac for philosophy. It has privileged the imagery of actors and institutions that favored rationalism while ignoring the many stories about belief in demons, miracles, hidden treasure troves or ancient sages who had saved knowledge from the time before the Flood – stories frequently told by later historical actors to legitimate and situate those sciences (van Koningsveld 1998; Pingree 2001, 2002–2003; Pingree and Steele 2014; Van Bladel 2009[CB]; Yücesoy 2009).

The problematic nature of this historiography becomes clear when one looks for serious efforts to define the supposed translation movement's properties beyond mere descriptions of examples. While there are many and diverse discussions about the historical events subsumed under the label, no historian of science, medicine or philosophy has tried to explain why such a label might be adequate and appropriate for theorizing about the manifold actual translations that took place in those centuries and what its consequences for our views on those activities might be. The most succinct statement about the properties thought to characterize the concept is A. I. Sabra's (1924–2013) formulation of 1987:

> The translation movement in the early ʿAbbāsid period was not a sideline affair conducted by a few individuals working in the dark under threat of being found out and thwarted. It was a massive movement which took place in broad daylight under the protection and active patronage of the ʿAbbāsid rulers. Indeed, in terms of intensity, scope, concentration and concertedness, it had had no precedent in the history of the Middle East or of the world. Large libraries for books on the "philosophical sciences" (*ḥikma*, or *al-ʿulūm al-ḥikmiyya*) were created, embassies were sent out in search of Greek manuscripts, and scholars (Christians and Sabians) were employed to perform the task of translation, all of this at the instigation and with the financial and moral support of the ʿAbbāsid caliphs,
>
> *(Sabra 1987[CB], 228).*

Numerous papers and a few books have been written since then that deal in different ways with the processes labeled 'translation movement' (Gutas 1998; Pingree 2000; Pingree 2001–2002; Rashed 2006, 2011, 2015; Sabra 1996; Saliba 1998, 2002, 2007; Ragab 2017; Vagelpohl 2009; Van Bladel 2014[CB]; Watt 2013; Yücesoy 2009). Several of them agree with the basic positions expressed in Sabra's quote, in particular his emphasis on Greco-Arabic translations. Others have challenged this focus on classical and late antique works directly (Pingree 2005; Yücesoy 2009) or indirectly (Gutas 1998). In general, however, most summaries of these translation activities tend to glorify, streamline, institutionalize and modernize the same set of events.

In fact, Sabra's claim that the Abbasid translation movement had "no precedent in the history of the Middle East or of the world" is historically false. Translations of Buddhist literature into Chinese and other East and Southeast Asian languages started earlier and were at least equal in cultural importance. By the same token, earlier translations of Sumerian knowledge into Akkadian had a long-term and long-distance impact on numerous fields of knowledge across the ancient Near East and beyond to Egypt, Greece, Rome, Iran, South and Central Asia (Xuanmin and Yunajian 2009; Crisostomo 2019).

The belief in a translation movement limited in time and place has created obstacles to investigating the existence of translation activities that were carried out in other circles and sponsored by other patrons with other goals. A few examples highlighting the existence of various kinds of translation projects from the 1st/7th to the early 5th/11th centuries are discussed in Chapters I.3.–I.4. In 1996, Sabra proposed considering the entire 3rd/9th century as a phase of diverse projects by diverse individuals (Sabra 1996, 655). Along the same lines, Morelon appraised Thābit ibn Qurra's (d. 288/901) astronomical works as carrying his very personal signature (Thābit ibn Qurra 1987[CB]). Sidoli and Isahaya underlined Thābit's active reworking of Euclid's text of the *Data* while translating it (Thābit ibn Qurra 2018[CB]). Rashed set the studies of the Banū Mūsā (3rd/9th century) apart from those of al-Kindī ([d. *c.* 256/870]; Rashed 2012, 2; Chapter I.3). Adamson proposed understanding the large group of translations made by Christian translators and Christian and Jewish scribes for al-Kindī as the result of al-Kindī's intellectual development as an autodidact (Adamson 2006, 10–12). And Quṣṭā ibn Lūqā's (d. *c.* 299/912) translations of mathematical texts for Aḥmad, a son of Caliph al-Muʿtaṣim bi-Llāh (r. 218–227/833–842), can be seen as a late product of al-Kindī's years as Aḥmad's teacher of philosophy and the mathematical sciences. All these cases undercut the idea of a single movement caused either by caliphal policies (Gutas 1998) or by the non-Arab, non-Muslim professional scribes as a response to their alleged loss of positions in the caliphal administration (Saliba 2007, 49–50, 52–64, 71 etc.).[2] To call these many different translation projects a movement requires evidence that they were somehow coordinated, rather than reflecting the desires, needs or interests of individuals or smaller circles. Such evidence is, however, either missing completely or unconvincing.

Earlier examples of individual translation projects can easily be found in the activities of courtiers and scholars of the 2nd/8th century. Ibn al-Muqaffaʿ (d. 142/759) is said to have translated texts on logic, politics and history. The Barmakid family's patronage of translations in astronomy, medicine and perhaps geometry provide other examples. Similarly, patrons like Ṭāhir ibn Ḥusayn (d. 207/822) sponsored translations of Pseudo-Aristotle's *On Virtues and Vices* and Nicomachus' (d. *c.* 120) *Introduction to Arithmetic*. The former belongs to a small set of texts that had been translated from Greek into Syriac as well as Armenian in the 6th and 7th centuries. This set also included Pseudo-Aristotle's *On the Cosmos*, Porphyry's (d. *c.* 305) *Introduction* [to the *Categories*] and Alexander of Aphrodisias' (late 2nd–early 3rd centuries) *On the Principles of the Universe According to the Opinion of Aristotle, the Philosopher* (*Maqālāt al-Iskandar al-Afrūdīsī fī l-qawl fī Mabādiʾ al-Kull bi-ḥasab raʾy Arisṭāṭālīs al-faylasūf*), now extant only in Arabic (King 2010; Alander of Aphrodisias 2001). A copy of some of those Syriac translations and adaptations is preserved from the 1st/7th century (MS London, British Library, Add. 14568; King 2010).

The Arabic rendering of Alexander of Aphrodisias's work on the universe was part of the body of translations made for al-Kindī and was connected to al-Kindī's own cosmological works in a manner that Fazzo and Wiesiner describe as a back and forth between translation and new composition (Fazzo and Wiesiner 1993). Before we consider amalgamating all these individual projects into a single societal process worthy of the label *movement* in a meaningful, nonreductionist manner, we need studies of scholarly and patronage networks that show who interacted in what manner with whom and pursued which kinds of goals with which kinds of means – studies, which are currently lacking. But even if future scholarship shows some coordination among some of these networks and actors, that will still leave many translations for which we do not know the translators, patrons, texts and circumstances. Hence, the existence of a *movement* in contrast to numerous separate activities will always be difficult to prove.

Beyond the problem of whether we are justified in thinking of the translations of the early Abbasid period in terms of a *movement*, other issues need to be given attention. In identifying the beginnings of the *movement* with the translation of Middle Persian texts at the court of Caliph al-Manṣūr (r. 136–158/754–775) and their possible anchoring in ideological practices of the Sasanian dynasty (r. 224–651) that ended more than a century earlier, Gutas (1998) operates on the basis of both explicit claims and silent assumptions. Due to space limitations, only one explicit claim can be discussed here. This concerns the existence of a royal ideology that promoted cross-cultural translation as a primordial cultural good in order to repair the damage inflicted by Alexander of Macedonia (r. 336–323 BCE) on knowledge that had been available in the Achaemenid Empire (550–330 BCE), which he destroyed (Gutas 1998, 34–45). The source for Gutas's belief in a (late) Sasanian ideology of this sort is the Middle Persian work *Dēnkard*, which was compiled and revised by Zoroastrian priests in the 3rd/9th century (*Dēnkard* 1874–1928; Rezania 2017, 337–8). However, the status of the *Dēnkard* as evidence for Gutas's thesis is problematic. Recent debates among historians of Sasanian Iran and Middle Persian literature suggest that not all of the *Dēnkard* reproduces pre-Islamic material: some of it reflects intellectual and ideological developments around the Abbasid court at the time of its compilation (Chapter I.16). For example, it includes answers to questions by Zoroastrian and Christian correspondents with one of the Zoroastrian compilers (Rezania 2017, 338). Rejecting earlier interpretations of the *Dēnkard* as a Zoroastrian encyclopedia, Rezania understands it as an apologetic work designed to serve interreligious polemics against Muslim theologians (Rezania 2017, 339). In contrast to Rezania, de Jong (2015) continues to accept the part of the *Dēnkard* (Book IV) reporting on the written codification of Zoroastrian religion as a product of the early 6th century (de Jong 2015, 99). On the other hand, he describes the relationship between the late Sasanian court and the Zoroastrian priests as repeatedly changing in response to external and internal religious, political and social challenges, with profound consequences for Zoroastrianism itself (de Jong 2015, 96–100, see also p. 94 for his interpretation of the negative Zoroastrian attitude toward Alexander). In other words, the existence of a single, clearly formulated ideology of cross-cultural translation under late Sasanian rule is at the very least doubtful.

I.1.3 Marginalized translation activities and results

The narrative of the translation movement has either marginalized or ignored whole classes of translations into Arabic and other languages. These include writings on occult matters, anonymous translations, translations of fragments, translations of historical information and translations of religious teachings. All these categories need to be considered important parts of scholarly, administrative, diplomatic and commercial life under Muslim rule.

Additionally, there are other translation activities that are rarely or never accounted for in the discussions on the translation movement. Scholarly communities established in late antiquity and later governed by Muslim rulers continued to be engaged in various kinds of translation. Christian clerics in Baghdad, Constantinople and several monasteries in the Abbasid realm exchanged historical and political reports, translating them for each other from Syriac or Greek into Arabic and vice versa

(Hoyland 1991, 213, 219–23, 226). Knowledge about the Qur'ān was shared with Syriac writers of history. Muslim converts who knew Syriac in addition to Arabic were famous for their knowledge of apocryphal literature such as the *Books of Daniel* and provided an important body of material for astrometeorological and magical literature in Arabic (Hoyland 1991, 226–7; Sezgin 1979, 7, 313–18). Christians and Jews of various denominations translated religious texts from Hebrew, Aramaic, Syriac, Coptic and Greek in territories under Muslim rule (Griffith 2013, 1, 127).

A further limitation of the discussions on the translation movement consists in the silent assumption that translators of ancient Greek texts, among them Ḥunayn ibn Isḥāq (d. 259/873), limited their work to medicine, philosophy or the mathematical sciences. But Ḥunayn ibn Isḥāq also translated the Pseudo-Aristotelian work on physiognomy and is believed to have translated or revised astrological texts or works with astrological material (Sezgin 1979, 7, 134 and 328; Chapter I.14). Indeed, Arabic historical and scientific sources considered it highly plausible that Ḥunayn had translated treatises on occult and astrological themes (Sezgin 1979, 7, 65). Weisser's studies of two such texts attributed to Bālīnās or Pseudo-Apollonius of Tyana confirm that Ḥunayn indeed participated in translation activity of this sort and quoted Pseudo-Apollonius in his own works (Sezgin 1979, 7, 66; Weisser 1980, 27), leading me to ask whether the same is true for other translators of the 3rd/9th century.

Translations of occult and astrological works pose a series of problems that apply less often or not at all to other fields of knowledge. The anonymity of most of their translators makes dating and contextualizing the works difficult. Some were ascribed to famous ancient scholars like Pythagoras, Hippocrates, Plato, Aristoteles or Ptolemy or to prophets like Zoroaster, Seth, Esra, Daniel and Hermes or to kings like Alexander of Macedonia, who clearly did not produce them (this means the works are *pseudepigraphs*). Some of their translations into Arabic have already been attributed to the 1st/7th, mid-2nd/8th or later 2nd/8th century. There are claims that translations into Middle Persian on royal command happened already in the 3rd century. Fantastic stories of genesis and placement appear in a number of those texts. And, finally, most of the extant copies of such works have received very scant academic study (Sezgin 1971, 4, 111–12, 1974, 7, 32–67, 313–18).

Some of these problems have a history of their own that long predates Islamicate societies. Their contradictory and incomplete histories, as well as often unreliable dating, complicate attempts to relate their contexts to concrete historical circumstances in Umayyad and early Abbasid times, meaning that our understanding of the translation activities between the second half of the 1st/7th and the first half of the 4th/10th centuries will always remain severely limited.

Notes

1 Consolidated bibliography.

2 Saliba's so-called alternative narrative suffers from much more serious flaws, which cannot be discussed here in detail. A few brief examples can highlight its problems. Saliba starts from stories told in historical sources of the later 3rd/9th and the 4th/10th centuries taking their claims at face value. But he does not merely abstain from applying even the classical methods of narrative analysis, let alone more recent ones, he also misstates the scholarly identities of historical actors and trivializes social processes as being in principle everywhere and always of the same kind (for some examples, see Saliba 2007, 56–7, 59–60). He calls Ibn al-Nadīm "the foremost theoretician of the early Islamic period". He claims that according to Ḥunayn ibn Isḥāq all Galenic works were translated for bureaucrats. But Ḥunayn either describes his clients as physicians or provides names of one Abbasid prince, three courtiers without specific responsibilities and four administrators of mostly very high ranks (vizier, governor; Saliba 2007, 30, 71; Ḥunayn ibn Isḥāq 1925, 3–10, 12–14, 16–38, 40–3). Saliba (2007, 49) also wishes us to believe that Abū Sahl al-Nawbakht told his story about the origin and subsequent history of the astral sciences "in order to secure a job for him and his descendants after him." A few pages later, he alludes to Abū Sahl as one of the allegedly "self-serving astrologers who were struggling to keep their position at the Abbāsid court" (Saliba 2007, 52). But we know nothing about the circumstances

of how Abū Sahl became one of Caliph al-Manṣūr's Persian court astrologers before or in 145/762, nor do we possess any information about him having been threatened with losing his position. Moreover, based on Ibn al-Nadīm's work, it is usually believed that he translated the Middle Persian book that contains this story into Arabic after he had been hired by the caliphal administration. There are many more such conflicts between Saliba's claims and the data found in the historical sources. Finally, Saliba explicitly admits that his interpretations of the historical sources are based on belief: "What could those communities do in response to those events? How could they awake from their first shock and try to reclaim their previous positions in the corridors of government? I think they did what most communities would do under such circumstances: go back and try to monopolize the government positions by other means. One such mean was to acquire the more advanced specializations in the very sciences that the government badly needed so that they would become once more indispensable to the running of the government. . . . In order to be able to compete with the new occupants of the *dīwāns*, and go back to monopolize the high positions of government, members of these communities of bureaucrats had to make use of their knowledge of both the Greek language and the elementary sciences that they used in the *dīwān*, and try to educate themselves or their children in the more advanced sciences, to which their elementary sciences referred for higher precisions and sophistication. They did all that in order to be able to deploy that new information and win their previous positions at the *dīwān*. Now that they had lost their jobs, they had an excellent motivation to go to Ptolemy's *Almagest*, that they knew only by name before, when they had no need for it, and to which they were referred by their co-religionists. Under the new conditions, and with the pain of unemployment, the bureaucratic communities would go back to teach their children and their co-religionists and to urge them to acquire the more advanced sciences about which they were well informed by the Greek as well as the Persian classical sources. And since Arabic had by then become the language of competition, they were obliged to demonstrate their competence both in the new bureaucratic language as well as in the sciences of the higher order. Again, all those difficulties had to be re-negotiated before they could reestablish the monopoly that they once had in the *dīwān*" (Saliba 2007, 59–61). Unfortunately, none of the men he names a moment later are known as sons or grandsons of bureaucratic families but as sons of doctors, apothecaries or perhaps astrologers (this third background is not attested in the historical sources; Saliba 2007, 61). They acquired their knowledge of "the sciences of the higher order" in their hometowns in Iran before they came to Baghdad or learned it from their medical co-religionists in Baghdad or abroad.

Bibliography

Sources

Alexander of Aphrodisias. ed. and tr. Genequand, J. 2001. *Alexander of Aphrodisias on the Cosmos. Arabic Text with English Translation, Introduction and Commentary* (Islamic Philosophy, Theology and Science. Texts and Studies 44) Leiden, Boston and Köln: Brill.

Dēnkard. ed. and tr. Sanjana, P. 1874–1928. *The Dînkard. The Original Pahlavi Text; the Same Transliterated in Zend Characters; Translations of the Text in the Gujarati and English Languages; a Commentary and a Glossary of Select Terms*. Vols. I–XIX. Bombay: Kegan Paul, Trench, Trübner & Co.

Euclid. ed., tr. and ann. Besthorn, R. O. and Heiberg, J. 1893. *Codex Leidensis 399,1. Euclidis Elementa ex interpretatione al-Hadschdschadschii cum commentariis al-Narizii*. Partis 1 Fasciculus 1. Copenhagen: In Libraria Gyldendaliana.

Ḥunayn ibn Isḥāq. ed. and tr. Bergsträsser, G. 1925. *Über die syrischen und arabischen Galen-Übersetzungen* [About the Syriac and Arabic Translations of Galen]. Abhandlungen für die Kunde des Morgenlandes 17.2.

Manuscript

MS London, British Library, Add. 14568.

Research literature

Adamson, P. 2006. *Al-Kindī*. Oxford: Oxford University Press.

Allsen, T. T. 1996. "Biography of a Cultural Broker: Bolad Ch'eng-Hsiang in China and Iran," in Raby, J. and Fitzherbert, T., eds. *The Court of the Ilkhans 1290–1340*. Oxford: Oxford University Press, 7–22.

Bondarev, D. 2014. "Old Kanembu and Kanuri in Arabic Script: Phonology Through Graphic System," in Mumin, M. and Versteegh, K., eds. *The Arabic Script in Africa*. Studies in the Use of a Writing System (Studies in Semitic Languages and Linguistics 71). Leiden and Boston: Brill, 107–40.

Brentjes, S. 2009. "Cartography in Islamic Societies," in Kitchin, R. and Thrift, N., eds. *International Encyclopedia of Human Geography*. 2 vols. Oxford: Elsevier, 1: 414–27.

Brentjes, S. 2013. "Giacomo Gastaldi's Maps of Anatolia: The Evolution of a Shared Venetian-Ottoman Cultural Space?" in Contadini, A. and Norton, C., eds. *The Renaissance and the Ottoman World*. Farnham, Surrey: Ashgate, 123–41.

Brentjes, S. 2014. "On Abū Bakr al-Dimashqī's (d. 1691) Hemispheric Map of the New World and the Representation of the Seas in His Maps of the World and the Continents," in Couto, D., Günergun, F. and Pedani, M. P., eds. *Seapower, Technology and Trade. Studies in Turkish Maritime History*. Istanbul: Piri Reis University Publications, 398–411.

Brentjes, S., Fidora, A. and Tischler, M. M. 2014. "Towards a New Approach to Medieval Cross-Cultural Exchanges," *Journal of Transcultural Medieval Studies* 1.1: 9–50.

Crisostomo, C. J. 2019. *Translation as Scholarship: Language, Writing, and Bilingual Education in Ancient Babylonia* (Studies in Ancient Near Eastern Records 22). Berlin: De Gruyter.

Fazzo, S. and Wiesiner, H. 1993. "Alexander of Aphrodisias in the Kindī-Circle and in al-Kindī's Cosmology," *Arabic Sciences and Philosophy* 3.1: 119–53.

García-Arenal, M. 2009. *Ahmad al-Mansur: The Beginnings of Modern Morocco*. Oxford: Oneworld Publications.

Goodrich, T. 1990. *The Ottoman Turks and the New World. A Study of Tarih-i Hind-i Garbi and Sixteenth-Century Ottoman Americana*. Wiesbaden: Otto Harrassowitz.

Griffith, S. H. 2013. *The Bible in Arabic. The Scriptures of the "People of the Book" in the Language of Islam*. Princeton, NJ: Princeton University Press.

Günergun, F. 2011. "The Ottoman Ambassador's Curiosity Coffer: Eclipse Prediction with De La Hire's 'Machine' Crafted by Bion in Paris," in Günergun, F. and Raina, Dh., eds. *Science Between Europe and Asia* (Boston Studies in the Philosophy of Science 275). New York: Springer, 103–23.

Haw, S. G. 2018. "The Persian Language in Yuan Dynasty China: A Reappraisal." http://www.eastasianhistory.org/39/haw. (Accessed 26 July 2018).

Herrera Casais, M. 2008. "The 1413–14 Sea Chart of Aḥmad al-Ṭanjī," in Calvo, E. *et al.*, eds. *A Shared Legacy: Islamic Science East and West. Homage to Prof. J. M. Millàs Vallicrosa*. Barcelona: Universitat de Barcelona, 283–307.

Hoyland, R. G. 1991. "Arabic, Syriac and Greek Historiography in the First Abbasid Century: An Inquiry into Inter-Cultural Traffic," *Aram* 3.1–2: 211–33.

Jong, A. de. 2015. "Religion and Politics in Pre-Islamic Iran," in Stausberg, M., Vevaina, Y. S.-D. and Tessmann, A., eds. *The Wiley Blackwell Companion to Zoroastrianism*. Oxford: Wiley Blackwell, 85–102.

King, D. 2010. "Alexander of Aphrodisias' On the Principles of the Universe in Its Syriac Adaptation," *Le Muséon* 123: 157–89.

Koningsveld, P. S. van. 1998. "Greek Manuscripts in the Early Abbasid Empire: Fiction and Facts about Their Origin, Translation and Destruction," *Bibliotheca Orientalis* 55.3–4: 345–72.

Mavroudi, M. 2013. "Translators from Greek into Arabic at the Court of Mehmet the Conqueror," in Ödekan, A., Necipoğlu, N. and Akyürek, E., eds. *The Byzantine Court: Source of Power and Culture*. Istanbul: Koç University Press, 195–207.

Morgan, D. O. 2012. "Persian as a Lingua Franca in the Mongol Empire," in Spooner, B. and Hanaway, W. L., eds. *Literacy in the Persianate World: Writing and the Social Order*. Philadelphia: University of Pennsylvania Press, 160–70.

Özervarlı, M. S. 2011. "Yanyalı Esad Efendi's Works on Philosophical Texts as Part of the Ottoman Translation Movement in the Early Eighteenth Century," in Schmidt-Haberkamp, B., ed. *Europa und die Türkei im 18. Jahrhundert/Europe and Turkey in the 18th Century*. Göttingen: V&R unipress, Bonn University Press, 457–72.

Pingree, D. 1975. "Al-Bīrūnī's Knowledge of Sanskrit Astronomical Texts," in Chelkowski, P. J., ed. *The Scholar and the Saint: Studies in Commemoration of Abu'l-Rayhan al-Biruni and Jalal al-Din al-Rumi*. New York: NYU Press, 67–81.

Pingree, D. 2000. "Cultures of Ancient Science: Some Historical Reflections. Hellenophilia Versus the History of Science," in Shank, M. H., ed. *The Scientific Enterprise in Antiquity and the Middle Ages. Readings from Isis*. Chicago: Chicago University Press, 30–9.

Pingree, D. 2001–2002. "From Alexandria to Baghdad to Byzantium: The Transmission of Astrology," *International Journal of the Classical Tradition* 8: 3–37.

Pingree, D. 2002–2003. "The Sabians of Harran and the Classical Tradition," *International Journal of the Classical Tradition* 9: 8–35.

Pingree, D. 2005. "Mashallah's Zoroastrian Historical Astrology," in Oestmann *et al.*, eds. [CB], 95–100.

Pingree, I. and Steele, J., eds. 2014. *Pathways into the Study of Ancient Sciences: Selected Essays by David Pingree*. Philadelphia: American Philosophical Society.

Pinto, K. 2011. "The Maps Are the Message: Mehmet II's Patronage of an 'Ottoman Cluster'," *Imago Mundi* 63.2: 155–79.

Ragab, A. 2017. "'In a Clear Arabic Tongue': Arabic and the Making of a Science-Language Regime," *Isis* 108.3: 612–20.

Rashed, R. 2006. "Greek into Arabic: Transmission and Translation," in Montgomery, J. E., ed. *Arabic Theology, Arabic Philosophy, From the Many to the One: Essays in Celebration of Richard M. Frank* (Orientalia Lovaniensia Analecta 152). Louvain: Peeters, 157–96.

Rashed, R. ed. el-Bizri, N. 2012. *Founding Figures and Commentators in Arabic Mathematics. A History of Arabic Sciences and Mathematics*. Vol. 1. London and New York: Routledge.

Rezania, K. 2017. "The *Dēnkard* Against Its Islamic Discourse," *Der Islam* 94.2: 336–62.

Ricci, R. 2010. "Islamic Literary Networks in South and Southeast Asia," *Journal of Islamic Studies* 21.1: 1–28.

Ricci, R. and van der Putten, J., eds. 2014. *Translation in Asia: Theories, Practices, Histories*. New York and London: Routledge.

Rothman, E. N. 2021. *The Dragoman Renaissance: Diplomatic Interpreters and the Routes of Orientalism*. Ithaca: Cornell University Press.

Sabra, A. I. 1996. "Situating Arabic Sciences: Locality Versus Essence," *Isis* 87: 654–70.

Saliba, G. 1998. *Al-Fikr al-ʿilmī al-ʿarabī; nash'atuhu wa-taṭawwaruhu*. Balamand: Balamand University.

Saliba, G. 2002. "Greek Astronomy and the Medieval Arabic Tradition," *American Scientist* 90.4: 360–7.

Sezgin, F. 1971. *Geschichte des arabischen Schrifttums*. Vol. 4: *Alchimie – Chemie – Botanik – Agrikultur bis ca. 430 H.* Leiden: Brill.

Sezgin, F. 1979. *Geschichte des arabischen Schrifttums*. Vol. 7: *Astrologie – Meteorologie und Verwandtes bis ca. 430 H.* Leiden: Brill.

Sobers-Khan, N. 2013. "East-West Knowledge Transfer in Mughal India," https://britishlibrary.typepad.co.uk/asian-and-african/2013/02/east-west-knowledge-transfer- in-mughal-india.html?_ga=2.94942940.1775040819.1562353033-405453899.1562353033 (Accessed 26 July 2018).

Speziale, F. 2018. *Culture persane et médecine ayurvédique en Asie du Sud*. Leiden: Brill.

Speziale, F. 2019. "Rasāyana and Rasaśāstra in the Persian Medical Culture of South Asia," *History of Science in South Asia* 7: 1–41.

Truschke, A. 2016. *Culture of Encounters: Sanskrit at the Mughal Court*. New York: Columbia University Press.

Vagelpohl, U. 2009. "The Abbasid Translation Movement in Context. Contemporary Voices on Translation," in Nawas, J., ed. *Abbasid Studies II. Occasional Papers of the School of Abbasid Studies*. Leuven, 28 June – 1 July, 2004. Leuven: Peeters, 245–67.

Watt, J. W. 2013. "The Syriac Aristotle between Alexandria and Baghdad," *Journal for Late Antique Religion and Culture* 7: 26–50.

Weisser, U. 1980. *Das "Buch über das Geheimnis der Schöpfung" von Pseudo-Apollonios von Tyana. Ars Medica: Abteilung 3, Arabische Medizin*, Band 2. Berlin and New York: De Gruyter.

Xuanmin, L. and Yunajian, H. 2009. *Translating China* (Topics in Translation). Bristol, North York, ON and Tonawanda, NY: Multilingual Matters.

Yücesoy, H. 2009. "Translation as Self-Consciousness: Ancient Sciences, Antediluvian Wisdom, and the Abbasid Translation Movement," *Journal of World History* 20.4: 523–57.

I.2
MULTIPLE TRANSLATION ACTIVITIES

Arash Zeini, Matteo Martelli, Paola Buzi, Alessandro Bausi and Peter Adamson

This second chapter contextualizes translation activities during the early Abbasid caliphate (132–early 5th century/750–early 11th century) by summarizing our knowledge about such practices elsewhere, as well as their thematic scope before, during and after the period of translations in the Abbasid capital Baghdad. It highlights the multiplicity of translation projects inside and outside Islamicate societies in Asia, Africa and Europe. It aims to clarify that translation was by no means a minor cultural practice in the chosen contexts and thus that translating from Greek into Arabic in Abbasid Baghdad was not as exceptional as has often been claimed. A more nuanced and more plausible historiography of translations carried out in Abbasid Baghdad depends on an improved knowledge not merely of the manifold local translation activities and their participants in the caliphal capital but also on a more extensive and substantial familiarity with translation activities in other regions, cultures, fields of knowledge and periods. This chapter can only offer brief summaries of five cases approached from two perspectives: (1) particular linguistic cultures related at times in substantive ways to translations into or from Arabic and at other times to one or more of the languages of importance during the translations in Abbasid Baghdad and (2) synchronic as well as diachronic comparisons within one specific field of knowledge. The goal is to elucidate in this manner similarities and differences between translation activities in different cultures and communities and thus prepare the ground for the two following chapters that are specifically dedicated to different views on translation activities in the mathematical sciences and medicine in Abbasid Baghdad. First, we talk about translations into Iranian languages (Arash Zeini), Syriac (Matteo Martelli), Coptic (Paola Buzi) and Gəʿəz (Alessandro Bausi), covering territories between what is today Iran, Egypt and Ethiopia. Then, translation activities in philosophy (Peter Adamson) are compared across time and regions in Asia, Africa and Europe. Moreover, these two perspectives are expected to help the reader to reflect on the questions raised in Chapter I.1. from cross-cultural and cross-disciplinary positions, in terms of both historical data and historiographical interpretation. They point out gaps and uncertainties in our knowledge and the obstacles we face in our constructions of interpretive narratives.

I.2.1 Translation in Iranian cultures

In the world of Iranian languages, translation, the act of making a text composed in one language available in another language, is strongly associated with the formation of writing systems, which seem to have been developed as an afterthought to translation. Of the two Old Iranian languages

DOI: 10.4324/9781315170718-4

to have survived the tribulations of history and antiquity, namely, Avestan and Old Persian, only the latter has a contemporary history of writing and translation. For millennia, the Avestan texts were transmitted orally by Zoroastrian priests. Although they constitute the oldest texts composed in an Iranian language, particularly the Old Avestan sections, which are commonly dated to the middle of the second millennium BCE on linguistic grounds, this corpus is only attested through manuscripts dating to the Islamic era. With the expansion of the Achaemenid dominion and the need for imperial administration the situation of writing and translation changes in Fars, the heartland of Achaemenid rule. Started in the year 520 BCE under the supervision of Darius I (r. 522–487 BCE), the Bīsotūn inscription is the oldest securely dated text in any Iranian or Indo-Iranian language (Huyse 2009). This trilingual inscription in Old Persian, Elamite and Babylonian was first carved in Elamite, although not composed by the king in this tongue, and then retranslated into Old Persian (DB §70: *ariyā*) for which a new cuneiform script was developed (Schmitt 2013). Its multilingualism was crucial to the decipherment of the inscription and the cuneiform writing system in general. However, in the Achaemenid era (550–330 BCE) translation was not limited to royal inscriptions but was also used '*ex tempore*' (Blois 2007, 1194–5). Scribes would translate texts spoken in Old Persian into Aramaic or Elamite for the purpose of written communication (letters) or accounting (tablets) across the vast empire. The receiving scribe would read out the letter in Old Persian, or in the vernacular language of the recipient, thus retranslating on the spot from Aramaic. This method of using one language for writing and another for reading out the same text, a phenomenon called 'alloglottography' by Gershevitch (1979), must have been the precursor of the heterographic elements in Middle Iranian writing systems (Henning 1958).

Centuries later, the early Sasanian kings still carved their inscriptions in Middle Persian, Parthian and Greek, although the latter two later fell into disuse as monumental languages. While translation played a pivotal role in the dissemination of Persian imperial culture in antiquity and late antiquity, through their reliance on oral transmission Avestan compositions lack the immediacy of a contemporaneous text of antiquity. The process of the codification of the Avestan compositions into distinct books in late antiquity (the so-called Sasanian Avesta as described in the Middle Persian compendium *Dēnkard*) cannot be traced with exactitude. According to Book Four of the *Dēnkard* (4.17–25), which relates an account of the transmission of the *Avesta*, the redaction of the texts started with the otherwise unknown Avestan king Wishtāsp and lasted until the reign of the Sasanian King Khōsraw I (r. 531–579). But we can discern two strands of scholarly activity by Sasanian priests as part of their codification project. First, there is the translation of the Avestan compositions into the vernacular of the Sasanian Empire and its priests, namely, Middle Persian. Second, there is an engagement with non-Iranian languages and cultures such as Greek, Syriac, Latin and Sanskrit. These internal translation activities resulted in the *Zand*, the Middle Persian translations of the Avestan corpus preserved with commentaries in bilingual manuscripts of Avestan texts. It is a commonly held view that the codification and the translation of the Avestan compositions into Middle Persian were completed by the late Sasanian era, presumably in the 6th century. Incidentally, this translation project resulted in the invention of another script, this time for writing down the Avestan language. The translation project of the Zoroastrian priests, which by no means was limited to the core Avestan texts, later continued into Sanskrit and Gujarati in India and into New Persian, the latter perhaps in Iran.

On the second strand of scholarly activity by Sasanian priests, we read in the *Dēnkard*:

> *Shābuhr*, the king of kings, son of *Ardashīr*, collected again also the writings, which were external to the religion, concerning medicine, astronomy, movement, time, space, essence, fortune, becoming, destruction, form, good conduct and rhetoric and

other arts and skills. These were dispersed in India, Rome and other lands, and he made them reckon again with the *Avesta*.

(For the Middle Persian text, see Madan 1911, 412–13, and Dresden 1966, 511. For a discussion of this passage, see Bailey 1943, 81–7)

According to this tradition, lost intellectual material was reimported into Sasanian–Zoroastrian culture and perhaps incorporated into the *Avesta*. The text is silent on the exact methods of the discovery and reintroduction, but translation could have been involved in this process. However, no translations from non-Iranian material such as Sanskrit or Greek into Middle Persian have come down to us. Most famously, it is claimed that the literary text *Kalīla wa-Dimna* was translated from Sanskrit into Middle Persian by the physician Burzōy (6th century) at the request of Khōsraw I (Blois 1990). This Middle Persian translation would have been the source of the text's extant Syriac and Arabic versions. Despite widespread references to the figure of Burzōy and his Middle Persian translation, such a text has not survived (Pietrăreanu 2018). Likewise, Middle Persian sources for Arabic texts presented as translations remain elusive. Two such texts stand out. First is the *Letter of Tansar*, which is a 7th/13th-century New Persian translation by Ibn Isfandiyār of Ibn al-Muqaffaʿ's (d. 142/759) lost Arabic version of an otherwise unknown Middle Persian letter, ascribed to a Zoroastrian priest presumably active under Ardashīr I ([224–242]; Zeini 2018a, 898). The second is the *Khwadāy Nāmag*, another elusive Middle Persian text, which is reputed to have been translated into Arabic, among others, also by Ibn al-Muqaffaʿ (Hämeen-Anttila 2018; Zeini 2018b, 1599). The assumption of Middle Persian translations of Greek philosophical material, either directly from Greek or via Syriac, cannot be substantiated through any surviving texts, although Zoroastrian exchanges with Greek philosophy seemingly occurred (König 2018; Chapter I.16).

Translation activities, however, were not limited to the Zoroastrian cultural sphere in late antiquity. A sizable corpus of translation literature has survived from Christians and Manichaeans. Syriac Christian texts were translated into Middle Persian (fragments of the Psalms have been found in Central Asia) and substantially into Sogdian, an Eastern Middle Iranian language.[1] Mani (d. 276/7), the founder of Manichaeism, is credited with linguistic as well as cultural translation. The latter is clearly attested in his *Shābuhragān*, a Middle Persian text composed for Shābuhr I (r. 239–270; MacKenzie 1979, 1980), since Mani explains there his religious teachings by leaning on Zoroastrian religious terminology. Mani's disciples later translated his works into Middle Persian, Parthian, Sogdian, Bactrian and other, also non-Iranian, languages.[2]

I.2.2 Syriac translations

Roughly between the 6th and the 7th/13th centuries, a variety of texts pertaining to various fields of ancient science and technology were translated into Syriac, originally the dialect of Aramaic spoken in Edessa, which became the official language of the Christians belonging to different churches settled between southeastern Anatolia, Syria and Mesopotamia: the Church of the West (also known as the Orthodox or Jacobite Church) and the Church of the East (often referred to as the Nestorian Church).

Translations reflect a multifaceted process of cultural transfer of ideas and practices, which featured various strategies of assimilation and readaptation. The inherited knowledge was not only translated into Syriac but was reshaped, digested and discussed in a variety of new writings: commentaries on Aristotle's (384–322 BCE) *Organon*; introductory works to Aristotelian logic; compendia on medicine, such as the Syriac *Book of Medicines*, whose first part reassembles passages from translations of a variety of Galenic works; astronomical treatises, such as Severus

Sēbōkht's (d. 46–47/666–667) treatises on the astrolabe and on constellations, which relied on writings by Ammonius of Alexandria (3rd century) and Ptolemy's (*c.* 100–170) *Handy Tables* (Villey 2014a, 149–90). Regrettably, a full understanding of this process is impaired by the lack of primary sources. On one hand, many Syriac translations – especially those produced during the Abbasid period – have been lost. On the other hand, different available sources are still poorly known or remain unpublished in manuscripts.

In the 6th century, drawing on the experience already gained in translating the Bible and Greek Christian literature, West Syriac scholars produced reader-oriented translations of a wide corpus of Greek scientific and secular works. The Syriac translation of Vindonius Anatolius's collection of agriculture practices (4th century) probably dates to this period (Guignard 2014, 215–42). In line with the curriculum of late antique Alexandrian schools, the study of philosophy (and theology in the new Christian context) was safely grounded in Aristotelian logic. The *Introduction* (~~*Eisagōg Isagoge*~~) by the Syrian philosopher Porphyry (d. *c.* 305) and various treatises belonging to Aristotle's *Organon* were among the first treatises to be translated into Syriac (Brock 1993; Hugonnard-Roche 2004, 23–101). A key figure was the scholar and physician Sergius of Rēsh ʿAynā (d. 536), who both wrote treatises on Aristotle's *Categories* and translated Pseudo-Aristotle's *On the Cosmos*, Alexander of Aphrodisias's (late 2nd–early 3rd centuries) *On the Principles of the Universe* (extant only in Arabic; Chapter I.1) and a wide selection of medical writings by Galen of Pergamon (129–216). The circulation of these translations of philosophical and medical texts in the East and their use in Sasanian Persia, for instance in the famous School of Nisibis (where both philosophy and medicine were taught), are still debated issues. Contemporary with Sergius is certainly Paul the Persian (6th century), who worked at the court of Khōsraw I and wrote a short treatise in Pahlavi about Aristotle's *On Interpretation*, a treatise that is available only in a Syriac version attributed to Severus Sēbōkht.

Between the 1st/7th and the early 2nd/8th century, many Syriac translations were revised, with a shift in translation pactice. The prestige and authority of the Greek language dramatically increased, and translators started aiming at a more literal, word-for-word rendering that could more faithfully mirror the original text (Brock 1983, 12–14). Monasteries played a key role in this process, with their libraries holding important Greek sources. Especially the West Syriac Monastery of Qenneshrē was a vital center of learning, where, under the guidance of Severus Sēbōkht, key translators and commentators of the Aristotelian *Organon* received their education: Jacob of Edessa (d. 89/708), Athanasius of Balad (d. 66/686), and George of the Arabs (d. 105/724). We must note that the late MS Paris, BnF, syr. 346 (dated 1309) preserves Severus's and George's astronomical writings as well as an anonymous Syriac translation of Ptolemy's *Tetrabiblos* (mutilated at the beginning), which is difficult to date and contextualize.

Similar difficulties arise from the study of other anonymous corpora preserved by late manuscripts, in particular the Syriac collections of alchemical treatises. The dates of these translations are as uncertain as the cultural contexts in which they were produced. Late chronicles point to a monastic milieu (such as the monastery of Qarṭmin), in which alchemy was known under the first Abbasids (Martelli 2014, 191–214). In this period, the interest of East Syriac scholars in logic and the sciences clearly emerges from the letters of the *Catholicos* Timothy I ([108–207/727–823]; head of the Church of the East since 163/780). He also refers to the medical school of Gundishapur, from where physicians of the Bukhtishūʿ family are said to have moved to Abbasid Baghdad (Debié 2014, 45–7; Takahashi 2015, 83–4). In the 3rd/9th century, Christian translators – such as Job of Edessa (d. after 217/832) and Ḥunayn ibn Isḥāq (d. 259/873) – took an active part in the so-called Arabic translation movement (Chapter I.1). Often bilingual or trilingual, they produced either Syriac or Arabic translations, depending on the creed or needs of their clients. In some cases, they revised earlier translations; in others, they translated Galenic and philosophical texts afresh. The scope of their work went beyond Aristotelian logic; Peripatetic works on natural philosophy were included as well, along with Theophrastus of Eresus's (*c.* 371–287 BCE) *Meteorology* (perhaps translated by Job of Edessa) and Nicolaus Damascenus's ([4th century?]; Fazzo and Zonta 2008, 682) *Compendium of Aristotelian Philosophy* (perhaps translated by Ḥunayn; Takahashi 2015, 72–3).

Many of these translations are only known through quotations in later authors, for instance, Jacob bar Shakko (d. 638/1241) and Bar Hebraeus (d. 685/1286). Arabic quickly became a reference language in the fields of science and philosophy, and during the so-called Syriac Renaissance (5th-7th/11th-13th centuries), Christian scholars packed their own encyclopedic writings with quotations from the masters of Arabic science and philosophy, such as Ibn Sīnā (d. 428/1037) to name but an undisputed authority.

I.2.3 Translations into Coptic and their contexts (3rd–5th/11th centuries)

In the last decade, increasing attention has been devoted to Greek–Coptic bilingualism in Egypt during late antiquity (Papaconstantinou 2014, 15–21; Camplani 2015, 129–53; Fournet 2009; Fournet 2014, 2, 599–607). There has also been new research on Coptic–Arabic bilingualism during the Middle Ages (Papaconstantinou 2007, 274–99, 2012, 58–76). Both lines of research show the difficulty of tracing precise boundaries in the use, competencies and finalities of these idioms. Unfortunately, only very few late antique booklists and inventories of the assets of monasteries and churches have survived, most of which date from the 6th century. Nevertheless, they are sufficient to confirm the frequent presence of bilingual books in their collections (Otranto 1997, 123–44; Dostálová 1994, 5–19). It has been clearly ascertained that, until the beginning of the 8th century, Greek represented the *high* language, normally used on any occasion outside communication within a local Egyptian community – whether secular, ecclesiastic or in the widest sense 'monastic' – but several aspects of this linguistic cohabitation still deserve careful analysis. However, during the 3rd to 5th centuries, Coptic was a language mainly used for translations. The most important sources on the development, contents and literary genres of Coptic literature are Orlandi (1997, 39–120) and Boud'hors (2012, 224–46). However, a complete and satisfactory history of Coptic literature remains to be written.

Although Coptic Egypt produced literature that, with very few exceptions, was Christian, there was also a sporadic reemergence of the classical tradition, especially in the early period, although sometimes this was unconscious and invariably revisited classical authors in the new Christian perspective. The influence of classical *paideia* on early Christian Egyptian culture is indisputable, to the point that classical texts were copied and classical philosophy was still studied in the schools of Alexandria in the 6th and 7th centuries (Larsen and Rubenson 2018). Roger Bagnall observed that "Antinoopolis maintained in the 4th century its active intellectual life, with instruction in rhetoric and law, medicine and the sciences," and the institution of the *gymnasium* "remained in use in the later fourth century and into the fifth – and perhaps later" (Bagnall 1993, 4).

As stated earlier, until the 5th century, Christian Egypt made use of Coptic almost exclusively as a medium of translation from Greek. The texts transmitted by the Nag Hammadi codices were translations, although this did not exclude a certain freedom in the redaction of the Coptic version of the texts. The Manichean texts found in the Fayyūm and in Kellis were also translations, not to mention the more obvious great number of biblical texts contained in codices that are among the oldest examples of manuscripts ever found. Many of the translations of the first phase of Coptic literature, however, belong to the patristic genre, an important witness to the complex relationship between Greek and Egyptian literary cultures.

While much still has to be understood about the relationship between translated works and original works written directly in Coptic, it is certain that a great deal of Greek literature was never translated into Coptic. This includes also scientific works, which were read, consulted and used directly in Greek, with the exception of medicine, probably due to practical needs. The medical literature was always transmitted anonymously. We should consider, however, that even the apparently simple act of including or excluding some works from the process of translation – and therefore of the creation of a literature in Coptic – is itself a creative act.

Monastic libraries preserve recipes for treatments of different forms of diseases. A parchment manuscript from the White Monastery of Shenoute (MS Naples, BNN, IB 14, fols 6–7), datable to the 10th–11th centuries, for instance, mentions impetigo and itching ailments, recommending a treatment based on swabbing with warm vinegar, attar of roses and water of juice of aloe mixed with the contents of a melon. Leprosy and efflorescence, two skin diseases, as well as diseases of the liver and of the kidneys, are mentioned in the same manuscript.

Although several recipes appear in magical manuscripts, the scientific description of the medical treatment to be applied appears serious and trustable. This is the case, for instance, of eight parchment leaves purchased in Madīnat al-Fayyūm (MS Ann Arbor, Michigan University, inv. 136; Worrell 1935, 1–37) from the 5th or 6th century, written by or for practicing doctors and dealing with various ailments, including diseases of the spleen and constipation, for which purgatives are recommended.

The previously mentioned cases show that the shelves of important monastic libraries also hosted this kind of literature, although not attributed to any specific author and therefore considered of lower dignity compared to other literary genres. Moreover, several other health problems, such as palpitations, sleeplessness and *hemoptysis* (coughing up blood), are dealt with by a great number of documentary papyri and ostraca (inscribed pottery), probably to be used as handbooks by doctors and pharmacists (*Ägyptische Urkunden* 1904, 24–31).

A recent and accurate assessment of the state of current knowledge of Coptic medical texts found in the area of western Thebes confirms that these works are normally translations from Greek, as their terminology attests, that monastic libraries owned codices dedicated to medicine and that in many cases, monasteries were "major providers of medical knowledge and treatment at a local scale" (Richter 2018, 160; see also Crislip 2005 and Schenke 2016).

I.2.4 Translations in late antique Ethiopia (2nd–9th centuries)

The historical area designated Ethiopia (corresponding to the highlands of the two present-day independent states of Eritrea and Ethiopia) is marked since the earliest linguistic phase in the first millennium BCE by phenomena of multilingualism (Bausi 2016a, 2018). While the non-Semitic, Agaw-Cushitic substratum is only documented in place-names, early Semitic settlers left the first written evidence in Ethiopia, with about 200 Sabaean inscriptions (in South Arabian script), which already betray the emergence of a local Ethiopian Semitic language. The earliest epigraphic documents in Gəʿəz (Ethiopic) language and script date from the 2nd/3rd century, during the flourishing of the kingdom of Aksum (1st–7th century), in a historical context strongly marked by the presence of Greek as the international language of culture, trade and diplomacy. The South Arabian script remained in use for reasons of prestige, along with the newly developed Ethiopic script (eventually vocalized) and Greek; this is well attested by the multilingual royal inscriptions from the kingdom of ʿEzānā (4th century), which are written in a bilingual combination of Greek and two versions of Gəʿəz, in Sabaean and Ethiopic script, respectively. These royal multilingual inscriptions presuppose processes of translation, the *Sitz im Leben* (life setting) of which is totally unknown to us.

The earliest evidence of translations of a literary character coincides with the Christianization of the country in the 4th century and with the body of literature that the process of Christianization involved (Bausi 2016a). In keeping with the epigraphic evidence that points to the use of Greek along with Ethiopic, there is a consensus that the early translations of Christian texts were also made from Greek between the 4th and the 6th centuries. This Aksumite corpus consisted of not only biblical, para-biblical, patristic, liturgical, hagiographic, homiletic and monastic but also historical texts, as has recently been shown. It has survived until today through complex, yet poorly understood, processes of transmission. This early corpus also comprises the *Physiologus*, the only Ethiopic text of a scientific genre that can be attributed to this early period of Ethiopic translations (Hommel 1877; Conti Rossini 1951). The only other passages with a scientific content translated in this period are the astronomical sections in the biblical book of Enoch (1 Enoch 72–82). Some parts of this once extensive corpus survived through direct copying, while others came down to us only in a heavily reshaped form, due to later partial retranslations from Arabic into Ethiopic. These Arabic translations are documented from the 12th century onward. Starting in the 10th century, if not earlier, Ethiopic copyists were already speaking a form of Ethiopic different from the language of the texts they were copying. Thus, in the following centuries, Gəʿəz gradually became a written language, exclusively employed for education and religious practice, for deeds and record keeping and for literature.

The analysis of this process is complicated to a great extent by two limiting factors: first, the scarcity of material sources and evidence and, second, the continuous and unceasing reworking of the manuscript and literary tradition over the centuries (Bausi 2016b). In fact, almost no written artifacts survive from the Aksumite period. Among the few remarkable exceptions, besides inscriptions and coins, are two Four-Gospel manuscripts from the monastery of ʾAbbā Garimā (the "Garimā Gospels"), which were carbon-14 dated to late antiquity (4th–7th centuries). Thus, almost all the Gəʿəz translations made in the Aksumite period survive only in much later manuscript copies, which do not contain dated colophons before the 12th/13th centuries.

The effects of these continuous revisions and rearrangements of older texts over the course of time resulted in dramatic changes of the format and arrangement of manuscripts. For some genres, the preserved manuscripts can be considered "corpus organizers"; their content is miscellaneous and is the result of long processes of transmission, organization and redistribution of previous corpora. While these corpora were at a certain point considered to be consistent and unitary, they actually included texts of very different origins in terms of time, place, provenance, source language (Greek, Arabic) and intermediary languages (Greek to Arabic or Greek to Coptic to Arabic or Greek to Syriac to Arabic and so on in various combinations).

The period between the Aksumite and post-Aksumite ages (8th–early 12th centuries) is obscure and scarcely documented. Written culture was probably affected by traumatic political, economic and military events with a consequent loss of whole bodies of texts. Besides material factors, changes in linguistic standards cannot be excluded from a major role. Translated texts, particularly very literal translations or versions based on nonstandard variants of language, became virtually unintelligible, even to the educated clergy. The combined effects of linguistic changes, the decline in literary and material culture, and the long textual transmission by copying manuscripts, all led to extensive textual corruptions. As a consequence, ancient Aksumite translations were either superficially revised or even transformed by complex and sometimes unexpected processes. In some cases, the older translations were abandoned entirely and replaced with completely new translations. A side effect of this process is the emergence of a literary phase that shows a deep discontinuity with the traditional teaching transmitted in ecclesiastical schools up to the present. This process of transmission strongly contributed to marking the Aksumite literature as a "classical literature" of a remote past, deeply rooted in the late antique context.

I.2.5 Philosophical translations

The history of philosophy is to no small extent the history of translations. This is true across the globe – one might think, for example, of the way that translations facilitated the passage of Buddhism from India to China and other Asian cultures. However, in this section, the focus is on those traditions rooted directly in Greek philosophical literature. Such works were produced well beyond the confines of Greece; for instance, Plotinus (204/5–270) wrote in Rome, and the most important school of late ancient Aristotelian commentators was at Alexandria. It is a well-known and well-studied fact that Greek works of philosophy and science were rendered into Latin by intellectuals of the Roman world, notably Cicero (106–143 BCE) and Boethius (d. between 524 and 526), and that Greek texts were also translated into Arabic, with Arabic works, in turn, being translated into Latin around the turn of the 13th century. These two processes of transmission were decisive in shaping the course of medieval philosophy. Indeed, the Arabic–Latin translations and Greek–Latin translations that followed in their wake serve as a convenient dividing point between early and later medieval philosophy in Europe, alongside the concurrent emergence of universities. In the Islamicate world, meanwhile, the work of renowned figures like al-Kindī (d. *c.* 256/870), al-Fārābī (d. 339/950–1), Ibn Sīnā (d. 428/1037), and Ibn Rushd (520–595/1126–1198) was a direct response to the availability of Aristotle and other Hellenic sources in Arabic (Adamson 2006, 2008, 2016; Gutas 2013; Adamson and Di Giovanni 2018; Rudolph *et al.* 2016[CB] [3]). Jewish authors too were involved. At first, Jewish medieval philosophy was always practiced in Arabic, but increasingly, scholars translated from this language into Hebrew, leading to numerous Hebrew versions of Ibn Rushd's commentaries on Aristotle.

However, the Greek–Latin, Greek–Arabic, Arabic–Latin and Arabic–Hebrew translation movements were only the most extensive and prominent cases of a broader cultural phenomenon. In fact, there were translations of philosophical literature involving pretty well every literary language used within striking distance of the Eastern Mediterranean. Thus, we have philosophical translations from Greek into Armenian, Gəʿəz, Georgian and Syriac; from Arabic into Gəʿəz and Syriac and Persian; and even between the less prominent languages. A 13th-century Armenian translation of Proclus's (412–485) *Elements of Theology* (*Stoicheiōsis theologikē*), for example, is not based directly on the original Greek but on the Georgian translation of Ioane Petritsi (11th century). Greek was, in some cases, the target language rather than the source language. Symeon Seth (11th century) rendered Arabic scientific works into Greek for a Byzantine readership, and Latin philosophy also made its way into the Eastern Greek realm with translations of such figures as Boethius (d. between 524 and 526), Thomas Aquinas (d. 1274), and Peter of Spain ([d. 1277]; for lists of medieval translations between various languages see Pasnau 2010, 2: Appendix B; Chapter V.14).

Of course, it does not follow that all intellectuals between, say, the 6th and 14th centuries were dealing with precisely the same body of sources, whatever their working language. Actually, very few texts existed in most or all of these languages. The most obvious example would be the group of Aristotelian logical treatises known as the *Organon*, which were so fundamental to education that they were at least partially available in nearly all the languages just mentioned. Other medieval "best sellers" are more surprising. Take the story of the "silent philosopher" Secundus (early 2nd century?), who vowed never to speak again after inadvertently causing his mother's suicide but agreed to set down his wisdom for a king in the form of wise sayings. He wrote originally in Greek and could later be read in Syriac, Latin, Armenian, Arabic and Gəʿəz (Brock 1978; Overwien 2016). This is a reminder that "wisdom literature," anecdotes and proverbs ascribed to famous philosophers, was a widespread literary genre across medieval cultures.

It would have loomed larger in the popular understanding of what philosophy was than, say, Aristotle's works on nature or Platonist metaphysics.

Obviously, one would like to understand the motivations underlying this cross-cultural process of translation. Just as obviously, however, the motivations were not always the same. Dimitri Gutas has provided a compelling account of the Greek–Arabic translations that situated them within the political context of the Abbasid caliphate. He pointed to a number of factors, including competition with the Byzantine Greeks, pragmatic considerations and such ideological tactics as the invocation of astrology to legitimate political rule (Gutas 1998[CB]; Chapter 1.1). Other factors will have been at work in 4th/10th-century Antioch, 12th-century Toledo, or 13th-century Armenia.

Nonetheless, it is possible to identify a few general features that apply to most, if not all, of the philosophical translation movements, both small and large. First, at a methodological level, it is striking that we often see a transition from overly literal or exact translations to more free, idiomatic versions. This is famously the case with Arabic and Latin. Already readers of the medieval Islamic world found some products of the circle around al-Kindī to be so slavishly close to their Greek sources that they were hard to understand and unpleasant to read (Daiber 1997). Far better were the more polished renderings of Isḥāq ibn Ḥunayn (d. 298/910–1), whose translations of Aristotle duly found a wider readership (Gutas 2016). Such complaints were also made by Latin readers who had to fight their way through texts written in a kind of "translationese" that even preserved grammatical elements from Arabic (Hasse 2013). The same tendency toward excessive exactness can also be observed in, for instance, Syriac translations of the 1st–2nd/7th–8th centuries and medieval Armenian translations, which have been described as being "in Greek, with Armenian words" (Mercier 1978–9, 62).

Another general, and perhaps even universal factor is that philosophical translations were often closely linked to religion. This may seem counterintuitive, given that we nowadays tend to draw a stark contrast between faith and reason, theology and philosophy. Some highly rationalist medieval thinkers, like al-Fārābī or Ibn Rushd, also adopted such a contrast. But, in fact, that attitude was vanishingly rare. Already in late antiquity philosophers like Iamblichus (*c.* 250–*c.* 325) and Proclus presented an explicitly pagan version of Platonism, while Jewish authors like Philo of Alexandria (d. after 40), and Christian church fathers writing in both Latin and Greek, borrowed terminology and argumentation from Hellenic philosophy for their theological purposes. The process was routinely compared to the Israelites "despoiling the Egyptians".

This famously led to a good bit of disquiet. A short list of figures attacked or condemned by religious authorities for their excessive devotion to philosophy might include Ibn Sīnā (posthumously accused of apostasy by al-Ghazālī; Chapter VI.3), John Italos (11th century) in Byzantium (anathematized in 1082) and whoever was being targeted in the condemnations handed down in Paris in 1277 (these probably took aim not only at arts masters of the university but also Aquinas; Wippel 1995). The more typical picture, however, is that religious education needed as a matter of course to include philosophical content. Mossman Rouché's observation regarding the Byzantine context holds more generally: "In so far as logic was a tool of philosophy and not a doctrine, its use by the Christian apologist was encouraged. Its application was inexorable and its utility common to all" (Roueché 1980, 72). Thus, one of the most widely used texts of Greek and Latin Christendom, John of Damascus's (d. after 132/749) *Fountain of Knowledge*, begins with a philosophical textbook giving the reader logical tools and other intellectual equipment drawn from those he calls the "outside" philosophers, that is, the pagans (Louth 2009). In John's case no translation was needed, since he was writing in Greek. But it was the same rationale that led scholars to translate, paraphrase and comment on Aristotle in a Syrian monastic context.

Thus, it would not be misleading to think of philosophical translation in the medieval era as being, in large part, an effort to acquire intellectual tools for understanding and explicating religious doctrine. To put the point more sharply, translation was like an arms race. Logic in particular was not just an instrument (*organon*) but also an intellectual weapon. As Gutas has observed, Aristotle's *Topics* – on dialectical argumentative technique – was translated into Arabic very early, perhaps because of its utility in interreligious disputation (Gutas 1998, 62–9). Certainly, in works of disputation that do survive, we see constant recourse to philosophical terminology and technique. For example, Christians, attempting to justify their belief in the Trinity against Muslim critics, explain the three Persons as individuals falling under the species that is divinity (Abū Rāʾiṭa 2006, 187). In the other direction, a refutation of the Trinity written by al-Kindī states explicitly that it deploys ideas from Porphyry's logical *Introduction* because it is familiar to the Christian opponents.

It is often said that logic, or philosophy more generally, offered a common ground shared by adherents of different religions. That is certainly true. An excellent example would be a polite epistolary exchange between the Christian philosopher Yaḥyā ibn ʿAdī (d. 363/974) and a Jewish scholar named Ibn Abī Saʿīd al-Mawṣilī (4th/10th century), who posed a serious of philosophical inquiries to him (Pines 1955). But more often material from translated texts was used for the sake of refutations and apologetics, the idea being that rational argumentation would have to be accepted by one's opponents. Thus, restricting our attention only to Christian writings, we can see philosophy being deployed in apologetic works written in Syriac, for instance the East Syrian patriarch Timothy I's account of his debate with the Caliph al-Mahdī (r. 158–169/775–78) or, later, in the writings of Bar Hebraeus (Mingana 2009; Griffith 1992). Similar works appear in Arabic, for example, Abū Rāʾiṭa's (2nd–3rd/8th–9th centuries) works on the Trinity (Abū Rāʾiṭa 2006). Peter Abelard (1079–1142) and Ramon Llull (1232–1316) both wrote dialogues in Latin pitting Christians against Jews and pagan philosophers (Payer 1979; Gomez 2018), and Eustratius's (d. *c.* 1120) Greek commentary on the *Nicomachean Ethics* (*Ēthika nikomacheia*) includes a diatribe against Islam (Trizio 2010). The same point could be made by listing examples from Islam or Judaism. For the latter, one might mention Saʿadia Gaʾon's (269–331/882–942) use of argumentation from John Philoponus (d. 574) in his Arabic treatise *Doctrines and Beliefs* (*Kitāb al-Amānāt wa-l-iʿtiqādāt*; Saʿadia Gaʾon 1942[CB]. In Hebrew, Levi ben Gershom (1288–1344; Gersonides) drew on Averroes (Ibn Rushd) to write his own rational justification of the faith, the *Wars of the Lord* (*Milḥamot ha-Shem*) (Levi ben Gershom 1984, 1987, 1999; Adamson 2014), and Shem Ṭov ben Falaquera (1225–*c.* 1295) wrote a fictional dialogue pitting a Jew against a philosopher (Harvey 1987; Adamson 2014).

A significant motive for translating philosophy into all these languages, then, was the facilitation of just such polemics. The goal was to defend one's own faith at a level of intellectual sophistication equal or superior to that found on the other side of the religious divide. But the various translation movements had their own momentum. Once a good portion of the Aristotelian corpus became available in Arabic, it seemed obviously desirable to have the complete set, and the same thing happened a few centuries further on with the Latin translations. Similarly, more aggressively rationalist and Aristotelian thinkers, like al-Fārābī in Baghdad, Ibn Rushd in al-Andalus, Petritsi in Georgia and the masters of the Arts Faculty in Paris, were arguably working not within but *against* the original spirit of the translation movements. Rather than deploying Hellenic thought for some political, ideological or religious ends, they became interested in Aristotelianism or Neoplatonism for its own sake, offering half-hearted religious justifications for this or no justification at all (Druart 2016; Leaman 2013; Gigineishvili 2007; Wei 2012).

It would be a mistake to take these extreme rationalists as definitive, or even characteristic, of medieval philosophy. They were exceptional figures who took a professional interest in material that was more typically adopted within other, more culturally mainstream, contexts. Those contexts might be theological but could also be literary, as in Arabic *adab* works and the French medieval poem *Romance of the Rose*, and philological, as with the Byzantine scholarship that preserved Plato for us more because of his fine Greek than for his ideas (Wilson 1996). A second context was scientific, as when Aristotle's visual theory was contested in treatises on optics and the theory of elements was taken up in medical textbooks (Lindberg 1996; Adamson and Pormann 2017[CB]). A third context was legal, as when logical tools were applied in the service of Islamic "principles of jurisprudence (*uṣūl al-fiqh*)" or philosophical ideas appeared in the Roman emperor Gratian's (d. 1144 or 5) *Decretum* (Emon 2010; Porter 2007). Once the translations were made and the texts of Antiquity had been made available, they could be used in any number of ways by authors of many different viewpoints. To reckon fully with medieval philosophy is, among other things, to study all the ways that translations were used.

Notes

1 Examples of Sogdian texts are given in Sims-Williams (2014) and his other publications.
2 For an example of the Manichaean material in Sogdian, see Benkato (2017).
3 Consolidated bibliography.

Bibliography

Sources

Abū Rā'iṭa. 2006. "The First Letter on the Holy Trinity," in Keating, S. T., ed. and tr. *Defending the 'People of Truth' in the Early Islamic Period: The Christian Apologies of Abū* Rā'iṭa. Leiden: Brill, 164–215.

Ägyptische Urkunden aus den Königlichen Museen zu Berlin, I. 1904. Berlin: Weidmannsche Buchhandlung.

Benkato, A. 2017. *Āzandnāmē: An Edition and Literary-Critical Study of the Manichaean Sogdian Parable-Book* (Beiträge zur Iranistik Band 42). Wiesbaden: Reichert Verlag.

Dresden, M. J., ed. 1966. *Dēnkart. A Pahlavi Text. Facsimile Edition of the Manuscript B of the K. R. Cama Oriental Institute Bombay*. Wiesbaden: Harrassowitz.

Hämeen-Anttila, J. 2018. *Khwadāynāmag: The Middle Persian Book of Kings* (Studies in Persian Cultural History 14). Leiden and Boston: Brill.

Levi ben Gershom. tr. Feldman, S. 1984. 1987. 1999. *Wars of the Lord*. 3 vols. Philadelphia: Jewish Publication Society.

Madan, D. M., ed. 1911. *The Complete Text of the Pahlavi Dinkard*. 2 vols. Bombay: The Society for the Promotion of Researches into the Zoroastrian Religion.

Mingana, A. 2009. *The Apology of Timothy the Patriarch before the Caliph Mahdī* (Woodbroke Studies 2). Piscataway, NJ: Gorgias Press.

Payer, P. 1979. *Peter Abelard: A Dialogue of a Philosopher with a Jew and a Christian*. Toronto: The Pontifical Institute of Mediaeval Studies Publications.

Research literature

Adamson, P. 2006. *Al-Kindī* (Great Medieval Thinkers). Oxford: Oxford University Press.

Adamson, P. 2008. *In the Age of Al-Farabi: Arabic Philosophy in the Fourth/Tenth Century* (Warburg Institute Colloquia 12). London: Warburg Institute.

Adamson, P. 2013. *Interpreting Avicenna. Critical Essays*. Cambridge: Cambridge University Press.

Adamson, P. 2014. "Podcast 164 – Man and Superman: Gersonides and the Jewish Reaction to Averroes," https://historyofphilosophy.net/gersonides (Accessed 12 August 2019).

Adamson, P. 2016. *Philosophy in the Islamic World. A History of Philosophy Without Any Gaps*. Vol. 3. Oxford. Oxford University Press.

Adamson, P. and Di Giovanni, M. 2018. *Interpreting Averroes. Critical Essays*. Cambridge: Cambridge University Press.

Bagnall, R. S. 1993. *Egypt in Late Antiquity*. Princeton: Princeton University Press.

Bailey, H. W. 1943. *Zoroastrian Problems in the Ninth-Century books*. Oxford: Clarendon Press.

Bausi, A. 2016a. "Ethiopic Literary Production Related to the Christian Egyptian Culture," in Buzi, P., Camplani, A. and Contardi, F., eds. *Coptic Society, Literature and Religion from Late Antiquity to Modern Times. Proceedings of the Tenth International Congress of Coptic Studies, Rome, September 17th–22nd, 2012, and Plenary Reports of the Ninth International Congress of Coptic Studies, Cairo, September 15th–19th, 2008*, I (Orientalia Lovaniensia Analecta 247). Leuven: Peeters, 503–71.

Bausi, A. 2016b. "On Editing and Normalizing Ethiopic Texts," in Bausi, A. and Sokolinski, E., eds. *150 Years after Dillmann's Lexicon: Perspectives and Challenges of Gəʿəz Studies* (Supplement to Aethiopica 5). Wiesbaden: Harrassowitz, 43–102.

Bausi, A. 2018. "Translations in Late Antique Ethiopia," in Crevatin, F., ed. *Egitto crocevia di traduzioni* (ΔΙΑΛΟΓΟΙ 1). Trieste: EUT Edizioni dell'Università di Trieste, 69–99.

Blois, F. de. 1990. *Burzōy's Voyage to India and the Origin of the Book of Kalīlah wa Dimnah*. London: Royal Asiatic Society.

Blois, F. de. 2007. "Translation in the Ancient Iranian World," in Kittel, H. *et al.*, eds. *Übersetzung-Translation-Traduction* (Handbücher zur Sprach- und Kommunikationswissenschaft 26.2). Berlin: De Gruyter, 1194–8.

Boud'hors, A. 2012. "The Coptic Tradition," in Fitzgerald, S. J., ed. *The Oxford Handbook of Late Antiquity*. Oxford: Oxford University Press, 224–46.

Bouriant, U. 1888. "Fragments d'un livre de médicin en copte thébain," *Académie des inscriptions et belles-lettres, Comptes rendus 1887*, ser. 4.15: 319–20, 374–9.

Brock, S. 1978. "Secundus the Silent Philosopher: Some Notes on the Syriac Tradition," *Rheinisches Museum für Philologie, Neue Folge* 121.1: 94–100.

Brock, S. 1983. "Towards a History of Syriac Translation Technique," in Lavenant, R., ed. *III Symposium Syriacum. Orientalia Christiana Analecta* 221. Rome: Pont. Institutum Orientale, 1–14.

Brock, S. 1993. "The Syriac Commentary Tradition," in Burnett, C., ed. *Glosses and Commentaries on Aristotelian Logical Texts*. London: Warburg Institute Surveys and Texts, 3–18.

Burnett, C. 2005. "Arabic into Latin: The Reception of Arabic Philosophy into Western Europe," in Adamson, P. and Taylor, R. C., eds. *The Cambridge Companion to Arabic Philosophy*. Cambridge: Cambridge University Press, 370–404.

Camplani, A. 2015. "Il copto e la chiesa copta. La lenta e inconclusa affermazione della lingua copta nello spazio pubblico della tarda antichità," in Nicelli, P., ed. *L'Africa, l'Oriente mediterraneo e l'Europa. Tradizioni e culture a confronto* (Africana ambrosiana 1). Milan: Biblioteca Ambrosiana-Bulzoni, 129–53.

Chassinat, E. 1921. "Un papyrus medical copte," *Mémoires publiés par le membres de l'Institut français d'archéologie orientale du Caire* 32.

Conti Rossini, C. 1951. "Il *Fisiologo* etiopico," *Rassegna di Studi Etiopici* 10: 5–51.

Crislip, A. 2005. *From Monastery to Hospital. Christian Monasticism and the Transformation of Health Care in Late Antiquity*. Ann Arbor: The University of Michigan Press.

Daiber, H. 1997. "Salient trends of the Arabic Aristotle," in Endress, G. and Kruk, R., eds. *The Ancient Tradition in Christian and Islamic Hellenism*. Leiden: Brill, 29–41.

Debié, M. 2014. "Sciences et savants syriaques : une histoire multiculturelle," in Villey, ed., 9–66.

Dostálová, R. 1994. "Gli inventari dei beni delle chiese e dei conventi su papiro [The Inventories of the Goods of the Church and the Convents on Papyrus]," *Analecta Papyrologica* 6: 5–19.

Druart, Th. 2016. "Al-Farabi," in *Stanford Encyclopedia of Philosophy*. https://plato.stanford.edu/entries/al-farabi/ (Accessed 12 August 2019).

Emon, A. M. 2010. *Islamic Natural Law Theories*. Oxford: Oxford University Press.

Fazzo, S. and Zonta, M. 2008. "Aristotle's Theory of Causes and the Holy Trinity. New Evidence About the Chronology and Religion of Nicolaus 'of Damascus'," *Laval théologique et philosophique* 64.3: 681–90.

Fournet, J.-L. 2009. "The Multilingual Environment of Late Antique Egypt: Greek, Latin, Coptic, and Persian Documentation," in Bagnall, R. S., ed. *Oxford Handbook of Papyrology*. Oxford and New York: Oxford University Press, 418–51.

Fournet, J.-L. 2014. "Alexandrie et la fin des hieroglyphs," in Méla, Ch. and Möri, F., eds. *Alexandrie la divine*. 2 vols. Neuchâtel: La Baconnière, 2: 599–607.

Gershevitch, I. 1979. "The Alloglottography of Old Persian," *Transactions of the Philological Society* 77.1: 114–90.

Gigineishvili, L. 2007. *The Platonic Theology of Ioane Petritsi* (Gorgias Eastern Christian Studies 4). Piscataway, NJ: Gorgias Press.

Gomez, N. U. 2018. *The Crusade of Ramon Llull: Apologetics and Evangelism to Muslims During the Thirteenth Century*. PhD dissertation. Louisville, KY: The Southern Baptist Theological Seminary.
Griffith, S. H. 1992. "Disputes with Muslims in Syriac Christian Texts: From Patriarch John (d. 648) to Bar Hebraeus (d. 1286)," in Niewöhner, F., ed. *Religionsgespräche im Mittelalter*. Wiesbaden: Harrassowitz, 251–73.
Guignard, Ch. 2014. "L'agriculture en syriaque : l'Anatolius Syriacus ('Géoponiques syriaques')," in Villey, ed., 215–42.
Gutas, D. 2013. "Avicenna's Philosophical Project," in Adamson, P., ed. *Interpreting Avicenna*. Cambridge: Cambridge University Press, 28–47.
Gutas, D. 2016. "The Rebirth of Philosophy and the Translations into Arabic," in Rudolph *et al.*, eds. [CB], 95–142.
Harvey, St. 1987. *Falaquera's Epistle of the Debate: An Introduction to Jewish Philosophy*. Cambridge, MA: Harvard University Press.
Hasse, D. N. 2013. "Die Überlieferung arabischer Philosophie im lateinischen Westen," in Eichner, H., Perkams, M. and Schäfer, C., eds. *Islamische Philosophie im Mittelalter: Ein Handbuch*. Darmstadt: Wissenschaftliche Buchgesellschaft, 377–400.
Henning, W. B. 1958. "Mitteliranisch," in Spuler, B., ed. *Handbuch der Orientalistik. Linguistik*. Abt. 1, Bd. 4, Abschn. 1, Lfg. 1. Leiden: E. J. Brill, 20–130.
Hommel, F. 1877. *Die aethiopische Uebersetzung des Physiologus nach je einer Londoner, Pariser und Wiener Handschrift*. Leipzig: J. C. Hinrichs'sche Buchhandlung.
Hugonnard-Roche, H. 2004. *La logique d'Aristote du grec au syriaque*. Paris: J. Vrin.
Huyse, Ph. 2009. "Inscriptional Literature in Old and Middle Iranian Languages," in Emmerick, R. E. and Macuch, M., eds. *The Literature of pre-Islamic Iran* (A History of Persian Literature 17). London: I. B. Tauris, 72–115.
Kolta, K. S. 1982. "Die Lepra im alten Ägypten in der koptischer Zeit," in Habrich, Ch., Williams, J. C. and Wolf, J. H., eds. *Aussatz-Lepra-Hansen-Krankheit. Ein Menschheitsproblem im Wandel. Katalog der Ausstellung im Deutschen Museum München*. Würzburg *et al.*: Deutsches Aussätzigen-Hilfswerk, 58–63.
Kolta, K. S. 1984. "Neue Ergebnisse zur Medizin der Kopten," *Sudhoffs Archiv* 68: 157–72.
Kolta, K. S. 1991. "Medicine, Coptic," in Atiya, A. S., ed. *Coptic Encyclopedia*. 8 vols. New York: Macmillan, 5: 1578–82.
Kolta, K. S. 2007. "Koptische Medizin," in Gerabek, W. E. *et al.*, eds. *Enzyklopädie Medizingeschichte*. Berlin: De Gruyter, 779–81.
König, G. 2018. "The Pahlavi Literature of the 9th Century and Greek Philosophy," *Iran and the Caucasus* 22: 8–37.
Larsen, L. I. and Rubenson, S. 2018. *Monastic Education in Late Antiquity: The Transformation of Classical Paideia*. Cambridge: Cambridge University Press.
Leaman, O. 2013. *Averroes and His Philosophy*. London and New York: Routledge.
Lindberg, D. C. 1976. *Theories of Vision from al-Kindi to Kepler*. Chicago: Chicago University Press.
Louth, A. 2009. *St. John Damascene: Tradition and Originality in Byzantine Theology* (Oxford Early Christian Studies). Oxford: Oxford University Press.
MacKenzie, D. N. 1979. "Mani's Šābuhragān," *Bulletin of the School of Oriental and African Studies* 42.3: 500–34.
MacKenzie, D. N. 1980. "Mani's Šābuhragān – ii," *Bulletin of the School of Oriental and African Studies* 43.2: 288–310.
Martelli, M. 2014. "L'alchimie syriaque et l'œuvre de Zosime," in Villey, eds., 191–214.
Mercier, C. 1978–1979. "L'École hellénistique dans la littérature arménienne," *Revue des études arméniennes* 8: 59–75.
Munier, H. 1918. "Deux recettes médicales coptes," *Annales du Service des Antiquités* 18: 284–6.
Orlandi, T. 1997. "Letteratura e cristianesimo nazionale egiziano," in Camplani, A., ed. *Egitto Cristiano. Aspetti eoyennei in età tardo antica*. Rome: Institutum patristicum Augustiniarum, 39–120.
Otranto, R. 1997. "*Alia tempora, alii libri*. Notizie ed elenchi di libri cristiani su papiro," *Aegyptus* 77: 123–44.
Overwien, O. 2016. "Secundus the Silent Philosopher in the Ancient and Eastern Tradition," in Cupane, C. and Krönung, B., eds. *Fictional Storytelling in the Medieval Eastern Mediterranean and Beyond* (Brill's Companions to the Byzantine World 1). Leiden: Brill, 338–64.
Papaconstantinou, A. 2007. "'They Shall Speak the Arabic Language and Take Pride in It': Reconsidering the Fate of Coptic after the Arab Conquest," *Le Muséon* 120: 273–99.

Papaconstantinou, A. 2012. "Why Did Coptic Fail Where Aramaic Succeeded? Linguistic Developments in Egypt and the Near East after the Arab Conquest," in Mullen, A. and James, P., eds. *Multilingualism in the Graeco-Roman Worlds*. Cambridge and New York: Cambridge University Press, 58–76.

Papaconstantinou, A. 2014. "Egyptians and 'Hellenists': Linguistic Diversity in the Early Pachomian Monasteries," in Tallet, G. and Zivie-Coche, Ch., eds. *Le myrte et la rose: mélanges offerts à Françoise Dunand par ses élèves, collègues et amis* (CENIM 9). Montpellier: Université Paul Valéry, 15–21.

Pasnau, P. ed. 2010. *The Cambridge History of Medieval Philosophy*. 2 vols. Cambridge: Cambridge University Press.

Pietrăreanu, O. 2018. "The Character of Barzawayh in an Anonymous Syriac Translation of Ibn al-Muqaffaʿ's *Kalīla and Dimna*," in Sitaru, L., ed. *Geographies of Arab and Muslim Identity Through the Eyes of Travelers*. Romano-Arabica 18. Bucharest: Center for Arab Studies, 125–44.

Pines, S. 1955. "A Tenth-Century Philosophical Correspondence," *Proceedings of the American Academy for Jewish Research* 24: 103–36.

Porter, J. 2007. "Custom, Ordinance and Natural Right in Gratian's *Decretum*," in Perreau- Saussine, A. and Murphy, J. B., eds. *The Nature of Customary Law*. Cambridge: Cambridge University Press, 79–100.

Richter, T. S. 2018. "Medical Care on the Theban Westbank in Late Antiquity," *Journal of Coptic Studies* 20: 151–63.

Roueché, M. 1980. "A Middle Byzantine Handbook of Logic Terminology," *Jahrbuch der Österreichischen Byzantinistik* 29: 71–98.

Schenke, G. 2016. "The Healing Shrines of St. Phoibammon: Evidence of Cultic Activity in Coptic Legal Documents," *Zeitschrift für Antike und Christendum* 20: 496–523.

Schmitt, R. 2013. "Bisotun iii. Darius's inscriptions," in *eIr*, https://iranicaonline.org/articles/bisotun-iii. (Accessed March 2019).

Sims-Williams, N. 2014. *Biblical and other Christian Sogdian Texts from the Turfan Collection* (Berliner Turfantexte 32). Turnhout: Brepols Publishers.

Sumner, C. 1974–1982. *Ethiopian Philosophy*. 5 vols. Addis Ababa: Addis Ababa University.

Takahashi, H. 2015. "Syriac as the Intermediary in Scientific Graeco-Arabica: Some Historical and Philological Observations," *Intellectual History of the Islamicate World* 3: 66–97.

Till, W. C. 1949. "Koptische Rezepte," *Bulletin de la Société d'archéologie copte* 12: 43–55.

Trizio, M. 2012. "A Neoplatonic Refutation of Islam from the Time of the Komnenoi," in Speer, A. and Steinkrüger, P., eds. *Knotenpunkt Byzanz. Wissensformen und Kulturelle Wechselbeziehungen*. Leiden: Brill, 145–66.

Uhlig, S. and Bausi, A. eds. 2003–2014. *Encyclopaedia Aethiopica* [Ethiopian Encyclopedia]. 5 vols. Wiesbaden: Harrassowitz.

Watt, J. W. 2010. *Rhetoric and Philosophy from Greek into Syriac*. Farnham: Variorum.

Wei, I. 2012. *Intellectual Culture in Medieval Paris: Theologians and the University, c.1100–1330*. Cambridge: Cambridge University Press.

Villey, É. 2014a. "Qennešre et l'astronomie aux VIe et VIIe siècles," in Villey, ed., 149–90.

Villey, É. ed. 2014b. *Les sciences en syriaque* (Études syriaques 11). Paris: Geuthner.

Wilson, N. G. 1996. *Scholars of Byzantium*. London: Duckworth.

Wippel, J. F. 1995. "Thomas Aquinas and the Condemnation of 1277," *Modern Schoolman* 72: 233–72.

Worrell, W. H. 1935. "Coptic Magical and Medical Texts," *Orientalia* 4: 1–37, 184–94.

Zeini, A. 2018a. "Letter of Tansar (Tosar)," in Nicholson, O., ed. *The Oxford Dictionary of Late Antiquity*. Oxford: Oxford University Press, 898.

Zeini, A. 2018b. "Xwaday Namag," in Nicholson, O., ed. *The Oxford Dictionary of Late Antiquity*. Oxford: Oxford University Press, 1599.

I.3
TRANSLATIONS IN THE MATHEMATICAL SCIENCES

Nathan Sidoli

I.3.1 Introduction

The delineations between the various mathematical, or exact, sciences in the ancient and medieval periods were different from what we might expect based on our own educational experience. In the first place, we do not find in our sources a clear distinction between pure and applied sciences – so that astronomy was understood to be just as much of a mathematical science as geometry, although they are obviously different sciences and were understood as such. Moreover, both astronomy and astrology were considered to be mathematical sciences, and the terminology did not always distinguish clearly between the two, although I am not aware of any text that actually confuses them – astronomy was the science of making claims and predictions about the arrangement of the heavenly bodies, and astrology was the science of making judgments, based on these, concerning earthly affairs. In general, these sciences are presented in different texts, or at least in different sections of the same text. Nevertheless, there was generally no institutional division between the practitioners of the various exact sciences, and often the same individuals wrote works in various fields of the exact sciences, as well as in the medical, religious and other sciences. The contents and categorizations of mathematical and astral sciences – astronomy and astrology – were understood somewhat differently in the various ancient traditions that Islamicate scholars adopted and developed, although by the end of the 3rd/9th century they seem to have become most influenced by Greek works in categorizing these sciences. Nevertheless, for the purposes of this chapter, I consider the mathematical, or exact, sciences to be any of those disciplines in which mathematics is applied or developed, particularly the astral sciences of astronomy and astrology – which follows an understanding of these sciences that can be gleaned from the sources themselves.

It is an undisputed historical fact that from the end of the 2nd/8th to the beginning of the 4th/10th century an unprecedented number of original treatises and translations of ancient treatises in the mathematical sciences were produced in Arabic, a language that had not previously been known for scientific works. Various motivations for this have been advanced as providing the social context in which this translation activity took place, such as administrative necessity on the part of the Umayyad caliphs, or imperial ideology on the part of the Abbasid caliphs (Gutas 1998[CB],[1] 11–104; Saliba 2007[CB], 1–72; Dallal 2010, 13–16). As well as any underlying social context that may have promoted translation, however, the study and translation of

DOI: 10.4324/9781315170718-5

the more theoretically obtuse texts, such as Archimedes's *Sphere and Cylinder* or Menelaus's *Spherics*, must have been principally motivated by the goals of the scholars themselves – the production and dissemination of new knowledge and the acquisition of the social status that accrued to this (Rashed 2006). Of course, the administrative goals and imperial ideology of the caliphates would have served as a social background and lent social prestige to the activities of those working in the mathematical sciences, but as we will see in the following discussion, the knowledge that was transmitted and produced went far beyond, or at least was tangential to, any such ends. In order to understand the practices of these mathematical scholars, we must also take into consideration the attitudes that they held toward the ancient sciences as expressed in the works that they produced.

In many ways, it is still premature to try to write the sort of survey or overview account that will be attempted in this chapter because we still lack critical editions of some of the most important texts and detailed critical studies of many, if not most, of the individuals involved. For example, even the *Catalog* (*Kitāb al-Fihrist*) of Ibn al-Nadīm (d. 380/990), one of our most important historical sources for scholarly activity in the early Islamicate world, is not available in a fully critical edition (Stewart 2006), and the various versions we do have are sometimes different in important details, such as a name, a crucial incident or the word used to describe a key concept (for examples, see Ibn al-Nadīm 1970[CB], 1: 263 no. 54, 2: 647 no. 43; 2: 827 no. 3). For the individuals discussed in this chapter, in many cases, the various medieval sources give differing, sometimes conflicting, accounts, which may themselves have their own rhetorical purposes. In only a few cases do we even have detailed scholarly discussions of all the known source passages (for examples, see al-Khwārazmī 2009, 15–24; Banū Mūsā 1979[CB], 3–6) so that it sometimes happens that the stories about these figures and their activity that circulate in the scholarly literature cannot be traced back to unambiguous and mutually consistent sources. In this chapter, I focus on reporting what the medieval authors say, but it should be borne in mind that some of the sources may have been written to serve goals other than the production of historical scholarship. Finally, we should recognize that some of the key terminology of our sources is used in different ways by different authors and sometimes has a rather unusual meaning from the perspective of our own practice and usage. In particular, one of the core topics of this chapter has to do with translation, which – following the usage of the medieval authors – must be understood to mean the general transmission of ideas and methods, as well as literary translation of texts. As an example, we may take the case of al-Kindī (d. *c.* 256/870), sometimes known as the "philosopher of the Arabs." According to the account of Ibn Abī Uṣaybiʿa (d. 668/1270), Ibn Juljul (d. after 383/994) credits al-Kindī with translating many philosophical works, and Abū Maʿshar (171–272/787–886) said that al-Kindī was one of the four great translators in Islam (Ibn Abī ʾUṣaybiʿa 2011[CB], 398, 2020[CB], 10:1.5[online translation]). Although this may strike us as strange, since al-Kindī did not read Greek, if we take a broader view of the concept of translation, it accords with al-Kindī's own statements that he was working to make the ideas of the ancients available to his fellow speakers of Arabic (Walzer 1945, 172–5). If we take this testimony seriously, it would indicate that when our sources tell us that someone was involved in "translation," or when they describe a "translation" project, we need not assume that they are always talking about literary translation from one written text to another, but they may also be describing a general process of transmitting ideas and methods. We must decide what kind of translation is meant on a case-by-case basis.

As discussed in the previous chapters, we should not think of translations of the exact sciences as taking place in isolation, and there is evidence for various types of translation being undertaken in the lands that the Muslims conquered before they arrived. Furthermore, the Syrian- and Persian-speaking scholars who came under Muslim rule doubtless advocated the importance of translation because it played an important role in narratives about learning in both Syrian and Persian. In this chapter, however, I focus on evidence for the transmission and translation of the exact sciences into Arabic. For more specific discussions of the practice and development of these sciences in Islamicate societies see Chapters I.5–I.7.

I.3.2 Obscure beginnings under the Umayyads

One of the first activities that involved the transmission of mathematical knowledge and practices for which we have reports concerns the *dīwāns* – that is, registers and offices of taxation and government accounting (Saliba 2007, 45–64; Rashed 2006, 160–2). We do not know the institutional details of these offices, which presumably differed from place to place, but they likely employed expert calculators and were certainly administered in different languages in the different regions that came under Arab Muslim rule. In the last of four accounts that Ibn al-Nadīm gives of the transmission of ancient sciences into Arabic, he presents a discussion of the translations of the *dīwān* during the Umayyad caliphate ([r. 41–132/661–750]; Ibn al-Nadīm 1970, 2, 581–3). This account makes it clear that the eastern *dīwān* was in Persian and the western *dīwān* in Greek and that expert knowledge was required to maintain these records. The reluctance of the administrator of the eastern *dīwān*, Zādānfarrūkh ibn Bīrī (d. *c.* 81/700) to have the records translated into Arabic, and his assertion, in this regard, that he was of more value to the governor than the governor was to him, makes it clear that running the office of the *dīwān* involved mathematical skills essential to the smooth functioning of the state. Ibn al-Nadīm then goes on to say that the *dīwān* in Damascus was administered by Sarjūn ibn Manṣūr (late 1st–early 2nd/7th–8th centuries) and his son, Yuḥannā ibn Sarjūn ibn Manṣūr, known in Greek as John of Damascus (d. after 132/749). Once again, the account makes it clear that the Greek-speaking administrator was reluctant to translate the records, and this translation was eventually carried out by someone else (Gutas 1998, 17, no.19). Both of these stories reinforce the impression that the methods of the *dīwān* constituted specialized knowledge, the possession of which bestowed status, and some political power, on their practitioners. The mathematical abilities of the officers of the *dīwān* are also asserted in a Greek source. In the hagiographic *Life of John of Damascus*; translation of an earlier Arabic version; not regarded as a historically reliable account), we are told that Yuḥannā and his cousin "trained in arithmetic proportions as skillfully as Pythagoras or Diophantus" (Diophantus Alexandrinus 1893–1895, 2, 36; Sahas 1972, 32–5). This passage is doubtless influenced by rhetorical hyperbole, but nevertheless, it is intended to convince the reader that Yuḥannā ibn Sarjūn was a competent calculator and that the methods at his disposal included those of solving problems involving unknown values as set out by Diophantus. It is likely that the sorts of mathematics being referred to here is that found in the scholia to mathematical problems in the late ancient *Palatine Anthology*, a number of which are solved through Diophantus's approach (Christianidis and Megremi 2019, 27–33). In the East as well, the officers of the *dīwān* must have had training in mathematics, which, in some Central Asian provinces, probably including computation using Indian methods – which were also known in Syria, as is attested by the remarks of Severus Sēbōkht (575–667) in his letter to Basil of Cyprus (7th century), in which he mentions that the Indians produced advanced astronomy and computed using a rational method involving nine symbols (Severus Sēbōkht 2000; Takahashi 2010, 21–4).

In both cases, since these were offices that oversaw the accounts and revenues of large territories, it is unlikely that the *dīwāns* simply consisted of a few individuals and some set of registers. Rather, they probably involved a large number of accountants, surveyors, engineers and other expert practitioners who could carry out arithmetic computations and solve basic problems in geometric mensuration and premodern algebra, as well as perform basic calendrical computations – as is later detailed by Ibn Qutayba (d. 276/889), who sets out the requirements of one seeking to be employed as a secretary (*kātib*) of the *dīwān* (Saliba 2007, 53–6). Hence, the translation of the *dīwān* discussed in our sources, would have involved the whole apparatus of this office, along with their accompanying technical vocabulary. Here, we do not need to take translation to mean that of individual texts, although some instruction manuals may have been translated, but rather to mean a transmission of the mathematical methods used in these offices.

In this way, a large part of the practical traditions of the mathematics of those regions that the Muslim armies had conquered would have been transmitted into Arabic and become naturalized in the Islamicate sphere. This view of the transmission activities of the *dīwān*s is supported by some of the earliest works written in Arabic that are devoted to mathematical subjects – such as al-Khwārazmī's (often al-Khwārizmī) work on Indian computation or his treatise on calculation through premodern algebraic methods, in which he assumes as well-known the process of naming sought values and operating on equalities, which had been fully described by Diophantus, before moving on to his new contribution of cataloging the types of equalities one encounters and detailing their solutions (Christianidis and Oaks 2013).

Astrology appears to have been given a central role in Arabic and Islamicate culture around the time of the shift of power from the Umayyads to the Abbasids (r. 132–656/750–1258). We have, however, only fragmentary evidence about this process, and the sources are subject to a range of interpretations. For example, al-Bīrūnī (362-d. after 444/973-d. after 1053) knew a text called the *Tables of the Arkhand* (*Zīj al-Arkhand*) that he regarded as having been rendered in a bad translation and which was related to material from the *Khaṇḍakhādyaka* of Brahmagupta (d. after 665; al-Bīrūnī 1910, 18). It is possible that this was an Arabic translation made from a Middle Persian *Arkhand* in 117/735 in Sind (today in Pakistan), but our sources do not make this certain (al-Hāshimī 1981, 207–11; Van Bladel 2014[CB], 60, no.13). On the other hand, one of the first certain translations of a work in the astral sciences into Arabic is the *Nativities* (*Kitāb al-mawālīd*), which was translated around 132/750, during the Abbasid revolution, from an Iranian source (Pingree 1997, 44–7; Van Bladel 2014, 273–4). This translation was apparently made in a period of intense political turmoil during the wars that resulted in the founding of the Abbasid caliphate – from which time forward we have clearer evidence for the presence of scholars in the mathematical sciences carrying out various projects and producing original work in an Islamic context.

Although it is possible that early translations were lost because they were superseded by later ones of better quality, the overall view that one draws from the sources is that during this period few works were translated in the sense that we would normally understand this term, but this would not have prevented knowledge of, or at least about, the ancient sciences from circulating in Arabic circles through scholars and practitioners who had available to them various scientific sources, or discussions of the sciences in the original languages. As examples, al-Hāshimī (*fl. c.* 277/890) relates that the Sasanian King of Kings Khōsraw I (r. 531–579) had a new astronomical handbook produced that was based on a comparison between Ptolemaic and Indian astronomy, and Severus Sēbōkht, although referring to Ptolemy's *Almagest*, at least in name, mentions that the Babylonians, Egyptians and Indians were also skilled in astronomy (al-Hāshimī 1981, fol. 95a; Takahashi 2010, 22–3). Hence, scholars and other practitioners working in Greek and Iranian languages within the Islamicate sphere understood the significance of the ancient sciences, were aware that there were different traditions of the sciences in the different ancient cultures and had, in some cases, familiarity with certain ancient texts.

I.3.3 The Abbasids

The first period of intense cultivation of the ancient sciences in an Islamic context is associated by the medieval historians with the reign of the second Abbasid caliph, al-Manṣūr (r. 136–158/754–775), who they tell us was interested in the religious and philosophical sciences, such as astrology – which itself must be regarded as a relatively new development in Arabic and Islamicate culture at that time (Gutas 1998, 28–60). Al-Manṣūr's motivation for his support of scholarship in the astral sciences was probably based on a number of different factors: an effort to model his authority on that of the late Sasanian kings who are reported to have supported the celestial sciences; the contacts of his court with T'ang China and of his companions with Central Asia, where both Chinese and Indian astrology was practiced; and his belief in, and reliance on, political, or historical, astrology (Borrut 2014). The sources make it clear that the caliph surrounded himself with scholars from various, but especially Persian and formerly Sasanian, backgrounds, who helped him determine the best date to found his new capital and served his state in various ways.

We are told by the historian al-Masʿūdī (d. 345/956), who was probably somewhat exaggerating to make his point, that translations from foreign languages began under al-Manṣūr, and a number of works in the mathematical sciences are named, such as a *Sindhind*, presumably that translated, or transmitted, by Abū Isḥāq al-Fazārī (2nd/8th century), Aristotle's (384–322 BCE) logical works, Ptolemy's (*c.* 100–170) *Almagest*, Nicomachus' (d. *c.* 120) *Introduction to Arithmetic*, and Euclid's *Elements* (Gutas 1998, 30). Of these early reported efforts, we now have only fragments of the *Sindhind*, so we cannot know what was meant by the concept of translation at this time. Nevertheless, the discussions of al-Fazārī's work on the *Sindhind* that are preserved as fragments and reports, mostly by al-Bīrūnī and al-Hāshimī, make it clear that the text that he produced was not a straightforward translation as we usually understand this term. Although al-Bīrūnī indicates that al-Fazārī and Yaʿqūb ibn Ṭāriq (d. c. 180 /796) learned Indian astronomy and computational methods from an Indian master who came to Baghdad as a member of an embassy from the Sind in 153/770, the surviving evidence concerning their work shows influences from various traditions, and the passages in our sources that discuss al-Fazārī's *Sindhind* indicate that it also incorporated elements of Greek and Persian material (Yaʿqūb ibn Ṭāriq 1968; al-Fazārī 1970). Moreover, the sources that mention this Indian scholar – an expert in the calculation of the *Sindhind* – do not name any particular Indian source, or sources, so that the astrologers gathered in al-Manṣūr's court may have learned what they knew about Indian methods directly from a master and not through the full translation of any definite text (Ṣāʿid al-Andalusī 1991[CB], 46; Van Bladel 2014, 260, no.10). Certainly, al-Bīrūnī found that the translations of Indian sources from the early Abbasid period that he read were imprecise and contained transliterated terms that were not clearly explained (al-Bīrūnī 1976, 189–90). Hence, still in the time of al-Manṣūr, the translations of ancient works that are reported in our sources may have resulted from various processes of transmission of ideas and methods and need not always be understood as a direct translation from one language into another that had literal accuracy as its goal.

What is clear from the sources, however, is that al-Manṣūr cultivated the activity of a number of individuals, who were experts in the mathematical sciences; who came from various linguistic, cultural and religious backgrounds; and who had some knowledge of the ancient sciences that had been developed in former times in the lands now under Muslim rule. This is especially clear regarding the founding of Baghdad – *Madīnat al-salām, The City of Peace.* The design of this new capital may have been overseen by the Abbasid official Khālid ibn Barmak (d.165/781–2), the scion of a prominent Central Asian Buddhist family, and based on the circular plan of a number of former Sasanian cites, as well as his ancestral home, the Nawbahār, a famous Buddhist monastery in Balkh, which had formally been a Sasanian palace (Beckwith 1984, 2009, 147). According to al-Yaʿqūbī (d. after 292/905), the technical work necessary to determine the best time for founding the city was carried out by the astrologers and masters of calculation, Nawbakht (d. *c.* 160/777), Māshā'allāh (d. *c.* 199/815), al-Fazārī and ʿUmar ibn al-Farrukhān al-Ṭabarī (d. *c.* 197/813), and al-Bīrūnī gives a horoscope for the construction, which can be dated to July 30, 762 (al-Fazārī 1970, 104). We do not know how these men learned their trade, but since some of them came from Iranian and formerly Sasanian cities, they may have been trained in the then current Sasanian and Indian traditions in the astral sciences.

The importance of Indian sources for the early development of the exact sciences in the early Abbasid period can be partly explained by the central role of the Barmakid clan, whose members, such as Khālid, were instrumental in the revolution itself and whose leaders served as viziers and companions to the early Abbasid caliphs and were patrons of scholarship until their downfall in 187/803, during the reign of Hārūn al-Rashīd (r. 170–193/786–809; Van Bladel 2012b). The Barmakids traced their lineage back for centuries in the Central Asian province of Bactria, where the head of the family had been the Barmak (from *pramukha*) and oversaw an important center of Buddhist learning (*vihāra*). This relation to Buddhism, and its connection to India, helps us understand the claim of Ibn al-Nadīm that it was Yaḥyā ibn Khālid ibn Barmak (d. 190/805), and the Barmakids in general, who most concerned themselves with India and summoned Indian physicians and philosophers to Baghdad (Ibn al-Nadīm 1970, 2: 827; Van Bladel 2011, 75).

Another possible influence on the significance given to Indian sciences in the early Abbasid period may have come from the diplomatic contacts between Abbasid emissaries and the court of the T'ang dynasty, in Chang'an. Chinese sources report embassies from Arabs in black robes arriving an average of more than once a year from 135/753 to 145/762, the year of the founding of Baghdad (Van Bladel 2014, 271). Chinese sources also report that the members of three Indian families, or schools, were working in the official Chinese astronomical service in 764, which had been reorganized in 758 as the Bureau of Astronomy 司天臺, and earlier in the same century, Indian computational methods had been used both in unofficial Indian Chinese calendars, as well as official Chinese calendars (Yabuuti with Yano 1979; Cullen 1982). It is unlikely that the Abbasid embassies learned any of the details of the Indian computational and astronomical methods, but it could not have escaped their notice that practitioners of the Indian tradition of the astral sciences were held in such high regard in Chang'an that they were given official positions in the bureaucracy of this wealthy and formidable state (Van Bladel 2014).

The Greek astral sciences were also studied and practiced at the early Abbasid court. The Chalcedonian Christian Theophilus of Edessa (d. 168/785), who witnessed the wars that brought the Abbasids to power, soon joined them in Baghdad, where he combined knowledge of Greek, Iranian and Indian astral sciences. He served under the first three caliphs of the dynasty and was appointed court astrologer to al-Manṣūr's son, al-Mahdī (r. 158–169/775–785). He is reported to have made translations from Greek into Syriac and to have read Middle Persian. In his extant Greek writings, he quotes the astronomical and astrological works of Ptolemy, Dorotheus of Sidon (1st century), Vettius Valens (2nd century) and Hephaestion of Thebes (mid-5th century) in essentially the same wording as our sources for the original Greek. Nevertheless, there are also Indian influences in his work, perhaps through Persian sources, such as the similarities between his work on military astrology and Varāhamihira's *Bṛhadyātrā* (Pingree 1976, 148, 2001, 13–17; Van Bladel 2014, 274–5). Hence, he presumably studied the astral sciences of these traditions as well, either through his own wide reading or directly from his colleagues in Baghdad.

The role of Persia in the transmission of the ancient sciences is promoted in the semi-mythical history of science written by Abū Sahl al-Faḍl ibn Nawbakht (*fl. c.* 153–193/770–809), as reported by Ibn al-Nadīm. The Nawbakhts traced their lineage from a distinguished, formerly Zoroastrian family, and Abū Sahl succeeded his father to become an astrologer in the court of al-Manṣūr (Pingree 1990, 293) and, later, al-Rashīd. According to Abū Sahl, who was building on and repurposing a story drawn from the Middle Persian sources of the near-contemporaneous *Dēnkard* (Rezania 2017; for the difficulties surrounding the dating and interpretation of this and other Middle Persian sources compiled during the Abbasid dynasty see Chapters I.1–I.2 and I.16), the sciences were a product of ancient Iranian culture that had then been scattered from Babylon to Egypt, Greece, India and China – particularly by that great enemy of the ancient Persians, Alexander III of Macedon (r. 336–323 BCE), who, we are told, destroyed the ancient buildings and plundered and burned the ancient books. While most of this allegedly Persian knowledge was sent to Egypt, some also found its way to China and India. Then, we are told that the first Sasanian rulers, Ardashīr I (r. 224–242) and Sābūr I (r. 240–270), engaged in a project of recalling the dispersed ancient knowledge from India, China and Byzantium and producing, once more, Persian compilations of the ancient sciences. Although much of this account is clearly fantasy, it ends with a list of the names of authors of books that we are meant to believe were then circulating in Middle Persian: Hermes the Babylonian, Dorotheus the Syrian, Fydrws the Athenian, Ptolemy the Alexandrian and Farmāsib the Indian (Van Bladel 2012a). Although we do not know who all these people were or what books Abū Sahl is referring to, this account, nevertheless, makes it clear that he wants us to understand that there were learned books in Middle Persian in circulation whose contents were held to reflect ancient knowledge from many parts of the world.

Although our sources for the early Abbasid activity in the mathematical sciences are rather fragmented and often seem to have been written with the goal of advancing certain political or cultural agendas, a few things do become clear. Under the first three Abbasid caliphs, Baghdad became an important center for the mathematical and astral sciences, and a fair number of astrologers, experts in calculation, worked there in the employ of the caliphs and their viziers. These men tended to practice, or at least advocate, the astral sciences of their own individual traditions, but, in fact, they almost always blended in elements from other traditions as well. This fusion is, indeed, one of the most distinctive features of the reports of the work that survive from this period. Indeed, in the texts of the following generations, which constitute almost the first treatises of the mathematical sciences in Arabic that have come down to us, we find already a mixed usage of Greek, Indian and Persian sources – if not in the same work, at least in different works by the same author.

I.3.4 The House of Wisdom

An institution that probably played some role in the transmission and preservation of the ancient sciences was the so-called *House of Wisdom* (*bayt al-ḥikma*) – also known as the "*Storehouse of Wisdom*" (*khizānat al-ḥikma*) and related terms (Chapter III.1). This institution was a caliphal library or housed such a library. We do not know anything about the architectural or organizational structure of the *bayt al-ḥikma*; we do not know when it was founded or when it ceased to exist; we do not know how it was funded or on what kind of budget it operated; we do not know what, if any, relation it had to other institutions of the Abbasid state. In fact, the modern views on the nature and function of this institution have varied considerably (Eche 1967, 9–65; Balty-Guesdon 1992; Gutas 1998, 53–60; Di Branco 2012; Janos 2014, 421–40; Richter-Bernburg 2016; Chapter III.1), so it may be best to simply report what is said in the sources – mainly, Ibn al-Nadīm and Ibn al-Qifṭī (568–646/1172–1248).

The *House of Wisdom* had a director (*ṣāḥib*), sometimes called a director of books, as well as other associates, or functionaries, whose positions are not specified in our sources. It is stated to have been operating during the time of, and for, both al-Rashīd and al-Maʾmūn. It had some association with two of the most influential Central Asian and Persian families of the early Abbasid period, the Barmakids and the Nawbakhts: ʿAllān al-Shuʿūbī (late 2nd–early 3rd/8th–9th centuries) is said in the same phrase to have been attached to the Barmakids and to have transcribed at the *bayt al-ḥikma*; Salm (or Salmān; 1st half 3rd/9th century) is reported to have overseen work carried out on behalf of Yaḥya ibn Khālid on Ptolemy's *Almagest* – for which activity we have no other evidence, and Abū Sahl al-Faḍl ibn Nawbakht is said to have translated Persian texts at the *bayt al-ḥikma* for al-Rashīd. The men who at one time or another were associated with the *bayt al-ḥikma* are known (a) for their work with manuscripts and books themselves, such as al-Shuʿubī, who transcribed; Ibn Abī l-Ḥarīsh (1st half 3rd/9th century), who bound books; and Salmān, who was part of a group sent to collect Greek books by al-Maʾmūn; (b) for belles lettres, such as Sahl ibn Hārūn (d. 215/830) and Saʿīd ibn Hārūn (or Hurayn; 1st half 3rd/9th century), both known for their eloquence; (c) for translations from Persian or from Greek, such as Abū Sahl, Salm and the Banū Mūsā (see the earlier discussion); or, (d) for their work in the exact and astral sciences, such as al-Khwārazmī, who wrote technical treatises in a number of the exact sciences; Yaḥyā ibn Abī Manṣūr (d. 215/830), who carried out observations and wrote technical works, of which the *Verified Tables* (*al-Zīj al-mumtaḥān*) was deposited in the library (Janos 2014, 432); and the Banū Mūsā, who, as well as translations, produced a number of original works in the ancient – especially, mathematical and astral – sciences (Gutas and Van Bladel 2009).

Although it is now not possible to know the details of how it functioned, it seems that the *House of Wisdom* played some role in the lives of a number of important scholars and increasingly during the reign of al-Maʾmūn in the lives of scholars who were producing original work in the exact and astral sciences, especially in those traditions that they traced back to Greek sources. In this sense, however, these scholars were probably simply following a general trend of new production in the mathematical sciences along with an emphasis on critical examination of the ancient sources that increasingly privileged Greek, over Indian and Iranian, texts.

I.3.5 New treatises, new translations

Throughout the 3rd/9th century, we can observe in our sources two parallel and complementary trends – on one hand, the production of new, sometimes highly original, treatises in the ancient sciences, often combining elements of formerly disparate traditions, and, on the other hand, the elaboration of new translations and editions of the most technically difficult works of the ancient sciences, often returning again and again to the same text and sometimes combining the various strengths of a number of different scholars. We may make the case for this characterization by considering the work of just a few of the known individuals from three generations of scholars working in the Abbasid sphere in the 3rd/9th century.

There is evidence of a strong interest in, and support of, the astral sciences during the reign of al-Maʾmūn. We are aware of the names, and have some of the work, of about 25 individuals – including one of al-Maʾmūn's wives – who were experts in the exact and astral sciences during his reign (Janos 2014, 406–13), although, of course, many of them had been active from before al-Maʾmūn assumed the caliph's cloak. Some of these individuals were attached to the caliph himself, or his viziers and courtiers, both in Marv and in Baghdad, and at least one of them followed him on his military campaigns. Some of them were powerful members of al-Maʾmūn's inner circle; some undertook observational programs at the caliph's command and, according to some of the sources, sometimes under his direct oversight; some wrote original treatises in the exact and astral sciences; and some carried out projects of transmission and translation of the ancient sciences into Arabic. A number of them engaged in all of these activities. From the perspective of the ancient and medieval history of the exact and astral sciences, this represents a significant number of known individuals active in a single social and political context.

One of the most famous men to work in the exact sciences during al-Ma'mūn's reign was al-Khwārazmī, who would later come to be regarded as one of the most important mathematicians of this period. He was attached to the *House of Wisdom* and wrote works on the art of calculation, premodern algebra, astronomical tables, calendrics, mathematical geography and astrological history, which incorporated various elements from the different traditions of the ancient sciences – especially those of India and Greece. His *Book of Indian Calculation* (*Kitāb al-ḥisāb al-hindī*), which is known only from Latin revisions, sets out rules for computing with Hindu–Arabic integers, as well as common and sexagesimal fractions (al-Khwārazmī 1990; al-Ḫwārizmī 1997, 2001; Chapter I.7). His *Algebra* (*Kitāb al-jabr wa-l-muqābala*) contains his new contribution to this ancient practice. He presents a standard set of six equations, with their solution, to which other equations can be reduced, along with proofs for the solution of the three composite equations. The rest of the book involves the solution of problems, mostly drawn from the Islamic science of inheritance (al-Khwārazmī 2009). This book exerted a profound influence on 3rd–4th/9th–10th-century mathematics and set off an intense development of premodern algebra, for example, by al-Ṣaydanānī (*fl. c.* 235/850), Abū Kāmil (*c.* 235–*c.* 317/*c.* 850–*c.* 930), Sinān ibn al-Fatḥ (1st half 4th/10th century) and al-Karajī ([d *c.* 419/1029]). Al-Khwārazmī's *Sindhind Tables* (*Zīj al-Sindhind*), which only survives in Latin and Hebrew versions, was apparently a reworking of al-Fazārī's astronomical handbook of the same name and preserved its Indo-Iranian framework but included a number of topics and elements from Ptolemaic, Greek astronomy (van Dalen 1996; King *et al.* 2001, 33–6). This handbook served as the basis of a number of commentaries and played an important role in the development of 3rd/9th-century astronomy in the Islamicate world. In all these texts, al-Khwārazmī sought to bring together various practices from the different ancient traditions and elements of Arabic and Islamic practice, such as finger reckoning or the divisions of inheritance according to Islamic law.

Yaḥyā ibn Abī Manṣūr was the son of one of al-Manṣūr's astrologers and had himself served al-Ma'mūn's vizier before converting to Islam and becoming a companion (*nadīm*) of the caliph. He was also attached to the *House of Wisdom* and oversaw a project carried out at the request of al-Ma'mūn to compare the astronomical works of the Greeks, Indians and Persians, which, at least according to Ḥabash al-Ḥāsib (d. after 255/869), came to the conclusion that Ptolemy's *Almagest* was the most correct of the ancient texts dealing with the astral sciences (Ḥabash al-Ḥāsib 1955, 142; Janos 2014, 436). He was involved in, and perhaps substantially carried out, the production of the *Verified Tables*, composed on the basis of new observations, in which he played a key role, that were made in Baghdad, allegedly under al-Ma'mūn's direction. While produced in a generally Ptolemaic framework, this work also incorporated material from Indian and Persian sources (Ḥabash al-Ḥāsib 1955, 142; van Dalen 2004; Chapters II.7 and III.5).

In this same generation, al-Ḥajjāj ibn Yūsuf ibn Maṭar (d. after 213/828) was engaged in various book-collection and translation efforts, dealing with Greek sources. He is said to have made two translations of Euclid's *Elements*, one for al-Rashīd and one for al-Ma'mūn, or more likely for their viziers (Brentjes 2008, 443–6; De Young 2016, 2-3). He apparently made a translation of the *Almagest*, which was neither the first nor the last. According to Ibn al-Nadīm, he was a member of an embassy, which included the director of the *House of Wisdom*, sent by al-Ma'mūn into Greek-speaking lands for the purpose of collecting books, some of which the caliph then sent to be translated (Ibn al-Nadīm 1970, 2, 584).

In the next generation, there were a number of influential men, from prominent Muslim Arab and Iranian families, who worked to advance the ancient sciences and turned their focus squarely on the Greek tradition. A key figure in this group was al-Kindī, who continued the late ancient project of trying to reconcile the philosophical and scientific ideas of authors that strike most modern readers as fundamentally incompatible, such as the philosophies of Plato (d. 348–347 BCE) and Aristotle, the cosmologies of Aristotle and Ptolemy and the optics of Aristotle and Euclid (Adamson 2005). In the course of his wide-ranging work, he and his colleagues produced a number of summaries and translations that, to our eyes, are closer to paraphrases meant to convey the ideas of an ancient text, with little attempt at textual fidelity – such as a loose summary of Ptolemy's *Almagest* (Rosenthal 1956; Gutas 1998, 145–7; Adamson 2005, 32–3). Al-Kindī wrote extensively in the exact and astral sciences (Ibn al-Nadīm 1970, 2, 615–26). In this work, he makes it clear that he saw himself as advancing the tradition of the ancient sciences, which he viewed as a cumulative project based on a full assessment and thorough critique of what had gone before and making this accessible to his fellow speakers of Arabic (Walzer 1945, 172–5; Rosenthal 1956, 445).

In the same generation, and apparently in fierce competition with al-Kindī, were Muḥammad, Aḥmad and al-Ḥasan ibn Mūsā – the sons of Mūsā ibn Shākir (early 3rd/9th century), a Central Asian warlord and practitioner of the astral sciences (*munajjim*) who became one of al-Ma'mūn's companions during his time in Marv. According to our sources, the brothers were raised at the caliph's order by a certain Isḥāq ibn Ibrāhīm al-Muṣʿabī (d. *c.* 235/850) and educated by Yaḥyā ibn Abī Manṣūr at the *House of Wisdom* (Banū Mūsā 1979[CB], 3–6). They administered various projects of scientific, engineering and political significance; became wealthy and influential; and entered into the dangerous politics of the palace. They produced original works in geometry, mechanics, music and the astral sciences, and they used their wealth and influence to support a number of scholars and translators – including financing book-collection trips in Byzantine, or formerly Byzantine, lands and the full-time support of translators (Ibn al-Nadīm 1970, 2; 584–5). For all their conflict with al-Kindī and his circle, however, the approach of the Banū Mūsā also had the effect of emphasizing the importance of the Greek tradition, in preference to that of India or Iran, for their work in the exact sciences.

Two other mathematical scholars in this generation should also be discussed: al-Farghānī (3rd/9th century) and the already mentioned Ḥabash al-Ḥāsib – for both of whom we have scant bibliographic information. They both worked in the exact and astral sciences, with particular attention to techniques and devices used for analog computation, such as the astrolabe. In the introduction of his treatise on the astrolabe, al-Farghānī states that he intends to give demonstrations of the correctness of this ancient device, which will provide a theoretical understanding of the instrument (al-Farghānī 2005, 24–5), and indeed when we compare his work with the much early descriptions of the astrolabe by John Philoponus (d. 574) or Severus Sēbōkht (John Philoponus 1839; Severus Sēbōkht 1899; Gunter 1932, 61–103), we find that his project is a sort of meeting ground between these practical instruction manuals and the mathematical approach of Ptolemy's *Planisphere* (*Kitāb . . . fī tastīḥ basīṭ al-kura*; Ptolemy 2007), which gives a geometrical method for modeling the sphere in a plane. In fact, al-Farghānī has gone much beyond either of these ancient traditions by including proofs for a number of mathematical facts that are simply assumed by Ptolemy, providing mathematical tables for the production of various lines and both mathematical and physical descriptions of the construction and usage of various aspects of the device.

The works that have come down to us from Ḥabash include a new astronomical handbook and a number of treatises on mathematico-astronomical instruments, all of which build on, but go considerably beyond, the work of his predecessors. In his astronomical handbook, Ḥabash continued Yaḥyā ibn Abī Manṣūr's project of producing a work in a Ptolemaic framework but included new parameters from the observational projects carried out under al-Ma'mūn and incorporated various techniques developed in the Indian and Persian traditions, for example, the use of the Indian trigonometric functions in the sections on spherical trigonometry (Debarnot 1987; King *et al.* 2001, 37–9). In two of his surviving works on astronomical instruments, Ḥabash describes the construction of novel devices for which we have no previous evidence of any kind, in the course of which he makes it clear that he had full mastery of certain aspects of both Indian and Greek mathematical traditions for which we now have little evidence in our surviving sources from these traditions (Kennedy *et al.* 1999[CB]; Ḥabash al-Ḥāsib 2001). In this regard, he explicitly asserts that he is working in an ancient tradition that is advanced through a critique of what has gone before by checking its results against new observations, with a willingness to constantly correct both one's own and one's predecessors' work. Indeed, in the introduction to his astronomical handbook, he quotes a passage from Ptolemy's *Almagest* to this effect and places himself squarely within this tradition (Ḥabash al-Ḥāsib 1955, 143–4).

Both of these scholars demonstrate considerable mastery of the mathematical traditions of India and Greece, including an understanding of what had already been done and what remained to do. Nevertheless, it is not necessary for us to suppose that they had at their disposal full translations of all the ancient texts with which they worked. Many texts were probably known only in summaries (Chapter III.5), some of the techniques were probably transmitted orally and it is sometimes possible for scholars to grasp the essentials of a work in their own field written in a language over which they do not have full control. We will see a clear example of such a scholarly project that involved direct study of a Greek source by scholars who were not known for their ability in Greek in the following section.

The generation of some of the most important translators of Greek texts in the mathematical and astral sciences into the Arabic language – namely, Isḥāq ibn Ḥunayn (d. 298/910–1), Qusṭā ibn Lūqā (d. *c.* 299/912) and Thābit ibn Qurra (d. 288/901) – followed three or four generations of scholars who worked intensively in these areas, producing their own original treatises, epitomes and summaries of past work, as well as scholarly translations of texts in the ancient sciences. Isḥāq was the son of Ḥunayn ibn Isḥāq (d. 260/873), the famous physician and medical translator whose work was supported by the Banū Mūsā, as well as wealthy Christian Syrian physicians in the milieu of the court (Watt 2014). Under his father's tutelage, Isḥāq would have learned the critical method of comparing multiple manuscripts to produce a single text before producing a scholarly translation (Ḥunayn ibn Isḥāq 2016, 10–11). Qusṭā was a Christian Greek physician, who brought Greek manuscripts with him from Baʿlabakk to Baghdad, and spent a number of decades in Baghdad and later Armenia writing original treatises in Arabic, mostly in medicine but also in the exact sciences, and translating Greek works, mostly in the exact sciences (Gabrieli 1912; Wilcox 1987). Thābit, a member of the pagan Sabian community of Ḥarrān, made an impression on Muḥammad ibn Mūsā with his linguistic abilities, was brought to Baghdad, possibly as a slave boy (*ghulām*) or charge (*ṣanīʿ*; Mimura 2020), and was educated by the Banū Mūsā in the most advanced mathematical and medical sciences of the time. He remained associated with them in various ways in the following years, even himself educating Muḥammad's children (al-Khwārazmī 2009, 15–24). These men produced a large number of original treatises in the various fields in which they worked, as well as translations and revisions of the translations of others, and in most cases, we can identify clear connections between their translation work and their own research activities and interests (Rashed 1989, 202).

It was in this generation that high-quality translations and editions of texts in the Greek mathematical sciences reached their greatest output. But in those cases in which we can compare these Arabic translations with Greek sources, there are often enough differences in the Arabic text that it is difficult for us to characterize them as straightforward scholarly translations of the Greek texts that we have received. For some of the most important treatises, such as Euclid's *Elements* and Ptolemy's *Almagest*, it may still be too early to render judgment, because the Arabic *Elements* is found in multiple versions, with blending and has only been partially edited, while the Arabic *Almagest* has yet to be critically edited. In most cases in which we can compare critically edited Greek sources with critical editions of the Arabic versions of these same texts, however, we find that the Arabic text is different in many places. Furthermore, it is often possible to explain many of these differences as the result of deliberate intervention on the part of the medieval scholars – either the original translators or correctors or copyists in the Arabic tradition. In other cases, there appear to have been changes made to the Greek texts by scholars working in that tradition, after the Arabic translations were made. Hence, we must consider each situation in its local circumstances. A few examples will suffice to make this point.

As a first example, we may consider the last of the introductory books of conic theory produced by Apollonius of Perga (*c.* 262–*c.* 190 BCE), *Conics* IV, which comes down to us in rather different versions in the Greek text edited by Eutocius of Ascalon (late 5th–mid-6th century) and in the Arabic version made by, and under the direction of, the Banū Mūsā. In particular, in the Arabic edition, the propositions have more detailed arguments, and the overall organization of the theory is superior from a mathematical perspective (Apollonius de Perge 2008–2010, 2.2, 12–21). Now, although it has been argued that this difference is due to the fact that the scholars working on the Banū Mūsā's project had access to better sources, closer to Apollonius's original, it is just as possible that the mathematical improvements to the texts are a result of the extensive editorial project carried out under the Banū Mūsā (for details of which, see the following section).

Another example can be drawn from the case of Euclid's *Data*. Thābit's version of Euclid's *treatise* is often somewhat different from either version of the text we find in the Greek manuscripts – some of these changes are probably due to changes introduced in the later traditions, both Greek and Arabic, but the majority of the substantial changes are most likely due to Thābit himself (Thābit ibn Qurra 2018[CB]). The arguments and diagrams are often somewhat different, but most of these changes are clear improvements. For example, there is a mathematical error in the Greek versions of *Data* 74 (Euclid 1896, 139, no. 1; Euclid 2003, 184–7). This mistake is not found in Thābit's version. There, the arguments for *Data* 74 and 75 are analogous, both sound, and both somewhat different from those in the Greek traditions (Thābit ibn Qurra 2018, 291–2). The most obvious explanation is that Thābit identified this error and reworked the text to improve the mathematical exposition.

The situation is even more striking in the case of Qusṭā's translation of books IV through VII of Diophantus's (3rd- or 4th-century) *Arithmetics*, which have been lost in Greek. Before his translation of the description of a third basic operation that can be carried out on equations, he uses the standard Arabic phrase for premodern algebra (*al-jabr wa-l-muqābala* = restitution and reduction), which originally denotes operations but does not correspond to the literal meaning of anything in the surviving parts of the Greek text – although the Greek text does fully describe these two operations (Diophantus 1982, 88, 284; Diophante 1984, 2–4; Diophantus Alexandrinus 1893–1895, 14). In fact, however, in the extant Greek text, when Diophantus, in the course of working out a problem, wants to apply both of these two operations, he consistently says, "Let a common, the lacking, be set out, and let the same be subtracted from the same" (*koinē proskeisthō hē leipsis kai aphērēsthō apo homoiōn homoia*; Diophantus Alexandrinus 1893–1895, 26, 28, 30, 90, 98, 257, 444), which states the same operations as the Arabic expression *al-jabr wa-l-muqābala*, and in the same order. It seems clear that when he translated this treatise Qusṭā disregarded the literal meaning of whatever he found in the Greek in favor of expressing the resulting Arabic text in the new terminology that had developed from the contributions of al-Khwārazmī and others to this type of problem-solving. Indeed, in general, the literal meaning of the Greek text has been replaced by the terminology introduced by al-Khwārazmī in his *Algebra*. Furthermore, the Arabic text contains material that completes or clarifies the argument, as well as verifications, solutions to the equation, and final statements, which are usually absent in the extant Greek text (Diophantus 1982, 29–33, 48–50). While it is sometimes claimed that this added material must have been in Qusṭā's Greek prototype (Diophantus 1982, 60–1), we have no certain confirmation of this position. It is equally possible that some of it, and in particular the detailed verifications of solutions, was introduced by Qusṭā himself. If this were the case, it would mean that Qusṭā was reorienting the text toward a focus on the solution of the equation, in line with al-Khwārazmī's reorientation of premodern algebra itself.

In these cases, we appear to have translations of the sort that have been called "reader-orientated" (Brock 1983, 4–5) – that is, these scholars appear to have felt free to alter the received text in order to make the meaning clearer, remove what they saw as mistakes and render the finished product more useful to mathematical scholars of their own time. The goal was most likely to produce a text of use in contemporary mathematical teaching and research, not historical scholarship.

Over the course of this century, mathematical scholars of the Muslim world articulated a clear concept of the mathematical sciences as the product of ancient cultures that progress through a constant, critical reevaluation of received knowledge, and they explicitly placed themselves within this tradition. We also perceive a distinct turn toward the Greek tradition. In the beginning of the century, Muḥammad al-Fazārī was writing poetry in the exact sciences in imitation of his Indian sources (Thomann 2014), but by the middle of the century, the competing groups around al-Kindī and the Banū Mūsā turned their attention to Greek sources and wrote original treatises emulating and advancing this material while still including elements of the Indian and Iranian traditions. Indeed, around the end of the 3rd/9th to the beginning of the 4th/10th century, the major works of the Greek tradition became the subject of critical studies and commentaries, and this work, along with the original treatises of the 3rd/9th century served as the basis of a flowering of the exact sciences in the 4th/10th century (Thomann 2014, 2017). Although Indian- and Persian-sourced methods and concepts were still used and discussed by mathematical scholars, it was not until the beginning of the 5th/11th century, with the work of al-Bīrūnī, that attention would again be directed to Indian texts as a source for translating treatises in the ancient mathematical sciences.

I.3.6 Apollonius's conic theory, a detailed example

For many of the specific sources and techniques of the ancient sciences, we know next to nothing about the process of their transmission into Islamicate scholarly circles, but for the theory of conic sections, as developed by Apollonius, we have a detailed firsthand narrative. Although it features a number of standard tropes and was probably written to emphasize the importance of the role played by the Banū Mūsā themselves in the overall project, because of the relevance of this story to the topic of this chapter it is worth going through it at some length.

In the introduction to their version of Apollonius's *Conics*, the Banū Mūsā give a fascinating account of their work on the treatise (Apollonius 1990, 620–9; Apollonius de Perge 2008–2010, 1.1.500–7). After making the claim that the ancients regarded the theory of conics as the apex of geometry and that Apollonius, who had mastered it, composed a treatise in eight books, they assert that this work had undergone corruption, both through the normal course of the manuscript transmission and because none of those copying it understood its contents. They then state that the situation was somewhat remedied by Eutocius, who produced a restoration (*iṣlāḥ*) of the text, both by collating manuscripts and by reworking the mathematical material in places where it no longer made sense. They then go on to claim that still in their own time very few understand geometry, and failing to comprehend the works of Euclid, some even put forward invalid proofs of false propositions.

Next, they give an account of the project that they directed and funded, namely, to produce a new translation and restoration of the *Conics*. At some point, they came into possession of a Greek manuscript of Books I through VII of the treatise, in its original form, but they could not understand it, due to the accumulation of errors. Then, al-Ḥasan ibn Mūsā made an investigation of the section of the cylinder, based on the mathematical characteristics of its diameters, axes and chords, including a theory of its area. This he compared with the closed section of a cone, the ellipse, and showed that the latter was the same curve as the section of the cylinder. After writing a treatise on his mathematical discoveries, al-Ḥasan passed away. The next breakthrough in this project came when Aḥmad ibn Mūsā took up a position as administrator of the postal service in Syria, where he was able to find a copy of the first four books of the *Conics* in the restoration by Eutocius. He studied these books and commented on them so that when he returned to Iraq, he was able to make this study the basis for his version of *Conics* I through VII, including explicit references to previous propositions where they are needed to justify steps in the argument. Finally, we are told that Aḥmad oversaw the translation, which was carried out by Hilāl al-Ḥimṣī (d. c. 266/880), for *Conics* I through IV, and Thābit ibn Qurra, for *Conics* V through VII.

Although this is just one area, indeed one text, of the mathematical sciences, and although other sources indicate that the situation with this text was probably more complicated and that there were once other Arabic versions of it in circulation (Apollonius de Perge 2008–2010, 29–44), the details of this episode give us insights into these processes that might not otherwise be clear. In the first place, the understanding and subsequent translation of technical works were long-term endeavors that could involve a number of different individuals. In the case of the *Conics*, it probably took at least some five, and perhaps as many as fifteen, years. Just as in Ḥunayn ibn Isḥāq's description of his repeated attempts, over some decades, to master and translate certain of the works of Galen into Syriac and Arabic (Ḥunayn ibn Isḥāq 2016), it is clear that the scholars working on the *Conics* returned to it and its subject matter again and again, and attempted, so far as possible, to collate manuscripts and apply their understanding of the technical material involved in order to rectify what they read in their sources.

Another key point is that, in this account, mathematical, as opposed to philological, scholarship is emphasized as the most crucial element in understanding the *Conics*. Of course, part of this may have been a desire on the part of the Banū Mūsā to underline their own contribution to the project. Nevertheless, it is clear that much of the work was carried out through direct study of Greek manuscripts before a complete translation – or at least a full and satisfactory translation – had been made. For example, when Aḥmad was in Syria, he studied and commented directly on a Greek manuscript. Perhaps this was done in consultation with colleagues who were proficient in Greek, but this is not mentioned. It is also possible that he knew enough Greek to make some sense of a text that he had been working on for years. This gives us the impression that in working on ancient mathematical sources a considerable amount was learned directly from the ancient manuscripts prior to the production of a complete translation. This is corroborated by Ibn al-Nadīm's description of the study of the *Almagest* that had been overseen by Yaḥyā ibn Khālid ibn Barmak, presumably during the reign of al-Rashīd (Ibn al-Nadīm 1970, 2, 639). Finally, the discussion of Eutocius's work in this passage makes it clear that the concept of a restoration (*iṣlāḥ*) of a text was here used to mean a rectification of the source based on both philological and technical considerations, the primary goal of which was the production of a text that would be useful for further scientific work, not one that was strictly faithful to the manuscript sources. This interpretation of the goal of editorial work is reinforced by studies of other Arabic versions of ancient Greek sources, such as al-Harawī's (d. between c. 380/990 and 390/990) version of Menelaus's *Spherics* or Thābit's restoration of Euclid's *Data* (Sidoli and Kusuba 2014; Menelaus 2017; Thābit ibn Qurra 2018). By their ascription of these practices to Eutocius, whether true or not, the Banū Mūsā articulate the idea that the rectification of an ancient source along both philological and technical lines is a necessary and proper continuation of an ancient practice – and by articulating their own work in these very terms they make the implicit claim that they are the true heirs, and only real current practitioners, of this ancient science.

I.3.7 Conclusion

By the end of the 2nd/8th century, the basic computational methods of the practical, administrative fields had been transmitted into Arabic, and a group of experts in the astral sciences from various cultures had begun to gather around the caliphal court. Although there was certainly knowledge about the ancient mathematical sciences, particularly in the Indian and Iranian traditions, it is not clear to what extent the ancient texts themselves were really mastered by the Christian, Jewish, Muslim and Zorastrian scholars of this generation. In the first half of the 3rd/9th century, however, there were major efforts to codify and organize the technical knowledge that was then in circulation, as well as to organize projects of translation – mostly focused on Syriac, Middle Persian and Greek sources. Important ancient texts were studied in detail and summaries and epitomes were produced. The different ancient traditions were compared against each other and against new observations. New treatises were composed synthesizing and extending this knowledge, which produced technical methods and genres of texts not found in the ancient sources. In this process, various failings of the ancient sources were identified and discussed, and some novel topics were addressed for the first time (Dallal 2010, 32–5). In the following generation, original treatises were composed that went well beyond any material that we find in the ancient sources, and which incorporate what were regarded as the best elements of the ancient traditions, particularly those of India and Greece. Many of the most important, theoretical treatises of the ancient sciences were mastered, translated and corrected – projects that involved both philological and technical command of difficult material. In this generation, there was a distinct shift toward Greek sources and Greek conceptions of the scientific enterprise – some of the more socially prestigious, Muslim members of this generation, such as al-Kindī and the Banū Mūsā, explicitly framed their activity as a continuation of the Greek tradition. Finally, a generation of scholars, trained in this rich intellectual milieu, produced both original treatises and scholarly translations and restorations of the ancient – that is, now primarily Greek – texts that they regarded as canonical. By the end of the century, we can see that the 3rd/9th-century scholars had, in fact, produced new styles of the exact and astral sciences through the process of further hybridizing and critiquing the ancient traditions (Dallal 2010, 26–43). They had set the foundations for the development of new genres of text, and indeed, new mathematical sciences that had been at best only adumbrated in ancient sources – such as separating out parts of what would later become trigonometry from its ancient context in astronomical writings, laying the basis for what would become the science of the configuration (of the universe; *ʿilm al-haʾya*), and clearly delineating algebra (*al-jabr*) as an independent science. By combining the interests, methods and goals of sources from the Greek, Indian and Iranian traditions, and subjecting these to critical scrutiny, these mathematical scholars were able to identify and articulate a concept of the ancient exact sciences as a critical, cumulative human endeavor and to position themselves as expert practitioners of this enterprise.

Note

1 Consolidated bibliography.

Bibliography

Sources

Apollonius. ed., tr. and ann. Toomer, G. J. 1990. *Conics Books V to VII: The Arabic Translation of the Lost Greek Original in the Version of the Banū Mūsā*. New York: Springer.

Apollonius de Perge. ed., tr. and ann. Rashed, R., Decorps-Foulquier, M., and Federspiel, M. 2008–2010. *Coniques*. 5 vols. Berlin: W. de Gruyter.

al-Bīrūnī. tr. and ann. Sachau, E. C. 1910. *Alberuni's India*. 2 vols. London: Kegan Paul, Trench, Trübner.

al-Bīrūnī. ed., tr. and ann. Kennedy, E. S. 1976. *The Exhaustive Treatise on Shadows by Abū al-Rayḥān Muḥammad b. Aḥmad al-Bīrūnī*. Aleppo: University of Aleppo.

Diophante. ed. Rashed, R. 1984. *Les Arithmétiques* [The Arithmetics]. Paris: Les Belles Lettres.

Diophantus. ed., tr. and com. Sesiano, J. 1982. *Books IV to VIII of Diophantus' Arithmetica in the Arabic Translation Attributed to Qustā ibn Lūqā*. New York: Springer.

Diophantus Alexandrinus. ed. Tannery, P. 1893–1895. *Opera omnia cum Graecis commentariis* [Complete Works with the Greek Commentaries]. 2 vols. Leipzig: Teubner.

Euclid. ed. Menge, H. 1896. *Euclidis* Data *cum commentario Marini et scholiis antiquis* [Euclid's *Data* with Marinus' Commentary and Ancient Glosses] (Euclidis opera omnia 6). Leipzig: Teubner.

Euclid. ed., tr. and ann. Taisbak, C. M. 2003. ΔΕΔΟΜΕΝΑ, *Euclid's Data: The Importance of Being Given*. Copenhagen: Museum Tusculanum Press.

al-Farghānī. ed., tr. and ann. Lorch, R. 2005. *On the Astrolabe*. Stuttgart: Franz Steiner.

al-Fazārī. Pingree, D. 1970. "The Fragments of the Works of al-Fazārī," *Journal of Near Eastern Studies* 29: 103–23.

Folkerts, M. 2001. "Early Texts on Hindu-Arabic Calculation," *Science in Context* 14: 13–38.

Ḥabash al-Ḥāsib. ed., tr. and com. Sayılı, A. 1955. "The Introductory Section of Habash's Astronomical Tables Known as the 'Damascene' Zīj," *Ankara Üniversitesi Dil ve Tarih-Coğrafya Fakültesi Dergisi* 13: 133–51.

Ḥabash al-Ḥāsib. ed., tr. and ann. Charette, F. and Schmidl, P. G. 2001. "A Universal Plate for Timekeeping by the Stars by Ḥabash al-Ḥāsib: Text, Translation and Preliminary Commentary," *Suhayl* 2: 107–59.

al-Hāshimī, ʿAlī ibn Sulaymān. intr. Haddad, F. I., Kennedy, E. S. and Pingree, D. 1981. *The Book of the Reasons Behind Astronomical Tables (Kitāb fi ʿilal al-zījāt)*. Delmar, New York: Scholars' Facsimilies & Reprints.

al-Ḫwārizmī. ed., tr. and com. Folkerts, M., with Kunitzsch, P. 1997. *Die älteste lateinische Schrift über das indische Rechnen nach al-Ḫwārizmī* [The Oldest Latin Work about the Indian Calulation According to al-Ḫwārizmī]. Munich: Bayerische Akademie der Wissenschaften.

al-Khwārazmī. tr. Crossley, J. N. and Henry, A. S. 1990. "Thus Spake al-Khwārizmī: A Translation of the Text of Cambridge University Library Ms. Ii.vi.5," *Historia Mathematica* 17: 103–31.

al-Khwārazmī. ed., tr. and ann. Rashed, R. 2009. *Al-Khwārizmī: The Beginnings of Algebra*. London: SAQI.

Menelaus. ed., tr. and ann. Rashed, R. and Papadopoulos, A. 2017. *Menelaus' Spherics: Early Translation and al-Māhānī/al-Harawī's Version*. Berlin: Walter de Gruyter.

Philoponus, John. ed. Codd, E. and Hase, H. 1839. "Joannis Alexandrini, cognomine Philoponi, de vsu astrolabii ejusque constructione libellus," *Rheinisches Museum für Philologie* 6: 127–71.

Ptolemy. ed., tr. and com. Sidoli, N. and Berggren, J. L. 2007. "The Arabic Version of Ptolemy's Planisphere or Flattening the Surface of the Sphere: Text, Translation, Commentary," *SCIAMVS* 8: 37–139.

Severus Sēbōkht. ed., tr. and ann. Nau, F. 1899. "Le traité sur l'astrolabe plan de Sévère Sabokt," *Journal asiatique* 9: 56–101, 238–303.

Severus Sēbōkht. tr. and com. Reich, E. 2000. "Ein Brief des Severus Sēḇōḵt," in Folkerts, M. and Lorch, R., eds. *Sic itur ad astra: Studien zur Geschichte der Mathematik und Naturwissenschaften. Festschrift für den Arabisten Paul Kunitzsch zum 70*. Wiesbaden: Harrassowitz, 478–789.

Yabuuti, K. with Yano, M. 1979. "Researches on the Chiuh-chih li: Indian Astronomy under the T'ang Dynasty," *Acta Asiatica* 36: 7–48.

Yaʿqūb ibn Ṭāriq. ed., tr. and com. Pingree, D. 1968. "The Fragments of the Works of Yaʿqūb ibn Ṭāriq," *Journal of Near Eastern Studies* 27: 97–125.

Research literature

Adamson, P. 2005. "Al-Kindī and the Reception of Greek Philosophy," in Adamson, P. and Taylor, R. C., eds. *The Cambridge Companion to Arabic Philosophy*. Cambridge: Cambridge University Press, 32–51.

Balty-Guesdon, M.-G. 1992. "Le bayt al-ḥikma de Baghdad," *Arabica* 39: 131–50.

Beckwith, C. I. 1984. "The Plan of the City of Peace: Central Asian Iranian Factors in Early ʿAbbâsid Design," *Acta Orientalia Academiae Scientiarum Hungaricae* 38: 143–64.

Beckwith, C. I. 2009. *Empires of the Silk Road*. Princeton: Princeton University Press.

Borrut, A. 2014. "Court Astrologers and Historical Writing in Early ʿAbbāsid Baghdād: An Appraisal," in Scheiner and Janos, eds. [CB], 455–501.

Brentjes, S. 2008. "Euclid's *Elements*, Courtly Patronage and Princely Education," *Iranian Studies* 41: 441–63.

Brock, S. 1983. "Towards a History of Syriac Translation Technique," *Orientalia Christiana Analecta* 221: 1–14.

Christianidis, J. and Oaks, J. A. 2013. "Practicing Algebra in Late Antiquity: The Problem Solving of Diophantus of Alexandria," *Historia Mathematica* 40: 127–63.

Christianidis, J. and Megremi, A. 2019. "Tracing the Early History of Algebra: Testimonies on Diophantus in the Greek-Speaking World (4th–7th Century CE)," *Historia Mathematica* 47: 16–38.

Cullen, C. 1982. "An Eighth Century Chinese Table of Tangents," *Chinese Science* 5: 1–33.

Dalen, B. van. 1996. "Al-Khwārizmī's Astronomical Tables Revisited: Analysis of the Equation of Time," in Casulleras and Samsó, eds. [CB], 195–252.

Dalen, B. van. 2004. "A Second Manuscript of the Mumtaḥan Zīj," *Suhayl* 4: 9–44.

Dallal, A. 2010. *Islam, Science, and the Challenge of History*. New Haven: Yale University Press.

De Young, G. 2016. "The Latin Translation of Euclid's /Elements/ Attributed to Adelard of Bath: Relation to the Arabic Transmission of al-Ḥajjāj," in Zack, M., and Landry, E., eds., /Research in History and Philosophy of Mathematics/. Cham: Springer, 1–13.

Debarnot, M.-T. 1987. "The Zīj of Ḥabash al-Ḥāsib: A Survey of MS Istanbul Yeni Cami 784/2," in King, D. A. and Saliba, G., eds. *From Deferent to Equant: A Volume of Studies in the History of Science in the Ancient and Medieval Near East in Honor of E. S. Kennedy*. New York: New York Academy of Sciences, 35–69.

Di Branco, M. 2012. "Un' istituzione sasanide? Il Bayt al-ḥikma e il movimento di traduzione," *Studia graeco-arabica* 2: 255–63.

Eche, Y. 1967. *Les bibliothèques Arabes publique et semi-publiques en Mésopotamie, en Syrie et en Égypte au Moyen Age*. Damascus: Institut français de Damas.

Gabrieli, G. 1912. "Nota bibliographica su Qusṭā ibn Lūqā," *Atti della Reale Accademia dei Lincei: Rendiconti, classe di scienze morali, storiche e filologiche* Ser. 5, 21: 341–82.

Gunter, R. T. 1932. *The Astrolabes of the World*. Oxford: Oxford University Press.

Gutas, D. and Van Bladel, K. 2009. "Bayt al-ḥikma," in *EI-3*, 2009–2: 133–7.

Janos, D. 214. "Al-Maʾmūn's Patronage of Astrology: Some Biographical and Institutional Considerations," in Scheiner and Janos, eds. [CB], 389–454.

Mimura, T. 2020. "*Ghulāms* (Slave Boys) and Scientific Research in the Abbasid Period: The Example of the Amājūr Family," *Historia Scientiarum* 29: 182–97.

Oaks, J. A. 2009. "Polynomials and Equations in Arabic Algebra," *Archive for History of Exact Sciences* 63: 169–203.

Pingree, D. 1976. "The Indian and Pseudo-Indian Passages in Greek and Latin Astronomical and Astrological Texts," *Viator* 7: 141–95.

Pingree, D. 1990. "Astrology," in Young, M. J. L., Latham, J. D. and Serjeant, R. B., eds. *Religion, Learning and Science in the ʿAbbasid Period*. Cambridge: Cambridge University Press, 290–300.

Pingree, D. 1997. *From Astral Omens to Astrology, From Babylon to Bīkāner*. Rome: Istituto Italiano per l'Africa e l'Oriente.

Pingree, D. 2001. "From Alexandria to Baghdād to Byzantium: The Transmission of Astrology," *International Journal of the Classical Tradition* 8: 3–37.

Rashed, R. 1989. "Problems of the Transmission of Greek Scientific Thought into Arabic: Examples from Mathematics and Optics," *History of Science* 27: 199–209.

Rashed, R. 2006. "Greek into Arabic: Transmission and Translation," in Montgomery, J. E., ed. *Arabic Theology, Arabic Philosophy: From the Many to the One, Essays in Celebration of R. M. Frank*. Leuven: Peeters, 157–96.

Rezania, K. 2017. "The Dēnkard Against Its Islamic Discourse," *Der Islam* 94: 336–62.

Richter-Bernburg, L. 2016. "Potemkin in Baghdad: The Abbasid 'House of Wisdom' as Constructed by 1001 Inventions," in Brentjes, Edis and Richter-Bernburg, eds. [CB], 121–31.

Rosenthal, F. 1956. "Al-Kindī and Ptolemy," in *Studi Orientalistici in onore di Giorgio Levi Della Vida*. 2 vols (Pubblicazioni dell'Istituto per l'Oriente 52). Roma: Istituto per l'Oriente, 2: 436–56.

Sahas, D. J. 1972. *John of Damascus on Islam. The "Heresies of the Ishmaelites"*. Leiden: Brill.

Sidoli, N. and Kusuba, T. 2014. "Al-Harawī's Version of Menelaus' Spherics," *Suhayl* 13: 149–212.

Stewart, D. J. 2006. "Scholarship on the Fihrist of Ibn al-Nadīm: The Work of Valeriy V. Polosin," *Al-ʿUsūr al-Wusṭā: Bulletin of Middle East Medievalists* 18.1: 8–13.

Takahashi, H. 2010. "Between Greek and Arabic: The Sciences in Syriac from Severus Sebokht to Barhebraeus," in Kobayashi, H. and Kato, M., eds. *Transmission of Sciences: Greek, Syriac, Arabic and Latin*. Tokyo: Organization for Islamic Area Studies, Waseda University, 16–39.

Thomann, J. 2014. "From Lyrics by al-Fazārī to Lectures by al-Fārābī: Teaching Astronomy in Baghdād," in Scheiner and Janos, eds. [CB], 503–25.

Thomann, J. 2017. "The Second Revival of Astronomy in the Tenth Century and the Establishment of Astronomy as an Element of Encyclopedic Education," *Asia* 71: 907–57.

Van Bladel, K. 2011. "The Bactrian Background of the Barmakids," in Akasoy, A., Burnett, C. and Yoeli-Tlalim, R., eds. *Islam and Tibet – Interactions along the Musk Routes*. Farnham: Ashgate, 43–88.

Van Bladel, K. 2012a. "The Arabic History of Science of Abū Sahl ibn Nawbaḫt (fl. ca 770–809) and Its Middle Persian Sources," in Opwis, F. and Reisman, D., eds. *Islamic Philosophy, Science, Culture, and Religion*. Leiden: Brill, 41–62.

Van Bladel, K. 2012b. "Barmakids," in *EI-3*, 2012–3: 32–8.

Walzer, R. 1945. "Arabic Transmission of Greek Thought to Medieval Europe," *Bulletin of the John Rylands Library* 29: 160–83.

Watt, J. 2014. "Why Did Ḥunayn, the Master Translator into Arabic, Make Translations into Syriac? On the Purpose of the Syriac Translations of Ḥunayn and his Circle," in Scheiner and Janos, eds. [CB], 363–88.

Wilcox, J. C. 1987. "Our Continuing Discovery of the Greek Science of the Arabs: The Example of Qusṭā ibn Lūqā," *Annals of Scholarship* 4: 57–74.

I.4

TRANSLATIONS OF MEDICAL AND OCCULT TEXTS INTO ARABIC AND SYRIAC AND THEIR CONTEXTS AFTER 80/700

Peter E. Pormann

I.4.1 Introduction

The study of Greco-Arabic translations has a long and distinguished history, in which scholars have endeavored to identify individual translators who, collectively, make up a movement. Yet the original confidence with which pioneers like Gotthelf Bergsträsser (1886–1933; 1913) attributed translations to Ḥunayn ibn Isḥāq (d. 260/873) or Ḥubaysh ibn al-Ḥasan al-Aʿsam al-Dimashqī (2nd half 3rd/9th century) has largely evaporated. No less a scholar than Gerhard Endreß[1] (1989) exclaimed in the face of these difficulties: '*δός μοι ποῦ στῶ*' (give me somewhere to stand) and argued that Greek–Arabic lexicographical studies could shed new light on authorship. Through a number of such studies, Manfred Ullmann (2002, 2006, 2007, 2009, 2011, 2012, 2018) has identified a number of translators such as al-Biṭrīq ([2nd half 2nd/8th century]; translator of the old versions of the Hippocratic *Aphorisms* and Galen's (128–216) *On Simple Drugs* [*Fī l-Adwiya al-mufrada*]);[2] Eustathius ([d. 201/817]; translator of the *Agronomy* [*Fī l-Filāḥa*] and part of the *Nicomachaean Ethics* [*Akhlāq Nīqūmākhūs*]); Stephen, son of Basil (Iṣṭifān ibn Basīl, [1st half 3rd/9th century]; translator of Dioscorides's (d. 90) *Medicinal Substances*; and Ḥunayn ibn Isḥāq. Yet, with the exception of Ḥunayn ibn Isḥāq, who penned an *Epistle* (*Risāla*) about his translations of Galen's works (Ḥunayn ibn Isḥāq 2016[CB][3]), we know very little about these translators, who largely remain names without much historical context. But even Ḥunayn's *Epistle* poses some problems, as the dispute about who translated Galen's *On the Affected Parts* (*Fī l-mawāḍiʿ al-ālima*) shows: Ullmann (2006, 28–31) argued for Ḥunayn as the translator, whereas Garofalo (1995) was convinced that it must have been Ḥubaysh.

The studies mentioned in the previous paragraph focus on Greco-Arabic translation technique and often ignore the fact that there were previous translations that may have served as intermediaries, notably Syriac versions by authors such as Sergius of Rēsh ʿAynā (d. 536), which are largely lost today. Gutas (2012) has shown the pitfalls of ignoring the Syriac intermediary in the case of the *Poetics*, and there is a growing endeavor to study the place of Syriac in Ḥunayn's school with reference to Galen's *Simple Drugs* (Pormann 2012; Bhayro *et al.* 2013). Yet scholars continue to compare Greek and Arabic terms and phrases by putting them side by side in order to determine a translation according to lexicon, syntax and style. If, however, the

DOI: 10.4324/9781315170718-6

Arabic target text is the result of a centuries-long process of recurrent translation into Syriac and then Arabic, and if the Arabic translations (as far as we know from Ḥunayn's *Epistle*) are often the result of teamwork, how does it make sense to try to identify individual Arabic translators and their technique?

We have grown accustomed of speaking of translation 'schools', 'circles' and 'workshops', such as the 'Kindī circle' or 'Ḥunayn's school' or 'workshop'. But even these groups of translators are, at least in some cases, not even collectively responsible for the whole translation process, because they drew on earlier versions. If this is the case, can we then really still speak of Arabic versions being authored by one translator or one group of translators? Moreover, can we really speak of the Greco-Arabic translation movement? What do the early translations of al-Biṭrīq, al-Ḥajjāj ibn Yūsuf ibn Maṭar (d. after 213/828) and others have in common with those of the 3rd/9th and 4th/10th centuries? Recent scholarship has shown that the translations from various ages of the 'translation movement' differ greatly in style, approach and vocabulary. Surely both the societal and institutional settings and the motivations, as far as they can be ascertained, were equally different. If that is the case, should we not abandon the term *translation movement*, which suggests a certain unity of purpose, and rather investigate the various phenomena involved in translations?

To illustrate the problematic nature of the concept of a translation movement, I shall review a number of case studies that show what a complex phenomenon these Greco-Arabic translations are, focusing on areas in which there has been some recent scholarly progress. Before doing so, it would be remiss not to mention two influential accounts of the Greco-Arabic translation movement, namely those by Dimitri Gutas (1998[CB][4]) and George Saliba (2007[CB]). Although both differ substantially in interpretation and analysis, they view the translation movement as a unified historical phenomenon, and both endeavor to explain the causes that animate it. Gutas focuses on the political ideology of the Abbasids, who strove to include Arab and non-Arab clients and incorporated other cultural traditions (including Sasanian court ritual) to a much larger extent than their Umayyad predecessors. Their cosmopolitanism, to use an anachronistic term, facilitated the sponsorship of Greco-Arabic translations, which were produced for the Abbasid elite on a massive scale. Saliba argued (rather strangely, it has to be said; see Pormann 2010) that the reforms of the Caliph ʿAbd al-Malik (r. 65–86/685–705) left Syriac-speaking Christians unemployed, and they subsequently turned to Greco-Arabic translation. Let us now turn to the case studies of individual translation phenomena that illustrate the problematic nature of matching individual translators to texts and explaining their motivations in a unified manner.

I.4.2 Hippocratic *Aphorisms*

When we deal with Greco-Arabic translations, we have to distinguish between external and internal evidence, that is, reports about various Syriac and Arabic translations of a specific text and the translations themselves, as far as they survive. In the case of the Hippocratic *Aphorisms*, the situation is further complicated by the fact that Galen's *Commentary on Hippocrates' 'Aphorisms'* played a major role in the transmission of the Hippocratic text, as it quoted it extensively in the form of so-called lemmas. Ḥunayn provides the following account of how Galen's commentary was rendered into Arabic:

> *The Commentary on the Book of Aphorisms.* He [sc. Galen] composed this book in seven sections [*maqālāt*]. Ayyūb [ar-Ruhāwī, that is Job of Edessa, (d. after 217/832)] had translated it badly. Jibrīl ibn Bukhtīshūʿ wanted to improve it, but he only corrupted it further. Then I collated the Greek with it, and corrected it in a way that amounted

to retranslating it. Then I added the lemmas of Hippocrates' text [*faṣṣ kalām Buqrāṭ*] on their own.

Aḥmad b. Muḥammad, known as Ibn al-Mudabbir, had asked me to translate it for him. I translated one section of it into Arabic. Then he asked me not to begin with the translation of another section before he had read the one that I had translated. Yet, the man was too busy, and therefore the translation was interrupted. When Muḥammad b. Mūsā saw this section, he asked me to complete [the translation] of the book. Therefore, I translated it completely.

(*Ḥunayn ibn Isḥāq 2016, 94–7*)

This entry in Ḥunayn's *Epistle* comprises two parts: he first describes the translations into Syriac produced by Job of Edessa and Jibrīl ibn Bukhtīshūʿ, two prominent physicians of the early 3rd/9th century. As is so often the case, Ḥunayn is rather scathing about his predecessors. The second part shows that the Abbasid elite commissioned the Arabic translations: both Ibn al-Mudabbir (d. 270/883–884) and Muḥammad ibn Mūsā (d. 259/873) were prominent figures at the court of the caliph, although, importantly, the caliph himself is never mentioned in the *Epistle* (Ḥunayn ibn Isḥāq 2016, 94–7).

Two Syriac and two Arabic translations of the *Aphorisms* have come down to us, at least partially (Pormann 2016). In recent years, there has been intense debate about the younger Syriac translation, extant in MS Paris, BnF, arabe 6734: Overwien (2015) argued for Ḥunayn's authorship, whereas Mimura (2017) and Barry (2018) argued against it, and this with excellent reasons that I cannot detail here. Likewise, there is an older and a newer Arabic version, the former thought to be by al-Biṭrīq and the latter by Ḥunayn. Al-Biṭrīq's version survives in the quotations from the *Aphorisms* in the *Book of History* (*Kitāb al-Taʾrīkh*) by al-Yaʿqūbī (d. after 292/905) and in the lemmas of a commentary previously thought to be by the late antique Greek physician Palladius ([5th century]; Biesterfeldt 2007) but in reality, an Arabic composition (Pormann *et al.* 2017).

I.4.3 Dioscorides's *Medicinal Substances* and Galen's *Simple Drugs*

In a truly groundbreaking study, Ullmann (2009) showed that previous scholarship about the transmission of Dioscorides's *Medicinal Substances* has to be completely revised. His conclusions can be summarized as follows. One manuscript (MS Istanbul, Sülemaniye Library, Aya Sofya 3704) preserves in large parts an anonymous old Arabic translation. Ullmann also demonstrated that the text contained in the edition of Dubler and Terés (Dioscorides 1953–1957) is Iṣṭifān ibn Basīl's translation and not that by Ḥunayn ibn Isḥāq as previously thought. Ḥunayn did, however, add a number of glosses explaining Iṣṭifān's text without making any fundamental changes. He notably failed to replace the many transliterated Greek terms for which Iṣṭifān did not know an Arabic equivalent but for which Ḥunayn had identified such equivalents in his other Arabic versions of Greek pharmacological texts. Furthermore, Ullmann discovered that Iṣṭifān's text was revised at the end of the 4th/10th century by the scholar al-Ḥusayn ibn Ibrāhīm al-Nātilī, about whom little else is known. Then, in the 6th/12th century, a certain Abū Sālim al-Malaṭī (middle 6th/12th century) produced an Arabic version on the basis of Ḥunayn's Syriac version, and Mihrān ibn Manṣūr al-Masīḥī (d. after 553/1158) produced yet another version, because that by al-Malaṭī was of such poor quality.

The major Greek work on medicinal substances is Galen's *Simple Drugs*. It can be divided into two parts: books 1 through 5 deal with the theory of simple drugs, whereas books 6 through 11 list various medicinal substances and detail their powers and qualities. For instance, books 6 to 8 contain an alphabetical inventory of medicinal plants. In this second part, we find quite a lot of overlap with Dioscorides's *Medicinal Substances*. For Galen's *Simple Drugs*, we have two Syriac and two Arabic translations that are at least partially extant. Ḥunayn also provides an account in his *Epistle* but only talks about the Syriac versions:

> Joseph al-Ḫūrī translated into Syriac the first part of this book (that is, the first five books), but poorly. Later, Job of Edessa translated it in a manner more correct than Joseph, although his corrective edition was not as thorough as it should have been. Later, I translated it for Salmawayh b. Bunān, and I tried my hardest to produce a corrective edition of it. Sergius of Rēsh ʿAynā translated the second part of this book. Yuḥannā b. Māsawayh asked me to collate for him the second part of this book and correct it, which I did, although it would have been more accurate to translate it [anew].
>
> *(Ḥunayn ibn Isḥāq 2016, 66–9)*

We have here a similar picture to that of the *Aphorisms*: a number of translators about whom little is known worked on various Syriac versions, which Ḥunayn criticizes for their quality. Sergius of Rēsh ʿAynā, however, is a well-known figure: he had studied medicine in Alexandria and was particularly influenced by the Alexandrian medical curriculum in his choice of translation. He rendered the so-called *Sixteen Books of Galen* (*al-Kutub al-sittata ʿashara*)[5] into Syriac; these were a group of works taught by iatrosophists in the amphitheaters of Alexandria (Pormann 2010). His translation of the second part of *Simple Drugs* (*ʿAl sammānē peshīṭē*) partially survives in two manuscripts (Afif *et al.* 2018a, 2018b), as well as a number of later quotations (Afif *et al.* 2018c, 2019). A tentative comparison based on a small sample of Sergius's Syriac version and that by Ḥunayn reveals that the latter is much more indebted to the former than one would gather from the latter's disparaging remarks in the *Epistle* (Bhayro *et al.* 2013, 139–43).

Through a curious quirk of fate, we also have two Arabic versions for book 6 of *Simple Drugs*, as one manuscript preserved an older translation (Ullmann 2002; Pormann 2011, 2012). Therefore, for both Dioscorides's *Medicinal Substances* and Galen's *Simple Drugs*, we can compare an older Arabic version with a younger one from Ḥunayn's workshop (Ullmann 2002, 2009; Pormann 2011, 2012). This allows us to trace the development of pharmacological terminology. The older period of the late 2nd/8th and early 3rd/9th centuries was characterized by hesitancy and a lack of sophistication. The translators often transliterated the Greek technical terms, such as drug names, without having a clear Arabic equivalent. Moreover, they simplified the text in their translation, and not infrequently misunderstood it. Iṣṭifān's version of Dioscorides marks progress in this respect, but he, too, lacked the astounding translational abilities of Ḥunayn ibn Isḥāq.

1.4.4 Paul of Aegina's *Childcare* and *Medical Handbook*; the *Alexandrian Summaries*; alchemical works

There are, to be sure, many other Greek medical texts not by Galen for which we do not have Ḥunayn's account about how they were rendered into Syriac and Arabic. One prominent example is the 1st/7th-century physician Paul of Aegina, about whom little is known beyond the fact that he probably witnessed the sack of Alexandria in 20/641. He wrote a monograph on childcare, which is lost in Greek but survives in Arabic quotations in later pediatric authors (Pormann 1999). Yet his greatest claim to fame is his medical handbook or encyclopedia in seven books, which partially survives in a single manuscript and in quotations by later authors. We know that this latter work was translated into both Syriac and Arabic, and Pormann (2004a) tentatively argued that the Arabic version had been made in Ḥunayn's workshop, although Barry (2018) has recently argued against this. If Barry is right, then Paul's Arabic version must remain anonymous for us, and a fortiori, we shall not be able to say anything about the milieu of the translators or their motivations.

The case is similar to another set of texts from late antique Alexandria, the so-called *Alexandrian Summaries* (*Jawāmiʿ al-Iskandarānīyīn*, also known by their Latin title *Summaria Alexandrinorum*). This is a very heterogeneous collection of abridgments of various Galenic texts that had become popular in the amphitheaters of the iatrosophists who taught medical students the basic texts such as Galen's *On the Sects for Beginners* (*Peri haireseōn tois eisagomenois*), *On the Elements according to Hippocrates for Beginners* (*Peri tōv kath' Hippokratēn stoicheiōn tois eisagomenois*) and so on (Pormann 2004b; Walbridge 2015; Overwien 2019).

Finally, there is the example of Ibn Sarābiyūn (2nd half 3rd/9th century), a physician who wrote a Syriac medical encyclopedia in seven books, known as the *Small Compendium* (*al-Kunnāsh al-ṣaghīr*), which was translated three times into Arabic, including by the famous glossographer Bar Bahlūl (4th/10th century; Pormann 2004c). Again, little is known about these translations and how they were produced; in particular, the intriguing question of the evidence for Greek–Syriac–Arabic word lists contained in Bar Bahlūl's glossary has hardly been broached (but see Pormann 2004a, 14–20; Barry 2018).

We find a similar picture in the area of alchemy: here, too, the translations are mostly anonymous, and we do not even know when they were produced. Moreover, the transmission into Arabic of Greek alchemical literature is further complicated by the fact that many of the works were pseudepigraphical: they circulated in Arabic under false names such as that of Zosimus (Chapters I.2. and I.12). Zosimus of Panoplis (late 3rd–early 4th century) in Egypt was the first major author on alchemy in Greek and dominated the early history of Arabic alchemy (Hallum 2008). For instance, the anonymous *Tome of Images* (*Muṣḥaf al-ṣuwar*), an Arabic alchemical compilation in dialogue form, probably dating to the 3rd/9th century, contains significant material from Zosimus's works (Hallum 2009; Martelli 2017). Yet, in most of these cases, we do not know who produced these translations and when. Some texts claim to be Arabic renderings from Coptic, but the role of this language in the transmission from Greek into Arabic has, again, hardly been broached (Richter 2009, 2015; Chapter I.2).

I.4.5 Conclusion

Over the last twenty years, Greco-Arabic scholarship has made a lot of progress and we have access to many more translations than before, especially in the area of medicine and the occult. And yet, scholars such as Ullmann still focus on individuals as authors of these translations. In most instances, however, the translations that have come down to us are the product of a long process of translation, retranslation and revision that involved physicians from different times and places. Where we do know about the modalities of translation, we see teamwork: older drafts are the starting point for new work, and in most cases, more than one person helps produce the final version. It is even harder to find the causes behind this translation activity: money will undoubtedly have played its part but can only be one reason among many for the fact that so much was rendered from Greek into Arabic. Given these uncertainties, we have to question whether it still makes sense to talk about this important historical phenomenon as a 'movement', because it suggests a unity of purpose that does not reflect the historical reality as we can discern it.

Notes

1 This is the spelling of this scholar's last name in German publications. In English publications the name is spelled as Endress.
2 Many of the Arabic titles provided here represent only one of the designations found in extant manuscripts.
3 Consolidated bibliography.
4 Consolidated bibliography.
5 This is a modern shorthand. The titles found in the manuscripts differ among each other.

Bibliography

Sources

Dioscorides. ed. Dubler, C. E. and Terés, E. 1953–1957. *La versión árabe de la 'Materia Medica' de Dioscórides (texto, varients e índices)*. Tetuán and Barcelona: Tipografía Emporium.

Manuscripts

MS Istanbul, Sülemaniye Library, Aya Sofya 3704.
MS Paris, BnF, Arabe 6734.

Research literature

Afif, N. *et al.*: Afif, N., Bhayro, S., Kessel, G., Pormann, P. E. and Sellers, W. I. 2018a. "The Syriac Galen Palimpsest: A Tale of Two Texts," *Manuscript Studies* 3.1: 110–54.

Afif, N. *et al.*: Afif, N., Bhayro, S., Kessel, G., Pormann, P. E. and Sellers, W. I. 2018b. "The Syriac Galen Palimpsest: A Tale of Two Texts," *Manuscript Studies* 3.1: 155–85.

Afif, N. *et al.*: Afif, N., Bhayro, S., Pormann, P. E., Sellers, W. I. And Smelova, N. 2018c. "On Digamma and the Armenian Earth," *Le Muséon* 131.3–4: 391–414.

Afif, N. *et al.*: Afif, N., Bhayro, S., Pormann, P. E., Sellers, W. I. And Smelova, N. 2019. "GalenQt: A New Software Tool for Recovering Lost Text," [forthcoming].

Barry, S. C. 2018. *Syriac Medicine and Hunayn ibn Ishaq's Arabic Translation of the Hippocratic Aphorisms* (Journal of Semitic Studies Supplement 39). Oxford: Oxford University Press.

Bergsträsser, G. 1913. *Ḥunain Ibn Isḥāk und seine Schule: sprach- und literargeschichtliche Untersuchungen zu den arabischen Hippokrates- und Galen-Übersetzungen*. Leiden: Brill.

Bhayro, S. 2017. "On the Problem of Syriac 'Influence' in the Transmission of Greek Science to the Arabs: The Cases of Astronomy, Philosophy, and Medicine," *Intellectual History of the Islamicate World* 5: 211–27.

Bhayro, S. *et al.*: Bayro, S., Hawley, R., Kessel, G. and Pormann, P. E., 2013. "The Syriac Galen Palimpsest: Progress, Prospects and Problems," *Journal of Semitic Studies* 58: 131–48.

Biesterfeldt, H. 2007. "Palladius on the Hippocratic Aphorisms," in D'Ancona, C., ed. *The Libraries of the Neoplatonists*. Leiden: Brill, 385–97.

Endreß, G. 1989 [1992]. "Die griechisch-arabischen Übersetzungen und die Sprache der arabischen Wissenschaften," in *Symposium Graeco-Arabicum II*. Amsterdam. [Reprinted as chapter "8.7 Die Entwicklung der Fachsprache," in Fischer, W., ed. *Grundriß der arabischen Philologie:Band III: Supplement*. Wiesbaden: Reichert Verlag, 3–23.]

Garofalo, I. 1995. "La traduzione araba del *De locis affectis*," *Studi classici e orientali* 45: 13–63.

Gutas, D. 2012. " The *Poetics* in Syriac and Arabic Transmission," in Tarán, L. and Gutas, D., eds. Aristotle. *Poetics* (Mnemosyne, Supplements, 338), Leiden and Boston: Brill, 77–128.

Hallum, B. C. 2008. *Zosimus Arabus: The Reception of Zosimos of Panopolis in the Arabic/Islamic World*. PhD thesis. London: University of London.

Hallum, B. C. 2009. "The Tome of Images: An Arabic Compilation of Texts by Zosimos of Panopolis and a Source of the *Turba Philosophorum*," *Ambix* 56: 76–88.

Martelli, M. 2017. "Translating Ancient Alchemy: Fragments of Graeco-Egyptian Alchemy in Arabic Compendia," *Ambix* 64: 326–42.

Mimura, T. 2017. "A Reconsideration of the Authorship of the Syriac Hippocratic Aphorisms: The Creation of the Syro-Arabic Bilingual Manuscript of the Aphorisms in the Tradition of Ḥunayn ibn Isḥāq's Arabic Translation," *Oriens* 45: 80–104.

Overwien, O. 2014. "Syriac and Arabic translators of Hippocratic texts," in Jouanna, J. and Zink, M., ed. *Hippocrate et les hippocratismes: médecine, religion, société: actes du XIVe Colloque international hippocratique.* Paris: Académie des Inscriptions et Belles-Lettres, 421–31.

Overwien, O. 2019. *Medizinische Lehrwerke aus dem spätantiken Alexandria: Die "Tabulae Vindobonenses" und "Summaria Alexandrinorum" zu Galens "De sectis"* (Scientia Graeco-Arabica 24). Berlin: De Gruyter.

Pormann, P. E. 1999. *The Greek and Arabic Fragments of Paul of Aegina's Therapy of Children.* MPhil thesis. University of Oxford. https://ora.ox.ac.uk/objects/ora:3106.

Pormann, P. E. 2004a. *The Oriental Tradition of Paul of Aegina's "Pragmateia"* (Studies in Ancient Medicine 29). Leiden: Brill.

Pormann, P. E. 2004b. "The Alexandrian Summary (*Jawāmiʿ*) of Galen's *On the Sects for Beginners*: Commentary or Abridgment?" in Adamson, P. *et al.*, eds. *Philosophy, Science and Exegesis in Greek, Arabic and Latin Commentaries. Bulletin of the Institute of Classical Studies. Supplement* 83.1–2. 2 vols. London: Institute of Classical Studies, School of Advanced Study, University of London, 2: 11–33.

Pormann, P. E. 2004c. "Yuḥannā ibn Sarābiyūn: Further Studies into the Transmission of His Works," *Arabic Sciences and Philosophy* 14: 233–62.

Pormann, P. E. 2010. "Arabic Astronomy and the Copernican 'Revolution'," *Annals of Science* 67: 243–8.

Pormann, P. E. 2011. "The Formation of the Arabic Pharmacology Between Tradition and Innovation," *Annals of Science* 68: 493–515.

Pormann, P. E. 2012. "The Development of Translation Techniques from Greek into Syriac and Arabic: The Case of Galen's *On the Faculties and Powers of Simple Drugs*, Book Six," in Hansberger, R., al-Akiti, M. A. and Burnett, C., eds. *Medieval Arabic Thought: Essays in Honour of Fritz Zimmermann.* London: Warburg Institute; Turin: Nino Aragno Editore, 143–62.

Pormann, P. E. 2016. "Al-Tarjamāt al-yūnāniyya al-suryāniyya al-ʿarabiyya li- l-nuṣūṣ al-ṭibbiyya fī awāʾil al-ʿaṣr al-ʿabbāsī," in Koetschet, P. and Pormann, P. E., eds. *Nashʾat al-ṭibb al-ʿarabī fī l-qurūn al-wusṭā.* Publications de l'Institut français de Damas. Beirut, Damascus and Cairo: Press de l'ifpo, 43–59.

Pormann, P. E., *et al.* 2017. "The Enigma of Arabic and Hebrew Palladius," *Intellectual History of the Islamicate World* 5.3: 252–310.

Richter, T. S. 2009. "What Kind of Alchemy is Attested by Tenth- Century Coptic Manuscripts?" *Ambix* 56: 23–35.

Richter, T. S. 2015. "The Master Spoke: Take One of the Sun and One Unit of Almulgam. Hitherto Unnoticed Coptic Papyrological Evidence for Early Arabic Alchemy," in Schubert, A. T. and Sijpesteijn, P. M., eds. *Documents and the History of the Early Islamic World.* Leiden and Boston: Brill, 158–94.

Ullmann, M. 2002. *Wörterbuch zu den griechisch-arabischen Übersetzungen des 9. Jahrhunderts.* Wiesbaden: Harrassowitz.

Ullmann, M. 2006. *Wörterbuch zu den griechisch-arabischen Übersetzungen des 9. Jahrhunderts. Supplement I:* A *bis* O. Wiesbaden: Harrassowitz.

Ullmann, M. 2007. *Wörterbuch zu den griechisch-arabischen Übersetzungen des 9. Jahrhunderts. Supplement II:* Π *bis* Ω. Wiesbaden: Harrassowitz.

Ullmann, M. 2009. *Untersuchungen zur arabischen Überlieferung der Materia medica des Dioskurides.* Wiesbaden: Harrassowitz.

Ullmann, M. 2011. *Die Nikomachische Ethik des Aristoteles in arabischer Übersetzung: Teil 1: Wortschatz.* Wiesbaden: Harrassowitz.

Ullmann, M. 2012. *Die Nikomachische Ethik des Aristoteles in arabischer Übersetzung: Teil 2: Überlieferung Textkritik Grammatik.* Wiesbaden: Harrassowitz.

Ullmann, M. 2018. *Wörterbuch zu den griechisch-arabischen Übersetzungen des 9. Jahrhunderts: Supplement Band III: Zur Agronomie.* Wiesbaden: Harrassowitz.

Walbridge, J. 2015. *The Alexandrian Epitomes of Galen.* 2 vols. Vol. 1: *On the Medical Sects for Beginners; The Small Art of Medicine; On the Elements according to Hippocrates.* Provo, UT: Brigham Young University Press.

I.5
GEOMETRY AND ITS BRANCHES

Glen Van Brummelen

Geometry is everywhere. It is such a fundamental part of the human experience that it seems impossible to exist without it. Inevitably, then, every culture thinks geometrically in one form or another. However, different groups have shared very different experiences. The dominant expression of geometrical thought in western culture, the logical form that manifests itself in axiomatic-deductive systems inspired by Euclid's (3rd century BCE) *Elements*, was also a major influence on scholars in Islamicate societies. But this is not to say that they were mere followers. Indeed, geometers in those societies took Euclidean geometry much further than Euclid. While the arsenal of geometric theorems was greatly expanded, the range of methods and approaches deepened and became more diverse.

Geometers also looked beyond their discipline, implementing their methods in new settings. For instance, astronomy had already been expressed as a form of applied geometry, but ritual needs such as the determination of the direction of Mecca triggered the spread of geometry into geography. Disciplines such as optics and architecture also participated in the geometrical project. They not only benefited from the Euclidean approach but also relied on the geometry of mensuration used in daily practice. The use of quantitative measurement in geometry helped produce physically and societally meaningful results that could not have been achieved with Euclid's teachings alone.

I.5.1 Euclidean geometry

Geometrical texts comprised a substantial portion of the scholarly translations in the 2nd/8th and 3rd/9th centuries. Not surprisingly, one of the first to find its way into Arabic was Euclid's *Elements*. Although this book contains topics like ratio and prime number theory, it was then and is now known mostly as the archetype for geometrical reasoning. The first translator, al-Ḥajjāj ibn Yūsuf ibn Maṭar (d. after 213/828), actually performed the task twice: first under the Abbasid Caliph Hārūn al-Rashīd (r. 170–193/786–809) no later than the first decade of the 3rd/9th century and the second about 25 years later under Caliph al-Ma'mūn (r. 198–218/813–833). Later in the 3rd/9th century, Isḥāq ibn Ḥunayn (d. 298/910–1) might have produced a new translation, edited by Thābit ibn Qurra (d. 288/901). Researchers are at the moment attempting to untangle the strands of the question of whether the Arabic passages

DOI: 10.4324/9781315170718-7

in our possession are due to al-Ḥajjāj, Isḥāq ibn Ḥunayn or Thābit ibn Qurra; the issue has not been resolved.

What is being found in these passages emphasizes that the word *transmission*, suggesting a passive download of information from one culture to another, is a misleading term.[1] For instance, the first proposition of *Elements* Book II states (Figure I.5.1):

> If there are two straight lines [A and BC], and one of them is cut into any number of segments whatever [BD, DE, and EC], then the rectangle contained by the two straight lines equals the sum of the rectangles contained by the uncut straight line and each of the segments.

We might interpret this statement to be a way to state that

$$a(x_1 + x_2 + \cdots + x_n) = ax_1 + ax_2 + \cdots + ax_n$$

(where $a = A = BG$, $x_1 = BD$, $x_2 = DE$, . . .). But for Euclid, a rectangle was a rectangle, and geometric magnitudes were entirely different types of objects from numbers. On the other hand, al-Ḥajjāj (or a later editor) expresses part of the passage corresponding to "rectangles" as

> that which results from the product (*ḍarb*) of one of the lines by the other.
>
> *(Djebbar 1996, 102)*

This difference reflects an implicit rejection of Euclid's distinction between geometric magnitudes and numerical quantities. This might have been done for pedagogical reasons or perhaps in response to deeper philosophical concerns. In any case, it is different, and the translation is no longer entirely the voice of Euclid (De Young 1991; Djebbar 1996; Brentjes 1996).

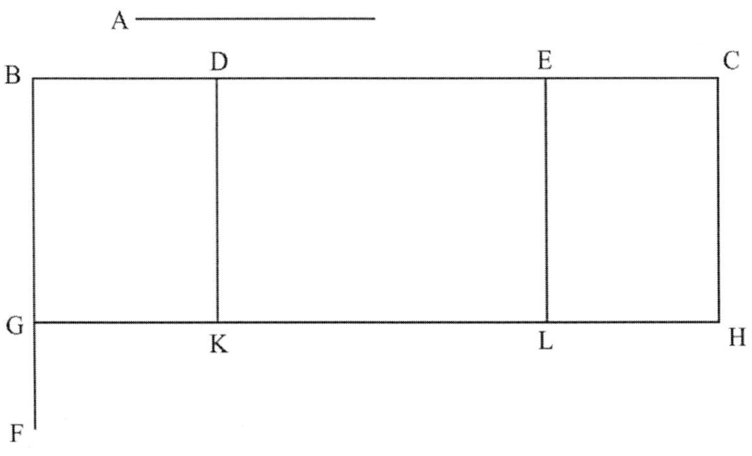

Figure I.5.1 Euclid's *Elements* II.1

Source: © Glen Van Brummelen

Another translation, until recently ascribed to Thābit ibn Qurra, is quite different. Here the same passage is translated in a way that Euclid might have recognized, although it causes more trouble for the modern reader:

> the surface with right angles that is enclosed by two rectilinear lines.
> When the word "rectangle" is needed, the translator employs *saṭḥ*, or "area".

The *Elements* continued to receive attention throughout the medieval period, both as a research tool and as a textbook. Most well-known is the *Taḥrīr* (redaction and commentary) of the *Elements* by polymath Naṣīr al-Dīn al-Ṭūsī (597–672/1201–1274). Part of his project to produce Arabic editions of many of the Greek classic works of mathematics and astronomy, the *Taḥrīr* may have been taught at the Maragha Observatory (Chapter II.7). It soon took on a life of its own as the preferred entry to geometry, translated into Persian and Sanskrit. Al-Ṭūsī's preferences (for instance, his lack of interest in the study of irrational magnitudes in Book X) were thus transmitted to many students of geometry in the late medieval period (De Young 2008–2009).

Scholars also engaged in rich dialogues with the *Elements*, leaving hardly any part of its contents or methods untouched. For instance, al-Nayrīzī's (d. *c.* 310/922) extensive early 4th/10th century commentary brings together both ancient Greek and then recent Arabic authors to question Euclid's approaches to the parallel postulate and the nature of ratio. With respect to the latter, al-Nayrīzī replaces Euclid's cumbersome formulation with the so-called anthyphairetic definition of ratio of his predecessor al-Māhānī (d. after 252/866; two ratios are said to be equal if the Euclidean algorithm is applied to both, and the same sequence of numbers emerges both times), an approach closer to our arithmetic understanding.[2] Expressed most clearly in ʿUmar al-Khayyām's (439–*c.* 517/1048–*c.* 1123) *Commentary on the Problems Posed by Certain Postulates of Euclid's Treatise* (*Sharḥ mā ashkala min muṣādarāt kitāb Uqlīdis*), the anthyphairetic definition contains such an arithmetic treatment of geometric magnitudes that it has been argued that one might as well equate it with the modern theory of real numbers (al-Khayyām 1999, 2002; Vahabzadeh 1997, 2004; Vitrac 2000).

Other authors, clearly not concerned by the issues that had motivated Euclid, reorganized parts of the *Elements* to align with their objectives. For instance, the geometer Abū Sahl al-Kūhī (2nd half 4th/10th century) composed several treatises recasting or augmenting certain books of the *Elements*. His take on Book I breaks with Euclid in two ways. First, he removes all of Euclid's constructions (example: "*to construct an equilateral triangle on a given finite straight line*"), leaving behind all the theorems. Presumably he felt that these were different activities and should be dealt with separately. Second, he reorganizes the logical structure. Book I had been organized to avoid using the problematic parallel postulate (an axiom equivalent to the assertion that the angles of a triangle sum to two right angles) as long as possible, but al-Kūhī had no such concerns (al-Kūhī 2005). Al-Kūhī's version of Book II is similarly transformative. His diagrams for the propositions contain only the line segments needed to state the results, not those needed to prove them; for instance, compare al-Kūhī's diagram for II.1 (Figure I.5.2) with Euclid's. The

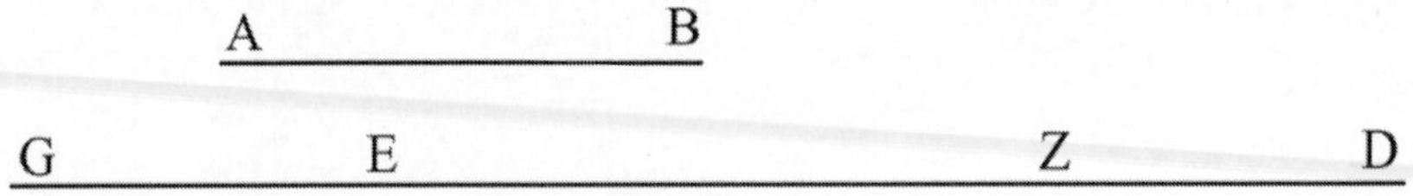

Figure I.5.2 al-Kūhī's diagram for Euclid's *Elements* II.1

Source: © Glen Van Brummelen

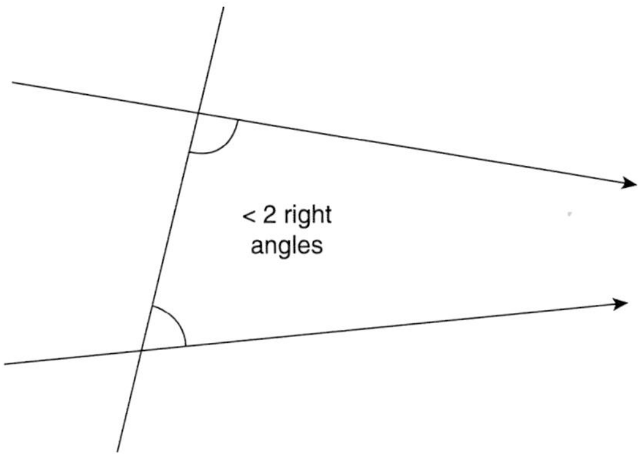

Figure I.5.3 The parallel postulate

Source: © Glen Van Brummelen

result is a proof that does not rely on a diagram, and while it is not exactly algebraic, it separates at least partly from purely geometric reasoning. Even so, his purpose is still geometric; he adds a number of propositions in order to fill gaps in proofs in Apollonius's (*c*. 262–*c*. 190 BCE) *Conics* (al-Kūhī 1992; 2002–2003).

Book I is especially important for its establishment of axioms as the logical starting point of geometric inquiry. It is also the most controversial for one of these axioms, known as the *parallel postulate* (Figure I.5.3):

> That, if a straight line falling on two straight lines make the interior angles on the same side less than two right angles, the two straight lines, if produced indefinitely, meet on that side on which the angles are less than the two right angles.

This statement, seemingly obvious and yet known today to be unprovable from Euclid's other axioms, was a source of discomfort. A number of scholars, as early as Thābit ibn Qurra, attempted to prove it in various ways. Thābit's proof worked by replacing Euclid's definition of parallels (lines that "do not meet each other") with a definition that requires parallel lines to be equidistant (Rashed and Houzel 2005; Sabra 1968). Introducing this notion changes the ground rules and renders the postulate provable. The most famous proof, by ʿUmar al-Khayyām (in the same treatise we mentioned earlier), claims that lines that converge must intersect, thereby also admitting a notion of distance into the argument (Rashed and Vahabzadeh 1999, 2002; for the theory of parallels in Islamicate societies in general, see Jaouiche 1986 and Rosenfeld 1988).

I.5.2 Conic sections

Although scholars in Islamicate societies went far beyond their Greek antecedents in their use of Euclidean methods (straightedge and compass), they showed particular interest in changing the tools that were permitted onto the geometric playing field. For instance, several authors (including al-Fārābī [d. 339/950–1] and Abū l-Wafāʾ [328–388/940–998]) explored what constructions are possible using only a *rusty compass*, one with a fixed rather than a variable opening (Berggren 2016, 104–11; Woepcke 1855). More attention was paid to so-called *solid problems*, those that can be solved using conic sections. The extension of the geometric tool set to include ellipses, parabolas and hyperbolas expands the domain of the subject considerably, rendering problems soluble that had been inaccessible to straightedge and compass. A tool for drawing conic sections called the *complete* or *perfect compass*, akin to the compass for drawing circles, was devised by Abū Sahl al-Kūhī, and its use was explored by al-Sijzī (d. *c.* 411/1020–1), as well as Abū l-Wafāʾ (Abgrall 1997, 2004; Rashed 2003, 2004).

The fundamental text for the study of conic sections was the *Conics* of Apollonius, but only the first four of its eight books survive in Greek. Books V through VII survive in Arabic translations by Thābit ibn Qurra and the three brothers known as the Banū Mūsā (3rd/9th century; Apollonius 1990; Apollonius de Perge 2008–2010). The eighth book is lost altogether and was reconstructed by Ibn al-Haytham (354–430/965–*c.* 1040) in his *Completion of "The Conics"* (*Maqāla fī tamām Kitāb al-Makhrūṭāt*; Ibn al-Haytham 1985, 2000). Ibn al-Haytham seems to have believed that Apollonius had been building toward problems involving drawing tangent lines to conic sections, and he goes about solving those problems in his work.

One of the favorite topics especially in the 3rd/9th to 5th/11th centuries was the construction of regular polygons that cannot be drawn with straightedge and compass, in particular, the heptagon and nonagon. Although it had not been a focus of special attention in ancient Greece, a heptagon construction exists in a treatise attributed to Archimedes, translated into Arabic by Thābit ibn Qurra. It resembles what is known as a *neusis*, or verging, construction. The heart of the construction (Figure I.5.4), from which Archimedes goes on to construct his heptagon (not drawn here), is as follows: draw square *ABGD*, extending *BA* to the right, and draw diagonal *BG*. Then draw a line segment from *D* upward and to the right so that it crosses *BA* at *Z* to the right of *A* so that triangles *AHZ* and *DTG* have the same area. Clearly there is such a point, but simply positing its existence and building a heptagon from it does not do much good, since the regular heptagon itself clearly also exists and we may as well just posit it instead. This was the objection of a number of 4th/10th- and 5th/11th-century authors who rejected such procedures as inadequate, belonging to "moving" rather than "fixed" geometry. (These terms reflect the intuition that one must "move" point *Z* back and forth along the extension of *BA* until the two triangles have equal areas.)

Instead, all of the dozen constructions of the regular heptagon that have come down to us employ conic sections, which – while not within the straightedge and compass tool set – were still considered to be part of fixed geometry. Several of these constructions show how point *Z* may be constructed on Archimedes's line *BA*. One such method, given by Kamāl al-Dīn ibn Yūnus (551–636/1156–1242) around the year 595/1200, works as follows. In Figure I.5.5, draw a square *AXUS* equal to *ABGD*, and draw a hyperbola through *U*, whose asymptotes are *AX* and

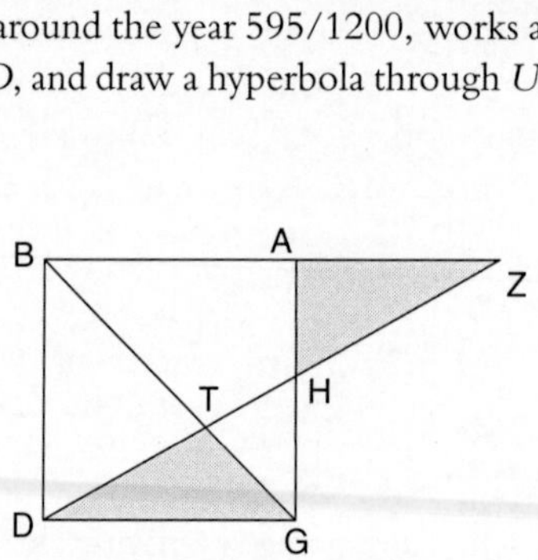

Figure I.5.4 The key step in Archimedes' construction of the regular heptagon

Source: © Glen Van Brummelen

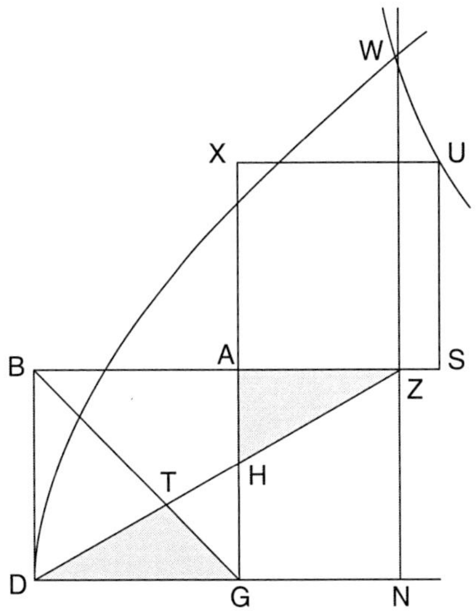

Figure I.5.5 Kamāl al-Dīn ibn Yūnus's solution to the regular heptagon problem

Source: © Glen Van Brummelen

AS. Then draw an equilateral hyperbola (one for which the asymptotes are perpendicular) *DW*, with axis *DG*. Drop a perpendicular from the intersection point *W* of the two hyperbolas onto the extension of *DG*. It turns out that the intersection of *WN* with the extension of *BA* is our sought point *Z*. So, Kamāl al-Dīn and his colleagues shared a criterion for acceptability: conic sections passed the test, but verging constructions did not (Hogendijk 1984; Berggren 2016; Rashed 1979; Ibn al-Haytham 2000; Anbouba 1997). We shall see, however, that this criterion did not apply to all contexts.

The construction of a regular nonagon has special interest because it happens to be a special case of the trisection of a given angle, one of the three classical unsolved problems of antiquity. As before, ancient verging constructions were rejected, and solutions using conics were generated by authors including Thābit ibn Qurra and al-Kūhī (Hogendijk 1979, 1981; Knorr 1989, 213–24 and 254–374). We shall revisit trisection in the section on trigonometry.

Trisecting the angle happens to be mathematically equivalent to solving a cubic equation, so it is not as surprising as it might sound that the study of conic sections spread into algebra. Several authors used conic sections to solve cubic equations, among them Abū Jaʿfar al-Khāzin (d. 350 or 360/961 or 971) and Ibn al-Layth ([4th/10th century]; al-Khayyām 1981, 12 and 18; Abū l-Jūd 2010). By far the best-known treatment is by ʿUmar al-Khayyām in his *Algebra*, where he deals systematically with the various possible cases (al-Khayyām 1981, 1999, 2000, 2008). What does it mean to solve an algebraic equation geometrically? We illustrate with al-Khayyām's solution (simplified slightly) to his first equation $x^3 + mx = n$. In Figure I.5.6., draw a semicircle with a diameter $BY = n/m$. Then draw a parabola with vertex *B*, axis *BC*, and parameter $\sqrt{m}$ (in modern terms, a parabola with equation $y = x^2/\sqrt{m}$, where the origin is at *B*). At the parabola's intersection *D* with the semicircle, draw perpendicular *DE*. Then it can be shown that the length *BE* satisfies our equation $x^3 + mx = n$ (Berggren 2016).

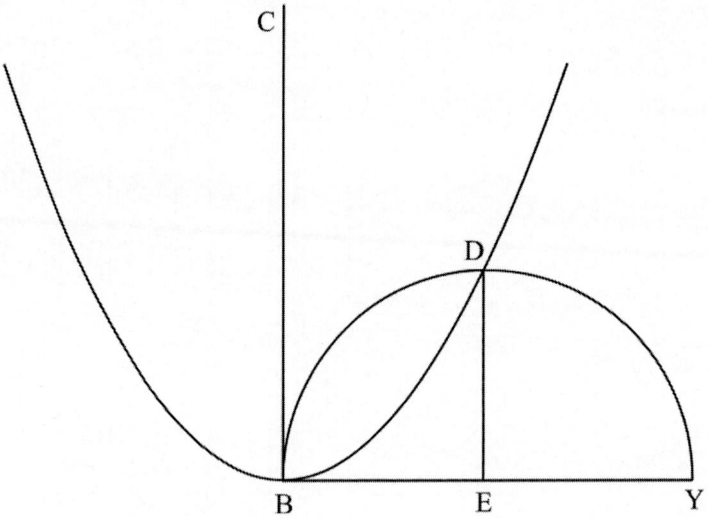

Figure I.5.6 al-Khayyām's solution to $x^3 + mx = n$

Source: © Glen Van Brummelen

I.5.3 Analysis and synthesis

In the late 4th/10th century, al-Sijzī composed a treatise on geometric problem solving, akin to George Polya's (1887–1985) modern classic *How to Solve It* (al-Sijzī 1996, 2004). He lays out seven methods, many of which are informal, such as being clear about what is given and what is sought or understanding the theorems relevant to the problem. However, one of his methods – *analysis and synthesis* – was a more formal procedure. It originated in ancient Greece, but was transformed to its own distinct method in the time of the Abbasid caliphate and became extremely popular (Bellosta 1991; Jaouiche 1988; Ibn al-Haytham 1991; Rashed 1991, 2017; Mawaldi 2000).

Suppose we are asked to construct an object A with a certain property (say, an equilateral triangle) from one or more given objects (e.g., a line segment), and we do not know how to do it. The analyst begins by assuming that he has object A in front of him, and he performs a *transformation* of it; that is, he constructs some other objects from it. At some point, he arrives at an object B that he feels he will be able to construct from the given objects. Then, using only the givens, he attempts to construct it, using theorems not from the *Elements* but from Euclid's lesser-known work, *Data*. This text contains statements asserting that if certain objects are "given", then some other object is also given: for instance, "if two lines given in position cut one another, their intersection is given in position" (*Data* 25). Once a path of "givens" has taken the geometer from the starting point to object B, he has completed the *resolution*. The transformation and the resolution combined are known as the *analysis* (Figure I.5.7).

From these two halves, the formal Euclidean proof, or the *synthesis*, may be assembled. First, one takes the path of the "givens" in the resolution and expands it to a formal, *Elements*-like *construction* of object B. Then, following the path of the transformation but in reverse (i.e., reasoning from B back to A), we achieve a demonstration that A satisfies the requirements of the problem (Hankel 1874, 137–50; Berggren and Van Brummelen 2000).

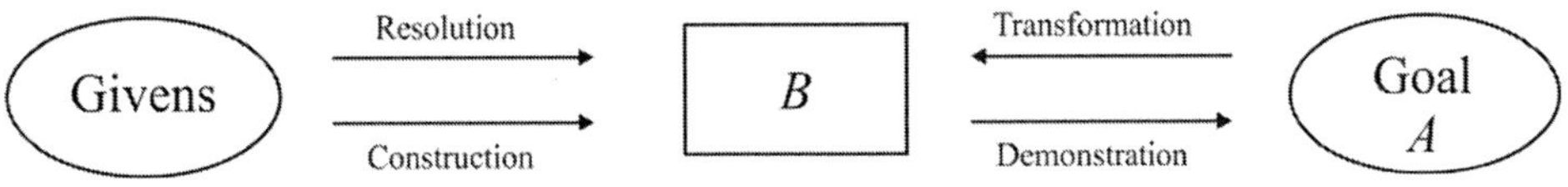

Figure I.5.7 The structure of analysis and synthesis

Source: © Glen Van Brummelen

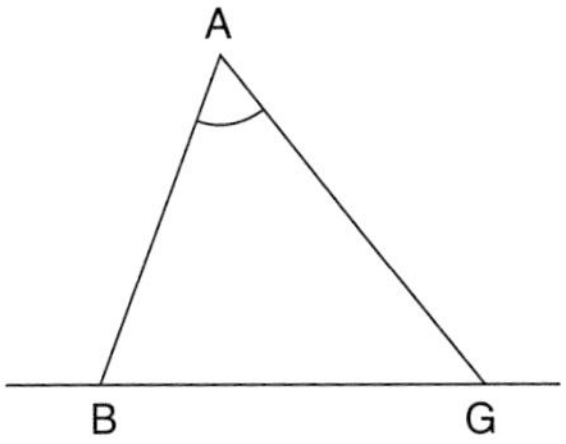

Figure I.5.8 An example of an analysis

Source: © Glen Van Brummelen

Effectively, this rather arcane process employs a reduction of the original problem to another (hopefully easier) one through the transformation. However, in medieval Islamicate societies, the rest of the process, especially the resolution, was recognized as having value on its own. Thus, many geometric works equated *A* and *B*, and the analysis came to consist only of the resolution. In fact, often the author would not include a synthesis at all, promising to complete it later if time permitted. Thus, a number of Arabic geometric texts consisted only of resolutions.

To give the reader the flavor of such resolutions, we present a late 4th/10th-century example from the work of al-Kūhī. The problem is to draw two lines *AB* and *AG* from *A* to the given line *BG*, containing a given angle, so that the ratio between the two segments is equal to some given ratio (Figure I.5.8). (Note that the term *known* is interchangeable with *given*.)

> Since the ratio of the line *BA* to *AG* is known,[3] and angle *BAG* is known, triangle *ABG* is known in form. Therefore angle *ABG* is known; but point *A* is known, so line *AB* is known in position. Therefore line *AG* is known in position, since angle *BAG* is known. And line *BG* is known in position, and therefore each of the two points *B* and *G* is known, and that is what we wanted to know.
>
> *(al-Kūhī 2001a, 66)*

This is the entirety of al-Kūhī's text. Clearly, although the ancient names of analysis and synthesis were still used, the process was no longer the same. The new structure became very common, applied both to straightedge and compass constructions and to conic sections.

I.5.4 Practical geometry

The reader will search in vain in the preceding for much of the heart of school geometry: for instance, formulas for areas, volumes, and perimeters. Such formulas are obviously useful in daily life, in land measurement, meting out inheritances and so forth. However, they do not fit well with the scholarly geometry we have seen so far. But alongside the theoretical approach was a practical discipline, often called *misāḥa*, where measures of geometric magnitudes were common. The word comes from "surveying" but seems to have taken on a more general meaning and is often translated in the context of this tradition as "measurement".

This knowledge seems to have existed prior to theoretical geometry and is found as early as the 3rd/9th century in the works of the great algebraists al-Khwārazmī (d. after 233/847) and Abū Kāmil (*c.* 235–*c.* 317/850–930). At the beginning of Abū Kāmil's *Book of Surveying* (*Kitāb al-Misāḥa*), he begins with prescriptions for finding the area of a square, the length of its diagonal, and the area of a rectangle. Next, for the area and circumference of a circle, we find an implicit use of the standard approximation $\pi \approx 3\frac{1}{7}$. Other topics include the areas of triangles and quadrilaterals, and volumes and surface areas of prisms, cylinders, pyramids and spheres. In the more advanced section of the treatise, he shows how to calculate the areas of various regular polygons inscribed in circles of known diameter. Abū Kāmil gives numerical examples throughout the book, directing readers who seek geometric proofs to Euclid. He may have intended his treatise to be used by landowners, but the approximate methods already in use by officials all inflated the magnitudes of their holdings. Thus, the landholders may not have appreciated the diminishment that Abū Kāmil's precise methods would have caused (Abū Kāmil 1996, 2014).

Some of these treatises were more extensive, for instance, the 4th/10th century *Epistle on Surface Measuring* (*Risāla fī l-taksīr*) by Ibn ʿAbdūn ([311–after 366/923–after 976]; Ibn ʿAbdūn 2005[CB], 2006[CB]; Berggren 2016, 113–17; Wiedemann 1909 for such a text by Ibn al-Haytham). This work, the oldest-known Arabic mathematical text from al-Andalus (the parts of the Iberian Peninsula ruled by Muslim dynasties), contains well over 100 problems, some of them more challenging: for instance, find the area of a rectangle when one knows the length of the diagonal and the difference of its side lengths. Some of his prescriptions concerning quantities in circles are approximate. For instance, to find the length of the arc q of a segment of a circle given its arrow a and its diameter D (Figure I.5.9), Ibn ʿAbdūn's rule corresponds to $q = 2a + a/7 + r$. This method guarantees q to be greater than r, which is clearly not the case when a is small.

One attempt to merge the two geometric traditions seems to have been very popular. Shams al-Dīn al-Samarqandī's (d. 701/1302) *Fundamental Propositions* (*Ashkāl al-Taʾsīs*) is an introductory work to the computational sciences, referring specifically to algebra and *misāḥa* (al-Samarqandī 2001; De Young 2002). However, its 35 propositions are taken from the geometric theorems in Euclid's *Elements*, including 29 from Book I. While the *Fundamental Propositions* is closer to the *Elements* than a practical work, we find occasional phrasings that help adapt the

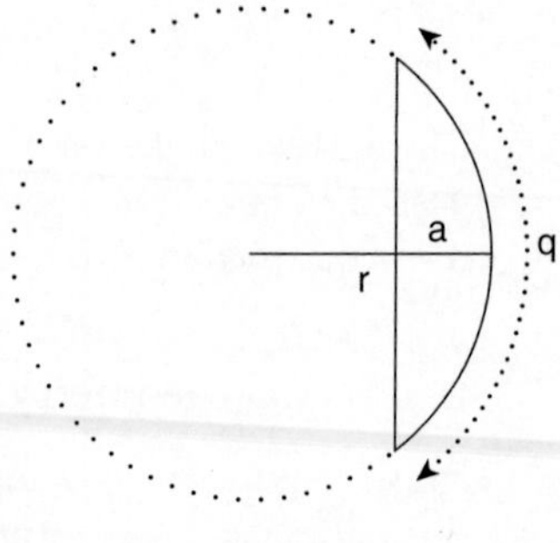

Figure I.5.9 finding the length q of a circular segment from its arrow a and radius r

Source: © Glen Van Brummelen

propositions for use in other contexts. For instance, its Proposition 31, taken from *Elements* II.1 (see the earlier discussion), is

> [t]he multiplication of something into something is equivalent to its multiplication into its parts.
>
> *(al-Samarqandī 2001, 109)*

This statement is a much more practical, or perhaps algebraic, statement of the theorem.

I.5.5 Geometry in other contexts

Geometry affects most areas of life, so it is not surprising that the methods chosen by geometers to practice their trade in Islamicate societies affected other disciplines. We describe several of them here.

Astronomy: Considered to be a form of applied geometry, mathematical astronomy relied heavily on geometric methods. Models to predict the positions of the celestial bodies were based on the motions of circles, mostly inspired by Ptolemy's (*c.* 100–170) *Almagest*. The use of quantitative observations and the demand for quantitative predictions of astronomical phenomena forced measurement into the geometry, but this was done by expanding the scope of theoretical geometry rather than employing tools from *misāḥa*. Indeed, theoretical methods such as analysis and synthesis often operated either implicitly or explicitly in astronomical work. An example of its use is al-Kūhī's method for finding the distance to the shooting stars (al-Kūhī 2001b, 2001c, 2002).

Trigonometry was the mathematical basis in astronomy; the primitive tool was the sine, taken from India as an improvement over the Greek chord. Geometrical methods prescribed what was considered acceptable in trigonometry, starting with the problem with which trigonometry begins: calculating values of sines of given arcs. Some of these values, for instance of the multiples of 3°, could be found using quantitative equivalents of straightedge and compass constructions. However, most values (including sin 1°) could not; they required procedures equivalent to trisecting the angle. Many authors followed Ptolemy by capturing these quantities between two bounds; for instance, in his *Handbook for the Khāqān* (*Zīj-i Khāqānī*), the Persian astronomer al-Kāshī (d. 832/1429) finds

$0.01745238 < \sin 1° < 0.01745244.$[4]

But these bounds do not allow us to conclude that sin 1° is given, so al-Kāshī (like his colleagues) is forced to approximate by taking the average of the bounds. Several astronomers objected, including Ibn al-Haytham and the Maghribi scholar al-Samawʾal ibn Yaḥyā (d. 570/1175); the latter went so far as to compose a sine table for a circle divided into 480 parts rather than 360 in order to bypass the problem (Ibn al-Haytham 1985, 12; Van Brummelen *et al.* 2012). It was not until al-Kāshī later found sin 1° as the solution of the cubic equation $\sin 3° = 3x - 4x^3$ that the restrictions of the geometrical method were removed by a transition to algebra (Van Brummelen forthcoming, 2009, 140–9).

Mathematical astronomy takes place on the celestial sphere, so the geometry of the sphere is fundamental to it. Greek authors of spherical geometry such as Theodosius (late 2nd century BCE) and Menelaus (*fl.* 100) carefully kept astronomical applications hidden just below the surface, but Arabic authors generally did not share such concerns (al-Ṭūsī 2003; Nadal *et al.* 2004). Through the 4th/10th century, spherical trigonometry relied on the same two propositions from Menelaus's *Spherics* that Ptolemy had used in the *Almagest*, but around the turn of the millennium, a number of new theorems such as the Rule of Four Quantities reinvented the subject in eastern Islam (Menelaus 2006; Van Brummelen 2009, 179–92; al-Bīrūnī 1985). The study of spherics was bolstered by interest coming from several directions. The first was various aspects of Islamic ritual that required astronomical methods, especially the determination of the direction of Mecca (Chapter VI.2; King 1986). The second came from demanding problems that arose in the mathematical study of astrology (Chapter I.6; Kennedy 1996; Casulleras 2008–2009; Casulleras and Hogendijk 2012). Finally, a substantial tradition rose around problems related to determining the time of day using the altitude of the sun (Chapter V.4; King 2004[CB], 2005[CB]).

Also on the interface between geometry and astronomy was the study of projections of the sphere onto a plane surface. This was motivated by the design of astronomical instruments, in particular the astrolabe, which is an embodiment of a stereographic projection of the celestial sphere (Chapter IV.3). The astrolabe was successful because circles on the sphere are projected to circles or lines on the plane, rendering the device relatively easy to build. The challenge to construct various curves on the astrolabe was taken on several times. Astrolabe variants based on other projections – in particular the azimuthal equidistant projection – were studied, although no such astrolabes are extant (Kennedy *et al.* 1999). Finally, a remarkable pair of 11th/17th-century world maps use a projection that places Mecca at its center and preserves the direction and distance to Mecca for all locations (King 1999[CB]).

Optics: In his *Catalogue of the Sciences*, early 4th/10th-century philosopher al-Fārābī lists optics among the mathematical sciences that should hardly be distinguished from geometry. Although the situation is more complicated than this (Chapter I.8), it is true that the study of optics in the Abbasid caliphate was inspired by Euclid's *Optics* (*Optika*), a work that relied heavily on geometrical methods (Kheirandish 1999, 2003). It has been argued that al-Kindī's commentary on Euclid's *Optics* shifted the study of the subject away from geometry and closer to physical reality (al-Kindī 1997). Indisputably the central Arabic work in this field was Ibn al-Haytham's *Optics* (*Kitāb al-Manāẓir*), which opted for the theory that rays travel from the object to the eye rather than from the eye to the object. Ibn al-Haytham's optical theory involved more than geometry and even included a discussion of psychological aspects of vision, but it still relied heavily on geometrical methods. For instance, the famous "Alhazen's problem" posits a spherical mirror, a point of sight and a point of radiation and asks for a construction of the point of reflection on the mirror's surface (Ibn al-Haytham 1989, 2002; Sabra 1983, 2003; Smith 2001, 2006, 2008).

Of special interest for geometry was the attention paid to "burning mirrors", surfaces that cause light rays to concentrate at a particular location. Inspired by the ancient geometer Diocles (*fl.* 190 BCE), a number of Muslim scholars from the 4th/10th century (including al-Kūhī, Abū l-Wafā' and Abū Saʿd al-ʿAlā' ibn Sahl) used the theory of conics to explore the properties of mirrors in the shapes of paraboloids and hyperboloids (Diocles 1976, 2000; Ibn al-Haytham 1993; Diocles and "Dtrūms" 1997; Rashed 1990, 2000, 2005; Bellosta 2002a; Neugebauer and Rashed 1999).

Architecture: Whether or not geometry played a direct role in architecture beyond the basic mensuration formulas is unclear (Chapter II.10). We do know that a number of geometers, including Abū l-Wafā', Ibn al-Haytham, al-Karajī (d. *c.* 419/1029) and al-Kāshī, wrote on problems directly inspired by architecture, although we may never know if these works were actually used by practitioners. Al-Kāshī's *Key to Arithmetic* (*Miftāḥ al-ḥisāb*) contains studies of arches, *qubbas* (domes) and the *muqarnas* (ornamented vaulting). For instance, al-Kāshī estimated the volumes and surface areas of architectural features that can be represented as surfaces of revolution about an axis. Relying on theorems and constructions from Euclid and Archimedes, his estimates were accurate to within a couple of percent (Dold-Samplonius 1992a, 1992b, 1993, 1996b, 2000, 2003; Sakkal 1995).

Applications of geometry extended beyond building design to ornamental decorations on walls. Artisans held conversations with geometers to help them construct the intricate patterns we find today in palaces like the Alhambra leading to works such as Abū l-Wafā'’s *On What the Artisan Requires of Geometrical Constructions* (*Kitāb fī mā yaḥtāju ilayhi al-ṣāniʿ min al-aʿmāl al-handasiyya*; Chapter II.10). Of special interest here is the choice of method: while (as we have seen) geometers preferred the use of conic sections over "verging" constructions for logical reasons, artisans favored verging constructions because they were easier to implement (Özdural 1996, 2000, 2015; Hogendijk 2013; Berggren 2007, 612–16; Nepçioğlu 2017).

These disciplines by no means exhaust the list of impacts that geometry had on various aspects of society. As one might expect for a subject so ubiquitous, the geometers' knowledge and the ways they chose to think and implement it helped to shape the ways in which Islamicate societies saw the world.

Notes

1 In a now classic article, Sabra (1987; see consolidated bibliography) argues that the notion of "reception" of foreign concepts should be replaced with "appropriation" and "naturalization" in order to better reflect the changes that occur in the text and its meanings in the new context.

2 Al-Nayrīzī's commentary is of interest to historians of both Greek and Arabic mathematics for its own sake but also because it preserves otherwise lost texts by other authors. A massive translation of the entire commentary is being undertaken by Anthony Lo Bello, so far encompassing the first four books of the *Elements* (al-Nayrīzī 2003, 2009).

3 The debate regarding the precise meaning of this term goes back to ancient Greece. For geometry in Islamicate societies, see Thābit ibn Qurra in Dold-Samplonius (1996a) and Bellosta (2002b) and Ibn al-Haytham (Ibn al-Haytham, Ibn Sahl and al-Qūhī 1993; Rashed 2017).

4 Medieval astronomers used a circle of radius 60 rather than the unit circle, so al-Kāshī's values were 60 times what is given here.

Bibliography

Sources

Abū Kāmil. ed. and ann. Sesiano, J. 1996. "Le *Kitāb al-Misāḥa* d'Abū Kāmil," *Centaurus* 38: 1–21.

Abū Kāmil. ed., tr. and ann. Sesiano, J. 2014. "Abū Kāmil's *Book on Mensuration*," in Sidoli, N. and Van Brummelen, G., eds. *From Alexandria, Through Baghdad*. New York: Springer, 359–408.

Abū l-Jūd. ed., tr. and ann. Rashed, R. 2010. "Les constructions géométriques entre géometrie et algèbre: L'épître d'Abū al-Jūd à al-Bīrūnī," *Arabic Sciences and Philosophy* 20: 1–51.

Apollonius. ed., tr. and ann. Toomer, G. 1990. *Conics, Books V to VII: The Arabic Translation of the Lost Greek Original in the Version of the Banū Mūsā*. 2 vols. New York: Springer.

Apollonius de Perge. ed., tr. and ann. Rashed, R. 2008–2010. *Apollonius de Perge, Coniques*. 5 vols. Berlin: De Gruyter.

al-Bīrūnī. ed., tr. and ann. Debarnot, M.-T. 1985. *Kitāb Maqālīd ʿilm al-Hay'a: La Trigonométrie Sphérique chez les Arabes de l'Est à la Fin du Xe Siècle*. Damascus: Institut Français de Damas.

Diocles and "Dtrūms". ed., tr. and ann. Rashed, R. 1997. "Dioclès et Dtrūms: Deux traités sur les miroirs ardents," *Mélanges Institut Dominicain d'Études Orientales du Caire* 23: 1–155.

Diocles. ed., tr. and ann. Toomer, G. 1976. *Diocles on Burning Mirrors: An Arabic Translation of the Lost Greek Original*. New York: Springer.

Diocles *et al.* ed., tr. and ann. Rashed, R. 2000. *Les Catoptriciens Grecs*]. Vol. 1. Paris: Société d'Édition Les Belles Lettres.

Euclid. ed., tr. and ann. Kheirandish, E. 1999. *The Arabic Version of Euclid's Optics*. New York: Springer.

Ibn al-Haytham. ed., tr. and ann. Hogendijk, J. 1985. *Ibn al-Haytham's Completion of the Conics*. New York: Springer.

Ibn al-Haytham, Ibn Sahl and al-Qūhī. ed., tr. and ann. Rashed, R. 1993. *Géometrie et Dioptrique au Xe Siècle: Ibn Sahl, al-Qūhī et Ibn al-Haytham*. Paris: Les Belles Lettres.

Ibn al-Haytham. ed., tr. and ann. Rashed, R. 1979. "La construction de l'heptagone régulier par Ibn al-Haytham," *Journal for the History of Arabic Science* 3: 309–87.

Ibn al-Haytham. ed., tr. and ann. Rashed, R. 1991. "La philosophie des mathématiques d'Ibn al-Haytham – L'analyse et la synthèse," *Mélanges d'Institut Dominicain d'Études Orientales du Caire* 20: 31–231.

Ibn al-Haytham. ed., tr. and ann. Rashed, R. 1993. "La philosophie des mathématiques d'Ibn al-Haytham – II. 'Les Connus'," *Mélanges d'Institut Dominicain d'Études Orientales du Caire* 21: 87–275.

Ibn al-Haytham. ed., tr. and ann. Rashed, R. 2000. *Les Mathématiques Infinitésimales du IXe au xIe Siècle*. Vol. 3: *Ibn al-Haytham: Théorie des Coniques, Constructions Géometriques et Géométrie Pratique*. London: Al-Furqān Islamic Heritage Foundation.

Ibn al-Haytham. ed., tr. and ann. Rashed, R. 2002. *Les Mathématiques Infinitésimales du IXe au XIe Siècle*. Vol. 4: *Ibn al-Haytham: Méthodes Géométriques, Transformations Ponctuelles et Philosophie des Mathématiques*. London: Al-Furqān Islamic Heritage Foundation.

Ibn al-Haytham. ed. Sabra, A. 1983. *Kitāb al-Manāẓir (The Optics of Ibn al-Haytham). Arabic Text of Books 1, 2, and 3 on Direct Vision*. Kuwait: National Council for Culture, Arts, and Letters.

Ibn al-Haytham. tr. and ann. Sabra, A. 1989. *The Optics of Ibn al-Haytham, Books I-III: On Direct Vision*. 2 parts. London: The Warburg Institute.

Ibn al-Haytham. ed., tr. and ann. Sabra, A. 2002. *Kitāb al-Manāẓir of al-Ḥasan ibn al-Haytham. Books IV-V. On Reflection, and Images Seen by Reflection*. Safat: National Council for Culture, Arts and Letters.

Ibn al-Haytham. ed., tr. and ann. Voss, D. 1985. *Ibn al-Haytham's Doubts Concerning Ptolemy: A Translation and Commentary*. PhD dissertation. University of Chicago.

al-Khayyām, tr. and ann. Djebbar, A. 2002. "Épitre d'Omar Khayyam *Sur l'explication des prémisses problématiques du livre d'Euclide*," *Farhang* 14: 79–136.

al-Khayyām. tr. Khalil, R. 2008. Algebra wa al-Muqabala: An Essay by the Uniquely Wise *'Ab[d]el Fath Omar bin al-Khayyam on Algebra and Equations*. Reading: Garnet.

al-Khayyām. ed., tr. and ann. Rashed, R. and Djebbar, A. 1981. *L'Œuvre Algébrique d'al-Khayyam, Établie, Traduite et Analysée par Roshdi Rashed et Ahmad Djebbar*. Aleppo: Institute for the History of Arabic Science.

al-Khayyām. ed., tr. and ann. Rashed, R. and Vahabzadeh, B. 1999. *Al-Khayyam Mathématicien*. Paris: Librairie Scientifique et Technique Albert Blanchard (English translation: 2000. *Omar Khayyam the Mathematician*. New York: Bibliotheca Persica Press).

al-Kindī. ed., tr. and ann. Rashed, R. 1997. *Œuvres Philosophiques et Scientifiques d'al-Kindī*. Vol. 1: *L'Optique et la Catoptrique*. Leiden: Brill.

al-Kūhī. ed., tr. and ann. De Young, G. 1992. "Abū Sahl's Additions to Book II of Euclid's *Elements*," *ZGAIW* 7: 73–135.

al-Kūhī. tr. and ann. Berggren, J. L. and Van Brummelen, G. 2001a. "Abū Sahl al-Kūhī's 'On Drawing Two Lines from a Point at a Known Angle'," *Suhayl* 2: 161–98.

al-Kūhī. ed., tr. and ann. Berggren, J. L. and Van Brummelen, G. 2001b. "Abū Sahl al-Kūhī on the Distance to the Shooting Stars," *Journal for History of Astronomy* 32: 137–51.

al-Kūhī. ed., tr. and ann. Rashed, R. 2001c. "Al-Qūhī: From Meteorology to Astronomy," *Arabic Sciences and Philosophy* 11: 157–204.

al-Kūhī. ed., tr. and ann. Rashed, R. 2002. "Al-Qūhī: De la méteorologie à l'astronomie," *Oriens Occidens* 4: 1–57.

al-Kūhī. ed., tr. and ann. Berggren, J. L. and Van Brummelen, G. 2002–2003. "From Euclid to Apollonius: al-Kuhi's Lemmas to the *Conics*," *ZGAIW* 15: 165–74.

al-Kūhī. ed., tr. and ann. Berggren, J. L. and Van Brummelen, G. 2005. "Al-Kūhī's Revision of Book I of Euclid's *Elements*," *Historia Mathematica* 32: 426–52.

Menelaus. ed., tr. and ann. Sidoli, N. 2006. "The Sector Theorem Attributed to Menelaus'," *SCIAMVS* 7: 43–79.

al-Nayrīzī. tr. Lo Bello, A. 2003. *The Commentary of al-Nayrizi on Book I of Euclid's Elements of Geometry*. Boston: Brill.

al-Nayrīzī. tr. Lo Bello, A. 2009. *The Commentary of al-Nayrizi on Books II-IV of Euclid's Elements of Geometry*. Boston: Brill.

al-Samarqandī. tr. and ann. De Young, G. 2001. "The *Ashkāl al-Ta'sīs* of al-Samarqandī: A Translation and Study," *Zeitschrift für Geschichte der Arabisch-Islamischen Wissenschaften* 14: 57–117.
al-Sijzī. ed., tr. and ann. Hogendijk, J. and Bagheri, M. 1996. *Al-Sijzi's Treatise on Geometric Problem Solving*. Tehran: Fatemi.
al-Sijzī. ed., tr. and ann. Rashed, R. 2004. *Œuvre Mathématique d'al-Sijzi*. Vol. 1: *Géometrie des Coniques et Théorie des Nombres au Xe Siècle*. Leuven: Peeters.
al-Ṭūsī. ed., tr. and ann. Pinel, P. and Taha, A. 2003. "Le travail d'al-Ṭūsī sur les *Sphériques* de Ménélaus: Établissement critique du texte, apport mathématique, interpretation astronomique," *Farhang* 15–6: 33–109.

Research literature

Abgrall, P. 1995. "Les cercles tangents d'al-Qūhī," *Arabic Sciences and Philosophy* 5: 263–95.
Abgrall, P. 1997. "Al-Qūhī et les courbes coniques," in Hasnawi, A., Elaramni-Jamal, A. and Aouad, M., eds. *Perspectives Arabes et Médiévales sur la Traditions Scientifique et Philosophique Grecque*. Leuven: Peeters, 21–9.
Abgrall, P. 2004. *Le développement de la géometrie au IXe-XIe siècles – Abū Sahl al-Qūhī*. Paris: Blanchard.
Anbouba, A. 1977. "Construction de l'heptagone régulier par les Arabes au 4e siècle de l'hégire [in Arabic]," *Journal for the History of Arabic Science* 1: 352–84.
Bellosta, H. 1991. "Ibrāhīm ibn Sinān: On Analysis and Synthesis," *Arabic Sciences and Philosophy* 1: 211–32.
Bellosta, H. 2002a. "Les instruments ardents dans la tradition arabe," *Matapli* 67: 73–88.
Bellosta, H. 2002b. "Uncomplément arabe aux *Données* d'Euclide: Le *Kitab al-Mafrudat de Tabit ibn Qurra*," in Ansari, S., ed. *Science and Technology in the Islamic World: Proceedings of the XXth International Congress of History of Science*. Turnhout: Brepols, 71–82.
Berggren, J. L. 2007. "Mathematics in Medieval Islam," in Katz, V., ed. *The Mathematics of Egypt, Mesopotamia, China, India, and Islam: A Sourcebook*. Princeton, NJ: Princeton University Press, 515–675.
Berggren, J. L. 2016². *Episodes in the Mathematics of Medieval Islam*. New York: Springer.
Berggren, J. L. and Van Brummelen, G. 2000. "The Role and Development of Geometric Analysis and Synthesis in Ancient Greece and Medieval Islam," in Suppes, P., Moravcsik, J. and Mendell, H., eds. *Ancient and Medieval Traditions in the Exact Sciences*. Chicago: University of Chicago Press, 1–31.
Brentjes, S. 1996. "Remarks about the Proof Sketches in Euclid's *Elements*, Book I as Transmitted by MS Paris, B. N. Fonds Latin 10257," in Folkerts, ed. [CB], 115–38.
Casulleras, J. 2008–2009. "Mathematical Astrology in the Medieval Islamic West," *ZGAIW* 18: 241–68.
Casulleras, J. and Hogendijk, J. 2012. "Progressions, Rays and Houses in Medieval Islamic Astrology: A Mathematical Classification," *Suhayl* 12: 33–102.
De Young, G. 1991. "New Traces of the Lost al-Ḥajjāj Arabic Translations of Euclid's *Elements*," *Physis* 38: 647–66.
De Young, G. 2002. "Ḳāḍīzāde al-Rūmī on Samarḳandī's *Ashkāl al-Ta'sīs*: A Mathematical Commentary," in Ansari, S., ed. *Science and Technology in the Islamic World: Proceedings of the XXth International Congress of History of Science*. Turnhout: Brepols, 83–90.
De Young, G. 2008–2009. "The *Taḥrīr Kitāb Uṣūl Uqlīdis* of Naṣīr al-Dīn al-Ṭūsī: Its Sources," *ZGAIW* 18: 1–71.
Djebbar, A. 1996. "Quelques commentaires sur les versions arabes des *Eléments* d'Euclide et sur leur transmission à l'Occident Musulman," in Folkerts, ed., 91–114.
Dold-Samplonius, Y. 1992a. "Practical Arabic Mathematics: Measuring the *muqarnas* by al-Kāshī," *Centaurus* 35: 193–242.
Dold-Samplonius, Y. 1992b. "The XVth Century Timurid Mathematician Ghiyāth al-Dīn Jamshīd al-Kāshī and his Computation of the *qubba*," in Demidov, S. *et al.*, eds. *Amphora: Festschrift for Hans Wussing on the Occasion of His 65th Birthday*. Basel: Birkhäuser, 171–81.
Dold-Samplonius, Y. 1993. "The Volume of Domes in Arabic Mathematics," in Folkerts, M. and Hogendijk, J., eds. *Vestigia Mathematica: Studies in Medieval and Early Modern Mathematics in Honour of H. L. L. Busard*. Amsterdam: Rodopi, 93–106.
Dold-Samplonius, Y. 1996a. "*The Book of Assumptions*, by Thābit ibn Qurra (836–901)," in Dauben, J. *et al.*, eds. *History of Mathematics: States of the Art*. San Diego: Academic Press, 207–22.
Dold-Samplonius, Y. 1996b. "How al-Kashi Measures the *muqarnas*: A Second Look," in Folkerts, M., ed., 56–90.
Dold-Samplonius, Y. 2000. "Calculation of Arches and Domes in 15th Century Samarkand," in Williams, K., ed. *Nexus III: Architecture and Mathematics*. Ospedaletto: Pacini Editore, 45–55 (Reprint: Williams,

K. and Ostwald, M. J., eds. 2015. *Architecture and Mathematics from Antiquity to the Future*. 2 vols. Cham: Birkhäuser, 1: 297–307).

Dold-Samplonius, Y. 2003. "Calculating Surface Areas and Volumes in Islamic Architecture," in Hogendijk and Sabra, eds., 235–65.

Hankel, H. 1874. *Beiträge zur Geschichte der Mathematik im Alterthum und Mittelalter*. Leipzig: Teubner [Reprinted Hildesheim: Olms 1965].

Hogendijk, J. 1979. "On the Trisection of an Angle and the Construction of a Regular Nonagon by Means of Conic Sections in Medieval Islamic Geometry," *Preprint 113*, University of Utrecht. www.jphogendijk.nl/publ.html.

Hogendijk, J. 1981. "How Trisections of the Angle Were Transmitted from Greek to Islamic Geometry," *Historia Mathematica* 8: 417–38.

Hogendijk, J. 1984. "Greek and Arabic Constructions of the Regular Heptagon," *Archive for History of Exact Sciences* 30: 197–330.

Hogendijk, J. 2012. "Mathematics and Geometric Ornamentation in the Medieval Islamic World," in Latała, R. *et al.*, eds. *European Congress of Mathematics: Proceedings of the 6th? Congress (6 ECM) Held at the Jagiellonian University, Kraków, July 2–7, 2012*. Zurich: European Mathematical Society, 727–41.

Hogendijk, J. and Sabra, A., eds. 2003. *The Enterprise of Science in Islam: New Perspectives*. Cambridge, MA: MIT Press.

Jaouiche, K. 1986. *La Théorie des Paralleles en Pays d'Islam*. Paris: Vrin.

Jaouiche, K. 1988. "L'analyse et la synthèse dans les mathématiques arabo-islamiques: Le livre d'Ibn al-Haytham," in Atik, Y., ed. *Histoire des mathématiques arabes*. Algiers: Maison des Livres, 106–24.

Kennedy, E. S. 1996. "The Astrological Houses as Defined by Medieval Islamic Astronomers," in Casulleras and Samsó, eds. [CB], 535–78 [Reprint: *Astronomy and Astrology in the Medieval Islamic World*. Aldershot: Variorum, XIX].

Kheirandish, E. 2003. "The Many Aspects of 'Appearances': Arabic Optics to 950 AD," in Hogendijk and Sabra, eds., 55–83.

King, D. 1986. "The Earliest Islamic Mathematical Methods and Tables for Finding the Direction of Mecca," *ZGAIW* 3: 82–149.

Knorr, W. R. 1989. *Textual Studies in Ancient and Medieval Geometry*. Boston: Birkhäuser.

Luther, I. 2002. "The Solution of Apollonius' Problem in the Medieval Arab East," in Ansari, S., ed. *Science and Technology in the Islamic World: Proceedings of the XXth International Congress of History of Science*. Turnhout: Brepols, 91–9.

Mawaldi, M. 2000. "Méthode de l'analyse et de la synthèse de Kamal al-Din al-Farisi," in İhsanoğlu, E. and Günergun, F., eds. *Science in Islamic Civilisation*. Istanbul: IRCICA, 193–9.

Nadal, R., Taha, A. and Pinel, P. 2004. "Le contenu astronomique des *Sphériques* de Ménélaos," *Archive for History of Exact Sciences* 58: 381–436.

Necipoğlu, G. ed. 2017. *The Arts of Ornamental Geometry. A Persian Compendium on Similar and Complementary Interlocking Figures*. Fī tadākhul al-ashkāl al-mutashābiha aw al-mutawāfiqa (*Bibliothèque nationale de France, Ms Persan 169, fols. 180r-199r*). Leiden and Boston: Brill.

Neugebauer, O. and Rashed, R. 1999. "Sur une construction du miroir parabolique par Abū al-Wafā' al-Būzjānī," *Arabic Sciences and Philosophy* 9: 261–77.

Özdural, A. 1996. "On Interlocking Similar or Corresponding Figures and Ornamental Patterns of Cubic Equations," *Muqarnas* 13: 191–211.

Özdural, A. 2000. "Mathematics and Arts: Connections between Theory and Practice in the Medieval Islamic World," *Historia Mathematica* 27: 171–201.

Özdural, A. 2015. "The Use of Cubic Equations in Islamic Art and Architecture," in Williams, K. and Ostwald, M., eds. *Architecture and Mathematics from Antiquity to the Future*. 2 vols. Basel: Birkhäuser, Cham, 1: 467–81.

Rashed, R. 1990. "A Pioneer in Anaclastics: Ibn Sahl on Burning Mirrors and Lenses," *Isis* 81: 464–91.

Rashed, R. 1991. "L'analyse et la synthèse selon Ibn al-Haytham," in Rashed, R., ed. *Mathématiques et Philosophie de l'Antiquité à l'Âge Classique: Hommage à Jules Vuillemin*. Paris: CNRS, 131–47.

Rashed, R. 2003. "Al-Qūhī et al-Sijzī: Sur le compas parfait et le tracé continu des sections coniques," *Arabic Sciences and Philosophy* 13: 9–43.

Rashed, R. 2005. *Geometry and Dioptrics in Classical Islam*. London: Al-Furqān Islamic Heritage Foundation.

Rashed, R. 2017. *Ibn al-Haytham's Geometrical Methods and the Philosophy of Mathematics*. New York: Routledge.

Rashed, R. and Houzel, C. 2005. "Thābit ibn Qurra et la théorie des parallèles," *Arabic Sciences and Philosophy* 15: 9–55.

Rosenfeld, B. 1988. *A History of Non-Euclidean Geometry: Evolution of the Concept of a Geometric Space*. New York: Springer.

Sabra, A. I. 1968. "Thabit ibn Qurra on Euclid's parallel postulate," *Journal of the Warburg and Courtauld Institutes* 31: 12–32.

Sabra, A. I. 2003. "Ibn al-Haytham's Revolutionary Project in Optics: The Achievement and the Obstacle," in Hogendijk, J. and Sabra, A., eds. 2003. *The Enterprise of Science in Islam: New Perspectives*. Cambridge, MA: MIT Press, 85–118.

Sakkal, M. 1995. "Geometry of Ribbed Domes in Spain and North Africa," *Journal for the History of Arabic Science* 11: 53–73.

Smith, A. M. 2001. *Alhacen's Theory of Visual Perception*. Philadelphia: American Philosophical Society.

Smith, A. M. 2006. *Alhacen on the Principles of Reflection*. Philadelphia: American Philosophical Society.

Smith, A. M. 2008. "Alhacen's Approach to 'Alhazen's Problem'," *Arabic Sciences and Philosophy* 18: 143–63.

Vahabzadeh, B. 1997. "Al-Khayyām's Conception of Ratio and Proportionality," *Arabic Sciences and Philosophy* 7: 247–63.

Vahabzadeh, B. 2004. "Umar al-Khayyām and the Concept of Irrational Number," in Morelon, R. and Hasnawi, A., eds. *De Zénon d'Élée à Poincaré: Recueil d'Études en Hommage à Roshdi Rashed*. Leuven: Peeters, 55–63.

Van Brummelen, G. 2009. *The Mathematics of the Heavens and the Earth: The Early History of Trigonometry*. Princeton, NJ: Princeton University Press.

Van Brummelen, G. forthcoming. "Crossing a Mathematical Rubicon: Al-Kāshī's Two Methods of Computing Sin 1°," in Chemla, K., ed. *Sciences in the Ancient World*.

Van Brummelen *et* al.: Van Brummelen, G., Mimura, T. and Kerai, Y. 2012. "Al-Samaw'al's Curious Approach to Trigonometry," *Suhayl* 11: 9–31.

Vitrac, B. 2000. "Omar Khayyām et Eutocius: Les antécédents grecs du troisième chapitre du commentaire *Sur certaines prémisses problématiques du Livre d'Euclide*," *Farhang* 12: 51–105.

Wiedemann, E. 1909. "Kleinere Arbeiten von Ibn al-Haiṯam," *Sitzungsberichte der physikalisch- medizinischen Sozietät zu Erlangen* 41: 16–24.

Woepcke, F. 1855. "Analyse et extrait d'un recueil de constructions géométriques par Aboûl Wafâ," *Journal Asiatique* 5.5: 218–359.

I.6

THE ASTRAL SCIENCES THROUGH THE 7TH/13TH CENTURY

Attitudes, experts and practices

Sonja Brentjes

This chapter discusses knowledge about the heavens as a whole, by attempting, as far as possible, to discuss astronomy and astrology together. It acknowledges that in certain periods the distance between astronomy and astrology was greater than in others. The chapter begins with a discussion of the classification of the astral sciences and then moves on to debates, mostly about theories and to a lesser extent about practices. It concludes with a section on mathematical practices. It thus pays attention to several themes that previous surveys of the astral sciences in Islamicate societies ignored or marginalized and abstains from repeating summaries of astronomical themes found in previous surveys (Rashed and Morelon 1996[CB];[1] Morrison 2013).

I.6.1 Introduction

Until recently, a more or less strict division between mathematical astronomy and predictive astrology was considered appropriate for the study of astral knowledge in all Islamicate societies after about 145/762. Saliba (1982; reprint 1994[CB], 66, 72), for example, suggested that the emergence of *the science of the configuration [of the universe]* (*ʿilm al-hayʾa*) was the result of scholars defending mathematical astronomy against attacks by religious scholars on astrology. The clearest expression of this trend is the separation of terminologies, concepts and practices in astronomy and astrology.

Aiming to avoid earlier overgeneralizations and essentializing judgments, Ragep (2009) identified the motivations for practicing the astral sciences in Islamicate societies as practical, religious, philosophical and political. He argued that the rejection of astrology by various religious scholars was only one among the many reasons for the emergence of *the science of the configuration [of the universe]* and that this rejection took various forms. Moreover, according to my experiences in manuscript libraries and with manuscript catalogues, numerous treatises presenting themselves as books on the *science of configuration* (*ʿilm al-hayʾa*) include chapters on astrological themes.

In this handbook, we are also making a serious effort to avoid such overgeneralizations and modern classifications. I have encouraged contributors to use the term *astral sciences* and render the actors' self-designation *munajjimūn* as *astral experts*. Due to differences in opinion,

DOI: 10.4324/9781315170718-8

not everyone follows this approach; some continue differentiating between astronomers and astrologers according to themes, methods and problems. The decision to treat astral knowledge as a unit despite the numerous differences between practitioners, periods and regions is based on five points: (1) the absence of a specialized actor-related terminology separating astronomers from astrologers, (2) the presence of shared theoretical and mathematical components in the two domains, (3) the representation of practices and goals by historical actors, (4) the engagement in astrology of many scholars considered by modern historians of sciences primarily as astronomers and (5) the existence of many works treating the astral sciences as a whole.

One example will suffice to document such overlaps. Abū l-Ṣaqr al-Qabīṣī (mid-4th/10th century) was a court astrologer of Emir Sayf al-Dawla (r. 333–356/945–967) in Aleppo. He studied Ptolemy's *Almagest* and probably other works on the mathematical sciences with another astrologer in his hometown of Mosul. Among his writings as a mature astral expert are a lost treatise, apparently on questions concerning mathematical astronomy, called *Doubts on the* Almagest (*Shukūk fī l-Majisṭī*), intended for testing people claiming astral expertise, a lost *Book on the Reasons behind the Astronomical Handbooks* (*Kitāb ʿilal al-zījāt*), the extant *Treatise on the Distances and (Heavenly) Bodies* (*Risāla fī l-abʿād wa-l-ajrām*), the extant *Book on the Introduction in the Science of the Judgment of the Stars* (*Kitāb al-Mudkhal ilā ṣināʿat aḥkām al-nujūm*) and the extant *Treatise on the Testing of Those Who Call Themselves Astral Experts* (*Risāla fī imtiḥān al-munajjimīn mimman huwa muttasim bi-hādha l-ism*). These books show that al-Qabīṣī was trained in all fields of the astral sciences and practiced them at the Hamdanid court. He insisted that the *complete* astral expert had to be solidly qualified in both species of the astral sciences; be able to solve all of their problems "through his own intelligence," not the mere repetition of texts or the use of an astrolabe; to be capable of rational demonstrations and proofs; and to be able to compile an astronomical handbook based on direct observations, not mere re-computations of data obtained from earlier scholars (Burnett 2002, 203–4; al-Qabīṣī 2004, 2–6).

The content, fields and texts of the predictive parts of the astral sciences in Islamicate societies have been expertly summarized in a recent article on astrology by Burnett (2007) in the third edition of the *Encyclopaedia of Islam*. Ragep (2009) published an analogous summary of some other parts of the astral sciences. The interested reader should consult both of them in conjunction with this chapter, although both articles contain several claims with which I disagree. A recent bibliographical survey of publications on the astral sciences in Islamicate societies until 1450 can be found on the website of the Ptolemaeus project at the Bavarian Academy of Science (https://ptolemaeus.badw.de/astrobibl/section/G).

I.6.2 Classifications of the astral sciences

The purpose of this section is to shed some light on the contributions of philosophers to the astral sciences, to point to some of the fundamental differences between philosophers and some of the astral experts and to stress that the adoption of Aristotelian natural philosophy and epistemology as the foundations of the astral sciences was neither a simple, straightforward process nor an uncontested commitment.

Classifications of the astral sciences were manifold, contradictory and in continuous flux. Two main positions existed: (1) astral knowledge is one discipline with different species, and (2) astral knowledge belongs to two different sciences marked by different epistemological values. The non-prognosticating parts were understood as forming a mathematical discipline, which we call today astronomy. The prognosticating parts, which we call astrology, were defined in two ways: (1) a second component of the astral sciences as a mathematical discipline and (2) a subdiscipline or branch of natural philosophy. The subsequent issue was how to define the relationship between the two species or sciences: Were both parts of equal standing, or were they in a hierarchical relationship, and if so, which one of the two should precede the other? A third question asked whether one or both should be based on an epistemically higher discipline from which their principles and foundational proofs had to be derived. If affirmed, the higher discipline was natural philosophy. A fourth issue concerned the reliability of methods and the soundness of theories.

The answers to the first two questions depended on views about the positions of the mathematical sciences in the entire system of knowledge, the role of philosophy regarding revelation or religious law, and the classification of the philosophical disciplines. The answers to the third question usually depended on how scholars saw the role of Aristotelian epistemology as elaborated in the *Posterior Analytics* in the system of the sciences. The positions taken in regard to the fourth question reflected religious beliefs, experiences with predicted results and the social status of practitioners (Burnett 2002). Behind these questions addressed to the astral sciences more general debates lurked that had emerged in antiquity: What were the status, objects and methods of the three theoretical parts of philosophy (metaphysics, natural philosophy, the mathematical sciences), their interrelations and the character of their results (conjectures or certain knowledge)? Among the ancient scholars who had grappled with these issues (Bowen 2007; Feke 2014, 2018), three were certainly known through translations of parts of their works: Geminus (*c.* 70 BCE), Hero (1st century) and Ptolemy (*c.* 100–170). They reinterpreted the relationship and ranking between the three theoretical disciplines of philosophy in favor of the mathematical sciences, on the basis of the certainty of their proofs and hence the knowledge they engendered. It is, however, still unclear whether their opinions guided some of the debates briefly summarized in this section or inspired some of the deviating positions attributed to scholars mentioned below.

The works of al-Kindī (d. *c.* 256/870) and his associate Abū Maʿshar (171–272/787–886) marked out an important position. They agreed that contrary to common belief the science of the stars was not based on conjecture and guesswork but was firmly grounded on first principles. It worked with analogical reasoning and demonstrations on the basis of reliable observations and trustworthy experience (Abū Maʿshar 2019[CB], 1: 46–7, 54–69). Abū Maʿshar believed that this science consisted of two species. The first was the science of the universe. This was apparently a mathematical science, since it contained the science of quality and quantity of the higher spheres, and because its arguments were derived from arithmetic, geometry and the science of measurement. The second, the science of the judgments of the stars, was the knowledge of the natures of the planets, the powers of their movements and their impacts on the sublunar world. Hence, it was not a mathematical science but connected to natural philosophy. These two species were not on an equal footing, since the second species received its information from the first (Abū Maʿshar 2019, 52–5). On the other hand, Abū Maʿshar affirmed that aspects of the second species that are not obvious and clear are "inferred by means of clear analogies (rational arguments) from the science of the natures of things (physics)" (Abū Maʿshar 2019, 1: 55). At the same time, the science of the stars was universal, like medicine, but higher and nobler than this discipline, because its objects belonged to the celestial realm (Abū Maʿshar 2019, 1: 68–71). According to Adamson, Abū Maʿshar defended this position through theological types of reasoning following a precedent set by al-Kindī (Adamson 2002, 249–51).

In the 4th/10th and 5th/11th centuries, a number of authors with different intellectual affiliations and standing wrote short texts on how to arrange, evaluate and define the various fields of knowledge. Most writers discussed the fields of knowledge as a whole. A few chose other approaches such as a classification by technical terminology or the practices of the people who engaged in such matters. Differences in goals and social contexts yielded differences in emphasis, style and evaluation. Al-Tawḥīdī (d. 414/1023), a secretary and litterateur, treated the astral sciences like the medical sciences, setting both up from the perspective of the practitioners and dividing both fields into two parts – knowledge (*ʿilm*) and action (*ʿamal*). While the description of medicine leaves little doubt that the practitioner should be well versed in both parts, the discussion of those who contemplate the stars is more equivocal. The beginning of the description implies that the practitioner applies himself to both parts, the first part being the conditions of the stars, the differences between their ways and stations, their rising and setting, and the second being the exploration of the judgments of the stars regarding future matters. But after the enumeration of these content elements it becomes clear that it would be best to deal only with the first part. Not only are the tasks of the second part more demanding and less certain in their implementation, but the gains are also poor and the impact on society more damaging than ignorance (which al-Tawḥīdī despised). Pursuing the first part leads directly to astonishment followed by an opening of the heart and an enlargement of the breast that reinforce belief in and veneration of God (al-Tawḥīdī 1963–1964, 267, Arabic Pl. 2).

In the late 3rd/9th and early 4th/10th centuries (Callataÿ 2013, 301), a work ascribed to a group called the Brethren of Purity (*Ikhwān al-ṣafāʾ*) began to take shape. This group allegedly consisted of well-educated religious scholars. Their affiliation within Islam remains puzzling, although numerous features of their encyclopedia of 51 or 52 individual treatises (Ikhwān al-Safāʾ 1957[CB]) were believed to suggest an inclination toward several Shīʿī denominations. Recently, De Vaulx d'Arcy proposed a different interpretation arguing for al-Kindī's student and later courtier Aḥmad ibn al-Ṭayyib al-Sarakhsī (d. 286/899) as the sole author of this collection of treatises (https://journals.openedition.org/mideo/3397?lang=ar). Whoever authored the collection took several stances in the epistles concerning classification, only one of which can be mentioned here as an example. At the beginning of their *Third Epistle: On Astronomy* (*al-Risāla al-thālitha fī asṭrunūmiyā*), they split the mathematical part of the science of stars into two species so that the whole discipline now encompassed three species: (1) the science of the configuration [of the universe] (*ʿilm al-hayʾa*), (2) the science of understanding astronomical handbooks, almanacs, calendars and (3) the science of judgments (Ikhwān al-Safāʾ 2015, 25). Within the epistle they only treated the first and third parts, mentioning the second part only in four brief sentences (Ikhwān al-Safāʾ 2015, 59).

The philosopher al-Fārābī (d. 339/950–1) went in a different direction. He divided the single science of the stars into two different sciences, placing the science of the judgments of the stars before the mathematical science of the stars (*ʿilm al-nujūm al-taʿlīmī*; al-Fārābī 1996[CB], 57). Their contents differed only in minor details from the views of his predecessors (al-Fārābī 1996, 58–61).

Ibn Sīnā (d. 428/1037), himself a famous and highly influential philosopher, mixed Neoplatonic and Aristotelian perspectives in his classifications of the sciences, including the astral sciences. His oeuvre is, however, too rich to summarize here. As an example, a brief summary from his early *Treatise on the Parts of the Rational Sciences* (*Risāla fī aqsām al-ʿulūm al-ʿaqliyya*) has to suffice. In this treatise, Ibn Sīnā presented his youthful understanding of the parts of Peripatetic philosophy as three theoretical sciences and three practical sciences. The astral sciences are discussed in the first two of the theoretical sciences: natural philosophy and the mathematical sciences. The second main part of the natural sciences encompasses knowledge about the main components of the universe such as the heavens, what they contain and the four elements, as established in Aristotle's *On the Heavens* (Ibn Sīnā 1298/1880, 75). Among the branches of the natural sciences, Ibn Sīnā lists first medicine, followed by the judgments of the stars (Ibn Sīnā 1298/1880, 75). According to the Neoplatonic quadrivium, the *science of the configuration [of the universe]* (*ʿilm al-hayʾa*) is the third of the fundamental mathematical sciences.

Its only subdiscipline concerns the compilation of astronomical handbooks and almanacs (Ibn Sīnā 1298/1880, 76). While the placement of the *judgments of the stars* by Ibn Sīnā has often been interpreted by modern historians as downgrading them, this is questionable. At least, it was not a new position when compared with the debates in the Hellenistic period about the question of whether Babylonian horoscopic astrology was a science and, if so, where it should be placed (Bowen 2007; Feke 2014, 297–8).

As seen in the brief survey of earlier classifications, even those scholars who considered the judgment of the stars as a species within one single science and equal in position and relevance to *the configuration [of the universe]*, rarely saw the *science of the judgments of the stars* a field of mathematical knowledge. The breaking apart of the two species and the equalization of medicine and the astral sciences as disciplines of equal rank and structural composition, while already considered by Abū Maʿshar, appear as classificatory principles only in the 4th/10th century.

In the late 6th/12th century, the influential religious and philosophical scholar Fakhr al-Dīn al-Rāzī (d. 606/1210) declared the judgment of the stars and other occult sciences to belong to the mathematical sciences (Melvin-Koushki 2017, 127). Under his impact and that of the Brethren of Purity, as well as Ibn Sīnā, the identification of the parts of the astral sciences shifted back and forth between natural philosophy and mathematics. This remained a constant feature of the debates in Islamicate societies until the 12th/18th century at the very least. A second significant feature of those debates consists in partitioning the astral sciences into an increasing number of disciplinary components, which were either portrayed as independent fields of knowledge or as subdisciplines. A third aspect is the relative instability of meaning and of relationships between such designations. Similarly, outside of treatises on classification, for instance in biographical dictionaries or historical chronicles, meanings and relationships between the content of subdisciplines remained unstable, as did the relationships to their practitioners.

I.6.3 Debates on astral theories and practices

I.6.3.1 Philosophical foundations

A good number of the many ancient Greek texts translated into Arabic during the 2nd/8th, 3rd/9th and 4th/10th centuries discussed the structure and properties of the universe. Among them were Plato's (d. 348–347 BCE) *Timaios* and Galen's (129–216) commentary on it, Aristotle's (384–322 BCE) *On the Heavens* and Alexander of Aphrodisias's (late 2nd–early 3rd centuries) commentary, John Philoponus's (d. 574) and Aristotle's *Meteorology* and Theophrastus's (c. 371–287 BCE) texts on weather phenomena (Lettinck 1999; Taub 2003). Combined with claims and arguments from Ptolemy's *Almagest* and *Planetary Hypotheses*, these translations provided the astral sciences in Islamicate societies with theoretical foundations that remained largely operative from the 3rd/9th until the 12th/18th century, although different authors occasionally questioned and critically discussed various components.

Practitioners of the astral sciences believed that the universe was spherical and without void, that the earth was spherical and immobile in its center and that the universe was divided into an immutable part and a mutable part. The immutable part was filled by the fifth element ether. It began with the orbs supporting the moon and ended with the outermost sphere. The mutable part was the constantly changing world below the moon. It was composed of earth, as the heaviest element, followed in ascending order by water, air and fire. The eternal bodies made up from ether, from the moon outward, moved uniformly and in a circle around a center, preferably the center of the universe. The bodies of the sublunar world were as a rule composed of different parts of the four elements and characterized by different degrees of the two qualities (hot, cold;

wet, dry) of each element that formed part of the mixture. They resulted from different processes of generation and decay. Their so-called natural motion (a motion not caused by compulsion) went in a straight line to their natural place according to the heaviness of each body, depending on the mixture of the two heavy (earth, water) and the two light elements (air, fire) that composed it. Heavy bodies moved downward, while light bodies moved upward.

So far, there is no overarching study of the process that led to the adoption of those natural philosophical assumptions among experts in the mathematical sciences and, in particular, those who focused partly or entirely on the astral sciences. Based on existing research on specific texts and scholars, it seems likely that the scholars who were primarily responsible for the acceptance of this set of beliefs during the 3rd/9th century were philosophers like al-Kindī, astrologers like Abū Maʿshar and physicians like Ḥunayn ibn Isḥāq (d. 260/873).

Al-Kindī addressed many questions concerning the universe, studying them from different perspectives. He was most interested in elucidating the connections between the celestial and terrestrial realms. He asked, for instance, how the planets can initiate transformations in the bodies of the sublunar world, if they consist of a different element, ether. How can they emit light and heat, like the element fire, when they consist of such a profoundly different element? Why is it proper to attribute specific effects to each planet when all of them consist of one and the same element? As his fundamental position, al-Kindī declared that God is the cause for all that happens in the universe. However, God does not act directly but is the remote agent cause. In order to bring about change, the proximate agent causes are needed. These proximate agent causes are the planets. Al-Kindī established this position by using Aristotle's theories of the four causes (material, agent, formal, final), the four qualities (hot, dry, cold, wet) and the four elements (Adamson and Pormann 2012, 153–4). On how they achieve these changes, he offered three different lines of argumentation in different treatises apparently written in different periods of his life. According to the earliest one, planets cause sublunar processes due to the heat and friction that their movements create according to their speeds, sizes and distances. In another account, perhaps formulated late in his life, al-Kindī argued that changes happen because both planets and elements emit rays. In a third account, he attributed the planetary effects to mathematical proportions and symbolic analogies (Adamson and Pormann 2012, 154–5, 162–4; Orthmann 2016, 730; Adamson 2020). The last treatise deals with weather forecasting, while the other two are cosmological and theological in orientation. These contradictory claims suggest that al-Kindī continued to reflect on the complex questions about the relationship between God, the celestial realm and the sublunar world, without achieving a coherent theory. But he did argue throughout that cosmological theories appropriated from ancient philosophy, in particular Aristotle, supported the fundamental Islamic beliefs in God's role in the world (Adamson and Pormann 2012, 155–6).

Al-Kindī also shared some Platonic views, for example, that "the planets were rational, . . . spiritual beings capable of intelligence and speech," and hence the causes and administrators of everything in the sublunar world. Their activities followed the command of the Prime Creator who exercises through them control and providence over the sublunar world (Saliba 1994, 55; Adamson 2020). In the 4th/10th century al-Nawbakhtī (d. *c.* 307/920) rejected such a position in a treatise specifically dedicated to this subject, *Physical Arguments, Drawn from the Books of Aristotle, Negating Those Who Have Claimed That the Heavenly Sphere is Living and Rational* (*Ḥujaj ṭabīʿiyya mustakhraja min kutub Arisṭāṭālīs fī l-radd ʿalā man zaʿama anna l-falak ḥayy nāṭiq*).

Al-Kindī's erstwhile opponent and later enthusiastic student Abū Maʿshar profited not only from his teacher's knowledge of astral matters but also from his studies of Aristotelian and Neoplatonic philosophy. Like him he explained planetary effects on the sublunar world as resulting from their differences in motion. He insisted on the existence of a First Cause that operates those motions like a distant force and which he identified with God (Orthmann 2016, 730–1; Abū Maʿshar 2019, 1: 12). The motions of the planets are the cause of the movement of the sublunar elements and allow their potentiality for change to actualize. This impact over a distance is only possible because the sublunar elements are inherently susceptible to it (Abū Maʿshar 2019, 1: 11). Abū Maʿshar combined these basically Aristotelian ideas with Platonic concepts such as the form that is the defining feature of each thing and operates like an artisan, while the four sublunary elements are its tools. The most powerful of all the planets is the sun (Abū Maʿshar 2019, 1: 12). While those and other concepts and claims serve to justify the reality and the scientific nature of the art of judgments (astrology), the cause and regulator of all planetary activities and effects is God alone.

Some scholars also challenged the philosophical assumptions summarized above either in part or as a whole. Among them were Thābit ibn Qurra (d. 288/901), al-Bīrūnī (362–d. after 444/973–d. after 1053), ʿUmar al-Khayyām (439–*c.* 517/1048–*c.* 1123) and perhaps also al-Kūhī (2nd half 4th/10th century) and al-Isfizārī ([late 5th–early 6th/11th–early 12th centuries;] Rashed 2009, 696–9). These challenges may have emerged (at least partly) from some of the epistemological views expressed by Hellenistic scholars, in particular Ptolemy. But there is not enough research to come to a definitive conclusion. Thābit ibn Qurra may have even questioned the entire physical doctrine of Aristotle (Rashed 2009, 676–7). But we cannot be certain because not enough of his treatises on these issues have survived. The surviving fragments indicate his rejection of Aristotle's theories of natural place, the nonexistence of the void, the definition of time as a measure of celestial movement and some of the properties of movement. He also seems to have been indecisive about the role of the Prime Mover in the movements of the spheres (Rashed 2009, 683–99). Al-Bīrūnī is well known for his rejection of several Aristotelian natural philosophical doctrines in his letters to Ibn Sīnā (al-Bīrūnī and Ibn Sīnā 1970). His letters express curiosity, challenge, criticism and dissatisfaction. But there are no studies yet of the philosophical positions he took in his later works. In general, this is also true for the overarching question of which philosophical positions astral experts held in different times and regions.

As a whole, however, according to our current knowledge, Aristotelian natural philosophy formed a stable theoretical foundation on which the astral experts practiced their crafts whether mathematical or more speculative. Aristotelian doctrines were incorporated into manuals or synopses of astronomical matters, usually following the geometrical propositions, mostly taken from Euclid's (3rd century BCE) *Elements*. An example is Athīr al-Dīn al-Abharī's (d. between 660/1263 and 663/1265) *Concise [Book] on the Science of the Configuration [of the Universe]* (*Mukhtaṣar fī ʿilm al-hayʾa*). He opened the second chapter with the simple acknowledgment that natural philosophy provides the basis for astronomical knowledge:

> It has been established in the natural sciences that the celestial bodies are equal in their parts and natures, that their figures are spherical, that they move in circular (forms), that there is no beginning in them for a straight inclination; that the place of the earth is in the middle of the world.
>
> *(MS Paris, BnF, Arabe 2515, fol. 5b,10–13)*

The above mentioned author/s of the *Treatises of the Bretren of Purity* (*Rasāʾil Ikhwān al-Ṣafāʾ*) treated natural philosophical doctrines as well-accepted underpinnings of the astral sciences. In their 3rd epistle titled *Astronomy* (*asṭrunūmiyā*) they confirm that the impact of the stars on the sublunar world is due to the movements of the planetary orbs and the planets, that the theory of the four qualities applies to the zodiacal signs and that all celestial orbs are spherical and move in circular motions. They also likened the planets to spirits and the zodiacal signs to bodies (Ikhwān al-Safā'2015, 25, 31–2). They presupposed more specific beliefs about the material nature of the planets as round and luminous bodies and the orbs as spherical, transparent and hollowed-out bodies (Ikhwān al-Safā'2015, 25–6).

Despite the substantial difference in mathematical content and overall religious outlook, this epistle has several features in common with al-Bīrūnī's *Book of Instruction on the Elements of the Art of Astrology* (*Kitāb al-Tafhīm li-awā'il ṣināʿat al-tanjīm*), dedicated in 420/1029 to Rayḥāna, a young woman probably from Khwarazm (al-Bīrūnī 1934). Both are teaching texts, both illustrate concepts and claims with diagrams, and both simplify the access to scholarly matters by using comparisons with things from everyday life. The sphere of the moon encloses the lightest element air "like the shell of an egg encloses its egg white"; the earth is compared to the yolk of the egg and the outermost, perpetually moving sphere is likened to a wheel (Ikhwān al-Safā' 2015, 26–7, 46).

Historians of astronomy often argue that natural philosophy played its most important role in the development of non-Ptolemaic planetary models (Chapter II.6). Aristotle and his followers ascribed the discussion of the "essence" (structure and properties) of the universe as a whole to natural philosophy and not to the astral sciences. Also, no text specifically devoted to the meaning of natural philosophical teachings for the astral sciences is known to have been written by scholars of the early Abbasid caliphate. However, Morelon emphasized that several of their extant writings are concerned with this topic (Morelon 1999, 115). He showed, for example, for the treatise of the Banū Mūsā (3rd/9th century) *On the Solar Year* (*Kitāb fī sanat al-shams*) – falsely attributed to Thābit ibn Qurra (d. 288/901) – that it interweaves natural philosophy and mathematical astronomy. An analogous point was made by Ragep in his discussion of al-Battānī's (244–317/858–929) description and criticisms of ancient and contemporary theories, observations and the philosophical groundings of celestial movements (Ragep 1996).

Independently from Morelon and Ragep, Orthmann, a historian of the Persianate world and the social impact of astrology, emphasized that models of the universe were questioned from philosophical, scientific and religious positions (Orthmann 2016). Major points concerned the structure and properties of the universe as a whole and its parts.

One of those discussions concerned the existence of spheres beyond the sphere of the ecliptic (the path of the mean sun, also called the sphere of the zodiac or the sphere of fixed stars), the eighth and perhaps outermost sphere. Following Aristotle, Ptolemy had declared the movements of the heavens to originate due to a Prime Mover situated in the sphere beyond the ecliptic. In the 3rd/9th century, the Banū Mūsā rejected the existence of such a ninth sphere in their *Book on the Motions of the Celestial Spheres* (*Kitāb Ḥarakāt al-aflāk*). The brothers explained the daily motion by the rotations of each single sphere from the moon to the fixed stars (Casulleras 2007, 1: 93). One of the them, Aḥmad, reportedly delivered a geometrical proof against the existence of the ninth sphere (Casulleras 2007, 1: 93). In the following centuries, arguments for more spheres than eight were also offered due to other observational results. An important issue was the precession of the equinoxes. According to Ptolemy this slow movement along the ecliptic of the two points of intersection between the ecliptic and the celestial equator was due to the slow eastward rotation of the sphere of the fixed stars around the ecliptic poles. Another phenomenon seemingly requiring additional external spheres was the apparently changing obliquity of the ecliptic (the angle of intersection between the ecliptic and the equatorial planes) which was explained by increasingly complex theories of trepidation (Ragep 1996). Identifying in the 4th/10th century the ancient divine Prime Mover or the First Cause with Allāh, facilitated the model's spread beyond the circles of astral experts. Two centuries later, scholars engaged in planetary theory defined the ninth sphere as a fixed coordinate system (Ragep 1996, 286).

The position of Mercury and Venus with regard to the sun was apparently first formulated as a question of symmetry. In a letter on weather forecasting, a subfield of the astral sciences, al-Kindī argued that four spheres existed on each side of the sun (al-Kindī 2000, 245–6). The upper spheres were those of Mars, Jupiter, Saturn and the fixed stars. The lower spheres consist, in addition to the sphere of the moon, of three single spheres for the following pairs: Venus and Mercury; fire and air; water and earth. Al-Kindī believed that Mercury and Venus were equidistant from the sun and hence should be placed in a single sphere together (al-Kindī 2000, 245–6). His justification for putting water and earth, as well as fire and air, in single spheres reflects Aristotelian physics: water and earth are both cold and heavy, move downward to the center and do not form clearly ordered spheres, as earth rises above water in many places (continents and islands). They were often treated as forming a composite sphere. Similarly, air and fire are both hot and light, move upward and may not be separated from each other by a simple geometrical boundary (al-Kindī 2000, 245).

Al-Kindī's statements resemble opinions already held in antiquity. They were particularly prominent among Roman writers (Eastwood 2007, 36–7, 49). We do not know if other scholars in Islamicate societies subscribed to al-Kindī's views. There are, however, some diagrams in Arabic and Persian texts that present at least the sublunar world in agreement with al-Kindī's model. A comment by Naṣīr al-Dīn al-Ṭūsī (597–672/1201–1274), in his *Recension of the Almagest* (*Taḥrīr al-Majisṭī*), also might point to a longer life of this view. He discussed earlier observations of the transits of Venus. Some of these are clearly spurious, but others are possibly correct. Using this evidence, he rejected the ideas that the two inferior planets are in the sphere of the sun and that the centers of their epicycles coincide with the center of the sun's body (Mozaffari 2016, 276, no. 40).

Al-Kindī's placement of Mercury and Venus is a special form of one of the three opinions held by ancient and medieval scholars about where they move: both below the sun, both above the sun or Mercury below and Venus above the sun. According to Ptolemy, deciding this question with certainty was impossible due to a lack of observations, the difficulty of observing the occultation of a large body (the sun) by a much smaller body (a planet) and the long periods between the occurrence of such occultations (as is the case for Venus; Goldstein 1972, 43). Because of the accepted Aristotelian principle of the unchangeable celestial world, the observation of any dark spot on the solar surface would have been (mis)taken for the transit of one of the two inner planets (Goldstein 1972, 44).

Although the worlds below and above the moon were rigidly separated doctrinally, natural philosophy delivered the basis for analogical arguments about the impact of the planets on the sublunar world. The sun was recognized as causing the seasons and the change between day and night, the moon as connected to rain and other forms of humidity, and the planets as contributing their shares to the temperatures during the seasons (Abū Maʿshar 2019, 9, 225, 257, 259). Discussions about tides linked them primarily to lunar movements. In particular, Abū Maʿshar's substantial deliberations in the *Great Introduction to Astrology* were picked up repeatedly by writers about terrestrial matters (Abū Maʿshar 2019, 5, 19, 263–86).

I.6.3.2 Debates on astronomical problems

Important theoretical work began as early as the 3rd/9th century dealing with three big topics – the structure of the universe and the movement of the celestial bodies, the constructions of instruments and the solution of the three tasks related to the religious duties of Muslims (first visibility of the crescent after New Moon, prayer times and prayer directions). These themes were picked up time and again until the end of the period covered in this chapter (and beyond).

Here I focus on the first topic as it belongs in a narrower sense to the theoretical debates (for the other two topics see Chapters IV.2, IV.3, V.4, VI.2).

Almost all issues of the first topic were related to studies of Ptolemy's *Almagest*. A set of them concerned the observed changes of the obliquity of the ecliptic, the equinoctial points and the solar eccentricity, the definition and the determination of the length of the solar year and the motion of the solar apogee. The most important works on those subjects were written in Baghdad, Ghazna and Toledo. Their authors evaluated observations, recomputed astronomical magnitudes and devised qualitative or quantitative geometrical models.

In the treatise *On the Solar Year*, the Banū Mūsā rejected Ptolemy's claim that the movements of the sun, the moon and the fixed stars were interconnected and hence his convention that the tropical year (the time in which the sun travels all the way around the ecliptic) could be taken as the standard year in astronomy and astrology. They argued for the sidereal year (the time of one revolution of the sun with regard to the fixed stars, in modern terms about 1 h 20 m longer than the tropical year), which they considered of the same length as the anomalistic solar year (the time of one revolution of the sun with regard to its apogee, in modern terms some 2 h 55 m shorter than the sidereal year). Their arguments went back and forth between natural philosophy, mathematical astronomy and observations. Their arguments made use of solar observations that the Banū Mūsā brothers had executed in Baghdad, together with what they perceived as the critical spirit of ancient Greek astronomy. They presented their own observations not merely as new observational results but as part of a history of observations, which included – as an important method – an explicit criticism of earlier observations by ancient scholars (Neugebauer 1962; Morelon 1994, 116; Chapter II.7).

The observed variability of precession and the decrease of the value of the obliquity of the ecliptic motivated some scholars of the 3rd/9th or 4th/10th centuries to assume a periodicity of the two phenomena (Neugebauer 1962, 290). In a treatise called *On the Motion of the Eighth Sphere* extant only in Latin translation (*Liber de motu de octaue sphere*) – also attributed falsely to Thābit ibn Qurra – these issues are represented in the first known qualitative geometrical model (Neugebauer 1962, 291–9; for possible authors of the Arabic treatise see Samsó 1994, I: 15, 2020[CB], 585). The model resulted from and contributed to the continued discussion of what was called trepidation or the motion of accession and recession. The assumption, already formulated in late antiquity in Greek and Indian works but without geometrical models, was that the equinoctial points move forward and backward at a rate of 1°/80 years along an arc of 8°. In the ancient mathematical model, the longitude of a star will decrease during the forward west-to-east movement (accession). During the backward east-to-west movement (recession) it will increase again until it reaches its original value.

Trepidation theory was significantly reinterpreted, although the beginnings are not well known (Ragep 1996, 283–8). Al-Battānī (244–317/858–929) described the results of this process of reformulation in his astronomical handbook as a physical and cosmological interpretation of trepidation (Ragep 1996, 271). According to this interpretation, the stars oscillate because of a movement of the ecliptic orb. This movement is described as occurring in addition to the west–east movement of the fixed stars (precession). Al-Battānī criticized trepidation because there are several consequences resulting from this version which are in conflict with Aristotelian celestial physics (for a discussion, see Ragep 1996).

Despite the work done in the East, trepidation theory became truly influential only in al-Andalus and the Maghrib from the 5th/11th to the 8th/14th century (Comes 2001; Samsó 1994, I). According to the historical records, the theory was picked up from a treatise written in Baghdad and further developed in the circle around Ṣāʿid al-Andalūsī (419–462/1029–1070) in Toledo. A later member, Ibn al-Zarqāllu (d. 493/1100) proposed three different trepidation models in his *Treatise on the Motion of the Fixed Stars* (*Maqāla fī ḥarakat al-iqbāl wa-l-idbār*). They substantially shaped the further development of astronomical theory and handbooks among his successors in the West. Samsó's analysis of the three models brings to light the fact that Ibn al-Zarqāllu privileged ancient and Abbasid scholars over Andalusians, and his own recomputed data over observational data (for the details of the models and Samsó's analysis, see Samsó 2020, 592–610). In the 7th/13th and later centuries, observations undertaken in the Maghrib contradicted values given in Andalusian astronomical handbooks based on a trepidation model. As a result, trepidation models were first criticized and then fell out of use in the Maghrib. This change was supported by the values for the obliquity of the ecliptic and the precession of the equinoxes found in material arriving in the Maghrib from Ayyubid and Mamluk Egypt and Syria (Samsó 2020, 610–41).

Underlying many of the debates mentioned here and those going beyond the available scope of this chapter are two more general issues: (1) the relationship between observation, calculation and theoretical interpretation and (2) the necessity (or not) to engage critically with the practical and theoretical results of previous scholars back to antiquity. Time and again, scholars of the period under discussion formulated opinions on the merits as well as errors of previous results and their causes. In response, they offered new mathematical solutions, theoretical insights and verifications of observations, reinforcing the identity of the astral sciences and at times even changing their direction (see, e.g., Morelon 2009, 613–17).

I.6.3.3 Is the science of the judgment of the stars *a science?*

The classification literature summarized earlier explained what kind of science astrology was – either mathematical or physical – and its status and rank. Other kinds of texts dealt with the larger question inherited from ancient scholars: whether it was a science at all, what characterized the true expert of the astral sciences and which people were acceptable practitioners as opposed to charlatans. Following Ptolemy, writers addressing these questions defended the scientific character of both species or fields of knowledge. They acknowledged epistemological differences between the mathematical and physical parts. The results of the mathematical parts were certain, due to infallible, rational proofs. But the conclusions of the physical part were at best conjectural, as they were drawn from observing the nature of planetary influences on the life on earth. In addition, some writers argued for the existence of possible knowledge (not only of necessary knowledge), as well as the creation of knowledge through methods of analogy and comparison (Burnett 2002, 206–10). They promoted the application of tools from logic, the execution of careful, individual observations, and the reliance on long-term observational results from competent astral experts of the past (Burnett 2002, 212).

Kūshyār ibn Labbān (late 4th–early 5th/10th–11th centuries) formulated these positions in favor of a nondeterministic science, also subscribed to by al-Kindī, Abū Maʿshar and al-Qabīṣī, as follows:

> The first division of the science of the stars is that for which there is geometrical proof . . . the second is the science that is grasped by experience . . . and analogy. . . . Most people reject the second science and claim that it concerns things that happen by accident and there is no demonstration in it. But we say that, so far as accidents are concerned, when it lasts long or occurs in most circumstances, then they have the force of demonstration.
>
> (*Burnett 2002, 202*)

The questions formulated by al-Qabīṣī for testing astral experts who approached his princely patron leave no doubt that the scientific character of both parts of the astral sciences depended on the relationship between them. An expert on the *science of the judgments of the stars* had to have mastered the *Almagest*, natural philosophy, instrumental techniques, geometry and arithmetic. In his refutation of ten kinds of criticisms against astrology, Abū Maʿshar explains that experts in mathematical astronomy are insufficiently trained if they do not know natural philosophy or practice the *science of the judgments of the stars*, because the latter is the fruit of astral knowledge (Burnett 2002, 209; al-Qabīṣī 2004, 6–7). This view on the connectedness of mathematical astronomy and astrology was still held in the early 9th/15th century, when either the Timurid prince Iskandar Sulṭān or possibly one of his court astrologers wrote in the preface to an otherwise lost scientific *summa* that the *science of the configuration [of the universe]* is the foundation of the noble *science of the stars* and the latter the fruit of all the parts of mathematical astronomy (Brentjes 2009, 465).

I.6.3.4 Methods for houses, rays and progressions

The most important discussions on astrological methods and their mathematical implementation took place in astronomical handbooks and specialized treatises (Kennedy and Krikorian 1972, 3–4; Casulleras and Hogendijk 2012). They concern three fundamental theories that historical actors considered interrelated: (1) the theory of *houses*, (2) the theory of *aspects* and (3) the theory of *rays*. Houses (also translated as *places*; not to be confused with the *domiciles* of the planets) resulted from the division of the ecliptic circle or the zodiacal belt into 12 parts. They could either have fixed or mobile limits, often defined by the projection of divisions from an auxiliary great circle like the equator onto the ecliptic (Kennedy 1996). Similarly, there were different views on how to determine the seven astrological *aspects*, significant points of *rays* allegedly cast by a planet, and predictions for any given date by progressions. Casulleras and Hogendijk have traced nine methods for projecting *rays* and identified six methods for predictions made by progression (Casulleras 2009; see Casulleras and Hogendijk 2012 for a survey of the older literature, editions and translations of primary sources and the details of the general themes, as well as their respective methods). They consider methods (also called systems) as different when their geometrical definitions are not mathematically equivalent (Casulleras and Hogendijk 2012, 37). As a result, the computations employed can also differ. Some of the systems are mathematically interrelated, while others are independent. The applied mathematical practices are nontrivial and should thus be acknowledged as valuable contributions to the history of the mathematical sciences in Islamicate societies (see, for instance, Hogendijk 2015, 285).

The fact that the different methods were unevenly distributed and evaluated in different areas of the Islamicate world points to locally specific developments and perspectives (Casulleras and Hogendijk 2012, 45, 73, 77–8, 82; Chapters V.3, V.4 and VI.2). The evaluation by practitioners concerned eight issues related to each other in different ways. The first four were the effectiveness, reliability and applicability of each of the methods for specific planetary and imagined celestial positions and the soundness of a method. These formed one complex. The relationship between methods and the possibility of extending a method to a more complex case reflect a second set of questions. The remaining two issues were the exact or approximate character of the computations and were related to a methodological problem also raised outside the astral sciences. This also applies to the question of which mistakes various authors had committed (Casulleras and Hogendijk 2012, 43, 45, 56–7, 59, 63–4, 70, 73, 81, 83, 91).

I.6.4 Mathematical practices

The main mathematical skills people working on astral matters needed to possess began with the abilities to add, subtract, multiply, divide and extract roots in the sexagesimal system (a positional system with basis 60). They needed to know the relevant great and small circles, arcs and angles on the celestial and terrestrial spheres and how to do geometrical constructions with a ruler and a compass. They had to be able to transform verbal statements into plane or spherical diagrams and vice versa. In more advanced mathematics, they needed to handle proportions between numbers or geometrical magnitudes, work with trigonometric magnitudes and relations, trisect arcs or angles, to project elements of a spherical surface or a plane unto another plane and make a linear interpolation between two values. More rarely quadratic interpolations were used. Last, they had to be able to translate geometrical problems or procedures into arithmetical or algebraic ones and thus recognize equivalences between problems from different fields of the mathematical sciences, which allowed them to search for arithmetical or algebraic solutions (Chapters I.5 and I.7).

The more proficient experts among them knew how to make theoretical claims and prove them with methods taken from Arabic translations of Euclid's *Elements* and the *Data* (*Dedomena*), Ptolemy's *Almagest*, the treatises on the sphere by Autolycus of Pitane (d. *c.* 250 BCE), Theodosius (late 2nd century BCE) and Menelaus (*fl.* 100), as well as new trigonometric tools derived from translations of Sanskrit astral texts and their further elaborations in the 3rd/9th and 4th/10th centuries (Chapters I.5, I.7, I.8). They executed complicated calculations and complex geometrical constructions. They were also able to check the methods applied by other practitioners and point out their mistakes and theoretical or practical shortcomings. Sometimes scathing remarks of this kind permeate the works of several scholars of the astral sciences, among them al-Bīrūnī and Ibn Muʿādh (Casulleras 2004, 387).

An entire literary genre of this kind of critical engagement with the foundations and practices of the scholarly field developed in medicine, philosophy, some of the religious sciences (Chapters I.11 and I.12) and in the astral sciences.

Difficult mathematical problems and complex procedures to solve them were applied not only in those parts of the astral sciences that lacked astrological applications, but also in procedures needed for producing horoscopes, such as the projection or casting of rays, or the division of a chosen great circle for determining the houses with fixed or mobile limits (Hogendijk 1989, 2005; Casulleras and Hogendijk 2012; Casulleras 2020–2021). The mathematical procedures used for such higher-level astrological practices were the same that the experts used for calculations and constructions in the mathematical part of the astral sciences (astronomy). The same holds for proofs and demonstrations or the preference of exact solutions over approximate ones. Exact solutions occur primarily in discussions of theoretical problems of the astral sciences. Casulleras assumes that more complex formulations of theories and their procedures led to an increase in scholarly reputation, which in turn inspired the creation of more difficult theories involving efforts to find more sophisticated mathematical solutions (Casulleras 2004, 385–6).

Approximate methods, on the other hand, seem to have dominated the daily practice of horoscope casting, computation of astronomical and astrological tables and the determination of prayer times and prayer directions. Arithmetical procedures that required only the use of tables seem to have been preferred over geometrical methods (Casulleras 2004, 388, 392–7; Casulleras and Hogendijk 2012, 76). Practicing astrologers could also work with astrolabes, which are effectively analog computers. Some specimens are known or described in texts that include special plates or plates with sets of lines for astrological tasks (Hogendijk 2005, 98–9; Casulleras 2013; Casulleras and Hogendijk 2012, 50–3, 55 *et al.*; see the latter for their reservations about the usability of some of those plates because of the agreement of the geometrical representation for several of the methods). The use of an armillary sphere, as well as a globe, was discussed by Jewish scholars at the court of Alfonso X of Castile ([r. 1252–1284]; Casulleras and Hogendijk 2012, 52).

I.6.5 Final remarks

This chapter has focused on themes that are only partly included in surveys of the history of astronomy in Islamicate societies. Astrology was excluded until very recently (Burnett 2007; Morrison 2013). A survey of the various complex and difficult mathematical methods used for predictions of the future still needs to be written. It is not possible to separate astrology from mathematical astronomy on the grounds that it is not considered a science today. If we did this, we would also have to abandon mathematical astronomy, since its natural philosophical foundations and its models are also faulty and equally contradict reality as understood today. The themes presented here indicate the fruitfulness of studying the different parts of the astral sciences as a unified field of knowledge interrelated by epistemological and natural philosophical issues of great importance, a broad range of mathematical methods and sociocultural contexts.

Note

1 Consolidated bibliography.

Bibliography

Sources

al-Bīrūnī, Abū l-Rayḥan. tr. Wright, R. R. 1934. *Kitāb al-Tafhīm li-awā'il ṣinā'at al-tanjīm* [The Book of Instruction in the Elements of the Art of Astrology] (Written in Ghaznah, 1029 A.D.). London: Luzac. Reproduced from British Museum MS. Or. 8349.

al-Bīrūnī and Ibn Sīnā. ed. Nasr, S. H. and Mohaghegh, M. 1970. *Al-as'ila wa-l-ajwiba* [Questions and Answers]. Tehran: Shūrā'ī-yi ʿālī-yi hunar u-farang.

Ikhwān al-Safā'. ed. and tr. Ragep. F. J. and Mimura, T. 2015. *On Astronomia: An Arabic Critical Edition and English Translation of Epistle 3. Epistles of the Brethren of Purity*. Oxford: Oxford University Press.

al-Kindī. ed., tr. and com. Bos, G. and Burnett, C. 2000. *Scientific Weather Forecasting in the Middle Ages: The Writings of Al-Kindi*. London and New York: Kegan Paul International.

al-Qabīṣī. eds. and trs. Burnett, C., Yamamoto, K. and Yano, M. 2004. *Al-Qabīṣī (Alcabitius): The Introduction to Astrology. Editions of the Arabic and Latin Texts and an English Translation*. London and Turin: The Warburg Institute and Nino Aragno Editore.

al-Tawḥīdī, Abū Ḥayyān. ed., tr. and ann. Bergé, M. 1963–1964. "Épitre sur les sciences (Risāla fī l-ʿulūm) d'Abū Ḥayyān al-Tawḥīdī (310/922 (?) – 414/1023). Introduction, Traduction, Glossaire Technique, Manuscrit et Édition Critique," *Bulletin d'études orientales* 241–77, 279–83, 285–98, 300.

Manuscripts

MS London, BL, Or. 1992.
MS New York, Pierpont Morgan Library, M. 788.
MS Paris, BnF, Arabe 2515.
MS Paris, BnF, Supplément Turc 242.

Research literature

Adamson, P. 2002. "Abū Maʿšar, al-Kindī, and the Philosophical Defense of Astrology," *Recherches de Théologie et Philosophie médiévales* 69: 245–70.

Adamson, P. 2006. *Al-Kindi*. New York: Oxford University Press.

Adamson, P. 2020. "al-Kindi," in Zalta, E. N., ed. *Stanford Encyclopedia of Philosophy*. Spring 2020 Edition. https://plato.stanford.edu/entries/al-kindi/#Cos.

Adamson, P. and Pormann, P. E., trs. 2012. *The Philosophical Works of al-Kindi*. Karachi: Oxford University Press.

Bevilacqua, A. 2018. *The Republic of Arabic Letters: Islam and the European Enlightenment*. Cambridge, MA: Belnap Press.

Bowen, A. C. 2007. "The Demarcation of Physical Theory and Astronomy by Geminus and Ptolemy," *Perspectives on Science*, 15.3: 327–58.

Brentjes, S. 2009. "The Interplay of Science, Art and Literature in Islamic Societies before 1700," in Dev, A., ed. *Science, Literature and Aesthetics. History of Science, Philosophy and Culture in Indian Civilization*. Vol. XV.3. New Delhi: PHISPC, Centre for Studies in Civilizations, 453–86.

Brentjes, S. 2010. *Travellers from Europe in the Ottoman and Safavid Empires, 16th–17th Centuries. Seeking, Transforming, Discarding Knowledge*. London: Routledge, Variorum.

Burnett, Ch. 2002. "The Certitude of Astrology: The Scientific Methodology by al-Qabīṣī and Abū Maʿshar," *Early Science and Medicine* 7.3: 198–213.

Burnett, Ch. 2007. "Astrology," in *EI-3*, 2007–2: 165–75.

Callataÿ, G. de. 2013. "Magia en al-Andalus*: Rasāʾil Ijwān al-Ṣafāʾ, Rutbat al-ḥakīm y Gāyat al-ḥakīm (Picatrix)*," *al-Qanṭara* 34.2: 297–344.

Casulleras, J. 2004. "Ibn Muʿādh on the Astrological Rays," *Suhayl* 4: 385–402.

Casulleras, J. 2007. "Banū Mūsā," in Hockey *et al*., eds. [CB], 1: 92–4.

Casulleras, J. 2009. "Métodos para determinar las casas del horóscopo en la astrología medieval árabe [Methods for Determining the Houses of the Horoscopes in Medieval Arab Astrology]," *Al-Qanṭara* 30.1: 41–67.

Casulleras, J. 2013. "The Instruments and the Exercise of Astrology in the Medieval Arabic Tradition," *Archives internationales d'histoire des sciences* 63.170–1: 517–40.

Casulleras, J. 2020. "Horoscope," in *EI-3*, 2020–1: 50–6.

Casulleras, J. and Hogendijk, J. P. 2012. "Progressions, Rays and Houses in Medieval Islamic Astrology: A Mathematical Classification," *Suhayl* 11: 33–102.

Comes, M. 2001. "Ibn al-Hāʾim's Trepidation Model," *Suhayl* 2: 291–408.

Eastwood, B. S. 2007. *Ordering the Heavens. Roman Astronomy and Cosmology in the Carolingian Renaissance* (History of Science and Medicine Library 4). Leiden and Boston: Brill.

Feke, J. 2014. "Meta-mathematical rhetoric: Hero and Ptolemy against the philosophers," *Historia Mathematica* 41.3: 216–76.

Feke, J. 2018. *Ptolemy's Philosophy. Mathematics as a Way of Life*. Princeton: Princeton University Press,.

Goldstein, B. R. 1972. "Theory and Observation in Medieval Astronomy," *Isis* 63.1: 39–47.

Hogendijk, J. P. 1989. "The Mathematical Structure of Two Islamic Astronomical Tables for 'Casting the Rays'," *Centaurus* 32: 171–202.

Hogendijk, J. P. 2005. "Applied Mathematics in Eleventh Century Al-Andalus: Ibn Muʿādh al-Jayyānī and Astrological Houses and Aspects," *Centaurus* 47: 87–114.

Hogendijk, J. P. 2015. "Al-Bīrūnī on the Computation of Primary Progression (*tasyīr*)," in Burnett, C. and Greenbaum, D. G., eds. *From Māshāʾallāh to Kepler: Theory and Practice in Medieval and Renaissance Astrology*. Lampeter: Sophia Centre Press, 279–308.

Kennedy, E. S. 1996. "The Astrological Houses as Defined by Medieval Islamic Astronomers," in Casulleras and Samsó, eds. [CB], 2: 535–78 (Reprint: *Astronomy and Astrology in the Medieval Islamic World*. Aldershot: Variorum, XIX).

Kennedy, E. S. and Krikorian, H. 1972. "The Astrological Doctrine of 'Projecting Rays'," *Al-Abḥāth* 25: 3–25 (Reprint: Kennedy, E. S. *et al.* 1983. *Studies in the Islamic Exact Sciences*. Beirut: American University of Beirut, 372–84).

Lettinck, P. 1999. *Aristotle's Meteorology and Its Reception in the Arab World: With an Edition and Translation of Ibn Suwār's Treatise on Meteorological Phenomena and Ibn Bājja's Commentary on the Meteorology*. Leiden: Brill.

Melvin-Koushki, M. 2017, "Powers of One: The Mathematicalization of the Occult Sciences in the High Persianate Tradition," *Intellectual History in the Islamicate World* 5: 127–99.

Morelon, R. 1994. "Ṯābit b. Qurra and Arab Astronomy in the 9th Century," *Arabic Sciences and Philosophy* 4.1: 111–39.

Morelon, R. 1999. "Astronomie physique et astronomie mathématique dans l'astronomie précopernicienne," in Rashed, R. and Biard, J., eds. *Les doctrines de la science dans l'antiquité à l'âge classique*. Leuven: Peeters, 105–29.

Morelon, R. 2009. "The Astronomy of Thābit ibn Qurra," in Rashed, R., ed. *Thābit ibn Qurra. Science and Philosophy in Ninth-Century Baghdad*. Berlin and New York: Walter de Gruyter, 601–18.

Morrison, R. G. 2013. "Islamic Astronomy," in Lindberg, D. C. and Shank, M. H., eds. *The Cambridge History of Science*. 6 vols. Vol. 2: *Medieval Science*. Cambridge: Cambridge University Press, 109–39.

Mozaffari, S. M. 2016. "A Forgotten Solar Model," *Archive for History of Exact Sciences* 70: 267–91.

Neugebauer, O. 1962. "Thâbit ben Qurra 'On the Solar Year' and 'On the Motion of the Eighth Sphere'," *Proceedings of the American Philosophical Society* 106.3: 264–99.

Orthmann, E. 2016. "Philosophy and Natural Science," in Rudolph *et al.*, eds. [CB], 727–33.

Plofker, K. 2007. "Fazārī: Muḥammad ibn Ibrāhīm al-Fazārī," in Hockey *et al.*, eds., 1: 362–3.`

Ptolemaeus Arabus et Latinus. "AstroBibl. 3. Middle Ages (500–1450). 3.3. Islam/Arabic World," https://ptolemaeus.badw.de/astrobibl/section/G

Ragep, F. J. 1996. "al-Battānī, Cosmology, and the Early History of Trepidation in Islam," in Casulleras and Samsó, eds., 1: 267–98.

Ragep, F. J. 2009. "Astronomy," in *EI-3*, 2009–1: 120–50.

Rashed, M. 2009. "Thābit ibn Qurra, la *Physique* d'Aristote et le meilleur des mondes," in Rashed, R., ed. *Thābit ibn Qurra. Science and Philosophy in Ninth-Century Baghdad*. Berlin: Walter de Gruyter, 675–714.

Russell, G. A., ed. 1994. *The 'Arabick' Interest of the Natural Philosophers in Seventeenth Century England*. Leiden: Brill.

Saliba, G. 1982. "Astronomy/Astrology, Islamic," in Strayer, R. J., ed. *Dictionary of the Middle Ages*. 13 vols. Vol. 1. New York: Scribner, 616–24 (Reprint in Saliba 1994 [CB], 66–84).

Samsó, J. 1994. "Andalusian Astronomy: Its Main Characteristics and Influence in the Latin West," in Samsó, J., *Islamic Astronomy and Medieval Spain*. Aldershot: Variorum, I.

Taub, L. 2003. *Ancient Meteorology*. New York: Routledge.

URL https://journals.openedition.org/mideo/3397?lang=ar

I.7
ARITHMETIC AND ALGEBRA

Jeffrey A. Oaks

I.7.1 Introduction

Techniques of arithmetical calculation arose among early civilizations in ancient Eurasia in response to the technical demands of government administration, trade, astronomy and other activities. Bureaucrats, merchants, judges, surveyors, astronomers and others practiced one form of arithmetic or another to solve a variety of problems. These include determining the areas of fields, designing and maintaining irrigation systems, measuring volumes and weights of commodities, calculating exchange rates, dividing estates, weather forecasting, and determining the positions of the stars and planets for timekeeping, divination or astrology.

People working in medieval Islamicate societies continued these practices, which are covered by the Arabic word *ḥisāb*.[1] This word can be translated as "arithmetic", "calculation" or "reckoning", and at its core, it was concerned with the different techniques for operating on numbers. Three main methods of calculation were already in use before the advent of Islam. Finger-reckoning was a technique common among the Greeks and Romans and across the ancient Near East, sexagesimal (base 60) calculation had been practiced by astronomers from Greece to India, and the Indian arithmetic that originated on the subcontinent was introduced to the Muslim world around the middle of the 2nd century/3rd quarter of the 8th century.

Ḥisāb also included the methods of finding numerical unknowns that were traditionally associated with finger reckoning. The rule of three, single false position, double false position and algebra were the most popular. Mensuration was also considered to be part of *ḥisāb*. Unlike the geometry we read in classical Greek books, the lines, surfaces and bodies of practical Arabic geometers all possessed numerical measure, so even here calculations belonged to arithmetic. It is because of geometrical and metrological applications that the numbers in practical Arabic arithmetic include any positive quantity that can arise through calculations, including fractions and irrational roots.

Classical Greek texts provided another source of arithmetical knowledge appropriated and explored by scholars writing in Arabic. The number theory in Books VII to IX of Euclid's (*c.* 3rd century BCE *Elements* and Nicomachus of Gerasa's (d. *c.* 120) *Arithmetial Introduction* inspired new developments, and despite the Greek restriction of numbers to positive integers, various of its elements were integrated into Arabic *ḥisāb*.

DOI: 10.4324/9781315170718-9

The sources of our knowledge of Arabic arithmetic (which includes algebra) consist largely of textbooks, together with some works with more theoretical leanings. Samples of actual calculations made by merchants, surveyors or other people are rare because most were performed mentally or were written on some temporary surface. Many of those we do possess are occasional calculations scribbled in blank areas of manuscripts.

The two main systems of writing numbers in Islamicate societies were *jummal* and Indian. *Jummal* notation is an adaptation to the Arabic alphabet of the corresponding system already in use in ancient Greek, Hebrew, and Syriac. The 28 letters of the Arabic alphabet are assigned numerical values: 1, 2, . . . , 9, 10, 20, . . . , 90, 100, 200, . . . , 900, and 1000. The number 542, for example, was shown with the letters signifying 500 ش, 40 م and 2 ب; hence, 542 is shown as ثمب. *Jummal* was also called *abjad*, after *alif*, *bā*, *jīm* and *dāl*, the first four letters of the Arabic alphabet, which were assigned the values 1, 2, 3 and 4, respectively. Indian notation consists of the nine numerals 1 through 9 together with the 0 designating an empty place. We call them "Arabic" numerals because Europeans learned them from Arabic sources. The shapes of the numerals have varied greatly over time and place. The forms we use today derive from the forms current in the western part of the Islamic world in the late medieval period.

I.7.2 Methods of calculating

Calculations were performed either mentally or were worked out in notation on a temporary surface like a dust-board or wax tablet. A dust-board is a board covered with dust or sand on which one wrote with a finger or stylus. The wax tablet functioned similarly, with wax instead of dust covering the board. These surfaces allowed for easy erasing and rewriting of digits.

Even with the proliferation of books that came with the introduction of paper, and the intense focus on scholarship beginning around the 2nd/8th century, the transmission of knowledge continued in large part to be conducted orally. In an environment in which recitation remained fundamental to instruction, physical copies of books were to a large extent treated as transcriptions of lectures. Notation serves no purpose to the listener, so the calculations in arithmetic books tend to be expressed all in words, including numbers. In many books, the samples of *jummal* or Indian notation are shown only as figures illustrating what the student should put down on the dust-board and are not part of the running text.

I.7.2.1 Finger reckoning

Finger reckoning, called by various names including hand arithmetic (*ḥisāb al-yad*), open arithmetic (*al-ḥisāb al-maftūḥ*) or aerial arithmetic (*al-ḥisāb al-hawāʾī*), was a method popular among government secretaries and merchants. Operations were performed mentally in base 10, with intermediate results from 1 to 9,999 being stored by positioning the fingers in particular ways. According to one 9th/15th-century poem, units are stored using the last three fingers on the right hand, tens are stored with the thumb and index finger on the same hand, hundreds by positioning the thumb and index finger of the left hand, and thousands with the last three fingers on the left hand. For units, a 1 is stored by extending the last three fingers on the right hand; a 2 by bending the last two fingers; a 3 by bending all three of these fingers and so on (Saidan 1968). In one example, Abū l-Wafāʾ (328–388/940–998) explains how to divide 4624 by 16. First 46 ÷ 16 gives 2 (i.e., 200) with a remainder of 14. Then 142 ÷ 16 gives 8 (i.e., 80) with a remainder of 14, and finally 144 ÷ 16 gives 9. The 200, then the 280, would be remembered with the fingers while continuing through the divisions. *Jummal* notation was typically used to record the results of these calculations in documents.

The most common way of expressing fractions for finger reckoners was to make some combination of the basic fractions 1/2, 1/3, . . . , 1/10. In this system, 1/12, for example, is "half of a sixth", and 11/15 is "three fifths and two thirds of a fifth".

The chapters of Arabic books on finger reckoning cover ratios (fractions), multiplication and division, often with further chapters on mensuration, business applications and problem-solving by algebra and other methods. The books also describe shortcuts for different situations. To pick just one example, to multiply a number by 37 one can multiply its third by 1, 10 and 100 and add the results. For example, to multiply 42 by 37 one takes a third of the 42, which is 14, and adds 14 + 140 + 1400 = 1554.

Rules in mensuration (*misāḥa*) explain how to calculate numerically the areas of rectangles, triangles, circles, and other shapes, as well as the surface areas and volumes of spheres, parallelepipeds, cones, and other bodies. Basic trigonometry is usually covered too for the arcs and chords of a circle.

I.7.2.2 Sexagesimal arithmetic

Calculations in astronomy, which included trigonometry, had been written in base 60 since Babylonian times. The system had spread to Greek, Sanskrit and Persian before the Arab conquests, so it was only natural for those working in the Caliphate to adopt it as well. In Arabic, the *jummal* forms for 1 through 59 were used together with a special sign for the zero or the empty place. The base 10 number 86,415, for example, is $24 \cdot 60^2 + 0 \cdot 60 + 15$, and was written with the *jummal* figures for 24, 0, 15. The system was also used to express fractions, where, for example, $2/60 + 25/60^2$ was written with the figures for 2, 25.

I.7.2.3 Indian arithmetic

Our Arabic numerals 1, 2, . . . , 9 and the 0 originated in India around the 1st century and were known in Syria by the mid-7th century. It appears, however, that it was only in the second half of the following century that "Indian arithmetic" (*al-ḥisāb al-hindī*), as they called it, became known to Arabic calculators.[2]

The earliest known Arabic book on the topic is the *Book on Indian Arithmetic (al-Kitāb fī l-ḥisāb al-hindī)* by al-Khwārazmī, a scholar working in Baghdad in the first half of the 3rd/9th century. The Arabic original is no longer extant, but we do possess a reworking of a medieval Latin translation. As with other Arabic books on Indian calculation, this one covers the shapes of the numerals and the rules for forming numbers, followed by rules for addition, subtraction, doubling, halving, multiplication, division and root extraction. Fractions in this system were usually borrowed from finger reckoning.

Al-ḥisāb al-hindī was also called "dust arithmetic" or "board arithmetic", from the dust-board used for making the calculations. In fact, the rules for operating on these numbers usually call for the shifting, erasing and rewriting of digits. In the mid-4th/mid-10th century, the Damascene mathematician al-Uqlīdisī wrote a book on Indian calculation in which he also gave rules for working out the operations with pen and paper, without erasing. He presented decimal fractions, too, in which one would write, for example, a version of 43.25 instead of $43\frac{1}{4}$. Despite al-Uqlīdisī's overtures, the dust-board and common fractions remained popular in various parts of the Islamicate world as late as the 13th/19th century.

I.7.2.4 *Approximation methods*

Algorithms for approximating square and cube roots of numbers, either in base 60 or base 10, were known in other civilizations before the Islamic conquests. Scholars writing in Arabic devised rules for extracting higher roots, and more generally the roots of polynomial equations, by techniques equivalent to what we call the Ruffini-Horner method. Important work was done in this area by people like ʿUmar al-Khayyām (439–*c.* 517/1048–*c.* 1123) and Sharaf al-Dīn al-Ṭūsī (d. *c.* 610/1213), but I will mention two calculations of al-Kāshī (d. 832/1429), who, since about 1415, had been a resident of Samarqand in Central Asia. In one treatise, he applied trigonometric identities and extracted the required square roots to obtain a sexagesimal approximation for *pi* that in base 10 is correct to 16 decimal places. In another treatise, he found an equally accurate sexagesimal approximation for sin 1° by taking the root of a cubic equation. Calculating sin 1° is required for the construction of trigonometric tables, and other approximation techniques, including sophisticated interpolation schemes, were devised with the aim of producing various astronomical tables efficiently and accurately (see Chapter I.6).

I.7.3 Numerical problem-solving

The three systems described earlier cover rules for calculating known numbers. Practitioners in Islamicate countries also solved problems asking for unknown numbers. For example, one might want to work out this problem posed by Ibn al-Yāsamīn (d. 600/1204) in the late 6th/12th century: "A quantity: you add its third and its fourth and get six. How much is the quantity?" (Ibn al-Yāsamīn 1993, 195.9). Today we find the answer to problems like this by algebra. We name the unknown x, then set up and solve an equation. But in Islamicate societies algebra was just one of several methods for finding unknown numbers. Ibn al-Yāsamīn, in fact, solved this problem by three different methods: single false position, algebra, and then double false position.

I.7.3.1 *The rule of three and single false position*

Given four proportional numbers $a : b :: c : d$, any one of them can be found from the other three by a multiplication and a division. Arabic authors called this the method of "the four proportional numbers", and it is known to us as "the rule of three". A simple example solved by the rule of three from al-Khwārazmī's *Book of Algebra* (*Mukhtaṣar fī ʿilm al-jabr wa-l-muqābala*) asks "ten [items] for six [dirhams]. How many will you get for four [dirhams]?" (al-Khwārazmī 2009, 198, quoted after Rashed's translation).

The method of single false position applies the rule of three. A convenient value is posited for the answer, and the correct value is found from the value and the error. Ibn al-Yāsamīn's first solution to the problem stated above is worked out by this method. He posits the answer as 12 (this is the "false position"), and he calculates its third and its fourth to be 7. Then the unknown is to 12 as 6 is to 7, so it is found by multiplying 12 by 6, then dividing the result by 7, to get $10\frac{2}{7}$.

I.7.3.2 *Double false position*

Another popular method that derives from proportion is double-false position, or, as it was called in Arabic, "the two errors" (*al-khaṭāʾān*). Two values are posited, and the answer is found by a rule using these values and their errors. Ibn al-Yāsamīn's third solution follows this method. He first supposes the answer is 8. Its third and its fourth are $4\frac{2}{3}$, which is short of the desired 6 by an error of $1\frac{1}{3}$. Then he supposes it is 12. The calculation gives 7, which is an error in excess of 6 by 1. The answer is found by calculating $\left(1\frac{1}{3}\cdot 12+1\cdot 8\right)\div\left(1\frac{1}{3}+1\right)=10\frac{2}{7}$. The rule varies slightly for cases in which the calculated values both fall short or they both exceed the desired outcome.

Qusṭā ibn Lūqā (d. *c.* 299/912) wrote a short treatise in the 3rd/9th century proving that this method gives the correct value by a geometric argument in the style of Euclid's *Data*. In the later 7th/13th century, the Persian scholar Kamāl al-Dīn al-Fārisī (d. 718/1319) gave a different proof in his *Foundation of Rules on Elements of Benefits* (*Asās al-qawāʿid fī uṣūl al-fawāʾid*), this time by arithmetic in the style of Euclid's number theory books. People using the method on the job would not have had much interest in Euclidean-style proofs, but mathematicians often sought to integrate ideas and approaches from different cultures.

I.7.3.3 Algebra

In the rule of three, single false position and double false position, all calculations are performed on known numbers. In algebra, by contrast, unknowns are given names, and these names are operated on to set up and solve equations. Algebra was called *al-jabr wa-l-muqābala* in Arabic, literally "restoration and confrontation", and this name was often shortened to just *al-jabr*. It is from this word, via medieval Latin and Italian transliteration, that our word *algebra* derives.

The name given to the first-degree unknown in Arabic algebra is a "thing" (*shayʾ*), or sometimes a "root" (*jidhr*). The word *jidhr* also takes the meaning of "square root" in arithmetic. Its square, the second-degree unknown, is called a *māl*, which ordinarily means "an amount of money".[3] The third-degree unknown is called a "cube" (*kaʿb*). Higher powers are expressed as a combination of *māl* and *kaʿb*, like *māl māl kaʿb* for the seventh power. As in arithmetic generally, units were counted in dirhams (a silver coin), or simply as "units" or "in number".

The solution to a problem by algebra is played out in three basic stages. In the first stage, an unknown is named in terms of the names of the powers, usually as "a thing", and the conditions of the enunciation are applied to this name to set up a polynomial equation (although sometimes divisions and roots were allowed). Here is a polynomial from the solution of problem (21) in al-Khwārazmī's book: "four ninths of a *māl* and nine dirhams less four roots" (al-Khwārazmī 2009, 181.13), which in modern notation corresponds to $\frac{4}{9}x^2+9-4x$.

In the second stage the equation is simplified. The central steps are "restoration" (*al-jabr*) and "confrontation" (*al-muqābala*). Take for example this equation from al-Karajī's (d. *c.* 419/1029) *[Book dedicated to] Fakhr al-Mulk on the Art of Algebra* (*[Kitāb] al-Fakhrī fī ṣināʿat al-jabr wa-l-muqābala*; 401/1011–2): "two thirds of a thing less three dirhams equal twenty dirhams" ($\frac{2}{3}x-3=20$). The left side was regarded as a deficient $\frac{2}{3}x$, which must be restored to a full $\frac{2}{3}x$. Then to balance the equation 3 must also be added to the other side: "So restore two thirds of a thing by three dirhams, and add its same to the twenty, so it becomes: twenty-three dirhams equal two thirds of a thing [$23=\frac{2}{3}x$]" (Saidan 1986, 170.9). "Confrontation" is invoked when there are like terms on opposite sides of the equation. The difference between the terms is taken, and the remainder is placed on the side of the greater. For example, al-Khwārazmī simplifies his rhetorical version of $3\frac{1}{3}+\frac{1}{3}x=x$ by writing: "So confront them. You cast away a third of a thing by a third of a thing, leaving: two thirds of a thing equal three dirhams and a third" (al-Khwārazmī 2009, 233.9). The Arabic name of algebra comes from these two steps in the simplification of equations, *al-jabr* and *al-muqābala*.

Today we have one standard form of the simplified quadratic equation, $ax^2 + bx + c = 0$. Because the rules for solving equations call for calculations on coefficients, and because only positive numbers were acknowledged, Arabic algebraists worked with six types of the simplified equation of the first and second degrees. Three are simple, with two terms, and three are composite, with three terms. Here they are in the order given by al-Khwārazmī, together with their modern notational versions:

1. some *māls* equal some roots	$ax^2 = bx$
2. some *māls* equal a number	$ax^2 = c$
3. some roots equal a number	$bx = c$
4. some *māls* and some roots equal a number	$ax^2 + bx = c$
5. some *māls* and a number equal some roots	$ax^2 + c = bx$
6. some roots and a number equal some *māls*	$bx + c = ax^2$

In the third stage, the simplified equation is solved. There was a specific rule for each type of equation. The simple equations require only a division, but several steps are necessary to solve the composite equations. Al-Karajī gives this example to explain the solution to the type 4 equation in his *al-Fakhrī*:

> Two *māls* and twenty things equal a hundred twelve dirhams [$2x^2 + 20x = 112$]. Return the *māls* to one *māl*: you take half of it and half of everything with it, so it becomes: a *māl* and ten things equal fifty-six dirhams [$x^2 + 10x = 56$]. Halve the [number of] things to get five, multiply it by itself, and add fifty-six to it to get eighty-one. Take its root to get nine, and cast away half the roots [i.e., half the number of things] from it, leaving four, which is a root of the *māl*, and the *māl* is sixteen.
>
> *(Saidan 1986, 149.14)*

The second solution Ibn al-Yāsamīn gives to his problem is "by *al-jabr*". He names the unknown quantity "a thing", and sets up the equation "three sixths of a thing and half a sixth of a thing equals six" [$\frac{7}{12}x = 6$]. Stage 2 is unnecessary because the equation is already simplified, and the solution is found by a simple division, so "the thing is ten and two sevenths".

Because algebra is not tied to proportion, it can be used to solve a wider variety of problems than the other methods surveyed earlier. Here is an example from Ibn Badr's (*c.* 7th/13th century) *Brief Book on Algebra* (*Kitāb fīhi ikhtaṣār fī l-jabr wa-l-muqābala*; Saidan 1986, 447.7):

> **[Enunciation]**
>
> Ten: you divide it into two parts. You multiply each part by itself, and you add the products to get eighty-two.
>
> **[Stage 1: setting up the equation]**
>
> To work this out you make one of the parts a thing, leaving the other as ten less a thing. Multiply the first part by itself to get a *māl*, and multiply the second part by itself to get a hundred and a *māl* less twenty things. Add them as stipulated, and your sum is a hundred dirhams and two *māls* less twenty things, [which] equal eighty-two [$100 + 2x^2 - 20x = 82$].

[Stage 2: simplifying the equation]

Restore and confront,[4] so you get: two *māls* and eighteen dirhams equal twenty things. [Restoration gives $100 + 2x^2 = 20x + 82$, and confrontation gives $2x^2 + 18 = 20x$.]

[Stage 3: solving the simplified equation]

Return all of what you have to one *māl*, so you get: a *māl* and nine dirhams equal ten things [$x^2 + 9 = 10x$]. Work this out according to what was shown in the fifth problem.[5] You find that the first part in one, leaving the other as nine.

I.7.4 Developments in algebra

The two earliest known Arabic texts dealing with algebra are the *Book of Algebra* by al-Khwārazmī, which we possess complete, and the book of almost the same title by Ibn Turk (1st half 3rd/9th century), of which only one part survives. Both books were written in Baghdad during the reign of the Caliph al-Maʾmūn (r. 198–213/813–833). From the language and structure of al-Khwārazmī's work, it is clear that he set down on paper a method that had already been circulating orally among people working in the trades (Oaks 2012a). In fact, the same method had been applied by Diophantus of Alexandria (*c.* 3rd century) in his *Arithmetic* in late antiquity (Christianidis and Oaks 2013), and we know from testimony in a 4th/10th-century Arabic source that the astronomer Hipparchus of Bithynia also wrote a book on *al-jabr* around the 2nd century BCE (Ibn al-Nadīm 1871–1872[CB],[6] 269.15, 283.15; Ibn al-Nadīm 1970[CB], 642.17, 668.6). None of these people invented algebra. Instead, all indications suggest that they composed books on the method already in use among practitioners.

Al-Khwārazmī may not have invented algebra, but Abū Kāmil (d. *c.* 317/930) later noted that it was he who "established its principles" (Abū Kāmil 2012, 245.3). This may mean that al-Khwārazmī was the one who decided on the classification of the six equations and who devised geometric proofs for the solutions to the composite types. In the proofs of both al-Khwārazmī and Ibn Turk, the unknown thing is represented by a line, and the *māl* is a square on that line. These proofs compare straight lines and rectangles with no appeal to either Euclid or to ratio and proportion.

Near the end of the 3rd/9th century, both Abū Kāmil and Thābit ibn Qurra (d. 288/901) wrote geometric proofs based on Propositions II.5 and II.6 of Euclid's *Elements* for these rules, and later Arabic algebraists devised variations on them. Some authors, beginning in the early 5th/11th century with al-Karajī, give proofs based instead in arithmetic. Ibn al-Bannāʾ (654–721/1256–1321), for example, gives three sets of distinct arithmetical proofs in two of his books. In one set of proofs, he completes the square in the context of the equation; in another, he appropriates a rule from finger reckoning as the basis for his proofs; and in the third set, he works with arithmetical versions of Euclid's *Elements* II.5 and II.6 (Oaks 2018).

Al-Fārisī, working around the same time, gave arithmetical proofs modeled on those in Euclid's number theory books. Just after presenting his proofs, he criticized others who prove results in arithmetic (which includes algebra) by geometry, maintaining that "it is more proper whenever possible to firmly establish [the propositions] in number, so they should not be demonstrated by means of lines" (al-Fārisī 1994, 524.17). Most Arabic algebraists, however, saw no problem with geometric proofs.

In the mid-3rd/9th century, Diophantus's (d. *c.* 350) *Arithmetic* was translated into Arabic by Qusṭā ibn Lūqā as a *Book on Algebra* (*Kitāb fī l-jabr wa-l-muqābala*), and certain of its elements were incorporated into later Arabic books and chapters on algebra. Diophantus's arithmetical derivations of the rules for solving composite equations were copied by al-Karajī into his *al-Fakhrī* and later became the basis for al-Karajī's arithmetical proofs in his short treatise *Causes of Calculation* (*ʿIlal al-ḥisāb*). Also, algebraists writing in Arabic copied Diophantus in extending the names of the powers to include their reciprocals. Diophantus's book is well known for its treatment of indeterminate problems. Such problems had been solved by algebra in Arabic before Diophantus became known, but it was Diophantus's book that provided inspiration for the most thorough examination of the techniques in al-Karajī's *Marvelous [Book] on Calculation* (*al-Badīʿ fī l-ḥisāb*).

In another direction, al-Karajī devised rules for operating on polynomials that mimic the rules for operating on numbers in Indian notation. He was able, for example, to give a rule for extracting the square roots of large degree polynomials. Later, in the mid-6th/12th century, al-Samawʾal ibn Yaḥyā (d. 570/1175) built on these ideas in his *Dazzling [Book] on the Science of Calculation* (*al-Bāhir fī ʿilm al-ḥisāb*).

Algebra was not only studied and developed by theoretically minded mathematicians, but it also served as a tool for classical geometrical problem-solving. Al-Khayyām wrote the best-known work on the topic, *Treatise on the Proofs of Algebra Problems* (*al-Risāla fī l-barāhīn ʿalā masāʾil al-jabr wa-l-muqābala*). In it he classified the 25 simplified equations of degree three and less, and he gave geometric solutions to the cubic equations as lines extending to the intersections of conic sections (circles, parabolas, hyperbolas). He had already applied this technique in another work to solve a problem of dividing a quadrant of a circle according to a particular ratio.

Around the 6th/12th century a notation for algebra developed in the western part of the Islamic world. This was an extension of Indian notation in which the first letter in the name of the power is placed above its "number" (coefficient). Like the notations in basic arithmetic, it is shown in snippets as textbook illustrations of what should be put down on the dust-board. It was not considered to be of any theoretical interest, and books continued to express algebra rhetorically. Previous to the development of this notation people had worked out algebraic solutions either mentally or by writing only the coefficients in their dust-board calculations (Oaks 2012b).

I.7.5 Greek number theory

In addition to the mathematics originating in practice that has been described so far, Arabic mathematicians also translated, studied, and expanded on Greek number theory. The two main sources translated into Arabic are Books VII to IX of Euclid's *Elements* and Nicomachus's *Arithmetical Introduction*. Between them they cover the classification of numbers into even, odd, and their subspecies; prime numbers and divisibility; amicable and perfect numbers; ratio and proportion; figurate numbers; and series summation.

Arabic mathematicians were keenly interested in each of these areas. I will give two examples of Arabic contributions. Thābit ibn Qurra derived a rule for generating pairs of amicable numbers and gave a proof following the style of Euclid's number theory books. The theorem states that if the three numbers $p = 3 \cdot 2^{n-1} - 1$, $q = 3 \cdot 2^{n} - 1$ and $r = 9 \cdot 2^{2n-1} - 1$ are prime, and if $n \geq 2$, then the pair $2^n \cdot pq$ and $2^n \cdot r$ is amicable (Hogendijk 1985). The pair 17,296 and 18,416 obtained from $n = 4$ was most likely unknown to the Greeks and was probably found by Thābit. In any case, this pair was later mentioned by Ibn Munʿim (d. 626/1228). Another important result is due to Ibn al-Haytham (354–*c.* 430/965–*c.* 1040), who stated what we now call Wilson's theorem, after the 18th-century mathematician John Wilson, who later rediscovered it: if p is prime, then $1 + (p - 1)!$ is divisible by p (Rashed 1980). As with practical arithmetic, these results were expressed in books rhetorically, without notation.

Aristotle provided the philosophical foundation for this Greek arithmetic. He divided the genus of "quantity" into two classes: continuous and discrete. Continuous quantities are lines, surfaces and bodies in geometry, and also time. Number, on the other hand, is discrete. As Euclid defined it, "[a] number is a multitude composed of units" (Euclid 1956, 2: 277). Because the unit (1) was regarded as being indivisible, fractional quantities were not considered to be numbers. And because it is not a multitude, 1 is not a number either. Numbers, then, consist only of the positive integers 2, 3, 4 and so on.

This notion of number is at odds with the fractions and roots that pervade Arabic arithmetic. Some effort was made by different mathematicians to reconcile the two, either by postulating two kinds of unit or by at least identifying fractions with ratios of integers (Oaks 2019, 109–10). One notable solution is found in al-Khayyām, who regarded the numbers of Arabic arithmetic as being the dimensionless measures of continuous magnitudes. Because he assumed a unit line, a number like $5\frac{2}{3}$ or $\sqrt[4]{11}$ can be the length of a line, the area of a surface or the volume of a body.

Scholars in Islamicate societies admired the logical structure and philosophical foundation of Greek mathematics, and they often sought to incorporate practical techniques into that framework. Books on finger reckoning and Indian calculation written after the 6th/12th century sometimes begin with Euclid's definition of number, even if that definition excludes the fractions and irrational numbers that follow. Two examples are Ibn al-Khawwām's (d. 724/1324) *Magnificent Benefits of the Rules of Calculation* (*al-Fawā'id al-bahā'iyya fī l-qawā'id al-ḥisābiyya*; 675/1277) and Ibn al-Bannā''s *Essays on Calculation* (*Maqālāt fī l-ḥisāb*). In addition to these theoretical trimmings, some number theory topics were incorporated into practical books. The classification of numbers is sometimes reviewed, the sieve of Eratosthenes for finding prime numbers was given to aid in calculations with fractions, and rules for series summation were applied to textbook problems. Mention should also be made of the numerical reinterpretation of incommensurable magnitudes from Euclid's Book X that would have given arithmetic students a stronger foundation in calculating with irrational roots. Some books went further by giving Greek-style proofs not just to the rules for solving equations in algebra but to the rules of arithmetic generally. Among the authors working in this direction were Abū Manṣūr al-Baghdādī (d. 429/1037–8), Ibn Munʿim and al-Fārisī.

I.7.6 Conclusion

It was an ongoing characteristic of Arabic arithmeticians to combine the best features of the various systems of calculation and problem-solving in circulation. One of the pleasures of reading through a hybrid work such as Ibn al-Bannā''s *Condensed [Book] on the Operations of Arithmetic* (*Talkhīṣ aʿmāl al-ḥisāb*), for example, is seeing how he explains Indian calculation together with multiplication shortcuts and problem-solving from finger reckoning, with some definitions and rules from Greek number theory scattered in between.

Among the problem-solving methods algebra stands out for its complexity and its applicability to a wide range of problems. It is because of these features that it attracted the attention of theoretical mathematicians, who extended it and reworked it in directions beyond the realm of practice. Mathematicians found ways to give its rules a foundation in Greek-style proofs; they extended the rules of calculation to polynomials of arbitrary degree, and the operations to include roots; and they found a way to apply algebra to theoretical geometry.

Notes

1 For overviews of Arabic arithmetic, see the introdutions to al-Uqlīdisī (1978) and Abdeljaouad and Oaks (2021).

2 See Folkerts (2001) for a more comprehensive overview of Indian calculation.

3 There is no good English equivalent of *māl*, so I leave it untranslated. Also, I write the plural with the English suffix: *māls*.
4 Ibn Badr uses the phrase "restore and confront" to mean "restore and/or confront".
5 That is, the fifth type of equation. Like many algebraists, Ibn Badr does not spell out the steps in stage 3.
6 Consolidated bibliography.

Bibliography

Sources

al-Hawārī. ed., tr. and com. Abdeljaouad, M. and Oaks, J. 2021. *Al-Hawārī's* Essential Commentary: *Arabic Arithmetic in the Fourteenth Century*. Berlin: Max-Planck-Gesellschaft zur Förderung der Wissenschaften. https://edition-open-sources.org/sources/14/.

Abū Kāmil. ed. and tr. Rashed, R. 2012. *Algèbre et Analyse Diophantienne*. Berlin: Walter de Gruyter.

Euclid. tr. Heath, T. 1956, *The Thirteen Books of the Elements*. 3 vols. New York: Dover Publication Inc.

al-Fārisī, Kamāl al-Dīn. ed. Mawāldī, M. 1994. *Asās al-qawāʿid fī uṣūl al-fawāʾid* [The Basis of the Rules on the Principles of the Uses]. Cairo: Maʿhad al-makhṭūṭāt al-ʿarabiyya.

Ibn al-Yāsamīn. ed. Zemouli, T. 1993. *Muʾallafāt Ibn al-Yāsamīn al-riyāḍiyya* [Mathematical Writings of Ibn al-Yāsamīn]. M.Sc. thesis in History of Mathematics. Algiers, E.N.S.

al-Khwārazmī. ed. and tr. Rashed, R. 2009. *Al-Khwārizmī: The beginnings of algebra*. London: SAQI.

Saidan, A. S. 1986. *Taʾrīkh ʿilm al-jabr fī l-ʿālam al-ʿarabī* [History of Algebra in Medieval Islam]. 2 vols. Kuwait: al-Majlis al-waṭanī li-l-thaqāfah wa-l-funūn wa-l-ādāb, Qism al-turāth al-ʿarabī.

al-Uqlīdisī. tr. Saidan, A. S. 1978. *The Arithmetic of al-Uqlīdisī: The Story of Hindu-Arabic Arithmetic as Told in Kitab al-fusul fi al-hisab al-Hindi, by Abu al-Hasan, Ahmad ibn Ibrahim al-Uqlidisi, Written in Damascus in the Year 341 (A.D. 952/3)*. Dordrecht: D. Reidel.

Research literature

Christianidis, J. and Oaks, J. 2013. "Practicing Algebra in Late Antiquity: The Problem-solving of Diophantus of Alexandria," *Historia Mathematica* 40: 127–63.

Folkerts, M. 2001. "Early Texts on Hindu-Arabic Calculation," *Science in Context* 14: 13–38.

Hogendijk, J. P. 1985. "Thābit ibn Qurra and the Pair of Amicable Numbers 17296, 18416," *Historia Mathematica* 12: 269–73.

Oaks, J. A. 2012a. "The Series of Problems in al-Khwārizmī's *Algebra*." Published in HASTEC's online Carnet de Recherche. http://problemata.hypotheses.org/157; https://www.uindy.edu/cas/mathematics/oaks/files/oakscarnet.pdf.

Oaks, J. A. 2012b. "Algebraic Symbolism in Medieval Arabic Algebra," *Philosophica* 87: 27–83.

Oaks, J. A. 2018. "Arithmetical Proofs in Arabic Algebra," in Laabid, E., ed. *Actes du 12è Colloque Maghrébin sur l'Histoire des Mathématiques Arabes: Marrakech, 26–27–28 mai 2016*. Marrakech: École Normale Supérieure, 215–38.

Oaks, J. A. 2019. "Proofs and Algebra in al-Fārisī's Commentary," *Historia Mathematica* 47: 106–21.

Rashed, R. 1980. "Ibn al-Haytham et le théorème de Wilson," *Archive for History of Exact Sciences* 22: 305–21.

Saidan, A. S. 1968. "Finger Reckoning in an Arabic Poem," *The Mathematics Teacher* 61: 707–8.

I.8
OPTICS: EXPERIMENTS AND APPLICATIONS

Johannes Thomann

I.8.1 Thought experiments or real experiments?

It is often said that the scientific art of experiment developed only after 1600 and that Galileo Galilei (1564–1642) and Evangelista Torricelli (1608–1647) were the first experimenters (Janich 1995, 622). Therefore, it has been asked whether the many test procedures described in premodern Arabic texts could indeed be called experiments in the modern sense, meaning "a controlled manipulation of events, designed to produce observations that confirm or disconfirm one or more rival theories or hypotheses. To experiment is to put questions to nature and the experimental method is contrasted with the passive acceptance of whatever observations happen along" (Blackburn 2016, 168). Matthias Schramm (1930–2005) was a strong advocate that medieval optics was at least well on the way to experimental physics in the modern sense in his book *Ibn al-Haythams Weg zur Physik* (Ibn al-Haytham's Path to Physics): "He is able by means of his experimental technique in a singular case to create the conditions, which in the first place enable the justification of a mathematically distinguished approach" (Schramm 1963, 288). According to this interpretation, Ibn al-Haytham (354–*c.* 430/965–*c.* 1040) not only performed experiments for testing an existing theory or hypothesis but also used them to discover principles in nature that he declared to be invisible and hidden from natural sense perception. For example, in his experiments in the camera obscura with color filters and colored objects, he discovered the decrease of color intensity with increasing distance of the object from the projected image (Schramm 1963, 244–74).

But there were also some more critical assessments of Ibn al-Haytham's reports of experiments. In the case of an experiment with refraction angles, Mark Smith concluded, from the fact that no concrete values were mentioned, that it was a mere thought experiment that was not carried out in practice (Smith 2015, 214). But this can hardly hold for all his reported experiments, as shown in the following cases.

Many treatises on optics are purely mathematical, and also in the *Book of Optics* (*Kitāb al-Manāẓir*) of Ibn al-Haytham most of the text consists of geometrical proofs (Ibn al-Haytham 1983, 1989, 2001, 2002, 2006, 2008, 2010; Rashed 2005, 184–223). But optics was also connected with other disciplines. A bundle of applied optical research fields existed, for instance, in astronomy, meteorology, physiology, military technology and aesthetics.

DOI: 10.4324/9781315170718-10

I.8.2 Optics in astronomy

Ibn al-Haytham observed a partial solar eclipse with the camera obscura (ed. and transl. Raynaud 2016). The sickle-shaped images obtained were presented in drawings and seem to be well preserved in the manuscript tradition. Dominique Raynaud was able to date and locate the eclipse. According to his thorough investigation, Ibn al-Haytham most likely observed the partial eclipse of 28 Rajab 380/21 October 990 in Baṣra (Raynaud 2016, 161–86). But his treatise *On the Shapes of the Eclipses* (*Maqāla fī ṣuwar al-kusūf*; Raynaud 2016) is significant for another reason; it is the first experimental study of the camera obscura, with extensive investigations of the size and form of the hole in order to obtain optimally sharp images. While his predecessors used the dark chamber as a simple observational tool, he succeeded in solving long-standing problems (Raynaud 2016, 96). It was a long way, both in time and in methodology, from the observation of a partial eclipse in the projected images produced by gaps between the leaves of a tree, which is reported in *Problemata* in the *Corpus Aristotelicum XV.11*, to the empirically tested and optimized device of Ibn al-Haytham. The first case is a simple observation of a natural phenomenon as it occurs spontaneously without human intervention (Aristoteles 1984, 2: 1419). In the second case, in order to obtain data of a precision not seen before, advanced technology is used to produce conditions that do not occur naturally.

I.8.3 Optics and astrophysics

Another similar case is Ibn al-Haytham's investigation of the physical quality of the moon's surface by means of an optical instrument. This is described in his treatise *On the Light of the Moon* (*Fī ḍaw' al-qamar*; German tr. Kohl 1923; Arabic ed. Ibn al-Haytham 1938, no. 8; partial Engl. tr. Zubairy 1969). A common opinion was that the moon is like a mirror that reflects the sun's light. Since it was known that the moon has a spherical shape, beams of light at any angle to the surface can be observed from the center of the moon to the periphery. With the naked eye, it is impossible to decide if a particular spot seen on the moon is seen by a beam of light actually coming from that spot on the moon's surface. In mirrors, objects are seen in places where they are not. Therefore, Ibn al-Haytham constructed an instrument by which the path of moonlight can be controlled (Schramm 1963, 152 and 168; Kohl 1923, 334).

It consists of a diopter (Figure I.8.1). The ocular plate has a round hole through which the eye is looking. The objective plate has a slit. Mounted on it is a moveable diaphragm with a slit turned by 90° to the slit of the objective plate (Figure I.8.2).

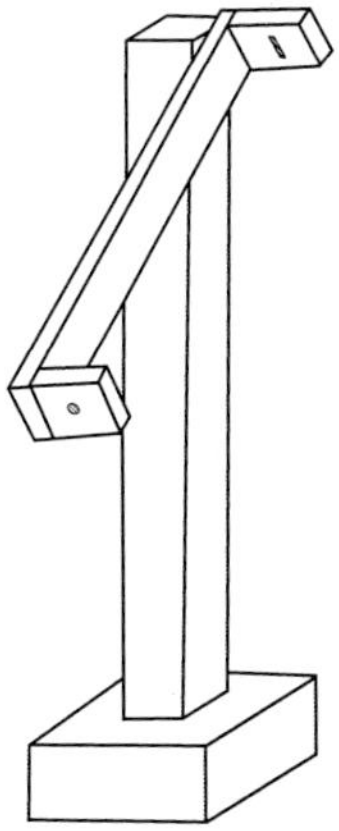

Figure I.8.1 Ibn al-Haytham, diopter. Source: © Johannes Thomann

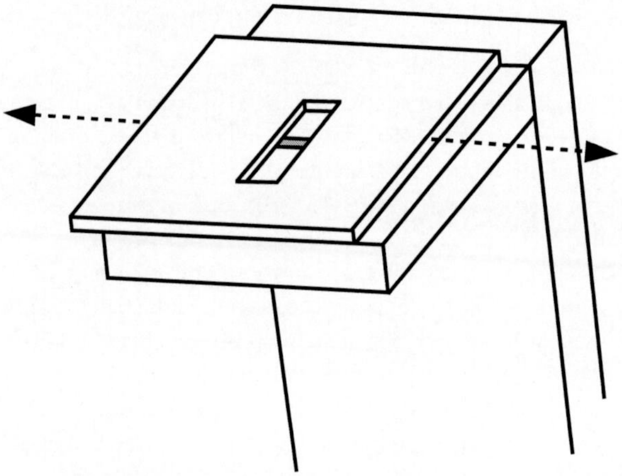

Figure I.8.2 Ibn al-Haytham, diaphragm on objectivve plate.

Source: © Johannes Thomann

This allows for an easy selection of areas of the moon's surface from the periphery to the center. Instead of looking directly through the hole of the ocular plate, a projected image can be observed on a translucent screen a short distance from the hole. The diaphragm is moved back and forth, and it is seen that the projected image does not change. Therefore, it is concluded that from every point on the lunar surface light is going out in all directions above the surface. In modern terms, this phenomenon is based on diffuse reflection. Ibn al-Haytham calls this a self-shining body. Some of his remarks speak strongly in favor of the actual performance of these experiments, for example, his statement that assistants are necessary in order to firmly stabilize the instrument when the diaphragm is moved (Kohl 1923, 338).

I.8.4 Optics and meteorology

Another natural phenomenon that became a debated topic in optics was the rainbow. In this case, Kamāl al-Dīn al-Fārisī (d. 718/1319) brought the experimental method a step further toward the replacement of natural objects by technical devices in his *Revision of Optics* (*Tanqīḥ al-manāẓir*, written in 708/1309; Kamāl al-Dīn al-Fārisī 1928–1929). In the chapter on the rainbow, he criticizes Ibn al-Haytham, who thought that the cause for a rainbow is a reflection in a cloud like in a burning mirror (Wiedemann 1910). Kamāl al-Dīn argued that this cannot be correct because it would exclude the empirical fact that two rainbows can occur at the same time and that a rainbow can occur in a clear sky. He revived Aristotle's explanation that a rainbow is produced by water droplets in the air. Since he assumed them to be spherical in shape, he could start with Ibn al-Haytham, who treated the problem in analogy to spherical burning mirrors. But Kamāl al-Dīn developed a more complicated explanation by assuming that the light beam split at the water–air border into a beam refracted into the air and a beam reflected back into water. This happens again to the second beam when it meets the water–air border. In that way, al-Fārisī was able to explain the secondary rainbow. He went on to make empirical tests of his geometrical analysis. He made an enlarged model of a water droplet, a spherical container of glass filled with water. With that he tested the reflection and refraction paths of light beams, as well as their path in double reflection at the surface of the glass container. He was able to determine which of the two reflections produced which circular shape in the sky (Figure I.8.3).

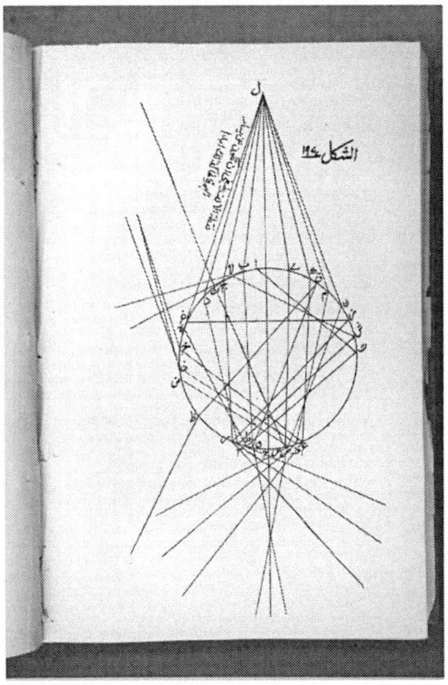

Figure I.8.3 The traces of the rays of light in the spherical vessel depicted in Kamāl al-Dīn al-Fārisī's autograph, MS Adilnor Collection, Sweden.

Source: © Adilnor Collection, Sweden

Almost at the same time, Dietrich of Freiberg (1240–1318) in Koblenz (today in Germany) came to similar results in his treatise *On the Rainbow and the Impressions Caused by Rays* (*Tractatus de iride et de radialibus impressionibus*). For him, like for Kamāl al-Dīn, the *Book of Optics* of Ibn al-Haytham was the starting point. Kamāl al-Dīn's experimentation was by no means naive. He addressed the problem of the different refractive indices of water and glass, but by measurement, he found empirically that the refractive index of his model as a whole was close to that of water. All experiments were made in a camera obscura to control conditions for the light rays.

I.8.5 Optics and physiology

Physiological optics has its root in antiquity. Ptolemy (*c.* 100–170) made an experiment to understand binocular viewing. His experiment consisted of putting a white and a black rod at different distances from the forehead. When one fixes one's gaze on the nearer white rod, the more distant black rod appears on both sides of the other. If one closes the left eye, the black image to the left disappears (Simon 2003, 134). This is what today is called diplopia. Ibn al-Haytham was inspired by this experiment but went a step further. He tried to understand what happens inside the viewer's body (Ibn al-Haytham 1989). He came to the (wrong) conclusion that the vitreous body is the sensory organ that sends the perceived forms of the objects to the optical nerve. According to Ibn al-Haytham this happens in both eyes and the two forms are united when they reach the chiasma, where he located the ultimate sensory faculty (Simon 2003, 139). From there, the united form is sent through the optical nerve to the supreme faculty in the brain. Ibn al-Haytham's entirely new model of binocular viewing was immensely influential until the 17th century, when Kepler discovered that the retina is the sensory organ, because there is no nerve connected to the vitreous body (Simon 2003, 214).

I.8.6 Optics and military technology

Applications of optics in the fields mentioned so far were entirely confined to knowledge production. But there is one application of optics that was supposed to serve human affairs in a dramatic and practical way. There are ancient accounts that Archimedes (287–212 BCE) defended Syracuse by setting the Roman fleet on fire (Rashed 2002, 317–19, 335–42). To achieve this, he ordered trained soldiers to reflect the sun's light by means of mirrors to one spot on a ship until fire broke out. The Romans, the legend reports, believed they were seeing a miracle executed by an angry god. It has been doubted that such an effect could have been achieved with ancient technology, on the grounds that it would have been extremely difficult to produce a burning mirror with the focal length necessary to reach the ships. The optimal shape of a burning mirror is a paraboloid, as was known in antiquity, but even a spherical mirror with a long focus is a technical challenge. However, the legend does not specify the form of the mirror. An array of plane mirrors might have been used. In 2005, David Wallace, professor of mechanical engineering at the Massachusetts Institute of Technology, made an experiment with his students, in which they reflected sunlight with 127 plane mirrors onto a model of a Roman ship at a distance of 100 feet (more space was not available). After a failed experiment, they finally succeeded in setting a small part of the ship on fire. Wallace concluded that such an effect could have been produced in Archimedes's time. But the FAQ posted on his website, describing the experiments in an attractive manner, highlights the various difficulties on the technical side that would have had to overcome in the case of an actual sea battle (http://web.mit.edu/2.009/www/experiments/deathray/10_ArchimedesResult.html).

The legend of Archimedes was translated into Arabic, and also the treatise on burning mirrors of Diocles (late 3rd–early 2nd century BCE), which had a great influence in the Islamicate world (Rashed 2002, 335–42, 3–151). Ibn al-Haytham alone wrote three treatises on the topic. Most works on burning mirrors are purely mathematical without any reference to military technology. But there is at least one exception to this rule. An Arabic work attributed to Alexander of Macedonia (r. 336–323 BCE) describes the use of burning mirrors in war. It recommends using 30 plane mirrors. With them it is possible, the text says, to set objects on fire in a distance up to 500 m (MS Gotha, Orient. A. 1348, fols 109b–124a). However, even if it is corroborated that Islamicate scholars believed that burning mirrors had military applications, it remains doubtful that this technique was ever used in war. Fire grenades thrown from catapults might have been more effective weapons. After this grim topic it seems appropriate to finish with a happier subject.

I.8.7 Optics and aesthetics

A further field in which optics was applied was visual aesthetics. While an immense literature on the aesthetics of linguistic artifacts exists, visual aesthetics is almost absent as a philosophical or scientific topic in the Islamicate world. But there is a prominent exception. Book II, chapter 3 of Ibn al-Haytham's *The Book of Optics* contains a treatise on beauty (Ibn al-Haytham 1989, 1: 200–6). This is again an innovation of Ibn al-Haytham. He presents for the first time a theory of visual beauty, defining ugliness as the absence of beauty. His idea is based on his theory of visual perception, in which he distinguishes 22 visual properties that vision is able to grasp (Ibn al-Haythama 1989, 1: 126–206). These are light, color, distance, position, solidity, shape, size, separation, continuity, number, motion, rest, roughness, smoothness, transparency, opacity, shadow, darkness, similarity, dissimilarity, beauty and ugliness (Puerta Vílchez 2017, 714). The property of beauty is produced by all those properties, except ugliness. He discusses one property after another and gives concrete examples. For instance:

> Motion produces beauty; hence the beauty of dancing, and of the movements of the dancer, and of many of the gestures and movements of man in speech and in action. Rest produces beauty, and therefore gravity and staidness appear beautiful. Roughness produces beauty. Thus many rough clothes and covers look beautiful; and for this reason many of the goldsmith's artifacts become beautiful by having their surfaces roughened and textured. Smoothness produces beauty, and therefore it is beautiful in cloth and utensils.
>
> *(Ibn al-Haytham 1989, 1: 202)*

In these, as in other discussions, two contrary properties can produce beauty in different ways. For example, similarity produces beauty in the paired organs of an animal. And dissimilarity creates ugliness in the nose, if it is not equally wide from beginning to end. The same is true for script. Furthermore, general principles of beauty, such as proportionality and harmony are discussed.

While Ibn al-Haytham's theory of visual aesthetics was not further developed in the Islamicate world, it was of crucial importance for the development of aesthetics in the Italian Renaissance. At least, this has been argued by influential art historians like Erwin Panofsky (Panofsky 1955, 89–90, n.63). David Summers especially pointed to the important role of Ibn al-Haytham's *Great Optics* in the formation of aesthetic theory (Puerta Vílchez 2017, 702, n. 708; Summers 1987, 74, 167–71). Graziella Federici Vescovini and Gérard Simon established the importance of Ibn al-Haytham for the development of central perspective in Renaissance art (Federici Vescovini 1998; Simon 2001, 2003). Ibn al-Haytham's work influenced not only painting but also the other visual arts, including architecture. Europe would literally not look as it does if Ibn al-Haytham's work had not reached western Europe.

Bibliography

Sources

Aristoteles. 1984. ed. Barnes, J. *The Complete Works: The Revised Oxford Translation* (Bollingen Series). 2 vols. Princeton, N.J.: Princeton University Press.

Ibn al-Haytham. 1938. *Rasā'il* [Treatises]. Ḥaydarābād: Dā'irat al-maʿārif al-ʿuthmāniyya.

Ibn al-Haytham. ed. Sabra, A. I. 1983. *Kitāb al-Manāẓir li-l-Ḥasan Ibn al-Haytham. al-Maqālāt 1, 2, 3 fī l-ibṣār al-istiqāma* [*Book of Optics* by al-Ḥasan Ibn al-Haytham. Books 1, 2, 3 on Direct Vision]. Al-Kuwayt: al-Majlis al-waṭanī li-l-thaqāfa wa-l-funūn wa-l-ādāb.

Ibn al-Haytham. ed. and tr. Sabra, A. I. 1989. *The Optics of Ibn al-Haytham*. London: The Warburg Institute.

Ibn al-Haytham. ed. Sabra, A. I. 2002. *Kitāb al-Manāẓir li-l-Ḥasan Ibn al-Haytham. al-Maqālāt al-rābiʿa wa-l-khāmisa fī nʿikās al-aḍwā' wa-mawāḍiʿ al-khayālāt al-mubṣara bi-l-inʿikās* [*Book of Optics* by al-Ḥasan Ibn al-Haytham. Books 4 and 5 on Reflection and Images Seen by Reflection]. Al-Kuwayt: al-Majlis al-waṭanī li-l-thaqāfa wa-l-funūn wa-l-ādāb.

Ibn al-Haytham. ed. and tr. Smith, A. M. 2001. *Alhacen's Theory of Visual Perception: A Critical Edition, with English Translation and Commentary, of the First Three Books of Alhacen's "De aspectibus", the Medieval Latin Version of Ibn al-Haytham's "Kitāb al-Manāẓir"*. Philadelphia: American Philosophical Society.

Ibn al-Haytham. ed. and tr. Smith, A. M. 2006. *Alhacen on the Principles of Reflection: A Critical Edition, with English Translation and Commentary, of Books 4 and 5 of Alhacen's "De Aspectibus", the Medieval Latin Version of Ibn al-Haytham's "Kitāb al-Manāẓir"*. Philadelphia: American Philosophical Society.

Ibn al-Haytham. ed. and tr. Smith, A. M. 2008. *Alhacen on Image-Formation and Distortion in Mirrors: A Critical Edition, with English Translation and Commentary, of Book 6 of Alhacen's "De Aspectibus", the Medieval Latin Version of Ibn al-Haytham's "Kitāb al-Manāẓir"*. Philadelphia: American Philosophical Society.

Ibn al-Haytham. ed. and tr. Smith, A. M. 2010. *Alhacen on Refraction: A Critical Edition, with English Translation and Commentary, of Book 7 of Alhacen's "De Aspectibus," the Medieval Latin Version of Ibn al-Haytham's "Kitāb al-Manāzir"*. Philadelphia: American Philosophical Society.

Kamāl al-Dīn al-Fārisī. 1928–9. *Tanqīh al-Manāzir* [Revision of *The Optics*]. Ḥaydarābād: Dā'irat al-maʿārif al-ʿuthmāniyya.

Kohl, K. 1923. "'Über das Licht des Mondes'. Eine Untersuchung von Ibn al-Haitham," *Sitzungsberichte der Physikalisch-Medizinischen Sozietät zu Erlangen* 55: 305–89.

Rashed, R. 2002. *Les catoptriciens grecs. Tome I. Le miroirs ardents*. Paris: Les Belles Letters.

Rashed, R. 2005. *Geometry and Dioptrics in Classical Islam*. London: Al-Furqan Islamic Heritage Foundation.

Raynaud, D. 2016. *A Critical Edition of Ibn al-Haytham's* On the Shape of the Eclipse. *The First Experimental Study of the Camera Obscura*. Cham, Switzerland: Springer.

Wiedemann, E. 1910. "Über die Brechung des Lichtes in Kugeln nach Ibn al-Haiṯam und Kamâl al Dîn al Fârisî," *Sitzungsberichte der Physikalisch-Medizinischen Sozietät zu Erlangen* 42: 15–58 [Reprint *Aufsätze zur arabischen Wissenschaftsgeschichte, Hildesheim*: Olms, 1: 596–640].

Zubairy, H. N. 1969. "Treatises of Ibn al-Haitham. The Light of the Moon," in *Ibn al-Haitham. Proceedings of the Celebrations of 1000th Anniversary Held under the Auspices of Hamdard National Foundation*. Karachi: Hamdard National Foundation, 203–14.

Manuscript

MS Gotha, Forschungsbibliothek, Orient. A. 1348.

Research literature

Belting, H. 2011. *Florence and Baghdad. Renaissance Art and Arab Science*. Cambridge, MA: Belknap Press of Harvard University Press.

Blackburn, S. 2016. *The Oxford Dictionary of Philosophy*. Oxford: Oxford University Press.

Federici Vescovini, G. 1998. "Alhazen vulgarisé: le 'De li aspecti' d' un manuscrit du Vatican (Moitié du XIVe siècle) et le Troisième Commentaire d'Optique de Lorenzo Ghiberti," *Arabic Science and Philosophy* 8: 67-96.

Janich, P. 1995. "Experiment," in *Enzyklopädie Philosophie und Wissenschaftstheorie. 1: A – E*. Stuttgart: Metzler, 621–2.

Panofsky, E. 1955. *Meaning in the Visual Arts. Papers in and on Art History*. Garden City, NY: Doubleday.

Puerta Vílchez, J. M. 2017. *Aesthetics in Arabic Thought. From Pre-Islamic through al-Andalus*. Leiden: Brill.

Rashed, R. 1970. "Le modèle de la sphère transparente et l'explication de l'arc-en-ciel: Ibn al-Haytham, al-Fārisī," *Revue d'histoire des sciences* 23: 109–40.

Schramm, M. 1963. *Ibn al-Haythams Weg zur Physik*. Stuttgart: Steiner.

Simon, G. 2001. "Optique et perspective: Ptolémée, Alhazen, Alberti," *Revue d'histoire des sciences* 54: 325–50.

Simon, G. 2003. *Archéologie de la vision: L'optique, le corps, la peinture*. Paris: Seuil.

Smith, A. M. 2015. *From Sight to Light: The Passage from Ancient to Modern Optics*. Chicago: University of Chicago.

Summers, D. 1987. *The Judgment of Sense: Renaissance Naturalism and the Rise of Aesthetics*. Cambridge: Cambridge University Press.

I.9
AUTOMATA AND BALANCES

Constantin Canavas

In this chapter, I discuss scholarly writings on automata and balances in Islamicate societies and current research on them, in the period from the 3rd/9th to the beginning of the 9th/15th centuries. I focus on Arabic texts, Arabic being the common scholarly language in the Islamicate world in this period. Persian or Arabic literary texts reporting legendary pre-Islamic Sasanian thrones and other Sasanian court marvels (automata) are not considered. The temporal limits also place Ottoman, as well as South Asian Islamicate (Mughal), societies beyond the scope of the chapter.

I.9.1 Introduction

The ancient Greek term *automaton* (pl. *automata*) denotes, in philosophical terms, "something acting by its own will." From a functional point of view, objects – commonly artificial mechanisms – would be called *automata* when they performed actions "on their own", "by themselves" (*auto-*), that is without obvious external intervention. In the Arabic tradition, such devices are commonly described by the term *ḥīla* (pl. *ḥiyal*), which has been often translated into English as "ingenious (mechanical) devices" (Canavas 2009a, 74) – a translation in which the term *ingenious* carries both connotations of genuine as well as of engineer in the broader sense of the French word *ingénieur*. Besides the technological context, the term *ḥīla* is encountered – mostly in the plural form *ḥiyal* – in several further semantic frames of the Arabic tradition. It is used as a technical term for stratagems in war, as a literary genre on the tricks practiced by beggars, conjurers, forgers, and like people, and – when applying Islamic law in practice – it describes the use of legal means for extra-legal ends (Schacht 1971, 511a). A linguistic-etymological review such as that provided by Muhammad Abattouy reveals some common notions in using the term in these different fields (Abattouy 2000a, 12–13); however, etymological and linguistic comparisons may conceal rather than elucidate the underlying *epistemological* problems – at least for the scientific and technological fields of the term's usage.

Abattouy has been working since the 1990s on the Arabic traditions of mechanics and their relations to the Greek traditions. His projects include both editing and epitomizing, interpreting and comparing Arabic texts on mechanics and related fields – which renders his publications central for the present exposition, both as material source and as interpreting perspective. His approach follows an epistemological perspective that perceives the Arabic

DOI: 10.4324/9781315170718-11

traditions concerning the science of mechanical devices (*ʿilm al-ḥiyal*) as ingenious applications of mathematics to natural bodies. In fact, philosophers in the Arabic Islamic tradition such as al-Fārābī (d. 339/950–1) have proposed theoretical systems of categorizing scientific disciplines in which *ʿilm al-ḥiyal* is mentioned in relation to the science of weights (or heavy bodies; *ʿilm al-athqāl*) and the doctrine of balances (*mīzān/mawāzīn*). According to the systematization of al-Fārābī, the science of mathematics comprises, among others, the science of weights (concerned with the doctrine of balances and the principles of the devices for moving heavy things) and the science of ingenious mechanical devices (*ʿilm al-ḥiyal*) which actualize and demonstrate the applicability of mathematical principles in the natural bodies (al-Fārābī 1948[CB],[1] 88–9; Abattouy 1997, 14). Following this classification, Abattouy argues that the term *ḥiyal* is generally used in Arabic texts – not only in philosophical ones and without any restriction of context, author or periodization – to denote the discipline or methods for applying theoretical (that is mathematical) mechanics (Abattouy 1997, 14–18, 2000a, 2000b). Abattouy indicates further that these ingenious mechanical devices are also referred to as moving devices (*al-ḥarakāt*), for example, in *The Catalog* (*al-Fihrist*) by Ibn al-Nadīm ([d. 380/990]; Ibn al-Nadīm 1871–1872[CB], 1: 265, 272, 285), as well as in *The Keys of the Sciences* (*Mafātiḥ al-ʿulūm*) by Abū ʿAbdallāh al-Khwārazmī ([d. 387/997]; al-Khwārazmī 1968, 246–9). He deduces from this fact a "double tradition" in the Arabic literature: one tradition, *al-ḥiyal*, following theoretical mechanics in the tradition of Pseudo-Aristotle, Hero (1st century) and Pappus (4th century), both from Alexandria, and another one, that of *al-ḥarakāt*, corresponding to concepts of "practical, technological" devices "imitating motions" (Abattouy 2000a, 14–16).

The suggestions of Abattouy regarding the Arabic terminological and semantic networks involving concepts and realization of ingenious mechanical devices are certainly informative and valuable with respect to the specific authors, treatises and periods concerned. His effort to cast the enormous amount of texts and traditions *into one comprehensive interpretative (epistemological) model* by applying a unique homogeneous terminology over several centuries and several authors acting in different and heterogeneous contextual circumstances, however, leads to a certain extent to oversimplifications, occasional inconsistencies and possible misunderstandings. The treatises by the Banū Mūsā (3rd/9th century) and al-Jazarī (d. *c.* 602/1206) explicitly treat and refer to *al-ḥiyal*. According to Abattouy's argument, they belong to the treatises of *al-ḥarakāt*. Translating the term *al-ḥiyal* generally by "mechanics", Abattouy postulates arbitrarily an apparently consistent theoretical background beyond specific historical reception contexts. This supposedly includes any Arabic treatise of a theoretical character, for example, Pseudo-Aristotelian works circulating in the 4th/10th century, and again presumably comprising the term *al-ḥiyal* (Abattouy 1997, 11), as well as the compendium of ingenious devices by al-Jazarī more than two centuries later. Since al-Jazarī does not include any theory, Abattouy deduces the theoretical affinity of the treatise by exalting its importance and locating it in a homogeneous and continuous tradition of "mechanical devices":

> Al-Jazarī's devices, apart from being practical machines, incorporate techniques and components of great importance in the development of machine technology.
>
> *(Abattouy 1997, 31)*

Another problem is the use of the practical label *al-ḥarakāt* by Ibn al-Nadīm for the theoretical *Treatise of Ingenious Devices* (*Kitāb al-ḥiyal*) by the Banū Mūsā. Abattouy tries to avoid this apparent counterexample by suggesting that it indicates the plurality of terminology, the dependence of the terms on the specific perspective (stressing the inner *principles* of functioning, *al-ḥiyal*, or the *phenomenon* of motion, *al-ḥarakāt*), as well as the lack of a consistent, generally accepted categorization in writing *practice* beyond the confinement of textual *genres* – even among authors of the same period (3rd/9th and 4th/10th centuries).

A diversified understanding of the polysemy and the different classification systems of sciences and practices in the Islamicate world concerning mechanics, balances and ingenious devices (automata) can be achieved when authors are regarded with respect to their *intentions*, their intended public and their embeddedness in the scholarly debates in each specific period and Islamicate societal environment – that is in a contextualized approach (Brentjes 2018[CB], 187–211). Ibn al-Nadīm offers a bibliographical classification for potential readers; al-Fārābī proposes a rigorous Aristotelian synthesis of scientific knowledge based on the theoretical potential–actual dichotomy and intended for scholars (theologians, philosophers). In contrast, al-Khwārazmī offers an overview of existing disciplines and practices addressing the Samanid administration and, further, a general, nonscholarly public. A similar effort that contextualizes the transmission of texts on ingenious mechanical devices, and including the transmission of knowledge on weights and balances, needs to be made for the heterogeneous collection of texts forming the corpus of al-Isfizārī (late 5th–early 6th/11th–early 12th centuries) on the sciences of weights and mechanical devices edited and translated into English by Abattouy and al-Hassani (al-Isfizārī 2013, 2015). It regards, after all, the question of calling a collection of heterogeneous texts "corpus of . . .". Brentjes and Renn have proposed a model for such an approach in the case of Thābit ibn Qurra's (d. 288/901) studies of the balance, focusing on his intellectual field and the specific circumstances under which balance studies *evolved* during the 3rd/9th century in the Islamicate world (Brentjes and Renn 2016, 67–99).

In the following, the history of research on Arabic texts on automata and balances will be reviewed in the context of the specific questioning. In a next step, the texts themselves will be presented in a hermeneutical approach that stresses their specific historical and transmission context.

I.9.2 Reconstructing the history of research on Arabic texts on automata and balances

Scholarly research on automata in the Islamicate world occurred in several phases with different intentions. At the end of the 19th century, scholars such as Bernard Carra de Vaux (1867–1953) were looking for Arabic manuscripts transmitting lost Greek scientific texts. Carra de Vaux discovered, studied, edited and translated such Arabic treatises into French, including the Arabic versions of Hero's *Mechanics* in 1894 (Hero 1894/1988) and Philo of Byzantium's (*c.* 280–*c.* 220 BCE) *Pneumatics* in 1903 (Carra de Vaux 1903).

The next phase was framed (mainly) by Eilhard Wiedemann (1852–1928), professor of physics at Erlangen University in Germany. Wiedemann – with the help of Arabic philologists – published a great number of translations of Arabic texts on theoretical physics (including the subjects of balances, levers and the steelyard [*al-qarasṭūn*], a balance with unequal arms), also technological objects (including compasses, mirrors, and automata), as well as mathematics, alchemy, medicine and other fields of medieval Islamicate sciences and technology. His works – short or long papers dedicated eclectically to a single object or issue, or a thematic chapter excerpted from an unpublished Arabic manuscript – were published in series such as *Sitzungen der Physikalisch-medizinischen Sozietät in Erlangen*, or *Abhandlungen der Kaiserlich Leopoldinisch-Carolinischen Deutschen Akademie der Naturforscher*. Later most of them were republished together in two volumes. Wiedemann's goal was to make topics in Arabic treatises accessible to scholars and, through his public lectures, to the broader public. Following the orientalist impetus of his time, he showcased the continuity from the Hellenistic period and Graeco-Roman late antiquity, with Arabic texts referring to Archimedes (d. 212 BCE), Hero and other well-known authors who appeared in Arabic Islamic scientific texts (Wiedemann 1910, 1913–1916a, 1913–1916b; Thābit ibn Qurra 1911–1912). The balances, hydraulics, pneumatics and the automata in the *Mechanics* of Hero were impressive reference points. The relevant Greek manuscripts had just been critically collated, translated, commented and published in 1899–1900 by Wilhelm Schmidt (1862–1905), including the Arabic recension of Hero's *Mechanics* (the Greek original is not extant). Wiedemann did not consider the historical and linguistic context in which the texts he treated had been produced. He was not interested in codicological questioning, nor was he ambitious in collating several manuscripts of the same text (in the cases he had access to them). When he considered several manuscripts on the same topic, he did it with the intention of "reconstructing" a plausible description of a technically functioning device – this goal was dominant in all his publications. In the same context, we should also mention Friedrich Hauser's (1883–1958) German translation of the *Kitāb al-ḥiyal* by the Banū Mūsā that was induced by similar intentions (Banū Mūsā 1922). It should be remarked that the huge effort of reconstructing the device's functionality from the point of view of a physicist (Wiedemann) solely on the basis of texts and illustrations was related to the lack of archaeological evidence for the ingenious devices (automata) described in the extant Arabic texts.

Another phase of modern scholar research on the Arabic literature on automata was set in the first half of the 20th century by art historians, who became interested in the illustrations of Arabic manuscripts – typically by the Banū Mūsā and al-Jazarī. Sadly, the illustrations studied and published were on folios previously removed from the corresponding manuscripts by Western or Middle Eastern collectors, traders and thieves. These folios eventually found their way to collections, galleries and museums, mostly in western Europe and the US, while the main textual part was left behind mutilated and was "discovered" later in libraries such as that of the Topkapı Palace or Hagia Sophia in Istanbul. Referring to MS Istanbul, Süleymaniye Library, Aya Sofya 3606, D. Hill indicates:

> It is well known to historians of art, because a number of its miniatures were abstracted and dissipated in the West. Much has been written about these miniatures. Of the 50 main illustrations 23 have been removed from the manuscript, and of these 15 have been traced [for example six in the Museum of Fines Arts in Boston and two in the Fogg Art Museum of Harvard University], leaving 8 unaccounted for. Some of the smaller illustrations are also missing.
>
> *(1981, 89–90)*

Manuscripts of mutilated or complete Arabic treatises were registered, compiled, edited, translated into English and published during the next phase of scholarship which focused on critical editions of works on automata, such as the *Book of Ingenious Devices* (*Kitāb al-ḥiyal*) by the Banū Mūsā (Banū Mūsā 1981; Banū Mūsā 1979[CB]) and the *Compendium on Knowledge/Science and Useful Practice on the Art of Ingenious Devices* (*al-Jāmiʿ bayna l-ʿilm wa-l-ʿamal al-nāfiʿ fī ṣināʿat al-ḥiyal*) by al-Jazarī (Banū Mūsā 1979; al-Jazarī 1974[CB]). These projects were conducted in the 1970s and 1980s by the engineer Donald Hill and scholars of the Institute for the History of Arabic Science at Aleppo University, Syria (founded in 1976), under Aḥmad Yūsuf al-Ḥassan (1925–2012). From this engagement resulted, among other studies, concise overviews of the history of Islamicate technology (Hill 1991; al-Hassan and Hill 1986[CB]). On the basis of their critical editions, questions on possible links between these treatises and earlier ones (including Arabic translations of texts from the Greek compiled in late antiquity and known under the names of Hero or Philo) could be addressed and, in many cases, answered through philological and codicological methods.

A substantial component of Hill's pioneering work was the construction of full-scale models of the automata, especially the huge hydraulic-mechanical clocks described by al-Jazarī. In doing so he followed texts and illustrations from the collated manuscripts. Such projects reveal the necessity of taking decisions regarding questions that the modern constructor cannot answer directly from the text. Comments on these texts and illustrations based on the experience with modern reconstructions are documented in the translation volumes of the Arabic treatises. Constructing physical models and animating three-dimensional simulations of the automata described by the Banū Mūsā and al-Jazarī have increasingly attracted the attention of scholars, academic institutions and museums. These include the Institute of the History of the Arab Islamic Science in Frankfurt-am-Main, Germany, the Istanbul Museum for the History of Science and Technology in Islam, both founded by Fuat Sezgin (1924–2018), as well as the Institute for the History of Arabic Science at Aleppo University, Syria, and the Institut du Monde Arabe in Paris. Contemporary mass media and the activities of institutions such as the Foundation for Science, Technology and Civilisation with its website in the UK feature Islamicate automata in ambitious and controversial projects, popularizing medieval Islamicate science and technology (al-Hassani 2012). In several cases, the latter oversimplify academic narratives or even ignore or distort established scientific and historical evidence (Brentjes *et al.* 2016[CB]).

An overview of existing works will provide answers to questions concerning the relationship between the various treatises, the development of approaches to specific problems, as well as the significance of the automata, the science of weights and balances, and the mechanics in general, in the Islamicate scientific tradition. Since 1990s, more Arabic texts on mechanics than the ones presented earlier have been identified and studied, a nucleus of research being the Max Planck Institute for the History of Science in Berlin (Abattouy 1997, 2000a, 2000b, 2001, 2002). In 2008, Abattouy published a list of 35 Arabic treatises on mechanics with reference to weighing and the theory of balances. Fifteen of these are from the 3rd–6th/9th–12th and 20 from the 7th–13th/13th–19th centuries, mainly from Syria and Egypt (Abattouy 2008, 92–8). As already mentioned, in 2013 and 2015, Abattouy and al-Hassani published a critical edition and an English translation of a collection of Arabic texts on mechanics, mainly from the corpus of al-Isfizārī. The two scholars consider the texts as complying with the epistemological categorizations by al-Fārābī and Abū ʿAbdallāh al-Khwārazmī, which they consider compatible to each other and valid for all extant Arabic texts involving the term *al-ḥiyal* (which they generally translate as "mechanics"), balances (*mīzān*), and devices that move "by themselves" (automata). Another approach followed by Brentjes and Renn focuses on historically contextualizing the transmission paths of specific Arabic texts on mechanics. Both approaches are discussed in the following.

It should be remarked that, with the exception of a coded lock mechanism, which is discussed later, there is no archaeological evidence for the ingenious devices (automata) described in the Arabic texts studied so far, although art historians have tried to indicate similarities between some of the described and depicted devices on one side and objects known in other contextual frames through archaeological or art historical research on the other. Islamicate and Byzantine historical evidence concerning automata in the Abbasid court is discussed in the next section.

In the following, I present research results by organizing the texts in two groups: Texts dealing explicitly with artificially constructed ingenious devices, and those dealing exclusively with theoretical considerations on mechanics.

I.9.3 Technological context: the tradition of automatic artificial devices

The evidence for descriptions of devices that perform actions "on their own" (automata) in Arabic texts is retrieved in this section without limitation to a certain literary *genre* or to a contemporary epistemological category mentioned earlier. Under the term *ḥiyal* – used in a technological sense – we find systematic descriptions of artificial mechanisms that are actuated mechanically, pneumatically or hydraulically or by some combination of these modes. The written sources call the field *art* or *knowledge/science of mechanical devices* (*ṣināʿat* or *ʿilm al-ḥiyal al-handasiyya*), thus indicating its status as an independent field of knowledge and of specific technological expertise.

A first group of texts is translations of treatises from the Hellenistic period or late antiquity from Greek (or Syriac) into Arabic. Genuine Arabic texts on automata include the treatises *Book of Ingenious Devices* by the Banū Mūsā and *Compendium on Knowledge/Science and Useful Practice on the Art of Ingenious Devices* (*al-Jāmiʿ bayna l-ʿilm wa-l-ʿamal al-nāfiʿ fī ṣināʿat al-ḥiyal*), also known under the title *Book of Knowledge of Ingenious Mechanical Devices* (*Kitāb fī maʿrifat al-ḥiyal al-handasiyya*) completed in 602/1206 by al-Jazarī.

The *Book of Ingenious Devices* by the Banū Mūsā contains in the critical edition of the Arabic text a sequence of 100 devices which are treated in several extant manuscripts with more or less established authenticity (Banū Mūsā 1981). From a modern point of view and terminology, these devices are mainly based on pneumatics and hydraulics, with special emphasis on automatic control. Most of the devices are various kinds of trick vessels (beakers, pitchers, jars), such as a pitcher that "pours out ritual washings only for believers and does not pour out ritual washings for heretics (*zindīq*) and the like" (Banū Mūsā 1979, 64). The rest are alternating fountains, self-filling and self-trimming lamps and a clamshell grab. Although the authors do not refer to predecessors, the marvelous devices and their descriptions could be compared to devices described by Hero (1st century) or Philo. However, we do not know whether Hero's *Pneumatica* or Philo's *Pneumatica* were known in Arabic translations to the Banū Mūsā. It is possible that more texts and illustrations in this field were circulating in that period, although no firm evidence for such a circulation is extant today. In contrast to Hero and Philo, the descriptions of the Banū Mūsā do not reveal any explicit theoretical ambitions, for instance by making connections with theoretical mechanics. Instead, they claim they have made highly sophisticated, complex combinations of elementary parts function by insisting on the accurate construction of novel technological devices like conical control valves or the use of specific connecting details. This is perhaps the reason why Ibn Khaldūn (732–808/1332–1406) remarked that the technical proofs of the Banū Mūsā are difficult to follow (Ibn Khaldūn 1967[CB], 3: 132).

During the 3rd/9th century we know that the three Banū Mūsā brothers, Muḥammad, Aḥmad and al-Ḥasan, were patrons of intellectual life, supporting Thābit ibn Qurra and Ḥunayn ibn Isḥāq (d. 259/873), and collected manuscripts. They are also known for their prevalent interest in mathematical, astronomical, mechanical and technological subjects and for their own translations from Greek and Syriac into Arabic. From the 20 works known to us by name, the *Book of Ingenious Devices* and a work on a musical automaton, are the only extant ones with technological content (Banū Mūsā 1979). Their *Book of Ingenious Devices* should be seen in the context of the intellectual debates on the relations between rationalism (inspired by the "foreign sciences" philosophy, logic, mathematics, physics and medicine), theology and political power in the Abbasid court during the turbulent 3rd/9th century. One could speculate on the affinity to the act of demonstrating the functionality of the automata in a *rationalist* context to certain Islamic movements (such as the Muʿtazila) with strong political implications in that period – but such speculations on the role, the inspirations and the political-theological aspirations of the Banū Mūsā brothers go beyond the scope of the present study. In the same speculative horizon belong considerations on the character of the automata descriptions: These could be pure thought experiments (e.g., in logical/rhetoric debates in the court or other circles), or they could imply a form of performative action – not necessarily engaging the whole collection of really constructed automatic devices. In fact, there is no supplementary evidence supporting the construction of automata described by the Banū Mūsā.

Al-Jazarī's material is organized in six categories (*naw'*). The devices treated comprise water-clocks and candle-clocks (Category I), vessels and pitchers suitable for drinking sessions (Category II), pitchers, basins and other devices for handwashing – even one supposed to control automatically the amount of blood drained during phlebotomy (Category III), fountains and musical automata (Category IV), machines for raising water (Category V) and finally miscellaneous constructions including a coded lock for a chest (Category VI). It constitutes an Arabic technological compendium. Al-Jazarī criticizes the treatise of the Banū Mūsā as inaccurate in some descriptions (al-Jazarī 1974, 157). While al-Jazarī's descriptions focus on construction and material details to ensure good functioning (an indication that he most probably *did* construct the devices), the descriptions of the Banū Mūsā appear to the reader rather as ingenious thought experiments in dealing with complexity in technological systems whose functioning depends on the perfect construction of a crucial component. In a different context, the text of Hero on pneumatic automata refers explicitly to physical theories – for example, when the author challenges Aristotle's claim that there is no void in nature by experimenting in thought with "ingenious" devices (pneumatic-hydraulic automata) that function precisely by means of an artificially created void (Canavas 2009b).

The most prominent devices in al-Jazarī's treatise are the castle-like clocks (Figure I.9.1) based on hydraulics, pneumatics and general mechanics. Further works dealing with hydraulic

Figure I.9.1 Castle Clock, copy of Ibn al-Razzāz al-Jazarī's *Compendium on Knowledge and Useful Practice on the Art of Ingenious Devices*, dated 755/1354. Museum for Fine Arts, Boston, accession number 14.533.

Source: © Museum for Fine Arts, Boston

and pneumatic clocks (time automata) include those of Ibn Khalaf al-Murādī, who lived in al-Andalus in the 5th/11th century (al-Murādī 2008), as well as the treatise *On the Construction of Clocks and Their Use* (*Kitāb ʿan ʿilm al-sāʿāt wa-l-ʿamal bihā*), dated 600/1203, by Riḍwān ibn al-Sāʿātī (a name that implies a family tradition of clockmakers; d. after 600/1203). The field of clocks in the Islamicate world is very large and lies beyond the scope of the present chapter, but for a concise overview, see Hill (1981).

In the Category VI of al-Jazarī's *Compendium*, Chapter 3 is dedicated to a combination lock for a chest using 12 letters of the Arabic alphabet (al-Jazarī 1974, 199–201, with valuable explanatory notes regarding the figures provided by al-Jazarī, 268–9). The lock on the chest slide is formed by two brass plates held together by four combination locks – each one consisting of a series of four concentric discs with three letters marked on each disc. By means of an elaborate system of cylinders and notches, the arrangement of the four cylindrical locks permits the opening of the lid only by a preset combination of the position of the four discs in each lock.

In the section on Islamic Art in the David Collection in Copenhagen under inv. no. 1/1984 there is a fragment of a 4.4 × 23.5 × 18.5 cm. box with a combination lock from Isfahan (Iran), made from cast and hammered brass, inlaid with silver and copper (Figure I.9.2). This combination lock is the work of the astrolabe-maker Muḥammad ibn Ḥāmid al-Isfahāni dated by its inscription to 597/1200–1, only five years before the appearance of al-Jazarī's *Compendium* in Āmid (today Diyarbakır). The four double dials, each of which can be set in 16 positions, allow for an enormous number of combinations. Entering the combination releases an inner metal plate, which attaches to a handle on the outside of the chest and to the locking mechanism on the inside. The four locks are set in a line, whereas in al-Jazarī's device, they are on the four sides of the chest. Remarkably, the lock on the upper disc dial has its pointer in the form of a bird's head – exactly as in al-Jazarī's description! A similar chest with four locks

Figure I.9.2 Fragment of a box with a combination lock, cast and hammered brass, with silver and copper inlay, perhaps made in Isfahan, Iran, 597/1200. Description by the holding institution: "The four double dials, each of which can be set in 16 positions, allows for 4,294,967,296 combinations. When the right combination is entered, it releases the inner metal plate, which is attached both to an external handle and to the locking mechanism itself." The David Collection, inv. no. 1/1984.

Source: © The David Collection, Copenhagen. Photographer: Pernille Klemp

in a line – each one provided with two-level dials – is found in the Museum of Fine Arts in Boston under the accession number 55.1113. Probably from Khurasan or Isfahan, it is dated 593/1197. It is made from leaded brass over wood and inlaid with silver, gold and copper. Its exterior dimensions are 18 × 22 × 16.7 cm, and it is also attributed to Muḥammad ibn Ḥāmid al-Isfahāni. Both objects correspond to al-Jazarī's description and yield archaeological evidence for a technology established at least between Anatolia and Iran by the end of the 6th/12th century.

A major feature of any kind of automaton is the maker's intention of impressing the observer. In the case of automata in the medieval Islamicate world this feature is exemplified by devices installed in the reception halls and gardens of palaces. The most famous of these is the artificial tree in the Palace of the Tree (*Dār al-Shajara*) in Baghdad, which is referred to in several historical sources, for example, al-Khaṭīb al-Baghdādī (d. 463/1071), in his description of the visit of a Byzantine delegation to Caliph al-Muqtadir (r. 295–320/908–932) in 305 /917 (Lassner 1970, 86–91; Le Strange 1897). According to al-Khaṭīb's report the tree was made of gold and silver and had singing golden and silver birds perched on it. Presumably, there were earlier versions of such artificial trees. The oldest known reference mentions one in the *burj* palace in Samarra, which was the Abbasid capital from 221–279/836–89 (Northedge 2001, 64).

These types of mechanisms are also found in contemporary reports about Byzantine ceremonies, such as in the reception hall of Magnaura, built after 360, which housed the senate in Constantinople. Early Arabic reports of automata in the Byzantine court, as well as Byzantine references that suggest Arabic influences on the installation of such devices in imperial palaces in Constantinople, imply that the rival powers Byzantium and the Abbasid caliphate competed during that period (3rd–4th/9th–10th centuries) in installing automata in areas used for the reception of foreign ambassadors (Canavas 2003). Although the original makers of the automata have not been established, it is certain that both Arabs and Byzantines were acquainted with representative texts on automata from late antiquity. Traces of court automata and the sense of wonder they excited can be found in the treatises on marvels (*ʿajāʾib*) and precious gifts. The most detailed description of the tree in *Dār al-Shajara*, for instance, is given in the *Book of Gifts and Rarities* (*Kitāb al-hadāyā wa-l-tuḥaf*, 5th/11th century). The author of the treatise is unknown according to Ghāda al-Ḥijjāwī al-Qaddūmī, editor of the critical edition (Anonymous 1996, 148–55). But following a different transmission line, some scholars, such as the editor of the first edition, Muḥammad Ḥamīdullāh, have attributed it under a slightly different title to a certain al Ibn al-Zubayr (al-Zubayr 1959).

The throne automata as well as the richly illustrated manuscripts containing the treatises of the Banū Mūsā and al-Jazarī have been the focus of a revival of the interest by art historians (Haase 2010; Müller-Wiener 2007). An indicator of a similar interest and awareness – beyond the conventional fields of history and archaeology – are public events such as the exhibition "Allah's Automata: Artifacts of the Arab-Islamic Renaissance (800–1200)" in ZKM, the Center for Art and Media Technology in Karlsruhe, Germany, from October 31, 2015, to September 4, 2016, curated by Ayhan Ayteş and George Saliba (catalogue edited by Zielinski and Weibel 2015). The exhibition was a hybrid between creative artistic reflection and path network analysis in art history, with annotated references to Arabic primary sources. Above all, it was seeking to improve public awareness of Islamicate cultures ("machines constructed . . . in praise of God the Almighty"). But it was also an example of how material from the history of medieval Islamicate science and technology (automata in form of clocks or automatic music instruments or texts on theoretical and practical mechanics) can be appropriated by media technology and exhibition practitioners for the sake of a modernized, recontextualized and "spectacular" projection upon "our present lives" with novel inspiring impact in arts and perception patterns. The focus of the exhibition on "the rich and fascinating world of the automata that were developed and built during the golden age of the Arabic- Islamic cultures" creates its own interpretative canvas. On this canvas, the exhibits – text facsimile, illustrations and device models – constitute a novel homogeneous historical ensemble, in which the existence of the "built" artifacts is taken for granted – "robotics" *avant la lettre*.

The actual use of mechanisms described in the *ḥiyal* literature, even of those that might have existed in reality like the devices described by al-Jazarī, is controversial. Several scholars, especially in publications and activities with the goal of popularizing medieval Islamicate science and technology, raise the claim of – possible – practical applications. Despite the surface plausibility of such claims, they are methodologically arbitrary since explicit textual evidence for the actual existence of large scale automata is rare, or – as in the case of court automata – such evidence is likely to be tainted by ideological bias and politically motivated exaggerations. A link between *ḥiyal* descriptions and everyday applications can be traced in evidence provided in descriptions of gardens (Tabaa 1992) or in geographic or cosmographic treatises, such as *The Most Remarkable Things in the Time as Marvels of the Earth and the Sea* (*Nukhbat al-dahr fī ʿajāʾib al-barr wa-l-bahr*) by Shams al-Dīn al-Dimashqī (d. 727/1327) (Mehren 1866, 188, 1874, 254–5), a compilation of geographical material (in the widest sense; Chapter 1.13) with narratives on marvels (*ʿajāʾib*) resembling the cosmography of al-Qazwīnī (d. 682/1283). In a section on Azerbaijan al-Dimashqī refers to a watermill in Merend (today in Iran) with a mechanism for raising the water needed for moving the waterwheel that functions like a *perpetuum mobile*; the flowing water is raised back to the height needed for driving the wheel by a device "powered from the water itself" in its downward flow – a marvel made by human hand! Difficulties in following the textual traces of such devices might arise from the author's inability to understand the mechanism or from his intention to impress the readers by obscuring the physical principles that govern the devices he describes – or from both reasons (Chapter IV.6). Such descriptions are often accompanied by illustrations; however, in most cases, they are no more explanatory than the text itself.

The fascination with automata is also underlined by evidence on importing such devices, conserving such gifts, or contracting craftsmen from abroad to construct such a device at the court of a sovereign. Such a fascination dominates the report of the Franciscan monk William of Rubruck (d. *c.* 1293) on his travel between 1253 and 1255 to the court of the sovereign of the Mongols in Karakorum. In this report, he describes an artificial tree that pours wine and other drinks automatically (Rubruck 1925, 111–12). Allegedly the constructor was a craftsman named William from Paris. A similar description, closer to the devices of Hero, the Banū Mūsā and al-Jazarī, is provided later by the Spanish ambassador Ruy Gonzalez de Clavijo (d. 1412), who visited Timur's court between 1403 and 1406 (Clavijo 1928, 270). To what extent these descriptions correspond to real objects in the Mongol or Timurid court, or to what extent they reproduce only Byzantine or Arabic narratives with reference to devices constructed by European craftsmen, is a question that cannot be answered at present.

I.9.4 Contextualizing automata in theoretical frames: balances, weights and mechanics

As mentioned earlier regarding Hero's *Pneumatics*, the epistemological context of the automata in the Greek tradition is related to questions concerned with the void, hydrostatics and pneumatics, as well as theoretical mechanics. The latter is a major issue in the Arabic tradition, attaching the automata to theoretical discourses about balances and weights. Texts include the corpus of mechanics of al-Isfizārī, a native of Isfizār in Khurasan, and of al-Khāzinī ([d. after 525/1130–1]; both dealing with "the sciences of weights and ingenious devices" (*ʿilmān al-athqāl wa-l-ḥiyal*). However, the use of the term *ḥiyal* refers more to mathematical concepts, such as the concept of balances with equal or unequal arms (*mīzān* or *qarasṭūn*), than to the construction of ingenious or marvelous devices (Figure I.9.3). The theory of the steelyard, a balance with unequal arms (*al-qarasṭūn*), for example, treats questions on the equilibrium of a beam suspended at a point that is not its center when various weights are suspended from different points on the beam (Abattouy 1997, 51, 2008, 86). These issues – especially in the *Book of the Balance of Wisdom* (*Kitāb mīzān al-ḥikma*) by al-Khāzinī (al-Khāzinī 1359/1940, 2008) – have links to Greek treatises on geometry and theoretical mechanics ascribed to Aristotle, Euclid, Archimedes, Hero and Pappus. However, they do not connect to the purely descriptive narratives regarding form, construction and performance of machines provided by the Banū Mūsā and al-Jazarī.

The writings on mechanics by al-Isfizārī, mentioned earlier, comprise seven texts: (1) an incomplete version of *Guiding People of Knowledge to the Art of the Steelyard* (*Irshād dhawī al-ʿirfān ilā ṣināʿat al-qaffān*), a treatise dealing with the theory of the balance with unequal arms (*al-qarasṭūn*); (2) a summarized version of *Guiding People of Knowledge* as it has been published by al-Khāzinī in his *Book of the Balance of Wisdom* including a part that preserves a (presumably) missing section of al-Isfizārī's treatise; (3) reproductions of the figures of machines in the *Book of Ingenious Devices* by the Banū Mūsā (*Ḥikāyāt ṣuwar Kitāb al-ḥiyal li-Banī Mūsā*); (4) reproductions of 52 machines out of the 78 pneumatic or hydraulic devices from Philo's *Pneumatics* (*Ḥikāyāt Kitāb Fīlūn al-mījānīqī fī l-ḥiyal*); (5) a summary of Hero's *Mechanics* titled *On Raising Heavy Objects by a Small Force* (*Maʿānī Kitāb Īrun al-mījānīqī fī rafʿ al-ashyāʾ al-thaqīla bi-l-quwwa l-yasīra*), a treatise on theoretical (mathematical) mechanics, extant only in the Arabic translation by Qusṭā ibn Lūqā (d. *c.* 299/912) done actually between 247–251/862–866; (6) *The Book of Apollonius on the Pulley* (*Kitāb Abulūniyūs fī l-bakara*), a work on the pulley by Apollonius of Perga (*c.* 262–*c.* 190 BCE); (7) a three-page fragment under the title *The Figure of the Clock Box* (*Ṣūrat ṣunduq al-sāʿā*),

Figure I.9.3 Balance, 2nd–3rd/8th–9th centuries; Arabic numbers, indicated weight: raṭl; religious inscription in Kufic: "In the name of the merciful and compassionate God / there is no god but God, He alone, He has no equal / Muhammad is the messenger of God." The David Collection, inv. no. 12/1994.

Source: © The David Collection, Copenhagen. Photographer Pernille Klemp

which describes a hydraulic organ known through the longer text of Mūrisṭus, a Greek author of works on organ construction that have only been preserved in Arabic and are mentioned under his name in Ibn al-Nadīm's *Catalog* (Figure I.9.4).

The method of al-Isfizārī consists in epitomizing theoretical treatises on mathematical mechanics designated *ḥiyal*, such as (3) and (4), next to compendia of ingenious mechanical devices, also known – although in a different text tradition – under the term *ḥiyal*. He also redraws the figures of the devices and corrects them if he considers them corrupted through the process of successive copying, as in the case of the treatise of the Banū Mūsā (3). It should be remarked that the collection (3) through (6) exists in three manuscript copies *separate* from the manuscripts reproducing (1) and (2). This collection has no title; Abattouy and al-Hassani refer to it as *Majmūʿ fī l-ḥiyal* (Codex of *ḥiyal*; al-Isfizārī 2015, 156). They introduce the collection as follows:

> We have collected in this book what has reached us from the books of the Ancients written about different machines, and those who came after them until our era, such as the book of Philon the inventor of machines, and the book of Heron the mechanician [*sic*] on the kinds of machines by which heavy objects are lifted with minimal force, and (the work of) Apollonius on the kinds of pulleys which are useful in moving wheels and mills and the machines of wringers.
>
> *(al-Isfizārī 2015, 158)*

The fragment (7) on the clock box – obviously an important copy for the transmission of knowledge on hydraulic organs – is mentioned neither in this introduction nor in the table of contents in the incipit of the collection, where the texts listed are attributed to al-Isfizārī. The fragment could have been added by a copyist or a compiler during the process of copying "similar" texts together. By editing the *practical* texts (3) through (6) and (7) together with the *theoretical* ones (1) and (2), Abattouy and al-Hassani suggest a homogeneity of theoretical mechanics, practical machines and ingenious (tricky) devices in the Arabic tradition over 500 years, but this homogeneity is not sufficiently supported by the provenance of the manuscripts and their hardly

Figure I.9.4 Pulley, 6th–7th/12th–13th centuries; Iran. bronze, inlaid with silver. The holding institution assumes that such a costly pulley might have been part of a clock. The David Collection, inv. no. 42/1981.

Source: © The David Collection, Copenhagen. Photographer: Pernille Klemp

traceable transmission paths. The annotating comments and the redrawing procedure of the figures by al-Isfizārī, author of the treatise on the balances (1), document forms of transmission of knowledge in Iran at the beginning of the 6th/12th century. When compared with the (textual) comments of al-Jazarī on the tradition of machines and on his predecessors some decades later, some differences in their practices come to the fore.

Al-Jazarī mentions that he had been told about the peculiarities of the constructions of other craftsmen. At the beginning of the description of a candle-clock he remarks:

> I have never come across the work by anyone on candle-clocks, . . . , I heard tell, however, of a candle-holder.
>
> *(al-Jazarī 1974, 63)*

He also reports about his various ways of seeking for written or material information. For example, concerning a perpetual flute, he wrote:

> I came across a well-known paper by Apollonius. . . . I also examined another old instrument about which I found no written report. . . . I also examined a paper written in Baghdad in year 517 A.H. by the eminent inventor Hibat Allāh ibn al-Ḥusayn al-Asṭrulābī.
>
> *(al-Jazarī 1974, 170)*

On several occasions he mentions that he is continuing an already existent craft tradition. For example, at the beginning of the chapter on the letter-coded lock for a chest he tells us:

> The earlier workers in this craft made locks for locking and opening by means of the letters.
>
> *(al-Jazarī 1974, 199)*

Al-Jazarī explicitly links his work with material, written, and oral traditions from the perspective of a practitioner. The introduction of al-Isfizārī's compilation prioritizes the textual and pictorial tradition, however, more information on the context of his activities is available from other sources. In one of the few sources on al-Isfizārī's life, Ẓahīr al-Dīn al-Bayhaqī's (d. 565/1169–70) mentions that Isfizari constructed an accurate balance that could be used to uncover forgery (al-Bayhaqī 1988, 125; Abattouy and al-Hassani 2015, 147). This remark brings the works on theoretical mechanics closer to practical issues in some Islamicate societies. It also offers another plausible link to important institutions of medieval Islamicate societies, the office of *ḥisba* that regulated weighing in the markets, including both the construction and the use of weighing instruments, and the office of the *muḥtasib* (controller) of the marketplace, an office that also included checking weights and balances (Abattouy 2002, 21–6). Both positions had presumably existed since the 2nd/8th century.

Brentjes and Renn (2016) have approached the entanglement of theoretical and practical knowledge on mechanics comprehensively on a historical level – for example, the theory of the balance and weighing or raising machines – by considering specific historical situations in which the balance was used in Islamicate societies. They have examined the written transmission of knowledge on a theoretical level in the case of Thābit ibn Qurra and his *Book on the Steelyard* (*Kitāb al-qarasṭūn*). The historical evidence provided by Brentjes and Renn enables the reconstruction of a network of texts and scholars in conjunction with the steelyard in an area extending from Iraq to the northeast of Iran (today southern Turkmenistan) for a period of nearly three centuries (mid-3rd–mid-6th/mid-9th–mid-12th centuries). In this network, Thābit ibn Qurra is portrayed as a collector of books under courtly patronage, as an author grappling again and again with the issue of the steelyard, as a compiler or editor of texts, as a student and as a teacher.

I.9.5 Teaching and other modes of transmitting knowledge on automata and balances

The possible function of the treatises presented above as teaching manuals can be approached in a first attempt by seeking relevant remarks by the authors themselves. In a second step, biographical references in various literary sources can be collected and the structure and the style of the treatises can be analyzed. In the case of the Banū Mūsā, we do not have explicit references, although we have historical evidence of their involvement in patronage (as patrons of Thābit ibn Qurra, perhaps involving teaching mathematics) and collecting earlier scientific manuscripts. The remark by Ibn Khaldūn characterizing their treatise as difficult to follow can be considered a negative indication of its value in teaching.

In the case of al-Jazarī's explanations, structure and style possess a didactic character appropriate for teaching. Although he occasionally comments on the ways *he* got access to existing knowledge (as discussed in the previous section of this chapter), he does not provide us with any hint of transmission by teaching. Besides, no information regarding his life is available beyond his own few remarks (Canavas 2017; al-Jazarī 1974).

We have biographical evidence about Thābit ibn Qurra provided by al-Muḥassin ibn Thābit (d. 400/1010) that the purpose of his various synopses and commentaries on mechanical treatises of other authors was to facilitate understanding. A later copyist (early 7th/13th century) judged Thābit's *Book on Compound Ratios* to be organized like a "textbook" (Brentjes 2018, 51–2). Moreover, in three copies of the *Book on the Steelyard*, it is mentioned that the text had been "dictated" by Thābit, or that it was the result of "listening" to him. Both terms are associated with situations of teaching and learning in a medieval Islamicate context under the guidance of a teacher (Brentjes 2018, 53). In later times, these activities became constitutive for the certificates (*ijāza*) authorizing students to transmit the text dictated by their master. These pieces of evidence, together with a morphological analysis of Thābit's compilations, support the claim that Thābit's texts had the character of teaching material in the service of the transmission of knowledge on the steelyard (Brentjes and Renn 2016, 90).

Regarding al-Isfizārī we have a remark of al-Bayhaqī who comments on al-Isfizārī's teaching activities. Al-Bayhaqī describes him as being tender to pupils, compared to ʿUmar Khayyām ([439–*c.* 517/1048–c. 1123]; Abattouy 2001,5; al-Isfizārī 2015, 147; al-Bayhaqī 1988, 125).

Teaching mechanics, at least in the form of the study of the balance and the steelyard, was occasionally chosen as a field of *madrasa* teaching in Mamluk Cairo during the 9th/15th century. These activities were apparently related to the increasing practical interest in the balance and counterweights; however, they were made institutionally possible only through the growth of the *madrasa* to an accepted teaching institution in Islamicate Egypt (Brentjes 2018, 90).

Having stressed the importance of Arabic compilations of treatises on mechanics composed by Iranian authors we should also indicate that similar activities are documented in Persian presumably from the 4th/10th century onward, although not to the same extent. These activities include translations into Persian from Greek, such as the book *On Raising Heavy Objects by a Small Force* belonging to Hero's *Mechanics*, and compilations of these works together with Arabic ones, such as the *Book of Ingenious Devices* by the Banū Mūsā (Ferriello 1998, 2005). These works are also documented in Persian encyclopedias of the sciences (Brentjes 2018, 212–15; Vesel 1986), and can be regarded as an expression of the strengthening of an Iranian self-consciousness, on both a political and a cultural level, from the 4th/10th century onward.

I.9.6 Concluding remarks

Reconsidering the question posed in the introduction on the relation between theoretical mechanics and practical knowledge in the field of automata and balances in the medieval Islamicate world, the various authors and transmission lines presented in this chapter yield a rather heterogeneous network of patterns. In this network, the compilations of al-Isfizārī in the early 6th/12th century constitute a crucial intersection of several traditions. At the first glance, al-Isfizārī expresses a combined formal interest in both traditions, mathematical mechanics and practical machines. However, he does not present any practical machines of his own – he just summarizes and comments on *al-ḥiyal* of his "predecessors", without linking his theoretical reflections with any practical issues embedded in the ingenious devices of the Banū Mūsā and al-Jazarī. Besides, he lacks the epistemological-cosmological interest of Hero for questions of physics. We still do not know whether al-Isfizārī had any ambition to synthesize earlier work or to contribute to the distinct tradition of mechanics. Perhaps more compendia with a similar or different compilation of heterogeneous Arabic texts on mechanics will be identified in the future. This would offer more textual and iconographic support for a comparative study of the history of the transmission of important issues of Islamicate mechanics. In this sense, the polysemy of *ʿilm al-ḥiyal*, *ʿilm al-athqāl*, *ʿilm al-mīzān* constitutes a texture with knots in which the paths of several traditions of Greek and Islamicate mechanics come together before departing on different routes.

Note

1 Consolidated bibliography.

Bibliography

Sources

Anonymus. tr. al-Qaddūmī, G. al-Ḥijjāwī. 1996. *Book of Gifts and Rarities – Kitāb al-hadāyā wa-l-tuḥaf.* Cambridge, MA: Harvard University Press.

Banū Mūsā. ed., tr. and ann. Hauser, F. 1922. *Über das kitāb al-ḥijal (Das Werk über die sinnreichen Anordnungen) der Banū Mūsā. Abhandlungen zur Geschichte der Naturwissenschaften und der Medizin* 1. Erlangen: Mencke.

Banū Mūsā. ed. al-Hassan, A. Y. 1981. *Kitāb al-Ḥiyal. "The Book of Ingenious Devices" by the Banū (sons of) Mūsā bin Shākir.* Aleppo: Institute for the History of Arabic Science.

al-Bayhaqī, Ẓahīr al-Dīn. ed. Kurd ʿAlī, M. 1988 [1946]. *Tārīkh ḥukamāʾ al-islām* [History of the Wise Men of Islam]. Damascus: Majmaʿ al-lugha al-ʿarabiyya.

Carra de Vaux, B. 1903. "Le livre des appareils pneumatiques et des machines hydrauliques, par Philon de Byzance, édité d'après les versions arabes d'Oxford et de Constantinople et traduit en français," *Notices et extraits des manuscrits de la Bibliothèque Nationale et autres bibliothèques* 38: 27–235.

Clavijo. tr. Le Strange, G. 1928. *Embassy to Tamerlane 1403–1406.* London: Routledge.

al-Dimashqī, Shams al-Dīn. ed. Mehren, A. F., ed. 1866. *Cosmographie de Chems-ed-Din Abou Abdallah Mohammed ed- Dimichqui (texte arabe).* Saint Petersburg: Eggers *et al.*

al-Dimashqī, Shams al-Dīn. tr. Mehren, A. F., ed. 1874. *Manuel de la cosmographie du moyen âge, traduit de l'arabe.* Copenhagen: C. A. Reitzel, Paris: E. Leroux, Leipzig: F. Brockhaus.

Hero of Alexandria. ed. and tr. Carra de Vaux, B., Hill, D. R. and Drachmann, A. G. 1894–1988. *Les Mécaniques ou l'élévateur des corps lourds*. Arab text by Qusṭā Ibn Lūqā. Paris: Les Belles Lettres (Reprint of the *editio princeps* with French translation published in the *Journal Asiatique* 1893).

al-Isfizārī. ed. Abattouy, M. and al-Hassani, S. 2013. *Matn al-Muẓaffar al-Isfizārī fī ʿilmay al-athqāl wa-l-ḥiyal. Taḥqīq naqdī wa-dirāsa tārīkhiyya li-nuṣūṣ jadīda fī taqlīd al-mīkānīkā al-ʿarabiyya*. [The Mechanical Corpus of al-Isfizārī in the Sciences of Weights and Ingenious Devices] London: Al-Furqan Islamic Heritage Foundation.

al-Isfizārī. tr. Abattouy, M. and al-Hassani, S. 2015. *The Corpus of Al-Isfizārī in the Sciences of Weights and Mechanical Devices: New Arabic Texts in Theoretical and Practical Mechanics from the Early XIIth Century. English Translation and Historical Commentaries*. London: Al-Furqan Islamic Heritage Foundation.

al-Khāzinī. 1359/1940. *Kitāb mīzān al-ḥikma* [The Book of the Balance of Wisdom]. Hyderabad: Matbuʿat Daʾirat al-maʿārif al-ʿuthmāniyya.

al-Khāzinī. ed., tr. and ann. Bancel, F. L. 2008. *Kitāb Mīzān Al-Ḥikma de ʿAbd al-Raḥmān al-Khāzinī* [The Book of the Balance of Wisdom by ʿAbd al-Raḥmān al-Khāzinī]. Carthago: Beït al-Ḥikma.

al-Khwārazmī, Abū ʿAbdallāh Muḥammad. ed. van Vloten, G. 1968^{2}. *Mafātiḥ al-ʿulūm* [The Keys of the Sciences]. Leiden: Brill.

al-Murādī, Aḥmad ibn Khalaf. 2008. *Kitāb al-asrār fī natāʾij al-afkār* [The Book of Secrets in the Results of Ideas]. Milan: Ed. Leonardo3.

Rubruck, W. von. ed. and tr. Herbst, H. 1925. *Der Bericht des Franziskaners Wilhelm von Rubruck über seine Reise in das Innere Asiens in den Jahren 1253/1255*. Leipzig: Griffel-Verlag.

Thābit ibn Qurra. tr. Wiedemann, E. 1911–1912. "Die Schrift über den *Qarastun*," *Biblioteca Mathematica* 12: 21–39 [German translation of *Kitāb fī l-qarastūn* by Thābit ibn Qurra].

al-Zubayr. ed. Ḥamīdullāh, M. 1959. *Kitāb al-dhakhāʾir wa-l-tuḥaf* [The Book of Treasures and Gifts]. Kuwait: Daʾirat al-maṭbuʿāt wa-l-nashr.

Research Literature

Abattouy, M. 1997. *The Arabic Tradition of Mechanics: General Survey and First Account on the Arabic Works on the Balance*. Berlin: Max Planck Institute for the History of Science, Preprint 76.

Abattouy, M. 2000a. *Mechané vs. ḥiyal: Essai d'analyse sémantique et conceptuelle*. Berlin: Max Planck Institute for the History of Science, Preprint 152.

Abattouy, M. 2000b. *Nutaf min al-ḥiyal: An Arabic Partial Version of Pseudo-Aristotle's Mechanica Problemata*. Berlin: Max Planck Institute for the History of Science, Preprint 153.

Abattouy, M. 2001. "Greek Mechanics in Arabic Context: Thābit Ibn Qurra, al-Isfizārī and the Arabic Traditions of Aristotelian and Euclidean Mechanics," *Science in Context* 14: 179–247.

Abattouy, M. 2002. *The Arabic Tradition of the Science of Weights and Balances*. Berlin: Max Planck Institute for the History of Science, Preprint 227.

Abattouy, M. 2008. "The Arabic Science of Weights (*ʿIlm al-Athkāl*): Textual Tradition and Significance in the History of Mechanics," in Calvo, E. *et al.*, eds. *A Shared Legacy – Islamic Science East and West*. Barcelona: Universitat de Barcelona, 83–116.

Abattouy, M. 2016. "The Corpus of Mechanics of al-Isfizārī: Its Structure and Signification in the Context of Arabic Mechanics," *Micrologus* 24: 121–69.

al-Hassani, S. T. S. 2012^{3}. *1001 Inventions: The Enduring Legacy of Muslim Civilization: Official Companion to the 1001 Inventions Exhibition*. Washington, DC: National Geographic.

Brentjes, S. and Renn, J. 2016. "Contexts and Content of Thābit ibn Qurra's (died 288/901) Construction of Knowledge on the Balance," in Brentjes, S. and Renn, J., eds. *Globalization of Knowledge in the Post-Antique Mediterranean, 700–1500*. London and New York: Routledge, 67–99.

Canavas, C. 2003. "Automaten in Byzanz. Der Thron von Magnaura," in Grubmüller, K. and Stock, M., eds. *Automaten in Kunst und Literatur des Mittelalters und der Frühen Neuzeit*. Wiesbaden: Harrassowitz, 49–72.

Canavas, C. 2009a. "Automata," in *EI-3*, 2009–4: 74–80.

Canavas, C. 2009b. "From Philon of Byzantium to al-Ǧazarī: Shifting of Perspective in Dealing with Complexity," *Proceedings of the 9th Symposium on the History of Arabic Science, 28–30 October 2008, Damascus*. Aleppo: Aleppo University Publications, Institute for the History of Arabic Science, 37–47.

Canavas, C. 2017. "Al-Jazarī's Compendium of Ingenious Devices: A Model of Representing and Communicating Technical Knowledge in a Medieval Islamicate Context," in Hilaire-Pérez, L. *et al.*, eds. *Le livre technique avant le XXe siècle à l'échelle du monde*. Paris: CNRS, 71–82.

Ferriello, G. 1998. *Il sapere tecnico-scientifico fra Iran e Occidente, una ricercar nelle fonti*. Napoli: Tesi di Dottorato in Studi Iranici, Istituto Universitario Orientale.
Ferriello, G. 2005. "'The Lifter of Heavy Bodies' of Heron of Alexandria in the Iranian World," *Nuncius* 2: 327–45.
Haase, C.-P. 2010. "Modest Variations – Theoretical Tradition and Practical Innovation in the Mechanical Arts from Antiquity to the Arab Middle Ages," in Zielinski, S. and Fürlus, E., eds. *Variantology* 4. Köln: W. König, 195–213.
Hill, D. R. 1981. *Arabic Water-Clocks*. Aleppo: Institute for the History of Arabic Sciences.
Hill, D. R. 1991. "Arabic Mechanical Engineering. Survey of the Historical Sources," *Arabic Sciences and Philosophy* 1: 167–86.
Hill, D. R. 1993. "Mūsā, Banū," in *EI-2*, 7: 640a–41a.
Lassner, J. 1970. *The Topography of Baghdad in the Early Middle Ages*. Detroit: Wayne State University Press.
Le Strange, G. 1897. "A Greek Embassy to Baghdad in 917 A.D. Translated from the Arabic MS of al-Khaṭīb in the British Museum Library," *Journal of the Royal Asiatic Society* 29: 35–45.
Müller-Wiener, M. 2007. "Vom irdischen Paradies zum höfischen Theater: Islamische Automaten und mechanische Konstruktionen des 9. bis 13. Jahrhunderts," *Eothen, Jahresheft der Gesellschaft der Freunde Islamischer Kunst und Kultur* 4: 143–62.
Northedge, A. 2001. "The Palaces of the Abbasids at Samarra," in Robinson, C. F., ed. *A Medieval City Reconsidered. An Interdisciplinary Approach to Samarra*. Oxford: Oxford University Press.
Schacht, J. 1971. "ḥiyal," in *EI-2*, 3: 511a–13a.
Tabaa, Y. 1992. "The Medieval Islamic Garden. Typology and Hydraulics," in Hunt, J. D., ed. *Garden History. Issues, Approaches, Methods*. Washington, DC: Dumbarton Oaks Research Library and Collection, 303–29.
Vesel, Ž. 1986. *Les encyclopédies persanes. Essaie de typologie et de classification de sciences*. Tehran and Paris: Éditions Recherche sur les Civilisations.
Wiedemann, E. 1910. "Über die Kenntnisse der Muslime auf dem Gebiet der Mechanik und Hydrostatik," *Archiv für Geschichte der Naturwissenschaften* 2: 394–8.
Wiedemann, E. 1913–1916a. "Al-Karastūn," in *EI-1*, 4: 757–60.
Wiedemann, E. 1913–1916b. "Al-Mīzān," in *EI-1*, 5: 530–9.
Zielinski, S. and Weibel, P. eds. 2015. *Allah's Automata: Artifacts of the Arab-Islamic Renaissance (800–1200)*. Ostfildern: Hatje Cantz.

I.10
MEDICINE

Peter E. Pormann

I.10.1 Introduction

Medicine evolved into a highly complex and variegated discipline from the 1st/7th to the 15th/21st century in the various lands of Islam (Pormann and Savage-Smith 2007[CB];[1] Pormann 2018). Medicine transcended the confines of country and creed, as physicians from diverse religious, linguistic and ethnic backgrounds shared in its scientific discourse (Pormann 2015). This medical tradition also had a profound impact on surrounding cultures, notably European university medicine as it developed from the 12th century onward. It survives today, in modified form, in many Muslim countries and among Muslim communities across the world. In this chapter, I highlight some aspects of this multifaceted process.

I.10.2 Late antique predecessors

Among the desert-dwelling population of the Arabian Peninsula, where Islam emerged, various medical techniques appear to have been known (Pormann 2007). Conditions such as coughing (*suʿāl*), ophthalmia (*ramad*) and various injuries (often caused by tribal warfare) all figure in the extant pre-Islamic and early Islamic poems, and the cures were often simple: camel urine and honey, for instance, had some prominence. In the two centuries before the emergence of Islam, the Arabs also came into contact with the two great empires, the Sasanian and the Byzantine, as well as the Syriac-speaking Christians who often had to flee from religious persecution at the hands of their coreligionists who declared them to be heretics. All these communities possessed quite a sophisticated medicine, with that of the Greeks clearly standing out among the others.

Syriac-speaking Christians and the medical schools of Alexandria played a crucial role in the translation of medical texts from Greek into Arabic (Chapter I.4): over the course of the 3rd/9th century, most available Greek medical texts were translated into Arabic, often via Syriac intermediary translations. They formed the basis on which the Islamicate medical tradition was to develop, although this also blended other elements and influences (be they Indian, Persian, Chinese), which were incorporated into a framework of humoral pathology.

DOI: 10.4324/9781315170718-12

I.10.3 Medical theories

The theory known as 'humoral pathology' dominated medical discourse in the Islamicate and the Christian worlds until the advent of germ theory in the second half of the 14th/19th century. According to this theory, which goes back to the Hippocratic text *On the Nature of Man* (*Peri phusios anthrōpou*; *Fī ṭabīʿat al-insān*), health consists in a balance (*iʿtidāl*) of the four humors, blood (*dam*), phlegm (*balgham*), yellow bile (*mirra ṣafrāʾ*) and black bile (*mirra sawdāʾ*; Pormann 2019). Each of these four humors has two of the four primary qualities, hot (*ḥārr*) or cold (*bārid*), and dry (*yābis*) or moist (*raṭb*). For instance, black bile is cold and dry, whereas blood is hot and moist. When an imbalance in the four humors occurs, disease ensues. The therapy then aims at restoring the balance by removing excessive humors – for instance through venesection (*faṣd*) and cupping (*ḥijāma*) – and regenerating deficient humors – for example, by eating a diet that produces the deficient humors. Because this humoral pathology did not have any religious connotations, it allowed physicians from different backgrounds to partake in one medical discourse, irrespective of country and creed.

Although this theory remained dominant in medicine, Savage-Smith (2013) has persuasively argued that in practice, it played far less of a role. In therapy, physicians aimed at restoring balance, but it was mainly the balance of the four primary qualities (hot and cold, dry and moist), not the four humors. In other words, physicians discussed humoral balance and often justified their treatment as a rebalancing of the humors, but in reality, they were far more likely to regulate temperature and levels of moisture.

Following in the footsteps of their Greek forebears, physicians in the medieval Islamic world took an acute interest in anatomy (*tashrīḥ*). Like the Greek term *anatomē*, the Arabic *tashrīḥ* was ambiguous, denoting both the study of human physiology (what we nowadays call 'anatomy' in English), and dissection, the 'cutting open' of human and animal bodies, either dead (dissection) or alive (vivisection). Anatomy in the modern sense was a greatly esteemed pursuit (Savage-Smith 1995). Not only did physicians repeatedly state that students must study it, but theologians such as al-Ghazālī (d. 505/1111) also prized it highly, since it made man understand God's providence (*ʿināyat Allāh*). In other words, the wonderful structure of the human body shows God's intelligent design. Therefore, 'if you occupy yourself with the science of anatomy, your faith in God increases', as the famous Andalusian physician and philosopher Ibn Rushd ([520–595/1126–1198]; Averroes) reportedly put it (Ibn Abī Uṣaybiʿa 1965[CB], 2: 77, 13–14, 2020[CB], 13.66). Although dissection was not regularly performed, there was no taboo against its practice on human bodies. We even have a number of famous cases in which Muslim physicians challenge Galenic anatomy. In his commentary on Ibn Sīnā's ([d. 428/1037]; Avicenna) *Canon of Medicine* (*al-Qānūn fī l-ṭibb*), for instance, the physician and philosopher-theologian Ibn al-Nafīs (d. 687/1288) discovered the pulmonary transit: the fact that blood does not pass from the right ventricle of the heart to the left via an opening (*manfadh*) in the septum but, rather, passes through the lungs (Fancy 2013[CB]).

I.10.4 Medical genres

Generally speaking, one can distinguish four broad categories of medical writing: (1) introductory works for students, (2) monographs on single topics, (3) medical encyclopedias and handbooks and (4) commentaries. For instance, Ḥunayn ibn Isḥāq (d. 260/873) wrote an *Introduction to Medicine* (*Mudkhal li-ʿilm al-ṭibb*) as well as *Questions on Medicine* (*Masāʾil fī l-ṭibb*), giving similar information in question-and-answer format. He also wrote a specialist work on ophthalmology (*al-ʿAshr maqālāt fi l-ʿayn*; see the following discussion). Al-Rāzī (d. 313/925 or 323/935), on the other hand, wrote a monograph on *Smallpox and Measles* (*Fī l-judarī wa-l-ḥaṣba*), a medical encyclopedia called *Book for al-Manṣūr* (*al-Kitāb al-manṣūrī*), and a commentary on the Hippocratic *Aphorisms* (now only extant in fragments). There are more than a dozen other authors who also wrote commentaries on the *Aphorisms*, which shows the vibrancy of the genre (Pormann and Karimullah 2017). Another text often commented on is the already mentioned *Canon of Medicine*, a medical encyclopedia in five books by Ibn Sīnā; he was arguably the most influential physician after Galen of Pergamum (*c.* 129–216). This book represents a true watershed in the writing of medical encyclopedias, and much medical instruction, whether in the East or the West, is subsequently based on the *Canon* and the many commentaries, super-commentaries and abridgments written on it. Most medical books were written in prose, but we also have a number of works in verse, such as the *Poem of Medicine* (*al-Urjūza fī l-ṭibb*) by the same Ibn Sīnā. This medical literature covered a huge array of topics, beginning with anatomy.

I.10.5 Medical practices

Already in late antiquity, physicians divided medical practice into prophylactics and therapeutics, and Arab physicians paid greater attention to the former than their Greek forebears. Diet or regimen (*tadbīr*) played a crucial role. Food obviously has a direct effect on one's well-being. The various foodstuffs were integrated into the system of humoral pathology and primary qualities. Some were seen to generate good humors such as blood, whereas others gave rise to disease. Exercise also served to preserve health. In this way, physicians manipulated the 'six nonnaturals' to prevent the patient from becoming ill. The 'six nonnaturals' (*al-ashyāʾ ghayr al-ṭabīʿiyya*), as they were known – namely, (1) the surrounding air, (2) food and drink, (3) sleeping and waking, (4) exercise and rest, (5) retention and evacuation, and (6) the mental state – also affected the health of a person. Too much exercise (under 3), for example, could cause excessive heat in the body, which had other physiological consequences; lack of sleep (under 4) could lead to health problems; and so on. Retention and evacuation refer to the bowel movement and urination of the patient but could also take other forms such as sexual intercourse, during which semen is evacuated (in both men and women). Sexual hygiene evolved into a separate subject with monographs by authors such as al-Kindī (d. *c.* 256/870), al-Rāzī and Ibn Sīnā.

The mental state is the last of the six 'nonnaturals', and the link between mental and physical states was a strong one. On one hand, sadness, sorrow, grief, fright, and fear could cause bodily reactions leading to disease. On the other hand, mental states were seen as the result of a person's mixture or temperament (*mizāj*, Greek *krāsis*). Galen had written a treatise with the programmatic title *That the Faculties of the Soul Follow the Mixtures of the Body*, which was translated into Arabic (*Fī anna quwā al-nafs tābiʿa li-mizāj al-badan*). For instance, melancholy (*malinkhūliyā*) is, as its name suggests, a disease caused by black bile (*al-mirra al-sawdāʾ*, Greek *melaina cholē*). Yet, it can be acquired in a variety of ways: not only could the wrong food lead to melancholy but also the wrong lifestyle and even mental activities, such as excessive thinking. Melancholy was only one of many mental disorders for which physicians in the Islamic world developed sophisticated categories and therapies, ranging from medication to social interaction; moreover, music played a particular role in the care of those suffering from mental diseases.

The regulation of the 'six nonnaturals' was important in preventing disease and curing it. Medication, however, occupied an even more prominent place. Here one has to distinguish between simple drugs (*adwiya mufrada*) and compound drugs (*adwiya murakkaba*). Simple drugs are single substances such as mint, honey, arsenic, or opium, which possess certain qualities, both primary (dry, moist; hot, cold) and others (e.g., styptic, purging). Following Galen, these qualities were often rated in degrees from one (lowest) to four (and occasionally higher). Compound drugs consist of more than one ingredient and could at times be very complicated. For instance, some recipes for theriac (*tiryāq*, from Greek *thēriakē*) – a drug originally made to counter the effect of snake bites and later used as a sort of panacea – contained dozens and dozens of different, and at times difficult to procure, ingredients. From a modern point of view, some ingredients seem highly effective (e.g., opium), whereas the usefulness of others is disputed.

Simple drugs are one of the areas where we see clear local differences: the same plants do not grow in Spain, Greece, the Levant and Fertile Crescent, the eastern parts of the Abbasid Empire, India, and China. Al-Bīrūnī (362–d. after 444/973–d. after 1053), for instance, waxed lyrical about Dioscorides, the Greek pharmacological writer of the 1st century ce, saying that "if Dioscorides had lived in our region and had directed his efforts at knowing the things in our hills and valleys, the plants would have all become remedies, and their fruits would have been turned through his expertise into medicaments" (al-Bīrūnī 1991, 12). Indeed, one of great achievements of Islamicate pharmacology is to incorporate drugs from different parts of the known world into the system of humoral pathology (Pormann 2011a). Yet this integration also led to problems. For instance, we have an anonymous 6th/12th-century commentary on Dioscorides written in al-Andalus that goes to great length to identify medicinal substances not found in that part of the world (Dietrich 1989).

Another means of therapy is surgery (*jirāḥa*). Surgery ranged from milder and simpler interventions such as bone-setting (*jabr*) to quite complex operations. For instance, excessive humors could be removed through both venesection (or phlebotomy, *faṣd*) and cupping (*ḥijāma*). In the former technique, one of the patient's veins was incised, and the blood would then run out. At times, blood was let in this way until the patient fainted. Two types of cupping existed: dry cupping and wet cupping. In both cases, cupping glasses were applied to suck disease matter and superfluities out of the body. In the latter case, small incisions on the skin were also made, and some blood would come out of them. Physicians and surgeons also frequently resorted to cauterization (*kayy*): a heated iron (or cautery, *mikwāh*) was put onto the skin so as to burn it; this would staunch bleeding and disinfect to some extent. Sometimes, extremely hazardous surgical procedures are explained in great detail, but it is doubtful that they were ever performed.

I.10.6 Medicine and the stars

Astronomy and astrology had been linked since classical Greek times, and authors in the Islamicate tradition heeded the Hippocratic utterance "that astronomy contributes not very little, but rather very much to medicine" (*Airs, Waters, Places*, book 1, chapter 2; 1839–61, 2: 14). For the environment, one of the six nonnaturals, determines health, and the environment is in its turn regulated by the course of the stars, in particular in terms of the four primary qualities. As Abū Maʿshar (171–272/787–886) put it in his *Great Introduction* (*al-Mudkhal al-kabīr*; Abū Maʿshar 1985[CB], 110), change in the environment "is only brought about by the movement of the stars, because the sun has the power to heat, whereas the moon has the power to moisten". Therefore, we find astrological elements in many medieval medical writings (Akasoy *et al.* 2008). One author, Yuḥannā ibn Ṣalt (*fl.* 256–298/870–910), who was acquainted with Ḥunayn ibn Isḥāq, even wrote a *Compendium on Astrological Medicine* (*Kunnāsh ṭibbī nujūmī*), comprising ten chapters dealing with topics such as zodiacal signs, the special quality of stars, how to pick the right time to take drugs or let blood, or other such activities.

I.10.7 Female practitioners?

The discourse about medicine in the premodern Islamicate world was dominated by men, and yet, we have to ask whether there were any female physicians, and how the conditions specifically affecting women were treated. Much of the standard medical care, the 'bodywork', was probably carried out by women (Pormann 2009, 2014). Whether as mothers, sisters, aunts, grandmothers, wise women, or nurses, women played a significant role in the medical marketplace. Yet, because the medical historiography was largely a male domain, and as the society as a whole was highly patriarchal, women's voices only reach us only faintly across the centuries. Still, we have indirect evidence that women practiced medicine in various guises. Women were not only practitioners but also patients. Even if women might at times feel shame to be treated by male physicians, it appears that in extreme cases, male doctors would even examine female genitalia. Such practices are justified by the Islamic legal principle of necessity (*ḍarūra*): the woman's welfare outweighs other considerations.

Gynecological conditions include not just menstruation but also pregnancy and breastfeeding, which also feature in manuals on pediatrics, which form a separate branch in the medical literature in Arabic (Giladi 2014). This literature is particularly rich and a testament to the care and attention paid to children by physicians in the medieval Islamic world. It provides advice about rearing children and some specific conditions that affected them.

I.10.8 Specializations

Some parts of the body require special attention, such as the eyes. Therefore, ophthalmology developed into another specialist subject. It generated its own genre of monographs by authors such as Ḥunayn ibn Isḥāq, who wrote the famous *Ten Treatises on the Eye* (*al-ʿAshar maqālāt fī l-ʿayn*), ʿAlī ibn ʿĪsā al-Kaḥḥāl (4th/10th century), ʿAmmār ibn ʿAlī al-Mawṣilī (*fl. c.* 390/1000), and Khalīfa ibn Abī l-Maḥāsin al-Ḥalabī (d. after 674/1275). Although physicians drew heavily on the Greek legacy in this area, they also made new discoveries and distinguished previously unknown ailments, as the example of *sabal* (pannus) shows (Savage-Smith 1980). This disease, in which blood vessels from the limbus invade the cornea, does not appear in the classical Greek medical works. Yuḥannā ibn Māsawayh (d. 243/857–8) and his pupil Ḥunayn ibn Isḥāq, however, included it in their ophthalmological works and advise on its treatment.

I.10.9 Hospitals

A prominent topic in recent scholarship is the Islamic hospital: What are the antecedents of the Islamic hospital and in what way was it original (Ragab 2015)? Certainly, Byzantine institutions and notions of Christian charity, as well as late antique Greek medicine played an important role. I have argued elsewhere, however, that five factors came together in Islamic hospitals which render them unique, and which together mark a significant departure from previous institutions (Pormann 2008, 2010). They are, briefly, (1) legal and financial security through the status of pious foundation (*waqf*) in Islamic law, (2) the 'secular' character of the medical therapy, (3) the presence of elite practitioners, (4) medical research, and (5) medical teaching. The combination of these factors certainly constitutes innovation. Moreover, only the institutional setting made it possible for physicians like al-Rāzī to carry out large-scale research or to encounter rare diseases.

Al-Rāzī is arguably the greatest clinician of the medieval period. In his major and highly influential treatise *On Smallpox and Measles*, for instance, he distinguishes between the two conditions and offers tools for differential diagnosis, a topic on which he also wrote a separate work with the title *What Differentiates [between Diseases]* (*Kitāb mā l-fāriq*). *On Smallpox and Measles* continued to be highly influential not only in the East but also in Europe, with Latin, English, and French translations appearing in the 12/18th and 13th/19th centuries (Pormann 2013, 41–9).

I.10.10 Controversies

A hotly debated topic among both physicians and theologians was contagion, and whether or not one should leave a locality infested by plague or other epidemic diseases (Stearns 2011). The Prophet Muḥammad reportedly had denied the existence of contagion in the following *ḥadīth*: 'There is no transmission, no augury, no owl and no (bad omen in the month of Ṣafar; *Lā ʿadwā wa-lā ṭiyarata wa-lā hāmata wa-lā ṣafara*)' [Bukhārī 1862–1908, 4: 55, no. 19)]. In other words, this utterance of the prophet appears to deny the existence of 'transmission' (*ʿadwā*). Yet there is another tradition, linked to the plague of Emmaus (ʿAmwās, located some 20 miles northwest of Jerusalem) that occurred in 17/638. Here, the faithful were enjoined not to enter a region affected by the plague if they were outside it, nor to leave if they were inside it. The presumed reason for this is an acknowledgment that transmission does exist. Medical sources, on the other hand, recognized contagion in certain cases, although, here, too, the theory of miasmas, inherited from Hippocratic works, remained one of the etiological explanations.

One should, however, reject the notion that religion hampered medical progress in the Islamicate world. Rather, there is a large body of religious scholarship that actively encourages the pursuit of medicine. After all, the Prophet reportedly said that "God did not send down any disease without also sending down a cure for it" (*mā anzala llāhu dāʾan illā wa-anzala lahū dawāʾan*). The genre of prophetic medicine (*al-ṭibb al-nabawī*), also known as 'Medicine of the Prophet' (*ṭibb al-nabī*), developed from the 4th/10th century onward (Perho 1995). Legal scholars drew on collections of *ḥadīth* (utterances of the Prophet) and *sunna* (reports about the behavior of the Prophet) to establish a religiously sound medical tradition. This genre gained greater prominence from the 13th century onward.

Religion also played a role in other ways. When faced with illness, many Muslims, Christians, and Jews reacted by praying to God and seeking His succor. But they went further: at times, they would, for instance, write certain verses (*sūras*) on a piece of paper that they would carry as a pendant or drink water from bowls inscribed with Qurʾānic verses. Here the line between licit religious practice and illicit use of magic (*siḥr*) is not always clear (Maddison and Savage-Smith 1997).

The distinction between licit and illicit practices featured prominently in works on medical ethics (Bürgel 2016[CB]). Elite physicians endeavored to distinguish themselves from other practitioners in the medical marketplace, with varying degrees of success. On one hand, they argued for a canon of medical knowledge that all physicians should master in order to have access to the profession. For instance, in a manual on market inspection (*ḥisba*) from the 6th/12th century, its author, the physician al-Shayzarī, demanded that physicians be tested according to the instructions given in Ḥunayn ibn Isḥāq's *On the Examination of the Physician* (*Fī Miḥnat al-ṭabīb*). Other manuals on medical ethics, such as those by Ayyūb al-Ruhāwī ([d. after 217/832]; Job of Edessa) and Ṣāʿid ibn al-Ḥasan (d. 464/1072), or on how to examine physicians, such as that by al-Sulamī (d. 604/1208), also refer to a canon of testable knowledge, largely based on Greek texts in Arabic translation. The famous physician and philosopher ʿAbd al-Laṭīf al-Baghdādī (d. 629/1231) even urged his readers to return to the example of Hippocrates and Galen. In this way, the medical canon of textbooks served as a touchstone. Yet, it is clear, too, from the same manuals on medical ethics and testing physicians that the medical elite rarely succeeded in excluding their competition. Moreover, there are injunctions to treat patients for free and not derive financial gain from exercising the medical profession.

I.10.11 Medical education and belles lettres (*adab*)

One should not think, however, that the literature on medical ethics reflected the situation on the ground. Let us take the example of how physicians are trained. We find instances of reports that students sought instructions from famous physicians: al-Rāzī, for instance, had a large cycle of students. Ṣāʿid ibn al-Ḥasan singled out the hospital as a good place to train because one can observe rare diseases there. We know from al-Rāzī's *Doubts about Galen* (*al-Shukūk ʿalā Jālīnūs*) that hospitals were important to his testing medical ideas; and his *Book of Experience* (*Kitāb al-Tajārib*) case notes collected by his students and published posthumously demonstrate that his pupils played an active role in recording cases. Confessional boundaries do not seem to have played a prominent role in medical apprenticeships, as Jews, Christians and Muslims could study with members of confessions other than their own. Finally, there are some medical autodidacts: physicians who learned by themselves through reading books; a famous example is ʿAlī ibn Riḍwān (388–453/998–1061 or later), who was later embroiled in a bitter dispute with Ibn Buṭlān (d. 458/1066) on a question of natural philosophy. In all this, however, the close link between medicine and philosophy is apparent: students of medicine need to master the basic principles of physics, logic, and biology to progress.

That the Best Physician Is Also a Philosopher (*Hoti ho aristos iatros kai philosophos*; *Fī anna l-ṭabīb al-fāḍil faylusūf*) is the title of an introductory work by Galen, which shows the close interaction between medicine and philosophy (Adamson and Pormann 2017[CB]). Many of the translators of Greek medical texts were often the same who also rendered philosophical works. Moreover, physicians in the Islamic world were often also philosophers, with famous examples including al-Rāzī and Ibn Sīnā. But some works such as the *Paradise of Wisdom* (*Firdaws al-ḥikma*) by ʿAlī ibn Sahl Rabban al-Ṭabarī (d. before 250/864) or the *Benefits of Bodies and Souls* (*Maṣāliḥ al-abdān wa-l-anfus*) by Abū Zayd al-Balkhī (d. 322/934) actually constitute works on both medicine and philosophy, mixing the two disciplines. Many medical ideas also entered philosophical discourse, not least in the interrelationship between mind and body.

Arabic literature (or 'belles lettres' [*adab*]) contains a number of medical anecdotes. The judge al-Tanūkhī (329–384/941–994), for instance, reported some extraordinary cases, such as that of the Siamese twins, joined at the hip, who had to do everything together or that of the girl at death's door because of a tick in her vagina (its removal caused her great shame; Bray 2006, 224–6 source). The relationship between literature and medicine has different aspects. One can ask, for instance, what rhetorical devices are employed in medical discourse or how physicians wrote literature. *The Famous Physician's Dinner-Party* (*Daʿwat al-aṭibbāʾ*) by Ibn Buṭlān, a doctor from Baghdad, is a work of *belles lettres* (*adab*), in which the author makes fun at the expense of his colleagues. Moreover, medical discourse also pervaded literature, both prose and poetry.

I.10.12 Translations and impact in Latin Europe

The Islamicate medical tradition had a tremendous impact on Europe during the Middle Ages and the Renaissance: in Italy, Spain, and Antioch, many Arabic medical texts were translated into Latin (Haase 2016). The two figures who excelled in these endeavors were Constantine the African (d. before 1099), and Gerard of Cremona (d. 1187). They translated not only the great encyclopedias by al-Rāzī,[2] ʿAlī ibn al-ʿAbbās al-Majūsī (d. *c.* 384/994),[3] and Ibn Sīnā,[4] but also many monographs such as that by Isḥāq ibn ʿImrān (d. *c.* 291/904) *On Melancholy*

(*De Melancholia*) or that by the Ibn al-Jazzār (d. 395/1004–5) *On Sexual Intercourse* (*De coitu*). *The Introduction to Medicine* (*Mudkhal li-ʿilm al-ṭibb*) by Ḥunayn ibn Isḥāq became known in Latin as *Isagoge Ioannitii* and was core curriculum in most of the nascent European universities from the 13th century onward. Likewise, during the European Renaissance, Ibn Sīnā's *Canon* was printed and reprinted dozens of times; it was also (together with the Qurʾān) the first book to be printed in Europe in Arabic for the Arabic market (Siraisi 1987). Even the great Renaissance anatomist Andreas Vesalius (d. 1564) wrote a *Paraphrase of al-Rāzī's Ninth 'Book for al-Manṣūr'* (*Paraphrasis in nonum librum Rhazae medici Arabis clarissimi ad regem Almansorem*). There can therefore be no doubt that Arabic medicine in Latin translation had a profound and lasting impact on the history of medicine in the West. Some physicians during the Renaissance, however, resented the prominent position of Arabic medicine, and fought vigorously to erase the Arab and Muslim contribution to medicine (Pormann 2011b). At times, they succeeded in sidelining and removing the Arabic and Islamic heritage from the history books, although this trend is now declining.

The exchange of medical ideas across the Mediterranean through translation continued into the early modern period. Two examples illustrate this. Dāwūd al-Anṭākī (d. 1007/1599), a physician from Syria, wrote the *Memorandum Book for Those Who Have Understanding and Collection of Wondrous Marvels* (*Tadhkirat ulī l-albāb wa-l-jāmiʿ li-l-ʿajab al-ʿujāb*). In it, he drew not only on the earlier Greco-Arabic tradition exemplified by Ibn Sīnā's *Canon*, but also incorporated descriptions of new diseases such as syphilis together with some European recipes for medications.

The court physician Ibn Sallūm (d. 1079/1669) commissioned the translation of a treatise titled *The New Chemical Medicine of Paracelsus* (*Kitāb aṭ-Ṭibb al-jadīd al-kīmiyāʾī taʾlīf Barākalsūs*), in which a Christian colleague, called Nicolas (11th/17th century), translated the work of two German followers of Paracelsus's chemical medicine (Bachour 2018; Chapter VI.6). This 'new chemical medicine' demonstrates that the exchange of ideas between East and West continued in the Ottoman Empire. Even the many encounters with colonial medicine throughout the 13th/19th century are not always ones of Western superiority.

I.10.13 The continued vitality of ancient practices

Islamic medicine is also a continuous tradition. In many Muslim countries, the texts of Ibn Sīnā are eagerly read, and in the souks, one can buy the ingredients necessary to compose the various drugs. On the Indian subcontinent, this medical tradition has developed into what is nowadays called *ṭibb-i yūnānī* (lit. 'Greek Medicine'; Chapters V.7; V.8). Next to Ayurveda, it is the major classical medical tradition and, together with Muslim migrant communities, has now reached most corners of the world. Likewise, the Medicine of the Prophet enjoys great popularity, and many of the works mentioned remain in print in numerous editions. Finally, there is also a large market for what one could call 'fusion medicine', syncretic collections of Greek humoral pathology and modern (Western) medicine that are commercially highly successful. Therefore, in many ways, the medical tradition that developed in the medieval Islamic world continues to thrive and grow in many different ways.

Notes

1 Consolidated bibliography.
2 It is known in Latin as the *Book for al-Manṣūr* (*Liber ad Almansorem*).
3 It is known in Latin as the *Royal Book* (*Liber regius*).
4 It is known in Latin as the *Canon of Medicine* (*Canon medicinae*).

Bibliography

Sources

al-Bīrūnī. ed. Zaryāb, ʿA. 1991. *Kitāb al-ṣaydana fī l-ṭibb* [Book on Pharmacy and Materia Medica]. Tehran: University Press.

Bukhārī, Muḥammad ibn Ismāʿīl. ed. Krehl, L. 1862–1908. *Le recueil des traditions mahométanes*. Leyden: E. J. Brill.

Research literature

Akasoy *et al.*: Akasoy, A. A., Burnett, C. and Yoeli-Tlalim, R., eds. 2008. *Astro-medicine: Medicine and Astrology, East and West*. Micrologus Library 25. Florence: Sismel.

Bachour, N. 2018. "Iatrochemistry and Paracelsism in the Ottoman Empire in the Sixteenth and Seventeenth Centuries," *Intellectual History of the Islamicate World* 6: 82–116.

Bray, J. 2006. "The Physical World and the Writer's Eye: al-Tānūkhī and Medicine," in Bray, J., ed. *Muslim Horizons: Writing and Representation in Medieval Islam*. London: Routledge, 215–50 (Reprint in Pormann 2011b, 1: 343–80).

Dietrich, A. 1989. *Dioscurides Triumphans: Ein anonymer Arabischer Kommentar (Ende 12. Jahrh. n. Chr.) zur Materia medica*. Wiesbaden: Harrassowitz.

Giladi, A. 2014. *Muslim Midwives: The Craft of Birthing in the Premodern Middle East*. Cambridge: Cambridge University Press.

Maddison, F. and Savage-Smith, E. 1997. *Science, Tools & Magic. Part One: Body and Spirit Mapping the Universe* (The Nasser D. Khalili Collection of Islamic Art 17). London and Oxford: The Nour Foundation in association with Azimuth Editions and Oxford University Press.

Perho, I. 1995. *The Prophet's Medicine: A Creation of the Muslim Traditionist Scholars*. Helsinki: Finnish Oriental Society.

Pormann, P. E. 2007. "Islamic Medicine Crosspollinated: A Multilingual and Multiconfessional Maze," in Akasoy, A. A., Montgomery, J. E. and Pormann, P. E., eds. *Islamic Crosspollinations. Interactions in the Medieval Middle East*. Oxford: Oxbow, 76–91.

Pormann, P. E. 2008. "Medical Methodology and Hospital Practice: The Case of Tenth-Century Baghdad," in Adamson, P., ed. *In the Age of al-Farabi: Arabic Philosophy in the 4th/10th Century* (Warburg Institute Colloquia 12). London: Warburg Institute, 95–118.

Pormann, P. E. 2009. "Female Patients and Practitioners in Medieval Islam," *The Lancet* 373: 1598–9.

Pormann, P. E. 2010. "Islamic Hospitals in the Time of al-Muqtadir," in Nawas, J., ed. *Abbasid Studies II: Occasional Papers of the School of ʿAbbasid Studies, Leuven, 28 June – 1 July 2004* (Orientalia Lovaniensia Analecta 177). Leuven and Dudley, MA: Peeters, 337–82.

Pormann, P. E. 2011a. "The Formation of the Arabic Pharmacology: Between Tradition and Innovation," *Annals of Science* 68: 493–515.

Pormann, P. E. 2011b. "The Dispute between the Philarabic and Philhellenic Physicians and the Forgotten Heritage of Arabic Medicine," in Pormann, P. E., ed. *Islamic Medical and Scientific Tradition. Critical Concepts in Islamic Studies*. 4 vols. London: Routledge, 2: 283–316.

Pormann, P. E. 2013. *Mirror of Health: Medical Science during the Golden Age of Islam*. London: Royal College of Physicians.

Pormann, P. E. 2014. "Female Patients, Patrons and Practitioners in the Medieval Islamic World," *Aspetar* 3.3: 656–60. www.aspetar.com/journal/upload/PDF/2014127143654.pdf (Accessed 19 September 2019).

Pormann, P. E. 2015. "Abrahamic Religions and the Classical Tradition," in Silverstein, A. and Stroumsa, G. G., eds. *Oxford Handbook of the Abrahamic Religions*. Oxford: Oxford University Press, 297–314.

Pormann, P. E. 2018. *1001 Cures: Contributions in Medicine and Healthcare from Muslim Civilisation*. Manchester: Foundation for Science, Technology and Civilisation.

Pormann, P. E. 2019. "Medical Conceptions of Health from Antiquity to the Renaissance," in Adamson, P., ed. *Health: A History*. Oxford: Oxford University Press, 43–74.

Pormann, P. E. and Karimullah, K., eds. 2017. The Arabic Commentaries on the Hippocratic 'Aphorisms'. *Oriens* 45.1–2.

Ragab, A. 2015. *The Medieval Islamic Hospital: Medicine, Religion and Charity*. Cambridge: Cambridge University Press.

Savage-Smith, E. 1980. "Ibn al-Nafīs's *Perfected Book on Ophthalmology* and His Treatment of Trachoma and Its Sequelae," *Journal for the History of Arabic Science* 4: 147–204.
Savage-Smith, E. 1995. "Attitudes Toward Dissection in Medieval Islam," *Journal of the History of Medicine* 50: 67–110 [Reprint in Pormann 201b, 1: 68–111].
Savage-Smith, E. 2013. "Were the Four Humours Fundamental to Medieval Islamic Medical Practice?" in Hsu, E. and Horden, P., eds. *The Body in Balance: Humoral Theory in Practice*. Oxford: Berghahn Books, 89–106.
Siraisi, N. 1987. *Avicenna in Renaissance Italy: The Canon and Medical Teaching in Italian Universities after 1500*. Princeton: Princeton University Press.
Stearns, J. 2011. *Infectious Ideas: Contagion in Premodern Islamic and Christian Thought in the Western Mediterranean*. Baltimore, MD: John Hopkins University Press.

I.11
NATURAL PHILOSOPHY, 100–700/700–1300

Andreas Lammer

Philosophy, as it was conceived in the Islamicate world, was "science" in its broadest terms. It was divided into theoretical sciences, practical sciences and applied sciences (Chapter VI.4). While applied sciences included, for example, astrology, agriculture and medicine, practical sciences comprised, for instance, ethics and politics. The theoretical sciences, in turn, split into four areas: logic, mathematics, natural philosophy and metaphysics. Of these, natural philosophy was further divided into seven "particular" investigations. These were concerned with the heavenly bodies, the elements, meteorological phenomena, minerals, plants, animals and the soul, respectively. Additionally, one fundamental or "common" science – called "physics" – covered all concepts immediately relevant not only for all the seven particular disciplines within natural philosophy but also for a number of the applied sciences such as medicine and astrology. It is important to understand that the term "natural philosophy" was not synonymous with "physics" but demarcated a wide scientific area concerned with the exploration of all the aspects and the inner governing structures of the corporeal world, whereas "physics" was but one – if the most essential – of its disciplines. The ancient and late ancient corpus on questions and themes intrinsically related to natural philosophy was vast. Likewise, the philosophical literature that was both translated and produced by intellectuals in the Islamicate world was, to a large extent, devoted to accounts and explanatory models of the physical world. Thus, in terms of sheer quantity, natural philosophy formed a considerable part of the scientific endeavor of exploring and understanding reality.

Besides philosophy, there were also the emerging indigenous Islamic sciences, such as jurisprudence, Qurʾānic exegesis, the study of *ḥadīth*, grammar and others. In addition, there was also theology (*kalām*), a science that aimed at providing a detailed theoretical account of the universe in the light of, and in support of, the revealed teaching of God and His creation in the Qurʾān. At the heart of Islamic theology was a particular theory that belonged to natural philosophy: atomism.

Atomism was a specific theory not only of the structure of the corporeal world but of its very foundation. Adopting an atomistic perspective on the corporeal world had far-reaching consequences and often preconfigured in various ways the potential answers considered and given in the quest of explaining and determining reality in all areas of scientific investigation, be that ontology, ethics, psychology or theology proper. This "natural philosophical" theory, although belonging to the so-called "subtle topics" of theology, bore greatly on the more substantial

DOI: 10.4324/9781315170718-13

"major topics." Moreover, this theory was successful even beyond the Islamic confession, as we know also of Jewish intellectuals who adopted atomism around the same time. Thus, in the Islamicate world, we are confronted with the rather peculiar situation that theological concerns in and of themselves often called for an occupation with natural philosophy and that, at the same time, positions adopted in the traditional topics of natural philosophy had an immediate potential significance for theological reasoning.

During the 2nd/8th and 3rd/9th centuries, we recognize two major intellectual developments in the Islamicate world: a "period of enormous creativity" in theology (*kalām*), as well as the onset and early phase of the "truly epoch-making" and "astounding achievement" of translating Greek philosophical literature into Arabic (van Ess 2002, 28; Gutas 1998[CB][1], 2, 8). In both regards, natural philosophical concerns proved to be central.

I.11.1 A "period of enormous creativity" in theology (*kalām*)

The origin of atomism in the Islamicate world is still unclear and it seems that we are confronted with a wealth of possibilities as well as an abundance of doctrinal influences that all contributed, some more than others, to the formation of this theory (as well as to other theories that in the beginning competed with atomism before the latter gained acceptance). Scholars have emphasized the importance of Epicurean and Stoic sources, pre-Islamic Iranian and Indian sources, and Galenic medical and pseudo-Galenic doxographical sources (Dhanani 1994; van Ess 2002; Langermann 2009; Pines 1997; Schwarb 2018; Wolfson 1976). Broadly conceived, we may distinguish two phases of early atomism: a phase from its beginning up to Abū ʿAlī al-Jubbāʾī (d. after 303/915) and a phase beginning with his son Abū Hāshim al-Jubbāʾī (d. 321/933). What they all agreed on is that a body is what has length, breadth and depth and that it is composed of a certain number of atoms. In contrast, atoms themselves are *not* thought of as bodies, because they lack the three dimensions. One reason for this is that atoms are indivisible: if they had length, breadth and depth, they would no longer be indivisible, as they would divide into precisely these three dimensions.

What they also seemed to have agreed upon is that atoms should be conceived as somehow cubical. Thus, all bodies consist of a large but still finite number of small indivisible cubes. These cubes not only make up bodies but are also the substrates in which various accidents can inhere. Examples of accidents include color, taste, heat and wetness, as well as will, life and pain. It is clear that atoms cannot *themselves* be qualified as hot or wet, for otherwise they would be divisible again (into an accidental property and its underlying bearer). Moreover, we perceive, and acquire knowledge of, bodies through the accidents that inhere in their atoms.

What the atomists up to and including Abū ʿAlī and those beginning with his son Abū Hāshim disagreed about, however, was the question of whether the atom was itself extended. The former believed that atoms were not extended, having no magnitude themselves (despite their belief that atoms are still somehow cubical), whereas the latter believed that atoms were themselves extended, having magnitude and taking up some space. (It remains controversial in what sense theologians up to and including Abū ʿAlī could consider the atom to be cubical while denying that it has magnitude or shape; Dhanani 1994, 134, 137 convincingly argues for minimal parts discrete geometry along the lines of Epicurean atomism.) What is interesting, and immediately relevant to the history of natural philosophy in the Islamicate world, is that the convictions of the atomists about the structure of bodies also influenced their views about the structure of other aspects of reality: If bodies were atomic, was space atomic, too? That is to say, is the universe as a whole structured by a three-dimensional grid of discrete cellular space-cubicals (i.e., cube-shaped singular amounts of space) subsequently filled by indivisible atoms?

If that is the case, then what about motion? If a body's motion from "here" to "there" must advance through several adjacent and indivisible cellular cubicals, one after another, then motion must seem to be atomic, too. Finally, time, with its intrinsic relation to motion, would seem to be atomic as well. What we as humans perceive as a continuous, smooth reality would actually be more like a movie on TV, which consists of 24 (or more) discrete frames per second, where each frame as a still configuration of colored squares or dots is slightly different from both the preceding and the succeeding frame. Thus, reality would consist in the succession of discrete static configurations of atoms and accidents that are always in their arrangements slightly different from one another. We may *perceive* it as consisting of continuous motions of continuous bodies, yet in truth, everything is divided up into small constituent parts and moments, each individually designed, configured and, thus, *created* by the omnipotent God.

A further disagreement divided the early theologians into two camps that corresponded to two theological schools prevalent in Basra and in Baghdad. This disagreement concerned the question of whether space could be empty, that is whether there could be some amount of space – some cellular cubicals – not occupied by atoms. In a word, it was about the possibility of a void or vacuum. This was already an ancient question that had been discussed for centuries, often in the context of perplexing phenomena like water being attracted into a closed vessel after one has sucked some air out and subsequently immersed it upside-down in water (Dhanani 1994, 78–80, 85–6; Lammer 2018a, 413–14). What attracted the water into the vessel?

The theologians from Basra, who allowed for the existence of void space, simply accepted the phenomenon as it is: by sucking air out of the vessel, we created a real void, that is, some empty space, inside the vessel. This empty space was subsequently filled with water once the vessel was immersed in water and its mouth no longer blocked. In contrast, their colleagues from Baghdad, who denied the possibility of empty space, claimed that through sucking, we merely warmed the air inside the vessel with our breath. After the vessel was immersed in water, the air – vehemently in motion through the warmth – rushed out and was replaced immediately by water. Thus, according to the theologians from Baghdad, it is actually the *impossibility* of a void that we perceive at work here; since the formation of empty space is impossible, the outgoing air had to be replaced immediately by something else – in this case the adjacent water.

A related well-known example involved the "clepsydra," an ancient device whose Greek and corresponding Arabic name translates as "water-thief." It is a hollow vessel with a narrow neck and a perforated base. Once immersed in water, it is filled with water. If the narrow hole on top is now blocked with a thumb and the device is raised above the water, we experience the puzzling phenomenon that the water, against its heavy nature, remains in the device and does not drip out until we remove the thumb from the top hole. To what precisely should we attribute the "attractive power" that keeps the water inside the device? The puzzling character of such phenomena was aggravated by the fact that not all liquids behave the same way: if, for example, one were to use mercury instead of water, it simply would flow out of the clepsydra (and a void would be formed inside, as some theologians hastened to add).

On one hand, such discussions about the possibility (and purported attractive power) of empty space were simply a corollary of the theologians' adherence to atomism. If atoms exist, they must exist somewhere, that is in some space, especially if they are also characterized, as the followers of Abū Hāshim did, as "*space*-occupying" (Dhanani 1994, 61–6). This view leads naturally to the questions of what this space essentially is and whether it could exist without being filled. On the other hand, this discussion also arose from experimentation in the natural sciences and was inspired by medical concerns, too. Some of the related phenomena under discussion include fractured bones that are set into place through vacua created by hot glass bowls or well-known medical treatments with hot cupping glasses that suck in the flesh once placed on the skin (Dhanani 1994, 76–8, 85–6; Chapter I.10). Consequently, it comes as no surprise that intellectuals from other disciplines, outside theological circles, contributed to the debate, in an attempt not merely to witness and utilize but to also understand the phenomena.

Among them was the philosopher al-Fārābī (d. 339/950–1), who apparently reacted to the theologians of his time, especially those hailing from Basra. He composed a treatise in which he rejected the claim that experiments with devices such as closed vessels confirmed the existence of the void (Daiber 1983; Lammer 2018a, 414–17). Moreover, he generally rejected atomism as the proper explanatory background for any such experiment. Neither an existing void nor some warmth is responsible for the attraction of water into the vessel. Instead, only the theory, inherited from the Aristotelian philosophical tradition, that bodies consist not of atoms but of "matter and form" can adequately account for these phenomena. In brief, al-Fārābī argued that by sucking air out of a vessel, the matter of the remaining air inside was unnaturally expanded so that, once the vessel was immersed into water and its mouth no longer blocked, the air counteracted this unnatural expansion by returning to its natural size, thus making room for water to enter. This, he explained, was a common behavior of natural bodies, as water, for example, analogously returns to its natural temperature after it has been heated up forcefully by fire. This behavior he attributed to the form of the air or the water that resided in the underlying matter, inducing an innate desire to remain in a state that is "natural" for the air or water. Here the term "state" is meant to cover qualities (like temperature) as well as quantities (like size) and other features such as places so that a stone would naturally fall through air and sink through water in a desire to return to its natural place somewhere down toward the center of the Earth (which was also considered the center of the universe).

This account was adopted and expanded by the most influential philosopher and physician of the Islamicate philosophical tradition: Ibn Sīnā (d. 428/1037). He was drawn into a technical correspondence with the scholar and historian al-Bīrūnī (362–after 444/973–after 1053), who asked him why water, against its heavy nature, rises inside a closed vessel after air has been sucked out of it if it were not for the void that was formed in the vessel and that, apparently, thus existed in the world (Lammer 2018a, 409–12; Hullmeine 2019). Ibn Sīnā, like al-Fārābī, a staunch critic of both atomism and the actual existence of the void, first sought to explain the phenomenon by taking recourse to a warmth that was created through our sucking on the vessel, leading to rarefaction and expansion of the air. Once the vessel was immersed in cold water, the air inside cooled down and, thus, condensed and contracted, thus making room for water to enter. In his more mature writings, Ibn Sīnā combines this account, which like the earlier explanation of the theologians from Baghdad relied on a change of temperature, with al-Fārābī's suggestion of an unnatural increase of the air's size even without any warmth being created through our sucking, thus providing an explanatory model that could account for all related experiments that had so far troubled the scholars (Lammer 2018a, 417–26).

I.11.2 The "truly epoch-making" Greco-Arabic translations

Overall, we observe an interesting development and refinement of natural philosophical concepts by the intellectual elites over decades and centuries as part of a continuous engagement with scientific experiments in various theological, medical and philosophical contexts. In addition, it is surely no coincidence that the theologians from Baghdad, in contradistinction to their colleagues from Basra, favored the rejection of the actual existence of the void just as al-Fārābī (and after him Ibn Sīnā) did. Al-Fārābī was likewise stationed in Baghdad – the very city in which most of the Greek scientific and philosophical works were translated between the 2nd/8th and 4th/10th century, including Aristotle's (384–322 BCE) philosophical corpus containing detailed arguments against atomism, against the void and in favor of "hylomorphism" (the theory that all bodies consist of matter and form, which was adopted by most philosophers at that time including al-Fārābī and Ibn Sīnā).

Yet, it is well known that not all theologians were atomists and that atomism and the experiments allegedly demonstrating the existence of the void were not the only reasons that called for an engagement with natural philosophy and with newly translated texts in that scientific area. An early instance was the theologian Hishām ibn al-Ḥakam (d. *c.* 179/795). He did not think that bodies were constituted by atoms but that everything that existed simply was corporeal. Extending his theory to God, he argued that God, too, was a body (van Ess 1991–97, 1: 355–64). There is good reason to believe that he worked with, and critically examined, what appears to be the earliest full translation of Aristotle's *Physics* into Arabic, produced by Sallām al-Abrash already during the reign of Hārūn al-Rashīd, the fifth Abbasid caliph at Baghdad (r. 169–193/786–809; Gutas 1998, 73; Lammer 2018a, 10–11). Only a little later, the influential anti-atomist theologian Ibrāhīm al-Naẓẓām (d. *c.* 230/845) – incidentally the nephew of the founding figure of Islamic atomism Abū l-Hudhayl (d. 226/840–1 or 235/849–50) – professed in the presence of the vizier Jaʿfar al-Barmakī (d. 187/803) that he had a thorough knowledge of Aristotle's *Physics* "from the beginning to the end" and "from the end to the beginning" and that he had "refuted Aristotle's book." (Dhanani [1994, 188] reports this anecdote on the authority of ʿAbd al-Jabbār al-Hamadhānī [d. 415/1024].) His motivation for refuting Aristotle, despite their shared anti-atomist stance, was doubtlessly their strong differences regarding other fundamental concepts such as the nature of body, motion and infinity.

Apparently within the first two generations after the Abbasid defeat of the Umayyad caliphate in 132/750, we find increasing evidence of thorough discussions of natural philosophy in and between various circles of scientific learning and practice (theological, scientific and philosophical alike). In the course of time, a diverse network emerged of interacting philosophers and theologians, working across fields and creedal boundaries. The common practice involved discussion and debate among scholars about scientific findings and the objectivity, or at least validity, of those findings. Debate was stimulated, in part, by textual materials freshly translated from Greek into Arabic (including first and foremost Aristotle's *Physics*). It was motivated by a common desire to grasp the inner structures of the corporeal world as a whole and, on that basis, understand the mechanisms *behind* the phenomena that were observed and exploited in scientific practices. There are two exceptionally vivid examples, described in what follows, that represent both the vigor and the entanglements of scholarly exchange among various social, religious, political and academic groups that are so characteristic of the period before Ibn Sīnā (Kraemer 1992). Examples such as these allow us a glimpse into what the scientific community in the Abbasid caliphate looked like and to learn about that community's research activities.

The first example consists of the contents of the only extant manuscript of a historical Arabic translation of Aristotle's *Physics* (MS Leiden, University Library, or. 583). This manuscript, which contains the translation by Isḥāq ibn Ḥunayn (d. 298/910–1), derives from the annotated copy produced by the theologian Abū l-Ḥusayn al-Baṣrī (d. 436/1044) while he was studying Aristotelian natural philosophy in Baghdad, which was subsequently copied by an unknown hand in 470/1077 and finally transcribed in 524/1129–30 by the physician and poet Abū l-Ḥakam al-Maghribī (d. 549/1155). That al-Baṣrī's own copy originated in a context of *teaching* Aristotelian natural philosophy is also apparent in the organization of its text as a series of "lessons," a practice already known from earlier Greek and Syriac teaching circles.

Al-Baṣrī received his philosophical education from the so-called "Baghdad Aristotelians," a circle that was prospering around the most famous philosopher of the 4th/10th century, Yaḥyā ibn ʿAdī (d. 363/974). Ibn ʿAdī himself was taught by Abū Bishr Mattā (d. 328/940) – the famous translator of Aristotle's logic and a colleague of Isḥāq ibn Ḥunayn (our translator of the *Physics*) – and probably also by al-Fārābī. In turn, he became the teacher of Abū ʿAlī ibn al-Samḥ (d. 418/1027) and al-Ḥasan ibn Suwār (d. after 407/1017). The latter, again, was the teacher of Abū l-Faraj ibn al-Ṭayyib (d. 435/1043) who, together with Ibn al-Samḥ, read the *Physics* with the theologian al-Baṣrī. Thus, our Arabic manuscript of Aristotle's *Physics* directly derives from, and through its many annotations and glosses bears witness to, the rich and sophisticated context of teaching and discussion thriving in 4th/10th-century Baghdad.

Moreover, with the exception of al-Maghribī, al-Baṣrī and al-Fārābī, the figures mentioned in this example were Christians, who collaborated with their Muslim colleagues, and many of them were, in one way or another, related to Ibn Sīnā, the towering figure of Arabic philosophy. Al-Fārābī was the author of a commentary on Aristotle's *Physics* that, now lost, was probably influential on the formation of Ibn Sīnā's own views on physics (as we could already see earlier in the case of their explanation of the attraction of water into a closed vessel). Ibn Suwār and Ibn al-Ṭayyib were vehemently criticized by Ibn Sīnā during the latter's stay in Hamadan. And al-Baṣrī would later exercise a strong influence on Ibn al-Malāḥimī (d. 536/1141), who, stationed in Khwarazm, was to become one of the earliest theological critics of Ibn Sīnā and a late representative of the early theological schools.

It is important to note that studying natural philosophy did not only mean translating, reading and discussing *Aristotle*; other texts were also appreciated. (For a survey of translated materials on physics see Lammer 2018a, 9–37; a convenient list of translated materials on philosophy is given by Gutas 2010.) First and foremost, there were the commentaries on Aristotle's various works on natural philosophy. In the case of the *Physics*, we know of translations of the commentaries of Alexander of Aphrodisias (late 2nd–early 3rd centuries), Porphyry (d. *c.* 305), Themistius (d. *c.* 385) and John Philoponus (d. 574). None of them is still extant in Arabic, and only the latter two are extant in Greek. (In fact, only the first half of Philoponus's commentary is extant. The margins of MS Leiden or. 583 provide some Arabic comments and excerpts of the lost commentaries of Alexander and Philoponus.) Yet, these commentaries were highly regarded by the scholarly community as we now learn from the next example, which represents the sophistication and the extent of the scholarly exchange at that time.

From an interesting report made by the historian Ibn al-Qifṭī (568–646/1172–1248), we learn that one of Ibn al-Ṭayyib's students, a physician named al-Yabrūdī (d. *c.* 442/1050), owned a large ten-volume Arabic manuscript of Philoponus's commentary on Aristotle's *Physics*. In its margins, he copied the Arabic text of Themistius's commentary, which had also originally been in Greek. Earlier, this very manuscript of Philoponus's commentary in the Arabic tongue had been owned by ʿĪsā ibn ʿAlī (d. 391/1001), who read it together with Ibn ʿAdī (the Christian head of the "Baghdad Aristotelians" mentioned earlier) and had added his own comments to the margins. Now, ʿĪsā ibn ʿAlī's father was Baghdad's "Good Vizier" ʿAlī ibn ʿĪsā ibn al-Jarrāḥ (d. 334/946), who at one point ordered his secretary Abū Rawḥ al-Ṣābiʾ to translate the first book of Alexander's commentary on Aristotle's *Physics* from Greek into Arabic. In turn, this translation of Alexander was subsequently revised and expanded by his son's teacher Ibn ʿAdī. All in all, Ibn al-Qifṭī's report demonstrates the positive aspects of the interweaving of politics with science as well as the political sponsoring of, and interest in, academic research in the 4th/10th century – an interest that was not merely peripherally directed toward natural philosophy.

What is more, this shared interest in natural philosophy appears to have been intrinsically cross-confessional (Chapter VI.1). Not only do we know of Jewish theologians who adopted atomism, but we are also in the possession of letters on various philosophical themes, such as time, space and body, between Ibn ʿAdī, the Christian head of the "Baghdad Aristotelians," and Ibn Abī Saʿīd al-Mawṣilī (4th/10th century), a Jewish scholar hailing from Mosul (Pines 1955).

In posing his questions, Ibn Abī Saʿīd not only refers to Aristotle and his commentators but also mentions other figures, particularly Galen (129–early 3rd century) and Hippocrates (d. *c.* 370 BCE). The natural philosophical works of these and other authors were absorbed in the wake of the Greco-Arabic translation activities, including the doxography of Pseudo-Plutarch (2nd century), Theophrastus's (c. 371–287 BCE) *Meteorology*, Galen's medical works (including his treatise *On the Elements according to Hippocrates*) and his paraphrase of Plato's (d. 348–347 BCE) *Timaeus*, Alexander's vindications of Aristotle's theories against Galen's criticisms, alongside his own account of the *Principles of the Universe*, materials from Plotinus's (204/5–270) *Enneads*, Proclus's (412–485) *Elements of Physics*, and Philoponus's detailed rebuttal of Proclus's 18 arguments for the eternity of the world. Besides the translated Greek materials, we have reports about various Arabic commentaries or expositions of the contents of Aristotle's works and about their own accounts of natural philosophy by figures from all parts of the Islamic world between the late 3rd/late 9th and the 7th/13th centuries. These culminated in the West with Ibn Bājja (d. 533/1139) and Ibn Rushd (520–595/1126–1198), and in the East with two new paradigms that profoundly shaped the conception of natural philosophy, in general, and of physics, in particular. These two new paradigms are considered next.

I.11.3 New paradigms in content and structure

Ibn Sīnā's main work *The Cure* (*al-Shifāʾ*) contains his most extensive treatment of natural philosophy, in eight volumes, the first of which is devoted to the foundational discipline of that scientific area: physics. Ibn Sīnā wrote *The Cure* at the request of his disciples, who urged him to start with natural philosophy, and so he began to work on his own systematic exposition of all the topics and concepts that Aristotle and his commentators had discussed under the common heading of "physics": matter and form, body, nature, causes, chance and luck, motion and rest, place, inclination, time and eternity, continuity, atomism, finitude, directionality and so on. That his work constituted a *new version of physics* – and not just another commentary on Aristotle's *Physics* – is crucial and may explain the otherwise perplexing fact that there is only *one* extant copy of an Arabic version of Aristotle's famous work but numerous manuscripts of Ibn Sīnā's new version. Ibn Sīnā actually *replaced* Aristotle, not only in natural philosophy but also in philosophy more generally, setting the new and most up-to-date standard of philosophical inquiry. Thus, Ibn Sīnā terminated the "Aristotelian tradition" (at least in the eastern parts of the Islamicate world) and initiated an "Avicennian tradition".

I.11.3.1 Content

In terms of content, Ibn Sīnā remained an Aristotelian at heart, developing and defending Aristotle's theories in light of the developments and criticism in the previous Neoplatonic scholarship down to his time. A good example of such a defense is his vindication of the Aristotelian account of place, which had been criticized – even ridiculed – with devastating objections and counterexamples (Lammer 2018a, 307–427). According to Aristotle, the place of something was the two-dimensional inner surface of the surrounding body containing it. The place of some water, for example, might be the adjacent glass surface of its bottle. Already Aristotle's ancient successors were disappointed, and their concerns grew into a potentially fatal critique of Aristotle's theory in the 6th century through Philoponus's commentary on the *Physics*. Among other things, Philoponus argued that it was absurd to claim that a *three-dimensional* body had a *two-dimensional* place and that the universe, as a whole, if one really wanted to apply Aristotle's theory, would have to have no place at all, because there was no surrounding body outside the universe containing it. In his response, Ibn Sīnā not only revealed Philoponus's line of argument as logically inconclusive and refined the concept of a two-dimensional surface (in part also by engaging with the theories of the theologians from Basra and Baghdad), but he also explained why the universe does not have a place and, indeed, does not *need* a place (McGinnis 2006). So, Ibn Sīnā allowed himself to deviate from earlier conceptions and interpretations and to refine or reject them with novel insights. In other words, he demonstrated his philosophical acumen by further developing already well-known concepts, such that they could withstand or overcome the criticism leveled in earlier centuries, and he displayed his capacity for innovation by taking original and surprising turns in his argumentation. The result was a new, *Avicennian* paradigm of Aristotelian natural philosophy successfully superseding all earlier, now outdated versions – and it was this paradigm that, in subsequent centuries, was both criticized and defended, rejected and further refined.

I.11.3.2 Structure

Second, Ibn Sīnā also provided a new structure for philosophy. Among his works, we find the intricate but influential *Pointers and Reminders* (*al-Ishārāt wa-l-tanbīhāt*). Two features about this work stand out. First, it generated a full-blown commentary and teaching tradition from the 6th/12th century until the modern period (Wisnovsky 2004, 2013). Second, it provided a new structure of philosophical investigation, as a whole, and of natural philosophy, in particular (McGinnis 2013). While most of his other works follow a traditional division of philosophy into logic, natural philosophy, mathematics and metaphysics (with their various subsections), Ibn Sīnā's *Pointers and Reminders* consist of only two main parts: logic and everything else. This abandonment of the traditional division of scientific investigation into distinct areas and disciplines was a considerable change in the *macrostructure* of philosophy (Chapter VI.4). At least one of the reasons for Ibn Sīnā's dissatisfaction with that traditional division was his realization that some philosophical topics transcended these categories or disciplines. For example, the soul was typically discussed in the discipline of psychology, which was one area within natural philosophy. But a full investigation showed that the soul was not only relevant for, or even at the core of, matters of natural philosophy, where the soul was considered the principle of life, movement and perception in animals. It was also needed in metaphysics and theology, where the soul was considered with regard to both its cosmological dimension in creation and its religious dimension in salvation (Gutas 2014[CB], 270–96; also Falcon 2005).

Ibn Sīnā's changes in the macrostructure of philosophical inquiry also provoked a new *microstructure* on the thematic level, which illustrates again – this time not in terms of content but in terms of structure – the onset of a new Avicennian tradition in philosophy itself and in natural philosophy in particular. Tables I.11.1 and I.11.2 below compare the broad thematic structure of a work on the science of physics, which remained the most fundamental discipline within natural philosophy. The first table focuses on the relevant works by Aristotle (his *Physics*), by Ibn Sīnā (the book on physics from *The Cure*) and by the influential Jewish-Muslim convert and philosopher Abū l-Barakāt al-Baghdādī ([d. *c.* 560/1164–5]; the part on physics from *The Considerations* [*al-Muʿtabar*]). The second table compares the structure of Ibn Sīnā's discussion of physical topics in his *Pointers and Reminders* with the parts on physics from *The Intimations* (*al-Talwīḥāt al-lawḥiyya wa-l-ʿarshiyya*) by al-Suhrawardī (d.

Table I.11.1 *Rough* outline of the "Aristotelian" structure of physics ending with Abū l-Barakāt al-Baghdādī

Aristotle	Book I		Book II			Book III		Book IV			Book V – VI	. . .
Physics:	method	matter/form	nature	causes	luck/chance	motion	infinity	place	void	time	continuity	. . .
Ibn Sīnā	Book I					Book II				Book III		. . .
The Cure, physics:	method	matter/form	nature	causes	luck/chance	motion	place	void	time	atomism/continuity/ infinity		. . .
Abū l-Barakāt												
The Considerations, physics:	method	nature	matter/form/ causes		luck/chance	motion	place	void	time	infinity		. . .

Table I.11.2 *Rough* outline of the "Avicennian" structure of physics beginning with Ibn Sīnā

Ibn Sīnā													
Pointers and Reminders:	atomism	matter/form I	infinity	matter/form II	void	directions	place	nature	motion	heavens	elements	. . . time	. . .
al-Suhrawardī													
The Intimations, physics:	atomism	matter/form I	infinity	matter/form II	directions	nature	heavens	place	void	motion	time	elements	. . .
al-Abharī													
The Guide to Philosophy, physics:	atomism	matter/form I	infinity	matter/form II	void	directions	place	nature	motion	time	heavens	elements	. . .

587/1191) and *The Guide to Philosophy* (*Hidāyat al-ḥikma*) by al-Abharī (d. between 660/1263 and 663/1265).

As can be seen in Table 1.11.1., Ibn Sīnā by and large adopted the thematic structure of Aristotle's physics. After a chapter on method, he focused on matter and form as the constituent principles of the subject matter of physics: natural things. The close relation between the form of a natural body and its inherent nature led to a discussion of the latter, which he conceived as a cause of motion. Already talking about causation, Ibn Sīnā proceeded with a full discussion of the four causes and their importance for natural philosophy, which also demanded a brief treatment of chance and luck (as some considered them to be causes, too). All in all, the first book of Ibn Sīnā's physics from *The Cure* deals with the principles and causes of natural things. Its second book turns to motion and what is closely related to it, in particular place and time. Here, we witness Ibn Sīnā skipping over Aristotle's discussion of infinity and finitude. This was certainly a deliberate decision, because the investigation of finite and infinite magnitudes has no relation to the phenomenon of motion as such while being essential to the discussion of continuity and the rejection of atomism. So, Ibn Sīnā moved this topic out of the second book and placed it at the beginning of his third book, which incidentally corresponded to the account of continuity in books five and six of Aristotle's *Physics* that follow immediately after the discussion of time in book four (Hasnawi 2002).

In the 6th/12th century, Abū l-Barakāt also followed the Aristotelian structure. Or maybe it would be more accurate to say that he actually followed Ibn Sīnā's *version* of the Aristotelian structure, as he, too, discussed finitude and infinity not after having dealt with motion but later, after his account of time. In addition, Abū l-Barakāt introduced some further changes of his own. For example, he moved the discussion of nature to the beginning as the second topic, presumably with two motives. On one hand, he intended to introduce the subject matter of physics – natural things – by referring to their inherent nature. On the other hand, to further clear up the picture, he proposed fusing the account of matter and form – which are two of the four Aristotelian causes – with the other two causes, agent or efficient cause and end or final cause. The most important point, however, is that there may have been no one in the Islamicate world after Abū l-Barakāt who followed the Aristotelian structure of physics; instead, they followed a novel *Avicennian* structure – a structure established by Ibn Sīnā in his *Pointers and Reminders*, as Table 1.11.2. shows.

The *Pointers and Reminders* have been mentioned already as one of those works in which Ibn Sīnā dissolved the traditional macrostructure of the various sciences; now they serve to show how this change also entailed a new thematic division of the topics of physics. Although there was no designated section on natural philosophy in the *Pointers and Reminders* anymore, Ibn Sīnā of course still treated natural philosophical topics – yet, as Table I.11.2. illustrates, he began with his rejection of atomism, arguing for matter and form as the constituent principles of corporeal reality. Moreover, bodies are essentially finite, as he briefly explained in the subsequent discussion of infinity and finitude, before he returned to other aspects of matter and form, followed by a rejection of the void and a brief account of direction and place. An elaboration on the concept of nature and of natural motion led to motion in general and the investigation of the supra-lunar motion of the heavenly bodies and the sublunary motions of the elements. The theory of time was excluded from this train of thought and discussed later as part of the account of creation.

Now, the prominent and highly influential thinker al-Suhrawardī did not follow Ibn Sīnā's changes in the macrostructure, and so his philosophical compendium *The Intimations* is divided into the traditional parts: logic, natural philosophy and metaphysics (with its various sections). Yet, as can be seen, he nonetheless adopted Ibn Sīnā's new microstructure of the thematic divisions *inside* the section on physics. There are only two minor differences. First, he moved the discussion of the void closer to the discussion of place. Second, he reintegrated the discussion of time, because, following the traditional macrostructure, his work contains a proper section on the discipline of physics, which would have been quite incomplete without a discussion of time as part of it. (Ibn Sīnā's discussion, in turn, was not incomplete, because his *Pointers and Reminders* did not contain a separate section on physics, as already mentioned.)

Al-Suhrawardī's novel layout for philosophical compendia was taken over by al-Abharī, the author of the most influential philosophical textbook in the Islamicate tradition, the *Guide to Philosophy*, with minor changes such as moving the discussion of the void back to where it had been placed by Ibn Sīnā. However, these changes are very subtle, because al-Abharī's treatment of the topics is rather superficial and often merges different topics with one another. For example, it could be questioned whether his investigation actually contains a distinction between the topic of place and that of the directions at all. (This is one of the reasons why I label these two tables "rough" outlines; see also McGinnis 2013).

In philosophical works, Ibn Sīnā's new structure remained in place way beyond the 7th/13th century, whereas in theological works, another structure became predominant that, although inspired by Ibn Sīnā's theoretical conception of the topics of metaphysics, was established by Fakhr al-Dīn al-Rāzī ([d. 606/1210]; Eichner 2007, 2009, 31–60). His highly influential works *Eastern Investigations* (*al-Mabāḥith al-mashriqiyya fī ʿilm al-ilāhiyyāt wa-l-ṭabīʿiyyāt*) and *The Compendium of Philosophy* (*al-Mulakhkhaṣ fī l-ḥikma*) divide all philosophical inquiry into three parts: (1) "common things", (2) accidents and substances and (3) theology. It is in the second division that we find most topics from natural philosophy neatly integrated into a novel structure: perceptible qualities (such as hot/cold, wet/dry, color, pain, and so on), causation, motion and time are all considered accidents and, thus, treated in the first subdivision of the second part (2.1 on accidents), while corporeality, atomism, heavenly bodies, the elements, plants, the soul and so on are considered substances and, consequently, treated in the second subdivision of the second part (2.2 on substances).

Like Ibn Sīnā, Fakhr al-Dīn dissolved and abandoned the traditional classification of the philosophical disciplines, providing a new order and context for the discussion of the themes of natural philosophy. Moreover, his new structure of theological inquiry was at least equally successful; it was taken up by ʿAbdallāh al-Bayḍāwī (d. *c.* 685/1286 or 716/1312) and ʿAḍud al-Dīn al-Ījī (d. 756/1355) in their main works *The Rising Lights* (*Ṭawāliʿ al-anwār min maṭāliʿ al-anẓār*) and *The Stations* (*al-Mawāqif*), respectively. The commentary tradition that these two works initiated, following the model established by Fakhr al-Dīn, lasted until the 12th/18th century in the former case and until the 14th/20th century in the latter. The structural changes that are so characteristic for the first periods of engagement with Ibn Sīnā also coincided with doctrinal shifts, which were the result of centuries full of debate and fruitful discussion on the contents and contentions of Ibn Sīnā's new natural philosophy. Some of these shifts have been described recently in the research literature, but it will be the topic for future investigation to determine them in more detail (some first results: Adamson 2017, 2018a, 2018b; Adamson and Lammer 2019; Ahmed 2016; Janssens 2010, 2018; Lammer 2018b; McGinnis 2018). At any rate, it seems clear that here, at the turn of the 6th/12th to the 7th/13th century, we witness the onset of a new phase within the Islamicate intellectual tradition – a phase that resulted from the interplay of early theology, the Greco-Arabic translations, Ibn Sīnā's contributions, and the developments in the Avicennian tradition in all areas of scientific practice, including not least natural philosophy.

Note

1 Consolidated bibliography.

Bibliography

Sources

al-Abharī, Athīr al-Dīn. tr. al-Attas, S. A. T. 2009. *A Guide to Philosophy: The Hidāyat al-Ḥikmah of Athīr al-Dīn al-Mufaḍḍal ibn ʿUmar al-Abharī al-Samarqandī*. Subang Jaya: Pelanduk Publications.

Aristotle. ed. Badawī, ʿA. R. 1964. *al-Ṭabīʿa: Tarjamat Isḥāq ibn Ḥunayn maʿa shurūḥ Ibn al-Samḥ wa-Ibn ʿAdī wa-Mattā ibn Yūnus wa-Abī Faraj ibn al-Ṭayyib* [Physics: The Translation by Isḥāq ibn Ḥunayn Together With the Commentaries by Samḥ, Ibn ʿAdī, Mattā ibn Yūnus and Abū l-Faraj ibn al-Ṭayyib]. 2 vols. Cairo: al-Dār al-qawmiyya li-l-ṭibāʿa wa-l-nashr.

Aristotle. tr. Waterfield, R. 1996. *Physics*. Oxford and New York: Oxford University Press.

Avicenna and Abū Rayḥān al-Bīrūnī. tr., Berjak, R. and Iqbal, M. 2003–2007. "Ibn Sīnā – al-Bīrūnī Correspondence I – VIII," *Islam & Science* 1.1–5.1.

Avicenna and Abū Rayḥān al-Bīrūnī. ed. Moḥaqqeq, M. and Nasr, S. H. 1973. *al-As'ila wa-l-ajwiba* [Questions and Answers]. Tehran: Chāpkhāna-yi Mu'assasa-yi intishārāt chāp-i dānishgāh-i Ṭihrān. Shūrā-yi ʿālī-yi farhang-u hunar. Markaz-i muṭālaʿāt-u hamāhangī-yi farhangī 9.

Avicenna. tr. Inati, S. C. 1984. *Remarks and Admonitions, Part One: Logic* (Medieval Sources in Translation 28). Toronto: Pontifical Institute of Medieval Studies.

Avicenna. tr. Inati, S. C. 1996. *Ibn Sīnā and Mysticism: Remarks and Admonitions, Part Four*. London and New York: Kegan Paul International.

Avicenna. tr. Inati, S. C. 2014. *Remarks and Admonitions: Physics and Metaphysics*. New York: Columbia University Press.

Avicenna. tr. McGinnis, J. 2009. *The Physics of The Healing: A Parallel English-Arabic Text* (Islamic Translation Series). Provo: Brigham Young University Press.

al-Baghdādī, Abū l-Barakāt. ed. al-Ḥaḍramī, ʿA. and al-Yamānī, A. b. M.1938–1939. *Kitāb al-Muʿtabar fī l-ḥikma* [The Book of Carefully Considered Teachings in Philosophy]. Hyderabad: Maṭbaʿat Dā'irat al-maʿārif al-ʿuthmāniyya.

al-Baydawi, ʿAbd Allah. tr. Calverley, E. E. and Pollock, J. W. 2002. "Ṭawāliʿ al-anwār min maṭāliʿ al-anẓār [The Rising Light from Far Horizons]," in Calverley, E. E. and Pollock, J. W., eds. *Nature, Man and God in Medieval Islam* (Islamic Philosophy, Theology and Science. Texts and Studies 45). Leiden: Brill.

al-Fārābī, Abū Naṣr. tr. Lugal, N. and Sayılı, A. 1951. *Halâ Üzerine Makalesi: Article on Vacuum* (Türk Tarih Kurumu Yayınlarından XV, Seri 1). Ankara: Türk Tarih Kurumu Basımevi.

al-Ījī, ʿAḍud al-Dīn. 1907. *Kitāb al-Mawāqif fī ʿilm al-kalām* [Book of the Stations on the Science of Kalām]. Cairo: Maṭbaʿat al-saʿāda.

al-Rāzī, Fakhr al-Dīn, 1923–1924. *Kitāb al-Mabāḥith al-mashriqiyya fī ʿilm al-ilāhiyyāt wa-l-ṭabīʿiyyāt* [Book of the Eastern Investigations on the Divine Science and Natural Philosophy]. Hyderabad: Dā'irat al-maʿārif al-niẓamiyya.

al-Suhrawardī, Shihāb al-Dīn. ed. Habibi, N. Q. 2009. *al-Talwīḥāt al-lawḥiyya wa-l-ʿarshiyya* [Book of Information of the Tablet and the Throne]. Tehran: Mu'assasa-yi pazhuhashī-yi ḥikmat-u falsafa-yi Īrān.

Manuscripts

MS Leiden, Universiteitsbibliotheek, or. 583.

MS or. oct. 623. Berlin: Preußischer Kulturbesitz, Staatsbibliothek.

Research literature

Adamson, P. 2017. "Fakhr al-Dīn al-Rāzī on Place," *Arabic Sciences and Philosophy* 27.2: 205–36.

Adamson, P. 2018a. "Faḫr al-Dīn al-Rāzī on Void," in Al Ghouz, A., ed. *Islamic Philosophy from the 12th Till the 14th Century*. Göttingen: Bonn University Press and V&R unipress, 307–24.

Adamson, P. 2018b. "The Existence of Time in Faḫr al-Dīn al-Rāzī's al-Maṭālib al-ʿāliya," in Hasse, D. N. and Bertolacci, A., eds. *The Arabic, Hebrew, and Latin Reception of Avicenna's Physics and Cosmology*. Berlin and Boston: Walter de Gruyter, 65–99.

Adamson, P. and Lammer, A. 2019. "Fakhr al-Dīn al-Rāzī's Platonist Account of the Essence of Time," in Shihadeh, A. and Thiele, J., eds. *Philosophical Theology in Medieval Islam: The Later Ashʿarite Tradition*. Leiden: Brill.

Ahmed, A. Q. 2016. "The Reception of Avicenna's Theory of Motion in the Twelfth Century," *Arabic Sciences and Philosophy* 26.2: 215–43.

Daiber, H. 1983. "Fārābīs Abhandlung über das Vakuum: Quellen und Stellung in der islamischen Wissenschaftsgeschichte," *Der Islam* 60: 37–47.

Dhanani, A. 1994. *The Physical Theory of Kalām: Atoms, Space, and Void in Basrian Muʿtazilī Cosmology*. Leiden: E. J. Brill.

Eichner, H. 2007. "Dissolving the Unity of Metaphysics: From Faḫr al-Dīn al-Rāzī to Mullā Ṣadrā al-Šīrāzī," *Medioevo: Rivista di Storia della Filosofia Medievale* 32: 139–97.

Eichner, H. 2009. *The Post-Avicennian Philosophical Tradition and Islamic Orthodoxy: Philosophical and Theological summae in Context*. Habilitationsschrift. Halle-Wittenberg: Martin-Luther-Universität.

Ess, J. van. 1991–7. *Theologie und Gesellschaft im 2. und 3. Jahrhundert Hidschra: Eine Geschichte des religiösen Denkens im frühen Islam*. 6 vols. Berlin: Walter de Gruyter.

Ess, J. van. 2002. "60 Years After: Shlomo Pines's *Beiträge* and Half a Century of Research on Atomism and Islamic Theology," in *Proceedings of the Israel Academy of Sciences and Humanities* 8.2. Jerusalem: Israel Academy of Sciences and Humanities, 19–41.

Falcon, A. 2005. *Aristotle and the Science of Nature: Unity without Uniformity*. Cambridge: Cambridge University Press.

Gutas, D. 2010. "Greek Philosophical Works Translated into Arabic," in Pasnau, R. and van Dyke, Chr., eds. *The Cambridge History of Medieval Philosophy*. 2 vols. Cambridge: Cambridge University Press, 2: 802–14.

Hasnawi, A. 2002. "La Physique du Šifāʾ. Aperçus sur sa structure et son contenu," in Janssens, J. and De Smet, D., eds. *Avicenna and his Heritage*. Leuven: Leuven University Press, 67–80.

Hasse, D. N. and Bertolacci, A., eds. *The Arabic, Hebrew, and Latin Reception of Avicenna's Physics and Cosmology*. Berlin: Walter de Gruyter.

Hullmeine, P. 2019. "al-Bīrūnī and Avicenna on the Existence of Void and the Plurality of Worlds," *Oriens* 47: 114–144.

Janssens, J. 2010. "The Reception of Ibn Sīnā's *Physics* in Later Islamic Thought," *Ilahiyat Studies* 1.1: 15–36.

Janssens, J. 2018. "Avicennian Elements in Faḫr al-Dīn al-Rāzī's Discussion of Place, Void and Directions in the *al-Mabāḥiṯ al-mašriqiyya*," in Hasse and Bertolacci, eds., 43–63.

Kraemer, J. L. 1992. *Humanism in the Renaissance of Islam: The Cultural Revival during the Buyid Age*. Leiden, New York and Cologne: E. J. Brill.

Lammer, A. 2018a. *The Elements of Avicenna's Physics: Greek Sources and Arabic Innovations*. Berlin: Walter de Gruyter.

Lammer, A. 2018b. "Time and Mind-Dependence in Sayf al-Dīn al-Āmidī's *Abkār al-afkār*," in Hasse and Bertolacci, eds., 101–61.

Langermann, Y. T. 2009. "Islamic Atomism and the Galenic Tradition," *History of Science* 47.3: 277–95.

McGinnis, J. 2006. "Positioning Heaven. The Infidelity of a Faithful Aristotelian," *Phronesis* 51.2: 140–61.

McGinnis, J. 2013. "Pointers, Guides, Founts and Gifts: The Reception of Avicennan Physics in the East," *Oriens* 41.3–4: 433–56.

McGinnis, J. 2018. "Changing Motion: The Place (and Misplace) of Avicenna's Theory of Motion in the Post-Classical Islamic World," in Hasse and Bertolacci, eds., 7–24.

Pines, S. 1955. "A Tenth Century Philosophical Correspondence," *Proceedings of the American Academy for Jewish Research* 24: 103–36.

Pines, S. 1997. *Studies in Islamic Atomism*. Jerusalem: Magnes Press.

Schwarb, G. 2018. "Early Kalām and the Medical Tradition," in Adamson, P. and Pormann, P. E., eds. *Philosophy and Medicine in the Formative Period of Islam*. London: Warburg Institute, 104–69.

Wisnovsky, R. 2004. "The Nature and Scope of Arabic Philosophical Commentary in Postclassical (ca. 1100–1900 AD) Islamic Intellectual History: Some Preliminary Observations," in Adamson, P.,

Baltussen, H., and Stone, M. W. F., eds. *Philosophy, Science and Exegesis in Greek, Arabic and Latin Commentaries*. 2 vols. London: Institute of Classical Studies, 2: 149–91.

Wisnovsky, R. 2013. "Avicenna's Islamic Reception," in Adamson, P., ed. *Interpreting Avicenna: Critical Essays*. Cambridge: Cambridge University Press, 190–213.

Wolfson, H. A. 1976. *The Philosophy of the Kalam*. Cambridge, MA and London: Harvard University Press.

I.12

ALCHEMY AND THE CHEMICAL CRAFTS[1]

Regula Forster

The chemical crafts are concerned with all kinds of chemical processes, from the dying of textiles and alloying of metals to the production of glass. The same or similar processes have been important to what is called alchemy, the art concerned with the transmutation of base metals, such as lead, into precious ones, especially gold or, sometimes, silver. Alchemy has often been characterized as "proto-chemistry" by historians of science. While the transmutation of metals may be seen as a primary goal of Islamicate alchemy, however, it was often pursued as a comprehensive natural philosophy more broadly, even as a part of (Neoplatonic) metaphysics (Ullmann 1972, 145, 257). As F. S. Taylor put it, "[a]lchemy is identical neither with mysticism nor with metallurgy. . . . The hall-mark of Alchemy is the combination of a spiritual and a practical aspect in the making of precious materials" (Taylor 1937, 30–1). This dual focus of alchemy, famously, led the Swiss psychiatrist and cofounder of psychoanalysis C. G. Jung to interpret European alchemical symbolism in purely psychological terms; while suggestive, such an interpretation runs the risk of becoming ahistorical.[2] Yet despite its importance to Western history of science and history of religion, Islamicate alchemy remains notoriously under-researched, even more so than its European counterparts, not least because the surviving texts are often extremely cryptic.

It might be somewhat ahistorical to divide chemistry and alchemy, but it seems important to stress that alchemy cannot exist without a solid background of philosophical ideas, while the chemical crafts as practices very often have not left us any theoretical framework we could consider (Porter 2011, 21–43; Chapter II.8). It therefore seems appropriate to at least mention both, even though alchemy has left many more traces, as indicated by the thousands of manuscripts extant, and will therefore receive more attention in this chapter.

I.12.1 Alchemy in the Islamicate world

Alchemy seems to be of ancient Egyptian origin: there, in the temple workshops, the craft was used to imitate gold and other precious metals (Taylor 1937). Two papyri found in Upper Egypt document this early stage of the development of alchemy (Halleux 1981). However, this Egyptian strand goes back, itself, to older, Babylonian traditions (Martelli and Rumor 2014). These craft traditions were then fused in Egypt with ideas taken from Gnosticism, Hermetism, Babylonian astrology, and Egyptian mythology (Schütt 2000, 15; Ullmann 1979, 110b; Vereno 1992, 8–9. For an introductory overview, see also Braun 2016, 13–14). But it was only through

DOI: 10.4324/9781315170718-14

this fusion of craft with Greek natural philosophy that alchemy emerged as a major science with a wide impact. The oldest alchemist we know much about is likewise an Egyptian: Zosimus (*fl.* 4th century ce), although he himself refers to Mary the Jewess (*Maria Prophetissima*) as an earlier important practitioner.

Islamicate alchemy, in turn, is based in the first place on this Hellenistic, Greco-Egyptian tradition, although Chinese and probably Indian influences can be evidenced; Persian traditions, too, seem to live on in its terminology.[3] Dating the beginning of alchemical activities and writings in the Islamicate world, however, is extremely difficult, as there are no clear indications regarding which texts might be oldest. Traditionally, the Umayyad prince Khālid ibn Yazīd (d. after 85/704) is credited with the introduction of alchemy into the Arabic-speaking world, which would mean that alchemical texts were among the first to be translated. But it has been shown that he most probably had no interest in this science (Ullmann 1978; Anawati 1996; see now also Dapsens 2016).[4] Be that as it may, we can assume that some works on alchemy were indeed translated from Greek, sometimes probably via Syriac, as happened in other scientific fields (Ullmann 1972, 148). It is quite likely that some of these translations date back as far as the 2nd/8th century, with the translation trend being established firmly by the 3rd/9th century. Besides translations, works were soon composed in Arabic and attributed to older, especially Greek authorities, often already popular in Greek alchemy, such as Hermes or Democritus (Ullmann 1972, 151).

A much earlier date, however, is found in connection with Zosimus: There is a gloss in an Arabic manuscript that suggests that one of Zosimus's works was translated into Arabic as early as 38/658. However, this gloss is almost certainly inauthentic (Hallum 2008, 159–60). In Islamicate alchemy, Zosimus is considered a very important authority, his name given in different forms: Ibn al-Qifṭī has Zūsīm and Rūsīm (Hallum 2008, 94). No less than 25 works attributed to him have been identified in Arabic, 22 of them extant in manuscripts (Hallum 2008, 106). Of these, some are translations of extant Greek texts, some can be proved to be translations from a Greek original that is either not extant or not yet found, and others include original material from Zosimus's works but were probably composed in Arabic. Yet others seem to be pseudepigraphical, although Hallum's term "forgeries" seems to be too harsh in my view (Hallum 2008, 111–2). The most disputed of these texts is the *Book of Pictures* (*Muṣḥaf al-ṣuwar*), surviving in a unique manuscript kept at Istanbul's Archeological Museum, an impressively illustrated exemplar dating to the 7th/13th century. The text presents a series of dialogues between Zosimus and his pupil Theosebeia. It remains to be conclusively shown whether these dialogues should be called a translation from Greek, as suggested by Abt, or rather an Arabic work incorporating original Zosimus material, as per Hallum (Abt 2007, esp. 21–68, against Hallum 2008, esp. 257).

I.12.2 Theories and principles of alchemy

The dual physical and metaphysical focus of Islamicate alchemy is already apparent in Ibn al-Nadīm's *Catalogue* (*Fihrist*), a bookseller's catalog of 4th/10th-century Baghdad. Ibn al-Nadīm explains:

> Persons interested in the art of alchemy, which is the manufacture of gold and silver not stemming from their mines, state that the first man who spoke about the science of this art was Hermes, the wise man and Babylonian . . .
>
> *(Ibn al-Nadīm 1393/1973[CB],[5] 417)*[6]

Ibn al-Nadīm thus defines alchemy as the art of acquiring gold and silver without having to mine them. But just after this, he adds the name Hermes, associating alchemy with the

Hermetic sciences. Hermes was probably already considered an alchemist in late antiquity (Taylor 1937, 46; Fowden 1986, esp. 89–91). Indeed, Hermes is one of the most important authorities in Islamicate alchemy, cited in almost all writings, and several pseudepigraphical works are attributed to him (Anawati 1996, 858; on Hermes in Arabic, see Van Bladel 2009[CB], esp. 121–233).

As transmutation is one of the central elements of Islamicate alchemy, we need to discuss how in theory the transmutation was supposed to work. First of all, the precondition of all methods of transmutation is that metals are considered to be species (*anwāʿ*, singular *nawʿ*) of a single genus (*jins*) – a concept accepted generally. Metals thus differ only in their accidents, not their essences. (Accidents are not an essential part of any substance but can be removed or changed without destroying the substance itself.) Furthermore, the metals were regarded as developing and changing even in a natural environment. They were not seen as already fixed in their final state as a substance from inception but as ripening toward that state. By this principle, then, every base metal would, if left alone for untold years, ripen to eventually become silver or gold.

In changing base metals into gold, the alchemist therefore sought not to change their natural evolution but only to accelerate it. This doctrine was crucial for the acceptance of alchemy by religious scholars, such as the 6th/12th-century philosopher, theologian and commentator of the Qurʾān, Fakhr al-Dīn al-Rāzī (d. 606/1210), who argued that natural products could also be produced artificially and that the specific color and weight of gold were only accidents (Ullmann 1972, 253–4).

Over time, three main methods of transmutation were identified. The most important method was focused on producing the so-called Philosopher's Stone. In order to make gold, the alchemist had to take a base metal, preferably a cheap one, and strip it of its accidents. This process was called "blackening" (*taswīd*), as its result was the prime matter, thought to be black, passive and free of accidental qualities. To this prime matter, the Philosopher's Stone – also called elixir, egg of the sages, and many other names – would be applied. The production of the stone came to be considered the core element of alchemy; many texts are mainly concerned with making the stone, not with the blackening and so on, although the stone would, in theory, not work on "normal" metals. However, many stories were told in which exactly this happened: the "stone" or elixir, usually depicted as a powder, was applied to any metal and produced gold and/or silver immediately. It was said to work like yeast in dough. What happened when it was applied was nothing out of the ordinary but only a radically accelerated ripening process. If the process was perfectly realized, gold would be produced (*taḥmīr*, "reddening") and, if imperfectly, silver (*tabyīḍ*, "whitening"; Ullmann 1972, 257–60).

The second prominent theory of transmutation is the mercury/sulfur theory already mentioned in the Jābir corpus (Ullmann 1972, 260–1; Hill 1990, 334–5; Hill 1993, 80; Artun 2013, 75–9). This is a large corpus of writings, mainly but not only on alchemy, that is attributed to one Jābir b. Ḥayyān, who is said to have lived in the 2nd/8th century and to have been, in alchemy, a pupil of the sixth Shīʿī *imām*, Jaʿfar al-Ṣādiq (d. 147/765). Whether Jābir should even be considered a historical figure, much less the author of this corpus, is one of the most disputed questions among scholars of Islamicate alchemy. Paul Kraus (1904–1944) challenged the very existence of Jābir. He dated the corpus to the 3rd–4th/9th–10th centuries and attributed it to a "school", based on three arguments: (1) that the corpus is too vast to have been written by a single author, (2) that the influence of Ismāʿīlī thought and the Greek-inspired cosmology were unlikely to occur in texts of the 2nd/8th century and (3) that the chronology offered was incoherent (see Kraus 1930, 1942, 1942–1943, vol. 1: xlviii–lvii). After a period in which Kraus's arguments were generally accepted, they were challenged by Sezgin (Sezgin 1971, 132–269) and Haq (Haq 1994, esp. 8–32). In fact, the corpus is not so vast that single authorship is impossible, and some of the so-called Ismāʿīlī influence need not necessarily be Ismāʿīlī at all but could also be explained as (proto-)Shīʿī (see Haq 1994, 21–4; Lory 1983, 47; Marquet 1970). A third position was adopted by Pierre Lory, who stipulated a development, reworking and ongoing interpolation of the corpus over time. The references to Greek philosophy could then be regarded as later additions to basically technical manuals (Lory 1988, 11–13). Whatever stance we might take, the Jābirian corpus seems to be quite early, if only in part going back to at least the 3rd/9th century.

According to the mercury/sulfur theory as presented in the Jābirian corpus, mercury and sulfur consist of opposing elements: mercury of water and earth, sulfur of fire and air. If it were possible to purify these to an extreme degree and then fuse them into a perfect mixture, gold would be produced after heating this mixture. If the process were conducted to an insufficient degree, another, lesser metal would result, depending on the primary quality that was dominant; if the process were hindered by coldness, silver would be produced; if too dry, copper; and so on.

The third method proposed for the transmutation was based on the so-called theory of the balance (*ʿilm al-mīzān*). This theory is associated in the first place with the Jābirian corpus (Anawati 1996, 865–7; Artun 2013, 94–9; Hill 1990, 335; Hill 1993, 80–2; Ullmann 1972, 261). For this method, the primary qualities (hotness, coldness, dryness, moistness) of the metals are to be studied and precisely measured. The alchemist would then attempt to produce a compound in which the qualities were perfectly balanced.

The Philosopher's Stone, the elixir, was sometimes seen as an elixir of life, a panacea that would cure all illnesses and extend life indefinitely (Ullmann 1972, 260). The functioning as a panacea is already present in the Jābirian corpus (Jābir ibn Ḥayyān 1935, 303–5). However, this medical aspect never became dominant in Islamicate alchemy. The panacea would always be primarily considered the objective of medicine, not of alchemy, which makes Islamicate alchemy quite different from early modern European alchemy, where the quest for eternal life was a central aspect.

I.11.3 Critiques of alchemy

Modern critiques of alchemy usually focus on the contention that alchemists never actually achieved the transmutation of metals into gold and silver, and hence must be considered deluded at best and charlatans at worst. Premodern critiques, however, usually focused on its theoretical underpinning. Such a tack is taken by Ibn Sīnā (d. 428/1037; Avicenna), who insists that metals are not actually one and the same substance, different only in their accidents, but are, in fact, different by specific, unchangeable qualities (Sezgin 1971, 7–9; Ullmann 1972, 249–52). His opposition to alchemy is most prominent in his *The Indication for Knowing the Fallacy of Judicial Astrology* (*al-Ishāra ilā ʿilm fasād aḥkām al-nujūm*), but he also takes this stance in his encyclopedic *The Book of Healing* (*Kitāb al-Shifāʾ*) (Ibn Sīnā 1385/1965, 22; cf. Allemann in al-Baghdādī 1988, 41–4). It is ironic, then, that several alchemical works have been attributed to Ibn Sīnā, which would therefore seem to be pseudepigraphical (Storey 1977, 436–7, no. 758). The authenticity of *The Epistle of the Elixir* (*Risālat al-iksīr*) attributed to Ibn Sīnā has been questioned (Gutas 2014, 459), but in any case, its author dismisses alchemy, mentioning that he found the books of the art to be empty of reasoning (*qiyās*), which should, however, be the basis (*ʿumda*) of every art (*ṣināʿa*). He goes on to explain that alchemists will never really change metals but only dye them. However, by heating, these dyes would disappear again, and it would become clear that no essential change had taken place (Ibn Sīnā 1953, 35–6).

Other prominent critics of alchemy were the famous translator Ḥunayn b. Isḥāq (d. 260/873) and the philosopher Abū Yūsuf Yaʿqūb al-Kindī (d. *c.* 256/870). Ḥunayn's negative attitude toward alchemy only becomes clear from citations by the alchemist ʿIzz al-Dīn Aydamir al-Jildakī ([d. *c.* 743/1342]; *The Pearls' Lights in the Explanation of the Stone*, MS Berlin, Staatsbibliothek, Landberg 606, fol. 28v and *The End of the Search in Commenting the 'Acquired'*, MS Berlin, Staatsbibliothek, Landberg 350, vol. 1, fol. 16r, cf. Ullmann 1972, 249). Al-Kindī – as far as we can judge from later citations – argued that men are not able to perform certain actions that are nature's alone, including the speeding up of natural processes (Ullmann 1972, 250).

The most interesting critique is perhaps the one by ʿAbd al-Laṭīf al-Baghdādī (d. 629/1232), who had been an adept of the art in his youth but then turned philosopher and adversary of alchemy. In his *Epistle on the Disputation of Two Sages, Alchemist and Theoretical Philosopher* (*Risāla fī mujādalat al-ḥakīmayn, al-kīmiyāʾī wa-l-naẓarī*), he depicts a kind of soliloquy between his old self, the alchemist, and his new, current self, the philosopher. The philosopher explains why he does not hold alchemy to be a valid science, casting those who believed in the existence of the elixir as naïve. Here again, the main argument pivots on the problem of accidents versus substance; ʿAbd al-Laṭīf's alter ego argues that for the transmutation to truly occur, a change of substance, not of accidents, must take place (al-Baghdādī 1988, §§ 59–68).

Conversely, practical alchemists often defended alchemy also on a theoretical level, as did Abū Bakr al-Rāzī (d. 313/925 or 323/935), who wrote an epistle against al-Kindī titled *Book in Refutation of al-Kindī's Refutation of the Art* (*Kitāb al-radd ʿalā l-Kindī fī raddihi ʿalā l-ṣināʿa*); both works are now lost (Ullmann 1972, 250).

I.12.4 Transmission of alchemical knowledge

Alchemical writings and instruction must be kept secret. This literary topos can be found in almost all works on alchemy, as well as other occult sciences generally. While the topos usually is just that, a topos, it has complicated scholarship by the use of a coded language: many alchemical works resort to enigmatic codes instead of straightforward language to safeguard their secrecy. Some of the codenames (Decknamen) are commonly used in alchemical writings and therefore easily identified; for example, *shams*, "sun", means "gold"; *qamar*, "moon", means "silver"; and *ʿuqāb*, "eagle", stands for sal ammoniac. Other identifications seem to be limited to specific works or authors. Lists of codenames are given in several works of alchemy, for example, in al-Jildakī's *Book of the Proof* (*Kitāb al-burhān*). The modern compilation of such alchemical codenames by Siggel should be used with caution, however, as its sources are not always clear (Siggel 1951).

It is in keeping with this idea of hiding and secrecy that alchemy is also presented as a prophetic mode of knowledge, an art revealed only to those chosen by God (Braun 2016, 35). The teacher must therefore choose his disciples carefully – as the disciples should the teacher. The adept of alchemy must aspire to deserve the mercy of God. This theme was an important aspect already in Syriac alchemy. The true alchemist was completely pure, pious, free of avarice, assiduous and righteous. In fact, only the adept who sought true knowledge, not the manufacture of gold, could hope to be successful (von Lippmann 1919–54, vol. 1, 77).

In the *Book of the Element of Foundation* (*Kitāb usṭuquss al-uss*), one of the works of the Jābirian corpus, the adept is described as being of "sound intellect (*raʾy ṣaḥīḥ*), thorough reasoning, and constant studying". His assiduity (*ṣabr*) and his purity are likewise stressed (Holmyard 1928, 71, 100, 109–10). Accordingly, the philosopher al-Fārābī (d. 339/950–1) explains that making gold is not the main goal of the alchemist. If it were, the civilized world could not last, as trade would become impossible if gold and silver were omnipresent. The alchemist's real goal, therefore, must be the training of his intellect:

> Therefore, they [the ancient alchemical authorities] have not spoken openly in their books about any of the works of this art (*ṣināʿa*) and their goal was not to teach it through their books and to disseminate it among the masses. Rather, they sought to awaken an understanding (*al-fiṭan*) of the science . . . And he who is of superior intelligence will attain perfect knowledge and happiness without him knowing from whence they come, and his joy in the philosophy that he has grasped will then be greater than his joy in what he has attained of this art.
>
> *(al-Fārābī 1951, 77)*

If intellectual training is of paramount importance for any student of the art, then the question becomes: How are they to be trained? As in other fields of learning, oral instruction was often seen as central. At the same time, however, the worth and importance of books are frequently alluded to in alchemical texts. Already in the Jābirian corpus, students of alchemy are reminded that only by studying books – namely, Jābir's books – will they ever succeed in the divine art (Holmyard 1928, 100).

Therefore, we should consider orality to be a literary topos, at least in part, which explains the very large number of alchemical dialogues, especially works on alchemy written in the form of a discussion between a master and his disciple. The most prominent dialogue on alchemy is between the Christian monk Maryānus and the Umayyad prince Khālid ibn Yazīd, the alleged founder of Arabic alchemy, titled *Khālid's Questions to the Monk Maryānus* (*Masāʾil Khālid li-Maryānus ar-rāhib*), the first Arabic text on alchemy ever translated into Latin (probably by Robert of Chester in 1144; Kahn 1990–1991; Lemay 1990–1991; Cardelle de Hartmann 2007, 79; Dapsens 2016). The work, perhaps dating to the 3rd–4th/9th–10th centuries (Forster 2016, 401), depicts how Khālid had been long searching for instruction in alchemy, but unsuccessfully, when he learns of the existence of the monk Maryānus and his superior knowledge of the art. The prince thereupon has him brought to his palace and tries to obtain his confidence in order to attain his alchemical goal. The monk, however, does not need to be persuaded to reveal his knowledge, realizing that Khālid is a worthy pupil (Forster 2017, esp. 76). Other dialogues give speaking roles to Aristotle, the late antique god Agathodaimon, the queen Cleopatra, Mary the Copt, and the early Islamic specialist of things "Nabatean", Ibn Waḥshiyya ([d. 318/930–1]; Müller 2 012; Braun 2016; Forster 2017).

Besides reading and oral instruction, learning in a laboratory seems to have been another option for getting acquainted with alchemy and the chemical crafts, but evidence for this remains scarce. One text that constructs such a background is *The Great Epistle on the Spheres* (*al-Risāla al-falakiyya al-kubrā*), a text of the late 3rd/9th or early 4th/10th century (Vereno 1992, § 36–191), in which Hermes asks the high priest Uwīrūs (Osiris) for counsel. Their dialogue is an instruction for laboratory processes, obscured by symbolic language.

Oral instruction in such a context is likely also reflected in the fact that recipes were usually written in a very much abbreviated form that would make sense only for those already initiated – not unlike, in fact, modern cookbooks (Braun 2016, 35). Often, recipes were added to manuscripts on spare leaves after the fact or written on flyleaves (Margoliouth and Holmyard 1931; Braun 2016, 53). Among the few larger collections of alchemical recipes, one, probably dating from the 4th–5th/10th–11th centuries, has been attributed to the Muʿtazilite *qāḍī* ʿAbd al-Jabbār al-Hamadhānī (d. 415/1024). While this attribution is probably incorrect, it emphasizes the fact that an interest in practical alchemy was considered to be acceptable even for a judge (Leube 2013).

While alchemical manuscripts are not usually illustrated, when they are, depictions of alchemical apparatus feature most prominently; these are often quite detailed. They show furnaces and stoves and very often different apparatus for distillation (Chapter IV.5). These are usually relatively simple pen drawings, only sometimes in color or with elements of perspective. These illustrations seem to serve the purpose of explaining utensils and processes. While images of apparatus are not always transmitted in all manuscripts of one text, diagrams and tables are important and therefore usually an indispensable part of a text. However, they often serve not alchemical purposes but illustrate general rules, such as the relationship of the primary qualities. A nice, although somewhat late example is Jalāl al-Naqqāsh's (*fl.* 9th/15th century) *Solarizing Full Moons: A Versified Commentary on 'The Splinters'* (*Tashmīs al-budūr fī takhmīs al-shudhūr*), a commentary on Ibn Arfaʿ Ra's's (d. 593/1197) *The Splinters of Gold* (*Shudhūr al-dhahab*), that contains a lengthy illustrated introduction to the general foundations of alchemy; see, for example, Halle, University Library, DMG, 65, f. 7r (probably 12th/18th century). Another function of tables is to give lists of synonyms. See, for example, the unusually heavily illuminated manuscripts of *The Seven Climes* (*al-Aqālim al-sabʿa*), attributed to Abū l-Qāsim al-ʿIrāqī al-Sīmāwī ([*fl.* 7th/13th century]; for example Riyadh, King Saud University, 3167, pp. 30–1, probably 10th/16th century; and Figure I.12.1).

Finally, there is a tradition of allegorical illustrations that starts in early texts like the *Book of Pictures* attributed to Zosimus, who was discussed earlier, or Ibn Umayl's (*fl.* 4th/10th century) *Book of the Silvery Water and the Starry Earth* (*Kitāb al-mā' al-waraqī wa-l-arḍ al-najmiyya*). The latter text contains an interesting frontispiece in one of its extant manuscripts dating from the

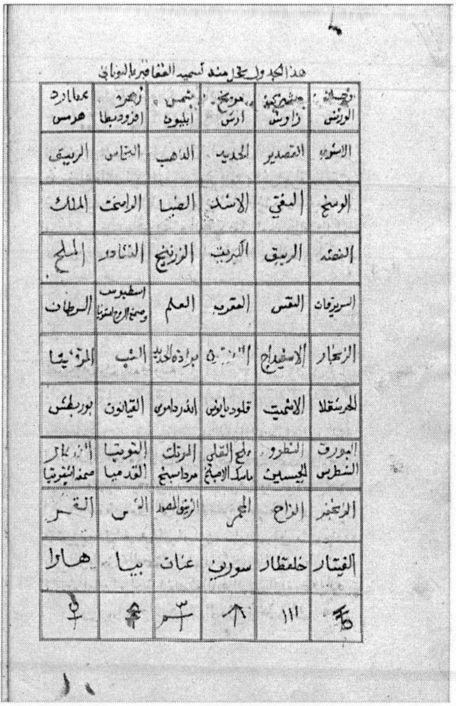

Figure I.12.1 A table giving the synonyms of various substances, provided in al-Sīmāwī al-ʿIrāqī's *The Seven Climes* (*al-Aqālīm al-sabʿa*); MS Gotha, Research Library, orient. A 1261, fol. 16b.

Source: © Research Library, Gotha

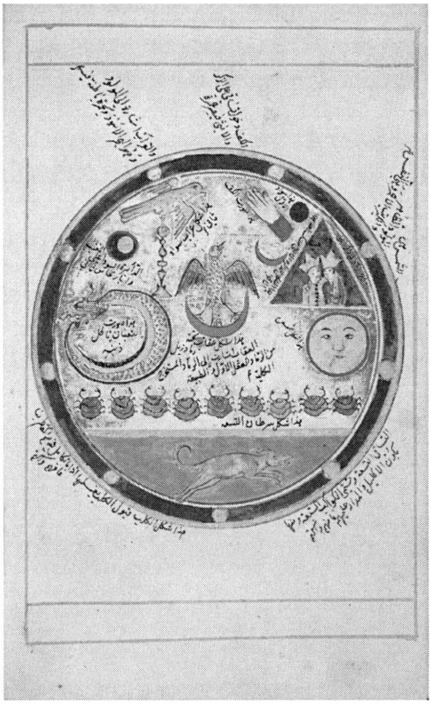

Figure I.12.2 A depiction of the mirror of wonders (*mir'āt al-ajā'ib*), a mysterious mirror that displays different symbols of alchemy, as contained in al-Jildakī's *The Shining of Thoughts* (*Lawāmiʿ al-afkār*); MS Oxford, Bodleian Library, Greaves 14, fol. 3b.

Source: © Bodleian Library, University of Oxford, Oxford

8th/14th century (see Abt 2003; Berlekamp 2003). Allegorical illustrations may also be found in later commentaries like al-Jildakī's *The Shining of Thoughts* (*Lawāmiʿ al-afkār*) that is illustrated allegorically in at least one case (Oxford, Bodleian Library, Greaves 14; Figure I.12.2) and especially in *The Seven Climes* (*al-Aqālīm al-sabʿa*), attributed to al-Sīmāwī, which features many allegorical illustrations.

I.12.5 (Al)chemical processes and products

While so far, due to the lack of sources, we have only discussed alchemy, the question about processes and products allows refocusing also on the chemical crafts. Alchemy as well as the chemical crafts took place in laboratories and workshops. However, we cannot say who exactly was involved, as the craftsmen did not sign their vessels or, for obvious reasons, their products. (For alchemical equipment, see Chapter IV.5.)

Abū Bakr al-Rāzī is perhaps the earliest author writing in Arabic to give a systematic overview of the processes (*tadābīr*, singular *tadbīr*) used in early chemistry and alchemy and their names in his *Book of Secrets* (*Kitāb al-asrār*; al-Hassan and Hill 1986 [CB], 134–5). The most important (al)chemical process is distillation (*taqṭīr*). Here, a liquid substance is heated in a vessel, and the distillate is then caught in an alembic and collected in a receiver. If solid substances were used, the process that served to purify substances was called sublimation (*taṣʿīd*). Unlike distillation, the substance to be sublimated would be put into an aludel (*uthāl*) instead of a cucurbit (*qarʿ*). Sublimation was a technique that served to purify the sublimated substances.

Distillation was not only the central process described in the quest for the transmutation, but it was also used in the production of alcohol, petroleum, oils and fats, acids and, most important, perhaps, in the production of rose water and essential oils used in the perfume industry. Al-Kindī describes the distillation of wine, and the renowned Andalusian physician Abū l-Qāsim al-Zahrāwī (fl. 1st half 5th/11th century) that of vinegar and of rose water. Most treatises on perfume production known to have existed are lost, but we have descriptions of the preparation of rose water by al-Kindī, al-Jawbarī (fl. 1st half 7th/13th century) and in manuals on agriculture. Distilled petroleum seems to have been used as a fuel and as a medicine (al-Hassan and Hill 1986[CB], 141–50). Although alchemy and medicine were strongly linked (Carusi 2003), it was only in early modern times, under the influence of Paracelsian iatrochemistry, that drugs were produced by alchemical processes, especially distillation and sublimation, and mainly from minerals (Bachour 2012).

(Al)chemical processes were sometimes described in great detail. The oxidation of mercury, an important innovation of Islamicate alchemy, is described by Abū l-Qāsim Maslama al-Qurṭubī (d. 353/964) in his *The Station of the Sage* (*Rutbat al-ḥakīm*), who remarks that the weights remained the same before and after the process (Holmyard 1931, 78; Ullmann 1979, 114).

Control of heat was essential in many of these processes. It was managed by the use of different furnaces and stoves, and heating via air (as in the process called *tashwiya*, "roasting") or via water (in baino marie). In the process of distillation, the tube was cooled with sponges or cloths, but systematic cooling by water cannot be proved to have existed. The fact that the instrument used to do so was called, in western Europe, a "Moor's Head" implies, however, an Islamicate origin (al-Hassan and Hill 1986 [CB], 138–9).

Other (al)chemical processes are calcination (*taklīs*, heating of a substance until it becomes a powder), ceration (reduction to a waxlike state), solution (*taḥlīl*, which seems to have included the use of acids), putrefaction (*taʿfīn*, a kind of decomposition aided by water), combination (of several substances into one powder or solution) and fixation (*tajmīd* or *taʿqīd*, in which the substance was put in a closed vessel and left to set).

Dying of metals, especially to make them look like silver or gold, seems to have been an important concern of medieval chemists. In his *Epistle of the Elixir*, Ibn Sīnā (or, possibly, pseudo-Ibn Sīnā) gives detailed explanations of permanent dyes, resistant to heat. The most important element of the process, he explains, is the use of mercury, which will allow for a permanent change of color, although for a red color he suggests the use of sulfur (*kibrīt*; Ibn Sīnā 1953, 37–46, 54).

Chemical methods were also used in the production of soap, glass (Djebbar 2001, 347–9) and, at least to some extent, ceramics (Porter 2011, 49–88) and metals, especially alloys. Chemical discussions of the production of swords may be found in al-Kindī's *Epistle to One of His Brethren about Swords* (*Risāla ilā baʿḍ ikhwānihi fī l-suyūf*; al-Hassan 1978; Djebbar 2001, 364). Finally, the production of inks and pigments was yet another field for practitioners of chemistry (al-Hassan and Hill 1986 [CB], 150–75, 246–55).

While alchemy has left a large written (and partially drawn) heritage that allows us to discuss its theories, principles, and actors, the chemical crafts are far more elusive, having left fewer textual traces that are accessible to us today. However, drawing a clear separation between these fields would be artificial, to some extent: Alchemy and the chemical crafts were based on similar or, in some cases, much the same substances, apparatus, and processes. It appears worthwhile to study alchemy from its different angles, as both the proto-chemistry (and therewith in the context of the chemical crafts) and the natural philosophy that it was, in order to get a clearer picture of the social and intellectual contexts of its practitioners, about whom little is known so far.

Notes

1 This chapter was written within the framework of the project "Between Religion and Alchemy. The scholar Ibn Arfaʿ Ra's (d. 1197) as a model for an integrative Arabic literary and cultural history" funded by the Swiss National Science Foundation.

2 This strand has been recently followed in the publications of the "Corpus Alchemicum Arabicum" series; see, for example, Abt 2003, 2007; Abt et al. 2003; Abt and Fuad 2011. For applications to Western science, see Dobbs 1983.

3 For the Chinese influence, see Hill 1990, 332–3 and Strohmaier 2016. Indian influence on Islamicate alchemy has not been extensively studied yet. See, however, Speziale 2010, 430–3, and the same author's chapter in this volume. Additionally, the prominent position of mercury in Indian alchemy and in the Islamicate tradition is striking (on the oxidation of mercury see Ullmann 1979, 114; Holmyard 1931, 78); on mercury in Indian alchemy Hellwig 2009, 12–13. For a skeptical view on the Persian connection, see Ullmann 1972, 148.

4 Anawati's chapter is largely based on Ullmann 1972 but is useful for an English-speaking audience.

5 Consolidated bibliography.

6 I adapt Dodge's translation who gives "which is the making of gold and silver from other metals" (Ibn al-Nadīm 1970[CB], 2: 843). For a slightly different rendering of this passage see Fück 1951, 88.

Bibliography

Sources

Abt, T., ed. 2007. *The Book of Pictures/Muṣḥaf aṣ-ṣuwar by Zosimos of Panopolis. Facsimile with an Introduction.* Zurich: Living Human Heritage Publications.

Abt, T. and Fuad, S., tr. 2011. *The Book of Pictures/Muṣḥaf aṣ-ṣuwar by Zosimos of Panopolis.* Zurich: Living Human Heritage Publications.

Abt, T. *et al.*; Abt, T., Madelung, W. and Hofmeier, T. eds. and Fuad, S. and Abt, T., tr. 2003. *Book of the Explanation of Symbols/Kitāb Ḥall ar-Rumūz by Muḥammad Ibn Umail.* Zurich: Living Human Heritage Publications.

al-Baghdādī, ʿAbd al-Laṭīf. ed. and tr. Allemann, F. 1988. *ʿAbd al-Laṭīf al-Baghdādī, Ris. fī Muǧādalat al-ḥakīmain al-kīmiyāʾī wan-naẓarī („Das Streitgespräch zwischen dem Alchemisten und dem theoretischen Philosophen"). Eine textkritische Bearbeitung der Handschrift: Bursa, Hüseyin Çelebi 823, fol. 100–123 mit Übersetzung und Kommentar.* PhD thesis. University of Berne.

al-Fārābī. ed. Sayılı, A. 1951. "Fârâbî'nin Simyanın Lüzûmu Hakkındaki Risâlesi [Fararbi's Treatise on the Necessity of Alchemy]," *Belleten* 15: 65–79.

Halleux, R., ed. 1981. *Papyrus de Leyde. Papyrus de Stockholm. Fragments de recettes.* Paris: Société d'Edition Les belles lettres.

Holmyard, E. J., ed. 1928. *The Arabic Works of Jâbir ibn Ḥayyân.* Vol. 1, part 1 (Arabic texts). Paris: Librairie Orientaliste Paul Geuthner.

Ibn Sīnā. ed. Ateş, A. 1953. "İbn Sīnā. Risâlat al-iksîr [The Epistle on the Elixir]," *Türkiyat Mecmuası* 10: 27–54.

Ibn Sīnā. ed. Madkūr, I. 1385/1965. *al-Shifāʾ [2]: al-Ṭabīʿiyyāt 5: al-Maʿādin wa-l-āthār al-ʿulwiyya = Les métaux et la météorologie.* Cairo: al-Hayʾa al-ʿāmma li-shuʾūn al-maṭābiʿ al-Amīriyya.

Jābir ibn Ḥayyān. ed. Kraus, P. ed. 1935. *Mukhtār rasāʾil Jābir b. Ḥayyān* [Selected Epistles of Jābir ibn Ḥayyān]. Paris: Maisonneuve; Cairo: al-Khānjī.

Leube, G. 2013. *Die Rezepte der Freiburger alchemistischen Handschrift des ʿAbd al-Ǧabbār al-Hamaḏānī. Edition, Übersetzung und Kommentar.* Berlin: Klaus Schwarz.

Vereno, I. 1992. *Studien zum ältesten alchemistischen Schrifttum. Auf der Grundlage zweier erstmals edierter arabischer Hermetica.* Berlin: Klaus Schwarz.

Manuscripts

MS Berlin, Staatsbibliothek Preußischer Kulturbesitz, Orientabteilung, Landberg 350.
MS Berlin, Staatsbibliothek Preußischer Kulturbesitz, Orientabteilung, Landberg 606 .
MS Gotha, Research Library, orient. A 1261.
MS Halle, University Library, DMG, 65.
MS Oxford, The Bodleian Libraries, Greaves 14.
MS Riyadh, King Saud University, 3167.

Research literature

Abt, T. 2003. *The Great Vision of Muḥammad Ibn Umail*. Los Angeles: C.G. Jung Institute of Los Angeles.

Anawati, G. C. 1996. "Arabic Alchemy," in Rashed, R., ed. *Encyclopedia of the History of Arabic Science*. Vol. 3. London and New York: Routledge, 853–85.

Artun, T. 2013. *Hearts of Gold and Silver: The Production of Alchemical Knowledge in the Early Modern Ottoman World*. PhD thesis, Princeton University.

Bachour, N. 2012. *Oswaldus Crollius und Daniel Sennert im frühneuzeitlichen Istanbul. Studien zur Rezeption des Paracelsismus im Werk des osmanischen Arztes Ṣāliḥ b. Naṣrullāh Ibn Sallūm al-Ḥalabī*. Freiburg: Centaurus.

Berlekamp, P. 2003. "Painting as Persuasion: A Visual Defense of Alchemy in an Islamic Manuscript of the Mongol Period," *Muqarnas* 20: 35–59.

Braun, C. 2016. *Das Kitāb Sidrat al-muntahā des Pseudo-Ibn Waḥšīya. Einleitung, Edition und Übersetzung eines hermetisch-allegorischen Traktats zur Alchemie*. Berlin: Klaus Schwarz.

Cardelle de Hartmann, C. 2007. *Lateinische Dialoge 1200–1400. Literaturhistorische Studie und Repertorium*. Leiden and Boston: Brill.

Carusi, P. 2003. "Il filosofo e il marinaio. Alchimia islamica e medicina alle prese con la natura," in Crisciani, C. and Paravicini Bagliani, A., eds. *Alchimia e medicina nel Medioevo*. Florence: SISMEL edizioni del Galluzzo, 19–31.

Dapsens, M. 2016. "De la *Risālat Maryānus* au *De Compositione alchemiae*. Quelques réflexions sur la tradition d'un traité d'alchimie," *Studia graeco-arabica* 6: 121–40.

Djebbar, A. 2001. *Une histoire de la science arabe. Introduction à la connaissance du patrimoine scientifique des pays d'Islam. Entretiens avec Jean Rosmordu*. Paris: Seuil.

Dobbs, B. J. T. 1983. *Foundations of Newton's Alchemy, or The Hunting of the Greene Lyon*. Cambridge: Cambridge University Press.

Forster, R. 2016. "The Transmission of Secret Knowledge: Three Arabic Dialogues on Alchemy," *Al-Qantara* 37: 399–422.

Forster, R. 2017. *Wissensvermittlung im Gespräch. Eine Studie zu klassisch-arabischen Dialogen*. Leiden and Boston: Brill.

Fowden, G. 1986. *The Egyptian Hermes. A Historical Approach to the Late Pagan Mind*. Cambridge: Cambridge University Press.

Fück, J. 1951. "Arabic Literature on Alchemy According to An-Nadīm," *Ambix* 4: 81–144.

Hallum, B. 2008. *Zosimus Arabus. The Reception of Zosimos of Panopolis in the Arabic/Islamic World*. PhD thesis, The Warburg Institute, London.

Haq, S. N. 1994. *Names, Natures and Things. The Alchemist Jābir Ibn Ḥayyān and His Kitāb al-Aḥjār (Book of Stones)*. Dordrecht: Kluwer.

al-Hassan, A. Y. 1978. "Iron and Steel Technology in Medieval Arabic Sources," *Journal for the History of Arabic Science* 2: 31–43.

Hellwig, O. 2009. *Wörterbuch der mittelalterlichen indischen Alchemie*. Eelde and Groningen: Barkhuis.

Hill, D. R. 1990. "The Literature of Arabic Alchemy," in Young, M. J. L., Latham, J. D. and Serjeant, R. B., eds. *Religion, Learning and Science in the ʿAbbasid Period*. Cambridge: Cambridge University Press, 328–41.

Hill, D. R. 1993. *Islamic Science and Engineering*. Edinburgh: Edinburgh University Press.

Holmyard, E. J. 1931. *Makers of Chemistry*. Oxford: Clarendon.

Kahn, D. 1990–1. "Note sur deux manuscrits du Prologue attribué à Robert de Chester," *Chrysopoeia* 4: 33–4.

Kraus, P. 1930. "Dschābir ibn Ḥajjān und die Ismāʿīlijja," in Ruska, J. and Kraus, P., eds. *Dritter Jahresbericht des Forschungs-Instituts für Geschichte der Naturwissenschaften in Berlin. Mit einer wissenschaftlichen Beilage: Der Zusammenbruch der Dschabir-Legende*. Berlin: Springer, 23–42.

Kraus, P. 1942. "Les dignitaires de la hiérarchie religieuse selon Ǧābir ibn Ḥayyān," *Bulletin de l'Institut français d'archéologie orientale* 41: 83–97.

Kraus, P. 1942–1943. *Jābir ibn Ḥayyān. Contribution à l'histoire des idées scientifiques dans l'Islam.* 2 vols. Cairo: Institut français d'archéologie orientale.

Lemay, R. 1990–1991. "L'authenticité de la préface de Robert de Chester à sa traduction du Morienus," *Chrysopoeia* 4: 3–32.

Lippmann, E. O. von. 1919–54. *Entstehung und Ausbreitung der Alchemie.* 3 vols. Berlin: Springer.

Lory, P. 1983. *Dix traités d'alchimie de Jâbir ibn Hayyân. Les dix premiers traités du Livre des Soixante-dix.* Paris: Sindbad.

Lory, P. 1988. *L'élaboration de l'élixir suprême: quatorze traités de Ǧâbir b. Ḥayyân sur le grand œuvre alchimique.* Damascus: Institut Français de Damas.

Margoliouth, D. S. and Holmyard, E. J. 1931. "Arabic Documents from the Monneret Collection," *Islamica* 4: 249–71.

Marquet, Y. 1970. "Ikhwān al-Ṣafā'," in *EI-2,* 3: 1071–6.

Martelli, M. and Rumor, M. 2014. "Near Eastern Origins of Graeco-Egyptian Alchemy," in Geus, K. and Geller, M., eds. *Esoteric Knowledge in Antiquity.* Berlin: Max-Planck-Institut für Wissenschaftsgeschichte, 37–62.

Müller, J. 2012. *Zwei arabische Dialoge zur Alchemie. Die Unterredung des Aristoteles mit dem Inder Yūhīn und das Lehrgespräch der Alchemisten Qaydarūs und Mītāwus mit dem König Marqūnus. Edition, Übersetzung, Kommentar.* Berlin: Klaus Schwarz.

Porter, Y. 2011. *Le prince, l'artiste et l'alchimiste. La céramique dans le monde iranien. Xe- XVIIIe siècles.* Paris: Hermann.

Schütt, H.-W. 2000. *Auf der Suche nach dem Stein der Weisen. Die Geschichte der Alchemie.* Munich: Beck.

Sezgin, F. 1971. *Geschichte des arabischen Schrifttums, Vol. 4: Alchimie, Chemie, Botanik, Agrikultur bis ca. 430 H.* Leiden: Brill.

Siggel, A. 1951. *Decknamen in der arabischen alchemistischen Literatur.* Berlin: Akademie-Verlag.

Speziale, F. 2010. "Les traités persans sur les sciences indiennes: médecine, zoologie, alchimie," in Herman, D. and Speziale, F., eds. *Muslim Cultures and the Indo-Iranian World during the Early-Modern and Modern Periods.* Berlin: Klaus Schwarz, 403–47.

Storey, C. A. 1977. *Persian Literature. A Bio-Bibliographical Survey, Vol. II, Part 3: F. Encyclopedias and Miscellanies. G. Arts and Crafts. H. Science. J. Occult Arts.* London: Royal Asiatic Society of Great Britain and Ireland.

Strohmaier, G. 2016. "Elixir, Alchemy and the Metamorphoses of Two Synonyms," *Al-Qanṭara* 37: 423–34.

Taylor, F. S. 1937. "The Origins of Greek Alchemy," *Ambix* 1: 30–47.

Ullmann, M. 1972. *Die Natur- und Geheimwissenschaften im Islam.* Leiden: Brill.

Ullmann, M. 1978. "Ḫālid ibn Yazīd und die Alchemie: Eine Legende," *Der Islam* 55: 181–218.

Ullmann, M. 1979. "al-Kīmiyā'," in EI-2, 5: 110–15.

I.13
GEOGRAPHY AND MAPMAKING UNTIL 700/1300

Michael Bonner

Geography and cartography together constituted an area of high achievement for the Islamicate world. They did not, however, enjoy full scientific status in the manner of, say, geometry and astronomy; for instance, they did not appear in the classification of the sciences by the philosopher al-Fārābī (d. 339/950–1). Of course, geography was not the only discipline with such ambivalent status; history in particular offers parallels (Rosenthal 1968, 30–53). All the same, Islamicate geography achieved a striking "style" and "look" of its own, with the visual beauty of its maps, the refinement of its texts, its love of precise observation and its far-ranging curiosity. It also presented an original blend of what we might call, in our modern terms, the Islamic sciences and humanities, raising interesting questions about each of these.

I.13.1 Origins

It would be difficult to say, at least in a brief summary, how the Arabs of the pre-Islamic era represented the physical world to themselves. To begin with, their most precious cultural resource, their poetry, required a vast knowledge of Arabian topography and place-names, from both poets and audiences. Beginning in the late 20s–early 30s/630s, however, the great conquests and the ensuing exodus from Arabia had the effect of blurring, if not erasing this knowledge. The Arabic philologists of the Umayyad and early Abbasid eras accordingly compiled massive amounts of information about Arabian toponyms, and much of this later found its way into the geographical literature (Kramers 1954a, 84). However, our fullest information in this area comes from the Qurʾān. Here the earth is flat, with the sky raised over it as a "guarded roof" (*saqf maḥfūẓ*, 21:32) and the mountains serving to keep it stable (13:3, 15:19, 50:7, 79:30, etc.). God has also fashioned the sky into seven heavens (*samāwāt*, 2:29, 41:12, etc.) or roads or tracts (*ṭarāʾiq*, 23:17). On the whole, this Qurʾānic cosmology accords with archaic Near Eastern mythology and tradition. In the following centuries, it received amplification in *ḥadīth* (Prophetical Tradition) and in *tafsīr* (scriptural exegesis). However, it remained mostly separate from the literature described here as 'geographical,' despite attempts at reconciling the two (Ibn al-Faqīh 1885, 3; Muqaddasī 1877, 10–19).

This geographical literature began in circumstances that remain obscure to us for several reasons, including the fact that none of it is extant from before the later 2nd/8th century, as is the case for Arabic literature more generally. We do have references to mapping during the

DOI: 10.4324/9781315170718-15

Umayyad and early Abbasid eras, apparently for practical purposes. We know, for example, about maps of Daylam (Ibn al-Faqīh 1885, 283) and Bukhara (Ṭabarī 2010, 2:1199), commissioned by the Umayyad governor al-Ḥajjāj (d. 95/714), and a map of the swamps of lower Iraq (Balādhurī 1866, 371; Tibbetts 1992a, 90) commissioned by the Abbasid Caliph al-Manṣūr (r. 136–158/754–775). All in all, however, two things stand out about this literature's early history. The first is its composite nature: it emerged from a variety of activities corresponding to a variety of needs. The second is that it began with acts of translation, not only from Greek but also from Sanskrit, Middle Persian (Pahlavi) and Syriac.

The first important encounter was with India. It appears that several Indians learned in astronomy, astrology and cosmography visited the court of Caliph al-Manṣūr (r. 136–158/754–775), known for his interest in astrology. As a result, we have some of the earliest translations (or adaptations) of foreign works into Arabic (Gutas 1998[CB],[1] 24–5). The Sanskrit word *siddhānta*, appearing in the titles of several astronomical works, entered Arabic as *sindhind*. Although the Arabic works in this tradition are now lost, Indian ideas retained a place in Islamicate geography and, afterward, in its Latin counterpart. These include the notion of a "cupola of the earth," a summit or center of the known and inhabited hemisphere of the world, located at a place in central India equally distant from the northern, southern, eastern and western limits of this hemisphere.

As the translators churned out more and more scientific texts, Arabic speakers interested in astronomy, astrology, geography and mapmaking shifted their attention to the Greek heritage in these fields, which Claudius Ptolemy of Alexandria (*c.* 100–170) came to represent practically all by himself. Ptolemy's *Megalē Syntaxis* was translated early on into Arabic, as *al-Majisṭī* (from which came its later Latin title, *Almagest*). In this astronomical work, Ptolemy had devoted a section to the division of the earth into "climes" ("inclinations," Greek *klimata*, sing. *klima*), which divided the northern hemisphere into zones by latitude. Arabic scholars now appropriated this term as *iqlīm*, pl. *aqālīm*, and made it a vital element of their own understanding of these matters. It is unclear, however, if the translators of the 3rd/9th and 4th/10th centuries ever produced an Arabic version of Ptolemy's work *Geography*, devoted entirely to this field of knowledge. Ibn Khurradādhbih (d. 299/911–2) claimed to have translated it, and translations were attributed to later scholars. In the end, however, it seems that if an Arabic translation of Ptolemy's *Geography* did ever exist, it arrived later than the events about to be described here.

I.13.2 al-Khwārazmī and mathematical geography

What appears to be a mere gap in the vast library of Arabic scientific translations acquires more importance when we turn to a foundational moment for Islamicate geography, the appearance of the *Image of the Earth* (*Ṣūrat al-arḍ*) by al-Khwārazmī ([d. *c.* 235/850;] see Nallino 1944; Kramers 1954a, 192–5; Krachkovskiĭ 1957, 91–2; Hopkins 1990; Tibbetts 1992a; Brentjes 2007). An astrologer with broad interests, Khwārazmī here embarked on a project best described as Ptolemaic. Its Arabic title, which means, as indicated earlier, "image, representation of the earth," is formed (calqued) after the Greek *geōgraphia*, which means exactly the same thing. Both works consist of tables, and al-Khwārazmī has taken over large amounts of data from his predecessor.

How did al-Khwārazmī get access to Ptolemy's *Geography*? It is unlikely that he knew Greek himself, although perhaps someone could have translated for him on the spot. Since the *Geography* had been known previously to Syriac authors (Krachkovskiĭ 1957, 20), including Bardesanes (d. *c.* 222), an anonymous translator (mid-5th century), and Jacob of Edessa (d. 89/708), Syriac could quite plausibly have served as an intermediary (Chapter I.2). In any case, while al-Khwārazmī replicated many of his predecessor's values for latitude and longitude, he changed many others, while rearranging the material's order and headings. He also added data relating to places that did not exist in Ptolemy's time and substituted Arabic toponyms for now-obsolete Greek ones.

Then where did al-Khwārazmī obtain his information, whenever he did not take it directly from Ptolemy? And on what basis did he "correct" many of Ptolemy's values? We have no evidence suggesting that he dispatched teams of collaborators to hundreds of locations throughout the world. Accordingly, the likely answer is that al-Khwārazmī tweaked Ptolemy's data on the basis of information from more recent sources, not all of them (or even most of them) necessarily mathematical or astronomical at all.

Tables of longitude and latitude were of little practical use at that time, although scholars would continue to compile them for centuries to come. Quite a few modern researchers have maintained, however, that al-Khwārazmī's project included drawing a world map, meant to fit within the dimensions of his book. Another idea, often proposed though not universally accepted, is that al-Khwārazmī was contributing here toward a large-scale world map, using coordinates of latitude and longitude, for Caliph al-Maʾmūn (r. 189–218/813–833). Such a project is indeed mentioned in some sources. In further support of this notion, it appears that Khwārazmī participated in the "House of Wisdom" (*bayt al-ḥikma*) in Baghdad, which many modern scholars have described as a research and/or teaching center supported by the caliphal court (Chapter III.1). However, problems arise. The *Image of the Earth* (of which only one manuscript has survived) does not mention any cartographical project, big or small. A few later sources do refer to a "Maʾmūnian image [or map]" (*ṣūra maʾmūniyya*), but these accounts seem inaccurate or ill informed, especially that of al-Masʿūdī ([d. 345/956]; 1894, 33), who apparently does not understand the Ptolemaic climes and who follows the Persian system of *kishvar*s (regions) in his own work. Similarly, a *bayt al-ḥikma* may have existed in Baghdad as early as the reign of Caliph al-Manṣūr, but its nature and role are more obscure to us today than they were several decades ago; in any case, it seems to have been more involved with Iranian learning than with Greek (Balty-Guesdon 1992; Gutas 1998, 53–60 [CB]; Richter-Bernburg 2016). Most significantly, difficulties arise regarding the process of mapmaking itself. Possibly because he did not have a complete translation of Ptolemy's *Geography*, al-Khwārazmī apparently did not consult the chapter on projections and therefore did not address the problem of how to represent the earth's spherical surface upon the flat surface of a map. All this makes it less likely that the Maʾmūnian world map was based on mathematical coordinates; there is reason to think instead that it used the Persian system of *kishvar*s. Looking ahead, we have a partial explanation here for why Islamicate cartography never dealt fully with the problem of projection (Tibbetts 1992a, 94–5), with the exception of al-Bīrūnī (362–d. after 444/973–d. after 1053) and possibly a few other scholars before 700/1300. Al-Khwārazmī's work in this area continued among later scholars including the astronomer al-Battānī (244–317/858–929), who developed further tables for geographical coordinates, and the enigmatic Suhrāb ([1st half 4th/10th century]; see Miquel 1967–88, 1: 79–80), who did the same while providing explicit instructions for preparing a map.

Most geographers considered the sphere of the earth as divided into two halves, only one of them known or even knowable. They divided the known half of the earth once more in half, with the northern part, or quadrant, constituting the inhabited world. They then further divided this northern quadrant with equally spaced latitudinal lines, reaching from the equator to the pole; the area between each two of these constituted one of the seven climes. Work also continued on questions going back to the Hellenistic geographers, including how to calculate the distance of a single degree of latitude. Meanwhile, from the beginning of Islam, scholars faced the problem of determining the direction of Mecca (the *qibla*) from any point in the world; their solutions, both approximate and accurate, constituted a great achievement for Islamicate geography over many centuries (King 1999[CB]; King and Lorch 1992[CB]; Chapter VI.3).

I.13.3 Ibn Khurradādhbih and administrative geography

Islamicate geography received input early on not only from Indian and Greek but also from Persian learning. Especially important was an Iranian tradition of representing the world as divided into seven *kishvars*, with one *kishvar* located centrally and the other six grouped around it. Islamic-era writers in this tradition referred to the central *kishvar* as *Īrānshahr*, the "domain of Iran," including not only modern-day Iran but also Iraq and some other places. This scheme appeared often in Islamicate geography, in effect as an alternative to the Ptolemaic system. However, since Arabic writers rendered Persian *kishvar* with the same term that they used for the Ptolemaic "clime," namely *iqlīm*, and since they assigned a total of seven units or regions to each system, confusion often prevailed.

Until now our story has connected scientific work with centers of political power. It has also involved a process of translation relating, though only in part, to those same centers of power. It has also been associated with the drawing of maps. Meanwhile, other things were happening in the vast area of the caliphate or Islamic Empire. In particular, armies, bureaucrats, merchants, seekers of religious merit and knowledge, criminals and other groups were on the move, often over long distances. This movement of so many people over so many routes created all sorts of material, organizational and intellectual needs.

Here again the caliphal government took the lead, through its chancery and bureaucracy. The management of long-distance communications led to the creation (or continuing) of a postal and intelligence bureau (*dīwān al-barīd*; Silverstein 2007). Heading this for a time in the mid-3rd/9th century was Ibn Khurradādhbih, a professional bureaucrat of Persian descent, whose *Routes and Realms* (*al-Masālik wal-mamālik*) can be considered, along with al-Khwārazmī's *Image of the Earth*, as one of the foundations of Islamicate geography. Although Ibn Khurradādhbih includes smatterings of Greek and Persian geography in his text, and even claims to have translated Ptolemy himself, he proceeds independently of astronomical and mathematical geography. Instead, and true to its title, *Routes and Realms* has a framework constituted by its wide-ranging itineraries. It indicates distances with stages (what a traveler can cover in a day), and with units of measurement such as the Persian *farsakh* (six kilometers) and the *mīl* or "Arab mile" (somewhat longer than the English mile). It seems obvious – although it is never stated outright – that these itineraries result from compilations by the *dīwān al-barīd* and its predecessors. At times Ibn Khurradādhbih adds landmarks for help in identifying place-names. He also gives fiscal information, less useful for travelers but much appreciated by later historians. At the same time, he cites poetry connected with particular places. He tells about wonders such as Sicilian volcanoes and the riches of Visigothic Spain, once seized by the Muslim conquerors. He also relays the purported accounts of travelers, such as the expedition of Sallām the Interpreter, sent by an Abbasid caliph to the land of Yājūj and Mājūj (Gog and Magog) in the (otherwise legendary) northeastern corner of the world. The overall result, while enjoyable for readers, lacks coherence, as some later geographers would point out. Meanwhile, Ibn Khurradādhbih makes no use whatsoever of maps. But he used schemes or diagrams, for instance for representing various *qibla*s (King 1999).

I.13.4 Geography under the unitary caliphate and its immediate successors

Two characteristics of Ibn Khurradādhbih's work proved essential for the next phase of geographical literature. The first was the participation of the imperial bureaucracy in compiling and disseminating useful knowledge. The second was the cultivation of a fine literary style, together with the display – indeed, the flaunting – of one's knowledge of many topics, including, but not restricted to, matters relating to imperial administration. These traits distinguished the *adīb*, or cultivated *littérateur*, while the relevant cultural domain was that of *adab*, which referred, on one hand, to polite, refined behavior and conduct and, on the other hand, to literary works that conveyed those values and showed readers how to acquire them.

In the following decades, several authors followed Ibn Khurradādhbih's example. These included al-Jayhānī (d. after 367/978), a vizier (chief administrator) for the principate of the Samanids in Transoxania (today Turkmenistan and Uzbekistan) in the early 4th/10th century, and the author of a book bearing the same title as Ibn Khurradādhbih's, *Routes and Realms*. We do not have this work, although many fragments have survived through direct or indirect quotation. It seems clear, in any case, that Jayhānī followed Ibn Khurradādhbih in his use of official archives and his emphasis on *adab*.

Somewhat earlier we have al-Yaʿqūbī (d. after 292/905), one of the most interesting writers in this entire literature. Yaʿqūbī worked in government service in various places, including the frontier region of Armenia. With his astute understanding of high politics, he comes across as a historian first and foremost. His surviving works on history and geography effectively treat the latter as the handmaiden of the former; geographical knowledge constitutes a prerequisite for the study of history. In style and presentation, Yaʿqūbī is direct and unornamented. Even if he subscribes to the values of *adab*, he does not foreground them.

We find a comparable trajectory in Masʿūdī, mentioned earlier. Iraqi by birth, Masʿūdī traveled far and wide (although not as far and wide as he claimed) and acquired a great stock of experiential knowledge, together with an impressive erudition derived from books. Like Yaʿqūbī, Masʿūdī made history his primary concern. He differed from Yaʿqūbī, however, in being unencumbered by government service, and in his cultivation of the literary and educational values of *adab*. Indeed, Masʿūdī imitated the prose style of the immortal al-Jāḥiẓ (d. 255/868), more successfully than anyone before or since. This "literarization" of the material, political and ethical worlds had already reached a peak in two geographers active around the turn of the 4th/10th century, Ibn al-Faqīh al-Hamadhānī and Ibn Rusta, both of western Iranian origin.

Qudāma ibn Jaʿfar is known to us as a loyal servant of the empire active in Baghdad in the earlier 4th/10th century and as the author of the *Book of the Land-tax and the Secretary's Art* (*Kitāb al-kharāj wa-ṣināʿat al-kitāba*). Geography constitutes only one of the disciplines forming the basis of this broad-ranging work. The *Book of the Land-tax* also differs from its predecessors in the presentation of its material. For instance, it isolates the itineraries within a chapter of their own. Yet it rightly belongs within our sequence of geographical works for several reasons, beginning with its treatment of the frontiers. In Qudāma's day, the unitary caliphate was fast becoming a fiction, as it lost control of its territories while successor states arose everywhere to supplant it. This fact underlies Qudāma's chapter on the frontiers of Islam (*thughūr al-Islām*), where he counterintuitively portrays an Islamic world where unity prevails, not only in culture and religion but also, and above all, in political and administrative matters. This unity expresses itself through engagement with the frontiers of Islam throughout the known world.

Another reason for including Qudāma within the geographical literature is his awareness of its three main components, "hard" science, bureaucratic administration and *adab*. Qudāma shows ample respect for the Ptolemaic tradition, and not surprisingly, he emphasizes the importance of administration. However, even though he advocates the cultivation of a fine Arabic style, he opposes the privileging of the values of *adab* characteristic of many of his contemporaries, and he "rejects any attempt to subordinate the geographical discipline to the literary goals of a cultural elite" (Heck 2002, 96).

We have seen both similarities and contrasts among the geographical writers active during the decades after al-Khwārazmī and Ibn Khurradādhbih. Some modern researchers group these into a putative 'Iraqi school,' but this seems to serve mainly for the purpose of contrasting them with the 'Balkhī school,' which we will take up next. One thing that does unite them, however, is their neglect of cartography, as they knew about maps but did not use them.

I.13.5 The 'Balkhī School'

Modern scholarship has identified a 'Balkhī school,' active throughout most of the 4th/10th century but including only three members, with a sort of honorary membership usually extended to a fourth. However, not only is this 'Balkhī school' a concept of modern scholarship, but the relations among these authors, especially the first three, are enigmatic. Perceptive scholars such as al-Idrīsī (d. after 560/1164–5) and Yāqūt al-Rūmī (d. 626/1229) showed perplexity over what these three took from one another and whether they were not really just one or two individuals. A solution came in 1871, when M. J. de Goeje (1836–1906) showed that these were indeed three distinct authors; that the geographical work of the first of them, Balkhī, was no longer extant; and that the other two composed distinct, although related, books. Afterward, scholars went on to associate this 'Balkhī school' with a collective project that they called 'the atlas of Islam' (Kramers 1932; Krachkovskiĭ 1957, 194–219; Miquel 1967–88, 1:267–330; Miller 1986 1.1.17, 5:109; Hopkins 1990, 312–15; Tibbetts 1992b).

The founding figure was the courtier and *littérateur* Abū Zayd al-Balkhī (d. 322/934), who composed a *Book of the Image of the Earth* (*Kitāb Ṣūrat al-arḍ*). This began with a map of the world and then presented twenty maps of provinces or "climes" with accompanying text. Balkhī's work was taken up afterward by al-Iṣṭakhrī (d. after 340/951), who traveled extensively and composed a *Book of Routes and Realms* (*Kitāb al-masālik wal-mamālik,*), revising Balkhī's maps while adding new text. We know little else about the connection between these two, since Balkhī's maps and text became absorbed into Iṣṭakhrī's or simply disappeared. Then in 331/943, a young man known as Ibn Ḥawqal departed from Baghdad, intending to map and describe as many of the regions of the world as he could. Some time later, he met Iṣṭakhrī (Ibn Ḥawqal 1938, 329–30). The encounter is often said to have taken place in Sind, since Ibn Ḥawqal describes it in his chapter on that province, but this is incorrect, since Ibn Ḥawqal almost certainly never went there (Miquel 1967–88, 1:299–300).

According to Ibn Ḥawqal, after they had shown each other some of their maps, Iṣṭakhrī asked the younger man to correct and continue his own work. Ibn Ḥawqal agreed, but after doing this for a while, he decided to take over Iṣṭakhrī's work on his own account and rewrite the text, redraw the maps and publish it all under his own name. In the wake of this apparent act of plagiarism we have, first of all, a book called *Routes and Realms* ascribed entirely to Iṣṭakhrī, and then we have Ibn Ḥawqal's work which exists in two main recensions, the first called *Routes and Realms*, and the second *Image of the Earth*. In both recensions, Ibn Ḥawqal incorporates Iṣṭakhrī's text while adding new material of his own. In other words, while the younger writer has devoured his predecessor whole, the predecessor's work has also survived, separately and unscathed.

The 'Balkhī School' is also usually considered to include the great work by Ibn Ḥawqal's younger contemporary al-Muqaddasī (2nd half 4th/10th century), *The Best Division for Knowledge of the Climes* (*Aḥsan al-taqāsīm li-maʿrifat al-aqālīm*). Muqaddasī knew about Balkhī and, apparently, about Iṣṭakhrī (De Goeje 1871, 47–8). However, even though he and Ibn Ḥawqal were engaged in similar projects, they show no awareness of one another, whether out of ignorance or jealousy.

Like other Islamicate geographers, the 'Balkhī' authors show respect for the Ptolemaic tradition. In practice, however, they follow different principles, especially by using the *iqlīm* as an organizing principle not in its Ptolemaic sense, but in a sense more akin to the Persian *kishvar*.

As a result, *iqlīm* for them means something like province. Accordingly, the question of where to draw the lines, how to delineate an *iqlīm*, ranks high among their concerns. We may recall that this was a time of political fragmentation, as the Abbasid caliphs lost all their territories and most of their authority. For the 'Balkhī' authors, the "climes" do not pertain to a caliphate (*khilāfa*), or even to a juridically constructed Abode of Islam (*dār al-Islām*) but rather to a "Realm of Islam" (*mamlakat al-Islām*). Like the individual provinces, this realm is defined not only by religious and political criteria but also by cultural and economic ones. The work of the 'Balkhī' authors has accordingly been described as the high point of an Islamicate 'human geography' (Miquel 1967–88, 1: 267–330).

All this may help us understand another difference between the 'Balkhī' authors and their predecessors. Beginning with Ibn Khurradādhbih, travelers' itineraries are an essential feature of the geographical literature, extending through individual provinces all the way to the end of the Islamic world, and constituting the communications network of a great empire. At times, Ibn Khurradādhbih and his successors also look beyond the frontiers, relaying information about India, China, Byzantium and so forth. Now, however, while the 'Balkhī' authors present itineraries as painstakingly detailed as their predecessors, they delineate them within the bounds of individual provinces (Figure I.13.1). At the same time, with regard to their overall

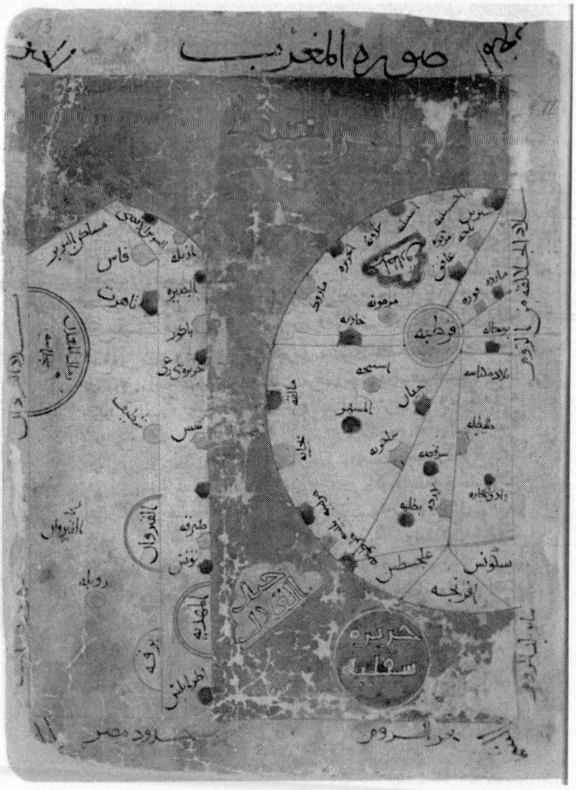

Figure I.13.1 Map of the Maghrib, from al-Iṣṭakhrī's *Book of Routes and Kingdoms*; MS Gotha, Research Library, orient. A 1521, fol. 13a.

Source: © Research Library, Gotha

coverage, the 'Balkhī' authors limit themselves (at least in principle) to the treatment of the Islamic world: a restriction which results logically from their deployment of the concepts of *iqlīm* and *mamlakat al-Islām*. For similar reasons, the frontier regions – where Islam comes up against other polities and civilizations – have less importance for them than for, say, Ibn Khurradādhbih or Qudāma.

With the exception of their eponymous founder, the 'Balkhī' authors are neither bureaucrats nor courtiers. They do share a need for autonomy, even privacy, in areas including their personal life, where they come across as solitary figures. In their religious life, Ibn Ḥawqal and perhaps Muqaddasī have Ismāʿīlī associations, which they mainly keep to themselves. For their professional activities, they conduct commerce, mostly on their own account. They show expertise in fiscal administration, commerce, agriculture, architecture, theology, law and just about everything else. What unites them above all is their critical acumen and power of observation. They articulate and follow research methods of remarkable sophistication; among the earlier (nonmathematical) geographers, it is al-Yaʿqūbī who most has this in common with them.

Finally, a defining and original characteristic of the 'Balkhī school' is its use of maps. We do not know what models Balkhī himself used for these, but it seems that in his work the maps held pride of place while the text provided commentary. In Iṣṭakhrī, however, the text can function without the maps.[2] The same applies to Ibn Ḥawqal, while Muqaddasī's maps may even seem superfluous at times. In later centuries, however, the maps came to enjoy more popularity than the texts and acquired complex transmission histories of their own. Some modern scholars have maintained that these 'Balkhī school' maps appeared in two distinctive varieties, the 'Iṣṭakhrī' and the 'Ibn Ḥawqal' types. However, this division is difficult to trace, especially in its earlier stages, since the earliest complete (or nearly complete) set of 'Balkhī school' maps that we have, accompanying an Ibn Ḥawqal manuscript, dates from around a century after the author's lifetime, while most extant examples date from considerably later than that. All the 'Balkhī school' maps have distinctive features in common, including a lack of indications of latitude and longitude, and a predilection for simple geometrical shapes and straight lines. They show the inhabited world enclosed by the *Encircling Sea* (*al-baḥr al-muḥīṭ*) or Ocean, with two large inlets extending inland from this sea, forming the *Sea of the Romans* (Mediterranean) and the *Sea of the Persians* (Indian Ocean, Arabian Sea, Persian Gulf, Red Sea). In the following centuries, these became shared features of most Islamicate world maps. Very recently, Karen Pinto (2016) has identified a symbolic vocabulary of meanings for these features and traits. She considers many historical sources for them and identifies Iranian, rather than Greco-Roman tradition, as their primary source. She also, less convincingly, deprives Balkhī and his 'school' of the creative role usually attributed to them in cartographical tradition.

I.13.6 Geography after the unitary caliphate

All the literature discussed so far has been in the Arabic language. But now a major work in Persian finally arrives, the *Regions of the World* (*Ḥudūd al-ʿālam*) by an unknown author, dated to 272/982 (Minorsky 1937). This comprehensive and well-organized work probably has the (now-lost) Arabic-language *Routes and Realms* of Jayhānī as a major source. It also makes extensive use of Iṣṭakhrī and perhaps Ibn Ḥawqal and lists no coordinates of longitude and latitude for its place-names. It accordingly has affinities with the 'Balkhī school.' On the other hand, it does not include maps, apparently even in its original form, although it does occasionally refer to a single map (*ṣūrat*), apparently of the world (Tibbetts 1992c, 138–9). In style and tone, it is direct and unornamented.

In 2002, the Bodleian Library of Oxford acquired an Arabic manuscript by an unknown author titled *The Book of Curiosities of the Sciences and Marvels for the Eyes* (*Kitāb gharā'ib al-funūn wa-mulaḥ al-ʿuyūn*). Online access was provided and in 2014 a full, illustrated edition and translation appeared (Rapoport and Savage-Smith 2014). Both manuscript and modern edition are indeed "marvels for the eyes." The editors date the composition of the work to the early to mid-5th/11th century, with Fatimid Egypt the likely place of origin. The manuscript itself was copied, according to the editors, probably around the turn of the 7th/13th century. While the text corresponds to the time in question (before 442/1050), the maps are almost certainly different from the originals, although it is difficult to say how much so since here, as with the 'Balkhī' maps, we have no direct knowledge of the first century or so of their existence. One of the world maps resembles the famous round map associated with al-Idrīsī. The manuscript also includes a rectangular world map and maps of three great seas, five rivers and various formations in the Mediterranean (Figure I.13.2), the area of greatest interest to the author and his readers.

The *Book of Curiosities* differs from the geographical literature discussed so far, in being divided into two parts (*maqālas*), the first on the heavens and their influence and the second on the earth. In the second part, the author claims to follow Ptolemy's *Geography*, but he actually proceeds somewhat, although not entirely, in the manner of the 'Balkhī' authors; the geographical authors on whom he relies most are Masʿūdī and Ibn Ḥawqal. The work concludes with a section on wonders (*ʿajā'ib*). The *Book of Curiosities* is of high interest for many reasons, including its context of courtly patronage under the Ismāʿīlī Shīʿī caliphate of the Fatimids, at a time when their old rivals, the Abbasids of Baghdad, no longer enjoyed much authority or even autonomy.

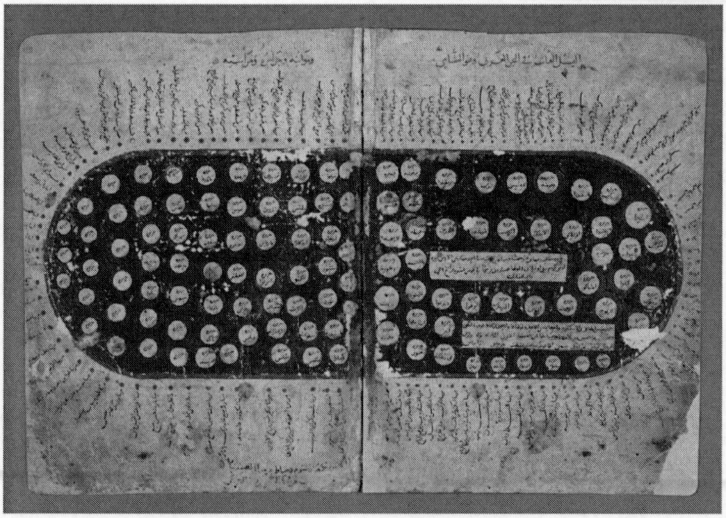

Figure I.13.2 Map of the Mediterranean Sea, from the *Book of Curiosities* written by an anonymous author from Fatimid Egypt; MS Oxford, Bodleian Library, Ar. 90 c., fols 30b–31a.

Source: © Bodleian Libraries, University of Oxford, Oxford

I.13.7 al-Bīrūnī

We have seen that a number of scholars after al-Khwārazmī continued to work on mathematical and astronomical geography, while their administratively – and humanistically – oriented colleagues received more attention. Now, however, at least from our perspective, mathematical geography recovers the spotlight with al-Bīrūnī, who spent much of his career in the service of the Ghaznavids, in the easternmost regions of the Islamicate world. Bīrūnī made a supreme contribution to 'human geography' with his book on India (Bīrūnī 2000). On the more strictly mathematical side, like his predecessor Ibn Yūnus (d. 399/1009), Bīrūnī combined geographical coordinates with tables of stellar coordinates. His *Handbook for al-Masʿūd* (*al-Qānūn al-masʿūdī*) provided the best tables of latitude and longitude available at the time, covering an unprecedented total of 600 places. In doing this, Bīrūnī addressed the old problem of the measurement of a single degree of latitude (Tibbetts 1992c, 141–2). As is well known, he arrived at a remarkably accurate figure for the circumference of the earth. He criticized the theory of projections that he attributed to Marinus of Tyre (*c.* 70–130) and Ptolemy of Alexandria and proposed instead the method that we know now as the equidistant azimuthal projection. In this projection, all points on a map are at proportionally correct distances from the center point; they are also at the correct direction (or *azimuth*) from the center point. This method was taken up afterward by Gerard Mercator (1512–1594) and other cartographers of the early modern era and remains prominent in contemporary mapmaking.

These advances had relatively little effect on Islamicate cartography overall, with one major exception. Until this time most geographers had thought that at the southern end of the world, the landmass that we now call Africa extended eastward as far as China, where it met the *Surrounding Sea*. The resulting confusion between "Ethiopia" and "India" pervaded medieval European travel literature. Bīrūnī proved that this was wrong and, in *his Book of Instruction in Astrology* (*Kitāb al-Tafhīm li-ṣināʿat al-tanjīm*), included a sketch map of the inhabited world that became influential and widely used.

I.13.8 al-Idrīsī

One cartographer and geographer who did not follow Bīrūnī in this last matter was one of the most remarkable figures in the entire tradition, al-Idrīsī. A scholar of noble descent from Ceuta, Morocco, al-Idrīsī received an invitation to the court of the Norman king Roger II of Sicily (r. 1130–1154), which had been conquered from the Muslims over half a century previously. There, he became involved in a project of constructing a large-scale world map that, in its surviving versions, is closely related to the world map accompanying the Fatimid-era *Book of Curiosities* (Rapoport and Savage-Smith 2014, 31). Al-Idrīsī also undertook the composition of a book combining maps with texts. In his introduction, he relates how it all began:

> [King Roger] ordered that a disk should be prepared for him of pure silver and divided into massive, bulky sections. . . . He ordered the workers to engrave upon it the likenesses of the seven climes with their countries, regions, shores, rural areas, gulfs, seas, and watercourses . . . inhabited and uninhabited places . . . well-trodden routes . . . and distances. . . . He also ordered them to compose a book [to which he] added a description of . . . the countries and lands, with regard to their people, places, locations, likenesses, seas, mountains, distances, crops and revenues; the types of their buildings and the characteristics of the [economic] activities performed there; the industries according with them; the merchandise carried to and from them; the marvels . . .

> associated with them; and [their] relation to the seven climes, all the while keeping account of their inhabitants' condition, appearance, mores, costume and speech. [King Roger] ordered that this book should be named *The Book of Pleasant Journeys into Faraway Lands* (*Kitāb nuzhat al-mushtāq fī khtirāq al-āfāq*). This took place in the first ten days of January, corresponding to Shawwāl of the year 548. Here I execute his order and command.
>
> *(al-Idrīsī 1970, 1: 6–7)*

Al-Idrīsī prudently attributes everything to his royal patron, but more important, we see a convergence between 'human geography' of the 'Balkhī' type and the mathematical-astronomical geography of Ptolemy and al-Khwārazmī. Later in his introduction, al-Idrīsī says that he began by drawing the usual seven latitudinal lines across the inhabited world and then drew ten longitudinal lines, thus achieving a grid with seventy rectangular units. He then provided each of these with a detailed map and description. This procedure allowed al-Idrīsī to combine the two dominant modes of Islamic cartography within a single format. Claiming to follow Ptolemy, al-Idrīsī gave coordinates of latitude and longitude. At the same time, and most remarkably, he provided a *summa* of Arabic geographical learning up to his time while concomitantly pursuing empirical knowledge wherever he could find it and from anyone who would provide it to him, not only in the Mediterranean environment (Figure I.13.3) but all the way to the northern coast of Europe, where at some time he apparently went himself.

Al-Idrīsī also composed a shorter book titled *The Gardens of Delight and the Diversion of Souls* (*Rawḍ al-faraj wa-nuzhat al-muhaj*), possibly an abridgment of *The Pleasure Excursion* but more likely an independent work (Maqbul 1992, 157–8, 163–7). Consisting largely of itineraries and distances, *The Gardens of Delight* includes an additional clime (*iqlīm*) just south of the equator, which it considers together with the "usual" first clime. It includes 73 sectional maps, smaller, less detailed, and less unified in format than the maps of *The Pleasure Excursion*.

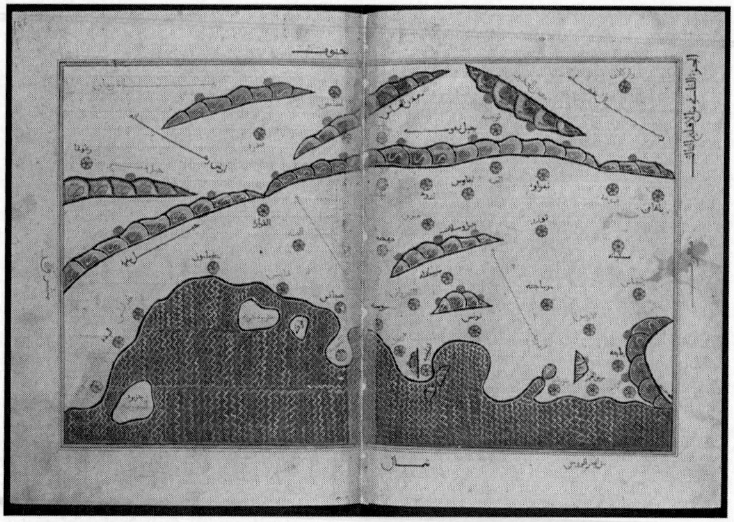

Figure I.13.3 Map of a part of the Mediterranean Sea and adjacent territory around Tunis, from al-Idrīsī's work *The Book of Pleasant Journeys into Faraway Lands* (*Kitāb nuzhat al-mushtāq fī khtirāq al-āfāq*); MS Oxford, Bodleian Library, Greaves 42, fols 115b–116a.

Source: © The Bodleian Libraries, University of Oxford, Oxford

Because of his service for a European Christian prince, al-Idrīsī had a problematic reputation in later Islamic scholarship. Nonetheless, he had considerable influence over later geographers and historians including the North Africans ʿAlī ibn Mūsā ibn Saʿīd al-Maghribī (d. 685/1286) and Ibn Khaldūn (732–808/1332–1406), and, in the Timurid East, Ḥāfiẓ-i Abrū (d. 883/1430). The round world map associated with al-Idrīsī dates from after his lifetime and is now the best-known image from all of Islamic cartography.

I.13.9 Other developments

The term *geographical* may seem inappropriate when applied to all the maps and texts discussed so far and those still remaining to be discussed. It certainly was not the term of choice for those authors and mapmakers themselves, for whom *jughrāfiyā* meant about the same thing as *ṣūrat al-arḍ*, image or map of the world. We still use it now, however, in part because of the lack of an alternative. Interestingly, when our geographical authors proclaimed their ambitions and projects, sometimes they did not identify their subject matter and discipline in a coherent or uniform way (e.g., Ibn Ḥawqal 1938, 3–4). Meanwhile, the broad category of geographical literature includes subgenres that we have not covered here, because of lack of space. These include the literature of travelers on land and sea that overlaps, notably, with the narratives of the 'Balkhī' authors. Another subgenre is that of "wonders of the world," which enjoyed popularity and became increasingly integrated into 'mainstream' geography, especially in the *Wonders of Creation* (*ʿAjāʾib al-makhlūqāt*) of al-Qazwīnī (d. 682/1283). And finally, we have the encyclopedias of place-names, of which the greatest is the *Dictionary of the Countries* (*Muʿjam al-buldān*) of Yāqūt al-Rūmī (d. 626/1229). This reader-friendly work presents thousands of place names in alphabetical order, providing ample literary, historical and indeed, geographical information. Today it remains an indispensable tool for students and scholars, enduring testimony to the fascination, utility and enjoyment that this geographical literature brings.

Notes

1 Consolidated bibliography.
2 The first printed edition of Iṣṭakhrī (Leiden, 1870) has no maps, since the editor had none available. The second edition (Cairo, 1961) does include maps.

Bibliography

Sources

al-Balādhurī, Aḥmad ibn Yaḥyā. ed. de Goeje, M. J. 1866. *Kitāb Futūḥ al-buldān* [The Book of the Conquest of the Countries]. Leiden: Brill.

al-Bīrūnī. tr. and ed. Sachau, E. 2000. *Alberuni's India*. London: Routledge.

Ibn al-Faqīh al-Hamadhānī. 1885. "Kitāb al-Buldān [The Book of the Countries]," in *BGA* 5.

Ibn Ḥawqal, Abu l-Qāsim al-Naṣībī. ed. Kramers, J. H. 1938. "Kitāb Ṣūrat al-arḍ [The Book of the Image of the Earth]," *BGA 2* [English tr. by Bonner, M. forthcoming. Reading: Garnet].

Ibn Khurradādhbih, Abū l-Qāsim. 1889. "Al-Masālik wa-l-mamālik [The Routes and the Realms]," in *BGA* 6.

Ibn Rusta, Abū ʿAlī Aḥmad ibn ʿUmar. 1892. "al-Aʿlāq al-nafīsa [The Precious Gems]," *BGA* 7.

al-Idrīsī, Abū ʿAbdallāh Muḥammad. 1970–84. *Opus geographicum* [. . .] = *Nuzhat al-mushtāq fī khtirāq al-āfāq* [Geographical Work (. . .) = The Book of Pleasant Journeys into Faraway Lands], eds. Bombaci, A., *et al.* Leiden: Brill.

al-Iṣṭakhrī, Abū Isḥāq Ibrāhīm. 1870. *Al-Masālik wa-l-mamālik* [The Routes and the Realms], ed. M. J. de Goeje, M. J. *BGA* 1. [newly ed. by M. Ḥīnī. 1961. Cairo: Wizārat al-Thaqāfa].
al-Masʿūdī. 1894. "Al-Tanbīh wa-l-ishrāf [The Notification and the Verification]," *BGA* 8.
Minorsky, V., ed. and tr. 1937. *Ḥudūd al-ʿālam* [The Regions of the World]. London: Luzac.
al-Muqaddasī, Shams al-Dīn Abū ʿAbdallāh Muḥammad. 1877. "Aḥsan al-taqāsīm fi maʿrifat al-aqālīm [The Best Divisions for Knowledge of the Regions]," in *BGA* 3 [Collins, B. A., tr. Reading: Garnet, 1994].
al-Qazwīnī, Zakariyyā' ibn Muḥammad. ed. Wüstenfeld, F. 1849. *ʿAjā'ib al-makhlūqāt* [The Wonders of the Created (Beings)]. Göttingen: Dieterichsche Buchhandlung.
Qudāma ibn Jaʿfar. ed. al-Zubaydī, M. H. 1981. *Kitāb al-kharāj wa-ṣināʿat al-kitāba* [The Book of Taxes and the Art of Calligraphy]. Baghdad: Dār al-Rashīd li-l-nashr.
Rapoport, Y. and Savage-Smith, E., trs. and eds. 2014. *An Eleventh-Century Egyptian Guide to the Universe: The Book of Curiosities*. Leiden: Brill.
al-Ṭabarī, Muḥammad ibn Jarīr. eds. de Goeje, M. J. *et al.* 1879–1901. *Ta'rīkh al-rusul wa-l-mulūk* [History of Prophets and Kings]. Leiden: Brill [Reprint: 2010].
al-ʿUmarī, Shihāb al-Dīn. 1924. *Masālik al-abṣār fi mamālik al-amṣār* [The Ways of Discernment into the Realms of the Capital Cities]. Cairo: Dār al-kutub al-miṣriyya.
al-Yaʿqūbī, Ibn Abī Yaʿqūb ibn Wāḍiḥ. 1892. "Kitāb al-Buldān [The Book of the Countries]," in *BGA* 7.
Yāqūt, Abū ʿAbdallāh Yaʿqūb al-Ḥamawī al-Rūmī. 1955. *Muʿjam al-buldān* [Dictionary of the Countries]. Beirut: Dār Ṣādir.

Research literature

Balty-Guesdon, M.-G. 1992. "Le *Bayt al-ḥikma* de Bagdad," *Arabica* 39: 131–50.
Brentjes, S. 2007. "Khwārizmī," in Hockey *et al.*, eds. [CB], 631–3.
De Goeje, M. J. 1871. "Die Istakhrī-Balkhī Frage," *ZDMG* 25: 41–58.
Heck, P. 2002. *The Construction of Knowledge in Islamic Civilization: Qudāma b. Jaʿfar and His* Kitāb al-kharāj wa-ṣināʿat al-kitāba. Leiden: Brill.
Hopkins, J. F. P. 1990. "Geographical and Navigational Literature," in Young, M. J. L., *et al.*, eds. *Religion, Learning and Science in the ʿAbbāsid Period*. Cambridge: Cambridge University Press, 301–27.
Kennedy, E. S. and Kennedy, M. H. 1987. *Geographical Coordinates of Localities from Islamic Sources*. Frankfurt am Main: Institut für Geschichte der Arabisch-Islamischen Wissenschaft, Johann-Wolfgang-Goethe-Universität.
Krachkovskiĭ, I. J. 1957. *Arabskaya geograficheskaya literatura. Izbranniye sochineniya*. Vol. 4. Moscow and Leningrad: Izdatel'stvo Akademii Nauk SSSR.
Kramers, J. H. 1932. "La question Balhī-Iṣṭahrī et l'atlas de l'Islam," *Acta Orientalia* 10: 9–30.
Kramers, J. H. 1954a. "La littérature géographique classique des Musulmans," in *Analecta Orientalia*. Leiden: Brill, 172–204.
Kramers, J. H. 1954b. "Al-Biruni's Determination of Geographical Longitude by Measuring the Distances," *Analecta Orientalia*. Leiden: Brill, 205–22.
Maqbul Ahmad, S. 1992. "Cartography of al-Sharif al-Idrisi," in *GTISAS*, 156–74.
Maqbul Ahmad, S. 1995. *A History of Arab-Islamic Geography, 9th–16th Century AD*. Mafraq: Al al-Bayt University.
Maqbul Ahmad, S. 1973–1978."Djughrāfiyā," in *EI-2*, 4: 575–87.
Miller, K. 1986[2] [1926–1931]. *Mappae arabicae. Arabische Welt- und Länderkarten des 9–13 Jahrhunderts*. Wiesbaden: L. Reichert.
Miquel, A. 1967–1988. *La géographie humaine du monde musulman jusqu'au milieu du 11e siècle*. Paris: Mouton, EHESS.
Nallino, C. A. 1944. "Al-Khuwârizmî e il suo rifacimento della Geografia di Tolomeo [al-Khwārizmī and His Rewriting of Ptolemy's *Geography*]," *Raccolta di scritti editi e inediti*. Rome: Istituto per l'Oriente, 5: 458–532.
Pinto, K. 2016. *Medieval Islamic Maps: An Exploration*. Chicago: University of Chicago Press.
Richter-Bernburg, L. 2016. "Potemkin in Baghdad: The Abbasid 'House of Wisdom' as Constructed by *1001 Inventions*," in Brentjes *et al.*, eds. [CB], 121–32.
Rosenthal, F. 1968. *A History of Muslim Historiography*. Leiden: Brill.
Silverstein, A. 2007. *Postal Systems in the Pre-Modern Islamic World*. Cambridge: Cambridge University Press.

Tibbetts, G. 1992a. "The Beginnings of a Cartographic Tradition," in *GTISAS*, 90–107.
Tibbetts, G. 1992b. "The Balkhī School of Geographers," in *GTISAS*, 108–36.
Tibbetts, G.1992c. "Later Cartographic Developments," in *GTISAS*, 137–55.
Zadeh, T. 2011. *Mapping Frontiers Across Medieval Islam: Geography, Translation and the 'Abbasid Empire.* London: Tauris.

I.14
PHYSIOGNOMY
Science of intuition

Liana Saif

'Ilm al-firāsa, physiognomy, which will be discussed in this chapter, was the science of reading physical features and appearances to discern character traits. Nowadays, it is generally understood to focus on the face; however, historically, it involved the entire body, stature, posture, movements, lines on hands, feet and the forehead, color and skin irregularities such as moles. In this chapter, I first look at the Greek and Indian backgrounds of the practice of physiognomy. In the following section, I argue that physiognomy is a science that systematizes and legitimizes the application of intuition (*ḥads*) making it an ally to medicine, especially since they share the focus on the body. The underlying premise of this survey is that occult practices were part and parcel of the scientific activities of medieval and early modern Islamicate societies. Adepts of such practices investigated phenomena that also concerned other sciences perceived by us retrospectively as "mainstream". For example, action at a distance was at the heart of astrological and magical theories, and the nature of intuition was the concept with which medicine, divination generally and physiognomy particularly grappled. The questions of action at a distance and intuition continue to be relevant to the fields of quantum mechanics, mathematics and medicine and constitute a major philosophical component still negotiated and contested in these sciences, as they were in medieval physiognomy, astrology and medicine.

To demonstrate intuition as a negotiated and contested concept, I also investigate a later trend that promoted physiognomy as an esoteric science, that is knowledge reserved to the spiritually elite, a skill gifted by God to those who seek and achieve proximity to the divine. The sphere of natural philosophy that contained the practice of physiognomy competed in its understanding with experiential and the mystical spheres of knowledge. Respectively, one centered on the discursive ability of deduction while the other sought to attribute extraordinary acts to revelation and inspiration in order to verify the privilege and stratification of those who achieve gnosis. This chapter demonstrates these epistemological maneuvers in the case of physiognomy by surveying texts, transmissions and classifications.

In the intellectual and scientific spheres of medieval Islamicate societies, physiognomy (*firāsa*) was linked to divination (*kihāna*), which was deemed as one of the "occult sciences" (*al-'ulūm al-khafiyya*) or "subtle sciences" (*al-'ulūm al-daqīqa*; Ikhwān al-Ṣafā' 2008[CB],[1] 4: 107). The Ikhwān al-Ṣafā' (The Brethren of Purity; [*fl.* 4th/10th century]),[2] in their short epistle on magic, enumerate five of these sciences: alchemy, astrology, magic/talismanry,

DOI: 10.4324/9781315170718-16

medicine and one they call *the science of abstracts* (*ʿilm al-tajrīd*) "whereby the soul knows itself" (Ikhwān al-Ṣafāʾ 2008, 4: 287). Divination does not appear in this list, but in the long version of the same epistle, *kihāna* (divination) is included as one of the "magical" sciences concerned with "telling what will happen before its occurrence" (Ikhwān al-Ṣafāʾ 2008, 4: 312–13). This division of the occult sciences is common; however, physiognomy (*firāsa*) and its subset chiromancy or palmistry, and separately pedomancy, occupy an ambiguous place in these classifications (Figure I.14.1).[3] The Ikhwān do briefly discuss physiognomy as they include it among practices, such as auguries (*zajr*), auspices (*al-faʾl*) and divination (*kihāna*), that indicate those things which are not apparent or have not yet occurred. For the Ikhwān, physiognomy is "the method of extracting morals (*akhlāq*) by examining features (*khalq*)" (Ikhwān al-Ṣafāʾ 2008, 4: 298). In this sense, it is a kind of "inductive divination", using Toufic Fahd's (1923–2009) term (Fahd 2012). These practices are distinguished from any other divinatory disciplines that predict via material or natural means such as astrology, which produces predictions through the study of the locations of the stars and planetary configurations (Ikhwān al-Ṣafāʾ 2008, 4: 190). This is made clear elsewhere in the *Epistles* (*Rasāʾil*), where we encounter physiognomy as one of the skills and sciences associated with the planet Mercury, attained through the mind's "estimative" faculty (*al-quwwa al-wahmiyya*). In addition to physiognomy, these skills include imagination, thought, analysis, conceptualization, distinction, inspiration, feeling and sensation (Ikhwān al-Ṣafāʾ 2008, 4: 222). The inclusion of physiognomy among these mental skills implies that it is without material, physical or astral mediation. Instead, it is a science reliant on intuition.

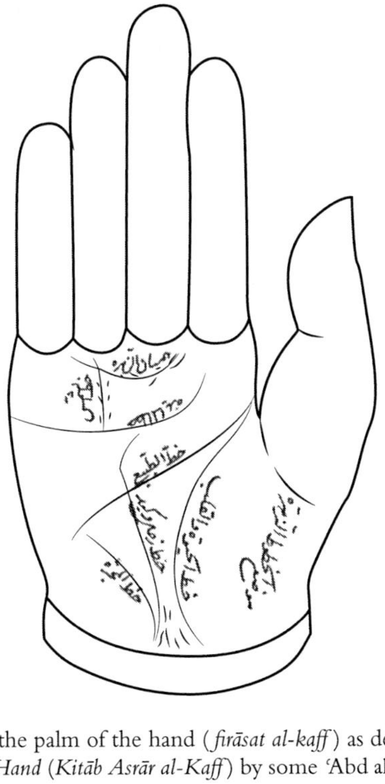

Figure I.14.1 Physiognomy of the palm of the hand (*firāsat al-kaff*) as described in *The Book of the Secrets of the Palm of the Hand* (*Kitāb Asrār al-Kaff*) by some ʿAbd al-Ḥasan, published in 1303/1886.

Source: @ photographer Liana Saif, Amsterdam

1.14.1 The Greek and Indian background of Islamic physiognomy

Physiognomy in Arabic sources represents Mesopotamian, Indian and Greek traditions of reading human marks (Akasoy 2008, 121–6). The Mesopotamian physiognomic omen collection known as *Šumma alamdimmû* (dated 11th century BCE) strikingly resembles in structure and content the Indian physiognomic omens, particularly the marks of men and women known as *puruṣalakṣaṇa* and *strīlakṣaṇa*. The Mesopotamian and Indian systems share a similar protasis and apodosis structure,[4] the male–female divisions and their inclusion of chiromancy and pedomancy. The first Indian system of bodily marks is found in one chapter of the *Astral Sciences According to Gārga* (*Gārgīyajyotiṣa*) from around the 1st century, although the precise date is contested (Mitchiner 1986, 82; Pingree 1981, 69–71; Pingree 1987, 95).

It is followed by two chapters in Varāhamihira's 6th-century *Great Collection of Verses* (*Bṛhatsaṃhitā*). In both works, physiognomy is embedded in the principal doctrines of the Brahmanical *jyotiḥśāstra*, meaning "the science of the stars". It contains two main sections on the marks of men and women, with the former's content applied to the upper three classes: Brahmans, *kṣatriya* (ruling/military class) and *vaiśya* (merchants, craftsmen and landowners), and the latter focused on adolescent girls. Both adopt the toe-to-head formula (Zysk 2014, I: 66–9). The *Great Collection of Verses* was known to the Khwarazmian scholar al-Bīrūnī (362–after 444/973–after 1052), and its root text was translated into Persian in the 8th/14th century (Zysk 2014, 1: 56).

Physiognomy in Arabic shares with the Indian system four major aspects. One of them is the inclusion of chiromancy and pedomancy as subsections. Chiromancy, for instance, is found in the *Astral Sciences According to Gārga* as part of the study of women's marks (Zysk 2014, 1: xi, 37–8, 55–6, 62–3). Thus, it is not surprising that Arabic texts that deal with chiromancy attribute this science to India (Ṭaşköprüzāde 1985[CB], 1: 327; al-Rāzī n. d., 145; al-Rāzī 1939, 11–12). In a widely cited statement, the Damascene geographer and Sufi Shams al-Dīn al-Anṣārī (727/1327) states in his *Epistle on the Science of Physiognomy* (*Risāla fī ʿilm al-firāsa*) that chiromancy "is [a part] of physiognomy, attributed to Ṭamṭam the Indian, Tankulūshā, and the scientists of India such as Sharāsīm the Indian" (MS Riyadh, University of Riyadh, 415, fol. 86b).

A second feature that Indian and Arabic teachings of physiognomy have in common is their treatment of race-based characteristics (Zysk 2014, 1: ix–x; Hoyland 2005, 381–3). These are not found in one of the two Greek physiognomic texts that were fundamental for the later Arabic literature by the politician and intellectual Polemon of Laodicea (*c.* 88–144). The author of the Pseudo-Aristotelian *Physiognomy*, the other main Greek source for such teachings in Arabic, acknowledges that there are some practitioners of physiognomy who take into account the characteristics of nations. Yet he rejects this method (Swain 2007, 639, 645; Ghersetti 1999, 20).

Another two features they share are the inclusion of animal similes and the protasis–apodosis form.These two points link the Indian and the Arabic systems of human marks with Greek physiognomy as manifested in the Pseudo-Aristotelian *Physiognomy*, dated from about 300 BCE and translated into Latin in the 13th century by Bartholomaeus de Messina at the court of Manfred I of Sicily (r. 1258–1266; Vogt 1999, 197; Foerster 1893, 4–91, vii–cxcii). Like the Indian sources, the two Greek texts use animal similes. In the Pseudo-Aristotelian physiognomy, the author questions the usefulness of these similes arguing that many animals share the same features, many of them have similar character traits, therefore establishing a correlation and resemblance with one animal is not possible. Furthermore, no human being looks like an animal but rather vaguely resembles one or another (Swain 2007, 639–41). Instead, he privileges human physical traits, although sometimes he compares them with those found in more than one animal to support a conclusion about the characteristic indication (Swain 2007, 641–3).

Nonetheless, a comparison of the Greek and Indian systems yields that the physiognomic marks in the Pseudo-Aristotelian text match well with those in the *Gārgīyajyotiṣa* (Zysk 2014, 2–3, 20, 25–6, 28–31, 39, 42–4, 46).

I.14.2 The emergence of physiognomy in Arabic

In comparison to astrology, alchemy and magic, Arabic physiognomy only appeared as the subject matter of entire treatises written by Muslim scholars relatively late (Hoyland 2007, 261–3). The earliest systematic Muslim-authored text dedicated solely to physiognomy is Fakhr al-Dīn al-Rāzī's (606/1210) *The Book on Physiognomy* (*Kitāb al-firāsa*). As a science, it does not appear in the classifications of al-Kindī (d. *c.* 256/870) and al-Farābī (d. 339/950–1), but we see it discussed by Ibn Sīnā (d. 428/1037), who considers it alongside medicine, astrology, oneiromancy, talismanry and alchemy as a natural science of a secondary order (Ibn Sīnā n.d., 110). This was taken up by three authors: the Cairene physician, herbalist and scholar Ibn al-Akfānī (d. 749/1348) in his *Guide for Those Aspiring to the Most Elevated Ends* (*Irshād al-qāṣid ilā asnā al-maqāṣid*), Ṭāshköprüzāde (d. 968/1561) in his *The Key to Felicity and the Lamp to Mastery on the Subject Matters of the Sciences* (*Miftāḥ al-saʿāda wa-miṣbāḥ al-siyāda fī mawḍūʿāt al-ʿulūm*) and Ḥajjī Khalīfa ([d. 1067/1657]; Kātib Chelebi) in his *Removal of Doubts on the Names of the Books and the Disciplines* (*Kashf al-ẓunūn ʿan asamī l-kutub wa-l-funūn*; Ghersetti 2007, 285–7; Ṭaşköprüzāde 1985, 1: 301–2; al-Shahrazūrī 1965, 1: 30).

However, in the form of translation, instruction on physiognomy could be acquired in Arabic already in the first half of the 3rd/9th century. The Pseudo-Aristotelian *Physiognomy* was the first Greek work on this subject to appear in Arabic. Its translator was Ḥunayn ibn Isḥāq (d. 260/873). A translation of Polemon's *Physiognomy* is the second oldest, from some time in the same period. The original translation is lost but formed the main source for the surviving Arabic renderings (Swain 2007, 2–4).

The racial interpretation of human marks evolved over time into a major theme of Arabic physiognomy, both on the scholarly and on the socioeconomic level. In the 6th/12th and 11th/16th centuries, Fakhr al-Dīn al-Rāzī and Ibn al-ʿUmarī (d. 965/1558), author of *The Human Delight in Human Physiognomy and the Satisfying Delight in Faith Physiognomy* (*al-Bahja al-insiyya fī l-firāsa al-insāniyya wa-l-bahja al-riḍḍiyya fī l-firāsa al-īmāniyya*), confirm physiognomy's major concern with races (*ajnās*). They count them among the strongest indicators of human character alongside natures, temperaments and ages since they constitute "essential qualities" (Ghersetti 1999, 8–9, 29–30; MSS London, British Library, Or. 8878, fols 9b–10a; Ankara, Milli Library, 4091, fols 4a–10a). This led to the production of physiognomies that primarily treat racial qualities to aid in selecting slaves such as a text known as *The Merits of Races* (*Maḥāsin al-ajnās*) dedicated to one of Ṣalāḥ al-Dīn al-Ayyūbī's (r. 564–589/1169–1193) descendants, and *The Correct Account of Choosing Slaves* (*al-Qawl al-sadīd fī khtiyār al-ʿabīd*) by the Egyptian scholar al-Amshāṭī (d. 902/1496). As a result of this ethno-stereotyping, physiognomy was integrated into Arabic economic literature, which primarily treated questions of how to run a household (Ghersetti 2007, 287). In relation to this, it has been shown that physiognomy played an administrative role as is evident in Ottoman manumission documents where the *hilye* (the physical appearance) of the slave was carefully described with the particularities of the texts on physiognomy discussed here. In addition to selecting and freeing slaves, physiognomy was widely employed in the forcible conscription of non-Muslim individuals from the Balkans (Sobers-Khan 2014, 98–9).

Animal traits as physiognomic marks became a second major theme in Arabic sources. Fakhr al-Dīn al-Rāzī insisted on their importance, arguing that if the resemblance runs across an entire species, and we can see in human beings with these resemblance similar internal traits, then we can legitimize this practice by the weight of the resemblances rather than differences. In addition, it is an accepted form of logic to connect the thing with that which it resembles, especially when it is corroborated by experience (al-Rāzī 1939, 21–2). A short time later, Shams al-Dīn al-Anṣārī included them in his *Epistle on the Science of Physiognomy* (*Risāla fī ʿilm al-firāsa*; MS Riyadh, University of Riyadh, 415, fols 7a–12a).

Thus, Arabic physiognomy was undoubtedly nourished by both Indian and Greek systems. But it also developed its own characteristics and orientations. Two of them, the central role of intuition and the emergence of a Sufi type of physiognomy, will be discussed in the next two sections.

I.14.3 Physiognomy as a science of intuition

Physiognomy was seen early on as based on intuition (*ḥads*). Defining *firāsa* in his dictionary *The Language of the Arabs* (*Lisān al-ʿArab*), Ibn Manẓūr (630–711/1233–1312) explains:

> on the *ḥadīth*: "beware the *firāsa* of the believer", Ibn al-Athīr said: "it [*firāsa*] is said with two meanings: one of which is what the manifest [sense] of the *ḥadīth* signifies and it is what God Almighty brings down into the hearts of his Friends so they know the states of some people by way of charisma (*karāmāt*), correctness of guesswork (*ẓann*) and intuition (*ḥads*). The second is the type learned by signs, experience, features, and morals whereby people's states are known".
>
> *(Ibn Manẓūr n. d., 6: 159–60)*

On the other hand, al-Zamakhsharī (467–538/1074–1143), in *The Foundation of Eloquence* (*Asās al-balāgha*), defines judgment by intuition as physiognomy, associating it with guesswork (*ẓann*; al-Zamakhsharī 1998, 1: 174). As we shall see in the following, this association permeates the discussion of physiognomy beyond those early sources.

Intuition is also a central constituent of Avicennan epistemology. He considers it the movement of the mind that seeks the middle term[5] in a syllogism. Intelligibles can thus be acquired by intuition. Some individuals achieve this faster than others, and those who are able to come upon the middle term spontaneously are characterized by acumen (*dhakāʾ*). In this early phase of conceptualizing intuition, Ibn Sīnā considered it responsible for bringing the dispositional intellect to the level of the acquired intellect, that is, to intellectual perfection. Later, he revised this theory, seeing intuition no longer as the movement of the mind, fast or slow, but a spontaneous action, whereas thinking constitutes the movement (Gutas 2001, 3–5). Gutas explains that this difference is significant

> in emotional appeal and philosophical clarity. Intuition, regardless of its technical definition, is a difficult concept upon which to build an entire epistemological system, which may also explain why it was never fully appreciated by both medieval and modern scholars. Thinking is not; it is the most human of our faculties and one which is expected to lead the way in any epistemology.
>
> *(Gutas 2001, 26)*

Physiognomy's affinity with intuition, assumption and estimation (*takhmīn*) made it in the eyes of many Muslim scholars and courtiers an inferior science or skill, as its findings were not deemed to be based on rational, evidence-based thinking. The sharp-tongued secretary and eloquent courtier *al-Tawḥīdī* (310–414/922–1023) established a kind of epistemic racial profiling by associating the Indians with intuition in contrast to the rationality of the Greeks: "deduction (*istinbāṭ*), in-depth study (*ghawṣ*), nuance (*tanqīr*), investigation (*baḥth*), exploration (*istikshāf*), examination (*istiqṣā'*) and thought (*fikr*) is for the Greeks, while imagination (*wahm*), intuition (*ḥads*), guesswork (*ẓann*), deception (*ḥīla*), trickery (*taḥayyul*) and sorcery (*sha'badha*) is for the Indians" (*al*-Tawḥīdī 2011, 147). The leading religious scholar of the 9th/11th century, al-Sharīf al-Jurjānī (740–816/1339–1413), took a general position independent of racial identities by juxtaposing intuition with thought in his *The Book of Definitions* (*Kitāb al-Ta'rīfāt*; al-Jurjānī 1983, 88).

Due to such cultural as well as scholarly positions, a lurking unease drove the codification of intuition in physiognomy. In the Pseudo-Aristotelian physiognomy, the author had already added a disclaimer: people who practice physiognomy need to be aware that many people experience several states of the soul yet have one demeanor, such as the brave and the imprudent. In fact, according to the Pseudo-Aristotelian physiognomy, there are only few people whose inner states can be detected through their exterior. The translator Ḥunayn ibn Isḥāq commented that Aristotle had only said this because he was surrounded by good people who were able to control the externalization of their inner states, but in Ḥunayn's time that was not the case, for those who could "control themselves" (*ḍabṭ al-nafs*) were the very few (Ghersetti 1999, 9). Pseudo-Aristotle continues and warns about another shortcoming of physiognomy and intuition:

> [I]t is possible for a person to imagine (*yatawahham*) that one of the signs is fixed (*thābit*) and verified and [yet is proven] false; what it indicates is correct except that this cannot always be so if this sign is not continuous and attached forever to that thing it indicates.
> *(Ghersetti 1999, 12)*

In a chapter on physiognomy in *The Book of Governance and Administration* (*Kitāb al-siyāsa fī tadbīr al-riyāsa*) – another Pseudo-Aristotelian text whose treatment of physiognomy became very influential – Alexander the Great, Aristotle's disciple, is advised by his master:

> O Alexander, do not be hasty in judgment based on a single sign, gather up all your indications. Whenever you are met with contradicting indications, lean towards the strongest and the most likely, and you shall be correct and succeed in your endeavours with the aid of God Almighty and His generosity.
> *(Badawī 1954, 124)*

Fakhr al-Dīn al-Rāzī expresses the same condition, adding that a physiognomer must know that not all physical signs are manifest equally on everyone; some of them are more subtle than others. Al-Rāzī admits the possibility of signs contradicting one another. In this case, one must exercise *tarjīḥ* (speculation) and give preponderance to signs that appear on the physical parts associated with the trait one is inferring. For example, inferring courage from the chest is better than from the eyes. If the signs are equally contradicting in quality, quantity and value, then one must desist altogether. Finally, he warns, as did the author of the Pseudo-Aristotelian physiognomy, that some people have various inner states but a single demeanor, such as brave and impudent individuals (al-Rāzī 1939, 26–30).

In *The Merits of Races*, the author confirms that enslaving men, women and boys based on their features is supported by "people of intuition and physiognomy" (*ahl al-ḥads wa-l-firāsa*; MS London, British Library, Or. 7592, fol. 2b). However, in physiognomy's defense, he distances it from the pitfalls of intuition. He writes: "[P]hysiognomy is not like intuition (*al-ḥads*) and assumptions, because intuition and assumptions do not result from consideration and observation (*naẓar wa-mushāhada*), and physiognomy results only from analysis and observation." Intuition, according to him, involves understanding a situation by relating it to other events or indications, such as telling whether a running woman is pregnant from the way she holds her belly. Notwithstanding, physiognomy and intuition are skills that can be found in one person since they both require intelligence (MS London, British Library, Or. 7592, fols 3b–4a).

Physiognomy is not the only science that is intuition-in-practice. Astrology and medicine are often presented as – or challenged for – being such practices. It is useful to situate physiognomy relative to these sciences in order to understand the epistemological class to which it belonged and to highlight the link physiognomy itself has had with these two sciences.

The anxiety surrounding intuitive knowledge is repeatedly expressed in astrological literature. In his magnum opus, *The Great Introduction to Astrology* (*Kitāb al-Madkhal al-kabīr ilā ʿilm aḥkām al-nujūm*), Abū Maʿshar (171–272/787–886) defends astrology's rationale, writing that

> [m]any people thought astrology is something stumbled upon by intuition and estimation (*ḥads wa-takhmīn*) without having a sound origin with which to work or from which syllogisms can be made . . . and so we composed our present book to establish the judgments [of astrology] with convincing arguments and evidence . . . and whatever that is not there can be deduced by those who know the foundations of this practice.
>
> *(Abū Maʿshar 1995–1996 [CB], 2: Bk. 1, Ch. 1, 3–4)*

The juxtaposition between evidence-based knowledge and intuition is explicit here. The word Abū Maʿshar uses for intuition is *ḥads*. For astrology, he favors the Aristotelian model of scientific substantiation: "Most of the science of the judgments [of the stars] is manifest, visible, and clear, and that part not manifest is inferred by clear syllogism from the science of the nature of things and from the powers of the planetary motions manifest in this world" (Abū Maʿshar 1995–1996, 2: Bk. 1, Ch. 2, 7). Following this assertion, he gives examples of non-astrological inferences and predictions that common people (*al-ʿāmma*) engage in: knowledge that can be described as physiognomic. For example, they know if a woman is pregnant when her eyes are hollow and her eyelids droopy, having pure-looking pupils and thick white sclera (Abū Maʿshar 1995–1996 [CB], 2: 11).

Contradicting Abū Maʿshar's statement, Ibn Khaldūn (732–808/1332–1406) undermines astrology by overemphasizing its reliance on intuition:

> Some people claim that there exist ways of perceiving the unknown without loss of sense. Among them are the astrologers who refer to astral indications . . . these astrologers have [attained] nothing of the unknown; it is merely intuitive guesswork and estimation (*ẓunūn ḥadsiyya wa-takhmīnāt*) based on astral influences . . . even if it were confirmed, the means is [still] intuition and guesswork.
>
> *(Ibn Khaldūn 2005[CB], 1: 226)*

Ibn Khaldūn was responding to a long and firm tradition established by Abū Maʿshar of an Aristotelianized and naturalized astrology according to which the stars signify sublunary events and influence minerals, plants, animals and humans, because they are the efficient causes of their generation and thus have a formal link to them (Saif 2015, 9–16).

As for medicine, according to the physician Ibn Abī Uṣaybiʿa (d. 668/1270), there are three diagnostic skills in medicine: inspiration (*ilhām*), intuition (*ḥads*) and inference (*istinbāṭ*) (Ibn

Abī Uṣaybiʿa n.d.[CB], 1: 158–160, 2020[CB], 1.2). In fact, he asserts, "[M]ost of the practice of medicine is intuition (*ḥads*) and estimation (*takhmīn*) and rarely is certitude involved" (Ibn Abī Uṣaybiʿa n. d., 4: 359, 2020, 15. 51. 8. 24). Error is particularly associated with intuition. Authors of physiognomic and medical literature agree in this point. But practitioners with intelligence, experience and experimentation can often avoid errors and establish probability (Ibn Abī Uṣaybiʿa n. d., 1: 156–62, 2020, 1.2). Other works confirm that the concept of intuition also played an important role in therapeutic practices. In his *The Compendium of Simple Medicaments and Foods* (*Jāmiʿ li-mufradāt al-adwiya wa-l-aghdhiya*), Ibn al-Bayṭār (593–646/1197–1248), a leading specialist of *materia medica* in his time, describes the administration of medicinal material as based on an intuitive impulse (Ibn al-Bayṭār 1992, 4: 354).

The integration of various fields of contested knowledge into the disciplinary canon of knowledge in different Islamicate societies was not achieved at the same time and through the same cultural and epistemic processes. Magic, for instance, gained its stable place as a natural science, especially between the 3rd/9th and the 6th/12th centuries, by incorporating concepts, theories and methods from astrology. Physiognomy, in contrast, solidified its position through medicalization. The main idea around which this process revolved is that "thoughts (*afkār*) and mindsets (*ʿuqūl*) follow bodily states" (Ghersetti 1999, 3). Through the practice of commenting, Ḥunayn refined this notion from the Pseudo-Aristotelian physiognomy. Citing Galen, he reflected on the body–soul interlink: the powers of the soul are affected by the temperaments (*mizāj*; Ghersetti 1999, 4). Galen, indeed, expressed such an opinion in the tract *That the Power of the Soul Follows the Temperament of the Body* (*Fī anna quwwat al-nafs tābiʿa li-mizāj al-badan*). Ḥunayn was familiar with this Galenic treatise, because he had translated it himself. Thus, various textual practices such as translating, reading, memorizing and commenting came together in setting the path for physiognomy to become a natural science of intuition. A further step was assured through the anonymous author of *The Book of Governance*. Ascribing the work to Aristotle, the author insists that an essential part of "spiritual medicine" – that complements physical medicine – is "sensing the soul through the manifest marks." He adds that such a practice is reliant on *ẓann* (guesswork), which is also instrumental in *kihāna* (divination) part of which is physiognomy (Badawī 1954, 116).

The most systematic medicalization of physiognomic intuition was undertaken by Fakhr al-Dīn al-Rāzī in his *Book on Physiognomy* (Akasoy 2008, 129–30). He asserts: "[T]he foundations of this sciences are based on the natural science and its branches as ascertained by experience. It is like medicine on all levels; therefore, any slander directed at this science is exactly that directed at medicine" (al-Rāzī 1939, 6). Furthermore, the physiognomer has to know all the things that make up the temperament including the Aristotelian four causes: material, formal, efficient, teleological. The material cause is composed of the organs, spirit, the four natures and the elements. The formal has effect on the temperaments and physical abilities; and the agent is the cause of health or illness. Concerning the causes of health and illness, there are six: air, food, sleep, wakefulness, motion and stillness, purging, congestion and psychological influences. The physiognomer should be able to observe and delineate the links between the external marks, the character and mindset of the individual and their physical state (al-Rāzī 1939, 8–9).

This alliance between medicine and physiognomy is understandable. In addition to being the only natural sciences that "have the human body as object and remain anchored to the present", they both operate within "the paradigm of semiotic inference" that through observing symptoms/marks reveals states hidden from the immediate senses (Ghersetti 2007, 285). The practical codification of physiognomy and its theoretical conceptualization as intuition-based medical practice secured a place for it in the natural sciences. However, for Fakhr al-Dīn, the diagnosis of the physiognomer and the physician must also be fortified with astrology, sharp senses and long experience (al-Rāzī 1939, 28).

I.14.4 Esotericization of physiognomy

The medical framework given to physiognomy did not detract from it being perceived as ultimately reflecting a type of insight that was associated with piety and devoutness. The Qurʾān appears to sanction such an association: "In this are signs for the *mutawassimūn* (the marked ones)" (Qurʾān15: 75). The last word is equated, in widely accepted interpretations, with *mutafarrisūn*, meaning those who discern by signs (Hoyland 2007, 240). Furthermore, in a *ḥadīth*, it is reported that the Prophet said: "Beware the *firāsa* of the believer, for he sees with the light of God" (Hoyland 2005, 363–4). The great legal scholar and imam al-Shāfiʿī (d. 204/820) was famed for his skill in physiognomy. The story goes that he collected many books on the subject during a trip to Yemen. It is said that he composed a text on the subject, but this is doubtful (Hoyland 2007, 241–3). Mystics and ascetics capitalized on this kind of physiognomy to distinguish it from a more mundane type based on mere knowledge of physical marks and signs (Hoyland 2005, 387).

The spiritual correlation of appearances and morals became so entrenched that it appeared in Ottoman art practices in the form of the *hilye-i şerīfe* (noble *hilye*), the verbal calligraphic portrait of the physical appearance and moral character of the Prophet Muḥammad. It is based on the content of certain *ḥadīths* that became popular in the Ottoman empire, in addition to a popular devotional poem in Ottoman Turkish, the *Hilye-i hakanī*, composed in the later 16th century. The earliest dated examples of the prophetic *hilye* belong to the later decades of the 17th century, usually credited to the calligrapher Ḥāfiẓ ʿOsmān ([1052–1110/1642–1698]; Stanley 2018, 559–60; Schick 2008; Derman 1998, 34–7). The most common text used in constructing the *hilye* is from a *ḥadīth* attributed to the Prophet's cousin ʿAlī ibn Abī Ṭālib (d. 40/661):

> He was neither very tall nor excessively short but was a man of medium size. He had neither very curly nor flowing hair but a mixture of the two. He was not obese. He did not have a very round face, but it was so to some extent. He was reddish-white. He had wide black eyes and long eyelashes. He had protruding joints and shoulder blades. He was not hairy but had some hair on his chest. The palms of his hands and his feet were calloused. When he walked, he raised his feet as though he were walking on a slope. When he turned around, he turned completely. Between his shoulders was the seal of prophecy, and he was the seal of the prophets.
>
> *(Stanley 2018, 562)*

Tim Stanley suggests that gazing at the *hilye* – an important devotional practice in the Ottoman period – may have been connected with Sufi practices such as *naẓar* (contemplating divine beauty by gazing at human beauty), in combination with the anti-image culture of the Naqshabandi Sufi order and others (Stanley 2018, 570). In this way, the Prophetic *hilye* becomes an object of contemplating the perfect alignment of *khalq* and *akhlāq* in the person of Muḥammad.

Prior to the rise of systematized and institutionalized Sufism in the 6th/12th and the 7th/13th centuries, the Abbasid translations and intellectual activities between the 2nd/8th and the 4th/10th centuries fostered and established a scientific episteme that integrated astrology, alchemy, magic and divination as part of the natural sciences that grapple with phenomena such as action at a distance and, in the case of divination, intuition. The mentioned changes in Sufism led to a shift in the perceived paradigms of legitimacy applied to the occult sciences. Now revelatory forms of knowing the hidden became privileged (Saif 2017). As a result, physiognomy was spoken of as a revelatory science practiced by an elite group of mystics. However, in that discourse, physiognomy seems to be less about intuiting through physical signs but rather simply an accurate, unerring and unmediated intuition. In his *The Ranks of Sufis* (*Ṭabaqāt al-ṣūfiyya*), the Sufi hagiographer al-Sulamī (325–412/937–1021) cites a certain Abū ʿUthmān Saʿīd al-Naysābūrī defining *firāsa* as "guesswork (*ẓann*) consistent with truth; for guesswork is hit and miss, if it is verified as *firāsa*, it is verified as judgment" (al-Sulamī 2003, 143). In *The Qushayriyan Epistle* (*al-Risāla al-qushayriyya*), the Sufi al-Qushayrī (d. 465/1072), who references al-Sulamī, dedicates an entire chapter to the subject giving many definitions of *firāsa* as a spontaneous divinely supported act, including "an occurrence of thought in the heart that refutes what negates it" (al-Qushayrī 1989, 398).

Ibn al-ʿArabī (558–638/1165–1240) subscribed to a similar view (Akasoy 2008, 119–29). In a chapter titled "On the Prestige of Physiognomy and its Secrets", in *The Meccan Disclosures* (*al-Futūḥāt al-makiyya*), Ibn al-ʿArabī tells us that the people who can be "read" by *firāsa* are the *shawārid*, meaning those whose consciousness strays from the divine and who are too attached to their corporeality. They fear the dispraise of the physiognomer who sees their spiritual flaws and vulnerabilities. Furthermore, there are two kinds of physiognomy: one described as natural (*ṭabīʿiyya*) and medical (*ḥikmiyya*), resulting from the physical temperament (*mizājiyya*), and another, spiritual (*rūḥāniyya*), of the soul (*nafsiyya*), founded on faith (*īmāniyya*) and divine matters (*ilāhiyya*). Natural physiognomy looks for physical marks on bodies that differ according to the diversity of natural temperaments due to the harmony (*ulfa*) decreed by God between the four natures and elements. A semi-substantial (medical) spirit mediates between these mixtures and the composite whole. As a result, the balance or imbalance of the temperament manifests on the body and affects the soul and character. Ibn al-ʿArabī elaborates on some of these physical marks and their significances. Divine physiognomy, however, is revelatory, resulting from a divine light pouring into "the eye of insight", revealing the inner states of individuals without recourse to physical marks. It is a talent attained by the faithful whose heart is illuminated by divine light as a result of fixating consciousness on divine attributes and names. So, if people of balanced temperament are ignorant of salvific means, they can consult the *ʿulamāʾ* (experts), including physiognomers for aid. The latter can also guide toward bliss and salvation those whose souls are astray (*munḥarif, shārid*) and whose temperaments are imbalanced (Ibn al-ʿArabī 2006, 1: 354–63). This exposition is also found in his *Divine Affairs* (*al-Tadbīrāt al-ilāhiyya*) in a chapter titled "On Religious and Medical Physiognomy" (Ibn al-ʿArabī 2003, 58–69).

Fakhr al-Dīn al-Rāzī, an aspiring mystic, also makes this distinction. In his *The Lofty Aspirations* and *The Book on Physiognomy* he speaks of "spiritual physiognomy" revealed to the heart without reliance on any physical signs. It is attained by souls characterized by transparency and luminescence due to their detachment from material things. Physiognomy in sleep belongs to this type. The second and natural type is inference through manifest states and marks that reveal hidden or inner states: "This is a science with reliable foundations, [but] assumptive in its branches (*ʿilm yaqīnī al-uṣūl, ẓannī al-furūʿ*)" (al-Rāzī 1939, 6–7). Al-Rāzī implies that, unlike its natural counterpart, spiritual physiognomy cannot err, being a divine inspiration. Thus, it is exempt from the tension present in trusting intuition from which natural physiognomy suffers. He writes:

> One of the Sufis was asked about the difference between these two parts. He said, "Guesswork (*ẓann*) occurs from the turning over of signs by the heart. [Faith-based] physiognomy happens by the disclosure of the light of the Lord of the Heavens."
>
> *(Fakhr al-Dīn al-Rāzī, MS London, British Library, Or. 7592, fol. 2b)*

This is the physiognomy of the prophets and the saints (*awliyāʾ*). Education and training are required for the natural-physical type (al-Rāzī 1939, 6–7; also see al-Rāzī n. d., VIII: 145).

As mentioned earlier, al-Anṣārī dedicated an entire treatise to physiognomy, and there its revelatory aspect is clearly articulated (Akasoy 2008, 131–4). His sources as he lists them are al-Shāfiʿī, Ibn al-ʿArabī, Fakhr al-Dīn al-Rāzī, Aristotle, Polemon, Hippocrates and Indian sources (MS Riyadh, University of Riyadh, 415, fol. 1a). The distinction he offers between natural and revelatory physiognomy is taken almost verbatim from Fakhr al-Dīn al-Rāzī (MS Riyadh, University of Riyadh, 415, fols. 3a–4a). In similar words as Fakhr al-Dīn, al-Anṣārī asserts that the epistemological foundations of natural physiognomy are identical to medicine, that is, intuiting inner moral and physical states from outer marks (MS Riyadh, University of Riyadh, 415, fols 2a–b).

Physiognomy set within a Sufi discourse is also found in the *Ranks of the Wayfarers* (*Madārij al-sālikīn*) by the theologian Ibn Qayyim al-Jawziyya (961–751/1292–1350). He opted for the literary technique of commentary to talk about physiognomy, although he is also known for his traditionalist stance discrediting occult practices. His text is mainly a commentary on *The Stations of the Wayfarers* (*Manāzil al-Sāʾirīn*) by the Persian Sufi al-Harawī (396–481/1005–1089). In it, al-Jawziyya speaks of three types of physiognomy. The first is called faith-based (*īmāniyya*). It is the "sharpest" since it is "a light cast down by God into the heart of the believer". Reports from the *salaf* (the predecessors) is given in support of this type, where we are told that the Rightly Guided Caliphs, Abū Bakr (r. 11–13/632–634), ʿUmar and ʿUthmān (r. 23–35/644–656), were among the best skilled in faith-based physiognomy. He cites the legendary ascetic Abū Jaʿfar al-Ḥaddād saying "physiognomy is the first thought without objection," which echoes al-Qushayrī's aforementioned definition (al-Jawziyya 2001, 2: 192–4). The second is physiognomy by asceticism (*firāsat al-riyāḍa wa-l-jūʿ wa-l-sahar wa-l-takhallī/* physiognomy by discipline, hunger, sleeplessness and abstinence). This is a universal way of attaining the skill "shared between the faithful and the infidel; it indicates neither faith nor allegiance" (al-Jawziyya 2001, 2: 194). Monks are known for it. Unlike the first type of physiognomy, this one gives partial information. Therefore, it is useless and leads astray. The third is physical physiognomy, which is adopted by physicians who observe external signs to learn about internal natures. Al-Jawziyya seems to approve of it (al-Jawziyya 2001, 2: 194–6). Based on the teachings of al-Harawī, al-Jawziyya, furthermore, mentions three degrees of physiognomy. The first is common physiognomy, which occurs very rarely to those mindless of God without any knowledge of how it took place. It could be an afflatus for the sake of warning or delivering good tidings, but it could also be demonic inspiration to undermine faith and cause fear. The second degree is the physiognomy of the faithful, and it is pure divine inspiration. Finally, there is "esoteric physiognomy" (*firāsa sirriyya*), which occurs to the noblest souls concerned with secrets and is articulated either explicitly or in symbols (al-Jawziyya 2001, 2: 197–200). In his exposition on physiognomy, al-Jawziyya incorporates physiognomy into the strata of spiritual development and thus stretches further the spectrum, which usually begins with the discursive-physical and ends with nondiscursive and revelatory (generally ideologically neutral) by distinguishing universal asceticism from a "mysticism" that is pronouncedly Islamic, privileging the physiognomy of the latter.

Esoteric physiognomy is also found in the aforementioned *Human Delight*, clearly inspired by Ibn al-ʿArabī's taxonomy and terminology. Ibn al-ʿUmarī cites Polemon, the two Rāzīs, in addition to Ibn al-ʿArabī and al-Qushayrī (London, British Library, Or. 8878, fol. 2a–b). Here, too, he speaks of physical physiognomy and faith-based physiognomy (*īmāniyya*), the latter being revelatory (*bi-l-mukāshafa*; MS London, British Library, Or. 8878, 2a–b). Ibn al-ʿUmarī begins with physical physiognomy, emphasizing its link with medicine, often borrowing verbatim from Fakhr al-Dīn al-Rāzī (MS London, British Library, Or. 8878, fols. 3b–6b). Curiously, Ibn al-ʿUmarī equates revelatory physiognomy with the inherent and inspired behavior of some animals, namely, bees and hoopoes. The first is honored by an entire *sūra* of the Qurʾān, and the other is depicted there as Solomon's trusty servant who brings him news of Balqīs in Saba' (Yemen; MS London, British Library, Or. 8878, f. 7b; Qurʾān 16 [The Bees]; Qurʾān 27: 20, 27–28). Being divinely guided, this type of physiognomy is not tarnished by *ẓann* and never fails (London, British Library, Or. 8878, fol. 8b). The treatise ends by exhorting the reader to employ what can be called reflective physiognomy. It is directing one's insight into the self, evaluating its preoccupations, values and behaviors. On judgment day, as the Qurʾān pronounces, one's own exterior – legs, hands, tongue and so on – shall stand witness to one's interior states; people's eternal destiny will manifest on their faces. It is the day of complete externalization of the inner state of being (MS London, British Library, Or. 8878, fols. 27a–28b).

I.12.5 Concluding Remarks

This chapter considered physiognomy as a set of practices based on intuition. As such, investment in speculative knowledge (*ḥads*, *ẓann*, *takhmīn*), no matter how systematized, required legitimization. This was carried out mostly through medicalization from the 3rd/9th to the 6th/12th centuries, and later, due to the systematization of Sufism, through its transformation into a part of esoteric doctrines and practices. The first process consolidated physiognomy's association with natural philosophy, contending at some point with an explicit esotericism that reclaimed physiognomy as a passively attained token of the spiritual elite. Recognizing the epistemological shifts that underlie these two practices frees us from the ossified perception that the occult sciences, including astrology and magic, occupy a marginal space contoured according to the value of their end results and claims, and anachronistically measured according to our own intellectual and epistemological borders. Simultaneously, by identifying the universal phenomenon physiognomy grapples with, namely, intuition, we begin to discern a scientific and philosophical concern that is still relevant today. In the field of medicine today, the role of intuition is acknowledged as a diagnostic factor that sometimes challenges evidence-based medicine yet plays a role in clinical decision-making (Greenhalg 2002, 395–400). Physiognomy was applied in economics, particularly in slave trading. It was valorized in Islamicate courts, especially among the Ottomans. It provided guidelines for selecting members of the ruling elite and a tool for imperial propaganda that was buttressed by racial profiling. Physiognomy as a science of intuition was utilized for decision-making pertaining to the management of body, society and state administration (Lelić 2017, 623–6; Chapter II.10).

Notes

1 Consoldated bibliography.
2 The dating and identity of those scholars is highly contested. For a survey on those problems, see Callataÿ 2013 and a new alternative interpretation by De Vaulx d'Arcy 2019.
3 Chiromancy is the divinatory practice of reading the features of the hands, including, but not restricted to, the lines of the palm. In Arabic, it is known as *ʿilm al-kaff* (the science of the palm) or *ʿilm sarāʾir al-kaff* (the science of the palm's lines). Pedomancy is examining the soles of the feet to divine the future and to diagnose personality traits and health conditions and tendencies.
4 Protasis and apodosis: the two basic constituents of a conditional sentence. Protasis (literally, what stands before) is the antecedent clause that expresses the condition, the "if clause". Apodosis (literally, what comes after) is the clause that expresses the consequence.
5 The term appearing in both premises of a syllogism but missing in the conclusion. For example, all men are human; Socrates is a man; therefore, Socrates is a human; here, *man/men* is the middle term.

Bibliography

Sources

Akasoy, A. 2008. "Arabic Physiognomy as a Link Between Astrology and Medicine," in Akasoy, A., Burnett, C., and Yoeli-Tlalim, R., eds. *Astro-Medicine: Astrology and Medicine, East and West*. Florence: SISMEL, 119–41.

Badawī, ʿA., ed. 1954. *al-Uṣūl al-yūnāniyya li-l-naẓariyyāt al-siyāsiyya fī l-islām* [The Greek Elements of the Political Theories in Islam]. Cairo: Maktabat al-nahḍa al-miṣriyya.

Devriese, L. ed. 2019, *Aristotle, Physiognomonica Translatio Bartholomaei de Messana*, Turnhout: Brepols.

Foerster, R. 1893. *Scriptores Physiognomonici, Graeci et Latini*. Vol. 1. Leipzig: B. G. Teubner.

Ghersetti, A., ed. 1999. *Il Kitāb Arisṭāṭalīs al-faylasūf fī al-firāsa nella traduzione di Ḥunayn ibn Isḥāq*. Venice: Herder Editrice.

Ibn al-ʿArabī, Muḥyī al-Dīn. ed. ʿal-Kayyalī, Ā. 2003. *al-Tadbīrāt al-ilāhiyya fī iṣlāḥ al-mamlaka al-insāniyya* [The Divine Affairs Dealing with the Rectification of the Human Kingdom]. Beirut: Dār al-kutub al-ʿilmiyya.

Ibn al-ʿArabī, Muḥyī al-Dīn. ed. Shams al-Dīn, A. 2006. *al-Futūḥāt al-makkiyya* [The Meccan Disclosures]. 9 vols. Beirut: Dār al-kutub al-ʿilmiyya.

Ibn al-Bayṭār. 1992. *al-Jāmiʿ li-mufradāt al-adwiya wa-l-aghdhiya* [The Compendium of Simple Medicaments and Foods]. 4 vols. Beirut: Dār al-kutub al-ʿilmiyya.

Ibn Manẓūr, Abū l-Faḍl. n.d. *Lisān al-ʿarab* [The Language of the Arabs]. 15 vols. Beirut: Dār Ṣāḍir.

al-Jawziyya, Abū ʿAbdallāh Muḥammad. ed. Riḍwān, R. J. 2001. *Madārij al-sālikīn* [The Ranks of the Wayfarers]. Cairo: Muʾassasat al-mukhtār.

Pack, R. 1972. "A Pseudo-Aristotelian Chiromancy," *Archives d'histoire doctrinale et littéraire du Moyen Âge* 36: 189–241.

al-Qushayrī, ʿAbd al-Karīm. ed. Ibn al-Sharīf, M. 1989. *al-Risāla al-qushayriyya* [The Qushayrian Epistle]. Cairo: Dār al-shaʿb.

al-Rāzī, Fakhr al-Dīn. ed. Ḥijāzī Saqqā, A. n.d. *al-Maṭālib al-ʿāliya min al-ʿilm al-ilāhī* [The Lofty Aspirations on Divine Science]. 9 vols. Beirut: Dār al-kutub al-ʿarabī.

al-Rāzī, Fakhr al-Dīn. ed. Mourad, Y. 1939. *La Physiognomie arabe et le* Kitāb al-Firāsa *de Fakhr al-Dīn al-Rāzī*. Paris: Libraire Orientaliste Paul Geuthner.

al-Shahrazūrī, Muhāmmad ibn Maḥmūd. 1965. *Rasāʾil al-shajara al-ilāhiyya fī ʿulūm al-ḥaqāʾiq al-rabbāniyya* [Epistles of the Sacred Tree on the Sciences of Divine Truths]. 3 vols. Tehran: The Iran Research Institute of Philosophy.

al-Sulamī, Abū ʿAbd al-Raḥmān. ed. ʿAbd al-Qādir ʿAtā, M. 2003. *Ṭabaqāt al-ṣūfiyya* [The Ranks of Sufis]. Beirut: Dār al-kutub al-ʿilmiyya.

al-Tawḥīdī, Abū Ḥayyān. ed. al-Ṭaʿīmī, H. 2011. *al-Imtāʿ wa-l-muʾānasa* [The Book of Delectation and Conviviality]. Beirut: al-Maktaba al-ʿaṣriyya.

Vogt, S. 1999. *Aristoteles: Opuscula VI: Physiognomonica* [Aristotle: Small Works VI: Physiognomic]. Berlin: Akademieverlag.

al-Zamakhsharī, Maḥmūd ibn ʿUmar. ed. ʿUyūn al-Sūd, M. B. 1998. *Asās al-balāgha* [The Foundation of Eloquence]. 2 vols. Beirut: Dār al-kutub al-ʿilmiyya.

Manuscripts

MS Ankara, Milli Library, 4091.

MS London, British Library, Or. 7592.

MS London, British Library, Or. 8878.

MS Riyadh, University of Riyadh, 415.

Research literature

Callataÿ, G. de. 2013. "Magia en al-Andalus: *Rasāʾil Ijwān al-Ṣafāʾ, Rutbat al-ḥakīm y Gāyat al-ḥakīm (Picatrix)*," *al-Qanṭara* 34.2: 297–344.

Derman, M. U. 1998. *Letters in Gold: Ottoman Calligraphy from the Sakıp Sabancı Collection, Istanbul*. New York: The Metropolitan Museum of Art.

Fahd, T. 1986 [2012²]. "Firāsa," in *EI-2*, 2: 916–7.
Ghersetti, A. 2007. "The Semiotic Paradigm: Physiognomy and Medicine in Islamic Culture," in Swain, S., ed. *Seeing the Face, Seeing the Soul: Polemon's Physiognomy from Classical Antiquity to Medieval Islam*. Oxford: Oxford University Press, 281–308.
Greenhalg, T. 2002. "Intuition and Evidence – Uneasy Bedfellows?" *British Journals of General Practice* 52: 395–400.
Gutas, D. 2001. "Intuition and Thinking: The Evolving Structure of Avicenna's Epistemology," in Wisnovsky, R., ed. *Aspects of Avicenna*. Princeton: Markus Wiener Publishers, 1–38.
Hoyland, R. 2005. "Physiognomy in Islam," *Jerusalem Studies in Arabic and Islam* 30: 361–402.
Hoyland, R. 2007. "The Islamic Background to Polemon's Treatise," in Swain, S., ed. *Seeing the Face, Seeing the Soul: Polemon's Physiognomy from Classical Antiquity to Medieval Islam*. Oxford: Oxford University Press, 227–80.
Isenman, L. D. 1997. "Toward an Understanding of Intuition and Its Importance in Scientific Endeavour," *Perspectives in Biology and Medicine* 40.3: 395–403.
Lelić, E. 2017. "Physiognomy (*ʿilm-i firāsat*) and Ottoman Statecraft: Discerning Morality and Justice," *Arabica* 64: 609–46.
Mitchiner, J. 1986. *The Yuga Purāṇa: Critical Edition with an English Translation and a Detailed Introduction* (Bibliotheca Indica 312). Calcutta: Asiatic Society.
Pack, R. 1972. "On the Greek Chiromantic Fragment," *Transactions and Proceedings of the American Philological Association* 103: 367–80.
Pack, R. 1978. "Aristotle's Chiromantic Principle and Its Influence," *Transactions of the American Philological Association* 108: 121–30.
Pingree, D. E. 1981. *Jyotiḥśāstra: Astral and Mathematical Literature*. Wiesbaden: Harrassowitz.
Pingree, D. E. 1987. "Babylonian Planetary Theory in Sanskrit Omen Texts," in Goldstein, B. R. and Berggren, J. L., eds. *From Ancient Omens to Statistical Mechanics: Essays on the Exact Sciences Presented to Asger Aaboe*. Copenhagen: University Library, 91–9.
Saif, L. 2015. *The Arabic Influences on Early Modern Occult Philosophy*. Hampshire: Palgrave Macmillan.
Saif, L. 2017. "From Ġāyat al-ḥakīm to Šams al-maʿārif: Ways of Knowing and Paths of Power in Medieval Islam," *Arabica* 64: 297–345.
Schick, İ. C. 2008. "The Iconicity of Islamic Calligraphy in Turkey," *RES: Anthropology and Aesthetics* 53.4: 211–24.
Sobers-Khan, N. 2014. "Firāsetle Naẓar Edesin: Recreating the Gaze of the Ottoman Slave Owner at the Confluence of Textual Genres," in Firges, P. *et al.*, eds. *Well Connected Domains*. Leiden: Brill, 89–109.
Stanley, T. 2018. "From Text to Art Form in the Ottoman Hilye," in *Studies on Islamic Art and Architecture in Honor of Filiz Çağman'a Armağan*. Istanbul: Lale Yayincilik, 559–70.
Swain, S. 2007. "Introduction," in Swain, S., ed. *Seeing the Face, Seeing the Soul: Polemon's Physiognomy from Classical Antiquity to Medieval Islam*. Oxford: Oxford University Press, 1–18.
Vaulx d'Arcy, G. de. 2019. The *Epistles of the Brethren of Purity* Edited by the Institute of Ismaili Studies, MIDÉO 34: 253-330. https://journals.openedition.org/mideo/3397?lang=ar. (Accessed 12 November 2021).
Zysk, K. G. 2014. *The Indian System of Human Marks*. 2 vols. Leiden: Brill.

I.15

THE HIEROGLYPHIC SCRIPT DECIPHERED? AN ARABIC TREATISE ON ANCIENT AND OCCULT ALPHABETS

Christopher Braun

In the middle of the 1st/7th century, Muslim troops conquered Egypt, a land that had seen the rise of one of the most advanced civilizations of antiquity. The ubiquitous legacy of the pharaohs such as the awe-inspiring pyramids, the extensive temple complexes and the soaring obelisks must have elicited questions on their long-forgotten origins in later times. Arab and Arabic-writing scholars began to inquire who the builders of these gigantic buildings were and whether the enigmatic inscriptions on the temple walls provided a key to the wisdom and the sciences of this bygone civilization. Spurred by curiosity, historians began to collect information on ancient Egypt from the 3rd/9th and 4th/10th century onward and recorded legendary accounts and tales on the pharaohs and their miraculous feats that circulated at the time (Haarmann 1996; Pettigrew 2004).

In their quest for knowledge, they tapped late antique Greek sources that had become available in Arabic from the 2nd/8th century onward (Gutas 1998[CB];[1] Sezgin 1993, 1994, 2001, 2002, 2004). A treatise on ancient and secret alphabets represents a case in point. Medieval tradition attributes it to Ibn Waḥshiyya (d. 318/930–1), an authority in the Arabic occult sciences (for the scholarly debates on Ibn Waḥshiyya and his presumed lifetime, see Hämeen-Anttila 2006, 3–9). Modern investigations have raised numerous questions about its authorship, content and context (see the following discussion). This concise work presents 93 alphabets, among them the ancient Egyptian hieroglyphs, and their supposed equivalents in Arabic. It comes as no surprise that the author did not successfully decipher the hieroglyphs' meaning almost a millennium earlier than the French scholar Jean-François Champollion (d. 1832), nor were his methods in any way comparable to those of Champollion. Nevertheless, his little treatise constitutes a fascinating and unique testimony to the scholarly interest in the ancient Egyptian and other ancient writing systems. A major element of the author's scholarly approach consists in his continuation of Hellenistic interpretations of the hieroglyphic script, which themselves relied on views formulated by ancient Egyptian priests (von Lieven 2010)

I.15.1 Arab encounters with ancient Egypt

Around 18–9/640, the Arab commander ʿAmr ibn al-ʿĀṣ (d. 42 or 43/662 or 664) and his retinue conquered Egypt on behalf of Caliph ʿUmar ibn al-Khaṭṭāb (r. 13–23/634–644; Kennedy 2006, 62). They did not set foot on *terra incognita*. Arabs had lived, worked, traded and

DOI: 10.4324/9781315170718-17

visited cultic centers in Egypt long before the advent of Islam. In fact, the ties between Egypt and Arab tribes reach back to the time of the Middle Kingdom (*c.* 2055 BCE–*c.* 1650 BCE). The Arab perception of Pharaonic culture during the pre-Islamic period, however, remains completely unknown (el-Daly 2003, 40–1). The first written records we have date back to the Abbasid period, when Islamicate scholars began to collect information on the history of ancient Egypt and put it down in writing. A diverse set of preserved Arabic sources deal with the Pharaonic civilization. These mostly legendary accounts are transmitted in chronicles, geographical surveys, books on *mirabilia*, popular epics (*siyar shaʿbiyya*), occult treatises and other literary genres.

The earliest known Arabic text depicting Egypt's legendary pharaonic past is a history of Egypt written in the 3rd/9th century. The Egyptian legal scholar and historian Ibn ʿAbd al-Ḥakam (d. 257/871) dedicated a part of his local history of Egypt titled *The Conquest of Egypt, North Africa and al-Andalus* (*Futūḥ Miṣr wa-l-Maghrib wa-l-Andalus*) to his homeland's pharaonic past. Around one century later, al-Masʿūdī's (d. 345/956) lengthy account on the ancient Egyptian rulers and their marvelous constructions in his universal history *The Book of the Meadows of Gold and the Mines of Gems* (*Kitāb Murūj al-dhahab wa-maʿādin al-jawhar*) contributed to the medieval Arab perception of ancient Egypt. The historian depicts the pharaonic civilization as a realm of wonders, miracles and magic. A similar image of ancient Egypt is presented by *The Book of Mirabilia* (*Kitāb al-ʿajāʾib*), a treatise falsely attributed to al-Masʿūdī. Its author was very likely a shadowy writer and authority on ancient Egyptian lore known as Ibrāhīm ibn Waṣīf Shāh or Ibn Waṣīf al-Ṣābiʾ. Recent studies have shown that these accounts of ancient Egypt are partly based on late antique material (Sezgin 1993, 1994, 2001, 2002, 2004).

Interest in Egypt's pre-Islamic past did not abate in the following centuries. In the 7th/13th century, the Upper Egyptian *ḥadīth* scholar Abū Jaʿfar al-Idrīsī (d. 649/1251) composed a whole monograph on the pyramids of Giza and other ancient Egyptian monuments. In this treatise, called *The Book of the Lights of the Supernal Bodies on the Unveiling of the Pyramids' Secrets* (*Kitāb Anwār ʿulwī al-ajrām fī l-kashf ʿan asrār al-ahrām*), he transmits an impressive collection of narratives on the millennia-old monuments on the Giza plateau and on ancient Egypt's history and culture in general.

The preserved texts suggest that some Egyptian scholars regarded their country's pre-Islamic monuments with awe and amazement, despite the frequent association of the ancient monuments with the Qurʾānic narrative of Pharaoh (*firʿawn*) as the antagonist of the prophet Moses. These scholars felt great admiration for the ancient civilization's architectural feats and expressed interest in the enigmatic scripts engraved on the pharaonic monuments (Haarmann 1996). Since knowledge of the hieroglyphic script had been lost much earlier, 'wild ideas' on these writings from the past and their potential meaning circulated among Arab and Arabic-writing scholars, as far as the surviving evidence tells us, from the Abbasid period onward. These ideas were either newly invented or the result of reading ancient texts in Arabic translation. Some scholars even claimed to have succeeded in unlocking their secrets.

1.15.2 Medieval Arabic treatises on the hieroglyphic script

Hieroglyphs are the pictographic, non-cursive system of ancient Egyptian writing (Figure I.15.1). The Greeks came to call it τὰ ἱερογλυφικὰ γράμματα (*tà hieroglyphikà grámmata*), 'the sacred engraved letters.' This script was already in use before 3000 BCE and continued to be used, as far as the evidence tells us, until around the end of the 4th century CE (Zauzich 1978). Its characters consist of signs that represent either words (logograms), sounds (phonograms), or determinatives. The latter identify certain semantic categories.

Figure I.15.1 Tomb of Seti I (r. 1290–1279 BCE), second ruler of the 19th dynasty (r. 1292–1189 BCE); Upper Egypt, Valley of the Kings, Thebes; fragment of a limestone relief with five vertical lines of hieroglyphic text; British Museum, London, museum no. EA 5610.

Source: © The Trustees of the British Museum, London

Knowledge of the hieroglyphic script dwindled dramatically in late antiquity. Although numerous Greek and Latin treatises deal with the hieroglyphs, the authors focused mainly on the allegedly symbolic use of these characters and hardly considered the phonetic aspects of the Egyptian script. The Christian theologian Clement of Alexandria (d. *c.* 215) is among the early proponents of this tendency (Zauzich 1978). In *The Enneads* the Greek philosopher Plotinus (204/5–270) expounded this late antique understanding of the hieroglyphs. He conceived of these characters as 'symbols' that convey the ancient Egyptians' wisdom and scientific knowledge:

> The wise men of Egypt seem to me to have grasped this, whether with precise understanding or innately. When they want to display their wisdom, they don't use forms of letters which spell out arguments and propositions, or imitate sounds and the verbalization of statements. Rather, they draw images, and carve them in their temples – each thing having one image, rather than a discursive description. So, then, each image is a bit of scientific understanding or wisdom: a substrate whose parts are taken altogether, not an act of discursive thinking or deliberation.
>
> *(Plotinus, The Enneads, V 8, 6; trans. Gerson 2017, 616)*

Later Greek and Roman authors followed Plotinus's way of interpretation. It comes as no surprise that the Arabs, who entered the world stage in the 1st/7th century and came into contact with late antique traditions in the newly conquered territories of the Middle East and North Africa, adopted the same understanding of the old Egyptian script.

Throughout the Middle Ages, scholars had at best a very limited knowledge of the hieroglyphic script. Al-Sukkarī (d. 275/888), a philologist and transmitter of Ibn al-Kalbī's (d. 204 or 206/819 or 821) book of genealogy, is one of the first Muslim authors to refer to it. He called it *musnad* (Ibn al-Kalbī 1966, 614). Muslim scholars used this term to refer to inscriptions in the ancient South Arabian script (Figure I.15.2). Since they had little or no knowledge of this alphabet and often provided fanciful depictions of the characters, it is less surprising that they

Figure I.15.2 Ancient South Arabian script in Minaean language on an incense burner made from limestone (Minaean period 3rd–2nd (?) centuries BCE); Haram, Yemen; 23.31 × 30.48 cm. British Museum, London, museum no. 125111.

Source: © The Trustees of the British Museum, London

sometimes indiscriminately used this term for the South Arabian as well as the hieroglyphic script (Beeston 1993).

Moreover, several Yemeni tribes migrated to Egypt after the Arab conquest and left their imprint on the Arabic tradition regarding the building of the pyramids (Fodor and Fóti 1976, 160, no. 13). This influence might have led to such an imprecise nomenclature.

The scholars, historians and geographers of the following centuries often used different terms for the hieroglyphic script. The Upper Egyptian author Abū Jaʿfar al-Idrīsī, mentioned earlier, called it sometimes 'the pharaonic temple-script' (*al-qalam al-birbāwī*), sometimes 'the script of auspices' (*qalam al-ṭayr*) or 'the divinatory script' (*al-qalam al-kāhinī*; Haarmann 1991, 60,9; 61,7; 62,11; 63,10; 95,9). Al-Idrīsī only refers to the hieroglyphs. The Egyptian historian al-Mubashshir ibn Fātik (5th/11th century) is the only author who refers to the ancient Egyptian script's three variants, Hieroglyphic, Hieratic and Demotic, in a passage that ultimately derives from Porphyry's (d. *c.* 305) *Life of Pythagoras* (Ibn Fātik 1958, 54).

In general, medieval Arab and Arabic-writing scholars agreed that knowledge of this ancient language had long been lost. They repeatedly mention in their writings that no one knew how to decipher them. ʿAbd al-Laṭīf al-Baghdādī (d. 629/1232), for example, writes that no one had an inkling of the meaning of the scripts in the temples and that he neither met nor heard of anyone able to decipher them (Zand *et al.* 1965, 119). More than hundred years later, the traveler Shams al-Dīn al-Lawātī al-Ṭanjī, widely known as Ibn Baṭṭūṭa (d. 770 or 779/1368 or 1377), encountered these mysterious writings while passing through Egypt. He asserts that no one understood the 'writings of the ancients,' which he had encountered in the then still standing temples of Akhmīm (ancient Panopolis; Ibn Baṭṭūṭa 1958, 1: 65). Given this complete lack of information on the history and the meaning of the script, it comes as no surprise that theories on its obscure origins and fanciful interpretations of its meaning soon began to spread.

I.15.3 Theories on the origins and meaning of the hieroglyphs

Medieval authors commonly conceived of the inscriptions on the ancient Egyptian temples as symbols conveying the former knowledge (mostly of an occult nature) of the ancient Egyptians or the late antique syncretistic deity Hermes Trismegistus. This idea appears very prominently in the so-called Arabic Hermetic tradition, that is to say the Arabic astrological, alchemical, magical and other occult treatises attributed or linked to Hermes Trismegistus. In these texts, Hermes was depicted as a prophet of science. The astrologer Abū Maʿshar (171–272/787–886) wrote a most influential account of him. In *The Book of the Thousands* (*Kitāb al-Ulūf*), he identifies three different Hermes. The first Hermes, Abū Maʿshar informs us, was the antediluvian prophet Idrīs who bequeathed his knowledge to later generations by inscribing it on Egyptian monuments. The second dwelt in Babylonia and invented astrology. The third Hermes lived in Egypt after the Flood. He was the author of magical and alchemical treatises (Van Bladel 2009[CB], 1–63; on the possible Egyptian roots of the three Hermes, see von Lieven 2007).

The idea that Hermes or other ancient philosophers engraved the hieroglyphic script on the ancient Egyptian temples and pyramids in order to safeguard their wisdom and their sciences from falling into oblivion proved quite alluring in the following centuries. It resurfaces time and again in Arabic occult writings. In the introduction to Abū l-Qāsim Maslama al-Qurṭubī's (d. 353/964) *The Goal of the Sage* (*Ghāyat al-ḥakīm*), a famous magical compendium known in the Latin West as *Picatrix*, the author claims that the ancient philosophers inscribed their wisdom in form of images (*nuqūshāt al-ṣuwar*) on the ancient Egyptian temples and monuments (*barābī*; al-Qurṭubī 1962, 2). In one of the works on magic attributed to the occultist al-Būnī (d. 622/1225), Idrīs is said to have saved his alchemical knowledge from the Deluge by inscribing it on the ancient temples (Ullmann 1972, 234). The Syrian cosmographer Shams al-Dīn al-Dimashqī (d. 727/1327) asserts that the images (*taṣāwīr*) in the temple of Akhmīm were believed to be signs (*rumūz*) referring to the sciences of the ancient Egyptians, such as the talismanic art, medicine, alchemy and astrology (al-Dimashqī 2013, 44).

Thus, Arab and Arabic-writing scholars comprehended the hieroglyphic script mainly as *rumūz* (signs, codes, ciphers) used by past sages to prevent the common masses and the ignorant from learning the precious secrets of their sciences and occult arts. Their interpretation of the ancient Egyptian script followed the late antique authors' approach who, as said earlier, interpreted the characters in a symbolical and allegorical way and hoped to retrieve ancient and esoteric wisdom from them. It is hardly surprising that the *kashf* or *fakk al-rumūz*, the 'unveiling' or 'deciphering of the symbols,' became a popular title of Arabic occult writings (Heinrichs 1995). Some authors even composed treatises in which they claimed to reveal the meaning of ancient and occult alphabets, including the hieroglyphic script. The most famous example is an Arabic treatise that was once hailed as the 'key' to the lost meaning of the hieroglyphs.

I.15.4 *The Yearning of the Love-Stricken* and the revelation of ancient and secret alphabets

The Arabic treatise *The Yearning of the Love-Stricken concerning the Knowledge of the Symbols of the Writing Systems* (*Shawq al-mustahām fī maʿrifat rumūz al-aqlām*) is an exposition of ancient and secret scripts that circulated under the name of the previously mentioned authority on occult matters, Ibn Waḥshiyya. The treatise consists of eight chapters (*bāb*), each divided into several sections (*fuṣūl*) and a conclusion (*khātima*) on the rites of the ancient Egyptian priests. The contents are as follows:

Chapter 1 (divided into 3 sections): Kufic, Maghribī Arabic, Indian
Chapter 2 (divided into 7 sections): Syriac, Nabatean, Hebrew, the alphabet of the ancient Egyptian temples (*birbā*, pl. *barābī*), *Lakamī* (?), *Musnad*, Greek
Chapter 3 (divided into 7 sections): contains the alphabets of the seven famous Greek philosophers and sages (Hermes, 'Aflīmūn' [Polemon?], Plato, Pythagoras, Asclepius, Socrates and Aristotle)
Chapter 4 (divided into 24 sections): contains the alphabets invented after the seven famous philosophers and named after their inventors (mostly Greek sages such as Apollonius of Tyana, Ptolemy, Dioscorides and so forth)
Chapter 5 (divided into 7 sections): alphabets of the seven "planets", from Saturn to Moon
Chapter 6 (divided into 12 sections): alphabets of the 12 zodiacal signs, from Aries to Pisces
Chapter 7 (divided into 10 sections): alphabets of the ancient Syrian, 'Hermesian', Pharaonic, Canaanaean, Chaldean, Nabatean, Kurdish, 'Kasdanian', Persian and 'Coptic kings'
Chapter 8: On the hieroglyphs
Conclusion: On the rites of the ancient Egyptian priests

Of these writing systems, only a small number correspond to existing alphabets, such as the Kufic and Maghribī Arabic scripts, the Greek, Syriac, Hebrew and South Arabian alphabets. These 'historic' alphabets resemble their authentic exemplars to greater or lesser degrees, whereas the remaining alphabets are fanciful inventions (Toral-Niehoff and Sundermeyer 2018, 252–3). In the preface, the author briefly explains his objective in composing this work, namely, to provide a collection of ancient alphabets for those interested in 'philosophical sciences (*al-ʿulūm al-ḥikmiyya*)' (Ibn Waḥshiyya 1806, 1).

Thus, the author composed his survey of ancient and magical alphabets with a didactic purpose in mind. The presentation of the different scripts shall enable students to read the ancient philosophers' books on *al-ʿulūm al-ḥikmiyya*, the 'philosophical sciences.' In the main text, the author provides for each character of these ancient alphabets an Arabic letter. Since he does not elaborate on the actual method of reading these scripts, he seems to imply that the deciphering of the ancient philosophers' texts simply consists of substituting an Arabic letter for each character of these different scripts. For *Kūfī*, it actually does work, since the *Kūfī* letters are just another variant of Arabic letters. You can substitute the *Kufic* letter with a *naskhī* letter, the text will still be in Arabic. But for Hebrew and Syriac the mere substitution of one letter with an Arabic letter only provides a transliteration of the text and not a translation into Arabic. He therefore conceives of these alphabets as ciphers and not as the writing systems of foreign languages.

The hieroglyphic script presented in this treatise is an exception to this rule. The author believes that different hieroglyphic scripts exist. He calls them 'the scripts of Hermesians' (*aqlām al-harāmisa*), since they were invented by Egypt's first royal dynasty, the Sage Hermes the Great (*al-ḥakīm Hirmis al-akbar*) and his descendants (Figure I.15.3). He asserts that the hieroglyphs are, in fact, signs, each conveying a single concept:

> Every one of these kings invented, according to his own genius and understanding, a particular alphabet, in order that none should know them but the sons of wisdom.

Figure I.15.3 Representation of hieroglyphs in a copy (dated 1165/1751) of *The Yearning of the Love-Stricken concerning the Knowledge of the Symbols of the Writing Systems* (*Shawq al-mustahām fī maʿrifat rumūz al-aqlām*) attributed to Ibn Waḥshiyya (d. 318/930–1). MS Paris, BnF, arabe 6805, fol. 50b.

Source: © BnF, Paris

> Few, therefore, are found who understand them in our time. They took the figures of different instruments, trees, plants, quadrupeds, birds, or their parts, and of planets, and fixed stars. In this manner these hieroglyphical alphabets became innumerable, like the alphabets of the Indians and Chinese. They were not arranged at all in the order of our letters *a*, *b*, *c*, *d*, but they had proper characters agreed upon by the inventors of these alphabets, . . .
>
> *(Ibn Waḥshiyya 1806, 14–15, trans. Hammer)*

In the section on the hieroglyphic alphabet attributed to the Great Sage Hermes *(al-ḥakīm Hirmis al-akbar*), the author points out that one is not dealing with an alphabet proper but with 'ciphers' (*rumūz*) and 'allusions' (*ishārāt*; Ibn Waḥshiyya 1806, 81). He is thus following the general idea Arab scholars had of the hieroglyphic script (Figure I.15.4).

I.15.5 Scholarly practices in *The Yearning of the Love-Stricken*

The unknown author of the Arabic treatise on hieroglyphs and other alphabets was not a mere dreamer or eccentric. He built his text according to standard rules of scholarly practices. He carefully arranged the presentation of the alphabets, giving each of them a title. He begins with a declaration of his objectives and goals. Then, he proceeds more or less systematically from one topic to the other, arranging his discussion of alphabets and scripts as appropriately combined parts. Together with this topical procedure, he follows an (admittedly not fully) systematic (sometimes speculative and sometimes realistic) chronology. When presenting his ideas about the various alphabets or scripts, he tells stories, many of which are the fruits of ancient speculations and stories told across numerous religious and intellectual communities in western Asia and the Mediterranean, but they, too, are systematically arranged in a sequence. It is obvious that he

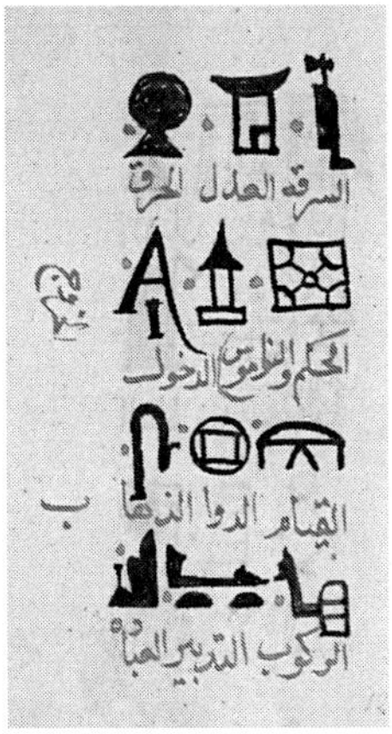

Figure I.15.4 Representation of hieroglyphs in a copy (dated 1165/1751) of *The Yearning of the Love-Stricken concerning the Knowledge of the Symbols of the Writing Systems* (*Shawq al-mustahām fī maʿrifat rumūz al-aqlām*) attributed to Ibn Waḥshiyya (d. 318/930–1). MS Paris, BnF, arabe 6805, fol. 56a.

Source: © BnF, Paris

wishes to be seen as a scholarly writer. That is why he strengthens his claims about types and forms of alphabets and scripts through two means: real, corrupt or invented examples in visual format and translations of collected knowledge. In addition to confronting the reader with material and intellectual 'proof' for the claimed multitude of alphabets and scripts, he finally turns to the core of 'scientific' work; he offers evidence. Not satisfied with one kind of evidence, the unknown author of the Arabic text selected three kinds to bolster his claims and win the trust of the reader. First are scholarly texts produced by divine or human authorities such as Agathodaimon (the Serpent God of Fortune Telling) or Jābir ibn Ḥayyān (d. 193/812?). Then he offers visits to temples and observations of samples with scripts, which include elements that can be verified and confirmed even today, such as the presence of script on the outside of pyramids or the use of clay tablets for writing. Finally, he gives a presentation of results of translations of texts written in other languages by past writers. His style of writing oscillates between the factual and the narrational, and his style of self-representation takes up standard scholarly values, still used by many today – humility and pride.

I.15.6 Scholarly reflections on *The Yearning of the Love-Stricken* in the Renaissance and at the turn to the 19th century

The German Jesuit Athanasius Kircher (1602–1680) was, as far as we know, the first scholar in Christian Europe who chanced upon a manuscript transmitting *The Yearning of the Love-Stricken* (Figure I.15.5). Kircher, who tried to decipher the hieroglyphs during his lifetime, discovered the manuscript on the island of Malta. He briefly mentions Ibn Waḥshiyya's treatise in one of his books (Ibn Waḥshiyya 1806, xvii–xix).

The Yearning of the Love-Stricken rose to greater prominence at the beginning of the 19th century when the Austrian diplomat and orientalist Joseph Hammer (after ennoblement: Joseph von Hammer-Purgstall [1774–1856]; Figure I.15.6) discovered another manuscript of this treatise

Figure I.15.5 Portrait of Athanasius Kircher at age 53 from *Mundus Subterraneus* (1664); the copper engraving was made in Rome, in 1655 by Cornelis Bloemaert. Size: 155 × 100 mm. Nuremberg, Germanisches Nationalmuseum, Graphische Sammlung (Paul Wolfgang Merkel'sche Familienstiftung), inv. no. MP 12693, Kapsel-Nr. 217.

Source: © German National Museum, Nuremberg

Figure I.15.6 Joseph von Hammer-Purgstall (1774–1856). Copper engraving (*c.* 1857) by Tommaso Bennedetti (1797–1863) after a drawing of Thomas Lawrence (1769–1830).

Source: https://commons.wikimedia.org/wiki/File:Joseph_von_Hammer-Purgstall._Benedetti_(um_1857).jpg

in Cairo. In 1806, he published an edition and English translation of *The Yearning of the Love-Stricken* in London under the title *Ancient Alphabets and Hieroglyphic Characters Explained; with an Account of the Egyptian Priests, Their Classes, Initiation, and Sacrifices, in the Arabic Language by Ahmad Bin Abubekr Bin Wahshih and in English by Joseph Hammer, Secretary to the Imperial Legation at Constantinople.* The publication coincided with the renewed interest in ancient Egypt and its history in the wake of Napoleon's Egyptian Campaign (1798–1801) and the nationalist "race" for the decipherment of the hieroglyphs.

In his introduction, Hammer explains that he discovered the manuscript in Cairo, despite the French scholars' efforts at the time to scour the Egyptian libraries for any valuable information on the meaning of the hieroglyphs (Ibn Waḥshiyya 1806, i). He is quite confident that the scripts in this treatise contain a kernel of truth. Hammer even believes that the specimens of the hieroglyphic script might eventually lead to their correct interpretation, despite the alphabets being in general, as he readily admits, reproduced in a garbled manner (Ibn Waḥshiyya 1806, iv–v). Hammer's conviction in the veracity of these alphabets prompted him to publish and translate this small booklet.

The earliest history of *Yearning* remains unknown. In a concluding note, which all five extant manuscripts preserve, the unknown author states that after spending 21 years on this work he donated it to the library (*khizāna*) of the (Umayyad) Caliph ʿAbd al-Malik (r. 65–86/685–705) on Thursday, 3 Ramadan 214/4 November 829. Not quite 200 years later, in 413/1022, a certain Ḥasan ibn Faraj ibn ʿAlī claims to have produced a copy of the original, which, in turn, became the sub-archetype of all preserved manuscripts. Ḥasan pretends descent from Thābit ibn Qurra (d. 288/901), the most prominent member of the well-known family of translators and scholars from Ḥarrān. The first claim is patently anachronistic, the Umayyad caliph ʿAbd al-Malik having died more than a century earlier – not to mention other exceptionable features of his phrasing. Hammer would not rule out the existence of a library bearing the caliph's name (Ibn Waḥshiyya 1806, xix). However, a copyist might simply have tried to raise the price for his manuscript by contriving such a noble origin (Fahd 1999, 36). Toral-Niehoff and Sundermeyer also reject Ḥasan ibn Faraj's claims. According to them, his etiological tale would be "all too fitting" the Nabatean context of the Ibn Waḥshiyya corpus (Toral-Niehoff and Sundermeyer 2018, 254). In fact, the two scholars argue that Ibn Waḥshiyya was not the author of the bulk, the eight chapters, of the treatise, and very likely not even of the last part, the *Khātima*, which had been a separate treatise before being bound together with the preceding exposition of the alphabets (Toral-Niehoff and Sundermeyer 2018, 252).

I.15.7 Ibn Waḥshiyya

Who is this enigmatic author to whom the treatise is attributed? Ibn Waḥshiyya is a 4th/10th-century authority on occult sciences and the author of a widely read Arabic agronomy manual (Fahd 1993). There is no reliable historical evidence that proves his existence. The biographical information that can be gleaned from his treatises is very scarce. In his most famous work, known as *The Nabatean Agriculture* (*Kitāb al-Filāḥa al-nabaṭiyya*), a sort of treatise on agronomy comprising magical and astrological passages, he presents himself as the translator of Syriac sources into Arabic. The Finnish Arabist Jaakko Hämeen-Anttila recently undertook to demonstrate that there is no reason to raise serious doubts as to Ibn Waḥshiyya's very existence. On the other hand, he confirms that Ibn Waḥshiyya did not compose all the treatises on alchemy, astrology, and magic that were later attributed to him (Hämeen-Anttila 2006, 6–8).

I.15.8 Academic research on *The Yearning of the Love-Stricken* since the 19th century

Scholarly discussions on *The Yearning of the Love-Stricken* began shortly after Hammer published the treatise. In 1810, the French orientalist Antoine-Isaac Silvestre de Sacy (d. 1838; henceforth: de Sacy) wrote a review of Hammer's edition and translation (de Sacy 1810). He drew attention to similar Arabic manuscripts on magical alphabets in what is now the Bibliothèque nationale de France in Paris. All these works have in common, de Sacy observed, that the revealed scripts shall serve the reader to produce talismans or amulets, to counter magical spells or to find hidden treasure. Some are attributed to famous authors such as the – similarly shadowy – alchemist Jābir ibn Ḥayyān. The French Orientalist doubted that *The Yearning of the Love-Stricken* really offers the actual meaning of the presented hieroglyphs and that all the presented characters can be found on ancient Egyptian monuments. To substantiate his claims, Silvestre points out that the historical alphabets (Hebrew, Nabatean, Greek, etc.) in *The Yearning of the Love-Stricken* only resemble to a small degree the actual alphabets and concludes that such discrepancies give rise to concern on the reliability of the reproduced material. He concludes that this work must be the product of some impostor who 'cashed in' on his contemporaries' gullibility (de Sacy 1810).

After de Sacy's scathing critique, the treatise fell into oblivion for decades. In 1861, the German orientalist Alfred von Gutschmid (1831–1887) dealt shortly with its rendition of 'Nabatean' script (von Gutschmid 1861). Thereafter, Western scholars again ignored it for more than a century.

In 1975 and 1999, Toufic Fahd published two articles on *The Yearning of the Love-Stricken* in which he argues that the treatise reflects Hellenistic traditions that can be traced back to the 3rd, 4th and 5th centuries and were revived in the form of alchemical speculations in Arabic between the 2nd/8th and 3rd/9th century (Fahd 1975, 1999). Some of his interpretations were rejected by Jaakko Hämeen-Anttila (2006). In 2015, the Egyptologist Okasha el-Daly argued that the preserved Arabic texts dealing with hieroglyphs show that some Arab scholars "succeeded in deciphering at least half of the Egyptian alphabetical signs" (el-Daly 2005, 57), but his approach has been strongly criticized by his colleagues in Egyptology as unscientific (Eyma 2005). Another unsubstantiated hypothesis has been put forward in 2017. Jean-Charles Coulon argued that this text emerged in Fatimid Egypt in an attempt to evoke Egypt's pre-Islamic past and to shape an 'Egyptian' identity in response to the claims of the Abbasid Empire (Coulon 2017, 129–33).

In the latest study on *The Yearning of the Love-Stricken*, Isabel Toral-Niehoff and Annette Sundermeyer focus on Ibn Waḥshiyya's alleged authorship of *The Yearning of the Love-Stricken*. They observed that Ibn Waḥshiyya is mentioned only once as the author (in the colophon). Moreover, the treatise conjures up an ancient Egyptian setting. Finally, the topics common to the Ibn Waḥshiyya corpus, such as the translation of 'Nabatean' lore, are missing. Only in the last part of the treatise, the so-called appendix or conclusion (*khātima*), are references to Babylonia to be found. The two scholars therefore suggest that the treatise in fact consists of two originally separate sections which were combined at a later stage (Toral-Niehoff and Sundermeyer 2018, 251). The author's 'Egyptianizing' tendency, they claim, implies his tract's Egyptian origin (Toral-Niehoff and Sundermeyer 2018, 257).

I.15.9 Egypt or Ḥarrān?

More than 200 years after the publication of *The Yearning of the Love-Stricken concerning the Knowledge of the Symbols of the Writing Systems*, the question of its provenance still remains unresolved. Although this short treatise has attracted the attention of Western scholars for almost 400 years, its history is still largely shrouded in dense fog. While an Egyptian local context seems likely, given the numerous 'Egyptian' references, Fahd's and Hämeen-Anttila's suggestion that Ḥasan ibn Faraj, a self-styled descendant of Thābit ibn Qurra, might have been the author appears plausible at first sight (Fahd 1975, 1999; Hämeen-Anttila 2006, 21).

Thābit ibn Qurra was born into the Sabian community of Ḥarrān who continued to adhere to Hermeticism and Neoplatonic and Neopythagorean doctrines in Islamic times. Thābit, who was fluent in Arabic, Syriac, and Greek, and his successors were thoroughly acquainted with the late antique tradition of Greek science and philosophy and instrumental in the transmission of Greek esoteric knowledge to Arabic (Green 1992). Given the late antique predilection for speculations on the hieroglyphs' symbolic value, a practice that seems to have originated already among the ancient Egyptian priesthood (von Lieven 2010), it appears convincing that the presentation of single hieroglyphs as 'mysterious images of condensed wisdom' in *Yearning* is an Arabic continuation of Hellenistic interpretations. Ḥarrān might have been a nexus where the Hellenistic tradition was kept alive and eventually influenced later speculations on the hieroglyphs among Arab and Arabic-writing scholars.

However, the treatise exhibits characteristics that rather contradict this hypothesis. The false *nisbas* al-Bābilī al-Nūqānī attributed to Thābit ibn Qurra in the colophon are suspicious. Furthermore, the use of vocabulary that gained its current meaning at a later period (e.g., *salṭana*) and the obvious grammatical errors in the *khātima* render Ibn Faraj's involvement highly unlikely. Rather, the treatise was composed at a much later date by someone whose skills in the Classical Arabic language were limited. More detailed research into the background of *Yearning* is still required in order to provide a final assessment of its origins. One of the obstacles to be overcome in this field of inquiry is the dearth of historical information on the Arabic appropriation of Hellenistic literature on ancient Egypt. Although the treatise's genesis remains nebulous, it is an interesting attempt to collect and organize information on ancient and invented alphabets and a unique testimony to a long-lasting fascination for ancient Egypt's writing system. Even though outside the perimeter of the accepted syllabus of learning among premodern Muslims, the transmission of, and search for, such 'occult' knowledge deserves more thorough investigation in the future.

Note

1 Consolidated bibliography.

Bibliography

Sources

Al-Dimashqī. ed. al-Nāṣīr, G. D., al-Ḥadathay, Ṭ. S. and Ayyūb, A. M. W. 2013. *Nukhbat al-dahr fī ʿajāʾib al-barr wa-l-baḥr* [The Most Remarkable Things in the Time as Marvels of the Earth and the Sea]. Damascus: Dār al-ʿArrāb li-l-dirāsāt wa-l-nashr wa- l-tarjama and Nūr.

Fahd, T. ed. 1993. *al- Filāḥa al-nabaṭiyya* [The Nabataean Agriculture]. 2 vols. Damascus: al-Maʿhad al-ʿilmī al-faransī li-l-dirāsāt al-ʿarabiyya.

Haarmann, U. 1991. *Das Pyramidenbuch des Abū Ǧaʿfar al-Idrīsī (st. 649/1251)* [The Book on the Pyramids by Abū Jaʿfar al-Idrīsī (d. 649/1251)]. Stuttgart: Steiner.

Ibn Baṭṭūṭa. tr. Gibb, H. A. R. 1958–1971. *The Travels of Ibn Battuta*. 3 vols. Cambridge: Cambridge University Press.

Ibn Fātik. ed. al-Badawī, ʿA. R. 1958. *Mukhtār al-ḥikam wa-maḥāsin al-kalim* [Book of Selected Maxims and Aphorisms]. Madrid: Maṭbaʿat al-Maʿhad al-miṣrī li-l-dirāsāt al-islāmiyya.

Ibn al-Kalbī. ed. Caskel, W. 1966. *Ǧamharat an-nasab: das genealogische Werk des Hišām Ibn Muhammad al-Kalbī*. 2 vols. Leiden: Brill.

Ibn Waḥshiyya. 1806. *Ancient Alphabets and Hieroglyphic Characters Explained: With An account of the Egyptian Priests, Their Classes, Initiation, and Sacrifices in the Arabic Language by Ahmad bin Abubekr bin Wahshih; and in English by Joseph Hammer*. London: W. Bulmer and Co.

Plotinus. ed. and tr. Gerson, L. P. 2017. *The Enneads*. Cambridge: Cambridge University Press.

al-Qurṭubī. ed. Ritter, H. 1962. *Das Ziel des Weisen von Pseudo-Maǧrīṭī* [The Aim oft he Sage by Pseudo-Majrīṭī]. Vol 1: *Der arabische Text* [The Arabic Text. London: Warburg Institute.

Zand, K. H., Videan, J. A. and Videan, I. E. 1965. *The Eastern Key. Kitāb al-ifādah wa-l-iʿtibār of ʿAbd al-Laṭīf al-Baġdādī* [Book of Utility and Verification. London: Allen and Unwin.

Research literature

Beeston, A. F. L. 1993. "Musnad. (A.) 1. As a Term Applied to the Ancient South Arabian Script," in *EI-2*, 7: 704b–5a.

Coulon, J.-C. 2017. *La Magie en terre d'Islam au Moyen Âge*. Paris: CTHS.

el-Daly, O. 2003. "Ancient Egypt in Medieval Arabic Writings," in Ucko, P. and Champion, T., eds. *The Wisdom of Egypt: Changing Visions through the Ages. Encounters with Ancient Egypt*. London: UCL Press, 39–63.

el-Daly, O. 2005. *Egyptology: the Missing Millennium: Ancient Egypt in Medieval Arabic Writings*. London: UCL Press.

Eyma, A. K. 2005. "Review of Okasha El Daly's *Egyptology: The Missing Millenium*," *Egyptologists' Electronic Forum*, 28 June 2005. www.egyptologyforum.org/reviews/Missing1000.html (Accessed 27 June 2018).

Fahd, T. 1975. "Sur une collection d'alphabets antiques réunis par Ibn Waḥšiyya," in Leclant, J., ed. *Le déchiffrement des écritures et des langues: colloque du XXIXe Congrès des Orientalistes, Paris, du 16 au 22 juillet 1973*. Paris: L'Asiathèque, 105–19.

Fahd, T. 1999. "Les écritures des sages hermétiques d'après Ibn Wahshiyya (Xe siècle)," *Politica Hermetica* 13: 33–67.

Fodor, A. and Fóti, L. 1976. "Haram and Hermes: Origin of the Arabic Word Haram Meaning Pyramid," *Studia Aegyptiaca* 2: 157–67.

Green, T. M. 1992. *The City of the Moon God. Religious Traditions of Harran*. Leiden and Boston: Brill.

Gutschmid, A. von. 1861. "Die Nabatäische Landwirtschaft und ihre Geschwister," *ZDMG* 15: 16–21.

Haarmann, U. 1996. "Medieval Muslim Perceptions of Ancient Egypt," in Loprieno, A., ed. *Ancient Egyptian Literature: History and Forms*. Leiden: Brill, 605–27.

Hämeen-Anttila, J. 2006. *The Last Pagans of Iraq. Ibn Waḥshiyya and His Nabatean Agriculture*. Leiden: Brill.

Heinrichs, W. P. 1995. "Ramz," in *EI-2*, 8: 426b–8b.

Kennedy, H. 2006. "Egypt as a Province in the Islamic Caliphate, 641–868," in Petry, C. F., ed. *Islamic Egypt: 640–1517*. Cambridge: Cambridge University Press, 62–85.

Lieven, A. von. 2007. "Thot selbdritt: Mögliche ägyptische Ursprünge der arabisch- lateinischen Tradition dreier Hermesgestalten," *Die Welt des Orient* 37: 69–77.

Lieven, A. von. 2010. "Wie töricht war Horapollo? Zur Ausdeutung von Schriftzeichen im Alten Ägypten," in Knuf, H., Leitz, C. and von Recklinghausen, D., eds. *Honi soit qui mal y pense. Studien zum pharaonischen, griechisch-römischen und spätantiken Ägypten zu Ehren von Heinz-Josef Thissen*. Leuven: Uitgeverij Peeters and Departement Oosterse Studies, 567–74.

Naguib, S.-A. 2008. "Survivals of Pharaonic Religious Practices in Contemporary Coptic Christianity," in Dieleman, J. and Wendrich, W., eds. *UCLA Encyclopedia of Egyptology*. Los Angeles. http://escholarship.org/uc/item/27v9z5m8 (Accessed 27 June 2018).

Nöldeke, T. 1875. "Noch Einiges über die 'nabatäische Landwirtschaft'," *ZDMG* 29: 445–55.

Pettigrew, M. F. 2004. *The Wonders of the Ancients: Arab-Islamic Representations of Ancient Egypt*. PhD "dissertation. Berkeley: University of California.

Quatremère, É. M. 1835. *Mémoire sur les Nabatéens. Journal Asiatique* 15: 5–5, 97–137, 209–71.

Silvestre de Sacy, S. 1810. "Review: Ancient Alphabets and Hieroglyphic Characters Explained," *Magasin encyclopédique* 6: 145–75.

Sezgin, U. 1993. "Al-Masʿūdī Ibrāhīm b. Waṣīfšāh und das Kitāb al-ʿAǧā'ib. Aigyptiaka in arabischen Texten des 10. Jahrhunderts n. Chr," *ZGAIW* 8: 1–70.

Sezgin, U. 1994. "Pharaonische Wunderwerke bei Ibn Waṣīf aṣ-Ṣābi' und al-Mas'ūdī. Einige Reminiszenzen an Ägyptens vergangene Größe und an Meisterwerke der Alexandrinischen Gelehrten in arabischen Texten des 10. Jahrhunderts n. Chr., Teil II," *ZGAIW* 9: 189–249.

Sezgin, U. 2001. "Pharaonische Wunderwerke bei Ibn Waṣīf aṣ-Ṣābi' und al-Mas'ūdī. Einige Reminiszenzen an Ägyptens vergangene Größe und an Meisterwerke der Alexandrinischen Gelehrten in arabischen Texten des 10. Jahrhunderts n. Chr., Teil III," *ZGAIW* 14: 217–56.

Sezgin, U. 2002. "Pharaonische Wunderwerke bei Ibn Waṣīf aṣ-Ṣābi' und al-Mas'ūdī. Einige Reminiszenzen an Ägyptens vergangene Größe und an Meisterwerke der Alexandrinischen Gelehrten in arabischen Texten des 10. Jahrhunderts n. Chr., Teil IV," *ZGAIW* 15: 281–311.

Sezgin, U. 2004. "Ein arabischer Text (4./10. Jahrhundert) über Könige von Ägypten gewährt Einblicke in das spätantike Ägypten [Pharaonische Wunderwerke, Teil V]," *ZGAIW* 16: 149–206.

Thissen, H. J. 2001. *Des Niloten Horapollon Hieroglyphenbuch. Band I. Text und Übersetzung.* Munich and Leipzig: K.G. Saur.

Toral-Niehoff, I. and Sundermeyer, A. 2018. "Going Egyptian in Medieval Arabic Culture," in Orthmann, E. and el-Bizri, N., eds. *The Occult Sciences in Pre-Modern Arabic Literature* (Beiruter Texte und Studien 138). Beirut: Orient Institut Beirut, 249–62.

Ullmann, M. 1972. *Die Natur- und Geheimwissenschaften im Islam. Handbuch der Orientalistik. Erste Abteilung: Der Nahe und der Mittlere Osten. Ergänzungsband VI. Zweiter Abschnitt.* Leiden: Brill.

Zauzich, K.-T. 1978. "Hieroglyphen," in Heck, W. and Westendorf, W., eds. *Lexikon der Ägyptologie.* 7 vols. Wiesbaden: Harrassowitz, 2: 1189–99.

I.16
PRACTICES OF ZOROASTRIAN SCHOLARS BEFORE AND AFTER THE ADVENT OF ISLAM

Götz König

I.16.1 Introduction

Zoroastrianism is the most important religion of pre-Islamic Iran (Widengren 1965; Boyce 1975, 1979, 1982; Boyce and Grenet 1991; Stausberg 2002, 2004). Its history covers the last 3000 years. Emerging probably in eastern Iran, Zoroastrianism reached western Iran in the Achaemenid period (6th–4th centuries BCE) and spread to Asia Minor in late antiquity. With the rise of Islam, the Zoroastrian communities finally took refuge in eastern Persis (southern Iran) and by migration in western India, especially Gujarat, as early as the late 3rd/9th century (West 1880, 1882a; Cereti 2007, 212–15). The Zoroastrian migration was described early in the last century (Paymaster 1915; Hodivala 1920; Modi 1934; Unvala 1940, 56–7, colophon M 50). More recent studies include Cereti and Sanjana (1991), Eduljee (1991) and Williams (2009).

The Zoroastrian Middle Persian (Pahlavi) literature of the 1st millennium CE is built on the Avesta, which was for a long time orally transmitted and consists of two layers. The first layer is the so-called *Zand* (knowledge), Middle Persian translations of and commentaries on the Avestan texts, mostly from the Sasanian period (Cantera 2004). Second is the Pahlavi literature of the 3rd–5th/9th–11th centuries, including religious literature of different genres, mostly written by high priests (West 1896–1904; Tavadia 1956; Klíma 1968, 1–67; Boyce 1968; De Menasce 1975, 1983; Macuch 2009; Table I.16.1).

These layers of Zoroastrian literature reflect different modes of religious thinking. While the oldest layer is dominated by texts of ritual and meta-ritual nature, in the later layer, theological texts prevail. The *Zand* has a kind of mediating position. This historical process of reflection has changed not only the religious contents but also the intellectual style of the scholars. Two further factors were of great importance for the development of Zoroastrian literature in the Islamic period: the emergence of a literate culture and the political integration of Iran into the caliphate. Because of this integration, Zoroastrian priests had to learn to argue for their doctrines, especially in the religious debates under Caliph al-Maʾmūn (r. 198–218/813–833). This meant in particular developing tools of logic (König 2020a). Although Greek philosophy was brought to pre-Islamic Iran repeatedly, it is striking that the *Zand* shows almost no Greek influence. Only some of the later Pahlavi literature uses philosophical terms and concepts that reflect Greek doctrines. This suggests that in the (late) 2nd/8th and 3rd/9th centuries some Zoroastrian priests

DOI: 10.4324/9781315170718-18

Table I.16.1 Premodern Zoroastrian literature

Avesta	*Old Avesta*	*Eastern Iran*	*2nd/1st mill.* BCE
	Younger Avesta		
Pahlavi Zand		Western Iran (Persia)	3rd–9th c. (?)
Pahlavi literature			9th(–14th) c.

engaged with Aristotelian and Neoplatonic philosophy in a manner that substantially changed the deeper structure of Zoroastrianism.

A major problem for evaluating the acquisition and transmission of philosophical, scientific and medical knowledge in Zoroastrian literature is the question when Zoroastrian teachings were codified in written form. Much of the extant textual material comes from after the 5th/11th century. It might thus reflect partly the exposure of Zoroastrian scholars to practices and ideas of other scholarly communities active in Islamicate societies since the 2nd/8th century. Moreover, the understanding of pre-Islamic Iranian access to ancient Greek and Syriac texts is complicated by the relatively widespread stories about the Sasanian recovery of knowledge destroyed centuries before by Alexander of Macedonia (r. 336–323 BCE), including the Avesta, and the scholarly activities of Syriac physicians in Gundishapur in western Iran. That is why this chapter does not proceed chronologically but is organized according to questions and problems faced in current research.

I.16.2 Designations of Zoroastrian and non-Zoroastrian scholars and their fields of knowledge before the 5th/11th century

The *Avesta* knows a good number of terms that can be translated as 'ritual priest', the (judicial/religious) 'decider', the 'bearer of the (magical) word' or the 'religious teacher', especially the Avestan *aēthra-pati* and the Middle Persian *hērbed*. Originally, *hērbed*s meant ritual instructors. After the Sasanian period, the word also applied to authors of *Handarz* (wisdom literature). For example, the Avesta-editor Ādurbād Mānsarspandān (4th century) was equally called *hērbed*, in this context translated as teacher. Despite the importance of the the concept of knowledge in the Avesta (for the concept of knowledge in the *Gāthās* see Colpe 2003) - the Avestan word *zaiṇti* (cf. Pahlavi *zand*) means 'knowledge' - words for 'theologian' or 'philosopher', however, are absent. In Pahlavi literature, most non-Zoroastrian scholars are called teacher (*kēsh-dārān*; see in particular *Dēnkard* 1973, 3: 8). Exceptions are Greek philosophers (*fīlasōfā*; Sundermann 1982, 14–38; König 2018), for whom a few sources also use what is perhaps a calque of philo-sophos (*dānāī-dōst*, friend of knowledge) and Iranian and Indian 'wise men' (*dānāg*):

> by the philosophers (*fīlasōfā*) of Hrōm [Rome] and the wise (*dānāg*) Indian men and other wise and knowing (*dānāg* <*ud*> *shnāsag*) men.
>
> *(Dēnkard 4: 108)*

Other terms are *frazānag* (from *fra-zan-*) and *śnāsag*. Compare the Avestan *zaiṇti*, which becomes *shnāsēnishn* when translated into Pahlavi. A similar ancient Greek distinction between the philosopher and the wise man appears in Diogenes Laertius (1487, 1: 4–5 and 12).

Their special field was seen as *dēn dānāgīh*, "knowledge from/about the *dēn* (religion)". Occasionally, they were also qualified according to their field of philosophy. Thus, a *gōhr-shnāsag* is an ontologist, while a *chihr-shnāsag* is a philosopher of nature (for further categories, see Bailey 1943, 86 no. 2; Shaki 1999, 177). While the *dānāg* in the anonymous *Judgments of the Spirit of Wisdom* (*Dādestān ī Mēnōg ī Khrad*; date unknown) acquires knowledge by an act of ritual gnosis, the *dānāg* in the introduction of Manūshchihr's *Religious Judgments* (*Dādestān ī dēnīg*; [*fl.* around 250 AY/881]; AY here and below means 'the year of Yazdegird' [d. 30/651], the last Sasanian king; it is one of the eras used in Islamicate societies, particularly by astrologers; it begins in 10/632) knows and decides through reasoning (*āsn-khrad*), ratio (*cim*) and the priestly family tradition. *Dēnkard* 3: 420 proves that the 'wise men' worked with philological arguments. Furthermore, the notion of knowledge usually has an ethical dimension, since the Zoroastrians in that period believed that no deed can be regarded as good without knowledge. This can be seen in the gnomes (*wāzag*) of the *Handarz* literature as collected in the *Pahlavi Texts* (manuscript MK) and in *Dēnkard* 6. According to *Dēnkard* 6.205, persons who will go to paradise are the wise, the friend of the wise and the non-foe of the wise. Thus, the concept of the scholar is that of a wise or good priest (*dēn-dastwar ī dānāg*).

I.16.3 Shifts and changes in Zoroastrian scholarship since the 5th/11th century

As said in the introduction, two major problems for evaluating the acquisition and transmission of philosophical, scientific and medical knowledge in Zoroastrian literature concern the question of when Zoroastrian teachings were codified in written form and the relatively late dates of such writings in the forms extant today. In the following, I briefly summarize what we currently know about these two issues.

Before the rise of Manichaeism, Iranian religious texts were mainly transmitted orally. It seems that the first codification of Zoroastrian, especially Avestan, texts did not take place before the Sasanian period (early 3rd to the middle of the 1st/7th centuries; but see also Pausanias 1918, 5: 27.5–6). However, the reports on the transmission of the Avesta and the *Dēnkard* seem to suggest that at least in the 6th century, and perhaps already in the 4th-century edition of the Avesta of Ādurbād Mahraspand, the 'written text' became the model of a text per se (*Dēnkard* 4; *Dēnkard* 3: 420).

After the conversion of many people in the former Sasanian territories to Islam beginning slowly in the 2nd/8th century, Zoroastrian intellectual activities fared differently under different Muslim dynasties. The Tahirids (r. 205–259/821–873) burned Zoroastrian books (Jackson 1906, chapter 23). In 346/957–958, the Samanids (r. *c.* 203–395/*c.* 819–1005) promoted four Zoroastrians to produce a historical narrative in New Persian prose, the *Book of Kings* (*Shāhnāme*; Nöldeke 1920, 16). The Buyids (r. 320–447/932–1055), in particular ʿAḍud al-Dawla (r. 338–372/948–983), were also interested in the Zoroastrian past of Iran (Mottahedeh 2012, 153–60).

Our knowledge about Zoroastrian intellectual activities, whether religious or philosophical and scientific, rests, on one hand, on surviving manuscripts and, on the other, on information copied from earlier, now lost versions. The oldest-known Zoroastrian manuscripts stem from the Mongol period and were produced by a family of eastern Iranian scribes in Gujarat. Its earliest representative consists of Avestan and Pahlavi material. A scribe named Rustam Mihrābān wrote it probably between 668–677/1269–1278 (Barr 1944; Cereti 2002, 214). This manuscript is known as K7. Its colophon gives as a date *637 pārsīg* (K7b, fols. 107r–108r, esp. fol. 107r,15; cf. Geldner 1886, vi–vii, xxxviii–xxxix). The oldest codices including only Pahlavi literature were written in India in the 720s/1320s by the scribe Mihrābān Kaykhusrō from Dizūg in Sistan, today eastern Iran (on his location, see colophon I of the manuscript K5, Unvala 1940,128–9 and Kotwal and Hintze 2008, 15; in general, see West 1896–1904, 113; Dresden 1967; Cereti

2007, 211–21; König 2019; those codices are called MK and K20/K20b, Christensen 1931). Mihrābān Kaykhusrō also wrote several exegetical Avestan manuscripts (Mills 1893; Sanjana 1896; Christensen 1937–1939; Christensen 1941–1942; Barr 1944). He often worked from manuscripts copied by his great-great-uncle Rustam Mihrābān mentioned earlier. Both scribes had access to different Iranian traditions (Katrak 1941, 245–6; on Mihrābān's colophons, see Westergaard 1852–1854, 3 no.1 and 11 no.1; Sanjana 1896, XXX-XXXV; Hodivala 1920, 118–33; Tavadia 1933, column 569; Tavadia 1944, 317ff). One tradition is linked to the southern Iranian region around Kazerun and Shiraz in the 5th/11th and 6th/12th centuries. The other tradition points to northeastern Iran around Nishapur between 494/1100 and 648/1250. Thus, their works prove that long before 882/1478, when the first collection of texts was brought from Iran to India by messengers (*rivāyat*), contacts existed between Sistan and Uca at the Indus in the Punjab (Unvala 1922; Dhabhar 1932; Vitalone 1987; Hodivala 1920, 276–349; West 1896–1904, 82). Such early contacts are also confirmed by other material. At the beginning of the 5th/11th century, for instance, another key figure in the transmission of Avestan and Pahlavi literature, the priestly scribe Māhwindād ī Naremāhān ī Wahrām ī Mihrābān, worked with material he had apparently received from partners in India. His networks covered a large territory, since in 410/1020 he also produced a copy of the *Dēnkard* from a manuscript he had found in Baghdad (Andreas 1882, 72–3; Geldner 1896–1904, 14; Christensen 1936; Unvala 1940, 152). Although we possess only this kind of scattered information about Zoroastrian studies of ancient and contemporary religious and secular knowledge, the *Rewāyat*s indicate that even in the 10th/16th century, Zoroastrian communities were still intellectually active in Khorasan, Kerman and Sistan.

Important intellectual activities among Zoroastrian priests took also place in India, mainly Gujarat. Between 551/1157 and 631/1234, an eminent priest called Nēryōsang was active (Geldner 1886, xxxiii–xxxiv; Geldner 1896–1904, 50; Damesteter 1892, CXII–CXIII; Eduljee 1991, 62; other examples: Anquetil 1771, 1.2: 74; West 1871, x [9th/15th century]; Cereti and Sanjana 1991 [7th/13th century]; Tavadia 1944, 304; Boyce 1979, 168–9 [5th–6th/11th–12th centuries]). Nēryōsang and members of his family seem to have produced Pāzand transcriptions of Pahlavi texts (Tavadia 1956, 14). It is debatable whether Pāzand, the vocalized writing of Pahlavi in Avestan characters, was invented by Nēryōsang. In contrast to manuscripts produced in India, Pāzand writings are only sporadically used in Iranian manuscripts. For the first time, those Zoroastrians also translated exegetical Avesta texts and Pahlavi literature into Sanskrit opening therewith new intellectual opportunities (Bharucha 1906–1933). This is proved by a note in the Paris manuscript P11 (= Suppl. Pers. 28/Blochet IX), a *Sanskrit Yasna*, which is a copy of the manuscript J3 (on Nēryōsang's *Yasna* manuscripts and translations see Geldner 1886, xxxiii–iv; Boyce 1979, 168; Humbach 2003, 2: 199–202).

Thus, between the 5th/11th and the early 8th/14th centuries, Zoroastrian scholarship in Iranian languages other than New Persian is characterized by two features. First, scholarly priests were no longer active as authors (see the following discussion) but, rather, as philologists and copyists. Second, a network of Zoroastrian scholars and communities continued to exist in Baghdad (Iraq), Kazerun (Fars), Nishapur (Khorasan), Dizūg (Sistan), Uca (Punjab) and Gujarat.

A new feature emerged in the same period. Pahlavi texts were translated into New Persian in versified form. In 676/1278, Zartusht Bahrām finished his *Book of Zarathustra* (*Zartusht-Nāma*), which was ordered by a high priest (*mobed ī mobedān*; Eastwick 1843, 477–522; Spiegel 1860, 2: 181–2; Rosenberg 1959). In his *Book of Ardā Wīrāz* (*Ardā Wīrāz Nāmag*) Zartusht Bahrām mentions his father Bahrām Pazhdu, a scribe (*dabīr*), writer (*adīb*) or perhaps physician (Āmūzgār 1989, 524), priest (*hīrbed*), astronomer and poet in Pahlavi and New Persian (Āmūzgār 1989,

524–5). 330 verses of his poem *Springtimes* (*Bahāriyat*) praising *Nōrūz*, the Zoroastrian kings and Zartusht, have survived in only one manuscript, finished in 626 AY/1257.

I.16.4 The authors of the Pahlavi literature (3rd/8th–8th/14th centuries)

Another problem when constructing a history of knowledge in Iranian languages is the widespread absence of dates in their sources. In Pahlavi literature, the focus of this section, only a few dates are transmitted. The most important of them is found in one of the letters written by the spiritual leader (*rad*) of *Pars* and *Kerman*, Manūshchihr, to his brother Zādsparam, the priest of the community of Sīrgān (southeastern Iran) and to the whole Zoroastrian community of Iran, dated 250 AY/881 (see West 1882b, III: 21). Manūshchihr criticizes his brother's shortening of the purification ritual, which is required in case of contact with corpses. It is assumed that Zādsparam's innovation was stimulated by his interest in anatomy and physiology (Sohn 1996). Zādsparam's *Selected Scripts* (*Wizīdagīhā ī Zādspram*) prove his empirical interests, his scientific work and his knowledge of Greek scientific texts. He mentions a lost zoological book *Taxonomy* (*Tōhmag-ōshmurishnīh*; Sohn 1996, 3.57) and seems to have been familiar with texts by Aristotle (384–322 BCE) and Hippocrates (d. *c.* 370 BCE), among them Aristotle's *Historia Animalium* (Aristotle 1979). The heading of Zādsparam's chapters 29 and 30, called *On the Structure of Human Beings* (*abar passāzishn ī mardōmān*), may reflect Hippocrates' *On the Nature of Man* (Sohn 1996, 32). The date 881 AY/1513 is the key for the dating of many other Pahlavi works.

The *Greater Bundahishn* 35a is an important account of the genealogy of the priestly families of Iran (Anklesaria 1956). Made allegedly on the basis of some version of the *Book of Kings* (*Khwadāyih-namag*), which has a very difficult and highly contested transmission history (Hämeen-Anttila 2018; Chapters I.1 and I.2), the genealogical lines recorded there lead back into the Achaemenid period. The claims made for pre-Sasanian times are not at all trustworthy, and the reliability of the Sasanian information is also contested (Hämeen-Anttila 2018). Farrbay, a cleric and editor of the *Greater Bundahishn*, lists about 20 priestly ancestors for himself (Anklesaria 1956, 35a6 and a1). The text also mentions contemporaneous scholars. Even if we cannot date each single author, on the basis of *Greater Bundahishn* 35a and by means of further information we are able at least to compile a chronological-genealogical table (Table I.16.2).

The table gives the names of nearly all known Pahlavi authors. It shows that most of the authors are related to each other and belong to the highest clergy. This genealogical and social density probably aimed to minimize doctrinal differences. However, it also enabled doctrinal changes 'from above'. A central figure of such a change in Zoroastrian theology after 800 CE was Ādurfarrbay ī Farrokhzādān. He is seen as the most important authority by modern Parsis (or Parsees), because he wrote two very important books: part or all of the *Dēnkard* and 147 questions of the *Riwāyats*. A brief note in the *Mādayān ī Gizistag Abālish* proves his participation in the interreligious discussions in Baghdad, stating that Caliph al-Ma'mūn:

> ordered that all his [Muslim] wise men (*dānāgān*) and the [wise men] of the Jews and of the Christians should come to him and should dispute with [the Zoroastrian apostate] Abālish [from the Arabic Abā Laith according to Schaeder 1972, 287, no. 2; see also De Menasce 1975, 544; Tafażżoli 1982, 58]. And because of the *amīr ī mūmenīn*'s ("leader of the believers") command [also] Ādurfarrbay ī Farrokhzādān – he was the 'leader of the good believers' (*hudēnān pēšōbāy*) – , the qadi, the vizier and Māmūn (*sic*) himself were in conversation with Abālish.
>
> *(Chacha 1936, 0.5–6)*

Table I.16.2 Zoroastrian scholars/priests between the late 2nd/8th and the early 5th/11th centuries

	Ādurbād ī *Jāwandān teacher (?) of	**Ādurfarrbay ī Farrokhzādān** time of al-Maʾmūn; first *hudēnān pēshōbāy* ("leader of the good believers"); author and editor of the *Dēnkard* and of other books	
	Rōshn <ī Ādurfarrbay> died before the middle of the 3rd/9th century; author of a "book" (**niβə.*)	**Zartusht ī Ādurfarrbay** *hudēnān pēshōbāy*; around 232/847 **Wāhramshād** *hudēnān pēshōbāy* (?) **Juwānjam ī Shābuhrān** (or **ī Wāhramshād**) *hudēnān pēshbāy*	
Zurwāndād	**Manushchihr ī Juwānjamān** *Pārs ud Kirmān rad*; wrote three epistles (260/881), *Dādestān ī dēnīg*	**Ashawahisht <ī J.>** (probably *hudēnān pēshbāy*)	**Zādsparam ī J.** *Hērbed* in Sīrgān; wrote the texts of the *Wizīdagīhā* and other (lost) works
Farrbay ī Ashawahisht editor of the GrBd	***Frāy-srōsh <ī Ashawahisht (?)>** **Ashawahisht ī Frāy-Srōsh <ī Ashawahisht (?)>** (contemporary of Farrbay)	**Ēmēd ī Ashawahishtān** *hudēnān pēshōbāy* (?); wrote a *Riwāyat* **Ādurbād ī Ēmēdān** *hudēnān* pēshōbāy; editor of the *Dēnkard* **Isfandyār ī Ādurbād ī Ēmēd** killed in Baghdad 324/936; *Mōwbed* of Iran	
Farrokhmard ī Wahrāmān compiled the *Mādayān ī Hazār Dādestān*	**Farrbay/Frīy-Srōsh ī Wahrāmān** *mōwbedān mōwbed*; wrote a *Riwāyat* (4 *Pursishn*, 419/1028		

The analysis of *Dēnkard* 3 points to an impact of Greek, especially Aristotelian and Neoplatonic philosophy (Josephson 2012, 541–52; Shaki 1970, 1973, 1998, 1999, 2003; König 2016, 2018, 2020b). In my view, the original structure of Ādurfarrbay's edition, the most important work of Pahlavi literature, is based on the Neoplatonic doctrine of emanation. Further analysis shows that the Zoroastrian position is formulated to demarcate it from the Islamic, Christian, Jewish, Manichaean and materialistic positions. Through these discussions, logical means and philosophical concepts were developed and tested (König 2020a).

Ādurfarrbay's work was continued by later Zoroastrian writers but especially by Mardānfarrokh (3rd/9th century?), the son of Ohrmazddād, author of the *Doubt-Dispelling Explanation* (*Shkand Gumānīg Wizār*), a key text for religious history in early Abbasid time (Sundermann 2001, 325; Schaeder 1925, 200–1, no. 3). Active after the death of Ādurfarrbay and of his son

Rōshn, Mardānfarrokh refers several times to the work of Rōshn (de Menasce 1945, 10.50–8). West (1885, 169 no. 4) identifies this Rōshn with the commentator Rōshn, who is mentioned for example in *Shāyest-nē-shāyest* and *Pahlavi Vīdēvdād*. He is mentioned 10 times in *Pahlavi Vīdēvdād* (Cantera 2004, 208), where at one time he is in dispute with Abarag, the last commentator from the school of Ādurfarrbay-Narsē (Ādurfarrbay-Narsē, Sōshans, Abarag). He pays special attention to those passages of the *Dēnkard* that are composed by Ādurfarrbay and of Ādurbād ī Jāwandān (or Ādurbād ī Jāwandād or Ādarpādiiāwaṇdą; see de Menasce 1945; 4.106–7, 9.1–3, 10.50–8, 1.38). Ādurbād ī Jāwandān was probably the teacher (*bun-spās*, West 1885, 162; de Menasce 1945, 109) of Ādurfarrbay (de Menasce 1945, 9.1–3) and perhaps also the author of the lost books *Dēnkard* 1 and 2 (West 1885, 138–9, no. 9). It seems that Mardānfarrokh did not belong to the very dense and close circle of Persian high priests. Early wanderings (de Menasce 1945, 1.36–7, 10.43–49; compare the wanderings of the sage (*dānāg*) in *Mēnōg ī Khrad* 1.34–60) led him, as he says, "to the borders of many countries and seas" (de Menasce 1945, 1.37; see also 10.47). Perhaps a Manichaean with intimate knowledge of Manichaeism (a sharp polemic against Mani ["Mānāe"] is found in *Shkand Gumānīg Wizār* 1.59–60), at this time he found his "salvation" (de Menasce 1945, 1.54, 10.58) especially in Zoroastrian scriptures but above all in the texts of Ādurbād ī Jāwandān and Ādurfarrbay ī Farrokhzādān. His intellectually important book, which was widespread in copies in Pāzand and in a Sanskrit translation in India was "not made and arranged for the wise and experienced men, but for pupils and novices" (de Menasce 1945, 1.40). Parts of Mardānfarrokh's text also survived in an older Pahlavi version (West 1885, xxviii; de Menasce 1945, 159; see also Jâmâsp-Âsânâ and West 1887, xxvii, xxxvii–xxxviii; Cereti 2014). Mardānfarrokh completed what began in Ādurfarrbay's work. *Dēnkard* 3 and *Shkand Gumānīg Wizār* are connected in two important points, the development of polemical/demarcating ideas that are the results of reflection and discussion on non-Zoroastrian positions based on logical arguments and the development of a philosophical terminology and system.

I.16.5 What can we know about Zoroastrian scholarship before the 3rd/9th century?

It is difficult, if not impossible, to provide trustworthy information about the centuries preceding Ādurfarrbay, because we need to rely on material of uncertain dates. One such source is the previously mentioned *Shkand Gumānīg Wizār*. It indicates that Ādurfarrbay's (supposed) teacher, Ādurbād ī Jāwandān, had already worked with some philosophical concepts known from *Dēnkard* 3. Another source attributed some of the terms in the transmission report *Dēnkard* 4 to Aristotelian influence. The latter was written in the time of Khōsraw I (on the age of the text manuscript B 322.4ff., see Bartholomae 1920, 9, no. 2 and Cantera 2004, 106; for the text see Zaehner 1955, 7–9, 31–4 and Cantera 2004, 110–11). Aristotelian influence is also apparent in the terms *nērōg ī mēnōyīg/paydāgīhēnīdārīh* <ī> *gētīyīg* (Sanjana 1874–1928, 4.25, manuscript B 323, 11–14), which seem to combine the Zoroastrian distinction of *mēnōg/gētīg* "non-material"/"material" with the potential/actual distinction.

The relation of the two recensions of the *Bundahishn*, that is the *Greater Bundahishn* (*Iranian Bundahishn*) and the *Indian Bundahishn*, is methodologically important. Those parts of the *Greater Bundahishn*, which stem from the priestly editor Farrbay (see the earlier discussion), share philosophical elements known from *Dēnkard* 3. In the *Indian Bundahishn*, however, these elements are missing. That gives rise to the assumption that parts of the transmitted Pahlavi materials were reworked according to the categories of Ādurfarrbay's 'philosophical theology'. Hence, it is likely that at least some important Pahlavi texts are neither simply "final versions of texts which evolved in oral transmission over long periods of time" (Kreyenbroek 2002, 42–3) nor the late product of a long and continuous tradition of Greek philosophy in Zoroastrianism (Shaki's position in 1999, 176 and 2003, 323). It rather seems that a faction of the Zoroastrian clergy of the late 2nd/8th and 3rd/9th centuries tried to reformulate Zoroastrian theology while in contact with the flourishing intellectual life of that period. This perspective receives support by the remarkable fact that there are no philosophical elements in the pre-Islamic parts of the whole *Zand* (*Pahlavi Vīdēvdād; Hērbedestān/Nērangestān*), nor do we find them in those parts that are probably later (*Pahlavi Yasna, Pahlavi Zand-e Khorde Avesta*). But for a different view on the dating of those two parts see Skjærvø (2008). The older parts of the *Zand* and of the juridical and ritualistic literature integrate commentaries of the priests (*Mādayān ī Hazār Dādestān; Shāyest-nē- shāyest, Zand ī Fragard ī Juddēwdād*; König 2010a). A nice (but little known) dispute between two scholars, Gōgushasp and Mēdyōmāh, on the meaning of Avestan phrases, is found in *Wizirgard ī dēnīg* 24. Since Tavadia (1930), efforts have been made to distinguish different schools. However, the time frame of these schools is still controversial. Gignoux (1995) and Secunda (2012) prefer a post-Sasanian date. Cantera (2004, 207–20) prefers a Sasanian dating; according to Tavadia (1930, 28–9, 1956, 41–4), the commentators could belong to both periods. In any case, none of the translations of and commentaries on the Avesta of these schools resembles the (philosophical) school of Ādurfarrbay.

The *wisdom literature* (*Handarz*) is also more or less unaffected by philosophical elements. It consists of a series of texts ascribed to authors, who lived in different centuries: Ādurbād ī Māraspandān and his teacher Mihr-Ohrmazd (4th century); Khusrōy ī Kawādān (Khōsraw Anōshīravān; r. 531–579) and his minister Wuzurg-Mihr (6th century); Ādurfarrbay Farrokhzādān (early 3rd/9th century). The concept of reason or wisdom (*khrad*) is a key term in *Dēnkard* 3 and is also discussed in *Pahlavi Text* 17 (König 2010b, 116–18). Future comparison with the concept's use and understanding in the older text *Handarz* by *Wehzād Farrokh Pērōz* (late Sasanid period?) may illuminate the development of the Zoroastrian 'philosophical theology'. In comparison, the Sasanian *Brih-Nask* (*Dēnkard* 8: 9.1) focuses its *handarz* on the term *āsn* <*ud*> *srūd khrad* "non-empirical and empirical reasoning". An eminent figure from the early 5th century was Ādurbād ī Zardushtān (Cantera 2004, 217). His text on the relation of the material and nonmaterial written for King Yazdegird ī Shābuhragān (r. 399–421) has survived (de Menasce 1973, 3.137), along with his and Bakhtāfrīd's (1st half 6th century) *Gnomes* (*Wāzagīhā*). Bakhtāfrīd took part in the struggle against the Mazdakites (*Zand ī Wahman Yasn* 2.2.; see Cantera 2004, 218–20). His colleagues are partly mentioned on seals, in the Pahlavi translation and in *Shāyest-nē-shāyest*, which is a part of the Dēnkard (Cereti 1995, 175–6). Bakhtāfrīd is quoted in *Dēnkard* 6: A4 (*Shāyest-nē-shāyest* 20.11), 6: E22a and *Dēnkard* 3: 117. In addition, he is also known from juridical texts (Macuch 1993, 8.66, 75; Tavadia 1930, 8.10). The range of his works resembles that of Ādurfarrbay's works from the 3rd/9th century. Even if we cannot be sure that Aristotelian and/or Neoplatonic thoughts were adopted by Zoroastrian priests of the middle and late Sasanian period, this similarity indicates that at least Ādurfarrbay's 'intellectual stature', defined by a scholarly work of one and the same priest in the fields of 'philosophical theology', *Handarz* and juridical literature, has had older parallels.

Bibliography

Sources

Afifi, R. ed. 1964. *Arda Viraf Nama-ye Manzum e Zartusht Bahram e Pazhdo* [The Book of Arda Viraf in verse by Zartusht Bahram Pazhdo]. Mashhad: Mashhad University Press.

Andreas, F. C. ed. 1882. *The Book of the Mainyo-i-Khard, also an Old Fragment of the Bundehesh*. Kiel: Lipsius and Tischer.

Anklesaria, B. T. 1956. *Zand-Ākāsīh. Iranian or Greater Bundahišn. Transliteration and Translation in English.* Bombay: Bunyād-i Farhang-i Īrān.

Anklesaria, E. T. D. ed. 1908. *The Bûndahishn. Being a Facsimile of the TD Manuscript No.2 Brought from Persia by Dastur Tîrandâz and Now Preserved in the Late Ervad Tahmuras' Library.* Bombay: K. R. Cama Oriental Institute.

Anquetil Duperron, A. H. tr. 1771. *Zend-Avesta: Ouvrage de Zoroastre I-II.* 3 vols. Paris: N. M. Tilliard.

Aristotle. tr. Peck, A. L. 1979. *Historia animalium.* Vol. 1 (Loeb Classical Library 437). Cambridge, MA: Harvard University Press.

Barr, K. 1944. "Preface," in *Selections from Codices K7 and K25 (Vispered and Frahang I Pahlavīk) and Tracings of the Avesta Codex K1* (*Codices Avestici et Pahlavici Bibliothecae Universitatis Hafniensis XII*). Copenhagen: Nordisk Forlag.

Bharucha, E. D., ed. 1906–1933. *Collected Sanskrit Writings of the Parsis: Old Translations of Avestâ and Pahlavi-Pâzend Books as Well as Other Original Compositions; with Various Readings and Notes.* 6 vols. Bombay: The Trustees of the Parsee Punchayet Funds and Properties.

Burnouf, E. 1833. *Commentaire sur le Yaçna l'un des livres lithurgiques des Parses.* Paris: s.n.

Cereti, C. G. 1995. *The Zand ī Wahman Yasn. A Zoroastrian Apocalypse.* Rome: Istituto italiano per il Medio ed Estremo Oriente.

Cereti, C. G. and Sanjana, M. Sh. 1991. *An 18th Century Account of Parsi History. The Qesse-ye Zartoštiān-e Hendustān.* Naples: Istituto universitario orientale, Dipartimento di Studi asiatici.

Chacha, H. F. 1936. *Gajastak Abālish. Pahlavi Text with Transliteration, English Translation, Notes and Glossary.* Bombay: The Trustees Parsi Punchayet Funds and Properties.

Christensen, A. 1931. *The Pahlavi Codices K20 & K20b. Containing Ardāgh Vīrāz-Nāmagh, Bundahishn, etc.* Facsimile (*Codices Avestici et Pahlavici Bibliothecae Universitatis Hafniensis* 1). Copenhagen: University Library.

Christensen, A., ed. 1936. *The Pahlavi Codex K 43. First part. Containing a Fragment of the Great Bundahishn, the Dādhastān ī Mēnōghēkhradh, some Parts of the Dēnkard, and the Vahman Yasht* (*Codices Avestici et Pahlavici Bibliothecae Universitatis Hafniensis* 5). Copenhagen: Levin-Munksgaard.

Christensen, A., ed. 1937–1939. *The Avesta Codex K5. Containing the Yasna with Its Pahlavi Translation and Commentary. 3 Parts. With an Introduction by K. Barr.* Facsimile (*Codices Avestici et Pahlavici Bibliothecae Universitatis Hafniensis* 7–9). Copenhagen: University Library.

Christensen, A. 1941–1942. *The Avesta codices K 3a, K 3b and K 1, containing portions of the Venidad with its Pahlavi translation and commentary. 2, Containing the facsimile of K 1 (Vendidad chapters 9.1–22.26 and the colophon)* (*Codices Avestici et Pahlavici Bibliothecae Universitatis Hafniensis* 11). Copenhagen: University Library.

Damesteter, J., tr. 1892–1893. *Le Zend-Avesta. Traduction nouvelle avec commentaire historique et philologique.* 3 vols. Paris: Musée Guimet.

de Menasce, J. 1945. *Une apologétique mazdéenne du 9. siecle. Škand-Gumānīk vičār. La solution décisive des doutes. Texte Pazend-pehlevi transcrit, traduit et commenté.* Fribourg: Libraire de l'Université Fribourg en Suisse.

Dhabhar, E. B. N. 1932. *The Persian Rivayats of Hormazyar Framarz and Others. Their version with Introduction and Notes.* Bombay: K. R. Cama Oriental Institute.

Diogenes Laertius. tr. Arretini, F. 1487. *Francisci Arretini ad Pium Pont. Maximum in Diogenes Epistolas Proemium.* Florence: Antonius Francisci Venetus.

Dresden, M. J. 1966. *Dēnkart. A Pahlavi Text. Facsimile Edition of the Manuscript B of the K. R. Cama Oriental Institute Bombay.* Wiesbaden: Harrassowitz.

Eastwick, E. B. 1843. "Zarthusht-Namah [The Book of Zarthusht]," in Wilson, J., ed. *The Parsi Religion.* Bombay: American Mission Press, 477–522.

Eduljee, H. E. 1991. *Kisseh-i Sanjan* [The Story of Sanjan]. Bombay: K. R. Cama Oriental Institute.

Geldner, K. Fr. 1886–1896. *Avesta. The Sacred Books of the Parsis.* Part I: *Yasna* (1886); Part II: *Vispered and Khorda Avesta* (1889); Part III: *Vendīdād* (1896). Stuttgart: Kohlhammer.

Hippocrates. 1931. "On the Nature of Man," in Jones W. H. S., tr. *Hippocrates* (Loeb Classical Library, No. 150). Cambridge, MA: Harvard University Press, 4: 1–41.

Jamasp, D. H. 1907. *Vendidād. Avesta Text with Pahlavi Translation and Commentary, and Glossarial Index.* With the Assistance of M. M. Gandevia. Vol. 1: *The Texts.* Bombay: Government Central Book Depot.

Jamaspasa, K. J. 1902. *Arda Viraf Nameh. The Original Pahlavi Text with an Introduction, Notes, Gujarati Translation, and Persian Version of Zartosht Behram in Verse.* Bombay: Education Society's Steam Press.

Jâmâsp-Âsânâ, H. J. and West, E. W. 1887. *Shikand-Gûmânîk Vijâr. The Pazand-Sanskrit Text Together with a Fragment of the Pahlavi.* Bombay: Government Central Book Depot.
Jamasp-Asana, J. M. 1897. *The Pahlavi Texts.* With an Introduction by B. G. Anklesaria and a Preface by M. Nawabi. Bombay: Fort Printing Press.
Kotwal, F. M. and Hintze, A. 2008. *The Khorda Avesta and Yašt Codex E1. Facsimile edition.* Wiesbaden: Harrassowitz.
Macuch, M. 1993. *Rechtskasuistik und Gerichtspraxis zu Beginn des siebenten Jahrhunderts in Iran. Die Rechtssammlung des Farrohmard i Wahrāmān.* Wiesbaden: Harrassowitz.
de Menasce, J. 1973. *Le troisième livre du Dēnkart. Traduit du pehlevi.* Paris: Librairie C. Klincksieck.
Mills, L. H. 1893. *The Ancient Manuscript of the Yasna with its Pahlavi Translation (A.D. 1323), Generally Quoted as J2 (Bodleian Library/Oxford).* Oxford: Clarendon Press.
Modi, J. J. 1934. *Qisseh-i Zartûshtiân-i Hindûstân va Bayān-i Âtash Behrâm-i Naosari* [The Story of the Zoroastrians of India and of the Behram Fire in Navsari]. Bombay: Fort Printing Press.
Pausanias. tr. Jones, W. H. S and Ormerod, H. A. 1918. *Pausanias' Description of Greece.* 4 vols. Cambridge, MA: Harvard University Press.
Paymaster, R. B. 1915. *Kisse-i Sanjân* [The Story of Sanjan]. Bombay: Frot Printing Press.
Rosenberg, F. 1959. *The Book of Zoroaster Zarātusht Nāma by Zartusht Bahrâm Son of Pajdû.* Tehran: Commissionnaires de l'Académie Impériale des Sciences.
Sanjana, P. 1874–1928. *The Dînkard. The Original Pahlavi Text; the Same Transliterated in Zend Characters; Translations of the Text in the Gujarati and English Languages; a Commentary and a Glossary of Select Terms.* Vols. I–XIX. Bombay: Kegan Paul, Trench, Trübner & Co.
Sanjana, P. 1896. *The Kârnâmê î Artakhshîr î Pâpakân, being the Oldest Surviving Records of the Zoroastrian Emperor Ardashîr Bâbakân, the Founder of the Sâsânian Dynasty in Irân. The Original Pahlavi Text edited for the first time with a Transliteration in Roman Characters, Translations into the English and Gujerati Languages, with Explanatory and Philological Notes, an Introduction, and Appendix.* Bombay: Duftur Ashkara and the Education Society's Steam Press.
Shaked, Sh. 1979. *The Wisdom of the Sasanian Sages. (Dēnkard VI) by Aturpāt-i Ēmētān.* Boulder: Westview Press.
Tavadia, J. C. 1930. *Šāyast-nē-šayast. A Pahlavi Text on Religious Customs, Edited, Transliterated and Translated with Introduction and Notes.* Hamburg: de Gruyter & Co.
Unvala, E. M. R. 1922. *Dârâb Hormazyâr's Rivâyat* [Darab Hormazyar's Text Collection]. With an introduction by J. J. Modi. 2 vols. Bombay: British India Press.
Unvala, J. M. 1940. *Collection of Colophons of Manuscripts Bearing on Zoroastrianism in some Libraries of Europe.* Bombay: The Trustees of the Funds and Properties of the Parsi Punchayet.
West, E. W. 1871. *The Book of the Mainyō-i-Khard or the Spirit of Wisdom. The Pazand and Sanskrit Texts as Arranged in the Fifteenth Century by Neriosengh Dhaval. Edited in Roman Characters from Various Manuscripts, with an English Translation, an Introduction, Notes, a Glossary of the Pazand Text, with Sanskrit, Persian and Pahlavi Equivalents, and a Sketch of Pazand Grammar.* Stuttgart and London: C. Grüninger.
West, E. W. 1880. "The Pahlavi Inscriptions at Kanheri," *India Antiquary* 9: 265–8.
West, E. W. 1882a. "An Engraved Stone with Pahlavi Inscription from Baghdad," *India Antiquary* 11: 223–6.
West, E. W. 1882b. *Pahlavi Texts.* Part 2: *The Dādistān-ī Dīnīk and the Epistles of Mānūškīhar* (Sacred Books of the East 18). Oxford: Oxford University Press.
West, E. W. 1885. *Pahlavi Texts.* Part 3: *Dīnā-ī Maīnōg-ī Khirad, Sikand-Gūmānīk Vigār, Sad Dar.* (Sacred Books of the East 24). Oxford: Oxford University Press.
Westergaard, N. L. 1852–1854. *Zendavesta or the Religious Books of the Zoroastrians.* Vol. 1: *The Zend Texts.* Copenhagen: Berling Brothers.

Research literature

Āmūzgār, Ž. 1989. "Bahrām Paždū," in *eIr*, 3: 524–5.
Bailey, H. W. 1943. *Zoroastrian Problems in the Ninth-Century Books.* Oxford: Clarendon Press.
Bartholomae, C. 1915. *Die Zendhandschriften der K. Hof- und Staatsbibliothek in München.* Munich: Akademie-Verlag.
Bartholomae, C. 1920. *Zur Kenntnis der mitteliranischen Mundarten III.* Heidelberg: Akademie-Verlag.
Boyce, M. 1968. "Middle Persian Literature," in *Handbuch der Orientalistik* 4.2.1. Leiden: Brill, 31–66.
Boyce, M. 1975. *A History of Zoroastrianism.* Vol. 1: *The Early Period.* Leiden: Brill.

Boyce, M. 1979. *Zoroastrians. Their Religious Beliefs and Practices*. London: Routledge and Kegan Paul.

Boyce, M. 1982. *A History of Zoroastrianism*. Vol. 2: *Under the Achaemenians*. Leiden: Brill.

Boyce, M. and Grenet, F. 1991. *A History of Zoroastrianism*. Vol. 3: *Zoroastrianism under Macedonian and Roman Rule*. Leiden: Brill.

Browne, E. G. 1902. *A Literary History of Persia*. Vol. 1: *From the Earliest Times Until Firdousí*. Cambridge: T. F. Unwin.

Cantera, A. 2004. *Studien zur Pahlavi-Übersetzung des Avesta*. Wiesbaden: Harrassowitz.

Cantera, A. and de Vaan, M. 2005. "Remarks on the Colophon of the Avestan Manuscripts Pt4 and Mf4," *Studia Iranica* 34: 31–42.

Cereti, C. G. 2001. *La Letteratura Pahlavi. Introduzione ai testi con riferimento alla storia degli studi e alla tradizione manoscritta*. Milano: Mimesis.

Cereti, C. G. 2004. "Sul codice M51 di Monaco [On the Codex M51 of Monaco]," in Cereti, C. G. *et al.*, eds. *Varia Iranica*. Rome: Istituto Italiano per l'Africa e l'Oriente, 119–30.

Cereti, C. G. 2007. "Some Primary Sources on the Early History of the Parsis in India," in Vahman, F. and Pedersen, C. V., eds. *Religious Texts in Iranian Languages. Symposium Held in Copenhagen May 2002*. Copenhagen: The Royal Danish Academy of Sciences and Letters, 211–21.

Cereti, C. G. 2014. "Škand Gumānīg Wizār," in *eIr*, online edition. www.iranicaonline.org/articles/shkand-gumanig-wizar (Accessed 01 Aug 2019).

Colpe, C. 2003. "Wissen und Erkennen in den Gathas," in Colpe, C., ed. *Iranier – Aramäer – Hebräer – Hellenen. Iranische Religionen und ihre Westbeziehungen. Einzelstudien und Versuch einer Zusammenschau*. Tübingen: Mohr Siebeck, 272–80.

de Menasce, J. 1975. "Zoroastrian Literature after the Muslim Conquest," in Frye, R. N., ed. *The Cambridge History of Iran*. Vol. 4: *The Period from the Arab Invasion to the Saljuks*. Cambridge: Cambridge University Press, 543–65.

de Menasce, J. 1983. "Zoroastrian Pahlavi Writings." In Yarshatar, E., ed. *The Cambridge History of Iran*. Vol. 3.2: *The Seleucid, Parthian and Sasanid Periods*. Cambridge: Cambridge University Press, 1166–95.

Dresden, M. J. 1967. "Pahlavi Manuscripts," in *Sir J. J. Zarthoshti Madressa Centenary Volume*. Bombay: The Trustees of the Parsi Punchayet Funds and Properties, 74–83.

Geldner, K. F. 1896–1904. "Awestaliteratur," in Geiger, W. and Kuhn, E., eds. *Grundriss der Iranischen Philologie*, 2.1. Strassburg: K. J. Trübner, 1–53.

Gignoux, P. 1995. "La controverse dans le mazdéisme tardif," in Le Boulluec, A., ed. *La controverse religioeuse et ses forms*. Paris: Cerf, 127–49.

Hämeen-Anttila, J. 2018. *Khwadāynāmag. The Middle Persian Book of Kings*. Leiden: Brill.

Hodivala, S. K. 1920. *Studies in Parsi History*. Bombay: S. K. Hodivala.

Humbach, H. 2003. "Neriosangh and His Sanskrit Translations of Avesta Texts," in Cereti, C. G. and Vajifdar, F., eds. *Ātaš-e dorun. The Fire Within. Jamshid Soroush Soroushian Memorial Volume 2*. Bloomington, IN: Authorhouse, 199–212.

Jackson, A. V. W. 1906. *Persia, Past and Present*. London: MacMillan Company.

Josephson, J. 2012. "The Evolution and Transmission of the Third Book of the Dēnkard," in Cantera, A., ed. *The Transmission of the Avesta*. Wiesbaden: Harrassowitz, 541–52.

Katrak, J. C. 1941. *Oriental Treasures Being a Condensed Tabular Descriptive Statement of Over a Thousand Manuscripts and of Their Colophons Written in Iranian and Indian Languages and Lying in Private Libraries of Parsis in Different Centres of Gujarat*. Bombay: J. C. Katrak.

Klíma, O. 1968. "Avesta. Ancient Persian Inscriptions. Middle Persian Literature," in Rypka, J., ed. *History of Iranian Literature*. Dordrecht: Reidel, 1–67.

König, G. 2010a. "Der Pahlavi-Text *Zand ī Fragard ī Juddēvdād*," in Macuch, M., Weber, D. and Durkin-Meisterernst, D., eds. *Ancient and Middle Iranian Studies. Proceedings of the 6th European Conference of Iranian Studies, Held in Vienna, 18–22 September 2007*. Wiesbaden: Harrassowitz, 115–32.

König, G. 2010b. "Didaktisches Erzählen in der neupersischen zoroastrischen Literatur," in Günthart, R. and Forster, R., eds. *Didaktisches Erzählen. Formen Literarischer Belehrung in Orient und Okzident (Interdisziplinäre Tagung Berlin 9./10. Oktober 2009)*. Frankfurt a. M.: Peter Lang Verlag, 109–32.

König, G. 2014. "Die Pahlawi-Literatur des 9./10. Jh. und ihre frühe Kodex-Überlieferung (II)," *Estudios de Iran y Turan* 1: 43–74.

König, G. 2016. "A Re-thinking of Mansour Shaki's Contributions to the Philosophical Writings in Pahlavi," *Quarterly Journal of Language and Inscription. Dedicated to Professor Mansour Shaki* 1: 4–26.

König, G. 2018. "The Pahlavi Literature of the 9th Century and Greek Philosophy," *Iran and the Caucasus*: 8–37.

König, G. 2019. "Die Pahlavi-Literatur des 9./10. Jh. und ihre frühe Kodex-Überlieferung (I)," in Hintze, A., Durkin-Meisterernst, D. and Naumann, C., eds. *A Thousand Judgements. Festschrift Maria Macuch*. Wiesbaden: Harrassowitz, 263–86.

König, G. 2020a. "Training in Thinking. Religious Criticism and the Use of Logic in Zoroastrian Theology," Ruani, F. and Timuş, M., eds. *Quand les dualistes polémiquaient Zoroastriens et Manichéens, Actes du colloque international, 12–13 juin 2015, Collège de France*. Leuven: Peeters Publishers.

König, G. 2020b. "On the question of neo-Platonic elements in the Zoroastrian literature of the 9th century," in Weltecke, D. and Echevarría Arsuaga, A., eds. *Religious Plurality and Interreligious Contacts in the Middle Ages*. (Workshop „Religious Plurality and Interreligious Contacts in the Middle Ages." Ein gemeinsames Spanisch-Deutsches Arbeitsgespräch der Herzog-August-Bibliothek Wolfenbüttel und der Fundación de Salas, 30.11.–02.12. 2015) (Wolfenbütteler Forschungen 161). Wiesbaden: Harrassowitz, 65–80.

Kreyenbroek, P. 2002. "Millennialism in the Zoroastrian Tradition," in Amanat, A. and Bernhardsson, M., eds. *Imaging the End: Visions of Apocalypse from the Ancient Middle East to Modern America*. London: I. B. Tauris, 33–55, 339–44.

Macuch, M. 2009. "Pahlavi Literature," in Emmerick, R. E. and Macuch, M., eds. *The Literature of Pre-Islamic Iran. Companion Volume 1 to A History of Persian Literature*. London: I. B. Tauris, 116–96.

Mottahedeh, R. P. 2012. "The Idea of Iran in the Buyid Dominions," in Herzig, E. and Stewart, S., eds. *Early Islamic Iran (Idea of Iran 5)*. London: I. B. Tauris, 153–60.

Nöldeke, T. 1920. *Das iranische Nationalepos*. Berlin: de Gruyter.

Nyberg, H. S. 1974. *A Manual of Pahlavi*. Part 2: *Ideograms, Glossary, Abbreviations, Index, Grammatical Survey, Corrigenda to Part 1*. Wiesbaden: Harrassowitz.

Schaeder, H. H. 1925. "Die islamische Lehre vom Vollkommenen Menschen, ihre Herkunft und ihre dichterische Gestaltung," *ZDMG* 79: 192–268.

Schaeder, H. H. 1972. *Iranische Beiträge I*. Halle 1930. Reprint Hildesheim: Olms.

Secunda, S. 2012. "On the Age of the Zoroastrian Authorities of the Zand," *Iranica Antiqua* 47: 317–49.

Shaki, M. 1970. "Some Basic Tenets of the Eclectic Metaphysics of the Dēnkart," *Archiv Orientálni* 38: 277–312.

Shaki, M. 1973. "A Few Philosophical and Cosmological Chapters of the *Denkart*," *Archiv Orientálni* 41: 133–64.

Shaki, M. 1975. "Two Middle Persian Philosophical Terms LYSTK' and M'TK'," in Gignoux, P. ed. *Iran ancien: Actes du XXIXe Congrès international*. Paris: L'Asiathèque, 52–7.

Shaki, M. 1981. "The Dēnkard Account of the History of the Zoroastrian Scriptures," *Archiv Orientálni* 49: 114–25.

Shaki, M. 1998. "Elements," in *eIr*, 8: 357–60.

Shaki, M. 1999. "Falsafa I. Pre-Islamic Philosophy," in *eIr*, 9: 176–82.

Shaki, M. 2003. "Greek Influence on Persian Thought," in *eIr*, 11: 321–6.

Skjærvø, P. O. 2008. "Review of Alberto Cantera, Studien zur Pahlavi-Übersetzung des Avesta. Wiesbaden: Harrassowitz, 2004 (Iranica 7)," *Kratylos* 53: 1–20.

Sohn, P. 1996. *Die Medizin des Zādsparam. Anatomie, Physiologie und Psychologie in den Wizīdagīhā ī Zādsparam, einer zoroastrisch-mittelperischen Anthologie aus dem frühislamischen Iran des neunten Jahrhunderts*. Wiesbaden: Harrassowitz.

Spiegel, F. von 1856–1860. *Einleitung in die traditionellen Schriften der Parsen. 2 Teile*. Wien: Verlag W. Engelmann (Leipzig).

Stausberg, M. 2002. *Die Religion Zarathushtras. Geschichte – Gegenwart – Rituale*. Vols. 1 and 2. Stuttgart: Kohlhammer.

Stausberg, M. 2004. *Die Religion Zarathushtras. Geschichte – Gegenwart – Rituale*. Vol. 3. Stuttgart: Kohlhammer.

Sundermann, W. 1982. "Soziale Typenbegriffe altgriechischen Ursprungs in der altiranischen Überlieferung," in Welskopf, E. Ch., ed. *Soziale Typenbegriffe im alten Griechenland VII*. Berlin: Akademie-Verlag, 14–38.

Sundermann, W. 2001. "Das Manichäerkapitel des Škand Gumānīg Wizār in der Darstellung und Deutung Jean de Menasces," in van Oort, J. *et al.*, eds. *Augustine and Manichaeism in the Latin West*. Leiden: Brill, 325–37.

Tafażżolī, A. 1373/1994. *Zabān-i Pahlawī adabīyāt wa dastūr-i ān*. Tehran: Tihrān Nashr-i Muʿīn.

Tafażżoli, A. 1982. "Abāliš," in *eIr*, 1: 58.

Tavadia, J. C. 1933. *Review* "Christensen, A., The Pahlavi Codices K20 & K20b [. . .]," *Orientalistische Literaturzeitung* 8–9: coll. 567–9.

Tavadia, J. C. 1944. "Zur Pflege des iranischen Schrifttums im Mittelalter," *ZDMG* 98: 294–339.
Tavadia, J. C. 1956. *Die mittelpersische Sprache und Literatur der Zarathustrier*. Leipzig: Harrassowitz.
Vitalone, M. 1987. *The Persian Revāyats. A Bibliographic Reconnaissance*. Napoli: Istituto universitario orientale, Dipartimento di studi asiatici.
West, E. W. 1896–1904. "Pahlavi Literature," in Geiger, W. and Kuhn, E., eds. *Grundriss der Iranischen Philologie*, II.3. Strassburg: K. J. Trübner, 75–129.
Widengren, G. 1965. *Die Religionen Irans*. Stuttgart: Kohlhammer.
Williams, A. 2009. *The Zoroastrian Myth of Migration from Iran and Settlement in the Indian Diaspora. Text, Translation and Analysis of the 16th Century Qesse-ye Sanjān 'The Story of Sanjan'*. Leiden: Brill.
Zaehner, R. C. 1955. *Zurvan. A Zoroastrian Dilemma*. Oxford: Clarendon Press.

I.17

EVALUATING THE PAST: SCHOLARLY VIEWS OF ANCIENT SOCIETIES AND THEIR SCIENCES

Ulrich Rudolph

From the very beginning, scientific activity in the Islamicate world was not considered something confined to the boundaries of the Muslim community. This applies, first of all, to the synchronic perspective, insofar as the actors in the field were Muslims, as well as Christians, Jews, Zoroastrians or adherents to other creeds. At the same time, however, it applies to the diachronic perspective, for all these scholars were well aware that their own activities were deeply anchored in ideas and practices developed before the rise of Islam.

The sciences, including philosophy, were thus understood to be a universal project of common interest to mankind. This seems to have been a widespread conviction shared by numerous scholars, among them the famous historian and leading intellectual of the 8th/14th century, Ibn Khaldūn (732–808/1332–1406). In his *Introduction* [*to the Study of History*] he writes:

> The rational sciences (*al-ʿulūm al-ʿaqliyya*) are natural to man, inasmuch as he is a thinking being. They are not restricted to any particular religious group but are studied by the people of all religious groups, who are all equally qualified to learn them and to do research in them.
>
> *(Ibn Khaldūn 1858[CB],*[1] *3: 86–7; tr. Rosenthal 1958[CB], 3: 111, with slight modifications)*

As a consequence, Ibn Khaldūn considers the rational sciences to "have existed (and been known) to the human species since civilization had its beginning in the world" (Ibn Khaldūn 1858, 3: 86–7; tr. Rosenthal 1958, 3: 111). This raises the question of how they developed in former societies and, more specifically, of which scientific traditions immediately preceded Islam and served Islamicate societies as a pattern for their own activities. Here again, Ibn Khaldūn's answer is very clear. As he says, two ancient nations have to be highlighted in this context: the Persians, that is the Persian dynasties from the time of the Achaemenid Empire, and the Greeks, whose most important political representative was Alexander of Macedonia (r. 336–323 BCE). Both nations cultivated the sciences extensively, each of them accumulating a treasure of knowledge. Yet, still according to him, it should be admitted that the routes of transmission were quite different in each case, the transmission of Greek knowledge having

DOI: 10.4324/9781315170718-19

been much more successful than its Persian counterpart (Ibn Khaldūn 1858, 3: 89–91; tr. Rosenthal 1958, 3: 113–16).

The picture emerging from these lines is not surprising. Actually, it fits very well within Ibn Khaldūn's general theory of history. As is well known, he reflected extensively on the connection between sociological factors, military success and the development of civilization, emphasizing that crafts and sciences can only flourish in powerful nations and sedentary societies (e.g., Ibn Khaldūn 1858, 2: 307–11; tr. Rosenthal 1958, 2: 347–51), such as the ancient Persians as well as the Greeks, including their Roman and Byzantine successors, had been.

One should keep in mind, however, that the 'power-focused' analysis presented by him is only one way in which Muslim scholars viewed ancient societies and their sciences. There have been different views on the topic, in particular by authors living in earlier periods and having more immediate access to the process of appropriating ancient science and philosophy as it happened from the 2nd/8th century onward. Four of these authors are discussed in the following. These are Isḥāq ibn Ḥunayn (d. 289/910–1) and Ibn al-Nadīm (d. 380/990) who both originated from the central (Eastern) regions of the Islamicate world, as well as Ibn Juljul (d. 384/994) and Ṣāʿid al-Andalusī (419–462/1029–1070) originating from its far West. Their respective individual experiences differed due to the various historical and cultural backgrounds they came from. However, they all reflected extensively on the question at stake here, namely, science in Islamicate societies and its relationship to earlier scientific endeavors, thereby offering us a stimulating variety of answers to it.

I.17.1 Isḥāq ibn Ḥunayn and his *History of the Physicians* (*Taʾrīkh al-aṭibbāʾ*)

Isḥāq ibn Ḥunayn is known mainly as an outstanding Greco-Arabic translator. He belonged to the celebrated group of translators working with his father, Ḥunayn ibn Isḥāq (d. 260/873), and contributed significantly to their tremendous output. Also, much like his father, Isḥāq was an excellent scholar. Both of them worked as physicians at the court of the Abbasids and composed scientific writings documenting the high level of their scholarship (Endress 2017, 426–8; for a detailed account of Ḥunayn's scientific achievements, see Gutas 2017a).

One of these writings is Isḥāq's *History of the Physicians* (*Taʾrīkh al-aṭibbāʾ*), a short history of medicine from its beginnings to the time of the author (Rosenthal 1954). The text is written in a terse style and, admittedly, tells us few historical details. Furthermore, Isḥāq seems to have followed his sources, among them Yaḥyā al-Naḥwī ([d. 574]; John Philoponus), very closely, making it sometimes difficult to decide whether he is speaking for himself or just reproducing another's voice (see Rosenthal 1954, 57–60). Despite these difficulties, however, his *History of the Physicians* can be read as a fascinating testimony of how an early Arabic scholar viewed the history of medicine and legitimated his own activities in the field.

Two aspects of his text are of particular interest to us (for the following see also Rudolph 2011, 291–3). The first is the way Isḥāq presents and organizes his subject matter: He subdivides the history of medicine into eight successive periods, emphasizing that each of them has been inaugurated by an eminent authority in the field. The eight coryphaei mentioned by him explicitly are Asclepius I, Ghūrūs, Mīnos, Parmenides (d. after 450 BCE), Plato the physician, Asclepius II, Hippocrates (d. *c.* 370 BCE) and Galen (129–early 3rd century; following the reading by Rosenthal 1954, 75). As it appears, the list includes several names that touch on the sphere of ancient religion (especially Asclepius I and II). The impression that medicine and religion are somehow connected is further corroborated by the fact that the *History of the Physicians* discusses at some length the question of whether medicine has existed from eternity (*qadīm*) or has been created in time (*muḥdath*; Rosenthal 1954, 62–4 [Arabic] and 73–5 [English]). The discussion as presented here is particularly intriguing, but it does not really concern the topic of our investigation; our focus is on Isḥāq's historiographical approach as expressed in the list of authoritative names. As far as these can be identified, they are all Greek, and this is the first point to be retained in our context: according to this testimony, medical science as practiced in the Islamicate world is entirely derived from Greek authorities.

The second aspect worth mentioning here is the particular role assigned in the *History of the Physicians* to Galen. Not only is he described as one of the eminent authorities inaugurating a new period in the history of medicine (in his case: the last period), but his knowledge is also said to have superseded the knowledge of all other physicians, making him the unsurpassable master of this discipline. The term used in this respect is *khātam al-aṭibbāʾ* (the seal of the physicians). Thus, Isḥāq ibn Ḥunayn refers once more to the sphere of religion. This time, however, his reference is not to ancient Greek religion but straightforwardly to Islam, for it is well known that, in a famous verse of the Qurʾān, Muḥammad (d. 11/632) is called *khātam al-nabiyyīn* (Q 33: 40), "the seal of the prophets".

This inevitably raises the question of how to understand the term *khātam* (seal) properly. Nowadays, it is mostly interpreted as meaning "the last" (prophet or so), that is the final element in a long series (thus Rosenthal 1954, 65, translating *khātam al-aṭibbāʾ* by "the last of the physicians"). Yet, this interpretation seems to be a reduction of its original semantic range. *Khātam al-nabiyyīn* has been understood in different ways within the Islamic tradition, the two major options being "the last prophet" and "the paradigmatic prophet", paradigmatic in the sense that Muḥammad realized prophethood in a perfect manner and thus can serve as a model confirming and certifying ("sealing") other prophets (Horovitz 1926; see now Sangaré 2016).

Apparently, the *History of the Physicians* uses the term *khātam* in the second meaning. For Galen was certainly not the last physician ever, but a perfect physician, whose knowledge and practice could serve as a pattern for all others. We may thus conclude that Isḥāq ibn Ḥunayn's view on the history of medicine is a shrewd presentation, combining some historical knowledge with an overall strategic goal: He openly acknowledges the overwhelming importance of the Greek tradition but, at the same time, tries to islamicize this tradition, legitimizing his own adherence to it by characterizing its outstanding representative (Galen) in a way similar to how Muḥammad was characterized in the Qurʾān.

I.17.2 Ibn al-Nadīm and his *Kitāb al-Fihrist*

The perspective changes considerably when we turn to our next author, Ibn al-Nadīm. Like Isḥāq ibn Ḥunayn, he spent most of his life in Baghdad, the capital of the Abbasid caliphate. Yet their respective approach to our topic and their professional background seem to have been different in several ways. In contrast to Isḥāq, Ibn al-Nadīm did not participate in the Greco-Arabic translation activities but only flourished when they had finished. He did not work as a physician but was a book dealer or stationer (*warrāq*). And far from being educated in a specific science (as Isḥāq was in medicine), he was a scientific amateur and generalist interested in a wide range of subjects and consulting as many books as possible (on his life, see Ibn al-Nadīm 1970[CB], 1: XV–XXIII).

The result of this widespread intellectual interest is Ibn al-Nadīm's most celebrated *Catalogue* (*Kitāb al-Fihrist*), in his own words a "catalogue of the books of all peoples, Arab and foreign, existing in the language of the Arabs, dealing with various sciences . . . from the beginning of the formation of each science to this our own time" (ed. Ibn al-Nadīm 1871–1872[CB]; tr. Ibn al-Nadīm 1970; the quotation is taken from Ibn al-Nadīm 1970, 1: 1). As has often been emphasized, the book is not just an exhaustive list of Arabic writings available in Baghdad at the end of the 4th/10th century but a veritable "encyclopaedia of medieval Islamic culture" (Ibn al-Nadīm 1970, 1: XIX) containing chapters on any kind of intellectual activity flourishing at that period, including literature, historiography and, most important, the sciences.

Chapter VII of the *Catalog* is devoted to the so-called ancient sciences, comprising first and foremost philosophy, the mathematical sciences[2] and medicine. Consequently, Ibn al-Nadīm discusses each of these disciplines extensively, characterizing its most important authors, their major works, their teaching (if possible), their impact and so on. Before going into such details, however, he offers a general introduction to the subject matter. Its main goal is to explain the origin and the historical development of the "ancient" sciences with a special focus on the question of how they have been transmitted to the Islamicate world.

The picture drawn by Ibn al-Nadīm on this occasion is manifold. Actually, it is not just one picture but a series of four accounts attributed to various authors whose writings must have been accessible to him. Each of these accounts is different regarding its length, the historical perspective and the approach to the topic but they all have one characteristic feature in common: their tendency to emphasize the role of ancient Persia in the process of developing the sciences and transmitting them to the Islamicate world.

This tendency is most obvious in the first account, attributed by Ibn al-Nadīm to Abū Sahl ibn Nawbakht (*fl. c.* 153–193/770–809). Abū Sahl served as an astrologer at the court of the Abbasid Caliph al-Manṣūr (r. 136–158/754–775) and composed a book on the history of the sciences. The book, probably titled *On Nativities* (*Kitāb al-Tohmagān [fī l-Mawālīd]*) and apparently depending on Middle Persian sources, has not been preserved on its own but is partially accessible thanks to an extended quotation found in the *Catalog* (Gutas 1998[CB], 33 and 38, 2017b, 100; Van Bladel 2012; concerning the title see Van Bladel 2012, 41 no.1). According to this fragment, Abū Sahl maintained that the sciences had been cultivated by several ancient nations, among them the Babylonians, the Egyptians (who learned them from the Babylonians) and the Indians. However, none of them could compete with the Persian nation, whose contribution to the development of the sciences, their transmission and long-term preservation superseded the contributions of all others. Supposedly, the Persians started their initiatives in the field very early, in the period of the legendary King Jamshīd. Later on, things became more complicated, as they experienced several vicissitudes and setbacks, for example, by Alexander the Great, seriously affecting the scientific culture of Iran. Yet, when the Sasanian kings took power, everything turned out for the best. Starting with Ardashīr ibn Bābak (r. 226–242) and culminating in Khōsraw Anūshirwān (r. 531–579), the Sasanians unconditionally promoted the sciences, collecting important writings from wherever they could get them (Persia, India, China, Babylon, Greece) and ordering them to be translated into Persian (Ibn al-Nadīm 1872, 2: 238.9–239.31, 1970, 2: 572–5; further Engl. translations by Gutas 1998, 39–40 [partial], and Van Bladel 2012, 44–7 [complete]; on the ideological background see Gutas 1998, 98–101, and Van Bladel 2012, 47–62).

The second account offered by Ibn al-Nadīm is taken from Abū Maʿshar (171–272/787–886), the famous astrologer (on his life and doctrine see Orthmann 2017, 727–33). He composed several books about astronomical tables (*zīj*), one of which must have contained the account about the transmission of the sciences quoted at length in the *Fihrist* (Ibn al-Nadīm 1872, 2: 240.5–241.6, 1970, 2: 576–8). The report opens with a statement stressing, just as much as Abū Sahl had done, the outstanding importance of the Persian tradition. Abū Maʿshar says:

> Because of their care in preserving [the books about] the sciences, their eagerness to make them endure throughout the ages, and their guarding them from celestial happenings and earthly damages, the kings of Persia actually chose for them the writing material which was the most durable in case of accident, the longest lasting in time, and the least prone to decay or effacement. . . . The peoples of India, China, and the neighboring countries imitated them.
>
> *(Ibn al-Nadīm 1872, 2: 240.5–9, 1970, 2: 576)*

Then follows a long report about the location in Iran where scientific books have been stored, including a reflection on the geographical and astrological reasons for selecting this particular region. All of this confirms that astronomy and astrology are supposed to have flourished nowhere else as much as in Persia, although, at the end of his text, Abū Maʿshar admits that both were also recognized as important sciences by the Indians and the Chaldeans (Ibn al-Nadīm 1872, 2: 240.31–241.3, 1970, 2: 578).

The last two accounts given by Ibn al-Nadīm are more difficult to assess, all the more so as they are reported anonymously, lacking information about the author and his background. Account number four contains the well-known story that the Umayyad prince Khālid ibn Yazīd (d. after 85/704) had had Greek books on alchemy and other occult sciences translated into Arabic (Ibn al-Nadīm 1872, 2: 242.7–11, 1970, 2: 581). This drew considerable attention in medieval sources, as well as in modern scholarship. As it turned out, however, we should not give too much credit to it since, despite its renown, the report seems to be a later fabrication (Ullmann 1978).

Account number three, in contrast, appears to be both more reliable and more instructive (Ibn al-Nadīm 1872, 2: 241.16–242.6, 1970, 2: 579–81). According to its anonymous author, philosophy emerged among the Greeks and the Romans long before the religious code of the Messiah (*qabla sharīʿat al-Masīḥ*) and was originally cultivated without restriction. In later times, however, things changed as a result of the rise of Christianity: The Romans and then the Byzantines (*al-rūm*) became adherents to the new religion and "prevented people from speaking about anything in philosophy which was opposed to the prophetic [Christian] doctrine" (Ibn al-Nadīm 1872, 2: 241.21–2, 1970, 2: 579).

This affirmation corresponds to the famous report, attributed to al-Fārābī (d. 339/950–1), as well as to several other authors, which has become notorious under the label of 'from Alexandria to Baghdad' (Gutas 1999). The report contains a similar passage telling us that the teaching of philosophy was restricted in late antiquity because Byzantine bishops were afraid of its allegedly harmful influence on Christian faith (Rudolph 2017, 51–3 with references to further secondary literature). The anonymous author quoted by Ibn al-Nadīm reproduces nearly the same story, thus sharing with al-Fārābī the critical tone toward both Byzantium and Christianity. Yet, when continuing his own account, he deviates from the Fārābian report, shifting instead to another topic: the Roman emperor Julian. As we are now told, Julian (r. 360–363), the 'Apostata' of the Christian tradition, returned to cultivating and encouraging philosophical studies. Furthermore, he had regular contact with the Sasanian Empire, mainly on the battlefield but also in terms of cultural exchange. How this exchange may have been operative remains unexplained. Instead, we are told some legendary tales, for example, the well-known story of Sābūr II (r. 309–379) traveling into the Byzantine Empire in disguise. However, the main goal of the narrative is unambiguous: The anonymous account not only criticizes the Byzantines as well as the

Christians but also actively promotes the role of ancient Persia in transmitting philosophy and the rational sciences, thereby fitting perfectly the overall tendency of Ibn al-Nadīm's presentation.

I.17.3 Ibn Juljul and his *Generations of the Physicians and the Sages*

In contrast to this, our next author betrays an unambiguously 'Western' perspective. I am talking here of Ibn Juljul (d. after 383/994), a physician from Córdoba who apparently spent all his life in Andalusia (Dietrich 1971). Ibn Juljul was deeply rooted in western Islamic civilization. As a matter of fact, he seems to have been the first Islamic scholar to quote well-known Latin authorities such as Orosius (d. *c.* 418), Hieronymus (d. 420) and Isidore of Seville (d. 636) in Arabic translation. Another proof for his explicitly 'Western' perspective is his famous book titled *The Generations of the Physicians and the Sages* (*Ṭabaqāt al-aṭibbā' wa-l-ḥukamā'*), which was meant to present a history of the rational sciences (medicine, philosophy, mathematics, astronomy) from their origin to the author's lifetime (Ibn Juljul 1955[CB], 1985[CB]).

As the title reveals, the book follows the so-called *ṭabaqāt*-structure originally developed within the realm of the Islamic sciences (*fiqh*, *kalām* and so forth). According to this structure, the history of a discipline can be understood as a succession of "generations" of scholars, each of them learning from the preceding "generation" and continuing their teaching and scholarly work.

Ibn Juljul uses the term *ṭabaqa* in a loose way, dividing the entire history of the rational sciences into nine "generations". The first consists of five religious figures, namely, Hermes I (having allegedly lived before the Flood), Hermes II (having lived in Babylon after the Flood), Hermes III (having transmitted the sciences to Egypt), Asclepius and Apollon (on the various incorporations of Hermes see now Van Bladel 2009[CB], 121–63). In other words, during the time posited as the first *ṭabaqa*, science appeared in the Near East and later on was transmitted via Egypt to Greece. The second generation corresponds to the classical period of Greek philosophy and science, including several well-known authorities such as Hippocrates, Socrates (d. 399 BCE), Plato (d. 348–347 BCE), Aristotle (384–322 BCE) and Dioscorides (d. 90). The third *ṭabaqa* refers to the period between Alexander the Great and Cleopatra (r. 51–30 BCE), including, among others, Ptolemy (*c.* 100–170) and Euclid (3rd century BCE). *Ṭabaqa* number four, corresponding chronologically to the Roman Empire from Augustus to its end, is completely dominated by one outstanding scholar, namely, Galen. The same Galen is still at the focus of *ṭabaqa* number five, devoted to the so-called *Summa Alexandrinorum*, that is the famous synopsis of Galenic writings composed in Alexandria during the 6th century (Chapter I.10).

Ibn Juljul locates the remaining four 'generations' within the Islamic era. Number six corresponds to the time starting with Muḥammad and ending with the fall of the Umayyad caliphate. Number seven equates to the early Abbasid period, while number eight comprises the scholars living at that period in the West of the Islamic world. *Ṭabaqa* number nine is devoted to scholars flourishing at the same period in Ibn Juljul's own region, that is al-Andalus.

All in all, *The Generations of the Physicians and the Sages* is an instructive piece of scholarship. It combines detailed information about individual physicians, philosophers, scholars of the mathematical sciences and astrologers with an overall historiographical perspective that is carefully composed. Two of its features are most striking: the tendency to 'islamicize' the history of the ancient sciences and the tendency to 'Westernize' them. The first goal is attained by applying the *ṭabaqāt* structure taken from the Islamic sciences to the entire presentation. The second becomes manifest when we consider the geographical regions, which Ibn Juljul has included and excluded from his book. Apart from a short note on the legendary oriental sage 'Hermes' (itself a Greek name!) he focuses exclusively on Greek scholars when presenting the pre-Islamic history of the sciences, whereas no mention is made of ancient Persia let alone the regions eastern to it.

I.17.4 Ṣāʿid al-Andalusī and his *Generations of the Nations*

One might thus suspect Andalusian scholars not to be interested in the East of the Islamic and pre-Islamic world. Yet, this suspicion turns out to be untenable when we take into account what our fourth and last author has to tell us. I am talking of Ṣāʿid al-Andalusī, who served as a *qāḍī* in Toledo at the time of the so-called party kings (*mulūk al-ṭawāʾif*; on his life see Blachère in Ṣāʿid al-Andalusī 1935[CB], 7–11). He authored a book on the history of the rational sciences titled *The Generations of the Nations* (*Ṭabaqāt al-umam*; Ṣāʿid al-Andalusī 1912[CB], 1935; see also Balty-Guesdon 1997). Therein he follows Ibn Juljul's *Generations of the Physicians and the Sages* quite closely whilst also extending the perspective and scope of his presentation in two important respects.

One of these extensions concerns the semantic level. Apparently, Ṣāʿid understands the term *ṭabaqa* in a much broader sense than any of his predecessors. According to his usage of the word, *ṭabaqa* means neither a single generation of scholars (as was usual in Islamic *ṭabaqāt* works) nor a certain time within a larger historical period (as in Ibn Juljul). Instead, it refers to the entire era in which a "nation" (*umma*) was flourishing and delivering its specific contribution to the development of the sciences and to cultural history.

The second extension applies to the geographical scope covered by his presentation. As it appears, Ṣāʿid was the first Islamic scholar to connect the essentially "Eastern" perspective on the history of science, promoted by authors such as Ibn al-Nadīm, with the "Western", that is, Andalusī perspective prevalent in the text of Ibn Juljul. As a result, his book can be considered one of the broadest and most perspicuous presentations of the subject. Its goal is to integrate the scientific activities of the Muslim community into a wider conceptual framework ranging chronologically from the first scientific efforts of humankind to the author's lifetime and geographically from India in the East to al-Andalus in the West.

The presentation is subdivided into eight chapters corresponding to the eight nations that, according to Ṣāʿid, excelled in the sciences. They all are supposed to have flourished at different times, delivering their respective contributions one after the other. Hence, their succession can be interpreted as the history of the sciences consisting of eight remarkable *ṭabaqāt*.

The first *ṭabaqa* identified by Ṣāʿid are the Indians. According to him, they already had some knowledge in politics, ethics and theology but were particularly successful in the fields of mathematics and medicine (Ṣāʿid 1912, 11–15, 1935, 43–8). The second generation is to be equated with the (ancient) Persians, who excelled in politics, medicine, religion and astronomy, including astrology (Ṣāʿid 1912, 15–17, 1935, 49–52). Next came the Chaldeans (i.e., the Babylonians) who knew much about mathematics, metaphysics, astronomy and astrology. Furthermore, they gave rise to Hermes who was the first sage and scholar to be known by his name (Ṣāʿid 1912, 17–19, 1935, 53–6; on Ṣāʿid's interpretation of Hermes see Van Bladel 2009, 129–30). Despite its importance, however, Hermes's contribution was soon superseded by the efforts of the fourth generation, that is, the Greeks. Their accomplishments are unparalleled in the history of the sciences, especially in philosophy but also in medicine, mathematics and astronomy. As a consequence, they are treated extensively in *The Generations of the Nations*, Ṣāʿid giving detailed information about the outstanding figures such as Empedocles (5th century BCE), Pythagoras (d. c. 490 BCE), Socrates, Plato, Aristotle, Hippocrates, Galen, Apollonius of Perga (*c.* 262–*c.* 190 BCE), Euclid, Archimedes (d. 212 BCE) and Ptolemy (Ṣāʿid 1912, 19–33, 1935, 57–76).

The following generations benefitted heavily from the Greeks but did not contribute themselves much to the development of the sciences. This applies in particular to *ṭabaqa* number five, the Romans (Ṣāʿid 1912, 33–7, 1935, 76–82), and *ṭabaqa* number six, the Egyptians, who are more or less equated with Egypt in late antiquity (Ṣāʿid 1912, 38–41, 1935, 83–7). Things changed, however, after the rise of Islam. The new society was open-minded and promoted all kinds of scientific activity. As a result, philosophy, medicine, the mathematical sciences and astrology started flourishing again in a way only comparable to their tremendous development among the Greeks. This development applies first and foremost to the Arabs, corresponding to the seventh *ṭabaqa* (Ṣāʿid 1912, 41–87, 1935, 88–154), but also includes the Jews, who constitute *ṭabaqa* number eight (Ṣāʿid 1912, 87–90, 1935, 155–60). As Ṣāʿid explains, both of them are still cultivating the sciences at an extremely high level, especially in al-Andalus, which, according to him, has meanwhile become the center of scientific activity within the Islamicate world.

In sum, the picture drawn in *The Generations of the Nations* is extremely broad and impressive. It covers the world, as far as a 4th/10th century Muslim could know it, neglecting only those regions which were out of his reach (like the Chinese and the Turks, both of whom Ṣāʿid mentions in his introduction [Ṣāʿid 1912, 5–11, 1935, 31–41] without treating them later on). Another point to note is the concept of science underlying his presentation. It emphasizes the universality of the rational sciences assuming that all 'civilized' nations were able to contribute to the same scientific project. By stressing this argument, Ṣāʿid practiced already what Ibn Khaldūn, many centuries later, expressed theoretically: the acknowledgment that the rational sciences were altogether "natural to man" and "not restricted to any particular religious group."

Notes

1 Consolidated bibliography.

2 According to Ibn al-Nadīm, the mathematical sciences include geometry, arithmetic, algebra, astronomy, astrology, music, optics, instruments, artificial devices (Chapter I.9) and pneumatics (Ibn al-Nadīm 1872, 2: 265–85; 1970, 2: 634–62).

Bibliography

Sources

Rosenthal, F. 1954. "Isḥāq b. Ḥunayn's Ta'rīkh al-aṭibbā' [Isḥāq b. Ḥunayn's *History of the Physicians*]," *Oriens* 7: 55–80.

Research literature

Balty-Guesdon, M. G. 1997. "Al-Andalus et l'héritage grec d'après les *Ṭabaqāt al-umam* de Ṣāʿid al-Andalusī," in Hasnawi, A., Elamrani-Jamal, A. and Aouad, M., eds. *Perspectives arabes et médiévales sur la tradition scientifique et philosophique grecque. Actes du colloque de la SIHSPAI, Paris, 31 mars-3 avril 1993*. Leuven: Peeters, 331–42.

Dietrich, A. 1971. "Ibn Djuldjul," in *EI-2*, 3: 778–9.

Endress, G. 2017. "Arabic Aristotelianism and the Transmission of the Organon in the 3rd/9th and 4th/10th Centuries," in Rudolph *et al.*, eds. [CB], 423–8.

Gutas, D. 1999. "The 'Alexandria to Baghdad' Complex of Narratives. A Contribution to the Study of Philosophical and Medical Historiography Among the Arabs," *Documenti e studi sulla tradizione filosofica medievale* 10: 155–93.

Gutas, D. 2017a. "Ḥunayn b. Isḥāq," in Rudolph *et al.*, eds. [CB], 680–704.

Gutas, D. 2017b. "The Rebirth of Philosophy and the Translations into Arabic," in Rudolph *et al.*, eds., 95–142.

Horovitz, J. 1926. *Koranische Untersuchungen*. Berlin: De Gruyter.

Orthmann, E. 2017. "Philosophy and Natural Science," in Rudolph *et al.*, eds., 727–33.
Rudolph, U. 2011. "Die Deutung des Erbes: Die Geschichte der antiken Philosophie und Wissenschaft aus der Sicht arabischer Autoren," in Goulet, R. and Rudolph, U., eds. *Entre Orient et Occident: La philosophie et la science gréco-romaines dans le monde arabe* (Entretiens sur l'antiquité classique LVII). Vandœuvres-Genève: Fondation Hardt, 179–320.
Rudolph, U. 2017. "The Late Ancient Background," in Rudolph *et al.*, eds., 29–73 (therein: Philosophy and the Sciences, 48–51).
Sangaré, Y. T. 2016. *La notion de khatm al-nubuwwa (scellement de la prophétie) en Islam: Genèse et évolution d'une doctrine*. Thèse doctorale. Strassbourg: Université de Strasbourg.
Ullmann, M. 1978. "Ḫālid ibn Yazīd und die Alchemie: Eine Legende," *Der Islam* 55: 181–218.
Van Bladel, K. 2012. "The Arabic History Science of Abū Sahl ibn Nawbakht (fl. ca. 770–809) and Its Middle Persian Sources," in Opwis, F. and Reisman, D., eds. *Islamic Philosophy, Science, Culture, and Religion: Studies in Honor of Dimitri Gutas*. Leiden: Brill, 41–62.

PART II

Scientific practices at courts, observatories and hospitals (2nd–13th/8th–19th centuries)

II.1
THE EMERGENCE OF PERSIAN AS A LANGUAGE OF SCIENCE

Hossein Kamaly

During the 3rd–5th/9th–11th centuries, the Persian language emerged as a distinct medium for literary expression, adopting and incorporating linguistic elements from multiple sources, including a myriad of Middle Iranian vernaculars as well as Arabic.[1] In addition to court poetry, several tracts on logic, philosophy, astronomy and medicine appeared in Persian. During this time, scientific writing in Persian became a serious activity, which ought to be studied on its own terms, not merely as a minor, peripheral and subordinate phenomenon ancillary to the history of Arabic science.[2] One way to account for the use, or lack of use, of Persian as a language of choice for science is to highlight the link between the practice of writing and the patronage networks that supported it. Writing on particular subjects occurred when and where potential authors found interested readers and supportive patrons, either inside or outside the courts. Drawing on the role of patronage networks to explain the emergence of the Persian as a language of scholarship counters commonplace ethno-nationalistic hypotheses and assumptions about the inherent superiority of Persian that have to date had the upper hand in explaining this major development.

The new form of the Persian language (New Persian) put down roots in the 3rd/9th century, beginning in Khurasan – the region that stretched from what is now northeastern Iran to the eastern limits of present-day Uzbekistan. Greater Khurasan also included parts of present-day Afghanistan, Pakistan, Turkmenistan, and Tajikistan (Rante 2015). Rapidly spreading from there, by the late 5th/11th century Persian was established as a major literary language of the vast Persianate cosmopolis (Pollock 2006), which extended as far east as the Indus Valley, as far west as present-day Azerbaijan, as far north as the Oxus River to the east and the Aras River to the west and as far south as Sistan to the east and Shiraz, in present-day Iran, to the west.

II.1.1 Interpretations of the rise of New Persian

The gradual emergence and the rapid spread of a new form of the Persian language have puzzled scholars for a long time. Writing at the turn of the 20th century, Edward G. Browne (1862–1926) tied this phenomenon to the rise in Khurasan of Persian dynasties operating autonomously or semi-independently from the Abbasid caliphate in Baghdad, writing that "[the Persian language] begins, under those semi-independent dynasties, the Saffarids and Samanids, and even under the earlier Tahirids, to be employed as a literary language" (Browne 1902, 445).

DOI: 10.4324/9781315170718-21

Browne combined linguistic and ethnic elements in his category of 'Persian,' and generations of authors have followed suit, including Western orientalists as well as native scholars (Levy 1923, 21–6; Arberry 2004, 17, 31). The Iranian litterateur Bahār (1886–1951) propounded a similar idea in his authoritative study, *Stylistics, or The History of Change in Persian Prose* (*Sabk-shināsī, yā tārīkh-i taṭawwur-i na<u>th</u>r-i fārsī*), which first appeared in 1942. Adopting Browne's and Bahār's time frame, Zarrīnkūb (1923–1999), of the University of Tehran, published a popular and influential book, *Two Centuries of Silence*, where the title referred to the period between the defeat of Sasanian armies at the Battle of Qadisiyya (*c.* 15/636–7) and the rise of the Tahirid dynasty in Khurasan (*c.* 215/830). In it he described the ethno-nationalistic origins of Persian:

> Having suffered under Arab injustice for long years, Sistan and [Khurasan] prepared to become autonomous. . . . Suzerainty and authority that used to be the exclusive prerogative of the Arabs now returned to Persian hands everywhere, including Baghdad. The language of the Persians, which had fallen into "two centuries of silence" . . . gradually broke the spell.
>
> *(Zarrīnkūb 1334 sh/1955, 291–2)*

This account of the origins of Persian as a literary language faces several challenges. First, ignoring early uses of Persian outside courtly milieus, it fails to explain how Persian evolved as a supra-regional vernacular long before becoming a literary court language. Nor does it account for the fact that the earliest of the semi-independent Persian dynasties, the Tahirids (r. 205–259/821–873), took no interest in Persian (Bosworth 1969; Ṣadīqī 1345 sh/1966, 79). In fact, evidence shows that the 'Persian' Tahirids thwarted the development of Persian literature by supporting Arabic belles lettres and withholding patronage from Persian writings. Browne, Bahār and Zarrīnkūb knew well that the Tahirid court was *not* Persian-speaking. They may also have noticed that the other powerful 'Persian' dynasty, the Buyids (r. 333–446/945–1055), scarcely extended patronage to Persian writings during their heyday in the 4th/10th century. Even the Samanid dynasty (r. 203–395/819–1005), the rulers of which presented themselves as heirs to the throne of the Sasanian kings of Iranshahr, seem to have preferred Arabic over Persian before the reign of Naṣr II (r. 301–331/914–943). Furthermore, the ethno-nationalist view oversimplifies the political, economic and cultural relations between local dynasts in Khurasan and the caliphate in Baghdad. The historical record clearly shows that the Tahirid, Samanids and the Ghaznavids (r. 366–582/977–1186) were not enemies of the caliphate. Finally, the ethno-nationalistic explanation that celebrates literary expression in Persian language as a means of liberation from presumed Arab suzerainty has at times undergirded baseless claims of the ethnic or racial supremacy of Persians over Turks, Indians and others. More than a century after Browne's pioneering work, and decades after Zarrīnkūb's, any updated appraisal of the emergence of Persian language and literature ought to incorporate significant examples of noncourtly early writings in Persian that have since come to light, and should draw on sounder theoretical perspectives.

Recently, the social historian Richard Bulliet has put forth an alternative socioeconomic hypothesis for the emergence of the New Persian language (Bulliet 2009, 140–5). Focusing on the growth of urban-based networks of trade and cultural interaction, Bulliet suggests that Muslim merchants' pressing need to communicate with their counterparts led to the reformation of Persian as *koinē* or *lingua franca*. During the 2nd/8th and 3rd/9th centuries, this vernacular evolved from multiple lineages, mixing words and phrases borrowed from Arabic with vocabulary from several Middle Iranian languages, including Parthian, Pahlavi, Sogdian, Bactrian, Khwarazmian, Tokharian and Yaghnobi. The streamlined *koinē* glossed over differences among the tongues used in various localities, forging a more simplified grammar and morphology. Unlike the ethno-nationalist hypothesis, this socio-economic view readily accounts for the linguistic features found in an early to mid-2nd/8th-century mercantile document that Aurel Stein (1862–1943) unearthed in 1901 at Dandān-Öiliq, in Chinese Turkestan (Moreen 1995, 71). In time, adopting the Arabic script provided a symbolic common denominator and a means for overriding the diversity of Middle Iranian writing systems.[3]

To highlight the role of patronage networks in the emergence and use of Persian as a language of scholarly writing is to acknowledge the contribution of multiple interrelated political, economic, linguistic and ideological factors. Assessing the existence and strength of patronage networks counteracts the simplistic reduction of historical developments to ethno-nationalist assumptions. Tracing the formation and functioning of author–patron networks allows for a better understanding of the process of integration that culminated in religious, political and linguistic unification in 2nd–5th/8th–11th century Khurasan. Moreover, it provides a consistent explanation for the widespread adoption of Persian as a literary language in non-Persian-speaking territories, most notably in India and in Anatolia. Meanwhile, the historical analysis of the rise and fall of patronage networks obviates the tendentious contentions of ethnic or linguistic superiority that continue to permeate some contemporary works.

II.1.2 The role of patronage networks in the emergence of Persian as a scholarly language

In this section, I focus on patronage networks in Khurasan during the early phase of the emergence of New Persian as a language of scholarship. Other studies are needed on moments in which the Persian language became a prominent medium for science writing, including episodes during the 7th–8th/13th–14th centuries in Anatolia, northwest Iran, and parts of the Levant; the 9th/15th-century revival in Herat and other Timurid cities; the 10th–11th/16th–17th centuries in Safavid Iran; and the 10th–12th/16th–18th centuries under the Mughals in India.

By the 4th/10th century, the growth of an urban-based Persian-speaking Muslim elite in Khurasan had paved the way for a decisive shift in the patterns of patronage by the ruling Samanids during this century. This shift reflected a confluence of literary and cultural interests and sensibilities on the part of the court and the local elite. By the 320s/930s, the Samanid court attracted pioneering Persian poets from towns or villages across Khurasan – including Shahīd (d. 324/935) of Balkh, Rūdakī (d. 329/940–941) of Samarqand and Abū Shakūr (*fl.* 4th/10th century) of Balkh. Samanid kings, princes and viziers commissioned works of prose and poetry in Persian.

But patronage during Samanid rule went beyond literature and the court. The late Ṣadīqī (1905–1992) of the University of Tehran published an annotated list of early works that were written in Persian or translated into this language before and during the 4th/10th century (Ṣadīqī 1345 sh/1966). A close study of Ṣadīqī's list indicates that some translations and compositions were made under the supervision of the Samanid court while others received patronage from outside the court. A Persian translation of the fables of the *Kalīla wa-Dimna* (Ṣadīqī 1345 sh/1966, 89–91), possibly from the original Sanskrit but more likely based on the 2nd/8th-century Arabic translation by Ibn al-Muqaffaʿ (d. 142/759), belongs to the first group. So does Muvaffaq ibn ʿAlī Hiravī's (*fl.* 4th/10th century) *Book of Structures: On the True Natures of Medicines* (*Kitāb al-anbiyāʾ ʿan ḥaqāʾiq al-adwiya*; al-Haravī 2009), which is a collection of descriptions of 561 simple and compound medicinal remedies dedicated to the Samanid ruler Manṣūr ibn Nūḥ (r. 350–366/ 961–976; Browne 1921, 93; Sarton 1927, 1: 678–9). Similarly, the oldest extant work on practical medicine in Persian verse, the *Compendium*, completed in 370/980 by Ḥakīm Maysarī ([324–367/935–977]; Zanjani and Muhaqqiq 1366 sh/1987), was dedicated to the Samanid general Sebüktigin (*c.* 331–387/942–997). In the second group, we find accounts of the teachings and actions of local heroes such as Bihāfarīd (d. 130/747), al-Muqannaʿ (d. 169/785) and Ḥamza, son of Āzarak ([d. 218/833]; Ṣadīqī 1345 sh/1966, 61–3, 72–4). Other works written outside the court include early compositions on the practical arts of healing, such as *The Gift of Marvels* (*Tuḥfat al-gharāʾib*), by an anonymous author from the year 306/918 (Ṣadīqī 1345 sh/1966, 92–6), and collections of wise sayings and moral advice, represented by Māturīdī's (d. 333/944) *Book of Advice* (*Pandnāma*). The last stands out as the earliest extant work of its kind in Persian (Māturīdī 1340 sh/1961), along with *The Guide for Students on Medicine* (*Hidāyat al-mutaʿallimīn fī l-ṭibb*), the first major work on medicine in Persian, completed by 371/981. These examples demonstrate that the uses of the Persian language in the early period went well beyond composing panegyrics and lyrical poetry.

Other translations and scholarly compositions reflect the predilections that the Samanid court and its secretaries shared with the local elite in Khurasan, who principally followed the Ḥanafī legal school. Notable among works that evidence this convergence of ideological interests are translations from Arabic into Persian of the voluminous chronicle of al-Ṭabarī (d. 310/922), an interpretive translation of the Qurʾān purportedly based on the latter's grand *Commentary* (*tafsīr*; Yaghmāʾī 1339 sh/1960),[4] a treatise on Ḥanafī law by al-Ḥakīm Isḥāq al-Samarqandī (1st half 4th/10th century; Lazard 1993, 628–30), and a translation of another Ḥanafī work, Samarqandī's *The Great Majority* (*al-Sawād al-aʿẓam*; Ṣadīqī 1345 sh/1966, 117–18).

Inside and outside the court, advocates of Persian for official record keeping and epistolary correspondence created another pillar of patronage for Persian writing. Besides having a sensitive ear for panegyric poetry, prose stylistics and wisdom literature, a bureaucrat or secretary (*kātib*) needed a vast range of knowledge. An ideal secretary was expected to cultivate encyclopedic learning, including mathematical sciences and medicine, as well as philosophy, music and more. In the words of Ibn Qutayba (d. 276/889) on the *Education of the Secretary* (*Adab al-kātib*):

> The Persians (*al-ʿajam*) used to say: 'Whoever lacks knowledge of the following subjects can hardly qualify as a secretary: extracting waters, digging canals and blocking disused shafts; variations in the length of days, revolutions of the sun and astronomical measurements such as what pertains to the rise and phases of the moon; weights and measures; assaying triangles, squares and polygonal shapes; construction of arched and straight bridges; erecting water wheels and irrigation machines; the tools artisans use and the precise computation'.
>
> *(Ibn Qutayba 1346/1928, 10)*

Already during the 4th/10th century, court secretaries spearheaded the composition of works in Persian on such topics as calendar calculation, *zīj* compilation and astrolabe construction (Chapters I.6, IV.2, V.4; Ṣadīqī 1345 sh/1966, 111). The geographical work, *Territories of the World from East to West* (*Ḥudūd al-ʿālam*; Bosworth 2004; Lazard 1963; Chapter I.13), which stands out as the earliest known work of its kind in Persian, was completed around 372/982 by an anonymous *kātib* at the court of a local ally of the Samanids in Gūzgānān, an eastern province of Khurasan. After the overthrow of the Samanid dynasty by Sultan Maḥmūd of Ghazna (r. 388–421/998–1030) at the turn of the 5th/11th century, Persian served as the principal language of bureaucracy until 401/1010. In that year, the pro-Persian vizier Faẓl ibn

Aḥmad Isfarā'īnī (d. 403/1012) was forced out of office and his successor Aḥmad Maymandī (d. 443/1052) decreed that court secretaries ought to avoid using Persian and return to issuing declarations and citing rhetorical expressions in Arabic (Ṣadīqī 1345 sh/1966, 82). Although Persian poetry continued to thrive in Sultan Maḥmūd's court, the writing of works of prose in Persian declined.

The change in policy to disfavor Persian writing may have played a part in Ibn Sīnā's (d. 428/1037) decision to move away from Khurasan. The timing coincides with the emergence of patronage for Persian writing outside Khurasan and with the spread of science writing in new territories. For example, it was in Isfahan and later in Hamadan that Ibn Sīnā found the most welcoming patrons for his Persian writings. The most outstanding title among these works, *The Book of Knowledge for ʿAlāʾ* (*Dānishnāma-yi ʿAlāʾī*; Ibn Sīnā 1353 sh/1974), belongs to this period. Considered the author's most extensive composition after his renowned Arabic *Book of Healing* (*Kitāb al-Shifāʾ*), this encyclopedic work discusses such topics as logic, metaphysics, music and astronomy. Other short treatises in Persian attributed to Ibn Sīnā treat psychology, knowledge of the veins and arteries in the human body and the concept of being.[5]

The ban on using Persian for bureaucratic matters continued at least until Maymandī's dismissal from the office of vizirate and his imprisonment in 415/1024. Meanwhile, Ghaznavid court patronage for works related to science in Arabic continued. Most notable among such writings are the four major works by Bīrūnī (362–d. after 444/973–d. after 1053), *The Book of Pharmacy in Medicine*, *The Comprehensive* [*Book*] *on Gemstones*, *The Remaining Traces of Past Centuries* [*Chronology*] and *The Verification of What Belongs to India* [*Book on India*] (*Kitāb al-Ṣaydala fīl-ṭibb*, *al-Jamāhir fī maʿrifat al-jawāhir*, *al-Āthār al-bāqiya ʿan qurūn al-Kẖāliya*, *Taḥqīq mā li-l-Hind*). It was during this period of vanishing patronage for Persian that Bīrūnī made his oft-cited pejorative remarks about the inadequacy of the Persian language for writing on science (Bīrūnī 1973, 12). Once the ban was lifted, Bīrūnī completed one of the most comprehensive works on the sciences of the stars in the Persian canon, *The Book of Understanding for the Art of Astrology* [*Introducing the Basics of the Art of Astrology*] (*Kitāb al-Tafhīm li-ṣināʿat al-tanjīm*), which he completed in 419/1029 and dedicated to Rayḥāna, presumably a young daughter of a hitherto unidentified patron named Ḥasan/Ḥusayn from Khwarazm (Bīrūnī 1318 sh/1939, 1). In and of itself, this well-structured, technically precise and cogently written work suffices to refute claims about the intrinsic weakness of Persian for science writing, even in its early phase of development.

As of the second half of the 5th/11th century, with the coming of the Seljuqs (r. 429–548/1037–1153), patronage patterns rapidly changed, and science writing in Persian dwindled for some time. The present chapter has confined itself to science-related works in Persian that were written before that time. But the main idea elaborated here holds there as well: any proper understanding of the use and lack of use of Persian for the writing of science needs to account for variations in networks of patronage.

Notes

1 For an up-to-date report on the state of research, see de Bruijn, *General Introduction to Persian Literature*. For other scholarship, see Lazard (1963, 1975, 1993).

2 For a different view, see Saliba (1998).

3 For more linguistic analysis, see Ṣādiqī (1357 sh/1978; Pisowicz (1985); Paul (2002); and Perry (2009, 43–70).

4 On the consultation with the local Ḥanafī *ʿulamāʾ*, see Yaghmāʾī (1339, sh/1960, 1: 5). ʿAbbās Zaryāb Khūʾī's detailed analysis shows discrepancies between the extant Persian translation and its putative Arabic source referenced in Ṣadīqī (1345 sh/1966, 113, note 1).

5 At least some of these works may have been completed not by him but by students and colleagues in his circle.

Bibliography

Sources

Bīrūnī, Abū Rayḥān Muḥammad b. Aḥmad Khwārazmī. ed. Humā'ī, J. 1318 sh/1939. *Kitāb al-Tafhīm li-awā'il ṣinā'at al-tanjīm* [The Book of Understanding for the Basics of the Art of Astrology]. Tehran: Anjuman-i Ās̱ār-i Millī.

Bīrūnī, Abū Rayḥān Muḥammad b. Aḥmad Khwārazmī. tr. Hakim Said, M. 1973. *Al-Biruni's Book on Pharmacy and Materia Medica*. Karachi: Hamdard Academy.

Harawī, Abū-Manṣūr Muwaffaq ibn ʿAlī. 2009. *Kitāb al-abniya ʿan ḥaqā'iq al-adwiya = Rawżat al-uns wa-manfaʿat al-nafs* [Book of Foundations: On the True Natures of Remedies = The Garden of Love and the Usefulness of the Soul]. Facsimile edition MS. A.F. 340-i Kitābkhāna-yi Millī-yi Uṭrīsh. Tehran: Markaz-i Pazhūhishī-yi Mīrās̱-i Maktūb.

Ibn Qutayba, ʿAbd Allāh ibn Muslim. ed. Khaṭīb, M. 1346/1928. *Adab al-kātib* [The Education of the Secretary]. Cairo: al-Maktaba al-tijāriyya al-kubrā. http://hdl.handle.net/2333.1/s1rn8s6b.

Ibn Sīnā. *Dānish-nāma-yi ʿAlā'ī* [The Book of Knowledge for ʿAlā' (al-Dīn)], eds. Mishkāt, S. M. and Muʿīn, M. 1353 sh/1974. Tehran: Dānishgāh-i Tihrān.

Māturīdī. ed. Afshar, Ī., ed. 1340 sh/1961. "Pandnāma-yi Māturīdī [Māturīdī's Book of Advice]," *Farhang-i Īrān-zamīn* 9: 46–67.

Minorsky, V. 1937. *Ḥudūd al-ʿĀlam, "The Regions of the World": A Persian Geography 372 AH – 982 AD* (Gibb Memorial Series, N.S. XI). Oxford: Oxford University Press.

Research literature

Arberry, A. J. 2004. *Classical Persian Literature*. Richmond: Curzon Press, 1994 (originally 1958, Macmillan).

Bosworth, C. E. 1969. "The Ṭāhirids and Persian Literature," *Iran* 7: 103–6.

Bosworth, C. E. 2004. "Ḥodūd al-ʿālam," *eIr* XII.4; 417–18. www.iranicaonline.org/articles/hodud-al-alam.

Browne, E. G. 1902. *A Literary History of Persia: From the Earliest Times to Firdawsí*. New York: Charles Scribner & Sons.

Browne, E. G. 1921. *Arabian Medicine*. Cambridge: Cambridge University Press.

Bruijn, J. T. P. de. ed. 2009. *General Introduction to Persian Literature*. London: I. B. Tauris.

Bulliet, R. W. 2009. *Cotton, Climate, and Camels in Early Islamic Iran: A Moment in World History*. New York: Columbia University Press.

Dabashi, H. 2012. *The World of Persian Literary Humanism*. Cambridge, MA: Harvard University Press.

Fragner, B. G. 1999. *Die "Persophonie": Regionalität, Identität und Sprachkontakt in der Geschichte Asiens*. Berlin: Das Arabische Buch.

Lazard, G. 1963. *La langue des plus anciens monuments de la prose persane*. Paris: C. Kincksieck.

Lazard, G. 1975. "The Rise of the New Persian Language," in Frye, R. N., ed., *Cambridge History of Iran*. 4 vols. Cambridge: Cambridge University Press, 4: 595–632.

Lazard, G. 1993. "Reconstructing the Development of New Persian," *al-ʿUṣūr al-Wusṭa* 5: 28–30.

Levy, R. 1923. *An Introduction to Persian Literature*. Oxford: Oxford University Press.

Moreen, V. B. 1995. "A Supplementary List of Judeo-Persian Manuscripts," *The British Library Journal* 21.1: 71–80.

Paul, L. 2002. "A Linguist's Fresh View on 'Classical Persian'," in Szuppe, M., ed. *Iran. Questions et connaissances. Actes du IV*[e] *Congrès Européen des Études Iraniennes, organisé par la Societas Iranologica Europaea, Paris, 6–10 Septembre 1999*. Vol. 2. *Périodes médiévale et moderne. Cahiers de Studia Iranica* 2 6: 21–34.

Perry, J. R. 2009. "The Origin and Development of Literary Persian," in Bruijn, J. T. P. de, ed. *A History of Persian Literature (Book 1). General Introduction to Persian Literature*. London: I. B. Tauris, 43–70.

Pisowicz, A. 1985. *Origins of the New and Middle Persian Phonological Systems*. Krakow: Nakladem Uniwersytetu Jagiellonskiego.

Pollock, Sh. I. 2006. *The Language of the Gods in the World of Men: Sanskrit, Culture, and Power in Premodern India*. Berkeley: University of California Press.

Rante, R., ed. 2015. *Greater Khurasan: History, Geography, Archaeology and Material Culture*. Berlin: De Gruyter.

Ṣadīqī, ʿA.-A. Tīr 1345 sh/June–July 1966. "Baʿżī az kuhantarīn ā<u>th</u>ār-i na<u>th</u>r-i fārsī tā pāyān-i qarn-i chahārum," *Majalla-yi Dānishkada-yi adabīyāt-i dānishgāh-i Tihrān*: 56–126.

Ṣādiqī, ʿA.-A. 1357 sh/1978. *Takwīn-i zabān-i fārsī*. Tihrān: Intishārāt-i dānishgāh-i āzād-i Īrān.
Saliba, G. 1998. "Persian Scientists in the Islamic World: Astronomy from Maragha to Samarqand," in Hovannisian, G. R. and Sabagh, G., eds. *The Persian Presence in the Islamic World*. Cambridge: Cambridge University Press, 126–46.
Sarton, G. 1927. *Introduction to the History of Science*. 3 vols. Baltimore: The Williams and Wilkins Co.
Yaghmāʾī, Ḥ., ed. 1339 sh/1960. *Tarjama-yi tafsīr-i Ṭabarī*. Tehran: Tehran University Press.
Zanjani, B. and Muhaqqiq, M., eds. 1366 sh/1987. *Dānish-nāma-yi Ḥakīm-i Maysarī dar ʿilm-i pizishkī: kuhantarīn majmūʿa-yi ṭibbī bi shiʿr-i fārsī*. Tehran: Muʾassasa-yi Muṭālaʿāt-i islāmī-yi Dānishgāh-i McGill, Shuʿba-yi Tihrān.
Zarrīnkūb, ʿA.-Ḥ. 1334[2] sh/1955. *Dū qarn-i sukūt*. Tehran: Sukhan.

II.2

THE EMERGENCE OF A NEW SCHOLARLY LANGUAGE: THE CASE OF OTTOMAN TURKISH

Ahmet Tunç Şen

In the early days of the 1560s, one of the renowned figures of Istanbul's literary circles, ʿĀshıq Chelebi (d. 982/1574), presented to his former master Ṭāshköprīzāde (d. 968/1561) the Turkish translation of the latter's sought-after biographical dictionary of scholars, originally written in Arabic. The story goes that when Ṭāshköprīzāde received his onetime student's gift, he teased him by saying "O, our dear fellow ʿĀshıq! We had already written it (as) Turkish, you bothered (to translate it) in vain!" (Nevizade Atai 2017, 1; 595). The way Ṭāshköprīzāde teases ʿĀshıq Chelebi raises several intriguing but little examined questions: What did it entail to translate a work from Arabic or Persian during the high point of the (Ottoman) Turkish language, which modern scholars sometimes characterize as a "hybrid" (İz 1976, 118) or "miscellaneous" (Küçük 2018, 320) language, which was already replete with words, grammatical forms and morphological units borrowed from the other two? How did the language that would ultimately be named Ottoman Turkish develop in the first place and what factors played a role in its increasing prestige in certain fields of scientific writing vis-à-vis Arabic, Persian and the earlier, supposedly simpler form of Anatolian Turkish?[1] In which particular fields of scientific activity did Ottoman Turkish gain wider prominence, and what were the reasons as well as the limits of this appeal?

Any study that aims at addressing such ambitious questions by exploring scientific texts in the Turkish language written during Ottoman times must first acknowledge that the topic is too vast to be covered by a single author in a limited space. The available sources are numerous and varied, deriving from a wide array of scientific disciplines, theoretical, as well as practical, pursued in the extensive temporal and spatial range of the Ottoman world. More important, these texts were evidently written in different registers of Turkish. Thus, one should first revisit the historical evolution of Turkish from the 8th/14th through at least the 10th/16th century, during which (Old) Anatolian Turkish was gradually replaced by a "high" literary form of Turkish that has traditionally been recognized as classical "Ottoman Turkish." What I do in the following is provide, first, a brief overview of the status of Turkish in scientific writing produced before the emergence and ultimate crystallization of the so-called Ottoman Turkish. I then follow this overview by a more detailed examination of texts written in different genres pertaining to the astral sciences, such as almanacs (*taqwīms*), manuals of instrument use and treatises on theoretical astronomy, as well as on astrological techniques. Taking the astral sciences as a case study, I hope to portray as concretely as possible the extent of the use of Turkish in the early modern Ottoman scientific landscape.

DOI: 10.4324/9781315170718-22

II.2.1 The rise of Ottoman Turkish

Although there is hardly a consensus over the number and nomenclature pertaining to the classification of written Turkish, the narrative that has received more scholarly recognition is the one drawing on Maḥmūd of Kashghar's (late 5th/11th century) *Compendium of the Languages of the Turks* (*Dīwān lughāt al-turk*), the oldest comprehensive dictionary of Turkish dialects in Arabic. In his compendium, Maḥmūd divides the Turkish language into two main branches: eastern and western Turkish. While the former was the language of the Qarakhanid state located in Central Asia, the latter, also known as Oghuz Turkish, spread through the mass migration of Oghuz tribes from the 4th/10th century onward into the region between the Caspian and the Adriatic Seas, including Anatolia. It is the western branch of the Turkish language from which the categories conventionally named in Turkish studies as "(Old) Anatolian"– and "Ottoman" Turkish–derived (Mansuroğlu 1954; İz 1976; Darling 2012).

Notwithstanding the growing Turkish population in Anatolia from at least the late 5th/11th-century onward, the earliest extant texts in Western (Oghuz) Turkish dialect that were written in Arabic script come down to us only from the late 7th/13th century. Semih Tezcan argues that in the first two centuries when Turkish tribes settled in Anatolia (5th/11th and 6th/12th centuries), they did not develop a written language and the Turkish language was only spoken (Tezcan 2012, 67). The lion's share of the responsibility for the lack of Turkish texts produced at the time should go to the Seljuqs of Anatolia (r. 473–641/1081–1243), who, despite originating from the Turkish speaking tribal peoples, retained Persian for administrative and literary purposes. The waning of Seljuq rule in Anatolia and the rise of Turkmen principalities, including the Ottomans, led by *ghāzī* warlords who only spoke Turkish, paved the way for the emergence of the first texts written in Anatolian Turkish. Thanks mostly to the patronage and population dynamics of late medieval Anatolia, the number of works produced in written representations of spoken Turkish significantly increased. These were either free translations from Persian and Arabic or original compositions. While catechistic texts, popular mystical treatises and literary works constituted the great majority of the Turkish corpus at the time, compendiums of medical knowledge and treatises about different branches of the mathematical sciences were not entirely absent. According to Ramazan Şeşen's estimate (1994), there were over 90 different works translated into Turkish in the 9th/15th century. While half of these translated works were on the religious sciences, Sufism and literature, there were at least 12 medical texts which constitute between 10% and 15% of the entire collection.

In the early decades and even centuries of (Old) Anatolian Turkish, when the language was still far from establishing its prestige vis-à-vis Arabic and Persian, the authors writing in Turkish often felt compelled to justify their choice of language. Some of the authors even say in an apologetic tone that although Turkish remains deficient in creating the intended meaning and delicacy due to its lack of grammatical sophistication and rhetorical eloquence, they still write in the language out of simple courtesy to render the contents of the work more accessible to their Turkish-speaking audience. They often add to these apologetic remarks their plea that more learned contemporaries should not chastise them for writing in such a "dry", "crude" and "indelicate" idiom (Yavuz 1983).

Although not necessarily written in an apologetic tone, the prologues of scientific treatises composed during the earlier phases of literary activity in Turkish often contain similar remarks about the urgent needs of their target audience (Fazlıoğlu 2003). For instance, in two of his translated works, Aḥmed-i Dāʿī (late 8th–early 9th/late 14th–early 15th centuries), an important man of letters traveling back and forth between different princely courts in western Anatolia in search of patrons, described in detail the production processes of his books. In his Turkish translation of an unnamed treatise in Arabic on prophetic medicine, he says that his dedicatee-patron Umur ibn Tīmūrtāsh (early 9th/15th century), one of the viziers of the Ottoman Sultan Murād II (r. 824–848/1421–1444, 850–854/1446–1451), kindly asked him to translate the work, which was, in Umur's words, "precious" in value and "felicitous" in view of its discourse, but not comprehensible to him due to its being in Arabic (Ahmed-i Dâî 1992, 405–6; Stanley 2004, 323–1). In the second book, which translates the 7th/13th-century polymath Naṣīr al-Dīn al-Ṭūsī's (597–672/1201–1274) *Treatise in Thirty Chapters* (*Risāla-yi sī faṣl*), a short textbook in Persian on almanac-making and foundational knowledge in astrology, Aḥmed-i Dāʿī states that since Ṭūsī's original text in Persian was difficult for beginners (*mübtedīler*) in the science of the stars, he sought to translate the work to make it easily comprehensible to them (Ahmed-i Dâî 1984, 14). The translation of Dāʿī, and many other translated works at the time, indeed, kept the key technical terms in their Arabic and/or Persian originals while changing the word order and verb structure according to the rules of Turkish grammar. No attempt was made to "Turkicize", for instance, the names of the planets, astrological signs or specific astral configurations.

By the mid-9th/15th-century, the (Old) Anatolian Turkish served as a medium venerable enough to communicate religious and nonreligious themes. But the century following the conquest of Constantinople in 857/1453 by the Ottomans, and attesting to the gradual transformation of the Ottoman enterprise from a relatively minor regional power to a bureaucratic world empire, enabled the formation of what would later be classified and classicized as "Ottoman Turkish" (Kuru 2012). The fashioning of this new, "imperial" idiom took place at the hands of the cultured elites of the empire and ran parallel to the establishment and growing maturation of Ottoman imperial culture. Grammatically speaking, Ottoman Turkish was not a separate language independent of Anatolian Turkish, nor was it an artificially constructed one like Esperanto. It was yet a new idiom "built" on vocabulary and syntactical elements borrowed extensively from Arabic and Persian. In the eyes of its most articulate champions, to be able to speak and write in this "astonishing language . . . was a meritorious act" (Fleischer 1986, 22). In order to qualify as a cultured Ottoman individual, one was expected to abandon the use of simple Turkish and cultivate the necessary rhetorical skills and eloquence, no matter in which discipline, genre, and form one wrote.

For linguists and literary historians, this "hybrid" language functioned for centuries as the "official jargon" of the empire. It prevailed to such an extent that some of the works already rendered into simpler Anatolian Turkish in the early part of the 9th/15th century were later "retranslated" into "Ottoman Turkish." An illuminating case is an early 12th/18th-century translation of *The Book by Qābūs (Qābūsnāma)* by Naẓmīzāde Murtażā (d. after 1140/1728), who says in the prologue of the work that while this important compendium of useful knowledge for rulers originally written in Persian in the 5th/11th century already had Turkish renditions made in the 9th/15th century, he still had to produce a new version for the grand vizier, because the Turkish in the previous translations was obsolete, unpleasant and inappropriate for the taste of contemporary elites and laypeople alike (Özdemir 2018; for early Turkish translations of the *Qābūsnāma* see Doğan 2012).

The flowering of Ottoman Turkish as a high literary medium from the mid-9th/15th century did not necessarily rule out the use of Arabic, Persian and Anatolian Turkish. Arabic, for instance, still functioned as the primary language of scholarly activities in the *madrasa* setting, although scattered anecdotal evidence suggests that even if the books studied in the scholastic tradition of the *madrasas* including theology, jurisprudence, or exegesis were primarily Arabic, in the daily interactions of the *madrasas* was Turkish also used (Yavuz 1983; Chapter III.2). Persian also retained its prestige, chiefly in poetry and literary pieces consumed among Sufi circles. In fact, as the Ottoman dynasty rose to prominence against the tumultuous post-Timurid changes in the Persianate world in the second half of the 9th/15th century, numerous émigré Persianate scholars, literati and practitioners of natural and mathematical sciences were lured to the flourishing Ottoman capital (Sohrweide 1970; Heiderzadeh 2005). This led to a transitory boom of Persian at the Ottoman court, not only in the fields of history writing and poetry but also in certain practical branches of learning, such as the science of the stars or medicine. The ascendancy of Persian in certain branches of literary-intellectual life in the late 9th/15th- and early 10th/16th -century Ottoman world is best captured by the voluminous inventory of the Ottoman palace library composed in the year 908/1502–3 (Necipoğlu *et al.* 2019). But (Old) Anatolian Turkish continued to feature in the textual lore of sciences written mostly, but not exclusively, for the consumption of laypeople.

II.2.2 The astral sciences as a case study

The main contours and changing trajectories of Ottoman Turkish as a new scholarly language can be vividly traced through the lens of the astronomical and astrological corpus. In this second part of the chapter, I focus on select texts from an array of genres pertaining to different branches of the astral sciences in order to provide a more precise description of the layers and limits of the use of Ottoman Turkish as a new scholarly language.

Among the relevant textual output, the *taqwīm* (almanac with prognostication) stands as the most characteristic genre of production in the astral sciences. Almanacs that were produced primarily for courtly consumption were composed by astral experts on a yearly basis before or around the time of the spring equinox (the beginning of the new solar year in March). As such, they offered their readers not only the calendrical information for the upcoming year but also tabulated astronomical data on planetary positions and corresponding astrological prognostications. Although the earliest available Ottoman *taqwīm* dates only from the early 820s/1420s, the genre itself was an older and seemingly ubiquitous practice in the medieval Islamicate world. The earliest references to *taqwīms* go as far back as the 3rd/9th century, and there are extant *taqwīms* produced in Persian at the courts of other principalities in 8th/14th-century Anatolia. The surviving Ottoman corpus of *taqwīms* from the 9th/15th down to the early 14th/20th centuries, however, outweighs all other sets of such works from different Islamicate societies (Şen 2016, ch. 4).

Exploring these *taqwīms* from the Ottoman realm that survive almost in a complete run from the mid-9th/15th to the early 14th/20th centuries can help us trace, and thus historicize, the changing constitutive elements of the genre pertaining to the contents, style, page layout and language. In terms of the language of *taqwīms*, there is a documented shift from the mid-9th/15th to the mid-10th/16th centuries. Whereas most of the extant *taqwīms* before the 880s/1480s were written in vernacular Anatolian Turkish, the overwhelming majority of surviving almanacs from the last two decades of the 9th/15th and the first two decades of the 10th/16th century were in Persian, likely reflecting the influence of Persianate émigré astronomers/astrologers populating the Ottoman court during the final years of the reign of Meḥmed II (r. 848–850 /1444–1446 and 854–886/1451–1481) and the entire sultanate of his son, Bāyezīd II (r. 886–918/1481–1512). Subsequently, the language of *taqwīms* shifted from Persian to (Ottoman) Turkish as a result of the ongoing process of Ottomanization in cultural and bureaucratic life over the course of the 10th/16th century, and the gradual emergence of a new, indigenous group of Ottoman experts who served as court astronomers/astrologers (Şen 2016, chapter 4). But this time the register of Turkish was distant from the plain Turkish of the early 9th/15th century, with numerous words and expressions borrowed from Persian and Arabic aligned according to Turkish syntax. Save for rare examples of *taqwīms* written in Arabic in the 11th/17th century, Ottoman Turkish dominated the *taqwīms* genre from the mid- 10th/16th century onward.

The systematic shift in the language of *taqwīms* was accompanied at the time by several individual attempts to produce works in Ottoman Turkish in the fields of both practical and theoretical astronomy, including timekeeping (*mīqāt*), the uşe of different astronomical instruments and the science of the configuration of the heavens (*ʿilmü l-hey'et*). In the early 10th/16th century, Meḥmed el-Qonevī ([d. *c.* 930/1524); al-Qūnawī], who worked as a timekeeper in the prominent mosque complexes of Istanbul or Edirne, for example, composed or rendered into Turkish at least four relevant works, including the sine tables of the famous Damascene timekeeper Shams al-Dīn al-Khalīlī ([d. *c.* 781/1379–80]; Fazlıoğlu 2007a). Qonevī often refers, in the prologue sections of his works in Turkish, to the *iḫvān* (brethren) and *evlād* (descendants) as the primary beneficiaries of his works. This implies that he had the urge to ensure the transmission of his astronomical expertise among Turkish-speaking members of learned circles (İhsanoğlu *et al.* 1997[CB],[2] 1: 84–9).

Qonevī's efforts seem to have been maintained in the mid-10th/16th century by another timekeeper-cum-court astrologer Muṣṭafā ibn ʿAlī (d. 958/1571), who composed a series of manuals in Ottoman Turkish for instructing the interested students (*ṭālib ve rāġıb olanlar*) in the use of different astronomical instruments (Fazlıoğlu 2007b). Decisive evidence is currently lacking for a master–student relationship between Qonevī and Muṣṭafā. Nevertheless, the abundance of the copies of both Qonevī's and Muṣṭafā's works, scattered around a vast geography between the 10th/16th and early 13th/19th centuries, clearly shows that they had a lasting influence on the proliferation of Turkish learning on the use of astronomical instruments and timekeeping manuals. For example, while compiling a list of surviving manuscripts of Muṣṭafā's works, İhsanoğlu *et al.* (1997, 1: 161–79) found more than 125 copies of his treatise on the astronomical instrument called the *ul muqanṭar* quadrant (*Kifāyetü l-vaqt*). Aside from practical astronomy, the rare examples of Ottoman Turkish texts in theoretical astronomy did also figure in the mid-10th/16th century, thanks mostly to the personal initiatives of different individuals including the Ottoman sea captain Seydī ʿAlī Reʾīs (d. 970/1562), who reworked in Turkish the well-esteemed *hay'at* book of ʿAlī Quşçu (Qūshjī; 805–[879/1403–4]). With around thirty extant copies, the latest of which dates to the 13th/19th century, Seydī ʿAlī's *hey'et* work in Turkish proved to be a useful yet unique addition to a rich body of Arabic treatises on theoretical astronomy some of which can be documented in the *madrasa* curriculum (İzgi 1997, 1: 361–412; Seydi Ali Reis 2010).

Compared to the growing use of (Ottoman) Turkish in *taqwīms*, manuals of astronomical instruments, or treatises on timekeeping, the established *zīj* canon (astronomical handbook of tables) in Persian, including the *Zīj-i Īlkhānī* (the Ilkhanid tables produced at the Maragha observatory in the second half of the 7th/13th century), the *Zīj-i Ulugh Beg* (Ulugh Beg's tables produced at the Samarqand observatory in the first half of the 9th/15th century) and various commentaries on the two remained largely untranslated until the 11th/17th and 12th/18th centuries. While the relative ease of using tabulated information in Persian and Arabic notation might be a reason for the indifference of the 9th/15th- and 10th/16th-century experts to translating these tables to Ottoman Turkish, I should also note that the most favored tables among Ottoman astronomers, the *Zīj-i Ulugh Beg*, did not go entirely untranslated until European tables superseded those of Islamicate origin in the later 11th/17th century. As far as the manuscript records are concerned, the *Zīj-i Ulugh Beg* was rendered into Turkish, first, by the 11th/17th-century court astrologer Meḥmed Chelebi (d. 1040/1631), and then by the 12th/18th-century physician ʿAbbās Vesīm (d. 1175/1761–1762). But if the number of extant copies of these Turkish translations is taken as an indication, they seem to have been slighted by the community of readers compared to the Persian original (İhsanoğlu *et al.* 1997, 1: 274, 445–6). As to the growing Turkish corpus on European astronomical tables, one should mention the translations from French or Latin of contemporary European ephemerides, including Jacques Cassini's (1677–1756) *Tables astronomiques* or Joseph Jérôme de Lalande's (1732–1807) *L'astronomie* (İhsanoğlu *et al.* 1997, 1: clxxii–clxxiv).

Intriguingly, the 12th/18th century witnessed court-sponsored (Ottoman) Turkish translations of a series of textbooks from the Islamicate astrological corpus. The chief reason behind this trend was the keen astrological interests of Sultan Muṣṭafā III (r. 1171–1187/1757–1774), who commissioned at least five translations from Persian and Arabic, including the canonical works of Pseudo-Ptolemy's *Centiloquium* (*Kitāb al-Thamara*) or ʿAlī-Shāh al-Bukhārī's (d. after 690/1291) *The Book of Fruits and Trees* (*Kitāb Athmar wa-ashjar fī l-nujūm*; İhsanoğlu *et al.* 2011, 1: 64, 70–1, 75–6). An immediate comparison, at this point, between the patronage and cultural dynamics of the 12th/18th-century Ottoman court and those at the time of Bāyezīd II will be illuminating. Like his distant grandson Muṣṭafā III, Bāyezīd II also had a special interest in the science of the stars (Şen 2017). However, among numerous astronomical and astrological books presented directly to Bāyezīd II himself there was not a single item in Turkish (Necipoğlu *et al.* 2019). This could be partially explained by the prevailing Persian cultural orientations of this ruler's court. However, one should also bear in mind that during Bāyezīd II's reign at the turn of the 10th/16th century, (Ottoman) Turkish was still not mature and the intellectual circles replete with émigré-scholars were not fully "Ottomanized."

II.2.3 Conclusion

To what extent could the patterns visible in the gradually developing use of (Ottoman) Turkish for the case of the astral sciences be extended to other disciplines? While recognizing that those sciences and texts taught consistently in Arabic in the *madrasas*, such as theology, jurisprudence or logic, do not follow this pattern, it seems plausible to explicate the dynamics in other natural or mathematical sciences using the pattern described here. The evolutionary passage from (Old) Anatolian Turkish in the 8th/14th and 9th/15th centuries to the refined Ottoman Turkish of the post-10th/16th century, with brief episodes of the prevalence of either Persian or Arabic, depended also on the personal proclivities of patrons. Therefore, any attempt at generalization should not overlook the needs of the target audience as well as the peculiarities of the aims and scope of specific scientific branches. For a particularly practical and vital science such as medicine, for instance, where both learned readers and laypeople alike might have preferred easy and immediate access, it would not have been a serious concern whether the language was refined enough or not. This must be the reason why over the centuries there was a steady stream of medical compendia composed in a language reminiscent of translations in (Old) Anatolian Turkish, where the original technical terms were retained and sentences were simply realigned in accordance with the rules of Turkish syntax (İhsanoğlu *et al.* 2008; İhsanoğlu 2004). For certain branches of the astral sciences too, the immediate pragmatic needs of individuals from all walks of life led them to seek methods for making simple calculations for calendrical purposes or quick prognostications for planning purposes, which gave rise in turn to a sizable scientific corpus in "simpler" registers of Turkish.

Notes

1 *Scientific writing* is defined here broadly, ranging from the so-called traditional sciences taught at established institutions of learning to the different kinds of inquiries into the natural as well as the supernatural world, which were often categorized as *al-ʿulūm al-gharība* (occult or unusual sciences).

2 Consolidated bibliography.

Bibliography

Sources

Ahmed-i Dâî. ed. Çağıran, Ö. 1992. *Tıbb-ı Nebevî* [Prophetic Medicine]. Unpublished PhD dissertation. Malatya: İnönü University.

Ahmed-i Dâ'î. ed. Gencan, T. N. and Dizer, M. 1984. *Muhtasar fi ilm el-tencim ve marifet el-takvim (Risale-i si fasl)* [Epitome on the Science of Astrology and the Knowledge of the Almanac (*Epistle in Thirty Sections*)]. Istanbul: Boğaziçi Üniversitesi Kandilli Rasathanesi.

Nevizade Atai. ed. Donuk, S. 2017. *Hadâ'iku'l-Hakâ'ik fî Tekmileti'ş-Şakâ'ik* [The Gardens of Truth Complementing (Ṭāshköprüzāde's Biographical Dictionary Entitled) Peonies]. Istanbul: Yazma Eserler Kurumu Başkanlığı.

Seydi Ali Reis. ed. Cengiz, M. 2010. *Ḫulāṣatü'l-Hey'e* [The Essence of the Configuration (of the Universe)]. MA thesis. Ankara: Hacettepe University.

Research literature

Darling, L. 2012. "Ottoman Turkish: Written Language and Scribal Practice, 13th to 20th Centuries," in Spooner, B. and Hanaway W. L., eds. *Literacy in the Persianate World: Writing and the Social Order*. Philadelphia: University of Pennsylvania Museum of Archaeology and Anthropology, 171–96.

Doğan, E. 2012. "On Translations of Qabus-nama During the Old Anatolian Turkish Period," *Uluslararası Sosyal Araştırmalar Dergisi* 5: 76–85.

Fazlıoğlu, İ. 2003. "Osmanlı Döneminde 'Bilim' Alanındaki Türkçe Telif ve Tercüme Eserlerin Türkçe Oluş Nedenleri ve Bu Eserlerin Dil Bilincinin Oluşmasındaki Yeri ve Önemi," *Kutadgubilig* 3: 151–84.

Fazlıoğlu, İ. 2007a. "Qunawī: Muḥammad ibn al-Kātib Sīnān," in Hockey *et al*., eds. [CB], 945–6.

Fazlıoğlu, İ. 2007b. "'Alī al-Muwaqqit: Muṣliḥ al-Dīn Muṣṭafā ibn 'Alī al Qusṭanṭīnī al Rūmī al-Ḥanafī al-Muwaqqit," in Hockey *et al*., eds., 33–4.

Fleischer, C. H. 1986. *Bureaucrat and Intellectual in the Ottoman Empire: The Historian Mustafa Âli (1541–1600)*. Princeton, NJ: Princeton University Press.

Heiderzadeh, T. 2005. "Patronage, Networks, and Migration: Turco-Persian Scholarly Exchanges in the 15th, 16th, 17th Centuries," *Archives internationales d'histoire des sciences* 155: 419–34.

İhsanoğlu, E. 2004. "Tıp Dilinin Türkçeleşmesi Meselesi," in Ekrem Kadri Unat *et al*., eds. *Osmanlıca Tıp Terimleri Sözlüğü* [Ottoman Turkish Dictionary of Medical Terminology]. Ankara: Türk Tarih Kurumu, xiii– xxviii.

İhsanoğlu, E. *et al*.: İhsanoğlu, E., Şeşen, R., Serdar Bekar, M., Gündüz, G. and Bulut, V. 2008. *Osmanlı Tıbbi Bilimler Literatürü Tarihi*. 4 vols. Istanbul: IRCICA.

İhsanoğlu, E. *et al*.: İhsanoğlu, E., Şeşen, R., Serdar Bekar, M., Gündüz, G. and Bulut, V. 2011. *Osmanlı Astroloji Literatürü Tarihi ve Osmanlı Astronomi Literatürü Tarihi Zeyli*. 2 vols. Istanbul: IRCICA.

İz, F. 1976. "Ottoman and Turkish," in Little, P., ed. *Essays on Islamic Civilization: Presented to Niyazi Berkes*. Leiden: Brill, 118–39.

İzgi, C. 1997. *Osmanlı Medreselerinde İlim*. 2 vols. Istanbul: İz.

Küçük, B. H. 2018. "Arabic into Turkish in the Seventeenth Century," *Isis* 109.2: 320–5.

Kuru, S. 2012. "The Literature of Rum: The Making of a Literary Tradition (1450–1600)," in Faroqhi, S. N. and Fleet, K., eds. *Cambridge History of Turkey*. 4 vols. Cambridge: Cambridge University Press, 2: 548–92.

Mansuroğlu, M. 1954. "The Rise and Development of Written Turkish in Anatolia," *Oriens* 7.2: 250–64.

Necipoğlu *et al*.: Necipoğlu, G., Kafadar, C. and Fleischer, C. H., eds. 2019. *Treasures of Knowledge: An Inventory of the Ottoman Palace Library (1502/3–1503/4)*. 2 vols. Leiden: Brill.

Özdemir, R. T. 2018. *Nazmî-zâde Hüseyin'in Kâbus-nâme Tercümesi*. MA thesis. Adnan: Menderes University.

Şen, A. T. 2016. *Astrology in the Service of the Empire: Knowledge, Prognostication, and Politics at the Ottoman Court, 1450s – 1550s*. PhD thesis. Chicago: University of Chicago.

Şen, A. T. 2017. "Reading the Stars at the Ottoman Court: Bāyezīd II (r. 886/1481–918/1512) and His Celestial Interests," *Arabica* 64: 557–608.

Şeşen, R. 1994. "Onbeşinci Yüzyılda Türkçe'ye Tercümeler," in *XI. Türk Tarih Kongresi*. Ankara: Türk Tarih Kurumu, 899–919.

Sohrweide, H. 1970. "Dichter und Gelehrten aus dem Osten im osmanischen Reich," *Der Islam* 46: 263–302.

Stanley, T. 2004. "The Books of Umur Bey," *Essays in Honor of J. M. Rogers: Muqarnas* 21: 323–31.

Tezcan, S. 2012. "Eski Anadolu Türkçesi ve Yunus Emre Şiirlerinin Dili Üzerine," in Ocak, A. Y., ed. *Yunus Emre*. Ankara: Kültür ve Turizm Bakanlığı, 67–82.

Yavuz, K. 1983. "XIII – XVI. Asır Dil Yadigarlarının Anadolu Sahasında Türkçe Yazılış Sebepleri ve Bu Devir Müelliflerinin Türkçe Hakkındaki Görüşleri," *Türk Dünyası Araştırmaları* 27: 9–57.

II.3
IMPERIAL DEMAND AND SUPPORT

Eva Orthmann

> The secretary, the poet, the astrologer and the physician belong to the people close to the king, and they are indispensable for him. The maintenance of the rule is by the secretary, the perpetuation of an eternal name by the poet, the ordering of affairs by the astrologer, and the health of the body by the physician.
>
> – *Niẓāmī ʿArūżī (1910, 11)*

In the introduction to his *Four Discourses* (*Chahār maqāla*), a Persian prose book from the 6th/12th century, Niẓāmī ʿArūżī (*fl.* 504–577/1111–1161) enumerates the four professional groups he considers indispensable for a prosperous and successful rule. The four main chapters of his book are dedicated to them. Besides the secretary and the poet, he mentions the physician and the astrologer. The quotation from his book points to the close relationship between rulers and scholars, and the anecdotes he recounts confirm the rulers' special relationship with members of these two professions.

During most of the Islamic centuries and in most Islamic dynasties, scholars were promoted by the rulers. In spite of this close relationship, however, most mirrors for princes do not refer to the need to promote and employ physicians, astrologers, scholars of the mathematical sciences or other experts at court. The perennial imperial demand for scholars was not something discussed or commented on in advice literature. It is however included in anecdotal material. In the *Four Discourses*, *The Table-Talk of a Mesopotamian Judge* (*Nishwār al-muḥāḍara*) by al-Muḥassin al-Tanūkhī (d. 384/994) or the *Relief of the Concerned about the History of Astrologers* (*Faraj al-mahmūm fī ta'rīḫ 'ulamā' al-nujūm*) by Ibn Ṭāwūs (d. 646/1266), we find quite a few allusions to the circumstances of scholars working the exact and natural sciences, especially astrologers. In these anecdotes, we encounter them not only at court, engaged by the caliph or a local ruler, but we also see them working as independent scholars, like, for example, physicians, who are called to sick persons in case of need. The anecdotes furthermore refer to street astrologers, primarily portraying them as ignorant charlatans. One more element captures attention in these sources: scholars are described as endangered persons, exposed to their patrons' caprices and sometimes even treated as a part of the booty in wars and conquests.

Looking at the relationship between scholars and rulers in more detail, two different perspectives can be adopted: first, that of the scholars and their careers and employment options, and

DOI: 10.4324/9781315170718-23

second, that of the ruler and his interest in employing scholars. The first perspective has been dealt with in articles by Sonja Brentjes, who has not only asked about the disciplines sponsored at court and the nature of relationships between ruler and scholar but also shown that with the spread of the *madrasa* in the Islamic world, a new opportunity for employment emerged, allowing not only religious scholars but also scholars working in the exact and natural sciences to get positions outside the court (Brentjes 2008a, 2008b).

The focus in what follows is the perspective of the court and the interest of the ruler in promoting science and scholars. These interests depended on the historical circumstances and were not the same for every region and period. We can discuss here only selected dynasties and their specific needs. The three dynasties and epochs dealt with in this chapter are (a) the Abbasids, (b) the Fatimids and (c) the Mughals.

II.3.1 The Abbasids

The early Abbasid period (2nd–4th/8th–10th century) was distinguished by a growing interest in what was often called the sciences of the ancients and less often the foreign sciences (Chapter I.3). This interest was reflected in the promotion and financial support of translations from other languages, along with an awakening of scholars' own scientific agendas. While translations from Greek into Arabic form the best-known and most prominent part of these activities, Pahlavi, Sanskrit and Syriac were also important source languages for translations. The exact reasons for this interest are still debated (Chapter I.1), but evidently, the ruling caliphs and several other high-ranking members of the Abbasid society, including some scholars themselves, promoted both the translation activities and the accruing advancement of their own research (Gutas 1998[CB],[1] 121–50).

The promotion of science had benefits on a practical and probably also an ideological level. Dimitri Gutas has argued that the interest in foreign sciences was triggered by the imperial need for astrology (Gutas 1998, 29–52). Texts on mundane astrology belong to the very early translated texts and were used to argue for the legitimacy of Abbasid rule. In utilizing mundane astrology for imperial ends, the Abbasid rulers may well have adopted and imitated earlier Sasanian practices. Astrology was also employed for practical ends, like determining the appropriate moment for specific undertakings, the most prominent case being the foundation of Baghdad (e.g., al-Yaʿqūbī 1892[CB], 238–41). Taking into consideration the extensive astrological literature produced in that period, we may wonder to what extent it reflects other everyday needs. The *Book of Questions* (*Kitāb al-Masāʾil*) by al-Qaṣrānī ([3rd/9th century]; Sezgin 1979, 138–9), for instance, covers all kinds of practical questions, like the seating order at a banquet or finding a runaway slave or a hidden treasure, a topic dealt with in other astrological treatises from that period, too (Burnett *et al.* 1997). We can assume that court astrologers had to advise their patrons in a variety of ways; they also accompanied them on travels and military campaigns.

The interest in astronomy was related to that in astrology and partially instigated by it. However, astronomy was also valued per se. The earliest known observatories, built in the time of Caliph al-Maʾmūn (r. 198–218/813–833), were probably constructed out of a genuine interest to verify the geometrical models and contradictory parameters found in the Greek, Iranian and Indian books used by the astronomers of Abbasid times (King *et al.* 2001[CB], 36–9; Sayılı 1960[CB], 50–87). The rulers' concern with the movement of the celestial bodies is equally reflected in the earliest translation of the *Almagest*, produced either for al-Maʾmūn or for Yaḥyā ibn Khālid ibn Barmak (d. 190/805; King *et al.* 2001, 35–6).

Astronomy was not the only mathematical science promoted by the rulers. There was a practical demand for geometry to improve land measurement, needed for administration and taxation.

Again, in the time of al-Ma'mūn and by order of the caliph, a complicated project to determine the length of one degree of the meridian was undertaken. Al-Bīrūnī also reports about other geodetic enterprises during this caliph's reign (al-Bīrūnī 1992, 212–14, 220, 234; Brentjes 2008b, 320–1). Arithmetic was not least required for the complicated Islamic inheritance law and for taxation and so considered essential for secretaries (Gutas 1998, 113; Rebstock 1992, 59–66, 81–5; Chapter I.7). Closely connected to the mathematical sciences were engineering and mechanics, with hydraulics especially needed for irrigation (Hill 1990, 260–73; Chapter I.9). For the development and functioning of the administration and infrastructure, it was therefore in the best interest of the ruler to promote these sciences. The need for physicians and their protection and promotion does not require much explanation. Theoretical medical knowledge in the Islamicate world was mostly based on Hellenistic traditions, especially on Galenic medicine (Chapters I.4, I.10). These traditions were not only received through translations but also from the physicians who worked at the hospital of Jundīsābūr. Abbasid caliphs usually had their own physicians at court but promoted also the construction of hospitals, at least from the time of Caliph al-Muqtadir (r. 295–320/908–932) onward (Pormann and Savage-Smith 2007[CB], 96–7; Chapter I.10).

Other sciences like alchemy and mineralogy deserve to be mentioned. Rulers had an interest in developing mines, exploiting natural resources and finding a way to fabricate gold (Chapter II.9), and there are quite a few other examples of imperial demand. Of course, many of the reasons given here for promoting science apply also to other periods in Islamic history. The specific challenge for the Abbasids lay in the fact that the knowledge they needed was already available to a large extent in the regions they ruled but not in the required language. So it was in their best interest to translate and develop it. In this regard, their situation was different from that of later Muslim rulers, who could all build on the rich Arabic scientific heritage of Abbasid times. This situation implicated that physicians and scholars of the mathematical sciences could work and earn a living as translators. Some of these activities might have been related to the *bayt al-ḥikma*, an institution of uncertain function mentioned in the context of translation (Gutas 1998, 53–60; Chapters I.3–I.4, III.1).

Scholars engaged in translation and research worked in manifold conditions. As the biographical data show, a clear-cut distinction between astrologers, astronomers, mathematicians, engineers, physicians and so on is often impossible, since numerous savants were experts in several fields of knowledge. Many of them worked for the caliphs; it is, however, not so easy to understand their exact working conditions, since most information is available only in later texts like the *Sources of Information on the Classes of Physicians* (*ʿUyūn al-anbāʾ fī ṭabaqāt al-aṭibbāʾ*) by Ibn Abī Uṣaybiʿa (d. 668/1270) or the *History of Physicians* (*Taʾrīkh al-ḥukamāʾ*) by Ibn al-Qifṭī (568–646/1172–1248), which might be influenced by the conditions of their authors' times. The verb mostly used in the context of physicians and astrologers and their service for rulers is *khadama*, to serve (e.g., Brentjes 2008a, 407; for a discussion of terminology, see Brentjes 2008b). To designate an exclusive relationship between patron and client, we also find "he was dedicated to him" (*kāna munqatiʿan ilayhi*; Ibn an-Nadīm 1415/1994CB], 335, 339) and "he was set aside for him" (*kāna khaṣīṣan bihī*; Ibn an-Nadīm 1415/1994, 345). Such an exclusive relationship might have implied that the client was not supposed to attend to any other patron.

At the caliphal court, several physicians and astrologers were employed at the same time, with one of them being at the head of the team. The physicians at court were supposed to monitor the caliph's health, including his diet, while the astrologers gave advice based on astral indications (al-Ṭabarī 1879–80, 317). At the *majlis*, the entertainment gatherings organized by Muslim rulers, they could engage in scientific debates and contests (Ibn al-Qifṭī 1903[CB], 443). The close relationship between physicians and patrons is very well described in the biographical data about Jibrīl ibn Bukhtīshūʿ (d. 212/827). He is said to have been the first to meet Caliph Hārun al-Rashīd (r. 170–193/786–809) in the morning; to have asked him about his whereabouts at night, his food, his drinking habits and so on; and to have generally been a very trusted person (al-Ṭabarī 1879–1880, 735–7). Evidently, he used to come to the *majlis* of the caliph (al-Ṭabarī 1879–1880, 667). A similar bond of trust is also reported for the astrologer-astronomer Thābit ibn Qurra, who is said to have been on such close terms with Caliph al-Muʿtaḍid (r. 279–289/892–902) that he was allowed to sit together with him alone, without the vizier or anybody else of the caliph's trusted people (*khāṣṣa*; Ibn al-Qifṭī 1903, 115–16). A very special relationship is recorded for al-Ma'mūn and the Banū Mūsā (3rd/9th century), who had allegedly been educated at the caliph's instance. They later became competent scholars not only in the field of astronomy but also in the fields of medicine and mechanics and promoted other scholars themselves (Ibn al-Qifṭī 1903, 441–2; Gutas 1998, 133–4; Sezgin 1979, 129–30).

Remunerations mentioned for physicians and astronomers included not only single payments of sometimes very high amounts of money for specific treatments or prognostications but also the conferment of administrative posts and land grants. While they could thus gain high esteem and become very rich, their position was precarious, since incorrect prognostications or inefficient treatment could cause severe punishment (al-Ṭabarī 1879–80, 317, 737; Ibn al-Nadīm 1415/1994, 337). Furthermore, all patronage relationships were personal. The loss of one's patron consequently forced a client to find a new one. Service to the wrong patron was therefore a risk when the latter died or lost his power, and the successor did not approve his predecessor's clients (Ibn Abī Uṣaybiʿa 1965[CB], 128–9, 2020[CB], 8.3.3.–8.3.7.). Since there was not yet any institutionalized form of learning, as later in the *madrasa*, patronage relationships were the most common way of life for scholars (Chapter III.2).

II.3.2 The Fatimids

The model institution of learning in the Ismāʿīli context was the House of Knowledge (*dār al-ʿilm*) or House of Wisdom (*dār al-ḥikma*), founded by the Fatimid Caliph al-Ḥākim (r. 386–411/996–1021) in 395/1005. This institute specialized in the teaching of religious and nonreligious sciences, imitating probably a similar Buyid institution in al-Karkh, a suburb of Baghdad. The purpose of the *dār al-ʿilm* was the promotion of scholarship. This goal was, on the one hand, pursued by employing scholars of different branches of science, like Qur'ān readers and jurists (*fuqahā'*), experts on tradition, grammarians, physicians and scholars of the mathematical sciences. On the other hand, the *dār al-ʿilm* held a huge and well-equipped library. The books were paid from the palace treasury and were meant to be read by anybody who wanted to come and study or copy them. Religious purposes were apparently not at stake, since Sunnis and Shīʿīs were employed alike, with no preference given to adherents of the Ismāʿīliyya. From 399/1009, the payment of the *dār al-ʿilm* was institutionalized by an endowment (*waqf*), freeing it from the direct control of the caliph. Since the details of the endowment have been preserved, we can understand which expenditures it covered, giving us unique insight into the running of such an institution (Halm 1997, 71–8; Walker 2009, 151–3). Renowned scholars could henceforth be employed at the palace or the *dār al-ʿilm*: Ibn Yūnus (d. 399/1009), the author of the famous *Astronomical Handbook for al-Ḥākim* (*Zīj al-ḥākimī*), worked there, while the position of court astrologer was taken by ʿAlī al-Ṭabarānī (d. after 411/1021; Walker 2017, 275–7; Halm 1997, 76). Nevertheless, scholars still had to fear the caliph's wrath: al-Ḥākim had at least two scholars of the *dār al-ʿilm* executed for unknown reasons, and two astrologers killed when the caliph turned against astrology for a while (Walker 2009, 152–3, 155).

The *dār al-ʿilm* certainly was a special case, since it was only in operation during the time of al-Ḥākim. The Fatimid court did generally promote natural philosophy, because the study of the natural order was perceived to be related to religion. Both were considered to be in mutual correspondence, with the former mirroring the latter. Due to a lack of biographical data, however, we know much less about the conditions of employment and funding of scholars outside the *dār al-ʿilm* and during other periods of Fatimid rule (Walker 2017, 273–4). Most of the information available concerns practicing physicians, since there was an evident need for them at court (Walker 2009, 155–6). In the case of Yūsuf ibn Ḥasadāy (d. *c.* 530/1136), who spent a large part of his time writing medical treatises, a decree (*manshūr*) indicating his duties and rights has been preserved (Walker 2017, 282–3). But even for such eminent scholars as the scholar of the mathematical sciences, physician and philosopher Ibn al-Haytham (354–*c.* 430/965–*c.* 1040), only questionable and rather anecdotal evidence is available (Halm 1997, 76; Walker 2017, 279–80).

If we inquire about specific Ismāʿīlī aspects in the promotion of science and the interests of Ismāʿīlī rulers and missionaries (*dāʿīs*), besides anything related to the propagation of Ismāʿīlī doctrine, astrology was at the core of their attention. Similar to their Abbasid predecessors, they used mundane astrology and conjunctional astrology for political and ideological ends. However, while the Abbasids needed mundane astrology to argue for an enduring predominance of Islam and Arab rule, the Ismāʿīlīs' expectation of the Mahdi was supported by astrological calculations. His coming was promised by the Ismāʿīlī *dāʿīs* and was inter alia determined on the basis of conjunctional astrology. For this reason, knowledge of the stars played an important role in early Ismāʿīlī propaganda (*daʿwa*). This is true too for the two other kinds of political organization in addition to the Fatimids that were underpinned by Ismāʿīlī doctrines: the Qarmaṭians (r. 281–end of 5th century/894–end of 11th century) and the Nizārites (r. 483–654/1090–1256). The Qarmaṭians established their rule in the east of the Arabian Peninsula and Bahrain and expected the imminent return of the Mahdi. Their attack on Mecca in 318/930 was probably inspired by a Saturn–Jupiter conjunction, which was interpreted as a sign of major political change: the coming of the Mahdi. Conjunction astrology was again used to explain the failure of this fatal prognostication and to postpone the date of the Mahdi's arrival (Madelung 1959, 77–81; Orthmann 2006, esp. 138–9). The Nizārites, who split off from the Fatimids in the 5th/11th century and had their center in northwestern Iran, promoted astrology and astronomy for similar reasons. One of the very few texts that survived the destruction of their main fortress Alamut by the Mongols in 654/1256 was the *Rules for the Astrologers* (*Dustūr al-munajjimīn*), a book that combines astronomy, astrology and historiography to explain the course of history (Figure II.3.1). In all likelihood, it was composed after the turn of the year 500/1106 and was meant to explain failed eschatological expectations associated with that date. Being mainly a compilation, the books referred to in the *Rules for the Astrologers* indicate the amount of astronomical and astrological literature stored in the castle's library (Karimi Zanjani Asl *et al.*, 2017, 81–2, 105). The continuing promotion of astral knowledge by the Nizārites is best testified to by the famous scholar Naṣīr al-Dīn Ṭūsī (597–672/1201–1274), who lived and worked from 633/1236 to 654/1256 in Alamut (Lane 2018; Daftary 2000, 59–63, 67; Dabashi 1996, 231–9) and who wrote there among other books his *Ship of (Astrological) Judgments* (*Safīnat al-aḥkām*), a manual on astrological practices (Madelung 2017).

Activities in the fields of astrology and astronomy are recorded for the Fatimid court as well. An early astrological text written at the Fatimid rulers' behest is the *Book of the Intervals of Time and the Conjunctions of the Tens* (*Kitāb al-Fatarāt wa-l-qirānāt al-ʿashara*), a treatise that testifies to successive adaptations of the date of salvation and the coming of the Mahdi under the reign of different Fatimid rulers (Halm 1975; Orthmann 2012). While astrology was briefly forbidden during the time of the Fatimid ruler al-Ḥākim (Walker 2017, 277–9), astronomy kept its importance;

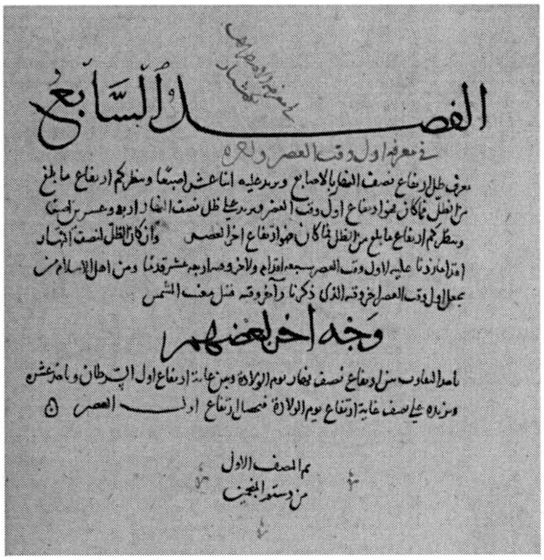

Figure II.3.1 Page from the *Rules of the Astrologers* (*Dustūr al-munajjimīn*) stating that the first half of the work has been completed; MS Paris, BnF, arabe 5968, fol. 188b.

Source: © BnF, Paris

one of the most significant astronomical handbooks of the Islamic world, the *Zīj al-ḥākimi*, was prepared for that caliph (Walker 2017, 275–7; Halm 1997, 76; Chapter III.8). Attempts to construct a large observatory failed, however (Sayılı 1960, 167–75; Halm 1997, 87–90).

II.3.3 The Mughal Empire

On the last day of his life, the Mughal emperor Humāyūn (r. 937–947, 962–963/1530–1540, 1555–1556) mounted the roof of his library to observe the rising of Venus. He called for his mathematicians and astronomers and intended to hold a *majlis* at that auspicious moment. When he descended from the roof, however, he slipped, fell on his head and died shortly afterward (Abū l-Fażl 2016, 498–9). These circumstances are indicative of the emperor's deep interest in both the occult and the rational sciences. Humāyūn's fondness for the mathematical sciences was so characteristic of him that on one of the dynastic paintings of the Mughal emperors, he is even represented with a pair of compasses in his hands (Goswamy and Fischer 1987, 91). Humāyūn had celestial and terrestrial globes prepared for him and was interested in all kinds of observational tools. He studied the nature of the elements; had a collection of old books, which he carried around with him (Firishta 1909,70–1; Abū l-Fażl 2015, 430–1); and liked to be in the company of educated people. Mawlānā Nūr al-Dīn Muḥammad Tarkhān (10th/16th century), a specialist in the mathematical sciences, was among his courtiers (Abū l-Fażl 2016, 100–1). He ordered the production of an encyclopedic work, the *Gemstones of the Sciences for Humāyūn* (*Jawāhir al-ʿulūm-i humāyūnī*), and wrote at least one short treatise on geometry. Toward the end of his life, he was eager to meet the Ottoman admiral Seydī ʿAlī Reʾīs (904–970/1498–1562), who was informed about the New World and the Strait of Magellan (Anooshahr 2017, 313–14). His interest in the rational and mathematical sciences went along with a strong regard for the occult sciences. He was himself very well versed in astrology and had several astrologers working for him. They were present not only at the birth of his son and successor Akbar (r. 963–1014/1556–1605; Figure II.3.2) and cast his horoscope (Orthmann 2005), but also at the foundation of Dīn-Panāh, the city built by him (Khwāndamīr

Figure II.3.2 The miniature shows in the upper register Ḥamīda Bānū Bīgum (d. 1013/1604) just after giving birth to Akbar and in the middle register Humāyūn in conversation with his court astrologers; it is an illustrates a copy of Abū l-Fażl ibn Mubārak's (958–1011/1551–1602), *Akbarnāma*, MS London, BL, Or. 12988, fol. 22a; the painters were Sānwala (10th-11th/16th-17th centuries) and Narsingh (d. after 1013/1604).

Source: © British Library Board, London

1940, 85). One of his astrologers was a Hindu named Mawlānā Chānd (10th/16th century); he accompanied the emperor even when he was in great trouble during the wars against his brothers and Shīr Khān (r. 947–952/1540–1545; Abū l-Fażl 2015, 552–3). Besides astrology, lettrism and all kinds of omens attracted his attention (Chapter II.9). He not only promoted scholars by employing them but intended his city, Dīn-Panāh, also to be a safe haven for intelligent and learned people.

This interest in the rational, mathematical and occult sciences continued in the time of Akbar and was broadened by the translation activities supported by the court (Figure II.3.3). Among the prominent translations in the field of mathematics was the *Līlāvatī* (allegedly the name of the author's daughter) of Bhāskarācārya (12th century), translated into Persian in 995/1587 by Fayżī (d. 1004/1595). One of the outstanding scholars at his court was Fatḥ Allāh Shīrāzī (d. 997/1589), a broadly educated expert in mathematics, astronomy, astrology, statecraft, accounting, finance and further fields of knowledge, who had come from Iran to the Mughal Empire via the ʿĀdil-Shāhī Sultanate of Bijapur (r. *c.* 895–1097/1490–1686). Fatḥ Allāh Shīrāzī was responsible for the introduction of the "divine" (*ilāhī*) calendar, an era based on the regnal years of Akbar, and served also as a tutor for some notables' children, among other duties. One of his main achievements was his participation in the reform of the *madrasa* curriculum, supporting medicine and the mathematical sciences instead of the religious sciences (Anooshahr 2014). Taking into consideration the emperor's emphasis on logical reasoning and the use of the intellect (*ʿaql*; Grobbel 2001, 70–1), we understand that one of the cornerstones of his imperial ideology was the promotion of the rational and mathematical sciences. Accordingly, scholars of these disciplines were very well received at court and patronized, while the influence of traditional religious learning and especially of those *ʿulamāʾ* who were opposed to Akbar's openness toward other religions was curbed.

Before looking in more detail at their situation, let us first briefly consider the situation in the time of Akbar's son and successor Jahāngīr (r. 1014–1037/1605–1627). This emperor is a unique case insofar as he was himself involved in the study of nature, especially animal life, and is famous for his detailed descriptions of birds. Jahāngīr furthermore undertook quite a few experiments, such as interbreeding goats or testing the medical effects of camel milk and saffron.

Figure II.3.3 Ceiling in the form of a horoscope in Akbar's palace Dawlatkhāna-yi khāṣṣ in Fātiḥpūr Sīkrī

Source: © Photographer Eva Orthmann, Göttingen

He was however working alone, without a team of scholars assisting him. These experiments and descriptions therefore testify to his inquisitive mind and can be interpreted as an element of Jahāngīr's self-representation as a second Solomon. The paradigm of just rule in the Qur'ān, Solomon, was associated with the idea of pacifying the animal kingdom and bringing peace to men and animals. Learning and the observation of nature were a part of this Solomonic approach and a means to mastery of the animated world, a notion underlined by the many portraits of Jahāngīr showing him with a globe in his hands (Koch 2009). For information about the situation of scholars, however, we must turn to other sources.

One of the best and most telling books on the working of patronage in Mughal times was written for ʿAbd al-Raḥīm Khān-i Khānān (d. 1036/1627), an outstanding notable in the time of Akbar and Jahāngīr. The last volume of this source, the *Illustrious Acts of Raḥīm* (*Ma'āthir-i raḥīmī*), is dedicated to the biographies of people patronized by ʿAbd al-Raḥīm. While a large part of it deals with men of letters and poets, quite a few biographies describe the lives of mathematicians, astronomers, physicians and other scholars. Furthermore, even the biographies of the poets contain relevant information on the general working of patronage in Mughal times.

The most common term used for patronage relationships was – as earlier – *khidma. Khidma* relationships were of a permanent nature, with the service period ending with the death of the protégé or his request to leave the court. Without permission by the patron, an employed person could not relinquish service. Employed persons received different types of payments, the two most typical being *jāgīr* and *ʿalūfa*. A *jāgīr* was a land grant, and *ʿalūfa* a regular payment in cash. Diverse other stipends and payments are also mentioned in the sources. Besides *khidma* relationships, patronage encompassed also remunerations and benefits for single services (Orthmann 2017). The boundaries between different types of services – administration, literature and culture, scholarship – were not fixed; both poets and scholars could obtain official posts and were remunerated in similar ways. In the *Illustrious Acts of Raḥīm*, six physicians are mentioned in total, two of them having come to the court for training and education (Nihāwandī 1931, 43–4, 65–6). Physicians are described as trusted and highly esteemed persons, who were well paid, with one of them spending more than 25 years in the service of ʿAbd al-Raḥīm (Nihāwandī 1931, 44–8). As for the other scholars, it is often not so easy to attribute to them a specific qualification, since many were polymaths, trained in both rational (*maʿqūl*) and traditional (*manqūl*) sciences, some of them with a *madrasa* career. Quite a few were widely traveled and of Iranian origin. ʿAbd al-Raḥīm was obviously interested in mathematics and astronomy and even had an astronomical handbook, the *Astronomical Tables for ʿAbd al-Raḥīm* (*Zīj-i raḥīmī*), prepared for his own use (Munzawī 1378, no. 2954). Mawlānā Farīd al-Dīn (d. 1039/1630), the author of this *zīj*, had a high standing at court. He is also described as an expert in talismans and magic and furthermore composed chronograms for his master (Nihāwandī 1931, 9–17). The other astronomer, who is specifically mentioned, designed an astrolabe for ʿAbd al-Raḥīm but was also proficient in astrology and composed a treatise on prognostications (Nihāwandī 1931, 62). The *Illustrious Acts of Raḥīm* inform us about other functions scholars fulfilled at court. They were employed as teachers for ʿAbd al-Raḥīm's sons and were responsible for financial concerns; one of them translated a book from Arabic into Persian and another worked in the library (Nihāwandī 1931, 32–5, 57–60). But more than anything, most likely, ʿAbd al-Raḥīm needed them for intellectual conversations, to spend time in an inspiring atmosphere, to satisfy his thirst for learning and to be surrounded by high-standing scholars in his *majlis*. His court certainly mirrored the needs and motives of the emperors' patronage policies. As delineated earlier, Humāyūn, Akbar and Jahāngīr were all depicted as eager to learn and of inquisitive mind. We also know about their need for physicians and their interest in promoting medical care (Sigaléa 1995, 425, 433–54).

II.3.4 Conclusion

The scope of this chapter has been limited to three dynasties, covering different epochs and regions of the Islamicate world. As all three cases have shown, it was attractive for rulers and other persons in prominent positions to have scholars working for them, be it to have a competent expert at hand or to adorn their court with the prestige of outstanding scholars; contests between their "own" experts and outsiders were counted among the entertainments in the *majlis* of rulers. In addition to this general demand, ideology and propaganda played a role in the specific values attributed to science and learning, whether to imitate the example of earlier dynasties, to support religious expectations or to counter traditional learning with a rational and supra-religious worldview.

Note

1 Consolidated bibliography.

Bibliography

Sources

ʿAllāmī, Abū l-Fażl. ed. and tr. Thackston, W. 2015. *The History of Akbar.* Vol. 1. Cambridge, MA: Harvard University Press.

ʿAllāmī, Abū l-Fażl. ed. and tr. Thackston, W. 2016. *The History of Akbar.* Vol. 2. Cambridge, MA: Harvard University Press.

al-Bīrūnī, Abū l-Rayḥān. ed. Bulgakow, P. and Aḥmad, I. I. 1962. *Kitāb Taḥdīd nihāyat al-amākin li-taṣḥīḥ masāfat al-masākin* [Book of the Determination of the Places for the Corrections of the Distances Between the Inhabited Localities]. Cairo: Maṭbaʿat lajnat al-taʾlīf wa l-tarjama wa-l-nashr (Reprint: 1992. Frankfurt am Main: Institute for the History of Arabic-Islamic Science at the Johann Wolfgang Goethe University).

Firishta, Muḥammad Qāsim. tr. Briggs, J. 1909. *Tarikh-i-Firistah or History of the Rise of the Mahomedan Power in India Till the Year A. D. 1612.* Vol. 2. Calcutta: R. Cambray and Co.

Khwāndamīr, Ghiyāth al-Dīn Muḥammad. ed. Ḥusayn, M. H. 1940. *Qānūn-i humāyūnī, nīz musammā bi Humāyūn-nāma* [The Canon of Humāyūn, also called Book of Humāyūn]. Calcutta: Royal Asiatic Society.

Nihāwandī, ʿAbd al-Bāqī. ed. Ḥusayn, M. H. 1931. *Maʾāthīr-i raḥīmī* [Illustrious Acts of Raḥīm]. Vol. 3. Calcutta: Baptist Mission Press.

Niẓāmī ʿArūżī, Aḥmad ibn ʿUmar. ed. Mīrzā Muḥammad. 1910. *Chahār maqāla* [Four Chapters]. Leiden: Brill.

al-Ṭabarī, Abū Jaʿfar Muḥammad ibn Jarīr. ed. De Goeje, M. J. 1879–1880. *Tārīkh al-rusūl wa al- mulūk* [History of the Prophets and Kings] (Tertia series). Leiden: Brill.

Research literature

Anooshahr, A. 2014. "Shirazi Scholars and the Political Culture of the Sixteenth-century Indo-Persian World," *The Indian Economic and Social History Review* 51.3: 331–52.

Anooshahr, A. 2017. "Science at the Court of the Cosmocrat: Mughal India, 1531–56," *The Indian Economic and Social History Review* 54: 295–316.

Brentjes, S. 2008a. "Courtly Patronage of the Ancient Sciences in Post-classical Islamic Societies," *Al-Qanṭara* 29: 403–36.

Brentjes, S. 2008b. "Patronage of the Mathematical Sciences in Islamic Societies," in Robson, E. and Stedall, J., eds. *The Oxford Handbook of the History of Mathematics.* Oxford: Oxford University Press, 301–27.

Burnett, C., Yamamoto, K. and Yano, M. 1997. "Al-Kindī on Finding Buried Treasure," *Arabic Sciences and Philosophy* 7: 57–90.

Dabashi, H. 1996. "The philosopher/vizier: Khwāja Naṣīr al-Dīn al-Ṭūsī and the Isma'ilis," in Daftary, F., ed. *Medieval Isma'ili History and Thought*. Cambridge: Cambridge University Press, 231–45.

Daftary, F. 2000. "Naṣīr ad-Dīn Ṭūsī and the Isma'ilis of the Alamut Period," in Pourjavady, N. and Vesel, Ž., eds. *Naṣīr ad-Dīn Ṭūsī. Philosophe et savant du XIIIe siècle*. Tehran: Presses universitaires d'Iran, 59–67.

Goswamy, B. N. and Fischer, E. 1987. *Wunder einer goldenen Zeit. Malerei am Hof der Moghul-Kaiser. Indische Kunst des 16. und 17. Jahrhunderts aus Schweizer Sammlungen*. Zürich: Museum Rietberg.

Grobbel, G. 2001. *Der Dichter Faiḍī und die Religion Akbars*. Berlin: Klaus Schwarz Verlag.

Halm, H. 1975. "Zur Datierung des ismā'īlitischen 'Buches der Zwischenzeiten und der zehn Konjunktionen' (kitāb al-fatarāt wa-l-qirānāt al-'ashara), HS Tübingen Ma VI 297," *Die Welt des Orients* 8: 91–107.

Halm, H. 1997. *The Fāṭimids and Their Traditions of Learning*. London and New York: I. B. Tauris.

Halm, H. 2014. *Kalifen und Assassinen. Ägypten und der Vordere Orient zur Zeit der ersten Kreuzzüge 1074–1171*. München: C. H. Beck.

Hill, D. R. 1990. "Mathematics and Applied Sciences," in Young, M. J. L., Latham, J. D. and Serjeant, R. B., eds. *The Cambridge History of Arabic Literature. Religion, Learning and Science in the 'Abbasid Period*. Cambridge: Cambridge University Press, 248–73.

Karimi Zanjani Asl *et* al.: Karimi Zanjani Asl, M., Orthmann, E. and Schmidl, P. 2017. "The Sources and the Composition of the Dustūr al-munajjimīn," in Orthmann and Schmidl, eds. 2017, 35–114.

Koch, E. 2009. "Jahangir as Francis Bacon's Ideal of the King as an Observer and Investigator of Nature," *Journal of the Royal Asiatic Society, Series 3* 19: 293–338.

Lane, G. E. 2018. "Ṭusi, Nāṣir-al-Dīn," in *eIr*, online edition, www.iranicaonline.org/articles/tusi-nasir-al-din-bio (Accessed 19 April 2018).

Madelung, W. 1959. "Fāṭimiden und Bahrainqarmaten," *Der Islam. Zeitschrift für Geschichte und Kultur des Islamischen Orients* 34: 34–88.

Madelung, W. 2017. "Ismaili Astrology in Naṣīr ad-Dīn Ṭūsī's Safīnat al-Aḥkām," in Orthmann and Schmidl, eds., 263–71.

Munzavī, Aḥmad. 2003². *Fihristvāra-yi kitābhā-yi fārsī* [Catalog of Persian Books]. 11 vols. Tehran: Center for the Great Islamic Encyclopedia, Vol. 4.

Orthmann, E. 2005. "Circular Motions: Private Pleasure and Public Prognostication in the Nativities of the Mughal Emperor Akbar," in Oestmann *et al.* [CB] , 101–14.

Orthmann, E. 2006. "Astrologie und Propaganda. Iranische Weltzyklusmodelle im Dienst der Fāṭimiden," *Die Welt des Orients* 36: 131–42.

Orthmann, E. 2012. "*Ẓāhir* und *bāṭin* in der Astrologie: Das *kitāb al-Fatarāt wa l-qirānāt al-'ašara*," in Klemm, V. and Biesterfeld, H., eds. *Differenz und Dynamik im Islam*. Würzburg: Ergon-Verlag, 337–58.

Orthmann, E. 2017. "'The Donor is Somebody Else.' Abdur Rahim Khan-i-Khanan as Patron and Benefactor," in Hossain, S. and Ray, D., eds. *Celebrating Rahim*. Ahmedabad: Interglobe Foundation, Aga Khan Trust for Culture, 72–86.

Orthmann, E. and Schmidl, P., eds. 2017. *Science in the City of Fortune. The Dustūr al-munajjimīn and Its World*. Berlin: EB Verlag.

Rebstock, U. 1992. *Rechnen im islamischen Orient. Die literarischen Spuren der praktischen Rechenkunst*. Darmstadt: Wissenschaftliche Buchgesellschaft.

Sezgin, F. 1979. *Geschichte des Arabischen Schrifttums*. Band VII: *Astrologie – Meteorologie und Verwandtes Bis ca. 43 H.* Leiden: Brill.

Sigaléa, R. 1995. *La Médicine traditionelle de l'Inde. Les doctrines prévédique, védique, âyurvédiques, yogique et tantrique. Les empereurs moghols, leurs maladies et leurs médicins*. Geneva: Éditions Olizane.

Walker, P. 2009. *Caliph of Cairo: Al-Hakim bi-Amr Allah, 996–1021*. Cairo: American University in Cairo Press.

Walker, P. 2017. "Science in the Service of the Fatimids and Their Ismaili *Da'wa*," in Orthmann and Schmidl, eds. 2017, 273–91.

II.4
THE PRACTICE OF PHARMACY IN LATER MEDIEVAL EGYPT

Leigh Chipman

How can we know how pharmacy and medicine were practiced in the past? Most of our sources are theoretical in nature and present the physician's point of view. They are books on medicine written by physicians for physicians. Even genres like casebooks, for example, Abū Bakr al-Rāzī's (d. 313/925 or 323/935) famous 33 cases (Meyerhof 1935; Álvarez-Millan 1999, 2000) and regimens such as Ibn Riḍwān's (d. 453/1061) *On the Prevention of Bodily Ills* (*Kitāb Dafʿ madārr al-abdān bi-arḍ Miṣr*; Dols 1984) can be suspected of being literary products, aiming more at prescription and increasing the author's renown, rather than description. Almost all our knowledge of the practice of pharmacy refers to the capital city of Cairo and nearby Fustat, and the different sources mean that we know about certain aspects of practice at particular times and not at others.

This chapter covers what we know (or think we know) about how pharmacy (and, to a certain extent, also medicine) was practiced in Egypt in the medieval period, under the Fatimid, Ayyubid and Mamluk dynasties (6th–10th/12th–16th centuries). It will be based mainly on the evidence for practice derived from a pharmacopoeia explicitly written for practitioners in the marketplace: *How to Run a Pharmacy* (*Minhāj al-dukkān*), a manual for apothecaries composed in 13th-century Cairo by the otherwise unknown al-Kūhīn al-ʿAṭṭār al-Isrāʾīlī, from his name a Jewish druggist (for a detailed investigation of this work, see Chipman 2010). That evidence is combined with material found in the Cairo Genizah, a storeroom at the Ben Ezra Synagogue in Fustat, which was a repository for documents that could not be destroyed because of their religious content. Medical prescriptions, and the chapters on pharmacy-related occupations from the manuals for the market inspector (*muḥtasib*), are considered here in order to discuss the practice of pharmacy in the marketplace. So far, archeological findings have been informative either about other periods, such as the early Islamic period in Fustat (Awad 2002; Hamarneh and Awad 2002a, 2002b), or about other medical specializations, such as surgery (Eger 2018). And while the *muḥtasib's* manuals portray a Muslim point of view, the Genizah material derives from a specific Jewish community. It is my assumption – shared by the vast majority of scholars of the Genizah – that the findings regarding the minority culture of the Jews of the Ben Ezra Synagogue in Fustat may be extrapolated to the larger society.

II.4.1 Pharmacy in the marketplace

The marketplace was the main location in which pharmaceutical practice took place in late medieval Cairo, and the importance of the pharmacists is indicated by the existence of a *sūq al-ʿaṭṭārīn* or *murabbaʿat al-ʿaṭṭārīn* (market or quarter of the perfumers/druggists; see Cambridge

DOI: 10.4324/9781315170718-24

University Library Or.1080 J23, Or.1080 J38). This is where the *dukkān* (literally, shop) referenced by al-Kūhīn al-ʿAṭṭār would have been located. *How to Run a Pharmacy* comprises 25 chapters, of which the majority are concerned with compound medicines, organized according to their formulation: a chapter on syrups, one on pills, one on ointments, and so on. At the end of the book, a number of chapters deal with practical aspects of being a pharmacist. These begin with chapters on substitute drugs, drug synonyms, and weights and measures and continue with chapters on the gathering, storage and preparation of simple drugs or the raw material to be used for prescriptions. Lists of substitute drugs and plant synonyms help the practitioner decide what materials to use. Chapters on the gathering and storage of local products, and basic rules for the preparation of certain forms of medicine, mark stages in the transition from simple to compound medicines. The lists of substitutes and synonyms in *How to Run a Pharmacy* are both longer and more concise than comparable lists, such as those by Abū Bakr al-Rāzī in the case of substitutes and by Maimonides (d. 601/1204) in the case of synonyms. The lists are also organized in Arabic alphabetical order, rather than the *abjad* (numerical) order, a change to alphabetical arrangement that occurred across many genres at this time (Chipman 2010, 78–88). Oliver Kahl has suggested that the weights and measures used by pharmacists can be divided into three groups: specific, semi-specific and nonspecific. Specific units can be converted to the modern metrical system; semi-specific units refer to a cupful, a handful and so on or to the weight/shape of a certain nut or lentil; and nonspecific units are terms such as *part* (*juzʾ*), amount (*miqdār*) and the like (Ibn at-Tilmīḏ 2007, 34–5). However, al-Kūhīn al-ʿAṭṭār's list shows that he does not differentiate between Kahl's specific and semi-specific terms. Precise equivalents are given for terms, which in Ibn al-Tilmīdh's (d. 560/1165) *Dispensatory* are more vaguely defined. This may reflect historical development, with terminology becoming more exact over time.

Advice on the proper methods of preparing various medicaments appears principally in the second half of chapter 23, otherwise devoted to moral advice for the pharmacist (ostensibly al-Kūhīn al-ʿAṭṭār's son), indicating that al-Kūhīn al-ʿAṭṭār saw no difference between the moral and practical sides of good practice. Al-Kūhīn al-ʿAṭṭār is anxious to make sure that the reader starts by preparing the raw materials correctly and, to this end, devotes a great deal of attention to the right way to grind drugs:

> It is incumbent upon you not to grind drugs that are hard to grind, like gum tragacanth, together with quickly ground drugs like bamboo sugar, for each of them will harm the other, especially the hard will damage the soft. . . . Do not crush myrobalan too finely and pulverize cardiac drugs or those that correct the temperament, so that they reach the heart quickly. . . . If you wish to crush spikenard, cut it with scissors before crushing it and it will be pulverized more quickly. If you wish to crush moss, moisten it with a little water and it will be pulverized more quickly.
>
> *(al-Kūhīn al-ʿAṭṭār 1992, 269, 271)*

Al-Kūhīn al-ʿAṭṭār proudly reveals his knowledge of craft secrets, as in this example:

> If you peel pumpkin seeds, soak them in water for an hour, and if you want to peel quince seeds, soak them in water and remove its mucilage with a rag several times until no mucilage remains, then peel it and it will peel quickly. This is a matter tenaciously kept secret, so know it!
>
> *(al-Kūhīn al-ʿAṭṭār 1992, 271)*

He also gives advice on the use of the correct tools for specific tasks: "Cut colocynth pulp with shears . . . when you extract pomegranate juice, cut the pomegranate with an ebony knife" (al-Kūhīn al-'Aṭṭār 1992, 269–70). The last comment reminds us that magic was often intertwined with medicine (although a wooden knife may simply have crushed the pomegranate most efficiently).

While the emphasis here is on preparation, the maturing and storage of prepared medicines is no less important. Al-Kūhīn al-ʿAṭṭār gives advice on how to treat oils and conserves:

> Place oils in the sun for forty days . . . when you take oils out of the sun, filter them and throw away the sediment, then return the oil to its vessels – in this way you will prevent spoilage. Do not fill jars with rose conserve or violet conserve or anemone conserve and in general leave [containers of] anything preserved with sugar or honey less than a third empty, lest they ferment and split and be squandered . . .
>
> *(al-Kūhīn al-ʿAṭṭār 1992, 270)*

Further advice appears throughout *How to Run a Pharmacy*. This advice can be divided into the following categories: (1) basic methods of preparation of oils; (2) preparation of materials used in many other medicines (burned scorpions, 'honey' of various fruits, colocynth pulp, whey, juices, barley soup); and (3) rules for the correct preparation of syrup (Chipman 2010, 103 and references there).

The rules bring us to the official in charge of preventing fraud in the marketplace, the *muḥtasib*, and the manuals devoted to guiding him in the execution of his duty. Under the Mamluks, there were three *muḥtasibs* at any one time: that of Cairo, whose jurisdiction was over all of Lower Egypt except for Alexandria, that of Alexandria, and that of Fusṭāṭ Miṣr, whose jurisdiction was over all of Upper Egypt. The person bearing the title of *muḥtasib* was not the one actually inspecting the markets – the areas of responsibility were too large for a single person to cover adequately. Rather, assistants would do so, some of whom had a general mandate and some of whom were responsible for controlling a particular trade. These last bore the title of *ʿarīf*, indicating detailed, perhaps inside, knowledge of a specific profession, which enabled the assistant to perform effectively (Buckley 1999, 2–11).

II.4.2 *Ḥisba* manuals

Ḥisba manuals, that is, manuals instructing the market inspector (*muḥtasib*) in his task of regulating public morals, are known from many places and periods of the Islamic world and are valuable representatives of their region's economic conditions (Shatzmiller 1994, 82). I discuss here three manuals from 6th–7th/12th to 9th/15th-century Egypt: ʿAbd al-Raḥmān ibn Naṣr al-Shayzarī's (d. 589/1193) *The Utmost Authority in the Pursuit of Ḥisba* (*Nihāyat al-rutba fī ṭalab al-ḥisba*), Ibn al-Ukhuwwa's (d. 729/1329) *Milestones on the Laws of Ḥisba* (*Maʿālim al-qurba fī aḥkām al-ḥisba*), and Ibn Bassām's (7th/13th century) *The Utmost Authority in the Pursuit of Ḥisba* (*Nihāyat al-rutba fī ṭalab al-ḥisba*).

An insistence on cleanliness and freshness of ingredients can be found throughout the *ḥisba* manuals. The adulteration of foodstuffs or the passing-off of shoddy goods as being of high quality were of concern to the *muḥtasib* in his dealings with other trades, but in no other chapters is such great detail supplied as in the chapters on the pharmacist, which are largely devoted to listing the commonest frauds and how to detect them – a subject which was of concern within the profession also. Ibn Bassām states very clearly at the beginning of these chapters that "a reliable *ʿarīf*" should be appointed, as there are very many kinds of medical plants, perfumes and syrups (Ibn Bassām 1968, 85, 96), while al-Shayzarī begins his chapters on pharmacists with the following words:

> In this chapter and others we have only mentioned the well-known and popular frauds, and said nothing about those which are not well known. . . . I will only mention those whose adulterations are well known and which are often prepared, and will not refer to those whose adulteration and preparation is a secret and which few of the pharmacists occupy themselves with.
>
> *(al-Shayzarī 1999, 69–70)*

Clearly, the authors of these manuals were concerned that there were more ways of adulterating drugs or ingredients and/or otherwise defrauding the customer than the layperson could be expected to master. The simples most commonly falsified seem to have been rhubarb, manna, saffron, musk, ambergris and camphor, among the commonest ingredients in the Arabic pharmacopoeia. In light of the commonplace that there are more ways of adulterating and falsifying *materia medica* than can be detected by the *muḥtasib* or his representative, it is interesting that of the 81 materials mentioned by al-Kūhīn al-ʿAṭṭār in chapter 25, 60 do not appear in the *ḥisba* manuals examined, while 19 other materials are mentioned in these manuals that are not mentioned in *How to Run a Pharmacy*. Furthermore, a mere 15 items are mentioned by all three *ḥisba* manuals (of which four do not appear in *How to Run a Pharmacy*). Not only does this show the accuracy of the previously mentioned saying, but it is also a concrete example of the wide range of *materia medica* at the disposal of the pharmacist in the medieval Islamic world.

How did the responsible pharmacist prove the genuineness and purity of a given raw material? The most common method was the minute observation of external color (usually when an inferior grade was suspected, for example, the substitution of Chinese for Tibetan musk), followed by tasting the substance and, finally, by examining its heaviness or lightness relative to its size. Smell and texture were also popular tests, as were observation of the color of a broken edge and dissolving the substance in water. Of course, a given drug could be tested in more than one way, and there was no substitute for experience. Al-Kūhīn al-ʿAṭṭār describes what a good specimen of a drug looks like and where it should come from, and whether he has personal knowledge of this substance (Chipman 2010, 97–9).

In contrast, the *ḥisba* manuals are far more concerned with adulteration than with good drugs: there are no descriptions of how to identify good drugs, only how to identify bad ones. In several cases, there is no mention of how to test whether a substance has been adulterated, only descriptions of how it may be adulterated (al-Shayzarī 1999, 68). For example, Ibn Bassām states: "Some adulterate tamarinds with the flesh of plums" (Ibn Bassām 1968, 87–8). It is important to point out that the various *ḥisba* manuals tended to recycle content from each other with slight changes. Thus, what may not be found in one may be found in another. Ibn al-Ukhuwwa, for example, describes another way that tamarind may be adulterated and knows how to test it: "Some adulterate tamarinds with wax and salt and vinegar and say that this is 'country paste.' Its adulteration is exposed when it becomes putrid, as 'country paste' does not putrefy" (Ibn al-Ukhuwwa 1972, 199–200). This is more characteristic of the manuals' approach: each drug is followed by a list of common methods of adulteration and then a method of testing.

As noted earlier, the range of simple drugs mentioned in the *ḥisba* manuals is far smaller than that appearing in *How to Run a Pharmacy*. Al-Shayzarī refers to 22 materials, Ibn al-Ukhuwwa to 31 and Ibn al-Bassām to 35. There is a great deal of overlap, particularly between al-Shayzarī and Ibn al-Bassām (who clearly wished to emulate and outdo his predecessor, even giving his book the same title). Regarding methods of testing, a similar phenomenon can be seen in the range of 'special' tests that are specific to particular drugs: *How to Run a Pharmacy* has 14 tests for 18 different materials (22% of all drugs have a special test), while the *ḥisba* manuals range from two to five special tests. However, when studying the standard tests, we can find similarities: Color is the test most commonly used in the *ḥisba* manuals, followed by taste and then by smell. The top three tests in *How to Run a Pharmacy* are the same. The *muḥtasib* is more willing than is the pharmacist to test the quality of drugs by dissolving in water or by burning, both essentially destructive methods – perhaps because he is not going to lose any money! The relative weight of a substance and the appearance of a broken edge are far more popular criteria for genuineness among pharmacists than among the *muḥtasib's* representatives. It is possible that these methods required a particularly high level of experience and practical knowledge of the raw materials involved to be accurate (Chipman 2010, 100–1).

The *muḥtasib's* main business was with simple drugs, but he was also concerned with compounds, particularly syrups, which are repeatedly shown to be the most common form of medicament. The chapter on syrups is by far the longest in *How to Run a Pharmacy* (al-Kūhīn al-ʿAṭṭār 1992, 17–56). Ibn al-Ukhuwwa gives a list of the most common syrups, mentioning 70 by name. He differentiates between the "active ingredient," as it were, which gives the syrup its name, and the julep (*jullāb*), which forms the base of all syrups. This has no role in healing *per se*, but forms a method of sweetening a bitter draught and getting the medications to the various organs quickly. He also gives a general rule for preparing syrups: "what appears in the *Medical Rule* (*Dustūr al-ṭibb*) is incumbent upon them, and that is for every ten *raṭls* of sugar, three and one third *raṭls* of fruit juice" (Ibn al-Ukhuwwa 1972, 185). In the case of syrups based on roots, flowers or herbs, a greater or lesser proportion of sugar was to be used, in all cases dependent on physicians' opinions. This rule appears in Ibn Abī l-Bayān's (d. 634/1236) hospital formulary, *The Hospital Rule* (*al-Dustūr al-bīmāristānī*; Ibn Abī l-Bayān 1932 3, 44).

The proportions given there are four *ūqiyya*s of fruit juice to one *ratl* of sugar. As there are 12 *ūqiyya*s to a *raṭl*, this is equivalent to the numbers cited by Ibn al-Ukhuwwa.

Consumer protection also means quality control and cleanliness. For example, in addition to the *muḥtasib* examining simple drugs for fraud every week, syrups should be examined at the beginning of every month. If the *muḥtasib* finds that some ingredients have gone bad, the syrup-maker may not use them (al-Shayzarī 1999, 77). According to Ibn Bassām, the syrup-maker, before preparing new batches, has to bring his ingredients to the *ʿarīf* for approval. He also must be sure to use only clean white sugar and bring syrups to the correct consistency before storing them and keep essential oils from Iraq separate from oils from Syria (Ibn Bassām 1972, 93, 95). At the end of his chapter on *sharābiyyīn* (preparers of syrups), Ibn al-Ukhuwwa states that the shopkeeper must have an adequate supply of clean water so that he can wash his utensils daily. Also, he must take precautions to keep flies away during the day and dogs out during the night. With regard to the *ʿaṭṭār*, he states that perfumes and drugs may be sold only by someone who combines professional knowledge and experience with religious trustworthiness and fear of God (Ibn al-Ukhuwwa 1972, 197–9).

While this last recommendation might indicate that only a Muslim pharmacist was acceptable to Ibn al-Ukhuwwa in the Mamluk period (for the changing attitude to non-Muslim medical professionals at that time, see Lewicka 2013), evidence from the Genizah indicates that under the Fatimids and Ayyubids, pharmacy was a very popular profession among Jews (Goitein 1971, 261–72), and it seems likely that Jewish pharmacists had non-Jewish as well as Jewish clients (see Chapter VI.1). The prescriptions of the Genizah – almost all of which date to the 5th/11th and 6th/12th centuries – form one of our major sources of knowledge of actual practice, in that these prescriptions are, in fact, autographs from the prescribing physicians themselves. Maimonides gave personal testimony to this practice in one of his most famous letters, where he wrote,

"I converse with them and prescribe for them while lying down from sheer fatigue" (Cohen 1995, 125). Prescriptions come about as close to the actual words of medieval physicians as we are likely to get. They reflect the medical reality that actually existed, which at times corresponds with that found in books.

II.4.3 Prescriptions

Let us look at a sample prescription, T-S Ar.42.60. It reads as follows:

1. Black spleenwort nine dirhams, lichen ten dirhams . . .
2. three dirhams of Iraqi basil and three dirhams of white tragacanth
3. and three dirhams of gum arabic and two dirhams of starch
4. and one dirham of lead ceruse and one dirham of Indian salt
5. and half a dirham of mineral sal-ammoniac and half a dirham of gold scoria
6. and half a dirham of silver scoria. Knead all of it in rue water
7. and three dirhams' weight of sugar-candy; also melt
8. in the rue water Isfahan [kohl?] to the weight [of all?]
9. Beneficial, if God wills.

(Lev and Chipman 2012, 116–18)

This prescription does not state what it is for, but the high level of mineral *materia medica* (ceruse, Indian salt, sal-ammoniac, gold scoria, silver scoria) and the reference to 'Isfahan', possibly meaning Isfahan kohl, a known ingredient of eye medicines, make it seem likely that this is an eye salve. The plants mentioned here (black spleenwort, lichen, basil, rue water) are also used for various eye ailments, although this is not their primary use – all are also good for the digestive tract. Gum arabic, tragacanth and starch may be carrier materials for the other ingredients. If this is a preparation for external use, the function of the sugar candy is unclear.

Eye medicines normally come in the form of powders or salves. The fact that this prescription contains gums and starch that are kneaded in rue water indicates that it is a salve. The final weight of the recipe comes to 90 grams, perhaps indicating long-term use. There is no mention of dosage or application – possibly, these instructions were given orally by the physician or the pharmacist. The chapters on eye medicines in *How to Run a Pharmacy* (al-Kūhīn al-ʿAṭṭār 1992, 135–52), the *Hospital Rule* (Ibn Abī l-Bayān 1932–33, 53–60) and Ibn al-Tilmīdh's *Dispensatory* (Ibn at-Tilmīḏ 2007, nos. 245–79) include many recipes containing some combination of the *materia medica* appearing in this prescription, but so far, an exact equivalent has not been found in any pharmacopoeia. This indicates the practicality and originality of these prescriptions and thus their value in reflecting reality on the ground (or, rather, in the market), in contrast to books, in which recipes were copied regardless of their actual usefulness or availability of substances.

This is one of 30 prescriptions of sufficient quality to be analyzed in detail, identified so far in the Cambridge University Library. In general, all the prescriptions in the Genizah are short, focused and make use of known and common drugs such as ginger, aniseed, cumin and rhubarb. The fact that very few prescriptions are similar, and none is identical, to recipes in contemporary medical books and pharmacopeias attests, on one hand, to their originality and practical characteristics and, on the other, to the way the Jewish physicians were thinking and working. As mentioned earlier, the assumption is that this was the general attitude and Jewish physicians represent practitioners of other religions as well. While medical books, and especially pharmacopeias, served as the basis of knowledge, the recipes that they describe were generic. Nevertheless, the actual healing process, which according to the Galenic system in the Islamicate world required a long meeting with the physician so that he might become acquainted not only with the current symptoms but also with the patient's unique usual temperament and qualities, would lead to a decision regarding the correct treatment. This included in most cases a recipe for a drug, instructions for its administration and diet – what today we would call holistic treatment. Sometimes, combined medical and dietetic advice extended to an entire household (Lev *et al.* 2008). Hence, in most cases, the prescription was unique to the patient and fitted their symptoms, the stage of the disease, their normal conditions of health, the weather, the season and the availability of medicinal substances in the markets or on the shelves of the local pharmacies (Lev and Chipman 2012, 144).

Symptoms should be distinguished from diseases, since conceptions of disease entities change over time. The same physical manifestations may be interpreted as symptoms of different diseases in different periods. For example, today, a high temperature (fever) is considered a symptom of a disease, whereas in the ancient and medieval world, fever was thought of as a diseased condition, or unhealthy state, in its own right. Prescriptions can teach us about the prevailing diseases and conditions, that members of the community actually suffered from. Unlike the information derived from books, which usually was copied from classical or contemporary medical sources, the prescriptions are clear-cut primary evidence of the medicinal substances actually used, the medical conditions that afflicted the members of the community and how they were treated. Investigation into the *materia medica* mentioned in Genizah material shows clearly that what was theoretically available was much greater than what was actually used (for a detailed study of the *materia medica* used by the Genizah people, see Lev and Amar 2007). Unfortunately, in most cases neither the symptoms nor the patient's name appears on the prescription. Perhaps these were not considered relevant, since in most cases the patient was present while the physician wrote the prescription. When there are names and/or symptoms on prescriptions, this might indicate that the patient was not present and was not checked by the physician. In this case, the patient's name and possibly symptoms on the prescription would be more important since a third party (likely a family member or an assistant) would take the prescription to the pharmacist to be prepared (Lev and Chipman 2012, 146). Letters preserved in the Genizah reveal that there were cases of prescriptions being written for a patient on the basis of their written description of their symptoms – see, for example, the letter and the prescription that Maimonides sent to his former pupil Tobias (Cambridge University Library T-S Ar.30.286).

II.4.4 Endowment documents

According to Hamarneh (1962), medicine and pharmacy became separate professions in Baghdad by the 3rd/9th century. Looking beyond the marketplace to the endowment documents (*waqfiyyāt*) of hospitals (Amīn 1976, 358–9; Hoffmann 2000, 314–15), we get the impression that the pharmacist was clearly subordinate to physicians, merely making up medicines according to the physician's instructions and under his supervision. The very fact that several pharmacopoeias intended for use in hospitals survive (Ibn at-Tilmīḏ 2007, 2008; Sbath 1932–33) shows that this was an important site of pharmaceutical practice. However, the *waqfiyyāt* are prescriptive, indicating the patron's plans for the hospital and not what occurred in practice. The Manṣūrī hospital in Cairo, founded by the Mamlūk Sultan al-Manṣūr Qalāwūn (r. 678–689/1279–1290) and still in existence today, has been the focus of recent studies (Sabra 2000; Northrup 2013; Ragab 2015), and in this case, we have the reports of the administrator and author al-Nuwayrī (677–733/1279–1333), who served as the hospital's administrator. Unfortunately, although the distribution of medicines to the public was stipulated in the *waqfiyya*, al-Nuwayrī does not refer to the practice of pharmacy as such (al-Nuwayrī 1395–1410/1975–1990[CB],[1] 31: 107–8). We can say, however, that the medicines available in the *bīmāristān* were probably more limited in number than those available in the *dukkān*. Fewer recipes appear in pharmacopeias intended for hospital use, like the *Hospital Rule* or the recension of Sābūr ibn Sahl's (d. 255/869) dispensatory for use in the ʿAḍudī Hospital in Baghdad, compared to pharmacopeias intended for use in the community, like *How to Run a Pharmacy* or other versions of Sābūr's dispensatory. Based on the medical indications appearing as headings in the recipes in *How to Run a Pharmacy* and other pharmacopeias, and the large quantities sometimes required there, I suspect that there were also people who were unable or unwilling to go to a doctor and who went straight to the pharmacist for diagnosis and treatment – just as many people do in countries across the world today.

Note

1 Consolidated bibliography.

Bibliography

Sources

Amīn, M. M. 1976. *Wathā'iq waqf al-Sulṭān Qalā'ūn ʿalā l-bīmāristān al-manṣūrī* [Endowment Deeds of Sultan Qalā'ūn for the Manṣūrī Hospital]. Cairo: Dār al-kutub al-miṣriyya.

Dols, M. W., tr. 1984. *Medieval Islamic Medicine: Ibn Ridwan's Treatise "On the Prevention of Bodily Ills in Egypt"*. Berkeley: University of California Press.

Ibn Abī l-Bayān. ed. Sbath, P., ed. 1932–33. "Le Formulaire des hôpitaux d'Ibn abil Bayan, médicin du bimaristan annacery au Caire au XIIIe siècle," *Bulletin de l'Institut d'Egypte* 15: 9–78.

Ibn Bassām, Muḥammad ibn Aḥmad. ed. Ḥuṣām al-Dīn al-Sāmarrā'ī. 1968. *Nihāyat al-rutba fī ṭalab al-ḥisba* [The Utmost Authority in the Pursuit of the *Ḥisba*]. Beirut: Maṭbaʿat al maʿārif

Ibn Sahl, Sābūr. ed., tr. and ann. Kahl, O. 2008. *Sābūr Ibn Sahl's Dispensatory in the Recension of the ʿAḍudī Hospital*. Leiden: Brill.

Ibn at-Tilmīḏ. ed., tr. and an. Kahl, O. 2007. *The Dispensatory of Ibn at-Tilmīḏ: Arabic Text, English Translation, Study and Glossaries*. Leiden: Brill.

Ibn al-Ukhuwwa, Muḥammad ibn Muḥammad ibn Aḥmad al-Qurashī. ed. M. M. Shaʿbān. 1972. *Maʿālim al-qurba fī aḥkām al-ḥisba* [Milestones on the Rules of *Ḥisba*]. Cairo: al-Hay'a al-miṣriyya al-ʿāmma li-l-kitāb.

al-Kūhīn al-ʿAṭṭār, Abū l-Munā Dāwūd ibn Abī Naṣr. ed. Ḥ. al-ʿĀṣī. 1992. *Minhāj al-dukkān wa-dustūr al-aʿyān fī aʿmāl wa-tarākīb al-adwiya al-nāfiʿa li-l-insān* [The Management of the (Pharmacist's) Shop and the Rule for the Notables on the Preparation and Composition of Medicines Beneficial to Man]. Beirut: Dār al-manāhil.

Meyerhof, M. 1935. "Thirty-Three Clinical Observations by Rhazes (Circa 900 A.D.)," *Isis* 23.2: 321–72.

al-Shayzarī, ʿAbd al-Raḥman b. Naṣr. tr. Buckley, R. P. 1999. *The Book of the Islamic Market Inspector: Nihāyat al-Rutba fī Ṭalab al-Ḥisba (The Utmost Authority in the Pursuit of the Ḥisba)*. Oxford: Oxford University Press.

Manuscripts

Cambridge, Cambridge University Library, Or.1080 J23.
Cambridge, Cambridge University Library, Or.1080 J38.
Cambridge, Cambridge University Library, T-S Ar.30.286.

Research literature

Álvarez-Millan, C. 1999. "Graeco-Roman Case Histories and Their Influence on Medieval Islamic Clinical Accounts," *Social History of Medicine* 12: 19–43.

Álvarez-Millán, C. 2000. "Practice Versus Theory: Tenth-century Case Histories from the Islamic Middle East," *Social History of Medicine* 13: 293–306.

Awad, H. A. 2002. "Medical Prescriptions," in Bacharach, J., ed. *Fustat Finds*. Cairo: AUC Press, 190–7.

Chipman, L. 2010. *The World of Pharmacy and Pharmacists in Mamlūk Cairo*. Leiden: Brill.

Cohen, Mark R. 1995. "The Burdensome Life of a Jewish Physician and Communal Leader," *Jerusalem Studies in Arabic and Islam* 16: 125–36.

Eger, A. 2018. "Bronze Instruments from Tüpraş Field and the Islamic-Byzantine Medical Trade," in Kozal, E. *et al.*, eds. *Questions, Approaches, and Dialogues in Eastern Mediterranean Archaeology: Studies in Honor of Marie-Henriette and Charles Gates*. Münster: Ugarit-Verlag, 735–59.

Goitein, S. D. 1971. *A Mediterranean Society*. Vol. 2*: The Community*. Berkeley: University of California Press.

Hamarneh, S. K. 1962. "The Rise of Professional Pharmacy in Islam," *Medical History* 6: 59–66.

Hamarneh, S. K. and Awad, H. A. 2002a. "Glass Vessel Stamp Data for *Materia Medica*," in Bacharach, J., ed. *Fustat Finds*. Cairo: AUC Press, 167–75.

Hamarneh, S. K. and Awad, H. A. 2002b. "Medical Instruments," in Bacharach, J., ed. *Fustat Finds*. Cairo: AUC Press, 176–89.

Hoffmann, B. 2000. *Waqf im mongolischen Iran: Rašīduddīns Sorge um Nachruhm und Seelenheil*. Stuttgart: Franz Steiner Verlag.

Lev, E. and Amar, Z. 2007. *The Practical Materia Medica of the Cairo Genizah*. Leiden: Brill.

Lev, E. and Chipman, L. 2012. *Medical Prescriptions in the Cambridge Genizah Collections: Practical Medicine and Pharmacology in Medieval Egypt*. Leiden: Brill.

Lev, E., Chipman, L. and Niessen, F. 2008. "Chicken and Chicory Are Good for You: A Unique Family Prescription from the Cairo Genizah (T-S NS 223.82–83)," *Jerusalem Studies in Arabic and Islam* 35: 335–52.

Lewicka, P. 2013. "Medicine for Muslims? Islamic Theologians, Non-Muslim Physicians and Medical Culture of the Mamluk Near East," in Conermann, S., ed. *History and Society during the Mamluk Period (1250–1517)*. Göttingen: V&R Unipress, 83–106.

Northrup, L. 2013. "Al-Bīmāristān al-Manṣūrī Explorations: The Interface Between Medicine, Politics and Culture in Early Mamluk Egypt," in Conermann, S., ed. *History and Society during the Mamluk Period (1250–1517)*. Göttingen: V&R Unipress, 107–42.

Ragab, A. 2015. *The Medieval Islamic Hospital: Medicine, Religion, and Charity*. Cambridge: Cambridge University Press.

Sabra, A. A. 2000. *Poverty and Charity in Medieval Islam: Mamluk Egypt, 1250–1517*. Cambridge: Cambridge University Press.

Shatzmiller, M. 1994. *Labour in the Medieval Islamic World*. Leiden: Brill.

II.5
OTTOMAN AND SAFAVID HEALTH PRACTICES AND INSTITUTIONS*

Miri Shefer-Mossensohn

What did an ill person do in the Middle East in the early modern period? What medical knowledge was accessible then? What was the available medical practice? How were medical services offered and by whom? In this chapter I deal with these kinds of questions by delineating the outlines of institutions and patterns of action open to the authorities and individuals in the realm of health. The discussion is framed by the two great Islamic empires of the Middle East in the early modern period, namely, the Ottomans and the Safavids. This is the first attempt to examine points of similarity and difference in these parallel health systems.

Discussing health and medicine in the Ottoman and Safavid Empires jointly is justified on several accounts. The Ottomans and the Safavids were neighboring dynasties and two of the last three major Muslim empires (the third being the Mughals in the Indian subcontinent). These were strong states, centralized on a scale previously unknown. In both of them, firearms played a significant role, albeit not as the only agent, in establishing an efficient central system. The empires were linked economically, socially and culturally. The cultural composition of each empire was a complex and unique indigenous mosaic, but there were also common cultural and social factors. The rulers and different parts of the populace were Muslims (even if of different types – Sunni in the Ottoman and Mughal regions and Twelver Shiʿi in the Safavid) and shared a Mongol-Turkic heritage. They relied on Persian culture and similar patterns of patronage of scholars and scholarship. Texts and people traveled widely between the Ottoman, Safavid and Mughal realms. They also shared the *madrasa* as an institution and worked with a quite similar set of medical works generated in the preceding period.[1]

The similar cultural context was also the basis for the related (but not identical) dynamics of scientific practice and application of knowledge. Furthermore, European travelers, Middle Eastern physicians, scientific texts, clinical practices and medical instruments passed back and forth between the Ottoman Empire and Persia (Brentjes 2010). The following two careers were quite ordinary: In the 9th/15th century, Ghiyāth al-Dīn Muḥammad, studied first in his native Isfahan, and then with the Ottoman physician Sheref el-Dīn Ṣābūnjuoghlu (d. after 872/1468) at Amasya in central Anatolia. In 895/1490 Ghiyāth al-Dīn Muḥammad dedicated a treatise in Persian to the reigning sultan Bāyezīd II (r. 885–917/1481–1512) (Elgood 1970, x–xii, 4). In the 11th/17th century, Shems el-Dīn ʿItāqī (d. *c.* 1049/1639–40), originally from Shirvan (an area in the Caucasus at the Caspian Sea, on the Ottoman–Safavid border), migrated to Istanbul, where he authored a tract on anatomy (Kâhya 2000, 63–8). Against this backdrop – the existence

DOI: 10.4324/9781315170718-25

of a strong central administration and similar cultural and scientific traditions – it is appropriate to compare the way in which medicine and health operated in these two neighboring Islamic empires.

The state of the fields, that is the history of medicine and health in the Ottoman Empire and the Safavid Empire, respectively, is far from even. The corpus of research in Western and Middle Eastern languages on physicians, health, health institutions and clinical realities in the Ottoman lands is much more extensive than what exists so far on the Safavids. A sizable group of scholars dealing with Ottoman health in very broad terms has formed, including myself. However, this is not the case for research on health practices in the Safavid lands. The overall number of scholars focusing on the Safavids is decidedly smaller than that of scholars who deal with the Persianate world in previous or later periods or with the Ottoman world. In addition, the group of scholars who focus on the Safavids have limited interest in medicine and health. Moreover, some of the more central contributions to medicine in the Safavid state continue to adhere to a version of the "decline thesis."[2]

Despite the differences in the state of the research, and perhaps precisely because of them, it is useful to compare health practices and institutions in the Ottoman and Safavid domains. This is a framework that allows us to obtain a broad picture of the common medical knowledge and procedures from the southeastern Balkans to Central Asia in the early modern era. Discussing Ottoman and Safavid medicines together may also convince more scholars to focus their attention on medicine during the Safavid period.

Before we delve into the contents of Ottoman and Safavid medicine and health, an explanation of the very usage of the terms *Ottoman medicine* and *Safavid medicine* is in order. Is a term that refers to a political unit valid for a discussion of medicine and health? My answer is in the affirmative, on two counts. First, these political units had unique cultural facets identified specifically with them that have a bearing on health. Elsewhere, I have argued that medical practice and learning in the Ottoman Empire, indeed science and technology as a whole, operated in uniquely Ottoman modes (Shefer-Mossensohn 2009, 181–96; Shefer-Mossensohn 2015, 159–69). It can be presumed that a similar reality perhaps existed also on the Safavid side. Second, the organization of health services was tied to the bureaucracy, even in the early modern period; hence, one cannot discuss medical practices without contextualizing them within the relevant political frameworks.

II.5.1 What was medicine?

Medicine in the Middle East in the early modern period was a preventive lifestyle and health regimen. It combined diet with the full range of human activity in relation to one's environment: motion and rest, emotions, intercourse, retention and so on. Like other traditional medical systems, medicine in the Middle East demonstrated plurality, with a variety of etiologies, practices, and legitimacies. The expression of this medical diversity is the coexistence of (at least) three medical traditions. While there was considerable and meaningful overlap in knowledge and clinical reality (Stearns 2011), the traditions presented themselves as independent and distinct.

One important medical tradition had arisen from the Arab (and Syriac) translation and adaptation, with considerable additions and transformations, of *Galenic humoralism*, brought from antiquity into Islamic science and culture via the patronage of the early Muslim urban elites (see Chapters I.4 and I.10). This medical tradition understood the animal body (human and nonhuman) as a microcosm of Nature, which – according to a physical and philosophical meta-theory – is a mixture of fire, earth, air and water. Accordingly, the body consists of four humors, or fluids, the physiological building blocks of the body: blood (air), phlegm (water), black bile (earth) and yellow bile (fire). Health is a state of balance (*mīzān*), and illness is a state of imbalance in the body, that is, a case in which one of the four humors was in excess or deficient. *Folklore* based on indigenous custom was a second tradition. *Religious medicine* known as "Prophetic medicine" or "The Prophet's medicine" (*al-ṭibb al-nabawī* or *ṭibb al-nabī* in Arabic), that situated medical thinking and practice in divine revelation was a third tradition.

This brief presentation of Ottoman medical traditions rests on Conrad (1995a), who first identified these three subsystems within the larger medieval Islamicate medical system. In the Ottoman case, in light of recent research, the triple division should be rethought as a four-segmented medical system. From about 1700, Paracelsian chemical medicine and philosophy gained some success also in Istanbul and some other urban centers, although it did not permeate Ottoman society nearly as deep as other medical traditions (Küçük 2020).

All these components operated in both Ottoman and Safavid medicines, yet the meaning and contents of each component, including the specific interpenetration of these traditions, were different. Several factors contributed to medical pluralism. One of them, already mentioned, is that the Middle East sustained diverse cultural and scientific traditions. As the Ottomans and Safavids were located in different places in the Middle East, they did not engage with the same regional traditions. The Ottoman Empire, for instance, interacted with Mediterranean and other European traditions in Latin, like the pre-Ottoman Byzantines and the Italians and Habsburgs, who were contemporaries (Dursteler 2011; Morrison 2014; Morrison 2016; Küçük 2016). The Safavids, however, were on the eastern border of the Ottoman empire, in the Asian hinterland far from the Mediterranean basin. Instead, Ayurvedic medicine and Indian physicians seem to have become quite known in Safavid Persia (Elgood 1951, 372–4; Speziale 2018; Chapter V.8). Although the interaction between the two medical systems occurred in Persian, led by Persian physicians like 10th/16th-century Muḥammad Qāsim Firishta, most of it occurred in Mughal India rather than making its way back into Safavid lands (Speziale 2014a, 2014b).

Another factor behind medical diversity was the environment. The area stretching from North Africa to the Iranian plateau contains several very different geographical zones and climates. Folklore that relies on local ingredients is one factor expected to yield differences. The Safavids, and even more so the Ottomans, maintained complex ecological systems that also differed from each other. The topography and the climate shaped a diversity of plants and minerals that in turn determined what *materia medica* would be common, available and affordable to sustain indigenous medicine. Also, the regions with which the Ottomans and the Safavids traded impacted greatly the availability of medicinal materials.

A third factor was religious differences. Ottoman Sunni and Safavid Twelver Shiʿi practitioners made efforts to anchor medicine in sacred texts: the Qurʾān and *ḥadīth*. The resemblance between Sunni and Shiʿi medical practices is substantial; both use natural cures and incorporate humoral understanding and anatomical knowledge, with a preference for revealed sources and emphasis on God's creation. The Shiites, however, add to their sources traditions about the twelve imams; hence, Shiʿi religious medicine is also called the *medicine of the Imams* (*ṭibb al-aʾimma*). The somewhat different sources are a factor in, and a symptom of, the nuances and variations in the medical theory and practice portrayed in Sunni sources (Elgood 1962a; Elgood 1962b; Bürgel 1976; Perho 1995) and in Shiʿi sources (Elgood 1970, 36–8; Newman 2003; Ispahany and Newman 1991).

II.5.2 Who was a healer?

The Ottoman and Safavid populations availed themselves of a host of medical caregivers. The variety of people who provided medical treatment was enormous in their background and training, in the type of treatment they offered, in the degree of their medical practice as an exclusive occupation, and more. Given the reality of considerable diversity among professionals, the broader term *healer* is an appropriate analytical category, replacing *physician*, which is a narrower term. References to healers appear in a variety of sources, ranging from literature like chronicles and biographical dictionaries to medical-scientific tracts, legal writing (court protocols and manuals regulating the markets) and state archival documents. Many of these sources are the products of the elite milieu and refer to the type of healers who were especially active in that specific social context. These healers tended to be associated with the humoral medical system. When humoral terminology is used, healers are classified according to the three disciplines acknowledged by humoral medicine. They may be physicians, that is, *ṭabībs*, *ḥakīms*, or *mutaṭabbib*. The first two terms seem to be popular ones in Ottoman sources more than the third one, which appears more frequently in Safavid sources. Humoral healers may also be ophthalmologists (*kaḥḥāls*) or surgeons (*jarrāḥs*). Anyone who did not fit this tripartite division was either omitted from the historical record or presented in a manner that was adapted to the humoral categories while blurring the more historically accurate medical association.

Women healers are usually missing from some of the sources but evidently were part of the available medical profession. Gynecology and female fertility and anatomy were discussed in learned tracts by men, in both the Ottoman and Safavid realms (Elgood 1970, 193–284; Elgood 1968; Aqajari *et al.* 2013; İtâkî 1996; Russell 1992). Yet there are recurring references to female healers. Women were not only involved in the traditional female healing occupations of midwives (*dāyas*) and wet nurses (*qābilas*). Indeed, evidence reveals that women healers enjoyed diverse professional training, qualifications, and expertise and dealt with male and female patients and colleagues alike. One female healer by the name of Ṣāliḥa Ḫātūn (Lady Ṣāliḥa) specialized in the surgical treatment of hernia (*fıtık* in Turkish, *fatq* in Arabic). Between Muḥarram 1032/November 1622 and Rajab 1033/April 1624, S Ṣāliḥa appears in the protocols (*sijill*) of the Muslim court of Üsküdar, the Asian neighborhood of Istanbul, as treating 21 patients. They were all male, some Muslim Turks and some Greek Christians (Sahillioğlu 1998, 64). The sources do not raise the claim that a woman treating a man (or vice versa) is something inherently wrong, even in cases that involved intimate body parts. Likewise, an 8th/14th-century prophetic medicine text clearly states that this is permissible. Elgood misidentified the author as Jalāl al-Dīn al-Suyūṭī (d. 911/1505); in fact, it is Shams al-Dīn al-Dhahabī ([d. 1348]; Elgood 1962, 133; Perho 1995, 36–40).

Multiple methods for acquiring medical knowledge existed in Ottoman and Safavid societies, in the theoretical and clinical sense: autodidactic learning, learning with a teacher at the *madrasa* (law school) and at the few existing medical schools, home learning with a private tutor, apprenticeship to a physician, and lectures that were available at hospitals and in public spaces, like markets (Brentjes 2018, 91–8 and 115–31; see also Chapter III.2). Travel was also a common pattern in learning (Brentjes 2018, 135–44). The phenomenon of families of physicians was quite common, and the family was sometimes the framework of learning and practical instruction (Elgood 1951, 353; Brentjes 2018, 131–5). Many blended a variety of ways of learning in accordance with their economic capability, the accessibility of various methods of training, their way of life or personal leanings and their outlook on medicine as either profession or an intellectual pursuit – all of which obviously change over time.

The tendency to value current and practical expertise over past theoretical studies is reflected in the numerous legal suits against physicians. Muslim court records throughout the Ottoman Empire include many that we would call today "medical malpractice." Disappointed patients sued their physicians, pleading that the physicians had not complied with the contract between them; the accused physicians received payment for full recovery and did not deliver the goods. In this procedure, the status of the defendant physician in the local professional community was evaluated. However, at least according to the court records, no explicit examination was conducted of the physician's precise training. This pattern is part of what seems to be a larger trend, namely, a concept of production standards and expertise (what we may call "quality assurance") anchored in the present, with earlier training not regarded as part of it (Cohen 2001; Yi 2004).

Plurality in medical education was feasible as there was no formal and binding process of licensing physicians. Instead, the Ottoman and Safavid medical systems were "open"; that is, medical knowledge was accessible to both laypersons and doctors (Porter 1985). There was no formal graduation procedure for medical practitioners, whether physicians, surgeons, oculists, pharmacists or other medical and para-medical personnel or indeed for those who dealt with other bodies of knowledge. *Madrasa* students of religious disciplines and logic often received a certificate (*ijāza*) from their teacher for having studied one or more texts with him (Barker 2017; Chapter III.2). Some such certificates are also preserved for students of medicine. However, a significant number of practitioners did not study at a *madrasa*.

A special genre in Islamic literature, *ḥisba*, was concerned with the enforcement of the standard codes of commercial law and the regulations for the safeguarding of public morality and faith, and for protecting Muslims against fraud, trickery and charlatanry, including in health-related issues (Levey 1963, 176). Because of the moral-religious context of this genre, the texts indicated 'Muslims' as the target of protection but they should be understood as referring more broadly to 'subjects'. Although, unfortunately, Ottoman and Safavid *ḥisba* manuals have not been studied till now in their medical context, we know that these treatises emphasized the importance of regular inquiries into the knowledge and capabilities of health practitioners (Elgood 1970, 32, 63, 130, 133, 146–7, 231–2; Levey 1963, 176–82; Levey 1967). This was one of the tasks entrusted to the imperial head physician (the *ḥakīmbāsh*). He could discipline physicians whose wages were paid by the state, promote, hire and fire them (Bayat 1999; Elgood 1970, 63; Tadjbakhsh 2003, 223). However, many more physicians, those working in the neighborhoods and markets outside the palaces and hospitals, were outside the reach of the formal medical institution, even in the Ottoman centralized state, prior to the 13th/19th-century reforms. Instead, such physicians seem to have been regulated by their patients and market inspectors.

Archival and literary sources reveal an Ottoman and Safavid medical professional hierarchy subordinated to and regulated by the state, embodied in the office of the head physician and other officials (Shefer-Mossensohn 2014; Newmann 1995, 783–5). However, most Ottoman medical practice in the early modern period was outside the state; healers rarely practiced medicine in imperial hospitals or at the court, and most were not associated with the *madrasa*s. In fact, there may have been little interaction between medical practitioners paid by the state and those in the open market; the latter being the majority in the Ottoman world (Küçük 2016, 224). Our knowledge regarding the Safavid medical scene is limited but the indications are that the overall picture may have been much the same.

Direct interaction may indeed have been minimal, but many practicing physicians in the market were using similar medical texts to the elite physicians, at least at the elementary level. The reality of some overlap sharpened the need to differentiate the two groups, which translated into competition between the people who followed different educational methods, about who possessed "real" medical knowledge. Ottoman doctors hurled the word *ignorant* (Arabic: *jāhil*) at each other when a personal disagreement between them, a theoretical medical argument or a professional struggle, led them to denounce rivals and competitors as frauds, even when the adversaries may have been skilled professionals (Conrad 1995b; Shefer 2002; Pormann 2005).

II.5.3 What was medical treatment?

Medical treatment provided both preventive and curative measures. Following the understanding of human beings as integrating body and soul, the material and the spiritual, diagnosis and then treatment were also integrative. This medical methodology built on medical treatments from earlier Islamicate periods and even late antiquity.

Food and beverages were the first course of action. For Ottomans and Safavids, the medical aspects of food were not just a matter of theory, they were very much a matter of concrete and practical daily routine. Food was a vital therapeutic tool. Yet, food and beverages were more than curative agents. They also acted as an insurance against illness, to be followed by more invasive and violent measures only if they failed. Therapeutic abilities were assigned to certain foodstuffs or dishes. Some of them included local ingredients, other include rare items that were not usually present in the kitchen for regular meals. In most cases, it is impossible to distinguish whether their primary importance was medical, gastronomic or both (Rogers 1994).

Medication was the next course of action. Middle Eastern pharmacology in the early modern period continued previous medieval Arabic practices, especially in two aspects. First, most drugs used were of plant origin, hence the uncertainty in some cases whether the recipes are for condiments, foods or medications. The archives at the Topkapı Palace include recipes that were written down for patients living there, and we also have examples of late Ottoman recipes and pharmaceutical equipment (Sandalcı 1997–2006). Narcotics were a special case of medication. From as early as the 3rd/9th century, hashish and opium (Arabic: *afyūn*) were prominent among the drugs used by Muslims in medical interventions, socially for recreation, or even as agents to maintain health (Tibi 2006). Opium is singled out here because it was an indispensable pharmaceutical agent for Ottomans and Safavids. It was a painkiller and a panacea for all sorts of diseases. It was relaxing and fortifying. During the Safavid period, it also became a very popular recreational drug among Persians, especially the elite, but it was quite popular and accessible to commoners as well (Matthee 2005, 97–116).

Early modern pharmacy also continued previous forms of pharmaceutical formularies. The range was surprisingly wide and included most forms familiar to the modern patient, such as infusions, decoctions, pomades, pills, syrups, pastilles, powders, emulsions, suppositories and enemas (rectal or urethral). Medications were taken by mouth, that is eaten, drunk or swallowed. There was also nasal medicine: fumes were inhaled and powders sniffed, sometimes with a blower used to open the nostrils. Baths, too, were used, so the patient would inhale the fumes from the drug more easily (the entire body was not necessarily immersed). Other medications were not ingested but rubbed on the outside, as in the case of ointments. Two forms seem to have been especially popular: syrup (*shurba* in Arabic) and doughy paste (Arabic *maʿjūn*) (Elgood 1970, 30–55).

Surgery was limited when compared to diet and medication. Very few surgical tracts exist. The number of anatomical tracts is also very limited. One notable tract was the 9th/15th-century *The Surgical Operation of the Khan* (*Jerrāhiyyet ül-ḫāniyye*) by Sheref el-Dīn Ṣābūnjuoghlu from Amasya in Anatolia, mentioned earlier as the teacher of Ghiyāth ibn Muḥammad. Another important work was the 11th/17th-century *The Anatomy of the Body Parts and Expounding the Role of the Philosophers* (*Teshrīḥ-i ebdān ve terjumān-i qābāle-yi faylasūfān*) by Shems el-Dīn ʿItāqī, which expounds the natural scientific aspects of anatomy (Shefer-Mossensohn 2015, 107; Elgood 1970, 121–89).

In the age before antiseptics, antibiotics and painkillers, the practice of surgery was obviously limited. Yet Emilie Savage-Smith's conclusion about surgical practices in the medieval Islamicate world seem apt also for the Ottomans and Safavids (Savage-Smith 1995; Savage-Smith 2000).

Based on surgical manuals, miniatures, observations by European travelers and cases described in the legal courts, surgery of a noninvasive nature was undertaken after all other avenues of treatment had been exhausted. Indeed, various surgical and dental instruments were employed, including cauteries, trocars, extractors, suture instruments, scalpels, pincers, pinchers, forceps and scissors (Chapter IV.4).

Circumcision and castration aside (as they are religious and cultural procedures rather than medical ones), two procedures seem to be mentioned more than others in Ottoman and Safavid sources: phlebotomy and cauterization. Bloodletting, that is the withdrawal of large quantities of blood from a patient whether by cutting a vein or via leeches, was a routine medical treatment for various aches and pains. The humoral reasoning was that removing blood alleviated problems arising from an excess or corruption of the blood in the body. On the popular front, it was a means of bringing a body into alertness as preventive medicine as well (Thevenot 1687, vol. 1, 37–8). Cauterization (Turkish *dağ*), namely, burning a body extremity with a hot iron, was a popular treatment for a variety of ailments in all body parts, ranging from headaches to fistulas and hemorrhoids. Cauterization treated also nonphysical problems, like forgetfulness or mental problems such as melancholy (Sabuncuoğlu 1992, vol. 2, 16r – 53r, 121r-124r; Özçelik 1998).

Amulets in every shape and size, pilgrimage to holy sites and visits to graves of holy figures were common means of treatment. Their popularity stemmed from the marriage of belief and medicine. One form was a general belief that patients have resources in themselves to endure and overcome pain and malady. Another form of patients' belief was in the healing powers of their healers. The marriage of medicine and religion also regarded Allah to be the ultimate healer of all maladies. The Healer (*al-shāfī* in Arabic) is one of His sacred 99 attributes, or Most Beautiful Names. Following this belief, amulets included written verses of the Qurʾan, which is believed to be His direct speech. Other amulets included words, letters, digits, geometrical drawings or special shapes, like the palm-shaped *khamsa* (five in Arabic).

Then, as today, medical services were not equally accessible. Financial resources and gender were among the factors that shaped medical options. Medication, like medical dietary plans, could be a status symbol, as wealthy people could afford medication prepared from rarer and more expensive ingredients (although the outcome was not necessarily more effective medically). The affluent patients in the Ottoman imperial palace, for example, received medication prepared from drugs like opium or hashish, pulverized gems and precious metals. This possibility did not exist for the patients in the hospitals or people who bought such ingredients on the open market at full price (Shefer-Mossensohn 2009, 34–45).

An extreme example of a drug available only to a few is theriac (a word loaned from Greek), the famous cure-all antidote that reached Muslims from antiquity. Theriac existed in many versions. They had various purposes, but the more famous were those intended as antidotes for poisons. Theriacs were among the most complex of Islamic pharmaceutical products from medieval times to the early modern period, as they contained huge numbers of ingredients. The most important (and perhaps the most unusual) of these were viper flesh and snake venom. Theriac recipes could also include opium as an antidote to snake venom. Moreover, the process of manufacturing was complex and time-consuming (Estes and Kuhnke 1984, 134–5).

Just as in our own time, people with means could enjoy a wider range of medical treatments although the effectiveness of expensive drugs, in comparison to the cheaper ones, is anyone's guess. However, high social status closed at least as many medical options as it opened, since many obligations and limitations were placed on the conduct of female members of the elite. Social conventions seem to have allowed for a broader range of possibilities below the upper crust of society. The accepted behavioral code in Ottoman and Safavid societies, as in other premodern Muslim societies, recommended physical and social separation between men and women who were not related to one another by marriage or blood ties. Women of the elite were more obliged than others to observe the separation of men and women, as they had the financial means to afford the costs incurred by seclusion (Dengler 1978, 232, 237). This medical, social and financial reality meant that when Ottoman imperial women fell sick, their experiences were very different from those of women from other social echelons. The social and cultural restrictions placed upon them – not approaching a male physician, for example – were stricter and more tightly observed, as documented by several observers (Bon 1996, 89–90; Hierosolimitano 2001, 36 n.4).

II.5.4 Where was medical treatment offered?

Medical treatment was dispensed at various sites: at the physician's house, at the pharmacist's store, in the marketplace, in the patient's house, in the open air and at the hospital. The Islamic hospital (referred to interchangeably in Arabic, Ottoman Turkish and Persian as *bīmāristān, bīmārḫāne, tīmāristān, tīmārḫāne,* and *dār al-shifā'*) was the outcome of several traditions, namely, early Islamic Arab, Hellenized Christian, Persian and Indian medical cultures. Although how each tradition might have contributed to its construction is not clear, the early Islamic hospital is now commonly regarded as a novel type of medical institution.

History reveals that Islamic hospitals were unique in several important respects. They were founded as urban charitable institutions, whose purpose was to heal sick patients (rather than serve as hostels or hospices caring for the dying or chronically ill), administered and financed as endowed institutions (Arabic *waqf*) and distinguished by the traditional medical practice of Galenic humoralism (in its Arabic-Muslim interpretation). The first Islamic hospital that became a lasting institution is associated with 3rd/9th-century Abbasid Baghdad. Hospitals spread out from Mesopotamia to Iran and Egypt during the 4th/10th century, and matured in the periods that followed, under the Zengids (r. 521–660/1127–1262), the Ayyubids (r. 567–658/1171–1260) and the Mamluks (r. 648–922/1250–1517). It also diffused beyond the urban centers of the Arabic-speaking Middle East, where it had started, into the Turkish principalities of Anatolia and Ilkhanid Iran.

Under the Ottomans and Safavids, the institution reached its peak. The absolute number of hospitals had never been large, and they were not intended to dispense medical aid to the entire public . However, in most large urban centers there was at least one hospital founded by elite patronage. Bursa, Edirne and Istanbul, the three Ottoman imperial capitals, had one or more hospitals associated with the sultans and their family members. Isfahan, Tabriz, Shiraz, Qazwin, Mashhad and other Safavid urban centers all had hospitals, likewise under the patronage of the Shahs and other Safavid notables and merchants (Tadjbakhsh 2012, 27–36; Floor 2012, 38–44).

In both regions, hospitals were funded by endowments. The Mughals, the third great dynasty in early modern Eurasia, for their part administered their hospitals differently – more directly by government funds rather than by private donation. The Ottoman and Safavid association of hospitals with *waqfs* was a choice and not a necessity (Speziale 2012, 5). Hospitals founded by the elite in Ottoman and Safavid domains were established as part of religious charitable complexes. In the Ottoman case, the complexes were usually composed of mosques, soup kitchens and schools, in addition to an occasional hospital (Inalcik 1990). The Safavid complexes seem to have been centered more around mosques and shrines (Speziale 2012, 2–5, 8–12).

Hospitals were initiated by the elite – the rulers and people of high rank. The actual patients, however, were of other social classes. In the Safavid case, they were vagabonds, pilgrims and the poor, who enjoyed not only medical care but also free meals (Speziale 2012, 6). Patients in the hospitals founded by the Ottoman elite may have come from somewhat different background. Lists of estates found in the Ottoman court archives include some deceased hospital patients. The very fact that people who died in hospitals appear on such lists testifies that they left some property when they died; otherwise, the Ottoman *qāżī* would not have been asked to intervene in his capacity as settler of inheritances. Yet the estates of such hospital patients were quite modest, and all were movable items (Öztürk 1995, 135–7, 327, 360, 369, 377, 442, 462, 466). Although the patients treated in Safavid and Ottoman hospitals were from social circles outside the elites, the low number of hospitals and the limited number of beds in them reveal that they were not meant to aid the population in its entirety. Perhaps this is evidence of many or even most patients' identity as visitors to the city. Being foreign and not necessarily poor, although decidedly not affluent, with no local support network, they may have needed institutional help when they fell ill.

There is some evidence of medical teaching in some Ottoman and Safavid hospitals, for instance in Tabriz, Isfahan and Istanbul (Brentjes 2018, 129). Yet, these pieces of information are scarce and scattered. Some hospitals had schools (*madrasas*) nearby, but they were not necessarily exclusively devoted to medicine (Speziale 2012, 3–4; Brentjes 2018, 91–8).

II.5.5 Some concluding remarks

Ottomans and Safavids were very much concerned with their medical (physical and mental) well-being. They subscribed to a variety of medical concepts and cultivated many different health practices. The written sources reveal that humoral theory, Prophetic medicine, folk medicine and magic and, later on, also chemical medicine were all available. Yet these subsystems were not independent of each other. None enjoyed complete hegemony or was regarded by all as superior to others, absolutely true and exceptionally efficient. Medical philosophy aside, the therapeutic reality on the ground was a fusion of medical options, that did not exclude each other, and that aimed at rebalancing the body in a broad sense (Savage-Smith 2013). Medicine, as a body of knowledge and a set of skills, diffused back and forth between all segments of society and between all sorts of practitioners. Medicine was not the exclusive domain of professional physicians or medical institutions. but rather formed a widespread activity, from the court and literati to commoners.

Notes

* The research for this chapter was supported by The Israel Science Foundation (Grant no. 290/16).

1 Some of the literature that looks at the three early modern Islamic empires as a single analytical unit are (Hodgson 1974; Robinson 1997; Dale 2010).

2 This is the thesis that after some moment in time, often the 5th/11th or 6th/12th centuries, all intellectual and even many cultural spheres in any Islamicate society withered due to reasons identified differently by different authors. Among them, the most prominent reasons are the Mongol invasion in the middle of the 7th/13th century; natural catastrophes, in particular the plague; hostile religious orthodoxy; and the rise of superstitious practices and beliefs (Huff 2011, 2017). This viewpoint has now been severely criticized (Brentjes 2011, 2018; Barker 2017).

Bibliography

Sources

Bon, O. tr. Withers, J. and Goodwin, G. 1996. *The Sultan's Seraglio: An Intimate Portrait of Life at the Ottoman Court.* London: Saqi Books.

Elgood, C. 1962a. "Tibb ul-Nabi or Medicine of the Prophet," *Osiris* 14: 33–192.

Hierosolimitano, D. tr. and intro. Austin, M., ed. Lewis, G. 2001. *Domenico's Istanbul*. Warminster: E. J. W. Gibb Memorial Trust.

Ispahany, B. and Newman, A. J. 1991. *Islamic Medical Wisdom: The Tibb al-A'imma*. Watford: Alif International.

İtaki, Ş. ed. Kâhya, E. 1996. *Şemseddîn-i İtâkî'nin Resimli Anatomi Kitabı* [Şemseddîn-i İtâkî's Book of Anatomic Drawings]. Ankara: Atatürk Kültür Merkezi Yayını.

Levey, M. 1967. "Medical Ethics of Medieval Islam with Special Reference to al-Ruhawī's 'Practical Ethics of the Physician'," *Transactions of the American Philosophical Society* 57.3: 1–100.

Sabuncuoğlu, Ş. ed. Uzel, İl. 1992. *Cerrahiyyetü'l-Ḫāniyye* [The Surgery of the Han]. 2 vols. Ankara: Türk Tarih Kurumu.

Sandalcı, M. 1997–2006. *Belgelerle Türk Eczacılığı, 1840–1948* [Turkish Pharmacology in Documents, 1840–1948]. 5 vols. Istanbul: Dr. Nejat F. Eczacıbaşı Vakfı.

Thevenot, J. 1687. *The Travells of Monsieur de Thevenot into the Levant* . . . London: Printed by Henry Clark, for John Taylor, at the Ship in St. Paul's Church-Yard.

Research literature

Aqajari *et al.*: Aqajari, S. H., Sadighi, B. and Karimi, B. 2013. "Women in Safavid Medical Discourse: A Case Study of General Medical Texts, Anatomical Writings, and Religious Medical Manuscripts," *Iran Nameh* 28.3: 4–23.

Barker, P. 2017. "The Social Structure of Islamicate Science," *Journal of World Philosophies* 3: 37–47.

Bayat, A. H. 1999. *Osmanlı Devleti'nde Hekimbaşılık Kurumu ve Hekimbaşılar*. Ankara: Atattürk Kültür Merkezi Yayınları.

Brentjes, S. 2010. *Travellers from Europe in the Ottoman and Safavid Empires, 16th–17th Centuries: Seeking, Transforming, Discarding Knowledge*. Farnham: Ashgate Variorum.

Brentjes, S. 2011. "The Prison of Categories: 'Decline' and its Company," in Opwis, F. and David Reisman, D., eds. *Islamic Philosophy, Science, Culture, and Religion: Studies in Honor of Dimitri Gutas*. Leiden: Brill, 131–56.

Brentjes, S. 2018. *Teaching and Learning in Islamicate Societies (800–1700)*. Turnhout: Brepols.

Bürgel, J. Ch. 1976. "Secular and Religious Features of Medieval Arabic Medicine," in Charles, L., ed. *Asian Medical Systems: A Comparative Study*. Berkeley: University of California Press, 44–62.

Cohen, A. 2001. *The Guilds of Ottoman Jerusalem*. Leiden: Brill.

Conrad, L. I. 1995a. "Medicine – Traditional Practice," in *The Oxford Encyclopedia of the Modern Islamic World*. New York: Oxford University Press, 3: 85–8.

Conrad, L. I. 1995b. "Scholarship and Social Context in the Near East," in Bates, D., ed. *Knowledge and the Scholarly Medical Traditions*. Cambridge: Cambridge University Press, 81–101.

Dale, S. F. 2010. *The Muslim Empires of the Ottomans, Safavids, and Mughals*. Cambridge: Cambridge University Press.

Dengler, I. C. 1978. "Turkish Women in the Ottoman Empire: The Classical Age," in Beck, L. and Keddie, N., eds. *Women in the Muslim World*. Cambridge, MA: Harvard University Press, 229–44.

Dursteler, E. R. 2011. "On Bazaars and Battlefields: Recent Scholarship on Mediterranean Cultural Contacts," *Journal of Early Modern History* 15: 413–34.

Elgood, C. 1951. *A Medical History of Persia and the Eastern Caliphate*. Cambridge: Cambridge University Press.

Elgood, C. 1962. "The Medicine of the Prophet," *Medical History* 6: 146–53.

Elgood, C. 1968. "Persian Gynaecology," *Medical History* 12: 408–12.

Elgood, C. 1970. *Safavid Medical Practice, or The Practice of Medicine, Surgery and Gynaecology in Persia between 1500 A.D. and 1750 A.D.* London: Luzac.

Estes, J. W. and Kuhnke, L. 1984. "French Observations of Disease and Drug Use in Late Eighteenth-Century Cairo," *Journal of the History of Medicine and Allied Sciences* 39: 121–52.

Floor, W. 2012. "Hospitals in Safavid and Qajar Iran: An Enquiry into Their Number, Growth and Importance," in Speziale, F., ed. *Hospitals in Iran and India, 1500–1950s*. Leiden: Brill, 37–116.

Hodgson, M. G. S. 1974. *The Venture of Islam: Conscience and History in a World Civilization*. 3 vols. Chicago: University of Chicago Press.

Huff, T. E. 2011. *Intellectual Curiosity and the Scientific Revolution: A Global Perspective*. Cambridge: Cambridge University Press.

Huff, T. E. 2017. *The Rise of Early Modern Science: Islam, China and the West*, 3rd edition. Cambridge: Cambridge University Press.

Inalcik, H. 1990. "Istanbul: An Islamic City," *Journal of Islamic Studies* 1: 1–23.

Kâhya, E. 2000. "One of the Samples of the Influences of Avicenna on the Ottoman Medicine, Shams al-Din Itaqi," *Belleten* 64.4: 63–8.

Küçük, B. H. 2016. "New Medicine and the *Ḥikmet-i Ṭabīʿiyye* Problematic in Eighteenth-Century Istanbul," in Langermann, Y. T. and Morrison, R. G., eds. *Texts in Transit in the Medieval Mediterranean*. University Park, PA: The Pennsylvania State University Press, 223–42.

Küçük, B. H. 2020. *Science without Leisure: Practical Naturalism in Istanbul, 1660–1732*. Pittsburgh, PA: University of Pittsburgh Press, Ch. 5 "The Recipe," 143–65.

Levey, M. 1963. "Fourteenth Century Muslim Medicine and *Ḥisba*," *Medical History* 7: 176–82.

Matthee, R. P. 2005. *The Pursuit of Pleasure: Drugs and Stimulants in Iranian History, 1500–1900*. Princeton: Princeton University Press.

Morrison, R. G. 2014. "A Scholarly Intermediary between the Ottoman Empire and Renaissance Europe," *Isis* 105: 32–57.

Morrison, R. G. 2016. "The Role of Oral Transmission for Astronomy among Romaniot Jews," in Langermann, Y. T. and Morrison, R. G., eds. *Texts in Transit in the Medieval Mediterranean*. University Park: The Pennsylvania State University Press, 10–28.

Newman, A. J. 1995. "Medicine in the Safawid Period," in "Ṣafawids," *EI-2*, 777–87.

Newman, A. J. 2003. "Bāqir al-Majlisī and Islamicate Medicine: Safavid Medical Theory and Practice Re-examined," in Newman, A. J., ed. *Society and Culture in the Early Modern Middle East: Studies on Iran in the Safavid Period*. Leiden and Boston: Brill, 371–96.

Özçelik, S. 1998. "Bāsur Hastalığı ve Tedavisiyle İlgili 15. Yüzyılda ait bir Metin," *Yeni Tıp Tarihi Araştıtmaları* 4: 207–24.

Öztürk, S. 1995. *Askeri Kassama ait Onyedinci Asır İstanbul Tereke Defterleri (Sosyo-Ekonomik Tahlil)*. Istanbul: Osmanlı Araştırmaları Vakfi.

Perho, I. 1995. *The Prophet's Medicine: A Creation of the Muslim Traditionalist Scholars*. Helsinki: Finnish Oriental Society.

Pormann, P. E. 2005. "The Physician and the Other: Images of the Charlatan in Medieval Islam," *Bulletin of the History of Medicine* 79: 189–207.

Porter, R., ed. 1985. *Patients and Practitioners: Lay Perceptions of Medicine in Pre-Industrial Society*. Cambridge: Cambridge University Press.

Robinson, F. 1997. "Ottomans-Safavids-Mughals: Shared Knowledge and Connective Systems," *Journal of Islamic Studies* 8: 151–84.

Rogers, J. M. 1994. "The Palace, Potions and the Public: Some Lists of Drugs in Mid–16th Century Ottoman Turkey," in Heywood, C. and Imber, C., eds. *Studies in Ottoman History in Honour of Professor V.L. Mènage*. Istanbul: Isis Press, 273–95.

Russell, G. 1992. "'The Owl and the Pussycat': The Process of Cultural Transmission in Anatomical Illustration," in Ihsanoglu, E., ed. *Transfer of Modern Science and Technology to the Muslim World*. Istanbul: IRCICA, 191–5.

Sahillioğlu, H. 1998. "Üsküdar'ın Mamure (Cedide) Mahallesi Fıtık Cerrahları," *Yeni Tıp Tarihi Araştırmaları* 4: 59–66.

Savage-Smith, E. 1995. "Attitudes Toward Dissection in Medieval Islam," *Journal of the History of Medicine and Allied Sciences* 50: 67–110.

Savage-Smith, E. 2000. "The Practice of Surgery in Islamic Lands: Myth and Reality," *Social History of Medicine* 13: 307–21.

Savage-Smith, E. 2013. "Were the Four Humors Fundamental to Medieval Islamic Medical Practice?" in Hsu, E. and Horden, P., eds. *The Body in Balance: Humoral Theory in Practice*. New York: Berghahn Books, 89–106.

Shefer, M. 2002. "Medical and Professional Ethics in Sixteenth-Century Istanbul: Towards an Understanding of the Relationships between the Ottoman State and the Medical Guilds," *Medicine and Law* 21: 307–19.

Shefer-Mossensohn, M. 2009. *Ottoman Medicine: Health and Medical Institutions, 1500–1700*. Albany: State University of New York Press.

Shefer-Mossensohn, M. 2014. "The Many Masters of Ottoman Hospitals," *Turkish Historical Review* 5: 94–114.

Shefer-Mossensohn, M. 2015. *Science Among the Ottomans: The Cultural Creation and Exchange of Knowledge*. Austin: University of Texas Press.

Speziale, F. 2012. "Introduction," in Speziale, F., ed. *Hospitals in Iran and India, 1500–1950s*. Leiden: Brill, 1–26.

Speziale, F. 2014a. "The Persian Translation of the *tridoṣa*: Lexical Analogies and Conceptual Incongruities," *Asiatische Studien-Études Asiatiques* 68: 783–96.

Speziale, F. 2014b. "A 14th Century Revision of the Avicennian and Ayurvedic Humoral Pathology: The Hybrid Model by Šihāb al-Dīn Nāgawrī," *Oriens* 42: 514–32.

Speziale, F. 2018. *Culture persane et médecine ayurvédique en Asie du Sud*. Leiden: Brill.

Stearns, J. K. 2011. *Infectious Ideas: Contagion in Premodern Islamic and Christian Thought in the Western Mediterranean*. Baltimore: John Hopkins University Press.

Tadjbakhsh, H. 2003. *History of Medicine and Veterinary Medicine in Iran*. Lyon: Fondation Mérieux.

Tadjbakhsh, H. 2012. "Hôpitaux et médecins avicenniens en Iran à l'époque Safavide," in Speziale, F., ed. *Hospitals in Iran and India, 1500–1950s*. Leiden: Brill, 27–36.

Tibi, S. 2006. *The Medicinal Use of Opium in Ninth-Century Baghdad*. Leiden: Brill.

Yi., E. 2004. *Guild Dynamics in Seventeenth-Century Istanbul: Fluidity and Leverage*. Leiden: Brill.

II.6
PLANETARY THEORY

Amir Mohammad Gamini

This chapter surveys works on theoretical and physical models of planetary movements beginning with Ibn al-Haytham's contributions in the early 5th/11th century. The main topics in this kind of theoretical astronomy were geometrical planetary models, Ptolemaic and non-Ptolemaic, the observational and physical reasons behind them, the centrality and immobility of the Earth in the universe, the distances and sizes of the planets and other problems related to Ptolemaic astronomy.

II.6.1 Physical planetary models

Planetary models served as a theoretical basis for computational astronomy. Those in Claudius Ptolemy's (*c.* 100–170) *Almagest* were purely mathematical, incorporating some observational results (Ptolemy 1998[CB[1]]). While Ptolemy had presented the physical interpretation of those models in his *Planetary Hypotheses*, which aimed at conformity with the principles of natural philosophy (Ptolemy 1967), the available Greek copies were apparently incomplete and difficult to understand (see below). The early phase of reflections on physical models of planetary movements is not yet well studied, as it used to be believed that the question of the physical existence of such models attracted interest only later. Recent work suggests that this view needs to be revised. In what follows, I survey major lines of research without aiming at comprehensiveness.

Indian and Iranian astral practices employed computational astronomy for the prediction of planetary positions, with at best a few remarks on physical properties of the universe. In Greek astral literature, various approaches can be found. In the *Almagest*, Ptolemy put forward several geometrical models but did not discuss their physical aspects. He used the models as geometrical devices for computing the true positions of the sun, moon and five planets without explicitly reflecting on a physical configuration of the celestial spheres (Pedersen 2010, 27). In the 3rd/9th century, the physical interpretation of his models in the *Planetary Hypotheses* seems to have seriously challenged the skills and knowledge of its translator, apparently neither a Greek nor an Arabic native speaker, since his translation is permeated by mistakes in both languages (Ptolemy 1967; Murschel 1995). This may have been one of the reasons for the subsequent historical development.

DOI: 10.4324/9781315170718-26

II.6.1.1 Ibn al-Haytham's on the configuration of the cosmos

The scholar often considered the first major discussant of the physical nature of geometrical models of the planets and the entire universe in Islamicate societies is Ibn al-Haytham (354–*c.* 430/965–*c.* 1040). In his book *On the Configuration of the Cosmos* (*Fī Hay'at al-ʿālam*), he argued for the physical reality of the celestial spheres. He put forward a series of physical planetary models containing nested celestial spheres that functioned and rotated according to Aristotelian natural laws, such as the denial of the void and the impossibility of all change, beyond physical motion, for celestial bodies (Sabra 1998a, 296–7; for images and a discussion of these models, see Section II.6.1.2 and Figure II.6.1). But this does not mean that nobody before him interpreted the Ptolemaic models physically (al-Ṭūsī 1993, 31). Examples can be found in (Pseudo-) Mashā'allāh's (d. *c.* 815) *Book of the Universe* (*Liber de orbe*), which recently has been identified with Dūnash ibn Tamīm's (d. 332/944) *Book on the Configuration [of the Universe]* (*Kitāb fī l-ha'ya*) and Muḥammad ibn Mūsā's (d. 259/873) *The Movement of the First Sphere* (*Ḥarakat al-falak al-ulā*). Both deal with the physical existence of the celestial spheres (Mimura 2015, Saliba 1994b).

Although works with several similar titles appeared before Ibn al-Haytham's treatise, these lacked any specific discussion of whether or how the planetary models could reflect physical reality. In the following decades, new generations of astronomers followed in Ibn al-Haytham's footsteps, bringing a new genre of astronomical texts into being under the older title, calling them books of *hay'at al-aflāk*, meaning "configuration of the heavenly spheres". In their standard form, these works usually have a four-chapter structure: (1) some preliminaries from geometry and natural philosophy, (2) planetary models, (3) mathematical geography and (4) sizes and distances of the heavenly bodies. In the abridged form of this genre, these topics were usually considered without reference to their observational or mathematical underpinnings.

Modern writers have raised doubts about accepting Abū ʿAlī Hasan ibn al-Haytham as the author of *On the Configuration*, proposing Muḥammad ibn al-Haytham as the author (Rashed 1993, 2: 13–14; Sabra 1998b rejected this suggestion). However, some medieval scholars held that the former was indeed the scholar who presented a physical interpretation of the planetary models. One important voice in this chorus was that of Quṭb al-Dīn al-Shīrāzī (633–710/1236–1311), who wrote in his *Preferences for Muẓaffar* (*Ikhtīyārāt-i muẓaffarī*):

> A group of the moderns [that is Islamicate scholars] have put forward the topic of the solidity of spheres and the representation of the causes of the motions, which they have found through observation, and have assumed for any motion a sphere as its mover; one such was Abū ʿAlī ibn al-Haytham, who was one of the great scholars of the science of mathematics. The [discipline] of the configuration of celestial spheres, considered according to the method of embodiment, has greatly benefited from his work and that of people similar to him.
>
> (*MS Tehran, Millī Library 1954, 91*)

Al-Shīrāzī attributed to Abū ʿAlī ibn al-Haytham not only the "topic of solid spheres" but also the "assumption for any motion [or moving point] (that it has) a sphere as its mover". It is likely that the former refers to *On the Configuration* and the latter to Ibn al-Haytham's *Doubts Concerning Ptolemy* (*al-Shukūk ʿalā Baṭlamīyūs*; Ibn al-Haytham 1971, 1985). Ibn al-Haytham had realized that the simple physical interpretation of the Ptolemaic models was not without difficulties, and his *Doubts Concerning Ptolemy*, discussed in Section II.6.2, reflects such a position.

II.6.1.2 Al-Kharaqī's new approach

Unsatisfied with the state of the art, al-Kharaqī (477–553/1084–1158), a major and innovative representative of the *hay'a* genre, took his predecessors to task in his *Ultimate Comprehension of the Subdivision of Celestial Spheres* (*Muntahā l-idrāk fī taqāsīm al-aflāk*) (526/1131):

> And they [the authors of *hay'a* books] did not include any demonstration or any of the reasons that led them to what they have imagined to be the position of the spheres in relation to one another, nor any of the anomalies of the planetary motions. . . . Abū ʿAlī ibn al-Haytham and some others represented the combination of the spheres according to how they imagined the revolutions of solid spheres carrying the stars, arranged in relation to one another. . . . So I decided to compile a book for my friends, including most of what is needed . . . [while] adding explanations and reasons . . . to spare the reader from mere imitation.
>
> *(Qalandari 2020, 2 & 3, parag. [3], [4])*

Al-Kharaqī appears to refer here to Ibn al-Haytham's *On the Configuration*, which, as he says, lacks any demonstration for any of the topics. Because it filled this gap, al-Kharaqī's book served as a comprehensive work useful even for experts. Nevertheless, al-Masʿūdī (d. after 613/1216), al-Jaghmīnī ([d. 618/1221–2]; S. Ragep 2016) and many others continued to produce abridged *hay'a* books for a general audience.

Figure II.6.1 shows two models. The one on the left represents the purely mathematical Ptolemaic model for the superior planets in the *Almagest*. The one on the right is the standard physical interpretation of this model as it appears in most of the *hay'a* books, replacing the circles with cross sections of three-dimensional ethereal spheres. These are either filled with colors for contrast – shown here in gray tones – or appear as ink line drawings only. Point P is the planet that uniformly rotates as an effect of the motion of a sphere with center C, which is called an

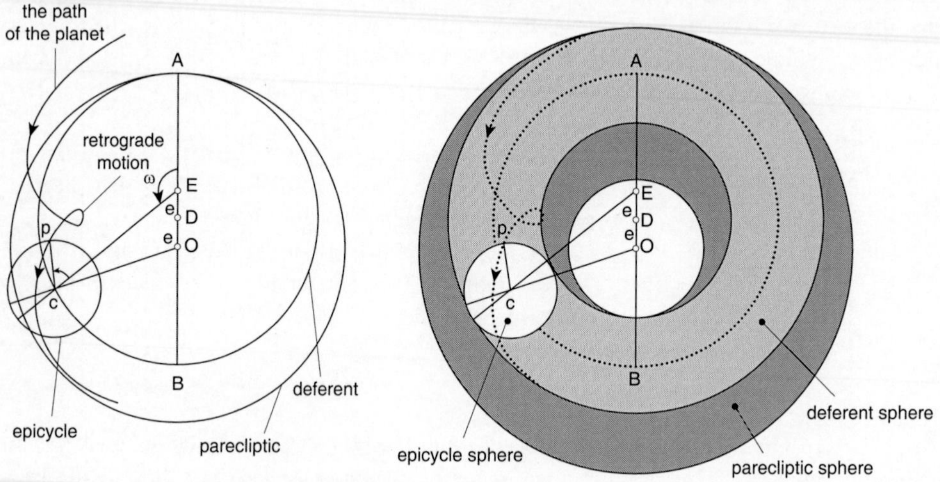

Figure II.6.1 The physical planetary model for the superior planets in *hay'a* books (right) compared to the two-dimensional *Almagest* model (left). In both diagrams, P, planet; C, center of the epicycle; D, center of the deferent; E, equant point; O, center of the world = position of the observer; e, eccentricity with OD = DE; *ω*, angular velocity of the deferent.

Source: © Amir Mohammad Gamini, Tehran

epicycle. The epicycle, in turn, is moved by the rotation of the deferent orb with center D. As a result, the motion of point P becomes predictable for any given time from the viewpoint of the observer on earth at the center of the universe O. The angular velocity of P on the epicycle and C on the deferent are tuned in such a manner that the planet P retrogrades in longitude in accordance with its period of conjunction (Pedersen 2010, 285). To explain the motions of the planets in latitude, Ptolemy supposed some oscillation for the spheres. The rotation of the deferent is not uniform around D. Rather it rotates in a manner that makes the angular velocity of C uniform around E. Point E is called the equant point and stands at the same distance to D as O. Ptolemy introduced this point to explain the observed variation of the sizes of the retrograde arcs of the planets at different points of the ecliptic (Gamini and Masoumi-Hamedani 2013; Swerdlow 2004). This point also designates the epoch of the epicycle.

II.6.2 Planetary models and Aristotelian physics

Almost all the authors of the *hay'a* books agree that the physical nature of the spheres conforms to Aristotelian physics: (1) the celestial spheres, like the planets and stars themselves, are composed of the invisible fifth element "ether"; (2) no vacuum exists in either the celestial realm or the sublunar world. Thus, for each planet, a third sphere, called the parecliptic (*mumaththal*), with center O, fills the spaces between deferent spheres. In Figure II.6.1, the parecliptic is shown as the two dark gray sections of the figure. Although the parts of a parecliptic were physically separated from each other by the intervening deferent orb, they were usually regarded as a single body. The parecliptics of the planets are nested within each other according to their order from the earth and cause the centers of the deferents to rotate slowly around the center of the universe, producing the observed motions of their apogees.

While a large-scale figure of the cosmos (Figure II.6.2) appears in many of the *hay'a* books, it is unclear when this step toward visualization was taken. No drawing of this kind is known from Greek or early Arabic manuscripts. While a late manuscript of Ibn al-Haytham's *On the Configuration* includes this figure, the two earlier surviving copies represent it only in verbal form. The figure, as found in later Arabic, Persian and Turkish texts, shows a schematic cross section through the parecliptics. No eccentrics and epicycles are given, and all the parecliptics are

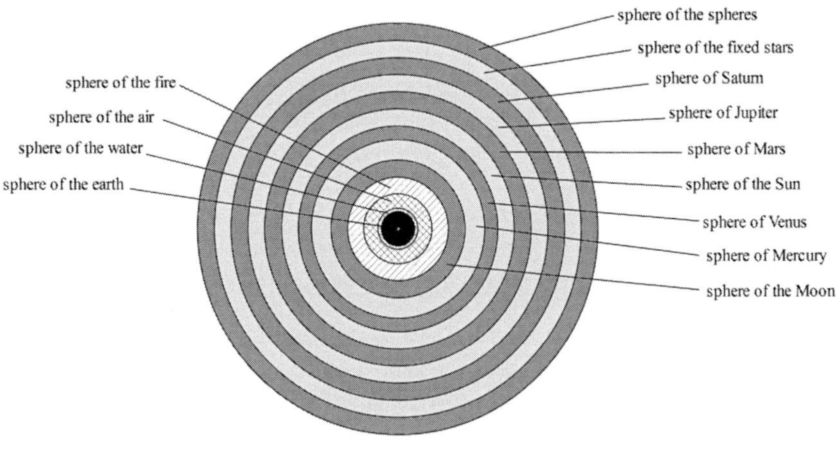

Figure II.6.2 The order of the parecliptic spheres and the elements.

Source: © Amir Mohammad Gamini, Tehran

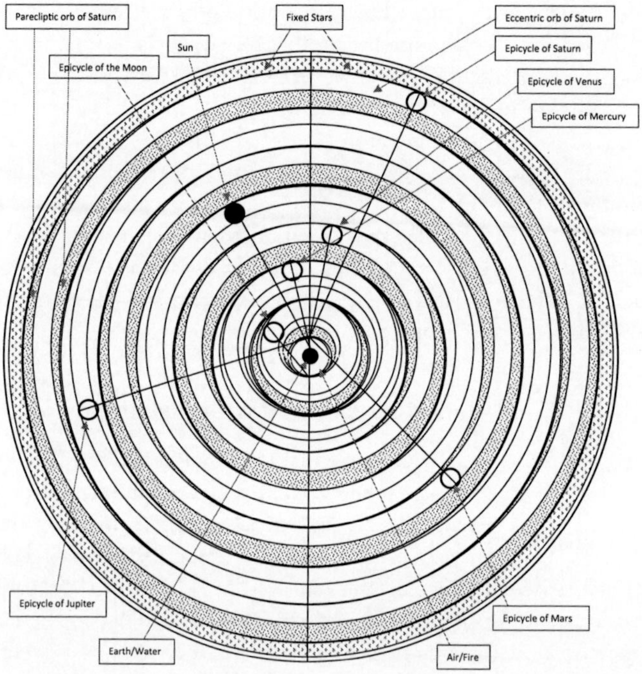

Figure II.6.3 The Ptolemaic cosmos with the spheres of all the planets according to Ibn al-Haytham's verbal depiction in his *Configuration* as reconstructed by Dirk Grupe, Munich, in a slightly simplified manner. The positions and sizes of the sun and the epicycles on their deferent circles are chosen arbitrarily.

Source: © Dirk Grupe, Munich.

represented as equal in depth, although the sizes of the epicycles vary substantially, with Venus and Mars being the largest, with correspondingly large parecliptic orbs. These size differences are treated in the fourth part of books on *hay'a*.

By way of comparison, Dirk Grupe (Munich) translated Ibn al-Haytham's verbal depiction in his *Configuration* into a slightly simplified diagram. The positions of the sun and the epicycles on their deferent circles, as well as the sizes and directions of the eccentricities, are chosen arbitrarily. The depths of all the parecliptics are shown as equal, but the eccentrics are also present (white); consequently, the parecliptic for each planet appears as a pair of elongated crescents, which are shaded in the cases of Saturn, Mars, Venus and the moon. The moon and Mercury have double sets of deferents, the innermost of which are not shaded.

At the beginning of his *Doubts on Ptolemy* (*al-Shukūk ʿalā Baṭlamyūs*), Ibn al-Haytham formulated four fundamental principles based on Aristotelian natural philosophy. They are also found in an appendix to *On the Configurations* and summarized here:

> (1) a natural body cannot have more than one natural movement; (2) the natural movement of a simple body can only be invariable [meaning: a rotation with constant speed]; (3) the celestial bodies cannot be violated [meaning: penetrated in depth]; (4) no vacuum can exist.
> *(Ibn al-Haytham 1985, 55–6, 1990, 7, 231; Saliba 1996, 1: 74)*

On this basis he declared that a simple physical interpretation of the Ptolemaic models, assuming a physical sphere corresponding to each one of the circles, as proposed in *On the Configuration*, contradicts Aristotle's physics, because the speed of a deferent is not constant around its center. Ptolemy required the point on the circumference of the deferent that carried the epicycle center to rotate uniformly in relationship to a point other than the deferent center, that is, the equant. According to Aristotle, however, the motion of a celestial sphere required that each point of its circumference rotate uniformly around the sphere's center. Ibn al-Haytham stated repeatedly that the lines and circles of the models provided by Ptolemy in the *Almagest* were merely imaginary and did not describe configurations of real bodies (Saliba 1996, 1: 74–80). Nevertheless, he went on to reject the physical models of spherical shells proposed in Ptolemy's *Planetary Hypotheses* and *On the Configuration*, whether written by him or by the other Ibn al-Haytham (Saliba 1996, 1: 80–2).

In the course of his discussion of a broad range of defects, contradictions and implausible claims in Ptolemy's works, Ibn al-Haytham emphasized that four features of his models violated the four principles listed earlier: (1) the nonuniform motion of the deferent around its geometric center (the equant problem) violated principle number 2, while (2) the oscillation of the lunar epicycle in longitude according to the *prosneusis* point; (3) the oscillations of the epicycles in latitude; and (4) the oscillation of the inferior planets' deferents, all violated principle (1) above (Ibn al-Haytham 1971, 16–18, 27, 36, 41). These points led to the establishment of a series of 16 difficulties (*ishkālāt*) in the models of the planets, which were investigated by the next generations of *hay'a* writers (al-Ṭūsī 1993, 1: 50; Saliba 1996, 1: 60–1). One may classify them into two categories: difficulties related to (1) the equant problem, and (2) the periodic oscillation of the epicycles and the deferents out of the plane of the ecliptic.[2]

Here we focus on the equant and similar problems, and one type of solution. An early example can be found in Abū ʿUbayd al-Jūzjānī's (d. 462/1070) *Summary of the Arrangement of the Spheres* (*Khilāṣ tarkīb al-aflāk*), where the author presented a solution for the equant problem that is not without difficulties (Saliba 1980). The issues regarding oscillation are more difficult and would go beyond the space available here. Surveys of both problems can be found in various Arabic and Persian treatises, one example being Muḥammad ibn Qāsim's (known as al-Akhawayn; d. *c.* 905/1500) *Problems in the Knowledge about the Configuration* [*of the Universe*] (*Ishkālāt fī ʿilm al-hay'a*; Saliba 1996, 1: 60–1). Some of the *hay'a* writers explicitly asserted that the motions depicted in Ptolemy's models were not compatible with the nature of the celestial spheres. Nonetheless, they never denied the validity of Ptolemaic models for computational applications.

Over the course of the 7th/13th century, astronomers discussed the problems of the standard models with greater interest than before. Among other topics, they sought to develop new configurations of the celestial spheres to replace the ones established by Ptolemy. Modern scholars refer to the initial contributors to this debate, Mu'ayyad al-Dīn al-ʿUrḍī (d. 645/1266), Naṣīr al-Dīn al-Ṭūsī (597–672/1201–1274) and Quṭb al-Dīn al-Shīrāzī, as the "Maragha School," because, at one point or another, they all worked in the Maragha observatory, built near the court of the Ilkhanid dynasty in northwestern Iran. Their works were not confined to proposing new planetary models. Rather, following al-Kharaqī's *Ultimate Comprehension*, they covered most of the topics and problems associated with the science of *hay'a*, including the observational and geometrical reasons behind the planetary models, the centrality and immobility of the earth, the sizes and distances of the heavenly bodies, the division of the inhabited parts of the earth into seven climes (Chapter I.13) and so forth. These comprehensive *hay'a* books, unlike abridged ones, go into the details of each problem, mentioning and criticizing the ideas of other experts.

Although al-Kharaqī's *Ultimate Comprehension* was a comprehensive *hay'a* book, it did not include non-Ptolemaic models. Al-ʿUrḍī, a court astrologer of the last Ayyubid ruler of Aleppo and Damascus, was the oldest member of the Maragha group. Shortly before the Mongol conquest of Syrian cities, he had finished his book on *hay'a* (*Kitāb al-hay'a*; 657/1259), in which he laid out new planetary models that avoided using deferents with nonuniform motion, introducing instead an additional epicycle and redefining the eccentricity of the deferent. In his physical models, all his spheres and orbs rotate uniformly about axes that pass through their geometrical centers, although many of these centers are distant from the center of the cosmos. Furthermore, his book does not follow the four-chapter structure of a standard *hay'a* book. Rather, it retains the order of the *Almagest*, starting with the position of the earth in the universe and then treating the configuration of the spheres for the Sun and the Moon, followed by the computation of their distances and sizes. The Ptolemaic and non-Ptolemaic planetary models of Mercury, Venus and the superior planets appear in their corresponding places. Whether this structural format reflects a rejection of al-Kharaqī's literary practice or a local continuation of an older tradition is unknown (al-ʿUrḍī 1990; Schmidl 2007).

The next scholar who proposed a series of new models was al-Ṭūsī. In his *Memoir on Astronomy* (*al-Tadhkira fī ʿilm al-hay'a*; 659/1261), al-Ṭūsī describes the difficulties of the Ptolemaic models and puts forward his ideas for solutions. Although the *Memoir* was written after the Mongol conquest, when al-Ṭūsī had already moved to the observatory he and al-ʿUrḍī established at Maragha, he had been busy with his new models already some time before but not earlier than ʿUrḍī's book on *hay'a*. In 643/1245, while in the Ismāʿīlī town of Tūn, he proposed some of them in a Persian treatise called *The Solution of the Difficulties for Muʿīn [al-Dīn]* (*Ḥall-i mushkilāt-i muʿīniyya*; Ragep 2017, 164). Muʿīn al-Dīn (1st half 7th/1st half 13th century) was one of the sons of the 26th Imam of a branch of the Ismāʿīlīs called Nīzārīs (Chapter II.3), for whom al-Ṭūsī had written a *hay'a* book called *Treatise for Muʿīn [al-Dīn]* (*al-Risāla al-muʿīniyya*). Four natural philosophical principles inspired by Aristotelian teachings precede his discussions of the theoretical and observational problems and their solutions in the *Memoir*, which I summarize here according to Ragep's translation in (al-Ṭūsī 1993, 1: 98–102):

> (1) a body is either simple in which case it has a single nature and what issues forth from that nature does so monoformly, or else compound, in which case it is composed of simple bodies and may turn out to be a species distinct from them. A simple body is either celestial or elemental. The celestial bodies are the orbs and the luminous bodies whose proper place is the orbs.. . . A void is impossible; (2) every motion has a principle (. . . self-moved . . . , or moved by something other than itself). . . . If the motion of a self-moved mobile is monoform, its principle of motion is called a nature whether the motion is natural and elemental or voluntary and celestial. . . . (3) motion due to a nature is divided into: (a) that which is toward the center;. . . (b) that which is away from the center; . . . (c) that which is about the center; this motion is in place and circular, and it is characteristic of the celestial bodies . . . ; (4) nothing having the principle of circular motion can undergo any rectilinear motion at all, and conversely, except by compulsion. Thus the celestial bodies neither tear nor mend, grow nor diminish, expand nor contract; neither does their motion intensify nor weaken. They do not reverse direction, turn, stop, depart from their confines, nor undergo any change of state except for their circular motion, which is uniform at all times.

Based on these principles, al-Ṭūsī proposed modified models for the moon, Venus and the three upper planets. To solve the outstanding problems, and especially to avoid the use of deferents with nonuniform motion, he employed a mathematical device he had devised himself consisting of two circles or orbs, which between them created a reciprocating, straight-line motion. (This is described in more detail in Section II.6.3.) For Mercury, he admits that "it has not yet been possible for [him] to conceive how it should be done" (al-Ṭūsī 1993, 1: 208).

Having received his advanced mathematical and astronomical training in the circle of al-ʿUrḍī and al-Ṭūsī at the Mongol court, al-Shīrāzī aimed to go beyond the works of his teachers, composing a series of three comprehensive *hay'a* books, in each of which he critically surveyed their non-Ptolemaic models, correcting their errors and subsequently his own. *The Limits of Attainment in the Understanding of the Heavens* (*Nihāyat al-idrāk fī dirāyat al-aflāk*) was written in Arabic in 680/1281. After only four months, its author published another book, in Persian, *Preferences for Muẓaffar* (*Ikhtīyārāt-i muẓaffarī*) (Gamini 2017, 183, no. 54). And four years later, recognizing that some of his own results were not without errors, he wrote, again in Arabic, *The Royal Gift* (*al-Tuḥfa al-shāhiyya*), which was recognized as his main achievement (Saliba 1979; Gamini 2017, 165). Al-Ṭūsī, al-ʿUrḍī and al-Shīrāzī were the leading scholars of the science of *hay'a* of their time. In what follows I present a short survey of one the models proposed by al-Ṭūsī to solve one of the difficulties (*ishkālāt*; see explanation above).

II.6.3 Al-Ṭūsī's model for the superior planets

In this section I present al-Ṭūsī's model for the superior planets as an example of a non-Ptolemaic model. As said earlier, when thinking in terms of physical, solid spheres, the two main issues of Ptolemy's models were the equant problem and the problem of the oscillation of epicycles. In the models for the superior planets the equant is a point the same distance from the center of the deferent as the center of the universe, but on the opposite side, and serves as the center of uniform rotation for the epicycle center. Because of this displacement of the center of rotation, none of these motions could be reconstructed as orbs or spheres that rotated uniformly about axes passing through their geometrical centers. To solve this difficulty, al-Ṭūsī designed a simple model of three spheres moving uniformly around their geometrical centers (Ragep 1999, 195–209). Two of these spheres constitute a device, which is now called the "Ṭūsī couple". Figure II.6.4 represents al-Ṭūsī's description of the planar case.

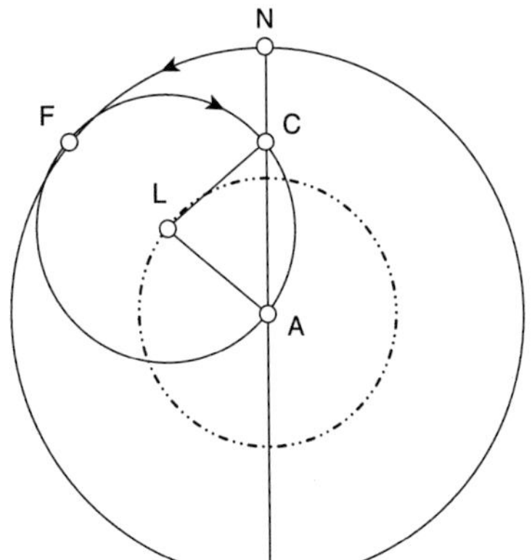

Figure 2.6.4 Ṭūsī couple in the planar case; it converts circular motion into linear motion: A, center of circle FMN; F, tangential point of the two circles FMN and ACF; L, center of circle ACF.

Source: © Amir Mohammad Gamini, Tehran.

> If two coplanar circles, the diameter of one of which is equal to half the diameter of the other, are taken to be internally tangent at a point, and if a point is taken on the smaller circle – and let it be at the point of tangency – and if the two circles move with simple motions in opposite direction in such a way that the motion of the smaller [circle] is twice that of the larger so the smaller completes two rotations for each rotation of the larger, then that point will be seen to move on the diameter of the larger circle that initially passes through the point of tangency, oscillating between the endpoints.
>
> *(al-Ṭūsī 1993, 1: 194)*

Thus, the whole device generates a repeating straight-line motion from two circular motions. The planar case may be converted into a three-dimensional object, suitable for incorporation into spherical models for planetary motion, by imagining the circles in Figure II.6.4 as the equators of two spheres that rotate about axes at A and L. The smaller sphere rotates freely inside a cavity in the larger one. Another version of the Ṭūsī couple deployed by Ṭūsī and his successors makes the two circles into lines of latitude on concentric rotating spheres (Ragep 2000, 2017). This version of the Ṭūsī couple may be applied in the models to solve other problems than the first version, which was developed to solve the equant problem, for example, the oscillation of the epicycles.

Ṭūsī explains that if one assumes a deferent circle with center E, carrying a Ṭūsī couple on its circumference with the angular velocity ω (the same as the velocity of the large circle), point C, the center of the epicycle carrying the planet, while describing a path close to the circumference of a circle with center D, moves uniformly with respect to point E (Figure II.6.5). This is what I call "al-Ṭūsī's simple model", which may be employed instead of the Ptolemaic deferent. In the physical version of the Ṭūsī couple, to accommodate the epicycle sphere carrying one of the superior planets (the white sphere with center C in Figure II.6.6), the epicycle sphere rotates inside a small orb (dark gray in Figure II.6.6), carried inside a larger orb (light gray in Figure II.6.6). But the diameter of the smaller orb is no longer half of the large one. It can be as large as the epicycle needs. In fact, the paths of the motions of the small and large spheres are equivalent to the circles of a Ṭūsī couple, as it is presented by dashed circles in Figure II.6.6. All these spheres and orbs rotate uniformly about axes that pass through their geometrical centers. (For the

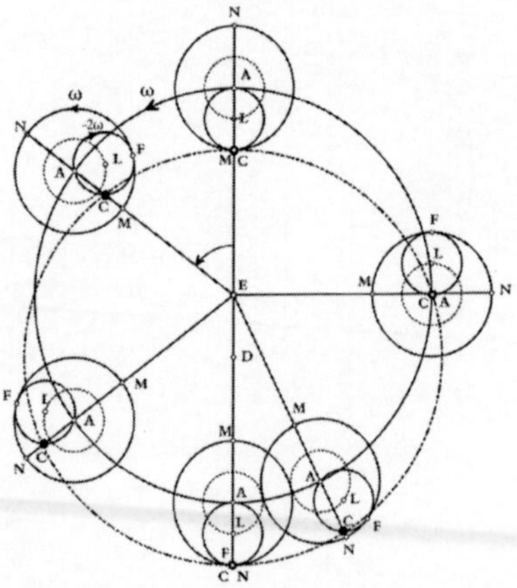

Figure II.6.5 Ṭūsī's simple model

Source: © Amir Mohammad Gamini, Tehran

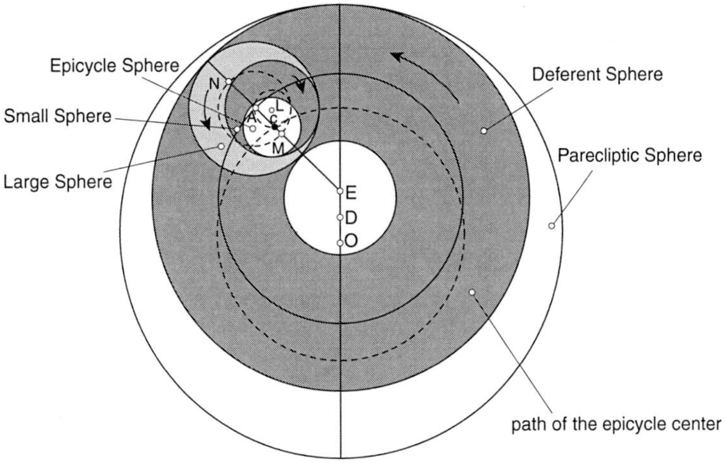

Figure II.6.6 Ṭūsī's model for superior planets

Source: © Amir Mohammad Gamini, Tehran

complete model, see al-Tūsī 1993, 1: 200–6 and esp. Fig. T13, p. 204.) Note that in contrast to Ptolemy, the deferent orb in al-Ṭūsī's model is concentric to the equant. For the moon, however, al-Ṭūsī continues to use eccentric deferents, since the universe center plays the role of the equant. He adds similar arrangements of epicycles and Ṭūsī couples to closely approximate the results of Ptolemy's equant, again using only spheres and orbs that rotate uniformly about axes passing through their geometrical centers.

II.6.4 Discussions of planetary theory after 700/1300

The scientific tradition established by al-Ṭūsī's, al-ʿUrḍī's and al-Shīrāzī's works continued, for example, through later authors' commentaries on al-Ṭūsī *Memoir* and the study of al-Shīrāzī's *The Royal Gift* (Figure II.6.7). There are many extant manuscripts of these two works produced in Herat, Samarqand, Isfahan, Mashhad and other cities during the 9th/15th century and afterward (Brentjes 2010, 331, 404). This continued interest in modifying previous models or establishing new ones addressed some of the inherent problems that the scholars from the Maragha School had not been able to solve.

Al-Shīrāzī, for instance, had hoped to propose his new models as the standard planetary models. But he was not successful with regard to a final model for Mercury, since it although his consisted of more spheres than al-ʿUrḍī's. None of the proposed models was better than the others in absolute terms. Finding criteria enabling a selection of models that were both efficient or minimalistic and true was one of the main issues that al-Shīrāzī struggled with. He emphasized, for example, that al-Ṭūsī's model for the moon includes more spheres than his own model. In his opinion, this shows the superiority of his model for the moon over al-Ṭūsī's model (MS Tehran, Millī Library 1954, 132–3). This criterion appears to be a specific case of al-Shīrāzī's principle of "the elimination of what is not needed". This principle, which appears several times in his text, brings to mind "Occam's Razor", also known as the principle of parsimony [= reject from two equally explanatory theories the one with the most variables] (Gamini and Sadrforati 2022). When he argues that there is no need to presuppose a sphere for each one of the fixed stars, he justifies this by saying that "in the [realm of] celestial bodies there is no addition that is not needed" (MS Tehran, Milli Library 1954, 34). Later scholars did not follow al-Shīrāzī's principles in their efforts to choose between

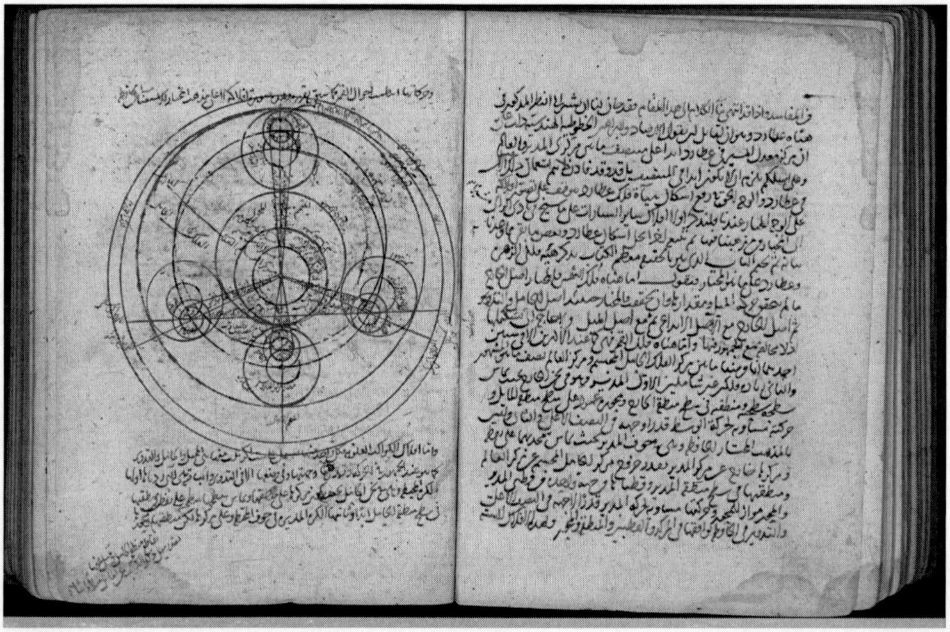

Figure II.6.7 Quṭb al-Dīn Shīrāzī's (633–710/1236–1311) non-Ptolemaic model for the superior planets according to his *The Royal Gift* (*al-Tuḥfa al-shāhiyya*); MS Tehran, Majlis Library, 3835, fol. 131b-132a.

Source: © Majlis Library, Tehran

models. In spite of various remarkable attempts and although al-Shīrāzī's efforts created a basis for later generations of scholars, the tradition of non-Ptolemaic modeling after al-ʿUrḍī, al-Ṭūsī and al-Shīrāzī did not converge to a series of standard models (Gamini 2017).

Currently, some of the best-known successors of the scholars from Maragha are Ṣadr al-Sharīʿa (d. 747/1346–1347), Ibn al-Shāṭir (705–777/1304–1375), ʿAlī Qūshjī (805–879/1403–1474) and Shams al-Khafrī (d. 956/1550). Most of these scholars also published religious work or worked in religious institutions. One may also add Niẓām al-Dīn al-Nīsābūrī (d. c. 729/1330) to the list of Marāgha's followers, even if he did not produce any non-Ptolemaic planetary models. He studied with al-Shīrāzī in Tabriz and worked at the Ilkhanid court there. Niẓām al-Dīn wrote a commentary (*Tawḍīḥ al-Tadhkira*) on Ṭūsī's *Memoir on Astronomy*, which was widely adopted in *madrasa*s in the Persianate world, including that of Ulugh Beg in Samarqand (Morrison 2007). Niẓām al-Dīn's near contemporary Ṣadr al-Sharīʿa lived in Bukhara (modern Uzbekistan). His major known work is *The Adjustment of the Configuration of the Celestial Spheres* (*Kitāb Taʿdīl hay'at al-aflāk*), which follows al-Shīrāzī in using al-ʿUrḍī's method for avoiding nonuniform motion of the deferent. Ṣadr al-Sharīʿ modified al-Shīrāzī's lunar model to deal with the *prosneusis* problem, which Shīrāzī had not addressed (Niazi 2014, 134–9). Ṣadr al-Sharīʿa solved it by adding another epicyclic sphere (Ṣadr al-Sharīʿa 1995, 283).

Ibn al-Shāṭir's professional life was spent as head timekeeper (*muwaqqit*) at the Umayyad Mosque in Damascus. He published at least two *zij*s and developed radically new planetary and lunar models building on al-ʿUrḍī. Ibn al-Shāṭir's models completely eliminated eccentrics and used only concentric deferents and epicycles, including double epicycles. His models for the moon and Mercury later appeared in the astronomical works of Nicholas Copernicus

(1473–1543) (Saliba 1987, 371; Nikfahm-Khubravan and Ragep 2019; see also the following discussion).

ʿAlī Qūshjī was born and educated in Samarqand, where he became director of the observatory founded by Ulugh Beg (r. 812–850/1409–1447 as governor of Samarqand; 850–853/1447–1449 as head of the dynasty). After Ulugh Beg's assassination Qūshjī served in various other princely courts, ending his life under the patronage of Sultan Meḥmed II (r. 848–849, 855–886/1444–1446, 1451–1481) in Constantinople ([c. 876–878/1472–1474]; today Istanbul; Barker and Heidarzadeh 2016). His children and students founded a dynasty of astrologers at the Ottoman court. ʿAlī Qūshjī both contributed to the *Zīj-i Ulugh Beg*, written at the Samarqand observatory, and corrected it after publication. He also produced original models for Mercury and Venus that differed from both Ptolemy's and al-Shīrāzī's (Fazlıoğlu 2007). In 2005 Jamil Ragep pointed out a striking similarity between a proof and diagram in a book written by Regiomontanus (1436–1476), published in Nuremberg in 1496, and work by ʿAlī Qūshjī. Like Copernicus's use of Ibn al-Shāṭir, this seems to be further evidence of advanced Islamicate planetary theory being available in Italy.

The connections between Copernicus, Ibn al-Shāṭir and earlier members of the Marāgha school have been intensely studied by historians of astronomy since their discovery in the 1950s. Despite some recent controversy, the main consensus is that Copernicus obtained information from works of al-Ṭūsī, al-ʿUrḍī, and Ibn al-Shāṭir, probably during his education in Italy, and incorporated their ideas in his own astronomical work (Swerdlow and Neugebauer 1984; Nikfahm-Khubravan and Ragep 2019). A central problem concerns the transmission of non-Ptolemaic planetary models from Arabic and Persian writers. Here, the focus was for a long time on texts and their presence in Italian cities. Less attention has been paid to the study of such works in centers of the Mamluk and Ottoman empires, and almost none in the Safavid and Mughal empires. It was not recognized that, for instance, al-Shīrāzī's models were copied and known in Mamluk Cairo and are also preserved in some Ottoman libraries. The Ottoman scholar Taqī al-Dīn Muḥammad ibn Maʿrūf (d. 985/1585) and director of the short-lived observatory in Istanbul knew several works of Ibn al-Shāṭir, in particular those describing his new instruments.

Furthermore, Langermann has shown that another resident of the Ottoman Empire, Mūsā Jālīnūs (d. after 1536), knew of Ibn al-Shāṭir's planetary theories (Langermann 2007, 290). He identified this Jewish merchant and physician, who was apparently also the holder of a military fief, as Moses Galeano (d. after 1536). Langermann emphasized that this is the first available evidence that someone who had visited the Veneto while Copernicus was there at the turn of the 16th century (especially 1500–1502) was familiar with four types of planetary models – the homocentric ones of al-Biṭrūjī (active 2nd half 6th/12th century; see the following discussion), the Ptolemaic models that used epicycles and eccentrics, those of Ibn al-Shāṭir that used only concentric circles carrying epicycles, and those of Levi ben Gershon ([1288–1344]; Langermann 2007, 288, 291). Morrison has followed Galeano's trail in greater detail and provides further evidence for the possible transfer of information on non-Ptolemaic planetary models from the Ottoman empire to Italy at the time Copernicus was there (Morrison 2011, 2014).

Other events to be considered in this complex history were highlighted by Dobrzycki, Kremer, Ragep, Barker and Heidarzadeh. Dobrzycki and Kremer have argued that knowledge of works from Maragha must have been already available in Europe before 1461, on the grounds that two ephemerides originating with Regiomontanus's teacher Georg Peurbach (1423–61) used a Maragha-like harmonic device to calculate planetary longitudes (Dobrzycki and Kremer 1996). As mentioned earlier, Ragep pointed to a connection between Regiomontanus and ʿAlī Qūshjī, who also discussed the rotation of the Earth and proved a proposition through which any Ptolemaic epicyclical model could be transformed into an eccentric model (Ragep 2005, 2007, 75), while Barker and Heidarzadeh emphasized the direct Venetian contacts with Muslim courts and their scholars in eastern Anatolia and western Iran (Barker and Heidarzadeh 2016). Going beyond the similarities of the mathematical models and the possibilities of transmission, Ragep stressed the much larger intellectual context in which the relationship between the debates in Islamicate and western Christian societies about the understanding of the universe needs to be seen. In addition to the rotation of the earth and the transformation of epicyclical into eccentric models, this larger context encompassed questions about comets and the epistemic relation between astronomy and natural philosophy (Ragep 2001a, 2001b, 2007, n. 24).

While Copernicus was working in Europe, Shams al-Dīn Khafrī (d. 942/1535) was adding to the Maragha tradition of non-Ptolemaic models in Shiraz, Persia. He constructed three new solutions for all 16 difficulties in the models of the planets initially pointed out by al-Haytham and codified Mercury (Saliba 1994a).

However, the non-Ptolemaic models never succeeded in entering the standard abridged *hay'a* books, the main teaching material for astronomy at *madrasa*s in Islamicate societies. Those abridged *hay'a* works continued to present Ptolemy's models as the standard arsenal. Manuscripts of several texts on *hay'a* are extant from the centuries following the Maragha tradition. Some of the most famous ones, in addition to the previously mentioned titles, are al-Kāshī's *The Quintessence for Iskandar [Ṣulṭān]* (*Lubāb-i Iskandarī*), al-Qūshjī's *Hay'a in Persian* (*Farsī-i hay'at*), Ghiyāth al-Dīn al-Dashtakī's (866–948/1462–1542) *Ambassador* (*al-Safīr*) and Bahā' al-Dīn al-ʿĀmilī's *Dissection of the Orbs* (*Tashrīḥ al-aflāk*). Bahā' al-Dīn al-ʿAmilī (953-1030 /1547-1621), as well as readers in Mughal India and the Ottoman Empire, kept the tradition alive and taught them at madrasas or in their homes. A number of other books called *A Treatise on Hay'a* are extant, whose authors are not well-known yet. Examples are: Abū Isḥāq Kūbanānī (d. after 886/1481; see Chapter II.10), Tāj al-Dīn Saʿīdī (d. 951/1544), Asadallāh Shāhmīr (10th/16th century), Aḥmad ibn ʿAbd al-Aḥad Fārūghī (971–1024/1564–1616) and Muḥammad Bāqir Najm Thānī, Masʿūd ibn ʿAbd al-Raḥīm Anṣārī, Tāj al-Dīn Munajjim Shīrāzī and Muḥammad Farmānī Amadī, whose dates are unknown.

II.6.5 Planetary theory in North Africa and al-Andalus

The tradition of *hay'a* books was not limited to Islamicate societies in the East. Al-Qaṭṭān's (4th/10th century) *Book of the Configuration [of the Universe]* (*Kitāb al-hay'a*), one of the first *hay'a* books in the western Islamic world, mixed Ptolemaic instructions on sizes and distances of the celestial bodies with Indian elements (Samsó 1994, 2), though other features of the genre of *hay'a*, such as models, are missing. In al-Andalus, although there are signs of familiarity with Ibn al-Haytham's *Doubts* and *On the Configuration*, criticism of Ptolemaic astronomy originated from the *Recapitulation in Regard to Ptolemy* (*al-Istidrāk ʿalā Baṭlamīyūs*) by an unknown writer (Samsó 1994, 5; Saliba 1991, 88). This criticism developed a different approach than in the eastern Islamicate world. Aristotelians of different varieties like Ibn al-Bājja (d. 533/1139), Ibn al-Ṭufayl (d. 581/1185), Ibn Rushd (520–595/1126–1198) and al-Biṭrūjī rejected not only the nonuniform motion of the spheres in contradiction with Aristotelian natural philosophy, but also epicycles and eccentric deferents, contending that the celestial motions had to go around the center of the world, not some hypothetical centers (Sabra 1984). Although Ibn Rushd accepted these in his *Epitome of the Almagest* (*Muktaṣar al-Majisṭī*; Lay 1996), in all his other writings he completely rejected them. The Andalusian approach to planetary models was therefore strictly homocentric, as its representatives attempted to revive and improve upon the approach to planetary models represented in antiquity by Eudoxus (d. 337 BCE) and his followers. Although they developed many interesting and elaborate schemes, they never reached the point of being able to calculate numerical values for planetary positions. From the viewpoint of practical astronomy, or astrology, their work was therefore of only philosophical interest.

Surviving works that presented new models trying to represent planetary motions without using eccentric or epicycles include al-Biṭrūjī's (2nd half 6th/12th century) *Book on the Configuration* (*Kitāb fī l-hay'a*; dated between 567–596/1185–1217; al-Bitrūjī 1971; Samsò 1994, 12) and *The Light of the World* (*Nūr al-ʿālam*), composed in Judeo-Arabic by Joseph ben Naḥmi'as, who lived in Christian Toledo around 1400 (Morrison 2016[CB]). The possible intellectual or personal connections between al-Biṭrūjī and the Andalusian Aristotelian philosophers and the kind of literature from which Ben Naḥmi'as appropriated his views of Aristotelian physics are not well understood yet. However, Ben Naḥmi'as's overall approach develops ideas found in treatises by Ibn al-Ṭufayl, whom al-Biṭrūjī mentions (al-Biṭrūjī 1971, 1: 61), by Maimonides, who himself connects them to Ibn Bājja (al-Biṭrūjī 1971, 4) and by Ibn Rushd (Sabra 1984; Samsó 2007). Joseph ben Naḥmi'as's work employs Ṭūsī couples, indicating a further infusion of eastern Islamicate astronomy in the intervening two centuries (Morrison 2016).

II.6.6 Encountering heliocentric astronomy

From the 11th/17th century on, Muslim scholars learned from English, Italian and possibly other European visitors about the incipient astronomical paradigm shift that involved replacing the geocentric by a heliocentric model of the universe and the many spheres by a few ellipses (Chapter VI.5). During the 12th/18th century, collaborative work under the Rajput local ruler Sawā'i Jai Singh (r. 1111–1156/1699–1743) led to the introduction of the telescope and further new instruments into the observational work for a new astronomical handbook for the Mughal Sultan Muḥammad Shāh ([r. 1131–1161/1719–1748]; Chapter II.7). New English and French texts on this new astronomy (and other scientific themes) were translated into or summarized in Persian. Muslim scholars began to respond to the new ideas, expressing, for instance, new doubts about planetary models, this time also including the non-Ptolemaic ones.

This new page in the history of the astral sciences in Islamicate societies is not yet well researched. One known example is the work of Abū Ṭālib Ḥusaynī (12th/18th century), a Persian scholar living in India and familiar with early modern astronomy. He criticized the whole process of non-Ptolemaic model building and the impossibility of finding a final, satisfactory solution:

> The first person who decided to solve the problems was Abū ʿUbayd al-Jūzjānī . . . and he proved a hundred and twenty spheres, but by deep consideration it is clear that his solution is erroneous. Afterwards, Ibn al-Haytham Miṣrī decided to solve some of these problems, but [his solution] is also not without difficulty. Later on, Khwāja Naṣīr al-Dīn al-Ṭūsī and Muḥyī al-Dīn al-Maghribī did so with difficulty and much toil, but their error is obvious to the intelligent people. And al-Shīrāzī proposed nine solutions to solve [the problem of] the equant point and then he confessed the invalidity of eight cases, describing seven of them and withholding the eighth to test the intellects of the intelligent people, as he says. And he supposed the ninth one to be satisfying, while it is not without failure as well. . . . It is worth noting that these problems are due to the existence of the spheres.
>
> *(Masoumi Hamedani 1363/1984, 162)*

In contrast, however, to Abū Ṭālib Ḥusaynī's fourth claim, there are no known non-Ptolemaic models attributed to al-Maghribī.

Based on oral knowledge from English settlers in India and his own study of books from Christian Europe, Ḥusaynī believed that he knew what the solution to all the difficulties of the Ptolemaic models should be. He proposed to omit the physical spheres as allegedly done by early modern European astronomers, since without spheres there would be no equant problem (Masoumi Hamedani 1363/1984, 162). Other scholars in various parts of India described their new understanding of the universe either in a comparison between non-Ptolemaic models and models learned from Latin, English or French books or by focusing on one or more of the new results from Christian Europe.

II.6.7 Conclusion

In this chapter, we dealt with the *hay'a* tradition as developed in Islamicate societies. Many of the astronomers, especially Ibn al-Haytham and those after him, worked to develop a satisfactory physical interpretation of the Ptolemaic models. Although they more or less accepted the mathematical and observational soundness of Ptolemy's models,[3] they created various non-Ptolemaic physical models to produce the same observed or theoretically conceived motions for the planets as the Ptolemaic models, employing spheres and orbs that rotated uniformly about their axes. At first this condition could not be met in the genre of *hay'a* launched by al-Kharaqī, with its emphasis on the physical interpretation of Ptolemy's models. These difficulties inspired the Maragha astronomers and their followers to develop new, non-Ptolemaic models. Most notably, al-Ṭūsī and al-ʿUrḍī devised separate solutions to the problems of constructing models that avoided nonuniform motion of deferents in planetary models as well as new models for the moon. The students and successors of al-Ṭūsī and al-ʿUrḍī continued to write *hay'a* and devise improvements on their non-Ptolemaic models as late as the 11th/17th century in Safavid Persia and 12th/18th century in Mughal India. The latest appearance of new work of this sort remains a subject for future research, although new commentaries on Bahāʾ al-Dīn al-ʿĀmilī's *Dissection of the Orbs* (*Tashrīḥ al-aflāk*) were actually printed for example in Lucknow in 1246/1849, in Iran in 1284/1867–8, in Calcutta in 1300/1883 (Mahdavi 2021).

The end of the *hay'a* traditions should probably be linked to the spread of European science during the period of western imperial expansion and its displacement of indigenous traditions. Persian texts, as well as observational activities written or undertaken in the 12th/18th and 13th/19th centuries in the Mughal Empire and other South Asian Muslim states confirm the growing interest in and impact of astronomical theories, books and instruments brought by missionaries and merchants from Portugal, England, France and the Netherlands to different regions in South Asia or acquired by Indian diplomatic and private visitors in Europe (Ansari 1985, 2015, 586–9). This process was part and parcel of a general shift in scientific attitudes and discussions across the Indian subcontinent (Chapter VI.7).

Note

1 Consolidated bibliography.
2 Saliba described the major problems with Ptolemy's models as seen by authors in Islamicate societies in a slightly different order and formulation: (1) *prosneusis* problem, (2) the problem of inclination and deviation of the spheres in the models for Mercury and Venus, (3) the equant problem for the superior planets and (4) the inconsistencies of the planetary distances (Saliba 1996, 1: 59).
3 An exception is Ptolemy's lunar model, in which the distance of the moon to the Earth varies much more than what is expected from observation of lunar apparent size. This problem was efficiently solved in Ibn al-Shāṭir's lunar model (Saliba 1996).

Bibliography

Sources

al-Biṭrūjī. ed. and tr. Goldstein, B. R. 1971. *Al-Biṭrūjī: On the Principles of Astronomy*. 2 vols. New Haven, CT: Yale University Press.

Ḥusaynī, Abū Ṭālib. ed. Masoumi Hamedani, H. 1363/1984. "A Book to Prove the Modern Astronomy," *Maʿarif* 2: 117–85.

Ibn al-Haytham. ed., tr. and com. Langermann, Y. T. 1990 [2017]. Ibn al-Haythams *On the Configuration of the World*. New York City: Garland Publishing Inc. [Reprint: London and New York: Routledge]

Ibn al-Haytham. ed., tr. and ann. Rashed, R. 1993. *Les Mathématiques infinitésimales du IXe au XIe Siècle*. Vol. 2. London: Furqan.

Ibn al-Haytham. ed. Sabra, A. 1971. *Doubts Concerning Ptolemy* (*Al-Shukūk ʿalā Baṭlamīyūs*). Cairo: Dār al-kutub al-miṣriyya.

Ibn al-Haytham. ed., tr. and ann. Voss, D. L. 1985. *Ibn Al-Haytham's Doubts Concerning Ptolemy: A Translation and Commentary*. PhD dissertation. Chicago: University of Chicago.

al-Kharaqī, ʿAbd al-Jabbār. ed. Qalandari, H. 2020. *Muntahā al-idrāk fī taqāsīm al-aflāk* [The Ultimate Comprehension of the Subdivision of Celestial Spheres]. Tehran: Miras Maktoob.

Ptolemy, C. ed., tr. and com. Goldstein, B. R. 1967. "The Arabic Version of Ptolemy's Planetary Hypotheses," *Transactions of the American Philosophical Society* 57.4.

Ṣadr al-Sharīʿa. ed., tr. and com. Dallal, A. S. 1995. *An Islamic Response to Greek Astronomy: Kitāb Taʿdīl hay'at al-aflāk of Ṣadr al-Sharīʿa*. Leiden: E. J. Brill.

al-Ṭūsī, Naṣīr al-Dīn. ed., tr. and com. Ragep, F. J. 1993. *Naṣīr al-Dīn al-Ṭūsī's Memoir on Astronomy (al-Tadhkira fī ʿilm al-hay'a)*. 2 vols. New York: Springer.

al-ʿUrḍī, Mu'ayyad al-Dīn. ed. Saliba, G. 1990. *The Astronomical Work of Mu'ayyad al-Dīn al-ʿUrḍī (Kitāb al-Hay'a): A Thirteenth Century Reform of Ptolemaic Astronomy*. Beirut: Markaz dirāsat al-waḥda al-ʿarabiyya. (In Arabic with an English introduction.)

Manuscripts

MS Tehran, Majlis Library, 3835.

MS Tehran: Milli Library, 1954.

Research literature

Ansari, S. M. R. 1985. "Introduction of Modern Astronomy in India during the 18th–19th Centuries," *Indian Journal of History of Science* 20: 363–402.

Ansari, S. M. R. 2015. "Survey of Zījes Written in the Subcontinent," *Indian Journal of History of Science* 50.4: 575–601.

Barker, P. and Heidarzadeh, T. 2016. "Copernicus, the Ṭūsī couple and East-West Exchange in the Fifteenth Century," in Grenada, M. A. and Bonner, P., eds. *Man and Cosmos*. Barcelona: University of Barcelona Press, 19–57.

Brentjes, S. 2010. "The Mathematical Sciences in the Safavid Empire: Questions and Perspectives," in Hermann, D. and Speziale, F., eds. *Muslim Cultures in the Indo-Iranian World during the Early-Modern and Modern Periods*. Berlin: Klaus Schwarz, 325–402.

Dobrzycki, J. and Kremer, R. L. 1996. "Peurbach and Marāgha Astronomy? The Ephemerides of Johannes Angelus and Their Implications," *Journal for the History of Astronomy* 27: 187–237.

Fazlioğlu, I. 2007. "Qūshjī: Abū al-Qāsim ʿAlā' al-Dīn ʿAlī ibn Muḥammad Qushčizāde," in Hockey *et al.*, eds. [CB], 946–8.

Gamini, A. M. 2017. "Quṭb al-Dīn al-Shīrāzī and the Development of Non-Ptolemaic Planetary Modeling in the 13th Century," *Arabic Sciences and Philosophy* 27.2: 165–203.

Gamini, A. M. and Masoumi-Hamedani, H. 2013. "Al-Shīrāzī and the Empirical Origin of Ptolemy's Equant in His Model of the Superior Planets," *Arabic Sciences and Philosophy* 23: 47–67.

Gamini, A. M. and Sadrforati, M. M. 2022. "The principle of simplicity for Quṭb al-Dīn Shīrāzī," *Studies in History and Philosophy of Science* 91: 60-65.

Langermann, Y. T. 2007. "From My Notebooks: A Compendium of Renaissance Science: *Taʿalumot ḥokmah* by Moses Galeano," *Aleph* 7: 285–318.

Lay, J. 1996. "L'*Abrégé de l'Almageste*: Un inédit d'Averroès en version hébraïque." *Arabic Sciences and Philosophy* 6.1: 23–61.

Mahdavi, Y. 2021. *Astronomy in Safavid Persia: Bahāʾ al-Dīn ʿĀmilī and the Patronage of Science*. PhD dissertation. University of Oklahoma.

Mimura, T. 2015. "A Glimpse of Non-Ptolemaic Astronomy in Early Hay'a Work – Planetary Models in ps. Mashā'allāh's *Liber de orbe*," *Suhayl* 14: 89–114.

Morelon, R. 1996. "Eastern Arabic Astronomy between the Eighth and Eleventh Centuries," in Rashed and Morelon 1996 [CB], 1–19.

Morrison, R. 2007. *Islam and Science: The Intellectual Career of Niẓām al-Dīn al-Nīsābūrī*. New York: Routledge.

Morrison, R. 2011. "An Astronomical Treatise by Mūsā Jālīnūs alias Moses Galeano," *Aleph* 11.2: 385–413.

Morrison, R. 2014. "A Scholarly Intermediary Between the Ottoman Empire and Renaissance Europe," *Isis* 105.1: 32–57.

Murschel, A. 1995. "The Structure and Function of Ptolemy's Physical Hypotheses of Planetary Motion," *Journal for the History of Astronomy* 26: 33–61.

Niazi, K. 2014. *Quṭb al-Dīn Shīrāzī and the Configuration of the Heavens*. New York: Springer.

Nikfahm-Khubravan, S. and Ragep, F. J. 2019. "The Mercury Models of Ibn al-Šāṭir and Copernicus," *Arabic Sciences and Philosophy* 29: 1–59.

Pedersen, O. 2010. *A Survey of the Almagest: With Annotation and New Commentary by Alexander Jones*. New York: Springer.

Ragep, F. J. 2000. "The Persian Context of the Ṭūsī Couple," in Pourjavadi, N. and Vesel, Ž., eds. *Naṣīr al-Din Ṭūsī: Philosophe et Savant du XIIIe Siecle*. Tehran: IFRI and Presses universitaires d'Iran, 113–30.

Ragep, F. J. 2001a. "Tusi and Copernicus: The Earth's Motion in Context," *Science in Context* 14: 145–63.

Ragep, F. J. 2001b. "Freeing Astronomy from Philosophy: An Aspect of Islamic Influence on Science," *Osiris* 16: 49–71.

Ragep, F. J. 2005. "Ali Qushji and Regiomontanus' Eccentric Transformations and Copernican Revolutions," *Journal for the History of Astronomy* 36: 359–71.

Ragep, F. J. 2007. "Copernicus and His Islamic Predecessors: Some Historical Remarks," *History of Science* 45: 65–81.

Ragep, F. J. 2017. "From Tūn to Torun: The Twists and Turns of the Ṭūsī Couple," in Feldhay, R. and Ragep, F. J., eds. *Before Copernicus*. Montreal and Kingston: McGill-Queens University Press, 161–97.

Ragep, S. 2016. *Jaghmīnī's Mulakhkhaṣ: An Islamic Introduction to Ptolemaic Astronomy*. Berlin: Springer.

Sabra, A. I. 1984. "The Andalusian Revolt Against Ptolemaic Astronomy: Averroes and al-Bitrûjî," in Mendelsohn, E., ed. *Transformation and Tradition in the Sciences: Essays in Honor of I. Bernard Cohen*. Cambridge: Cambridge University Press, 233–53.

Sabra, A. I. 1998a. "Configuring the Universe: Aporetic, Problem Solving, and Kinematic Modeling as Themes of Arabic Astronomy," *Perspectives on Science* 6.3: 288–330.

Sabra, A. I. 1998b. "One Ibn al-Haytham or Two? An Exercise in Reading the Bio- Bibliographical Sources," *ZGAIW* 12: 1–50.

Saliba, G. 1979. "The Original Source of Quṭb al-Dīn al-Shīrāzī's Planetary Model." *Journal for the History of Arabic Science* 3: 3–18.

Saliba, G. 1980. "Ibn Sīnā and Abū ʿUbayd al-Jūzjānī: The Problem of the Ptolemaic Equant," *Journal for the History of Arabic Science* 4: 376–403.

Saliba, G. 1986. "The Determination of New Planetary Parameters at the Maragha Observatory," *Centaurus* 29.4: 249–71.

Saliba, G. 1987. "The Role of Maragha in the Development of Islamic Astronomy: A Scientific Revolution Before the Renaissance," *Revue de synthèse* 108.3: 361–73.

Saliba, G. 1991. "The Astronomical Tradition of Maragha: A Historical Survey and Prospects for Future Research," *Arabic Sciences and Philosophy* 1.1: 67–99.

Saliba, G. 1994a. "A Sixteenth-Century Arabic Critique of Ptolemaic Astronomy: The Work of Shams al-Din al-Khafri," *Journal for the History of Astronomy* 25.1: 15–38.

Saliba, G. 1994b. "Early Arabic Critique of Ptolemaic Cosmology: A Ninth-Century Text on the Motion of the Celestial Spheres," *Journal for the History of Astronomy* 25: 115–41.

Saliba, G. 1996. "Arabic Planetary Theories after the Eleventh Century AD," in Rashed and Morelon [CB], 1: 58–127.

Samsó, J. 1994. "On al-Biṭrūjī and the Hayʾa Tradition in al-Andalus," in Samsó, J., ed. *Islamic Astronomy and Medieval Spain*. Aldershot, Hampshire: Variorum.

Samsó, J. 2007. "Biṭrūjī: Nūr al-Dīn Abū Isḥāq [Abū Jaʿfar] Ibrāhīm ibn Yūsuf al-Biṭrūjī," in Hockey *et* al., eds. 2007 [CB], 133–4.

Schmidl, P. G. 2007. "ʿUrḍī: Muʾayyad (al-Milla wa-) al-Dīn (Muʾayyad ibn Barīk [Burayk]) al-ʿUrḍī (al-ʿĀmirī al-Dimashqī)," in Hockey *et al.*, eds., 1161–2.

Swerdlow, N. M. 2004. "The Empirical Foundations of Ptolemy's Planetary Theory," *Journal for the History of Astronomy* 35: 249–71.

Swerdlow, N M. and Neugebauer, O. 1984. *Mathematical Astronomy in Copernicus's De revolutionibus*. 2 vols. Berlin: Springer.

II.7

PRACTICES OF CELESTIAL OBSERVATION IN THE ISLAMICATE WORLD[1]

Amir Mohammad Gamini and Sonja Brentjes

This chapter surveys practices of celestial observations in the Islamicate world. They consisted of a two-stage process. The first is the actual observation of the chosen celestial object. Although the reference points may change, the main purpose of the observation is almost always to measure the position of a celestial object which is determined via pinpointing two coordinates and its aparent size. If the reference is the horizon, it is the altitude or azimuth; when it is measured from the ecliptic, it is called the ecliptic latitude and longitude; and if it is measured from the celestial equator, then it is the declination and the right ascension of the celestial object. The second stage is to use the obtained values for the calculation needed for various applications. In addition to the immediate act of observing, observational practices include social, technical and scholarly aspects. An observer had to find and convince wealthy sponsors, select collaborators and witnesses and, finally, narrate his activities to patrons and other interested audiences. Scholarly practices in a narrower sense encompassed various forms of knowledge production and reproduction such as the training of young men (no reliable records on female observers are known) and the acquisition of relevant literature or acts of recording, interpreting, appraising and verifying observational results. The preparation of observational activities by acquiring or building instruments, auxiliary tools and the necessary buildings required cooperating with craftsmen of different professions, selecting suitable locations and possibly undertaking excursions, adding a number of technical procedures to these social and scholarly activities.

The two main scholarly genres providing information about such practices are compendium-like handbooks with astronomical and other tables (*zīj*es, which will be referred to here as astronomical handbooks) and treatises about instruments (Chapters III.8, IV.2, IV.3). Most, but not all, celestial observations were undertaken with the goal of improving the parameters needed for the compilation of such handbooks. This was considered necessary because compilers repeatedly discovered differences between received data and celestial conditions, as well as between handbooks or calendars by different authors.

Treatises about instruments taught how to construct a single instrument, and how to use it, or both. Another, but quantitatively smaller, group within this genre describes several observational instruments often but not always in a summary fashion. Outside these genres, some treatises or quotes in other works deliver further information on observational activities. An example of such an additional source outside the two main literary genres is an unusual commentary on the *Almagest*. Muḥyī l-Dīn al-Maghribī (d. 682/1283) wrote it in Maragha, calling it *Compendium*

DOI: 10.4324/9781315170718-27

of the Almagest (*Talkhīṣ al-Majisṭī*). This treatise describes the methodological foundations and mathematical treatment of systematic observations (Saliba 1983, 1985; Mozaffari 2014a).

In addition to scholarly literature on astral themes, some information can be gleaned from philosophical, biographical or historical literature. Such sources occasionally report comets, eclipses, conjunctions and other natural events (Chapter V.9). Beyond texts, instruments also provide evidence for practical activities. Numerous craftsmen inscribed their names and date of fabrication on them, and technical analysis reveals the chemical and physical processes mastered by instrument makers (Sarma 2019, 31–2).

The dominant institutional settings of such practical activities were homes of scholars, artisanal workshops in towns or palaces, buildings used on a short-term basis for celestial observations, and architectural complexes built for long-term observations (Sayili 1960[CB];[2] King 1995, 2005[CB], 31; Sarma 2019, 30–1, 60). Pavilions in courtly gardens, the roofs of private buildings or palaces and even mosques served for observing the skies (Sayili 1960, 169–70; Chapter II.3). Calculations were executed in many of those settings, but mostly in private homes, observatories, *madrasas* and special houses created for timekeepers (Sayili 1960; Chapter V.4).

A full survey of all those sources and activities goes beyond the space allocated to this chapter. Instruments and their related practices are described in later chapters (Chapters IV.2, IV.3). This chapter focuses primarily on individual and collective observations and their literary and institutional contexts.

II.7.1 Individual and collective observations

Most recorded observations served for preparing tables that provided the information necessary for casting horoscopes. They were also undertaken to correct observational errors of previous scholars and to check astronomical parameters from translated Sanskrit, Middle Persian and Greek texts and tables. They supported debates about which solar year should be taken as a yardstick and for comparing planetary models. Our knowledge of Islamicate observational activities rests mainly on two types of information: (1) explicit statements in handbooks and related texts and (2) the mathematical analysis of the data provided in tables.

Observations were carried out by individuals, for a single event or as part of a systematic, long-term program. They were also undertaken by groups working together in specific or more general and long-term projects. Observations were carried out with the naked eye, often supported by a sighting device mounted on either portable or fixed instruments. Surveys of astronomical handbooks by Kennedy (1956), King *et al.* (2001[CB]) and van Dalen (unpublished) as well as the studies of Mozaffari (2014a, 2015, 2016, 2017, 2019a, 2019b) reveal that new observations were undertaken from the 2nd/8th century every century over about a millennium, with a good number of them continuing over many years. Thus, many scholars and instrument makers are now known to have carried out their own short-term or longer-term observations.

Until the late 6th/12th century, the most important observations were carried out by small groups of scholars and artisans or individually. The names of many of their participants have been transmitted in preserved texts and instruments.[3] In later centuries, observations were mostly executed at observatories supported by courtly patronage. These will be treated in the following sections.

An important source for early observations is Ibn Yūnus's *Zīj* for the Fatimid Caliph al-Ḥākim (r. 386–411/996–1021). He attests as one example that during 12 years (240–252/854–866) al-Māhānī in Baghdad observed four lunar eclipses, one solar eclipse, and conjunctions between Saturn and Venus, Mercury and Venus, and Mars and Venus, as well as the longitude of Regulus.

Al-Battānī, on the other hand, in his *Zīj* reports his own observations, among them those of the sun at equinoxes and solstices used to obtain the eccentricity of the solar deferent and the position of its apogee (al-Battānī 1899–1907[CB], 66–7). In his measurements, he followed Ptolemy's approach in the *Almagest* obtaining a result of 2;4,45ᵖ for a deferent radius of 60ᵖ, a better value than that of Ptolemy (2;30ᵖ; Ptolemy 1998[CB], 156). The numerous observations of Ibn al-Aʿlam and the Banū Amājūr were extraordinarily accurate and highly appreciated (Mozaffari 2019a, 59–60, 2019b, 518).

Almost all authors of the 3rd/9th century made their observations in Baghdad – except for those who participated in the activities commanded by Caliph al-Maʾmūn (r. 198–218/813–833) in Damascus and some astral experts in Central Asia. Among the latter, Muḥammad ibn Aḥmad al-Samarqandī (d. after 252/866) observed the sun and Regulus in Samarqand, and Ibn ʿIṣma (d. after 278/891) observed the obliquity of the ecliptic, the equinoxes and the solstices in Balkh. Like Yaḥyā ibn Abī Manṣūr, al-Marwarrūdhī or later al-Khujandī, Ibn ʿIṣma built an instrument several meters high (in his case a mural quadrant) in order to achieve the needed precision. From the 4th/10th century onward, a growing number of scholars and instrument makers made observations outside of Baghdad, following the new patronage opportunities at courts first in Iran, Syria, Egypt and al-Andalus and later also in North Africa, Anatolia, Central Asia and South Asia.

II.7.2 Astronomical handbooks in context

Compiling astronomical handbooks was a central practice in the context of celestial observations. Improving earlier data was the main goal, primarily achieved by recalculations. But in more cases than previously thought, new observations were executed in the process of producing a new handbook (see Section II.7.3). Although mostly claimed by a single author, the production of a handbook was a profoundly collective work. It depended and built on earlier books, often in multiple, mixed strands. In some cases, this composite nature was clearly intended by the compilers, as for instance in the *Astronomical Handbook for Ashraf* (*al-Zīj al-ashrafī*) by al-Kamālī ([d. after *c.* 703/1303]; known as Sayf, the astrologer from Yazd). In other cases, such mixtures reflect the limitations and instabilities of the book market.

Today extant manuscripts often contain texts without beginnings or ends, including many sets of tables. Those astronomical handbooks that possess a beginning and name their authors and titles indicate that courts and princely (or other) patronage were the dominant social frames of their production. Titles often include a reference to a specific patron and introductions may link them in a more elaborate manner to him. This connection to the ruling elites corresponds well to the ultimate purposes of astronomical handbooks – the casting of horoscopes and the production of calendars. A third purpose consisted in improving the reliability of the data needed for the astrological predictions (Samsó 2020[CB], 13, 174).

Numerous authors compiled not merely one or two astronomical handbooks but three or more. Sāʿid al-Andalusī (419–462/1029–1070) suggested that Ḥabash al-Ḥāsib's three handbooks corresponded to successive stages of access to different types of sources (Indian, Sasanian, Ptolemaic), as well as to the execution of new observations (Sāʿid al-Andalusī 1993[CB], 74, 1935[CB]). Remarks in several handbooks indicate that new observational results or access to new source material about observations motivated the repeated production of handbooks by the same scholar. In later centuries, this access to new source material about observations became a major reason for undertaking new observations and producing more than one astronomical handbook. Other reasons for such prolific activities were patronage but also the absence of sufficient instructions for the use of the tabulated data (Samsó 2020, 12).

II.7.3 Astronomical handbooks with new observations

Although a primary practice when compiling a handbook was recalculating data for a new location, numerous scholars considered improving their astronomical data a necessary part of their duties. They approached this task by critically analyzing the works of their predecessors for reliability, accuracy and consistency in comparison with contemporary celestial positions, thus providing us with valuable historical information (Mozaffari 2019a). Focusing their analysis mostly on previous handbooks, at times they discussed an impressive number of them.

In the context of the *zīj*es, observation meant measuring various positions of the sun, the moon, the planets and the fixed stars in longitude and latitude, as well as their times of rising, culminating and setting. Depending on the financial means of the observer either smaller, portable instruments were used (astrolabes, different kinds of rulers, quadrants, globes, sundials and time-measuring devices) or these were employed together with large fixed instruments (mural quadrants, mural sextants, armillary spheres, hemispheres and spherical segments, globes, sundials, water clocks or hourglasses, rulers, different kinds of rings and so on; Chapters IV.2, IV.3). The observations were usually undertaken at or near specific times such as oppositions, conjunctions, equinoxes, solstices and solar or lunar eclipses. Although there are references to continuous observations over several years (sometimes up to 40), they do not provide sufficiently detailed data to know how they were arranged (Mozaffari and Zotti 2013, 53–4; Samsó 2020, 578, 654, 903).

The resulting data were used to compute the geometrical parameters of the models, such as eccentricities, radii of the epicycles, positions of the apogees and mean motions. In numerous cases, the observations led to corrections of Ptolemaic parameters, the rejection of previous theoretical assumptions and the discovery of new celestial motions, such as the proper motion of the solar apogee (Morelon 1996, 30, 48; Mozaffari 2017, 2019a), as well as the acknowledgment of phenomena such as annular solar eclipses and the possible variability of the obliquity of the ecliptic and the rate of precession of the equinoxes (King *et al.* 2001, 39). The two last questions contributed to reflections on instrumental deficiencies and observational errors and were important factors for theoretical discussions about models of celestial motions (on trepidation and on the sun; Mozaffari 2016, 269; Samsó 2020, 577–681).

For astral experts, observing and calculating lunar and solar eclipses was an important component of their work. Small inaccuracies in the tables led to incorrect predictions. The errors in eclipse prediction usually did not exceed one hour, and many predictions were made with an accuracy of half an hour or even 20 minutes. But sometimes an error occurred regarding the magnitude or the zone of the eclipse (GiahiYazdi 2008). The latter happened, for instance, in Mamluk Cairo and caused much unrest, as well as anger against the scholars. This is not the only known example of a prediction that caused serious disturbances among the population and upset the elites. The social preparations against the allegedly dangerous effects of eclipses included special prayers in Friday mosques and the distribution of grain and other essential goods to the population.

The study of eclipses also contributed to correcting Ptolemy's view that the minimum angular diameter of the moon, occurring at its apogee, was equal to the constant angular diameter of the sun. If Ptolemy were correct, annular solar eclipses could not occur (Mozaffari 2015, 119). Although famous practitioners in the Islamicate world, for instance al-Ṭūsī, followed Ptolemy's position, others happened to live in times and regions where annular solar eclipses occurred. Statements about annular eclipses are found in a number of extant Arabic and Persian handbooks and other sources, among them historical chronicles (Mozaffari 2014b, 40). A report about such an observation in 259/873 in Khurasan is quoted in al-Bīrūnī's *Masʿudic Canon* (*al-Qānūn al-mas'ūdī*). Another quite detailed report about an observation in 681/1283 in Mughan (northwestern Iran) appears in *The Corrected Tables for the Sultan Based on the Principles of the Ilkhanid Tables* (*al-Zīj al-muḥaqqaq al-sulṭānī 'alā uṣūl al-Īlkhānī*) of al-Wābkanawī (d. after 720/1320), a leading mathematical practitioner at the observatory in Maragha after al-Ṭūsī (Mozaffari 2015, 213–14, Figure II.7.1). But critical voices against Ptolemy's position were raised also without actually observing an annular eclipse in the 3rd/9th and 6th/12th centuries by al-Battānī and al-Khāzinī (Mozaffari 2015, 215). Mozaffari argues that the development of views on solar eclipses among Muslim astral experts depended on detailed observations, the use of Indian hypotheses on angular diameters together with Ptolemaic assumptions, the elaboration of new methods and values, and Aristotelian beliefs about celestial physics. These came together to different degrees, shaping the choices and interpretations of the experts who participated in efforts to determine lunar and solar diameters, planetary distances and magnitudes of eclipses (Mozaffari 2015, 126–30).

Al-Wābkanawī's careful observational report is presented together with calculations of the times, the magnitude and further parameters of an eclipse. In those calculations, he depended on Muḥyī l-Dīn al-Maghribī's systematic, long-term observational program carried out in the 670s/1270s at the observatory in Maragha (see the following discussion) and their interpretations in his *Everlasting Circles of Light* (*Adwār al-anwār*) and *Compendium of the* Almagest (*Talkhīṣ al-Majisṭī*; Mozaffari and Zotti 2013, 52).

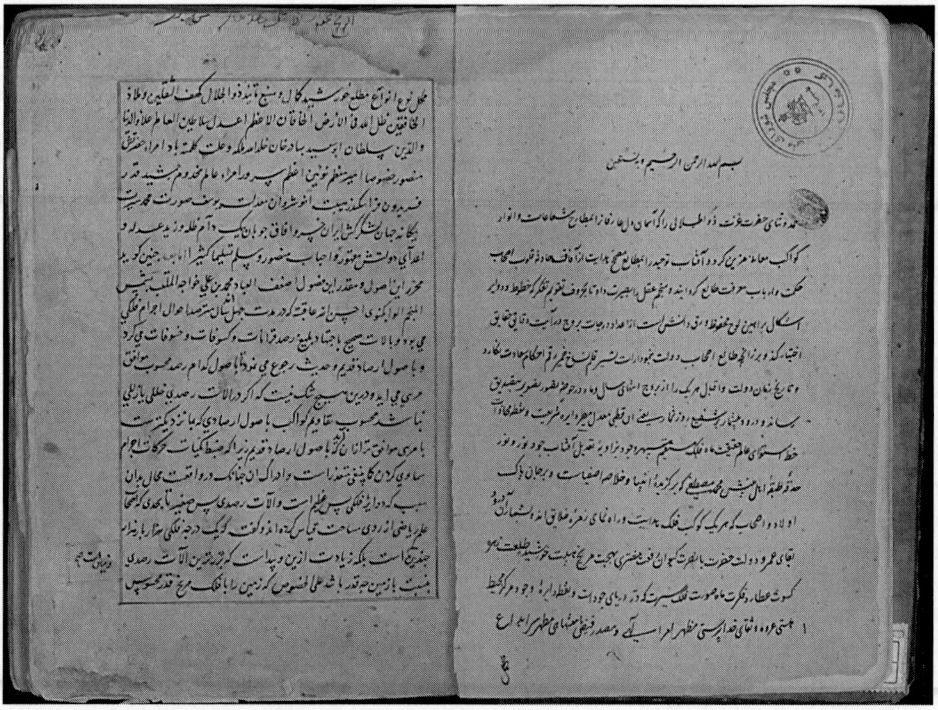

Figure II.7.1 Shams al-Munajjim al-Wābkanawī explains his observations on the first pages of *The Corrected Tables for the Sultan Based on the Principles of the Ilkhanid Tables* (*al-Zīj al-muḥaqqaq al-sulṭānī 'alā uṣūl al-Īlkhānī*); MS Tehran, Majlis Library, 6435, fols 1b–2a.

Source: © Majlis Library, Tehran.

Although too few tables in handbooks have been systematically investigated so far, those studied reveal the use of more new data than previously assumed. Kennedy showed that new observations were carried out for the *Ilkhanid Tables*, at least for the positions of the fixed stars (Kennedy 1956, 169–70). The planetary equations in al-Maghribī's *Everlasting Cycles of Light* were calculated from new observational values for the moon (eccentricity and radius of the epicycle), Mars (epicycle radius) and Saturn (eccentricity). Al-Wābkanawī executed the third systematic observational program at Maragha in order to compare and evaluate the numerical data in the *Ilkhanid Tables* with those in the *Everlasting Cycles of Light* and his own observations (Mozaffari and Zotti 2013, 51–2, 63). In China, observations by Muslim astrologers in the capital of the Yuan dynasty (r. 1271–1368) yielded numerous new planetary parameters and coordinates for fixed stars that were different from those determined by Ptolemy and his successors in Islamicate societies. Some of al-Kāshī's *Tables for the Khāqān* (*al-Zīj al-khāqānī*) were calculated for new values of the epicycle radii of Mars and of the moon measured during lunar eclipses in 808–809/1406–1407. Ulugh Beg's *Tables* built, for all planets except Mercury, on the new observational results achieved by al-Kāshī, Qāḍīzāde al-Rūmī (d. after 844/1440) and ʿAlī Qūshjī (805–879/1403–1474).

These examples could be easily extended to other scholars and their handbooks. So far the large-scale observatories have attracted more attention than the work of individual scholars in popular surveys of the history of observations in Islamicate societies. But the new research results of van Dalen, Mozaffari, Samsó and other historians of science leave no doubt that observations of solar, lunar and, at times, planetary positions, were much more common than previously realized. The recognition that individuals often made systematic and long-term observations invites us to abandon the older view that astronomical handbooks were first and foremost the outcome of computational practice, with little new observational input.

II.7.4 Instruments

In many cases, the preparation of long-term observations included the invention of new instruments and the modification of existing ones. Modifications consisted above all in increases in size, allowing more graduations within the same angle on a scale, plus the addition of further parts such as a ruler or a sighting tube. But changes in size could create serious problems for artisans and observers. A highly instructive case occurred during the efforts to build an observational site in early 6th/12th-century Fatimid Cairo. The first site chosen for the observations was considered too far from the city and thus impractical and costly for the transport of the construction material. The big instruments and their accessories were produced on-site; the construction material (metal, wood, and stone) had to be brought and processed there. Ten furnaces were built for smelting the metal and the patron participated in the construction process by commanding the simultaneous opening of the hot furnaces. The first casting did not succeed fully and had to be repeated. A second casting was successful and the instrument was placed in front of the mosque chosen for the observations, but the morning rays of the sun could not be observed there. Hence the instrument and its accessories had to be moved to a new site, again a mosque, but the transportation proved extraordinarily challenging. After the observations of the sun had started it was discovered that the big metal ring was so heavy that it sagged toward the ground by more than a minute. Corrections were needed but did not fully succeed. Then the first patron, the Fatimid vizier, died; his successor in the office continued the work but ran into difficulties with his master, the caliph. Rumors spread that the vizier intended to communicate with Saturn, perform magic, become a prophet and commit other evil deeds. The caliph executed his vizier and demolished the observational mosque (Sayili 1960, 168–71). As this example shows, building an observatory and the necessary large-scale instruments was challenging technically, astronomically, environmentally and politically, and needed a strong, healthy and long-living patron.

The number of newly invented or modified instruments made for specific observational purposes is as impressive as the number of observations made over time. They are too many to be described here, but a few at least may be mentioned. In Raqqa, al-Battānī made a globe for solar observations. In Baghdad, al-Kūhī oversaw the fixed installation of a segment of a hollow sphere with a diameter of 24m for observing solar altitudes. In Rayy, al-Khujandī constructed his famous 20m radius stone sextant, called the Fakhrī sextant, to measure the obliquity of the ecliptic and Rayy's latitude. In Hamadan, Ibn Sīnā (d. 426/1037) built an instrument 7m in diameter for measuring the azimuth and altitude of a celestial body. A modification of this instrument is Mu'ayyad al-ʿUrḍī's altitude-azimuth instrument built for the Maragha observatory. Another variant of the altitude-azimuth instrument that allowed the simultaneous measurement of the coordinates of two stars is attributed to Ulugh Beg (Bruin 1974, 274–6). After constructing an armillary sphere and a mural quadrant for the Istanbul observatory, Taqī al-Dīn ibn Maʿrūf invented three new instruments, one for determining equinoxes, a mechanical clock with a train of cogwheels, and a sextant (Sayili 1960, 300; Bir and Kaçar 2009, 69, 74–6).

II.7.5 Institutions and patrons

The geographical distribution of the observations indicates the diversity of their organizational contexts and patrons. The pioneering study of such contexts, but in particular of observatories, published in 1960 by Aydın Sayılı (1913–1993) has not yet been superseded by a book-length study, but the bibliography of this chapter presents much new research. The first observations patronized by a caliph occurred in the early 3rd/9th century. Sponsored by Caliph al-Ma'mūn, modern historians disagree about the interpretations of the motives for these observations, as well as their execution, possible institutional properties and surviving historical sources (Sayili 1960; Mercier 1992; King 2000[CB] ; Brentjes 2014). In the 4th/10th century, several Buyid emirs supported astronomical observations, their instruments and their settings. ʿAḍud al-Dawla (r. 338–372/949–983) financed observations in Rayy and Shiraz. After him, in 378/989, Sharaf al-Dawla (r. 362–368/983–989) established a small observatory in the garden of his palace in Baghdad under the supervision of al-Kūhī. Two observations of the exact times of the summer solstice and autumn equinox of 367/988 were witnessed by religious dignitaries. Thanks to a report of al-Bīrūnī, we know that the observational instrument was a semicircle with a radius of 12.5 m. Although there is no historical document confirming the existence of an observatory in Cairo in the time of Ibn Yūnus's observations, we know that he received 100 dinars per day from Caliph al-Ḥākim for his scientific activities (King 1999[CB], 164). It is not clear whether there were similar dedicated sites in al-Andalus, although many observations took place there, as well as in various North African cities (Samsó 2007, 2020).

The best-known observational sites after the 4th/10th century were the observatories of ʿUmar al-Khayyām (439–*c.* 517/1048–*c.* 1123) at the Seljuq court of Malik-Shāh (r. 465–485/1072–1092) in Isfahan, the observatories in the Ilkhanid capitals of Maragha, Tabriz and Sultaniyya, Ulugh Beg's observatory near Samarqand, Taqī al-Dīn ibn Maʿrūf's observatory in Istanbul and Jai Singh's five observatories in Mughal India (Samsó 2007; Mozaffari and Zotto 2013, 53, 62–3). The financial investment provided for buildings for making observations and instruments was substantial. Hülägü Khān (r. 654–663/1256–1265), the founder of the Ilkhanid dynasty, is believed to have spent 30,000 dinars for the instruments at Maragha alone (Brambilla

2012, 3). The architectural ambition and skills involved in producing these sites likewise were often extraordinary.

The observatory at Maragha covered an expanse of about 150 m × 350 m and consisted of six observational sites, a substantial library, a place for metal workers and housing units. The main building was a four-story stone structure with a diameter of 22 m. The mural quadrant installed inside this building had a radius of 40m; today, its remains measure 5.5m (Varjāvand 1987; Figure II.7.2). Other instruments included armillary spheres, armillary rings and globes, some of which were made by Mu'ayyad al-Dīn al-ʿUrḍī (d. 664/1266). A globe manufactured by al-ʿUrḍī's son Muḥammad is preserved at the Mathematical Salon in Dresden (Chapter IV.2). The entire complex was financed like a madrasa or a mosque, that is by making it a religious bequest (*waqf*). This was the first time, as far as we know, that an observatory was financed in such a manner. The second observatory financed in this way was that of Ghāzān Khān (r. 694–703/1295–1304; Sayili 1960, 228). The other major observatories seem to have received their funding either from the private purse of the patron or from the royal treasury.

Figure II.7.2 Remains of the mural quadrant at the observatory in Maragha.

Source: Photographer Behrad. This file is licensed under the Creative Commons Attribution-Share Alike 4.0 International license.

The observatory in Tabriz was part of the massive suburban development planned by Ghāzān Khān and his experts, around a 45m-high dodecagonal tomb tower erected for his everlasting reputation, emulating and outdoing the building achievements of the Seljuq ruler Sanjar (r. 511–552/1117–1157) in Marv. This development included a mosque, madrasas, hospitals, Sufi lodges, caravanserais and private houses. The ideological importance of the astral sciences and thus of the observatory for Ghāzān Khān was represented by the twelve zodiacal signs that decorated the walls of his tomb (Brambilla 2012, 6–8, 15). The Seljuq and Ilkhanid observatories were built in less than two years. At least for the one in Tabriz some factors for this speedy completion are known: careful planning of the entire project, hiring a huge workforce and highly skilled architects called *muhandisūn* (also meaning builders of canals or geometers) from Iran, Armenia and Georgia, as well as prefabricated architectural units (Brambilla 2012 16; Jahn 1940).

Ulugh Beg ordered his observatory to be built on a rocky hill north of Samarqand. The founder of the Mughal dynasty, the Timurid prince Bābur (r. 932–937/1526–1530), described it as a three-storied building (*Babur* 2006, 32). The main observational instrument was partly built underground and had a radius of 40m configured as a sextant (Kennedy 1998; Figure II.7.3). It was fixed on the meridian and was used to measure the meridian altitudes of the sun, the moon and stars. Further positions were measured with other instruments, such as armillary spheres and parallactic rulers (Chapter IV.2). The precision achieved at the observatory was very high (up to tens of seconds) and reduced observational errors considerably (Krisciunas 1992; Barthold 1963, 2: 132). The report of the historian ʿAbd al-Razzāq al-Samarqandī (816–887/1413–1482) was interpreted by W. W. Barthold (1869–1930) to say that the walls of the observatory were richly decorated by paintings of the nine heavenly spheres (seven spheres for the moon, the sun and the five planets visible to the naked eye, plus the zodiac carrying the stars and the sphere of the spheres = sphere of the Prime Mover). There were also images of planetary models with epicycles and the earth divided into the seven climes showing mountains, seas, deserts and so forth (Barthold 1963, 2: 132). But Sayılı suggested that some of those celestial and terrestrial representations were rather globes and other instruments (Sayili 1960, 282).

Figure II.7.3 Subterranean part of the sextant of Ulugh Beg's observatory North of Samarqand.

Source: Photographer Alexis.

Taqī al-Dīn ibn Maʿrūf took an interesting path of learning and social networking that led him to become the head astrologer of Sultan Selīm II (r. 974–982/1566–1574) in 987/1571. One important feature of the patronage elements in his career was access to literature and instruments that linked him to the scientific tradition in Samarqand and its continuation in Istanbul, as well as to printed books and mechanical clocks from the Netherlands and other European countries. Having made individual celestial observations in Egypt, as well as Istanbul (where they started in 980/1573), he had petitioned Murād III in 983/1575 to construct an observatory in Istanbul. In 985/1577 he finally became its director.[4] One of a series of famous miniatures depicts him as the head of a team of 15 members working with standard instruments of Islamicate observatories enriched by a few novel instruments from Europe (Figure II.7.4). As in previous cases, Taqī al-Dīn ordered large-scale instruments (a wooden mural quadrant, an armillary sphere, an azimuthal semicircle, rulers and rings) and built his own modified or newly introduced instruments (a movable wooden sextant with three rulers, a mechanical clock indicating hours, minutes and seconds). In addition to observations of the sun, the moon and the five planets, he observed the comet of Shaʿbān 985/November 1577 and thought it predicted Ottoman victory

Figure II.7.4 Miniature of Taqī al-Dīn ibn Maʿrūf's observatory in Istanbul. MS Istanbul, Istanbul University, Rare Works Library, F 1404, fol. 57a.

Source: © Istanbul University, Rare Works Library, Istanbul

in the military conflict with Safavid Iran, which had commenced already in 1576. In 986/1578, the Ottoman court begann a full-fleged war against its neighbour, which lasted until 998/1590. But the war was opposed by grand vizier Meḥmed Sokollu Pāshā (d. 987/1579), whose long-term influence on Ottoman foreign policy declined, however, rapidly under Murād III. A biography and a few other sources suggest that Taqī al-Dīn fell victim to this struggle against the grand vizier and a second supporter of his research program, the sultan's former teacher. The highest religious authority of the state and other religious scholars rejected astrology and in particular observations of the heavens as heresy and made them responsible for the outbreak of the plague in Istanbul in 986/1578. Thus, the observatory was closed in late 987/1579 and demolished on 4 Dhū l-Ḥijja 987/22 January 1580 (Fazlioğlu 2007; Şenel 2009; Sayılı 1956; Aydüz and Şen 2015). However, precious astrological and historical manuscripts produced at the Ottoman court for the sultan, as well as gifts commanded by the sultan for two of his daughters, leave no doubt that the sultan and his close advisors continued to support and cherish the astral sciences (MSS Paris, BnF, Supplement Turc 242; New York, Pierpont Morgan Library, M. 788).

The most famous Mughal observatories in India were built between 1136/1724 and 1145/1733 by the Rajput prince Jai Singh after the Mughal ruler Muḥammad Shāh (r. 1131–1161/1719–1748) had given his permission and support. In the preface to the resulting *Tables for Muḥammad Shāh*, Jai Singh justified this huge investment of time, money, material and manpower by pointing to the substantial differences between the data in the available Persian, Sanskrit, French and other handbooks. They differed among each other and with current observations of the sky, above all for the new moon. Consequently, new tables were needed for religious and administrative purposes, and for astrology (Sharma 1995, 19–20). Inspired by the description of Ulugh Beg's instruments and his *Tables*, he increased the size of his instruments beyond that already achieved by Ulugh Beg, hoping for still better accuracy. These included sundials for day time measurements, quadrants and other fixed instruments, and at least one newly invented cylindrical instrument called Rām Yantra (Figure II.7.5). In addition, Jai Singh invited Jesuits

Figure II.7.5 Rām Yantra, New Delhi, invented by Sawā'ī Jai Singh in about 1135/1723. It encompasses two circular structures of the same format with a pillar enabling continued observations of risings and settings of stars. The walls and the floor are graduated for reading azimuth and altitude angles. The pillars are marked with vertical stripes of 6° width (Kaye 1917, 43–4).

Source: © Photographer Roxana Ashtari, Madrid

to his court asking for information about European methods, instruments and theories. At least two Portuguese, five French and two Bavarian Jesuits worked for the Rajput prince between 1139/1727 to 1156/1743, while others came for some time from the Mughal capital (Sharma 1995, 294). They brought him telescopes and new astronomical tables. In 1141/1729, Jai Singh sent a group of dignitaries and scholars to Europe to learn how observatories were built and run there (Sharma 1995, 295). Together with the European tables that these emissaries brought back, Hindu and Muslim scholars at Jai Singh's court in Jaipur translated the *Ilkhanid Tables*, Ptolemy's *Almagest* and Euclid's (3rd century BCE) *Elements* from Persian into Sanskrit. Two or more people cooperated in measuring solar, lunar and stellar altitudes and declinations with the quadrant and the large sundial. Other instruments served for measurements of zenith distances, hour angles, azimuths and diameters of celestial bodies (Sharma 1995; Mukherji 2010). Thanks to the telescopes, for the first time at an observatory in the Islamicate world, Jai Singh's scholars observed the phases of Venus, the moons of Jupiter and the rings of Saturn, pictures of which can be found in some copies of the *Tables for Muḥammad Shāh* (Ansari 2015).

It is often stated that the observatories in Islamicate societies lasted only for a short time (Huff 2017, 162). But this is neither the full truth nor the proper perspective. These institutions began to appear in the 3rd/9th century and continued to appear until the 12th/18th century. They were meant to last for a full cycle of planetary observations, that is, for up to approximately 30 years (one complete revolution of Saturn), or serve shorter, specific observations. By comparison, observatories working for more than 30 years or more than one generation of observers in western Europe started only with Greenwich in 1675. Observations at observatories in Islamicate societies often extended beyond the death of famous scholars and rulers who founded them. In the case of the Samarqand observatory, for instance, new evidence shows that observational activities continued after Ulugh Beg's death (Richard 2011). Observations in Samarqand began in 810/1408, but the work at the observatory apparently began in the 1420s. It may have operated for about half a century.

The longest continuously functioning observatory headed by a Muslim before the modern era was created in 1271 in China, with altogether 37 employees. The Islamic Astronomical Bureau at the Yuan capital near Beijing, and its revived form in the Ming capital Nanjing, existed for about four centuries with their observational sites in the palace compounds. The *Islamic Astronomical System* (*Huihui li*) is a Chinese translation of a Persian handbook that was most likely produced by the Muslim astrologer and head of the Islamic Astronomical Bureau, Jamāl al-Dīn (d. *c.* 688/1289; Chapter V.12). Using new observational data for almost all relevant parameters and stellar coordinates, its tables were newly computed and sometimes set up in a novel format (van Dalen 2002, 20–2; Li 2016).

Furthermore, observatories in Islamicate societies were mainly objects of passion, fame and politics for individual rulers, not dynasties or governments. Despite their importance for the astral sciences and the great role that celestial predictions played in those societies, policies to promote observatories or other scientific practices were largely designed by and for individuals. This began to change only in the 13th/19th century with reforms of the armies, educational institutions, hospitals and governmental apparatus.

2.7.6 In lieu of a conclusion

Observatories in Islamicate societies have attracted a good deal of scholarly attention; however, we have shown in this chapter that there were also substantial programs of observation carried out by individuals, across a wide range of times and places, and that these frequently contributed to the production of new *zījes*. As indicated in this chapter, practices of celestial observations included a broad array of activities, although, for entire regions and periods, no such information has survived. The limited length of this chapter necessitated decisions on what to report. Because many less technical practices connected to observations have often been omitted in the traditional history of science in Islamicate societies, we chose to focus on surveying them. They are not only easier to grasp for the nonexpert reader but equally important to any historical account of celestial observations. We have avoided discussions of technicalities, because they demand the explanation of too many details and interest only a small circle of readers. The reader who wishes to learn more will have to turn to the literature provided in the bibliography. Samsó's new book (2020) is particularly useful for this purpose, although the information is scattered across many chapters and does not always explain every single step of an observation.

Notes

1 We are profoundly grateful to Benno van Dalen, Munich, for allowing us to use his unpublished survey of astronomical handbooks and his continuous help with unresolved problems. We thank S. M. Razaullah Ansari, Aligarh, for information on Jai Singh and Taha Yasin Arslan, Istanbul for his help with sources and technicalities.

2 Consolidated bibliography.

3 Yaḥyā ibn Abī Manṣūr (d. 215/830), al-Marwarrūdhī (d. after 216/832), the Banū Mūsā (3rd/9th century), Ḥabash al-Ḥāsib (d. after 255/869), al-Māhānī (d. after 252/866), the Banū Amājūr (late 3rd-early 4th/9th-10th centuries), al-Battānī (244–317/858–929), Ibn al-Aʿlam (d. after 374/985), al-Ṣūfī (291–376/903–986), al-Kūhī (2nd half 4th/2nd half 10th century), al-Khujandī (324–390/945–1000), Maslama al-Majrīṭī (d. before 397/1007), Ibn Yūnus (d. 399/1009), al-Bīrūnī (362-d. after 444/973-d. after 1053), ʿUmar al-Khayyām, Ibn al-Zarqāllu (d. 493/1100) al-Khāzinī (d. after 525/1130–1) and Ibn al-Fahhād (6th/12th century).

4 The observatory was only partly finished at that time. Hence, the start of its operation is often given only as early 987/1579, after all construction was finalized.

Bibliography

Sources

Babur. trs. Beveridge, A. S. and Hiro, D. 2006. *Babur Nama. Journal of Emperor Babur*. New Delhi: Penguin Books.

Manuscripts

MS New York, Pierpont Morgan Library, M. 788.
MS Paris, BnF, Supplement Turc 242.
MS Tehran, Majlis Library, 6435.
MS Tehran, Malik Library, 6037.

Research literature

Ansari, S. M. R. 2015. "Survey of Zījes Written in the Subcontinent," *Indian Journal of History of Science* 50.4: 575–601.

Aydüz, S. and Şen, H. 2015. "Galata'da Rasathane Kuran Osmanlı alimi Takiyüddin Efendi Öldürüldü mü?," *Derin Tarih* 35: 108–11.

Barthold, V., Minorsky, V. and Minorsky, T., trs. 1963. *Four Studies on the History of Central Asia*. 2 vols. Vol. 2: *Ulugh-Beg*. Leiden: Brill.

Bir, A. and Kaçar, M. 2009. *Zamanın Görünen Yüzü. Saatler. Visible Faces of the Time. Timepieces*. Istanbul: YKY.

Blochet, E. 1910. *Introduction à l'histoire des Mongols de Fadl Allah Rashid ed-din*. Leiden: Brill.

Brambilla, M. G. 2012. "Large Scale Building Techniques in Ilkhanid Iran," Paper presented at *Masons at Work*, a conference sponsored by the Center for Ancient Studies, University of Pennsylvania.
Brentjes, S. 2014. "Sanctioning knowledge.," Al-Qantara: revista de estudios árabes, 35.1: 277–309.
Bruin, F. 1974. "Astronomical Observations and Instruments of Islam," *Nederlands Tijdschrift voor Natuurkunde* 40.19: 271–5.
Burnett, C. 2002. "The Certitude of Astrology: The Scientific Methodology of al-Qabīṣī and Abū Mashar," *Early Science and Medicine* 7.3: 198–213.
Dalen, B. van. 2002. "Islamic Astronomical Tables in China. The Sources for the Huihui Li," in Ansari, S. M. R., ed. *History of Oriental Astronomy*. Dordrecht: Springer, 19–32.
Fazlioğlu, I. 2007. "Taqī al-Dīn Abū Bakr Muḥammad ibn Zayn al-Dīn Maʿrūf al-Dimashqī al-Ḥanafī," in Hockey *et al.*, eds. [CB], 1122–3.
GiahiYazdi, H.-R. 2008. "Solar Eclipses in Medieval Islamic Civilization: A Note on Cultural and Social Aspects," *Tarikh-e Elm: Iranian Journal for the History of Science* 6: 75–82.
Huff, T. 2017. *The Rise of Modern Science: Islam, China and the West*. 3rd ed. Cambridge: Cambridge University Press.
Jahn, K., ed. 1940. *Geschichte Ġāzān-Ḫān's aus dem Ta'rīḫ-i-Mubārak-i-Ġāzān al-Dīn Faḍlallāh b. ʿImād al-Daula Abū l-Ḫair* (E. J. W. Gibb Memorial Series, New Series 14). London: Luzac and Co.
Kaye, G. R. 1917. *The Astronomical Observatories of Jai Singh*. Calcutta: Superintendent Government Printing, India.
Kennedy, E. S. 1956. "A Survey of Islamic Astronomical Tables," *Transactions of the American Philosophical Society* 46.2.
Kennedy, E. S. 1998. "The Heritage of Ulugh Beg," in Kennedy, E. S., ed. *Astronomy and Astrology in the Medieval Islamic World*. Aldershot: Ashgate.
King, D. A. 1995. "Early Islamic Astronomical Instruments in Kuwaiti Collections," in Fullerton, A. and Fehérvári, G., eds. *Kuwait: Arts and Architecture: A Collection of Essays*. Kuwait: s. n., 76–96.
Krisciunas, K. 1992. "The Legacy of Ulugh Beg," in Paksoy, H. B., ed. *Central Asian Monuments*. Istanbul: Isis Press, 95–103.
Li Liang. "Arabic Astronomical Tables in China," *East Asian Science, Technology, and Medicine* 44: 21–68.
Mercier, R. P. 1992. "Geodesy," in *GTISAS* [CB], 175–88.
Morelon, R. 1996. "Eastern Arabic Astronomy between the Eighth and Eleventh Centuries," in Rashed and Morelon [CB], 20–57.
Mozaffari, S. M. 2014a. "Muḥyī al-Dīn al-Maghribī's Lunar Measurements at the Maragha Observatory," *Archive for the History of Exact Sciences* 68: 67–120.
Mozaffari, S. M. 2014b. "A Case Study of How Natural Phenomena Were Justified in Medieval Science: The Situation of Annular Eclipses in Medieval Astronomy," *Science in Context* 27.1: 33–47.
Mozaffari, S. M. 2015. "Annular Eclipses and Considerations About Solar and Lunar Angular Diameters in Medieval Astronomy," in Orchiston, W., Green, D. A. and Strom, R., eds. *New Insights from Recent Studies in Historical Astronomy: Following in the Footsteps of F. Richard Stephenson. A Meeting to Honor F. Richard Stephenson on His 70th Birthday*. Berlin: Springer, 119–42.
Mozaffari, S. M. 2016. "A Forgotten Solar Model," *Archive for History of Exact Sciences* 70: 267–91.
Mozaffari, S. M. 2017. "Holding or Breaking with Ptolemy's Generalization: Considerations about the Motion of the Planetary Apsidal Lines in Medieval Islamic Astronomy," *Science in Context* 30.1: 1–32.
Mozaffari, S. M. 2019a. "The Orbital Elements of Venus in Medieval Islamic Astronomy: Interaction Between Traditions and the Accuracy of Observations," *Journal for the History of Astronomy* 50.1: 46–81.
Mozaffari, S. M. 2019b. "Ibn al-Fahhād and the Great Conjunction of 1166 AD," *Archive for History of Exact Sciences* 73: 517–49.
Mozaffari, S. M. and Zotto, G. 2013. "The Observational Instruments at the Maragha Observatory After AD 1300," *Suhayl* 12: 45–179.
Mukherji, A. Sh. 2010. *Jantar Mantar. Maharaja Sawai Jai Singh's Observatory in Delhi*. Delhi: Ambi Knowledge Resource.
Pingree, D. E. 1970. *Census of the Exact Sciences in Sanskrit*. Series A. Vol. 1 (Memoirs of the American Philosophical Society 81). Philadelphia: American Philosophical Society.
Ragep, F. J. 2009. "Astronomy," in *EI-3*, 2009–1,120–50.
Richard, F. 2011. "Un témoignage sur l'activité de l'observatoire de Samarqand après la morte d'Ulug Beg," in Kerr, R. M. and Milo, Th., eds. *Writings and Writing from Another World and Another Era* (Festschrift for J. J. Witkam). Cambridge: Archetype Press, 39–47.

Saliba, G. 1983. "An Observational Notebook of a Thirteenth-Century Astronomer," *Isis* 74: 388–401.
Saliba, G. 1985. "Planetary observations at the Maraghah Observatory before 1275: A new set of parameters, " *Journal for the History of Astronomy* 16: 113–22.
Samsó, J. 2007. "Marṣad," in *EI-2*, 6: 599–602.
Sarma, Ś. R. 2019. *A Descriptive Catalogue of Indian Astronomical Instruments – Abridged Version*. Open Access eBook. http://crossasia-repository.ub.uni-heidelberg.de/4167/.
Sayılı, A. 1956. "Ala al-din Mansur's Poems on the Istanbul Observatory," *Belleten* 20 (1956): 429-484. https://belleten.gov.tr/tam-metin-pdf/1224/eng
Sezgin, G. 1974. *Geschichte des arabischen Schrifttums*. Vol. 5: *Mathematik bis ca. 430 H.* Leiden: Brill.
Sharma, V. N. 1995. *Sawai Jai Singh and His Astronomy*. New Delhi: Motilal Banarsidass Publishers Private Limited.
Şenel, C. tr. 2009. "Nevizade Ataî'nin Hadaikü'l-Hakaik'inden Takiyyüddin'in Biyografisi," *Osmanlı Bilimi Araştırmaları / Studies in Ottoman Science* 10.2: 130–33.
Varjāvand, P. 1987. *Kāvush-i raṣad khāna-yi Marāgha u negāhī pīshīnah-yi dānish-i sitāra- shināsī dar Īrān*. Tehran: Amirkabir.

II.8
THE PRACTICAL ASPECTS OF OTTOMAN MAPS

Gottfried Hagen

II.8.1 Introduction

On August 13, 1787, the Polish nobleman and diplomat Stanisław Tarnowski received a present from his dragoman, Joseph d'Alexandre, as a souvenir of his stay in Constantinople: a watercolor showing the most important venue of Ottoman diplomacy, the office of the *re'īsü l-küttāb*, in his *köşk* overlooking the Bosporus. D'Alexandre showed the reis seated on a divan in the far-left corner, facing a dragoman of a foreign power, with several other officials looking on. On the wall to the right hangs a large map, about 1.5 m by 1 m, showing (presumably) the Black Sea, as the most important theater of political conflict for the Ottomans at the time (Sotheby's 2018, Lot 111; Figure II.8.1).[1] The map is held up by wooden rods along the top and bottom edges and is suspended from a single hook on the wall, as if it was designed to be rolled up and swapped out for another map in a moment. This unassuming painting appears to be, to the best of my knowledge, the only extant representation of a map in use in an Ottoman context. Like many other state formations, the Ottoman enterprise, having assumed the mantle of empire with the conquest of the imperial city of Constantinople, was based on the two interrelated activities of surplus extraction and warfare. Increasingly stretched out over vast territories and a diverse manifold of populations, the imperial dynasty and its household, which formed the core of the empire, were in need of efficient administrative tools for purposes of recording and planning, to maintain control and keep direct exercise of violence at a minimum. By the same token, maps would help them to efficiently plan the deployment of military force. That maps as sophisticated displays of critical knowledge have a special place in the toolbox of the early modern and modern state is considered commonplace in the history of cartography since the seminal works of John B. Harley (1932–1991; 1988a, 1988b) or Joseph Konvitz (1987). D'Alexandre's watercolor appears to confirm that the Ottoman state, too, used maps as a tool for planning and decision-making, and for military and foreign policy purposes in particular.

However, assumptions regarding the institutionalization of cartographic practices in the service of the state between the mid-9th/15th and the late 12th/18th centuries deserve closer scrutiny. In this chapter, I discuss in detail the way the Ottoman state, and its civil, fiscal and military bureaucracies used maps as records and as planning devices, demonstrating that cartography remained an exceptional and rarefied technology that was never adopted in a systematic and institutionalized way at the state level before the 13th/19th century and the accelerated

DOI: 10.4324/9781315170718-28

Figure II.8.1 Painting by Joseph d'Alexandre from 1787 for the Polish nobleman Stanisław Tarnowski. On the wall to the right hangs a large map, about 1.5 m by 1 m, showing (presumably) the Black Sea, as the most important theater of political conflict for the Ottomans at the time.

Source: © Sotheby's

reforms of the periods known as *Niẓām-i jedīd* ("New Order"; after 1203/1789) and *Tanẓīmāt* ("Reorganization"; after 1255/1839).

II.8.2 Cartographic traditions

Historians of Ottoman cartography face the fundamental obstacle that use of maps is, with the exception of the earlier example, not documented, so that any inference in this regard has to be drawn from the object itself, today preserved in a library or an archive. This chapter builds on previously published maps; more detailed research especially in the Prime Minister's Archive is likely to add significantly to the corpus (Şakul 2009). So far, most attention has been paid to the collections in the library and archives of the Topkapı Palace (Goodrich 1993). The extant record is complicated and variegated, and there are almost no instances that speak directly to the practical application of the knowledge gathered and organized in texts and maps. Therefore, a quick overview of cartographic genres and their uses is in order.

In the early modern period maps were not theorized by Ottoman scholars as specialized technology to display information of spatial significance. The available corpus of maps shows that the Ottomans were familiar with a number of discrete cartographic and geographical traditions. Several of these are direct continuations of earlier Islamic traditions. The most popular of them is the genre of cosmography, building on collections of *mirabilia* (natural wonders) primarily for the sake of pious edification. Ottomans read the medieval texts of Muḥammad Ṭūsī

(d. after 562/1166), al-Qazwīnī (d. 682/1283) and the Pseudo-Ibn al-Wardī (*fl.* 860/1456) in Arabic, as well as Turkish, translations and abridgments (Kut 1985; Kut ed. 2012). The latter often contained highly stylized world maps. A much smaller number of manuscripts belongs to the so-called *Atlas al-Islam* tradition, which by the 10th/16th century was thoroughly obsolete in terms of political and cultural geography, but maps in these late versions – while retaining the same degree of abstraction – were frequently transformed into works of art, featuring figural decoration and elaborate calligraphy (Pinto 2011; Chapter I.13).

By contrast, we find only narrowly circumscribed interests in mathematical geography. A translation of Ptolemy's (*c.* 100–170) *Geography* into Arabic, prepared for Meḥmed II (r. 848–850/1444–1446 and 854–886/1451–1481), remained an outlier that never circulated outside of the palace (Mavroudi 2013). Older geographical works such as the tables of Abū l-Fidāʾ (672–732/1273–1331) included compilations of locational data, but Ottoman contributions are limited to lists of locations with latitude and the angle of the *qibla*, which were relevant for the construction of mosques. Muṣṭafā ibn ʿAlī (d. 979/1571), known as the Timekeeper (*al-muvaqqit*), knew astronomy well, but copied most coordinates from earlier works (İhsanoğlu *et al.* 1997[CB],[2] 161–79; İhsanoğlu *et al.* 2000[CB], 49–52). The mathematical problems of projection, which were defining for Ptolemy's maps, apparently did not concern Ottoman mapmakers (Brentjes 2005; Brentjes 2011).

The palace also houses the most significant collection of maritime maps known from Ottoman contexts. These include a number of portolan charts originating in the western Mediterranean as well as original Ottoman portolans and maritime atlases, showing that Ottoman sailors were active parts of what Şakul calls a pool of military (and nautical) expertise, that is, the Mediterranean (Şakul 2013, 183). The two world maps created by Pīrī Reʾīs (d. 961/1554), only preserved as fragments, show him aware of the discovery of the Americas; in a narrative text on the earlier one, he claims to have used a map by Columbus ([1451–1506]; Soucek 1996, 50). Pīrī Reʾīs also produced a description of the Mediterranean for which he adopted the format of the Italian island books (*isolario*), which combine multiple small-scale maps with textual description, interspersing critical information on resources or danger spots with the occasional personal narrative (Soucek 1973). Several other maritime atlases from the later 10th/16th century display supreme painterly skills, combining technical aspects of the portolan, such as loxodromes (lines that cross all circles of longitude at a constant angle), with knowledge of western maps (Özdemir 1992; Robles Macías 2012; Casale 2012).

The palace also sought to obtain recent European maps, and individual printed maps and atlases of European provenance probably circulated in Istanbul in the first half of the 11th/17th century (Arbel 2002; Agoston 2007; see Hagen 2003[CB] on maps available to Kātib Chelebi [1017–1067/1609–1657]). The traveler and raconteur Evliyā Chelebi (1020–after 1098/1611–after 1687) included the mapmakers among his list of guilds of Istanbul, claiming they used Western maps as sources, but it is impossible to separate fact from fiction in his account (Hagen 2012). Evliyā himself claimed to have learned how to draw maps from a master, possible in the employ of the palace (Dankoff *et al.* 2018, 245); the map of the Nile kept in the Vatican today and associated with his name is not in his hand. His contemporary Kātib Chelebi certainly did not have a high opinion of whoever was making maps in his days (Dankoff *et al.* 2018, 4). However, later illustrators were clearly capable of adapting maps from many different traditions for their purposes (in addition to Brentjes 2005 and Brentjes 2011, see multiple examples in Pinto 2016).

Only in the mid-11th/17th century did an Ottoman scholar for the first time attempt to integrate what he saw as valid and practically useful geographical knowledge from all these disparate genres and traditions into one compilation, as one pillar of a set of encyclopedic works, which together constructed a unified image of the world. Remarkably, Kātib Chelebi started out from cosmography, which was the least "practical" but also the most capacious genre at hand.

Experimenting with different types of maps and sources, he ultimately upended the Islamicate traditions altogether, and ushered in a new era in which Ottoman geographers translated European works and used Islamic sources only to supplement the new framework. Like all his encyclopedic works, his *Mirror of the World* (*Jihānnümā*) is devoid of literary ambitions, presenting its content in plain and easily accessible form. Kātib Chelebi also explicitly proclaimed that he intended to provide critical knowledge useful to political and military decision-makers (Kātib Çelebi 1729, 17). It is safe to assume that Kātib Chelebi was not thinking narrowly of a reference work for the sultan or grand vizier but envisaged an elite in which this critical knowledge was shared widely so that the elite could articulate expectations and critiques of their leaders in an informal system of checks and balances. Kātib Chelebi's cartographic work began with sketches that betray their autodidactic nature and thus the absence of an Ottoman tradition within reach even for one of the most erudite men of the century. His later work entirely relies on models of western cartography, showing outlines, rivers and mountain ranges in addition to cities, all placed in frames indicating latitude and longitude. Later manuscripts of his work feature many different cartographic styles, meaning that copyists and illustrators had access to quite different models in addition to his own. Hagen (2006) describes Kātib Chelebi's first attempts. Noteworthy later developments are the full-page regional maps in a *Jihānnümā* manuscript in the Topkapı Library (e.g., shown in Gökbilgin 1956), and later versions of maps of the Caspian Sea based on the Delisle–Homann map of 1720 in *Jihānnümā* manuscripts (Sotheby's 2013, lot 141, previously in the possession of Hamid Sadi Selen; see also Hagen 2003, 425).

II.8.3 Functions of maps: contemplation and representation

The utilitarian approach to geography, and by extension mapping, manifested in the *Jihānnümā*, however, was a novelty, as was its expression and justification in the work itself. For the earlier production of maps and texts surveyed earlier, the question about their social and practical function is more difficult to answer since there are no explicit statements in the works themselves. It has been generally accepted that cosmographical literature was primarily appreciated for lessons derived from the manifold marvels of this world, which lent themselves to the contemplation of God's omnipotence. A similar contemplative reading, although possibly with less theological implications, may be suggested for the cluster of beautifully illustrated copies of al-Iṣṭakhrī (d. after 340/951) *Book of the Routes and Kingdoms* (*Kitāb al-masālik wa l-mamālik*) produced in the orbit of Meḥmed II, inspired by a precious copy received from the Aqqoyunlu sultan (Pinto 2016, chs. 10–12). Meḥmed II had, as mentioned, received an Arabic translation of Ptolemy; a versified Italian rendering of the *Geography* was dedicated to him by the Florentine scholar Francesco Berlinghieri (1440–1501). We also find in the conqueror's possession maps of the Veneto, and of southeastern Europe, attached to a treatise on the art of war. Brotton has speculated that the sultan used those maps to move armies in his imagination like pawns on a chessboard, but surely, this would have been a mostly academic exercise linked to the study of history, not the practical planning of military campaigns (Roberts 2013; Brotton 1997, 103–4; Banfi 1954).

Even works that pretend to speak to practical purposes often assumed a different meaning with their elite audiences. The captain and litterateur Seydī ʿAlī Reʾīs (d. 970/1562), stranded in Gujarat after a botched naval campaign in 1558, produced a Turkish version of a pilot manual for the Indian Ocean, which probably served more the entertainment of the elite than the guidance of a captain at sea. Tibbetts dismissed it as the product of a "literary connoisseur or an antiquary whiling away his time in Gujarat with something topical", but without "real knowledge of what his texts were dealing with" (Tibbetts 1971, 45). Pīrī Reʾīs's world map depicts the marvelous beasts of Islamic cosmological lore in South America, and his maritime handbook contains legends along with trivialities no sailor would need to have explained to him.

In the case of these and other nautical maps, which are often precious and superbly decorated, scholars have debated if they were prepared for actual navigation or were appreciated by high-brow owners primarily for their aesthetic appeal. Scholars have attributed existing specimens, of Pīrī Reʾīs' *Book of the Sea* (*Kitāb-i baḥriyye*) in particular, to one or the other purpose based on the amount of decoration applied. It is worth remembering, however, that these attributions are usually not based on any documented usage or reading but are no more than scholarly assumptions about how a book may have been used or read by an Ottoman reader. Meanwhile, scholars have overlooked Pīrī Reʾīs's statement that he submitted his work to the sultan as a document of the achievements of navigators in the sultan's service (the passage is translated in Soucek 1996, 86. See also Soucek 1994; Emiralioğlu 2014, 26, 93). He did so in analogy to presentations to the sultan from guilds and craftsmen, showing that professional skills were cultivated by the men of the navy, without being incentivized as a matter of state policy, in the absence of an Ottoman counterpart to the Spanish *Casa de Contratación*. Speculation that the admiral Ḫādım Süleymān Pāshā (d. 954/1547) curated an archive of maps in Egypt that was still in use half a century later originates from a misreading of an account by Selanikī (Casale 2007, 187, 2012, 220). The only possible exception is the captain and author of a maritime atlas ʿAlī Majar Reʾīs (2nd half 10th/16th century), who was also registered among the imperial ateliers (*naqqāshkhāne*; Soucek 1971; Özdemir 1992).

Harley's conceptualization of maps as tools of state power and vehicles of royal propaganda has been adopted by scholars of Ottoman maps, despite the fact that many maps do not easily align with a presumed propagandistic goal and often were not accessible except to a select small audience in the imperial palace. For example, Pinto says:

> Specific information regarding sponsorship of cartographers is often difficult to determine. Because of the surreptitious nature of political propaganda, official patronage is frequently concealed from public view and not documented.
>
> *(Pinto 2016, 219; see also Emiralioğlu 2014)*

This is very different from the dissemination of printed maps in early modern Europe, which served as theater for 'cartographic battles' between nascent nations (for one example among many, see Duchhardt 2015). In any case, it is difficult to pinpoint anything like an imperial self-image in cartographic and geographical production up to Kātib Chelebi. Ottoman imperial identity was anchored in the imperial city of Constantinople and embodied in the persona of the sultan, but arguably no specific territory is ever identified in the geographical record as defining for the Ottoman Empire, very different from the agenda of Abbasid geographers and cartographers to depict a 'Realm of Islam' (Chapter I.13). To assume otherwise would risk projecting paradigms of the nation-state back onto premodern identities and ideologies (Kafadar 2007). Arguably, the idea of universal sovereignty, much discussed in recent scholarship, concerns the position of the Ottoman sultan among other rulers more than a specific territory under his immediate administration. Territorial boundaries are absent from the representation, and the mental maps are primarily organized in radial paths of movement from the imperial capital outward, and in concentric zones of familiarity, from an Islamic cultural space at the center, through a zone of the exotic to the periphery of utter alien-ness (Bonner and Hagen 2010). The bizarre creatures shown on Pīrī Reʾīs's world map testify to the tenaciousness of these tropes (Emiralioğlu 2014; Pinto 2016, chs. 5–9). Tellingly, the one 10th/16th-century world map that is framed by a flattering narrative in Turkish about the power of the Ottomans and their role in world history was not an Ottoman production at all but designed in Venice for an Ottoman market, although Casale (2013) has attempted to interpret it as an Ottoman artifact. Venetian authorities shut the enterprise down before any of the maps reached the empire (Ménage 1958).

Much attention has been paid to Maṭraqchı Nāṣūḥ's (d. 971/1564) *Explication of the Stations of the Campaign of Iraq* (*Beyān-i menāzil-i sefer-i ʿIrāqeyn*), which true to its title depicts every stop on Süleymān the Magnificent's (r. 926–974/1520–1566) march to conquer Baghdad in 940/1534. Strikingly, military action is entirely absent from the illustrations. While it has been interpreted as a depiction of an Ottoman frontier, the context of the Ottoman–Safavid sectarian rivalry and the focus on the sacred sites of Najaf and Karbala rather suggest an interpretation as an imperial pilgrimage narrative. Antrim has rightly pointed to the similarities of the illustrations with contemporary pilgrimage manuals (Antrim 2018, 78–96).

At the core of this uncertainty is the ambivalent concept of patronage, which, first of all, should be seen to relate to persons rather than ideas or books (Brentjes 2009). Patronage of geographical or cartographical work, like that of many other sciences, can be understood as a moral economy with patron and client and a wider public as main actors, in which sustenance, loyalty, knowledge and prestige are the units of currency. Patrons may commission works from their clients or reward works presented to them. Since typically both share a cultural formation, some degree of congruence between the ideas of a patron and those expressed by a client is to be expected. Yet, works dedicated to a ruler may also serve to articulate normative expectations, or contain aspects the dedicatee does not agree with, nor can the existence of a text with a dedication be taken as assurance that the client fulfilled all expectations of his patron. Therefore, to ignore the difference between commissioned and dedicated works, and to attribute the ideas expressed in dedicated works to the recipient, is a fallacy, if a common one; in an ideal situation, it is exactly the discrepancy between expectations and delivered goods (from both sides) that is most revealing to the historian.

Unfortunately, such information is mostly lacking in the case of Ottoman maps. Moreover, while patronage of knowledge production *per se* carries social prestige, it is not to be equated with the adoption and application by the patron of the knowledge produced. It remains a fact that for all the variegated geographical and cartographic production in the orbit of the Ottoman court it was only in 1067/1675 that the first work of this type was expressly commissioned by the sultan. Joan Blaeu's (1596–1673) *Atlas Maior* had been presented to the court by the Dutch envoy Justinus Colyer (1624–1682) in 1079/1668, and the translation was executed under the aegis of the scholar Ebū Bekr el-Dimeshqī ([d. 1102/1691]; Wurm 1971, 39–47; see also the following discussion). Thus, the first commissioned work falls in a period where space is no longer represented in cultural categories in the mental maps mentioned earlier but homogenized under the Ptolemaic grid of coordinates.

II.8.4 Administrative needs

So far, this survey has given us an idea about the mapping traditions, genres and techniques that Ottoman officials could draw on when they sought to harness cartography as a visual technology to display information critical for their service to the Ottoman state. Outside of a canon of literary production, or demonstrations of professional prowess, there remains a body of maps, and a number of documented instances, that shed more light on their practical use at the hands of the state and its officials, leading up to and expanding on the instance shown in d'Alexandre's painting for his Polish patron. Harley states:

> It is generally accepted that mapping is an activity designed to promote state efficiency and that with good maps the writ of centralized power can be made to run more uniformly of a country as a whole.
>
> *(Harley 1988b, 66)*

However, Scott cautions:

> The premodern state was, in many crucial respects, partially blind: it knew precious little about its subjects, their wealth, their landholdings and yields, their location, their very identity. It lacked anything like a detailed "map" of its terrain and its people. It lacked, for the most part, a measure, a metric, that would allow it to "translate" what it knew into a common standard necessary for a synoptic view.
>
> *(Scott 1998, 2)*

In comparison to these standards, the sophisticated Ottoman fiscal administration maintained uniquely detailed records of taxpaying households across the empire. Modern historians have not only used such information to construct maps but also found that the organization of the records themselves do not reflect geographical realities (see the maps in Hütteroth and Göyünç 1997 and Kołodziejczyk 2004). Contemporary observers noted that Ottoman dignitaries expressed specific interest in information about population size and resources, when it came to the translation of the Blaeu *Atlas*, meaning that the state was content with a representation of the empire that was based not on their own records but on external sources. It appears that the Ottomans themselves never attempted to translate the vast amount of spatial information contained especially in tax registers into visual terms and to create maps of their territory for their administrators. The most obvious explanation is that there was no purpose for such records and that decisions that required locational information were taken locally by officials who were able to verify such information as needed. This is most evident in instances where the boundaries of a property are described in the form of a circular path along landmarks, which can only be verified on-site (Singer 2002; see also Sarıcaoğlu 2020a, 2020d).

A similar preference for inscribing boundaries on the ground rather than recording them in maps manifested itself in the 12th/18th century in boundary disputes with Austria and Russia. When in 1126/1714 Russian and Ottoman commissioners set out to draw the new boundary between the two empires north of Crimea, it was the Ottomans who insisted on markers being erected on the ground, because recording them only on a map would be insufficient; they agreed that the 'boundary record' would be drawn up by the Russians and shared with all commissioners. The Russian commission, well furnished with all kinds of records, repeatedly scorned their Ottoman colleagues for a lack of preparation, including maps of the area (Bazarova 2015). In his memoir, an Ottoman scholar and member of the commission charged with registering the Austrian–Ottoman border along the Danube after the Treaty of Belgrade in 1152/1739 describes how he had to use all kinds of tricks to keep up with the surveying techniques of his Austrian counterparts, including secretly replicating their equipment, indicating that surveying for the purpose of a cartographic record was an unknown or at least unusual practice in the Ottoman military administration (Ebū Sehl Nuʿmān Efendi 1972; see also Sarıcaoğlu 2020b).

By contrast with the empire-wide scale of taxation records or the remoteness of boundary delimitations, the administration of water resources was local and crucial for the well-being of a significant number of subjects. By far, the most elaborate and complex water supply system was that of Istanbul, based on Byzantine beginnings and upgraded repeatedly under Ottoman rule. Sinān (d. 996/1588), the famed imperial architect, left a few construction drawings, including a sketch of the Qırqcheshme supply line, which he constructed as the largest of the Ottoman additions to the system; this sketch shows some measurements for the aqueducts (Çeçen 1992, 156–7). The Istanbul waterways have been glorified in literary works as well, sometimes with illustrations, which, however, do not serve any practical use. More relevant here is a copy in the Topkapı Palace library of a plan prepared in response to an inquiry of Sultan Murād III (r. 982–1003/1574–1595) to show him how the water in the city was distributed. While not an administrative document in itself, it goes to show how visualization, in this case within a well-established cartographic genre, was used to convey information to the ruler (Dursun 2013, no. 40; for two versions of the original see Karamustafa 1992, appx. 11.1). Even more interesting, however, is another map of the Qırqcheshme supply system for Istanbul dated 1127/1715, which shows all branches of this line within the city walls and records the shares of every resident or institution that had a claim to the water. The plan appears to follow Sinān's water distribution book of 976/1568 and was drawn as the result of an inspection. This map appears to be lost: Çeçen (1992) published an older photograph, acknowledging that he was unable to locate the map itself. Its functionality certainly had to do with the fact that it must have been stored in the office of the water administrator (*ṣu emīnī*) close to the object it helped govern. The invocation of a blessing on the anonymous maker inscribed on it goes to show that such a plan was a special accomplishment and by no means a regular tool in the administration of Istanbul's water. It remained an isolated document, as the 34 volumes of published records of Istanbul's water administration from the beginnings up to the 13th/19th century do not include a single map (Kal'a 1997).

II.8.5 Military maps

The crucial area for the use of maps in the service of the state, however, was warfare (Sarıcaoğlu 2020c). We have already seen the pride of Ottoman mariners in their navigational skills, celebrated in precious and artistic maps but have also noted that the state did not move to institutionalize the collection and development of nautical knowledge but left this task to informal networks of captains and navigators. As we will see, the institutionalization of mapping in military contexts took a different and more circuitous route.

In contrast to maritime cartography, military mapping on land was rare and unsystematic in the 10th/16th century, and later as well. The Greek historian Doukas (d. after 1462) describes how Meḥmed II, while still in Edirne, planned the siege of Constantinople.

> [He] traced the City's fortifications and designated to those skilled in siege warfare where and how to place the cannon, the breastworks, the trenches, the entrance into the fosse, and on which wall the scaling ladders were to be placed. In other words, he staged every operation during the night and in the morning his orders were executed.
> *(Doukas 1975, 202–3)*

In premodern military culture, in which sieges of fortified cities and castles were critical, such sketches of topographical situations would appear an obvious form of recording, but just as obviously such sketches would tend to be discarded after the action. In fact, most extant siege maps, as one typical genre of premodern Ottoman military mapping, appear to be narrative maps, probably designed to report to the sultan about the state of a campaign, for example, in the sieges of Belgrade 927/1521 and Malta 963/1565, or to explain its failure, as was the case after the Vienna siege in 1094/1683 (Dursun 2013, no. 48, 50, 51; Kreutel 1953). A rough sketch of the Dniepr and the fortress of Kiev, dated to around 906/1500, appears to be the exceptional case of a campaign proposal, submitted by a certain Ilyās the Pilot (or Reconnoiterer) from Morea (Dursun 2013, no. 47; Karamustafa 1992). The influence of maritime mapping is remarkable; out of these maps, two, Dniepr and Malta, show the hands of sailors. In stark contrast to the idiosyncratic style of these examples is a map of the Battle of Mezőkeresztes (1004/1596), which shows minimal topography, but its indication of troop contingents is a modern reimagination (TSMA. E-5539, see Uzunçarşılı 1982–88, 3: 1, pl. XVII). On the other hand, a peculiar anecdote narrated by the historian Muṣṭafā ʿĀlī (d. 1008/1600) suggests that maps may have been used to visualize the Hungarian frontier against the Ottomans. However, not only does this reference occur embedded in a dream narrative, we again do not have any surviving evidence for such a map (Hagen 1999, 121–2). Ottoman military campaigns required long preparations and relied on an efficient system of roads and supply routes. Information, for example, about water supplies, was recorded in campaign journals, but it is quite plausible, in analogy to the administration of tax revenues, that this system did not require visual records to function, as on-the-ground knowledge was obtained from spies and scouts (Murphey 1999, 67–8).

It was naval warfare again that provided the first impetus to rethink the uses of maps. Observing Ottoman struggles during the Crete campaign (begun in 1055/1645) and then spooked by the specter of an imminent Venetian attack on the straits and Constantinople in 1066/1656, Kātib Chelebi launched an urgent plea to adopt cartography as a matter of military technology:

> For the man who is in charge of affairs of state, the science of geography is one of the matters of which knowledge is necessary. If he is not familiar with what the entire earth's sphere is like, he should at least know the map of the Ottoman domains and that of the states adjoining it, so that when there is a campaign and military forces have to be sent, he can proceed on the basis of knowledge, and so that the invasion of the enemy's land and also the protection and defense of the frontiers becomes an easier task. Taking counsel with individuals who are ignorant of that science is no satisfactory substitute, not even when such men are of the region. Most locals are entirely unable to sketch the map of their own home region. Sufficient and compelling proof of the necessity for [learning] this science is the fact that the unbelievers, by their application to and their esteem for those branches of learning, have discovered the New World and have overrun the ports of India and the East Indies.
>
> *(Itzkowitz 1972, 106)*

The urgency of this appeal is underscored by four endorsements from leading religious authorities of the time (included in the editions Kātib Çelebi 1913, 2008). The immediate addressee was grand vizier Köprülü Meḥmed Pāshā (d. 1071/1661), first in a lineage of statesmen who patronized the pragmatic adoption of Western cartography. His son Köprülüzāde Fāżıl Aḥmed Pāshā (d. 1087/1676) initiated the translation of the Blaeu *Atlas*; an excerpt of that work may have been made in preparation for the campaign against Vienna in 1094/1683 (Hagen 2003, 154). As late as 1164/1751, a scion of the family patronized another translation, of Bernhardus Varenius's (1622–1650/1) *General Geography* (*Geographia Generalis*; İhsanoğlu *et al.* 2000, 157–61). Kātib Chelebi's reworking of Gerard Mercator (1512–1594) and Henricus Hondius's (1591–1651) *Atlas Minor* and Ebū Bekr el-Dimeshqī's translation of the Blaeu *Atlas* presented regional maps copied from European models, very different from what we saw previously (Hagen 2003, 2012). This development was carried on by the pioneer of printing and cartography, İbrāhīm Müteferriqa (d. 1160/1745), who edited Kātib Chelebi's cartographical work, *Mirror of the World* and *Gift for the Great (Men) on Naval Campaigns* (*Tuḥfetü l-kibār fī esfāri l-biḥār*), and also produced a printed map of the Black Sea, as well as a huge manuscript wall map of the Asian parts of the

Ottoman Empire and of Iran, of which two copies are known (Karamustafa 1992; on Müteferriqa see Sabev 2006).

Müteferriqa was a Hungarian convert to Islam who used his humanistic erudition and linguistic skills to make a unique career at the Ottoman court. He was, however, by no means exceptional as a cultural broker. The pool of technological expertise composed of the Mediterranean and the Danube region mentioned earlier continued to function, although its modes of operation – always historically contingent – were different from the earlier periods. Military technology was making major strides in Europe from the mid-17th century in many different ways, transforming the construction of fortifications and siege warfare and introducing light and mobile artillery for field battles or highly disciplined battalions trained to deliver volley fire in rapidly changing battle situations (Aksan 2007, 134; Pedley 2012). These innovations intersected with other needs of the emerging modern state to record, categorize and control territories, resources and populations. Consequently, mapping and surveying became essential tools in the service of the state, and the military in particular, and military and mathematical academies proliferated across Europe (Pedley 2012, 224; Sarıcaoğlu 2020c). The new experts trained in these sciences were also highly mobile and helped to disseminate them across the region. While in earlier centuries Ottomans had adopted new technologies for themselves, in this period they relied increasingly on 'innovation through patronage' (Şakul 2013, 194).

French cartographers active in the Ottoman lands already in the late 11th/17th century were interested in the fortifications at the Dardanelles and the Bosporus and operated in secrecy (Pérouse and Günergun 2016). Antiquarianism, however, should not be discounted as another important incentive (Pedley 2012). In the 12th/18th century, in the course of a French–Ottoman rapprochement, these men shared their products with Ottoman authorities. Beyond such general statements, however, it is difficult to assess how (and how early) Ottoman officers were exposed to new methods of military cartography. Two maps of the Battle of Pruth (1123/1711) offer an example by virtue of their stark difference in style. One, a separate sheet in the Topkapı Archive, is executed with all the details of European maps of the period, showing terrain, vegetation and roads, in addition to the Russian and Ottoman trenches. It also includes a scale and a wind rose (Karamustafa 1992, fig. 11.8). The inscriptions are by a professional Ottoman hand, but the date is given as 1711, in Christian era notation. A similarly 'modern' example, dated to 1095/1684, shows the fortress of Buda (Karamustafa 1992, fig. 11.7). The second map of the Battle of Pruth appears much less 'professional', lacking proportion and orientation, while it shows terrain and bridges on the river, unusual for Ottoman maps. An innovative, if probably improvised feature, indicates the strength of troops by little circles and the trajectories of troops by dotted lines. The map is part of an extensive narrative by a bureaucrat and eyewitness of the battle who lacked training as cartographer but must have had a good sense of what kind of information would be expected from a 'modern' map (Kurat 1968; Figure II.8.2). Other examples of individual copyists' innovations in maps of 12th/18th-century include copies of Kātib Chelebi's *Jihānnümā*, mentioned earlier.

In the course of the 12th/18th century, the Ottomans repeatedly accepted the services of French officers with the express desire to introduce new military technology and methods. The first prominent example was the Comte de Bonneval (1675–1747), who arrived in 1729. His Turkish name, Ḫumbaracı Aḥmed Pāshā, points to his expertise in fortification and mining. The curriculum for officers he introduced certainly included surveying and mapping, although it is doubtful that a separate 'College of Engineering' (*hendese-khāne*) was, in fact, established at the time (Şakul 2013). The officer and geometer Jean Lafitte-Clavé (1750–1793) was tasked with planning an assault on Ochakov using his expertise in 1200/1787. François Kauffer (d. 1801), mostly remembered for the first survey map of Istanbul, was asked to create maps of crucial

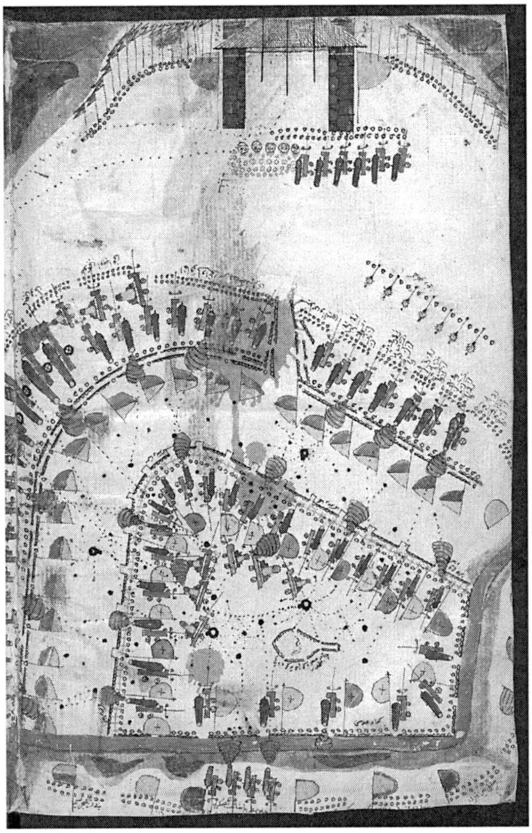

Figure II.8.2 Aḥmed ibn Maḥmūd, sketch of the Battle of Pruth, 1711, accompanying his untitled account of the campaign. MS Berlin, Preußischer Kulturbesitz, Staatsbibliothek, Orientabteilung, or. quart. 1209, fol. 305b.

Source: © STAATSBIBLIOTHEK ZU BERLIN – Preußischer Kulturbesitz, Orientabteilung

moments in the French occupation of Egypt (Pedley 2012; his map of Abukir is shown in Dursun 2013, no. 36). The English-born Ressām (or Enderūnlu) Muṣṭafā has been identified as the author of a number of maps of various parts of Europe beginning in 1181/1768 (van de Waal 1969, nos. 1447–1451, 1455; Karamustafa 1992, fig. 11.9 and 11.19; on his English origins see Sarıcaoğlu 2020c). These are the kind of maps d'Alexandre must have seen in the office of the *reʾīsü l-küttāb*.

It was, after all, the Ottoman rivalry with Russia and Austria in southeastern Europe that drove the adoption of new technologies, including military cartography, and created an opening for the French; it is only to be expected that the Black Sea and the Danube basin are the most frequently represented areas in the maps of the period. Still, the Ottomans were unable to keep up especially with progress in artillery and infantry, with the result that they lost almost every major field battle after the Vienna campaign of 1094/1683 (Şakul 2013, 193). After a relatively long period of peace, the Ottomans entered a war with Russia in 1182/1768, with disastrous outcomes both on land and at sea, the naval defeat at Çeşme 1184/1770 being particularly humiliating. In reaction, the School of Naval Engineering was founded in 1189/1775, followed by a School of Military Engineering in 1209/1795. With those institutions, engineering methods and cartography finally became central parts of the curriculum, which was supported by an in-house printing press that produced textbooks and maps. The multiple inventories surveyed by Beydilli show how the school constantly acquired maps (and the products of the Müteferriqa press), although a separate statute on how to access and archive them was issued only in 1208/1794 (Beydilli 1995, 292). When Abbé Toderini (1728–1799) visited the school during his research for his *Letteratura Turchesca*, he noted walls of classrooms hung with Turkish and French maps in manuscript and print (Toderini 1787, 1: 176–8). A department of cartography and geography was established in the School of Naval Engineering in 1212/1797, but maps produced by the school itself can only be traced after the 1830s (Pérouse and Günergun, eds. 2016, 150–1). The one notable exception is the *Translation of the New Atlas* (*Aṭlas-i cedīd tercümesi*) apparently based on one of William Faden's (d. 1836) atlases, with an introductory essay by Rā'if Maḥmūd Efendi (d. 1222/1807), of which 50 copies were printed in 1218/1803. An extremely expensive project, it did not find its way into the classroom of the academy, but, as Beydilli argues, served to demonstrate the viability of printing maps for the sultan (Beydilli 1995, 169–72).

The intense and systematic training in the engineering schools, which thoroughly transformed Ottoman visual culture, was not spared serious setbacks. Toderini (1787, 1: 174) cited scathing criticism of navigational ignorance that cost the Ottomans hundreds of ships. The historian Aḥmed Vāṣıf (d. 1221/1807) scoffed that many Ottoman statesmen were unable to understand how the Russian Baltic fleet appeared off Çeşme in the Aegean without passing through the Bosporus in 1184/1770 (Lewis 1982, 154, with more examples). As late as 1245/1829 negotiations with Russia sent Ottoman officials scrambling to identify a place that their Russian counterparts had mentioned and that they could not find on any map (Beydilli 1995, 395). It appeared that İbrāhīm Müteferriqa had been exceedingly optimistic when he praised the new medium of the map as 'easy to comprehend and to remember' on his map of the Black Sea. Obviously, his main purpose was to promote the medium and its dissemination through printing (Dursun 2013, no. 35). Even when the medium was available and made legible, it also required a new cultural approach that valued empirical data and planning over moral rectitude and zeal. This process, in which among others the aforementioned Aḥmed Vāṣıf played a prominent role (Menchinger 2016), extended far into the 13th/19th century and thus took much longer and was much less natural or self-evident than Müteferriqa envisaged.

Notes

1 I wish to thank the representatives of Sotheby's for making a high-resolution scan available. I am also grateful to Sonja Brentjes, Erdem Çıpa and Mary Pedley, who shared their thoughts with me.

2 Consolidated bibliography.

Bibliography

Sources

Blaeu, Joan. 1665. *Atlas Maior sive Cosmographica Blaviana*. 11 vols. Amsterdam: J. Blaeu.

Doukas. tr. Magoulias, H. J. 1975. *Decline and Fall of Byzantium to the Ottoman Turks*. Detroit: Wayne State University Press.

Kal'a, A., ed. 1997. *İstanbul Su Külliyâtı* [Collected Documents on the Istanbul Water Supply]. 34 vols. Istanbul: Istanbul Araştırmaları Merkezi.

Kātib Çelebi, Muṣṭafā ibn ʿAbdallāh. 1729. *Cihānnümā* [Cosmography]. Istanbul: Müteferriqa.

Kātib Çelebi, Muṣṭafā ibn ʿAbdallāh. 1913. *Tuḥfetü l-kibār fī esfāri l-biḥār* [Gift of Nobles: Naval Campaigns]. Istanbul: Maṭbaʿa-i baḥrīye.

Kātib Çelebi, Muṣṭafā b. ʿAbdullāh. ed. Bostan, İ. 2008. *Tuhfetü'l-Kibâr fî Esfâri'l-Bihâr (Deniz seferleri hakkında büyüklere armağan)* [Gift of Nobles: Naval Campaigns]. Ankara: T.C. Başbakanlık Denizcilik Müsteşarlığı.

Kolodziejczyk, D. 2004. *The Ottoman Survey Registers of Podolia (ca. 1681): Defter-i Mufassal-i Eyalet-i Kamaniçe*. Cambridge, MA: Harvard Ukrainian Research Institute.

Mercator, G. 1628. *Atlas minor*. Amsterdam: J. Jansz.

Nuʿmān Efendi, Ebū Sehl. ed. Savaş, A. İ. 1999. *Tedbīrāt-ı Pesendīde* [Appreciated Measures]. Ankara: Türk Tarih Kurumu.

Pīrī Reʾīs. tr. and ann. Soucek, S. 1996. *Piri Reis & Turkish Mapmaking after Columbus. The Khalili Portolan Atlas*. London: Nour Foundation, Azimuth Editions and Oxford University Press.

Tibbetts, G. R. 1971. *Arab navigation in the Indian Ocean before the coming of the Portuguese: Being a translation of Kitāb al-Fawāʾid fī uṣūl al-baḥr waʾl-qawāʾid of Aḥmad b. Mājid al-Najdī; Together with an Introduction on the History of Arab Navigation, Notes on the Navigational Techniques and on the Topography of the Indian Ocean and a Glossary of Navigational Terms*. London: Royal Asiatic Society of Great Britain and Ireland; Luzac.

Toderini, G. 1787. *Letteratura turchesca*. Venezia: Presso G. Storti.

Research literature

Agoston, G. 2007. "Information, Ideology, and the Limits of Imperial Policy: Ottoman Grand Strategy in the Context of Ottoman-Habsburg Rivalry," in Aksan, V. and Goffman, D., eds. *The Early Modern Ottomans: Remapping the Empire*. Cambridge: Cambridge University Press, 75–103.

Aksan, V. 2007. *Ottoman Wars, 1700–1870: An Empire Besieged*. Harlow: Pearson.

Antrim, Z. 2018. *Mapping the Middle East*. London: Reaktion Books.

Arbel, B. 2002. "Maps of the World for Ottoman Princes? Further Evidence and Questions Concerning 'The Mappamondo of Hajji Ahmed'," *Imago Mundi* 54: 19–29.

Banfi, F. 1954. "Two Italian Maps of the Balkan Peninsula," *Imago Mundi* 11: 17–34.

Baramova, M., Boykov, M. G. and Parvev, I., eds. 2015. *Bordering Early Modern Europe*. Wiesbaden: Harrassowitz.

Bazarova, T. 2015. "The Process of Establishing the Border between Russia and the Ottoman Empire in the Peace Treaty of Adrianople (1713)," in Baramova, Boykov and Parvev, eds., 121–32.

Beydilli, K. 1995. *Türk İlim ve Matbaacılık Tarihinde Mühendishâne, Mühendishâne Matbaası ve Kütüphânesi (1776–1826)*. Istanbul: Eren.

Bonner, M. D. and Hagen, H. 2010. "Ottoman Accounts of the Dār al-Ḥarb," in Irwin, R., ed. *The New Cambridge History of Islam*. Vol. 4: *Islamic Cultures and Societies to the End of the Eighteenth Century*. Cambridge: Cambridge University Press, 474–94.

Brentjes, S. 2005. "Mapmaking in Ottoman Istanbul between 1650 and 1750: A Domain of Painters, Calligraphers, or Cartographers?" in Imber, C., Kiyotaki, K. and Murphy, R., eds. *Frontiers of Ottoman Studies: State, Province, and the West*. 2 vols. London: I. B. Tauris, 2: 125–56.

Brentjes, S. 2009. "Patronage of the Mathematical Sciences in Islamic Societies," in Robson, E. and Stedall, J., eds. *The Oxford Handbook of the History of Mathematics*. Oxford: Oxford University Press, 301–28.

Brentjes, S. 2011. "Patchwork – The Norm of Mapmaking Practices for Western Asia in Catholic and Protestant Europe as well as in Istanbul between 1550 and 1750?" in Günergun, F. and Raina, Dh., eds. *Science between Europe and Asia*. Dordrecht: Springer, 77–102.

Brotton, J. 1997. *Trading Territories: Mapping the Early Modern World*. London: Reaktion Books.

Casale, G. 2007. "An Ottoman Intelligence Report from the Mid Sixteenth Century Indian Ocean," *Journal of Turkish Studies* 31.1: 181–8.

Casale, G. 2012. "From Hungary to Southeast Asia: The Ali Macar Reis Atlas in a Global Context," *Osmanlı Araştırmaları* 39: 55–62.

Casale, G. 2013. "Seeing the Past: Maps and Ottoman Historical Consciousness," in Çıpa, E. and Fetvacı, E. F., eds. *Writing History at the Ottoman Court: Editing the Past, Fashioning the Future*. Bloomington: Indiana University Press, 80–99.

Çeçen, K. 1992. *Sinan's Water Supply System in İstanbul*. Istanbul: T.C., İstanbul Büyük Şehir Belediyesi.

Dankoff, R., Tezcan, N. and Sheridan, M. D. 2018. *Ottoman Explorations of the Nile*. London: Gingko Library.

Duchhardt, H. 2015. "The Cartographic 'Battle of the Rhine' in the Eighteenth Century," in Baramova, Boykov and Parvev, eds., 3–13.

Dursun, H. A. ed. 2013. *Piri Reis'ten önce ve sonra Topkapı Sarayı'nda Haritalar/Before and After Pîrî Reis: Maps at the Topkapı Palace*. Istanbul: Topkapı Sarayı Müdürlügü.

Edney, M. H. and Pedley, M. S. eds. 2020. *The History of Cartography*. Volume Four: *Cartography in the European Enlightenment*. Chicago: University of Chicago Press.

Emiralioğlu, M. P. 2014. *Geographical Knowledge and Imperial Culture in the Early Modern Ottoman Empire*. Farnham, Surrey: Ashgate.

Gökbilgin, M. T. 1956. "Kanunî Sultan Süleyman devri başlarında Rumeli eyaleti, livaları, şehir ve kasabaları," *Belleten* 20: 247–85.

Goodrich, T. D. 1993. "Old Maps in the Library of Topkapi Palace in Istanbul," *Imago Mundi* 45.1: 120–33.

Hagen, G. 1999. "Träume als Sinnstiftung – Überlegungen zu Traum und historischem Denken bei den Osmanen (zu Gotha, Ms. T. 17/1)," in Stein, H., ed. *Wilhelm Pertsch – Orientalist und Bibliothekar. Zum 100. Todestag*. Gotha: Forschungs- und Landesbibliothek, 101–22.

Hagen, G. 2006. "Kâtib Çelebi and Sipahizade," in Kaçar, M. and Durukal, Z., eds. *Essays in Honour of Ekmeleddin İhsanoğlu*. Istanbul: IRCICA, 525–42.

Hagen, G. 2012. "Atlas and Papamonta as Sources of Knowledge and Power," in Tezcan, N. and Karateke, H., eds. *Evliya Çelebi Seyahatnamesi'nin Yazılı Kaynakları*. Ankara: Türk Tarih Kurumu, 104–29.

Harley, J. B. 1988a. "Maps, Knowledge, and Power," in Cosgrove, D. and Daniels, S., eds. *The Iconography of Landscape: Essays on the Symbolic Representation, Design, and Use of Past Environments*. Cambridge: Cambridge University Press, 277–312.

Harley, J. B. 1988b. "Silence and Secrecy: The Hidden Agenda of Cartography in Early Modern Europe," *Imago Mundi* 40: 57–76.

Hütteroth, W. D. and Göyünç, N. 1997. *Land an der Grenze: osmanische Verwaltung im heutigen türkisch-syrisch-irakischen Grenzgebiet im 16. Jahrhundert*. Istanbul: Eren.

Itzkowitz, N. 1972. *Ottoman Empire and Islamic Tradition*. New York: Knopf.

Kafadar, C. 2007. "A Rome of One's Own: Reflections on Cultural Geography and Identity in the Lands of Rum," *Muqarnas* 24.1: 7–25.

Karamustafa, A. T. 1992. "Military, Administrative, and Scholarly Maps and Plans," in *GTISAS* [CB] , 209–27.

Konvitz, J. W. 1987. *Cartography in France, 1660–1848: Science, Engineering, and Statecraft*. Chicago: University of Chicago Press.

Kreutel, R. F. 1953. "Ein zeitgenössischer türkischer Plan zur zweiten Belagerung Wiens," *Wiener Zeitschrift für die Kunde des Morgenlandes* 52: 212–29.

Kurat, A. N. 1968. "Hazine-i Bîrun Kâtibi Ahmed bin Mahmud'un Prut Seferine ait defteri," *Tarih Araştırmaları Dergisi* 4.6–7: 261–426.

Kut, G. 1985. "Türk Edebiyatında Acâibü'l-mahlükât Tercümeleri Üzerine," in *Beşinci Milletlerarası Türkoloji Kongresi Bildiriler*. Istanbul: İstanbul Üniversitesi Edebiyat Fakültesi Türkiyat Araştırma Merkezi, 183–93.

Kut, G.. ed. 2012. *'Acâyibü'l-Mahlûkât ve Garâyibü'l-Mevcûdât (İnceleme-Tıpkıbasım). Süleymaniye Yazma Eser Kütüphanesi, Nuri Arlasez Koleksiyonu, No. 128'deki Nüshanın Tıpkıbasımı*. Istanbul: Türkiye Yazma Eserler Kurumu Başkanlığı, 5–26.

Lewis, B. 1982. *The Muslim Discovery of Europe*. New York: W.W. Norton.

Mavroudi, M. 2013. "Translations from Greek into Arabic at the Court of Mehmed the Conqueror," in Ödekan, A., Necipoğlu, N. and Akyürek, E., eds. *The Byzantine Court: Source of Power and Culture. Papers from the Second International Sevgi Gönül Byzantine Studies Symposium, Istanbul 21–23 June 2010*. Istanbul: Koç University Press, 95–207.

Ménage, V. L. 1958. "The Map of Hajji Ahmet and Its Maker," *Bulletin of the School of Oriental and African Studies* 20: 291–314.

Menchinger, E. L. 2016. "Free Will, Predestination, and the Fate of the Ottoman Empire," *Journal of the History of Ideas* 77.3: 445–66.

Murphey, R. 1999. *Ottoman Warfare, 1500–1700*. New Brunswick: Rutgers University Press.

Özdemir, K. 1992. *Ottoman Nautical Charts and the Atlas of Ali Macar Reis*. Teşvikiye, İstanbul: Marmara Bank, Creative Yayıncılık ve Tanıtım.

Pedley, M. 2012. "Enlightenment Cartography at the Sublime Porte: François Kauffer and the Survey of Constantinople," *Osmanlı Araştırmaları* 39: 29–53.

Pérouse, J.-F. and Günergun, F., eds. 2016. *Üç denizin arasında. Osmanlı ve Fransız Boğaz haritaları/ Entre trois mers. Cartographie ottomane et française des Dardanelles et du Bosphore.* İzmir: Arkas.
Pinto, K. C. 2011. "The Maps Are the Message: Mehmet II's Patronage of an 'Ottoman Cluster'," *Imago Mundi* 63.2: 155–79.
Pinto, K. C. 2016. *Medieval Islamic Maps. An Exploration.* Chicago: University of Chicago Press.
Roberts, S. 2013. *Printing a Mediterranean World: Florence, Constantinople, and the Renaissance of Geography.* Cambridge, MA: Harvard University Press.
Robles Macías, L. A. 2012. "Zodiac on Earth: The Ecliptic on Two Sixteenth-Century Ottoman World Maps," *Osmanlı Araştırmaları* 39: 95–120.
Sabev, Orlin. 2006. *İbrahim Müteferrika ya da ilk Osmanlı matbaa serüveni (1726–1746). Yeniden değerlendirme.* Istanbul: Yeditepe Yayınevi.
Şakul, K. 2009. *An Ottoman Global Moment: War of Second Coalition in the Levant.* PhD dissertation. Washington, DC: Georgetown University.
Şakul, K. 2013. "Military engineering in the Ottoman Empire," in Lenman, B. P., ed. *Military Engineers and the Development of the Early Modern European State.* Dundee: Dundee University Press, 179–99.
Sarıcaoğlu, F. 1997. "Harita," in *Türkiye Diyanet Vakfı İslam Ansiklopedisi.* Istanbul: Türkiye Diyanet Vakfı, 16: 210–16.
Sarıcaoğlu, F. 2020a. "Administrative cartography in the Ottoman Empire," in Edney and Pedley, eds., 64–9.
Sarıcaoğlu, F. 2020b. "Boundary Surveying in the Ottoman Empire," in Edney and Pedley, eds., 199–203.
Sarıcaoğlu, F. 2020c. "Military Cartography by the Ottoman Empire," in Edney and Pedley, eds., 993–9.
Sarıcaoğlu, F. 2020d. "Property Mapping in the Ottoman Empire," in Edney and Pedleuy, eds., 1190–2.
Scott, J. C. 1998. *Seeing Like a State: How Certain Schemes to Improve the Human Condition Have Failed.* New Haven: Yale University Press.
Singer, A. 2002. "Transcrire les frontières de village," in Pouillon, F., ed. *Lucette Valensi à l'oeuvre: une histoire anthropologique de l'Islam méditerranéen.* Paris: éditions Bouchène, 133–43.
Sotheby's. 2013. *Travel, Atlases, Maps, and Natural History, May 14, 2013.* Lot 141. https://www.sothebys.com/en/auctions/ecatalogue/2013/travel-atlases-maps-natural-history-l13401/lot.141.html
Sotheby's. 2018. *Arts of the Islamic World, Oct. 24, 2018.* Lot 111. https://www.sothebys.com/en/auctions/ecatalogue/2018/arts-of-the-islamic-world-l18223/lot.111.html
Soucek, S. 1971. "The 'Ali Macar Reis Atlas' and the 'Deniz Kitabı'. Their Place in the Genre of Portolan Charts and Atlases," *Imago Mundi* 25: 17–27.
Soucek, S. 1973. "Tunisia in the *Kitāb-i baḥrīye* by Pīrī Re'īs," *Archivum Ottomanicum* 5: 129–296.
Soucek, S. 1994. "Piri Reis and Ottoman Discovery of the Great Discoveries," *Studia Islamica* 79: 121–42.
Uzunçarşılı, İ. H. 1982–8. *Osmanlı Tarihi* [Ottoman History]. Ankara: Türk Tarih Kurumu.
Waal, E. H. van de. 1969. "Manuscript Maps in the Topkapı Saray Library, Istanbul," *Imago Mundi* 23.1: 81–95.
Wurm, Heidrun. 1971. *Der osmanische Historiker Ḥüseyn b. Ǧaʿfer, gen. Hezārfenn, und die Istanbuler Gesellschaft in der zweiten Hälfte des 17. Jahrhunderts.* Freiburg i.Br.: Klaus Schwarz.

II.9
ANOTHER SCIENTIFIC REVOLUTION
The occult sciences in theory and experimentalist practice

Matthew Melvin-Koushki

The Scientific Revolution, a problematic 20th-century label that nevertheless faithfully conveys the ethos of many 16th- to 19th-century sources, is central to our narrative of the emergence of Western modernity. That revolution was multidisciplinary and progressive: first in astronomy and physics (Copernicus to Newton), then in chemistry and biology (Boyle to Darwin). In all fields, it was defined by a new empiricist and experimentalist program, whereby even the authority of Aristotle could now be set aside, albeit respectfully, when it contradicted direct observation (Galilei and Bacon). And while its protagonists were not revolutionary in the counterintuitive modern sense of the term, instead championing their empirical science as the revival ("the Renaissance") and extension of ancient, perennial wisdom, most shared a millenarian identity as the self-styled architects of a new age (Shapin 1996; McKnight 1998).

To that end, many of our Heroes of Science were also deeply invested in the occult sciences, from astrology and alchemy to geomancy and magic, as integral to their natural philosophy. Yet this simple fact long horrified modern historians, who until recently sought to winnow the chaff of such "medieval superstition" from the grain of True Science, engine of the Rise of the West. Such a heresiographical and teleological approach is no longer considered legitimate by historians of science. We now emphasize rather the messy, meandering, orthodoxizing and often abortive nature of "scientific progress," present and past, which was historically achieved by diehard occultists (like Newton and Bacon) as often as not (see, e.g., Yates 1964; Langman 2010; Newman 2019).

The same principle applies to the contemporary Islamicate and especially Persianate world, culturally just as "Western" as the Latinate world in its dual focus on Hellenic philosophy and natural science and Abrahamic monotheism, whose imperial states (Mughal, Safavid, Uzbek, Ottoman) encompassed a full third of the human race (Melvin-Koushki 2018a, 2018b). That Islamdom was far wealthier and more populous and cosmopolitan than Christendom throughout the early modern period, moreover, meant that the sciences – especially the occult sciences – flourished there to an even greater degree. Astrology, lettrism, geomancy, oneiromancy and physiognomy in particular enjoyed heavy royal patronage (see, e.g., Şen 2016; Lelić 2017; Melvin-Koushki 2018d). And no wonder: for these and their allied divinatory and magical disciplines promised control, especially *imperial* control, of minds and bodies, time and space. First you must predict the future; then you can change it.

DOI: 10.4324/9781315170718-29

The explosion in patronage and popularity of such disciplines in courtly and scholarly circles around the turn of the 9th/15th century is reflected by the sudden emergence of a new Arabo-Persian genre: the imperial grimoire (Melvin-Koushki 2018c). Unparalleled in the Latinate world, where political and economic units were smaller, these occult-scientific manuals are explicitly modeled on the vastly popular 4th/10th-century Pseudo-Aristotelian *Secret of Secrets* (*Sirr al-asrār*, Lat. *Secretum secretorum*), wherein the Greek sage teaches his protégé Alexander the Great that world empire can only be achieved and maintained by means of astrology, alchemy, physiognomy and magic. Similarly, some early modern Muslim sages sought to turn their royal patrons into mages by writing highly accessible manuals on these and other occult sciences – and sometimes succeeded in this project. In the Ottoman case, indeed, one such manual, ʿAbd al-Raḥmān al-Bisṭāmī's (d. 858/1454) *Key to the Comprehensive Prognosticon* (*Miftāḥ al-jafr al-jāmiʿ*), even came to function in the 10th/16th century as a *dynastic handbook*, whereby the political fate of the House of ʿOsmān could be effectively managed over the generations, even to the Last Hour itself (Fleischer 2009). Astrology, too, was so thoroughly institutionalized at the Ottoman court that the office of court astrologer only ceased with the fall of the Ottoman Empire in 1922, almost four and a half centuries since its institution (Şen 2016). Competing Islamicate empires used similar strategies to jockey for precedence. Early modern Islamicate imperialism, in short, was pinned to this specifically occult-scientific model of the Alexandrian Empire, synthesized with Mongol, Persian and Islamic precedent, and hence often pursued as an occult science in its own right (Melvin-Koushki 2018a, 2018c).

But the occult sciences were not of merely political, economic and military interest. Just as in West Europe, they were held up by some of the greatest Muslim philosophers and saints of the early modern era – who, again, often functioned as court mages and architects of Platonic Empire – as the basis for what we might call a *scientific revolution*. For all their perennialist respect for the ancients, these ambitious thinkers consciously achieved an unprecedented and fundamental break with Hellenic antiquity, while avidly using its tools. That is, they consciously engineered a paradigm shift, in Kuhnian terms. And they too thought of themselves as *modern* (Melvin-Koushki 2018b).

This new cosmological imaginary was born in the 6th/12th-century western Mediterranean, a beneficiary of climate change; began to crystallize in 7th/13th-century Mamluk Egypt and Ilkhanid Iran, where it was heavily supported by new forms of post-caliphal empire; then was given wings by the 8th/14th-century apocalypse that was the Black Death. Like other crisis-born cosmologies before and since, it was remarkably long-lived and would enjoy hegemony throughout the Islamicate world down to the colonial period, at all levels of society. The sheer profusion of relevant surviving manuscripts and artifacts aside, the surest index of the ubiquity of occult science in post-Mongol Islamicate societies is the fact that market inspectors (*muḥtasib*) had to be appointed to regulate this booming industry (Mauder 2019). Unlike our own, however, the cosmology that underpinned it was emanationist and analogist, which is to say, Neoplatonic – and often just as millenarian as Newtonian and Baconian science. As Noah Gardiner recently put it:

> In this new regnant vision of the cosmos, the stars, planets, plants, minerals, bodily humors, colors, scents, letters of the alphabet, numbers, nations, and countless other entities and accidents of creation were enmeshed in a web of sympathetic or antipathetic relationships that vibrated along the 'great chain of being' between the godhead and the manifest world.
>
> *(Gardiner 2019, 557)*

To a majority of early modern Muslim thinkers, occult sciences thus were simply those natural and mathematical sciences that could best penetrate and manipulate the structure of the cosmos to theoretical and practical advantage – and so help to divinize man, to achieve Neoplatonic *theosis*.

Most crucially for the history of science, this paradigm shift was also a specifically Neopythagorean one: the world would now be read in exclusively *mathematical-linguistic* – which is to say, *talismanic* – terms. This is a major source of the Two Books doctrine, peculiar to the bibliomaniac West, whereby the natural philosopher (Muslim, Jewish or Christian) must treat nature and scripture as twin expressions of the same divine revelatory mind. Henceforth the world was to be read, and magically rewritten, as a *mathematical text* (Håkansson 2001; Harrison 2001; Howell 2001; Killeen and Forshaw 2007).

Now this *mathesis*, of course, has been hailed by historians as the very heart of the Scientific Revolution. After all, did not Galilei declare that "the book of nature is written in the language of mathematics"? And were not Copernicus and Kepler ardent Pythagoreans (Hallyn 1990)? (While not consciously Pythagorean themselves, modern physicists still embrace both men as exemplars; and the new physics of information is curiously Neopythagorean in tenor [Wheeler 1990, Vlatko 2010, Melvin-Koushki 2016].) The overtly occult-scientific program that drove this shift, which fundamentally reorganized the human relationship to the cosmos, and which gained traction in the Islamicate world well before the Christianate, was thus no less important to Western intellectual and cultural history than the Copernican shift of a few centuries later – and even, recent scholarship suggests, helped make the latter possible (Saliba 2007[CB];[1] Ragep 2007; Westman 2011; Morrison 2014; Feldhay and Ragep 2017; Melvin-Koushki 2017). True, Muslim scholars remained formally committed to geocentric theory throughout the early modern period, out of deference to Aristotle and Ptolemy, or simply ignored the question (at least as far as we know). But the boom in Neoplatonic and especially Neopythagorean philosophy, the natural epistemological basis for heliocentrism since antiquity, emboldened reflective Westerners to finally break with Aristotle's physics. It must be emphasized that the Copernican theory was seen as essentially Neopythagorean by its proponents and opponents alike, including Copernicus himself. The same epistemology also helped inspire them to begin to read the Book of Nature systematically as a second scripture. And this reading flourished as the Platonopolis flourished: for science, occult or otherwise, both drives and is driven by empire.

II.9.1 Putting the "occult" in the occult sciences

Why was this earlier, Islamic Scientific Revolution specifically Occult? What was it about the occult sciences that made them so compelling to a majority of early modern Muslim scholars and rulers, not to mention their Jewish and Christian contemporaries? Why is their surge in popularity our best index for the beginning of Western *mathesis*, and hence modernity itself? It was not simply the status of astrology or alchemy as inductive natural sciences, according to the venerable Avicennan classification – although this feature alone would have ensured their persistence in post-Avicennan scholarly culture. Nor was it merely their obvious utility as technologies of empire; astrology, for example, had already been a vaunted imperial discipline in the Near East for millennia. Rather, as sciences concerned with extrapolating from visible data to invisible, from *ẓāhir* to *bāṭin*, they acquired an epistemological mobility not granted to any other rational (*ʿaqlī*) or traditional (*naqlī*), that is, Greek or Islamic, sciences.

The occult sciences, in short, became the only sciences in the Islamicate tradition to transcend the strict epistemic and generic divide between the natural and mathematical sciences, on one hand, and the religious and literary sciences, on the other. By the 10th/16th century, lettrism (*ʿilm al-ḥurūf*) and alchemy (*ʿilm al-kīmiyāʾ*) in particular were sciences classified and practiced as natural, mathematical and religious *simultaneously*. As such, they were embraced by philosophers as an especially effective means of reconciling the otherwise irreconcilable Neoplatonic emanationist and Quranic creationist accounts of the origin and destiny of humanity and the cosmos, thereby marrying Greek philosophy to its Islamic actualization (Melvin-Koushki 2019). So compelling was this concordist project, in fact, that European Renaissance humanists, from Giovanni Pico della Mirandola (d. 1494) onward, began to seize upon *kabbalah*, lettrism's coeval Hebrew twin, to much the same end. Thenceforth lettrism and kabbalah would be the primary driver of and vehicle for Neopythagoreanism throughout the early modern West (Melvin-Koushki 2016).

This transcendent quality is captured in the very term by which these sciences were routinely identified as an epistemologically coherent group, both in Arabic and in Latin: the *occult* sciences. As in modern English usage, *occult* (Ar. *khafī*, Lat. *occultus*), simply meaning "hidden," originally conveyed an air of mystery; the sciences so designated were often taught and practiced by individuals or within closed groups – "esotericist reading communities" – who jealously guarded them against profanation by non-initiates (Gardiner 2014). Unlike the current English term, however, *occult* was not meant as a slur, synonymous with *spooky* or, better, *paranormal*, but rather as an assertion of social and spiritual superiority and exclusivity – while also embracing the frequent, wonder-inducing *weirdness* of human experience more generally. As such, it was above all a simultaneously epistemological and technological designation: the sciences so named are those that allow one to penetrate and harness the unseen, subtler, weirder, more wondrous levels of existence beyond ordinary human ken; hence the common appellations *al-ʿulūm al-laṭīfa*, the subtle sciences, and *ʿulūm al-ghayb*, the sciences of the unseen. That we now prefer instead the neologistic prefixes *dark*, *sub-*, *super-* and *para-* (as in *dark matter*, the *subconscious*, the *supernatural*, *parapsychology*) to designate the same category merely testifies to our continued belief in the magical power of naming!

Because these sciences required both rigorous intellectual training and spiritual purity, however, few scholars achieved full mastery of one or more of them. But some of those who did became renowned as true Renaissance men, celebrated in biographical dictionaries and chronicles as the greatest scholars of their age. By the early modern period, philosophers famed as both jurists and occult scientists, like the Safavid sage-mage Shaykh Bahāʾī (d.1030/1621), could thus be officially eulogized in Persian as being 'possessed of eternal forms of knowledge (*ʿārif-i maʿārif-i azalī*), a master of sciences occult and manifest (*vāqif-i ʿulūm-i khafī u jalī*)' (Iskandar Beg Munshī 2003, 2: 967). Likewise, Mīr Fatḥ Allāh Shīrāzī (d. 997/1589), the celebrated Safavid polymath and inventor credited with bringing Islamic philosophy to Mughal India, earned this biographical encomium:

> He was a very learned scholar and superior to all the scholars of Khurasan, Iraq and India in his mastery of all types of rational and traditional sciences; during his lifetime he had no peer in all the world. He was also skilled in the occult sciences (*ʿulūm-i gharība*), particularly various types of talismans and spellcasting (*ṭilismāt, nīranjāt*).
>
> *(Niẓām al-Dīn Aḥmad 1931–1935, 2: 457)*

In view of their elite status, the term used in the earlier passage became standard in Arabic, Persian and Turkish for these sciences as a distinctive unit: *al-ʿulūm al-gharība*, the rare, difficult, paranormal and wondrous sciences, which may be read as a sociological descriptor as much as an epistemological one. Thus only 2% to 15% of scholars captured in Arabic and Persian biographical dictionaries of the 9th/15th to the 13th/19th centuries achieved fame specifically as occultists – roughly the same percentage, that is, that achieved fame as physicians or *ḥadīth* specialists (Murphy 2010, 90). Occult science, quite simply, had come to be a universal marker of prestige. Alchemy is here an exception, given the threat it represented to economic and hence political stability; a majority of Muslim alchemists labored in anonymity (Harris 2022; Artun 2013). And yet the historical and artifactual record shows that a much broader spectrum of society was involved in medical practices and hadith studies, precisely in view of their prestige. The same principle held true for the occult sciences (including medicine), which were practiced more or less competently by saints and charlatans alike; occult-scientific amateurs abounded. Elite sources (like biographical dictionaries) were thus concerned with *quality control*.

That the Arabic adjective *gharīb* means "elite" and not "strange" in this context is confirmed already in the 6th/12th century by no less a luminary than the philosopher-theologian Fakhr al-Dīn Rāzī (d. 606/1210), who in his *Compendium of the Sciences* (*Jāmiʿ al-ʿulūm*), the first comprehensive Persian encyclopedia of the rational and religious sciences, introduces his eastern scholarly readership to the new occult science of geomancy (*ʿilm al-raml*) by specifically styling it *gharīb*. In his entry on physiognomy (*ʿilm al-firāsa*), an occult science well known from Hellenic antiquity, he further glosses *gharīb* as *sharīf* – precisely meaning "elite" (Fakhr al-Dīn Rāzī 2003, 268, 432; on the many nuances of this descriptor see Zadeh 2020, 2023; on Rāzī as occultist, see Noble 2021). As a trenchant critique of the prevailing Aristotelian–Avicennan classification of all occult sciences as applied natural sciences only, and hence less capable of scientific precision, Rāzī here makes a revolutionary move: he reclassifies certain occult sciences, including judicial astrology (*ʿilm al-aḥkām*), geomancy and jinn magic (*ʿilm al-ʿazāyim*), as *mathematical* sciences, for the first time in the Western tradition as a whole. It is in this Rāzian revolution that we can see the seeds of the Occult-Scientific Revolution that was to follow, whereby Aristotle was increasingly if respectfully abandoned in favor of Plato and Pythagoras. In other words: Muslim occult scientists, like their Jewish and Christian peers, found they could best marry Plato to Aristotle by means of Pythagoras, sage, mage and saint in equal measure, on the basis of their personal experience. Significantly for the history of science, the history of religion and the history of empire alike, Pythagoras's Islamic reincarnation was widely held to be Imam ʿAlī – fountainhead of imperial sacral power (*walāya*) and mathematical occult science for Sunni and Shiʿi alike (Moin 2012; Melvin-Koushki 2018a).

II.9.2 The occult sciences in classifications of the sciences

The epochal process by which the occult sciences became simultaneously natural, mathematical and religious sciences, and hence ubiquitous in early modern Islamicate societies, may best be traced in those Arabic and Persian classifications of the sciences (*taṣnīf al-ʿulūm*) produced between the 4th/10th and the 11th/17th century. In such works, quite simply, the occult sciences (often formally designated as such) constituted roughly *half* of the Islamicate natural and mathematical sciences for a full millennium, right down to, and in some cases through, the colonial rupture (Melvin-Koushki 2017). Naturally, this percentage varies widely from century to century, from region to region and from author to author. But even manifestly anti-occultist encyclopedias were obliged to include many occult sciences as a matter of course; and the most influential classifications of the sciences in the later Arabo-Persian tradition are, without exception, heavily pro-occultist.

We have already seen how Fakhr al-Dīn Rāzī inaugurated this process by reclassifying several occult sciences, most notably astrology, from the natural to the mathematical. And to be a mathematical science, according to early modern Platonic-Aristotelian cosmology, is to have celestial, even spiritual, potency – hence Imam ʿAlī's new status as a mathematical genius. But even the natural sciences were increasingly valorized in their own right. Here it must be emphasized that

Maslama al-Qurṭubī's (d. 353/964) *Goal of the Sage* (*Ghāyat al-ḥakīm*), famed in Latinate Europe as the *Picatrix*, and among the most influential of all grimoires, was sometimes used as a primer in Aristotelian natural (occult) science till at least the 11th/17th century. Indeed, together with equally landmark works on astrology by Abū Maʿshar ([171–272/787–886]; Latin: Albumasar), it was a vector by which Aristotle entered Latin scholarship (Lemay 1962, 1993).

While occult sciences were perennially popular at Islamicate courts from the Abbasids onward, the 6th/12th to 8th/14th centuries, as noted, stand as a major inflection point in the history of Islamicate science and empire in tandem. For it was then that occultism, together with Sufism and Alidism, was made central to the new, experimentalist modes of millennial sovereignty and philosopher-kingship being developed in Ilkhanid Iran and Mamluk Egypt in particular. These modes would be fully realized over the next three centuries in the form of competing Timurid, Aqquyunlu, Safavid, Mughal and Ottoman imperialisms, each pressed into ever more radical avenues of ideological experimentation by their common Turko-Mongol Perso-Islamic rivalry (Melvin-Koushki 2018a). To reiterate: the post-Mongol, plague-wracked, intensely cosmopolitan early modern era was the golden age of Islamicate occultism, and the occult-scientific tenor of early modern Islamicate empire without precedent and parallel in Western intellectual-imperial history (Melvin-Koushki 2018b; Fleischer 2009). That the same era was also the golden age of Judeo-Christianate occultism is not an accident. But the latter was decidedly less imperial, and the Christian colonialism of later centuries was violently antimagical (Hanegraaff 2012).

This sea change, this earlier Scientific Revolution, is both heralded by and enshrined in the two most influential classifications of the sciences ever produced in the Islamicate encyclopedic tradition as a whole, Mamluk and Ilkhanid-Injuid, respectively, which served as models for most subsequent Islamicate such works:

1. *Guidance for the Seeker of the Sublimest of Goals* (*Irshād al qāṣid ilā asnā l-maqāṣid*), in Arabic, by Ibn al-Akfānī (d. 749/1348), a Cairene polymath who perished in the plague; and
2. *Jewels of Sciences Delightful to Behold* (*Nafāyis al-funūn fī ʿarāyis al-ʿuyūn*), in Persian, by Shams al-Dīn Muḥammad Āmulī (d. 752/1352), professor (*mudarris*) at Sultaniyya under the Mongol rulers Öljeitü (r. 704–716/1304–1316) and Abū Saʿīd (r. 716–736/1316–1335), then patronee of the Injuid Shaykh Abū Isḥāq (r. 743–54/1343–53) in Shiraz.

Following a general Avicennan schema, the first classifies all occult sciences as derivative natural sciences, a category they dominate. Here is his list of the applied natural sciences in full:

> Medicine (*ʿilm al-ṭibb*), veterinary medicine and falconry (*ʿilm al-bayṭara wa-l-bayzara*), physiognomy (*ʿilm al-firāsa*), oneiromancy (*ʿilm taʿbīr al-ruʾyā*), judicial astrology (*ʿilm aḥkām al-nujūm*), magic (*ʿilm al-siḥr*), talismans (*ʿilm al-ṭilasmāt*), illusionism (*ʿilm al-sīmiyāʾ*), alchemy (*ʿilm al-kīmiyāʾ*), agriculture (*ʿilm al-filāḥa*).

As Ibn al-Akfānī clarifies, these natural sciences concern either simple bodies or composite bodies or both. Thus, astrology concerns celestial simple bodies, and talismans terrestrial or elemental simple bodies. As for composite bodies, illusionism concerns those that have no constitution (*mizāj*), while those that have constitutions are further subdivided into those that do not have souls (alchemy) and those that do have souls. Here there are either insensible souls (agriculture) or sensible (veterinary medicine for animals, medicine for humans). Physiognomy then inquires into the inner states of rational human souls while they are conscious, and oneiromancy does the same while they are unconscious. That science that concerns both simple and composite bodies is magic (Melvin-Koushki 2017, 172–3).

Ibn al-Akfānī posits, in short, an *astrology–talismans–magic* continuum as the very backbone of natural philosophy and psychology alike, running the epistemological-ontological gamut from celestial simple bodies to terrestrial or elemental composite bodies and allowing the competent philosopher-scientist experiential control of the cosmos. Even *agriculture* is here conceptualized as an occult science. On the basis of such evidence, Gutas (2018) suggests that occult science represents the continuation of Avicennan natural science, even though Ibn Sīnā (d. 428/1037) himself, the second Aristotle, was no booster of the occult sciences. That Ibn al-Akfānī could be not only an Avicennan but also such an enthusiastic occultist testifies to the conceptual shift that intervened between the 5th/11th and 8th/14th centuries.

Where this Mamluk offering is highly succinct and dry, however, its Ilkhanid-Injuid counterpart is expansive, gorgeous and epistemologically revolutionary. Shams al-Dīn Muḥammad Āmulī also employs an Avicennan schema, synthesizing it with the then standard basic division between the Islamic or latter-day sciences (*ʿulūm-i avākhir*) and the philosophical or ancient sciences (*ʿulūm-i avāʾil*), and covering an unprecedented number of both. Yet he also seeks to transcend both schemas to move toward a new, more universal conception of knowledge – one in which, crucially, occult sciences like geomancy are to be reclassified as *mathematical*, following Fakhr al-Dīn Rāzī's radical move of a century and a half before. The various branches of the natural and mathematical sciences are as follows:

> Branches of natural science (*furūʿ-i ṭabīʿī*): medicine (*ʿilm-i ṭibb*), alchemy (*ʿilm-i kīmiyā*), letter magic (*sīmiyā*), oneiromancy (*ʿilm-i taʿbīr*), physiognomy (*ʿilm-i firāsat*), judicial astrology (*ʿilm-i aḥkām-i nujūm*), active properties (*ʿilm-i khavāṣṣ*), physiological professions (*ʿilm-i ḥiraf al-ṭabīʿa*) (these including the two divinatory techniques of scapulomancy (*aktāf*) and spasmatomancy (*ikhtilāj-i aʿżā*)), breath control (*ʿilm-i dam*), meditation and visualization (*ʿilm-i vahm*) (*dam* and *vahm* being Indian yogic sciences).
>
> Branches of mathematical science (*furūʿ-i riyāżī*): planetary theory (*ʿilm-i hayʾat*); optics (*ʿilm-i manāẓir*); the Middle Books (*mutavassiṭāt*); calculation (*ʿilm-i ḥisāb*); algebra (*ʿilm-i jabr u muqābala*); surveying (*ʿilm-i masāḥa*); constellations (*ʿilm-i ṣuvar-i kavākib*); star tables, ephemerides and astrolabes and other observational instruments (*ʿilm-i arqām u aʿmāl-i zīj u taqvīm u asṭurlāb u ālāt-i raṣadī*); administrative geography (*ʿilm-i masālik u mamālik*); magic squares (*ʿilm-i vafq-i aʿdād*); geomancy (*ʿilm-i raml*); mechanics and wondrous devices (*ʿilm-i ḥiyal*); games of strategy (*ʿilm-i malāʿib*).

Here, too, natural science is almost entirely occult, but the *Jewels of Sciences* further testifies to the progressive mathematization of the occult sciences, that is, their formal reclassification from the natural (*ṭabīʿī*) to the mathematical (*riyāḍī*) sciences. Again, this epochal development, peculiar to the Persianate East, must be considered central to the *mathesis* narrative going forward.

Equally epochally, the *Jewels of Sciences* is also the first encyclopedia to register the contemporary *sanctification* of occult science, and lettrism in particular. Āmulī designates the science of Sufism (*ʿilm-i taṣavvuf*) the supreme Islamic science, equal in importance to all the other religious and traditional sciences combined, and lettrism the supreme Sufi science, that is, *the* science of *walāya* (Melvin-Koushki 2017, 148–51; on this sanctification process generally see Gardiner 2014; Saif 2017).

Building on this Egyptian–Iranian precedent, the leading occultists of the next century – Timurid and Ottoman scholars radiating from Mamluk Cairo, and calling themselves, significantly, the New Brethren of Purity, after the shadowy 4th/10th-century group of Neopythagorean occultists of the same name – synthesized the parallel processes of mathematization and sanctification to reconceptualize lettrism as the supreme *Alid-Pythagorean universal-imperial* science, a total physics-metaphysics allowing the scholarly or royal adept comprehensive access to and experiential control over the various levels of being (Melvin-Koushki 2017; Fleischer 2009). Also in Mamluk Cairo, alchemy too was rendered a universal *mathematical-natural-religious* science by Aydemir al-Jildakī (d. *c.* 743/1342), the most important and prolific Muslim alchemist after (pseudo-)Jābir ibn Ḥayyān; all subsequent Arabic alchemy is Jildakian (Harris 2022; Artun 2013). The mathematization-sanctification of occult science generally and lettrism and alchemy specifically continued apace in the Persian-speaking world over the next centuries. Thus Safavid encyclopedias – even more heavily Neopythagorean-occultist in tenor – seize on these two sciences as both the primary mode of philosophical praxis and the best means of harnessing *walāya* – which is to say, of *magically incarnating the Imams*. It was then that ʿAlī ibn Abī Ṭālib, inventor of letter divination (*jafr*), was hailed the chief mathematician of Islam (Melvin-Koushki 2017).

For some early modern encyclopedists, moreover, the sanctification of certain natural-mathematical occult sciences was so total as to justify their reclassification exclusively as religious sciences. A case in point is the Ottoman polymath Ṭāshköprīzāde Aḥmed (d. 968/1561), who in his comprehensive *Key to Felicity and Lamp to Mastery* (*Miftāḥ al-saʿāda wa-miṣbāḥ al-siyāda*), greatly expanding on Ibn al-Akfānī, structures the totality of Islamic scientific and religious knowing according to an explicitly *lettrist* schema, for the first time in the Arabo-Persian tradition. At the same time, he classes lettrism proper not as a mathematical science but as a religious science and specifically as a subset of Quranic exegesis (*tafsīr*); the surest route to the heart of scripture is *magic* (Melvin-Koushki 2016). The Two Books doctrine remains intact; the emphasis, however, is on the primacy of the Quran as model of reality rather than on that model's call to mathematize the cosmos. Embodying the scientific mainstream, Ṭāshköprīzāde's offering became the primary model for all subsequent Ottoman encyclopedias, particularly Ḥājjī Khalīfa's ([d. 1067/1657]; Kātib Chelebi) landmark *Removal of Doubts About the Names of Books and Disciplines* (*Kashf al-ẓunūn ʿan asāmī l-kutub wa-l-funūn*).

This shift of emphasis would seem to reflect a larger Ottoman scholarly tendency to "de-mathematize" astronomy and to return occult sciences like astrology and geomancy to the natural sciences, apparently as a means of differentiating Ottoman learning from its Timurid-*cum*-Safavid model and rival, committed to an equally sacralizing, magical *mathesis* (cf. Fazlıoğlu 2014, 35; Brentjes 2010). For Ṭāshköprīzade's Safavid peers went to the opposite extreme: as noted, they singled out lettrism and alchemy as the two ultimate mathematical and Islamic-Imamic, which is to say universal, sciences (Melvin-Koushki 2017). In both Persianate cases, however, the first Sunni and the second Shiʿi, the moral cannot be more explicit – nor more divergent from contemporary developments in Inquisition-Reformation Europe, which was plagued by witch hunts: *Islam is to be lived as Magic.*

While much more research into the Islamicate encyclopedic tradition remains to be done, and the history of the occult sciences in Islam is still mostly unwritten, the overall trend from the medieval to early modern periods is clear. The natural, the mathematical and even the religious sciences became increasingly and disproportionately occult, and the greatest metaphysicians and physicians of early modern Islamdom were often also its greatest mages.

II.9.3 Occult-scientific experimentalism as early modernity

The historical study of Islamicate occult science is thus an especially efficient means of exploding colonialist-orientalist myths as to the eternal unmodernity of Islam: it is a project that cuts to the quick of what it means to be modern – and scientific.

For Muslim occult scientists constantly testify to the *empirical* nature of their operations. That is to say, Islamicate early modernity, like its Christianate parallel, is characterized by an increasingly heavy emphasis on the experimentally verifiable nature of science, occult and otherwise. Here the central Arabic terms are *taḥqīq*, "independent verification," and *tajriba*, "experiential essaying"; the latter's semantic evolution closely parallels that of its Latin cognate *experientia* (Janssens 2004; Langermann 2014; Melvin-Koushki 2018b; cf. Gutas 2018; Brentjes 2018[CB], 10, 74–7). This, it must be emphasized, represents a significant departure from Aristotelian precedent, wherein *empeiria* is a mode of general knowledge considerably inferior to *epistēmē*, "science." The empirical nature of occult science was already asserted by such authorities as Abū Maʿshar, systematizer of conjunction astrology, who influentially declared his discipline to be a science founded on analogical reasoning (*qiyās*) and experience (*tajriba*), the latter entailing careful observation (*muʿāyana*) and direct discovery (*al-wujūd al-ḥāḍir*; Burnett 2002, 210, 212). An identical argument is made for medicine (often classified as an occult science) by Ibn Hindū (d. 409/1019 or 420/1029), who rather uses the terms *raṣad* and *mushāhada* for observation and personal inspection respectively (Gutas 2002, 279–80). And Ibn Sīnā himself asserted *tajriba* to be superior to Aristotelian inductive reasoning (McGinnis 2010, 47–52; Gutas 2012).

By the early modern period, such categorizations became standard for the occult sciences as an epistemological unit. Geomancers, lettrists, astrologers and alchemists now insisted without exception on the *mujarrab* nature – the *reproducibility* – of their findings and procedures, now laid equal claim to *taḥqīq*. Representing the state of the art at the turn of the 10th/16th century, for example, ʿAlī Ṣafī (d. 939/1533), son of the great Timurid preacher, polymath and occultist Ḥusayn Vāʿiẓ Kāshifī (d. 910/1505), asserts in his programmatic lettrist manual *Amulet of Protection from the Trials of Time* (*Ḥirz al-amān min fitan al-zamān*): "Experiential essaying aimed at manifesting the effects and properties of the letters has often been successfully repeated; and experience is among the forms of proof that produce certainty (*tajriba yakī az iṣbāt-i yaqīn ast*)" (Melvin-Koushki 2017, 157 no. 103). And as we saw earlier, in a major departure from medieval precedent, the same period saw an explosion in the production of Arabic and Persian occult-scientific manuals, imperial grimoires, that explicitly invite the reader to master their contents independently of a teacher (cf. el-Rouayheb 2015[CB]). Personal experience of the seen and the unseen alike became an imperative.

Crucially for the history of science, such an occultist-experientialist ethos is highly reminiscent of Baconian experimentalism, that other basis of scientific modernity – wherein natural-mathematical science was just as frequently occult. Francis Bacon (d. 1626), the patron saint of Royal Society experimentalism, was himself a strong advocate of divinatory sciences like geomancy, although he did not go so far as to classify them as mathematical like his Muslim peers (Langman 2010). In this Persianate context, it is historically ironic that Bacon pursued the rehabilitation of what he imagined to be ancient *Persian* magic, as well as alchemy, astrology and various forms of divination, as the core of his new explicitly Protestant form of natural philosophy. He invoked magic not as superstition but as superstition's cure (Josephson-Storm 2017, 44–51).

In sum, as in Latinate Europe, Persianate occult science was widely and emphatically embraced as a means of demonstrating the validity of Neoplatonic and especially Neopythagorean physics-metaphysics, and hence a primary driver of *mathesis*. But unlike in Europe, it also became foundational to early modern Islamic, Mongol, Persian and Alexandrian-universalist *imperialism*. That is to say, Bacon's famous dictum – *knowledge is power* – would seem to have been taken more seriously, and to much greater political effect, by early modern Muslim ruling elites and the occultists that served them. Contemporary Christian imperial ideologies could hardly scientifically or materially compete (Melvin-Koushki 2018c). While it should be noted that Bacon's famous phrase has been routinely misunderstood – in context, it means rather "[God's] knowledge is itself power" (Josephson-Storm 2017, 46–7) – it accords neatly with the primary objective of our early modern Muslim occult philosophers: the attainment of *theosis*, or *theomimesis*, which drove in turn their projects to help imperially realize the Pythagorean One in time and space.

II.9.4 Conclusion

The Western cosmology now regnant, materialist though it may be, still descends in some part from that established first in the Islamicate world and then paralleled in the Christianate world during the early modern period, whereby the cosmos is to be scientifically understood, and imperially and technologically controlled, as a mathematical text. For what are Google and the Utah Data Center but imperial Comprehensive Prognosticons? And even Ronald Reagan (r. 1981–1989) had his court astrologer. So while the various occult sciences are no longer patronized by state and society to the great degree they once were, and we now believe Modernity to be the antithesis of Magic (Hanegraaff 2012), the Occult-Scientific Revolution achieved by our Muslim, Jewish and Christian forebears continues to inform our cosmology and hence our practice of science and empire.

Is occult science Science, then, in early modern Islamicate societies specifically and Western societies generally? Yes, naturally – and mathematically. We cannot historically explain *mathesis*, a project associated with Pythagoras from late antiquity to the present, or the rise of early modern experimentalism, the very matrix of scientific modernity, without it. That Magic is often simultaneously Religion – which is to say, concerned as much with mind as with matter – as well as Empire – which is to say, concerned as much with power as with knowledge – does not exclude it from history of science. Likewise, that Islam as imagined and lived was (and still is) often tantamount to Magic does not exclude it either from Western civilization. As a host of early modern Muslim scholarly and ruling elites declared, following Aristotle's advice to Alexander: *Islam is magic; magic is science; science is empire.*

Note

1 Consolidated bibliography.

Bibliography

Sources

Fakhr al-Dīn Rāzī. ed. Āl-i Dāvūd, S. ʿA. 2003. *Jāmiʿ al-ʿulūm* ("*Sittīnī*") [Compendium of the Sciences ("The Sixty")]. Tehran: Thurayyā.

Iskandar Beg Munshī. ed. Afshār, Ī. 2003. *Tārīkh-i ʿĀlam-ārā-yi ʿAbbāsī* [The World-Adorning History of ʿAbbās]. 2 vols. Tehran: Amīr Kabīr.

Niẓām al-Dīn Aḥmad. ed. De, B. 1931–1935. *Ṭabaqāt-i Akbarī* [The Akbarian Generations]. 3 vols. Calcutta: Asiatic Society of Bengal.

Research literature

Artun, T. 2013. *Hearts of Gold and Silver: Production of Alchemical Knowledge in the Early Modern Ottoman World*. PhD dissertation. Princeton: Princeton University Press.

Brentjes, S. 2010. "The Mathematical Sciences in Safavid Iran: Questions and Perspectives," in Speziale, F. and Hermann, D., eds. *Muslim Cultures in the Indo-Iranian World during the Early Modern and Modern Periods*. Berlin: Klaus Schwarz, 1–77.

Burnett, C. 2002. "The Certitude of Astrology: The Scientific Methodology of al-Qabīṣī and Abū Maʿshar," *Early Science and Medicine* 7: 198–213.

Fazlıoğlu, İ. 2014. "Between Reality and Mentality: Fifteenth Century Mathematics and Natural Philosophy Reconsidered," *Nazariyat* 1: 1–39.

Feldhay, R. and Ragep, F. J., eds. 2017. *Before Copernicus: The Cultures and Contexts of Scientific Learning in the Fifteenth Century*. Montreal and Kingston: McGill-Queen's University Press.

Fleischer, C. 2009. "Ancient Wisdom and New Sciences: Prophecies at the Ottoman Court in the Fifteenth and Early Sixteenth Centuries," in Farhad, M. and Bağcı, S., eds. *Falnama: Book of Omens*. London: Thames & Hudson, 232–43.

Gardiner, N. 2014. *Esotericism in a Manuscript Culture: Aḥmad al-Būnī and His Readers through the Mamlūk Period*. PhD dissertation. Ann Arbor: University of Michigan.

Gardiner, N. 2019. "Books on Occult Sciences," in Necipoğlu, G., Kafadar, C. and Fleischer, C. H., eds. *Treasures of Knowledge: An Inventory of the Ottoman Palace Library (1502/3–1503/4)*. 2 vols. Leiden: Brill, 1: 557–87.

Gutas, D. 2002. "Certainty, Doubt, Error: Comments on the Epistemological Foundations of Medieval Arabic Science," *Early Science and Medicine* 7: 276–89.

Gutas, D. 2012. "The Empiricism of Avicenna," *Oriens* 40: 391–436.

Gutas, D. 2018. "Avicenna and After: The Development of Paraphilosophy. A History of Science Approach," in Al Ghouz, A., ed. *Islamic Philosophy from the 12th to the 14th Century*. Göttingen: V&R unipress and Bonn University Press, 19–71.

Håkansson, H. 2001. *Seeing the Word: John Dee and Renaissance Occultism*. Lund: Lunds Universitet.

Hallyn, F., tr. Leslie, D. M. 1990. *The Poetic Structure of the World: Copernicus and Kepler*. New York: Zone Books.

Hanegraaff, W. J. 1995. "Empirical Method in the Study of Esotericism," *Method & Theory in the Study of Religion* 7.2: 99–129.

Hanegraaff, W. J. 2012. *Esotericism and the Academy: Rejected Knowledge in Western Culture*. Cambridge: Cambridge University Press.

Harris, N. 2022. *Better Religion through Chemistry: Aydemir al-Jildakī and Alchemy under the Mamluks*. PhD dissertation. Philadelphia: University of Pennsylvania.

Harrison, P. 2001. *The Bible, Protestantism and the Rise of Natural Science*. Cambridge: Cambridge University Press.

Howell, K. J. 2001. *God's Two Books: Copernican Cosmology and Biblical Interpretation in Early Modern Science*. Note Dame: Notre Dame Press.

Janssens, J. 2004. "'Experience' (*tajriba*) in Classical Arabic Philosophy (al-Fārābī – Avicenna)," *Quaestio* 4: 45–62.

Josephson-Storm, J. Ā. 2017. *The Myth of Disenchantment: Magic, Modernity, and the Birth of the Human Sciences*. Chicago: University of Chicago Press.

Killeen, K. and Forshaw, P. J., eds. 2007. *The Word and the World: Biblical Exegesis and Early Modern Science*. Houndsmill, Basingstoke: Palgrave Macmillan.

Langermann, T. 2014. "From My Notebooks: On *Tajriba/Nissayon* ('Experience'): Texts in Hebrew, Judeo-Arabic, and Arabic," *Aleph* 14: 147–76.

Langman, A. P. 2010. "The Future Now: Chance, Time and Natural Divination in the Thought of Francis Bacon," in Brady, A. and Butterworth, E., eds. *The Uses of the Future in Early Modern Europe*. New York: Routledge, 142–58.

Lelić, E. 2017. *Ottoman Physiognomy ('Ilm-i Firâset): A Window into the Soul of an Empire*. PhD dissertation. Chicago: University of Chicago.

Lemay, R. 1962. *Abu Ma'shar and Latin Aristotelianism in the Twelfth Century: The Recovery of Aristotle's Natural Philosophy through Arabic Astrology*. Beirut: American University of Beirut Press.

Lemay, R. 1993. "L'Islam historique et les sciences occultes," *Bulletin d'Études Orientales* 44: 19–37.

Mauder, C. 2019. "Nur hinter verschlossenen Türen? Das Amt des *muḥtasib* und die Öffentlichkeit von Astrologie, Wahrsagerei, Zauberei und Amulettgebrauch," in Günther, S. and Pielow, D., eds. *Die Geheimnisse der oberen und der unteren Welt: Magie im Islam zwischen Glaube und Wissenschaft*. Leiden: Brill, 319–43.

McGinnis, J. 2010. *Avicenna*. Oxford: Oxford University Press.

McKnight, S. A. 1998. "The Wisdom of the Ancients and Francis Bacon's *New Atlantis*," in Debus, A. G. and Walton, M. T., eds. *Reading the Book of Nature: The Other Side of the Scientific Revolution*. St. Louis: Sixteenth Century Journal Publishers, 91–110.

Melvin-Koushki, M. 2016. "Of Islamic Grammatology: Ibn Turka's Lettrist Metaphysics of Light," *Al-'Uṣūr al-Wusṭā: Journal of Middle East Medievalists* 24: 42–113.

Melvin-Koushki, M. 2017. "Powers of One: The Mathematicalization of the Occult Sciences in the High Persianate Tradition," *Intellectual History of the Islamicate World* 5.1: 127–99.

Melvin-Koushki, M. 2018a. "Early Modern Islamicate Empire: New Forms of Religiopolitical Legitimacy," in Salvatore, A., Tottoli, R. and Rahimi, B., eds. *The Wiley-Blackwell History of Islam*. Hoboken: Wiley-Blackwell, 353–75.

Melvin-Koushki, M. 2018b. "*Taḥqīq* vs. *Taqlīd* in the Renaissances of Western Early Modernity," *Philological Encounters* 3: 193–249.

Melvin-Koushki, M. 2018c. "*How to Rule the World*: Occult-Scientific Manuals of the Early Modern Persian Cosmopolis," *Journal of Persianate Studies* 11: 140–54.

Melvin-Koushki, M. 2018d. "Persianate Geomancy from Ṭūsī to the Millennium: A Preliminary Survey," in El-Bizri, N. and Orthmann, E., eds. *Occult Sciences in Premodern Islamic Culture*. Beirut: Orient-Institut Beirut, 151–99.

Melvin-Koushki, M. 2019. "World as (Arabic) Text: Mīr Dāmād and the Neopythagoreanization of Philosophy in Safavid Iran." *Studia Islamica* 114.2: 378–431.

Moin, A. A. 2012. *The Millennial Sovereign: Sacred Kingship and Sainthood in Islam*. New York: Columbia University Press.

Morrison, R. 2014. "A Scholarly Intermediary between the Ottoman Empire and Renaissance Europe," *Isis* 105: 32–57.

Murphy, J. 2010. "Aḥmad al-Damanhūrī (1689–1778) and the Utility of Expertise in Early Modern Ottoman Egypt," *Osiris* 25: 85–103.

Newman, W. R. 2019. *Newton the Alchemist: Science, Enigma, and the Quest for Nature's "Secret Fire"*. Princeton: Princeton University Press.

Noble, M. 2021. *Philosophising the Occult: Avicennan Psychology and 'The Hidden Secret' of Fakhr al-Dīn al-Rāzī*. Berlin: de Gruyter.

Ragep, F. J. 2007. "Copernicus and His Islamic Predecessors: Some Historical Remarks," *History of Science* 45: 65–81.

el-Rouayheb, K. 2015. "The Rise of 'Deep Reading' in Early Modern Ottoman Scholarly Culture," in Pollock, S., Elman, B. and Chang, K.K., eds. *World Philology*. Cambridge, MA: Harvard University Press, 201–24.

Saif, L. 2017. "From *Ġāyat al-ḥakīm* to *Šams al-maʿārif*: Ways of Knowing and Paths of Power in Medieval Islam," in Melvin-Koushki, M. and Gardiner, N. eds. *Islamicate Occultism: New Perspectives*, special double issue of *Arabica* 64.3–4: 297–345.

Şen, A. T. 2016. *Astrology in the Service of the Empire: Knowledge, Prognostication, and Politics at the Ottoman Court, 1450s-1550s*. PhD dissertation. Chicago: University of Chicago.

Shapin, S. 1996. *The Scientific Revolution*. Chicago: University of Chicago Press.

Vlatko, V. 2010. *Decoding Reality: The University as Quantum Information*. Oxford: Oxford University Press.

Westman, R. S. 2011. *The Copernican Question: Prognostication, Skepticism and Celestial Order*. Berkeley: University of California Press.

Wheeler, J. D. 1990. "Information, Physics, Quantum: The Search for Links," in Zurek, W. H., ed. *Complexity, Entropy, and the Physics of Information*. Redwood City, CA: Addison-Wesley, 3–28.

Yates, F. A. 1964. *Giordano Bruno and the Hermetic Tradition*. London: Routledge and Kegan Paul.

Zadeh, T. 2020. "Postscript: Cutting Ariadne's Thread, or How to Think Otherwise in the Maze," in Saif, L., Leoni, F., Melvin-Koushki, M. and Yahya, F., eds. *Islamicate Occult Sciences in Theory and Practice*. Leiden: Brill, 607–50.

Zadeh, T. 2023. *Wonders and Rarities: The Marvelous Book That Traveled the World and Mapped the Cosmos*. Cambridge, MA: Harvard University Press.

II.10
ARTS, SCIENCES, AND PRINCELY PATRONAGE AT ISLAMICATE COURTS (4TH/10TH–11TH/17TH CENTURIES)

Yves Porter

Art objects, including architecture, are often clear proof of princely patronage. The dedication inscriptions, displaying the names and titles of sultans or their courtiers on artifacts and monuments, attest to this fact. However, the theoretical knowledge of chemical formulae, technical methods or geometric calculations that made these creations possible is rarely acknowledged, although this kind of knowledge may suggest the intervention of scholars (whom we might today call engineers, chemists, or mathematicians) in the creative process. The use of costly materials, together with experimentation for new processes, implies the necessity of important financial support. Thus, it is possible to figure out a kind of triangular relation between the prince, the artist and the man of science or alchemist.[1]

The interest of the ruling elites in scientific domains can be observed, in the sphere of Islamic arts, as early as the Umayyad caliphate (41–132/661–750). The cupola of the caldarium in the little palace of Quṣayr ʿAmrāʾ (present-day Jordan), dating back to the late 1st/early 8th century, illustrates this interest. It shows a painting that is probably the oldest surviving semi-spherical depiction of the constellations of the northern hemisphere made for a Muslim prince and displays clear references to an iconography derived from the Ptolemaic descriptions (Saxl 1932, 289–303). Later, under the Abbasids (r. 132–656/750–1258), caliphs such as Hārūn al-Rashīd (r. 170–193/766–809) and his son al-Maʾmūn (r. 198–218/813–833), were known for encouraging scientific activities (al-Hassan and Hill 1986[CB],[2] 10). In parallel, some reputed scholars such as Abū l-Wafāʾ (328–388/940–998) or Bīrūnī (362–after 444/973–after 1053) made the greater part of their scientific careers under the patronage of emirs and sultans.

Unfortunately, contrary to the case of princes and scholars, the names of the artists and craftsmen of these early centuries are rarely known to us. Besides, the connection between arts and sciences is not universally recognized. Thus, in a recent essay, Gülru Necipoğlu writes:

> With a few exceptions, however, historians of Islamic art and science have tended to resist entertaining the possibility that geometer-astronomers, artisan-designers, and master builder-architects could meaningfully communicate with or learn from one another through their engagement in interconnected epistemic activities.
>
> *(Necipoğlu 2017a, 12)*

DOI: 10.4324/9781315170718-30

This chapter focuses on some examples of such triangular interactions between scholars, artisans and patrons, mostly coming from the lands of the eastern Islamicate world and dating from the 4th/10th to 9th/15th centuries. They document the skillful and very often highly sophisticated mastery of builders, potters, painters and producers of manuscripts, about whom we often know very little beyond the names, periods of life and patrons. But lack of information does not allow us to write a coherent narrative, which could explain the kinds of knowledge that those artisans (and in several cases also scholars) had acquired before and during their creation of splendid works of art and technology.

2.10.1 Early period: 4th–7th/10th–mid-13th century

During the 4th/10th century, Abbasid rule was challenged by the emergence of new power centers in the form of emirates. This was especially the case with the Buyid dynasty (r. 320–454/932–1062). Under its rule, the caliph became a powerless puppet. However, these changes did not have a negative effect on the sciences and the arts. Rather, the new emirates played an important role in attracting scholars and artists to their courts, thus supporting a wider geographical spread of intellectual and artistic activities, which enhanced their international prestige. The activities of Turkish dynasties in the 5th/early 11th century started a new phase in the Islamicate East. They adopted Persian as their language of culture and patronized, besides Arabic scientific works, also those written in Persian (Chapter II.1). The invasions of the Mongols represent the end of the first period surveyed here. Starting in Central Asia by the 610s/1220s, they culminated in the sack of Baghdad in 656/1258 and the assassination of the last Abbasid caliph. In the following section, six dynasties will be introduced whose members patronized technologically challenging architectural endeavors. They were also the patrons of scholars who wrote for craftsmen or about practical matters regarding ceramics, mineralogy and architecture, and lavishly illustrated and illuminated manuscripts were produced for them.

II.10.1.1 The Ziyarids

In the early decades of the 4th/10th century, Mardāvīj ibn Ziyār (r. 315–323/927–935) founded the Ziyarid dynasty (r. 315–*c.* 483/928–*c.* 1090), extending his domains from the southwestern corner of the Caspian Sea to the south of Isfahan, in present-day Iran. One of his successors, Qābūs ibn Vushmgīr (r. 367–402/978–1012), was a ruler with considerable literary taste and ability. His tomb, in the town now known as Gunbad-i Kāvūs, is an extraordinary brick-built monument. It is 51 meters high and has an external diameter of 17 meters; the height of the tower is three times its diameter. Above a circular plinth, the tower forms a decagon or ten-pointed star, a geometric figure requiring the golden section to build up the double pentagon. It also has a conic dome. Although nothing is known about the architects who designed it and the members of their team, the geometric sophistication of this tomb-tower makes it unique and the first masterpiece of Iranian Islamic architecture.

Bīrūnī, a prolific polymath of Central Asian origin, was one of the scholars who worked for a time at the Ziyarid court. He was acquainted with other celebrities of his time such as Ibn Sīnā (d. 428/1037). In 388/998, Bīrūnī was welcomed by Qābūs in Gorgan. There, probably around 390/1000, he wrote his first important work, *The Remaining Traces of Past Centuries* (*al-Āthār al-bāqiya ʿan al-qurūn al-khāliyya*; sometimes translated as "Chronology of Ancient Nations" or "Vestiges of the Past") on historical and scientific chronology, although he later made some amendments to the book (Pingree 1987 [2011], 906–9). Likewise, before going to the Buyid court of Rayy, Ibn Sīnā spent some time in Gurgan, although by that time Qābūs, whose patronage he had expected, had been killed.

II.10.1.2 The Buyids

The Buyid dynasty, also originating from the southern Caspian region, followed the Shīʿī doctrine. They ruled in Iraq and western Iran and played an important part as patrons of many excellent scholars in the mathematical sciences, astrology, philosophy and medicine. Their main capital cities included Baghdad, Rey and Hamadan. One of the astrologers at their courts was ʿAbd al-Raḥmān al-Ṣūfī (290–376/903–986). He is the author of the *Book on Star Constellations* (*Kitāb ṣuwar al-kawākib al-thābita*), which he dedicated to the Buyid ruler ʿAḍud al-Dawla ([r. 338–372/949–983]; Chapter III.8). Until recently, it was thought that an illustrated version of this text, with the name of Ṣūfī's son and allegedly dated 399/1009, was the oldest known copy of this text. However, it is now clear that the colophon was added later (Savage-Smith 2013[CB], 122–55). Other early copies of this work leave no doubt that Ṣūfī himself had illustrated his astronomical treatise and that at least one of his students had produced another illustrated copy. Indeed, Ṣūfī provided not only the figures of the constellations viewed from earth but also the constellations as depicted on a globe. This decision made the use of illustrations compulsory, since the general arrangement of the stars composing the constellations changes according to the place they are observed from.

Ṣūfī also created a celestial globe made of solid gold for ʿAḍud al-Dawla. It is unfortunately now lost, although it was still preserved in the 6th/12th century in the treasury of the Fatimid dynasty (297–567/909–1171) in Cairo (Destombes 1958, 308–9). Astronomical instruments, together with illustrated manuscripts, are obvious examples of the fruitful relations between the arts, the sciences and princely patronage. Unfortunately, most of the examples from these early centuries have disappeared.

Abū l-Wafāʾ ~~(328–388/940–998)~~, a mathematician and astronomer originally from Persia, is a further scholar patronized by members of the Buyid dynasty. He wrote, among other works, a treatise in Arabic on applied geometry called *Book on the Geometrical Constructions Necessary for the Artisan* (*Kitāb fī mā yaḥtāju ilayhi al-ṣāniʿ min aʿmāl al-handasa*), which had an enormous impact on artists and craftspeople (Aghayani-Chavoshi 1998, 95–116). In its preface, Abū l-Wafāʾ explains that he completed this work upon the order of his master, the Buyid emir Bahāʾ al-Dawla (r. 388–403/998–1012), who

> requested that I substantiate the notions that were discussed in the presence of His Excellency, concerning the geometrical constructions most commonly employed by artisans, leaving out all the demonstrations and proofs, in order to make these easy for artisans to use and to familiarize them with their method.
>
> *(quoted after Necipoglu 2017, 31)*

Abū l-Wafāʾ's *Treatise* was translated into Persian at least twice. The older translation is not precisely dated (5th/11th century?), but the later one is the work of Abū Isḥāq Kubanānī Yazdī, living in the 8th/15th century (see the following and Aghayani-Chavoshi 1998, 114–5).

II.10.1.3 The Ghaznavids

The Ghaznavid dynasty (r. 366–582/976–1186) takes its name from the town of Ghazni, now in Afghanistan. Its founder, Sebüktigin (r. 366–387/977–997), was originally a slave of Turkish stock, who rose to the rank of emir (general or governor). The reign of his son Maḥmūd

(r. 388–421/998–1030) is considered the apex of the dynasty. When he conquered Rayy and the possessions of the Maʾmunids (r. 385–408/995–1017), Maḥmūd of Ghazni took most scholars, including Bīrūnī, with him to Ghazna, although Ibn Sīnā refused the invitation of the Ghaznavid ruler.

One of Bīrūnī's major works, the *Book on the Knowledge of Stones* (*Kitāb al-jamāhir fī maʿrifat al-jawāhir*; literally: *Book of Gatherings on the Knowledge of the Gemstones*), was made possible because he was for some time in charge of the inspection of the royal gifts presented to Sultan Maḥmūd (Biruni 1989, 47). This position allowed him to examine all kinds of precious stones and metals, but also their imitations. It was an incentive for working on the relative weights and volumes of gems, specific gravities and relative hardness. Indeed, it was only through the knowledge of hardness and of the specific gravity of stones that imitation can be detected, and precious stones thus distinguished from fake ones. Bīrūnī acknowledges that he made use of two older texts to compose his own treatise: one was a now-lost Arabic work by al-Kindī (d. *c.* 256/*c.* 870) on gems and precious stones, and the other – written in Persian – was by Naṣr al-Dīnawārī ([205–283/820–896]; Biruni 1989, 243). In this book, dedicated to the grandson of Maḥmūd (Sultan Mawdūd [r. 432–440/1040–1048]), Bīrūnī described the varieties of precious stones, their mines, their specific weights and their prices. Concerning metals and ores, he gave a vivid account of their origins, uses and ways of refining them, some of which he determined experimentally. He also used the occasion to deal with Chinese porcelain, enamels and glass of diverse hues – particularly the ruby-red glass named *adhrak* – besides referring to artifacts found in ancient graves. Some hints in his book indicate that he used to sell precious items and was thus aware of their prices.

II.10.1.4 The Great Seljuqs

The Seljuqs (r. 429–552/1037–1157) belonged originally to a clan from the Oghuz Turkish people. They converted gradually to Islam by the end of 4th/10th century and extended their domains steadily after taking Khurasan from the Ghaznavids. Settling first in Nishapur, the Seljuqs undertook a Sunni revival, aiming to wipe out the Shīʿī doctrine of the Buyids with the aid of grand viziers such as Niẓām al-Mulk (409–485/1018–1092). Under his supervision, the institution of the madrasa as an independent structure for higher studies was developed. These madrasas, known as Niẓāmiyya, spread over Seljuq dominions, from Khurasan to Baghdad and Mosul (Chapter III.2). If the first goal of these learning centers was the diffusion of the Sunni doctrine, other branches of knowledge also developed. Thus, a library and a hospital were, for instance, attached to Nishapur's *Niẓāmiyya madrasa* (Finster 1994, 19). Significantly, the almost 30-meter-high dome built in front of the main *miḥrāb* of Isfahan's Great Mosque is ascribed to Niẓām al-Mulk and represents the largest cupola of its time. The mathematician and astronomer ʿUmar Khayyām (439–*c.* 517/1048–*c.* 1123), also famous for his *Quatrains* (*Rubāʿiyyāt*), was first in the service of Niẓām al-Mulk at the court of Malik-Shāh (r. 465–485/1072–1092) in Isfahan. Sultan Sanjar (r. 511–552/1117–1157) later invited him to his newly built capital in Marv (now in Turkmenistan).

In this city, the monument known as Sanjar's mausoleum was originally part of a vast complex with religious and secular buildings. It was probably built shortly after the sultan's death (552/1157). It has often been ascribed to the architect Muḥammad ibn ʿAzīz of Sarakhs, although nothing is known on the biography of this architect. Sultan Sanjar is considered the last Great Seljuq, since after his long reign, the dynasty started disintegrating. His mausoleum (a square with a side of 27 meters, the inner dome rising above 30 meters), is among the most impressive tombs of his time and served as a model for future monumental tombs.

2.10.1.5 The Khwarazmshahs

The Khwarazmshah dynasty (*c.* 470–628/*c.* 1077–1231) was first established as governors of the Khwarazm province (roughly between the Caspian and Aral seas) for the Seljuqs. After the weakening of the Seljuqs, the power of the Khwarazm shahs increased, and they formed a lineage that was only swept away by the Mongols in 628/1231.

A case similar to that of Bīrūnī, although much less famous, concerns Muḥammad Jawharī Nīshāpūrī, the author of a Persian *Treatise on Mineralogy* (*Jawhar-nāma-yi Niẓāmī*) dated 592/1196. Little is known of his life. We understand from his surname (*Jawharī*) that he was a jeweler and a goldsmith. Moreover, he wrote in his *Treatise* that he used to oversee the royal gifts at the court of the Khwarazmshah Sultan Tekesh (r. 568–596/1172–1199). His text, dedicated to the grand vizier, deals with precious stones, pearls, metals and minerals, but he also added chapters on the imitations of gems and on the making of glass-based enamels together with recipes for luster painting on glass and ceramics (Porter 2004a).

Luster painted ceramics are among the most amazing inventions of Muslim potters. The first examples, dating from the 2nd/8th century, were actually painted on glass. They were mostly made in Egypt. During the 3rd/9th century, the technique was applied to faience in the Abbasid workshops of the Basra area in Iraq. The process consisted in applying luster in several hues (thus called "polychrome lusterware"). Making polychrome lusterware is extremely complex, since each hue is made according to a specific formula. This might be the reason why, for the later Abbasid vessels, dating from the 4th/10th century, monochrome luster was used instead (Figure II.10.1). With rare exceptions, all the later lusterware productions, from al-Andalus to Iran, used monochrome luster only (Figure II.10.2). The luster technique reached Iran by the second half of the 6th/12th century. It is not known if Nīshāpūrī's text had any impact on this process. But it is clear that he reacted to it since, on several occasions, he specifies that he has experimented with the formulae and compares them with the search for the Elixir, thus situating his work in the frame of alchemy (Chapter I.12).

Figure II.10.1 Sherds, luster: upper exemplar: Egypt 5th/11th century; lower exemplars: left Kashan late 6th/12th century, right Kashan early 7th/13th century.

Source: © Photographer and owner Yves Porter, Aix-en-Provence

Figure II.10.2 Sherd, luster, horse in a star; 8th/14th century, Iran.

Source: © Photographer and owner Yves Porter, Aix-en-Provence

The recipes for luster painting described in his book are able to furnish no fewer than 24 hues and are thus reminiscent of the first Abbasid polychrome lusterwares. But these Abbasid practices and the redaction of the treatise are separated by 300 years. Moreover, the region where it was written (in Khwarazm or in Khurasan) never produced any lusterware. So this case is puzzling; it shows that the theoretical knowledge expounded by an author enjoying royal patronage does not necessarily imply that the techniques described were being employed for production.

Many other domains of experimentation we nowadays call "applied chemistry." They concern metalwork, textile dyes and mordants, tanning of leather and even ink-making. Many of these crafts involve simple chemical reactions. Black inks, for instance, can easily be obtained by mixing gallnuts with a metallic salt, the result being a black precipitate. A prince named Muʿizz ibn Bādīs (406–453/1016–1061) reigned in what is now Tunisia – Ifriqiya in medieval times. Although contested, he is often considered the author of a book on the arts of bookmaking, papermaking, inks and bookbinding, dated circa 416/1025. This work may not qualify, strictly speaking, as scientific, but it displays an amazing number of elements and reactions that are useful for a variety of chemical experiments. Consequently, the English translation of this text by Martin Levey was entitled *Chemical Technology in Medieval Arabic Bookmaking*. Interestingly, a Persian translation was made far away from Tunisia 500 years later. This incomplete and not very accurate Persian version was made at the request of the Bahmanid Sultan of the Deccan Maḥmūd Shāh II (r. 887–924/1482–1518). It testifies to the success of this text across the lands of Islam and reflects the prestige attached to bookmaking (Porter 1989, 61–7).

II.10.1.6 The Artuqids

The Artuqids (r. 491–629/1098–1231), a dynasty of Turkish origin, settled in eastern Anatolia by the late 5th/11th century. It was at their court that al-Jazarī (d. *c.* 602/1206), the author of one of the best-known treatises on automata, spent most of his life and career. Mechanics were considered part of either the mathematical sciences or natural philosophy (Chapter I.9).

Al-Jazarī leaned toward the second view. Practical applications were discussed as parts of the science of ingenious devices (*ʿilm al-ḥiyal*). Al-Jazarī's book on mechanical devices *The Book of Knowledge of Ingenious Mechanical Devices* (*Kitāb fī maʿrifat al-hiyal al-handasiyya*), describes engines for elevating water, water-clocks and machines designed for serving drinks. It continues a long-established tradition of automaton-making that started in ancient Greece and continued in Baghdad under the Abbasids. The attention paid to automata raises an interesting issue in the frame of the so-called Islamic arts. Not only do these figures imitate living beings, as sculpture does, but automata are also capable of motion. However, sculpture is perceived as the artistic form that approaches God's creative power the most; therefore, sculpture is rarely found in the Islamicate lands.

Al-Jazarī's treatise also contains the description of a pair of doors he designed for the Artuqid prince's palace. These were made of cast metal in an intricated geometric pattern and with a pair of doorknockers bearing interlaced dragons. They represent the only artifacts designed by al-Jazarī that have come down to us (von Folsach 1996, 385).[3] But the designs of several of his creations are illustrated in many manuscripts that still exist. Some of these were copied in northern Syria and western Anatolia. But at least one other illustrated copy was made in India, in the city of Mandu, the capital of the Malwa sultanate, in the early years of the 10th/16th century (Losty 1982, 68). The technology involved in the design for water-raising machines and irrigation devices had a tremendous impact in the whole Islamicate world, not only for obvious agricultural purposes but also for garden design (Chapter IV.5).

II.10.2 From the Mongols to the Mughals (7th/13th to 11th/17th centuries)

After the dramatic years following the Mongol invasions, which culminated in the sacking of Baghdad in 656/1258, the Mongol dynasty that settled in Iran became known as the Ilkhanids (r. 654–736/1256–1336). In the following sections, I survey some of the scholarly and artistic activities commissioned or honored by this new dynasty as well as some of those patronized by three other dynasties ruling mainly in Central Asia and India.

2.10.2.1 The Ilkhanids

Once established firmly in western Iran, the Ilkhanids allowed for a more peaceful era. A plethora of eminent scholars were patronized by them, among them Naṣīr al-Dīn Ṭūsī (597–673/1201–1274), now chiefly remembered as an astronomer and, more important, as a founder of Shīʿī *kalām*. Like several of the figures we have already considered, he was also the author of a book on precious stones (*Tansūḫ-nāma-yi īlkhānī*), partly inspired by the work of Muḥammad Jawharī of Nishapur. Like his predecessors Bīrūnī and Jawharī, Ṭūsī was probably in charge of the inspection of the royal treasury (Vesel *et al.* 2000, 147); compared with previous authors, he added some new data on specific subjects such as turquoise mines in the Kerman area (Vesel *et al.* 2000, 148).

With the support of the first Ilkhanid ruler, Hülägü (r. 654–663/1256–1265), Ṭūsī supervised a project that combined practical and theoretical skills in an exemplary way, the construction of the huge observatory near the Khān's capital at Maragha. The central building of the complex was a cylinder of 28 meters in diameter, which also served as a library. It contained a built-in astronomical quadrant more than four stories high. A variety of other large instruments for observing the stars was placed around the grounds. Hülägü's patronage and *Ṭūsī's* reputation attracted other famous scholars. Others came as prisoners of war such as the astrologer and instrument maker Mu'ayyad al-Dīn al-ʿUrḍī (d. 645/1266). Al-ʿUrḍī assisted both in constructing the instruments and making observations. The main result of the work there was a new astronomical handbook, the *Zīj-i īlkhānī*, named after the observatory's patrons (Ragep 1993, 3–20; Jorati 2014, 137–209; Chapters II.6, II.7, III.5).

Another scholar, patronized by the new dynasty, was Abū l-Qāsim of Kashan. He belonged to a well known family of potters but made most of his career as a historian and chronicler at the Ilkhanid court. In 699/1300, he finished a treatise on mineralogy, precious stones and other matters, written in Persian, and also partly inspired by Muḥammad Jawharī's work. But going beyond his predecessors, Abū l-Qāsim's text also contains a short account on the art of ceramic-making, which was considered, before the discovery of Muḥammad Jawharī's text, the only technical text dealing with ceramics in the entire Islamicate premodern world. It accurately explains the materials used by potters, such as how to make stone-paste (an artificial paste made with powdered flint, alkaline glaze and white clay), together with the different ores used for painting the ceramics. It also contains a single formula for luster-painting. Members of his family – particularly his father and brother – specialized in architectural décor for mosques and shrines, signing monumental tiled *miḥrāb*s painted with luster. Presumably, Abū l-Qāsim's place at the Ilkhanid court played a role in providing his family's workshop with commissions for producing costly architectural decorations.

As already mentioned, the names of the architects of most of the architectural masterpieces built in premodern eastern Islamicate lands are seldom known to us. This is also the case with the most magnificent building of the Ilkhanid period, the mausoleum of Öljeitü (r. 703–716/1304–1316; Figure II.10.3). Built on an octagonal basis, and with very thick walls, it rises 49 meters

Figure II.10.3 Öljeitü's dome, Sultaniyya, Iran; 8th/14th century.

Source: © Photographer Yves Porter, Aix-en-Provence

above ground and is thus the highest dome in Iran, and among the largest brick-built domes in the world. It is most probable that experienced geometers worked on this project, although no contemporary records exist on this matter.

Texts on applied geometry, then called the science of surveying (*ʿilm al-misāḥa*), taught knowledge about a series of planar figures such as triangles, squares, rectangles, polygons, circles, segments, sectors or egg- and drum-like shapes and occasionally included some simple constructions based on symmetry (Chapter I.5). Furthermore, the basic principle of harmony between the parts and the whole found multiple fields of application. Thus, it is not surprising that geometry was used, for instance, in bookmaking, to determine the ratios for formatting papers, and thus the size and shape of finished books, but also the layout and template of the pages, their ruling (*misṭara*) and, of course, the geometric ornaments of their illuminations and bookbindings (Porter 2004b). Ruling was a mechanical operation consisting in tracing lines for the written space, intended to help the copyist to write text on horizontal straight lines, in a prescribed space. The grid formed by the ruled lines could also serve painters as guidelines for illustrations. The templates used in making manuscripts allowed the rationalization of serial production in some cases, since the grid of ruling facilitated the reproduction of illustrations. This "serial production" is known for some specific workshops such as the foundation established by Rashīd al-Dīn Fażl Allāh (645–718/1247–1318), the vizier of the Ilkhanid ruler Öljeitü (Porter 2004b, 60–4). Rashīd al-Dīn was a polymath scholar in the service of various Ilkhanid rulers. Among his major works is the *Universal History* (*Jāmiʿ al-tavārīkh*), often considered the first attempt at a "global" history. Only one volume of the original text is preserved, and its illustrations show that they were built up with the help of the ruled grid (Rashīd al-Dīn 1995). Later illustrated manuscripts also used sophisticated calculations in formats, ratios, and ruling to create a harmony between the written pages and their illustrations.

II.10.2.2 The Timurids

In the last quarter of the 8th/14th century, Tīmūr ([r. 771–807/1370–1405]; Tamerlane) conquered a vast territory centered in modern Uzbekistan and founded the dynasty known as the Timurids (r. 771–912/1370–1506). The imprint of Tīmūr and his successors has been made manifest through their architectural achievements.

In addition to the lack of documents referring to architects, the fact that very little was written in premodern Islamic lands about architecture and building technologies is another obstacle to writing a coherent narrative of the triangular relationship between princes, scholars and craftsmen. An exception to this rule happened under the Timurid dynasty. Al-Kāshī (d.832/1429) wrote one of the rare texts relating to architectural technology. His explanations of how to produce prefabricated modules for domes and decorations appear in a chapter of the mathematical treatise *Key to Arithmetic* (*Miftāḥ al-ḥisāb*), which he dedicated to the Timurid prince Ulugh Beg (r. 812–850/1409–1447 as governor of Samarqand; 850–853/1447–1449 as head of the dynasty). Indeed, al-Kāshī, who was a renowned mathematician and astronomer born in Kashan (Iran), had first served at the court of another Timurid prince, Iskandar Sulṭān (r. 805–818/1403–1415), in Shiraz and Isfahan, before he was invited to Samarqand by Ulugh Beg, to teach at his newly founded madrasa in the Registan square (Figure II.10.4). Al-Kāshī's short text mainly explains the making of arches and vaulting based on Euclidean geometry. It deals particularly with the way of drawing multicentered arcs, cupolas of diverse profiles, together with squinches and *muqarnas* vaultings[4] (al-Kāshī 1987; Dold-Samplonius 1992, 1996, 2003).

Ulugh Beg was the grandson of Tīmūr. His reign as a Timurid sultan was very short, but he ruled Samarqand and its environs for many years as a governor. In that time, he was able

Figure II.10.4 Ulugh Beg's *madrasa*, Samarqand, Registan Square.

Source: © Photographer Yves Porter, Aix-en-Provence

to dedicate financial, administrative and intellectual support to his passion for astronomy. He built an observatory in Samarqand – the ruins of which can still be seen – with buildings and instruments similar to Naṣīr al-Dīn Ṭūsī's observatory in Maragha during the Ilkhanid period (Chapter II.7). Ulugh Beg also had an enviable library. Among his books was a magnificent and richly illustrated copy of Ṣūfī's *Book on the Star Constellations* (now in the BnF; Richard 1997, 78; Chapters III.8 and IV.7). Another manuscript from his library is a deluxe Arabic copy of Abū l-Wafāʾ's *Treatise* (MS Istanbul, Süleymaniye Library, Ayasofya 2753). Although it remains uncertain whether this text was directly used in the royal workshop of Samarqand, its influence in works of art, not only on architectural decoration but also on other domains such as book-binding and manuscript illumination, was undeniably important.

Qawwām al-Dīn ibn Zayn al-Dīn of Shiraz (d. 842 or 844/1438 or 1440) belongs to the few architects named in the historical sources of the Timurid period. Some authors qualify him as an engineer (or geometer, *muhandis*) and draughtsman (*ṭarrāḥ*; O'Kane 1994, 70). Although several buildings in Mashhad and Herat go to his credit, not much is known of his theoretical bases, although he possibly had a knowledge of theoretical books on geometry (Golombek and Wilber 1988, 1: 189–94).

II.10.2.3 The Aq Qoyunlus

The Aq Qoyunlus (780–908/1378–1502) were a dynasty formed by the heads of the White Sheep confederation of Oghuz tribes. They began their rise to power in eastern Anatolia (Diyar Bakr) and extended their dominions over time into Iraq, Iran, Azerbaijan and Armenia, repeatedly even threatening Timurid possessions as far as Herat (modern Afghanistan). They were linked by marriage to the Greek rulers of Trebizond and entertained relationships with Venice. They patronized the production of splendidly illustrated historical and literary manuscripts as well as architecture. One of the leading scholars of Ulugh Beg's court, ʿAlī Qūshjī (805–879/1403–1474), spent some time at their court in Tabriz before moving on to Ottoman Istanbul, acting as a diplomatic envoy between the two dynasties, and carrying Ulugh Beg's astronomy to the two courts (Barker and Heidarzadeh 2016, 48–52).

A fascinating architectural pattern book seems to have been produced at the dynasty's court in Tabriz, called today the *Topkapı Scroll* (Necipoğlu 1995). Contemporary with Kubanānī's 9th/15th-century translation of Abū l-Wafā"s *Treatise* mentioned earlier, the scroll is an almost unique example of geometric designs to be applied to tile-working and *muqarnas* vaulting. Indeed, the geometrical constructions, both bi-dimensional, for use in tilework or stucco, and three-dimensional, including "stalactite" or *muqarnas* vaulting designs, display amazing figures of an extraordinary complexity.

Another text follows Kubanānī's translation in the manuscript preserved today in the National Library of France and was written by the same scribe. It is an anonymous treatise on applied geometry titled *On Similar and Complementary Interlocking Figures* (*Fī Tadākhul al-ashkāl al-mutashābiha aw al-mutawāfiqa*; MS Paris, BnF, Persan 169, fols 180–99; Necipoğlu 2017b). Craftspeople using practical geometry did not necessarily have to know its theoretical aspects and could do their work by applying elementary constructions in a repetitive manner. Indeed, as Gülru Necipoğlu writes,

> correct and approximate constructions are juxtaposed, sometimes offering alternative methods, suggesting that this document was initially put together in a milieu where individuals with varying degrees of mathematical knowledge and artisanal expertise commingled, or at least became aware of each other's differing yet partly intersecting geometrical construction methods.
>
> *(Necipoğlu 2017a, 57)*

2.10.2.4 The Adilshahis

The Adilshahid dynasty (895–1097/1489–1686) ruled one of the so-called Deccan sultanates, the Sultanate of Bijapur in the western part of what is today the State of Karnataka in India. The founder of the dynasty was previously a governor of the Bahmanid dynasty mentioned earlier but managed to carve out his own, independent realm. Between 940/1534 and 1036/1627, the rulers carried out a major architectural program providing the city with a new wall, water canals, palaces, mosques and tombs. They are ranked today among the best examples of Indo-Muslim architecture.

One of their outstanding architectural masterpieces is the Gul Gunbad, which was built as the tomb of Muḥammad ʿĀdil Shāh (r. 1035–1067/1626–1656). It is one of the biggest single-chamber-domed rooms in the world and the most technically advanced construction found in any of the Deccan sultanates (Michell and Zebrowski 1999, 15). It has even been called "one of the supreme structural triumphs of Muslim craftsmen anywhere" (Burton-Page 1986, 75). But in contrast to other Muslim rulers elsewhere its patrons and builders favored size and technical excellence over the decorative arts and thus the tomb has almost no ornamentation. As in the case of the Tāj Maḥal in Agra (started in 1041/1632), almost nothing is known about the architects of this and other such wonders, neither of the theoretical geometrical bases nor of the statics and related domains of knowledge used by their builders.[5]

II.10.3 Conclusion

In the 8th/14th century, the Arab historian Ibn Khaldūn (732–808/1332–1406) wrote:

> The monuments of a given dynasty are proportionate to its original power. The reason for this is that monuments owe their origin to the power that brought the dynasty into being. The impression the dynasty leaves is proportionate to (that power).
>
> *(Ibn Khaldūn 1958[CB], 356)*

As lamented so often before, Ibn Khaldūn's quotation, too, makes no reference to the builders, designers or geometers who designed these monuments or to the artists and craftspeople who ornamented them. However, it was the combined skills of scholars, whether alchemists, geometers or students of mechanical works, and artists and other craftspeople that made these works possible, thanks to the support of the ruling elites.

Ibn Khaldūn's silence might seem strange when we compare it with what we know about the building of Constantinople's Aya Sofya. We know that Justinian (r. 527–565) commissioned Anthemius of Tralles (d. before 558) and Isidore of Miletus (442–537), both reputed geometers (but not architects) for this project. But some of this silence may reflect simply the genres of the sources that have survived – scholarly histories as in the case of Ibn Khaldūn or mathematical treatises as in the case of Abū l-Wafāʾ and Kāshī. Architects of the Ottoman empire are often better known, at least from the 10th/16th century onward, due to documents preserved in the palace archives and stories told by foreign visitors.

The situation is sometimes different where illustrated manuscripts are concerned (Porter 2009; Vesel *et al.* 2009). In some cases, we know the names of the people who drew the mathematical diagrams or painted the human figures of the star constellations. This also applies to astronomical instruments such as astrolabes or globes. Hence, in the future, more extended research on the artistic features of scientific products and their relationship to patrons has the potential to improve our knowledge about the people who were involved in those activities and their practices.

Further promising avenues to a better understanding of the triangular relationship between patrons, scholars and craftsmen consist in the scientific and technical analysis of manuscripts, paintings, instruments and architectural objects and the systematic inclusion of the results of archaeological excavations and investigations, as hinted at in this chapter with regard to pottery, ceramics and architectural modules.

Notes

1 This concept is reflected in the title of my book on Iranian ceramics (Porter 2011).

2 Consolidated bibliography.

3 This pair of door-knockers is unfortunately divided now, as one is in Copenhagen's David collection, while the other is in Istanbul's TİEM (Turkish and Islamic Art Museum).

4 Squinches are support beams placed diagonally across the corners of a square room; the first squinches added to the walls form an octagon. Successive additions create polygons that approximate a circular structure and more easily support a dome. Muqarnas, one of the most distinctive features of Islamic architecture, are a form of decoration for the interior of domes, arches and similar spaces. They consist of a honeycomb of small geometrical cavities conforming to the shape of the structure and spanning the space. In addition to the beauty of their inherent pattern, the appearance of muqarnas changes dramatically as the light illuminating them shifts with the time of day.

5 The evidence supporting the identification of Ustādh Aḥmad Lahūrī as the architect of the Tāj Maḥal is only circumstantial. No actual official document is asserting this fact. Other possible candidates would be Mīr ʿAbd al-Karīm and Makramat Khān, mentioned as supervisors in Mughal sources. On this subject, see Begley and Desai (1989, 261–75).

Bibliography

Sources

Biruni, Abu Raihan. tr. Said, H. M. 1989. *The Book Most Comprehensive in Knowledge on Precious Stones*. Islamabad: Pakistan Hijra Council.

al-Kāshī, Ghīyāth al-Dīn. tr. and ann. Jadhbī, S. ʿA.-R. 1392 sh/1987. *Risāla-yi ṭāq wa azaj* [The Epistle of the Vault and the Arch]. Tehran: Soroush Press.

Rashīd al-Dīn. ed., tr. and anno. Blair, S. 1995. *A Compendium of Chronicles: Rashid al-Din's Illustrated History of the World*. Vol. 27. London: The Nasser D. Khalili Collection of Islamic Art.

Research literature

Aghayani-Chavoshi, J. 1998. "Abu al-Wafa, innovateur de la géométrie pratique dans le monde islamique," in Vesel, Z., Beikbaghban, H. and Thierry de Crussol, B., eds. *La science dans le monde iranien*. Tehran: IFRI, 95–116.

Barker, P. and Heidarzadeh, T. 2016. "Copernicus, the Ṭūsī couple and East-West Exchange in the Fifteenth Century," in Grenada, M. A. and Boner, P., eds. *Man and Cosmos*. Barcelona: University of Barcelona Press, 19–57.

Begley, W. E. and Desai, Z. A. 1989. Taj Mahal. *The Illumined Tomb. An Anthology of Seventeenth Century Mughal and European Documentary Sources*. Cambridge, MA.: The Aga Khan Program for Islamic Architecture.

Burton-Page, J. 1986. "Bijapur," in Michell, G., ed. *Islamic Heritage of the Deccan*. Bombay: Marg.

Destombes, M. 1958. "Un globe céleste arabe du XIIe siècle," *Comptes rendus de l'Académie des Inscriptions et Belles Lettres* 102–3: 308–9.

Dold-Samplonius, Y. 1992. "Practical Arabic Mathematics: Measuring the Muqarnas by al-Kashi," *Centaurus* 35: 193–242.

Dold-Samplonius, Y. 1996. "How al-Kashi Measures the Muqarnas: A Second Look," in Folkerts, ed. [CB], 56–90.

Dold-Samplonius, Y. 2003. "Calculating Surface Areas and Volumes in Islamic Architecture," in Hogendijk, J. P. and Sabra, A. I., eds. *The Enterprise of Science in Islam, New Perspectives*. Boston: MIT Press, 235–65.

Finster, B. 1994. "The Saljuqs as Patrons," in Hillenbrand, R., ed. *The Art of the Saljuqs in Iran and Anatolia*. Costa Mesa, CA: Mazda, 17–28.

Folsach, K. von, ed. 1996. *Sultan, Shah, and Great Mughal. The History and Culture of the Islamic World*. Copenhagen: National Museum.

Golombek, L. and Wilber, D. 1988. *The Timurid Architecture of Iran and Turan*. Princeton: Princeton University Press.

Jorati, H. 2014. *Science and Society in Medieval Islam: Nasir al-Din Tusi and the Politics of Patronage*. PhD thesis. New Haven: Yale University Press.

Losty, J. P. 1982. *The Art of the Book in India*. London: British Library.

Michell, G. and Zebrowski, M. 1999. *Architecture and Art of the Deccan Sultanates* (The New Cambridge History of India). Cambridge: Cambridge University Press.

Necipoğlu, G. 1995. *The Topkapi Scroll: Geometry and Ornament in Islamic Architecture. Topkapi Palace Museum Library MS H. 1956*. Santa Monica: Getty Center for History.

Necipoğlu, G. 2017a. "Ornamental Geometries: A Persian Compendium at the Intersection of the Visual Arts and Mathematical Sciences," in Necipoğlu, ed. 2017b, 11–78.

Necipoğlu, G., ed. 2017b. *The Arts of Ornamental Geometry. A Persian Compendium on Similar and Complementary Interlocking Figures*. Leiden: Brill.

O'Kane, B. 1994. "Iran and Central Asia," in Frishman, M. and Khan, H., eds. *The Mosque, History, Architectural Development and Regional Diversity*. London: Thames and Hudson, 1994, 118–39.

Pingree, D. 1987 [2011]. "Al-Āṯār al-bāqīā ʿan al-qorūn al-ḵālīa," *EIr* II.8: 906–9. www.iranicaonline.org/articles/atar-al-baqia-an-al-qorun-al-kalia-the-chronology-of-ancient-nations-by-biruni (Acessed 19 March 2021).

Porter, Y. 1989. "Une traduction persane du traité d'Ibn Bâdis: *'Umdat al-Kuttâb (ca. 1025)*," in Déroche, F., ed. *Les Manuscrits du Moyen-Orient, Actes du Colloque d'Istanbul (1986)*. Istanbul and Paris: IFEA, 61–7.

Porter, Y. 2004a. "Le quatrième chapitre du *Jawâhar-nâma-i Nizâmi*: le plus ancien texte persan sur la céramique," in Pourjavady, N. and Vesel, Ž., eds. *Sciences, techniques et instruments dans le monde iranien (Xe-XIXe s.)*. Tehran: IFRI and Presses universitaires d'Iran, 341–60.

Porter, Y. 2004b. "La réglure (*mastar*): de la "formule d'atelier" aux jeux de l'esprit," *Studia Islamica* 96: 55–74.

Porter, Y. 2009. "Scientific and Technical Illustrations in the Eastern Islamic Arts," in Vesel, Tourkin and Porter, eds., 45–66.

Porter, Y. 2011. *Le prince, l'artiste et l'alchimiste. La céramique dans le monde iranien, Xe- XVIIe siècle*. Paris: Hermann.

Ragep, F. J. 1993. *Naṣīr al-Dīn al-Ṭūsī's Memoir on Astronomy*. Berlin: Springer.

Richard, F. 1997. *Splendeurs persanes. Manuscrits du XIIe au XVIIe siècle*. Paris: BnF.

Saxl, F. 1932. "The zodiac of Qusayr ʿAmra," in Creswell, K. A. C., ed. *Early Muslim Architecture, vol. I: Umayyads (A.D. 622–750), with a Contribution on the Mosaics of the Dome of the Rock and of the Great Mosque at Damascus by Marguerite Van Berchem*. Oxford: Clarendon Press, 289–303.

Vesel *et al.* 2000: Vesel, Ž., Afshar, I. and Mohebbi, P. 2000. "Le Livre des Pierres pour Neẓām [al-Molk] (*Javāher-nāme-ye Neẓāmī*) (592/1195–6): la source présumée du *Tansūkh-nāme-ye Īlkhānī* de Ṭūsī," in Pourjavady, N. and Vesel, Ž., eds. *Naṣīr al-Dīn Ṭūsī, philosophe et savant du XIIIe siècle*. Tehran: IFRI and Presses universitaires d'Iran, 145–74.

Vesel *et al.* 2009: Vesel, Ž., Tourkin, S. and Porter, Y., eds. 2009. *Images of Islamic Science: Illustrated Manuscripts from the Iranian World*. Teheran: IFRI.

II.11
PHYSIOGNOMY (*ʿILM-I FIRĀSET*) AND POLITICS AT THE OTTOMAN COURT

Emin Lelić

Physiognomy, the art of examining human nature, was known in the Ottoman world as the science of discernment (Arabic: *ʿilm al-firāsa;* Turkish: *ʿilm-i firāset*) or the science of following traces (Arabic: *ʿilm al-qiyāfa*; Turkish: *ʿilm-i qıyāfet*).[1] Islamicate physiognomy, which was deeply rooted in Hellenic physiognomy, developed a systematized method for discerning human nature based on corporeal signs embedded in the human face and body (Chapter I.14). Ghersetti (2007) has given a historical overview of physiognomy's scientific classification, confirming that it was considered a legitimate natural science (*ʿilm*). In the Ottoman world, physiognomy was cultivated and developed by scholars, Sufis and poets. Although in many ways an extension of the broader Islamicate physiognomical corpus, Ottoman physiognomy is distinguished by its linguistic adaptation into the Ottoman language and adjustments to specific Ottoman circumstances.

Physiognomy's injunctions reverberated across socio-economic and political divides, eliciting potential interest from anyone dealing with other human beings. Its claim to being a method for evaluating potential partners in business, marriage, politics or crime made it applicable to all. Even ascetics shunning human society used it as a guide on their path to self-knowledge.[2] Physiognomy was patronized, and likely practiced too, by those who held temporal power. It was thus also a royal science. This chapter focuses on physiognomy's role as a royal science in the Ottoman world.

II.11.1 Royal science

Physiognomy's main purpose as a royal science was to provide a reliable and scientific selection mechanism for the ruler, through which he could control entry into his retinue and ruling elite. Its association with rulers in the Islamicate world can be traced back to the Pseudo-Aristotelian *Secret of Secrets* (*Sirr al-asrār* or *Secretum Secretorum*), most likely composed, not by Aristotle nor in Greek, but in Arabic in the 4th/10th century (Chapter I.14). At the time, it was seen as the definitive translation of Aristotle's treatise on governance and became one of the most popular works on statecraft in both the Muslim and Christian worlds with translations into Ottoman, Latin and over fifteen other Middle Eastern and European languages.[3]

In this text, Pseudo-Aristotle counseled his world-conquering protégé, Alexander of Macedonia (r. 336–323 BCE), on rulership and governance. It included sections on hygiene, exercise, ethics, just rule, principles of governance, talismans, astrology and physiognomy. Alexander

DOI: 10.4324/9781315170718-31

is told that the science of physiognomy "is as much necessary for thee as those other sciences which rest upon conjecture. It is a great science, and the ancients knew it and practiced it, and prided themselves upon possessing it. It is a true science" (*Secretum Secretorum* 1920, 218). Physiognomy encompassed advice on everything from selecting ministers to the proper diet and physical exercises.

Among other things, Pseudo-Aristotle encouraged his student to learn physiognomy so that through it he might choose the best possible candidates for his retinue:

> O Alexander, since the science of physiognomy is one of the subtle and speculative and intellectual sciences which it is necessary for thee to know and to understand, because of the great need in which thou standest when appointing men to stand before thee, I will, therefore, put down for thee in this chapter all the tokens of physiognomy which are proved true and known in the days gone by, and which we have tested in sooth from olden times.
>
> (*Secretum Secretorum 1920, 218)*

Physiognomy's utility for a ruler, particularly a world-conquering ruler such as Alexander, was its claim to providing a scientific basis for the discernment of human character. This ability was of especial significance in the selection of royal servants since the king's direct representatives and members of his royal retinue were evaluated according to the same lofty characteristics that were expected of the king himself (see the lengthy section "On Ministers," *Secretum Secretorum* 1920, 227–41). In fact, the connection between the morality of the ruling elite and the health and survival of the polity stood at the heart of Islamicate and Ottoman political thought.

Following a list of major physiognomical characteristics and their respective significations in terms of human character, Pseudo-Aristotle describes the "best of men," worthy of the ruler's company in the following words:

> The best of men is one having a moderate-sized mouth, soft and moist flesh, neither too thin nor too fat, neither too tall nor too short, in colour either white inclining to red, or a clear brown colour, oval in face, and of even features, hair long – neither too thick nor too thin – of a colour between red and black, moderate-sized eyes, somewhat deep-set, moderate-sized head, straight neck, square shoulders inclined to sloping, moderately broad chest, back and thighs not too full, a clear and moderate voice, smooth palms, long fingers inclined to thinness, grave, thoughtful, amiable, cheerful so as to inspire others with his cheerfulness, and high minded.
>
> (*Secretum Secretorum 1920, 223)*

If indeed such a man could be found, concludes Pseudo-Aristotle, he should be recruited into the ruler's royal company and assigned a position of governance: "Therefore, O Alexander, whenever thou findest such a man choose him for thy company and for governing thy people and for serving thyself" (*Secretum Secretorum* 1920, 224). In short, this is the physical appearance of the *type* fit for a king's company and fit to rule as the king's representative.

Of course, it is an ideal type, which may not have been encountered too often in reality. Pseudo-Aristotle further added, perhaps alluding to the scarcity of "best men," that all character traits must be weighed together and men judged "on the whole." He alerts Alexander to the need for being flexible and even accepting contrary signs, in which case one should "lean towards those [signs] that are stronger and more conclusive so that thou mayest be rightly guided and achieve thy objects, by the help of God" (*Secretum Secretorum* 1920, 224).

II.11.2 An Ottoman royal science

The *Secret of Secrets* and other Islamic *Mirror for Princes* texts (*naṣīḥat*) in Arabic, Persian and Ottoman Turkish counseled the cultivation of physiognomy at royal courts to ensure the selection of qualified courtiers and, more broadly, the ruling elite. The classical Islamicate *Mirror for Princes* texts, many of which referenced physiognomy, met an enthusiastic reception among the Ottomans (Yılmaz 2018, 57–63). The emergence of a specifically Ottoman Turkish physiognomical tradition, in fact, dates back to the early stages of Ottoman court literature during the first half of the 9th/15th century. The first Ottoman writings on physiognomy were mainly – although not exclusively – located within the context of an emerging Ottoman *naṣīḥat* literature.[4] Early manifestations of Ottoman physiognomy were connected to the first Ottoman attempts at centralizing power and laying imperial foundations, which involved the development of a distinct Ottoman court language to unify the new polity, as detailed later. Not surprisingly, Ottoman rulers and their elites expressed a particular interest in guides and manuals on statesmanship. Knowledge and application of physiognomy were considered part of good statesmanship (Yılmaz 2018, 22–45).

Although physiognomic knowledge may have circulated in other languages (Arabic, Persian, Greek), both in written form as well as orally, some of the earliest physiognomy in the Ottoman language appears in translations of Kay Kā'ūs ibn Iskandar's (d. 475/1082) *Letter to Qābūs* (*Qābūsnāma*).[5] Here physiognomy appears as only one part of a larger text dealing with various subjects, as opposed to a text or treatise solely dedicated to physiognomy, which is discussed later. Widely read in the Ottoman world, it was a *Mirror for Princes* treatise, written by a king for his son and successor, which anticipates the challenges the young prince would face, both in terms of personal maturation as well as governing the kingdom. It gives advice accordingly. Given the provenance and nature of the text, its attraction to Ottoman rulers is hardly surprising, particularly at a time when they were rebuilding their polity following the destructive Interregnum Period (804–816/1402–1413).[6] The *Letter to Qābūs*'s section on physiognomy is particularly telling of physiognomy's role at the 9th/15th-century Ottoman court. Sultan Murād II (r. 824–848, 850–855/1421–1444, 1446–1451), for whom the *Letter to Qābūs* was translated, was laying the foundations for an expanded royal household made up of loyal slaves (*qūl*). They would compete with and ultimately subdue rival grandee and lordly households in the kingdom, which had gained great power during the political fragmentation of the Interregnum Period.

The twenty-third chapter of the *Letter to Qābūs*, titled "On the Purchase of Slaves," advised the young prince to learn the science of physiognomy if he was to know exactly how to choose the best slaves for purchase (Bedr-i Dilşād 1997, 2: 878; Mercimek Ahmet 1974, 144; Kaykāvūs ibn Eskandar 1951, 100):

> Whoever it may be that inspects the slave must first look at the face, which is always open to view, whereas the body can only be seen as occasion offers. Then look at eyes and eyebrows, followed by nose, lips, and teeth, and lastly at the hair.
>
> *(Kaykāvūs ibn Eskandar 1951, 100)*

Echoing Pseudo-Aristotle's *Secret of Secrets*, the author of the *Letter to Qābūs* was deeply interested in laying out the set of physical qualities that signaled positive character traits relevant to the expectations of court life:

> The learned say that one must know the indications and the signs by which to buy the slaves suited for particular duties. The slave that you buy for your private service and conviviality should be of middle proportions, neither tall nor short, fat nor lean,

> pale nor florid, thickset nor slender, curly-haired nor with hair over-straight. When you see a slave soft-fleshed, fine-skinned, with regular bones and wine-colored hair, black eye-lashes, dark eyes, black eyebrows, open-eyed, long-nosed, slender-waisted, round-shinned, red-lipped, with white regular teeth, and all his members such as I have described, such a slave will be decorative and companionable, loyal, of delicate character and dignified.
>
> *(Kaykāvūs ibn Eskandar 1951, 100)*

On the other hand, the physiognomical markers optimal for services requiring learning and intellectual aptitude are somewhat different:

> The mark of the slave who is clever and may be expected to improve is this: he must be of erect stature, medium in hair and in flesh, broad of hand and with the middle of the fingers lengthy, in complexion dark though ruddy, dark-eyed, open-faced and unsmiling. A slave of this kind would be competent to acquire learning, to act as treasurer or for any other [such] employment.
>
> *(Kaykāvūs ibn Eskandar 1951, 101)*

In contrast to the potential treasurer, a successful warrior could be recognized by the following physical characteristics:

> Hair is thick, his body tall and erect, his build powerful, his flesh hard, his bones thick, his skin coarse and his limbs straight, the joints being firm. The tendons should be tight and the sinews and blood-vessels prominent and visible on the body. Shoulder must be broad, the chest deep, the neck thick and the head round; also for preference he should be bald. The belly should be concave, the buttocks drawn in and the legs in walking well extended. And the eyes should be black.
>
> *(Kaykāvūs ibn Eskandar 1951, 101–2)*

Kay Kāʾūs ibn Iskandar furthermore provided physiognomically encoded markers that indicate suitability for various other services, such as musical entertainment, employment in the women's quarters, herdsmen and domestic servants, including cooks.

In other words, the text charted the physiognomy of the royal household, which stood at the heart of the Ottoman polity. Many of the corporeal markers that characterize particular services seem rather intuitive – that is, physical strength for the person bearing arms, a "clean" face and body for the domestic servant, lengthy fingers for the musician – while others make little to no immediate sense without the help of a physiognomy treatise.

II.11.3 Creating an imperial elite

Physiognomy was deeply connected to the selection of slaves, both in the pre-Ottoman, as well as the Ottoman, period (Müller 1980, 9–12). The institution of slavery, in turn, was a matter that surpassed mere household management and was fundamentally woven into Ottoman state (*devlet*) policy through the *devshirme*, a complex recruitment system for the elite janissary corps and the Ottoman ruling elite.[7] During the reign of Sultan Murād II, the slaves of the imperial household (*qūl*) were becoming increasingly influential. Indeed, during the reign of his son and successor Sultan Meḥmed II (r. 848–850/1444–1446; 855–886/1451–1481), the imperial *qūls* came to dominate state politics at the expense of the old aristocratic families. The first grand viziers of *qūl* origins were appointed by Mehmed II, who thus introduced a new tradition.[8] Henceforth, it became a tradition to appoint imperial *qūl*s to the grand vizierate and eventually staff the whole imperial ruling apparatus with members of the imperial household.[9]

The very notion of *devlet* during the latter half of the 9th/15th century and afterward came to represent "the decision-making power of the legitimate head of state as well as of those to whom he has delegated this power," who were increasingly *qūl*s or servile members of the imperial household (Abou-El-Haj 1991, 19). This was especially true during the 9th–10th/15th–16th century when the governance of the Ottoman polity was almost fully placed in the hands of the servile members of the imperial household. Thus, a mechanism for the selection of *qūl*s became an integral part of managing the Ottoman imperial venture. Physiognomy responded to this need by claiming to provide a scientific sifting mechanism for discerning talent and ability in potential recruits. Other considerations (background, relations and connections to serving members of the elite, skills and talents, and so on) also influenced selections.

Muṣṭafā ʿĀlī (d. 1008/1600), the highly prolific 10th/16th-century Ottoman historian and man of letters, recorded in many of his writings the use of physiognomy in elite Ottoman circles, particularly at court, and its indispensable role in monitoring access to the corridors of power. In his great history work, *Essence of History* (*Künhü l-akhbār*), he explained physiognomy's function in selecting palace pages at the court of Sultan Meḥmed II. In order for the Ottoman government to function properly, Muṣṭafā ʿĀlī maintained in his treatise *Tables of Delicacies* (*Mevā'idü n-nefā'is fī kavā'idi'l-mecālis*) that the *devshirme* recruits, who were its foundation, had to be examined individually and very discerningly for appropriate intrinsic qualities (ʿĀli 2003, 15–17). It was indeed the same sentiment that motivated the composition of many Ottoman physiognomy treatises, particularly during the 9th–10th/15th–16th century. Muṣṭafā ʿĀlī was describing what he saw as the exceptionally successful – and thus exemplary – functioning of the Ottoman state during the reign of Meḥmed the Conqueror, over a century prior to his own lifetime. It is worth noting that he does not record the presence of a court physiognomer during his own time and, indeed, seems to equate such an absence with the "twilight" of the "Golden Age."

According to Muṣṭafā ʿĀlī, during the exemplary reign of Sultan Meḥmed II,

> physiognomy was regarded highly. A noble face, which is a divine gift, was looked upon favorably, and guiding signs were decreed, by which the particular characteristics of noble and illustrious men were revealed and made known. This has always been the intention, through divine aid, of the customs of the Ottoman sultans and regulations of the kingdom-conquering rulers.
>
> *(ʿĀli 1277, 5: 14)*

Muṣṭafā ʿĀlī believed very firmly that the Ottoman tradition was built on – and owed its success to – the laws and customs that ensured a tried and very meticulous selection process for entry into the ruling elite. The insights of physiognomy, he claimed, were integrated into Ottoman custom to help distinguish men of noble character from their baser counterparts and to order society accordingly.

According to Muṣṭafā ʿĀlī, this process functioned seamlessly during the reign of Sultan Meḥmed II, who applied physiognomy to develop a strong royal household. By staffing state offices with loyal household servants, he was able to enforce an intense centralization of power and resources and thus lay the foundations for building an empire:

> During the time of Sultan Meḥmed Han, the children who came as war captives and non-Muslim children who were collected as tribute, in accord with the imperial law,

> were brought to the imperial threshold. In the presence of the most illustrious of men, the White Eunuch, they were divided, based on a perfect examination, into [1] those seen most fitting for and worthy of the highest service in the most honorable privy chamber, [2] gatekeepers, [3] Janissary corps, and [4] outer service.
>
> (*Āli 1277, 5: 15*)[10]

The new recruits, according to Muṣṭafā ʿĀlī, were examined by the White Eunuch and the Imperial Palace Teacher, whose physiognomical eye

> was capable of discerning every person's innermost heart. . . . Every boy on whose forehead were manifested signs of felicity was taken inside [to the inner palace service]. Those with signs of an evil disposition and depravity on their foreheads were deemed suitable for outer services.
>
> (*Āli 1277, 5: 15*)

Muṣṭafā ʿĀlī insisted that the process for selecting imperial pages, whose career track led to the highest imperial positions, must be conducted by a skilled physiognomer. According to his rationale, unless those of unworthy disposition were, from the very beginning, cut off from that illustrious career path that began with palace service, the rise of an oppressive elite was inevitable. It took, however, the great skill of one versed in the science of physiognomy to recognize the budding potential of an inherently unjust disposition in the appearance of a *devshirme* boy and nip the rise of haughty oppressors and intriguers in the bud by restraining them from entering "the highest service in the most honorable privy chamber" (ʿĀli 1277, 5: 15). He continues his argument thus:

> If a person of reprehensible moral character enters service in the imperial privy chamber and, by way of the regulated way of promotion, is promoted to the rank of officer, chieftain, brigade general, or even regional governor, he will become conniving, oppressive and overbearing. In order to prevent such a person from burning the ruled subject with the fire of oppression, he is assigned to outer service, where he is confined to being a common soldier. He is prevented from rising to the ranks of the elite corps, which would allow him to treat the people of the world with severity and torment.
>
> (*ʿĀli 1277, 5: 15*)

This type of application was propagated in contemporaneous Ottoman physiognomy treatises, which proliferated at the Ottoman court during the 9th–10th/15th–16th century. Numerous treatises were composed in Ottoman Turkish for reigning sultans and occasionally for incumbent viziers (Lelić 2017).

Some Ottoman physiognomy treatises were quite explicit about physiognomy's role in contemporary politics. Muṣṭafā ibn Evrenos, who lived during the reign of Sultan Selīm II (r. 974–982/1566–1574), to whom he dedicated his physiognomy treatise, not only commented on the current state of affairs in that same treatise but also gave very precise advice on the role that physiognomy could play in contemporary state politics. He presented the problems of his age as an infiltration by men of the lowest character into the ruling elite. Predictably, such men were accused of abusing their power and authority in pursuit of their desires. According to Muṣṭafā ibn Evrenos, this caused factionalism and infighting amongst the elite and paralyzed effective governance – a very concrete problem in Ottoman society at the time, which other contemporaries were also noticing and recording (Tezcan 2010, 183). As a solution to the very root of the problem, Mustafa ibn Evrenos proposed physiognomy.

"In short," Ibn Evrenos wrote, the sultan "will have knowledge of every person's life, what has befallen them since the day they were born and what will befall them until the day they die; he will have knowledge of all their qualities and characteristics" (MS Çankırı, Il Halk Library, 8 Hk 321, fol. 2b). The appeal of such knowledge – precise awareness of every man's disposition and destiny – to the ruler of an enormous empire requiring an extensive governing apparatus must have been great. With the help of physiognomy, Ibn Evrenos claimed, the sultan will instantaneously

> recognize those who are suitable and qualified to occupy the offices of the vizierate, provincial offices, high state offices, offices of trust, and religious offices. Everyone will be appointed precisely according to their level. The subjects, too, will be employed depending on their individual state. The kingdom will be safe and secure. For the sake of justice, the subjects must be kept safe; only thus can sovereignty endure permanently.
> *(MS Çankırı, Il Halk Library, 8 Hk 321, fol. 3a)*

The sultan in charge of the highly centralized 9th–10th/15th–16th-century Ottoman state was expected to oversee the distribution of imperial offices with a physiognomical eye. Proper selection to imperial offices guaranteed the perpetuation of imperial justice, which in turn was the ultimate *raison d'être* for the very existence of the Ottoman state. A kingdom without justice – according to Ottoman political thought as well as Ottoman physiognomy treatises – was doomed.

II.11.4 The physiognomy treatise (*Firāsetnāme/Qıyāfetnāme*)

The precise physiognomical significations of various corporeal markers were laid out systematically in physiognomy treatises, variously referred to in Ottoman as *Qıyāfetnāme* or *Firāsetnāme*. They matched particular corporeal signs – often starting with the face and working their way down the body to the feet – with character traits: "X physical marker signifies Y character trait."

The physiognomical entries in the long *Mirror for Princes* tradition, which are generally interested in vocational suitability, such as those listed earlier, differ from physiognomy treatises, which instead focus on specific character traits, such as courage/fear, stupidity/intelligence, honesty/duplicity and so on. For example, the explicit connection made in the *Letter to Qābūs* between a person's complexion or eye color and financial acumen is an interpretation of physiognomy's pairing of a "dark though ruddy" complexion with good judgment and dark eyes with intelligence.

The first Ottoman Turkish physiognomy treatise was written by Ḥamdullāh Ḥamdī (d. 908/1503), most likely to satisfy an increasing demand among Ottomans who could not read the readily available Arabic and Persian treatises (Çelebioğlu 1998).[11] Ḥamdī was a highly accomplished poet, who produced impressive Ottoman Turkish renditions of some of the great Persian epics. His fame rests even more on those accomplishments than on his *Physiognomy Treatise* (*Qıyāfetnāme*), which was well known in its own right (Gibb 1902, 2: 138–225; İz 2012; Öztürk 2001; Özyıldırım 1999).

A very telling reference to Ḥamdullāh Ḥamdī's *Physiognomy Treatise* is made by Evliyā Chelebi (1020–after 1098/1611–after 1687) – the man who has become the symbol of an "Ottoman mentality" thanks to his extensive travels and meticulous recording of his experiences (Dankoff 2004). He offers some insight into physiognomy's place among Ottoman elites. In the ninth volume of his grand *Book of Travels* (*Seyāḥatnāme*), Evliyā Chelebi took note of the inhabitants of Mecca, particularly the descendants of the Prophet and gave a physiognomically encoded description of their appearance.

> Eulogy of the people of Mecca and description of the noble descendants of the Prophet of the city of "Bakka": Based on what has been recorded in the *Physiognomy Treatise* (*Qıyāfetnāme*) of Ḥamdī Chelebi, which is not concealed from the people of intellect, the figure and face of the people of Mecca are of a ruddy hue and swarthy face, with gazelle-like eyes, sweet words, and light-filled faces, their secrets hidden deep within. The people of the House of Hāshim belong by virtue of both birth and actions to the class of true gentlemen.
>
> *(Evliya Çelebi Seyaḥatnāmesi 2007, 399)*

Evliyā Chelebi's reference to Ḥamdullāh Ḥamdī's *Physiognomy Treatise*, almost two centuries later, attests to its continued popularity. This popularity is confirmed not only by the sheer quantity of available manuscript copies in central, as well as provincial, Ottoman archival collections but also by the fact that it forms the basis of some later Ottoman physiognomy treatises. An example is Muṣṭafā ibn Bālī's (d. 1027/1618) *Treatise on Physiognomy* (*Risāle-i qiyāset-i firāset*), which incorporates Ḥamdī's treatise to the letter but also greatly expands on it.

One of the reasons Ḥamdī's treatise was so popular might be its rhymed and metered composition in fairly simple and not inelegant 9th/15th-century Ottoman Turkish. In addition to its relative brevity, these characteristics predisposed Ḥamdī's treatise to easy memorization and widespread inclusion in miscellaneous collections (*mecmūʿa*), presumably for those who failed to memorize it but nonetheless wished to have it readily available. In a stratified, hierarchical society, physiognomical injunctions offered a map for navigating professional and social networks – manipulating superiors by exploiting their weaknesses on the one hand and choosing and controlling inferiors on the other, both in the immediate household (marriage partners, children) and extended household (slaves, servants, clients).

In essence, Ḥamdullāh Ḥamdī's protasis–apodosis (if x, then y) pairing of corporeal features and character traits embodies the basis of traditional physiognomy. Although many later Ottoman treatises expanded physiognomy to include additional "methods" (*ṭuruq*), such as sonic physiognomy, zoological physiognomy, ethnic and social/estate-based typologies (Muṣṭafā ibn Bālī, *Treatise on Physiognomy*, cited earlier; Taʿlīqīzāde Meḥmed Ṣubḥī, *Physiognomy Treatise* [*Firāsatnāme*], MS Paris, BnF, Turc 1055) and gender physiognomy (Lelić 2018), among others, the constituent part or method was always the correspondence between corporeal features and character.

According to Ḥamdullāh Ḥamdī, God created man by giving every person his own unique form, and in turn, "He [God] made the (human) form an indicator of character" (MS Istanbul, SK, Bağdatlı Vehbi, 2162, fol. 1b). In the parlance of later Ottoman physiognomy treatises, which were addressed to a more learned audience, physiognomy "consists of inferring the inner character (*akhlāq-ı bāṭine*) from the exterior state (*aḥvāl-ı ẓāhire*)" (MSS Istanbul, SK, Nuruosmaniye, 4100, fol. 7a; Nuruosmaniye, 4099, fol. 4a).

An Ottoman who was exposed to the types of physiognomically encoded corporeal markers encountered in the previously quoted extracts from various *Mirrors for Princes* texts and Evliya Chelebi's *Book of Travels* could turn to Ḥamdullāh Ḥamdī's *Physiognomy Treatise* for answers. Thus, the "ruddy hue and swarthy face" of the Meccans encountered by Evliyā Chelebi is revealed by Ḥamdullāh Ḥamdī to signify quick-bloodedness and good judgment: "A reddish hue is a sign of quick-bloodedness / A dusky tint signifies good judgment" (MS Istanbul, Süleymaniye Library, Bağdatlı Vehbi, 2162, fol. 1b). In Evliyā's case, the physiognomical reference was likely a literary device to further reinforce his already positive impression of the members of the House of Hāshim in Mecca by adding an extra layer of signification, which could be decoded through reference to a physiognomy treatise.

Physiognomical references also permeated other Ottoman genres such as historical chronicles, *belles lettres*, poetry and even legal records, particularly manumission documents (Sobers-Khan 2014). Physiognomy was, in fact, an integral part of Ottoman intertextuality. This is further reinforced by the fact that Ḥamdullāh Ḥamdī wrote a physiognomy treatise as part of his extensive literary oeuvre but also that, as a leading Ottoman littérateur, he felt the need to provide a map for navigating the abundance of physiognomical references in Ottoman writings.

Physiognomy's intertextual potential is illustrated by pairing Ḥamdī's physiognomy treatise with Kay Kā'ūs's physiognomically laden references to produce new layers of meaning. According to the *Letter to Qābūs*, a slave that is "competent to acquire learning, to act as treasurer or for any other [such] employment" must be "of erect stature, medium in hair and in flesh, broad of hand and with the middle of the fingers lengthy, in complexion dark though ruddy, dark-eyed, open-faced and unsmiling" (Kaykāvūs ibn Eskandar 1951, 101).

Erect stature, according to Ḥamdullāh Ḥamdī's *Physiognomy Treatise* is a sign of honesty: "Whosoever is in stature tall / His heart is pure, without duplicity" (MS Istanbul, Süleymaniye Library, Bağdatlı Vehbi, 2162, fol. 1b). "Medium in hair" is not listed as a separate category; instead, the reader must deduce its meaning from descriptions of thick and thin hair: "Thickness of hair is proof of bravery/It brings good tidings for soundness of intellect // If soft then it is a witness of fear/ With a stupid mind and a cool brain" (MS Istanbul, Süleymaniye Library, Bağdatlı Vehbi, 2162, fol. 2a). A person "medium in hair" is thus neither brave nor stupid. Discerning the meaning of "medium in flesh" is also rather complicated because here, too, Ḥamdullāh Ḥamdī only lists the extremes: "A person with soft flesh possesses / Graceful nature and sharp-mindedness in abundance // Sturdiness, however, is a sign of strong body / But equally of foulness and stupidity" (MS Istanbul, Süleymaniye Library, Bağdatlı Vehbi, 2162, fol. 2a). Thus, the ideal budding treasurer is neither particularly graceful nor foul nor especially sharp-minded – although that is overturned by his dark eyes, which signify wit (MS Istanbul, Süleymaniye Library, Bağdatlı Vehbi, 2162, fol. 3a).

The possibility of a "broad hand" is not listed, but its meaning might perhaps be deduced from a wide foot, which signifies caution (MS Istanbul, Süleymaniye Library, Bağdatlı Vehbi, 2162, fol. 6a). Hands and feet are both bodily extremities and thus share certain parallels. Long fingers are a sign of mastery in every profession (MS Istanbul, Süleymaniye Library, Bağdatlı Vehbi, 2162, fol. 5b). A "dark though ruddy complexion," as already seen in Evliyā Chelebi's description, denotes quick-bloodedness and good judgment, while an open face is the mark of a "pure spirit" and maturity (MS Istanbul, Süleymaniye Library, Bağdatlı Vehbi, 2162, fol. 3b). The set of characteristics that emerge are those of a cautious, mature, capable, moderately witty, doubly pure and thus trustworthy person with good judgment – someone indeed cut out to be a good treasurer or "any other [such] employment" (Kaykāvūs ibn Eskandar 1951, 101).

Physiognomy treatises were the key that unlocked physiognomical signs in human appearances and textual descriptions thereof. The 9th–10th/15th–16th century witnessed a relative efflorescence of Ottoman physiognomy, particularly during the last quarter of the century, which produced treatises of increasing linguistic and theoretical complexity, as well as physiognomical entries in various compilations. Many of the 9th–10th/15th–16th-century treatises were specifically designated for royal use – they were dedicated to reigning sultans and incumbent grand viziers, themselves members of the imperial household. They purported to train the sultan and his household in developing a physiognomical eye and thus a scientifically proven method for implementing royal justice. Royal justice meant giving every person their due, which physiognomy claimed to discern based on inborn character and vocational capacity.

II.11.5 Conclusion

Physiognomy treatises, particularly those that propagated physiognomy's role as a royal science, far from speaking in a vacuum, seem to have been preaching to the choir of Ottoman intellectual-political thought. Muṣṭafā ʿĀlī, who has already been quoted, spoke of the necessity for cultivating physiognomy at court in numerous texts (Mustafa Āli 1979, 1: 17; Mustafa Āli 2003, 15–16, 167–8). On a more tangible level, physiognomical knowledge seems to have been incorporated into the broader *devshirme* collection procedure by alerting recruiting parties to particular corporeal characteristics and their physiognomical significations (*Qavānīn-i Yeniçeriyān* 1990, 9: 228–9). Even European diplomats and visitors to the Ottoman Empire noticed the presence of physiognomy in the process through which Ottoman elites were formed (Rycaut 1686, 45). The classical Ottoman *akhlāq*, or ethics treatises, drew directly and openly on physiognomical knowledge in their section on household management (Qınalızāde 2007, 389–97). Perhaps most importantly, physiognomy was duly classified as a legitimate natural science in Ottoman scientific taxonomies (Taşköprizade 1910–1, 272–4; Ḥāccī Khalīfa (Kātip Çelebi), 1310/1892–3, 2: 72).

Physiognomy's role as a royal science was recognized by another well-known 9th/16th- century Ottoman scholar, Ṭāshköprüzāde (d. 967/1560). In his *Treatise on the Secrets of the Human Caliphate and the Spiritual Sultanate* (*Risālā fī asrār al-khilāfa al-insānīya wal-salṭana al-manʿawīya*), which was inspired by Pseudo-Aristotle's *Secret of Secrets*, Ṭāshköprüzāde outlined ten "laws of sovereignty, . . . without which a sultan will not be delivered from torment" (MS Istanbul, SK, Veliyuddin 3275, fol. 117a). The first nine laws follow the general injunctions developed by Islamic political thought,[12] but the tenth, which is of particular interest here, enjoins the true ruler to,

> [o]bserve with a physiognomical eye the unfolding of events and to watch with the eye of discernment their manifestations and their consequences; and to render judgment on them based on the holy law if they are lucid and based on physiognomy if they are enigmatically concealed.
>
> *(MS Istanbul, SK, Veliyuddin 3275, fol. 118b)*

In short, Ṭāshköprüzāde advised rulers to cultivate physiognomical intuition – even beyond the specific process of selecting a ruling elite, such as Kay Kāʾūs ibn Iskandar or Muṣṭafā ʿĀlī were concerned with – in order to govern with true justice. Specifically, he presented physiognomical intuition as a method for resolving issues that were not encompassed by the parameters of the Ottoman legal system. The Ottoman ruler's most fundamental duty on which his very existence hinged – to guarantee and enforce justice in the realm – was, according to many early modern Ottoman voices, best facilitated through the cultivation of physiognomy. It ensured that he rendered just judgment and that he selected a fitting ruling apparatus that would do the same. Physiognomy treatises, particularly those composed for royal use, were crucial in developing physiognomy's role as a royal science by offering a manual for cultivating physiognomical intuition.

Notes

1 From these names the two interchangeable Ottoman designations for a physiognomy treatise: *Qıyāfetnāme* and *Firāsetnāme* were derived.

2 For more on Ottoman physiognomy as a method for the acquisition of self-knowledge see (Lelić 2017, 32–6).

3 For a detailed discussion of extant manuscripts, the dating issues, contents and contextualization in the Hellenistic and Arabic traditions of the *Sirr al-asrār* see (Manzalaoui 1974); for another discussion of origins and translations see (Grignaschi 1976); for an analysis of the Arabic and German renditions see (Forster 2006). For a critical Arabic edition see (Yuḥanna ibn al-Biṭriq 1954); for a 10th/16th-century Ottoman translation prepared for the grand vizier Sokollu Meḥmed Pāshā (d. 987/1579), see (Nevālī, *Translation of Aristotle's Advice* [Tercüme-i Neṣā'ih-i Arisṭaṭālīs]; Yılmaz, 2018, 63–4).

4 For a thorough list of Ottoman, Persian and Arabic physiognomy treatises, see Mustafa ibn Bâlî (2014, 46–51).

5 There are a number of Ottoman translations of the *Qābūsnāma*. The best-known are Mercimek Aḥmed's and Bedr-i Dilşād's translations; the latter is known as the *Murād-nāme*. Both translations were composed for and dedicated to Sultan Murād II, see (Mercimek Ahmet 1974, 144–56; Bedr-i Dilşād 1997, 2: 878–90). For an overview of Ottoman translations of the *Qābūsnāma*, see (Doğan 2012). For an English translation of the original Persian rendition see (Kaykāvūs ibn Eskandar 1951).

6 Following Sultan Bāyezīd I's (r. 791–804/1389–1402) disastrous defeat by Tīmūr at the Battle of Ankara in 804/1402, the defeated Ottoman polity was cut down in size and divided by the victor between Bāyezīd's four sons. This resulted in a bloody civil war, until the last contender standing, Sultan Meḥmed I (r. 816–824/1413–1421), emerged as victor over his brothers in 816/1413. The Ottoman translations of the *Letter to Qābūs* were commissioned by his successor, Sultan Murād II, who was dealing with the consequences of the Interregnum Period.

7 For more on Ottoman slavery, including a historiographical review, with a particular focus on the confluence of slavery and physiognomy, see (Sobers-Khan 2014). For an introductory study of the *devshirme* (Ménage 1991, 210–3). For the role of *qūl*s in the early development of the Ottoman polity, see (Inalcik 1954, 120–2).

8 For a defining transitional moment charting the rise of the *qūl* elite, see (Stavrides 2001).

9 It must be noted that this trend was occasionally interrupted with appointments to the grand vizierate and other elite positions from old Turkic aristocratic families and the household networks of the frontier warlords (Kafadar 1995, 110–13).

10 The numbering is mine, added for clarification. For more detailed information on the palace school see (Miller 1941; Uzunçarşılı 1943).

11 For the most part, Ottoman physiognomy treatises tend to be translations or reconfigurations of older Arabic and Persian treatises, with varying degrees of creativity, license and adaption to the Ottoman context. For example, Ḥamdullāh Ḥamdī's treatise was presented as an Ottoman Turkish translation of Imām Shāfiʿī's physiognomy treatise; although, most modern scholars find this questionable.

12 These are: 1) To think of himself, when it comes to rules, as one of the subjects (*raʿāyā*); 2) to provide for the needs of Muslims, which is the most excellent [form of] submission; 3) to follow the example of the Righteous Caliphs in food and dress; 4) to persuade with gentle words, without showing his rough side unless there is cause, and to patiently hear out petitions, especially from the poor and the aggrieved; 5) to avoid gentleness when passing judgment, nor to neglect it for the sake of pleasing the people, nor to contradict the holy law for anyone's pleasure; 6) to not be heedless of the weightiness of governance and sovereignty, because sovereignty is either an instrument of felicity or an instrument of misery; 7) to desire the company of scholars and the righteous ones; 8) to not alienate the people through tyranny and haughtiness; 9) to not neglect gathering information about oppressive agents and treacherous representatives, and to not release the wolves on the subjects (MS Istanbul, Süleymaniye Library, Veliyuddin 3275, fols. 117a-118b).

Bibliography

Sources

[Anonymous]. 1990. "Qavānīn-i Yeniçeriyān-i Dergāh-i Āli [The Regulations for the Janissaries of the Sublime Porte]," in Akgündüz, A., ed. *Osmanlı kanunnāmeleri ve hukukī tahlilleri* [Ottoman Lawbooks and Legal Rulings]. Vol. IX. İstanbul: Fey Vakfı.

Bedr-i Dilşād. ed. Ceyhan, A. 1997. *Murād-nāme* [The Letter to Murād]. Istanbul: Milli Eğitim Basımevi.

Evliya Çelebi. ed. Gökhan, O. Ş. 2007. *Evliya Çelebi Seyahatnāmesi* [Evliya Çelebi's Travel Account]. Vol. IX. Istanbul: Yapı Kredi Yayınları.

Ḥājjī Khalīfa (Kātib Chelebi). 1310/1892–3. *Kitāb Kashf al-ẓunūn ʿan asāmī al-kutub wa- al-funūn* [The Book of the Removal of Doubts on the Names of Books and Disciplines]. Istanbul: Maṭbaʿat al-ʿālam.

Kaykāvūs ibn Eskandar. tr. Levy, R. 1951. *A Mirror for Princes: The Qābūs-nāma by Kai Kā'ūs b. Iskandar Prince of Gurgan*. New York: E. P. Dutton.

Mercimek Ahmet. ed. Gökyay, O. Ş. 1974. *Keykavus, Kabusname* [The Letter to Qābūs by Kai Kā'ūs]. Istanbul: Milli Eğitim Basımevi.

Mustafa Āli. 1277/1860. *Künh ül-ahbar* [Essence of History]. Istanbul: Takvimhane-i Amire.

Mustafa Ālī. tr. Brookes, D. 2003. *Mevā'idü'n-nefā'is fī kavā'idi'l-mecālis* [Tables of Delicacies Concerning the Rules of Social Gatherings]. Cambridge, MA: Dept. of Near Eastern Languages and Civilizations, Harvard University.

Mustafa bin Bâlî. ed. Sarıçiçek, R. 2014. *Risâle-i Kiyâset-i Firâset/Ilm-i Firâset. Yüzler Hâli Söyler* [Epistle on the Measure of Physiognomy/The Science of Physiognomy. Character is Written in the Face]. Istanbul: Büyüyenay Yayınları.

Mustafa Āli. ed. and tr. Tietze, A. 1979. *Nüṣḥat üs-salāṭīn, Muṣṭafā 'Âlî's Counsel for Sultans of 1581*. Vols. 137, 158. Wien: Verlag der Österreichischen Akademie der Wissenschaften

Pseudo-Aristotle. tr. Ali, I. 1920. "Secretum Secretorum [The Secret of Secrets]," in Steele, R., ed. *Opera Hactenus Inedita Rogeri Baconi* [The Previously Unpublished Works of Roger Bacon]. Vol. 5. Oxford: Clarendon Press.

Qınalızāde, 'Alī Çelebi. ed. Koç, M. 2007. *Akhlāq-ı Alā'ī* ['Alā'īd Ethics]. Istanbul: Klasik.

Rycaut, Paul. 1686. *The History of the Present State of the Ottoman Empire*. London: Printed for C. Brome.

Taşköprizade. A. 1910–1911. *Miftāḥ al-sa'āda wa-miṣbāḥ al-siyāda* [The Key to Happiness and the Light of Nobility]. Hyderabad: Dā'irat al-ma'ārif al-niẓāmiyya.

Yuḥanna ibn al-Biṭriq. 1954. "*Kitāb al-Siyāsa fī tadbīr al-riyāsa (al-ma'rūf bi-*Sirr al-Asrār) [The Book of Politics on the Management of Leadership]." Part 2 of Badawī, 'A., ed. *al-Usūl al-yūnāniyya li-l-naẓariyyāt al-siyāsiyya fī l-islām* [The Greek Elements in the Political Theories in Islam]. Cairo: Maktabat al-Nahḍa al-miṣriyya.

Manuscripts

MS Istanbul, Süleymaniye Library, Veliyuddin 3275.
MS Çankırı, Il Halk Library, 8 Hk 321.
MS Istanbul, Süleymaniye Library, Bağdatlı Vehbi, 2162.
MS Istanbul, Süleymaniye Library, Hafid Efendi 253.
MS Istanbul, Süleymaniye Libnrary, Nuruosmaniye, 4099.
MS Istanbul, Süleymaniye Library, Nuruosmaniye, 4100.
MS Paris, BnF, Turc 1055.

Research literature

Abou-El-Haj, R. 1991. *Formation of the Modern State: The Ottoman Empire Sixteenth to Eighteenth Centuries*. Albany, N.Y: SUNY Press.

Çelebioğlu, A. 1998. "Kiyafe(t) ilmi ve Akşemseddinzade Hamdullah Hamdi ile Erzurumlu Ibrahim Hakkı'nın Kıyafetnameleri," in *Eski Türk Edebiyatı Araştırmaları: Prof. Dr. Amil Çelebioğlu*. Istanbul: MEB Yayınları, 225–62.

Dankoff, R. 2004. *An Ottoman Mentality: The World of Evliya Çelebi*. Leiden: Brill.

Doğan, E. 2012. "On Translations of Qabus-nama During the Old Anatolian Turkish Period," *Uluslararası Sosyal Araştırmalar Dergisi* 5.21: 76–85.

Forster, R. 2006. *Das Geheimnis der Geheimnisse: Die arabischen und deutschen Fassungen des pseudo-aristotelischen Sirr al-asrār/Secretum Secretorum*. Wiesbaden: Dr. Ludwig Reichert Verlag.

Ghersetti, A. 2007. "Physiognomy and Medicine in Islamic Culture," in Swain, S., ed. *Seeing the Face, Seeing the Soul: Polemon's Physiognomy from Classical Antiquity to Medieval Islam*. Oxford: Oxford University Press, 285–7.

Gibb, E. J. W. ed. Browne, E. 1902. *A History of Ottoman Poetry*. London: Luzac & Co.

Grignaschi, M. 1976. "L'origine et les metamorphoses du 'Sirr-al-asrār'," *Archives d'histoire doctrinale et littéraire du Moyen Âge* 43: 7–112.

Inalcik, H. 1954. "Ottoman Methods of Conquest," *Studia Islamica* 2: 103–29.

İz, F. 2012. "Ḥamdī, Ḥamd Allāh," in *EI-2*, 3: 131–2.

Kafadar, C. 1995. *Between Two Worlds: The Construction of the Ottoman State*. Berkeley: University of California Press.

Lelić, E. 2017. "Physiognomy (*ʿilm-i firāset*) and Ottoman Statecraft: Discerning Morality and Justice," *Arabica* 64: 1–37.

Lelić, E. 2018. "'The Greatest of Tribulations:' Constructions of Femininity in 16th Century Ottoman Physiognomy," in Karateke, H. *et al.*, eds. *Disliking Others: Loathing, Hostility, and Distrust in Premodern Ottoman Lands*. Brighton, MA: Academic Studies Press, 264–95.

Manzalaoui, M. 1974. "The Pseudo-Aristotelian 'Kitāb Sirr al-asrār'. Facts and Problems," *Oriens* 23–4: 147–257.

Ménage, V. L. 1991. "Devshirme," in *EI-2*, 2: 210–3.

Miller, B. 1941. *The Palace School of Muhammad the Conqueror*. Cambridge, MA: Harvard University Press.

Müller, H. 1980. *Die Kunst des Sklavenkaufs in arabischen, persischen und türkischen Ratgebern vom 10. bis zum 18. Jahrhundert*. Freiburg: Klaus Schwarz Verlag.

Öztürk, Z. 2001. *Hamdullah Hamdi'nin Yusuf ve Zeliha Mesnevisi*. Cambridge, MA: Harvard University Press.

Özyıldırım, A. E. 1999. *Hamdullah Hamdi ve Divanı*. Ankara: T. C. Kültür Bakanlığı Yayınları.

Sobers-Khan, N. 2014. "*Firāsatle naẓar edesin*: Recreating the Gaze of the Ottoman Slave-Owner at the Confluence of Textual Genres," in Firges, P., ed. *Well-Connected Domains: Towards an Entangled Ottoman History* (The Ottoman Empire and Its Heritage 57). Leiden: Brill.

Stavrides, T. 2001. *The Sultan of Vezirs: The Life and Times of the Ottoman Grand Vezir Mahmud Pasha Angelović (1453–1474)*. Leiden: Brill.

Tezcan, B. 2010. *The Second Ottoman Empire: Political and Social Transformation in the Early Modern World*. Cambridge: Cambridge University Press.

Uzunçarşılı, I. H. 1943. *Osmanlı devleti teşkilâtından Kapukulu ocakları*. Ankara: Türk Tarih Kurumu Basımevi.

Yılmaz, H. 2018. *Caliphate Redefined: The Mystical Turn in Ottoman Political Thought*. Princeton, NJ: Princeton University Press.

PART III

Learning and collecting institutions – debates and methods (3rd–13th/9th–19th centuries)

III.1

LIBRARIES – BEGINNINGS, DIFFUSION AND CONSOLIDATION

Lutz Richter-Bernburg

Scripture was foundational to Islam, yet in *literature*, broadly defined, the transition from orality – and *au*rality – to literacy took roughly two centuries (Schoeler and Toorawa 2009; Schoeler 2010; Touati 2013).[1]

In less pronouncedly religious fields the process went faster, in poetry and religious scholarship perceptibly more slowly. Actually, in the transmission of *ḥadīth*, the alleged sayings and actions of the Prophet (and his companions), the fiction of orality was maintained even after in fact, it had come to rest entirely on written records (cf. Lecker 2007).[2]

Extant monuments prove that as early as Muḥammad's lifetime, Arabian culture was not entirely oral (e.g., Munt 2017; Imbert 2013, 2012a, 2012b, 2011; Jones 1998; Gruendler 1993), and the exigencies of administering an ever-expanding empire lead to a rapid expansion of the production of written documents (van Berkel 2013; Meier 2012; Lewis 1954). It stands to reason that some sort of record keeping or archivism developed in tandem (Touati 2003; Heffening and Pearson 1987; Eche 1967); furthermore, it is no more than to be expected that those civilizations which had, before Islam, flourished in caliphal dominions exerted a tangible influence on this as on so many other areas of social and cultural practice. Libraries and archives had existed in the Roman or Christianized-Roman, so-called Byzantine, Empire as well as in the realm of its great rival, the Sasanian dynasty of Iran. Yet considering that the production of proper 'books' did not massively begin until the 3rd/9th century, whatever storage of written materials did exist cannot seriously be claimed as a library (Déroche 1987–89). Additionally, narrative sources are extant only from the 3rd/9th century and have to be scrutinized for a tendency to retroject then current conditions into the, as it were, heroic past. Still, the name frequently given to early sovereign repositories of precious materials, 'cabinet(s) of wisdom,' is by and large restricted to the pre-library period; as early as the first half of the 3rd/9th century, when notable private libraries began to be celebrated by contemporaries, usage changed. Now it was not a vaguely defined 'wisdom' that was collected and stored but simply 'books.'

III.1.1 Cabinets of wisdom

Unsurprisingly, the historiographic and biographic tradition has most, if still little enough, to say about the caliphs' collection of memorabilia or *bayt al-ḥikma* ('cabinet of wisdom'),[3] which was traced back as far as the Umayyad Caliph Muʿāwiya ([r. 41–60/661–680]; Eche 1967, 11–12).[4]

DOI: 10.4324/9781315170718-33

Muʿāwiya's care for the preservation of written documents allegedly included, in addition to *ḥadīth*, notebooks (*dafātir*) in – loosely defined – the genre of mirrors of princes (Eche, 1967 12; al-Masʿūdī 1965–1979, 3: 222, #1836).[5] Circumstantial evidence for Umayyad period archivism is provided by the relative wealth of preserved administrative documents – from as early as 22/643 onward. 'Cabinets of wisdom' were also attributed to pre-, or in any case, non-Islamic rulers in legendarily transformed locales, from far-flung Ceylon (Ikhwān al-Ṣafā' 2008 [CB], 4: 324,3, also note the context, 315–27)[6] to Iran (Ḥamza al-Iṣbahānī 2009, 64; Gutas 1998[CB],[7] 54–5), Egypt (Ibn al-Nadīm 2014[CB], 2: 466, 2–4, cf. also 2: 445,10–15 on 'temples' [*barābi'*]),[8] Greece (Ibn Isḥāq 1406/1, 41, 10–12, 48, 3–5, 49, -5-ult., 51, 3–6, penult., 52, 6, 56, 10),[9] and late Visigothic Hispania (al-Maqqarī 1968, 1: 242, 8–9, 243, 4–9).[10] They are featured as repositories of 'wisdom' in various media, more often than not in pictorial or sculptural rather than verbal, instantiation and as venues for the meetings of savants (al-Maqqarī 1968; al-Bīrūnī 1936, 166, 4–9[al-Bīrūnī 1995, 268, 12–269, 2]).[11] In any case, the notion of 'wisdom' which the pertinent texts suggest comprises everything worth knowing, particularly for princes.[12] It is not yet defined along the lines of later disciplinary demarcations.

III.1.2 The first libraries

The transition from a multifarious 'cabinet of wisdom' to a proper library appears to have taken place under the early Abbasid caliphs, up until the reign of al-Ma'mūn (r. 197–218/813–833). While the narrative sources give but a lacunary account of their *bayt al-ḥikma*, instead focusing on the individuals connected with it, it enjoyed quite a remarkable career in 20th-century scholarship by being all too readily identified as *the* venue of 'scientific' activity in the caliphal capital during the 3rd/9th century, as some sort of academy of science *avant la lettre*. Only during the past quarter century have such exaggerated claims been reduced to more realistic proportions (Richter-Bernburg 2016; Gutas 1998, 53–60; Balty-Guesdon 1992).[13] The Abbasid 'cabinet of wisdom' served as a storehouse of chancery documents and scholarly writings useful for the governance of the empire. Apart from the artisanal staff, those holding office in it served the caliphs as littérateurs, translators and 'scientists' – often in personal union. As indicated earlier, the composition of proper 'books' instead of pamphlets picked up speed from the end of the 2nd/early 9th century. The concurrent transformation of the caliphal 'cabinet of wisdom' into, more properly, a library consequently involved some professionalization of bookmaking, as the mention of specialized labor and, if later accounts be trusted, of librarianship suggests (Miskawayh 1952, 21,14–15).[14] As a library above all, al-Ma'mūn's 'cabinet of wisdom' was remembered by posterity; its corresponding designation as 'repository' (*khizāna*) – in variant combinations, such as 'of books' ([*khizānat*] *al-kutub*), 'of books of wisdom' (*kutub al-ḥikma*), 'of wisdom' (*al-ḥikma*) – continued to be used in bibliographic contexts.

If the Abbasid caliphal library was accorded a measure of immortality, it can, from as early as its flourishing under al-Ma'mūn onward, in no way claim uniqueness. Interest in learning and its written materialization was not a caliphal monopoly. In addition to high-placed bureaucrats such as the proverbial Barmakid family and perhaps even preceding them, scholars of means indulged in building up their own collections. The great historian of early Islam Muḥammad ibn 'Umar al-Wāqidī (130–207/747 or 748–822) is said to have kept busy two slaves as copyists 'day and night;' at his death, his personal library amounted to 600 chests, each requiring two men to lift, but already during his lifetime, a sale of books of his had fetched 2,000 *dīnār*s (Leder 2000; Ibn al-Nadīm 2014, 1: 308, 6–8).[15]

During the following, the 3rd/9th, century the production of and trade in books increased immeasurably – in tandem with burgeoning intellectual and artistic creativity (Kennedy 2017,

93–6). The profession of copyist in combination with publishing and marketing of books, or 'stationer' (*warrāq*), firmly established itself (Gruendler 2016, 34–40, 2011, 32; Montgomery 2013, 39–40, 472, no. 32). In addition to the relative ease of producing entire 'editions' by way of reciting and dictating an authorial text (Bloom 2017, 111–12), bookmaking was further facilitated by a fundamental material change, the introduction of paper (Bloom 2017, 113–4, 2001, 42–5).[16] Paper, eventually driving both papyrus and parchment from the market, was to affect all processes, intellectual as well as artisanal and commercial, involved in the writing and marketing of books.

The proliferation of libraries during the centuries to come owed its dynamic to yet another momentous innovation, which took place during the 3rd/9th century. The legal institution of mortmain (*waqf*, pl. *awqāf*), which counted as religiously sanctioned charity, was extended to cover, beyond real estate and its appurtenances, Qurʾān codices and other books, thus encouraging their donation for community use (Eche 1967, 68–74). All over the Islamicate world, mosques, hospitals, madrasas – as of the 5th/11th century – and Sufi convents were the recipients of such bequests and in spite of the ravages of time and of men, which did not spare the legally inviolable, perpetually established *awqāf* either, emerged as prime repositories of the literary heritage (Touati 2003, 54–7; Hirschler 2016[CB], 492a: 'endowed'; Chapter III.2).

Returning to the 3rd/9th century (Touati 2003, 21–30 and *passim*), it was then that beyond the needs of scholars for a reference collection, books began to be appreciated and collected for their aesthetic value as well as for symbolic capital. Bibliophily in the proper sense emerged too. And even before the donation of books as charity became widespread, some outstanding scholars opened up their private collections to all interested comers. The narrative sources usually focus on the extraordinary and momentous, nor can they be trusted for numbers, quantities and magnitudes, such as for library holdings or the sale prices of books. Institutional history does not usually capture the narrators' attention either. The following examples will illustrate these points.

As seen previously, passionate collecting of books by scholars began before the age of paper – and then accelerated. From the 3rd/9th to the early 4th/10th century, when *waqf*-bequests of books were still rare, remarkable acts of munificence toward interested readers are related about some well-placed gentleman littérateur-scholars and book collectors, but at the same time, the fate of their holdings after their deaths or even during their lifetime shows the precarious impermanence of such initiatives. The earliest one of those here to be mentioned, a Baghdadi scholar of inherited wealth (Abū ʿAbd al-Raḥmān al-Naysābūrī [d. 236/850–1]) housed his considerable collection in a separate building and opened it to the public, providing room and board and covering all expenses incurred in copying, such as for paper (Touati 2003, 49–50;[17] al-Azharī 1384/1964, 1: I 24). However, after his death, his heirs sold his books, allegedly for the enormous sum of 400,000 dirham; the figure, a multiple of the proverbially 'uncountable' forty, cannot be given credence. Somewhat later, an erudite member of the caliphal court (ʿAlī ibn Yaḥyā ibn Abī Manṣūr al-Munajjim [200–275/815 or 6–888 or 9]) made his 'scientific' library, which was housed in his sumptuous mansion, accessible to eager scholars for extended periods of study and provided for their every need (Touati 2003, 56). Nothing is known of the library's further destiny. Another few decades on, in Mosul, a supremely well-connected and brilliant local notable (Jaʿfar ibn Muḥammad ibn Ḥamdān [240–324/855–936]) established, now already as a charity, a 'house of knowledge' (*dār al-ʿilm*) with a library encompassing all disciplines; every prospective student or curious transient was welcome to it and if needy, given the necessary wherewithal. He himself held courses there too. However, an intrigue instigated against him in Mosul drove him to Baghdad into exile (under Caliph al-Muʿtaḍid [r. 279–289/892–902]); the fate of his library was left unrecorded (Touati 2003, 56).

III.1.3 Proliferation of book collections

The following, 4th/10th, century was a period of abundant productivity, commercial, artistic and intellectual. As for libraries, it saw a proliferation of elite private along with princely collections, but they, just like their antecedents, fell victim to ideological and political upheavals. The notion of public library took hold too, even though, again, the respective founders' political prominence virtually ensured their destruction during periods of civil unrest and war. Before the modern age, it was religious rather than 'secular' institutions – *madrasa*s, mosques, shrines – whose collections were granted permanence.

Some of the sovereign or court-related libraries founded during the 4th/10th century gained well-nigh proverbial fame for the numerical and qualitative richness of their holdings. The cultural unity of the Islamicate world is mirrored in their geographical distribution, from Hispano-Umayyad Córdoba to Fatimid Cairo to Buyid Shiraz, with Buyid-Abbasid Baghdad and Samanid Bukhara following closely behind. The Umayyad Caliph al-Ḥakam II al-Mustanṣir (r. 350–366/961–976) was known as a great collector of books, generous sponsor of writers from as far afield as Baghdad and respectable scholar himself (Wasserstein 1993). His library was said to number 400,000 volumes, but again, the figure is merely conventional rhetoric (cf. Hirschler 2016; Necipoğlu 2019, 17, 23). Alas, this splendor shone but briefly. First al-Ḥakam's collection was despoiled of allegedly offensive material in a gamble for power, trying to appease narrow-minded 'orthodoxy;' in 403/1013, plundering Berber troops put an end to what remained.

In Fatimid Cairo, the notorious Imam-Caliph al-Ḥākim (r. 386–411/996–1021) founded in 394/1003 – as a charity – an institution of learning, which included, but was not limited to, a seminary to train Fatimid-Ismāʿīlī missionaries; the library of this 'house of knowledge' (*dār al-ʿilm*), or apparently in official usage of 'wisdom' (*al-ḥikma*), did not remain unaffected by the subsequent political and religio-ideological instability of the Fatimid régime and disappeared with the dynasty in 567/1171 (Halm 2003, 206–9, 412–13, 2014, 145). Of lasting fame even long after its end was the imam-caliphs' own library; Ṣalāḥ al-Dīn ([r. 564–589/1169–1193]; 'Saladin'), leading the Sunnite 'reconquest' of Egypt, had it liquidated (Halm 2014, 293–4; Bora 2015, with dubious conclusions).

Further East, the most powerful and most ambitious member of the North Iranian dynasty of the Buyids, ʿAḍud al-Dawla (r. 338–372/949–983) strove to emulate great princes of the past also by establishing a palace library at his residence in Shiraz. It survived somewhat unsettled conditions after its founder's death and still flourished under his eventual successor Bahāʾ al-Dawla (r. 379–403/989–1012). In 375/985, the Arabic geographer Abū ʿAbdallāh Muḥammad al-Muqaddasī (2nd half 4th/10th century) enjoyed the privilege to be shown around the entire palace and given access to the library. He describes it as holding everything until then composed in any given discipline and as carefully catalogued; the volumes were shelved by subject in cabinets along the walls of a long, vaulted hall and its lateral recesses, and only respectable visitors were allowed in. Of 'staff' al-Muqaddasī mentions but three, presumably top echelon, positions, an executive (*wakīl*), a librarian (*khāzin*) and a controller (*mushrif*). ʿAḍud al-Dawla's library may well have been the most lavishly appointed in Buyid dominions, but it was by no means unique, neither within their borders nor without. Notable book lovers include the eminent Buyid administrators Ibn al-ʿAmīd (c. 300–360/*c.* 912–970) and al-Ṣāḥib Ibn ʿAbbād (326–385/938–995). Comparable conditions obtained in the rival Samanid dynasty's (*c.* 205–389/c. 820–999) territories. Princes and the wealthy civil élite alike eagerly collected books, and while they may not have thought of their holdings as public institutions, they did grant access to select individuals.

The Fatimid al-Ḥākim's foundation in Cairo was not the only 'public library' to be established during the 4th/10th century. In Buyid-dominated Baghdad, the vizier Sābūr ibn Ardashīr (336–416/948–1025 or 6) set up in charity a 'house of science' (*dār al-ʿilm*) in 381/991 or 992 for public benefit and donated '10,000 volumes' to it; its operation was entrusted to a librarian and a superintendent. In 451/1059, at a moment of foreign aggression and civil rioting, it was terminally ravaged by a conflagration and concomitant plundering by the mob and an avaricious vizier. As early as during the founder's lifetime, the reversal of his political fortunes had led to spoliation of the library.

The examples hitherto listed are really no more than what the word suggests: a few exemplary cases highlighting practices that were by no means a privilege of society's topmost élite; the literate classes shared both 'bibliophily' and the philanthropic impulse to give society at large access to learning and scholarship. Thus libraries attached to 'public', religiously connoted institutions, including hospitals, steadily grew through the centuries, in the process rivaling princely collections. Nevertheless, palace libraries continued to exist among late medieval and early modern dynasties, as their extant successors in Istanbul, Tehran and a number of former princely capitals in India amply demonstrate (Necipoğlu 2019, 14–17). A new locus of libraries comes into focus from the 7th/13th century onward – the observatory (Chapter II.7) – but may have existed earlier. Small to medium-size private libraries of scholars ranging from about tens to hundreds of volumes are documented in historical sources. Occasionally, they have survived to this day (Chapter VI.6). The storage of texts took different forms – stacked as loose leaves and wrapped up in cloth or placed in boxes or sewn together and bound in plain or sumptuous covers made from paper, cardboard or leather (Chapter IV.1). Maintenance – and replenishment – of libraries was often in the hands of a librarian, or it fell to the administrator or the owner of the respective institution and their families. Their tasks included the repair of damages caused by climate, water or insects and building up the collection through purchases from the book market or commissioning new copies.

Notes

1 The following references are not meant to be exhaustive, especially since many of the cited works in their turn have ample bibliographies. Continuing scholarly debate on the gradual adoption of literacy in 'literary' production is reflected in the just mentioned three titles. The present sketch is no more than a brief introduction to the subject of libraries; Eche 1967 remains indispensable as reference and guide to sources (now to be supplemented by internet resources).

2 Lecker's subject, al-Zuhrī (51–124[?]/671–742), was anecdotically – apocryphally? – reported to have irritated his wife by ensconcing himself among his 'books' (*kutub-h*) (Ibn Khallikān 1971 4: 177–9, no. 563, esp. 177,19–178,2; cf. the identical anecdote about al-Zubayr ibn Bakkār [174–256/790–870]; 2: 311–2, no. 240, esp. 312,10–2).

3 The indeterminacy of *bayt* which by no means only denotes separate structures has here, tentatively, been rendered as 'cabinet.' Similarly indeterminate is *ḥikma*, here, hewing to convention, translated as 'wisdom.' Certainly, in this early period, it did not have the later disciplinary meaning of philosophy and even less of mystical or experiential knowledge. Also see the following discussion and the following two notes.

4 Eche's source, a rebuttal of Bishr al-Marīsī (d. 218/833) by 'Uthmān b. Saʿīd al-Dārimī (d. Dhū l-ḥijja 280/894 or 282/895–96), does not focus on Muʿāwiya's 'so-called cabinet of wisdom' (*bayt yusammā bayta l-ḥikma*) as such but on the reported record keeping of *ḥadīth*s in it; at any rate, the transmitted wording suggests distance from, and possibly disapproval of, the said cabinet's designation (al-Dārimī 1939, 135).

5 *Dafātir*, referring to unbound quires or even entire books, were the common form of book manuscripts through the 4th/10th century at least – regardless of size (e.g., 'weighty notebooks', *dafātir ʿiẓām*, on 'the caliphs' customs and policies', *siyar al-khulafāʾ*; Abū l-Faraj al-Iṣfahānī 2002, 16: 12,5f [under

al-Muʿtamid, 256–279/870–892]). It would seem that *kitāb* frequently denoted the written work, whether letter or book, qua work, whereas *daftar* was the physical object as bearer of such work – but this distinction does not always hold.

6 Here 'house of wisdom' may be a more appropriate translation since reference is to a philosophical school akin to an order and possibly even the timely manifestation of a perennial entity.

7 Consolidated bibliography.

8 Even if in some transmissions *dār*, not *bayt*, *al-ḥikma* figures in the legend of ancient Egyptian temples (*barābi'*) as repositories of wisdom, the variant does not affect the content (cf. al-Nuwayrī 1342–1418/1923–1998[CB], I 394:8–18/I² 380:10–20).

9 The severly contaminated transmission of (Pseudo-?)Ḥunayn's text and the narrative of the 'golden house' in Pseudo-Aristotle and elsewhere have given rise to lively debate (e.g., Zakeri 2008; Overwien [2003]; Gutas 1986, esp. 30–1 with n. 60).

10 Here and further on al-Maqqarī is and will be quoted on the understanding that he merely relays much earlier authors, who will, however, not be traced and identified at present. The Visigothic King Roderic (Rodrigo in Spanish, Ludhrīq in Arabic), who in 711 lost his kingdom and his life to the Arab invaders, is said to have been forewarned of them by paintings in the 'cabinet of wisdom in our city' (*bi-bayt al-ḥikma bi-baladinā*); further according to legend, this establishment of – clearly including occult – wisdom was founded by Greeks – the 'party' renowned for sagacity – at the time of their exile because of Persian occupation, before Alexander the Great.

11 To al-Bīrūnī or to his source, the 'wisdom' of the temple at Akhmīm, which, moreover, is attributed to the Greeks, was most likely represented by the rich relief decoration typical of pharaonic buildings (cf. Richter-Bernburg 2009, esp. 137–40).

12 Thus, the tradition recorded by Ibn Juljul (wrote 377/987) about 'Umar b. 'Abd al-'Azīz (r. 99–101/717–720) finding a medical pandect in the 'repositories of books' might have a factual circumstantial background regardless of doubtful concrete details (in al-Qifṭī 1903[CB], 324, 19–325, 4; the mentioned *khazā'in al-kutub* would appear to have been 'armoires, cases,' rather than rooms). Admittedly, the existence of such 'repositories' under the Marwānids might still be rejected as retrojection along with the early dating of Māsarjawaih, the supposed translator of the pandect in question (al-Qifṭī conscientiously lists the variant traditions; the alternative date, roughly a century later, would appear to be more likely, al-Qifṭī 1903, 325, 7–326, 10).

13 It is to be hoped that present knowledge will, for whatever reasons: 'identitarian' and other ideological agendas or simply lazy ignorance, not fall into oblivion again.

14 In Miskawayh's (c. 320/932–421/1030 [?]) quotation from 'The superiority of understanding' (*Istiṭālat al-fahm*), a most likely pseudepigraphic work under al-Jāḥiẓ's name (c. 160/776–255/868–69), a catalogue of al-Ma'mūn's books figures as an entirely unremarkable accessory. If Miskawayh's account were taken with the utmost scepticism, his attribution of such professionalism to al-Ma'mūn's library would merely be a retrospective projection from a distance of roughly two centuries. Even at that it would have some value, expressing as it does, the notion that a well-organized library could not do without a reliable catalogue.

15 Ibn al-Nadīm's report in *al-Fihrist* is based on an 'antique' manuscript (. . . *bi-khaṭṭ 'atīq*), which suggests near-contemporaneousness with al-Wāqidī.

16 Bloom is not alone in taking a sceptical attitude to the often-quoted account of the introduction of papermaking to Islam. In this tradition (in al-Thaʿālibī [350–429/961–1038] 2003, 436, #892), which does seem to conform to certain well-worn patterns of constructing history, papermaking is said to have been brought to Samarqand by Chinese artisans taken prisoner by the victorious Muslim forces in the battle of Talas (Arabic: Ṭarāz) in July 133/751 (Bosworth 1998, 222–3). Once it is accepted that 'Chinese' in the given context covered every ally and subject of the Chinese regardless of ethnic or linguistic identity, it is entirely plausible to relate the report to Sogdian papermakers. But whether before or after Talas, their craft became established in the Sogdian central place Samarqand and in time gained the city near-proverbial prestige for their product; along with it, the local, Sogdian, term for paper, *kāγaδā*, passed into Persian, Arabic (*kāgha/idh*) and other languages (Norman 2015, 310–14; Porter 1999).

17 Here and elsewhere Touati's brilliantly conceived account is haunted by the proverbial devil lurking in the detail – which is to say references are not always exact, nor do the textual witnesses called on always support the author's conclusions. The present three references to Touati do not obviate consultation of the sources.

Bibliography

Sources

al-Azharī, Abū Manṣūr Muḥammad ibn Aḥmad. eds. Hārūn, ʿAbd al-S. M. and al-Najjār, M. ʿA. 1384–1396/1964–1975. *Tahdhīb al-lugha* [Refining the Lexicon]. 17 vols. Cairo: al-Mu'assasa al-miṣriyya al-ʿāmma li-l-taʾlīf wa-l-tarjama.

al-Bīrūnī, Abū l-Rayḥān Muḥammad ibn Aḥmad. ed. al-Hādī, Y. 1374/1995. *Kitāb al-Jamāhir fī maʿrifat al-jawāhir* [Book of the Choicest Parts on the Knowledge of the Gemstones]. Tehran: Sherkat-e Enteshārāt-e ʿelmī va-farhangī (*Mīrās-e maktūb*; 16: *ʿOlūm-ow fonūn*; 6).

al-Bīrūnī, Abū l-Rayḥān Muḥammad ibn Aḥmad. ed. Krenkow, F. 1355/1936. *Kitāb al-Jamāhir fī maʿrifat al-jawāhir*. Hyderabad: Dāʾirat al-maʿārif al-ʿuthmāniyya.

al-Dārimī, ʿUthmān ibn Saʿīd. ed. al-Fiqī, M. Ḥ. 1358/1939. *Radd al-Imām al-Dārimī ʿUthmān ibn Saʿīd ʿalā Bishr al-Marīsī l-ʿanīd* [Rebuttal by al-Imām ʿUthmān ibn Saʿīd al-Dārimī of the Obdurate Bishr al-Marīsī]. Beirut: Dār al-kutub al-ʿilmiyya.

Ḥamza ibn al-Ḥasan al-Iṣbahānī. ed. al-Ḍubayb, A. b. M. 2009. *Kitāb al-Amthāl al-ṣādira ʿan buyūt al-shiʿr* [Book of Proverbs Originating from Poetic Verses]. Beirut: Dār al-madār al-islāmī.

Ḥunayn ibn Isḥāq [attributed to] and Muḥammad ibn ʿAlī ibn Ibrāhīm ibn Aḥmad al-Anṣārī. ed. Badawī, ʿA. 1406/1985. *[Mukhtaṣar] Ādāb al-falāsifa* [(Abridgement of) The Wise Sayings of the Philosophers]. Kuwayt: Maʿhad al-makhṭūṭāt al-ʿarabiyya.

Ibn Khallikān, Aḥmad ibn Muḥammad ibn Ibrāhīm. ed. ʿAbbās, I. 1968–72. *Wafayāt al-aʿyān* [Eminent Lives]. 8 vols. Beirut: Dār Ṣādir [Reprint: 1414/1994].

al-Iṣfahānī, Abū l-Faraj ʿAlī b. al-Ḥusayn. eds. ʿAbbās, I., al-Saʿāfīn, I., ʿAbbās, B. 1423/2002. *Kitāb al-Aghānī* [Book of Songs]. 25 vols. Beirut: Dār Ṣādir [Reprints: 1426, 1429/2005, 2008].

al-Maqqarī, Aḥmad ibn Muḥammad. ed. ʿAbbās, I. 1388/1968. *Nafḥ al-ṭīb min ghuṣn al-Andalus al-raṭīb* [The Waft of Fragrance from the Verdant Branch of al-Andalus]. 8 vols. Beirut: Dār Ṣādir.

al-Masʿūdī, Abū l-Ḥasan ʿAlī ibn al-Ḥusayn. ed. Pellat, Ch. 1965-79. *Murūj al-dhahab wa-maʿādin al-jawhar* [Meadows of Gold and Mines of Gemstones] (Manshūrāt al-Jāmiʿa al-Lubnāniyya: Qism al-dirāsāt al-taʾrīkhiyya 10-11). 6 vols. in 7 pts. Beirut: al-Jāmiʿa al-Lubnāniyya.

Miskawayh, Abū ʿAlī Aḥmad ibn Muḥammad. ed. Badawī, ʿA. 1952. *Jāwīdānkhirad: Al-ḥikma al-khālida* [Perennial Wisdom]. Cairo: Maktabat al-nahḍa al-miṣriyya.

al-Muqaddasī, Muḥammad ibn Aḥmad ibn Abī Bakr. tr. Collins, B. 1994. *The Best Divisions for Knowledge of the Regions: Aḥsan al-Taqāsīm fī Maʿrifat al-Aqālīm*. Reading: Garnet [Reprint: 2001].

al-Thaʿālibī, ʿAbd al-Malik ibn Muḥammad. ed. Ibrāhīm, M. A. 1424/2003. *Thimār al-qulūb fī l-muḍāf wa-l-mansūb* [Fruits for Hearts on Genitive and Adjectival-relational Attributes]. Ṣaidā and Beirut: al-Maktaba al-ʿaṣriyya.

Yāqūt al-Rūmī. ed. ʿAbbās, I. 1993. *Muʿjam al-udabāʾ* [Dictionary of the Littérateurs]. 7 vols. Beirut: Dār al-gharb al-islāmī.

Research literature

Balty-Guesdon, M.-G. 1992. "Le *Bayt al-ḥikma* de Baghdad," *Arabica* 39: 131–50.

Berkel, M. L. M. van. 2013. "Archives and Chanceries: Pre-1500, in Arabic," in *EI-3*, 2013–2: 24–32.

Bloom, J. [M.] 2001. *Paper before Print: the History and Impact of Paper in the Islamic World*. New Haven and London: Yale University Press.

Bloom, J. [M.] 2017. "How Paper Changed the Literary and Visual Culture of the Islamic Lands," in Blair, Sh. and Bloom, J. [M.], eds. *By the Pen and What they Write – Writing in Islamic Art and Culture* (Biennial Hamad bin Khalifa Symposium on Islamic Art and Culture 6). New Haven and London: Yale University Press, 105–27.

Bora, F. 2015. "Did Ṣalāḥ al-Dīn Destroy the Fatimids' Books? An Historiographical Enquiry," *JRAS, Series* 3 25.1: 21–39.

Bosworth, C. E. 1998. "Ṭarāz," in *EI-2*, 10: 222–3.

Déroche, F. 1987–89. "Les manuscrits arabes datés du IIIe/IXe siècle," *Revue des études islamiques* 55–57: 343–80.

Eche [ʿUshsh], Y. 1967. *Les bibliothèques arabes publiques et semi-publiques en Mésopotamie, en Syrie et en Égypte au Moyen Age*. Damascus: IFÉAD.

Gruendler, B. 1993. *The Development of the Arabic scripts: from the Nabatean Era to the First Islamic Century According to Dated Texts* (Harvard Semitic Series 43). Atlanta, GA: Scholars' Press.

Gruendler, B. 2011. *Book Culture before Print: the Early History of Arabic Media* (The Margaret Weyerhaeuser Jewett Chair of Arabic Occasional Papers). Beirut: American University.

Gruendler, B. 2016. "Aspects of Crafts in the Arabic Book Revolution," in Brentjes, S. and Renn, J., eds. *Globalization of Knowledge in the Post-Antique Mediterranean, 700–1500*. London and New York: Routledge, 31–66.

Gutas, D. 1986. "The Spurious and the Authentic in the Arabic Lives of Aristotle," in Kraye, J., Ryan, W. F. and Schmitt, Ch. B., eds. *Pseudo-Aristotle in the Middle Ages: The Theology and other Texts*. London: The Warburg Institute, 15–36.

Halm, H. 2003. *Die Kalifen von Kairo*. München: Beck.

Halm, H. 2014 [2016²]. *Kalifen und Assassinen*. München: Beck.

Heffening, W. and Pearson, J. D. 1987. "Maktaba," in *EI-2*, 6: 197b–200b.

Imbert, F. 2011. "L'Islam des pierres: l'expression de la foi dans les graffiti arabes des premiers siècles," *Revue des mondes musulmans et de la Méditerranée [REMMM]* 129: 57–78.

Imbert, F. 2012a. "Le Coran des pierres. Graffiti sur les routes de pèlerinage," *Le monde de la Bible* 201: Aux origines du Coran 24–7.

Imbert, F. 2012b. "Réflexions sur les formes de l'écrit à l'Aube de l'Islam," *Proceedings of the Seminar for Arabian Studies* 42: 119–27.

Imbert, F. 2013. "Le Coran des pierres: statistiques et premières analyses," in Azaiez, M. and Mervin, S., eds. *Le Coran: nouvelles approches*. Paris: CNRS Édition, 99–124.

Jones, A. 1998. "The Dotting of a Script and the Dating of an Era: The Strange Neglect of PERF 558," *Islamic Culture* 72.4: 95–103.

Kennedy, H. 2017. "Baghdad as a Center of Learning and Book Production," in Blair, Sh. and Bloom, J. [M.], eds. *By the Pen and What They Write – Writing in Islamic Art and Culture* (Biennial Hamad bin Khalifa Symposium on Islamic Art and Culture 6). New Haven and London: Yale University Press, 89–103.

Lecker, M. 2007. "al-Zuhrī," in *EI-2*, 11: 565a–b.

Leder, S. 2000. "Al-Wāḳidī," in *EI-2*, 11: 101b–103a.

Lewis, B. 1954. "Daftar," in *EI-2*, 2: 77b–81b.

Meier, A. 2012. "Archives and Chanceries: Arab World," in *EI-3*, 2012–4: 17–22.

Montgomery, J. E. 2013. *Al-Jāḥiẓ: In Praise of Books*. Edinburgh: Edinburgh University Press.

Munt, H. 2017. "What Did Conversion to Islam Mean in Seventh-century Arabia?" in Peacock, A., ed. *Islamisation: Comparative Perspectives from History*. Edinburgh: Edinburgh University Press, 83–101.

Necipoğlu, G. 2019. "The Spatial Organization of Knowledge in the Ottoman Palace Library: An Encyclopedic Collection and its Inventory," in Necipoğlu, G., Kafadar, C. and Fleischer, C., eds. *Treasures of Knowledge – An Inventory of the Ottoman Palace Library* (1502/3–1503/4). 2 vols (Studies and Sources in Islamic Art and Architecture; XIV.1–2). Leiden and Boston: Brill, 1: 1–78.

Norman, J.†, Mei, T. and South Coblin, W., eds. 2015. "Inner Asian Words for Paper and Silk," *Journal of the American Oriental Society* 135: 309–17.

Overwien, O. 2003. "Ḥunayn b. Isḥāq, *Ādāb al-falāsifa*: Griechische Inhalte in einer arabischen Spruchsammlung," in Piccione, R. M. and Perkams, M., eds. *Selecta colligere*, 1.: Akten des Kolloquiums "Sammeln, Neuordnen, Neues Schaffen: Methoden der Überlieferung von Texten in der Spätantike und in Byzanz (Jena 21.–23. November 2002)," *Alessandria: Edizioni dell'Orso, Hellenica* 11: 95–115.

Porter, Y. 1999. "Notes sur la fabrication du papier dans le monde iranien médiéval (VIIIᵉ – XVIᵉ siècle)," in Zerdoun Bat-Yehouda, M., ed. *Le papier au Moyen Âge: histoire et techniques – Actes . . .* 1998 (Bibliologia 19). Turnhout: Brepols, 19–30.

Richter-Bernburg, L. 2009. "Between Marvel and Trial: al-Harawī and Ibn Ǧubayr on Architecture," in Vermeulen, U. and D'hulster, K., eds. *Egypt and Syria in the Fatimid, Ayyubid and Mamluk eras* VI (Analecta Orientalia Lovaniensia 183). Leuven, Paris and Dudley, MA: Peeters, 111–41.

Richter-Bernburg, L. 2016. "Potemkin in Baghdad: The Abbasid 'House of Wisdom' as Constructed by *1001 Inventions'* Companion Book," in Brentjes, S., Edis, T. and Richter-Bernburg, L., eds. *1001 Distortions: How (Not) to Narrate History of Science, Medicine and Technology in Non-Western Cultures* (Bibliotheca Academica, Reihe Orientalistik 25). Würzburg: Ergon, 121–31.

Schoeler, G. 2010. "The Relationship of Literacy and Memory in the Second/Eighth Century," *Proceedings of the Seminar for Arabian Studies* 40, Supplement: The Development of Arabic as a Written Language. Papers from the Special Session of the Seminar for Arabian Studies held on 24 July, 2009, 121–9.

Schoeler, G. and Toorawa, Sh. M. 2009. *The Genesis of Literature in Islam: From the Aural to the Read*. Edinburgh: Edinburgh University Press (revised English edition of Schoeler, G. 2002. *Écrire et transmettre dans les débuts de l'islam*. Paris: Presses universitaires de France).

Touati, H. 2003. *L'armoire à sagesse: bibliothèques et collections en Islam*. Paris: Flammarion and Aubier.

Touati, H. 2013. "Book," in *EI-3*, 2013–2: 66–73.

Wasserstein, D. J. 1990–1991 [1993]. "The Library of al-Ḥakam II al-Mustanṣir and the Culture of Islamic Spain," *Manuscripts of the Middle East* 5: 99–105.

Zakeri, M. 2008. "Before Aristotle Became Aristotle: Pseudo-Aristotelian Aphorisms in *Ādāb al-falāsifa*," in Akasoy, A. and Raven, W., eds. *Islamic Thought in the Middle Ages: Studies in Text, Transmission and Translation, in Honour of Hans Daiber*. Leiden: Brill, 649–96.

III.2
MADRASAS AND THE SCIENCES

Sonja Brentjes and Abdelmalek Bouzari

Madrasa is an Arabic noun connected to different forms of a verb that means, among other things, to read, to read repeatedly, to make someone read (repeatedly in order to remember), to teach someone to read, to lecture and hence to instruct, and to teach (*darrasa*; Lane 1968, 3: 870–1). Although the use of buildings called *madrasa* in historical sources ranged from places of reading or learning and prayer to prisons and offices for mercantile use, the word *madrasa* is often translated as school or college and thought of as a building or a social institution. It is usually regarded as the main sociocultural space where Muslims taught the doctrines and argumentative, demonstrative and arithmetical practices of their various legal schools that came into being and sometimes died out over time.

Opinions on the origin of the term and of its application to spaces of learning and teaching vary (Bulliet 1972, 47; Günther 2017). According to Bulliet, the word was used for a place of this sort already in the late 3rd/9th century by a jurist who apparently lived in a rural environment close to Nishapur (today in Iran; Bulliet 1972, 249). Other *madrasa*s were founded during the 4th/10th century in Nishapur, mostly by men who taught law according to different Sunni law schools but, in some instances, also by teachers of Sufi doctrines or of *ḥadīth* (sayings and deeds of the Prophet; Bulliet 1972, 249–55). Other early instances of *madrasa*s are known from Bukhara (today in Uzbekistan; Günther 2017).

A Muslim community called *Karrāmiyya*, which had its intellectual centers in Nishapur and Herat, but spread through merchants and migrants as far as Egypt, organized itself in *khānaqāh*s, which are buildings around a courtyard with numerous small rooms for male members of the movement. In other contexts, *khānaqāh*s also served to accommodate students of law and other fields of knowledge, as well as merchants and visitors from abroad. In the case of the *Karrāmiyya*, these accommodation complexes were apparently also called '*madrasa*' and may have been locations where the founder of the movement held sermons and perhaps other kinds of lectures (Zysow 2012).

At the latest during the 4th/10th and 5th/11th centuries, buildings called *madrasa* also arose in other cities such as Bukhara or Baghdad. They were usually identified as such by a scholar and teacher during his lifetime or as part of his legacy or built by a patron for a particularly cherished scholar. They often provided accommodation and stipends for a few students and were maintained thanks to an endowment (*waqf*; Günther 2017). A major boost in the establishment of *madrasa*s is linked to the Persian grand vizier Niẓām al-Mulk (409–485/1018–1092) of the

DOI: 10.4324/9781315170718-34

recently established Oghuz Turkish dynasty known as the Great Seljuqs (r. 429–552/1037–1157). He founded schools in various cities of the Abbasid Caliphate for the Shāfiʿī legal rite, among them Baghdad, the capital of the caliphate, Mosul north of the capital, Isfahan (today in southwestern Iran), Nishapur, Balkh and Herat (the last two today in Afghanistan). All those establishments were physical buildings that offered spaces for students and teachers to live or teach and learn there supported by financial allotments and donations in kind such as oil, bread, water and clothing. Niẓām al-Mulk also provided for service personnel like porters, water carriers, cleaners, librarians or readers of the Qurʾān from his vast landed properties. The schools he founded were named after him *Niẓāmiyya madrasa*s (Makdisi 1981). Hiring leading religious scholars as teachers (*mudarrisūn*), the *Niẓāmiyya madrasa*s quickly acquired fame and set a pattern to be emulated by rulers, their wives and daughters, courtiers and military leaders, as well as individual scholars and their family members. Their legal status followed stipulations of Islamic law. The sponsor of such a teaching establishment had to sign a legal donation (*waqf*) in which they specified the various properties to be used for housing the teacher, the students and the service personnel or for generating the funds needed to pay them, as well as the maintaining of the school and its courtyard.

The size and provisions for *madrasa*s varied considerably, as did their public recognition, administrative organization, teaching profiles and existence over time. This historical information is found in administrative documents, narrative sources like chronicles or biographical dictionaries, colophons at the end of texts or inscriptions on the exterior walls of the building. The impact that an appointment at a well-endowed and highly regarded *madrasa* could have on the career and financial status of a scholar induced many of them to compete for access to the best ones and to amass several such appointments. Scholars with more than one teaching position were often unable and sometimes also unwilling to fulfill all the obligations coming with them. They appointed deputies from among their disciples or advanced students to teach at the lower-ranking schools. They often gave their daughters in marriage to these deputies or arranged marriages of their sons with daughters of other influential scholars to build a stable social network in defense of their status and access to profitable positions. Intense and severe fights over teaching positions at *madrasa*s are particularly well known for major Syrian and Egyptian cities under the rule of the Ayyubids (r. 564–658/1169–1260) and the Mamluks (r. 648–923/1250–1517; Chamberlain 1995).

The later careers of *madrasa* students were apparently very varied. They included teaching at *madrasa*s and elementary schools, and serving as preachers, timekeepers, muezzins, legal witnesses, physicians and merchants. Students who passed through Mamluk *madrasa*s during the 9th/15th century seldom pursued careers corresponding to their fields of study (Berkey 1992), but this was different in the Ottoman Empire from about the early 10th/16th century on, because the state tried to regulate the *madrasa* system by legal, administrative and financial means (Imber 2009; Beyazıt 2012–2013; al-Muḥibbī 1966).

The spread of *madrasa*s as an option for young men wishing to acquire knowledge instead of, or in addition to, private tutoring did not follow the same path all over the Islamicate world. In North Africa, west of Egypt, the first *madrasa*s were only founded in the 7th/13th century. It took another century before such schools seem to have come into being in al-Andalus, the territories of the Iberian Peninsula ruled by Muslim dynasties. Abdelmalek Bouzari surveys current knowledge about the presence of the mathematical sciences in these North African, mostly royal institutions in Sections III.2.3 and III.2.4. In Sections III.2.1 and III.2.2., Sonja Brentjes discusses the introduction of the mathematical sciences, medicine, logic, philosophy and the predictive (or occult) disciplines into *madrasa*s in Iraq, Syria, Egypt and Iran and their subsequent presence at *madrasa*s in Central Asia and Anatolia. Brentjes also considers mosques as sites of teaching, since the boundaries between *madrasa*s and mosques were often fluid. In contrast, other types of teaching and learning institutions are excluded, such as the Houses for Reading the Qurʾān or *ḥadīth*, Sufi lodges (*khānaqāh*), *zāwiya*s (originally corners, mostly in mosques, later also lodges) or *ḥawża*s, either because they did not teach the fields of knowledge just listed or because we are not yet well informed about such activities.

III.2.1 The introduction of scientific disciplines to *madrasa*s and mosques

In 519/1125 and 553/1158 at the *Niẓāmiyya madrasa*s in Baghdad and Mosul, some anonymous and some named scribes copied ʿAbd al-Raḥmān al-Ṣūfī's (291–376/903–986) *Book on the Images of the Fixed Stars* (*Kitāb ṣuwar al-kawākib al-thābita*; usually rendered as *Book on the Constellations* or abbreviated as *Star Catalog*) and Theodosius of Bithynia's (late 2nd century BCE) *Spherics* (*Sphairika*; Theodosius 2010, 4; Savage-Smith 2013[CB],[1] 135–6). Three years later, further mathematical texts were copied in the same *madrasa* in Baghdad. These are the earliest known references to astronomical and geometrical texts produced within the confines of a *madrasa*. For this early time, there is only rare proof that copying such manuscripts in *Niẓāmiyya madrasa*s was undertaken by specialists in the mathematical sciences. A brief extract from a once larger text on arithmetic is the only witness to the presence of a mathematical expert at the *Niẓāmiyya madrasa* in Baghdad known so far. Its colophon states that in Ṣafar 590/February 1194 the calculator Muḥammad al-Baghdādī copied the first part of a work by the Andalusian scholar al-Ḥaṣṣār (alive in 552/1157) called *The Explanation* (*al-Bayān*) at the *Niẓāmiyya madrasa* in Baghdad (MS Philadelphia, University of Pennsylvania, lsj 293, fol. 87a).

A second copy of Theodosius's *Spherics* was the work of Muḥammad ibn Abī Bakr, a scribe who also copied Thābit ibn Qurra's textbook on compound ratios in the summer of 625/1228. Muḥammad also copied Arabic translations of ancient Greek geometrical texts such as Theodosius's treatises on *The Habitations* and *On Days and Nights*, Euclid's *Phenomena* and Autolycus's *The Risings and Settings* (Theodosius 2010, 3; Thābit ibn Qurra 2018[CB], 22). But early copying activities in *Niẓāmiyya madrasa*s were not confined to texts in the mathematical sciences. Dioscorides' (d. *c.* 90) pharmaceutical book called *On Medicinal Substances* (*De materia medica*), for instance, was copied in one of them in 637/1240 (Savage-Smith 2011, 43).

These and other manuscripts copied at early *madrasa*s, covering medicine, natural philosophy, logic and at least some of the occult sciences, suggest that teaching texts in such fields of knowledge spread through *madrasa*s and mosques in major urban centers in the Middle East during the later 6th/12th and throughout the 7th/13th centuries. From the 7th/13th century onward, in larger cities, as well as in smaller provincial towns, *madrasa* teachers taught other than religious disciplines, if they wanted to and had students interested in reading texts on such topics. This was particularly true for dynasties that made family members governors in important regional towns and in border fortresses. An important example is the Ayyubid dynasty, which ruled over Egypt, Syria, Yemen, the western Arabian Peninsula and northern Iraq. Although, in the first half of the 7th/13th century, the main intellectual debates continued to take place at courts in Cairo, Damascus, Hama, Aleppo or Karak, scholars at *madrasa*s are known to have taught medicine, algebra, arithmetic, geometry, logic and philosophy.

An example of such a *madrasa* teacher is Ibn Fallūs (590–636 or 649/1194–1239 or 1252). He was a scholar of Ḥanafī law, who also taught medicine and the mathematical sciences in Damascus and Cairo. Several short teaching texts of his have survived. They indicate that he taught mathematical knowledge in the form of rules to be learned by heart and applied to standard word problems, some types of which are very similar to word problems already taught in ancient Mesopotamia. In arithmetic he taught the Indian positional decimal system as it had evolved in Arabic presentations since the early 3rd/9th century, as well as oral rules of calculation. In algebra, he taught calculation with higher positive powers and with what we call today quadratic and cubic irrational numbers (Brentjes 1993, 6–7; Chapter I.7). In one of his treatises, Ibn Fallūs makes clear that he also knew ʿUmar al-Khayyām's (439–*c.* 517/1048–*c.* 1123) and Sharaf al-Dīn al-Ṭūsī's (d. *c.* 606/1210) treatments of cubic equations (Brentjes 1993, 8). In his text on number theory, he presents a survey on Euclidean and Nicomachean themes, among them perfect and amicable numbers, and gives rules for the first sequences of the binomial triangle (Brentjes 1988, 1990).

This range of topics and methods highlights that teaching mathematical skills at Ayyubid *madrasas* was not limited to practical knowledge needed for the determination of inheritance shares as taught by the different law schools. Students who took such classes were exposed to a much broader range of knowledge, including theoretical components of geometry and number theory and quite recent developments in algebra. In the following centuries, spherical geometry and trigonometry as well as the construction of astronomical instruments became further standard parts of mathematical education offered at various *madrasas* from North Africa to India, even though the institutional organization of this knowledge and its specific content could differ substantively (King 1988, 1990; Brentjes 2008a, 2008b, 2014, 2018[CB]).

A century earlier than the mathematical sciences, philosophy, in particular as codified by Ibn Sīnā (d. 428/1037), entered *madrasas*, at least as a manuscript shelved in a library or a collection of texts studied by interested individuals. Not surprisingly, this happened in the *Niẓāmiyya madrasa* of Nishapur not too far from Ibn Sīnā's hometown Bukhara (Brentjes 2018, 70).

In 626/1229, after the Ayyubid prince al-Ashraf Mūsā (r. 607–635/1210–1237, in different towns of Syria and Iraq) had taken Damascus away from his nephew Naṣīr al-Dīn Dāwūd (r. in Damascus 624–626/1227–1228 and in Karak 626–646/1228–1248), he allegedly sent a messenger through the city forbidding any other teaching at the city's *madrasas* than that of specific religious disciplines (law, exegesis). This idea that a ruler of Damascus (or any other city or town east of North Africa) could enforce a unified curriculum on all schools in a town, including many, perhaps a majority, founded by private donors, is very implausible, although it was taken at face value by modern historians for more than a century.

The scholar who is believed to have suffered direct repercussions of this alleged decree was Sayf al-Dīn al-Āmidī (551–631/1156–1233). Chroniclers of the 7th/13th and later centuries report widely different stories about the reasons for his dismissal from his position as the teacher at the *ʿAzīziyya madrasa*. One of them names his teaching of logic and philosophy, the other his intention to desert the Ayyubid rulers of Egypt and Damascus in favor of the ruler of his home town Āmida (today in Turkey and called Diyarbakır; Brentjes 2017, 155–64). Al-Āmidī was an important scholar of *kalām* (rational theology; Chapter III.3) and theoretical jurisprudence (*uṣūl al-fiqh*), using the methods of the latter in the former. He also was an important philosopher and logician, who wrote substantial commentaries on Ibn Sīnā's philosophical encyclopedia *Pointers and Reminders* (*Ishārāt wa-l-tanbihāt*) and Fakhr al-Dīn al-Rāzī's (d. 606/1210) critical commentary on Ibn Sīnā's text, as well as two extensive summaries of logic and philosophy (MS Princeton, Firestone Library, Garrett Collection, 42B; al-Āmidī 2001; Chapters III.3, III.4 and V.6).

Al-Āmidī's logical writings were read during his lifetime, but eclipsed by other scholars of the 7th/13th century, who shaped the lively discussions in this field of knowledge for several centuries (el-Rouayheb 2010, 49–51). His summary of philosophy (four out of five volumes exist) follows Ibn Sīnā's philosophical teachings structurally and according to Lammer also in spirit and content (Lammer 2017, 434–5). Al-Āmidī may have written this work in his early career in Cairo and perhaps taught it there. A reference to teaching can be found in the part of that work where he talks about his view that the world is not eternal but has an origin in time. In this quote, al-Āmidī praises Ibn Sīnā as "the greatest philosopher" and describes some of Ibn Sīnā's philosophical ideas about God as being "appropriate for the teaching of reasoning" (Lammer 2017, 437).

Thus, al-Āmidī was well prepared for teaching logic and philosophy at the *ʿAzīziyya madrasa* in Damascus and for the philosophical conflicts that awaited him at the Ayyubid court in the 620s/1220s with students of Fakhr al-Dīn al-Rāzī, which may have contributed to his later fate. Most likely, then, al-Āmidī was dismissed for reasons other than his teaching of philosophy. In general, however, debating philosophical issues in different intellectual contexts and writing summaries and other kinds of textbooks on logic and philosophy were a major feature of the intellectual life in Syria, Iran, Central Asia and apparently also in northern India during the 7th/13th century.

III.2.2 Scientific disciplines at madrasas and mosques after the Ayyubids

The replacement of the Ayyubid dynasty by members of their slave guard, the Mamluks, in 648/1250 in Egypt and in 658/1260 in Syria was caused by internal difficulties of the Ayyubid family, as well as by the way they had exercised their political and military power in their territories. Another major factor was the destruction wrought by the advancing Mongol armies, who devasted Aleppo, Damascus and other parts of Ayyubid Syria, only to be stopped in 658/1260, early in the newly established Mamluk reign. This shift in political power, ethnic identity, geographical origin and the structure of rulership changed the relationship of the rulers to the local intellectual, landowning and commercial elites. This in turn had a positive impact on the spreading of *madrasas*, mosques, hospitals and other religious and educational institutions. Sultans, their wives, sons and daughters, high-ranking officers and descendants of the Mamluk military leadership donated funds for new *madrasas* and mosques. One Mamluk sultan even appointed a teacher to the only chair for a mathematical science, the science of timekeeping, ever installed at a *madrasa*-mosque (Berkey 1992, 69).

Similar trends, although less well studied, enabled the spread of the sciences at *madrasas* in Iran, Central Asia and Anatolia. In 625/1227, the penultimate Abbasid caliph founded a big *madrasa* in Baghdad, the *Mustanṣiriyya*, which opened its doors about six years later. It included a chair for medicine and regular office hours for treating the sick (Conermann 2004, 61). Its name lives on in the Mustanṣiriyya University reestablished in Baghdad in 1963. The lack of specialized chairs for the various sciences did not impede the teaching of texts belonging to them. Their specialists were appointed as teachers of law, the standard type of teaching chair at a *madrasa* or a mosque. But no institutions existed before the 13th/19th century in most Islamicate societies that regulated and controlled the appointment of teaching and the content of classes, although the Ottoman and Mughal Empires were exceptions.

Hence donors, teachers, students, copyists, book traders and their networks had a great degree of leeway to decide whom to appoint, support or recommend and what to teach, read, copy and trade. Traveling students and teachers, as well as people who migrated over hundreds and even thousands of kilometers, forced by wars, shifts in patronage and other obstructions of teaching opportunities in their hometowns, carried books and ideas with them that enriched the teaching and learning climate and opportunities in the cities they passed through or decided to settle in. The religious obligation of pilgrimage offered a further incentive to travel and to spread as well as acquire knowledge (Brentjes 2018, 135–44). In the sciences, students of famous scholars in Iran, Central Asia and northern South Asia came in several waves to Syria, Egypt and Anatolia bringing mathematical, philosophical and medical treatises and long books with them that shaped the intellectual lives in those areas and to a lesser degree also those in North Africa and al-Andalus. Classical textbooks for *madrasa* teaching were composed during the 7th/13th centuries in philosophy and elementary astronomy in Central Asia, Iran and Iraq. Ancient textbooks of the mathematical sciences, translated during the 3rd/9th century in Baghdad, were revised and newly edited on the basis of older versions at Ayyubid courts in Syria and two courts in western Iran, the Ismāʿīlī court in Alamut and the Ilkhanid court in Maragha. The broadest and most influential set of revised ancient mathematical textbooks was produced by the Shīʿī scholar Naṣīr al-Dīn al-Ṭūsī (597–672/1201–1274). Hundreds of copies were produced in classroom work at *madrasas* between Morocco, India and Central Asia through the early 15th/20th century (Heidarzadeh 2019, 144–5).

Several major processes took place in the context of teaching the sciences at *madrasas* after the 7th/13th century. The textbooks written or revised during the previous century became, as already indicated, the core teaching material supplemented by a number of texts written directly in the context of *madrasa* teaching. Famous *madrasa* teachers such as Ibn al-Majdī (767–850/1366–1447) in Cairo, Ibn al-Hāʾim (753-d. after 815/1352-d. after 1412) in Jerusalem or al-Qūshjī (805–879/1403–1474) in Istanbul, to name only a few, either wrote texts directly addressing their students or used texts dedicated to princely patrons when teaching at *madrasas*. Unnamed compilers and teachers produced textual collections of ancient texts on geometry such as Euclid's *Elements* in Arabic and the commentaries written by scholars between the 3rd/9th and 8th/14th centuries in Baghdad, Maragha and perhaps Rayy. They taught these collections and commentaries, comparing them with several Arabic translations and editions of the *Elements* following basic epistemological rules for teaching in philosophy, *kalām* and the theoretical foundations of law and faith. This indicates that new developments shaping critical writing about other scholars' works, that is new approaches to commentary writing, found their place even in the mathematical sciences (Brentjes 2019; Wisnovsky 2013).

In philosophy and logic, these new rules governing how to dispute and which methods of analysis and refutation to apply were first formulated in Khurasan and Central Asia in the late 6th/12th and throughout the 7th/13th centuries. They spread during the following two centuries to India, Syria, Egypt, Anatolia and, with some delay, at least for logic, North Africa. In this period, changes in the literary organization of logical and philosophical knowledge took place in close connection with *madrasa* teaching. Instead of summaries and textbooks, commentaries now became the main form of expression (el-Rouayheb 2010). Sets of philosophical texts taught at *madrasas* were time and again organized in single manuscripts as portable libraries (Endreß 2001[CB]; Endress 2006[CB]). Metaphysics had place of pride. But writings on natural philosophy were also included, as were treatises on cosmological and more technical astronomical topics.

Teaching medicine, pharmacology and knowledge about plants moved from hospitals to *madrasas*, whether or not they were affiliated to a hospital. Family teaching in those fields remained a second major form of training for future druggists or physicians (Brentjes 2018, 131–5). *Madrasa* teaching of medical and related knowledge led to a trend in which such knowledge often became limited to textual studies; practical experience with patients was lost. Historical sources time and again report men well educated in medical books, who abstained from treating the sick (Brentjes 2002, 59; Brentjes 2018, 95–7).

Commentary writing also became a dominant literary form in medicine, although summaries continued to be composed apparently until at least the late 10th/16th century. Ibn Sīnā's *Canon of Medicine* (*al-Qānūn fī l-ṭibb*) remained a standard source for medical classes at *madrasas* as did several treatises by Hippocrates (d. *c.* 370 BCE) and Galen (129–216) together with the commentaries on them by Christian, Jewish and Muslim authors. A more recent author, Ibn al-Nafīs (d. 687/1288) became a widely taught medical authority (Fancy 2013[CB]). Regional differentiation went hand in hand with the development of new scientific languages in Islamicate societies. This applied in particular to Persian and Ottoman Turkish. With the spread of Muslim dynasties to new regions in South and Southeast Asia, medical texts were also composed in Urdu and other languages (Chapters II.1, II.2, V.7–V.8).

For all fields of knowledge already mentioned, two major methods of teaching emerged at the latest in the 8th/14th century. One consisted in repeated studies of a single text with several different teachers either in different cities or towns or in one major urban center. The other was expressed in the reduction of fully studied texts in favor of classes that taught only a few chapters of one or more texts. These developments reflect the methods applied in the religious disciplines called the disciplines of transmission, and hence the increasing importance of teacher chains as providers of scholarly authority and reputation in addition to the religious disciplines also in the sciences (Figure III.2.1). Other teaching and learning practices in the sciences were learning texts by heart (again reflecting methods used in the religious sciences), question–answer exchanges and visual presentations of problems and methods (used above all in the mathematical sciences; Brentjes 2018, 147–85).

Changes also occurred in the occult disciplines. More teachers taught texts on the magic properties of letters and numbers, the discipline of sand divination called *raml* (geomancy) and alchemy (Chapters I.12 and II.9). Already in the late 6th/12th and 7th/13th centuries *raml* was an important predictive practice employed before major battles, for instance, by Ayyubid princes. In the subsequent centuries it spread across major parts of the Islamicate world and was practiced before marriage ceremonies, births and battles and perhaps even before commencing

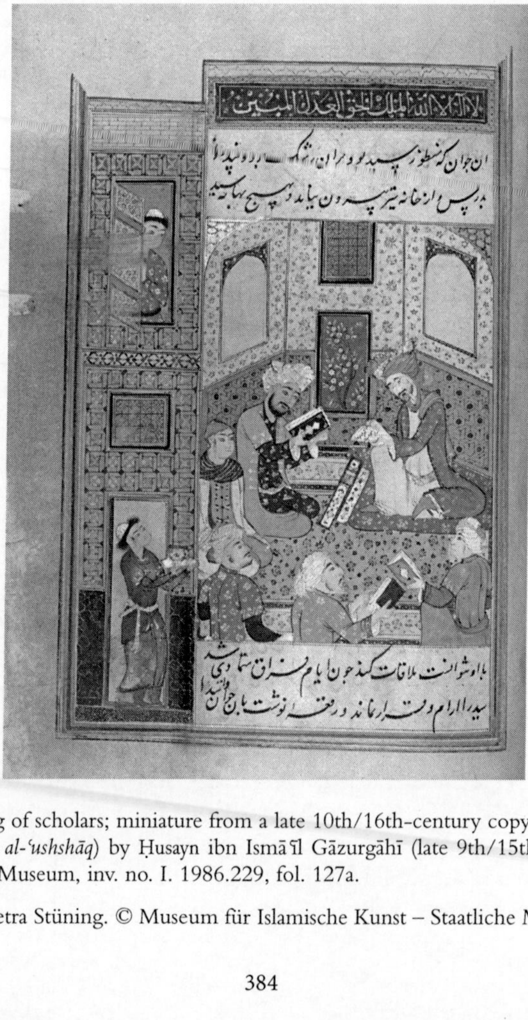

Figure III.2.1 Meeting of scholars; miniature from a late 10th/16th-century copy of *Sessions of the Lovers* (*Majālis al-ʿushshāq*) by Ḥusayn ibn Ismāʿīl Gāzurgāhī (late 9th/15th century); MS Berlin, Islamic Museum, inv. no. I. 1986.229, fol. 127a.

Source: Photographer Petra Stüning. © Museum für Islamische Kunst – Staatliche Museen zu Berlin

to build secular and religious buildings (although we have little evidence for such an application). Numerous Timurid, Mamluk, Safavid, Mughal and Ottoman *madrasa* scholars wrote survey or introductory texts on *raml* (Melvin-Koushki 2016). The rise of letter magic combined Ibn al-ʿArabī's (558–638/1165–1240) Sufi doctrines with a long tradition of mathematical studies of magic squares. It also incorporated extensive writings by Maghribī and Timurid authors on this field of knowledge, which they presented as a new religious philosophy that would enable universal rule (Melvin-Koushki 2014, 2018; Gardiner 2017a, 2017b).

Modern historians of science and philosophy have described the sociocultural character of these changes as a shift from the physician-philosopher to the jurist-physician and the jurist-philosopher, as well as the *mutakallim*-philosopher (Sabra 1987[CB], 19, 24; Behrens-Abouseif 1989; Endress 2006, 127–8; Brentjes 2018, 206). As so often happens, such labels cannot embrace all the nuances and complexities of shifts in intellectual practices. Indeed, the loss of philosophical knowledge as suggested by the shift from the physician-philosopher to the jurist-philosopher cannot be confirmed for those medical writers after the 6th/12th century whose biographies and works have been analyzed in greater depth (Fancy 2013, 2018; Chapter III.7). Moreover, some of the most prominent teachers of medical and mathematical texts among the *madrasa* scholars did not teach applied law. Their main fields of teaching were the theoretical foundations of faith and law (Brentjes 2018, 97, 165).

III.2.3 *Madrasas* in the Maghrib

The Greater Maghrib (North Africa west of Egypt), became part of the Islamicate world during the period of conquest from 11/632 to 133/751. It is divided into three zones: the Far Maghrib (today Morocco), the central Maghrib (today Algeria and the northern part of Mauritania) and the eastern Maghrib or Ifrīqiya (today Tunisia and Libya; Talbi 1971, 1073–6). In this section, I survey exclusively *madrasas* built in major urban centers. I primarily focus on the mathematical sciences, which included directly or as branches (Chapter VI.5) the astral sciences and the science of inheritance calculations (Chapters I.6 and I.7), although sometimes other disciplines will also be mentioned.

All mathematical activities in the Maghrib were essentially expressed in Arabic. The scholars I consider either came from one of the three parts of the Maghrib or had a clear impact on the mathematical practices in those three regions. I focus on the period after the Almohades (r. *c.* 524–667/*c.* 1130–1269), also called the post-imperial period. Three dynasties, all Berbers like their predecessors, ruled in the three parts of the Greater Maghrib in that time: the Merinids (r. *c.* 591–955/*c.* 1195–1548) in the Far Maghrib, the Zayyanids or ʿAbd al-Wadids (r. *c.* 633–796/*c.* 1235–1393) in the central Maghrib and the Hafsids (r. *c.* 625–982*c.* 1228–1574) in the eastern Maghrib. Two themes will be pursued in the following: a discussion of the most important *madrasas* in the Greater Maghrib and the presentation of the works of some of the main scholars of the mathematical sciences who worked in the great urban centers such as Qayrawan, Tunis, Constantine, Bejaïa, Tlemcen, Fez and Marrakesh.

Similarly to eastern Islamicate societies, the first spaces of religious and general education in the cities of the Maghrib were the elementary schools (*kuttāb* or *m'sid*). The second, vernacular word seems to be derived from *masjid* (mosque). In rural regions those schools were more often called *ribāṭ* (a kind of monastic compound). They are first known to have existed in the last quarter of the 3rd/9th century in Qayrawan. They taught the Qur'ān and the beginnings of Muslim religious duties. In the *ribāṭ* religious devotees, ascetics and later Sufis, as well as occasionally scholars of the mathematical sciences, taught basic religious knowledge and Arabic as a preparation for later studies at mosques in the cities. Mosques were the place of learning for adults. Ibn Tamīmī (d.

332/944), a writer of biographies of scholars from Qayrawan, reports that ʿIkrima (d. 105/723), a successor of the companions of the Prophet, taught there the first classes on *ḥadīth* and exegesis (Lamrabet 2014, 123). Following the example of the Abbasids (r. 132–656/750–1258), the erudite ruler of Ifrīqiya, Ibrāhīm ibn Aḥmad ibn al-Aghlab (r. *c.* 260–288/*c.* 874–901), opened a *House of Wisdom* (*bayt al-ḥikma*) in the newly built city of *Raqqāda*. He may have been inspired by the Abbasid institution of the same name (Chapters I.3 and III.1), installing in it astronomical instruments and a library described as rich in scientific works. Some Maghribi biographers attribute the beginnings of scientific activities in that area to the 3rd/9th and 4th/10th centuries (Brunshvig 1931, 267). Three scholars of that period are Mūsā ibn Yāsīn (3rd/9th century), Dūnash ibn Tamīm ([d. *c.* 349/960]; Chapter VI.1) and al-Ḥufrī ([d. 237/852]; Lamrabet 2014, 128–9; Abū l-ʿArab 1920, 174). Mūsā ibn Yāsīn and al-Ḥufrī are remembered as authors of works on inheritance calculations and other mathematical topics. Ibn Tamīm, a Jewish scholar, wrote on medicine, astronomy, arithmetic and philosophy, including the construction of the universe, the armillary sphere (Chapter IV.2) and the Indian decimal positional system. But information about teaching from the 3rd/9th to the 6th/12th centuries is rare and fragmented.

The first madrasas were founded in the 7th/13th century in several cities across the Greater Maghrib: Qayrawan and Tunis (eastern Maghrib), Tlemcen, Bejaïa and Constantine (central Maghrib) and Fez, Ceuta and Marrakesh (Far Maghrib; al-Manūnī 1985, 77–105). They all responded to two aims of the new powers: distributing the message and doctrines of Ibn Tūmart (d. 524/1130), the Mahdī (the Messias), and training an administrative and financial elite for the dynasties. Scholars of the mathematical sciences participated in the new institution, because one part of the training of judges consisted of the science of inheritance calculations, for which knowledge of arithmetic and algebra was needed. Although the mosques remained open to all social groups, the *madrasas* apparently were destined for students of modest means, because the well-off often had their own tutors. This feature drew various criticisms by jurists as well as teachers of the mathematical sciences.

We possess no descriptions of the pedagogical system employed at the *madrasa*s, but it seems that classes were taught in a *ḥalqa* (circle) as at mosques. Students sat in concentric circles around a teacher who commented on a text read out loudly by a reader whom he chose mostly among his disciples. Since the educational style was shared among most disciplines, the mathematical sciences surely will not have been an exception (Lamrabet 2014, 124). The teacher provided a pedagogical certificate called an *ijāza* (licence) to a student who had followed, worked hard and mastered specific books written or taught by the master. This licence granted its owner the right to teach one or more books, which he had learned with the master. In the *ijāza*, the master had to name his *isnād* (chain of transmitters), the teachers with whom he had studied, the teachers of his teachers and so on until the (true or imagined) founder of the discipline or the author of the foundational book of that discipline. With this document, the master should confirm that the recipient had the necessary skills and knowledge to merit the licnce and possessed all competences required for either replacing him in the same *madrasa* or to take another teaching position somewhere else. In the religious sciences, this licence existed in different formats: the general licence concerning not only the command of several disciplines but also the licence for a special theme often given in the form of a special poem and addressed to a master somewhere else who eventually would give the general licence to teach (Berkey 1992, 204). In the sciences treated in this book, the general form has not been attested so far. Even individual licences are fairly rare.

The *madrasa*s in the Great Maghrib, built by rulers and their relatives, were erected in close proximity to mosques and *zāwiya*s without supplanting them. *Zāwiya*s were often run by the head of a local Sufi order. In the 13th/19th century, some of them also taught astronomy and astrology (Sobieroj 2016, 197, 200–1). But little is known about their teaching of science, although this included logic taught in Islamicate societies east of the Greater Maghrib more or less regularly (el-Rouayheb 2010, 10, 72, 120–1). *Madrasas* were established less often than mosques. But their teaching of the sciences was more prominent. *Madrasas* in which scholars taught who left a mathematical corpus include, for instance, *al-Manṣūriyya* in Marrakesh, where Aḥmad ibn Munʿim (d. 626/1228) taught, who contributed major new results to combinatorics. At the so-called *New Madrasa* or *Tashfīniyya madrasa* in Tlemcen the widely read Saʿīd al-ʿUqbānī (d. 811/1408) taught logic and inheritance calculations. Other examples are two madrasas in Tunis, founded by princesses of the Hafsid dynasty in the middle of the 7th/13th century (*al-Madrasa al-Tawfīqiyya*) and at the first half of the 8th/14th century (*al-Madrasa al-ʿUnqiyya*). The first head teacher at this school was al-Hawwārī (d. 749/1348 or 9), one of Ibn Khaldūn's teachers. Another teacher of Ibn Khaldūn was al-Ābilī (681–757/1282–1356) from Tlemcen who had studied the mathematical sciences with Ibn al-Bannāʾ (654–721/1256–1321) in Marrakesh and the logical, philosophical and theological works of Ibn Sīnā, Fakhr al-Dīn al-Rāzī and al-Ṭūsī at *madrasas* in Syria and Iraq (Nasser 1964, 107–9). Ibn al-Bannāʾ himself had studied Euclid's *Elements* with al-Shārifī (d. 682/1283), a former disciple of Ibn Munʿim, at the *Shārifiyya Madrasa* in Tlemcen. It was still operating in the 9th/15th century, when al-Qalaṣādī (d. 891/1486) from the Nasrid kingdom of Granada studied there for several years (Marin 2004; Djebbar 1990a, 12–23).

Several other large as well as smaller madrasas were created across the Greater Maghrib between the 7th/13th and 9th/15th centuries (Figure III.2.2). At several of them, law professors also

Figure III.2.2 *Madrasa* in Marrakesh called after ʿAlī ibn Yūsuf (r. 851–861/1448–1458) the Wattasid ruler who held the city for the Merinid dynasty (r. 614–869/1217–1465). The school was, however, built a century later, in 969/1562 by the Saʿdī Sultan ʿAbdallāh al-Ghālib (r. 964–981/1557–1574). It was built with a *qibla* orientation of 146° instead of 91°21′. With more than 100 dormitory rooms for students it was the largest madrasa in town. Its classes continued until 1960, when it was transformed into a museum.

Source: Photographer Marc Moyon, Limoges

taught parts of the mathematical sciences (Harbili 1997, 318–28, 331; Lamrabet 2014, 157). With the extension of Ottoman rule to the eastern and central parts of the Maghrib after the liberation of Tunis from Spanish occupation in 982/1574, *madrasas* now began to teach Ḥanafī law. With the renewal of logic in the Greater Maghrib in the 11th/17th century, logic was again taught there too (el-Rouayheb 2010, 10, 120–1).

III.2.4 Mathematical works produced by *madrasa* teachers in the Maghrib

Despite serious efforts by historians of mathematics from the Greater Maghrib, we know very little about the organization of mathematical teaching at Maghribī *madrasas*. We can, however, draw conclusions about the content of teaching from the writings of *madrasa* teachers. In geometry, a mathematical discipline that is not well documented in North African sources, Euclid's *Elements* seem to have constituted the main source. It was taught in Tlemcen and Fez (Lamrabet 2014, 57). Algebra and the science of inheritance calculation attracted much more attention. As in other knowledge fields, didactic poems such as Ibn al-Yāsamīn's (d. 600/1204) *Poem on Algebra* (*Urjūza fī l-jabr*) were preferred teaching materials (Sobieroj 2016). The first ten lines praise God, the Prophet Muḥammad and Ibn al-Yāsamīn's teachers. The remaining 43 lines present an elementary introduction to algebra as defined in the early 3rd/9th century by al-Khwārazmī (d. *c.* 235/850; Chapter I.7). This poem gained celebrity as a basic tool for teaching algebraic methods. At least 15 commentaries meant to be used in class were composed on it. Among their authors were famous scholars of the mathematical sciences, all *madrasa* teachers in the Maghrib or al-Andalus, such as Ibn Qunfidh (740–810/1339–1407), Saʿīd al-ʿUqbānī and al-Qalaṣādī (Ibn Maryam 1908, 106, 296; Harbili 1997; Guergour 1988, 184; Djebbar 1990b, 11–12; Zemouli 1993, 197). A second didactic poem by Ibn al-Yāsamīn, also used in classes at *madrasas* in the Maghrib, is his 54-verse long *Poem on the Square Roots* (*al-Urjūza fī l-judhūr*). He also authored a very lengthy textbook on arithmetic with an appendix on geometry (Chapter I.7). It deals with natural numbers, fractions and roots in the standard operations of medieval arithmetic: multiplication, division, addition, subtraction, extraction of square and cubic roots, as well as with the solution of quadratic equations (Zemouli 1988, 100). Moreover, it confirms the use of symbols in some arithmetic and algebraic problems in teaching. These three works of Ibn al-Yāsamīn reflect the level of mathematical teaching at Maghribi *madrasas* in the late 6th/12th century.

Early in the 8th/14th century, the works of Ibn Munʿim dominated mathematical teaching at Maghribī *madrasas*. We know almost nothing about his life except that he was a practicing physician and a *madrasa* teacher in Marrakesh, as well as an expert in number theory and geometry. Although he composed several mathematical treatises, only one is partly extant, *The Book on the Knowledge of Arithmetic* (*Kitāb fī fiqh al-ḥisāb*). This text teaches in its first part the basic arithmetical operations for natural numbers and the extractions of any kind of root followed by sums of natural numbers, perfect, amicable and figurative numbers and concludes with an application of combinatorial rules on lexicographic problems. The second part treats fractions, the approximation of square and cubic roots and word problems (Lambrabet 2005, 154).

The last set of mathematical treatises that exercised a major influence on the teaching of mathematics at North African *madrasas* were those by Ibn al-Bannāʾ. He taught most of his life in Marrakesh but undertook also a few travels to Fez teaching there at the Qayrāwiyyīn Mosque and participating in disputations with local scholars. The biographers ascribe numerous treatises to him covering several of the mathematical sciences (arithmetic, algebra, astronomy), astrology and other occult sciences, law and the science of inheritance calculations (Ibn al-Bannāʾ 1992, 1994; Aballagh 1988, 1992, 2002; Djebbar and Aballagh 1995; Calvo 1997, 404). His most often and widely used work in *madrasa* teaching, also outside of the Maghrib, was *The Abridgement of the Operations of Arithmetic* (*Talkhīṣ aʿmāl al-ḥisāb*). It teaches in a concise manner arithmetical and algebraic operations for natural numbers, fractions and roots without proofs or examples. The second work emerged as a teaching commentary on the *Abridgement*. It is called *The Lifting of the Veil from the Classes of Operations of Arithmetic* (*Rafʿ al-ḥijāb ʿan wujūh aʿmāl al-ḥisāb*). While it was written for classroom work, it also contained some new ideas, among them purely algebraic justifications of al-Khwārazmī's algorithms for solving quadratic equations (Ibn al-Bannāʾ 1994). His third book taught in *madrasas* was entirely dedicated to algebra and followed the work of Abū Kāmil (*c.* 235–*c.* 318/*c.* 850–*c.* 930; Djebbar 1990b, 111–12; Chapter I.7).

Following in the abovementioned practice of teaching Ibn al-Yāsamīn's poems through commentaries, Ibn al-Bannā's *Abridgement* became the most commented upon introductory teaching text on elementary arithmetic and algebra in the Greater Maghrib from the 8th/14th to the 10th/16th centuries. This method of teaching elementary mathematical knowledge, starting from brief introductory surveys in combination with commentaries. It was occasionally complemented by higher-level texts and was shared across many Islamicate societies at the least until the 12th/18th century. It responded to the skills of the participating students. Older students in the position of teaching assistants often prepared a class by reading and conducting exercises with different materials in smaller groups of less advanced students.

In Ibn al-Bannāʾ's times, however, this method was still looked at with suspicion. In order to rebut the many criticisms leveled against him and his concise presentation of arithmetical and algebraic knowledge, he composed the following distich

> I have sought coniseness in my expositions,
> knowing that precision resides in conciseness . . .
> My attitude is that of experienced scholars.
> Explaining lengthily is convenient when teaching children.
> *(Ibn al-Bannāʾ 1969, 18)*

Beginning in the 10th/16th century, the level of mathematical teaching seems to have declined, focusing more and more on purely practical aspects, in particular those connected to problems of inheritance. Proofs for more complex problems and new research results can be found less and less in the surviving school texts. As in other Islamicate societies, older, well-known treatises and commentaries now were versified in order to simplify memorizing elementary knowledge. Examples are the poems on Ibn al-Bannāʾ's *Abridgement* written by two authors of the 10th/16th century: al-Wansharīsī (d. 956/1549) and al-Akhḍarī (d. *c.* 983/*c.* 1575). Al-Akhḍarī had great impact on *madrasa* teaching in logic in the Greater Maghrib as well as in Cairo (el-Rouayheb 2017, 510; Sobieroj 2016).

III.2.5 Instead of a conclusion

This chapter focuses on one of the most important educational institutions that arose in many Islamicate societies. While it served in the 5th/11th century for teaching one or the other legal system, a century later texts of other disciplines, religious as well as nonreligious, were also taught. Despite the importance of the educational institutions for the stabilization of nonreligious knowledge beyond the courts and outside the ruling elites, traditional history of science was and is little interested in their systematic study. It prefers studies of new knowledge to the investigation of institutionalized knowledge, its geographical and social spread and reproduction or other contextually embedded themes. As a result, too little is known about teaching patterns in different regions or times, the applied teaching methods, the literature used in class or the relationship between the various disciplines taught at *madrasas* and other types of schools, to name only a few aspects that characterize teaching as a social institution. For this reason, important regions such as Central Asia, the Indian subcontinent, Anatolia, the Balkans or sub-Saharan Africa could not be covered in this chapter.

Note

1 Consolidated bibliography.

Bibliography

Sources

Abū l-ʿArab. ed. and tr. Ben Cheneb, M. 1920. *Ṭabaqāt ʿulamā' Ifrīqiya wa-l-Qayrawān* [Classes of Scholars From Ifrīqiya and al-Qayrawān]. Algiers: Publications de la Faculté des Lettres d'Alger.

al-Āmidī, Sayf al-Dīn. facs. ed. Sezgin, F. 2001. *Al-Nūr al-bāhir fī l-ḥikam al-zawāhir* [Splendorous Light on Many-Flowered Wisdom]. Frankfurt am Main: Institute for the History of Arabic-Islamic Science.

Ibn al-Bannā'. ed., tr. and ann. Aballagh, M. 1994. *Rafʿ al-ḥijāb ʿan wujūh aʿmāl al-ḥisāb li-Ibn al-Bannā' al-Marrākushī* [The Lifting of the Veil from the Classes of Operations in Arithmetic by Ibn al-Bannā' al-Marrākushī]. Fez: Publications de la Faculté des Lettres et Sciences Humaines.

Ibn al-Bannā' al-Marrākushī. ed., tr. and com. Soussi, M. 1969. *Talkhīṣ Aʿmāl al-ḥisāb* [The Abridgement of the Operations of Arithmetic]. Tunis: Publications de l'Université de Tunis.

Ibn Maryam. ed. Ben Cheneb, M. 1908. *Al-Bustān fī dhikr al-awliyā' wa-l-ʿulamā' bi-Ṭilimsān* [The Garden Recording of the Saints and Scholars of Tlemcen]. Algiers: Office des Publications Universitaires.

Ibn al-Yāsamīn. ed., tr. and com. Zemouli, T. 1993. *Al-Aʿmāl ar-riyāḍiyya li-Ibn al-Yāsamīn* [The Mathematical Works of Ibn al-Yāsamīn]. MA Thesis. Algiers: E.N.S.

Lane, E. W. 1968. *An Arabic-English Lexicon*. 8 vols. Beirut: Librairie du Liban, 3: 870–1.

al-Muḥibbī, M. A. 1966. *Khulāṣat al-athar fī aʿyān al-qarn al-hādī ʿashar* [Essence of the Report on the Nobles of the 11th Century]. 4 vols. Beirut.

Theodosius. ed. Kunitzsch, P. and Lorch, R. 2010. *Sphaerica. Arabic and Latin Translations*. Stuttgart: Franz Steiner Verlag.

Manuscripts

MS Philadelphia, University of Pennsylvania Library, lsj 293.

MS Princeton, Firestone Library, Garrett Collection, 42B.

Research literature

Aballagh, M. 1988. "Les fondements des mathématiques à travers le *Rafʿ al-ḥijāb* d'Ibn al-Bannā' (1256–1321)," in *Actes du Premier Colloque International d'Alger sur l'Histoire des Mathématiques Arabes (Alger, 1–3 Décembre 1986)*. Alger: Maison du livre, 11–23.

Aballagh, M. 2002. "Iktishāf kitāb riyāḍī jadīd li-Ibn al-Bannā'," *Daʿwat al-ḥaqq* 363: 126–32.

Behrens-Abouseif, D. 1989. "The Image of the Physician in Arab Biographies of the Post-Classical Age," *Der Islam* 66: 331–43.

Ben Khouja, M. 1945. *Al-Āthār al-ḥafsiyya fī inshā' al-madāris bi-l-Maghrib*. Tunis: al-Thurayyā.

Berkey, J. 1992. *The Transmission of Knowledge in Medieval Cairo: A Social History of Islamic Education*. Princeton: Princeton University Press.

Beyazıt, Y. 2012–2013. "Efforts to Reform Entry into the Ottoman *İlmiyye* Career Towards the End of the 16th Century: The 1598 Ottoman *İlmiyye Kanunnamesi*," *Turcica* 44: 201–18.

Bouzari, A. 2003. "Procédure et circulation des 'nombres pensés' de l'Orient à l'Occident musulmans," in Spiesser, M. Y. and Guillemot, M., eds. *Actes du Colloque De la Chine à l'Occitanie, Chemins entre arithmétique et algèbre* (Toulouse, 22–24 Septembre 2000). Toulouse: C.I.H.S.O, Université de Toulouse II, 15–27.

Bouzari, A. 2009. "Les coniques en Occident Musulman entre le XIe et le XIVe siècle," *Llull: Revista de la Sociedad Española de Historia de las Ciencias y de las Técnicas* 32: 59–72.

Bouzari, A. 2015. "Les sections coniques d'Apollonius dans la tradition mathématique arabe," in Barbin, E. and Maltrel, J. L., eds. *Les mathématiques méditerranéennes d'une rive et de l'autre*. Marseille: Ellipses, 43–55.

Brentjes, S. 1988. "The First Seven Perfect Numbers and Three Types of Amicable Numbers in a Manuscript on Elementary Number Theory by Ibn Fallūs," *Erdem* 4.11: 467–83.

Brentjes, S. 1990. "Sur quelques travaux mathématiques d'Ibn Fallūs," *Archives Internationales d'histoire des sciences* 40: 239–57.

Brentjes, S. 1993. "An Algebraical Textbook Text: The 'Kitāb niṣāb al-ḫabr fī ḥisāb al-ǧabr' of Ibn Fallūs (590 h/1194–637 h/1239)," *Mitteilungen der Mathematischen Gesellschaft, Hamburg* 13: 5–11.

Brentjes, S. 2002. "On the Location of the Ancient or 'Rational' Sciences in Muslim Educational Landscapes (AH 500–1100)," *Bulletin of the Royal Institute of the Inter-Faith Studies* 4: 47–71.

Brentjes, S. 2008a. "The Study of Geometry According to al-Sakhāwī (Cairo, 15th c) and al-Muḥibbī (Damascus, 17th c)," *Acta Historica Leopoldina* 54: 323–41.

Brentjes, S. 2008b. "Shams al-Dīn al-Sakhāwī on 'Muwaqqits', 'Muʾadhdhins', and the Teachers of Various Astronomical Disciplines in Mamluk Cities in the Fifteenth Century," in Calvo, E. *et al*., eds. *A Shared Legacy: Islamic Science East and West: Homage to Professor J. M. Millàs Vallicrosa*. Barcelona: University of Barcelona, 129–50.

Brentjes, S. 2014. "Teaching the Mathematical Sciences in Islamic Societies. Eighth-Seventeenth Centuries," in Karp, A. and Schubring, G., eds. *Handbook on the History of Mathematics Education*. New York: Springer, 85–108.

Brentjes, S. 2017. "On Four Sciences and Their Audiences in Ayyubid and Mamluk Societies," in Hees, S. von, ed. *Inḥiṭāṭ – The Decline Paradigm: Its Influence and Persistence in the Writing of Arab Cultural History* (Arabische Literatur und Rhetorik – Elfhundert bis Achtzehnhundert [ALEA] 2). Würzburg: Ergon, 139–72.

Brentjes, S. 2019. "MS Munich, Bayerische Staatsbibliothek, Codex Arab. 2697 and Its Properties," *Micrologus* 27: 443–66.

Brunshvig, R. 1931. "Quelques remarques historiques sur les Medersas de Tunis," *Revue tunisienne* 6: 261–85.

Bulliet, R. W. 1972. *The Patricians of Nishapur. A Study in Medieval Islamic Social History*. Cambridge, MA: Harvard University Press.

Calvo, E. 1997. "Ibn al-Bannāʾ," in Selin, ed.[CB], 404.

Chamberlain, M. 1995. *Knowledge and Social Practice in Medieval Damascus, 1190–1350*. Cambridge: Cambridge University Press.

Conermann, S. 2004. "Die Einnahme Bagdads durch die Mongolen im Jahre 1258. Zerstörung-Rezeption-Wiederaufbau," in Ranft, A. and Selzer, S., eds. *Städte aus Trümmern: Katastrophenbewältigung zwischen Antike und Moderne*. Göttingen: Vandenhoeck and Ruprecht, 54–100.

Djebbar, A. 1980. *Enseignement et recherche mathématiques au Maghreb des XIIIe-XIVe siècles*. Paris: Université Paris Sud, Publications Mathématiques d'Orsay, 81–02.

Djebbar, A. 1987. "Les Mathématiques au Maghreb à l'époque d'Ibn al-Bannāʾ," in *Actes du Colloque International de la Société de Philosophie au Maroc sur "Mathématiques et Philosophie"* (Rabat, 1–4 Avril 1982). Paris: l'Harmattan and Rabat: Okad, 31–46.

Djebbar, A. 1990a. "Al-Qalaṣādī, ʿālim andalusī maghāribī min al-qarn al-khāmis ʿashar," *Revue Arabe des Technologies* 9: 12–23.

Djebbar, A. 1990b. "Les activités mathématiques dans le Maghreb central (XIIe– XIXe siècles)," in *Actes du troisième Colloque maghrébin sur l'histoire des mathématiques arabes (Alger, 1–3 décembre 1990)*. Algiers: Office des Publications Univérsitaires, 73–115.

Djebbar, A. 1990c. "Quelques éléments nouveaux sur les activités mathématiques arabes dans le Maghreb oriental (IXe–XVIe siècles)," in *Actes du deuxième Colloque maghrébin d'histoire des mathématiques arabes (Tunis, 1–3 Décembre 1988)*. Tunis: Université de Tunis I, I.S.E.F.C.-A.T.S.M., 53–73.

Djebbar, A. 1990d. *Mathématiques et Mathématiciens du Maghreb médiéval (IXᵉ– XVIᵉ siècles): Contribution à l'étude des activités scientifiques de l'Occident musulman.* Thèse de Doctorat. Nantes: Université de Nantes.

Djebbar, A. and Aballagh, M. 1995. *Ḥayāt wa-mu'allafāt Ibn al-Bannāʾ al-Marrākushī* [Life and Work of Ibn al-Bannāʾ al-Marrākushī]. Rabat: Publications de la Faculté des Lettres et Sciences Humaines.

Fancy, N. 2018. "Post-Avicennan Physics in the Medical Commentaries of the Mamluk Period," *Intellectual History of the Islamicate World* 6.1–2: 58–81.

Gardiner, N. 2017a. "Esotericist Reading Communities and the Early Circulation of the Ṣūfī Occultist Aḥmad al-Būnī's Works," *Arabica* 64.3–4: 405–41.

Gardiner, N. 2017b. "Stars and Saints: The Esotericist Astrology of the Ṣūfī Occultist Aḥmad al-Būnī," *Journal of Magic, Ritual, and Witchcraft* 12.1: 39–65.

Günther, S. 2017. "Education, general (up to 1500)," in *EI–3*, 29–48.

Guergour, Y. 1988. "Un mathématicien du Maghreb, Ibn Qunfudh al-Qusanṭīnī (740–809/1339–1406)," in *Actes du Premier Colloque International d'Alger sur l'Histoire des Mathématiques Arabes (Alger, 1–3 Décembre 1986).* Algiers: Maison du livre, 156–79.

Harbili, A. 1997. *L'enseignement des mathématiques à Tlemcen au XIVe siècle à travers le commentaire d'al-ʿUqbānī (m. 1408) au Talkhīṣ d'Ibn al-Bannāʾ (m. 1321).* MS Thesis. Algiers: Ecole Normale Supérieure de Kouba.

Harbili, A. 2017. "Saʿīd al-ʿUqbānī al-Tilimsānī: wa-musāhamatihī fī tadrīs al-riyāḍiyyāt fī gharb al-Islām," *Revue Dirassate* 57: 311–18.

Heidarzadeh, T. 2019. "The Marāgheh School and its Impact on Post-Mongol Science in the Islamic World," in Babaie, S., ed. *Iran After the Mongols* (The Idea of Iran 8). London: I. B. Tauris, 143–58.

Imber, C. 2009. *The Ottoman Empire, 1300–1650. The Structure of Power.* Basingstoke and New York: Palgrave Macmillan.

King, D. A. 1988. "Universal Solutions to Problems of Spherical Astronomy from Mamluk Egypt and Syria," in Kazemi, F. and McChesney, R. D., eds. *A Way Prepared: Essays on Islamic Culture in Honor of Richard Bayly Winder.* New York: New York University Press, 153–84 [Reprint: King 1993b, VII].

King, D. A. 1993a. "Mīḳāt: astronomical timekeeping," in *EI-2*, 7: 27–32. (Reprint: King 1993b, V).

King, D. A. 1993b. *Astronomy in the Service of Islam.* Aldershot: Variorum Ashgate Publishing Ltd.

Lammer, A. 2017. "Eternity and Origination in the Works of Sayf al-Dīn al-Āmidī and Athīr al-Dīn al-Abharī: Two Discussions from the Seventh/Thirteenth Century," *The Muslim World* 107: 432–81.

Lamrabet, D. 1994. *Introduction à l'histoire des mathématiques maghrébines.* Rabat: Imprimerie al-Maʿārif al-jadīda.

Lambrabet, D. 2005. *Fiqh al-ḥisāb.* Rabat: Imprimerie al-Karāma.

Makdisi, G. 1981. *The Rise of Colleges: Institutions of Learning in Islam and the West.* Edinburgh: Edinburgh University Press.

al-Manūnī, M. 1985. "Nashāṭ al-dirāsāt al-riyāḍiyya fī Maghrib al-ʿaṣr al-waṣīṭ al-rābiʿ," *al-Manāhil* 33: 77–115.

Marin, M. 2014. "The Making of a Mathematician: al-Qalaṣādī (d. 891/1486) and His Riḥla," *Suhayl* 4: 295–310.

Melvin-Koushki, M. 2014. "The Occult Challenge to Messianism and Philosophy in Early Timurid Iran: Ibn Turka's Lettrism as a New Metaphysics," in Mir-Kasimov, O., ed. *Unity in Diversity: Mysticism, Messianism and the Construction of Religious Authority in Islam.* Leiden: Brill, 247–76.

Melvin-Koushki, M. 2016. "Astrology, Lettrism, Geomancy: The Occult-Scientific Methods of Post-Mongol Islamicate Imperialism," *The Medieval History Journal* 19.1: 142–50.

Melvin-Koushki, M. 2018. "How to Rule the World: Occult-Scientific Manuals of the Early Modern Persian Cosmopolis," *Journal of Persianate Studies* 11.2: 127–99.

Nasser, N. 1964. "Le Maitre d'Ibn Khaldūn: Al-Ābilī," *Studia Islamica* 20: 103–14.

el-Rouayheb, K. 2010. *Relational Syllogisms and the History of Arabic Logic, 900–1900* (Islamic Philosophy, Theology and Science. Texts and Studies 80). Leiden and Boston: Brill.

el-Rouayheb, K. 2017. "Aḥmad al-Mallawī (d. 1767)," in el-Rouayheb and Schmidtke, eds. [CB], 509–34.

Savage-Smith, E. 2011. *A New Catalogue of Arabic Manuscripts in the Bodleian Library, Oxford.* Vol. 1: *Medicine.* Oxford: Oxford University Press.

Sobieroj, F. 2016. *Variance in Arabic Manuscripts. Arabic Didactic Poems from the Eleventh to the Seventeenth Centuries – Analysis of Textual Variance and Its Control in the Manuscripts* (Studies in Manuscript Cultures 5). Berlin: De Gruyter.

Talbi, M. 1971. "Ifrīḳaya," in *EI-2*, 3: 1073–6 [Reprint: 1986, 3: 1047–50].

Wisnovsky, R. 2013. "Avicennnism and Exegetical Practice in the Early Commentaries on the *Ishārāt*," *Oriens* 41: 349–78.

Zemouli, T. 1988. "Le poème d'Ibn al-Yāsamīn sur les nombres irrationnels quadratiques," in *Actes du Premier Colloque International d'Alger sur l'Histoire des Mathématiques Arabes (Alger, 1–3 Décembre 1986)*. Algiers: Maison du livre, 178–91.

Zysow, A. 2012. "Karrāmiya," in *eIr*, 15.6: 590–601.

III.3
SCIENTIFIC MATTERS IN *KALĀM* (THEOLOGY)

Ulrich Rudolph

In Islamicate societies, scientific matters have always been a subject of theological investigation. In a certain way, discussions of these issues can even be considered one of the basic elements of *kalām*. As the overall structure of early *kalām* compendia reveals, theological speculation was supposed to treat, before addressing God's essence and His attributes, three introductory topics: (a) the epistemological foundation of theological reasoning, (b) the ontological structure of the world (i.e., its physical components) and (c) proceeding from these two elements, a proof of the existence of God (Rudolph 2015, 201–3; Frank 1992, 12).

As a consequence, the *mutakallimūn* (experts in *kalām*) regularly reflected on physical matters. Yet, one might wonder whether they did so because they were genuinely interested in such questions, or just borrowed some elements from scientific discourse in order to pursue their own theological goals. As it appears, there are good arguments on both sides. Van Ess, for instance, assumes that the atomism developed by early Muʿtazilites such as Abū l-Hudhayl (d. 226/840–1 or 235/849–50) was not part of a physical theory but a theologoumenon (van Ess 1991–1997, 4: 460). In a similar way, Ibn Khaldūn (732–808/1332–1406), the famous historian of the 8th/14th century, had already characterized the 'physical' part of *kalām* treatises as purely functional. In his *Introduction* [*to the Study of History*] (Muqaddima) he writes:

> It should be known that the theologians most often deduced the existence and attributes of the Creator from the existing things and their conditions. As a rule, this was their line of argument. The physical bodies form part of the existing things, and they are the subject of the philosophical study of physics. However, the philosophical study of them differs from the theological. The philosophers study bodies in so far as they move or are stationary. The theologians, on the other hand, study them in so far as they serve as an argument for the Maker.
>
> *(Ibn Khaldūn 1858[CB],*[1] *3: 41–2; tr. Rosenthal 1958[CB], 3: 52–3).*

One might object, however, that the distinction introduced by Ibn Khaldūn is less evident than he assumed it to be. As a rule, not only theologians but also philosophers made use of reflections on the physical structure of beings in order to derive from them metaphysical conclusions. Actually, they were well aware of this fact. Ibn Rushd (520–595/1126–1198), for instance, in his *Decisive Treatise* (*Faṣl al-maqāl*), characterized philosophy as a mental activity

DOI: 10.4324/9781315170718-35

speculating about existing things (*al-naẓar fī l-mawjūdāt*) and considering them insofar as they refer to the Creator (*wa-iʿtibāruhā min jihati dalālatihā ʿalā l-ṣāniʿ*; Ibn Rushd 2001, 1.7–8 of the Arabic text). It thus appears that both camps had their own goals in mind when they treated scientific questions. We cannot exclude the possibility that the theological approach to the latter, as compared to the philosophical, was less 'genuine' and more 'functional,' but we have no reason to construe the *mutakallimūn*'s interest in these matters not to have been serious. As R. M. Frank put it, the second introductory part of *kalām* compendia, as mentioned earlier (= b),

> embraces a discussion of the modes of being and the basic ontological classes of being in general, . . . and comprises a lengthy exposition concerning the being of the things. . . . It consists fundamentally, thus, of a theoretical treatment of what counts as reality and of the basic classes of contingent beings and their properties.
>
> *(Frank 1992, 14; see also Sabra 2006, 2009)*

This should be kept in mind when, in the following, some of the scientific matters discussed in *kalām* will be presented. Due to restrictions of space and the limits of our knowledge, the presentation will be terse and selective. Yet, hopefully, it will show that the contributions made by Muslim theologians to these issues were considerable and are still worth exploring.

III.3.1 Atomism or the structure of the physical world

As mentioned before, one of the topics raised very early in *kalām* was the question of how to describe the ontological structure of the world. As a matter of fact, various theories were developed in this regard, the selection primarily coming down to three models. The first of these came from the 2nd/8th century and can be traced back to Ḍirār ibn ʿAmr (d. 180/796). According to this model, the world is made up of individual components, the so-called accidents (*ʿaraḍ*, pl. *aʿrāḍ*). Ḍirār understands them as all qualitative phenomena, that is, that which he considers to be perceivable to the senses. As for bodies, they play only a secondary role in his system. They have no self-subsistence, being nothing more than clusters of accidents. If they change, this is explained consequently as the reconfiguration of one or more constitutive accidents. In addition, he distinguished between accidents that form bodies, and those which only emerge in previously existing bodies. To the first kind, such as heat and cold, lightness and heaviness, he attributed a certain independence and called them parts (*abʿāḍ*). The second type, such as lust and pain, in his opinion, were not able to persist independently. Being not constitutive of bodies, they are only named accidents, in a more restricted sense (van Ess 1967, 251–4, 1991–7, 3: 37–42; Schöck 2016, 68–75).

The second model was diametrically opposed to this idea. It had various advocates, among whom al-Naẓẓām (d. before 232/847) was a leading figure. According to the latter, the material world is not constituted of accidents, but bodies. This means, then, that all the qualities that Ḍirār characterizes as merely accidental were defined by al-Naẓẓām as corporeal. They are not static, however, and can actually change, because bodies are constantly in a state of mixing. They penetrate each other (*mudākhala* or *tadākhul*) and may be concealed in one another (*kāmin*). These can become visible, however, as soon as a physical process effects a change. To illustrate this, al-Naẓẓām liked to name wood as an example. When wood burns, fire is freed from within, and in fire, the previously latent substances of heat and light show themselves. Thus, the world is presented as a single commixture of bodies that are outwardly perceivable in various portions (van Ess 1967: 246–50, 1991–7, 3: 331–52).

The third model functions at first like a synthesis of the first two, since it classifies both bodies and accidents as the foundational components of the physical world. But, in reality, a more radical change of perspective is at hand. This is because corporeal parts are conceived of here as the smallest indivisible pieces that exist (*al-juzʾ alladhī lā yatajazzaʾ* or *al-jawhar alladhī lā yanqasim*), that is as atoms. Atomistic teachings were professed by various thinkers of the 3rd/9th century. But it was the model created by Abū l-Hudhayl, which prevailed in the long run. According to this conceptualization, every created thing (*shayʾ*) that possesses existence (*wujūd*) must either be corporeal or an accident. The corporeal is defined as whatever occupies space (*mutaḥayyiz*), carries accidents (*ḥāmil* or *muhtamil li-l-aʿrāḍ*) and occasionally as that which can subsist through itself (*qāʾim bi-nafsihī*). Accidents are described with the opposite qualities. They cannot occupy space and can only reside in something else (*qāʾim bi-ghayrihī*). Apart for some exceptional cases (like will or time), they constantly require a substrate (*maḥall*), and this substrate by definition can only be a corporeal substance (*jism*). This raises the question of how many atoms were necessary for the formation of a *jism*. Abū l-Hudhayl said six, thinking three-dimensionally, but later scholars such as al-Iskāfī (d. 240/854) modified his model on this point. He advocated a minimum of two atoms, which finally led to the definition of a body as that which is composed (*muʾallaf/muʾtalif* or *mujtamiʿ*) of two parts (Pines 1936, 3–10; Dhanani 1994, 43–7; van Ess 1991–7, 3: 224–9, 4: 87).

Atomism was able to supplant the other two models. It began its way to dominance around the turn of the 3rd/9th to the 4th/10th century and from that point on left its long imprint on the physical worldview of most theologians, at first the Muʿtazilites and the Ashʿarites but, later on, the Māturīdites, too (Rudolph 2015, 248–50). Atomism thus became a common feature of *kalām* in general. This is confirmed by the fact that it constantly caused offense in the rival camp of the philosophers. Many of them composed refutations of atomism, irrespectively of their own religious affiliation. Among the most prominent critics we find authors such as Yaḥyā ibn ʿAdī (d. 363/974), a Miaphysite Christian (Endress 1984); Ibn Sīnā (d. 427/1037), who discussed the topic extensively in his *Buch of Healing* (*Kitāb al-Shifāʾ*) and in other writings (Ibn Sīnā 2009, 2: 273–310; Lettinck 1999; McGinnis 2013: 78–81); and Maimonides (d. 601/1204), who, in his *Guide of the Perplexed* (*Dalālat al-ḥāʾirīn*), gave a critical report on atomism as part of the famous 'twelve premises' attributed by him to the *mutakallimūn* (Maimonides 1856[CB], 1: 134–50; Maimonides 1963[CB], 1: 194–214; see Schwarz 1991–2).

Yet, when Maimonides composed his critical report, the theological discourse on these matters was already about to change. Starting with Fakhr al-Dīn al-Rāzī (d. 606/1210), the *mutakallimūn* integrated, more and more, philosophical concepts and arguments into their discussions about the ontological structure of the world. Instead of writing some pages, they filled whole volumes by exploring this topic. Following a template that was established by Rāzī in his *Summary on Philosophy and Logic* (*al-Mulakhkhaṣ fī l-ḥikma wa-l-manṭiq*), this exploration usually consisted of three parts: (1) a long discussion about existence, essence, the modalities of being, unity and multiplicity, eternity and temporality under the heading "On General Matters" (*al-umūr al-ʿāmma*); (2) a discussion about the *aʿrāḍ*; and (3) a discussion about the *jawāhir* (general overview by Gardet-Anawati 1981, 160–9; for a detailed analysis see Eichner 2009, 46–61).

As a consequence, the sections on the *aʿrāḍ* and the *jawāhir* contained much philosophical teaching, interpreting *ʿaraḍ* as "accident" and *jawhar* as "substance" in line with the Aristotelian tradition. In his early works, Fakhr al-Dīn al-Rāzī even adopted the Aristotelian ontology, thus abandoning the atomism of traditional *kalām* (Setia 2006, 116–17; Eichner 2009, 52–9). In contrast, he affirmed atomism in his later writings, such as *The Higher Issues on the Divine Science* (*al-Maṭālib al-ʿaliyya fī l-ʿilm al-ilāhī*), now adopting a critical stance toward Aristotelian hylomorphism (Setia 2006, 118–34). The same position is to be found in ʿAḍud *al-Dīn al-Ījī*'s (d. 756/1355) *Stations on the Science of kalām* (*Mawāqif fī ʿilm al-kalām*), who, after discussing the philosophical usage of *jawhar* and *ʿaraḍ* at length in this book, explains why he himself adheres to the theological understanding of both terms, that is, as "atom" and "accident" (Sabra 1994, 15–19 ; Dhanani 2017, 384–7). Thus, atomism continued to play a role in post-Avicennan theology, although this role was more contested and certainly less prominent than in early *kalām* (Dhanani 2015).

III.3.2 Occasionalism or the negation of natural causality

Another characteristic feature of *kalām* falling under the realm of scientific matters is occasionalism. It can be defined as a theory that stresses God's absolute power by negating any kind of natural causality and attributing every causal effect immediately to Him (Chapter III.4). Like atomism, occasionalism was the result of theological discussions that took place during the 3rd/9th century. In a certain way, the discussions that resulted in both theories were interconnected, which has sometimes led modern scholars to consider them something like theoretical twins (see, e.g., Setia 2006, 118). This assessment, however, seems to be inadequate. Occasionalism can be affirmed without affirming atomism and vice versa, as many examples within the Islamic tradition and beyond have demonstrated. It thus seems reasonable to consider the theory in its own right and from two perspectives: historically, by contextualizing it within the framework of early *kalām* and, systematically, by opposing it to other causal theories usually affirmed by philosophers.

As to the historical perspective, this takes us into the midst of the theological debates running through the second half of the 3rd/9th century. Actually, we are able to identify several theological positions affirmed at that time which, despite emerging independently from each other, contributed altogether to shaping the occasionalist theory (for a detailed discussion of the points following now see Perler and Rudolph 2000, 31–51, and Rudolph 2016, 350–4). These positions can be enumerated as follows:

(1) A conviction concerning God's omnipotence, first expressed by Abū l-Hudhayl. According to him, God is, in principle, in the position to have a heavy stone float in the air without the stone falling or to bring a piece of cotton into contact with fire without the cotton being burnt (Ashʿarī 1929–33, 2: 312.10–313.2 with further examples; tr. (German) van Ess 1991–7, 5: 400–1). This does not mean that God really does so. Yet, being omnipotent, he *could* do everything and *could* produce any kind of effect, including the unexpected, if only he wanted to.

(2) A doctrine concerning man's ability to act attributed in our sources to a theologian (nick-)named Ṣāliḥ Qubba (d. probably 245/860). According to him, man can only act on himself, which implies that he is neither able to act on other persons nor to produce any secondary effect. As a result, none of the events happening 'outside' a human being is actually done by him. Whatever occurs (e.g., the movement of a stone) does not occur *because* of his acts but only *when* he is 'acting' (*ʿinda fiʿlihī*). The real cause of 'his' effects is always God who, in the example of the stone, creates its movement spontaneously when the stone has been pushed by a human being (Ashʿarī 1929–33, 2: 406.6–15; tr. van Ess 1991–7, 6: 208).

(3) An idea about causality within the physical world, first promoted by Abū ʿAlī al-Jubbāʾī (d. after 303/915): According to him, nothing which happens in the created world can be considered the effect of 'natural' causes (in the sense of 'intramundane' causes). Instead, all

these events are directly caused by the Creator himself. He has established the 'habit' (*ʿāda*) of connecting certain things in a regular manner. So, at each moment when we are 'acting' (e.g., eating), He creates the particular event (e.g., satiety) corresponding by habit to our act (ʿAbd al-Jabbār 1956–65, 9: 109.4–20; 11: 43.1).

(4) A further development of atomism, attributed to an otherwise unknown Muʿtazilite theologian named al-Shaṭawī (d. 297/910). He applied the idea of minimal discrete unities, originally meant to explain the structure of material bodies, to the realm of time. According to him, God has to act at every moment as creator and as a cause in his creation because the accidents, which were supposed to be indispensable elements of the physical world, are by the sheer fact of being an 'accident', transitory, and cannot endure (*tabqā*) for two moments (*waqtayn*; Ashʿarī 1929–33, 358.5; see van Ess 1991–7, 4: 475–6).

All these positions strengthened the idea that the created world does not follow its ʿnaturalʾ course and cannot exist by itself even for one moment. As such, they all paved the way for the occasionalist theory, which was first formulated at the beginning of the 4th/10th by al-Ashʿarī (d. 324/935). He drew the conclusions from the ideas of his predecessors by generalizing them and connecting them to a coherent argumentative framework. As a result, he conceptualized occasionalism, teaching that God, in his omnipotence, produces every casual effect in this world immediately, the allegedly 'natural' course of the events being nothing but his habit (*ʿāda*). As is well known, the theory was extremely successful and became, in the long run, one of the characteristic features of Islamic theology (Perler and Rudolph 2000, 51–62; see Rudolph 2016, 354–7).

So far, the historical report. As it reveals, occasionalism was deeply rooted in the conceptual framework of early *kalām*. It was motivated by theological concerns and emerged from discussions taking place among theologians who wanted to elaborate a coherent doctrine about God and his relation to the creation. Notwithstanding this origin, however, occasionalism can be considered a generally valuable theory on causality. It responds to questions that have been raised at various times by many scholars, be they theologians, philosophers or scholars of other fields of knowledge. This leads us to the systematic aspect of the theory which can be explained best by turning to al-Ghazālī (d. 505/1111).

As is well known, al-Ghazālī wrote a long discourse on occasionalism in the famous 17th chapter of his *Incoherence of the Philosophers* (*Tahāfut al-falāsīfa*; Ghazālī 2000, 166–77). It is devoted to an extensive and completely innovative discussion of causality, comparing and weighing up different positions for the first time, such as (1) occasionalism, (2) the idea of natural forces (*ṭabāʾiʿ*) acting autonomously in the world (which may be ascribed to some kind of 'naturalists' or 'Dahrites') and (3) the idea of a complex interaction between effects emanating from ʿthe principles of temporary eventsʾ (*mabādiʾ al-ḥawādith*) and various dispositions (*istiʿdād*) existing in this world (which was the position of Ibn Sīnā and other philosophers). The chapter has been subject to varying scholarly comments and interpretations over the years, in particular on the question of what precisely Ghazālī's own position was (see, besides many others, Goodman 1978; Marmura 1981, 1995; Perler and Rudolph 2000, 68–105; Griffel 2009[CB], 147–73; see Rudolph 2016, 358–9). This point aside, however, all scholars agree on the fact that the discussion elaborated by him was extremely innovative. Far from being a simple defense of God's omnipotence, it launched theoretical reflections that shed new light on the problem of causality and had a far-reaching impact on later philosophical and scientific debates on the topic.

One of these innovative elements was his argument that empirical data and sensual observation can never establish causal connections. It inaugurated a skeptical trend in investigations about causality, leaving a long impression on the history of philosophy, in the Islamicate world as well as in Europe. In order to show this, it may suffice to mention just the skeptical arguments of David Hume (Perler and Rudolph 2000, 255, 257; Kukkonen 2010, 29–30, 42–3, 54). Another important element is the fact that Ghazālī was able to raise occasionalism to a higher theoretical level. As he explains it, it is a valuable causal theory that can be considered one possible way of explaining the events occurring in the world. In this respect, too, he was followed by numerous scholars, among them not only Muslim theologians but also 17th-century Cartesian philosophers such as Nicolas Malebranche (1638–1715; see Perler and Rudolph 2000, 14–20, 214–58).

III.3.3 Further scientific matters discussed in *kalām*

Without any doubt, occasionalism and atomism were the most original and most specific theories that theologians contributed to the history of science in the Islamicate world. In comparison, further doctrines and arguments raised in theological discussions had much less impact on the scientific discourse. As a consequence, it seems justified to mention them only briefly, that is by selecting just some of these matters and without going into the details. One topic discussed by many *mutakallimūn* was the question of how to conceive of 'space'. It was closely connected to the concept of atomism, given the fact that Abū l-Hudhayl and other promoters of this doctrine had defined atoms and bodies as "whatever occupies space" (*mutaḥayyiz*; see the earlier discussion). Apparently, there was some dispute on 'space' from the 3rd/9th century onward. As the sources tell us, two positions were prevailing, one of them ascribed to the Baghdādī Muʿtazilites and the other one to their colleagues in Basra. The Baghdādīs denied the existence of the void and believed that 'space' was a two-dimensional container enveloping a body (a position akin to the Aristotelian view). The Basrians upheld the existence of void spaces within the universe. Furthermore, they distinguished between space that is occupied by an extended object (*ḥayyiz*) and empty unoccupied space (*makān*). This seems to suggest that they may have believed in the idea of absolute space (Dhanani 1994, 66–8). However, this idea is not clearly stated in our sources, and all in all, the whole debate seems to have been quite vague on the theoretical level. As Dhanani puts it "it is sufficient to note that the theory of space . . . is implicit in most of the *kalām* texts of this period [= the 3rd/9th and 4th/10th centuries] and does not constitute a subject worthy of discussion in its own right" (Dhanani 1994, 67).

This stands in contrast with later *kalām* works in which the theory of space is discussed separately and much more elaborately. The example best explored so far is Fakhr al-Dīn al-Rāzī, who, in one of his later books, *The Higher Issues*, offers an interesting discussion of the topic. According to him, three major competing positions can be identified. One of them defines place as the inner surface of the containing body (= the Aristotelian position as understood by Ibn Sīnā). The other two agree on the idea that place is "space" (*faḍāʾ*) that can be pervaded by body but interpret it differently: One faction says that space is "pure and unmixed" (*maḥd, ṣirf*) and actually lacks existence (= the position of the *mutakallimūn*). The other one, identified as "Plato and many philosophers" in *The Higher Issues*, teaches that space does indeed exist. They define space as a self-subsisting "extension" (*buʿd*) becoming a place when it is occupied by bodies (Adamson 2017, 208–9). Having compared these positions and weighed up their arguments, Rāzī himself decides to join the third position. He is convinced that space is a subsistent three-dimensional extension, the existence of which is immediately obvious to us (Adamson 2017, 207, 213–19), introducing thereby a specifically Platonic view into his general adherence to atomism.

Another topic often raised in *kalām* was the question of how to conceive of 'time'. Here again, the conceptual framework of atomism played an important role in the discussion. As Abū l-Hudhayl had taught, every created thing must either be corporeal or an accident. Consequently, time could not be but an accident. Actually, he defined time as a series of transitory moments (*waqt*) separating one act (*ʿamal*) from the other (van Ess 1991–7, 3: 241–2; 5: 382).

As in other cases mentioned earlier, this led to a controversial exchange between philosophers and theologians. Ibn Sīnā (Ibn Sīnā 2009, 1: 223, 229) opposed the concept of time as developed in early *kalām*. Later *mutakallimūn* reacted to his opposition weighing up the different positions in order to formulate their own solution. Once more, Fakhr al-Dīn al-Rāzī is a striking example of this process. In his *Higher Issues*, he devoted a lengthy discussion to the concept of time, which deserves careful investigation (Adamson 2017, 207). So far, we can only say that, here again, he seems to have drawn on theoretical elements taken partially from early *kalām* and partially from the Platonic tradition. For he defines time as an "objective indeterminate substance or physical dimension . . . marked out by the accidents of motion and repose which inhere in it" and composed of consecutive, successive adjoining moments (Setia 2008. 407).

Apart from Rāzī's elaboration, however, there is another interesting reflection on time to be found in later *kalām*, this time in al-Ghazālī's *Incoherence*. It is part of the first chapter of this book which argues extensively against the doctrine of the world's past eternity. In the course of his argument against the philosophers al-Ghazālī denies that time is something (*shayʾ*) self-subsisting. Instead, it should be considered "a relation necessary with respect to us [only]" (*nisba lāzima bi-l-iḍāfa ilaynā*; Ghazālī 2000, 32.15). Thus, taken at face value, he seems to contest the objective reality of time, pleading for a kind of subjectivist approach to the matter, which foreshadows, in some respect, concepts developed much later in European contexts (see the interpretation by Obermann 1921, which is still worth considering).

Finally, it should be mentioned that several later *mutakallimūn* included discussions about astronomy in their writings. The case usually quoted in this regard is al-Ījī's *Stations* (Sabra 1994, 34–41), but, as it appears, elaborations on this topic are found in other *kalām* books, too (Morrison 2014, 218–26). The questions raised in this context are numerous ranging from Ījī's specific approach to astronomical issues (was it an instrumentalist approach?) to the general debate about the role of astronomy within philosophy versus within *kalām* and the other Islamic disciplines. All this is currently under investigation, the scholarly debate promising instructive insights (a survey of recent literature is given by Morrison 2014, 201–6; see also Ragep 2001). To all appearances, this will lead us to detect further scientific matters discussed in *kalām*, in particular in the later period, which certainly deserves more scholarly attention than it has received so far.

Note

1 Consolidated bibliography.

Bibliography

Sources

ʿAbd al-Jabbār ibn Aḥmad al-Qāḍī. ed. al-Ḥilmī, M. M. 1958–65. *Al-Mughnī fī abwāb al-tawḥīd wa-l-ʿadl* [Summa on the Matters of Divine Unity and Divine Justice]. 16 vols. Cairo: al-Muʾassasa al-miṣriyya al-ʿāmma li-l-taʾlīf wa-l-anbāʾ wa-l-nashr.

al-Ashʿarī, Abū l-Ḥasan. ed. Ritter, H. 1929–1933. *Maqālāt al-Islāmiyyīn wa-khtilāf al-muṣallīn* [The Claims of the Muslims and the Differences in Opinion between Those Who Pray]. 2 vols. and indexes. Istanbul and Beirut: Brockhaus.

Ibn Rushd. ed. and tr. Butterworth, Ch. E. 2001. *Faṣl al-maqāl. The Book of the Decisive Treatise Determining the Connection between the Law and Wisdom*. Provo, UT: Brigham Young University.
Ibn Sīnā. ed. and tr. McGinnis, J. 2009. *The Physics of the Healing*. Provo, UT: Brigham Young University.

Research literature

Adamson, P. 2017. "Fakhr al-Dīn al-Rāzī on Place," *Arabic Sciences and Philosophy* 27: 205–36.
Dhanani, A. 1994. *The Physical Theory of Kalām: Atoms, Space, and Void in Basrian Muʿtazilī Cosmology*. Leiden: Brill.
Dhanani, A. 2015. "The Impact of Ibn Sīnā's Critique of Atomism on Subsequent *Kalām* Discussions on Atomism," *Arabic Sciences and Philosophy* 25: 79–104.
Dhanani, A. 2017. "*Al-Mawāqif fī ʿilm al-kalām* by ʿAḍud al-Dīn al-Ījī (d. 1355), and Its Commentaries," in el-Rouayheb and Schmidtke, eds. [CB], 375–96.
Eichner, H. 2009. *The Post-Avicennan Philosophical Tradition and Islamic Orthodoxy: Philosophical and Theological Summae in Context*. Habilitation thesis. Halle/Saale-Wittenberg: Martin Luther University (unpublished).
Endress, G. 1984. "Yaḥyā Ibn ʿAdī's Critique of Atomism: Three Treatises on the Indivisible Part, Edited with an Introduction and Notes," *ZGAIW* 1: 155–79.
Ess, J. van. 1967. "Ḍirār b. ʿAmr und die „Cahmīya". Biographie einer vergessenen Schule," (I) *Der Islam* 43: 241–79.
Ess, J. van. 1991–1997. *Theologie und Gesellschaft im 2. und 3. Jahrhundert Hidschra. Eine Geschichte des religiösen Denkens im frühen Islam*. 6 vols. Berlin and New York: De Gruyter.
Frank, R. 1992. "The Science of *Kalām*," *Arabic Sciences and Philosophy* 2: 7–37.
Goodman, L. E. 1978. "Did Al-Ghazālī Deny Causality?" *Studia Islamica* 47: 83–120.
Kukkonen, T. 2010. "Al-Ghazālī's Skepticism Revisited," in Lagerlund, H., ed. *Rethinking the History of Skepticism: The Missing Medieval Background*. Leiden: Brill, 29–59.
Lettinck, P. 1999. "Ibn Sīnā on Atomism: Translation of Ibn Sīnā's *Al-Shifāʾ, Al-Ṭabīʿiyyāt 1: al-Samāʿ al-Ṭabīʿī*, Third Treatise, Chapters 3–5," *Al-Shajarah* 4: 1–51.
Marmura, M. 1981. "Al-Ghazālī's Second Causal Theory in the 17th Discussion of the Tahāfut," in Morewedge, P., ed. *Islamic Philosophy and Mysticism*. Delmar, NY: Caravan Books, 85–112.
Marmura, M. 1995. "Ghazālian Causes and Intermediaries," *Journal of the American Oriental Society* 115: 89–100.
McGinnis, J. 2013. "Avicenna's Natural Philosophy," in Adamson, P., ed. *Interpreting Avicenna: Critical Essays*. Cambridge: Cambridge University Press.
Morrison, R. 2014. "What Was the Purpose of Astronomy in Ījī's *Kitāb al-Mawāqif fī ʿilm al-kalām*?" in Pfeiffer, J., ed. *Politics, Patronage and the Transmission of Knowledge in 13th-15th Century Tabriz*. Leiden: Brill, 201–29.
Obermann, J. 1921. *Der philosophische und religiöse Subjektivismus Ghazālīs: Ein Beitrag zum Problem der Religion*. Wien and Leipzig: W. Braunmüller.
Perler, D. and Rudolph, U. 2000. *Occasionalismus: Theorien der Kausalität im arabisch-islamischen und im europäischen Denken*. Göttingen: Vandenhoeck & Ruprecht.
Pines, S. 1936. *Beiträge zur islamischen Atomenlehre*. Berlin: Hein.
Ragep, F. J. 2001. "Freeing Astronomy from Philosophy: An Aspect of Islamic Influence on Science," *Osiris* 16: 49–71.
Rudolph, U. 2015. *Al-Māturīdī and the Development of Sunnī Theology in Samarqand*. Leiden and Boston: Brill.
Rudolph, U. 2016. "Occasionalism," in Schmidtke, S., ed. *The Oxford Handbook of Islamic Theology*. Oxford: Oxford University Press, 347–63.
Sabra, A. I. 1994. "Science and Philosophy in Medieval Islamic Theology: The Evidence of the Fourteenth Century," *ZGAIW* 9: 1–42.
Sabra, A. I. 2006. "Kalam Atomism as an Alternative Philosophy to Hellenizing Falsafa," in Montgomery, J., ed. *Arabic Theology, Arabic Philosophy, from the Many to the One: Essays in Celebration of Richard M. Frank*. Leuven: Peeters, 199–272.
Sabra, A. I. 2009. "The Simple Ontology of *Kalām* Atomism: An Outline," *Early Science and Medicine* 14: 68–78.

Schöck, C. 2016. "Jahm b. Ṣafwān (d.128/745–6) and the 'Jahmiyya' and Ḍirār b. ʿAmr (d. 200/815)," in Schmidtke, S., ed. *The Oxford Handbook of Islamic Theology*. Oxford: Oxford University Press, 55–80.

Schwarz, M. 1991–2. "Who Were Maimonides' Mutakallimûn? Some Remarks on *Guide on the Perplexed*, Part 1, Chapter 73," *Maimonidean Studies* 2: 159–209 and 3: 143–72.

Setia, A. 2006. "Atomism Versus Hylemorphism in the *Kalām* of Fakhr al-Dīn al-Rāzī: A Preliminary Survey of the *Maṭālib al-ʿāliyya*," *Islam & Science* 4: 113–40.

Setia, A. 2008. "Time, Motion, Distance, and Change in the *Kalām* of Fakhr al-Dīn al-Rāzī: A Preliminary Survey with Special Reference to the *Maṭālib ʿAliyah* [sic]," *Islam & Science* 6: 13–29.

III.4
ASHʿARITE OCCASIONALIST COSMOLOGY, AL-GHAZĀLĪ AND THE PURSUIT OF THE NATURAL SCIENCES IN ISLAMICATE SOCIETIES

Frank Griffel

One of the most important books in Islamic philosophy is the *Tahāfut al-falāsifa* by the influential Ashʿarite theologian al-Ghazālī (d. 505/1111). Its title is often translated as *The Incoherence of the Philosophers*, but recent research has shown that it should rather be rendered as *The Precipitance of the Avicennans* (on "precipitance", which here means something like "incautious haste"; see Treiger 2012, 108–15). Al-Ghazālī's book discusses in twenty chapters teachings of the Aristotelian Muslim philosopher Ibn Sīnā (Avicenna, d. 428/1037) that al-Ghazālī found either objectionable from the point of view of Muslim theology or that he thought led to wrong impressions about the accomplishments of the scholarly tradition of philosophy. *The Precipitance of the Avicennans* marks a watershed moment in the history of philosophy in the Islamic world, when Muslim theologians (*mutakallimūn*) began to engage seriously with Ibn Sīnā's philosophy and when they started to distinguish between those elements of Avicennan philosophy (*falsafa*) that they could productively integrate into their theology (*kalām*) and those that they had to counter and refute. The book also triggered the beginning of a new tradition in Islamic philosophy, when scholars trained in *kalām* began to discuss, evaluate and eventually improve Avicennan philosophy. These scholars avoided the label *falāsifa* as a self-characterization but chose *ḥukamāʾ* instead, a word connected to Arabic *ḥikma* (wisdom, philosophy) that has Qurʾānic connotations (e.g., Qurʾān 2:129, 151, 231) and had already been used by al-Fārābī (d. 339/950–1) and others for denoting philosophy. Following al-Ghazālī, the word *falāsifa*, which before him was understood as a reference to any school of thought within the discipline of philosophy, now became a label for the purist followers of Ibn Sīnā. Al-Ghazālī's *Precipitance* introduced an understanding of *falsafa* as the original philosophical system of Ibn Sīnā. *Ḥikma*, on the other hand, came to mean the presentation, critical discussion and the improvement of philosophical teachings produced often by *mutakallimūn*.

In his introduction to *The Precipitance of the Avicennans*, al-Ghazālī explains that the most problematic part of the *falāsifa*'s views are their teachings in "*ilāhiyyāt*", meaning metaphysics and philosophical theology (al-Ghazālī 2000[CB],[1] 3–4). Only four of the twenty discussions

DOI: 10.4324/9781315170718-36

in that book deal with Ibn Sīnā's teachings in the natural sciences (*ṭabīʿiyyāt*). In a preamble to these four chapters, al-Ghazālī explains that the majority of teachings in the natural sciences do not clash with religion (al-Ghazālī 2000, 161). The most problematic aspect in this field is the *falāsifa*'s view of causality, which prompts them to deny some of the prophetic miracles that are soundly transmitted in books (al-Ghazālī 2000, 163). This leads al-Ghazālī into the 17th discussion where he addresses the *falāsifa*'s teachings on causality. Teachings about causality belong to what is referred to as cosmology, that is, a set of ideas that explains the physical situation and changes that humans find themselves in and that explains whether and how those changes are connected to one another. Al-Ghazālī begins this chapter with a highly programmatic statement:

> The connection (*iqtirān*) between what is habitually believed to be a cause and what is habitually believed to be an effect is not necessary (*ḍarūrī*) according to us. But [with] any two things that are not identical and which do not imply one another it is not necessary that the existence or the nonexistence of one follows necessarily (*min ḍarūra*) out of the existence or the nonexistence of the other. . . . Their connection is due to the prior decree (*taqdīr*) of God who creates them side by side (*ʿalā al-tasāwuq*), not to its being necessary by itself, incapable of separation.
>
> *(al-Ghazālī 2000, 166, adopted from Marmura's translation)*

When western researchers of the history of philosophy read this sentence in the 19th century, they understood it to mean that here al-Ghazālī denies that any given connection between a cause and its effect is stable and predictable and that he argues for the possibility that God could and would disrupt and suspend that connection and either hinder the effect from coming about or create a different one. Solomon Munk (1803–1867), for instance, who stands at the beginning of the European academic study of Arabic philosophy, wrote in 1844 that according to al-Ghazālī, "the philosophers' theory of causality is wrong" (Munk 1857–1859, 377). Al-Ghazālī argues, according to Munk, that the philosophers are mistaken when they say that the effects cannot come about without their causes (Munk 1857–1859, 377–9). Munk was also the first Western scholar who connected al-Ghazālī's critique of causality with the occasionalist cosmology developed by early Ashʿarite *mutakallimūn*. Munk was familiar with that cosmology from a detailed report by the influential Jewish thinker Moses Maimonides (d. 601/1204), who includes in his *Guide of the Perplexed* (*Dalālat al-ḥāʾirīn*) six chapters where he reports and critically evaluates the approach of Muslim *mutakallimūn* toward the question of the world's creation in time (Maimonides 5691/1930–1[CB], 121–62; Maimonides 1963[CB], 1: 175–231).

III.4.1 Occasionalism

The Hebrew and the Latin translations of Maimonides's *Guide* introduced occasionalism to a wide readership of Jews and Christians in Europe and elsewhere and made it a part of the history of Western philosophy. Originally, occasionalism was developed by Muslim theologians of the early Ashʿarite school. Al-Ashʿarī himself (d. 324/935–6), the founder of this tradition, played the most important role. During the 14th/20th century, a number of Arabic texts by early Ashʿarite theologians were discovered that allow us to reconstruct their occasionalist cosmology from their own presentations rather than from hostile reports, such as that of Maimonides. Particularly fertile is a report of the teachings of the master by Ibn Fūrak (d. 406/1015), one of al-Ashʿarī's early followers, as well as the theological works of early Ashʿarite theologians such as al-Bāqillānī (d. 403/1013) and Abū l-Maʿālī al-Juwaynī (d. 478/1085).

In his early career, al-Ashʿarī was a proponent of the Muʿtazilite school of *kalām*. His own occasionalist cosmology is built on many Muʿtazilite elements, among them atomism and the understanding that the laws of nature should be understood as "God's habit" (*ʿādat Allāh*). Muʿtazilite *mutakallimūn* taught that bodies are not infinitely divisible but are composed of "parts that cannot be divided" (Chapters I.11 and III.3). The atomist theory developed in early *kalām* is different from modern ideas about the atom, however, because it assumes that atoms are by themselves completely powerless and have no predetermined way of reacting to other atoms. The *mutakallimūn* describe all attributes that a body has other than its shape, as "accidents" (sing. *ʿaraḍ*). For them the created world consists only of two kinds of beings: atoms and accidents. Every nonmaterial feature, like a color, an odor, an impression or an idea, is understood as an accident of a material body that is composed of atoms. The *mutakallimūn* taught that when a human believes in God's existence, the atoms of her heart carry the accident of "belief in God." When an architect has a plan for a building, the atoms of his brain carry the accident of that plan. Both the atoms and the accidents are by themselves devoid of all power and need to be combined in order to create bodies, be they animated or lifeless. Atoms are empty building blocks, so to speak, and they only constitute the shape of a body. All other characteristics are formed by the accidents that inhere in the body (Dhanani 1994, 38–54, 90–140; Rudolph 2016, 348–54). This kind of atomism appealed to al-Ashʿarī because it does not assume that potentialities in things limit how they will develop in the future. Such potentialities would limit God's actions. Al-Ashʿarī insists on the nonexistence of any true potentiality outside of God (Frank 1966, 21, 29). In principle, any atom can adopt any kind of accident as long as God has created the association of this particular atom with that particular accident.

Al-Ashʿarī combined the atomism he had inherited from Muʿtazilism with a particular understanding of change and of time. Some Muʿtazilites had already speculated that movements are not continuous processes but consist of smaller leaps (singl. *ṭafra*) that our senses cannot detect and whose sum we perceive as a continuously flowing movement (Dhanani 1994, 138). This theory, in turn, led other Muʿtazilite thinkers to assume that time itself is not a continuous flow but is rather a procession of "moments" (sing. *waqt*), in a succession that is concealed from our senses. Al-Ashʿarī combined all these ideas and formulated what became known as occasionalism. Its main components are the atomism of the earlier *mutakallimūn* plus the idea that time is a sequence of moments. The latter idea is sometimes called an "atomism of time" (Perler and Rudolph 2000, 46–51). Al-Ashʿarī insisted that accidents cannot subsist from one moment to another. They need to be created every moment anew. And since bodies cannot exist without accidents, bodies exist from one moment to the next only because God creates their accidents anew in every moment. In order for an atom to exist from one moment to another, God has to create the accident of "subsistence" (*baqāʾ*) every moment He wants the atom to persist. This leads to a cosmology where in each moment, God must assign the accidents to the atoms and to the bodies they form. When one moment ends, God creates new accidents, and through these new accidents He ensures that the atoms persist. None of the accidents created in the second moment has any intrinsic causal relation to the ones in the earlier moment. If a body has a certain attribute from one moment to another, then God created two identical accidents that inhere in that body in two different moments. Movement and development occur when God decides to deviate from the arrangement of the moment before. A ball is moved, for example, when in the second moment of two, the atoms of the ball are created at a specific distance from the locus of the first moment. The distance determines the speed of the movement. The ball thus jumps in leaps over the playing field, as do the players' limbs and their whole bodies. This also applies to the atoms of the air if there is some wind. In every moment, God rearranges all the atoms of this world and creates their accidents anew – thus creating a new world every moment (Perler and Rudolph 2000, 51–6; Rudolph 2016, 354–7; Gimaret 1990, 43–130).

Occasionalism was conceived out of a strong desire to grant God control over each and every element of His creation at every point in time. This desire is connected to the Ashʿarites' dispute with the Muʿtazilites over the character of human actions. If God is omnipotent then He must create *everything*, including human actions. Ashʿarites taught that humans have no free will that decides over their actions; rather, God creates each and every human action and predetermines it. Created beings have no power over themselves nor over any other beings. There is no causal efficacy among God's creation: a ball on a playing field appears to be moved by a player, but in fact, it is moved immediately by God, and so are the corresponding motions of the player. There is only one single cause for all events in the universe, which is God. He has the most direct effect on all His creatures and no being other than He has any effect on others:

> The fact that the stone moves when it is pushed is not an act of him who pushes, but a direct act of God (*ikhtirāʿ min Allāh*). It would be perfectly possible that one of us pushed it without it being moved because God did not produce its movement, or that there is no-one who pushes it and it still moves because God directly produces its movement.
>
> *(Ibn Fūrak 1986, 132–3)*

At the heart of al-Ashʿarī's occasionalist ontology lies the denial of any unrealized potentialities in the created world. Al-Ashʿarī rejected the idea that created beings are compelled to act according to their nature (sing. *ṭabʿ*). The concept of "nature" plays an important role in Aristotelian explanations of causal change and hence also in Ibn Sīnā's. He considers the "nature" of a thing an internal cause of action that belongs to the genus of the thing's essence (Lammer 2018, 300–6). The rejection of natures was part and parcel of early Ashʿarite theology. We usually assume that if a date stone, for instance, is planted and fed, it can only develop into a date palm and not into an apple tree. While this may be true for all practical purposes, the early Ashʿarites taught that in theology this assumption unduly limits God's freedom to act. After discussing where such natures would be located in his cosmology, al-Ashʿarī determined that they can neither be classified as atoms nor as accidents. Thus, he concluded, the word *nature* (*ṭabʿ*) is empty of any comprehensible meaning. Those who use it wish to indicate that there is some regularity in the production of accidents in certain bodies, nothing more (Ibn Fūrak 1986, 131–2; Gimaret 1990, 403–9).

III.4.2 The 19th and 20th centuries: al-Ghazālī as destroyer of the sciences in Islam

Al-Ghazālī grew up in an Ashʿarite milieu in his hometown of Ṭābarān – Ṭūs. He studied at an Ashʿarite *madrasa* in the nearby bigger city of Nishapur, in what is today Northeast Iran. At that time, the *madrasa* was a novel institution of higher learning in Islamic law and theology (Chapter III.2). Al-Ghazālī's teacher was the renowned Ashʿarite al-Juwaynī, who taught occasionalism. Al-Ghazālī's own *kalām*-compendium *The Balanced-Book on What-to-Believe* (*al-Iqtiṣād fī l-iʿtiqād*), which he composed when he was himself a highly respected teacher at an Ashʿarite *madrasa* in Baghdad, can be read as a manual of Islamic theology based on an occasionalist cosmology. Hence, Solomon Munk and, following him, a long list of Western interpreters got the impression that in the 17th discussion of his *Tahāfut*, al-Ghazālī proposed Ashʿarite occasionalism as a way to explain physical change and to maintain the possibility of prophetical miracles. Whereas

Aristotelianism commits itself to the existence of natures or essences that things have and that allow physical change only within the range of what that thing's nature or essence determines, Ashʿarites had no problem explaining, for instance, how Moses's staff turned into a serpent. The change of Moses's staff into a serpent in front of Pharaoh and his magicians, a miracle reported both in the Qurʾān (7.103–18; 20.65–9; 26.32; 26.45), as well as in the Hebrew Bible (*Exodus*, chapter 4), became a stock example that according to al-Ghazālī any acceptable physical theory should be able to explain. In the 17th chapter of the *Tahāfut*, al-Ghazālī put forward five conditions that any explanation of physical change acceptable to him must fulfill (Griffel 2009[CB], 183–6). Among these five conditions is the view that God is the only true "agent" or "efficient cause" (*fāʿil*) in the world and that all events are created by Him. If a cotton ball catches fire from a spark, another standard example, al-Ghazālī maintained that the agent or efficient cause (*fāʿil*) of the cotton ball's combustion, of making it black and turning it into ashes is God (al-Ghazālī 2000, 167).

Prior to the 21st century, most Western readers of al-Ghazālī agreed that in the 17th discussion of his *The Precipitance of the Avicennans*, he denied the working of efficient causality, of natures and, hence, the validity of laws of nature. Rather, they said, he put forward an occasionalist cosmology that denied any connection between two created objects and taught that the only cause in this world is God, producing change by direct intervention. Al-Ghazālī's arguments were regarded as similar to those which were later developed by empiricists such as David Hume (1711–1776). Al-Ghazālī, for instance, pointed to the fact that observation (*mushāhada*) only witnesses the concomitant occurrence (*al-ḥuṣūl ʿindahu*) of causes and their effects not their combined occurrence (*al-ḥuṣūl bihi*). Observation also can never exclude the involvement of causes additional to those that are visible (al-Ghazālī 2000, 167), a point already made by earlier Ashʿarite *mutakallimūn* such as al-Bāqillānī. This interpretation of the 17th discussion led to the view that here al-Ghazālī attacked the *falāsifa*'s approach in the natural sciences. Although al-Ghazālī nowhere says so, it was assumed that he devalued the natural sciences, implicitly arguing that if laws of nature do not exist or can be easily suspended, then what use is there in studying them? This understanding was reinforced by Ibn Rushd's ([520–595/1126–1198]; Averroes) attack on al-Ghazālī in his response to the 17th discussion within his *Precipitance of "The Precipitance."* There, Ibn Rushd accuses al-Ghazālī of making skeptical (*safsaṭānī*) arguments against "the existence of efficient causes," which, in turn, lead to the denial of knowledge (*tabṭīl li-l-ʿilm*). Ibn Rushd's comments culminate in the conclusion that, "whoever rejects causes rejects rationality (*al-ʿaql*)" (Ibn Rushd 1987, 519–22, 1954 Engl. trans. 2: 318–9).

During the mid-19th century, the understanding of al-Ghazālī as a denier of causality served as an explanation for the assumed decline that Muslim societies entered into after the 6th/12th century. Whereas before the 6th/12th century, Muslim societies were considered leaders in philosophy, mathematics and the natural sciences, Western scholars mistakenly concluded that these disciplines were no longer practiced in Islamic societies after the 6th/12th and 7th/13th centuries, or if they were, they did not produce anything worthy of note. In 1852, the influential French religious historian Ernest Renan (1823–1892) portrayed al-Ghazālī as a pious Sufi who undertook to prove the radical incapacity of reason:

> And with a manoeuvre that has always seduced minds more fervent than wise, he founded religion on skepticism. In this fight he fielded an astonishing sharpness of mind. He opened his attack against rationalism especially through his critique of the causal principle. Hume had said the very same thing: We only perceive simultaneousness, never causality. Causality is only that the will of God creates two things ordinarily in sequence. Laws of nature do not exist, rather they express a mere habitual

> course. God alone is unchanging. This was, as one can see, the negation of all science. Al-Ghazālī was one of those bizarre minds who only embraced religion as a manner to challenge reason.
>
> *(Renan 1852, 73–4)*

Influenced by the European experience of a conflict between Enlightenment thought and church dogma, Renan saw in al-Ghazālī an enforcer of Islamic religious dogma, one may even say an inquisitor, who used both the power of his pen as well as institutionalized religious and political power to crack down on the freethinking *falāsifa*. According to Renan, these endeavors were successful and led to the abandonment of the sciences in the Islamicate world. At a time when medieval European Christians and Jews translated Ibn Rushd and other Muslim philosophers from Arabic into Latin or Hebrew and thus laid the foundation for a thriving philosophy in parts of Christian Europe that would culminate in the Enlightenment, Muslims abandoned these thinkers at their own peril (Renan 1852, 28–9). In a public lecture in 1883, Renan presented the view that as a religion, Islam is hostile to the pursuit of the sciences and poses an obstacle to their study and development: "for human reason, Islam has only been harmful" (Renan 1883, 9).

Renan's view of Islam's relationship to the sciences and its basis in al-Ghazālī's assumed denial of causality was repeated in numerous academic publications of the 19th and 20th centuries. Ignaz Goldziher (1850–1921), one of the most influential scholars of Islam during the 20th century, for instance, argued in a widely read article of 1915 on the "Attitude of Orthodox Islam Toward the Ancient Sciences" that most Muslim scholars were opposed to the rational sciences they had inherited from antiquity (*ʿulūm al-awāʾil*; Goldziher 1981b, originally 1915). In his popular textbook on Islam he wrote that "orthodox Islamic theology, built on Ashʿarite foundations, demands the rejection of the concept of causality, in any form whatever" (Goldziher 1981a, 114, originally 1910). Today, this idea of Islam's hostility to the sciences is rarely expressed in publications by academics who study Islam and Islamicate societies. Yet it still plays a very important role in the public perception of Islam and in debates outside of the field of Islamic studies. As recent as 2007, for instance, the American physicist and Nobel laureate Steven Weinberg (1933–2021) wrote that

> Islam turned against science in the twelfth century. The most influential figure was the philosopher Abu Hamid al-Ghazzali, who argued in *The Incoherence of the Philosophers* against the very idea of laws of nature, on the ground that any such laws would put God's hands in chains. According to al-Ghazzali, a piece of cotton placed in a flame does not darken or smoulder because of the heat, but because God wants it to darken and smoulder. After al-Ghazzali, there was no more science worth mentioning in Islamic countries.
>
> *(Weinberg 2007, 3)*

Views like this are hardly different from those of Ernest Renan in the 19th century. They are also the hallmark of highly popular lectures by the American physicist and educator Neil deGrasse Tyson (b. 1958), which proliferate widely on the Internet. But whereas Renan's and Goldziher's works must be read against the backdrop of European colonialism and its attempt to demonstrate the superiority of Europe's culture over the colonized ones, contemporary expressions of this notion happen in the context of religious, mostly Christian fundamentalist opposition to Darwinism. Weinberg and deGrasse Tyson tap into and reinforce popular stereotypes about Islam to illustrate their larger point – which is equally based on popular prejudice – that religions are hostile to the pursuit of the sciences.

III.4.3 The 21st century: the compossibility of occasionalism and secondary causality

The 19th-century interpretation of al-Ghazālī as a stern defender of occasionalism against the causal theories of Aristotelian philosophers, however, found its first doubters in the early 21th century, when more of his works became available (Perler and Rudolph 2000; Griffel 2011, 47–52) and has by now been abandoned by specialist scholars. A close reading of al-Ghazālī's works shows that with the exception of metaphysics (*ilāhiyyāt*), he endorses the philosophical sciences, particularly logic and mathematics as well as most natural sciences, and requires that even religious scholars study them. At various points in Muslim intellectual history, those who supported the study of the rational sciences relied on passages from al-Ghazālī – particularly from the introduction to *The Precipitance of the Avicennans* – to support their position (Riexinger 2015). Whereas the last decade of the 20th century saw two opposing interpretations of al-Ghazālī's views on causality, one by Michael E. Marmura (1929–2009) and another by Richard M. Frank (1927–2009), there is now a widespread consensus that al-Ghazālī developed his own, highly original theory of causality, which "had been misunderstood among scholars as a criticism of causality" (Daiber 2015, 21).

Although Marmura and Frank taught opposing views on al-Ghazālī's position on causality, both were able to provide solid documentation for their interpretations of the works of al-Ghazālī. Marmura saw in al-Ghazālī an occasionalist, who in some of his works adopted the philosophical language of Ibn Sīnā to express his occasionalist views. Frank, in turn, argued that al-Ghazālī's views on causality were largely similar to those of Ibn Sīnā and that he sometimes expressed these Avicennan views in occasionalist language (see the report in Griffel 2009, 179–82). Influenced by Aristotle and his Greek Neoplatonist interpreters, Ibn Sīnā had taught that the existence of every thing and event in the created world is the effect of a sufficient efficient cause yet that none of these causes can exist on its own but all rather depend for their own existence on other, "higher" efficient causes. Hence, causal chains unfold that all have their starting point in the "First Cause," a totally simple and immaterial intellect that Ibn Sīnā calls "the cause of (all) causes" (*sabab al-asbāb*). Ibn Sīnā also taught that in any given causal chain only the first cause is the absolute cause (*ʿilla muṭlaqa*), whereas all middle elements simply mediate the First's efficient causal activity to the effect (Griffel 2009, 144–5). The middle elements are understood as intermediaries or "secondary causes," which are subject to the First. Hence, Ibn Sīnā argued that God is the cause (*sabab*) of all things and all events in the world and that all are subject to His government (*tadbīr*) and predetermination (*taqdīr*). Ibn Sīnā also argued that God Himself has no choice about His creation and that creation unfolds with necessity from His being as an emanation that has no beginning in time (Griffel 2009, 133–43; McGinnis 2010, 163–72).

On this view, al-Ghazālī adapted Ibn Sīnā's theory of secondary causality so that it could become an alternative interpretation to the then existing Ashʿarite cosmology. Whereas Ashʿarites before al-Ghazālī all subscribed to an occasionalist model, al-Ghazālī made the point that secondary causality can be an equally possible explanation of God's creative activity as long as we exclude Ibn Sīnā's claim that God's actions are necessary consequences of His essence. In Ashʿarite theology, God is a free actor, who exercises a free choice (*ikhtiyār*) about what to create and when to create it. He actually created the world at one point in time. According to al-Ghazālī, God's way of creating can be either described in an occasionalist manner or through secondary causality. Al-Ghazālī's adapted model of the *falāsifa*'s secondary causality, which is different only insofar as it insists on God's free choice and the world's creation in time, fulfills the five conditions that al-Ghazālī set for causal theories just as his revised form of occasionalism does. In both theories, cause and effect are "created side by side" and this connection of concomitance between the two is "willed by God." The effects are also always "created by God." The beginning sentence of the 17th discussion in his *Tahāfut*, with its distinction between mere correlation and necessitation, has created much confusion among Western interpreters. It should, however, be read as a critique of Ibn Sīnā's views on the necessity of God's actions and not of his views on secondary causality. The connection of the cause to its effect is not necessary; that connection would be different if God had chosen to create a world different from this. For Ibn Sīnā any given causal connection is the result of God's necessity and His inability to create a world any different from the one He creates. Al-Ghazālī rejects this and maintains that God could create a different world in which the laws of nature would be different from those in our world. This is what he means when he points out that the connection between a cause and its effect is not necessary (Griffel 2009, 172–3).

A prophetic miracle would be the only event that would allow us to distinguish whether God creates by way of secondary causality or in an occasionalist way. Only in an occasionalist universe can God break His habit and create an effect without a cause. The model of secondary causality allows for no direct intervention of God once the process of creation unfolds. Here, however, al-Ghazālī sides with the *falāsifa* in the sense that he no longer thinks miracles are breaks in God's "habits." Rather, miracles are rare events in nature that are nevertheless the effect of natural causes, many of them unknown to us. In the case of Moses's staff turning into a serpent, the presence of a catalyst could have rapidly accelerated the slow natural transition from decaying wood, through fertile earth, plants and herbivores into the carnivore serpent (al-Ghazālī 2000, 172). In opposition to Ibn Sīnā, al-Ghazālī maintains that God could still break His "habits" – a phrase that stands in for the laws of nature – if He wanted so. Based on the Qur'ānic phrase: "You will not find any change in God's habit" (Qur'ān 33:62, 48:23), al-Ghazālī concludes, however, that God will never suspend the laws of nature. Given that God intends humans to pursue the natural sciences, according to al-Ghazālī, breaking His habits would be detrimental to their development and success (Griffel 2009, 194–201). If miracles are not understood as disruptions of God's habits or the laws of nature, a universe created by secondary causality remains indistinguishable from one created in an occasionalist way. Given also that God does not touch on this question in revelation, humans must, according to al-Ghazālī, suspend judgment on which of the two possible ways of how God relates to his creation is true. These two models are compossible, meaning two equally possible ways of explaining God's actions (Griffel 2009, 264–74).

One of al-Ghazālī's requirements toward causal theories is that they must account for our coherent experience of the universe and must allow predictions of future events, meaning that they must account for the successful pursuit of the natural sciences. At various points in his oeuvre he polemicizes against theologians who believe God would suspend His habits or the laws of nature. Hence it has been argued that even if al-Ghazālī does not regard the connections between causes and effects as necessary, he teaches that human knowledge of the connections and human judgments about causes and effects are indeed necessary (Griffel 2009, 201). There is nothing in al-Ghazālī's ontology or his epistemology that can be interpreted as an obstacle to the study of nature or the universe. Quite the opposite, al-Ghazālī was a champion of the rational sciences and he supported them in his works.

Future research will have to determine in what way al-Ghazālī's suggestion of the compossibility of an occasionalist cosmology and secondary causality was influential. It seems that during the first century after al-Ghazālī his view on the compossibility of explanations in *kalām* and in philosophical literature was picked up and led to a more general distinction between works of *kalām* and works of *ḥikma*. In the former, authors seem to have favored occasionalist explanations of physical change, whereas the latter genre remained committed to secondary causality.

The fact that some authors such as Fakhr al-Dīn al-Rāzī (d. 606/1210) or Sayf al-Dīn al-Āmidī (551–631/1156–1233) wrote books in both genres leads to the conclusion that they accepted al-Ghazālī's suggestion of a compossibility of these two cosmological models yet interpreted it in the context of a wider compossibility between explanations in *kalām* and *ḥikma*. This led to a situation in which books on *ḥikma* use Aristotelian hylemorphism (the doctrine that things consist of matter plus form) together with secondary causality, whereas books on *kalām* explain God's creation in terms of atomism and occasionalism. Most major textbooks of *madrasa* education in *kalām* in the Islamicate world's post-classical period, among them ʿAḍud al-Dīn al-Ījī's *Book of Stations* (*Kitāb al-Mawāqif*), ʿUmar al-Taftazānī's (d. 793/1390) *Commentary on the Creed* (*ʿAqāʾid*) of al-Nasafī (d. 508/1114) or the works of Ibn Yūsuf al-Sanūsī (d. 895/1490) on *kalām* and their commentaries, teach atomism and an occasionalist cosmology (Gardet and Anawati 1948, 327). Future studies will have to address the relationship between their cosmological models and their epistemologies. There are, however, no indications that any of these works of *madrasa* education ever argued against studying the natural sciences or against the interpretation of God's habits as the laws of nature.

Note

1 Consolidated bibliography.

Bibliography

Sources

Ibn Fūrak. ed. Gimaret, D. 1986. *Mujarrad maqālāt al-Ashʿarī* [The Bare Presentation of al-Ashʿarī's Teachings]. Beirut: Dār al-Mashriq.

Ibn Rushd. tr. van den Bergh, S. 1954. *Averroes' Tahafut Al-Tahafut (The Incoherence of the Incoherence)*. 2 vols. Cambridge: E. J. W. Gibb Memorial Trust.

Ibn Rushd. ed. Bouyges, M. 1987². *Tahāfut al-tahāfut* [Precipitance of the Precipitance]. Beirut: Dār al-Mashriq.

Research literature

Daiber, H. 2015. "God versus Causality. Al-Ghazālī's Solution and Its Historical Background," in Tamer, G., ed. *Islam and Rationality. The Impact of al-Ghazālī*. 2 vols. Leiden: Brill, 1: 1–34.

Dhanani, A. 1994. *The Physical Theory of Kalām. Atoms, Space, and Void in Basrian Muʿtazilī Cosmology*. Leiden: Brill.

Frank, R. M. 1966. "The Structure of Created Causality According to al-Ashʿarī. An Analysis of the *Kitâb al-Lumaʿ*, §§ 82–164," *Studia Islamica* 25: 13–75.

Gardet, L. and Anawati, G. 1948. *Introduction à la théologie musulmane*. Paris: J. Vrin.

Gimaret, D. 1990. *La doctrine d'al-Ashʿarī*. Paris: Les éditions du Cerf.

Goldziher, I. trs. Hamori, A. and Hamori, R. 1981a. *Introduction to Islamic Theology and Law*. Princeton: Princeton University Press (Originally published as *Vorlesungen über den Islam*. Heidelberg: Carl Winter, 1910).

Goldziher, I. tr. Swartz, M. L. 1981b [1915]. "The Attitude of Orthodox Islam Toward the Ancient Sciences," in Swartz, M. L., ed. *Studies on Islam*. New York: Oxford University Press, 185–215.

Griffel, F. 2011. "The Western Reception of al-Ghazālī's Cosmology from the Middle Ages to the 21st Century," *Dîvân. Disiplinlerarasi Çalişmalar Dergisi* (Istanbul) 16: 33–62.

Lammer, A. 2018. *The Elements of Avicenna's Physics. Greek Sources and Arabic Innovations*. Berlin: de Gruyter.

McGinnis, J. 2010. *Avicenna*. Oxford: Oxford University Press.

Munk, S. 1857–1859. *Mélanges de philosophie juive et arabe*. Paris: A. Franck.

Perler, D. and Rudolph, U. 2000. *Occasionalismus. Theorien der Kausalität im arabisch-islamischen und im europäischen Denken*. Göttingen: Vandenhoeck und Ruprecht.

Renan, E. 1852. *Averroes et l'averroïsme. Essai historique*. Paris: A. Durand.

Renan, E. 1883. tr. Ragep, S. P. and Wallis, F. 2011. *Islam and Science: A Lecture Presented at La Sorbonne, 29 March 1883*. www.mcgill.ca/islamicstudies/files/islamicstudies/renan_islamism_cversion.pdf (Accessed 2 June 2022).

Riexinger, M. 2015. "Al-Ghazālī's 'Demarcation of Science.' A Commonplace Apology in the Muslim Reception of Modern Science – and Its Limitation," in Griffel, F., ed. *Islam and Rationality. The Impact of al-Ghazālī*. 2 vols. Leiden: Brill, 2: 283–309.

Rudolph, U. 2016. "Occasionalism," in Schmidtke, S., ed. *The Oxford Handbook of Islamic Theology*. Oxford: Oxford University Press, 347–63.

Treiger, A. 2012. *Inspired Knowledge in Islamic Thought. Al-Ghazālī's Theory of Mystical Cognition and Its Avicennian Foundations*. London and New York: Routledge.

Weinberg, S. 2007. "A Deadly Certitude," *Times Literary Supplement*, 19 January 2007.

III.5
THE ROLE OF SENSE PERCEPTION AND EXPERIENCE (*TAJRIBA*) IN ARABIC THEORIES OF SCIENCE

Frank Griffel

If today a pharmaceutical company develops a new drug that it claims heals a certain disease and if it wishes to introduce the new medication into the market, it has to provide evidence for its effectiveness. In most cases, a clinical trial is the only kind of evidence governmental agencies, who approve the use of the new medication, will accept. In a clinical trial, a sufficiently large group of people affected by the disease is divided into two subgroups. One is given the medication and the other is given a placebo, a pill, for instance, that appears to be the drug but contains no medication. After some time, both groups are checked, and approval is granted only if a significantly larger number of people from the subgroup who were given the medication shows signs of improvement over those who were given the placebo. Clinical trials are often called the "gold standard" of contemporary research in medicine. They can tell us whether a drug is effective, whether a certain behavior or eating habit is harmful to our health, or whether a certain test procedure, such as mammography, truly extends life expectancy. The results of such trials are often surprising as they reveal causal connections that were often unknown and even unsuspected before the results of those trials. It should be noted, however, that clinical trials themselves are, strictly speaking, not concerned with causal connections. Clinical trials only find correlations. They show that the use of a certain drug usually or often *correlates* with the disappearance of a disease. The clinical trial itself says nothing about *how* the drug achieves or contributes to such a correlation. The trial only notes that two events, the taking of a drug, for instance, and the disappearance of a disease, tend to be concomitant. We may indeed imagine a situation – and this is not simply an imagination but part of medical research – that patients are given a certain kind of medication or told to act in a certain way because a clinical trial showed its correlation with a desired effect, without us knowing how that effect is achieved.

Universal causal connections are no longer the prime focus of today's medical research or today's natural sciences. Causal connections are universal (and, according to the Aristotelian understanding, necessary) when the effect *always* appears once the cause becomes complete. If fire reaches a ball of cotton, so the standard example discussed in Arabic literature on this subject, the cotton always combusts. If it does not, then the cause is either not complete, meaning the fire has not fully reached the cotton, or there are some external hindering forces, such as water

DOI: 10.4324/9781315170718-37

that has penetrated the cotton ball. Aristotle (384–322 BCE), who for many laid the foundations for premodern scientific inquiries into nature, required that a scientific explanation of any given phenomenon in the natural world must fulfill two conditions: it must be universal and necessary in the sense that the explanation always applies when the phenomenon appears, and it must be explanatory, meaning that it informs us of the true cause and that the effect cannot be achieved by a different one (Aristotle, *Analytica posteriora*, 71b.9–12). But how is such knowledge acquired? Muslim theologians who were critical of Aristotelian natural sciences pointed out that human sense perception only provides knowledge about the correlation of two events and not about their causal connection (al-Bāqillānī 1957, 43; al-Ghazālī 2000[CB],[1] 167). Our senses only tell us that every time a cotton ball is touched by fire, the next event is the combustion of the cotton ball. Sense perception does not tell us that the touching with fire is the cause for the cotton ball's combustion or that fire ignites cotton. In fact, Muslim theologians of the Ashʿarite school developed a different explanation in which fire is not the cause for the cotton ball's combustion (Chapters III.3 and III.4).

The opposition from Muslim theologians forced Aristotelian philosophers in Islamicate societies to come up with robust theories of scientific knowledge that fulfilled Aristotle's two demands of being universal and explanatory. Ibn Sīnā ([d. 428/1037]; Avicenna) made the most important and influential attempt in this regard. He taught that all human knowledge begins with sense perception. At the beginning of their life, the human is a blank slate without any knowledge. With this statement, Ibn Sīnā opposed Platonic theories of knowledge as recollection, as well as those who assume we have *a priori* knowledge that determines all our judgments. Ibn Sīnā concedes that humans have certain "first intelligibles" (*awwaliyyāt*) that are necessarily true under all circumstances, but these are not given to us *a priori*. Rather, we develop them during childhood based on our sense perceptions. Together with sense perception, the human faculty of estimation (*wahm*) perceives certain general rules that seem to apply to all objects. One such rule is, for instance, that the whole is greater than its parts. Another is "Everything that exists is spatially extended (*mutaḥayyiz*)." For Ibn Sīnā, these judgments of the faculty of estimation (*wahmiyyāt*) can be either true or false. Many of them are false, as the second example illustrates, because there are many things, such as the human soul or God, which are not spatially extended. All judgments of the faculty of estimation are examined by the rational faculty (*ʿaql*) and only when the latter judges them true are they accepted as first intelligibles (Ibn Sīnā 1985, 115–7, 121–3; Griffel 2011, 16–23). "The whole is greater than its parts," is an example for a first intelligible. The first intelligibles are general rules about the world that surrounds us and usually do not inform us of causal connections. In Ibn Sīnā's epistemology, judgments about causal connections must also be triggered by sense perception but they acquire their universality – meaning that the particular causal connection *always* applies – not from sense perception but from a different source that lies outside the human mind. Sense perception alone can never lead to universal and necessary judgments (Ibn Sīnā 1966, 44; McGinnis 2003, 313). Rather, Ibn Sīnā says that scientific knowledge is "abstracted" (*mujarrad*) from sense perception. What he truly means by that is disputed among modern interpreters. All agree that for Ibn Sīnā sense perception triggers a process of "abstraction" (*tajrīd*) at whose end the human has scientific knowledge of the perceived thing that includes knowledge about the causal connections that this thing engages in due to its essence. In premodern Aristotelian science, people did not speak of "causal laws" or "laws of nature." Rather, at the center of inquires in the natural sciences was the "nature" (*ṭabīʿa*) of a certain thing. "Fire" has a nature just as "cotton" has a nature. Part of fire's nature is to ignite combustible objects and part of cotton's nature is to be combustible. The latter is due to the fact that it belongs to the genus of plants. All plants are combustible due to their essence. Hence, fire has a certain active causal power (*quwwa fāʿiliyya*) and cotton has a certain passive causal power (*quwwa munfaʿila*). Once the two come together, there is a causal reaction where both the active and the passive causal powers are actualized. This is a necessary reaction that must come about every time these powers meet (McGinnis 2006, 442, 446).

How do humans know about the active and passive causal powers of things in the world? If they are part of the thing's essence, says Ibn Sīnā, they are abstracted from sense perception with the help of the Active Intellect. Since sense perception itself can only provide repeated observation and no universal judgments, scientific knowledge about the essences of things must come from something other than sense perception. Ibn Sīnā follows a long tradition of Aristotle interpretation when he says that this knowledge comes from the "Active Intellect," an incorporeal intellect, meaning a mind, that is separate from human minds and connected to one of the celestial spheres that revolve around the earth. The Active Intellect is the mind that governs the lowest of all spheres, the so-called sublunar one that includes the earth itself. Ibn Sīnā compares the process of concept acquisition with the visual perception of color. We perceive colors in the outside world only when sunlight shines on it. The sun is here compared to the Active Intellect. Just like colors on a meadow, for instance, can only be perceived if sunlight shines on them, so the concepts – and with them their essential active and passive causal powers – can only be acquired "through the mediation of illumination by the Active Intellect" (Ibn Sīnā 1959, 235; McGinnis 2007, 173–6, 180).

True scientific knowledge (*ʿilm*) about causal connections is fully universal and explanatory and can, according to Ibn Sīna, only come from the Active Intellect even if its acquisition must be triggered by sense perception (Davidson 1992, 83–94; McGinnis 2003, 312–3). This is in line with Ibn Sīnā's conviction that there is an eternal and an invariant formal level of being that shape both the individual things in the outside world as well as our knowledge of them. Ibn Sīnā argued that individual things are what they are because of real existing universal forms that determine these things. This determination includes active and passive causal powers. The Active Intellect can be understood as the repository of those universal forms – Ibn Sīnā says, it is "the giver of forms" (*wāhib al-ṣuwar*) – that provides them to the individual beings on earth as well as to our knowledge. The consistency of our knowledge with the outside world is due to the ontological coherence between the two. For Ibn Sīnā, like for many Aristotelians, the source of our knowledge of the essential active and passive causal powers of things is not nature and its observation but the separate Active Intellect.

Yet while we know many causal connections in a truly scientific way, there are, according to Ibn Sīnā, also some that we know neither in a fully universal nor explanatory manner. Most causal connections in medicine belong to this second group, as well as observations about celestial movements (Ibn Sīnā 1985, 113–4). The standard example Ibn Sīnā uses is the causal connection between the consumption of scammony, a weed native to the eastern Mediterranean, and the purging of the gallbladder. This, Ibn Sīnā teaches, is a causal connection that is not part of the essence of scammony; rather, it is an "inseparable accident" (*ʿaraḍ lāzim*) or a "proprium" (*khāṣṣ*) that is still considered part of scammony's nature (Ibn Sīnā 1966, 45–6, Janssens 2004, 55–6). Given that it is an accident and not part of the universal form of "scammony," this causal connection is not accessible to us via the Active Intellect. Rather, we acquire knowledge about this connection through repeated sense perception. If we repeatedly observe that people who take scammony purge their gallbladder, or if this happens just for the most parts (*aktharìyyan*), then we form the universal judgment that scammony purges the bile. The judgment, however, is universal only insofar as it is restricted by the circumstances in which it has been observed (*ʿilm kullī bi-sharṭ*; Ibn Sīnā 1966, 46). It is valid in our climate zone, for instance, but not in other ones, such as the colder Central Asia (Janssens 2004, 57–9). It is also not explanatory because we cannot say what in scammony causes the purging (McGinnis 2003, 320–1; Janssens 2004, 55). Still, Ibn Sīnā says, the judgment "scammony purges the bile" is valid, and it is the result of a syllogism based on repeated sense perception. This syllogism "appears in the mind while not being perceived as such" (Ibn Sīnā 1988, 134). Ibn Sīnā calls this kind of perception "experience" (*tajriba*), and he says it is certain knowledge, albeit of a limited kind (Janssens 2004, 59; Griffel 2009[CB], 209–10).

Jon McGinnis has argued that with Ibn Sīnā's observations on *tajriba* he moves away from a purely Aristotelian position on how we have knowledge of causal connections toward the direction of a more modern epistemology, where causal connections are not learned from universal forms or the Active Intellect (McGinnis 2003, 308, 325–7). Later thinkers went much further along this way. Ibn Sīnā's epistemological theories were foundational for all later debates about scientific knowledge in Arabic, even if his followers disagreed with him on some of his major assumptions. Most problematic was Ibn Sīnā's suggestion that there is a real formal level of being that underlies both the individuals in the outside world, as well as our knowledge of them. This assumption is called "epistemological realism" and many thinkers, who followed Ibn Sīnā, challenged it. Yet despite their rejection of Ibn Sīnā's realism, many of those, who came after him accepted important elements of his teachings about scientific knowledge and adapted them toward their purposes (Fazlıoğlu 2014). One early example is the Ashʿarite theologian and *mutakallim* al-Ghazālī (d. 505/1111). Like many of those, who followed Ibn Sīnā, he didn't accept his ideas about the real existence of essences and forms and neither did he accept the role of the Active Intellect for the acquisition of human knowledge. Despite this, al-Ghazālī's explanation of how humans acquire knowledge about causal connections should be regarded as Avicennian, because he elevates Ibn Sīnā's mode of experience (*tajriba*) to truly scientific knowledge and assumes that all causal connections can be known through repeated sense perception.

Al-Ghazālī's critique of Ibn Sīnā is nominalist in the sense that he rejects the assumption of universal forms in the Active Intellect that are the basis of our scientific knowledge. In another critique of Ibn Sīnā, al-Ghazālī assumes that causal connections are merely the repeated conjunctions of two events. Al-Ghazālī remains agnostic whether those two events are truly connected with one another. It is quite possible that God simply makes them appear one after the other, and that He does this out of habit and always (Griffel 2009, 175–201). Hence, the connection between cause and effect is for al-Ghazālī not necessary (al-Ghazālī 2000, 166; Griffel 2009, 147–73). Human knowledge about causal connections, however, is necessary even for al-Ghazālī (Griffel 2009, 201–8). All our judgments about causal connections – the "causal laws or "the laws of nature," as we would say – rely on repeated sense perception, where the connection has been observed either constantly or for the most part. This applies to all causal connections in a similar manner. Al-Ghazālī makes no distinction between fire igniting cotton and scammony purging the bile. Whereas Ibn Sīnā ultimately denies that repeated sense perception can lead to truly universal judgments, al-Ghazālī posits the opposite and claims that there is a "hidden syllogistic force" (*quwwa qiyāsiyya khafiyya*) that brings the individual observations together and merges them into one. Every individual observation strengthens the causal judgment like an additional eyewitness strengthens knowledge of a distant or historical event. If enough observations come together, "so that their number is so large that it cannot be determined . . . the soul grants assent (to the judgment)" (al-Ghazālī 1927, 122–3). Unlike Ibn Sīnā, al-Ghazālī does not limit the validity of these judgments to the place, where they were made and the circumstances. For him they are fully universal (*kullī*) and certain (*yaqīn*). Al-Ghazālī, however, no longer requires that scientific knowledge is also explanatory. Human judgments about causal connections rely on the regular appearance of an event together with another. On the level of ontology, al-Ghazālī pleads ignorant whether scammony truly has a causal effect on the human body. On the level of human knowledge, however, he assumes that the two events are always connected. (Griffel 2009, 211–3).

Al-Ghazālī's critique became typical for postclassical approaches to Ibn Sīnā's epistemology. With Abū l-Barakāt al-Baghdādī (d. *c.* 560/1164–5) and Fakhr al-Dīn al-Rāzī (d. 606/1210) – two thinkers who were heavily influenced by al-Ghazālī – began a period of nominalist challenges to Ibn Sīnā's epistemology (Ibrahim 2013; Falaturi 1969). These challenges led, in turn, to defenses by Naṣīr al-Dīn al-Ṭūsī (597–672/1201–1274) and his numerous students, which then triggered countercriticism by *mutakallimūn* of the 8th/14th century like ʿAḍud al-Dīn al-Ījī (d. 756/1355) and al-Jurjānī (740–816/1340–1413). Although we know as of yet too little about these numerous debates about scientific judgments during the postclassical period in Islamicate societies, our impression is that the general tendency was toward a nominalist critique of Avicennan epistemology, similar to the kind that al-Ghazālī produced. Authors point out that truly scientific knowledge in the Aristotelian sense of being universal and explicatory lies outside of human epistemological capacities. Instead, they focus increasingly to determine the object of scientific knowledge, which in Arabic was often referred to as *nafs al-amr*, "the thing itself" (Fazlıoğlu 2014).

Note

1 Consolidated bibliography.

Bibliography

Sources

al-Bāqillānī. ed. McCarthy, R. J. 1957. *al-Tamhīd fī-l-radd ʿalā l-mulḥida wa-l-muʿaṭṭila* [Arrangement on the Rebuttal of the Herectics and Deniers]. Beirut: Librairie Orientale.

al-Ghazālī. ed. Ṣabrī al-Kurdī, M. 1927. *Miʿyār al-ʿilm fī fann al-manṭiq* [Criterion of Knowledge in the Art of Logic]. Cairo: al-Maṭbaʿa al-ʿarabiyya.

Ibn Sīnā. ed. Badawī, ʿA. 1966. *al-Burhān* [The Proof/Analytica Posteriora]. Cairo: Dār al-Nahḍa al-ʿarabiyya.

Ibn Sīnā. ed. Dānishpazhūh, M. T. 1985. *al-Najāt min al-gharq fī baḥr al-ḍalālāt* [Salvation from the Immersion in the Sea of Errors]. Tehran: Intishārāt-i dānishgāh-i Ṭihrān.

Ibn Sīnā. ed. Muḥaqqiq, M. 1988. "Risālat al-Ḥukūma fī-l-ḥujaj al-muthbitīn li-l-māḍī mabdaʾan zamāniyyan [The Appraisal of the Arguments of Those Who Hold That the World Has a Beginning in Time]," in Ibn Ghaylān al-Balkhī, ed. *Ḥudūth al-ʿālam* [The Creation of the World in Time]. Tehran: Muʾassasat-i muṭālaʿāt-i islāmī, 131–53.

Ibn Sīnā. ed. Rahman, F. 1959. *Avicenna's De Anima (Arabic Text) Being the Psychological Part of Kitāb al-Shifāʾ*. London: Oxford University Press.

Research literature

Davidson, H. A. 1992. *Alfarabi, Avicenna, and Averroes, on Intellect. Their Cosmologies, Theories of the Active Intellect, and Theories of Human Intellect.* New York: Oxford University Press.

Falaturi, A. 1969. "Fakhr al-Din al-Râzî's Critical Logic," in Minuvi, M. and Afshar, I., eds. *Yādnāmah-yi Īrānī-yi Minyūrskī*. Tehran: Intishārāt-i dānishgāh-i Ṭihrān, 51–79.

Fazlıoğlu, İ. 2014. "Between Reality and Mentality – Fifteenth Century Mathematics and Natural Philosophy Reconsidered," *Nazariyat* 1: 1–39.

Griffel, F. 2011. "Al-Ghazālī's Use of 'Original Human Disposition' (*fiṭra*) and Its Background in the Teachings of al-Fārābī and Avicenna," *The Muslim World* 102: 1–32.

Ibrahim, B. 2013. "Faḫr al-Dīn ar-Rāzī, Ibn al-Hayṯam and Aristotelian Science: Essentialism Versus Phenomenalism in Post-Classical Islamic Thought," *Oriens* 41: 379–431.

Janssens, J. 2004. "'Experience' (*tajriba*) in Classical Arabic Philosophy (al-Fārābī – Avicenna)," *Quaestio* 4: 45–62.

McGinnis, J. 2003. "Scientific Methodologies in Medieval Islam," *Journal of the History of Philosophy* 41: 307–27.

McGinnis, J. 2006. "Occasionalism, Natural Causation and Science in al-Ghazālī," in Montgomery, J. E., ed. *Arabic Theology, Arabic Philosophy. From the Many to the One. Essays in Celebration of Richard M. Frank.* Leuven: Peeters, 441–63.

McGinnis, J. 2007. "Making Abstraction Less Abstract: The Logical, Psychological, and Metaphysical Dimensions of Avicenna's Theory of Abstraction," *Proceedings of the American Philosophical Association* 80: 169–83.

III.6
LOGIC
Didactics and visual representations

Johannes Thomann

This chapter does not contain a general history of logical literature in the Islamicate world. Excellent surveys of Arabic texts on logic exist (Endreß[1] 1992, 47–57; el-Rouayheb 2011, 2019; Street 2016; Rudolph *et al.* 2016[CB][2]). Instead, following the aim of this volume, practical aspects of logic are addressed. Some of these aspects have not been studied well or at all. The topics chosen are learning logic, controversies on logic and integration of logic into other scientific disciplines. This chapter consists therefore of a set of examples and unavoidably lacks historical continuity.

III.6.1 Learning Logic

Learning logic is a difficult task for any individual. At least in the beginning, the help of an instructor is necessary. Therefore, the living tradition of teaching logic among Christian scholars from late antiquity to the early Abbasid epoch was crucially important to the development of logic in the Islamicate world, even if its scope varied over time. The texts studied were Porphyry's (d. *c.* 305) *Isagoge*, and Aristotle's (384–322 BCE) *Categories*, *Hermeneutics* and the first seven chapters of his *Prior Analytics*. The earliest surviving Arabic work on logic by Ibn al-Muqaffaʿ (d. 139/756 or 142/759; or by his son), shows traces of the Syriac tradition of logic. (Hugonnard-Roche 1991, 203–4; Watt 2015, 21 n. 48), and covers the contents of these books. Ibn al-Muqaffaʿ might have translated it from an intermediate Middle Persian version (Hermans 2018). The original work might have been in Greek written in the Alexandrian school (Street 2004, 531). The topics in the four ancient works are five fundamental notions (genus, species, difference, proprium, accidens), the categories of the expression "to be," the relation between language and thought and the classes of statements and syllogistics (how to draw conclusions from propositions). There is no detailed information on philosophical education for the early Abbasid time as there is for Alexandria in late antiquity (Rudolph *et al.* 2017, 40–7). But we have a later account of how to learn logic by one the most influential logicians of all time.

Ibn Sīnā (d. 428/1037) wrote an autobiography that is, despite its subjective style, an idealizing account of how philosophy should be studied, and it is closely modeled after the ancient biographical accounts on the life of Aristotle (Ibn Sina 1974; Gutas 2014 [CB], 223). He writes that in his early youth, after having been trained in Indian arithmetic, he occupied himself with jurisprudence and became familiar "with the methods of posing questions and ways of raising objections to a respondent in the manner customary with these people" (Gutas 2014, 13–14).

DOI: 10.4324/9781315170718-38

After that, he studied the *Isagoge* with his teacher al-Nātilī (late 4th/10th century), but soon he went on to study Aristotle's books on logic and their commentaries on his own. Next, he studied Euclid's *Elements* and Ptolemy's *Almagest*, and again he soon surpassed his teacher in these fields. After having occupied himself with physics and metaphysics, he read books on medicine and cared for the sick. At the same time, when he was sixteen years old, he went back to studying jurisprudence and engaged in legal debates. In the next one and a half years, he devoted himself again to logic and the other parts of philosophy, but this time more intensively:

> I put together in front of me [sheaves of] scratch paper, and for each argument that I examined, I recorded the syllogistic premises it contained, the way in which they were composed, and the conclusions which they might yield, and I would also take into account the conditions of its premises [i.e., their modalities] until I had ascertained that particular problem.
>
> *(Gutas 2014, 16–17)*

According to this account, Ibn Sīnā's study of logic was based on Aristotle's works and commentaries on them. However, Ibn Sīnā's own works on logic demonstrate that he must have studied other sources besides Aristotle. Indeed, he had the possibility to do that, when he was enrolled as a physician in the service of Nūḥ II ibn Manṣūr (r. 366–387/976–997), the governor of Bukhara:

> One day I asked his permission to enter their library, look through it, and read its contents. He gave me permission and I was admitted to a building with many rooms; in each room there were chests of books piled one on top of the other. In one of the rooms were books on the Arabic language and poetry, in another jurisprudence, and so on, in each room a separate science. I looked through the catalogue of books by the ancients and requested those which I needed. I saw books whose very names are unknown to many and which I had never seen before nor have I seen since.
>
> *(Gutas 2014, 18)*

Obviously, access to these treasures of learning was limited to a small circle of employees at the court. A few libraries offered free access to everybody, or offered free book loans without deposit fees (Mez 1922, 169; Wüstenfeld 1866 1928, 4: 509–10; Chapter III.1). In general, scientific books were accessible to everybody, but some less-current works were only kept in exclusive court libraries, as is seen from Ibn Sīnā's account. Ibn Sīnā's way to learn logic must have been rather exceptional and only manageable for a highly gifted youth, who was able to work through Aristotle's books twice by the age of eighteen. For the less gifted, didactic help was necessary. One aspect of it will be the topic of the next section.

III.6.2 Visualization

Visualization was always a crucial didactic element in scientific traditions and diagrams were used early on, especially in the mathematical disciplines. Ancient Greek geometry and astronomy books contained large numbers of figures, but diagrams were also used in arithmetic and musical theory. One would assume that they would have been particularly helpful in logic. But with one exception, which will be discussed shortly, the surviving Greek manuscripts of Aristotle's works do not contain diagrams, nor are they found in Arabic manuscripts containing their translations. However, there are references to figures in Aristotle's text that indicate the use of diagrams together with the text. But in contrast to mathematics, in logic they did not become

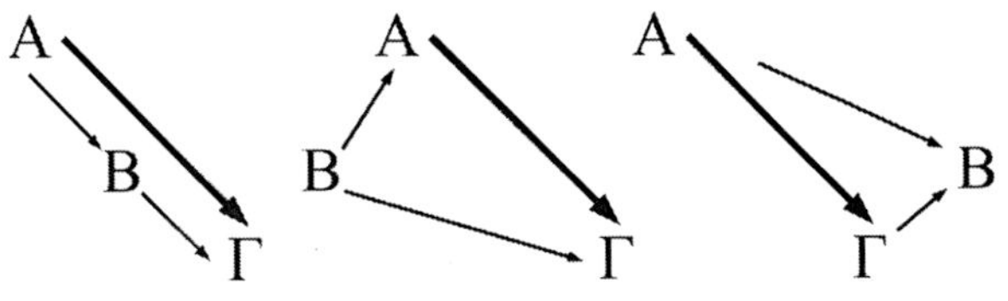

Figure III.6.1 Aristotle's syllogistic figures

Source: © Johannes Thomann, Zurich

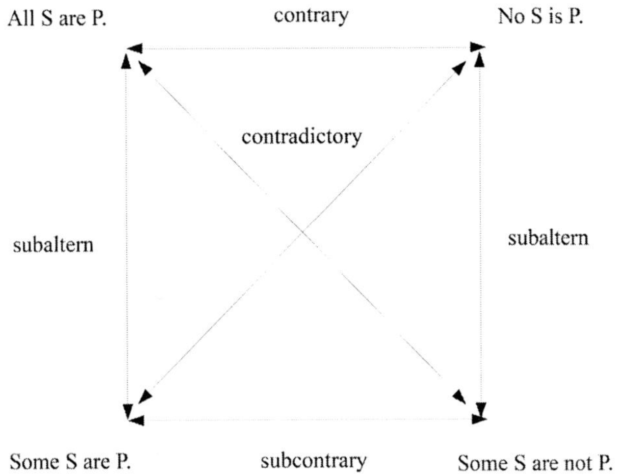

Figure III.6.2 Square of opposition

Source: © Johannes Thomann, Zurich

an integral part of the manuscript tradition. In some manuscripts of Ammonius's (d. after 517) commentary on the *Prior Analytics*, there are diagrams representing syllogisms. They became the model for diagrams in scholastic and Renaissance works on logic, but they do not seem to represent Aristotle's intention of how to visualize syllogistics (Gätje 1979). According to a recent reconstruction, they must have had the following form (Wesoły 2012; Figure III.6.1).

The letters A, B and Γ represent the terms of the syllogism, the fine arrows indicate the direction from predicate to subject in the premises, and the bold arrows indicate the same in the conclusions. Without having them in front of you, several passages in Aristotle's *Prior Analytics* remain incomprehensible, and the names of the two premises, "minor" and "major", are meaningless.

The most famous diagram in logic is the square of opposition (Figure III.6.2).

It summarizes the relationship between the four forms of subject–predicate proposition. In antiquity, it appeared in the Latin tradition only and continued to be reproduced in scholastic works on logic (Schepers 1989, 7: 1733–6). But in Aristotle's *Hermeneutics*, a square is matched to the text, which contains the opposition of propositions with the modes "necessary", "admissible", "possible" and "impossible". It is found in the oldest Greek manuscript (MS Milan, Bibliotheca Ambrosiana L 93, 9th century) in the margin (Aristotle 1949, 66). In the Arabic translation, it has the same square form as in the Greek tradition (Figure III.6.3).

But in Ibn al-Muqaffaʿ's work, the same diagram appears in a circular form (Figure III.6.4).

Furthermore, the mode of "admissible" is left out, both in the text and in the diagram. The classical square of oppositions appears in modified form too (Table III.6.1).

Figure III.6.3 Arabic translation of Aristotle's *Hermeneutics*; MS Paris, BnF, arabe 2346, fol. 188b.

Source: © BnF, Paris (gallica.bnf.fr)

Table III.6.1 Table of oppositions in Ibn al-Muqaffaʿ's treatise

general opposition	every man lives	none of the men lives
special opposition	some men live	not all men live
contradictory opposition	some men live	none of the men lives
indefinite opposition	the man lives	the man does not live
specified opposition	N.N. lives	N.N. does not live
connected by affirmation	every man lives	some men live
connected by negation	none of the men lives	not all men live

Besides the forms of opposition in the classical square, two more are included in this table: A proposition without a quantifier is called "indefinite" (*muhmal*), and a proposition with an individual as the subject is called "specified" (*makhṣūṣ*; Ibn al-Muqāffaʿ and Ibn Bahrīz 1978, 31).

Another kind of diagram occasionally appears in Arabic manuscripts: trees of definitions. This was a form of visual presentation of scientific content for general works on the sciences, including introductions to medicine. It is found in a manuscript of one of the earliest Arabic works on logic, the *Definitions in Logic* (*Ḥudūd al-manṭiq*) of Ibn Bahrīz (or Bihrīz; [late 2nd–early 3rd/late 8th–early 9th centuries]; Ibn al-Muqāffaʿ and Ibn Bahrīz 1978; Gutas 1993, 45 no. 81). It is also found in the margin of the famous manuscript of the *Organon*, written in 418/1027 (MS Paris, BnF, arabe 2346; Figure III.6.5).

Besides these simple diagrams, a much more complex type emerged in later times. Ibn Sīnā started a process in which Aristotelian logic was considerably extended. This was continued by a group of scholars who were working at the observatory at Maragha in the 7th/13th century

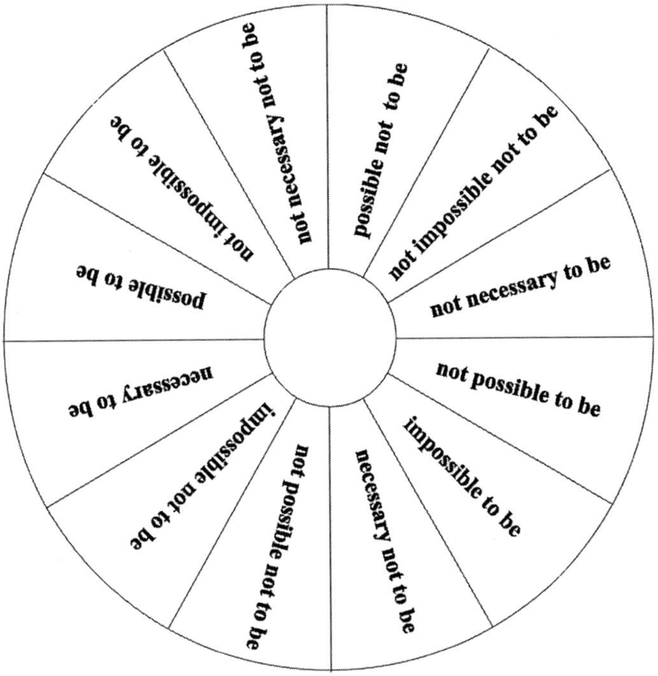

Figure III.6.4 Circular diagram of modal oppositions in Ibn al-Muqaffaʿ's treatise on logic

Source: © Johannes Thomann, Zurich

(Street 2013, Chapter I.5.2). The most famous was Nāṣir al-Dīn al-Ṭūsī (597–672/1201–1274). He wrote a commentary on Ibn Sīnā's *Pointers and Reminders* (*al-Ishārāt wa-l-tanbīhāt*), which contains a section on logic. But two of his colleagues were more influential authors for the later history of logic in the Islamicate world. Athīr al-Dīn al-Abharī (d. between 660/1263 and 663/1265) wrote an *Isagoge* (*Īsāghūjī*) and Najm al-Dīn al-Kātibī (d. 675/1276 or 693/1294) wrote his *Epistle for Shams al-Dīn* (*al-Risāla al-shamsiyya*), dedicated to the statesman Shams al-Dīn Muḥammad ibn Muḥammad Juwaynī (d. 683/1284). Both became standard textbooks in the *madrasa* teaching of logic. A third book was equally successful: *The Dawning of Lights* (*Maṭāliʿ al-anwār*) by Sirāj al-Dīn al-Urmawī (d. 682/1283), a handbook on logic and metaphysics.

It is a characteristic of this group that they distinguished more modes of propositions than Ibn Sīnā. They analyzed propositions such as the "absolute continuing" (*ḥīniyya muṭlaqa*): *All A's are always B's while A's [exist]*, or the "possible continuing" (*ḥīniyya mumkina*): *All A's are possibly B's while A's [exist]* (Street 2000, 217). The three textbooks mentioned are very brief and condensed. They were probably learned by heart, which was an important part in *madrasa* education, as the many didactic poems indicate. The versification of logical works was produced up to the 12th/18th century (el-Rouayheb 2016b). The short compendia needed explaining by an expert, and so from oral teachings written commentaries were created. Probably the most frequently read author in this field was Quṭb al-Dīn al-Rāzī al-Taḥtānī (d. 966/1364). His commentaries on the *Epistle for Shams al-Dīn* and on *The Dawning of Lights* exist in numerous copies. In Iranian libraries alone, there are 532 manuscripts of the first and 177 of the second (Dirayatī 2013, 7: 170–203;

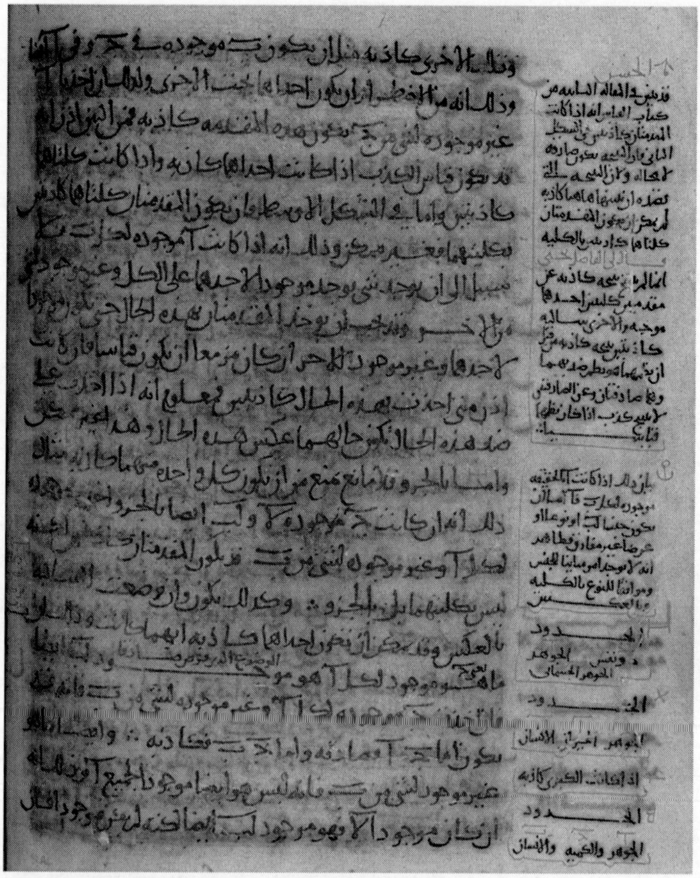

Figure III.6.5 Definitions in logic, MS Paris, BnF, arabe 2346, fol. 207b.

Source: © BnF, Paris (gallica.bnf.fr)

27: 508–21). One topic in syllogistics was the dependency of the modes of the conclusion on the modes of the premises, in the four syllogistic figures. In the system of the Maragha logic, the modes of propositions were many more than in classical logic. Time became a key element. A predicate could apply to a subject "always", "sometimes", or "at particular times" (like eclipses). In one of the oldest manuscripts of al-Taḥtānī's commentary on *The Dawning of Lights* (MS Zurich, ZB, or. 120), these dependencies are visualized in six diagrams. Their columns indicate a first premise (major), and rows represent a second premise (minor). In the diagram for the first figure, two words are written diagonally across a number of cells (Figure III.6.6).

The upper "like the major" (*ka-l-kubrā*) indicates the mode of the conclusion, while the lower "sterile" indicates that no conclusion is possible. Both words define as diagonals entire rectangular areas of cells. The large area with the label "like the major" corresponds to the rule "The conclusion here [that is in the first figure] is the same as the major [in modality], if it (that is the minor) is other than one of the two conditionals and the two conventionals" (Rescher 1967, 44). As an example, if the major is temporal and the minor is necessary, the conclusion must be temporal (like the major): "If all A's are necessarily B's at a given time, but

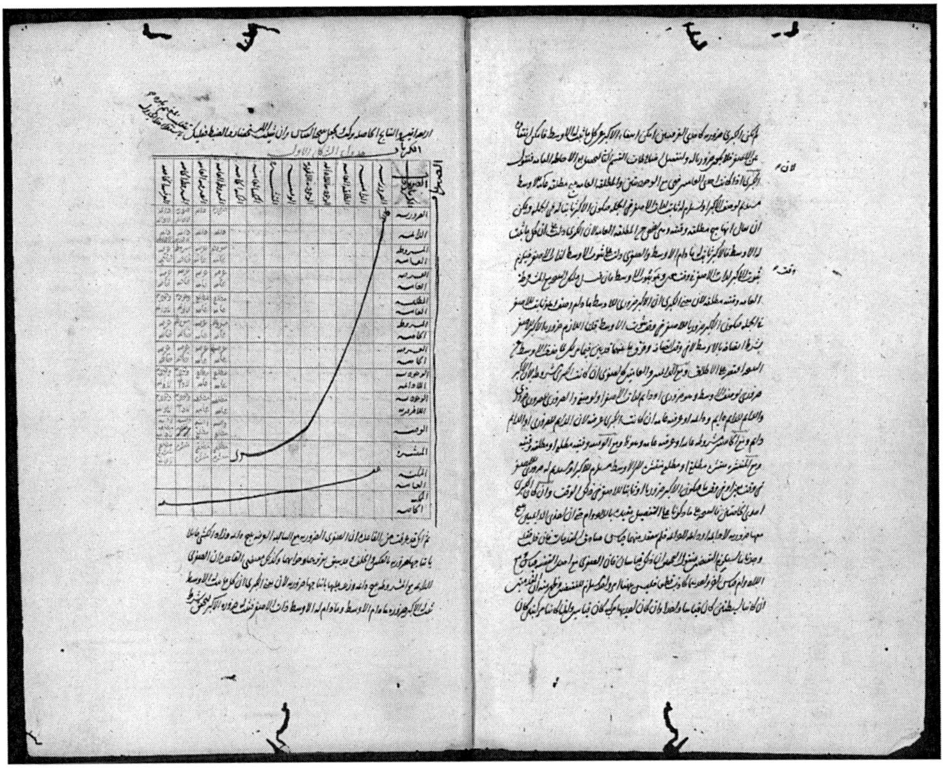

Figure III.6.6 Modal syllogisms according to Quṭb al-Dīn al-Taḥtānī (d. 766/1364); MS Zurich, ZB, Or. 120, fol. 169a.

Source: © ZB, Zurich

not always, and all B's are necessarily C's, then all B's are necessarily C's at a given time, but not always." In the upper-left area, the modes of the conclusion are written individually in each cell. In the other diagrams, larger areas are left with no indication. It is unclear if these were omitted by the scribe of the manuscript or if al-Taḥtānī did not determine all conclusions (Table III.6.2).

The majority of manuscripts do not contain these diagrams at all. In one manuscript of al-Taḥtānī's commentary (MS Tehran, Majlis Library, 540 s), the tables are drawn, but the headers and the fields contain no writing. A further peculiar feature is that the sequence of the headings of lines and columns changes in each diagram. Obviously, al-Taḥtānī sorted them in a way that allowed him to build large areas of equal indications for the conclusions. Similar diagrams are also found in his other commentary the *Redacting the Rules of Logic* (*Taḥrīr al-qawāʿid al-manṭiqiyya*). They are less complex and contain only a selection of eleven modes (Figure III.6.7).

Al-Taḥtānī's diagrams were promising as a didactic approach, but they do not seem to have had consequences for later works. Textbooks on logic remained for the most part void of illustrations. Thus, it has to remain an open question to what extent such diagrams were used in teaching.

Table III.6.2 Modal syllogisms in Quṭb al-Dīn al-Taḥtānī's treatise

		1	*2*	*5*	*8*	*9*	*10*	*11*	*12*	*13*	*3 gen. cond.*	*4 gen. conv.*	*6 spe. cond.*	*7 spe. conv.*
1	*necessary*										necessary	perpetual	n.-per. nec.	per. n.-per.
2	*perpetual*										perpetual	perpetual	per. n.-per.	per. n.-per.
3	*general conditional*										gen. cond.	gen. conv.	spe. cond.	spe. conv.
4	*general conventional*										gen. conv.	gen. conv.	spe. conv.	spe. conv.
5	*general absolute*										gen. abs.	gen. abs.	n.-per. ess.	n.-per. ess.
6	*special conditional*										gen. cond.	gen. conv.	spe. cond.	spe. conv.
7	*special conventional*					like the major					gen. conv.	gen. conv.	spe. conv.	spe. conv.
8	*nonperpetual existential*										gen. abs.	gen. abs.	n.-per. ess.	n.-per. ess.
9	*nonnecessary existential*										gen. abs.	gen. abs.	n.-per. ess.	n.-per. ess.
10	*temporal*										abs. temp.	abs. temp.	abs. n.-p. temp.	abs. n.-p. temp.
11	*spread*										spr. abs.	spr. abs.	cond.	spr. n.-p. cond.
12	*general possible*										Sterile			
13	*special possible*													

Figure III.6.7 Modal syllogisms according to al-Taḥtānī; MS Berlin, State Library, Landberg 954, fol. 79b.

Source: © STAATSBIBLIOTHEK ZU BERLIN – Preußischer Kulturbesitz, Orientabteilung

III.6.3 Controversies and criticism

There were problems in logic that led to controversies in philosophy. In his commentary on Aristotle's *Prior Analytics*, al-Fārābī (d. 339/950–1) criticized Galen for neglecting the hypothetical and the mixed syllogisms despite the fact that they are most useful in medicine and Galen as a physician should have directed most of his attention to them (quoted in Maimonides [d. 601/1204] 2015, 171). The best-known case is the question of the number of existing syllogistic figures. The three Aristotelian figures and their diagrams were already mentioned in the last section. A fourth figure ascribed to Galen is dealt with in Arabic logical works. It can be derived from the first figure, when subjects and predicates are exchanged in the premises. All earlier philosophers up to the epoch of Ibn Sīnā did not regard it as an independent figure but, rather, as a variant of the first figure. But Ibn al-Sarī, also called Ibn al-Salāḥ (d. 548/1154), wrote a treatise "On the Fourth Figure of the Assertoric Figures Attributed to Galen", in which he defends the independent status of the fourth figure and criticizes those who disapprove of it (Sabra 1965; Rescher 1965). Ibn al-Salāḥ refers to a book on "The Fourth Figure of Galen" written by Denḥā ([*fl. c.* 200/800]; Rescher 1966, 50; Rudolph *et al.* 2017, 87). The treatise had no impact on the philosophers in the western part of the Islamicate world, but in the East its ideas were a breakthrough. They were accepted among the Maragha logicians and the fourth figure found a place in the handbooks and commentaries.

The debate on the number of figures took place among philosophers. But there was also criticism from outside philosophy. The most famous case was the frontal attack against logic launched by the eminent theologian Ibn Taymiyya (d. 728/1328). From biographical contexts, it becomes clear that the logicians were not his primary target but, rather, thinkers who defended a theosophic system of theology following Ibn al-ʿArabī (558–638/1165–1240), using Ibn Sīnā's logic to argue for it. In Ibn Taymiyya's view, this kind of theological thinking went astray from sound Islamic belief. However, he attacked the logicians not from a standpoint of Islamic orthodoxy. Rather he attempted to beat them with their own weapons. His first attack was directed toward the assumptions that notions like "genus" or "species" are believed to actually exist in the external world. He claimed that they are mere conventions of some human individuals, thereby pulling away the safe ground for definition, one of the backbones of logic. He also declared the difference between essential and accidental properties/predicates as entirely arbitrary. Syllogisms require universal propositions. But in the external world, only individual particulars exist and nothing corresponding to universal propositions. Hence, syllogisms are useless, he argued, and their conclusions cannot extend knowledge about particulars in the real world (Hallaq 1993, xxxiii). This was a radical turn from philosophical realism to skeptical antirealism, in which it was denied that universal terms had any correspondence in reality (Chapter V.7).

An entirely different critique came from the side of mathematics. Ibn ʿIrāq (d. *c.* 427/1035-6), the teacher of al-Bīrūnī (362–d. after 444/973–d. after 1053) wrote a treatise against the methods of the Ismāʿīlīs to regulate the calendar and to calculate the first visibility of the crescent moon. In his debate with a high-ranking Ismāʿīlī official, he accused his adversary of not arguing mathematically but only logically. For a mathematician, he argued, the notion of "appromimation" has definite limits within which the true value must lie, while for a logician a value can only be either true or false, which is not sufficient for discussing the problem in question (Thomann 2013, 499; Ikhwān al-Ṣafāʾ 2010, 104). Ibn ʿIrāq's preference of mathematics over logic recalls al-Kindī, who admittedly included logic in his philosophical works but gave mathematics much more room.

Yet another line of critique came from a famous grammarian, al-Sīrāfī (d. 368/979). In 932, he was involved in a controversy with one of the prominent logicians of his time, Mattā ibn Yūnus (d. 328/940). In al-Sīrāfī's view, grammar is superior to logic as an instrument for deciding between truth and falsehood. Logic, as used by the philosophers, is based on the Greek language, and therefore, it cannot claim to provide universal rules of thinking. As far as propositions in Arabic are concerned, they must be analyzed by Arabic grammar and not by logic rooted in Greek grammar (Margoliouth 1905).

Another field of controversy was the question of whether a physician should study logic before studying medicine. Galen wrote in his treatise *That the Best Physician is Also a Philosopher* (Galen 1884–93, 2: 1–8) that according to Hippocrates, a physician must be trained in logic but stated that in his time this was rarely the case. Galen's treatise was translated into Arabic and widely read, and there is evidence that physicians indeed studied logic (Galen 1985). The most prominent case is of course Ibn Sīnā, whose method of studying has already been described. Another prominent physician and prolific author of medical works, Ibn Riḍwān (388–453/998–1061 or later), wrote also logical works, one on Aristotle's *Prior Analytics* and one on the application of logic in the sciences and arts (Schacht and Meyerhof 1937, 47–8, no. 76 and n. 91; Aouad 1997, 177). He was also involved into a controversy that took place in Cairo in 441/1049 (Ibn Bukhtishūʿ 1977; Schacht and Meyerhof 1937, 12). This occurred when Ibn Butlān (d. 458/1066), a Christian physician from Baghdad, who had arrived in Cairo, wrote a polemic against some Egyptian physicians, who maintained the traditional medical doctrine that the young of a bird has a warmer complexion than an adult bird. Ibn Riḍwān took up the gauntlet and wrote a fierce reply, firing angry salvoes at Ibn Butlān, reproaching him with a lack of logic and deconstructing his arguments word by word (Schacht and Meyerhof 1937, 77). For Ibn Riḍwān, medicine as such was not an independent science, but rather an art or a craft (Chapter III.5). Only in unity with philosophy could it be called a science, and thus, only those physicians who were also philosophers should be called physicians (*ṭabīb*). Practitioners without thorough philosophical education should be called "medical practitioners" (*mutaṭabbib*; Schacht and Meyerhof 1937, 77). At his time, besides practical training by a physician, much of the teaching of medicine consisted of the study of ancient texts of a theoretical nature. This was particularly true for Ibn Riḍwān, who could not afford the apprentice's fee demanded by medical practitioners of those days and thus had to learn everything from books (Schacht and Meyerhof 1937, 12).

Another controversy on the relationship of philosophy and medicine had taken place some years earlier in Basra. In 429/1037, an unnamed scholar argued that medicine is not a science of its own but a mere craft that gains its theoretical legitimacy only through philosophy. The critical case by which the methodological primacy of philosophy and medicine was discussed became a key issue. The adherents of the primacy of philosophy insisted on the separation of body and soul and refused to interpret fierce love as a bodily disease, in terms of medical concepts. Adherents of the primacy of medicine assumed close interactions between body and soul (Ibn Bukhtishūʿ 1977, 13). After the discussions, the latter were not satisfied with how their arguments were received in public and asked Ibn Bukhtishūʿ (d. after 450/1058) to write a statement. In his treatise, Ibn Bukhtishūʿ discussed not only the body–soul problem in line with Galen but also the role of logic in medicine. In the relevant chapter, he refuted the oft-repeated opinion that logic is a universal instrument of understanding and arguing (Ibn Bukhtishūʿ 1977, 54–6). His argument was that every art has its idiosyncratic terminology, which can only be learned within each practical discipline, especially among physicians. They deal with topics that cannot be understood without practical experience and that can only be acquired by joint training and usage. In his view, the otherwise privileged position of logic as the prerequisite of scientific study was thoroughly reduced, if not entirely negated. The impact of Ibn Bukhtishūʿ's treatise might have been minimal, but it gave a voice to many practitioners of medicine who remained silent but might have been the majority among physicians.

III.6.4 The significance of logic in other sciences

In the previous section, the discussions on the significance of logic for medicine have been outlined, and the examples given show that most authors of medical works regarded logical training as indispensable for physicians. Similar views are found in other disciplines where it may be less expected. In magic, for instance, which is often regarded as the most irrational product of the human mind, a similar attitude can be noted as in medicine. The earliest comprehensive Arabic book on magic, *The Aim of the Sage* (*Ghāyat al-Ḥakīm*), a real encyclopedia of its different branches, was probably written by Maslama ibn Qāsim al-Qurṭubī (d. 353/964). At the beginning of the book, after a general praise of philosophy, the author goes on to expose the importance of logic for the discipline of magic, which he names with the logical term "conclusion" (*natīja*). He argues that magic is the conclusion, the quintessence of all fields of philosophy. He elaborates on that by explaining what the elements of the Greek syllogism are. Furthermore, he explains the correspondence between logic and grammar and other forms that are not the topic of the logicians. The chapter might be short, but its prominent position at the beginning of the work directs the reader to see magic in as part and parcel of philosophical studies, all of which begin with logic (Ritter and Plessner 1962, 4–7).

A connection to logic exists also in the discipline of physiognomy. In Aristotle's *Prior Analytics*, it served as an example of a syllogism based on signs (*enthymēma*). In Arabic logical texts, it was called "the physiognomical syllogism" (*al-qiyās al-firāsī*), and the physiognomical method was discussed by Ibn Sīnā on several occasions (Thomann 1996). In Arabic physiognomical works, logic is rarely considered (Chapters I.14 and II.11). A rare case is Fakhr al-Dīn al-Rāzī's (d. 606/1210) *Book on Physiognomy* (*Kitāb al-firāsa*; Mourad 1939, 79–80, 133–4, Arabic Text 8–10; Chapter I.14). Al-Rāzī discussed in it the logical composition of arguments in physiognomy.

Another discipline with an undeserved reputation to be irrational was alchemy (Chapter I.12). But the praise of logic was also sung by alchemist authors. The author of the treatise *The Transformation of What is in Potentiality into Actuality* (*al-Ikhrāj mā fī l-quwwa ilā l-fiʿl*), which is a part of the *Jābir Corpus* (Kraus 1943), explains the structure of the syllogism in his own way (Kraus 1942, 312, no. 1):

Knowledge is light
Understanding is light
– – – – – – – – – – – – –
Knowledge is understanding

In classical logic, this would be a fallacy. A syllogism of the second figure, in which the middle term is predicate in both premises, one of the premises must be a negative proposition. The slogan of this logic was obviously "anything goes". There are 13 titles of works on logic mentioned in the *Jābir Corpus*, but none of them is preserved (Kraus 1943, 161–6).

The attitude of theologians toward logic varied (el-Rouayheb 2016a). The previous section referred to the attack on logic by Ibn Taymiyya. According to al-Fārābī, Christian theologians accepted logic to the extent it was outlined in *Isagoge*, *Categories*, *Hermeneutics* and the first seven chapters of the *Prior Analytics*, but they regarded the *Posterior Analytics*, in which the theory of scientific proof is explained, as dangerous for religious beliefs (al-Fārābī 1993, 128–30; Stroumsa 1991, 266–8). The historicity of this account has been questioned (Stroumsa 1991, 272). The case is still controversial, and statements maintaining a restricted study of the *Organon* (Gutas 2012, 86–7 no. 18) and denying such a restriction (Watt 2015, 13 no. 22) have been published in recent years. As we already saw, the part of logic that was considered acceptable was the content of the earliest Muslim work on logic by Ibn al-Muqaffaʿ. Indeed, logical elements were used by theological authors, yet they did not use Aristotelian logic. In most cases, their terminology goes back to Stoic logic (van Ess 1970, 27). It might have been transmitted by Christian or Manichaean theologians. Instead of "proof" (*burhān*), they used "sign" (*dalīl*), a literal translation of the Stoic term *sēmeion* as the key concept of the deduction of a conclusion. The resistance of the theologians against the Aristotelian system of proof was contrasted by the development in philosophy. With the work of al-Fārābī, the focus shifted from the *Prior Analytics* to *Posterior Analytics* and from syllogistics to the theory of scientific proof (el-Rouayheb 2011, 687). This process culminated in Ibn Sīnā, who based his theology on logical theory. This led to a change in the attitude of theologians toward Aristotelian, or better Avicennian, logic. Al-Ghazālī (d. 505/1111) was responsible for the long-lasting success of this new attitude. Otherwise known for his criticism of a number of philosophical positions, he defended the validity of logic in its philosophical form. Al-Ghazālī insisted on the use of philosophical logic in theology. In his view, real understanding is only possible if the syllogistic structure of an argument is understood, since only syllogisms assume knowledge to be conscious in the human mind, which otherwise would remain unconscious (Rudolph 2005, 81). Al-Ghazālī's plea led to a change in trend among the majority of theologians. It had a lasting impact on teaching theology.

In an earlier section, the use of handbooks and commentaries in the madrasas was already mentioned. Compendia for teaching theology contained chapters on logic, and the use of logic in theology became a part of theology, an example for which can be seen in the works of the very influential theologian ʿUmar al-Taftazānī (722–793/1322–1390; Walbridge 2011, 139–41). He wrote a commentary on Najm al-Dīn al-Kātibī's *Epistle for Shams al-Dīn* and included a section on logic in his introduction to theology for beginners (Madelung 2005, 232–3). The comprehensive study of Jamal Malik on the education of theologians in Lucknow contains many details on the place of logic in the curriculum (Malik 1997, 596 and below). At first, Kātibī's *Epistle for Shams al-Dīn* and the commentary of al-Taḥtānī were the textbooks (Malik 1997, 75). Later, in the 10th/16th century, they were replaced by al-Urmawī's *The Dawn of the Lights*, together with al-Taḥtānī's commentary (Malik 1997, 79). In the 12th/18th century, a reform of the curriculum took place in the *Farhangi Maḥall*, a traditional madrasa in Lucknow. Now the number of books on logic to be studied was increased to eight, some of them by more recent authors (Malik 1997, 526). The trend to give logic more room in the curriculum was continued in the 13th/19th century, despite some criticism and attempts to reduce it (Malik 1997, 215, 216, 235).

In the second half of the 14th/20th century, there were also attempts to modernize logic in the education of Islamic theologians. The most prominent author in this regard was the Iraqi Shiʿite Ayatollah Muḥammad Bāqir al-Ṣadr (d. 1980), with his book *The Logical Base of Induction*, in which he rejected philosophical realism, and proposed a system which he called "subjectivism". His important authorities are the British empiricists, and Bertrand Russell (1872–1970; Agha 2017). Nevertheless, even today, traditional logic is still a part of the theological curriculum in many religious institutions worldwide. A prominent example is al-Azhar University in Cairo, where al-Taftazānī's textbook is still used (Würtz, 2016, 2).

Much can be said about the practice of logic in jurisprudence. However, since it was closely linked to the use of logic in theology, and excellent surveys already exist on this topic (Brunschvig 1970; Hallaq 1990), a few remarks may suffice here. The principles of (arguing in) jurisprudence (*uṣūl al-fiqh*) were developed by al-Shāfiʿī (d. 204/820) before Aristotle's works became available. When the theory of the divine origin of law became triumphant, logic was suspect, and arguments were of a linguistic nature, serving the interpretation of the sacred texts (Hallaq 1990, 315–16). There were only a few jurists who accepted elements of Greek logic. These were al-Jaṣṣāṣ (d. 370/980), Ibn Ḥazm (384–456/994–1064) and Imām al-Ḥaramayn al-Juwaynī ([d. 478/1085]; Hallaq 1990, 318 n. 8). Like in theology, the turn to logic in jurisprudence came with al-Ghazālī (Hallaq 1990, 318–21). Henceforth, despite Ibn Taymiyya's intervention, logic remained an integral part of jurisprudence, as represented in the influential works of Ibn Qudāma (d. 620/1223), Ibn al-Hājib (d. 646/1249), Ibn al-Humām (d. 861/1456) and others (Hallaq 1990, 318, 321–7, 330–5). The jurist Ibn Khaldūn (732–808/1332–1406), who spent most of his life in North Africa, explicitly followed al-Ghazālī on the status of logic and theology. He regarded logic as the fundamental science, as it provides knowledge about the essences of things (Ibn Khaldun 1958[CB], 3: 137–8; Dale 2015). Similarly, from the 10th/16th to the 12th/18th centuries, the works of another North African, Ibn Yūsuf al-Sanūsī (d. 895/1490), came to dominate the teaching of theology and logic at the al-Azhar Madrasa in Cairo. The tradition that developed there and spread through North Africa offered a rational theology to both scholars and ordinary people that included logical proofs of the existence of God (el-Rouayheb 2015[CB]).

Another field with points of contact with logic is literature, and particularly poetry. Today, traditional logic is defined by the topics in the six books of Aristotle's *Organon* in its modern definition (*Categories, On Interpretation, Prior Analytics, Posterior Analytics, Topics, Sophistical Refutation*;

Blackburn 2016, 343). However, in late antiquity, logic was defined by an extended *Organon*, which included also Aristotle's *Rhetoric* and *Poetics*, and philosophers in the Islamicate world received logic in this way (Black 1990, 1–16). For them, tragedy, the main topic in Aristotle's *Poetics*, was an unknown literary genre, and they focused on Arabic poetry in their treatment of the poetic syllogism (*qiyās shiʿrī*); Gelder and Hammond 2008, 24–5; Schoeler 1983). While the Aristotelian concept of *mimēsis* refers to dramatic narrative or plot imitating human actions, the concept of Arabic *takhyīl* (imagination) refers rather to lyrical still images. Al-Fārābī used the metaphor of the mirror image (Arberry 1938, 274). A clear explanation of the poetic syllogism was given by Ibn Sīnā (Schoeler 1983, 45). It is analyzed as a syllogism of the first figure, and the example "so-and-so is a moon" is analyzed as follows:

> So-and-so is handsome
> Every handsome is a moon
> Therefore, so-and-so is a moon.

The purpose in this case is to evoke approval in the mind of the listener for the person who is praised. Another example is a verse by Ibn al-Rūmī (d. 283/896): "A rose is an anus of a mule with dung in the middle". In this case the aim is to provoke loathing for the thing being spoken about (Schoeler 1983, 65). Truth is not seen as the goal of a poetic syllogism. In poetry, often just the conclusion is expressed, while the premises are only implied. Ibn Sīnā called this the implicit form of the poetic syllogism (Gelder and Hammond 2008, 26; Schoeler 1983, 49–51). The theory of the poetic syllogism had hardly any significance for the genuine Arabic theory of poetry (Schoeler 1983, 78–82).

Scientific practices of logic, as referred to so far, were practices in which logical concepts appeared explicitly, but the focus could be opened to cases in which logical concepts were implicitly at work. There is hardly a more rigorous approach to that then the one undertaken in Benedikt Reinert's book on Afḍal al-Dīn Khāqānī (d. 595/1199) with the subtitle *Poetic Logic and Imagination* (*Poetische Logik und Phantasie*). An example may suffice to demonstrate how logical structure in the seemingly irrational poetic imagination can be analyzed. The rhetorical figure of hyperbole was defined by Ibn ʿAbd Rabbih (246–328/860–940) as "the egression from the real (*mawjūd*) into the unreal (*maʿdūm*)" (Reinert 1972, 51). It is important that the exaggeration concerns always particular aspects of subject and object. If it concerns the same aspect of subject and object it is called a primary hyperbolic relation. Furthermore, the relation must contain a distortion. This can be expressed as a distortion of an aspect of the subject, which is compared to the same aspect of the object, or it can be expressed by identification of an aspect of object, which itself is hyperbolically distorted, with the same aspect of the subject. Reinert refers to the fact that the form this primary hyperbolic relationship was conceptualized in was from one of the distinguished theoreticians of premodern Arabic rhetoric. ʿAbd al-Qādir al-Jurjānī (d. no later than 474/1074) insisted that in a hyperbole the aspect of the base subject and the predicate must be the same. He discusses this with the example "With my embellishment of some people with praise, I am similar to a man who is putting (*muʿalliq*) pearls on a sow" (Jurjānī 1959, 219). Al-Jurjānī insists that the poet compares his actions with the analogous actions of another person (Jurjānī 1959, 221). This means that both persons are compared with respect to the same aspect, which is characteristic for the primary hyperbolic relationship.

Reinert's analysis of logical structure in poetic creation is limited to the Persian poet al-Khāqānī. There is a wide field of unexplored material of implicit logic in Arabic, Persian and Turkish poetry and prose waiting for similar studies.

Notes

1 The spelling of this name in German publications is Endreß, in English publications Endress.
2 Consolidated bibliography.

Bibliography

Sources

Aristotle. ed. Minio-Paluello, L. 1949. *Categoriae et liber de interpretatione* [Categories and Book of Interpretation]. Oxford: Clarendon Press.
l-Fārābī. Rescher, N. 1993. *Al-Fārābī's short commentary on Aristotle's "Prior analytics."* Pittsburgh: University of Pittsburgh Press.
Galen. tr. Bachmann, P. 1965. *Galens Abhandlung darüber, dass der vorzügliche Arzt Philosoph sein muss*. Göttingen: Vandenhoeck.
Galen. ed. Mueller, I. 1884–1893. *Claudii Galeni Pergameni scripta minora* [The Minor Writings of Claudius Galenus Pergamenus]. Leipzig: Teubner.
Ibn al-Muqaffaʿ and Ibn Bahrīz. Dānish-Pazhūh, M. T. 1978. *Al-Manṭiq li-Ibn al-Muqāffaʿ. Ḥudūd al-manṭiq li-Ibn Bahrīz* [Logic by Ibn Muqaffaʿ. Definitions of Logic by Ibn Bahrīz]. Tehran: Anjuman-i shāhinshāhī-yi falsafa-yi Īrān.
Ibn Bukhtīšūʿ. tr. and ann. Klein-Franke, F. 1977. *Über die Heilung der Krankheiten der Seele und des Körpers, Abū Saʿīd Ibn Baḫtīšūʿ*. Beyrouth: Dar El-Machreq.
Ibn Sina. ed. and tr. Gohlman, W. E. 1974. *The Life of Ibn Sina: A Critical Edition and Annotated Translation*. Albany, NY: State University of New York Press.
Ikhwān al-Ṣafāʾ. tr. and ann. Baffioni, C. 2010. *Epistles of the Brethren of Purity: On Logic*. Oxford: Oxford University Press.
al-Jurjānī, ʿAbd al-Qāhir. tr. Ritter, H. 1959. *Die Geheimnisse der Wortkunst (Asrār al-balāġa)*. Wiesbaden: Steiner, 1959.
Maimonides. ed. and tr. Bos, G. 2015. *Medical Aphorisms. Treatise 22–25*. Provo, UT: Brigham Young University Press.
Mourad, Y. , *La physiognomie et le* Kitab al-Firasa *de Fakhr al-din (sic) al-Razi*. Paris: Geuzhner.
Ritter, H. and Plessner, M., trs. 1962. *Picatrix: das Ziel des Weisen von Pseudo-Maǧrīṭī*. London: Warburg Institute.

Manuscripts

MS Berlin, Staatsbibliothek, Landberg 954.
MS Milan, Bibliotheca Ambrosiana L 93.
MS Paris, BnF, arabe 2346.
MS Tehran, Majlis Library, 540s.
MS Zurich, ZB, or. 120.

Research literature

Agha, S. J. 2017. "Muhammad Baqir al-Sadr (d. 1979) on the Logical Foundations of Induction," in el-Rouayheb and Schmidtke, eds. [CB], 629–54.
Aouad, M. 1997. "La doctrine rhétorique d'Ibn Riḍwān et la Didascalia in rheoricam Aristotelis ex glosa Alpharabii," *Arabic Sciences and Philosophy* 7: 163–245.
Arberry, A. J. 1938. "Fārābī's Canons of Poetry," *Rivista degli Studi Orientali* 17: 266–78.
Black, D. L. 1990. *Logic and Aristotle's Rhetoric and Poetics in Medieval Arabic Philosophy* (Islamic Philosophy, Theology and Science. Texts and Studies 7). Leiden: Brill.
Blackburn, S. 2016[3]. *Oxford Dictionary of Philosophy*. Oxford: Oxford University Press.
Brunschvig, R. 1970. "Logic and Law in Classical Islam," in von Grunebaum, G. E., ed. *Logic in Classical Islamic Culture*. Wiesbaden: Harrassowitz, 9–20.
Dale, S. F. 2015. *The Orange Trees of Marrakesh: Ibn Khaldun and the Science of Man*. Cambridge, MA: Harvard University Press.

Dirayatī, M. 2013. *Fihristgān-i nuskhahā-yi khaṭṭī-yi Īrān* [Union Catalog of Iranian Manuscripts]. Tehran: National Library and Archives of the Islamic Republic of Iran.

Endreß, G. 1986. "Grammatik und Logik. Arabische Philologie und griechische Philosophie im Widerstreit," in Mojsisch, B., ed. *Sprachphilosophie in Antike und Mittelalter*. Amsterdam: Grüner, 163–299.

Endreß, G. 1992. "Die wissenschaftliche Literatur," in Fischer, W., ed. *Grundriß der Arabischen Philologie*, Bd. III: Supplement. Wiesbaden: Reichert Verlag, 3–152.

Ess, J. van. 1970. "The Logical Structure of Islamic Theology," in von Grunebaum, G. E., ed. *Logic in Classical Islamic Culture*. Wiesbaden: Harrassowitz, 21–50.

Gätje, H. 1979. *Logische Symbolik in lateinischen Aristoteles-Averroes-Fassungen*. Saarbrücken: Universität des Saarlandes.

Gelder, G. J. V. and Hammond, M. 2008. *Takhyīl: The Imaginary in Classical Arabic Poetics*. Cambridge: Gibb Memorial Trust.

Gutas, D. 1993. "Aspects of Literary Form and Genre in Arabic Logical Works," in Burnett, C., ed. *Glosses and Commentaries on Aristotelian Logical Texts: The Syriac, Arabic and Medieval Latin Traditions*. London: The Warburg Institute, 29–76.

Gutas, D. 2012. "The Poetics in Syriac and Arabic Transmission," in Tarán, L. and Gutas, D., eds. *Aristotle Poetics: Editio Maior of the Greek Text with Historical Introductions and Philological Commentaries*. Leiden: Brill, 77–128.

Hallaq, W. B. 1990. "Logic, Formal Arguments and Formalization of Arguments in Sunnī Jurisprudence," *Arabica* 37.3: 315–58.

Hallaq, W. B. 1993. *Ibn Taymiyya Against the Greek Logicians*. Oxford: Clarendon Press.

Hermans, E. 2018. "A Persian Origin of the Arabic Aristotle? The Debate on the Circumstantial Evidence of the Manteq Revisited," *Journal of Persianate Studies* 11: 72–88.

Hugonnard-Roche, H. 1991. "L'intermédiaire syriaque dans la transmission de la philosophie grecque à l'arabe," *Arabic Sciences and Philosophy* 1: 187–209.

Hugonnard-Roche, H. 1993. "Remarques sur la tradition arabe de l'Organon d'après leoyennet Paris, Bibliothèque nationale, ar. 2346," in Burnett, C., ed. *Glosses and Commentaries on Aristotelian Logical Texts: The Syriac, Arabic and Medieval Latin Traditions*. London: The Warburg Institute, 19–28.

Kraus, P. 1942. *Jābir et la science grecque*. Cairo: Institut français d'archéologie orientale.

Kraus, P. 1943. *Le corpus des écrits jābiriens*. Cairo: Institut français d'archéologie orientale.

Madelung, W. 2005. "At-Taftazānī und die Philosophie," in Rudolph, U. and Perler, D., eds. *Logik und Philosophie: Das Organon im arabischen und lateinischen Mittelalter*. Leiden: Brill, 227–38.

Malik, J. 1997. *Islamische Gelehrtenkultur in Nordindien: Entwicklungsgeschichte und Tendenzen am Beispiel von Lucknow*. Leiden: Brill.

Margoliouth, D. S. 1905. "The Discussion between Abu Bishr Matta and Abu Sa'id al-Sirafi on the Merits of Logic and Grammar," *The Journal of the Royal Asiatic Society of Great Britain and Ireland*: 79–129.

Mez, A. 1922. *Die Renaissance des Islâms*. Heidelberg: Winter.

Reinert, B. 1972. *Ḫāqānī als Dichter. Poetische Logik und Phantasie*. Berlin: de Gruyter.

Rescher, N. 1965. "New Light from Arabic Sources on Galen and the Fourth Figure of the Syllogism," *Journal of the History of Philosophy* 3.1: 27–41.

Rescher, N. 1966. *Galen and the Syllogism: An Examination of the Thesis, that Galen Originated the Fourth Figure of the Syllogism in the Light of New Data from Arabic Sources. Including an Arabic Text ed. and Annotated tr. of Ibn al-Salah's Treatise "On the Fourth Figure of the Categorical Syllogism"*. Pittsburgh, PA: University of Pittsburgh Press.

Rescher, N. 1967. *Temporal Modalities in Arabic Logic*. Dodrecht: Reidel.

el-Rouayheb, K. 2011. "Logic in the Arabic and Islamic World," in Lagerlund, H., ed. *Encyclopedia of Medieval Philosophy*. Dodrecht: Springer, 686–92.

el-Rouayheb, K. 2016a. "Theology and Logic," in el-Rouayheb and Schmidtke, eds. [CB], 408–31.

el-Rouayheb, K. 2016b. "Aḥmad al-Mallawī (d. 1767): Commentary on the Versification of the Immediate Implications of Hypothetical Propositions," in el-Rouayheb and Schmidtke, eds., 509–34.

el-Rouayheb, K. 2019. *The development of Arabic logic (1200–1800)* (Medieval and Early Modern Philosophy 2). Basel: Schwabe.

Rudolph, U. 2005. "Die Neubewertung der Logik durch al-Ġazālī," in Rudolph, U. and Perler, D., eds. *Logik und Theologie. Das Organon im arabischen und lateinischen Mittelalter*. Leiden: Brill, 73–97.

Sabra, A. I. 1965. "A Twelfth-Century Defence of the Fourth Figure of the Syllogism," *Journal of the Warburg and Courtauld Institutes* 28: 14–28.

Schacht, J. and Meyerhof, M. 1937. *The Medico-Philosophical Controversy Between Ibn Butlan of Baghdad and Ibn Ridwan of Cairo: A Contribution to the History of Greek Learning Among the Arabs*. Cairo: The Egyptian University.

Schepers, H. 1989. "Quadrat, Logisches," in Ritter, J. and Gründer, K., eds. *Historisches Wörterbuch der Philosophie*. Basel: Schwabe, 7: 1733–6.

Schoeler, G. 1983. "Der poetische Syllogismus: Ein Beitrag zum Verständnis der 'logischen' Poetik der Araber," *ZDMG* 133.1: 43–92.

Street, T. 2000. "On Arabic Modal Logic," *Journal of Islamic Studies* 11.2: 209–28.

Street, T. 2004. "Arabic Logic," in Gabbay, D. M. and Woods, J., eds. *Handbook of the History of Logic. Volume 1: Greek, Indian and Arabic Logic*. Amsterdam: Elsevier, 523–96.

Street, T. 2013. "Arabic and Islamic Philosophy of Language and Logic," in Zalta, E. N., ed. *The Stanford Encyclopedia of Philosophy*. Palo Alto: Stanford University. https://plato.stanford.edu/entries/arabic-islamic-language/ (Accessed 8 May 2018).

Street, T. 2016. "Kātibī (d. 1277), Taḥtānī (d. 1365), and the Shamsiyya," in el-Rouayheb and Schmidtke, eds., 348–74.

Stroumsa, S. 1991. "Al-Fārābī and Maimonides on the Christian Philosophical Tradition: A Re-evaluation," *Der Islam* 68: 263–87.

Thomann, J. 1996. "Avicenna über die physiognomische Methode," in Campe, R. and Schneider, M., eds. *Geschichten der Physiognomik: Text, Bild, Wissen*. Freiburg im Br.: Rombach, 47–63.

Thomann, J. 2013. "A Mathematician's Manifesto on Scientific Reasoning Against Religious Convictions," in Klemm, V. and N. al-Shaʿar, N., eds. *Sources and Approaches Across Disciplines in Near Eastern Studies*. Leuven: Peeters, 491–501.

Walbridge, J. 2011. *God and Logic in Islam: The Caliphate of Reason*. Cambridge: Cambridge University Press.

Watt, J. W. 2015. "The Syriac Aristotelian Tradition and the Syro-Arabic Baghdad Philosophers," in Janos, D., ed. *Ideas in Motion in Baghdad and Beyond Philosophical and Theological Exchanges between Christians and Muslims in the Third/Ninth and Fourth/Tenth Centuries*. Leiden: Brill: 7–43.

Wesoły, M. 2012. "ΑΝΑΛΥΣΙΣ ΠΕΡΙ ΤΑ ΣΧΗΜΑΤΑ: Restoring Aristotle's Lost Diagrams of the Syllogistic Figures," *Peitho: Examina Antiqua* 1.3: 83–114.

Würtz, T. 2016. *Islamische Theologie im 14. Jahrhundert: Auferstehungslehre, Handlungstheorie und Schöpfungsvor stellungen im Werk von Saʿd ad-Dīn at-Taftāzānī*. Berlin: de Gruyter.

Wüstenfeld, F. 1866–1928. *Muʿǧam el-buldān = Jacut's geographisches Wörterbuch*. Leipzig: F. A. Brockhaus.

III.7
MEDICAL COMMENTARIES

Nahyan Fancy

Chapters I.4 and I.10 have already addressed the translation and appropriation of Greco-Roman medical texts that led to the development of the learned tradition of medicine (*ṭibb*) over the course of the 2nd/8th and 3rd/9th centuries. This learned tradition of medicine was not merely a continuation of the Hellenistic medical tradition as found in places like Alexandria and other centers of Greek learning in the 6th century. It included influences from other medical traditions (Indian, Persian and others), as well as knowledge gleaned from local practices, trade networks and so forth. Historians of medicine have long recognized the novel applications of Galenic-Hippocratic humoral theory by physicians in Islamicate societies, such as al-Rāzī's (d. 313/925 or 323/935) application of humoral theory to smallpox and measles. They have recognized the emergence of specialist medical tracts on ophthalmology, pharmacology and surgery over time too. Physicians of premodern Islamic societies have also been credited for the systematization of medicine into large compendia, such as the *Book on Medicine for al-Manṣūr* (*al-Kitāb al-Manṣūrī fī l-ṭibb*) of al-Rāzī, *The Royal Book* (*al-Kitāb al-Malakī*) of ʿAli ibn al-ʿAbbās al-Majūsī (d. *c.* 384/994) and, the most famous one, *The Canon of Medicine* (*Kitāb al-Qānūn fī al-ṭibb*) of Ibn Sīnā (lat. Avicenna; d. 428/1037).[1] Nevertheless, most of the historical work has tended to ignore the Arabic medical commentary tradition.

III.7.1 Introduction

The commentary tradition dominated the learned medical scene from the end of the 5th/11th century well into the 12th/18th century across the entire expanse of Muslim-dominated regions lying between the Atlantic Ocean and the Oxus River (and beyond). Consequently, this chapter focuses on two distinct, yet interrelated sets of Arabic medical commentaries: the commentaries on the Hippocratic *Aphorisms* and commentaries on *The Canon of Medicine* and its abridgment, *The Epitome* (*Kitāb al-Mūjaz fī al-ṭibb*/*Kitāb Mūjaz al-Qānūn*). Commentaries were composed on other works too, both earlier Greek works, such as those of Hippocrates and Galen, and later Islamicate works, such as those of Najīb al-Dīn al-Samarqandī (d. 618/1222) and Maḥmūd al-Jāghmīnī (d. *c.* 618/1221–2). Commentaries were also composed in Persian, Hebrew and Ottoman Turkish, but the Arabic commentaries on the *Aphorisms* and *Canon* have received the most attention to date; that is why this chapter focuses on them.

DOI: 10.4324/9781315170718-39

Although commentaries have often been seen as a sign of intellectual stagnation, historians of astronomy, mathematics, philosophy and, increasingly, medicine have argued that commentaries "are sometimes a mine of their own to be exploited for the original ideas they contain" (Saliba 1996, 714). Medical commentaries of the post- 600/1200 period are concerned primarily with "verification" (*taḥqīq*; Fancy 2013b, 2017, 2018a; Karimullah 2019). As Khaled el-Rouayheb, Robert Wisnovsky and Asad Ahmed, amongst others, have shown, verification was the central tenet of philosophical commentaries, and also played a significant role in many other intellectual disciplines (el-Rouayheb 2006, 2015[CB];[2] Wisnovsky 2013; Ahmed 2013). In the philosophical texts, verification of the source text (*matn*) could (and often did) encompass both of the following activities: lexical analysis (definitions, alternative readings, identifications of texts) and theoretical analysis (completing proofs, reordering arguments). Verification could also lead commentators to raise objections against a given theory, refute it and even develop alternative ones in its stead (Wisnovsky 2013, 354–7). This latter part has attracted the most attention of historians for its revolutionary underpinnings, but it is important to recognize that "originality and dynamism can be found in [other] sections of commentaries" too (Wisnovsky 2013, 357).

By examining commentaries through the lens of verification, we can uncover three key features of learned medicine in the post-600/1200 period. First, there was a slow shift away from the Galenic practice of writing commentaries to a more scholastic, and philosophical practice of exegetical verification associated with scholars such as Fakhr al-Dīn al-Rāzī (d. 606/1210). Some Islamicate medical writers were looking beyond the Galenic corpus already in the 5th/11th century. By the mid-8th/14th century, both Galen's texts and commentarial method were abandoned in favor of other authors and exegetical practices, chief among them Fakhr al-Dīn's method of verification. Second, and on a related note, the penetration of philosophical verification practices into medical commentaries allowed medical writers to subject medical theory and natural scientific claims to careful scrutiny within their works. Philosophers, theologians and physicians thus all engaged in verifying theoretical claims within the ambit of medical texts. Finally, this verification of theoretical claims led to new developments in natural science and, more important, medical theory. The most famous case here is that of Ibn al-Nafīs (d. 687/1288), but it is important to recognize that subsequent medical authors continued to debate, modify and refine his claims by subjecting them to verification in later commentaries.

III.7.2 Commentaries on the *Aphorisms*

In a recently concluded project, Peter Pormann and his team at the University of Manchester edited and began analyzing the entire extant corpus of Arabic medical commentaries on Hippocrates's *Aphorisms*. The texts are now available online (Pormann and Karimullah 2017).[3] As these texts are further analyzed, we will undoubtedly learn more about the practice of medical composition, medical teaching (of both theory and practice) and even the place of medicine in intellectual and social circles of premodern Islamicate societies over time. Nonetheless, early investigations of this corpus have revealed some interesting insights into the evolution of this discourse, both in style and in content, over five centuries.

As Pormann and Joosse have stated, "[f]ew secular texts have had such an impact on subsequent generations as the Hippocratic *Aphorisms*. They influenced not only medical theory and practice, but also affected popular culture" (2012, 211). Moses Maimonides (d. 601/1204), the famous Jewish, Andalusian philosopher-theologian and physician to the Ayyubid ruler, Saladin (r. 570/1174–589/1193), stated that it was not uncommon for schoolchildren to know some aphorisms by heart. Similarly, the philosopher and physician, ʿAbd al-Laṭīf al-Baghdādī (d. 629/1231), stated that the *Aphorisms* "constituted the most important Hippocratic text for medical teaching; therefore, he, like many others, penned a commentary on them" (Pormann and Joosse 2012, 211). Many other well-known figures also composed commentaries on this work, including the Persian physician al-Rāzī, the Egyptian physician ʿAlī ibn Riḍwān (388–453/998–1061 or later), the Syrian-Egyptian physician-jurist Ibn al-Nafīs and the Syrian Christian physician-surgeon, Ibn al-Quff (d. 685/1286). In fact, apart from Ibn Bājja (d. 533/1139), the Andalusian philosopher, the only surviving commentaries on the *Aphorisms* are from physicians. This fact demonstrates the continued importance of the text for medical teaching in premodern Islamic societies, even though, as will be clear shortly, there is a significant drop in the production of commentaries on this text after 700/1300. Moreover, various features (lexical and theoretical) of the subsequent manuscript tradition of the *Aphorisms* and *The Canon of Medicine* commentaries suggest that, over time, the *Canon* and its abridgments came to share the central place that the Hippocratic *Aphorisms* (and other Hippocratic texts) had occupied in the medical curriculum in the pre-700/1300 era.

The early Arabic commentators on the *Aphorisms* saw themselves as continuing and reenergizing the legacy of the pre-Islamic Alexandrian medical curriculum. Both al-Nīlī (d. 420/1029) and Ibn Riḍwān were physicians who explicitly stated that their goal in composing their works on the *Aphorisms* was to reverse the "decline" in quality of medical education from the purported heyday of pre-Islamic era Alexandria. Although they both partly blame this decline on the rise of abridgments and commentaries, their own works are precisely in those genres. Both works continue the practice of Hellenistic Greek commentators in extracting, summarizing and distilling content from Galen's commentary on the *Aphorisms* for either personal benefit (in the case of Ibn Riḍwān), or that of contemporary physicians and students (in the case of al-Nīlī). As Kamran Karimullah states, "in examining Ibn Riḍwān's and Nīlī's abridgment strategies, it is apparent that the text of Galen's commentary exerted a near exclusive hold over how these authors understood the Hippocratic lemmata" (2017, 320).

Galen's hold over the medical commentators of the *Aphorisms* slackened over the next few centuries.[4] The Central Asian physician, Ibn Abī Ṣādiq's (d. after 460/1068) *Commentary on the Aphorisms* marks an important threshold because, judging from the surviving manuscripts, it was the most widely used Arabic commentary on the *Aphorisms*. Ibn Abī Ṣādiq followed Galen on most accounts and even went so far as to compose a separate response to al-Rāzī's *Doubts Against Galen* defending Galen's views. Yet, Ibn Abī Ṣādiq also brought into his *Commentary on the Aphorisms* works of other medical writers, including Greek commentators on the *Aphorisms* other than Galen, such as Rufus of Ephesus, and earlier Islamicate authors, such as al-Rāzī and Ibn Sīnā (Karimullah 2017, 330; Hauter 2020).

This displacement of Galen's textual authority is complete by the 7th/13th century. In the commentary of Ibn al-Nafīs, for example, Galen is rarely cited as an authority, and much of the discussion of individual aphorisms moves beyond the confines of Galen's commentary, often even challenging Galen's interpretation itself. Thus, in commenting on Aphorism V.48 ("when the child is male, it is more common for it to be generated on the right side"), Ibn al-Nafīs takes the anti-Galenic (and anti-Avicennan) position that semen generated in the right male testicle is not warmer than the left, implying that the heat of the right side of the womb is sufficient to generate a male fetus regardless of whether the semen is drawn from the right or the left testicle (Fancy 2017). Similarly, when discussing critical days for medical illnesses in his commentary on Aphorism II.24, Ibn al-Nafīs sides with Ibn Sīnā against Galen in calculating the critical days purely based on the effective synodic lunar month (29½ days minus three days) rather than an average of the sidereal month (27⅓ days) and the effective synodic month as Galen had employed (Cooper 2018).

In fact, it would be fair to say that Ibn Sīnā's *Canon* and the commentary tradition displaced Galen and took center stage in the *Aphorisms* commentaries from the 13th century onward. Both Ibn al-Nafīs and Ibn al-Quff composed commentaries on the *Aphorisms* and the *Canon*. Their commentaries on the *Aphorisms* often bring in material from the latter tradition. Ibn al-Quff, for example, employs the exegetical method of verification in his *Commentary on the Aphorisms*, a method that was deployed so effectively by Fakhr al-Dīn al-Rāzī in his commentaries on the Avicennan corpus, including the *Canon*. Moreover, when he discusses the concept of pain, Ibn al-Quff quickly summarizes Galen's position that pain is caused exclusively by the dissolution of continuity before proceeding to assess and analyze the positions on pain of Ibn Sīnā, Fakhr al-Dīn (found in his *Commentary on the Canon*) and Ibn Rushd (lat. Averroes; 520–595/1126–1198), the Andalusian philosopher-physician. By the end of this long investigation on the different positions on pain, Ibn al-Quff finally concurs with the position of Averroes "that noxious irregular mixture [i.e., a temperament that irregularly inclines toward cold, hot, moist or dry] causes pain intrinsically and that dissolution of continuity causes pain by means of it" (Karimullah 2019). This displacement of Galen's centrality in the *Aphorisms* corpus is so thorough that the discussion of pain in the 14th-century physician, ʿAbd al-Raḥīm al-Ṭabīb's (d. *c.* 787/1385) *Commentary on the Epitome of the Aphorisms* is entirely based on Ibn Sīnā's *Canon* (Karimullah 2017, 341–7).

To summarize, over the course of five centuries, Galen's hold over the interpretation and understanding of the Hippocratic *Aphorisms* was slowly relinquished. Not only did physicians start looking for sources outside of Galen's corpus to further their understanding of the *Aphorisms*, but they also employed non-Galenic commentarial strategies. By the 8th/14th century, they often did not even include Galen's interpretation of a particular aphorism. In the few cases they did, the commentators relegated Galen's interpretation to one of many options. The later commentaries on the *Aphorisms* also reveal the growing importance in medical teaching and practice of Ibn Sīnā's *Canon* and the commentaries and abridgments it spawned. This shifting of medical authority away from the Galenic corpus thus parallels to some degree the move away from Aristotle in philosophy (i.e., the Avicennan revolution) that has been highlighted by historians of philosophy for some time (e.g., Adamson 2011).

III.7.3 Commentaries on the *Canon* and its abridgments

Although a handful of commentaries were composed on the *Canon* in the first century after Ibn Sīnā, including those by his student, al-Jūzjānī (d. 462/1070), and one by al-Īlāqī (*fl.* 440/1068; Rahman 1986), there is a veritable explosion in commentaries on the *Canon* from the second half of the 6th/12th century onward (Savage-Smith 2013, 150). These latter commentaries include those by the Egyptian Jewish physician, Ibn Jumayʿ (d. 594/1198), and the philosopher-theologian, Fakhr al-Dīn al-Rāzī. Ibn Jumayʿ's *The Explication of the Obscure Through the Revision of the* Canon (*al-Taṣrīḥ bi-l-maknūn fī tanqīḥ al-Qānūn*) selects passages from the first three books of the *Canon* for further comment,[5] while Fakhr al-Dīn's commentary focuses exclusively on book one, *Kulliyyāt* (Universal Principles). Interestingly, the non-physician Fakhr al-Dīn's *Commentary on the Kulliyyāt* played a greater role in the development of the *Canon* commentary tradition over the next century.

Fakhr al-Dīn's importance to postclassical philosophy and theology in Islamicate societies cannot be overstated. The fact that he left his mark on medical writings of the 7th/13th century needs some explanation as he was neither a physician nor very well versed in medicine. His commentary focuses heavily on issues of logic (definitions, proper argumentation, consistency and so forth) and topics in natural science (Aristotelian physics) of importance to philosophers and theologians (motion, elements, sensation and so forth). This leads Fakhr al-Dīn to posit new opinions of importance to physicians sometimes, such as his claim that pain is only caused by irregular noxious mixtures within temperaments and not by the dissolution of continuity as maintained by Galen (Karimullah 2019). However, more often Fakhr al-Dīn seeks to tighten the argumentation of passages in the *Canon*. Or he tries to reconcile its claims with the natural scientific or logical principles found in Ibn Sīnā's philosophical work, *The Book of Healing* (*Kitāb al-Shifāʾ*), while upholding the validity of the main theoretical claims found in the *Canon*.

Nonetheless, Fakhr al-Dīn had a significant impact on subsequent medical commentaries in both content and style of exegesis. Primarily, this is due to the fact that many of his students – physicians, philosophers and theologians – composed their own commentaries on Book I, which were, in turn, read and responded to by scholars across the Mongol, Ayyubid and Mamluk realms. For example, al-Nakhjawānī (d. 651/1253), a philosopher in Tabriz and colleague of Naṣīr al-Dīn al-Ṭūsī (597–672/1201–1274), wrote a response to the doubts raised by Fakhr al-Dīn in his *Commentary on the Kulliyyāt*. A generation later, Quṭb al-Dīn al-Shīrāzī (633–710/1236–1311) wrote that when he started reading the *Canon* as part of his medical education, he found Fakhr al-Dīn's commentary to be the most helpful (Iskander 1967, 49). Fakhr al-Dīn's most famous pupils with regard to teaching and commenting on the *Kulliyyāt* include Quṭb al-Dīn al-Miṣrī (d. 618/1221), a Maghribī physician who died during the Mongol invasion of Nishapur; al-Khūnajī (d. 649/1249), a Persian philosopher who moved to and died in Egypt; and Khusrawshāhī (d. 652/1254), a Persian philosopher who entered into the service of the Ayyubid Sultan al-Malik al-Nāṣir Dāwūd (r. in Damascus 624–626/1227–1228 and in Karak 626–646/1228–1248). Through their written commentaries (al-Miṣrī and al-Khūnajī) and their direct teaching of the *Kulliyyāt* (al-Khūnajī and Khusrawshāhī), these students brought Fakhr al-Dīn's method of exegetical verification and substantive critiques of the *Kulliyyāt's* content (and their resolutions) to physicians in Syria, Egypt and beyond. Moreover, since the *Kulliyyāt* presented a "welcome exposition of natural and medical sciences" without the metaphysical or theological obstacles found in Ibn Sīnā's *Healing* (Endress 2006[CB], 391), Fakhr al-Dīn and his students' commentaries of the *Kulliyyāt* were also taught to jurists and theologians in *madrasa*s, where they spawned further commentaries on this text from philosophers and theologians.

The penetration into the Ayyubid intellectual scene of Fakhr al-Dīn's commentary is so thorough that every major physician associated with the famous Damascene physician and founder of the medical madrasa (Chapter III.2), al-Dakhwār (d. 628/1230), incorporated the content or style of Fakhr al-Dīn's exegetical verification to some degree. This is even true of the generation of students who studied under al-Dakhwār's students, such as Ibn al-Quff (Karimullah 2019) and al-Sāmirī (d. 681/1282). Al-Sāmirī states explicitly in the introduction of his commentary on the *Kulliyyāt* that he has followed the model of Fakhr al-Dīn and his student al-Khūnajī and added to it the critique and responses of al-Dakhwār's student, the physician Najm al-Dīn ibn Minfākh (d. 652/1254; MS Oxford, Bodleian Library, MS Marsh 464, fol. 1a).

A vital consequence of the use of Fakhr al-Dīn's *Commentary*, in particular its exegetical style, for later commentators is that medical writers were now free to test and verify Ibn Sīnā's natural scientific claims within the ambit of the medical commentary itself. In a famous passage from the *Canon*, Ibn Sīnā places certain restrictions on the kinds of inquiries physicians can conduct into theoretical matters insofar as they are physicians and not philosophers. Dimitri Gutas has cited this passage as a key reason for why "Arabic medicine . . . ultimately never went beyond Galenism and Avicennism" (Gutas 2003, 160). In actual fact, by adopting Fakhr al-Dīn's verification method and engaging with his critiques of the medical theory and physics found in the *Kulliyyāt*, subsequent medical commentators revised not only aspects of Avicennan medical theory but even Avicennan physics.

An excellent example of revisions in Avicennan physics can be found in the medical commentaries' discussions on the definition of pulse that is found in part two of the *Kulliyyāt*. This part is on the universal discourse on diseases, their causes and symptoms, which, according to Gutas, falls under the theory of medicine, and thus outside of the purview of dispute for a physician (Gutas 2003, 153). Fakhr al-Dīn's commentary on the definition accuses Ibn Sīnā of providing an improper definition, for Ibn Sīnā never prescribes the category of motion that pulse falls under nor its efficient cause. Fakhr al-Dīn ultimately provides a complete definition, which agrees entirely with Ibn Sīnā's own understanding, "Pulse is a movement of the receptacles of the spirit in the category of place, issued from its vital faculties [and] composed of expansion and contraction in order to temper the spirit with fresh air" (MS Oxford, Bodleian Library, MS Arch. Seld. A64, fol. 158a). However, by engaging with the different categories of motion and arguing for why pulse falls in the category of place, Fakhr al-Dīn set a precedent for subsequent medical commentators to engage in the *verification* of this claim rather than accepting it as an established natural scientific claim from within the philosophical corpus. This process of verification is what led Ibn al-Nafīs to challenge this categorization and, ultimately, led medical commentators to engage with and arrive at new understandings of motion and space (for details, see Fancy 2018a).

In brief, Ibn al-Nafīs recognized that arteries technically never leave the place that they occupy while they pulse. Based on the Aristotelian and Avicennan conception of place (*makān*) as the exterior surface of a contained body that is touching the interior surface of the containing body, Ibn al-Nafīs rightly remarked that the arteries are never *dis*placed – they never actually stop touching the surfaces they touch while they pulse. He thus placed arterial motion in the category of positional motion, akin to the rotation of a sphere. Yet, al-Shīrāzī in his own commentary on the *Kulliyyāt*, which he entitled *The Gift to Saʿd* (*al-Tuḥfa al-Saʿdiyya*), and following him the Anatolian philosopher-theologian, al-Aqsarāʾī (d. 779/1379), in his commentary on the *Epitome*, rightly pointed out that, according to Ibn al-Nafīs's strict understanding of motion in the category of place, water in a cup that is moved from one house to the next would not be considered to have moved with respect to place. This seemed to these two esteemed commentators to be patently absurd. Therefore, al-Shīrāzī and al-Aqsarāʾī recognized that understanding place with respect to the limits of contiguous bodies is problematic. Thus, they shifted to talking about place more metaphorically as where something is (*ayn*, pl. *uyūn*). This change in conception led the Persian physicians Nafīs al-Kirmānī (d. 853/1449–50) and Ḥakīm Shāh ibn al-Mubārak al-Qazwīnī (d. 928/1521) to develop distinctions between real and metaphorical place and between *per se* (in itself) and *per accidens* (by accident) motion to further refine their understanding of motion in the categories of place and position. As al-Qazwīnī concludes in his *Commentary on the Epitome*,

> However, I say that this objection [of al-Shīrāzī] is to be rejected, for the author (Ibn al-Nafīs) meant to say that the *per se* motion in the category of place necessarily requires that the moving body leave its real place (*makānihi al-ḥaqīqī*). The motion of the water due to the motion of the cup, on the other hand, is not *per se*, but rather *per accidens* such as the motion of a person sitting on a boat.
>
> *(Fancy 2018a, 75)*

It is worth emphasizing that none of the discussions surrounding pulse's category of motion or the proper understanding of place ("space" for us) has any bearing on medical theory or practice. Yet, medical commentators devoted considerable ink (a number of folios in the case of al-Shīrāzī and al-Qazwīnī) to this natural scientific topic that was (and was expected to be) treated in philosophical books, such as Ibn Sīnā's *Healing* or *Remarks and Admonitions* (*al-Ishārāt wa-l-tanbīhāt*), and the Persian philosopher Athīr al-Dīn al-Abharī's (d. between 660/1263 and 663/1265) *The Guide* (*al-Hidāya*) – a text that was widely used in madrasas across Islamicate societies from the 8th/14th century onward. This fact supports the findings of earlier scholars that theoretical medicine was part "of the repertoire of an erudite man of the time." Medicine was "taught in the circles of *ʿulamāʾ* and *šuyūḫ* in religious institutions or even privately" (Behrens-Abouseif 1989, 334). The disputative form and scholastic commentary style of the *Canon* and *Epitome* commentaries complemented the pedagogical priorities of madrasa education.

Nonetheless, some of these commentaries, especially those on the *Epitome*, were directly geared toward medical practitioners (Fancy 2013b, 531). Thus, the fact that these *Epitome* commentaries *still* continued to address and investigate natural scientific topics opposes the generally held conviction that the post-Ayyubid period (566/1171–658/1260) witnessed a separation of theoretical medicine and philosophy from medical practice to the extent that the "medical craft was . . . deprived of the scholarly ground on which it once flourished" (Behrens-Abouseif 1989, 341). Many of the commentators described above were at par with the best erudite scholars of an earlier generation and were practicing physicians. For example, al-Shīrāzī worked in a hospital setting and in courts as a physician while also being a philosopher-astronomer. Similarly, Ḥakīm Shāh al-Qazwīnī was not only an Ottoman court physician but was also trained in philosophy by one of the leading philosopher-theologians of Shiraz, al-Davānī (830–908/1426–1502). The greatest of them all, at least in terms of impact on subsequent medical writers, was Ibn al-Nafīs, who was a practicing jurist and physician, *and* composed works on philosophy and logic (Fancy 2013[CB]). Moreover, he modified and transformed Galenic-Avicennan medical theory within the confines of his exegetical commentary on the *Canon* – a commentary that was used extensively by later physicians.

III.7.4 Ibn al-Nafīs, verification and transcending Galen's physiology

Ibn al-Nafīs has long been recognized as the first person to seriously challenge Galenic cardiovascular anatomy. Whereas Galen had posited *in*visible pores in the septum wall to explain the (natural) movement of blood from the right to the left side of the heart, Ibn al-Nafīs rejected the existence of *in*visible pores. He instead argued that the venous blood of the right side is first carried to the lungs in the artery-like vein (pulmonary artery for us) where its fine parts seep out of the coats of the arteries and mix with air in the lumen (i.e. inner space) of the lungs. From there, this blood–air mixture is taken up by the vein-like artery (pulmonary vein for us), which carries it to the left side of the heart to generate fresh spirit (Fancy 2013[CB], 101–5). The real questions then to address are (a) How did Ibn al-Nafīs arrive at his pulmonary transit result? and (b) Why was he so convinced about the nonexistence of *in*visible pores?

The earlier responses to these questions either undercut Ibn al-Nafīs's ingenuity by suggesting that he had made a "happy guess" (Meyerhof 1935, 118) or speculated on the possibility that Ibn al-Nafīs had performed dissections (Chéhadé 1955). The problem with the latter claim should be obvious: it is impossible to disprove the existence of invisible pores by cutting open and observing dead animals. As far as cutting open live animals is concerned (vivisection), Ibn al-Nafīs explicitly denies doing so, even when he himself claims that it could provide the final proof for his new theory of pulse (see the following discussion).

Those who claim Ibn al-Nafīs made a "happy guess" deny the possibility that Ibn al-Nafīs challenged or modified Galenic physiology (Ullmann 1978[CB], 69). There are two problems with this claim. First, it incorrectly assumes that any physiology that is based on the concept of faculties, spirits and chief organs has to be Galenic, without paying attention to the subtle (and not-so-subtle) differences in the definitions and even number of faculties, chief organs and spirits between authors. Second, it misconstrues how Galen, Ibn Sīnā, Ibn al-Nafīs and their successors would have characterized systematic physiological investigation. Nowadays, undertaking physiological research means conducting experiments or inferring (and testing) the processes of the body through dissection, particularly of living things. For our historical actors, physiology (or medical theory) was the study of the natural part of medicine. The natural constituents of the body included the elements and faculties – concepts that were also discussed in the sections on natural science in philosophical compendia. Philosophical arguments thus played *the* dominant role in establishing physiology. Physiology, as Cunningham puts it succinctly, was a "thinking and talking discipline – a discourse" (Cunningham 2002, 645).

The fundamental starting point for physiology was an understanding of the soul and its relationship to the body. This was as true for Galen (with his tripartite soul as the source for all faculties of a living human being) as it was for Aristotle, Ibn Sīnā and Ibn al-Nafīs (Fancy 2013). Ibn Sīnā had asserted the superiority of his philosophy (natural science and metaphysics) and chided Galen (and physicians) for engaging in physiological disputes based on an incorrect metaphysics (such as that of a tripartite soul). When he asked physicians to accept physiological principles as given from the superior science of philosophy in Book I of the *Canon*, this is really what he had in mind. However, as we have seen, once Fakhr al-Dīn broke open the barriers to investigating philosophical content in his *Canon* commentary, it provided enough room for Ibn al-Nafīs to put Aristotle's, Galen's and Ibn Sīnā's medical theories to the test. This included a thorough investigation of the soul–body relationship, the definitions, numbers and kinds of faculties, spirits and chief organs and so forth. There is nothing inherently theological (or metaphysical) about Ibn al-Nafīs's investigation when compared to those of his predecessors. He engaged in the best practices of philosophical argumentation and verification to establish a new physiology on his revised understanding of the soul and the soul–body relationship. The fruits of this exercise in verification were the positing of the pulmonary transit of blood, a new understanding of fetal generation and his new theory of pulse (Fancy 2013[CB]).

In the *Commentary on the Canon*, Ibn al-Nafīs concurred with philosophers that the soul is unitary and uniform but rejected the notion that it can only be connected to a single body part. Instead, he posited that the soul is connected to the entire body, which is why it can provide every body part with the natural faculties directly. This led him to modify the definition of chief organs and remove the liver and generative organs from that category (contrary to Galen and Ibn Sīnā). For that reason, he also had no reason to posit the existence of a vital faculty as a preparatory faculty (as in Ibn Sīnā), nor as one necessary for passions or the pulse (as in Galen and Ibn Sīnā). He instead claimed that all members that move due to muscle fibers possess a kind of volitional faculty, including the heart, stomach, esophagus, intestines and uterus, even if in these latter cases we are not always conscious of their motion.[6] He calls this kind of unconscious volitional faculty, the natural volitional faculty (MS London, Wellcome Library, WMS Arabic 51, fol. 106b). This faculty is first emanated to the spirit from the soul when the former is generated in the heart. It is for this reason that the thick blood of the right ventricle needs to be filtered and mixed with air before arriving in the left side so as to prevent the spirit from being ruined and becoming incapable of receiving this faculty. Finally, the fineness of the arterial spirit is maintained by the pulse, which is a motion caused by the movement of the heart. When the heart expands, it sucks the remaining spirit of the arteries back into itself which forces their contraction, and when the heart contracts, the arteries return to their normal, distended state (Fancy 2013[CB]).

Ibn al-Nafīs defends every single one of his claims with rational arguments akin to those provided by Galen and Ibn Sīnā. In the case of his new theory of pulse, he devotes approximately 8000 words to argue for his position, taking into consideration (and rejecting) not only Galen's claim that the pulse is caused by the vital faculty but also other ancient and modern opinions. He even raises a possible empirical objection to his theory, namely that if the arteries are forced to contract due to the expansion of the heart, the arteries would contract at different times depending on their distance from the heart and this would be observed. His response here is that this is indeed the case, but the rapidity of arterial expansion and contraction does not permit us to actually observe these differences based on distances from the heart (MS London, Wellcome Library, WMS Arabic 51, fol. 106b-107a). He also raises one possible empirical verification of his new pulse teaching, which he claims would be definitive. This brief passage is worth quoting in full if only to highlight the fact that Ibn al-Nafīs's deep commitment to verification, both rational and empirical, does have its limits:

> As for the dissection of the heart and arteries and diaphragm and lungs, etc., one [must] be informed about the manner of their movements and *whether the motion of the arteries is synchronous with the movement of the heart or is different*, and similarly, the movement of the lungs along with that of the diaphragm. *It is a given fact that it can be learned only through dissection (tashrīḥ) of the living, but that is difficult because of the disturbance of the living due to its feeling of pain.*
>
> *(Ibn al-Nafīs,* Commentary on the Anatomy *as translated in Savage-Smith 1995, 101, my emphasis)*

No subsequent commentator on the *Canon* that I know performed this vivisection either. Yet, that did not prevent Ibn al-Nafīs's new account of pulse from being vetted by subsequent commentators. Al-Shīrāzī rejected his account entirely, preferring to stick to Galen's original account. He found the assigning of a volitional faculty to the heart absurd. However, others did not think so. Ḥakīm Shāh al-Qazwīnī defended Ibn al-Nafīs's entire account. Others, such as al-Kirmānī did not commit to the volitional faculty but otherwise followed Ibn al-Nafīs in accepting (a) that pulse only signifies the motion of the arteries without that of the heart and (b) that the contraction of the arteries is caused by the expansion of the heart and vice versa (Fancy 2019, 269). In fact, what we see in subsequent commentaries is that the process of verification leads authors after Ibn al-Nafīs to struggle with his novel claims, regardless of whether they ultimately accept them. And in this process, Galen becomes one of several options and, for many, the incorrect option. This is true about pulse, generation (Fancy 2018b), vital faculties (Fancy 2013[CB]) and more.

IIII.7.5 Conclusion

In a panel on commentaries on the *Canon* at the annual meeting of the Middle East Studies Association (2017), Peter Pormann correctly surmised that what we have so far are "islands of knowledge" in a vast "sea of ignorance!"[7] Although medical commentaries have been largely dismissed by historians of medicine over the past century as being derivative and unworthy of academic study, this chapter has shown that this blanket dismissal was premature. Worse still, in ignoring this vast corpus, historians of medicine have falsely asserted that medical thought "declined" and failed to move beyond Galen and Ibn Sīnā after 600/1200. Instead, even this cursory examination of small sections from the vast corpus of medical commentaries reveals that medical thought not only moved beyond Galen and Ibn Sīnā in the post-600/1200 period, but it continued to evolve well into the 9th/15th and 10th/16th centuries. Much more work still needs to be done on hundreds of these manuscripts to uncover the real trajectory of medical theory and practice in postclassical Islamicate societies.

Notes

1 For an overview of medicine in Islamicate societies see Savage-Smith (2013).
2 Consolidated bibliography.
3 A listing of all the editions, along with links to the electronic additions can be found here: http://hummedia.manchester.ac.uk/schools/salc/subjects/clah/projects/arabiccommentaries/List-of-Scholarly-Editions.pdf (Accessed January 11, 2018).
4 Al-Rāzī may have been an earlier exception as he was very critical of Galen. His Commentary on the Aphorisms is not extant, but his critique of Galen's commentary can be found in his Doubts against Galen. Ibn Riḍwān, amongst others, responded to al-Rāzī's criticisms and attempted to defend Galen but, in Ibn Riḍwān's case, these responses to al-Rāzī's critique were relegated to a separate text and are not found in his abridgment of Galen's commentary on the Aphorisms (Karimullah 2017, 322).
5 The Princeton manuscript, Handlist (New Series), no. 1526, is available online: http://arks.princeton.edu/ark:/88435/pz50gw161 (Accessed January 30, 2018).
6 In modern medicine, these muscles are also singled out as being outside of conscious control. The term used to encapsulate them all is *involuntary muscle*, although they are technically referred to as smooth muscles (stomach, esophagus, intestines and uterus) and cardiac muscle (heart).
7 The best overview of the extant commentaries on the Canon and its abridgments in Arabic, Persian and Turkish is the work in Urdu by Rahman (1986).

Bibliography

Manuscripts

MS London, Wellcome Library, Arabic 51. https://wellcomelibrary.org/item/b20294979#?c=0&m=0&s=0&cv=3&z=-0.6389%2C-0.0788%2C2.2778%2C1.5751 (Accessed 4 October 2018).
MS Oxford, Bodleian Library, Arch. Seld. A64.
MS Oxford, Bodleian Library, Marsh 464.

Research literature

Adamson, P., ed. 2011. *In the Age of Averroes: Arabic Philosophy in the Sixth/Twelfth Century*. London: The Warburg Institute.
Ahmed, A. Q. 2013. "Post-Classical Philosophical Commentaries/Glosses: Innovation in the Margins," *Oriens* 41: 317–48.
Behrens-Abouseif, D. 1989. "The Image of the Physician in Arab Biographies of the Post-classical Age," *Der Islam* 66: 331–43.
Cooper, G. 2018. "Medical Crises and Critical Days in Avicenna and After: Insights from the Commentary Tradition," *Intellectual History of the Islamicate World* 6: 27–54.
Chéhadé, A. 1955. *Ibn al-Nafīs et la découverte de la circulation pulmonaire*. Damascus: Institut français de Damas.
Cunningham, A. 2002. "The Pen and the Sword: Recovering the Disciplinary Identity of Physiology and Anatomy before 1800; I: Old Physiology—the Pen," *Studies in History and Philosophy of Biological and Biomedical Sciences* 33: 631–65.
Fancy, N. 2013. "Medical Commentaries: A Preliminary Examination of Ibn al-Nafīs's *Shurūḥ*, the *Mūjaz* and Commentaries on the *Mūjaz*," *Oriens* 41: 525–45.
Fancy, N. 2017. "Womb Heat Versus Sperm Heat: Hippocrates Against Galen and Ibn Sīnā in Ibn al-Nafīs's Commentaries," *Oriens* 45: 150–75.
Fancy, N. 2018a. "Post-Avicennan Physics in the Medical Commentaries of the Mamluk Period," *Intellectual History of the Islamicate World* 6: 55–81.
Fancy, N. 2018b. "Generation in Medieval Islamic Medicine," in Hopwood, N., Flemming, R. and Kassell, L., eds. *Reproduction: Antiquity to the Present Day*. Cambridge: Cambridge University Press, 129–40.
Fancy, N. 2019. "Galen and Ibn al-Nafīs," in Bouras-Vallianatos, P. and Zipser, B., eds. *Brill's Companion to the Reception of Galen*. Leiden: Brill, 263–78.

Gutas, D. 2003. "Medical Theory and Scientific Method in the Age of Avicenna," in Reisman, David, ed. *Before and After Avicenna: Proceedings of the First Conference of the Avicenna Study Group*. With the assistance of Ahmed H. al-Rahim. Leiden: Brill, 145–63.

Hauter, A. 2020. "Madness, Pain and *Ikhtilāṭ al-ʿaql*: Conceptualizing Ibn Abī Ṣādiq's Medico-Philosophical Psychology," *Early Science and Medicine* 25: 453–79.

Iskander, A. I. 1967. *Catalogue of Arabic Manuscripts on Medicine and Science in the Wellcome Historical Medical Library*. London: Wellcome Historical Medical Library.

Karimullah, K. I. 2017. "Transformation of Galen's Textual Legacy from Classical to Post-Classical Islamic Medicine: Commentaries on the Hippocratic *Aphorisms*," *Intellectual History of the Islamicate World* 5: 311–58.

Karimullah, K. I. 2019. "The Emergence of Verification (*taḥqīq*) in Islamic Medicine: The Exegetical Legacy of Faḫr al-Dīn ar-Rāzī's (d. 1210) Commentary on Avicenna's (d. 1037) *Canon of Medicine*," *Oriens* 47.1–2: 1–113.

Meyerhof, M. 1935. "Ibn an-Nafīs (XIIIth cent.) and His Theory of the Lesser Circulation," *Isis* 23: 100–120.

Pormann, P. E. and Joosse, N. P. 2012. "Commentaries on the Hippocratic Aphorisms in the Arabic Tradition: The Example of Melancholy," in Pormann, P., ed. *Epidemics in Context: Greek Commentaries on Hippocrates in the Arabic Tradition*. Berlin: De Gruyter, 211–49.

Pormann, P. E. and Karimullah, K. I. 2017. "The Arabic Commentaries on the Hippocratic *Aphorisms*: Introduction," *Oriens* 45: 1–52.

Rahman, Z. 1986. *Qānūn-i-Ibn-i-Sīnā aur us ke shāriḥīn-o-mutarjimīn* [The *Qānūn* of Ibn Sīnā and its Commentators and Translators]. Aligarh: Aligarh Muslim University Publication Division.

el-Rouayheb, Kh. 2006. "Opening the Gate of Verification: The Forgotten Arab-Islamic Florescence of the 17th Century," *International Journal of Middle East Studies* 38: 263–81.

Saliba, G. 1996. "Writing the History of Arabic Astronomy: Problems and Differing Perspectives," *Journal of the American Oriental Society* 116: 709–18.

Savage-Smith, E. 1995. "Attitudes toward Dissection in Medieval Islam," *Journal of the History of Medicine and Allied Sciences* 50: 67–110.

Savage-Smith, E. 2013. "Medicine in Medieval Islam," in Lindberg, D. and Shank, M., eds. *The Cambridge History of Science*. 6 vols. 2: *Medieval Science*. Cambridge: Cambridge University Press, 140–67.

Wisnovsky, R. 2013. "Avicennism and Exegetical Practice in Early Commentaries on the *Ishārāt*," *Oriens* 41: 349–78.

III.8

TEXTUAL GENRES AND VISUAL REPRESENTATIONS IN THE ASTRAL SCIENCES[1]

Sonja Brentjes

Textual and visual forms for imparting knowledge reflect two fundamental sets of practices, both of which aim to communicate knowledge. Various types of text, or genres, and various types of visualization (diagrams, tables, figures, calligraphy and so on) inform us in different ways about the audiences addressed, the intended purposes of their creators, and the reputation of the knowledge they imparted. Their study helps determine the place of the relevant forms of knowledge in their specific contexts, which may include, for instance, education, military endeavors, or prognostication for individuals or dynasties relating to political agendas, family life or elite entertainment. The organization of knowledge in specific literary formats is understudied when it comes to the sciences in Islamicate societies. The same holds true for its visual representation in the form of diagrams, tables, models of the universe or figurative depictions of the planets, the sun and the constellations. To date, planetary movements are the only topic whose visual modeling has been seriously analyzed (Chapter II.6).

This chapter cannot compensate for these gaps in previous research, and its goal is correspondingly modest. First of all, it summarizes current knowledge about nine textual genres, only one of which, the *zīj*, belongs exclusively to the astral sciences, while the other eight are shared with either a few other disciplines (notably philosophy and medicine) or with most of the fields of knowledge practiced in Islamicate societies. The second part of this chapter surveys the main forms in which astral knowledge was visually presented. While some of those forms were assimilated from other iconographic traditions in the process of textual translation, others were newly invented or elaborated with respect to their types of execution, complexity, functionality or interdisciplinary use, to use a modern expression. The ubiquitous presence of tables and diagrams in many texts, scientific and non-scientific, establishes them as important aspects of scholarly practice, meaning that, in addition to their contents, their layout, functions and cross-disciplinary distribution deserve serious investigation. These elements convey information about how and why formats, themes and genres were durable and successful, while others were marginalized either regionally or over time.

III.8.1 Textual genres in the astral sciences

III.8.1.1 Astronomical handbooks

Zījes contain astronomical, trigonometric, astrological and chronological tables and tables for calculating the dates of religious events, together with introductory surveys and explanatory sections (King *et al.* 2001[CB][2]). Compiling such handbooks primarily involved collecting and

DOI: 10.4324/9781315170718-40

calculating data. In other words, *zījes* are books that contain tables copied from older books or newly computed from previous, but not necessarily new, observations. Nonetheless, an impressive number of handbooks contain or analyze new observations (Chapter II.7). The practice of compiling astronomical handbooks continued over almost a millennium, because each such handbook had to be adapted to a specific location. Most of them were produced in urban centers and many were dedicated to courtly patrons.

Astronomical handbooks begin with one or more chapters on chronology, presenting different eras and calendars and their respective tables. The goal was to teach the conversion between calendars, the determination of the first weekday of a given year and month and the various eras and calendars which differed between different regions of the Islamicate world. The eras most often used were the ancient Greek and Roman, pre-Islamic Iranian and Muslim eras, and the calendars were solar, lunar or soli-lunar. The tables, of which these handbooks for the most part consist, usually deal with epoch positions, which are reference positions of celestial objects at a given time and specific to each handbook. These are followed by tables of mean motions, true motions, various angular values called equations, planetary apogees and latitudes and longitudes, as well as the stations and retrogradations of each planet. Other tables treat solar and lunar eclipses, solar and lunar parallax, the visibility of the lunar crescent, the zodiacal signs, fixed stars, astrological implications, prayer times (Chapter V.4), prayer directions (*qibla*; Chapter VI.2), trigonometric functions and other mathematical topics.

The user of a *zīj* followed various methods to select specific values from the tables in order to find the position of a planet required for further calculations involved in creating birth and other horoscopes, for example, or predicting the time of an eclipse. The mean position of a celestial body, for instance, is determined by additively combining a specific epoch position with the number of years, months, days and hours elapsed since that position. To find the true position of that body on the ecliptic, the mean position is corrected by the so-called equation (*taʿdīl*), which here has the technical meaning of an additive correction, rather than the modern meaning of a relation between two quantities. These equations can be determined by various methods, among them successive applications of trigonometric functions or the tabulation of a set of auxiliary functions in double-argument tables (King *et al.* 2001[CB], 24, 79; for examples of such functions and tables see, for instance, King 1973, 2004[CB]). Double-argument tables were one of the innovations that occurred in the context of astronomical handbook production. They significantly simplified the complicated Ptolemaic methods. For example, consulting parallax tables was a preparatory step for calculating and hence predicting eclipses: several parameters had to be tabulated, among them the times of equinoxes and solstices, apparent solar and lunar radii, as well as true solar and lunar motions in specific time intervals.

The earliest astronomical handbooks were compiled during the 2nd/8th century on the basis of Indic and Middle Persian predecessors. Traces of Arabic versions of Indic handbooks from the first half of the 2nd/8th century are scant and their ancestors debated (Kennedy 1956, 138). But methods and parameters of likely or certain Indian provenance can be found in numerous Arabic texts and tables from the late 2nd/8th century onward (Kennedy 1956, 138; King *et al.* 2001, 37–40, 55, 63, 77). The surviving traces of a famous Middle Persian *Royal Handbook* (*Zīk-i Shahriyārān* = *Zīj al-Shāh*) also include specific parameter values and computational methods, as well as specific doctrines or concepts such as the World Year or the political importance of Saturn–Jupiter conjunctions (Kennedy 1956, 7–8; King *et al.* 2001, 32; Chapter I.3). Examples of relevant parameters include those for the mean motions of Jupiter, Saturn and the lunar nodes (the points of intersection between the ecliptic and the lunar path), as well as the eccentricity and longitude of the apogee of Venus (King *et al.* 2001, 30–2; Mozaffari 2019). Before its possible translation into Arabic, the *Royal Handbook* was apparently used in 144/762 by a group of Iranian Zoroastrian and Jewish astrologers, together with one Muslim expert, to draw up a horoscope in order to choose an auspicious day for starting work on the new Abbasid capital at Baghdad (Chapter I.1).

Hints about astronomical handbooks newly compiled in Arabic in the late 2nd/8th and early 3rd/9th centuries survive in later material, together with a very small number of fragments (Sezgin 1974, 5: 218–19; Ragep 2009, 127). The two most famous exemplars are the largely lost handbook called *Sindhind* and the handbook of al-Khwārazmī (d. *c.* 235/850). The former seems to have been the result of cooperation between two Muslim astrologers, Yaʿqūb ibn Ṭāriq and Ibrāhīm al-Fazārī (both 2nd half 2nd/8th century), and an astrologer from Sind (today in Pakistan). Al-Khwārazmī's handbook, also called *Sindhind*, survives in Latin translations with substantial revisions made in Córdoba (King *et al.* 2001, 57). Before 187/803, Barmakid viziers had sponsored the translation of Ptolemy's *Almagest* and the *Handy Tables* in addition to Indian material. Following this, Ptolemaic parameters and methods made their way into the handbooks. Handbooks primarily based on Indian or Iranian material continued to be written in the first half of the 3rd/9th century. In the later 3rd/9th century the shift toward practices taken from Ptolemy's *Almagest* and the *Handy Tables* accelerated. Nonetheless, various Indian or pre-Islamic Iranian parameters, assumptions and methods continued to be used for centuries in several regions of the Islamicate world such as Iran, and not only in al-Andalus, as often claimed in older literature (Bruin 1974, 274; Mozaffari 2019).

The first two *zīj*es that document the transition to largely Ptolemaic parameters and methods are Yaḥyā ibn Abī Manṣūr's (d. 215/830) *Verified Tables* (*al-Zīj al-mumtaḥan*) and Ḥabash al-Ḥāsib's (d. after 255/869) handbook variously named *The Damascene Zīj* (*al-Zīj al-dimashqī*), *The Zīj Based on the Arabic [Calendar]* (*al-Zīj al-ʿarabī*), *The Verified Tables* (*al-Zīj al-mumtaḥan*) and *the Zīj [Based on the Observations] for al-Ma'mūn* (*al-Zīj al-ma'mūnī*; Kennedy 1956, 10; King *et al.* 2001, 33–8). As reflected in their names, these tables are based on observations carried out during the rule of Caliph al-Ma'mūn (r. 198–218/813–833) and codified in the *Verified Tables*, begun by Yaḥyā ibn Abī Manṣūr and continued by his collaborators. Both handbooks also introduced the first critical comments on Ptolemy and deviations from Ptolemaic parameters or models, brought about by the new observations carried out in Baghdad and Damascus in the 210s/820s–830s.

Another astronomical handbook, important mostly for the transmission of Ptolemaic astronomy to Christian societies in Europe, was the *Sabean Zīj* (*al-Zīj al-Ṣābi'*) compiled by the convert al-Battānī (244–317/858–929) around 290/902 in Raqqa (today in Syria). His many observations allowed him to critically evaluate earlier measurements and their suitability for planetary models (van Dalen 2007, 1: 102–3). While praised for their accuracy by contemporaries and modern scholars, scholars between the 4th/10th and 8th/14th centuries repeatedly criticized his observations for their errors, as well as their deviations from later observable conditions.

Among many other Ptolemaic handbooks, we consider in the following those with a long and far-reaching impact on other scholars. They crossed centuries and regions, thus pointing to important cultural practices that enabled and sustained scholarly activities in the astral sciences. They also provide evidence for personal or literary cooperation, patronage, travel and cross-community interaction. In addition to al-Battānī, the handbooks of the following scholars were widely read, taught and used: Abū l-Wafā' (328–388/940–998) from Buzjan in eastern Iran, Ibn Yūnus (d. 400/1009) from Egypt, al-Fahhād (*fl. c.* 545–572/1150–1176) from Shirvan (today in Azerbaijan), Ibn Isḥāq al-Tamīmī (*fl, c.* 589–617/1193–1220) from Tunis, Muḥammad ibn Abī Bakr al-Fārisī (d. *c.* 677/1278–9) from Yemen, Muḥyī l-Dīn al-Maghribī from Morocco but working in Damascus and Maragha, Ibn al-Shāṭir (d. *c.* 777/1375), from Damascus and, of course, the well-known representatives of cooperative work, including Naṣīr al-Dīn al-Ṭūsī in Maragha (597–672/1201–1274; for other members of this group, see Chapter II.6) and the group of scholars (see below) at the court of the Timurid prince Ulugh Beg (r. 812–850/1409–1447 as governor of Samarqand; 850–853/1447–1449 as head of the dynasty; King *et al.* 2001, 43–5, 49–50, 53, 60–1). Important Ptolemaic handbooks were created in al-Andalus, for example, by al-Zarqāllu ([d. 493/1100]; Latin: Arzachel) of Toledo. While influential in Europe and North Africa these had limited impact in the eastern Islamicate world.

III.8.1.2 Introductions and synopses

Several major texts on the astral sciences present themselves as introductions or synopses. Books with those designations share naming customs and typologies with books in medicine and logic in antiquity and the early Abbasid period. In those two disciplines texts with titles such as introduction, or synopsis, often served as teaching material. It is, however, unclear, whether this applied to the astral sciences as well (for the understanding of the terms, see Chapter I.6).

The most important texts of this kind were al-Kindī's *Book on the Great Art* (*Kitāb fī l-ṣināʿa al-ʿuẓmā*); al-Farghānī's (3rd/9th century) *Book on the Principles of the Science of the Stars* (*Kitāb fī uṣūl ʿilm al-nujūm*); Abū Maʿshar's *Great Introduction to the Judgments of the Stars* (*al-Mudkhal al-kabīr ilā aḥkām al-nujūm*) and its abbreviation, the *Small Introduction* (*al-Mudkhal al-ṣaghīr*); al-Qabīṣī's (d. 356/967) *Introduction into the Science of the Stars* (*al-Mudkhal ilā ṣināʿat ʿilm al-nujūm*); Abū Naṣr al-Ḥasan ibn Alī, the astrologer's (*fl. c.* 357/968) *Introduction to the Science of the Judgments of the Stars* (*al-Mudkhal ilā ilm aḥkām al-nujūm*); al-Ṣūfī's (291–376/903–986) *Book of the Introduction to the Science of the Stars and Their Judgments* (*Kitāb al-Mudkhal ilā ʿilm al-nujūm wa-aḥkāmihi*) and Kūshyār ibn Labbān's (2nd half 4th–early 5th/2nd half 10th–early 11th centuries) *Book of the Introduction to the Art of the Judgment of the Stars* (*Kitāb al-Mudkhal ilā ṣināʿat aḥkām al-nujūm*), also called *Compendium of the Principles of the Judgments of the Stars* (*Mujmal al-uṣūl fī aḥkām al-nujūm*). Other such introductions are lost but known from references or quotes in Ibn al-Nadīm's (d. 380/990) *Catalog* (*Kitāb al-Fihrist*) and several works of al-Bīrūnī (Ibn al-Nadīm 1970[CB]). Important authors were al-Kindī's student al-Sarakhsī (d. 286/899) and Abū Jaʿfar al-Khāzin (d. 350 or 360/961 or 971). All these introductions except the first two deal with the predictive part of the astral sciences.

Most of these introductory works have only a partial formal organization such as a table of contents or some other kind of referencing that allows the reader to navigate the text. Exceptions are, for example, al-Farghānī's *Book on the Principles of the Science of the Stars* and Abū Maʿshar's *Great Introduction*. The latter serves here as an example of the genre (Chapter I.6). Possibly the oldest extant copy of an Arabic manuscript of the mathematical sciences is the manuscript of Abū Maʿshar's book from 325/936–937 preserved in the National Library of France in Paris (MS Paris, BnF, Arabe 5902). Its scribe, ʿAlī al-Muṭarriz (1st half 4th/10th century), provided Abū Maʿshar's structural organization with parts, chapters and sections in red, calligraphic embellishments, indentations, floral forms of punctuation and other marks helping the reader access its content (III.8.1). Abū Maʿshar himself proposed to adopt the seven parts into which ancient scholars had divided their books for defining their position in the order of teaching: (1) aim of the book, (2) its benefit, (3) its author, (4) its title, (5) texts before or after which the book should be read, (6) whether the book belongs to the theoretical or to the practical part of the discipline and (7) the division of the book into chapters and sections (Abū Maʿshar 2019[CB], 1: 44–7). Remarkably, only some of these ancient rules for organizing a scientific text were followed by scholars in Islamicate societies. Not even points (3) and (4) were always adopted.

The development of the formal aspects of literary practice in the sciences took centuries to settle. There is no systematic study of such features available yet for scientific texts, and no effort has been made so far to define regional specifics. Tables of contents had already appeared in the

3rd/9th century but did not become a widespread convention in the sciences until the 6th/12th or 7th/13th centuries. Embellished forms characterize Timurid, Safavid and Ottoman multiple-text manuscripts produced for princes or high-ranking courtiers from the 10th/16th century on. Until very late in time, there were no page numbers, but the order of pages in such books was often, but not always, signaled by catchwords, which were inserted in the lower margin of each page and which provided the first words of the following page. Punctuation appeared irregularly, took different forms and mostly separated larger textual units. The marking of keywords, titles or headings in red, green or blue or by a change of calligraphy appeared in scientific texts in the 4th/10th century at the latest.

It is unclear whether the few extant manuscripts from the early centuries with such features were connected to courts or elite households. When used as artistic elements in scientific manuscripts from the 6th/12th century onward, such features mark the inclusion of the sciences in the courtly art of the book. Lavishly decorated and colored texts with beautiful calligraphic headings or tables appeared first in the mathematical sciences and medicine, but in later centuries, any kind of scientific work could appear in highly embellished copies, as copyists added to the earlier colorful decoration of titles and calligraphic scripts small-scale images, miniatures, full-page decorative designs, naturalistic scenes or human figures of scholars, painters, artisans, princes or religious men (mostly Sufis; Chapters II.10 and V.7). These developments differed among regions. From al-Andalus and North Africa very few scientific manuscripts of this kind

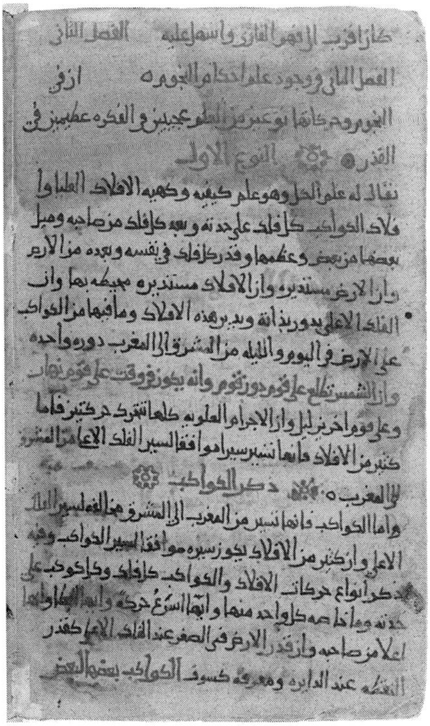

Figure III.8.1 ʿAlī al-Muṭarriz's presentation of Abū Maʿshar's *Great Introduction to the Judgments of the Stars*. MS Paris, BnF, arabe 5902, fol. 4b.

Source: © BnF, Paris

are known, among them a copy of ʿAbd al-Raḥmān al-Ṣūfī's *Book of the Images of the Fixed Stars* (*Kitāb ṣuwar al-kawākib al-thābita*), abbreviated from here on as *Book of the Constellations*, made in Ceuta.

By way of contrast, a surprising number of high-quality paintings in floral, geometric or figurative styles can be found in manuscripts with texts on planetary theory, instruments, star constellations or astrological themes, as well as astronomical handbooks produced under several dynasties in Iran, Central Asia, India or Anatolia from the 8th/14th to the 13th/19th centuries. Before the Ottoman conquests, there were fewer artistically produced texts on the astral sciences in Iraq, Syria and Egypt than in eastern regions, and their execution was less sophisticated.

III.8.1.3 Commentaries, refutations and aporias

Since antiquity, commentaries and refutations were standard textual means to engage with opinions and doctrines an author wished to explain, to expand or to challenge. In Islamicate societies, the genre of texts called aporias (puzzles) was a product of scholarly debates in the philosophical, medical and astral sciences. In this type of commentary, the author challenged a part of a disciplinary paradigm by raising doubts (*shukūk*) or objections (*iʿtirāḍāt*; see the following examples). Refutations, a format particularly prominent in the religious disciplines, went beyond a critical discussion of doubts to reject wholeheartedly another author's position. They were used in the sciences, but less often.

Until the early 8th/14th century, most exegetical commentaries in the astral sciences, so identified in their titles (*tafsīr*; *sharḥ*), engaged Ptolemy's *Almagest*. Major contributors to this literary genre were al-Nayrīzī (d. *c.* 310/922), al-Fārābī (d. 339/950–1) and Abū l-Wafāʾ. Despite the generic title, not all of them commented on the entire *Almagest* but picked and chose among its themes, geometrical constructions or calculations. In addition, there are a number of extant texts that do not make such a general claim but indicate in their titles that their authors aim to comment on, explain, analyze or prove single problems, methods or claims from the *Almagest*. One example of the latter is the anonymous *Proof of What Ptolemy Has Said About the Fourth Proposition of the Twelfth Book* [*of the Almagest*] (*Burhān mā qāla Baṭlamyūs fī l-shakl al-rābiʿ min al-maqāla al-thāniya ʿashra [min al-Majisṭī*]; MS Oxford, Bodleian Library, Thurston 3, fol. 107a). Both kinds of commentaries show close similarities to the types used for teaching or presenting research results associated with Euclid's *Elements*. The short treatises written by Thābit ibn Qurra, for example, served as teaching material for the *Almagest* (Thābit ibn Qurra 1987[CB]). In addition, a few commentaries were written on treatises about constructing scientific instruments and on didactic poems on astrology.

Direct refutations of Ptolemy's *Almagest* are known from only one author – al-Nawbakhtī (d. 300/912 or later), mentioned in Chapter I.6. It is unclear whether his *Book of the Refutation of Ptolemy on the Form of the Heaven and the Earth* [*Kitāb al-radd ʿalā Baṭlamyūs fī hayʾat al-falak wa-l-arḍ*] should be seen as a forerunner of the texts expressing doubt. As in the case of the commentaries, texts about doubts could tackle specific problems, such as the parallax of the moon and the various opinions held about it, or address one entire text, or even a set of texts on astral themes by an author. An example of the first type is al-Maʿdān's (1st half 4th/10th century) text on differing opinions about the parallax of the moon, which is only extant in an anonymous reply (*Answer to the Doubt Concerning the Parallax of the Moon from Abū l-Qāsim ʿAlī ibn al-Ḥasan al-Maʿdān's Doubts* [*Jawāb shakk fī khtilāf manẓar al-qamar min shukūk Abī l-Qāsim ʿAlī ibn al-Ḥasan al-Maʿdān*]; MS Oxford, Bodleian Library, Thurston 3, fols 101a–b). The best-known example of the latter type is Ibn al-Haytham's *Doubts on Ptolemy* (*Shukūk ʿalā*

Baṭlamyūs), although it is by no means the first: about a century earlier, al-Qabīṣī, primarily known for his astrological works, claimed to have written such a work, now lost, when the first aporetic commentaries surfaced in medicine and philosophy (Saliba 1994[CB], 20; Shihadeh 2017, 302). Another topic of aporetic commentary was linked to Aristotelian views on optics and comets. For example, ʿAlī ibn Sulaymān al-Hāshimī (*fl. c.* 277/890) wrote an *Enumeration of Doubts in Aristotle's Discourse on Vision and Enumeration of Doubts on Comets* (*Taʿdīd shukūk talzamu maqālat Arisṭūṭālis fī l-baṣar wa-taʿdīd shukūk fī kawākib al-dhanab*; Rosenfeld and Ihsanoğlu 2003, 117[3]).

According to the philosopher and physician al-Rāzī (d. 313/925 or 323/935), doubts could be dispelled either by offering a new solution or by abandoning the view against which the doubt was raised, thereby opening new avenues for theorizing (Shihadeh 2017, 302). Both approaches are at work in Ibn al-Haytham's aporetic commentary, in which he critically analyzed the *Almagest*, the *Planetary Hypotheses* and the *Optics* of Ptolemy (Ibn al-Haytham 1971, 1985[CB]; for a survey of the work's content see Goldstein 1973, 138–9; Chapter II.6). Saliba added texts named recapitulations (*istidrāk*) to the same genre, which he considered highly significant for the development of astronomical theory in Islamicate societies (Saliba 1994, 21).

III.8.1.4 Epistles, treatises, abbreviations and other textual formats

Before the early 8th/14th century, the largest body of texts written on some topic of the astral sciences fell under the rubrics of epistle (*risāla*), treatise or book(let) (*maqāla* or *kitāb*), abbreviations or paraphrases (*mukhtaṣar*) and works with thematic descriptions as titles. This raises the question of whether these terms identify single genres or cover several genres. The naming of works has its own history, which may be shared with other disciplines. On the other hand, this history may also have followed specific lines of development in the mathematical sciences, to which the astral sciences belonged (at least in part). Further research on such scholarly practices would be necessary to determine which of these alternatives prevailed.

Texts from all those categories can be short (as little as one or two pages) or can contain an exposition of considerable length and depth. A few arbitrary examples have to suffice: the works by ʿAlī ibn ʿĪsā, the astrolabe maker (d. after 216/832), *Epistle on the Astrolabe* (*Risāla ʿalā l-asṭurlāb*; MS Oxford, Bodleian Library, Huntington 193, fols 186a–199b); by Thābit ibn Qurra, *Epistle on the Movement of the Two Luminous* [*Bodies*] (*Risāla fī ḥarakat al-nayyirayn*; MS Oxford, Bodleian Library, Thurston 3, fols 102b–103a); by Ibn al-Haytham, *On the Movement of the Moon* (*Fī ḥarakat al-qamar*; MS Oxford, Bodleian Library, Arch. Seld. A.32, fols 100b–107a); and by an anonymous author, *Question on the Meaning of the Days of the World* (*Masʾala fī maʿnā ayyām al-ʿālam*; MS Oxford, Bodleian Library, Thurston 3, fol. 75a), are all short, the last consisting of a single folio. By contrast, Abū Maʿshar's *Book of the Conjunctions on the Judgments of the Stars* (*Kitāb al-Qirānāt fī aḥkām al-nujūm*; MS Oxford, Bodleian Library, Hyde 32, fols 1a–54a); and Muḥyī l-Dīn al-Maghribī's *Epistle on the Judgments According to the Revolution of the World Years* (*Risāla fī l-aḥkām ʿalā taḥwīl sinī al-ʿālam*; MS Oxford, Bodleian Library, Marsh 548, fols 42b–124b) are long and detailed.

Only a few texts in the astral sciences are called abbreviations, paraphrases, concise presentations (*mukhtaṣar*, *ikhtiṣār*) or corrections (*iṣlāḥ*). Examples are the *Abridgement of the Almagest* (*Mukhtaṣar al-Majisṭī*) by al-Ḥāzimī al-Saʿīdī (*fl. c.* 390/1000) and a treatise of the same name by Ibn Sīnā, which was a part of his philosophical summa *Book of Healing*, or *The Cure* (*Kitāb al-Shifāʾ*; Chapter I.11), as well as by Ibn Rushd (520–595/1126–1198), Ibn al-Haytham's

Concise Book on the Azimuth of the Qibla (*al-Maqāla al-mukhtaṣara fi samt al-qibla*) or Athīr al-Dīn al-Abharī's (d. between 660/1263 and 663/1265) *Concise* [*Book*] *on the Science of the Configuration* [*of the Universe*] (*Mukhtaṣar fī 'ilm al-hay'a*). Additional textual formats in the astral sciences used only in certain periods before the early 8th/14th century were open letters to students, friends or patrons about a specific problem or some larger problem, answers to colleagues or competitors and new editions of both the *Almagest* and the so-called *Middle Books*, which included treatises by ancient Greek scholars, as well as by writers from the 3rd/9th and the 4th/10th centuries.

The differences in survival and prominence between all these genres in the astral sciences reflect changes in at least three contextual factors: (1) the type of education involved and its institutional format; (2) the functions and duties of astral experts at various institutions; and (3) the sociocultural reputation of specific parts of astral knowledge. But much more research is needed, quantitative and qualitative, before the specific effects of those, and possibly other, factors can be evaluated in their concrete settings.

III.8.2 Representing the heavens visually

Forms of visual representation of astral knowledge appear in three formats – as parts of texts, as independent collections of diagrams, tables or figures and as parts of nontextual material objects. While diagrams in geometrical texts have increasingly attracted research in the history of mathematics, and figurative presentations are a standard research topic in art history, research on the visual organization of tables, on the content and functions of tabular structures tables and on the relationship between the different forms of visualizing knowledge is still in its infancy.

The most often visualized celestial objects were constellations, planets and lunar mansions. Constellations and planets were depicted in different parts of the astral sciences and in a broad variety of cultural objects. They could appear in treatises on the fixed stars, on instruments, horoscopes, lunar mansions or astral magic. They can also be found in introductions to Ptolemaic astronomy or in treatises on non-Ptolemaic models, calendars, astronomical handbooks and books about wonders and curiosities. In addition to texts, constellations and planets were depicted on coins, medals, ceramic bowls, metal inkwells, pen cases, ewers, basins or bowls, miniature paintings and buildings (Chapters II.10 and IV.7). In later periods, they became such familiar and widely known forms of knowledge that they were included in historical chronicles and religious texts.

From the 6th/12th century onwards, depictions of planets and zodiacal constellations as human, animal or mythological figures are extant in manuscripts and on other material objects. Many of them follow a fairly standard set of representations that in most cases seem to go back to Mesopotamian and Mediterranean forms. Examples are the depictions of Venus as a musician (although the instrument she plays varies between a kind of lute, a harp or a flute); Mercury as a scribe; Mars as a warrior in chain mail with a sword, holding a bleeding severed head; and Saturn as a dark-skinned, often half-nude old man with a scythe. Among the constellations, Capricorn is usually shown as a goatfish. Sagittarius is sometimes portrayed as a centaur with a bow and arrow shooting forward, but alternatively may appear as a hybrid of a big feline (or exceptionally a thick snake) and a male torso, shooting backward toward a dragon head at the end of the long tail of the tiger, lion or leopard. The depiction of Sagittarius as a centaur remained typical for copies of al-Ṣūfī's *Book of the Constellations*, while its feline form is primarily found in astrological works and on coins, medals, mirrors, ewers, basins or bowls made from metal or ceramics. It is also found on bridges, gates of fortresses or walls of observatories in

Iraq, eastern Anatolia, Syria, Egypt, Iran, Central Asia, Afghanistan or northern India between the 6th/12th and the 11th/17th centuries (for a survey, see Caiozzo 2003a; for the Anatolian case of the Rum-Seljuqs, see Peacock 2020). These modified iconographic features incorporate elements of Turkic, Indic, Iranian and possibly East Asian mythology or religious narratives. It is, however, largely unclear where and when such narratives were translated into pictures and how they moved westwards. One reason for this state of affairs is the lack of pictorial evidence before the 6th/12th century.

In texts on talismans, geomancy and similar topics, fascinating deviations from this standard pictorial repertoire of the heavens appear. An example is a multiple-text manuscript with several, partly incomplete magical and astrological texts in prose and verse, two of which are dated to 670/1272 and 671/1273. The author of some of this material, Naṣīr al-Dīn al-Rammāl al-Muʿazzim al-Sāʿatī al-Haykalī, fabricated geomantic predictive schemes, invocatory spells, clocks and astral talismans. Peacock interprets the term *al-Sāʿatī* differently as someone who was an expert for auspicious or inauspicious times or who had knowledge of the Day of Judgment (Peacock 2020). Naṣīr al-Dīn al-Rammāl may have intended his collection as a gift to the Rum-Seljuq ruler of parts of Anatolia, Ghiyāth al-Dīn Kay Khōsraw III (r. 663–683/1265–1284), then about ten to twelve years old (MS Paris, BnF, Persan 174; Caiozzo 2003b; Peacock 2020). On six folios, the planets and the sun are depicted in a kind of Indian style (MS Paris, BnF, Persan 174, fols 108a-110b). Painted in watercolors, all except Saturn have four arms (Figure III.8.2), and they carry additional symbols beyond those included in standard representations of the planets. Saturn has an

Figure III.8.2 Venus depicted with four arms playing a string instrument and holding a flute and a tambourine. MS Paris, BnF, Persan 174, fol. 108a.

Source: © BnF Paris.

uneven number of arms holding objects, which is highly unusual in Indian divine imagery. While the multi-armed depictions of the sun, Venus, Mercury, Mars and Jupiter remained rare, the seven-armed Saturn became widespread in manuscripts on the astral sciences produced in the Persianate world after the 7th/13th century, as well as in later Ottoman historical chronicles. The manuscripts and the many other objects with astrological symbols produced for the many smaller dynasties in Iraq, Anatolia, Syria and Egypt in the 7th/13th century (and elsewhere in later times) point to the political functions of the astral and other predictive sciences at their courts.

The most successful book with visual representations of constellations was ʿAbd al-Raḥmān al-Ṣūfī's *Book of the Star Constellations*. In about 353/964, al-Ṣūfī compiled it for his patron ʿAḍud al-Dawla (r. 338–372/948–983), appropriating the names, coordinates and other data of both the constellations and the zodiacal signs from Ptolemy's star catalogue in the *Almagest*. He apparently chose their pictorial forms from globes made by craftsmen from Harran in the 2nd/8th and 3rd/9th centuries and from a lost illustrated treatise on this topic by an author or craftsman of the 3rd or 4th/9th or 10th century, who was named after Mercury (ʿUṭārid). An Arabic translation of Aratos of Soli's (d. 245 BCE) Greek poem on the stars, of which no clear traces are known to exist beyond very few references to the work, may have served as a third, possibly only narrative, source for the stellar images.

Al-Ṣūfī's work was copied many times by scholars, calligraphers and painters and by interested laypeople. It was translated fully or partially into Persian, Ottoman Turkish, Castilian, Latin and Hebrew. Preparing this work, al-Ṣūfī studied Ptolemy's star catalogue and determined anew the coordinates of some of its stars as well as the magnitudes and brightness of many of them. He collected information about constellations he attributed to unnamed Bedouin tribes of the Arabian Peninsula, thus profiting from, or perhaps even participating in, the cultural conflicts about the merits of the Arabs and the Persians, which occurred in the 3rd/9th and 4th/10th centuries. One of those constellations, the Camel, made its way into the pictorial program of some of the later copies well into the 13th/19th century.

According to his long and impressive preface, which culminates in the work's dedication to ʿAḍud al-Dawla, al-Ṣūfī studied texts and globes by his predecessors and contemporaries, several of whom he sharply criticized for their lack of reliability and accuracy. According to colophons of some of the extant manuscripts, he drew the figures of the configurations himself and taught his work to students of the astral sciences. This information seems to be plausible, because in his preface al-Ṣūfī stresses the connection between precise astral knowledge acquired through observations and reliable visual depictions of the observed stars against the limited usefulness of numerical data learned from books and faulty depictions made by instrument makers, comments that bespeak personal experience.

This attractive book continued to be copied over the span of almost one thousand years. It was translated and compared with similar works in other languages. It was augmented with new observations of some of the stars and corrections of the numerical values in the tables. It was taught at courts, in private houses and perhaps also *madrasa*s (evidence for the latter is scarce). It became the object of philological studies, which commented on the terms it used, including their meaning and origin. Craftsmen used the figures when constructing globes. Finally, the configurations were redesigned to respond to new habits of dress, to fit new explanatory narratives, to reflect the aesthetics of new painting styles and to accommodate political expediency (Figure III.8.3). Several copies testify to the later use of the book among Christian and provincial, possibly non-Muslim, communities, as well as to the knowledge of alternative depictions of several configurations not found in the earlier copies

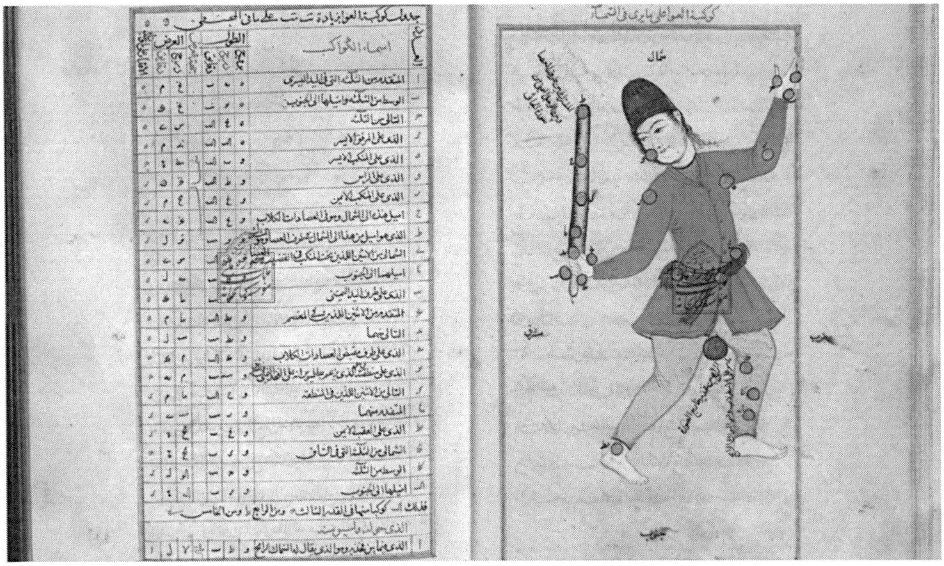

Figure III.8.3 Figure and table of stellar coordinates of the constellation Bootes in ʿAbd al-Raḥmān al-Ṣūfī's (291–376/903–986) *The Book of the Constellations* (MS Tehran, Malik Library, 6037, fols 59b–60a).

Source: © Malik Library, Tehran

of al-Ṣūfī's book. This suggests that Greek or Latin manuscripts (possibly in translation) or globes so far unknown to us, continued to circulate in later centuries. All in all, al-Ṣūfī's book was not only a scientific and cultural bestseller; it represented knowledge and taste considered epistemologically, culturally and politically relevant in many communities and social strata.

The visual representations of planets in introductions to Ptolemaic astronomy or books on wonders or curiosities also included lunar phases either for every night of a month or for some of the moon's main positions, lunar and solar eclipses and models of the universe, as well as the movements of each planet (Figure III.8.4). These depictions were usually two-dimensional, typically ink drawings in black or brown and red or ochre in manuscripts extant from before the early 8th/14th century, although watercolor images of planetary models in some copies of al-Jaghmīnī's *Epitome on the Science of the Elementary Configuration [of the Universe]* ([d. 618/1221–2]; *al-Mulakhkhaṣ fī ʿilm al-hay'a al-basīṭa*) introduce the appearance of depth. A similar use of color in planetary models can be found in copies of Zakariyyā' ibn Muḥammad al-Qazwīnī's *Wonders of Creation* (*ʿAjā'ib al-makhlūqāt fī gharā'ib al-mawjūdāt*) produced in courtly contexts (Chapter IV.7).

While the asterisms that defined the lunar mansions were often tabulated as dots or sets of dots, on three astrolabes produced in the early 7th/13th century in Damascus, and probably Mosul or Mayyafarikin, they appear as human, animal or hybrid figures (King 1997; Ward 2004). In later periods other images of this sort were depicted in manuscripts produced in Baghdad and Istanbul (Chapter IV.7). The origins of all those kinds of depictions are unknown.

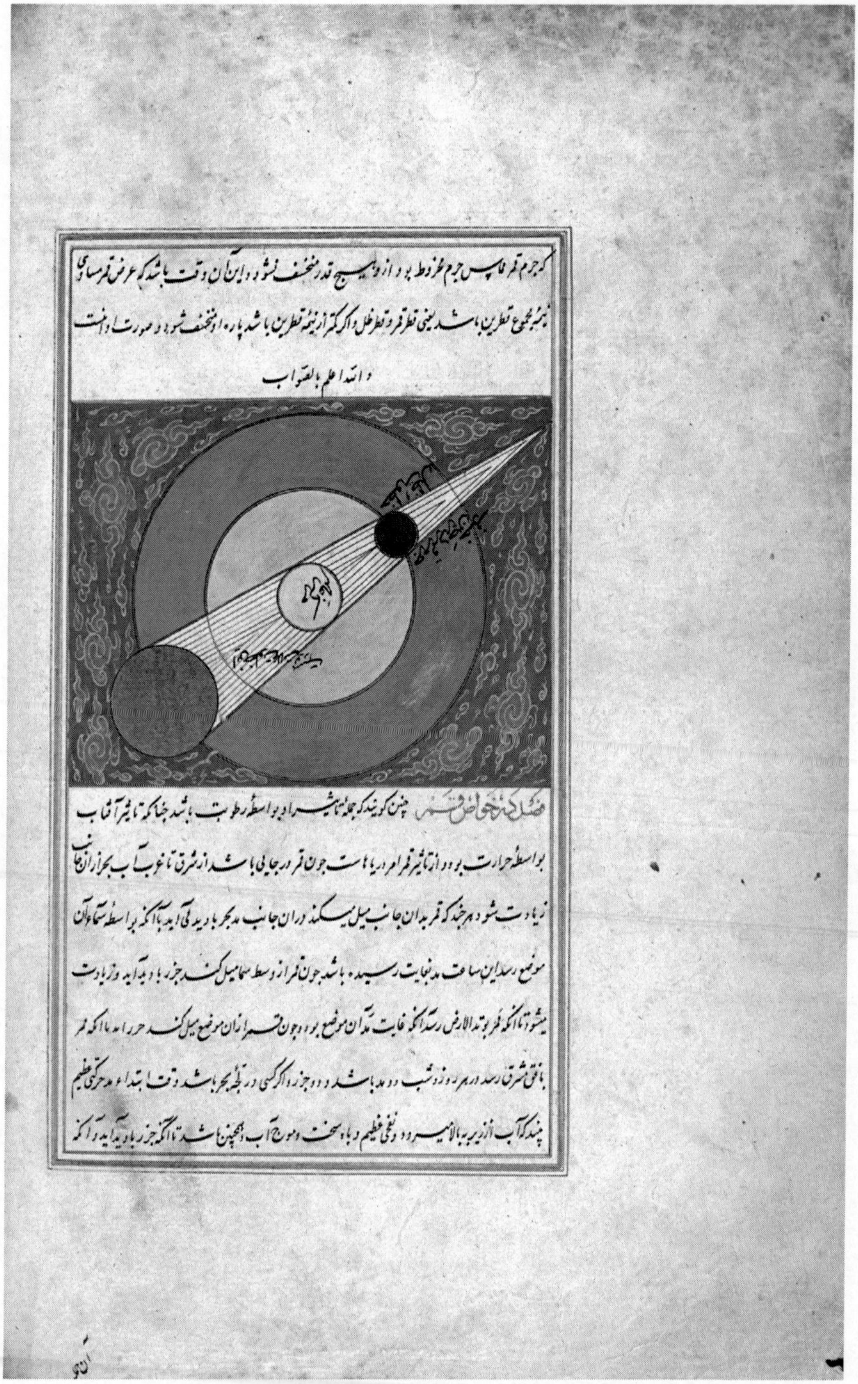

Figure III.8.4 Diagram of the principle of a solar eclipse in a copy of Zakariyyā' ibn Muḥammad al-Qazwīnī's *Wonders of Creation*. MS Cambridge, Cambridge University Library, Nn. 3.74, fol. 15b.

Source: © Reprodcued by kind permission of the Syndics of Cambridge University Library

III.8.3 Preliminary reflections instead of conclusions

For several decades, genres of scientific texts and visual representations of knowledge have been themes of growing importance in other fields of the history of science. They straddle the boundary between content and context analysis, belonging to both. With relation to content, their study informs us about practices of organizing and presenting knowledge in verbal, numerical and visual forms and about the relationship between textual and nontextual knowledge objects. It also provides us with insights into the preferred styles of argumentation, demonstration and verification of particular fields of knowledge. Given that in Islamicate societies many of the texts that have been preserved come from teaching institutions and libraries, the genres that dominate such collections and the layout and illustration of their texts give us data about teaching methods, audiences and the spread of the knowledge they deliver beyond narrow circles of scholars and patrons.

The comparative analysis of genres and illustrative styles across disciplines, periods, regions or localities increases our knowledge about the reputation not only of entire disciplines but of specific topics or styles, local specificities or geographical, as well as social distribution of such knowledge. Such themes deserve our attention and research, because their exploration helps us understand in a much more nuanced manner the factors that supported or hindered scholarly activities under concrete conditions. The study of technical content, as important as it is, is insufficient to answer such larger questions about what scientific knowledge meant in concrete Islamicate societies to whom, how it was generated, maintained and disseminated and which obstacles it encountered.

Notes

1 I am profoundly grateful to Benno van Dalen, Munich for allowing me to use his unpublished survey of astronomical handbooks and his continuous help with unresolved problems.
2 Consolidated bibliography.
3 The death date and identification of al-Hāshimī provided by the authors is wrong.

Bibliography

Sources

Ibn al-Haytham. ed. Sabra, A. H. and Shehaby, N. 1971. *Shukūk ʿalā Baṭlamyūs* (*Dubitationes in Ptolemaeum*). Cairo: The National Library Press.

Manuscripts

MS Oxford, Bodleian Library, Arch. Seld. A.32.
MS Oxford, Bodleian Library, Huntington 193.
MS Oxford, Bodleian Library, Hyde 32.
MS Oxford, Bodleian Library, Marsh 548.
MS Oxford, Bodleian Library, Thurston 3.
MS Paris, BnF, Arabe 5902.
MS Paris, BnF, Persan 174.

Research literature

Bruin, F. 1974. "Astronomical Observations and Instruments of Islam," *Nederlands Tijdschrift voor Natuurkunde* 40.19: 271–5.

Caiozzo, A. 2003a. "Une conception originale des cieux: planètes et zodiaque d'une cosmographie jalayride," *Annales islamologiques* 37: 59–78.

Caiozzo, A. 2003b. "La représentation d'al-Mirrīḫ et d'al-Zuḥal, planètes maléfiques et apotropaia," *Annales islamologiques* 37: 23–58.

Dalen, B. van. 2007. "Battānī: Abū ʿAbd Allāh Muḥammad ibn Jābir ibn Sinān al-Battānī al-Ḥarrānī al-Ṣābiʾ," in Hockey *et al.*, eds. [CB], 1: 102–3.

Goldstein, B. 1973. "Alhazen on Ptolemy," *Journal for the History of Astronomy* 4: 138–9.

Kennedy, E. S. 1956. "A Survey of Islamic Astronomical Tables," *Transactions of the American Philosophical Society* 46.2: 123–77.

King, D. A. 1973. "Al-Khalīlī's Auxiliary Tables for Solving Problems of Spherical Astronomy," *Journal for the History of Astronomy* 4: 99–110.

King, D. A. 1997. "The Monumental Syrian Astrolabe in the Maritime Museum, Istanbul," *Erdem* 9.27: 729–35.

Mozaffari, S. M. 2019. "The Orbital Elements of Venus in Medieval Islamic Astronomy: Interaction Between Traditions and the Accuracy of Observations," *Journal for the History of Astronomy* 50.1: 46–81.

Peacock, A. 2020. "A Seljuq Occult Manuscript and Its World: MS Paris Persan 174," in Canby, S. *et al.*, eds. *The Seljuqs and Their Successors*. Edinburgh: Edinburgh University Press, 163–79.

Ragep, F. J. 2009. "Astronomy," in *EI-3*, 2009–1: 120–50.

Rosenfeld, B. A. and Ihsanoğlu, E. 2003. *Mathematicians, Astronomers and Other Scholars of Islamic Civilization and Their Works (7th-19th c.)*. Istanbul: IRCICA.

Sezgin, G. 1974. *Geschichte des arabischen Schrifttums*. Vol. 5: *Mathematik bis ca. 430 H.* Leiden: Brill.

Shihadeh, A. 2017. "Al-Rāzī's (d. 1210) Commentary on Avicenna's Pointers [Sharḥ al-Ishārāt wa-l-Tanbīhāt]: The Confluence of Exegesis and Aporetics," in el-Rouayheb and Schmidtke, eds. [CB], 296–325.

Ward, R. 2004. "The Inscription on the Astrolabe by ʿAbd al-Karim in the British Museum," *Muqarnas* 21: 345–57.

PART IV

The materiality of the sciences (3rd–13th/9th–19th centuries)

IV.1
THE MATERIALITY OF SCHOLARSHIP

Konrad Hirschler

This chapter concerns the material dimensions of the production of scientific knowledge and its transmission. The focus will be on the codex as the prime carrier of written scientific knowledge. With the codex at its center, this chapter presents the materiality of the production and transmission of scientific knowledge starting with the support material, moving on to the textual format and finally discussing layout and writing implements.

IV.1.1 Writing materials

In Islamicate societies, written information was registered on the region's three standard materials, namely, papyrus (*qirṭās*), parchment (*raqq*/*riqq* or *jild*), and paper (*kāghid* or *kāghad* among other terms). This is not only evident from the extant material but also from the earliest advice texts on bookmaking, such as the *The Ornament of the Scribes* (*Zīnat al-kataba*), by the physician al-Rāzī ([d. 313 or 323/925 or 935]; Zaki 2011, 223–34). The importance of each of these materials did not remain static but between the 2nd/8th and the 4th/10th centuries, Islamicate societies underwent a drastic transformation with the adoption of paper as the preferred writing material. Papyrus, produced from fibers of the papyrus plant, played a particularly prominent role in Egypt, where some regions in the Nile Delta have the best climatic conditions for producing high-quality fibers. Papyrus was used outside Egypt to a lesser extent as users faced the choice between lower-quality local papyri or expensive imported papyri. Our knowledge of the use of papyrus in regions outside Egypt is, however, limited as more humid climatic conditions meant that survival rates of papyri have been significantly lower than in Egypt. Papyrus was the prime writing material during the first Islamic centuries and served as the support material for most of the documents written in this period.

Papyri come in the form of a (folded) single sheet or a scroll. In order to record longer texts, such as scientific texts, the scroll was the most feasible option, as either volume (unfolding horizontally) or *rotulus* (unfolding vertically). Papyrus was more rarely used to produce a codex (or simply 'book'), that is a collection of sheets (single 'folio'/plural 'folia'), gathered (and generally sewn) into quires, which are themselves in most cases stitched together and then provided with a protective cover. With the explosion of written information in Islamicate societies from the 2nd/8th century onward, the codex started to be the main format and pushed the scroll to the margins. In consequence, papyrus was used less and less so that by the 4th/10th century, it had

DOI: 10.4324/9781315170718-42

ceased to be a significant writing material, mainly limited to some specific forms of pragmatic literacy (especially documents and letters) and ritual literacy (sacred texts).

Parchment, animal skins prepared for writing, had been widely used in parallel with papyrus in the pre-Islamic and early Islamic periods. Compared to papyrus, parchment had the advantage of being more flexible in its uses, especially for producing codices, but it had the very distinct disadvantage of significantly higher costs. Parchment could thus be used for producing prestige items, such as lavishly produced copies of the Qurʾān. The best example of this conspicuous consumption is probably the majestic *Blue Koran* (D'Ottone Rambach 2017), a luxurious manuscript on blue vellum (high-quality parchment) most likely dating from the 3rd/9th century. However, the high costs of parchment made it an unattractive option for more mundane forms of writerly culture. While parchment always retained a role as a niche luxury item and for ritual purposes (e.g., talismans), the massive adoption of paper in Islamicate societies in the course of the 3rd/9th century marginalized it. Paper, produced from cellulose pulp fibers derived from wood, rags or grasses, started to be used from the mid-2nd/8th century onward but came into its own during the 3rd/9th century. In contrast to papyrus, it was versatile and could be easily used to produce codices; in contrast to parchment, it was cheap (Bloom 2001). Commodity prices are difficult to trace for the medieval periods, but one attempt to do so has argued that rising purchasing power and decreasing costs from the 2nd/8th century onward meant that paper and books became increasingly affordable for larger sections of society, at least in Iraq and Egypt (Shatzmiller 2015).

The rise of the paper codex helped to enable the increasingly writerly culture of Islamicate societies to undergo an expansion of pragmatic and scholarly literacy. In the scholarly field, the enormous number of texts produced from the 3rd/9th century onward reflected this development. This lively written culture found its monument in the *Catalog* (*Fihrist*) of Ibn al-Nadīm (d. 380/990), one of the most important sources for the intellectual life of the early Islamic centuries. In his *Catalog* Ibn al-Nadīm lists thematically the titles of books known to him, some 7,000, and it comes as no surprise that scientific subjects, such as alchemy, medicine, mathematical sciences and astronomy are well represented (Stewart 2011; Ibn al-Nadīm 1970[CB];[1] Ibn al-Nadīm 2014[CB]). The link between the adoption of paper and the expansion of written scientific knowledge was especially close, as one of the iconic elements of the explosion of written information, the so-called translation movement, focused on the sciences. This development meant that numerous works on philosophical, astronomical, mathematical and medical knowledge in Pahlavi, Sanskrit, Syriac and Greek were translated into Arabic between the late 2nd/8th and late 4th/10th century in Baghdad – traditionally, but perhaps not reliably, ascribed to the *Bayt al-Ḥikma* or House of Wisdom (Brentjes and Morrison 2010[CB], 564–72).

While we are informed that numerous works were produced in the first four Islamic centuries, many have been lost; it is estimated, for instance, that only 2 percent of the titles Ibn al-Nadīm mentioned are extant (Stewart 2011, 129). In addition, those that are extant generally survive in later copies. In the scientific fields, a typical example is the *Book of the Images of the Stars* (*Kitāb Ṣuwar al-kawākib al-thābita*) by al-Ṣūfī (291–376/903–986), one of the most important treatises on constellation iconography to be produced in Arabic. Al-Ṣūfī wrote this work in *c.* 355/965, but the earliest extant parts of this work, preserved in MS Doha, Museum of Islamic Art, 2.1998, were produced some 150 years later in 519/1125 (Savage-Smith 2013[CB]). In the same vein, one of the most important early *zīj*s (astronomical handbooks) was written by al-Khwārazmī (d. *c.* 235/850) in the first half of the 3rd/9th century, yet his work is known to us only indirectly from the subsequent Arabic tradition and from Latin translations of the 6th/12th century (King *et al.* 2001[CB], 33–5).

IV.1.2 Textual formats

While virtually all scientific texts have come down to us as codices, the textual format in these codices is far from uniform. For scientific texts, the most relevant differentiation is that between single-text manuscripts (one codex with one text) and those manuscripts containing several texts. The latter, in turn, come in the form of either multiple-text manuscripts (one codex with several texts resulting from one production process) or composite manuscripts (one codex with several texts that had previously been independent units; Friedrich and Schwarke 2016). Modern research on manuscript cultures has tended to prioritize the single-text manuscript as the prime textual format. This inclination goes back to the fact that this textual format tends to transmit the 'grand' texts with identifiable authors, while multiple-text manuscripts and composite manuscripts often have anonymous material. The preference for single-text manuscripts has been compounded by practical issues, as multiple-text manuscripts and composite manuscripts are often the last items dealt with by libraries on account of the time that it takes to catalogue them. In consequence, they have been less easily identifiable and accessible for research. However, multiple-text manuscripts and composite manuscripts have increasingly come into scholarly focus and scientific texts are no exception to this development. We are still far from having an overview of the relative distribution of single-text manuscripts, multiple-text manuscripts and composite manuscripts across the extant manuscript tradition or even in specific fields. It is however possible to venture the hypothesis that in most scholarly fields only a minority of texts were transmitted in the format of single-text manuscripts. Moreover, in some fields, such as the transmission of the Prophetic tradition (*ḥadīth*), single-text manuscripts were rather the exception.

The multiple-text manuscript, generally produced by one single scribe, is, in most cases, datable and thus gives fascinating insights into the intellectual preoccupations of (sometimes identifiable) individuals. Franz Rosenthal ([1914–2003], 1955) labeled these manuscripts as 'one-volume libraries'. For instance, in the mid-6th/12th century, an anonymous writer assembled a whopping 80 texts in Baghdad, and this multiple-text manuscript is today MS Damascus, Syrian National Library ('Ẓāhiriyya'), 4871. Of the original texts, only 43 remain, but the mixture of philosophic and scientific treatises from the fields of astronomy and geometry gives a taste of the writer's intellectual world (Ragep and Kennedy 1981). Some 60 years later, a certain al-Ḥasan ibn al-Ḥasan produced a manuscript in the same city, which contains numerous texts on Euclidean geometry, most of them works of the well-known geometer Aḥmad al-Sijzī (d. *c.* 411/1020–1). Al-Ḥasan ibn al-Ḥasan traced his descent back to the Seljuk vizier Niẓām al-Mulk (409–485/1018–1092) and this claim was certainly linked to the fact that he wrote his multiple-text manuscript in the Niẓāmiyya Madrasa, which this vizier founded over a century earlier (Endress 2016). This manuscript is today MS Dublin, Chester Beatty Library, Ar. 3652, listed in the Bibliography as *Collection* (*Majmūʿ*). MS Paris, BnF, arabe 2457, also a *Collection*, brings us to the textual practices of Aḥmad al-Sijzī himself; in this work, he collected 51 mathematical and astronomical texts by various authors. His choice was still meaningful enough to be copied over 500 years later as the Paris manuscript was produced between 969/1562 and 972/1565 by a scribe in southern Persia (Endress 2016).

Scientific texts have also come down to us in the shape of composite manuscripts, that is, previously independent codicological units bound together. While the multiple-text manuscript often allows us to pin down the moment of production and thus the manuscript's historical context, this is much more complicated in the case of composite manuscripts. Here we are often only able to give a *terminus post quem* dating on the basis of the last dateable text combined with a rough estimate on the basis of the binding. In addition to the problem of contextualizing these

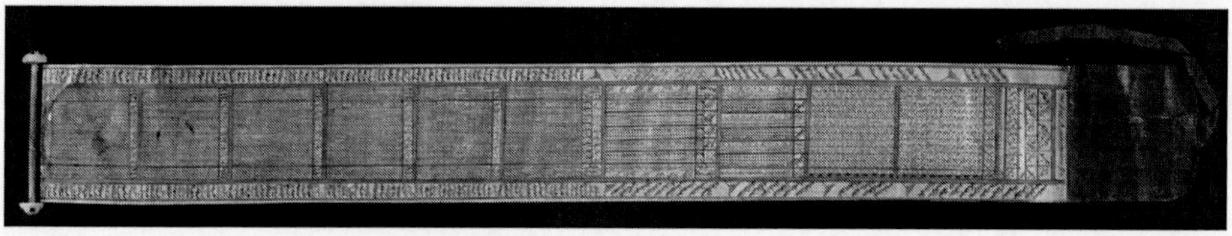

Figure IV.1.1 Ottoman calendar scroll for the years 1198–1282/1784–1866, acquired by Heinrich Friedrich von Diez (1751–1817) during his stay in Istanbul (1784–1790). MS Diez A duodez. 9a.

Source: © STAATSBIBLIOTHEK ZU BERLIN – Preußischer Kulturbesitz, Orientabteilung

manuscripts, they pose a particular problem in terms of accessibility: Compared to multiple-text manuscripts, the composite manuscripts are even more unwieldy for cataloguers with a wide range of scripts, formats and often themes. A typical example of such a complex composite manuscript is MS Leiden, University Library, Or. 644, another *Collection* (Schmidt 2016). The binder, more or less carefully, had to trim larger codicological units to the format of the new composite manuscript. Users later pasted in individual leaves and many of the texts only made it into the manuscript in fragmentary form. The thematic range of its 77 texts takes the reader on quite a ride: Qurʾānic sciences, Prophetic traditions, jurisprudence, prayers, official letters, documents, poems, magic formulas, almanacs, mysticism, alchemy, medicine, astronomy and, finally, geometry. This manuscript was bound in the mid-11th/17th century and gives an insight into the Ottoman literary culture of the previous 100 years or so. While we cannot say too much about who put it together and for what purposes, it is at least clear that the independent codicological units had their origin among what Jan Schmidt describes "as the (lower) middle class of Istanbul: soldiers, academics employed at the less prestigious schools, and a sheik" (Schmidt 2016, 215).

In addition to the various formats in which codices transmitted scientific texts, other formats did not entirely disappear from usage. Scrolls, for example, retained a role for specific texts in which this format offered distinct advantages. For instance, MS Hamburg, Museum am Rothenbaum Kulturen und Künste der Welt (Museum of Ethnology), no. 13.172:24, is a parchment scroll written in 1211/1796–7 in Istanbul. It is a *rūznāme*, a table for calculating the respective weekday and prayer times for any given date in the solar or lunar calendar. These tables were refined in the course of the 13th/18th century in the Ottoman lands and the grandly embellished nature of the Hamburg scroll shows that its producer intended it for elite usage (Haase *et al.* 2016, 67–9). With a width of merely 8.6 cm, it was of very compact size and its owner could carry it around more easily than a codex. Another example is the much shorter Ottoman calendar today in Berlin, Preußischer Kulturbesitz, Staatsbibliothek (Figure IV.1.1) At the same time the scroll format indicates the text's pragmatic purpose, a text to be used and not to be studied, as the user could quickly gain access to the required information without having to leaf through multiple folia.

IV.1.3 Layout

The organization of the text in any given manuscript, the layout of the page, is a crucial element of its materiality, but scholarship on Islamicate manuscript cultures has only recently started to engage with this topic. The seminal work on this issue is Daub (2016). On account of the often highly technical nature of the information contained in them, scientific manuscripts have one distinctive feature in page layout: the systematic and sophisticated use of tables and illustrations (such as diagrams, maps and drawings) to organize knowledge. None of these tools to organize the page is unique to scientific manuscripts, but they do appear in these texts much more commonly. It is clear that authors employed them in a uniquely systematic fashion, arguably also as a didactic device. For example, the table layout was widely used in medical fields to present rules for food and drug use and for listing diseases with associated therapies. One of the most influential treatises on medical substances was Ibn Buṭlān's (d. 458/1066) *Almanac of Health* (*Kitāb Taqwīm al-ṣiḥḥa*), which he organized in tabular format using 40 tables (Ibn Buṭlān 1990). Subsequent texts on the same topic by other authors show that this approach to page layout became paradigmatic in books on medical substances and led to an established tradition of tabular presentation. The text by the Andalusian author Ibn Baklārish, written at the turn of the 5th/11th and 6th/12th centuries, included 121 tables (Savage-Smith 2008).

The tabular layout was even more popular in the field of astronomy where one of the field's most important genres, the *zīj*, is basically a compendium of numerical tables (Chapters I.6 and III.8). These tables allowed determining the positions of the sun, the moon and the planets at a given day and time. They were mostly used for astronomical and astrological purposes such as horoscopes and could also be employed for ritual purposes such as identifying the direction of Mecca and determining the times of the five daily prayers, like the Ottoman almanac (*rūznāme*). They usually contain between about 50 and 400 tables and have only limited running text, mostly instructions for use (Van Brummelen 2014). Because the tables are a standard element of scientific manuscripts, they became increasingly user-friendly to spare the user complicated calculations. From the 7th/13th century, we have, for instance, an increasing number of tables that were organized in a way to simplify complicated Ptolemaic calculations (King *et. al.* 2001, 79). Astronomical tables were able to organize large sets of data, and individual tables often became so substantial that they circulated on their own. For instance, sexagesimal multiplication tables for products of numbers between 0 and 60 could come in tables with up to 216,000 entries and a table for timekeeping by the sun and the stars from early 10th/16th-century Istanbul carried some 250,000 entries. Probably the largest set of tables ever produced in the premodern period is the *Tables of the Time-Arc for all Latitudes* (*Jadāwil al-dā'ir al-āfāqī*) by the Egyptian author Najm al-Dīn al-Miṣrī (1st half 8th/14th century). This landmark work in the field of astronomical timekeeping contained some 400,000 entries in a wide array of differently organized tables (Charette 1998). The tabular layout became so popular in the scientific fields that scholars occasionally tabulated rather surprising functions, such as the oblique ascensions of the ascendant at the time when the muezzin announces a blessing of the Prophet (King *et al.* 2001, 96).

Another feature that is as prominent as tables in scientific manuscripts is branch-diagrams (*tashjīr*), which we find in the work of Ibn Baklārish. These branch-diagrams visually organized information in categories, divisions and subdivisions, as an *aide-memoire*. We find a similar layout in other fields such as poetry, where picture-poems became a popular way to organize the page. However, the sheer number of such branch-diagrams and their variety clearly indicates that their center of gravity was within the scientific fields of knowledge. Together with branch-diagrams, other types of diagrams were a further striking feature of scientific manuscripts. For instance, medical texts regularly described anatomical features such as the bones, nerves, muscles and organs (including the eye, liver, heart and brain) and routinely illustrated the textual descriptions with diagrams. The physician al-Rāzī, mentioned earlier, uses a diagram of the brain in his influential medical encyclopedia *The Book on Medicine for al-Manṣūr* (*al-Kitāb al-Manṣūrī fī al-ṭibb*), reproduced in Savage-Smith (2007).

A final distinctive feature of scientific manuscripts in an Islamicate context is the high number of painted illustrations. While painted illustrations were standard in Persianate and Turkic manuscript cultures, they remained marginal in Arabic manuscripts. The only exception in the nonscientific fields is a brief flowering between the middle of the 7th/13th century and the end of the 8th/14th century in the Iraqi, Egyptian and Syrian lands when there was a tendency to illustrate literary texts (Gacek 2009, 182–3). However, scientific manuscripts continued to carry illustrations after the 8th/14th century, and they were the largest group of Arabic manuscripts to do so, especially those on botany, pharmacology, medicine, zoology, astronomy, astrology and mechanics. These included the Arabic translations of the pharmacological *On Medicinal Substances* (*De materia medica*) by Dioscorides ([d. 90]; e.g., the illustration of two physicians preparing medicine, MS Baltimore, Walters Museum, W.675 produced in Baghdad in 621/1224), the *Book of the Theriaq* (*Kitāb al-Diryāq*) on antidotes by Pseudo-Galen (Pancaroğlu 2001), the *Book of the Star Constellations* by al-Ṣūfī mentioned earlier and al-Jazarī's (d. *c.* 602/1206) *The Book of Knowledge of Ingenious Mechanical Devices* (*al-Jāmiʿ bayna l-ʿilm wa-l-ʿamal al-nāfiʿ fī ṣināʿat al-ḥiyal*) on engineering (al-Jazarī 1974[CB]).

Although no complete illustrated manuscripts survive from before the 6th/12th century, fragments of Dioscorides's work show that they existed (Brentjes and Morrison 2010, 575–6).

IV.1.4 Storing manuscripts

Scholars in scientific fields needed access to physical copies of the works they did not own themselves – there were few who could afford to have a copy of all relevant works in their own private book collections. We do not yet have studies specifically on the storage of scientific manuscripts, but it is clear by now that the scientific fields were not separate from other fields of knowledge. Most importantly, they were part of *madrasa* teaching and two of the works considered so far neatly illustrate this. As mentioned previously, the multiple-text manuscript MS Dublin, Chester Beatty Library, Ar. 3652 was written in a *madrasa*, the Niẓāmiyya Madrasa in Baghdad. The Doha copy of al-Ṣūfī's *Book of the Constellations of the Stars*, in turn, was copied from a manuscript that was endowed to the same *madrasa*. It thus comes as no surprise that in the earliest surviving actual library catalogue of a *madrasa*, the late 7th/13th-century Ashrafiyya in Damascus, two of the 15 thematic categories are specifically for the sciences (Hirschler 2016[CB]). The Damascene Ibn Abī Uṣaybiʿa (d. 668/1270) wrote a particularly informative biographical dictionary of those active in scientific fields during this period (Brentjes 2010; Ibn Abī Uṣaybiʿa 1965[CB], 2020[CB]). His description gives a vivid picture of the availability of scientific texts across the urban topography and in various social settings.

In addition, sites of scientific activities clearly had their dedicated book collections. For example, the illustration of the Ottoman observatory in Istanbul in the late 10th/16th century depicts part of a library. This was seemingly the private library of its director, Taqī al-Dīn (932–993/1526–1585), and some books bearing the mark of his ownership are in Leiden today (King 1995; Chapter II.7). Finally, the lavish illustrations and the pragmatic purposes of many scientific manuscripts indicate that their buyers were also patrons at court or rich customers beyond the court. In the field of astronomy, such wealthy 'armchair astronomers', wanted to identify the constellations on their globes and simply enjoy a beautiful book (Savage-Smith 2013).

IV.1.5 Ink and writing implements

We have ample advice literature on making ink (*midād* and *ḥibr*) of which the best known are the *Ornament of the Scribes* by al-Rāzī, mentioned earlier, the *Staff of the Scribes* (*ʿUmdat al-kuttāb*) attributed to al-Muʿizz ibn Bādīs (r. 406–454/1016–1062), *Flowers on the Making of Ink* (*al-Azhār fī ʿamal al-aḥbār*) by Muḥammad ibn Maymūn al-Marrākushī (7th/13th century), and the *Treasure of the Elect* (*Tuḥfat al-khawāṣṣ*) by al-Qalalūsī (d. 707/1308). Ink recipes were of particular importance in books on bookmaking in general; the *Ornament of the Scribes* dedicates roughly a third of its text to the topic of ink-making (Zaki 2011). The multitude of recipes we find is partly down to the fact that writers had to deal with many different materials. Even after paper had become dominant, different inks were required to deal with the wide range of paper qualities. While the advice literature mentions different kinds of ink, including carbon ink with soot as the main ingredient, they most often discuss iron-gall ink. Although we find a wide degree of variation in recipes for iron-gall ink, also depending on factors such as whether the ink was intended for immediate use or long-term storage, most of them agree that the basic ingredients are gall nuts, water, vitriol and gum arabic. In the same vein, there is broad agreement that the processing includes crushing the gall nuts, boiling them in water, then straining them and leaving the mixture in the sun (Raggetti 2016).

Because of the relatively high number of illustrations, colored inks are of specific importance for discussing scientific manuscripts. Recipes for colored ink show as much variation as those for black ink, but there are some pertinent works, such as the *Ornament of the Scribes*, that do not have a single recipe for colored ink. The *Staff of the Scribes*, by contrast, lists more than 30 such recipes including those for red, golden, yellow, silver and azure blue ink. A particularly popular topic was invisible inks: The *Ornament of the Scribes* describes the preparation of 11 different such inks, including one that reappears only by night and another that disappears gradually. These invisible inks subsequently took another path, as they also became part of the trickster literature associated with the underworld and street life. The *Book of the Flowering Gardens* (*Kitāb zahr al-basātīn*) by al-Zarkhūrī, probably composed in the early 9th/15th century, is a good example of how these recipes were put to new uses and further developed to produce 'fire ink' or 'dust ink' (Raggetti 2016; al-Zarkhūrī 2012).

As well as multitudes of ink recipes, advice literature intensively discussed writing tools. Islamicate manuscripts were written almost without exception by using the *qalam*, a split reed pen, while the use of the brush was limited to illustrations. The reed had to be trimmed and the intended use of the resulting *qalam* determined its main feature, the shape of its nib. In consequence, we find a sophisticated vocabulary to describe the nib, such as the *qalam* with a straight (as opposed to oblique) nib, *jazm;* the inner side of the nib, *wajh al-qalam;* the right side of the half-nib, *ḥarf al-qalam*; the place where the cutting of the nib begins, *ḥadd al-qalam*; and so on. The main factor in producing a pen was whether the nib was cut straight or obliquely in order to produce thicker or finer lines. In addition to the pen, the writer's main implements were a knife for trimming the reed, an inkwell, a pen box, and a *milqaṭ/milqāṭ*, an implement for collecting eraser shavings (Gacek 2009; Déroche 2005; Gacek 2001, 129; Figures IV.1.2 and IV.1.3). While most of our knowledge of scribal practices relies on advice literature, fascinating additional insights can be gained from documents that emerged out of the scribal milieu. For instance, an estate inventory from the Cairo Geniza collection, probably from the 7th/13th century, enumerates the possessions of a deceased scribe, among them different inkwells, a copper ruler, parchment rolls and a copper nibbing block (Sadan 1977).

The production of Arabic scientific and nonscientific manuscripts took place in two ways. On one hand, we have a large number of manuscripts with '*li-nafsihi*' (for himself) notations. These notations mean that the manuscript was written by a scholar for his own personal use. On the other hand, we have manuscripts that were produced for wealthy (often royal) patrons. This is especially true for illustrated manuscripts that play a relatively important role within the corpus of scientific manuscripts. A question that remains open for Arabic manuscript cultures in general is to what extent professional workshops of scribes existed.

Figure IV.1.2 Inkwell, Iran, copper alloy with silver inlay, mid-7th/13th century; inv. no. 1890,431. Photographer Johannes Kramer.

Source: © Museum für Islamische Kunst – Staatliche Museen zu Berlin

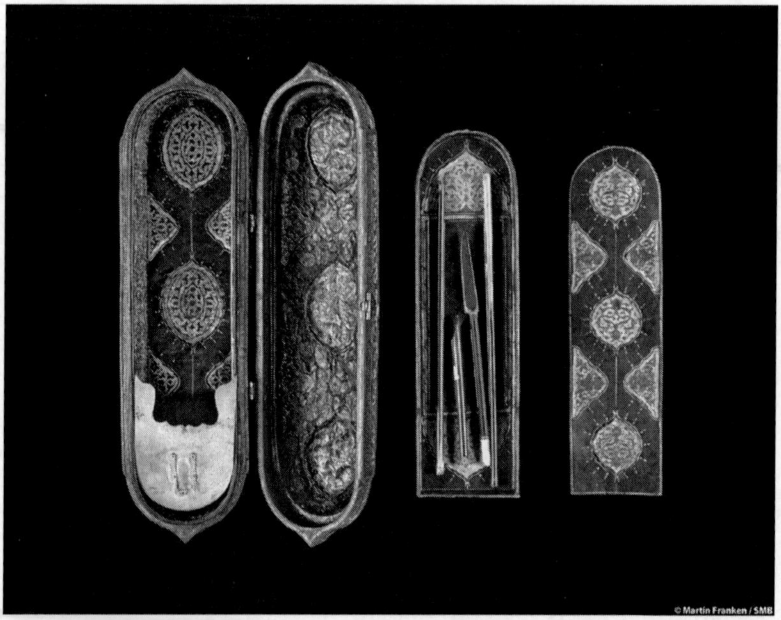

Figure IV.1.3 Pen box with writing tools qalam, knife, inkwell, mirror, ruler), Istanbul, late 11th/17th century; papier-mâché, goat leather, wood, bone, iron, glass and paint. The decoration indicates that those pen boxes and their tools were produced in workshops of bookbinders (Kröger 2004, 195); inv. no. I B 100a–i. Photographer Martin Franken.

Source: © Ethnologisches Museum der Staatlichen Museen zu Berlin – Preußischer Kulturbesitz

Irrespective of the production setting, a manuscript became at some point an object that was traded on the market. Middle Eastern cities routinely had specialized markets for paper, books, bindings and scrap paper (Figure IV.1.4). Ibn al-Nadīm was enabled to write his *Catalog*, mentioned earlier, because he himself was a book merchant. Even manuscripts that had at one point in time been endowed to a library routinely made their way back to the market at some later point. *Theft* would be too one-dimensional a term to describe the factors that drove these processes. The best overview of the book trade is Behrens-Abouseif (2018).

Figure IV.1.4 Miniature from a copy (c. 998/1590) of Ḥusayn ibn Ismāʿīl Gāzurgāhī's (d. after 908/1502–3) *Sessions of the Lovers* (*Majālis al-ʿushshāq*), portraying Khwāja ʿUbaydallāh Aḥrār (806–895/1404–1490), the shaykh of the Naqshbandī Sufis, sitting in front of a bookbinder's workshop. On the right are street peddlers with a balance; inv. no. I. 1986.229, fol. 149b. Photographer Petra Stüning.

Source: © Museum für Islamische Kunst – Staatliche Museen zu Berlin

Note

1 Consolidated bibliography.

Bibliography

Primary sources

Ibn Buṭlān. ed. Elkhadem, H. 1990. *Kitāb taqwīm al-ṣiḥḥa. Le Taqwīm al-ṣiḥḥa (Tacuini sanitatis) d'Ibn Buṭlān.* Leuven: Peeters.

al-Marrākushī, Muḥammad ibn Maymūn. ed. Shabbūḥ, I. 2001. "Kitāb al-azhār fī ʿamal al-aḥbār," [The Most Beautiful Flowers on the Production of Inks] *ZGAIW* 14: 41–133.

al-Qalalūsī, Muḥammad ibn Muḥammad. ed. al-ʿAbbādī, Ḥ. A. 2007. *Tuḥaf al-khawāṣṣ fī ṭuraf al-khawāṣṣ* [The Rarities of the Elite, On the Curiosities of the Elite]. Alexandria: Biblioteca Alexandrina.

al-Zarkhūrī, Muḥammad ibn Abī Bakr. ed. Gari, L. 2012. *Zahr al-basātīn fī ʿilm al-mashshātīn: Garden's Flowers on Sleight of Hand Knowledge. A Rare Medieval Islamic Text on Technology and Crafts*. Cairo: Maktabat al-Imām al-Bukhārī.

Manuscripts

MS Baltimore, Walters Museum, W.675.

MS Damascus, Syrian National Library ('Ẓāhiriyya'), 4871.

MS Doha, Museum of Islamic Art, 2.1998.

MS Dublin, Chester Beatty Library, Ar. 3652.

MS Hamburg, Museum of Ethnology, no. 13.172:24.

MS Leiden, University Library, Or. 644.

MS Paris, BnF, arabe 2457.

Research literature

Behrens-Abouseif, D. 2018. *The Book in Medieval Egypt and Syria (1250–1517). Patronage, Production and Market*. Leiden: Brill.

Bloom, J. M. 2001. *Paper before Print: The History and Impact of Paper in the Islamic World*. New Haven: Yale University Press.

Brentjes, S. 2010. "Ayyubid Princes and Their Scholarly Clients from the Ancient Sciences," in Fuess, A. and Hartung, J.-P., eds. *Court Cultures in the Muslim World: Seventh to Nineteenth Centuries*. London: I. B. Tauris, 326–56.

Charette, F. 1998. "A Monumental Medieval Table for Solving the Problems of Spherical Astronomy for All Latitudes," *Archives internationales d'histoire des sciences* 48: 11–64.

Daub, F.-W. 2016. *Formen und Funktionen des Layouts in arabischen Manuskripten anhand von Abschriften religiöser Texte: al-Būṣīrīs Burda, al-Ǧazūlīs Dalāʾil und die Šifāʾ von Qāḍī ʿiyāḍ*. Wiesbaden: Harrassowitz.

Déroche, F. 2005. *Islamic Codicology: An Introduction to the Study of Manuscripts in Arabic Script*. London: al-Fūrqān Foundation.

D'Ottone Rambach, A. 2017. "The Blue Koran. A Contribution to the Debate on its Possible Origin and Date," *Journal of Islamic Manuscripts* 8: 127–43.

Endress, G. 2016. "'One-Volume Libraries' and the Traditions of Learning in Medieval Arabic Islamic Culture," in Friedrich, M. and Schwarke, C., eds. *One-Volume Libraries: Composite and Multiple-Text Manuscripts*. Berlin: Walter de Gruyter, 171–205.

Gacek, A. 2001. *The Arabic Manuscript Tradition: A Glossary of Technical Terms and Bibliography*. Leiden: Brill.

Gacek, A. 2009. *Arabic Manuscripts: A Vademecum for Readers*. Leiden: Brill.

Friedrich, M. and Schwarke, C. eds. *One-Volume Libraries: Composite and Multiple-Text Manuscripts*. Berlin: Walter de Gruyter

Haase, C.-P. *et al.*, eds. 2016. "Herstellung und Gestaltung von Manuskripten/Production and Design of Manuscripts," *Manuscript Cultures* 9: 59–126.

King, D. A. 1992. "Some Remarks on Islamic Astronomical Instruments," *Scientiarum Historia* 18.1: 5–23.

King, D. A. 1995. "Illustrations in Islamic Scientific Manuscripts," in Atiyeh, G. N., ed. *The Book in the Islamic World: The Written Word and Communication in the Middle East*. Albany, NY: State University of New York Press, 149–77.

Pancaroğlu, O. 2001. "Socializing Medicine: Illustrations of the *Kitāb al-diryāq*," *Muqarnas* 18: 155–72.

Ragep, F. J. and Kennedy, E. S. 1981. "A Description of *Ẓāhiriyya* (Damascus) MS 4871: A Philosophical and Scientific Collection," *Journal of the History of Arabic Science* 5.1–2: 85–108.

Raggetti, L. 2016. "*Cum grano salis*. Some Arabic Ink Recipes in Their Historical and Literary Context," *Journal of Islamic Manuscripts* 7.3: 294–338.

Rosenthal, F. 1955. "From Arabic Books and Manuscripts V: A One-Volume Library of Arabic Scientific and Philosophical Texts in Istanbul," *Journal of the American Oriental Society* 75.1: 14–23.

Sadan, J. 1977. "Nouveaux documents sur scribes et copistes," *Revue des études islamiques* 45: 41–87.

Savage-Smith, E. 2007. "Anatomical Illustration in Arabic Manuscripts," in Contadini, A., ed. *Arab Painting: Text and Image in Illustrated Arabic Manuscripts* (Handbuch der Orientalistik I.90). Leiden: Brill, 147–59.

Savage-Smith, E. 2008. "Ibn Baklarish in the Arabic Tradition of Synonymatic Texts and Tabular Presentations," in Burnett, C., ed. *Ibn Baklarish's Book of Simples: Medical Remedies between Three Faiths in Twelfth-Century Spain*. London: The Warburg Institute, 113–31.

Schmidt, J. 2016. "From 'One-Volume-Libraries' to Scrapbooks. Ottoman Multiple-Text and Composite Manuscripts in the Early Modern Age (1400–1800)," in Friedrich, M. and Schwarke, C., eds. *One-Volume Libraries: Composite and Multiple-Text Manuscripts*. Berlin: De Gruyter, 207–31.

Shatzmiller, M. 2015. *An Early Knowledge Economy: The Adoption of Paper, Human Capital and Economic Change in the Medieval Islamic Middle East, 700–1300 AD* (CGEH Working Paper Series 64). Utrecht: Universiteit Utrecht.

Stewart, D. 2011. "Ibn al-Nadim," in Young, T. D. and St. Germain, M., eds. *Essays in Arabic Literary Biography I, 950–1350*. Wiesbaden: Harrassowitz, 129–42.

Van Brummelen, G. 2014. "The Travels of Astronomical Tables within Medieval Islam: A Summary," *Suhayl* 13: 11–21.

Zaki, M. 2011. "Early Arabic Bookmaking Techniques as Described by al-Rāzī in His Recently Rediscovered *Zīnat al-Katabah*," *Journal of Islamic Manuscripts* 2.2: 223–34.

IV.2
THREE-DIMENSIONAL ASTRONOMY
Celestial globes and armillary spheres

Taha Yasin Arslan

Astronomical instruments are physical manifestations of knowledge about the heavens to which various disciplines contributed. They provide information on the development of three of the four fundamental mathematical sciences – number theory, geometry, and astronomy, and their branches. They are also evidence of scholarly influences and transmission, mastery of craftsmanship, artistic and religious preferences, and the social acknowledgment of astronomy and astrology as well as the patronage and institutionalization of knowledge. This chapter deals with two of these instruments: celestial globes and armillary spheres. Celestial globes carry evidence of arts and craftsmanship in their time and region through the depictions of the constellations, the wording of the maker's signature, the names of stars, constellations, lunar mansions, and the nature of the scales included. Their popularity, as well as records of dedications, reveals non-scholarly interests and the patronage of different elite circles for globe making, while records of dates and places may provide information on the transition of knowledge between regions and centuries. In contrast, only a few examples of armillary spheres survive, preventing us from making a broad assessment of the instrument's role in society. However, extant manuals on its construction and use shed some light on the development of the instrument.[1]

IV.2.1 Celestial globes

The celestial globe is an instrument that imitates the daily motion of the sky and is used for observations, calculations, or demonstrations. It is usually made of metal and occasionally of wood and papier-mâché. More than 200 globes are extant and almost all have a hollow metal body (for a comprehensive description and a list of surviving celestial globes, see Savage-Smith 1990–1991). Their sizes vary between 5cm and 35cm in diameter. This is one of the largest classes of surviving astronomical instruments made in the Islamicate world. During the construction process, the first problem that had to be dealt with was how to produce a sphere. Unfortunately, no known text discusses this. A careful investigation of the extant instruments, however, gives us credible information on two types of metal spheres used as globes. First are two-piece globes, which consist of two identical hemispheres cast, joined by seams, and smoothed. Second are one-piece globes that were made by the lost-wax method, which involved casting a complete sphere and smoothing its surface, although this usually required plugs on the body to cover the

DOI: 10.4324/9781315170718-43

hole required in the casting process (Savage-Smith 1985, 83–6). Depending on the craftsmanship, construction markings, plugs, or traces of seams may hardly be visible.

Almost all surviving globes from the Islamicate world share major similarities. A standard globe has the celestial equator and ecliptic drawn. The equator carries a scale of 360°, while the ecliptic has twelve scales of 30°. Both begin at the vernal equinox and are divided into intervals of 5°, 6°, or 10°, where the divisions are labeled with Arabic alphanumeric notation (*abjad*) and subdivided into intervals of 1° or 2°. The twelve scales of the ecliptic are inscribed with the corresponding names of the signs of the zodiac. The poles of the equator and the ecliptic are usually marked with circles and occasionally labeled. The globe is divided into twelve sections by six great circles drawn at intervals of 30° between the ecliptic poles. One of these is a solstitial colure, which is a great circle through the celestial poles, the poles of the ecliptic, and the summer and winter solstices. None of the standard globes carries circles of ever-visible or ever-invisible stars. Depending on the type of globe, stars and the constellation figures may or may not be inscribed. When they are, the stars are usually shown in different sizes and shapes based on their brightness (Figure IV.2.1).

A standard globe has two rings, one for the meridian and another for the horizon. The globe is mounted in the meridian ring from the celestial poles and the ring is fixed within the horizon ring at the north–south axis, making it the prime meridian. The horizon ring sits on a stand. Because of this arrangement, manuscripts in the Islamicate world refer to the instrument as the *sphere with a stand* (*al-kura dhāt al-kursī*). Both rings carry either a scale of 360°, two scales of 180°, or four scales of 90° and are divided into intervals of 5°, 6°, or 10°, which are subdivided into intervals of 1° or 2°. Although standard globes do not have an attached ring or a gnomon

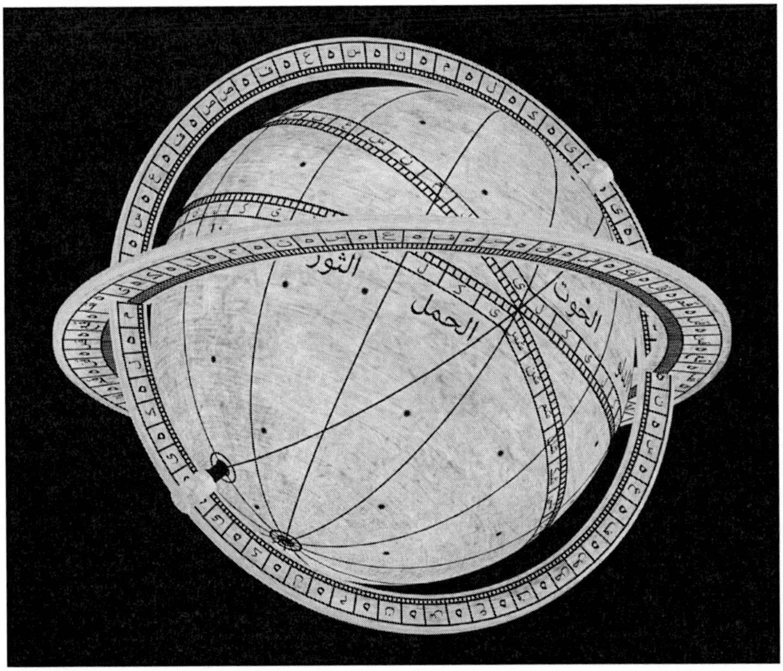

Figure IV.2.1 Model of a standard globe

Source: © Taha Yasin Arslan, Istanbul

for measuring the altitude of the Sun and the stars, the manuals on the use of globes propose using a suitable object as gnomon and a detached quadrant with a 90° scale to find the altitude of celestial objects. Standard globes are divided into three groups for the number of stars they represent. Globes of the first group have around a thousand stars usually distributed among the images of the 48 classical constellations. Globes of the second group display only selected stars, usually those found on astrolabes. The third group contains specimens that carry only the great circles but no stars (Savage-Smith 1985, 61–2).

In addition to those standard types, the works of Ptolemy (*c.* 100–180) and Battānī (d. 317/929–930) introduced two other designs to the Islamicate world. In the eighth book of the *Almagest*, Ptolemy describes a globe with a rotatable ecliptic ring (Ptolemy 1998[CB],[2] 404–7). The purpose of this modification is to allow for the deviation caused by the precession of the equinoxes over the years (approximately 1° in 72.2 years in modern value). Al-Battānī introduced a globe called an egg (*bayḍa*), which was a common term for globes, with five rings (al-Battānī 1977[CB], 210–13 [fols 142a–44b]). The celestial equator and the ecliptic are drawn on the globe as usual. Additionally, there are two meridian rings, one is fixed and the other rotates, a horizon ring, a zenith ring, and the outer ring, which carries a movable and removable gnomon. Unlike the other examples, al-Battānī's egg does not sit on a stand but is suspended. The second rotatable meridian ring serves for marking the star positions and making calculations accurately. The movable gnomon is used to mark the solar longitude and the altitude of the sun along with many other applications. Unfortunately, no example of these two ingenious devices has survived.

IV.2.1.1 Instructions for globe making

Instructions of how to draw on globes are very limited in extant astronomical writings. This is particularly surprising given the large number of surviving objects. ʿAbd al-Raḥmān al-Ṣūfī's (291–376/903–986) *Book of the Constellations* (*Kitāb ṣuwar al-kawākib al-thābita*) seems to be the only treatise that was used extensively all over the Islamicate world. Construction manuals for astrolabes and sundials, however, are much more common. They usually give precise values in ratios as pointers for engravings such as almucantars, azimuths, and seasonal hours on astrolabes and parabolas and hyperbolas of shadow lengths of various signs of the zodiac on sundials. On the other hand, globes require their users to consider only two parameters: (1) the obliquity of the ecliptic and (2) the precession of the equinoxes. While the value of the obliquity was calculated and given differently by various astronomers, the globe makers seem to have used approximate values such as 23.5° or 24°. Considering the small size of the instrument one should not expect a precision of more than half or a quarter of a degree. The value of the precession is more important since the makers usually used older catalogues for star positions (i.e., al-Ṣūfī's *Book of the Constellations*) and needed to consider the positional changes of stars during the period between the preparation of the star catalogue and the construction of the globe.

The scales, circles, and inscriptions all depend on the size of the instrument as well as the maker's initiative. The production process includes marking opposite points as poles, drawing great circles for the ecliptic and celestial equators and the three great circles through the poles of the ecliptic, as well as constructing scales with various intervals on some of these circles, and inscribing their values. Unlike the ones on an astrolabe, these circles do not require a table of radii (*jadwal anṣāf al-aqṭār*) but simple angular calculations since no type of projection is required on globes. For those tasks, the globe makers usually employed various tools, such as a quadrant with a scale of 90°; compasses; and engraving tools.

There are two important works with detailed descriptions on how to draw on globes. The first one was introduced by al-Battānī in the 7th chapter of his *Book of Astronomical Tables* (*Kitāb al-Zīj*), where he proposed the egg (al-Battānī 1977, 210–13 [fols 142a–44b]). According to al-Battānī, the maker needs to determine two points that are exactly opposite of each other on the globe, which will act as the celestial poles. However, he doesn't mention how this is done. Al-Battānī's summary of the different steps for construction reflects the scholar's effort for providing globe makers (craftsmen as well as scholars) with the necessary skills. Using the poles, the globe maker is told to draw a great circle as the celestial equator and then to mark four equidistant points on it. The steady leg of a pair of compasses has to be fixed at one of these points, and the drawing leg is used to draw another great circle that passes through two points on the equator and the poles, making it the solstitial colure. In order to determine the positions of the ecliptic poles and to draw the ecliptic, the globe maker has to inscribe a quarter of the equator with a scale of 90°. Learning from al-Battānī's text that the obliquity of the ecliptic is 23° 35′, the globe maker, using the scale, can now open the legs of the compasses as wide as the value of the obliquity. Then he fixes the steady leg at one of the celestial poles and rotates the drawing leg until it intersects with the solstitial colure. Wherever the tip of the leg touches on the colure he marks as an ecliptic pole. The same procedure is carried out for the second pole. Once the poles are marked, he can draw the ecliptic.

Determining and marking the stars' positions is only possible after the ecliptic circle is divided into twelve scales of 30°. First, the globe maker needs to obtain the values for a star's longitude and latitude in ecliptic coordinates from either catalogues or observations. Using the scale of 90° that was previously prepared for finding the ecliptic poles, he then opens the compasses as wide as the value of the latitude of the star. Keeping the width fixed, the globe maker has now to place the steady leg on the corresponding degree of the zodiacal sign as its longitude. He is instructed to use the drawing leg to draw a small, concave arc for an approximate position. Depending on the direction of the star, he draws the arc either on the north or the south of the ecliptic but always eastward. To make the position more precise, he has to open the compasses exactly 90° and to place the steady leg on a point that is 90° west of the longitude of the star on the ecliptic circle. Now he can draw a second arc by using the drawing leg. The intersection of the two arcs gives the position of the star. All designations are approximate since the scales have only 1° intervals.

The second example of instructions for how to make a globe is found in Abū ʿAlī al-Ḥasan al-Marrākushī's (2nd half of the 7th/13th century) *From A to Z: A Compendium of Timekeeping* (*Jāmiʿ al-mabādī wa-l-ghāyāt fī ʿilm al-mīqāt*).[3] In the fifth chapter of the second book, he gives details on how to construct celestial globes and armillary spheres (MS Oxford, Bodleian Library, Marsh 154, fols 2b–5a). In the section on the globe, he first describes finding the diameter of a globe and proposes to make a quadrant with a scale of 90° and the same radius as the globe. He then introduces a method for precision drawing. According to this, one picks any point on the globe as the center, then places the steady leg of the compass on this center and draws a small circle with four equidistant points marked on it. After this, the quadrant, that was introduced while finding the diameter of the globe, is required. The quadrant is placed on the globe so that its rim touches any one of the four equidistant points while one of its ends remains fixed at the original center. Starting from the center, an arc is drawn along the rim of the quadrant. The other end of the arc marks one of the poles. Then the quadrant is rotated until its rim touches the opposite one of the equidistant points. A secondary arc is drawn to mark the second pole. After this, al-Marrākushī proposes that one should use a lathe to engrave the celestial equator. To do that, the globe should be held axially in the lathe so that it can be rotated about its poles. Then, using a point equidistant from each pole as the reference, one can engrave a great circle that also passes through the center and the two remaining points of the small circle.

Al-Marrākushī's introduction of the quadrant, which carries a scale of 90°, makes marking the star positions rather easy in comparison to al-Battānī's method. One simply fixes one end of the quadrant to the north or south ecliptic pole and rotates it to the corresponding degree of the ecliptic for the longitude. Then one measures the latitude of the star and marks its position. Although none of the extant globes resembles fully the types introduced by al-Battānī or al-Marrākushī, determining whether the producers of the extant globes had indeed used any of their methods for drawing the circles and marking the star positions requires further research.

ʿAbd al-Raḥmān al-Ṣūfī was perhaps the scholar whose work had the greatest impact on the construction of globes. In his famous work, *Book of the Star Constellations*, he illustrates 48 constellations and marks 1025 stars from Ptolemy's star catalogue. He also offers the ecliptic coordinates (longitude and latitude) and the brightness of the stars in tables and gives detailed descriptions of the positions of the stars relative to their constellations. Each constellation is depicted as it is seen in the sky and on the globe. Al-Ṣūfī's catalogue relies heavily on Ptolemy's values, but he adds 12° 42′ to the longitude of each star to eliminate the difference caused by the precession of the equinoxes and occasionally changes some of the values to the ones he obtained through his own observations (Dekker 2012, 29–47). Al-Ṣūfī's catalogue and, more importantly, his illustrations were helpful for preparing star catalogues and constructing globes. It was used until the 13th/19th century.

IV.2.1.2 Uses of celestial globes

Once the globe is constructed, one can use it in any latitude.[4] To use it in a specific locality, the meridian ring is rotated until the angular difference between the celestial pole and horizon agrees with the local latitude. This is rather easy since the meridian ring is attached to the globe at the celestial poles and carries a scale. Depending on the globe's type, the number of applications that can be performed with it differs. Moreover, the celestial globes, with the exception of those using an additional gnomon and a detached quadrant, are not destined for observations but for calculations and demonstrations. Only a few examples of these additional tools survive (Figure IV.2.2).

Therefore, we do not know whether many of the extant globes were indeed used for observations. Before one uses the globe, one needs to obtain values such as the degree of the zodiacal

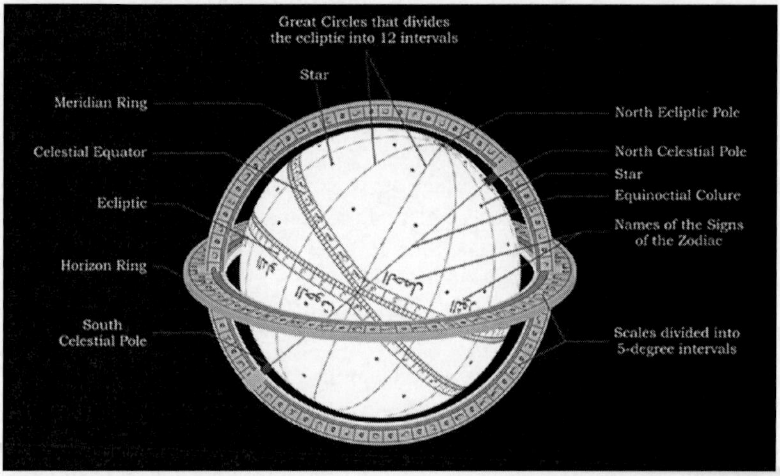

Figure IV.2.2 Muḥammad ibn Muʾayyad al-Dīn al-ʿUrḍī's (7th/13th century) celestial globe (*c.* 687/1288) on a wooden stand, made from brass, with gold and silver inlay; Mathematisch-Physikalischer Salon, Dresden, inv. no. E II 1.

Source: © Mathematisch-Physikalischer Salon, Staatliche Kunstsammlungen, Dresden sign for the intended day, the instantaneous altitudes of the sun, the moon and the stars and the ascendant. These values can be found in astronomical tables or through observations made with the globe or different instruments. When such parameters are known, a globe can be used for numerous calculations. Some are

- finding the longest and shortest days and nights in a given latitude.
- finding the maximum and minimum meridian altitude of the sun in a given latitude.
- finding the difference between the lengths of the different days in a given latitude.
- finding the daytime arc of any star marked on the globe.
- finding the azimuth of any star when it is rising or setting.
- finding the declination of any star on the globe.
- finding the distance between two stars on the globe.
- finding the time of rising and setting of any star on the globe.
- finding ever-visible and ever-invisible stars for a given latitude.
- finding the azimuth of the direction of Mecca (*qibla*).

As a consequence of its wide array of use, several manuals on this topic were written, copied, and used for centuries. Of those, two were particularly influential: Qusṭā ibn Lūqā's (d. *c.* 300/912–913) *Treatise on the Use of the Sphere with a Stand* (*Risālat al-ʿAmal bi-l-kura dhāt al-kursī*) and al-Ṣūfī's *Commentary on the Use of the Globe* (*Fī Sharḥ al-ʿAmal bi-l-kura*). These two works, compiled in the 3rd/9th and 4th/10th centuries, respectively, are the most comprehensive treatises on the subject. Qusṭā's work, on present evidence the more popular one of the two, consists of an introduction and 65 chapters.[5] First, it gives information on the celestial sphere followed by a brief description of a globe with those stars that can be found on astrolabes (chapters 1–5). It continues with the instructions on many different applications both for observations and calculations (chapters 6–65). Al-Ṣūfī's treatise is even more comprehensive. It deals with the problems of the use of globes in 157 chapters. Ḥabash al-Ḥāsib (d. after 255/869), al-Marrākushī, and Amīn al-Dīn al-Abharī (d. 733/1332–3) are among the few other scholars who wrote on the use of the celestial globe.

IV.2.1.3 A profile of the globe-making tradition

For the surviving globes, knowing their makers, dates of construction, and places of origin allows us to create a profile for the globe-making tradition. For instance, most of the extant globes are of either Persian or Mughal Indian origin. Only a handful of globes are dated from the 3rd/9th to the 10th/16th centuries. The number of dated globes rises exponentially in the 11th/14th century and the provenances shifted from the Middle East to Mughal India. The dates and signatures on the instruments provide proof of the continuous activity of families and individual specialists as instrument-making artisans. In comparison to other instruments such as astrolabes, quadrants, and sundials, relatively more non-scholarly artisans seem to be involved in the construction of globes. There were two noteworthy cities for instrument makers: Ḥarrān (today in Turkey) and Lahore (today in Pakistan). Although little of their production has survived, the historical record indicates that Ḥarrān was a major center for the construction of astronomical instruments between the 3rd/9th and the 5th/11th centuries. ʿAbd al-Raḥmān al-Ṣūfī states, for instance, that he has seen many globes made in Ḥarrān and that he disapproved of a globe made by Alī ibn Īṣā from Ḥarrān (MS Paris, BnF, Arabe 5036, fols 4b, 108b). Unfortunately, in the case of Ḥarrān, we lack details such as the list of makers, the sale records, and the number of workshops. We are better informed about a family who ran a workshop for instruments in Mughal Lahore during the 11th/17th century over four generations: Shaykh Allāhdād Aṣṭurlābī Humāyūnī Lāhūrī, his son Mullā ʿĪsā, the latter's two sons Qāʾim Muḥammad and Muḥammad Muqīm, Qāʾim's son Ḍiyāʾ al-Dīn Muḥammad and Muqīm's sons Ḥāmid and Jamāl al-Dīn (Sarma 2019). Of the astrolabes and globes they produced, 134 and 33, respectively, are extant. They made these 33 globes in the seven decades between 1032/1623 and 1102/1691, using the lost-wax method exclusively and successfully. They produced very accurate globes of every type. Although they were not the only instrument makers in Mughal India, this Lahore family constituted the frame of the Mughal globe-making tradition (Figure IV.2.3).

Figure IV.2.3 Cast and engraved celestial globe on a stand from the workshop in Lahore, Mughal India, made in 1036/1626–7. Height: 23 cm, Width: 17.5 cm. Inscriptions: (1) The work of the least of the servants Qāʾim Muḥammad ibn ʿĪsā ibn Allāhdād Aṣṭurlābī Humāyūnī Lahūrī. (2) Twenty-second year of the reign of Jahāngīr. Museum no. M.828PART/1–1928.

Source: © Victoria and Albert Museum, London

Celestial globe production in the Islamicate world continued for over a millennium. While the values for positioning the stars mostly remained true to the ones given by al-Ṣūfī, the style of wording and the images of the constellations varied from region to region during this period. Globe makers from Iran and Mughal India excelled at the lost-wax method around the 10th/16th century. Afterward, it became the preferred method for globe construction. Almost none of the extant globes has its original rings (the meridian ring and the horizon ring), but some have replacements. Noncontemporary replacements, in particular, indicate that these globes were in use for centuries.

IV.2.2 Armillary spheres

Like celestial globes, armillary spheres imitate the daily motion of the sky. But, unlike globes, they do not have a solid spherical body to indicate the stars and the constellations. The circles that are astronomically significant, such as the meridian and the ecliptic, are represented by nested rings. They can be used both for demonstrations and observations. However, it seems that they were exclusively employed for observations in the Islamicate world. This is rather different from Europe where armillary spheres were often employed in demonstrations and teachings.

Armillary spheres are made of wood or metal and constructed as large as possible in order to reach high accuracy in observations. Depending on the material and the craftsmanship, instruments were known to have been made in sizes between 1 and 6 meters in diameter. There were, however, certain limitations to size since parts can be deformed by expansion or bending. Unfortunately, only very few late examples survive from the 12th/18th and the 13th/19th centuries (Sarma 2019, 3261–80).

IV.2.2.1 The construction of an armillary sphere

The armillary sphere is a Greek invention and is known in Arabic as *dhāt al-ḥalaq* (the [instrument] with rings). The earliest record in the Islamicate world is ascribed to the 2nd/8th-century astrologer Muḥammad ibn Ibrāhīm al-Fazārī because of his lost work *Book on the Use of the Astrolabe Known as the Armillary Sphere* (*Kitāb al-ʿAmal bi-l-asṭurlāb wa-huwa dhāt al-ḥalaq*; Ibn al-Nadīm 1871–1872[CB], 2: 273). Although al-Fazārī categorized the armillary sphere as a type of astrolabe, this was not followed by others, and it was usually treated as a different instrument. The armillary sphere became popular very early in the history of astronomy in the Islamicate world through the translations of Ptolemy's *Almagest* in the 2nd/8th and 3rd/9th centuries. In the fifth book of the *Almagest*, Ptolemy introduces an armillary sphere with six rings, as follows (Ptolemy 1998, 217–18):

Two rings of the same size, representing the ecliptic and the solstitial colure, are joined perpendicularly. Then the ecliptic poles are located on the colure and two pegs, one for the inner surface and the other for the outer surface, are attached at each pole. Two rotatable rings are fixed at these pegs. The outer one is the large longitude ring, and the inner one is the small longitude ring. The small longitude ring houses another ring with two sighting vanes. This 'observation' ring is fitted so that it can only rotate parallel to and simultaneously with the small longitude ring. After measuring the angular difference between the ecliptic and celestial poles, positions of the latter are marked on the large longitude ring. One peg is attached to its outer surface at each pole. The outermost ring, which represents the meridian, is fixed at these pegs. The meridian ring is then fixed inside the horizontal plane as we have seen in globe making. The entire arrangement is carried by columns fitted under the horizon plane (Figure IV.2.4).

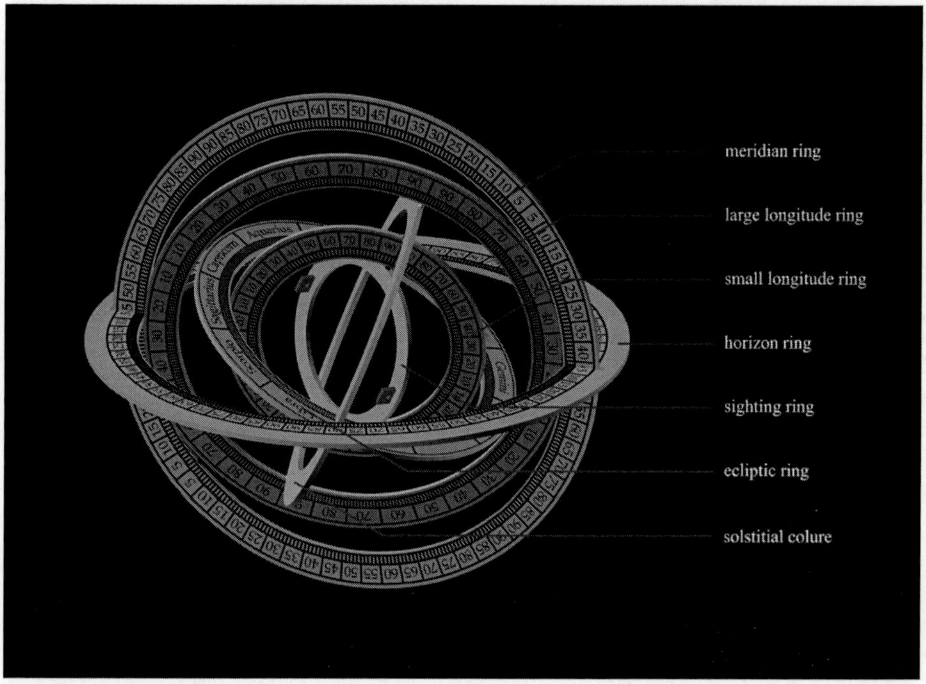

Figure IV.2.4 Ptolemy's model of the armillary sphere

Source: © Taha Yasin Arslan, Istanbul

The types of this instrument used for demonstration and for observation are distinctively different. An armillary sphere for demonstration contains a small globe, representing the Earth at the center, and does not require sighting vanes or an observation ring. The armillary sphere for making observations does not have a globe at its center. It was rather more popular in the Islamicate world and arguably had more practical value. It could be used to find the latitudes and longitudes of the sun, the moon, the five planets and the stars, their meridian altitude, the time, and the angular differences between two or more celestial objects, all according to the ecliptic coordinate system.

Ptolemy's description of an armillary sphere for making observations was used for centuries with some changes to the number and the function of the rings. For instance, the instrument maker and scholar Muʾayyad al-Dīn al-ʿUrḍī (d. 664/1266) switched the observation ring with an alidade (a sighting unit) by converting the small longitude ring into a hollowed disk so that the alidade could be fixed at its center. The Ottoman scholar Taqī al-Dīn ibn Maʿrūf (d. 993/1585), who headed the short-lived Istanbul Observatory between 984–987/1577–1580, constructed an armillary sphere with six rings but removed the observation ring while adding another for the celestial equator (MS Istanbul, Topkapı Library, Hazine 452). The new ring is of the same size as the meridian ring and was connected to it perpendicularly. In this model, the sighting vanes are on the celestial equator ring (Figure IV.2.5).

The construction of an armillary sphere does not necessitate sophisticated astronomical knowledge. Other than determining or obtaining the value of the obliquity of the ecliptic, one needs to execute only simple geometrical calculations to determine the sizes of the rings. There

Figure IV.2.5 A miniature of Taqī al-Dīn ibn Maʿrūf's armillary sphere. MS Istanbul, Istanbul University, Rare Works Library, F 1404, fol. 56b.

Source: © Istanbul University, Rare Works Library, Istanbul

is no astronomical significance in the proportions of the rings, the choice is based on aesthetics. However, constructing a flawless instrument, especially in large sizes as it was often done in the Islamicate world, is extremely hard. Metal rings may expand and contract due to temperature changes and both wood and metal rings may bend over time due to their own weight. Unskillful craftsmanship can also prevent free rotation of the rings. These become major problems since gathering systematic observations requires many years of application. However, making a large instrument creates advantages in terms of precision in the observations. Hence, instrument makers in the Islamicate world tried to perfect their designs with modifications of Ptolemy's model and constructed instruments as large as possible. In comparison to portable instruments like globes and astrolabes, these armillary spheres were rather heavy and quite expensive. That is why they were employed almost exclusively in observatories. Moreover, political struggles, wars, and the financial costs of maintenance shortened the lifespans of the observatories in the

Islamicate world. For instance, the Samarqand observatory may have been active no longer than 30 years and the Istanbul observatory survived only four. These relatively short working lives did not help the survival of large instruments.

IV.2.2.2 The use of the armillary sphere

The earliest known armillary spheres used in the Islamicate world are the two instruments made for the observatories in Baghdad and Damascus, respectively, between 213–14/828–829 and 216–217/831–832. Both were made by Khālid ibn ʿAbd al-Malik al-Marwarrūdhī (3rd/9th century), an instrument maker of the Abbasid Caliph al-Maʾmūn (r. 198–218/813–833; Sayili 1960 [1988][CB], 57). No further information about these instruments seems to have survived. Several centuries later, two firsthand records were written on armillary spheres used in the two most important observatories in the Islamicate world: Maragha and Samarqand. Al-ʿUrḍī, Maragha's instrument maker, compiled *The Epistle on How (to Make) Observations* (*al-Risāla fī kayfiyyat al-arṣād*). In it, he deals with several instruments, some of which were his own inventions. He also wrote a comprehensive section on the construction of a wooden armillary sphere with five rings and a diameter of approximately 160cm (MS Istanbul, Süleymaniye Library, Nuruosmaniye 2971, fols 76b–82a). Al-ʿUrḍī gives measurements for the diameter, width, and thickness of each ring and detailed instructions on how to attach the rings, as well as how to preserve the instrument's condition. He shows why Ptolemy's 'observation' ring is inconvenient and argues that an alidade is a better option for observations. In another text, Giyāth al-Dīn al-Jamshīd al-Kāshī (d. 832/1429), who was the first director of the observatory in Samarqand, mentions an armillary sphere with seven rings that were employed in the observations (Kennedy 1961, 100). It should be noted that both texts are derivations from Ptolemy's descriptions in the *Almagest*.

IV.2.3 Conclusion

Celestial globes and armillary spheres are miniaturized versions of the celestial sphere as it is understood in classical astronomy and, depending on construction, could be used for demonstrations and observations. They are helpful educational tools when used for demonstrations, because they offer a direct representation of the sky, unlike projection-based instruments such as astrolabes and quadrants. They are universal in nature; they can be used anywhere in the world. Due to challenges with their construction, globes were relatively small and portable, whereas armillary spheres for stationary use were built as large as possible for accuracy. Both instruments provide ample evidence for scholarly activities, astronomical interests in the corresponding communities, and the transmission of astronomical knowledge through time. The many unique surviving instruments and the manuals on their construction and use deserve thorough investigation. Most of the dated and authored globes were produced for the courtly elites such as Ayyubids, Safavids, or Mughals. Armillary spheres, on the other hand, were occasionally depicted and illustrated in manuscripts like the *Shahinshahnāma* by Sayyid Loqmān dedicated to Murād III (see Figure IV.2.5). Moreover, the social aspects of the existence and use of these instruments within the Islamicate world have been ignored for too long and are in need of comprehensive study.

Notes

1 For more on their purpose see Charette (2006, 123-38); and formore on the patronage, see Brentjes (2008, 403–36).

2 Consolidated bibliography.

3 For more on Marrākushī's work, see Charette (2003).
4 For examples of its use, see Evans (1998, 85–87).
5 For more details, see Samsó (2015, 68–72).

Bibliography

Manuscripts

MS Istanbul, Süleymaniye Library, Hamidiye 1453, fols 103b–123a.
MS Istanbul, Süleymaniye Library, Nuruosmaniye 2971, fols 74b–102a.
MS Istanbul, Topkapı Library, Hazine 452.
MS Oxford, Bodleian Library, Marsh 154.
MS Paris, BnF, Arabe 5036.

Research literature

Brentjes, S. 2008. "Courtly Patronage of the Ancient Sciences in Post-Classical Islamic Societies," *Al-Qanṭara* 29.2: 403–36.
Brentjes, S. 2015. "Sanctioning Knowledge," *Al-Qanṭara* 35.1: 277–309.
Dekker, E. 1999. *Globes at Greenwich: A Catalogue of the Globes and Armillary Spheres in the National Maritime Museum, Greenwich.* New York: Oxford University Press.
Dekker, E. 2012. *Illustrating the Phaenomena.* Oxford: Oxford University Press. http://doi.org/10.1093/acprof:oso/9780199609697.003.0004
Dekker, E. and Kunitzsch, P. 2008. "An Early Islamic Tradition in Globe Making," *ZGAIW* 18: 155–211.
Evans, J. 1998. *The History and Practice of Ancient Astronomy.* Oxford: Oxford University Press.
Gillispie, C. C. ed. 1981. *Dictionary of Scientific Biography.* New York: Charles Scribner's Sons.
Kennedy, E. S. 1961. "Al-Kāshī's Treatise on Astronomical Observational Instruments," *Journal of Near Eastern Studies* 20.2: 98–108.
Kunitzsch, P. 1959. *Arabische Sternnamen In Europa.* Wiesbaden: Otto Harrassowitz.
Samsó, J. 2015. "Qusṭa ibn Lūqā and Alfonso X on the celestial globe," *Suhayl* 5: 63–79.
Sarma, S. R. 2019. *A Descriptive Catalogue of Indian Astronomical Instruments.*
Edmonton: University of Alberta, Canada. https://srsarma.in/catalogue.php
Savage-Smith, E. 1985. *Islamicate Celestial Globes: Their History, Construction, and Use, with a*
Chapter on Iconography by Andrea P. A. Belloli. Washington, DC: Smithsonian Institution Press.
Savage-Smith, E. 1990–1991. "The Classification of Islamic Celestial Globes
in the Light of Recent Evidence," *Der Globusfreund* 38–9: 23–9.
Savage-Smith, E. 1992. "Celestial Mapping," in *GTISAS* [CB], 12–70.
Savage-Smith, E. 1997. "Islamic Celestial Globes and Related Instruments," in Maddison, F. and Savage-Smith, E., eds. *Science, Tools and Magic, Part One: Body and Spirit, Mapping the Universe* (Khalili Collection XII). London and Oxford: Nour Foundation, Azimuth Editions, Oxford University Press, 168–84.
Worrel, W. H. 1944. "Qusta ibn Luqa on the Use of the Celestial Globe," *Isis* 35.4: 285–93.

IV.3
PROJECTING THE HEAVENS
Astrolabes[1]

Taha Yasin Arslan

IV.3.1 Introduction

Instruments made for and used by practitioners of the astral sciences in Islamicate societies cover a fairly broad range of types either as standard or particular, often innovative, models. In Toledo, a famous and influential instrument maker and author of scholarly texts and tables, Abū Isḥāq Ibrāhīm ibn Yaḥyā al-Naqqāsh ibn al-Zarqālluh (d. 493/1100) grouped them into two main sets (instruments usable only during daytime; instruments also usable at nighttime) as far as they were known to him in al-Andalus. The first group depends on the shadow of the Sun projected by a gnomon and includes four types of sundials (horizontal, vertical, cylindrical, conical). The second group operates by using the rays of the Sun during daytime and of stars at night. It consists of plane and spherical astrolabes, which will be treated in this chapter, quadrants, briefly mentioned here, armillary spheres and globes, discussed in the previous chapter, and armillary rings and several instruments called alidades (from the Arabic *al-ʿiḍāda*).

Astrolabes visualize the motions of the sky above a terrestrial observer and present therefore a hand-held model of the universe. Of Greek origin, these multifunctional astronomical instruments are used for elementary observations, calculations, education and demonstrations. Most of the astrolabes in the Islamicate world can be classified in four categories: plane, universal, linear and spherical. The main purpose of astrolabes, no matter what type they are, is to find time by measuring the altitude of the Sun during daytime or of stars at night. Although there is no concrete evidence when the knowledge of this instrument arrived in the Islamicate world, both the historical accounts and surviving texts indicate that it was already known around the end of the 2nd/8th and beginning of the 3rd/9th centuries. Al-Nadīm (d. 380/990), a historian and bookseller, mentions Ibrāhīm ibn Ḥabīb al-Fazārī (2nd/8th century) as the very first astrolabe maker and the author of a lost treatise on the use of plane astrolabes (Ibn al-Nadīm 1871–2[CB],[2] 2: 273). Several treatises were written on the construction and use of astrolabes in the 3rd/9th century by prominent scholars such as Muḥammad ibn Mūsā al-Khwārazmī (d. *c.* 235/850), Ḥabash al-Ḥāsib (d. after 255/869) and al-Farghānī (3rd/9th century). One of the oldest surviving astrolabes is not dated but signed by an instrument maker known only as Khafīf (d. before 300/912–913), who described himself as an apprentice of another astrolabe maker called ʿAlī ibn ʿĪsā (3rd/9th century). The oldest extant astrolabe that is dated was made by Nastulus (3rd–4th/9th–10th centuries) in 315/927–8. Al-Nadīm states that many plane astrolabes were

DOI: 10.4324/9781315170718-44

made in Harran as early as the time of Caliph al-Maʾmūn (r. 198–218/813–833; Ibn al-Nadīm 1871–1872, 2: 284). Altogether, about a dozen astrolabes from early Abbasid times are preserved.

IV.3.2 Types of astrolabes

The astrolabes made by the craftsmen and most of those described by the scholars belong to the standard category of the plane astrolabe. Its components will be described in the following section. For the short survey here, it suffices to say that its construction worked almost always with the stereographic projection. This choice necessitates the fabrication of several plates for specific geographical latitudes, because the distance between the projection point (celestial pole) and the point above the head of the observer in the plane of the horizon (*zenith*) depends on the geographical latitude of the observer's location. This limits the usability of the standard astrolabe to the places represented by the plates and their immediate environments, except for some parts engraved on the back such as the shadow squares. Muslim instrument makers and scholars soon began to look for solutions to tackle this limitation. Ḥabash al-Ḥāsib, mentioned earlier, is remembered as the first to make such efforts (King 1999[CB], 330). He described, for instance, an astrolabe plate with half arcs of horizons for each latitude between the equator and the pole.

Plane astrolabes were made in a variety of forms by modifying some of their parts. An example is the so-called melon-shaped astrolabe described allegedly and perhaps even constructed already in the late 2nd/8th century by Muḥammad ibn Ibrāhīm al-Fazārī (d. 180 or 190/796 or 806). The earliest extant text on this instrument was written about half a century later by Ḥabash al-Ḥāsib (Kennedy *et al.* 1999[CB]). Instead of a stereographic projection, it is believed to have used a conic projection. But the reports about its curves and its functionality by numerous Muslim writers over the centuries are contradictory and often hostile (Kennedy *et al.* 1999, 4–7). Ḥabash al-Ḥāsib's text on the instrument, however, is filled with unusual and even complicated geometrical, trigonometric and arithmetic procedures for determining the various arcs and other kinds of markings to be inscribed onto the various parts of the astrolabe (Kennedy *et al.* 1999, 11–12).

In addition to the multiplate plane astrolabe, instrument makers or scholars from the Islamicate world invented two other kinds of astrolabes: the linear (*khaṭṭī*) and the spherical (*kurī*) astrolabe (King 2018). It is, however, unclear whether the spherical astrolabe is a modification of the plane astrolabe or the globe (Chapter IV.2). In comparison to plane astrolabes, spherical astrolabes require no projection. They carry evenly distanced altitude circles that are parallel to the horizon and number of great circles for azimuths. They can be used anywhere in the northern hemisphere by simply fixing the axis of the rete to the corresponding degree of the locality.

The fourth type of astrolabes, the universal astrolabe, was invented as another attempt to tackle the limitation caused by plates of plane astrolabes. The first extant instrumental solutions are known from al-Andalus. In Toledo during the 5th/11th century, two instrument makers turned scholars of the astral sciences, the just mentioned ibn al-Zarqālluh and Ibn Khalaf al-Shajjār (5th/11th century), created universally usable plates by applying the stereographic projection to the plane passing through the celestial poles and the solstices and choosing the equinoctial points (Aries 0° and Libra 0°) as the points of projection (Puig 2007a, 34–5, 2007b, 1258–60; Samsó 2020[CB], 441). As a consequence, the horizon is projected as a straight line and a ruler or the diameter of the *rete* rotating on the universal plate represents any needed horizon (Figure IV.3.1). Interestingly enough, at least Ibn al-Zarqālluh did not intend to fully replace the individual plates but only to complement them for places that they did not fit, in order to improve the accuracy of measurements and to relieve the user from cumbersome approximate calculations (Samsó 2020, 440).

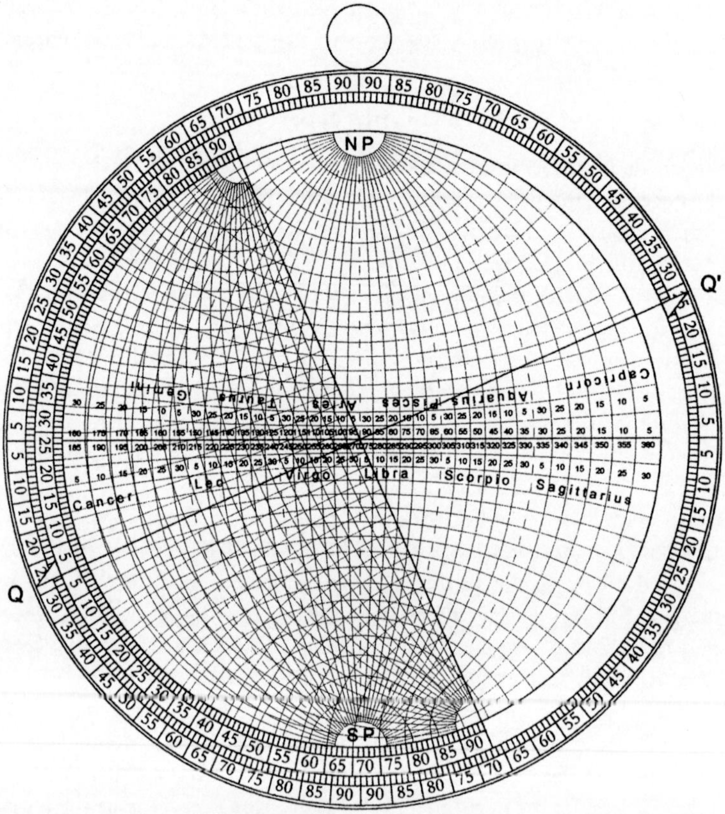

Figure IV.3.1 Reconstruction of ʿAlī ibn Khalaf al-Shajjār's (5th/11th century) universal plate by Emilia Calvo and Roser Puig Aguilar. First published in *Suhayl* 6, 155, figure 8. The rete is placed over the mater showing the obliquity of the ecliptic as 23° 3'. The signs between Aries and Virgo are shown in the southern hemisphere.

Source: © Josep Casulleras, editor of *Suhayl*

A third kind of universal plate invented in al-Andalus was that made by the timekeeper (*muwaqqit*), Ḥasan (or Ḥusayn) ibn Muḥammad ibn Bāṣo (d. 716/1316; Samsó 2020, 466–9; Chapter V.4). Universal plates were also known in eastern regions of the Islamicate world but seem to be related to their Andalusian predecessors. Ibn Khalaf's plate, for instance, was modified in Mamluk Syria by Shihāb al-Dīn Aḥmad ibn Abī Bakr, known as Ibn al-Sarrāj (8th/14th century) from Aleppo. Extant instruments rarely carry only a universal plate but combine specific plates stored in the *mater* with a universal plate engraved on the back. A small number of instruments that have or had a universal plate are extant in museums, among them the Oxford Museum of History of Science, the Benaki Museum in Athens and the Museum for Islamic Art in Cairo.

Similar to the efforts by al-Zarqālluh and Ibn Khalaf, but several decades before them, al-Bīrūnī (362–d. after 444/973–d. after 1053) replaced the plane of the equator by that of the solstitial colure when describing the cylindrical (today called orthographic) projection. In addition, he also explained non-stereographic projection methods (today called azimuthal equidistant and globular projections; Samsó 2020, 442–3).

We should briefly mention here the quadrants, astronomical instruments which were often incorporated in astrolabes or derived from them. The first kinds of quadrant were developed in the 3rd/9th century by al-Khwārazmī (horary quadrant for a specific latitude; sine quadrant) and Ḥabash al-Ḥasib (universal horary quadrant; King 1996, 20–1; Charette 2003, 209). The universal horary quadrant might have been derived from the sine quadrant (Lorch 1981). Two centuries later in Ghazna (today Afghanistan), al-Bīrūnī discussed the horary quadrant as a well-known device, either marked on the back of an astrolabe or constructed as an independent instrument with a plumb bob attached to a silk thread, a movable bead and sights. The earliest extant such quadrants on the back of astrolabes come from the 4th/10th century (Charette 2003[CB], 116). In the 7th/13th and 8th/14th centuries, several new types of horary quadrants were thought up in North Africa or al-Andalus and in Mamluk Egypt and Syria (Charette 2003, 117–39). In the same period, numerous versions of trigonometric quadrants were invented, ranging from simple to highly sophisticated, nonstandard formats (Charette 2003, 210–20). Astrolabic quadrants, a trifold version of the front of a plane astrolabe, were also a design of this period. They became and remained popular in Mamluk Egypt and Syria, as well as Ottoman realm until the 14th/20th century. Charette suggests that the thriving culture of astronomical instrumentation under the Mamluks and their predecessors, the Ayyubids, may have been the direct consequence of the emergence of a reorganization of astronomical knowledge and its practices in the new discipline of timekeeping (Charette 2003, 8–9; Chapter V.4). In some regions, such as the Ottoman realm and parts of North Africa, quadrants replaced the use of the astrolabe almost completely. Moreover, they also spread far to the East, including even the Malay Archipelago, although we have very little information otherwise about scientific practices there (Zaynalal and Ismail 2010).

IV.3.3 Main features of astrolabes

The standard astrolabe consists of four essential pieces: *mater* (mother) with an attachment for hanging it up, *rete* (net or grid), plates and alidade, with a pin and horse holding them together (Figure IV.3.2). Spherical astrolabes have all these pieces, excluding plates, but in different shapes (e.g., a spherical *rete*). The linear astrolabe consists of a rod, which represents the meridian. It has no plates, alidade, pin or horse. A plumb line can be attached.

The part representing the heavens, reflecting the geocentric view of the universe (Chapter II.6), is the star map called *rete* (grid or net) or spider (*ʿankabūt*; Figure IV.3.2). Its main elements are the ecliptic, an eccentric circle depicting the path of the Sun around the Earth through the zodiacal signs and the star pointers. The ecliptic is divided into 12 units of 30° each representing the signs of the zodiac which are usually further partitioned through intervals of 3°, 5°, or 6°. The rim of the *rete* corresponds to almost the entire Tropic of Capricorn. Between the ecliptic and the Tropic of Capricorn there is sometimes a third circle, in most cases reduced to an arched bar. It depicts the celestial equator. Additional circular and linear links stabilize the *rete*. Together with the Tropic of Capricorn and the equator they carry so-called star-pointers marking bright stars, which give the *rete* the character of a star map. Their number varies between different instruments, approximately between 15 and 40. Both the names of the signs of the zodiac and the stars that are shown by the star-pointers are usually inscribed on the *rete*. Due to the precession of the equinoxes, which causes the ecliptic to move approximately one degree in every 72.2 years (modern value), the positions of the fixed stars change significantly over long periods, thus making the *rete* incorrect over time.

The terrestrial part of the plane astrolabe is constituted by the plates, also called *tympans*. On standard astrolabe plates constructed with the polar stereographic projection, the celestial equator and the Tropics of Cancer and Capricorn are engraved as circles. The plates of earlier

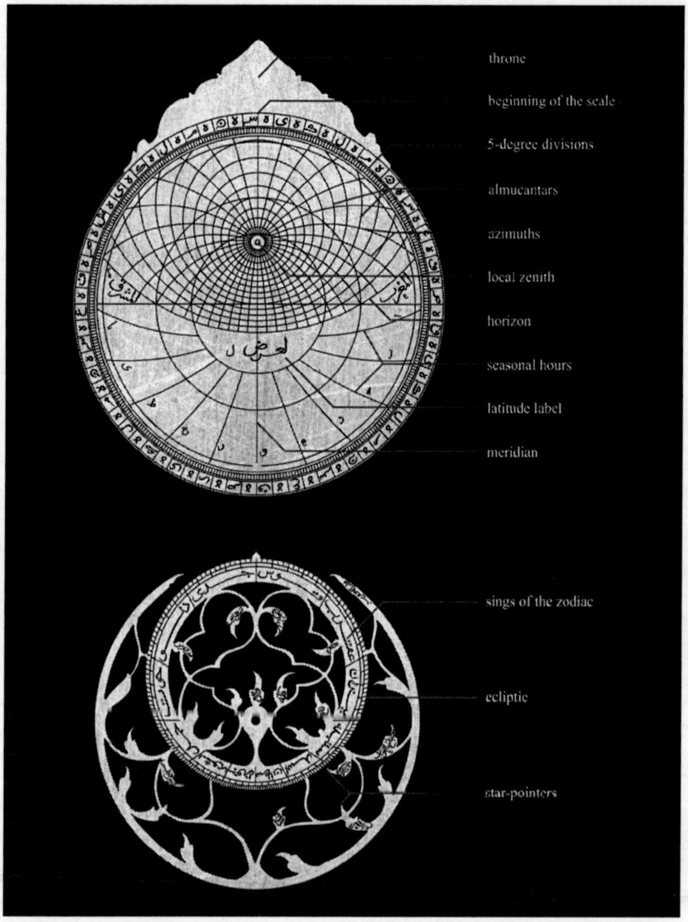

Figure IV.3.2 Mater and rete of an astrolabe

Source: © Taha Yasin Arslan, Istanbul

astrolabes display one magnitude: altitude. The plates of later astrolabes display two magnitudes: altitude and *azimuth*. The altitude refers to the position of any celestial object in the sky above the horizon. The *azimuth* (from the Arabic word *al-sumūt* [directions]) marks a distance on the horizon from a chosen reference point, for instance the horizon's south point. Altitudes are read off from circles (*almucantar*s, also derived from Arabic) drawn in regular intervals, depending on the size of the astrolabe in steps of 1, 2, 3, 5, 6, or 10°, concentrically around the *zenith*, the point 90° from the horizon and directly above the observer's head. Arcs of great circles radiate in regular angular distances from the *zenith* down to the horizon. With their help, the *azimuth*s are determined. Below the horizon, so-called temporal or seasonal hour lines are marked. Moreover, the latitude of each plate is usually given, very often also the length of the longest day and commonly the name of the cities that suit this latitude.

The *rete* and the plates are placed into a circular container called the *mater* (Latin for mother, corresponding to the Arabic *umm*). The latter are fixed in it with a little peg that fits in the small notches in all of them which prevents them from rotating. This allows only the *rete* to rotate for adjustment to the movements of the sun and the stars. The rim of the *mater* carries a scale of 360° usually divided in 5° intervals, which are subdivided in corresponding parts of 1°, or in hours, either two times 12 or 24 equal time units; sometimes both divisions are marked. The sky may be imagined to revolve 360° around the Earth once in 24 hours; hence, 1° represents 4 minutes and 15°, 1 hour. On the top of the *mater* are the throne (*al-kursī*), a shackle and a ring with a string, allowing an observer to suspend the astrolabe when measuring altitudes. The throne may be inscribed with a poem or a Qurʾānic quote.

The *mater* can be made from two pieces that are rivetted together. Both the *mater*'s front and back can accommodate further astronomical, astrological or geographical information such as shadow squares for calculating the tangent and cotangent of an angle and vice versa, circular scales for a calendar year, the signs of the zodiac, lunar mansions, a sine, horary, or qibla quadrant or a circular geographical gazetteer with the angles that describe the prayer directions toward the Kaʿba. The names of the signs of the zodiac and the lunar mansions are often inscribed. On some astrolabes, they also appear in pictorial form. The front can serve as a further latitude plate. On the rim of one upper quadrant of the back, at least one altitude scale from 0° to 90° is marked off. While early astrolabes carried very few such additional markings, over the centuries, more and more information was added. Workshops followed regional and even special local customs. Astrolabes from the Islamic East, for instance, often carry double shadow squares at the center of the lower half of the back and cotangent scales on the lower half of the rim. Astrolabes from al-Andalus and the Maghrib can include circular scales for the Julian calendar and the signs of the zodiac. All this information is usually arranged in two main structural designs: a division of the back in four quadrants or the inscription of circular scales and tables between the altitude scales and the center.

IV.3.4 Uses of astrolabes

The first and most basic operation to conduct with an astrolabe is the measurement of the altitude of a celestial object above the horizon. Subsequent tasks are the determination of time since sunrise, until midday, sunset, or sunrise, or of the rising, culmination and setting of a celestial object. The standard uses of astrolabes are subjects of many modern academic and popular texts, as well as many videos on the internet.

At the beginning of the 3rd/9th century, in his treatise on the astrolabe, the oldest preserved text on this instrument, al-Khwārazmī lists more than 30 different problems that can be solved by it. One century later, ʿAbd al-Raḥmān al-Ṣūfī (291–376/903–986) wrote arguably the most extensive treatise on the astrolabe. In its longest version, it includes 1760 chapters, less than half of them preserved, each one describing a problem that may be solved with an astrolabe.

An application unique to the Islamicate world is the determination of the prayer direction (*qibla*; Chapter VI.2). As an astronomical task this means to calculate the *azimuth* of the Sun when it reaches the *zenith* of Mecca. This happens twice a year, at Gemini 8° (May 28) and at Cancer 23° (July 16). Before the calculation, the user needs to know the longitudinal difference in degrees between Mecca and the intended locality. Today, longitudes are measured from Greenwich. Before the 12th/18th century, other reference points such as the Fortunate Islands (today the Canary Islands), the Atlantic coast of Morocco or the so-called meridian of water (27° west of Córdoba) were used in the Islamicate world.

The modern value for Mecca's longitude is approximately 40° East. We will use Istanbul for this example, and its longitude is approximately 29° East. Therefore, the longitudinal difference will be 11°. Using a plate that is engraved with markings specific to the intended locality, in this case for 41° latitude (Istanbul), the *rete* is rotated until Gemini 8° coincides with the meridian on the plate. At this position, the user reads the degree of the altitude circle that Gemini 8° intersects with. For the plate of 41° (Istanbul), this value will be approximately 71°. The longitudinal difference, 11°, is subtracted from this value, giving 60°. Then the *rete* is rotated counterclockwise by 11° with the help of the scale on the rim. At this point, the user reads off the *azimuth* of Gemini 8°. This value will be equal to the angle of the *qibla* from East. For Istanbul, it is approximately 61°. When measured from North, it is 151° (for further details, see Morrison 2007, 138–43).

IV.3.5 Skills and knowledge practices

How can we know which skills and what kinds of knowledge men and sometimes also women had to acquire who wished to produce an astrolabe or write about its construction or usage? How do we know where, when and with whom they had learned and whether their learning was mainly textually based or learning-by-doing? What can be said about division of labor between theoretically trained people and craftsmen? Did only theoreticians write texts about astrolabes but never really built one? Could all craftsmen read and write and produce their own manuals, or did they merely copy inscriptions of whatever content from lines written for them on a piece of paper by an educated person? How did craftsmen learn about the latest artistic style and its individual elements? Where did authors of texts compose their treatises, and how did they do it – in solitude at home surrounded by manuscripts they had bought or copied themselves, in special rooms for scholars in courtly settings with a scribe at hand and a student as helpmate, or at *madrasa*s, mosques and later in special buildings for timekeepers? Where scholarly authors and instrument makers neatly separated professional groups of different social standing? Did all of them work only in the capitals or great provincial centers?

Only a few of those questions can be answered with any degree of reliability. There is no material that would allow us to derive some kind of statistical survey for a period, a larger region or even only a city. Texts on astrolabes rarely talk about the technicalities and skills involved in making astrolabes from metal or wood, the two main materials used in the production process. In a few cases even cardboard was used; more often, however, for quadrants. Individual inscriptions, anecdotes or references to patrons, partners, students or buyers provide occasional information in sources such as teaching certificates (*ijāzāt*), colophons, remarks in the margins of a scientific text, entries of a biographical dictionary, a travel account or in catalogs.

The most significant insight that can be derived from theoretical texts and instruments is that numerous instrument makers were highly skilled in the theoretical aspects of constructing and using astrolabes. Surprisingly, many of them, as will be shown, acted both as producers of instruments and authors of texts. Some of them solved demanding geometrical and astronomical problems as pointed out above. Others taught their craft, progressed to become skilled observers, compiled astronomical tables with new data and even rose to head positions in expeditions or series of observations. Other combinations of skills, mostly documented by extant instruments and related metalwork, were imparted by artistically sophisticated metalworkers and craftsmen competent in the production of locks, pulleys and most likely further mechanical devices (Chapter I.9). Texts written by scholars in different centuries reveal that the division of labor also played an important role in the practices involved in producing astrolabes, above all technically sophisticated ones. In what follows, examples of such skills and relationships will be presented, beginning with what seems to have been the first occurrence of a process in which craftsmen became recognized as authors of scholarly texts. This development is sometimes expressed in a byname such as *al-ḥāsib* (the calculator). Early historical, geographical and mathematical sources imply that this professional designation describes an astrologer and in particular someone who calculated astronomical tables and knew operations like multiplication and division (Isḥāq ibn Ḥusayn al-Munajjim 1408/1988, 33–4; Jaouiche 1976, 164–5). Al-Yaʿqūbī already equated this byname with *munajjim* (astrologer) for the first maker of an astrolabe under the Abbasids, Muḥammad ibn Ibrāhīm al-Fazārī, (al-Yaʿqūbī 1849, 13). One of the most proficient people named *al-ḥāsib* was the already mentioned Ḥabash. He came from the city of Marv (today in Turkmenistan) and worked most of his life apparently in Baghdad and Samarra, the Abbasid capital in the second half of the 3rd/9th century. Although he invented new devices for the astrolabe (universal horary quadrant, plate of horizons) and wrote about variants of the instrument, it is unknown whether he actually constructed an astrolabe. Kennedy, Kunitzsch and Lorch commented that his treatise was not sufficiently detailed and precise for a craftsman to produce the melon-shaped astrolabe he describes (Kennedy *et al.* 1999, 4). They evaluated Ḥabash's skills as that of an "astronomer first and last", having been "both a practical observer and a theoretician" and "in full command of the branches of mathematics applicable to astronomy in his time" (Kennedy *et al.* 1999, 10). This means that in addition to Ptolemaic theory, Ḥabash used ancient Greek and Indian geometrical, trigonometric and iterative numerical methods in his various works. In the case of the treatise on the melon-shaped astrolabe, he calculated tables for altitude and azimuth circles before plotting the necessary arcs on paper. With a ruler and a compass, he drew complex diagrams, from which he derived trigonometric solutions. Afterward, he provided a numerical example (Kennedy *et al.* 1999, 10–2). The analysis of those diagrams shows that Ḥabash established them through two main methods: a piecemeal transformation of the three-dimensional case into two-dimensional components, which he then reassembled and a derivation through arithmetical procedures. According to Kennedy and Lorch, in particular, the second approach was applied by other writers on the mathematical sciences beyond astrolabe texts and might have been learned from Indian or Persian sources (Kennedy *et al.* 1999, 11–2). A more general method, used by Ḥabash, is his reliance on texts by other authors, which he mined for problems, examples and procedures without naming either the author nor the text's title (Kennedy *et al.* 1999, 12). This last method was standard for many centuries.

Cooperation between authors and craftsmen is also known from other regions and times. The scholar wrote a text and the craftsman translated it into the desired instrument, among them astrolabes, quadrants, armillary spheres, sundials and compound instruments. An example for such a cooperation in the case of astrolabes seems to be found in a work on the construction of such instruments by the Rasulid prince and later ruler al-Ashraf ʿUmar (r. 694–695/1295–1296). The teaching certificates (*ijāza*) by two scholars at the end of the treatise include their inspection of astrolabes produced on the basis of his work. They are described as faultless "'except a little from the craftsman turner' which al-Ashraf [ʿUmar] 'knew about and could correct'" (King 2005[CB], 646, see also 644). The craftsmen who worked for al-Ashraf ʿUmar with copper alloys also cast and struck the individual components of the instrument (King 2005, 645). Artisans in the Maghrib, Mamluk Egypt and Syria or the Ottoman Empire also produced astrolabes and quadrants from wood or papier mâché. They had to master the technical skills needed for those materials. In addition, craftsmen who aspired to sell their instruments to wealthy elites had to be informed about different decorative forms then in fashion.

An example of cooperation for an azimuthal quadrant is attested in the legend to the drawings of a specimen invented by the Mamluk timekeeper Najm al-Dīn al-Miṣrī (1st half of the 8th/14th century). An otherwise unknown instrument maker called Shaykh Muḥammad ibn al-Sāʾiḥ (7th/13th century) had produced several brass realizations of this quadrant that were sold after the master's death (Charette 2003, 137). Since the table for the *azimuth*s and the time-arcs provided by Najm al-Dīn lacked "a few entries which would have greatly facilitated the construction of the markings," the instrument maker must have been capable of compensating. This suggests that the better-skilled artisans could not merely read, write and calculate trigonometric and other numerical data, as well as draw and trace geometrical figures, they also could handle missing information and other idiosyncratic features in the scholarly texts, even if they had not made a full transition to scholarship.

Instrument makers produced more than one type of instrument. But families or individual artisans could become famous for the quality of their astrolabes (and globes). In such cases, their skills could be recognized through the byname "the astrolabe maker" (*al-asṭurlābī*). Already in the 4th/10th century, al-Nadīm mentioned among the group of instrument makers one who carried this byname (Ibn al-Nadīm 1970[CB], 2: 671). Other carriers of this byname indicate that al-Nadīm's strict separation between authors of texts, including those on instruments, and the makers of instruments wasmuch less rigid than he impiled. Instrument makers in the early 3rd/9th century participated in astronomical observations as producers of specific instruments, as well as observers and calculators and writers of the resulting new tables (van Dalen 2007, 1249–50; Bolt 2007, 740). In terrestrial measurements, they served as producers of instruments and surveyors who reported to the caliph about the final results (King 2000). They also trained artisanal apprentices, progressed into the ranks of scholars and in some cases, they were the first of several generations of scholarly students of the heavens (Ibn al-Nadīm 1970, 2: 671; Bolt 2007, 740; Chapter I.6).

ʿAlī ibn ʿĪsā (d. after 216/832), the astrolabe maker, started out as an apprentice of Khālid ibn ʿAbd al-Malik (d. after 216/832), an instrument maker, observer, author and founder of a dynasty of astrologers. In the desert of Sinjar, he participated with his master in the terrestrial measurement for newly determining the size of the earth, followed by astronomical observations in Baghdad and Damascus. As part of his task there, he improved a mural quadrant for verifying previous observational results. He also wrote a treatise on the astrolabe (Bolt 2007, 34). His educational path as an apprentice is repeated by al-Nadīm for each instrument maker he lists (Ibn al-Nadīm 1970, 2: 671). It is also confirmed by the inscription on one of the oldest extant Arabic astrolabe from Baghdad, which states that it was made by Khafīf (d. before 300/912–3), apprentice of ʿAlī ibn ʿĪsā. Two groups in al-Nadīm's list transmitted their knowledge within their families from father to son to grandson (Ibn al-Nadīm 1970, 2: 671). One of the extant astrolabes was even made as a joint work by two sons of an astrolabe maker from Isfahan, who had moved to Baghdad (Ibn al-Nadīm 1970, 2: 671). Analogous cases are known from Toledo and Valencia in the 5th/11th century and Lahore in the 11th/17th and 12th/18th centuries (see also the following discussion). Both kinds of training for future instrument makers remained active over the centuries in many, albeit not all regions of the Islamicate world.

A study of 40 mostly Moroccan astrolabes with 47 *retes* and 176 plates, produced from the 5th/11th to the 13th/19th centuries, has shown that their makers adapted their grids systematically to the changes due to precession and paid attention to specific prescriptions concerning prayer times (Mercier 2018, 218, 223, 226). There are continuous errors in the positions of eight stars, but the only errors in prayer times concern the curves for the midday prayer as used in al-Andalus and the afternoon prayer (Mercier 2018, 25–6). The most accurate *retes* were produced in the 7th/13th century, the worst ones in the 13th/19th century (Mercier 2018, 27). Mercier concluded on this basis that their makers did not simply copy older instruments but followed quite closely changes in the sky and the discussions in the discipline of timekeeping.

Not all astrolabes are newly designed but rather copied from earlier specimens. The results, however, depended heavily on the knowledge and skills of the maker – most obviously in late astrolabes made in the 13th/19th and 14th/20th centuries as souvenirs for tourists. In premodern times, instruments were used as templates, as depictions in surviving manuscripts of astrolabes illustrate. They often went beyond common features in design and data. Evidence of copying appears in an early undated and unsigned European astrolabe preserved in the History of Science Museum in Oxford whose *rete* is a nearly exact image of the astrolabes made by Ibrāhīm ibn al-Sahlī (5th/11th century), a member of the family in Valencia mentioned earlier.

On the other hand, extant texts and preserved instruments suggest that more complex and complicated specimens and problems were of interest only to a specialized, highly educated few. Despite all the variations in engravings and problems solved in texts over regions and times, standard astrolabes served for dealing with basic problems – determining times, heights, distances, or checking astrological data. Major alterations concern shadow scales and *azimuth* curves (not present on early Abbasid and Andalusī instruments), solar and calendrical scales (first appearing on early Andalusī instruments), tables of the lunar mansions (appearing first under Buyid rule) and *qibla* quadrants and gazetteers with dozens of entries (prominent in Safavid Iran). This evolution went hand in hand with the development of other disciplines, especially spherical trigonometry. Therefore, the essential knowledge and skills to make astrolabes had to conform to their sophisticated mathematical background and complicated craftmanship. In addition to astrologers and timekeepers, an important clientele for the works of astrolabe makers were rulers, their family members and high-ranking court officials. Many famous instruments in al-Andalus, the Abbasid and Fatimid Caliphates or the Ottoman, Safavid and Mughal Empires were dedicated to such lofty patrons. But the many extant specimens of simple kinds of astrolabes and quadrants highlight their fairly wide distribution among more modest owners, probably mostly in cities.

IV.3.6 Novelties and changes

All parts of the astrolabe attracted scholars as well as instrument makers to improve, modify and beautify them. On the front, the *rete* was designed in many different shapes. The throne and the shape of star-pointers varied with the mastery of the craftmanship. On the back, the four quadrants could house a number of different sets of engravings in order to increase the usefulness of the instrument (see IV.3.3). This potential created three noticeable highlights in the evolution of astrolabes. The first one is the culmination of early astrolabe making in the 4th/10th and 5th/11th centuries. The second one was the result of the ingenuity of Mamluk-era scholarly instrument makers between the 7th/13th and 9th/15th centuries. The third one consists in the artistic enrichment of instruments in the Persian and Indian astrolabe-making traditions during the 11th–12th/17th–18th centuries.

The 4th/10th and 5th/11th centuries witnessed the emergence of regional specifics in astrolabe making, particularly noticeable in the different rete designs common in the western and eastern parts of the Islamicate realm. In the East, many attempts were made in this time to unfold the potential of the instrument and a consensus became established of basic features to be represented on an astrolabe as summarized earlier. Two prominent scholars, al-Sijzī (d. *c.* 411/1020–1) and al-Bīrūnī, gave accounts of many different astrolabes with unusual *rete* configurations. Some of these were their own inventions, while others were already known in the 4th/10th century. Bīrūnī's work *The Book on Understanding All Possible Ways of Constructing the Astrolabe* (*Kitāb al-Istīʿāb al-wujūh al-mumkina fī ṣanʿat al-asṭurlāb*) is particularly important for it provides accounts about some unknown instrument makers and instruments (Figure IV.3.3). It is also one of the most comprehensive books on astrolabe making.

The period (4th–5th/10th–11th centuries) of flourishing instrument making did not just result in a kind of consensus in different regions as mentioned earlier but also encouraged novel designs, especially in the Islamicate East. Innovation was encouraged by the active and rich transmission of knowledge via the circulation of texts accompanying scholars, merchants and

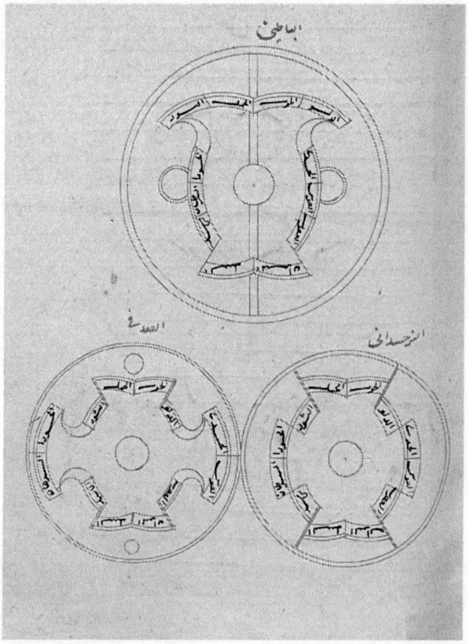

Figure IV.3.3 An extract showing different rete styles from al-Bīrūnī's *The Book on Understanding All Possible Ways of Constructing the Astrolabe* (*Kitāb Istīʿāb al-wujūh al-mumkina fī ṣanʿat al-asṭurlāb*). MS Istanbul, Süleymaniye Library, Aya Sofya 2576, fol. 31a

Source: © Süleymaniye Library, Istanbul

pilgrims, who traveled within the Islamicate world. Abū ʿAlī al-Ḥasan al-Marrākushī's monumental work (2nd half of the 7th/13th century) *From A to Z: A Compendium of Timekeeping* (*Jāmīʿ al-mabādī wa-l-ghāyāt fī ʿilm al-mīqāt*; written in Cairo) played a pivotal role in transmitting knowledge of instrumentation. The work consists of four books: (1) mathematical applications related to timekeeping and instrumentation, (2) instructions for the construction of dozens of instruments for observation and calculation, (3) explanations of the use of several astronomical instruments and (4) a list of questions with answers to 101 technical problems related to what was explained in the work. Several chapters on the construction or use of instruments are extensively revised versions of other scholars' treatises both from the East and West. Considering the popularity of the work, which survives in dozens of copies around the world, it was arguably one of the most influential texts on instruments, including different types of astrolabes.

Mastery of astronomical instrumentation peaked in Mamluk Egypt, Palestine and Syria in the 8th–9th/14th–15th centuries. Although all branches of astronomy were studied, most of the Mamluk scholars who worked on astronomy had a special interest in timekeeping and subsequently instrumentation. With the establishment of the office of timekeeper (*muwaqqit*) around the second half of the 7th/13th century, the main focus of interest became knowledge of timekeeping (*ʿilm al-mīqāt*), which deals with practical applications of astronomy such as determining the daily prayer times, finding time during day and night, calculating the *qibla* and preparing yearly prayer timetables (King 2004[CB], 2005; Chapters V.4, VI.2). This special attention to timekeeping created momentum for the advancement of many related tools such as sundials and astrolabes.

The idea of going beyond the standard shapes and markings of astrolabes led to the creation of ingenious designs. Prominent scholar-instrument makers such as Najm al-Dīn al-Misrī, ʿAlāʾ al-Dīn ʿAlī ibn Ibrāhīm known as Ibn al-Shāṭir (d. 776/1374–1375), Ibn al-Sarrāj, Shams al-Dīn al-Mizzī (750/1349–1350) and Jamāl al-Dīn al-Māridīnī (9th/15th century), focused on both enhancing and simplifying the astrolabe. For instance, Ibn al-Shāṭir designed several instruments, including a universal astrolabe and two quadratic trigonometric instruments. His most remarkable design was perhaps the instrument called the complete quadrant (*al-rubʿ al-tāmm*). It can be used for all functions of both faces of the standard quadrant. It contains two perpendicular sets of parallel lines for latitudes and longitudes, forming a triangular grid inside several scales with equal and unequal divisions at the rim and two radii. Since there are no altitude circles for a specific latitude, the instrument is universal in nature.

Ibn al-Shāṭir's out-of-the-box vision may have been surpassed only by his contemporary Ibn al-Sarrāj who also was a scholar and instrument maker. Very little is known about his life, but his unmatched ingenuity is documented by his extant instruments and treatises. His universal astrolabe, named *sarrājiyya* after himself, is arguably the most sophisticated astronomical instrument ever made in the Islamicate world (Figure IV.3.4). The previously mentioned astrolabe of Ibn Khalaf appears to be a similar instrument.

The third highlight in the evolution of astrolabes occurred in Iran and Mughal India during the 11th–12th/17th–18th centuries. Isfahan and Lahore became the epicenters of astrolabe-making and astrolabes flourished as art objects. The fusion between science and art in particular in the case

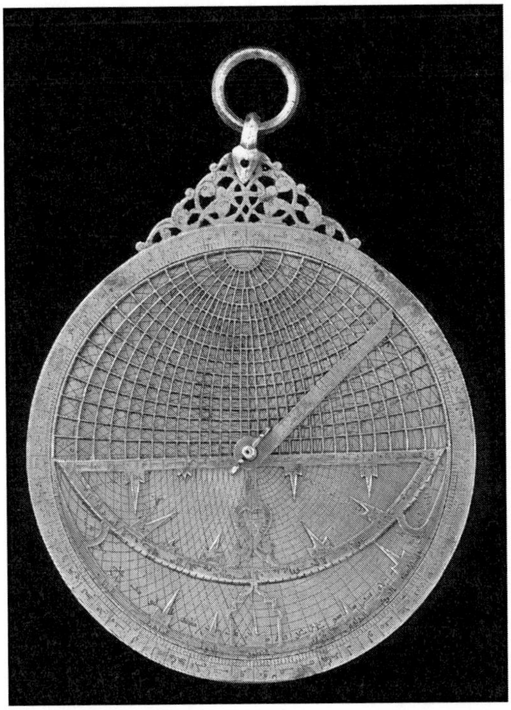

Figure IV.3.4 Ibn al-Sarrāj's universal astrolabe dated 729/1328–9. Athens, Benaki Museum, Greece, inv. no. ΓΕ 13178, side A.

Source: © Benaki Museum, Athens

of astrolabes had already become visible in the 4th/10th century, when the quatrefoil decoration and zoomorphic star pointers made their entrance on *retes*. A magnificent astrolabe of this early time is the one fabricated in Baghdad in 374/984–5 by another instrument maker turned scholar, al-Khujandī (*c.* 324–390/945–1000), today in the Museum for Islamic Art in Doha, Qatar. But Safavid and Mughal instrument makers took the artistic aspect of astrolabe making to an unprecedented level. The *mater*, the throne and the *rete* were more elaborately and artistically decorated than ever. In Iran, the instrument makers Muḥammad Mahdī al-Yazdī (11th/17th century) and ʿAbd al-Aʾimma (late 11th–early 12th/17th–18th centuries) are two of the most prolific artisans who made and/or decorated dozens of extant astrolabes (Figure IV.3.5). The inscriptions indicate further division of labor, naming a craftsman who constructed the instrument and one who had added the beautiful designs. The borderlines between these two responsibilities remained fluid, since an artisan could be named the maker on one instrument and the decorator on the next.

In Mughal India, a famous Lahore workshop run by a family of craftsmen and scholars dominated the astrolabe-making tradition. This family stretched over four generations from the second half of the 10th/16th century through the entire 11th/17th century (for its members, see Chapter IV.2). The family signatures appear on 134 astrolabes and 33 globes. Muḥammad Muqīm (*c.* 1030–1070/1621–1660) and his nephew Ḍiyāʾ al-Dīn Muḥammad (*c.* 1047–1091/1637–1681) stand out the most with 33 and 34 astrolabes, respectively (Sarma 2018).

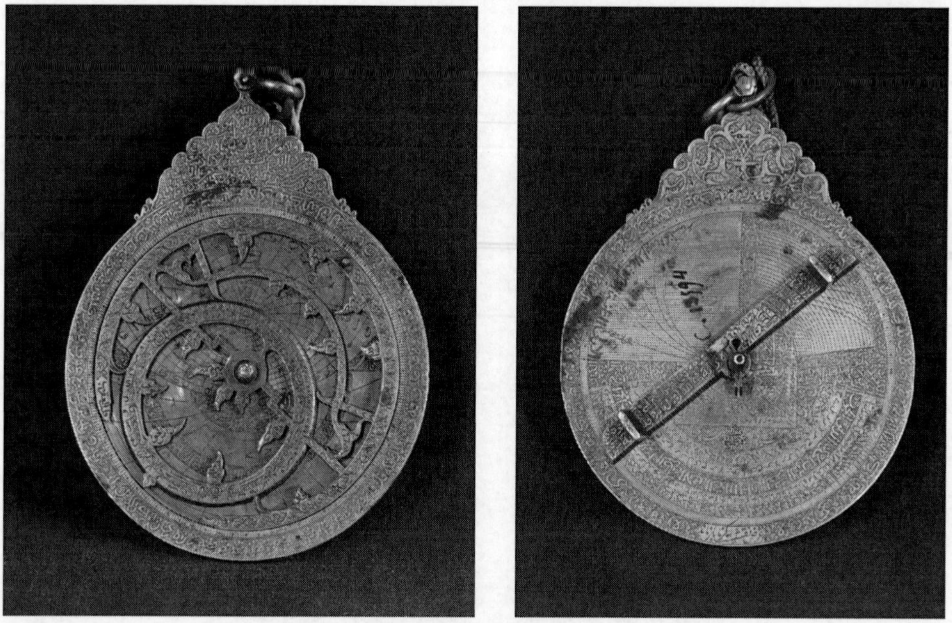

Figure IV.3.5a and IV.3.5b Astrolabe decorated (*namaqa*) by Muḥammad Mahdī al-Khādim al-Yazdī *c.* 1070/1659–60. Copper 12 × 9 cm. 5 plates, one of them for Isfahan (32°) and another one for Mecca (22°). On the lower rim of the back a hemistich from Abū Muḥammad Musharrif al-Dīn Saʿdī's (d. 690/1291) work *The Rosegarden* (*Gūlistān*) says: "The goal of the design is that it should endure." The instrument shares both inscriptions, as well as other elements with an instrument sold by Christies (Art of the Islamic and Indian Worlds. 21 April 2016, 54). Paris, BnF, Department Maps and Plans, inv. no. Ge A 327.

Source: © BnF, Paris

Astrolabe-making traditions in Iran and Mughal India differ significantly from each other and compared to earlier periods. They differ both in their makers and the related texts. Compared to Mamluk scholarly instrument makers, there are far more non-scholarly artisans known in this period. There is no doubt that they had adequate knowledge of the theory of the astrolabe. Although no technically unique type of astrolabe survives from this period and most of the texts on astrolabes are copies of earlier works, some Safavid instruments show particular or improved features. This concerns above all the Mecca-centered world maps (King 1999), and some of the *qibla* quadrants. On the other hand, the number of extant astrolabes from Safavid Iran and Mughal India, about 200, is higher than from any other region or period. This may reflect an increase in demand for astrolabes as artistic, precious objects rather than astronomical instruments. From the 3rd/9th century onward, astrolabes and celestial globes were commissioned by rulers, courtiers, rich families and individuals as symbols of knowledge and power, as well as an expression of social status. In the 11th–12th/17th–18th centuries this motivation seems to have achieved its highest level. Although astrolabes continued to be used for timekeeping applications, many of the surviving examples from Iran and Mughal India are almost in pristine conditions and show no signs of tear and wear. This can also be said for the extant Ottoman astrolabes from the 12th/18th and 13th/19th centuries. There was no need for astrolabes any longer, since most of the timekeeping applications could be done by either mechanical clocks or cheap-to-build and easy-to-use wooden astrolabe quadrants.

The period between the 7th/13th and 9th/15th centuries can be considered the peak of the tradition in the sense of the instrument's technical development. Although making astrolabes continued until the early 14th/20th century all around the Islamicate world, no further substantial changes were made.

IV.3.7 Conclusion

Astrolabes are mirrors of the societies they were produced in. For instance, the shape of a throne may reveal the artistic fashion of a time and region, or a set of markings on the back of an astrolabe may show the level of knowledge of geometry and trigonometry possessed by its maker. A dedication to a person may point out the owner's interests in astronomy. Worn-out plates and replacement pieces may indicate long-term use of the instrument. In short, an extant astrolabe provides a good deal of data for us to create patterns of connections between regions and periods, development stages in the mathematical sciences and the extent of patronage in the Islamicate world. This is also a valid argument for texts on the construction and use of astrolabes. For instance, the high number of inventions that are only available to us via texts from the Mamluk era may be the result of the establishment of the office of timekeeper. More than any other within the Islamicate world, this institutionally sponsored position arguably became a medium for scholars and *madrasa* teachers to engage with astronomical applications. Uninterrupted scholarly activity in Egypt, Palestine and Syria under Mamluk rule over two centuries, with no noteworthy discouragement from working on astronomy by the dynasty and its administration, allowed scholars and craftsmen to flourish in instrument making (for Mamluk patronage for the mathematical sciences, see Brentjes 2008a, 2008b). Comparable conditions applied to other regions in the Islamicate world in other periods. Bīrūnī's excellent scholarly activities under the patronage of Maḥmūd of Ghazna (r. 388–421/998–1030), systematic observation activities in Maragha and Samarkand in the 7th/13th and 9th/15th centuries, respectively, and Taqī al-Dīn ibn Maʿrūf's (d. 993/1585) activities in the short-lived Istanbul Observatory between 984–987/1577–1580 are some of the brief but fruitful periods for instrument making.

Notes

1 My thank goes to Sonja Brentjes, in particular for Section IV.3.3, and Petra G. Schmidl for their support and help with this chapter.
2 Consolidated bibliography.

Bibliography

Sources

al-Farghānī. ed. with tr. and comm. by Lorch, R. 2005. *On the Astrolabe*. Stuttgart: Steiner.
Ḥabash al-Ḥāsib. ed., tr. and com. Charette, F. and Schmidl, P. G. 2001. "A Universal Plate for Timekeeping by the Stars by Ḥabash al-Ḥāsib: Text, Translation, and Preliminary Commentary," *Suhayl* 2: 107–59.
Isḥāq ibn al-Ḥusayn al-Munajjim. ed. Saʿid, F. 1408/1988. *Ākām al-marjān fī dhikr al-madāʾin al-mashhūra fī kull makān* [Mounds of Pearls: Account of the Well-Known Cities in Each Place]. Beirut: ʿĀlam al-kutub.

Manuscripts

MS Doha, Qatar National Library, Or 5593.
MS Istanbul, Süleymaniye Library, Hamidiye 1453.
MS Istanbul, Topkapı Sarayı Müzesi Library, Ahmet III 3342. Ms Oxford, Bodleian Library, Marsh 154.
MS Paris, BnF, Arabe 5098.

Research literature

Bolt, M. 2007. "ʿAlī ibn ʿĪsā al-Asṭurlābī," in Hockey *et al*., eds. [CB], 34.
Bolt, M. 2007b. "Marwarrūdhī: Khālid ibn ʿAbd al-Malik al-Marwarrūdhī," in Hockey *et al*., eds. [CB], 740.
Brentjes, S. 2008a. "Courtly Patronage of the Ancient Sciences in Post-Classical Islamic Societies," *Al-Qanṭara* 29.2: 403–36.
Brentjes, S. 2008b. "Shams al-Dīn al-Sakhāwī on *Muwaqqits*, *Muʾadhdhins*, and the Teachers of Various Astronomical Disciplines in Mamluk Cities in the Fifteenth Century," in *A Shared Legacy — Islamic Science East and West: Homage to Professor J.M. Millàs Vallicrosa*. Barcelona: Edicions de la Universitat de Barcelona, 129–50.
Calvo, E. 1992. "Ibn Bāṣo's Universal Plate and Its Influence on European Astronomy," *Scientiarum Historia* 18: 61–70.
Calvo, E. and Puig, R. 2006. "The Universal Plate Revisited," *Suhayl* 6: 113–57.
Dalen, B. van. 2007. "Battānī: Abū ʿAbd Allāh Muḥammad ibn Jābir ibn Sinān al-Battānī al-Ḥarrānī al-Ṣābiʾ," in Hockey *et al*., eds., 101–3.
Dalen, B. van. 2007b. "Yaḥyā ibn Abī Manṣūr: Abū ʿAlī Yaḥyā ibn Abī Manṣūr al-Munajjim," in Hockey *et al*., eds., 1249–50.
Gunther, R. T. 1932 [1976]. *The Astrolabes of the World*, 2 vols. Oxford: University Press (Reprinted in one volume 1976: London: The Holland Press).
Hernández Pérez, A. 2018. *Catálogo razonado de los astrolabios de la España medieval*. Madrid: La Ergástula.
İzgi, C. 1997. *Osmanlı Medreselerinde İlim*. 2 vols. Istanbul: İz Yayıncılık.
Jaouiche, Kh. 1976. *Le livre du qarasṭūn du Ṯābit ibn Qurra*. Leiden: Brill.
Kennedy, E. S. and Destombes, M. 1967. *Introduction to Kitab al ʿamal bil asturlab*. Hyderabad: Dāʾirat al-Maʿārif al-ʿUthmāniyya (Osmania Oriental Publications Bureau).
King, D. A. 1987. *Islamic Astronomical Instruments*. London: Variorum Reprints.
King, D. A. 1996. *A Catalogue of Medieval Astronomical Instruments*. Parts 1.1 to 2.3. Islamic Astrolabes to *ca*. 1550 with Additional 'Old-Style' Astrolabes to *ca*. 1800. unpublished manuscripts. https://davidaking.academia.edu.
King, D. A. 2017a. "European Astrolabes to ca. 1500. An Ordered List," *Medieval Encounters* 23: 355–64. (Repr. in Rodríguez-Arribas *et al*.)
King, D. A. 2017b. "What is an Astrolabe, and What is an Astrolabe Not," www.davidaking.academia.edu.
King, D. A. 2018. "Spherical Astrolabes in Circulation: From Baghdad to Toledo and to Tunis & Istanbul," www.davidaking.academia.edu.

Mayer, L. A. 1956. *Islamic Astrolabists and Their Works*. Geneva: A. Kundig.

Mercier, E. 2018. "La qualité scientifique des instruments gnomoniques maghrébo-andalous (XI-XIX ème siècle)," in Abdeljaouab, M. and Hmida, H., eds. *Actes du XIIIe Colloque Maghrébin sur l'Histoire des Mathématiques Arabes* (Tunis, 30–31 mars et 1er avril 2018) (COMHISMA13). Tunis: Graphika Imprimerie, 213–34.

Mimura, T. 2017. "Too Many Arabic Treatises on the Operation of the Astrolabe in the Medieval Islamic World: Athīr Al-Dīn Al-Abharī's *Treatise* on Knowing the Astrolabe and His Editorial Method," *Medieval Encounters* 23: 365–403.

Morrison, J. E. 2007. *The Astrolabe*. Rehoboth Beach, DE: Janus.

Puig, R. 2007a. "ʿAlī ibn Khalaf: Abū al-Ḥasan ibn Aḥmar al-Ṣaydalānī," in Hockey *et al.*, eds. [CB], 34–5.

Puig, R. 2007b. "Zarqālī: Abū Isḥāq Ibrāhīm ibn Yaḥyā al-Naqqāsh al-Tujībī al-Zarqālī," in Hockey *et al.*, eds. [CB], 1258–60.

Rodríguez-Arribas, J., Burnett, C., Ackermann, S. and Szpiech, R., eds. 2018. *Astrolabes in Medieval Cultures*. Leiden: Brill.

Samsó, J. 1994. *Islamic Astronomy and Medieval Spain*. Aldershot: Variorum.

Sarma, S. R. 2018. *A Descriptive Catalogue of Indian Astronomical Instruments*. Edmonton: University of Alberta, Canada. https://srsarma.in/catalogue.php.

Schmidl, Petra G. 2017. "Knowledge in Motion: An Early European Astrolabe and Its Possible Medieval Itinerary," *Medieval Encounters* 23: 149–97 (Repr. in Rodriguez-Arribas *et al.*, eds.).

Schmidl, Petra G. 2018. "Using Astrolabes For Astrological Purposes: The Earliest Evidence Revisited," in Dunn, R., Ackermann, S. and Strano, G., eds. *Heaven and Earth United. Instruments in Astrological Contexts*. Leiden: Brill, 4–23.

Solla Price, D. de. 1955, "An International Checklist of Astrolabes," *Archives internationales d'histoire des sciences* 8: 243–9 and 363–81 (Repr. in Sezgin, F. *et al.*, eds. 1998. *Astronomical Instruments and Observatories in the Islamic World. Texts and Studies* 12. Islamic Mathematics and Astronomy 96. Frankfurt: Institut für Geschichte der arabisch-islamischen Wissenschaften, 93–119).

Solla Price, D. de, *et al.* 1973. *A Computerized Checklist of Astrolabes*. New Haven, CT: Self-Publishing Company.

Thomann, J. 2018. "Astrolabes as Eclipse Computers: Four Early Arabic Texts on Construction and Use of the Ṣafīḥa Kusūfiyya," *Medieval Encounters* 23: 8–41. (Repr. in Rodriguez-Arribas *et al.*, eds.).

Van Brummelen, G. 2013. *Heavenly Mathematics: The Forgotten Art of Spherical Trigonometry*. Princeton and Oxford: Princeton University Press.

Zaynalal, B. and Ismail, M. R. 2010. "Trigonometric Solutions Using Sine Quadrant," *Procedia, Social and Behavorial Sciences* 8: 721–8.

IV.4
MEDICAL INSTRUMENTS

Fabian Käs

The medical instruments described, illustrated or produced in Islamicate societies are of greatest interest for the history of surgery. Many chirurgical tools still in use today have, despite all further development, precursors in early modern Europe. These have, in turn, two forerunners, namely the instruments mentioned by ancient Greek authors and the descriptions and drawings in medieval Latin books, many of which were translations from Arabic, or depended on these translations. The exact extent of this influence cannot be determined, since theoretical medicine in medieval Islamicate countries was itself an adaption of that of antiquity. Almost every instrument mentioned in the present article has, for example, a Greek counterpart. What made Arabic medical texts attractive for the Christian scholars and patrons in Europe, was not their originality and innovativeness, but their clear systematic arrangement. This also counts for the most important Arabic monograph on surgery, *The Arrangement for One Who is Not Able to Compile a Book for Himself* (*al-Taṣrīf li-man ʿajiza ʿan taʾlīf*; in the following: *The Arrangement*), by al-Zahrāwī ([*fl.* 1st half 5th/11th century]; Figure IV.4.1), which was highly esteemed in Europe well into the early modern period.

IV.4.1 Sources

The body of source material for the study of medical instruments in Islamicate societies is significantly different from that of Greco-Roman surgical apparatus. The main problem is that virtually no original instruments exist dating from Islamicate societies before 700/1300. In comparison, great numbers of Roman chirurgical and ophthalmological toolsets have been excavated, especially in graves of physicians (Künzl 1996; Milne 1907; Bliquez 2015; see also Delpont 1996, 108–13). The reasons for this lack of Islamicate instruments are, mainly, the underdeveloped state of medieval archaeology in the Middle East and the fact that the monotheist religions forbade grave furnishings. It should also be stressed that most medical practitioners before the modern era actually used only a few instruments that can easily be confused with ordinary knives, saws or tongs. Interesting exceptions are several tools discovered in the ruins of al-Fusṭāṭ (Old Cairo) destroyed at the end of the 6th/12th century. Their interpretation is, however, difficult and some of them were apparently, contrary to earlier identifications, not designed for medical purposes (Hamarneh 1977; ʿĪsā Bāy 1925, plate 6; Sezgin 2011, 92–4; Delpont 1996, 165–6; see also Zozaya 1984).

DOI: 10.4324/9781315170718-45

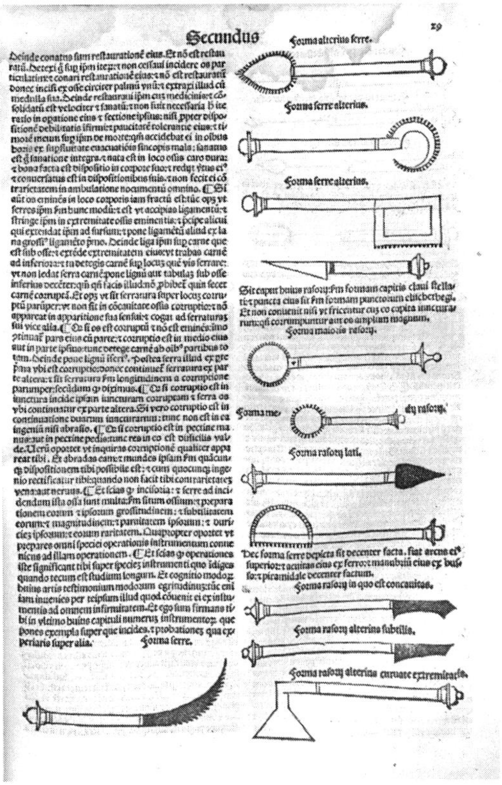

Figure IV.4.1 Saws in a printed edition of Gerard's translation of Abū l-Qāsim Khalaf ibn ʿAbbās al-Zahrāwī's work on surgery. *Cyrurgia parua Guidonis, Cyrurgia Albucasis cum cauteriis et aliis instrumentis* . . . Venice: Ottaviano Scoto 1500, fol. 29r.

Source: © BSB, Munich

The most important sources for our topic are, therefore, medical texts written between the 4th/10th and the 9th/15th centuries, and especially those containing sketches of instruments. Such drawings have no precursors in Greek books on medicine (Stückelberger 1994, 87–93; Künzl 1996, 2443–7; Savage-Smith 2000, 309) and may indeed represent an invention of scholars in Islamicate societies (Hamarneh 1961, 85; Leclerc 1861, VI). They can be found for the first time in Arabic books on ophthalmology and surgery written in the late 4th/10th and early 5th/11th centuries, such as ʿAlī ibn ʿĪsā's (d. after 400/1010) *Memorandum for Oculists* (*Tadhkirat al-kaḥḥālīn*; Savage-Smith 2011, 416, plate XXXIV), ʿAmmār al-Mawṣilī's (d. after 400/1010) *Concise Book on the Therapy of the Eye* (*Muntakhab fī ʿilāj al-ʿayn*; Hirschberg *et al.* 1904, 2: 63, 125, 131), and al-Zahrāwī's *Surgery* (see the following discussion). The latter is our main source for the reconstruction of the surgical apparatus in Islamicate countries. It also had a great impact on medieval medicine in Europe, via Gerard of Cremona's (d. 1187) Latin translation, the *Chirurgia*. Decorative miniatures depicting scenes of operations, which date back to antiquity, are often found in the Latin codices of al-Zahrāwī's *Chirurgia* (Stückelberger 1994, 88–90; Sudhoff 1918; al-Zahrāwī 2012). But these are missing from the extant Arabic manuscripts, probably due to their little practical value. However they appear in the expanded Turkish translation of the *Surgery* of 870/1465, perhaps picking up Byzantine or Latin influences (Huard and Grmek 1960).

Compared to the vast Arabic literature on medicine, monographs on surgery are extremely rare. This phenomenon might reflect the fact that *cheirourgia* or *al-ʿamal bi-l-yad* – both meaning 'handwork' – was not considered an integral part of the "medical art" in its strict sense by Greco-Islamicate physicians, but the specialty of craftsmen (see below). The *Hippocratic Oath* already contained the well-known sentence: "I will not use the knife, not even, verily, on sufferers from stone, but I will give place to such as are craftsmen therein" (Hippocrates 1957, 1: 299). Alongside the physicians (*ṭabīb*, *ṭabāʾiʿī*), who treated patients mainly with diets and drugs, there existed several medical professions. Authors on the ethics of medicine and the inspection of markets (Bürgel 2016[CB], 171–7, 203–4) mention oculists (*kaḥḥāl*), bonesetters (*mujabbir*), bloodletters (*fāṣid*, *faṣṣād*), cuppers (*ḥajjām*) and surgeons (*jarrāḥ*, *jarāʾiḥī*). A catalogue of questions for the examination of the professional skills of ophthalmologists, surgeons, and bonesetters was compiled by al-Sulamī (d. 604/1208), the head of physicians in Cairo (al-Sulamī 2004, 89–111).

In most Greek and Islamicate sources, surgical interventions were only described in the context of the respective diseases as a last resort, when other therapies seemed to be ineffective. The only preserved systematic presentation of surgery in Greek is the 6th book of Paul of Aegina's (7th century) medical compendium (Paul of Aegina 1846, 2: 247–511), which became available to Islamicate physicians via an Arabic translation by Ḥunayn ibn Isḥāq (d. 260/873). Although only fragments of this translation are still extant today, Paul's book was highly appreciated by early authors on medicine until the 5th/11th century (Pormann 2004, 300–2; Sezgin 1970, 168–70). It served as a model for the popular genre of handbooks (*kunnāsh*) and was an important source for authors dealing with surgery and ophthalmology. Ḥunayn's son Isḥāq (d. 298/910–1) authored one of the rare Arabic monographs on this topic, entitled *The Craft of the Therapy with the Iron* (*Ṣanʿat al-ʿilāj bi-l-ḥadīd*), which is only known from a bibliographical note (Sezgin 1970, 266). Abū Bakr al-Rāzī's (d. 313/925 or 323/935) *Book on Working with the Iron* (*Kitāb al-ʿAmal bi-l-ḥadīd*. Sezgin 1970, 292; De Koning 1896, 52) is also completely lost. Al-Rāzī also treated surgery in his famous handbooks *On Medicine for [the Samanid prince] al-Manṣūr* (*Kitāb al-Manṣūrī fī l-ṭibb*) and *The Comprehensive Book* (*Kitāb al-Ḥāwī*), especially in the chapters on tumors and fractures (al-Rāzī 1987, 301–36; al-Rāzī 1955, 12: 25–248; 13: 126–252). Several passages describing surgical interventions are also to be found in Ibn Sīnā's renowned *Canon* (Sanagustin 1986; Gurlt 1898, 1: 650–9). The only important compendium containing a long systematic chapter dedicated to surgery is *The Complete [Book] of the Medical Art* (*Kāmil aṣ-ṣināʿa al-ṭibbiyya*) by ʿAlī ibn al-ʿAbbās al-Majūsī (d. *c.* 384/994), which probably depends on Paul's handbook (al-Majūsī 1294, 2: 454–516; see also Gurlt 1898, 1: 615–18; Pormann 2004, 303–5). All these sources give more or less detailed accounts of the operations and the instruments used. Exact descriptions of the scalpels, cauteries and other tools employed are, however, rare and sketches appear for the first time in the books on ophthalmology mentioned above.

The most important source for our knowledge of medical instruments (Figure IV.4.2) is the chapter on surgery of al-Zahrāwī's medical encyclopedia *The Arrangement* (Savage-Smith 2011, 208–16; Sezgin 1970, 323–5). Not much about the author is known, except for the fact that he hailed from Madīnat al-Zahrāʾ, near Córdoba. According to new pieces of evidence, he cannot have died around the year 400/1010, as often maintained. Instead, he was still alive in the second quarter of the 5th/11th century (Käs 2010, 1:73–4). There has been some controversy about the question whether of he was a practicing surgeon, but this seems likely, since in a few cases he tells about his own experiences (al-Zahrāwī 1973, 2–4, 676). His *Arrangement* is divided into thirty chapters mainly dedicated to pharmacology and diet. The thirtieth – and by far the longest – chapter is his famous *Surgery* (*al-ʿamal bi-l-yad*), especially renowned for its great number of drawings of instruments, which are analyzed later. Although al-Zahrāwī did not often name his sources explicitly, it is clear that his text mainly depends on the Arabic version of Paul of Aegina

Figure IV.4.2 Cephalotribe and forceps from al-Zahrāwī's *Taṣrīf.* MS University of Pennsylvania Libraries, Schoenberg Institute for Manuscript Studies, LJS 435, fol. 35b.

Source: © Lawrence J. Schoenberg Collection of Manuscripts, Kislak Center for Special Collections, Rare Books and Manuscripts, University of Pennsylvania, Philadelphia

(al-Zahrāwī 1973, 841, index entries). Some instruments and operations having no parallels in the extant Greek and Arabic texts may, in fact, represent his own inventions (al-Zahrāwī 1973, IX). The *Surgery* including the sketches was translated into Latin by Gerard of Cremona (Figure IV.4.1), into Hebrew in 1261–1264 by Shem Ṭov of Tortosa ([d. after 1267]; Bos *et al.* 2011; Delpont 1996, 306), and into Turkish, with some additions, in 1465 by Sheref el-Dīn Sabunjuoghlu ([d. after 872/1468]; Huard and Grmek 1960; Shefer-Mossensohn 2009, 45–61).

Because of its importance for Western medicine and because of its unique character as virtually the only important Arabic monograph on surgery, al-Zahrāwī's book has been of the greatest interest for historians of medicine. John Channing (1703–1775) had already edited the Arabic and Latin texts in the 18th century (Figure IV.4.3; al-Zahrāwī 1778). In 1861, Lucien Leclerc (1816–1893) translated it into French with a special emphasis on the graphic reconstruction of the instruments (Gurlt 1898, 1: 648, plates IV–V; Hamarneh 1961; Figure IV.4.4). In 1918, Karl Sudhoff (1853–1938) analyzed the drawings in the Latin translation in the framework of his classic study on medieval surgery. In 1973, M. S. Spink and G. L. Lewis published an edition of the Arabic text with an English translation and facsimiles of the drawings from two Oxford manuscripts (Degen 1976). A facsimile of an illuminated Istanbul manuscript was published in 1986.

The second important Arabic monograph is *The Basics in the Craft of Surgery* (*Kitāb al-ʿUmda fī ṣināʿat al-jirāḥa*) by the Syrian Christian Ibn al-Quff ([d. 685/1286]; Ullmann 1970[CB], 176–7). His focuses were on anatomy, drugs, the treatment of wounds, and bone setting. Surgery in a stricter sense was only described at the end of the second volume depending largely on al-Zahrāwī, whom the author mentions from time to time. Although descriptions of the tools

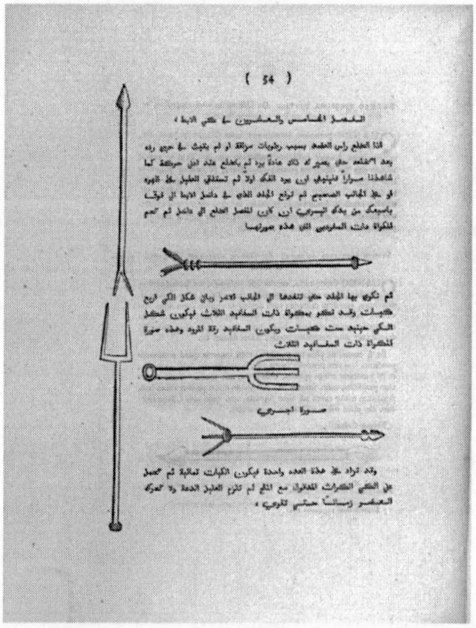
(54)

Figure IV.4.3 Cauteries in Channing's edition of al-Zahrāwī's work (1778, 54).

Source: © BSB, Munich

mentioned are almost completely missing, the book originally contained a few drawings of instruments as well (Ibn al-Quff 1937, 2: 160–73), which were unfortunately not depicted in the Hyderabad edition. A contemporary of his, Khalīfa ibn Abī l-Maḥāsin al-Ḥalabī ([d. after 674/1275]; Ullmann 1970, 212), wrote a manual, titled *The Sufficient Book on Ophthalmology* (*Kitāb al-Kāfī fī l-kuḥl*), which includes several sketches of cauteries, scalpels, cataract needles, scissors and other instruments (Hirschberg *et al.* 1904, 2: 153–94; Sezgin 2011, 6, 43–53).

Another monograph on surgery dating from the late centuries of Muslim Spain is Muḥammad al-Shafra al-Qirbilyānī's (d. 722/1322) *Minute Investigation on the Treatment of Wounds and Tumours* (*K. al-Istiqṣā' wa-l-ibrām fī 'ilāj al-jirāḥāt wa-l-awrām*; Ullmann 1970, 177; Franco and Sol 2004). Depending on al-Zahrāwī and focusing on superficial wounds, it is not very productive for our context. At least, it contains an interesting word of warning to practitioners not to trust in the descriptions of surgical interventions found in manuals (Shafra 1988, 60; Pormann and Savage-Smith 2007[CB], 130).

IV.4.2 Materials

Ancient medical instruments were in general made from copper alloys and especially bronze. Iron was only used for the blades of scalpels and cauteries (Bliquez 2015, 16–20; Milne 1907, 10–23; Gurlt 1898, 1: 506, 511). Al-Zahrāwī states, by contrast, that even probes should be made of steel (1973, 347). Some authors believed that golden cauteries were more efficient than the usual ones made from iron (Ibn al-Quff 1937, 2: 161). Gold or bronze hollow cataract needles were recommended by al-Mawṣilī (Hirschberg *et al.* 1904, 2: 131). Silver or ivory was mentioned as materials for syringes (al-Zahrāwī 1973, 407). Lead tubes or goose quills were used as cannulas (Gurlt 1898, 1: 314; 'Īsā Bāy 1925, 257). Cupping vessels were made from metal, horn, wood, or glass (Maddison and Savage-Smith 1997, 1: 42–7) and clysters (enemas) from leather or the bladders of animals. Wood was used for specula, splints, and benches for the setting of joints. A remarkable instrument was described in the *Book on Stones* (*Kitāb fī l-aḥjār*) by Pseudo-Aristotle (1912, 149). According to its unknown late antique author, one should fix a diamond on an iron probe and insert it into the urethra to destroy vesical calculus stuck there. However, there is no evidence that this instrument was actually used.

IV.4.3 Types of instruments

In what follows, I give brief accounts of the principal types of medical instruments mentioned in sources written before the 7th/13th century and their basic applications. The focus of the references is on sources in Arabic and especially al-Zahrāwī. Subtypes of the tools and further information can be found in the indices of the edition by Spink and Lewis (al-Zahrāwī 1973, 842–50). Sezgin (2011) depicted physical reconstructions, not all of which are totally plausible. The Greek equivalents of the instruments and the sources mentioning them can easily be located in studies on this topic (Bliquez 2015; Milne 1907; Durling 1993; cf. Künzl 1996; Gurlt 1898, 1: 314, 511). Almost all the types of instruments have Greek precursors. New inventions are rather rare (al-Zahrāwī 1973, IX), but further developments can be observed, for example in the case of cauteries. While the Greek tradition only knows about ten types (Milne 1907, 116–20; Bliquez 2015, 157–73), al-Zahrāwī describes no less than twenty-eight (Figure IV.4.4).

The different types of probes, Gr. *mēlē, spathomēlē*/Ar. *midass*, *barīd*, *misbār*, and *mirwad* (al-Zahrāwī 1973, 345–6), were used for various purposes, such as exploring wounds, removing foreign objects, piercing through tissue, or – especially those with a broad end – for applying remedies. Sharp spatulas (*spatē*/*mijdaʿ*) could be used like a scalpel (al-Zahrāwī 1973, 360; ʿĪsā

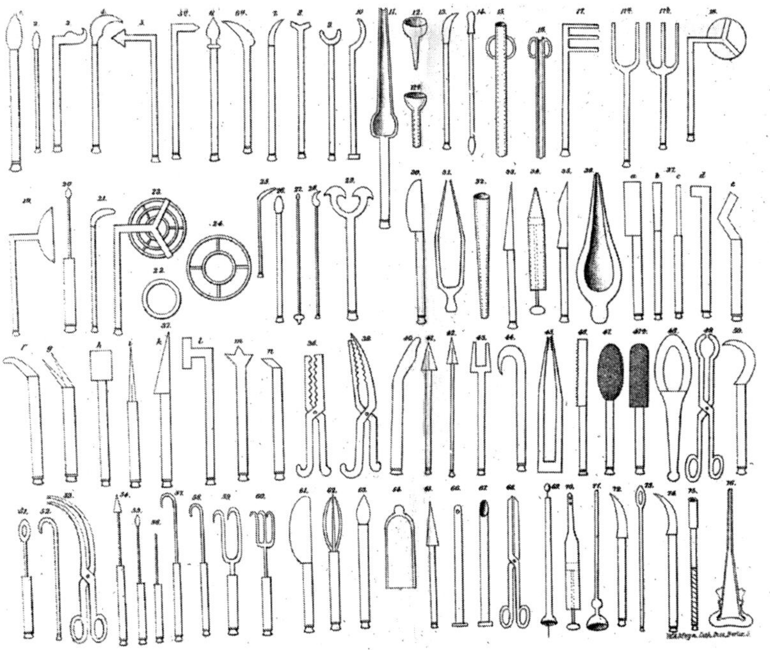

Figure IV.4.4 Instruments of Albucasis. Gurlt 1898, 1: 648, table IV.

Source: © Wellcome Collection, London

Bāy 1925, 264). Simple sharp or blunt hooks, or double- and triple-pronged hooks (*ankistron/ṣinnāra*) served for seizing blood vessels, raising small pieces of tissue for excision, and fixing and retracting the edges of wounds (Bliquez 2015, 173–83; Milne 1907, 86; al-Zahrāwī 1973, 350–5). Needles (*belonē/ibra*) were used for suturing wounds and operations on superficial aneurisms (al-Zahrāwī 1973, 371) as well as for ophthalmological purposes (see the following discussion). Al-Zahrāwī mentions, besides linen threads and animal gut, also ant heads used as pincers, a suturing material unknown to the Greeks (al-Zahrāwī 1973, 538, 550; Milne 1907, 161–2; Bliquez 2015, 148–9).

The cautery (*kautēr, kautērion/mikwāt*; Figures IV.4.3, IV.4.4), a glowing iron, "was employed for almost every possible purpose, as a 'counterirritant', as a haemostatic, as a bloodless knife, as a means of destroying tumors, etc." (Milne 1907, 116). The ancient tradition noted several forms of the head of this pencil-shaped instrument. Authors and practitioners from Islamicate societies enlarged this inventory considerably by new types for special purposes (see Section IV.4.1.; Sezgin 2011, 36–60, 67, 81).

Both classical and Arabic language sources mention a broad variety of knives and scalpels (*machaira, smilē/mibḍaʿ*) used by surgeons (al-Zahrāwī 1973, 354–60, 634; Sezgin 2011, 57, 83, 85). Special types are the lithotomy scalpel (*mibḍaʿ al-nashl*), as well as the scarifying scalpel (*mishraṭ*), and the lancet (*phlebotomon/faʾs*) required for wet cupping and bloodletting (al-Zahrāwī 1973, 354, 356, 413, 624, 629; Sezgin 2011, 50). The main instrument of the cupper was the vessel simply called *sikua/miḥjama*, but the application of medicinal leeches was also practiced (al-Zahrāwī 1973, 669–75; al-Rāzī 1987, 332). In his book on mechanical devices, al-Jazarī (d. *c.* 602/1206) described a sophisticated instrument for measuring the quantity of blood after bloodletting, which could be considered a gimmick with little practical value, but speaks for the zeal to achieve virtuosity in a broad range of imaginable applications of technical devices (Sezgin 2011, 35; Delpont 1996, 169).

Bone surgery required a special toolset, consisting of saws (*priōn/minshār*; Figure IV.4.1), chisels (*ekkopeus/miqṭaʿ*), and raspatories (*xustēr/mijrad*. See al-Zahrāwī 1973, 554, 556, 558, 565, 678; Sezgin 2011, 88–91). Trepanations of the skull were carried out with drills (*trupanon/mithqab*. See al-Zahrāwī 1973, 608, 682) or chisels (*miqṭaʿ*; al-Zahrāwī 1973, 608, 682, 703). The cephalotribe (*piestron/mishdākh, midfaʿ*; Figure IV.4.2) was a rare specialized instrument for crushing the skull of dead fetuses (al-Zahrāwī 1973, 491; Sezgin 2011, 78–9; Milne 1907, 154–5; Bliquez 2015, 42–3; Savage-Smith 2000, 315–6). A few types of specula (*dioptra/lawlab*) for investigating the vagina or the rectum were described by al-Zahrāwī (1973, 484; cf. Sezgin 2011, 74–7). In cases of urinary retention, the use of catheters was well known (*kathetēr/qāthāṭīr, mibwala*; al-Zahrāwī 1973, 403; Ibn al-Quff 1937, 2: 208,6; De Koning 1896, 44). Cannulas (*auliskos/unbūba*) served, among other things, for the drainage of the abdomen in cases of dropsy (al-Zahrāwī 1973, 196, 384, 406). Remedies could be injected into the bladder with syringes (*euthutrētos kathetēr/zarrāqa*) and into the rectum with clysters (*klustēr/ḥuqna*; al-Zahrāwī 1973, 406; Milne 1907, 106–8; Bliquez 2015, 208–13).

Many of the ophthalmological instruments, such as hooks, knives, and cauteries, differed from the ordinary tools only in size (Sezgin 2011, 42–53; Künzl 1996, 2612–7). However, al-Ḥalabī described several scalpels for individual operations, which he called the 'rose leaf' (*warda*), the 'spear' (*ḥarba*), the 'myrtle leaf', and the 'axe' (*ṭabar*; Hirschberg *et al.* 1904, 2: 165–8). He also mentioned special scissors (*miqaṣṣ, kāz, miqrāḍ*), one use of which was cutting off the growths on the eye known as pterygium. There were two techniques for couching or removing cataracts: The clouded lens could either be depressed with a solid needle (*miqdaḥ*) inserted into the eye, or the oculist could suck it out with a hollow needle (*miqdaḥ manfūdh*; *mikhaṭṭ mujawwaf*; al-Zahrāwī 1973, 257; Hirschberg *et al.* 1904–5, 1: 228, 2: 169, 2: 125, 2: 131; Künzl 1996, 2613).

Modifications of everyday objects, such as scissors, tweezers (*mudion, labidion/jift*), tongs, and forceps (*labis/kalālīb*), were used for various purposes, including the extraction of teeth (al-Zahrāwī 1973, 190, 279–81, 618). Other dental instruments were raspatories for the removal of tartar (*xustērion/mijrad*) and hooks for levering out broken teeth (al-Zahrāwī 1973, 272, 280–5; Milne 1907, 138; Sezgin 2011, 61–6; Demeisi 1999).

Bonesetters did not need many instruments, except for splints (*jabīra*) and ropes (al-Zahrāwī 1973, 688; ʿĪsā Bāy 1925, 258). The 'Hippocratic bench', an orthopedic bench (*dukkān*) for the treatment of dislocations of the dorsal vertebrae described by Hippocrates in his book on joints, was also known to scholars in Islamicate societies (*De articulis* 72; Hippocrates 1844, 4: 297, 4: 40–7; al-Zahrāwī 1973, 817; Sezgin 2011, 82).

Various utensils for the preparation of remedies, such as mortars, sieves, pots, and vessels, were mentioned in Arabic dispensatories (Kahl 2007, 34–6; Maddison and Savage-Smith 1997, 2: 290–319). Detailed descriptions are missing from the medical texts. They are more often found in alchemical treatises (Siggel 1950, 95–100; Chapter IV.5).

The great number of highly specialized instruments, as described by al-Zahrāwī and his followers, is somewhat misleading (Savage-Smith 2000; Pormann and Savage-Smith 2007, 130–5). In fact, medieval practitioners were equipped with a much smaller range. Most books on medicine do not differentiate between subtypes, for example, of scalpels or cauteries. An interesting piece of information was furnished by Ibn al-Ukhuwwa (d. 729/1329), who stated in his book on the duties of a market inspector that a surgeon (*jarāʾiḥī*) should own a set of scalpels (*mibḍaʿ*, *ḥarba*), a lancet (*faʾs*), a saw (*minshār*) and a 'rose leaf' (scalpel) for cutting off sebaceous cysts (*ward al-silaʿ*). An oculist should have hooks (*ṣanānīr*) and scalpels (*mabāḍiʿ*), as well as pencils for the application of remedies at his disposal, and a bloodletter should own several different scalpels (Ibn al-Ukhuwwa 1937, 169, 168, 161). With some caution, the instruments excavated at al-Fusṭāṭ can be identified as probes, cauteries, hooks, and tweezers (Hamarneh 1977; Sezgin 2011, 92–4). Similar probes or pencils were also found in al-Andalus (Zozaya 1984). The tool set of the oriental barber surgeons of the 19th century (Figure IV.4.5) actually consisted only of a razor, a lancet, a cautery, a single pair of tongs, and cupping glasses (Gurlt 1898, 2 :175, 189, 1: 18–21, 173–201; Lane 1908, 223).

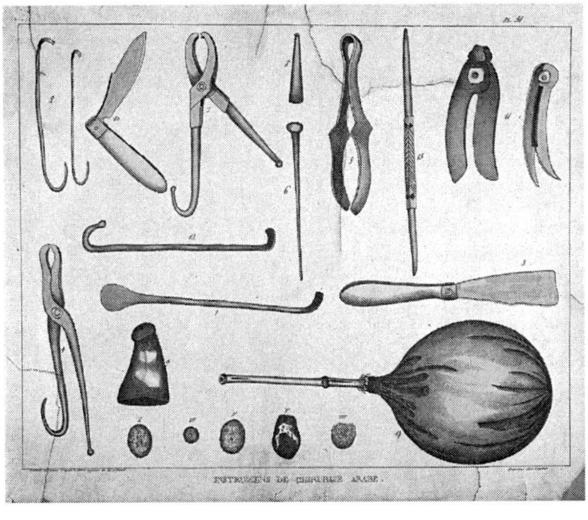

Figure IV.4.5 Arabic surgical instruments. Colored lithograph by F. Cazanave after J. J. Rifaud, *c.* 1830.

Source: © Wellcome Collection, London

IV.4.4 Conclusion

The preceding pages have tried to show that the study of medical instruments and surgery in premodern Islamicate societies is based on a small number of literary and material sources. The picture of skilled medieval surgeons disposing of a highly specialized and innovative tool set drawn in some popular scientific publications is certainly fallacious. The main source of such assumptions, al-Zahrāwī, did not describe the actual state of the art in his time but a utopian ideal concept. Other monographs on ophthalmology, surgery, and the treatment of wounds, as well as scattered accounts found elsewhere, show, however, that the main types were indeed in practical use.

Bibliography

Sources

Hippocrates. tr. Jones, W. H. S. 1957. *Hippocrates with an English Translation*. 4 vols. London and Cambridge, MA: Loeb Classical Library.

Hippocrates. ed. and tr. Littré, E. 1844. *Œuvres complètes d'Hippocrate*. 10 vols. Paris: Baillière.

Ibn al-Quff, Yaʿqūb ibn Isḥāq. 1356/[1937]. *Kitāb al-ʿUmda fī l-jirāḥa* [The Basics in Surgery]. Hyderabad: Dā'irat al-maʿārif al-ʿuthmāniyya.

Ibn al-Ukhuwwa, Muḥammad ibn Muḥammad. 1937. ed. Levy, R. *Maʿālim al-qurba fī aḥkām al-ḥisba* [Signs of Pious Deeds, on the Rules of *ḥisba*]. Cambridge: Cambridge University Press.

al-Majūsī, ʿAlī ibn al-ʿAbbās. 1294/[1877]. *Kitāb Kāmil al-ṣināʿa al-ṭibbiyya* [The Complete Book of the Medical Art]. 2 vols. Būlāq: al-Maṭbaʿa al-kubrā.

Paul of Aegina. tr. Adams, F. 1846. *The Seven Books of Paulus Aeginota*, 3 vols. London: Sydenham Society.

Pseudo-Aristotle. ed. Ruska, J. 1912. *Das Steinbuch des Aristoteles*. Heidelberg: Carl Winter.

al-Rāzī, Abū Bakr Muḥammad ibn Zakariyyā'. 1955–1970. *Kitāb al-Ḥāwī fī l-ṭibb* [The Comprehensive Book on Medicine]. 23 vols. Hyderabad: Dā'irat al-maʿārif al-ʿuthmāniyya.

al-Rāzī, Abū Bakr Muḥammad ibn Zakariyyā'. 1987. ed. al-Bakrī al-Ṣiddīqī, Ḥ. *al-Manṣūrī fī l-ṭibb* [The Book of Medicine for Manṣūr]. Kuwait: Manshūrāt Maʿhad al-makhṭūṭāt al-ʿarabiyya.

al-Shafra, Muḥammad. ed. Llavero Ruiz, E. 1988. *Un tratado de cirugía hispanoárabe del siglo XIV: el Kitāb al-istiqṣā'*. Granada: Servicio de Publicaciones, Universidad de Granada.

al-Sulamī, ʿAbd al-ʿAzīz. eds. Leiser, G. and al-Khaledy, N. 2004. *Questions and Answers for Physicians*. Leiden: Brill.

al-Zahrāwī, Abū l-Qāsim Khalaf ibn ʿAbbās. 1986. *Kitāb al-Taṣrīf li-man ʿajiza ʿan al-taʾlīf* [The Arrangement of Medical Knowledge for One Who is Not Able to Compile a Book for Himself], Facsimile edition. 2 vols. Frankfurt am Main: Institut für Geschichte der Arabisch-Islamischen Wissenschaften.

al-Zahrāwī, Abū l-Qāsim Khalaf ibn ʿAbbās. ed. Channing, J. 1778. *Albucasis De chirurgia. Arabice et latine*. Oxford: Clarendon.

al-Zahrāwī, Abū l-Qāsim Khalaf ibn ʿAbbās. ed. and com. Irblich, E. 2012. *Chirurgia. Cod. Ser. n. 2641 der Österreichischen Nationalbibliothek, Wien*, Facsimile edition. Graz: Akademische Druck- und Verlagsanstalt.

al-Zahrāwī, Abū l-Qāsim Khalaf ibn ʿAbbās. ed. and tr. Spink, M. S. and Lewis, G. L. 1973. *Albucasis. On Surgery and Instruments. A Definitive Edition of the Arabic Text with English Translation and Commentary*. London: Wellcome Institute.

Research literature

Bliquez, L. J. 2015. *The Tools of Asclepius. Surgical Instruments in Greek and Roman Times*. Leiden: Brill.

Bos, G. *et al.*: Bos, G., Hussein, M., Mensching, G. and Savelsberg, F., eds. 2011. *Medical Synonym Lists from Medieval Provence: Shem Tov ben Isaac of Tortosa* (Sefer ha-Shimmush, *Book 29. Part 1*). Leiden: Brill.

Degen, R. 1976. "Review of 'Albucasis. On Surgery and Instruments, eds. Spink and Lewis'," *Sudhoffs Archiv* 60: 301–2.

De Koning, P. 1896. *Traité sur le calcul dans les reins et dans la vessie par Abū Bekr al-Rāzī*. Leiden: Brill.

Delpont, É. 1996. *La médecine au temps des califes. À l'ombre d'Avicenne*. Gand: Snoeck- Ducaju and Zoon.

Demeisi, U. 1999. *Zur Geschichte der Erforschung von Leben und Werk des Abu l-Qāsim az- Zahrāwī unter besonderer Berücksichtigung der Zahnheilkunde*. Berlin: Mensch und Buch.

Durling, R. J. 1993. *A Dictionary of Medical Terms in Galen*. Leiden: Brill.

Franco Sánchez, F. and Sol Cabello, M. 2004. Muḥammad *aš-Šafra. El médico y su época*. Alicante: Secretariado de Publicaciones. Universidad de Alicante.

Gurlt, E. J. 1898. *Geschichte der Chirurgie und ihrer Ausübung. Volkschirurgie, Altertum, Mittelalter, Renaissance*. 3 vols. Berlin: Hirschwald.

Hamarneh, S. 1961. "Drawings and Pharmacy in al-Zahrāwī's 10th-century Surgical Treatise," *United States National Museum Bulletin* 228: 81–94 (paper 22).

Hamarneh, S. 1977. "Excavated Surgical Instruments from Old Cairo, Egypt," *Annali dell'Istituto e Museo di Storia della Scienza di Firenze* 2: 1–14.

Hirschberg *et* al.: Hirschberg, J., Lippert, J. and Mittwoch, E. 1904–1905. *Die arabischen Augenärzte*. 2 vols. Leipzig: Veith and Comp.

Huard, P. and Grmek, M. D. 1960. *Le premier manuscrit chirurgical turc rédigé par Charaf ed-Din (1465)*. Paris: Roger Dacosta.

ʿĪsā Bāy, A. 1925. "Ālāt al-ṭibb wa-l-jirāḥa wa-l-kaḥḥāla ʿind al-ʿArab," *Majallat al-Majmaʿ al-ʿilmī al-ʿarabī bi-Dimashq* 5: 253–74.

Kahl, O. 2007. *The Dispensatory of Ibn at-Tilmīḏ. Arabic Text, English Translation, Study and Glossaries*. Leiden: Brill.

Käs, F. 2010. *Die Mineralien in der arabischen Pharmakognosie*. Wiesbaden: Harrassowitz.

Künzl, E. 1996. "Forschungsbericht zu den antiken medizinischen Instrumenten," in Temporini, H. and Haase, W., eds. *Aufstieg und Niedergang der römischen Welt, Teil II: Principat, Band 37/3. Teilband Philosophie, Wissenschaften, Technik. Wissenschaften (Medizin und Biologie [Forts.])*. Berlin and New York: De Gruyter, 2433–639.

Lane, E. W. 1908. *The Manners and Customs of the Modern Egyptians*. London: Dent and New York: Dutton.

Leclerc, L. 1861. *La chirurgie d'Abulcasis*. Paris: Baillière.

Maddison, F. and Savage-Smith, E. 1997. *Science, Tools and Magic (The Nasser D. Khalili Collection of Islamic Art, XII)*. 2 vols. London: The Nour Foundation.

Milne, J. S. 1907. *Surgical Instruments in Greek and Roman Times*. Oxford: Clarendon Press.

Pormann, P. E. 2004. *The Oriental Tradition of Paul of Aegina's Pragmateia*. Leiden: Brill.

Sanagustin, F. 1986. "La chirurgie dans le Canon de la médecine (Al-Qânûn fi-ṭ-ṭibb) d'Avicenne (Ibn Sînâ)," *Arabica* 33: 84–122.

Savage-Smith, E. 2000. "The Practice of Surgery in Islamic Lands: Myth and Reality," *Social History of Medicine* 13: 307–21.

Savage-Smith, E. 2011. *A New Catalogue of Arabic Manuscripts in the Bodleian Library, University of Oxford*. Vol. 1: *Medicine*. Oxford: Oxford University Press.

Sezgin, F. 1970. *Geschichte des arabischen Schrifttums*. Vol. 3: *Medizin, Pharmazie, Zoologie, Tierheilkunde*. Leiden: Brill.

Sezgin, F. 2011. *Science and Technology in Islam. Catalogue of the Collection of Instruments of the Institute for the History of Arabic and Islamic Sciences*. Vol. 4: *Medicine, Navigation, Mineralogy*. Frankfurt am Main: Institut für Geschichte der Arabisch-Islamischen Wissenschaften.

Shefer-Mossensohn, M. 2009. *Ottoman Medicine. Healing and Medical Institutions, 1500–1700*. Albany: SUNY Press.

Siggel, A. 1950. *Arabisch-Deutsches Wörterbuch der Stoffe aus den drei Naturreichen, die in arabischen alchemistischen Handschriften vorkommen, nebst Anhang: Verzeichnis der Geräte*. Berlin: Akademie-Verlag.

Stückelberger, A. 1994. *Bild und Wort. Das illustrierte Fachbuch in der antiken Naturwissenschaft, Medizin und Technik*. Mainz: Zabern.

Sudhoff, K. 1918. "Die Instrumentenabbildungen der lateinischen Abulqâsim-Handschriften des Mittelalters," in Sudhoff, K., ed. *Beiträge zur Geschichte der Chirurgie im Mittelalter, Zweiter Teil*. Leipzig: Barth, 16–84 and plates II–XXIII.

Zozaya, J. 1984. "Instrumentos quirúrgicos de al-Andalus [Surgical Instruments from al-Andalus]," *Boletín de la Asociación Española de Orientalistas* 20: 255–9.

IV.5
ALCHEMICAL EQUIPMENT

Sébastien Moureau

IV.5.1 Introduction

From the very beginning, alchemy has been characterized as an art. Indeed, alchemists have always been aware that, along with the knowledge that was the foundation of their field, practice was a central element. And to practice their art, alchemists have always needed instruments. However, despite this insight, alchemists have not given as much focus to their instruments as to their doctrines and the recipes in their texts. The main problem when studying alchemical equipment is the paucity of sources at hand. It also requires caution and care, since most of the available material is open to differing interpretations.

As a prelude, it is important to highlight the links and differences between alchemy and craft. Most alchemists were neither pure theorists nor strict practitioners, but most often needed both theory and practice, personal as well as borrowed from others, to elaborate their systems. If alchemy is distinct from craft when considering its theoretical and nonexperimental sides, we must pay attention to the strong connections it has with craft when considering its practices and techniques. These links are striking when looking at alchemical recipes, but they become more obvious when we turn to instruments. Most alchemists' tools were craftsmens' tools. In fact, a sharper distinction can be drawn between alchemists and craftsmen than between alchemy and craft, yet this is not the subject of the present chapter.

As another prerequisite, it is worth underlining a certain stability of alchemical tools and instruments in the various time periods up to the industrial revolution. With this in mind, it is often interesting to look at later periods to study techniques in the Middle Ages, although this must be done carefully in order to avoid anachronisms.

Alchemical instruments have not received a lot of attention from scholars up to now, especially alchemical instruments in the Islamicate world. There are only a few articles and chapters on the subject. Early works on the topic were published by Henry Ernest Stapleton (1878–1962) when studying texts oriented toward practice with R. F. Azo in 1905 (Stapleton and Azo 1905), and with Azo and M. Hidāyat Ḥusayn in 1927 (Stapleton *et al.* 1927). Others were by Julius Ruska (Ruska 1923) and Wiedemann (Wiedemann 1909). More recent papers have focused on specific apparatus for alchemists (Zosimos of Panopolis 1995, CXIII – CLXIX; Savage-Smith 1997; Kurzmann 2009; Moureau and Thomas 2015). However, when looking at the publications on craftsmens' tools, the material becomes much more abundant, such as studies dealing

DOI: 10.4324/9781315170718-46

with tools that are common between craft and alchemy, especially when enlarging the geographical area and including the Latin Middle Ages. Among these, distillation has received particular attention (see, for instance, Forbes 1970), as well as mining technologies and metallurgy (Cressier and Canto García 2008).

IV.5.2 Sources

We can distinguish among four groups of sources of information on this topic: textual sources, iconographic sources, archaeological sources and experimental sources.

Textual sources on alchemical equipment divide into several types. Short accounts and lists appear in both non-alchemical and alchemical texts. We can find a shining example of an inventory of alchemical apparatus with concise descriptions in the *Keys of Wisdom* (*Mafātīḥ al-ʿulūm*), a handbook written by Abū ʿAbdallāh al-Khwārazmī (d. 387/997), a state officer at the court of the Samanid ruler Nūḥ II ibn Manṣūr (r. 366–387/976–997). In this work, the author presents the best-known alchemical instruments and describes them briefly (Khwārazmī 1968, 255–8; English translation in Stapleton *et al.* 1927, 362–3). However, most of the lists lack definitions, as in the *Book of the Beloved* (*Kitāb al-ḥabīb*; Berthelot *et al.* 1893, Arabic 3: 35–6, French translation 3: 77). Much more interesting are the texts that contain in-depth descriptions of instruments and, above all, descriptions of how to make the instruments, often placed within or between recipes where the tools necessary for the recipe are considered. There are three important writings in this category.

First, al-Rāzī's (d. *c.* 313/925 or 323/935) *Instructive Introduction* (*Madkhal taʿlīmī*) and *Book of Secrets* (*Kitāb al-asrār*) are two technical alchemical treatises. They are certainly the most valuable and precise source for studying alchemical instruments. In these texts, Rāzī provides us with a long list of tools for the alchemist, along with descriptions and recipes for making them. The *Book of Secrets* has been edited (Rāzī 1964), partially translated into English (Stapleton *et al.* 1927) and translated into German (Ruska 1937). The *Instructive Introduction* has been edited and translated into English (Stapleton *et al.* 1927, 412–7, translation 345–61).

Second, the *Source of the Art and Help of the Art* (*ʿAyn al-ṣanʿa wa- ʿawn al-ṣanʿa*) is a technical text written in Bagdad by a certain Abū l-Ḥakīm al-Khwārazmī al-Kāthī in 426/1034. This text is very much indebted to Rāzī's works. It has been edited and translated into English (Stapleton and Azo 1905). There is also an edited Persian translation (Maqbūl 1929).

Last, the pseudo-Avicennian *On the Soul* (*De anima*) is a Latin translation and compilation of three Arabic texts that have not been preserved (the largest part was composed between the third quarter of the 5th/11th century and the middle of the 7th/13th century, and the text might have been translated around 1226 or 1235). Although this very long text is quite unclear, it contains descriptions of how to make various instruments. It was edited and translated into French (Moureau 2016a, 2016b).

Some craft treatises also include accounts of instruments and their manufacture, as al-Ḥasan ibn Aḥmad al-Hamdānī's (280–334/893–945 or 946) *Book of the Two Substances* (*Kitāb al-jawharatayn*) (Moureau and Thomas 2016, 108–15). Iconographic sources are a second way of obtaining information about alchemical equipment. Unfortunately, iconographic elements in the alchemical Arabic manuscript tradition are very few. We very rarely find illustrations in manuscripts, especially in older (medieval) volumes. Most of the illustrations that we see usually come from later manuscripts and are small sketches of vessels or furnaces, such as in MS Rabat, al-Khizāna al-ḥasaniyya, 1393, fol. 153a (aludels and alembic; Figure IV.5.1), MS Rampur, Raza Library, Kīmiyāʾ 12, fols 132b–133b (9th/15th century; pots and aludels; Figure IV.5.2) and others. Some contain more delicate and elaborate drawings such as the distillation instruments and furnaces in MS Bethesda,

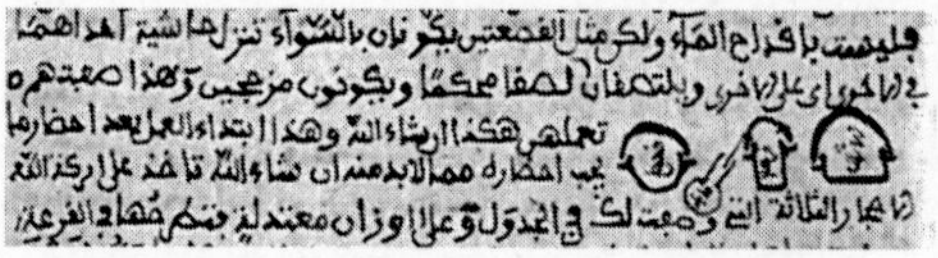

Figure IV.5.1 A "two cups" device, alembic and aludel; MS Rabat, al-Khizāna al-ḥasaniyya, 1393, fol. 153a.

Source: © al-Khizāna al-ḥasaniyya, Rabat

Figure IV.5.2 Images for making an aludel from the *Source of the Art and Help of the Art*; MS Rampur, Raza Library, Kīmiyāʾ 12, fol. 133b.

Source: © Raza Library, Rampur

National Library of Medicine, A65, fols 80b–81b, 83b–84a (1123/1712). Two treatises are exceptions to this general trend. First, the *Tome of Images* (*Muṣḥaf al-ṣuwar*), a pseudepigraph attributed to Zosimos of Panopolis (3rd–4th centuries), is an Arabic text which includes many illustrations. Almost all are allegorical, but one of them represents a furnace with a distillation device on top. A facsimile of the unique manuscript (Istanbul, Arkeoloji Müzeleri Kütüphanesi, 1574) was published (Abt 2007) and translated into English (Abt and Fuad 2011), but it is to be read with caution (see the remarks in Hallum 2009). Second, the *Seven Climes* (*ʿAqālīm sabʿa*)

Figure IV.5.3 Alembic and furnace in Abū l-Qāsim al-ʿIrāqī's *Seven Climes*; MS London, British Library, Add. 25724, fol. 56a, 12th/18th century.

Source: © British Library, London

by Abū al-Qāsim al-ʿIrāqī (mid-7th/13th century) is an alchemical treatise illuminated with numerous illustrations, mostly allegorical; some of these pictures depict alchemists at work or alchemical devices (Figure IV.5.3). The text has not been edited yet (for this study, I have used MS London, British Library, Add. 25724).

However, it is important to underline here that manuscript illustrations can often be misleading, since they were usually drawn by copyists who were most often not practitioners. These images are subject to interpretation. And when these illustrations are allegorical, such as in the *Tome of Images* and the *Seven Climes*, they must be treated with even more caution.

Archaeology brings a kind of information that is not available from texts or iconography: the material remains of instruments. Unfortunately, here again, for the Islamicate world, the field is almost *terra incognita*. When dealing with archaeological remains, two situations appear. Either the remains come from a (hopefully well) documented excavation, in which case their context is at least partially preserved, or they are kept in a museum, inherited from early and undocumented excavations, in which case their context is lost. The first case is obviously the better situation for studying any object. Disappointingly, up to the present, no firmly identified alchemical apparatus of the first variety, found in well-documented digs, is available for the medieval Islamicate world. One possible alchemical device was excavated in Ramla, in modern Israel, as proposed in (Gorzalczany 2010), but the proposal is extremely hypothetical and not very convincing. All the preserved instruments are in museums, and their context is usually known poorly or not at all. This causes significant problems and confusion for dating, localizing and even interpreting the actual nature of the instrument. For instance, medical cupping glasses for bloodletting are often mistaken for alembic heads (as in Kurzmann 2009). A beautiful glass alembic dated from the 4th–5th/10th–11th centuries in Iran (without any certainty) is kept at the Institut du monde arabe in Paris (object AI 90–08: *À l'ombre d'Avicenne* 1996, 143), and another, that is dated (but again without certainty) from the 4th/10th to the 7th/13th centuries and that might come from the Middle East, is kept at the Science Museum in London (object 1978–219; Figure IV.5.4). Savage-Smith (1997) also presents a few glass alchemical instruments,

Figure IV.5.4 Glass alembic, Science Museum, London, obj 1978–219.

Source: © Science Museum, London

but all of them are from unknown sources. Because of this lack of reliable sources, scholars often need to search for elements of information on alchemical apparatus in medieval Latin archaeology, which is more plentiful. For instance, remains of Latin medieval aludels allow us to better understand aludels in the medieval Islamicate world (again, with caution; Moureau and Thomas 2015; Thomas 2009). As for the other kinds of sources, looking at craft objects may also provide further information about alchemical instruments, although such information is usually very hypothetical (Shaddoud 2016).

As a final source, we need to mention experimentation, which currently remains quite marginal. Experimentation always relies on other sources and must therefore be carried out with caution. But reproducing instruments and actually using them always sheds new light on our understanding of tools and recipes. It often allows for the confirmation or invalidation of assumptions made when reading a text or examining an illustration. Experimentation also lets us understand practical know-how that is impossible to obtain from other sources. For an example of how experimentation enables a better understanding of medieval texts, see Moureau and Thomas (2016).

Given the sources presented here, the following descriptions apply primarily to alchemical instruments in the medieval Arabic-speaking Islamicate world but are also valid for early medieval Latin alchemy.

IV.5.3 Alchemical instruments

What is an alchemical instrument? At first glance, we could give two answers to this question: a broader one, namely, any instrument used by an alchemist, and a narrower one, namely, instruments that are specific to alchemists. In fact, the second definition is not valid for the very reason that this category is almost empty in the Middle Ages in the Islamicate and Latin worlds. Apart from some furnaces and several complicated and particular devices described by alchemists for very specific processes, all the instruments of alchemists are also used by craftsmen (or are simply instruments of day-to-day life, many of them being borrowed from the kitchen). These range from hammers and tongs, required in numerous crafts, to alembics and aludels, respectively utilized for the distillation of rose water and other products, and also for extraction of various substances, such as mercury from cinnabar.

For the sake of clarity, we can define three categories of instruments: tools and accessories, vessels and apparatus, and furnaces. Tools and accessories are certainly the least specific to alchemy. This category includes common tools such as hammers and ladles, tongs and files, and less common instruments such as molds, sieves or filters. As for vessels, a great diversity appears starting with cups, bottles, and cucurbits. Alchemists multiply technical terms for the various vessels they use. For instance, cups, bowls and plates may be called *qadaḥ*, *kūz*, *jām*, *bātiya*, *sukurruja*. or *riṭl*, each word being used depending on the context or the alchemist. These instruments are most usually made of clay with additional ingredients, such as grog (crushed, fired clay) or vegetable fiber, but can also sometimes be made of glass or metal. This category includes the more complex devices that are linked to particular operations and that I describe in more detail.

The alembic (*al-anbīq* or *al-ambīq*) (Figures IV.5.1, IV.5.3–IV.5.5) is the instrument for distillation *per ascensum*, namely distillation in the contemporary meaning, the separation of the components of a material by vaporization and recondensation. It is made up of a head (the actual *anbīq*) and a cucurbit (*qarʿ*) placed underneath. Alchemists place the material to be distilled (usually liquid) in the cucurbit, the alembic is set on the top of the cucurbit and sealed with lute (*ṭīn*), a mix of clay and other ingredients. The device is gently heated, usually in a water bath or in ashes or sand, in order to avoid thermal shocks and high temperatures. Water limits the temperature to 100 °C. The liquid vaporizes and rises into the head, where it condenses on its inner side and flows into the gutter of the *anbīq*; then, from the gutter, it flows into the beak of the alembic and thence into the receptacle (*qābila*). An alembic without a beak is called a blind alembic (*anbīq aʿmā*). Examples of alembic descriptions can be found in Rāzī's *Madkhal taʿlīmī* (Stapleton *et al.* 1927, 414, English translation 355–6), or, a more complicated one, in Pseudo-Plato's *Kitāb al-rawābiʿ* (Badawī 1955, 183–4).

The aludel (*al-uthāl*) is another complex device (Figures IV.5.1, IV.5.2, IV.5.6). It is used in the process of sublimation. In its medieval sense, sublimation (*taṣʿīd*) means vaporization, usually

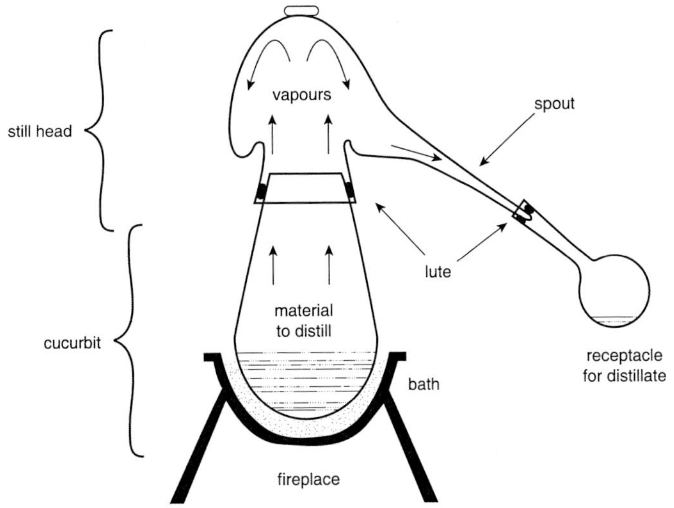

Figure IV.5.5 Alembic (Thomas 2009, 36).

Source: © Nicolas Thomas

of a solid, and its recondensation into a solid, as opposed to distillation where the material, usually liquid, condenses in a liquid form. The modern term means vaporization of a solid and its recondensation directly back into a solid without an intermediate liquid phase. The features of the aludel are quite similar to those of the alembic. It is composed of a head (the actual *uthāl*) set on the top of a pot (*burma* or other) and sealed with lute.

Rāzī's *uthāl* is made of a perforated clay disk attached to the pot, on which is laid a *mikabba*, a cone-shaped lid with a hole in its top to let the pressure out (similar to the lid of a tajine cooking pot); others kinds and shapes are described in several Arabic and Latin medieval texts (see Moureau and Thomas 2015). The device is heated (the fire can be hotter than for distillation) on a specific furnace, usually a *mustawqad* (see the following discussion), and the material sublimes into the head, where it condenses on the inner side. At the end of the process, the lid is removed, and the condensed matter gently scraped out.

The *būṭ bar būṭ* (Persian; literally crucible over crucible) is a device made up of two crucibles, one on top of the other, the bottom of the upper crucible being perforated with small holes. It is also called descensory. The alchemists put a metal to purify in the upper crucible and heat the device; the metal melts and flows into the lower crucible, while impurities remain in the upper crucible. This operation must not be confused with the distillation *per descensum*. The apparatus for this distillation also consists of two pots, one pot with a perforated bottom on the top of another pot, but the process is very different; it is used to extract vegetable oils and never to purify metals, and these two processes are clearly distinct in alchemists' minds (Thomas and Claude 2011). It is important to remember that neither the alembic nor the aludel or the *būṭ bar būṭ* is specifically alchemical, since they are all used in other crafts.

The variety of furnaces is also extremely rich. There are no less than seven different names of furnaces in Arabic (*kūr*, *tannūr*, *atūn*, *mustawqad* or *mawqid*, *ṭābadshān*, *kānūn*, *nāfikh nafsi-hi*). Furnaces are usually made of clay or earth (sometimes with other ingredients added, such as vegetal fiber), or bricks or stones. Furnaces are related to specific processes; a furnace for calcination is not the same as a furnace for distillation. For instance, the *mustawqad* is a cylindric stove ordinarily used in the process of sublimation (Figure IV.5.6). They are often very similar

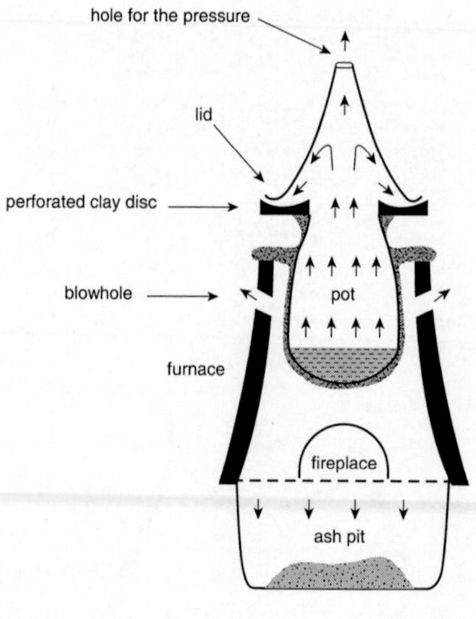

Figure IV.5.6 Aludel (Moureau and Thomas 2016, 244).

Source: © Nicolas Thomas

to craft furnaces, with some modifications when needed. Most alchemical furnaces are natural draught furnaces, but when a higher temperature is needed, bellows can be added (*minfakh*, *ziqq*, *kīr* or, in the case of cupellation, *rawbāṣ*). Furnace descriptions are quite frequent in alchemical literature, usually in recipes, to give details over specific shapes or measures (see, e.g., the *Risālat al-iksīr*, attributed, falsely in all likelihood, to Ibn Sīnā [1953, 39–40]). The fuel is usually wood or charcoal (charcoal allows the fire to reach a higher temperature), but other fuels, such as dried dung, may also be used in regions where wood is rare and too expensive.

IV.5.4 How did alchemists produce their instruments?

The question of how alchemists of the Islamicate and Latin medieval worlds made their instruments may be asked at two levels. In a historical perspective, we may ask how alchemists came to imagine such instruments. And in a more literal and practical sense, we may ask how alchemists made their instruments in their day-to-day life.

In the historical perspective, the question is very hard to answer, and will certainly not be answered by looking at Islamicate sources. Alchemists in the Islamicate world mostly inherited their tools from older civilizations. The invention of furnaces cannot be dated and goes back to early prehistory. Metallurgy appeared in the 4th millennium BCE in the Near East. Instruments for distillation and sublimation were already in use for centuries when the Islamicate civilization rose. The earliest known apparatus of this kind, which might have been used for sublimation or distillation, is a device found in Tepe Gawra, in northwest Iraq, possibly dating from 3500 BCE (Levey 1955). The form of this device is very close to that of alembics and aludels. Thus, this field is not characterized by major technological changes.

When studying this question from the perspective of the practical making of instruments, the sources provide more information. It is by comparing the different kinds of sources that we can learn how alchemists made their apparatus. Indeed, in the case of alchemical equipment, each kind of source when taken alone lacks sufficiently accurate information. An example of this problem is the precise description of the making of an aludel in Rāzī's *Book of Secrets*:

> Take a pot in the shape of a cooking pot (*burma*) that is one cubit long and two hands wide. Put it upside down on a flat thing. Scatter sifted ashes around the pot up to the distance of one hand and a half from the pot. Then, remove the pot. Make around it, in the sifted ashes, a disk of lute of wisdom, and leave it to dry. Then, remove it, polish its surface, cover it with white lead and egg white, and polish it a second time. Make a drill running on the edge of the disk, where you will leave a space from which you will take the sublimates. Leave it to dry. Then, turn the pot of the aludel upside down, cover it with a plain layer of lute, neither thick nor thin, and leave it to dry. Then, turn the disk upside down, place the pot on it and lute the joint on both sides. Make wings to the pot, at the distance of one hand beneath the disk, in order to prevent the flame of the fire from reaching and burning the disk and therefore damaging and ruining what is on the disk. Place the lid (*mikabba*) on the aludel. This lid has a secret that I will mention when speaking about sublimates [this secret is a hole at the top of the lid to allow pressure to exit].
>
> *(Rāzī 1964, 11, slightly modified with readings from MS San Lorenzo de El Escorial, árabe 700, fols 9a–b)*

This description, when read without any other source, is not sufficient to build a proper aludel (Figure IV.5.6). Even with the illustrations that, in some very rare cases, accompany texts in manuscripts, it is impossible to build this kind of device without further information (see for instance the pictures to help making an aludel in the *'Ayn al-ṣan'a* in manuscript Rampur, Raza Library, Kīmiyā' 12, Figure IV.5.2). Alchemists did not rely only on texts. In fact, they had two different kinds of sources: books and practice. And practice itself divides into two categories: the observation of other practitioners, craftsmen or alchemists and personal practical experience. Not only making alchemical instruments but also, more generally, practicing alchemy required a subtle mix of these three sources. The same goes for us: looking at the archaeological, iconographic and experimental material, we can grasp many details that are not explained in texts (and the same goes for each kind of source).

This recipe gives us another important piece of information: alchemists of this time and place made some of their devices themselves. These kinds of 'do it yourself' instructions are not very frequent in texts, except in relation to furnaces, which were probably most of time made by alchemists (they require less practical experience than vessels). On the other hand, we know that at least one kind of vessel was in all likelihood not made by alchemists themselves: glass vessels. Indeed, alchemists often mention glass containers, but shaping glass vessels requires long experience. Alchemists would have needed to be glassmakers to make them themselves. The same goes for iron tools (pincers, hammers and others), which would have obliged alchemists to be blacksmiths at the same time. Although clay vessels are easier to make, alchemists might also have asked potters to make specific instruments for them. A sort of middle way is visible in the description of the aludel quoted earlier. In this description, Rāzī explains how to make an aludel from an earthen cooking pot and a tajine lid (*mikabba*). We might infer that many alchemical vessels could also have been everyday pots diverted from their normal use. But whatever the way alchemists made (or had others make) their instruments, they had to manage the practices for conducting their experiments, such as luting vessels, building furnaces and customizing instruments.

IV.5.5 Conclusion

Among scientific instruments, alchemical equipment remains poorly studied up to now. The main reason for this is a patent lack of available material of all kinds. But the sources at our disposal do allow us to catch a glimpse of the alchemist's practices in the Islamicate and Latin medieval worlds.

Alchemists had two options for obtaining instruments: have craftsmen make them or make them themselves. Making instruments requires not only knowledge, which can be acquired through books and observation of practitioners, but also know-how that is not accessible through the other sources, but only through repeated personal practice. However, having others make their instruments was complicated, too, and certainly more expensive. But another, intermediate, way can also be glimpsed in the sources: alchemists could make new instruments from other common tools so that they might have been more familiar with apparatus customization than with their fabrication.[1]

Note

1 I am very grateful to Lawrence Principe and Nicolas Thomas for their accurate revision and helpful suggestions. I also thank Peter Barker, Sonja Brentjes, Marion Dapsens, Regula Forster and Bink Hallum for their help. Research for this article benefited from the support of the FNRS and the ERC project 'The Origin and Early Development of Philosophy in tenth-century al-Andalus: The impact of ill-defined materials and channels of transmission' (ERC 2016, AdG 740618, PI Godefroid de Callataÿ) at the University of Louvain (Université catholique de Louvain), from 2017 to 2022.

Bibliography

Sources

Abt, T. 2007. *The Book of Pictures. Muṣḥaf Aṣ-Ṣuwar by Zosimos of Panopolis. Facsimile with an Introduction.* Corpus Alchemicum Arabicum, II.1. Zurich: Living Human Heritage Publications.

Abt, T. and Salwa, F. 2011. *The Book of Pictures. Muṣḥaf Aṣ-Ṣuwar by Zosimos of Panopolis. Introduction and Translation.* Corpus Alchemicum Arabicum, II.2. Zurich: Living Human Heritage Publications.

al-Khwārazmī, Muḥammad ibn Aḥmad al-Kātib. ed. Vloten, G. van. 1968 [1895]. *Liber Mafātīḥ al-ʿulūm: explicans vocabula technica scientiarum tam Arabum quam peregrinorum* [Book of the Keys of the Sciences: It Explains the Technical Vocabulary of the Sciences both of the Arabs and the Strangers]. Leiden: Brill.

al-Rāzī, Abū Bakr ibn Zakariyyā'. ed. Dānish-Pazhūh, M. T. 1964. *Kitāb al-asrār wa-sirr al-asrār* [The Book of the Secrets and the Secret of the Secrets]. Tehran: Commission Nationale Iranienne pour l'UNESCO.

Badawī, ʿA. 1955. *Al-Aflāṭūniyya al-muḥdatha ʿinda al-ʿarab/Neoplatonici apud Arabes.* Cairo: Maktabat al-Nahḍa al-miṣriyya.

Berthelot, M., Houdas, O. V. and Duval, R. 1893. *Histoire des sciences. La chimie au Moyen Âge.* 3 vols. Paris: Imprimerie Nationale.

Ibn Sīnā. ed. Ateş, A. 1953. "*Risālat al-iksīr* [Epistle of the Elixir]." *Türkiyat Mecmuası* 10: 27–54.

Maqbūl, A. 1929. "A Persian Translation of the Eleventh Century Arabic Alchemical Treatise *ʿAin aṣ-Ṣanʿah wa ʿAun aṣ-Ṣanʿah.*" *Memoirs of the Asiatic Society of Bengal* 8.7: 419–60.

Moureau, S. 2016a. *Le* De anima *alchimique du pseudo-Avicenne.* Vol. 1 (Étude. Micrologus' Library, 76, Alchemica Latina, 1). Firenze: Sismel – Edizioni del Galluzzo.

Moureau, S. 2016b. *Le* De anima *alchimique du pseudo-Avicenne. Volume 2. Édition critique et traduction annotée* (Micrologus' Library, 76, Alchemica Latina, 1). Firenze: Sismel – Edizioni del Galluzzo.

Zosimos of Panopolis. ed. and tr. Mertens, M. 1995. *Les Alchimistes grecs. T. IV, 1. Zosime de Panopolis. Mémoires authentiques* (Collection des universités de France). Paris: Les Belles Lettres.

Manuscripts

MS Bethesda, National Library of Medicine, A65.

MS Istanbul, Arkeoloji Müzeleri Kütüphanesi 1574.

MS Rabat, Khizāna Ḥasaniyya, 1393.

MS Rampur, Raza Library, Kīmiyā' 12.

MS San Lorenzo de El Escorial, árabe 700.

Research literature

À l'ombre d'Avicenne 1996: À l'ombre d'Avicenne. La médecine au temps des califes. Exposition présentée du 18 novembre 1996 au 2 mars 1997. Gand, Paris: Snoeck-Ducaju and Zoon, Institut du monde arabe.

Cressier, P. and Canto García, A., eds. 2008. *Minas y metalurgia en al-Andalus y Magreb occidental. Explotación y poblamiento.* Collection de la Casa de Velázquez 102. Madrid: Casa de Velázquez.

Forbes, R. J. 1970. *A Short History of the Art of Distillation: From the Beginnings up to the Death of Cellier Blumenthal.* Leiden: Brill.

Gorzalczany, A. 2010. "A Possible Alchemist Apparatus from the Early Islamic Period Excavated at Ramla, Israel," *Antiguo Oriente: Cuadernos del Centro de Estudios de Historia del Antiguo Oriente* 8: 161–82.

Hallum, B. 2009. "The *Tome of Images*: An Arabic Compilation of Texts by Zosimos of Panopolis and a Source of the *Turba Philosophorum*," *Ambix* 56.1: 76–88.

Kurzmann, P. 2009. "Einige Glasgeräte der arabischen Alchemie," *Sudhoffs Archiv* 93.2: 184–200.

Levey, M. 1955. "Some Chemical Apparatus of Ancient Mesopotamia," *Journal of Chemical Education* 32.4: 180.

Moureau, S. and Thomas, N. 2015. "L'aludel : savoir et savoir-faire transmis du monde arabe à l'Occident médiéval?" in Gayraud, R.-P., Poisson, J.-M. and Richarté, C., eds. *Héritages arabo-islamiques dans l'Europe méditerranéenne.* Paris: La Découverte, 239–52.

Moureau, S. and Thomas, N. 2016. "Understanding Texts with the Help of Experimentation: The Example of Cupellation in Arabic Scientific Literature," *Ambix* 63.2: 98–117.

Ruska, J. 1923. "Chemische Apparatur bei den Arabern und Persern und im Abendland am Ausgang des Mittelalters," *Chemische Apparatur* 10: 137–9.

Ruska, J. 1937. "Al-Rāzī's Buch *Geheimnis der Geheimnisse*, mit Einleitung und Erläuterungen in deutscher Übersetzung," *Quellen und Studien zur Geschichte der Naturwissenschaften und der Medizin* 6: 1–246.

Savage-Smith, E. 1997. "Glass Alchemical Equipment," in Maddison, F. and Savage-Smith, E., eds. *Science, Tools and Magic.* The Nasser D. Khalili Collection of Islamic Art 12. 2 vols. London: Nour Foundation in Association with Azimuth Editions and Oxford, 1: 48–58.

Shaddoud, I. 2017. "Vaisselier de santé dans le monde arabe (VIIIᵉ–XVᵉ siècles): une restitution possible des usages grâce au croisement des sources," in Bocharov, S., François, V. and Sitdikov, A., eds. *Glazed Pottery of the Mediterranean and the Black Sea Region, 10th–18th Centuries* (Archaeological Records of Eastern Europe). 2 vols. Kazan: Stratum Publishing House, 2:189–205.

Stapleton, H. E. and Azo, R. F. 1905. "Alchemical Equipment in the Eleventh Century A.D.," *Memoirs of the Asiatic Society of Bengal* 1.4: 47–70.

Stapleton, H. E., Azo, R. F. and Hidāyat Ḥusain, M.1927. "Chemistry in 'Irāq and Persia in the Tenth Century A. D.," *Memoirs of the Asiatic Society of Bengal* 8.6: 317–418.

Thomas, N. 2009. "L'alambic dans la cuisine ?" in Ravoire, F. and Dietrich, A., eds. *La cuisine et la table dans la France de la fin du Moyen Âge. Contenus et contenants du XIVᵉ au XVIᵉ siècle. Actes du colloque de Sens (2004)* (Publications du Centre de Recherches Archéologiques et Historiques Médiévales [CRAHM]). Turnhout: Brepols, 35–50.

Thomas, N. and Claude, C. 2011. "Les vases à fond percé: Pratique de la distillation *per descensum* au bas Moyen Âge en Île-de-France," *Revue archéologique d'Île-de-France* 4: 267–88.

Wiedemann, E. 1909. "Über chemische Apparate bei den Arabern," in Diergart, P., ed. *Beiträge aus der Geschichte der Chemie.* Leipzig: Deuticke, 234–52.

IV.6
WATER AND TECHNOLOGY IN THE ISLAMICATE WORLD[1]

Charlotte Schriwer

IV.6.1 Introduction

Attempts to efficiently harness water for irrigation and energy have been ongoing for thousands of years in the Middle East; waterwheels are depicted on ancient Sumerian stone carvings of Mesopotamia, and the technological improvements of irrigation systems by the Nabateans and later the Romans in ancient Jordan, for example, are well known. These systems were readily adopted and adapted by early Islamic craftsmen and further developed over the centuries, with traditional water systems, such *qanāt* and *falaj* (subterranean channel systems)*, norias* (vertical water wheels), and water-driven mills (al-Hassan and Hill 1986[CB],[2] 80–1), forming ingenious and complex systems of irrigation and agricultural production. These technologies aided in transforming arid areas into regions that sustained cultivated crops, human settlements and rural economies (Glick 2005, 264). This chapter attempts to provide an overview of the principal types of water systems of Syria, Jordan, Iran and Oman through a broad overview of the environmental, technical, historical and sociopolitical aspects around their development and existence, from the 6th–12th/11th–17th centuries.

IV.6.2 Mill, *noria* and *qanāt* systems: technical and environmental considerations

The Middle East has a long history of settlement and land use, centered on the availability of water. West-central Syria and the areas around Homs and Hama developed along the banks of the Orontes River, while further south, the Hawran, the area of southern Syria and the desert steppe of northern Jordan, lacks such perennial rivers but is instead served by springs, aquifers and temporary water sources (Braemer *et al.* 2009). The fertile highlands of north Jordan are watered by perennial rivers and springs fed by seasonal precipitation. The construction of water systems was dependent on the geological nature of these landscapes, such as slope gradients, as well as the availability of building materials and the viability of agricultural production, in addition to the proximity and depth of water resources.

Water systems of West Central Syria include *qanāt, norias* and water mills fed by the Orontes River, with structures dating as far back as the Ayyubid and Mamluk periods still extant in the Hama and Homs regions (McPhillips 2016, 144–66). The Hawran has extensive networks

DOI: 10.4324/9781315170718-47

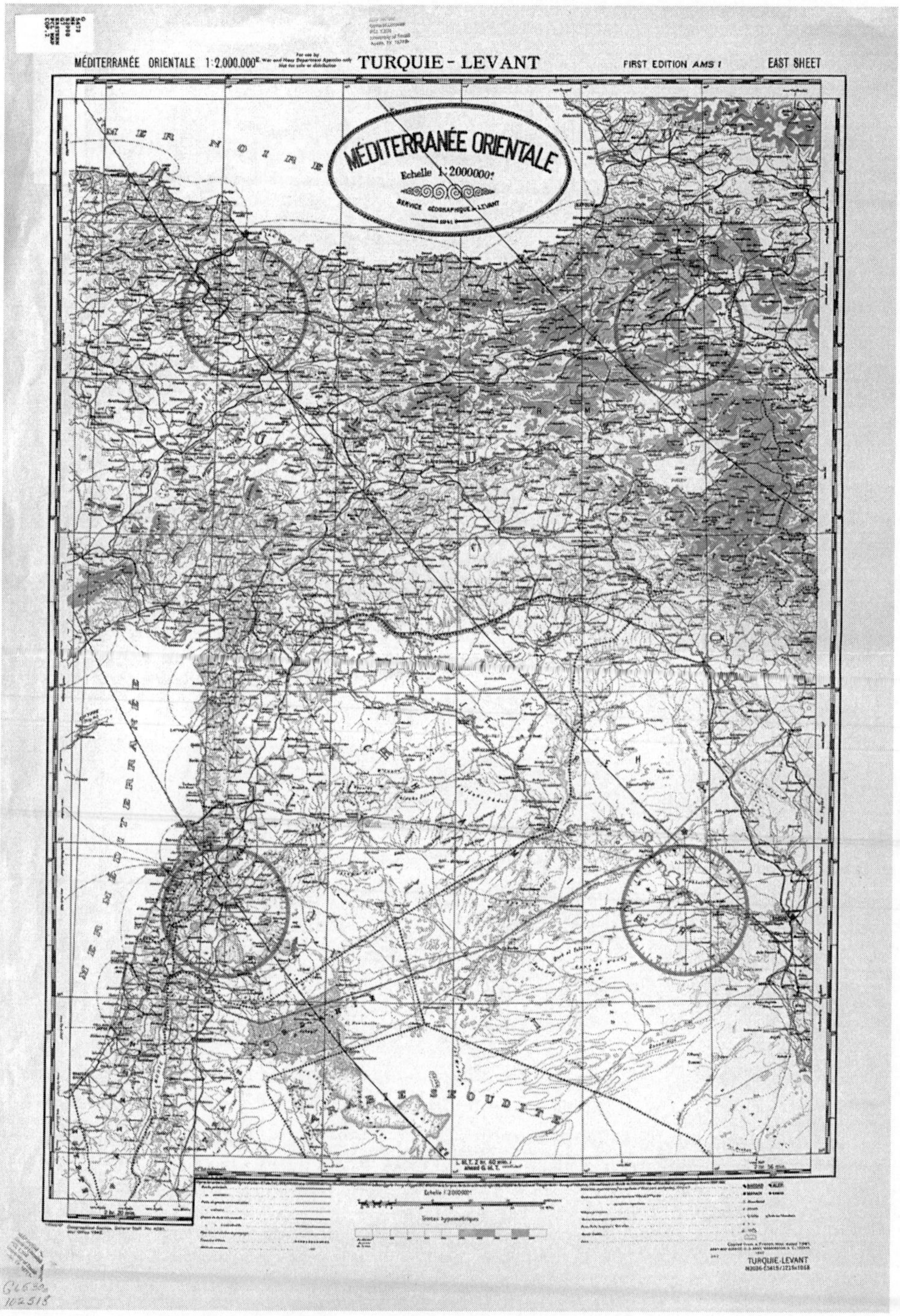

Figure IV.6.1 Map of the eastern Mediterranean, copied from a French map from 1941, War Office 1942; University of Texas, General Libraries, Ge6530s102518.

Source: © University of Texas, Austin

of ancient water channel systems, which capture rainwater and lead them into reservoirs or feed water mills (Braemer *et al.* 2009, 36–57), the majority of which date from the Ayyubid (r. 571–660/1171–1260), Mamluk (r. 750–917/1250–1517) and Ottoman periods (917–1322/1517–1922; Braemer *et al.* 2009). Extensive open and subterranean channel networks are also a common feature in the highland regions of Jordan, as well as the Jordan Valley, where they are often a part of a wider irrigation network that supplied water to power sugar and grain mills from the 8th/14th to 13th/19th centuries (Pringle 1997, 2000).

IV.6.3 Qanāt

Qanāt are ubiquitous across the Fertile Crescent, the northern and southern Arabian Peninsula and Iran. *Qanāt* were invented out of a necessity to find a system of water supply that could accommodate settlements in a vast arid climate where the water table was either too deep to access by other means or settlements were too remote from a water source (Bonine 1982, 2). Although the term predominantly refers to a subterranean aqueduct that moves groundwater along a sloping gradient (Lightfoot 1997, 433; Figure IV.6.1), they are also open, above-surface water channels (Figure IV.6.2). Geologically, the necessity of a gradient indicates that the aquifer or source of water originates at either a higher elevation, such as a mountainous region, or in lower ground where the water is carried over a great distance through a gradual slope (Lightfoot 1997). The depth of the *qanāt* can range from 10m to 90m, depending on its length, and vertical boreholes are dug at accessible intervals in order to facilitate maintenance and repair of the *qanāt*, which are excavated by professional *muqannī*, or *qanāt* excavators (Bonine 1982, 145–6).

The subterranean *qanāt* are most commonly found in Iran and Syria, while *qanāt* in Jordan appear to be mostly constructed above ground. *Qanāt* tend to follow the line of main or secondary roads and are often the main method of supply for crop irrigation. In some instances, where settlements are very small and scattered, the nearest river provides the water supply for crops, where it is diverted into a number of channels that irrigate nearby fields. Subterranean and surface *qanāt* are both known to also have supplied water to power mills (Harverson 1993, 149–77).

Figure IV.6.2 Surface *qanāt* in Hisban, Jordan. Note the dark line paralleling the road to its left.

Source: Photographer Charlotte Schriwer

IV.6.4 Grain and sugar mills

Vertical-wheeled and horizontal *arubah* (Hebrew *chimney*) penstock mills are the most prevalent type of water mill in Syria and Jordan (Hodge 1992; McQuitty 1995; Figures IV.6.3.a–c). The predominant use of horizontal mills was grinding or pressing either grain or sugar. As well as varying in wheel type, water mills also vary in size and capacity, ranging from structures with single apertures to those with multiple apertures that control the flow of the water. These apertures are called *penstocks*.

Figure IV.6.3a View of penstock and mill chamber from a water mill in Wadi Haydan, Jordan

Source: Photographer Charlotte Schriwer

Figure IV.6.3b Double-penstock grain mill in Rachaya, Lebanon

Source: Photographer Ali Kadri

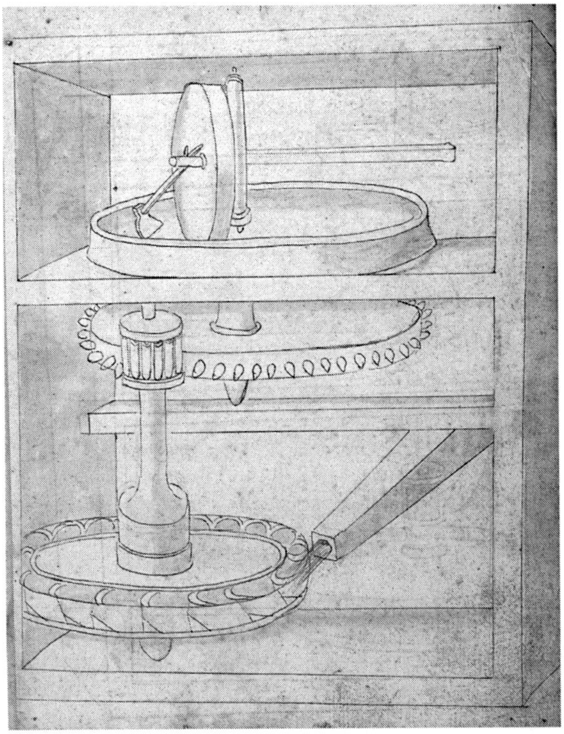

Figure IV.6.3c Mill with horizontal waterwheel. Drawing by Francesco di Giorgio Martini, from his *Trattato di Architettura*, between 1475 and 1480.

Source: © Getty's Open Content Program

In the horizontal mill, the penstock is connected to a water channel, or *millrace* (*qanāt*), which leads the water down the vertical shaft. At the bottom is an outflow through which the water can escape. This leads into the wheel chamber housing a horizontal wheel (*rāḥa* or *dawlab*). The wheel is attached by a spindle through a series of gears to the millstones (*ḥajar al-ṭāḥūn*) in a chamber above (Gardiner and McQuitty 1987). As the force of the water gains momentum in the penstock, it hits the wooden wheel with force, setting the wheel and milling gears into motion. The upper millstone, or *runner stone* (*hajar al-ṭaḥan al-fawqānī*), can be adjusted in relation to the lower millstone, or *bed stone* (*hajar al-ṭaḥan al-taḥtānī*), to control the degree of coarseness to which the flour is ground. An arched opening in the wheel room enables the water to leave the chamber as it turns the wheel; the water either returns to the river or empties into connected irrigation channels. Millraces rarely continue for any great length – usually no more than 25–30m – and are normally connected to a *qanāt* further along which follows the contours of the physical landscape.

Vertical waterwheel mills are most commonly operated through an undershot, breast-shot or overshot wheel, the difference being the angle at which the water hits the wheel's paddles. Fixed externally to the building, they are attached to the milling mechanism through a series of gears. Vertical mills are mostly associated with large rivers, such as the Orontes and Euphrates in Syria, but a unique example was found in Jordan, in the Wādī al-Lajjūn, which was used for grinding grain during World War I (Gardiner and McQuitty 1987).

Sugar mills were also operated by waterpower, using similar basic technical principles as the grain mill. However, the wheel is attached to a series of complex gears, which are joined to the *edge–runner stone*, which moves along the edge of the *bed stone* in a vertical position to crush the sugar. Some variations include two vertical millstones, activated in a similar manner to the horizontal millstones of the grain mill.

The majority of grain mills were built against steep terraced hills that made additional use of their position to catch precipitation and run-off. Sugar mills tend to be located at a lower altitude, closer to or even in the valley, where the warmer temperatures and humidity are suitable for sugar cane plantations. Sugar mills required a strong flow of water to power the larger grinding stones and hence are more frequently connected to long aqueducts.

IV.6.5 Norias

The *noria*, or water-lifting wheel, is predominantly found in Syria (Figure IV.6.4). *Norias* are located along the Orontes and Euphrates Rivers, and their tributaries in eastern Syria (Berthier 2001), and have been dated to as far back as the Roman period (de Miranda 2007, 79–93); however, the earliest extant examples date from the Ayyubid period. Constructed of giant wooden wheels with water-lifting compartments, they use the force of the river to gather its water, which pours from the compartments into open surface *qanāt*s that irrigate the surrounding lands through the use of aqueducts (de Miranda 2007, 28). Due to their heavy machinery, *norias* need a strong, predictable current of water for the wheels to rotate efficiently. In a number of places along the Orontes River, mill–*noria* complexes can be found through the use of dykes (Delpech *et al.* 1997). A *waqf* document of the early 10th/16th century describes *norias* and water mills as interdependent

Figure IV.6.4 Noria complex in Guadalquivir, Spain.

Source: © Graham Beards at English Wikipedia (Creative Commons Attribution 3.0 Unported)

in many cases (Sauvan 1975, 231–58). *Norias* were also installed to resolve irrigation problems. A study by French archaeologists conducted on Islamic settlements from the 1st/7th to the 9th/15th centuries around the Euphrates Valley in northeastern Syria (Berthier *et al.* 2001) showed that a lack of control of irrigation water channeled from the Khabūr River posed a problem for the settlements within the floodplain on the west bank of the Euphrates. The installation of a number of *norias* solved this problem (Berthier *et al.* 2001, 355), enabling the settlements to survive economically and continue their agricultural activities into the Mamluk period.

IV.6.6 Economies, politics and laws

Local and regional economic prosperity relied heavily on agricultural production and innovation during the Ayyubid, Mamluk and Ottoman periods. In the Ayyubid and Mamluk periods, settlements in the wider region of northern Jordan, with fortified outposts between Damascus and the estates of Montréal and al-Karak, as well as Belvoir (Kawkab) to the east, indicate steady occupation and consistent agricultural activity, by settled, semi-nomadic and nomadic communities (McQuitty 2005). This continued into the early Mamluk period, with areas for sugar cane production for trade located in the central and southern parts of Jordan and Palestine, with historically documented processing facilities located in Tawahīn al-Sukkar, Tell el-Tahūna and Kurdāna (Pringle 1997).

Urban centers such as Cairo and Damascus experienced profitable trade, particularly through the export of sugar. Various sultans took advantage of this and imposed monopolies on sugar, not only to stabilize and improve the economy but also to increase their own wealth (Sobernheim 1914, 75–85). Grain monopolies were also frequently imposed by the state, and these had a greater impact on merchants and millers, who were compelled to purchase grain from the warehouses of emirs and sultans at elevated prices (Lapidus 1969, 10). The grain economy was further sustained by keeping warehouses to store large provisions of grain in the empire's provinces, in particular at the citadels of Damascus, al-Karak and Shawbak (Ibn Mammātī 1943, 305). These warehouses, called *shuwān*, often had mills and bakeries attached to them to dispense bread and flour (Lapidus 1969, 6). The presence of water mills in the close proximity of these citadel towns furthermore suggests that there may have been an organized system of food storage and processing (Walker 2003), in which irrigation played an integral part.

Documents from Ottoman Palestine have brought to light the financial complexities concerning the use of irrigation systems and water mills. Hütteroth and Abdelfatteh provide a detailed study of the tax registers for Palestine at the end of the 10th/16th century (Hütteroth and Abdelfatteh 1977; Hütteroth 1978), where tax (*taḥrīr*) registers reveal that water mills were taxed at various rates. A water-sharing system was often employed in rural areas to ensure all farmers had access to water for irrigation purposes, enabling mills to produce a larger volume of flour, which could be sold on a local or a state level, depending on the energy capacity of the mill. In Ottoman Damascus, wheat millers also held a monopoly on milling commodities (Rafeq 1981, 657), resulting in the small farmers being pushed into poverty, as milling became an increasingly expensive endeavor (Rafeq 1981, 669).

Ownership of water systems had a direct impact on the survival of irrigation communities in Jordan and Syria. Sato (1997) demonstrated that mills were frequently owned by Mamluk rulers and that sugar mills were almost exclusively in the possession of a sultan due to sugar's lucrative nature and expense of production (Sobernheim 1914). This may further point to a state initiative of constructing water mills for specific economic purposes (Beazley and Harverson 1982, 83). Taxes were levied on the use of water mills, as well as for the annual repair of the irrigation canals and dams; in addition, tenants of perpetually irrigated lands paid annual rents at a fixed rate, on top of which local rulers levied a tax on sugar-cane plantations (Poliak 1939; Sato 1997; Frantz-Murphy 1986).

The system of state ownership became decentralized in Ottoman Jordan and Syria, although mill ownership by individual families was not common due to the cost of running and maintaining the irrigation systems and water rights (Given 2000; Rogan 1995). Mills were thus often either owned by a particularly wealthy sheikh, or by a consortium of shareholders, as suggested by Rogan for Ottoman Jordan (Rogan 1995).

Associations between mills and religious institutions in Jordan and Syria are more difficult to establish. Elisséeff frequently places the mills of Damascus near mosques and monasteries (Elisséef 1956, 77–9), such as the *Khānqāh al-Ṭaḥūn*, or the "Convent of the Mill", located outside the old city walls and attributed to the Zengid ruler Nūr al-Dīn (r. 540–569/1146–1174), who is said to have commissioned its construction in 560/1165–1166 (Elisséeff 1951, 24). In Jordan, water mills were endowments belonging to the Mamluk sultans of Egypt, which appears to have been a common practice there during the 7th/13th and 8th/14th centuries (Walker 2007). This tradition continued into the Ottoman period, where the establishment and maintenance of religious institutions frequently included irrigation systems. In Syria, for example, this can be seen in the *waqf* of Selīm II (r. 974–982/1566–1574), where the mills and *norias* were religious endowments left by the owner during his lifetime or upon his death (Sauvan 1975).

IV.6.7 Traditional irrigation methods in Iran and Oman

Historically, organized irrigation and water distribution systems were documented in Iran as early as the 4th/10th century, as an example from Marv, Khurasan shows. It states that "the superintendent of the irrigation systems based on the River Murghab had more authority than the prefect of Marv, and supervised 10,000 workers, each with a specific task" (al-Hassan and Hill 1986, 86). The geographer al-Muqaddasī (2nd half 4th/10th century) also refers to *norias* and mills in Khuzestan, built by the Buyid emir ʿAdūd al-Dawla (r. 337–372/949–983), which irrigated 300 villages (al-Muqaddasi 1994).

As was the case in Ayyubid and Mamluk Syria and Jordan, the expense of building *qanāt* meant that this activity was mostly confined to the ruling political and social elite, as suggested by Lambton (1992) in her investigations of *qanāt* in Yazd from the 8th/14th to the 13th/19th centuries. A thriving system of *qanāt* shareholding and trading existed. *Qanāt*, which were often a part of larger building complexes that included mosques, *khānqāh*s and *madrasas* (Lambton 1992, 29), were frequently established as *waqf*, or endowments, that were investments by the ruling elite (Lambton 1992). These have been documented for the Ilkhanid viziers Shams al-Dīn Juwaynī (d. 683/1284) and Rashīd al-Dīn Hamadānī (d. 718/ 1318), and the Timurid governor of Yazd, Amīr Chaqmaq ([d. *c.* 846/1446]; Lambton 1992, 26–8).

Lambton further states that the endower often had additional motives for following the tradition of naming *qanāt* as *waqf* to support charitable and religious foundations. These were to prevent the seizure of property, and to ensure that any revenues would remain in the possession of the endower's descendants (Lambton 1992, 33). Construction and maintenance of *qanāt* by the ruling elite continued in Yazd into the Safavid (r. 901–1134/1501–1722) and Qajar (r. 1189–1325/1789–1925) periods (Lambton 1992, 30). The scholar ʿAbd al-Wahhāb Taraz documented one of the later examples dating from 1241–1242/1841–1842 (Lambton 1992, 33–4).

Archaeological excavations and historical sources have both confirmed the association of *qanāt* with water mills. Integrating these as a source of energy was a useful system, as it helped to extend the use of the *qanāt* beyond its primary function as an irrigation mechanism (Figure IV.6.5). Water mills have been documented, amongst others, in the Yazd, Khurasan and

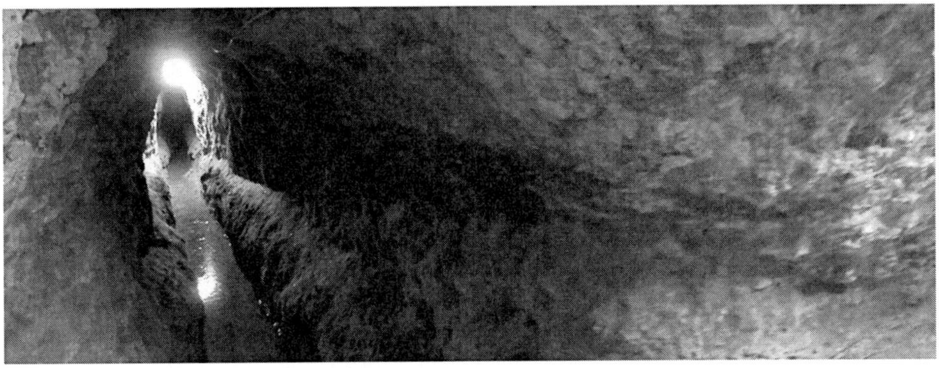

Figure IV.6.5 Example of subterranean *qanāt* in Iran

Source: © NAEINSUN – Own work, CC BY-SA 3.0, https://commons.wikimedia.org/w/index.php?curid=27037762

Isfahan provinces, but here, the mills are subterranean (Harverson 2000, 153–4; Bonine 1982, 148). Vertical-wheeled mills, usually of the breast-shot type, were utilized for grinding wheat. Mills located beside dams to make use of the constant availability of water also existed as integrated mill and water-raising devices (Harverson 2000, 8). Today, modern water pumping technology and food processing facilities have almost entirely replaced *qanāt* and mills, and the majority of these have fallen into disuse (Seyf 2006, 659–73).

In Oman, a *falaj* (pl. *aflāj*) is constructed following principles similar to a *qanāt*. This complex system of water channeling and distribution is used both for irrigation as well as household supply. In fact, as *aflāj* supply clean drinking water, the domestic use is the primary use of the *aflāj*, while irrigation is the secondary use (al-Ghafri 2004).

The *falaj* system is designed to be a perennial system supplying a continuous flow of water 24 hours a day all year long. The technological ingenuity of the *falaj* lies in the fact that it is engineered to effectively collect groundwater and transport it to the ground surface through the advantage of gravity. This is done first through the construction of a 'mother well' (*umm al-falaj*) which is dug up to 2m deep. Channels are constructed in parallel to the direction of the groundwater flow enabling the water to flow to the *falaj* outlet. Covered *falaj* are made up of a series of vertical wells, approximately 15m to 30m apart, with a diameter of 1.5m, accessed through a series of tunnels with diameters varying from half to one meter, through which the water can flow freely.

Depending on the quality and quantity of the flowing water and the geological nature of its path, a *falaj* can fall into three different categories: *ghaylī*, *dawūdī* and *'aynī* (UNEP 2001). The *ghaylī aflāj* are mostly open channels of a seasonal nature that have a water flow drawing on rainfall and shallow groundwater infiltration from *wādī* beds and mountain slopes (Mershen 2002). The *dawūdī aflāj* usually derive their water from regional aquifer systems providing them with a perennial flow of water through long tunnels, ranging from 3km to 12km, that are fed by extensive networks of tributaries, similar to the *qanāt* in Iran. Here also, to aid maintenance and ventilation, vertical shafts are dug at varying intervals (Costa 1983, 276). The *'aynī falaj* originates at the source of the spring (*'ayn*) and is channeled to its destination through an open water channel, or a combination of subterranean and above-ground *qanāt*. Approximately 3000 *aflāj* are still functioning in Oman today (UNESCO 1207).

The landscape of Oman is littered with complex *falaj* systems, both in urban as well as more remote rural areas. In desert palace-fort cities like the UNESCO World Heritage site of Bahlā'

(6th–9th/12th–15th centuries), in the interior of Oman, the water channels were engineered to supply every household with water from the fort's surrounding oasis springs. The *falaj* system also remained an integral part of solitary fort structures, such as al-Ḥazm Fort (1111/ 1711) in Rustāq, and formed part of a wider network of channels that fed surrounding villages. Despite the often challenging terrain of Oman's karst rock landscape, even remote interior oases were able to engineer a *falaj* system to feed villages that are otherwise cut off from the wider communication network. In Wādī Mayḥ near Muscat, for example, an 12th–13th/18th–19th-century farmstead was supplied by a system that was constructed for several kilometers along the *wādī* banks, and even in the villages of the Wādī Ṭīwī, located deep into the rocky hillsides now accessible only by four-wheel drive vehicles, clever construction and engineering has provided villagers with running water.

In areas such as Ṭīwī oasis, this ancient technology is essential as the landscape has made it difficult to install more modern forms of irrigation and water supply. This is also evident in the abandoned village of Tanūf, where a long and extensive network of still functioning *aflāj* have been built into the hillside on which the parts of the old village were constructed. Oman's higher mountain regions of Jebel Akhḍar also make use of this ingenious technological invention, where living villages, such as al-ʿAyn, and abandoned villages, such as Wādī Ibn Ḥammād, are still served by *aflāj* networks (Figure IV.6.6).

Written historical documentation regarding *aflāj* in Oman is rare, while oral histories account for most of the historical information available (Mershen 2002,107–11), in addition to archaeological investigations. A survey of the Wādī Dank region in 2006 associated a number of *aflāj*

Figure IV.6.6 A *falaj* in al-ʿAyn, Jebel Akhḍar mountains, Oman

Source: Photographer Charlotte Schriwer

with the medieval mining town of ʿArjāʾ, which also contained a number of water mills (Costa 2005), suggesting the irrigation and milling systems were an integral part of the region's copper industry well into the Abbasid period (150–658/750–1258; Costa 2005, 148). Based on interviews conducted with villagers in an archaeological study of Wādī Fanja and Wādī al-Khod and its *aflāj* systems, published in 2002, it was suggested that one of the *falaj*, Falaj al-Zuwayrī, was commissioned in the 12th/18th century by a member of the ruling elite belonging either to the Yaʿāriba dynasty (1033–1153/1624–1741) or the Āl Bū-Saʿīd dynasty (1158–present/1746–present), in order to develop a prominent agricultural estate (Mershen 2002, 110). This was met by considerable local opposition in Fanja village, as the addition of a private *falaj* would minimize the water supply available for existing plantations owned by the villagers (Mershen 2002).

Although not as commonly found as in Jordan or Syria, examples of water mills associated with *falaj* have been found in various locations, such as on the Bāṭina Coast in Oman (Costa 2005, 281–5; Wilkinson 1980), where they have been built either above or below ground.

IV.6.7 Technical and agricultural treatises

Irrigation and agricultural systems were documented in technical and agricultural treatises as early as the 4th/10th century. An important early work on water engineering, *The Search for Hidden Waters* (*Inbāṭ al-miyāh al-khafiyya*), written by the 5th/11th-century scholar of the mathematical sciences al-Karajī, contains detailed information and drawings on extracting groundwater (Rashed 2008). A 6th/12th-century document concerning irrigation, the *Book of Government Works and Records of the Accounts of the Treasury* (*Kitāb al-ḥāwī li-l-aʿmāl al-sulṭāniyya wa-rusūm al-ḥisāb ad-dīwāniyya*) is a work on the technical, economic and administrative aspects of irrigation in Buyid Iraq (Cahen 1951, 149–73). It is divided into three sections that describe irrigation machinery, developments in the field and potential problems during the planning and execution of irrigation works. One of the most famous treatises, by the craftsman and scholar al-Jazarī (d. c. 602/1206), the *Book of Knowledge of Ingenious Mechanical Devices* (*al-Jāmiʿ bayna l-ʿilm wa-l-ʿamal al-nāfiʿ fī ṣināʿat al-ḥiyal*; al-Hassan and Hill 1986), includes technical explanations and drawings of various hydraulic devices, such as water-raising machines (Figure IV.6.7). These were later elaborated by scholars such as the Ottoman polymath Taqī al-Dīn Muḥammad ibn Maʿrūf ([932–993/1526–1585]; de Miranda 2007, 84).

In his work *The Ordinances of Government Offices* (*Kitāb Qawanīn al-dawāwīn*), the Ayyubid scholar Ibn Mammātī (d. 609/1209) provides information on the government offices related to agriculture and irrigation (Ibn Mammātī 1942, 453). Among the duties of the administrator (*amīn*), he assisted the governor (*nāʾib*) with the inspection of waterworks. Ibn Mammātī described the cleaning of canals, confirming that there was a state program of maintaining waterworks. In his handbook *The Ultimate Ambition in the Art of Erudition* (*Nihāyat al-arab fī funūn al-adab*), the Mamluk scholar al-Nuwayrī (677–733/1279–1333) discussed regulations regarding irrigation, including detailed descriptions of the sugar manufacturing process, as well as the types of irrigation in place for maintaining sugar crops (al-Nuwayrī 1395–1410/1975–1990[CB]).

In addition to official documents, works regarding *ḥisba*, or market inspection, also contain information on the administration of irrigation systems. Two of the most renowned handbooks were those of Ibn Taymiyya (d. 728/1328) and al-Sunāmī (7th–8th/13th–14th centuries), both scholars from the Mamluk period. Al-Sunāmī explores questions regarding the use of water and mills, such as disputes between neighbors and the building of mills on one's own property, which were usually referred to the *muḥtasib* to be resolved (Izzie Dien 1997). Ibn Taymiyya describes the office of the *muḥtasib* as one that encompasses most public services, his duties being to "look

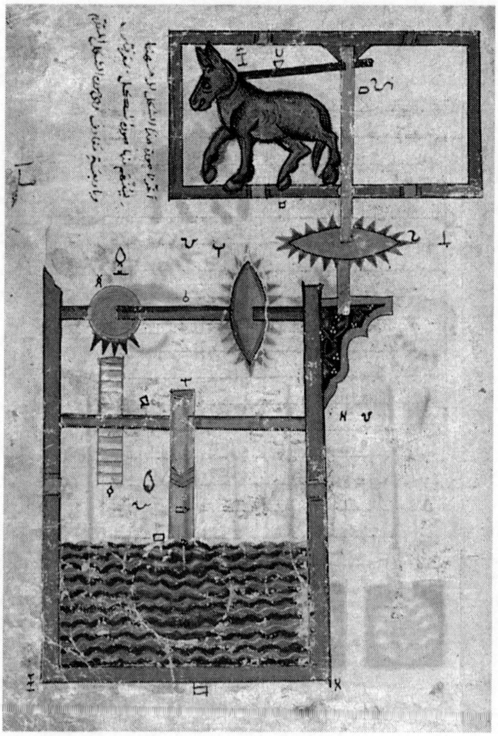

Figure IV.6.7 "Design on Each Side for Waterwheel Worked by Donkey Power," folio from *Book of the Knowledge of Ingenious Mechanical Devices* by al-Jazarī, dated 715/1315; acc. no. 55.121.11. Rogers Fund, 1955.

Source: © The Metropolitan Museum of Arts, New York

into the entire municipal administration such as street lighting, removal of garbage, architectural design of buildings, water supply and antipollution sanctions" (Ibn Taymiyya 1982, 141).

The office of *muḥtasib* continued into the Ottoman period, although the administration of irrigation systems was usually referred to in the *Mejelle*, or the Civil Law, or the Ottoman administrative law, known as the *Kānūn-nama* (Hammer 1963). The *Mejelle* contains detailed regulations for the administration of irrigation systems, landownership, water rights, maintenance of water channels and the building of mills.

IV.6.8 Conclusion

Historical and archaeological research has provided evidence of the economic, social and even political value of water, and the importance of efficient irrigation systems, since the early Islamic period. This is reflected in the numerous existing remains of such systems in the wider region today. Local craftsmen understood how to supply water for irrigation as well as how to use it as a source of energy. The links between watermills, archaeology, settlement and landscape enjoyed a continuously active economy and show that the rural and urban Islamicate world was productive and technologically enterprising. The ongoing importance of human-made water systems from the 4th/10th century through the recent past is a good example of the technological and scientific ingenuity of communities in the Islamicate world.

Notes

1 This chapter is a modified excerpt from a previously published work, Schrijwer, C. 2015. *Water and Technology in Levantine Society 1300–1900: An Historical, Archaeological and Architectural Analysis*. Archaeopress – British Archaeological Reports Publishing: Oxford, and reproduced with their kind permission.
2 Consolidated bibliography.

Bibliography

Sources

Ibn Mammātī. ed. ʿAtiyah, ʿA. S. 1943. *Kitāb qawānīn al-dawāwīn* [The Book of the Rules of Administration]. Cairo: Jamaʿiyyat al-zirāʿiyyat al-mālikiyya.
Ibn Taymīya. tr. Holland, M. and Ahmad, Kh. 1983. *Public Duties in Islam: The Institution of the Ḥisba* (Islamic economic series 3). Leicester: Islamic Foundation.
al-Muqaddasi. tr. Collins, B. A. 1994. *The Best Divisions for Knowledge of the Regions*. Reading: Centre for Muslim Contribution to Civilisation, Garnet.

Research literature

Beaumont, P., Blake, G. H. and Wagstaff, J. M. 1976. *The Middle East: A Geographical Study*. London: Wiley.
Beazley, E. and Harverson, M. 1982. *Living with the Desert. Working Buildings of the Iranian Plateau*. Warminster, UK: Aris and Phillips.
Berthier, S., ed. 2001. *Peuplement rural et aménagements hydroagricoles dans la moyenne vallée de l'Euphrate, fin VIIe-XIXe siècle: region de Deir ez-Zor-Abu Kemal (Syrie)*. Damascus: Institut Français de Damas.
Bienert, H.-D. and Häser, J., eds. 2004. *Men of Dikes and Canals. The Archaeology of Water in the Middle East. International Symposium held at Petra, Wadi Musa [H. K. of Jordan] 15–20 June, 1999* (Orient-Archäologie 13). Rahden, Westf.: Verlag Marie Leidorf.
Bonine, M. 1982. "From Qanat to Kort: Traditional Irrigation Terminology and Practices in Central Iran," *Iran* 20: 145–59.
Braemer, F. *et al.*: Braemer, F., Genequand, C, Dumond Maridat, C., Blanc, P.-M., Dentzer, J.-M., Gazagne, D. and Wech, P. 2009. "Long-Term Management of Water in the Central Levant: The Hawran Case (Syria)," *World Archaeology* 41.1: 36–57.
Cahen, C. 1951. "Service d'irrigation dans l'Iraq," *Bulletin d'Etudes Orientales* 13: 149–73.
Costa, P. M. 1983. "Notes on Traditional Hydraulics and Agriculture in Oman," *World Archaeology* 14.3: 273–95.
Costa, P. M. 2005. "Dank Archaeological Project: A Preliminary Report," *Proceedings of the Seminar for Arabian Studies* 36: 139–49.
Delpech, A. *et al.*: Delpech, A., Girard, F., Robine,G. And Roumi, M. 1997. *Les Norias de l'Oronte. Analyse technologique d'un element du patrimoine syrien*. Damascus: Institut français de Damas.
Elisséeff, N. 1951. "Les Monuments de Nur al-Din: Inventaire, Notes Archéologiques et Bibliographiques," *Bulletin d'études orientales* 13: 5–43.
Elisséeff, N. 1956. "Corporations de Damas sous Nur al-Din. Matériaux pour une topographie économique de Damas au XIIe siècle," *Arabica* 3: 61–79.
Frantz-Murphy, G. 1986. *The Agrarian Administration of Egypt from the Arabs to the Ottomans* (Supplément aux Annales Islamologiques 9). Cairo: Institut français d'archéologie orientale.
Gardiner, M. and McQuitty, A. 1987. "A Water Mill in the Wadi el-ʿArab: North Jordan and Water Mill Development," *Palestine Exploration Quarterly* 119.1: 24–32.
al-Ghafri, A. S. 2004. *Water Distribution Management of Aflaj Irrigation Systems of Oman*. PhD thesis. Sapporo: Hokkaido University.
Al Ghafri, A. S. 2008. "Traditional Water Distribution in Aflaj Irrigation Systems: Case Study of Oman," in Zafar, A., Schuster, B. and Bigas, H., eds. *What Makes Traditional Technologies Tick? A Review of Traditional Approaches for Water Management in Drylands*. Hamilton, Ontario: United Nations University, International Network on Water, Environment and Health, Desertification Series 8: 74–85
Given, M. 2000. "Agriculture, Settlement and Landscape in Ottoman Cyprus," *Levant* 32: 215–36.
Glick, T. 2005. *Islamic and Christian Spain in the Early Middle Ages*. Leiden: Brill.

Hammer, J. von. 1963. *Des Osmanischen Reichs Staatsverfassung und Staatsverwaltung*. Hildesheim: Olms.

Harverson, M. 1993. "Water Mills in Iran," *Iran* 31: 149–77.

Harverson, M. 2000. *Mills of the Muslim World*. London: Society for the Protection of Ancient Buildings, Mills Section.

Hodge, A. T. 1992. *Roman Aqueducts and Water Supply*. London: Duckworth.

Hütteroth, W.-D. . 1978. *Palästina und Transjordanien im 16. Jahrhundert: Wirtschaftsstruktur ländlicher Siedlungen nach osmanischen Steuerregistern*. Wiesbaden: Reichert.

Hütteroth, W.-D. and Abdelfatteh, K. 1977. *Historical Geography of Palestine, Transjordan and Southern Syria in the late 16th [sixteenth] Century*. Erlanger Geographisch Arbeiten (Sonderband 5) Erlangen: Fränkische Geographische Gesellschaft and Palm und Enke.

Izzi Dien, M. 1997. The Theory and the Practice of Market Law in Medieval Islam: A Study of Kitāb Niṣāb al-Iḥtisāb of ʿUmar b. Muḥammad al-Sunāmī (fl. 13th-14th century). Warminster, UK: E. J. W. Gibb Memorial Trust.

Lambton, A. K. S. 1992. "The Qanāts of Yazd," *Journal of the Royal Asiatic Society*, Third Series 2.1: 21–35.

Lapidus, I. 1969. "The Grain Economy of Mamluk Egypt," *Journal of the Economic and Social History of the Orient* 12: 1–15.

Lightfoot, D. 1997. "Qanats in the Levant: Hydraulic Technology at the Periphery of Early Empires," *Technology and Culture* 38.1: 432–51.

McPhillips, S. 2016. "Harnessing Hydraulic Power in Ottoman Syria: Water Mills and the Rural Economy of the Upper Orontes Valley," in McPhillips, S. and Wordsworth, P., eds. *Landscapes of the Islamic World: Archaeology, History, and Ethnograph*. Philadelphia: University of Pennsylvania Press, 143–66.

McQuitty, A. 1995. "Water-Mills in Jordan: Technology, Typology, Dating and Development," *Studies in the History and Archaeology of Jordan* 5: 745–53.

McQuitty, A. 2004. "Harnessing the Power of Water: Water Mills in Jordan," in Bienert, H. D. and Häser, J. eds., *Men of Dikes and Canals: The Archaeology of Water in the Middle East. International Symposium held at Petra, Wadi Musa [H. K. of Jordan] 15–20 June, 1999* (Orient-Archäologie 13). Rahden, Westf.: Verlag Marie Leidorf, 261–72.

McQuitty, A. 2005. "The Rural Landscape of Jordan in the Seventh-Nineteenth Centuries: The Kerak Plateau," *Antiquity* 79.304: 327–39.

Mershen, B. 2002. "'Let the Water Run Back.' Archaeological and Ethnohistorical Observations from Wadi Fanja/Wadi Khod, Oman: The Case of a Failed 18th/19th Century Agro-Economic Enterprise," *Proceedings of the Seminar of Arabian Studies* 32: 99–116.

Miranda, A. de. 2007. *Water Architecture in the Lands of Syria: The Water Wheel* (Studia Archaeologica 156). Rome: L'Erma di Bretschneider.

Poliak, A. N. 1939. *Feudalism in Egypt, Syria, Palestine, and the Lebanon, 1250–1900*. London: Royal Asiatic Society.

Pringle, D. 1997. *Secular Buildings in the Crusader Settlement of Jerusalem*, Cambridge: Cambridge University Press.

Pringle, D. 2000. *Fortification and Settlement in Crusader Palestine*. Aldershot: Ashgate.

Rafeq, A. K. 1981. "Economic Relations between Damascus and the Dependent Countryside, 1743–71," in Udovitch, A. L., ed. *The Islamic Middle East: Studies in Economic and Social History*. Princeton: Darwin Press, 653–87.

Rashed, R. 2008. "The Philosophy of Mathematics," in Rahman, S., Street, T. and Tahiri, H., eds. *The Unity of Science in the Arabic Tradition: Science, Logic, Epistemology and Their Interactions*. Dordrecht: Springer, 153–82.

Rogan, E. 1995. "Reconstructing Water Mills in Late Ottoman Transjordan," *Studies in the History and Archaeology of Jordan* 5: 753–7.

Sato, T. 1997. *State and Rural Society in Medieval Islam: Sultans, Muqta's, and Fallahun*. Leiden: Brill.

Sauvan, Y. 1975. "Une liste de fondations pieuses (*waqfiyya*) au temps de Selim II," *Bulletin d'études orientales* 28: 231–58.

Schrijwer, C. 2015. *Water and Technology in Levantine Society 1300–1900: An Historical, Archaeological and Architectural Analysis*. Archaeopress – British Archaeological Reports Publishing: Oxford.

Seyf, A. 2006. "On the Importance of Irrigation in Iranian Agriculture," *Middle Eastern Studies* 42.4: 659–73.

Sobernheim, M. 1914. "Das Zuckermonopol unter Sultan Barsbai," *Zeitschrift der Assyriologie* 27–8: 75–85.
UNEP. 2001. Sourcebook for Alternative Technologies for Freshwater Augmentation in West Asia, 2001 (Technical Publications Series 8. Osaka; Shiga, Japan: UNEP-IETC.
UNESCO. 1207. "*Aflaj* Irrigation Systems of Oman," https://whc.unesco.org/en/list/1207 (Accessed 10 June 2019).
Walker, B. J. 2003. "Mamluk Investment in the Southern Bilad al-Sham: The Case of Hisban," *Journal of Near Eastern Studies* 62.4: 241–61.
Walker, B. J. 2007. "Sowing the Seeds of 'Rural Decline'? Agriculture as an Economic Barometer for Late Mamluk Jordan," *Mamluk Studies Review* 11.1: 173–99.
Wilkinson, T. J. 1980. "Water Mills of the Batinah Coast of Oman," *Proceedings of the Seminar for Arabian Studies* 10: 127–32.

IV.7

ARTS AND SCIENCES IN THE ISLAMICATE WORLD

Anna Caiozzo

In this chapter, I discuss the relationship between the sciences and the arts in various regions of the Islamicate world from the 2nd/8th to the 13th/19th centuries. While I pay particular attention to illustrated or illuminated scientific texts, I also include other objects with scientifically relevant illustrations.

IV.7.1 The emergence of illuminated scientific manuscripts

Although not always reflected in written sources, the connection between the arts and sciences was very fruitful in many Islamicate societies. It dates back to early Umayyad art in Syria, when Caliph al-Walīd II (d. 125–126/743–744) ordered the building in Jordan of the small palace and bath of Quṣayr ʿAmrah (Vibert-Guigue and Ghazi-Bisheh 2007). Its caldarium (hot room) boasted a cupola painted with celestial bodies inspired by the representations of Greek constellations in the tradition of ancient celestial globes. In the art of the book, too, an important area in which the arts and the sciences interacted, the history of expert knowledge and its illustration stems in part from the introduction of Greek manuscripts to the Abbasid court in the 2nd/8th and 3rd/9th centuries.

The sciences in Islamicate societies fed on the ancient sciences reinvigorated by new discoveries, and their manuscripts provided a basis not only for translations but also for paintings (Gutas 1998[CB];[1] Collins 2000, 116–7). In many fields, such as geometry, arithmetic and philosophy, scholars worked mostly with diagrams or, in later periods, decorated tables (Burnett 2007). In others, however, including astronomy, astrology, pharmacy, botany, veterinary medicine and zoology, they worked with calligraphers or painters to produce beautiful images.

As far as we know, the insertion of imagery took place long after the translations of the ancient works themselves (Chapters I.2 and I.3): the images in question date back to the late Buyid emirate (r. 320–447/932–1055) or even to the Seljuq sultanate of the 6th/12th and 7th/13th centuries, when the visual arts began to flourish in Islamicate lands. It is untrue that the artistic representation of human forms was generally forbidden in Muslim, as in Jewish, circles, as is commonly claimed (Baer 1999, 2004, Contadini 2010, Vesel *et al.* 2009). Reluctance concerning that practice was widespread, particularly in Sunni circles, but it concerned mainly religious artwork, and it did not prevent the widespread use of the figurative arts in the sciences, poetry and history. Art historians have pointed out that among Shīʿītes and among the new conquerors

DOI: 10.4324/9781315170718-48

of the East, who were recently converted Turks, resistance to the representation of human beings was considerably weaker (Baer 1981). Additionally, in the 4th/10th and 5th/11th centuries, under the Samanids (r. 204–395/819–1005), Buyids, and Ghaznavids (r. 366–582/976–1186), the easternmost regions of the Abbasid Empire were home to a Persian revival. The participants in this so-called *shuʿubiyya* movement cultivated the Persian language and fueled nostalgia for Sasanian Persian rule, where figurative arts had held center stage (Chapter II.1 and II.9).

Additionally, images played an important didactic part in the Manichaean tradition, which was influential in Iraq and Central Asia from the 3rd century CE to the 4th/10th century (Gulácsi 2015). In Egypt, too, the artworks extant from the Fatimid caliphate (r. 297–567/909–1171), whose rulers were Ismaʿīlī Shīʿīs, bear the imprint of Coptic traditions as well as more ancient Egyptian art and emphasized figurative representations (Milstein and Brosch 1984, 23), sharply differing in this respect from works in other North African artistic traditions. The lack of local pictorial traditions in that region, combined with either rigorous or stern religious ideologies (Almoravids [*Murabiṭun*], Almohads [*Muwaḥiddun*]), meant that the western Islamicate world (*al-Maghrib al-aqṣā'* and al-Andalus) produced very few illuminated scientific manuscripts (*Sciences arabes* 2005, 169), with the exception of the images of surgical instruments in al-Zahrāwī's (1st half 5th/11th) *Book of the Medical Method* (*Kitāb al-Tasrīf*) (Chapter IV.4).[2]

In the 6th/12th and 7th/13th centuries, the so-called School of Baghdad began to produce illustrated scientific texts that addressed the treatment of horses; the structure of the cosmos and the earth; and the animal, plant, and mineral kingdoms (Contadini 2012; Vesel 2009). These Arabic texts, subsequently translated into Persian between the 7th/13th and the 11th/17th centuries, show the vitality of the sciences in Islamicate regions from Egypt to Central Asia from the 3rd/9th and 4th/10th centuries onward. They also evidence the spread of those parts of learning that were more accessible to the literati and upper-middle-class milieus or high secular culture, since such themes and their visualizations had already been included in poetry and prose literature (*adab*).

IV.7.2 Ancient models and Asian innovations

The iconography of extant scientific manuscripts has two main features. On the one hand, it draws on preexisting models, and on the other, it was profoundly innovative. Some of its models were inherited from antiquity or Byzantium. This applies in particular to illustrated manuscripts and metal artworks. The astrologer al-Ṣūfī (291–376/903–986), then in the service of the Buyid Emir ʿAḍud al-Dawla (r. 338–372/949–983), drew constellations illustrating his star tables (Chapter III.8). After correcting some data given in Ptolemy's *Almagest* and criticizing severely the works of his immediate predecessors, Abū Ḥanīfa al-Dīnawarī (d. after 282/895), al-Battānī (244–317/858–929) and ʿUṭārid (3rd–4th/9th–10th centuries), al-Ṣūfī added a set of images meant to serve a didactic purpose. While the black ink figures in certain extant copies, for instance the *Codex Vossianus* (MS Leiden, Leiden University Library, q69), a Byzantine copy from the 4th century, resemble those in some Greek manuscripts of the *Aratea*, al-Ṣūfī is thought to have followed a celestial globe rather than a manuscript (Savage-Smith 1992, 44–5). According to later manuscripts and colophons, he drew the constellations in two sets – one as seen on the sky and the other as seen on the globe. He indicated the names of larger-scale stars in red and marked some clusters of stars or lunar mansions with their Bedouin names, thereby initiating the pictorial tradition of the fixed stars in Islamicate societies.

This tradition exists in several variants, which differ with regard to both their astronomical content and their artistic form. One widespread version shows the figures corresponding to the constellations with eastern clothing – the men wear turbans and the women ankle bracelets, diadems, and dresses with many folds – and alternative names: the Giant Orion is called *al-ʿAwwā'*, "the Howler," Hercules is named *al-Jathī*, the Dancer, and so on (Caiozzo 2009). Other versions adapted their artistic conventions to the arts and tastes of the time and place. Occasionally astronomical attributes were reinterpreted, and a few constellations were borrowed from the Bedouin skies (Caiozzo 2020).

The finest copies seem to have been primarily made by scholars or for princes, who patronized astrologers (Savage-Smith 2013[CB]). For instance, a manuscript preserved in Doha, produced by a scholar in Baghdad in 519/1125 (MS Doha, IMA, 2.1998), has a closing colophon indicating that he wrote and illustrated it for his own use. Another example, MS Paris, BnF, Arabe 5036 (Figure IV.7.1), belonged to the Timurid prince Ulugh Beg (r. 812–850/1409–1447 as governor of Samarqand; 850–853/1447–1449 as head of the dynasty). The invention of a variety of astronomical instruments in princely observatories or by scholars engaged with timekeeping (Chapters IV.2–IV.3 and V.4), together with the longevity of this manuscript tradition, which lasted to the 13th/19th century, shows the durability of these interests. Another example is Taqī al-Dīn Muḥammad ibn Maʿrūf 's (d. 985/1585) *Observational Instruments for the Royal Astronomical Handbook* (*Ālāt al-raṣadiyya li-l-zīj al-shāhinshāhī*), created in Istanbul around 988/1580 (MS Paris, BnF, Suppl. Turc 1126).

A second group of illustrated manuscripts drawing on ancient sources directly imitated Byzantine models. Arabic manuscripts of Dioscorides' (d. 90) *On Medicinal Substances* (*De materia medica*) are a good example. This treatise on botany was translated more than once, including under the rule of Caliph al-Maʾmūn (r. 197–218/813–833) (Sadek 1983; Collins 2000, 117–47; Chapter I.4). The model of the oldest extant Arabic exemplars, MSS Leiden, University Library, or. 289, Samarqand, 475/1082–3, and Paris, BnF, Arabe 4947, Syria or Anatolia, 6th/12th

Figure IV.7.1 Astronomy: Andromeda, ʿAbd al-Raḥmān al-Ṣūfī (291–376/903–986), *Book of the Constellations* (*Kitāb ṣuwar al-kawākib al-thābita*); MS Paris, BnF, arabe 5036, fol. 89a. Probably made in Samarqand, before the middle of the 9th/15th century.

Source: © BnF, Paris

Figure IV.7.2 Pharmacy: Mint, Dioscorides (d. 90), *On Medicinal Substances*; MS Paris, BnF, arabe 4947, fol. 56b.

Source: © BnF, Paris

century, Figure IV.7.2, goes back to the Anicia Julia Codex (MS Vienna, ÖNB, Greek 1). Many of the artists who illustrated these manuscripts aimed to imitate the ancient iconography, an approach also evident in later examples (e.g., MS Istanbul, Topkapı Palace Library, Ahmet III, 2127, Syria (?), 625/1228). In contrast, other illustrators seem to have taken their inspiration from Coptic or Syriac copies of the Gospels (e.g., MS Oxford, Bodleian Library, Arab. d. 138, Baghdad, 637/1239–140). Despite variations, however, one distinctive feature of Arab painting, including in copies of Dioscorides' book, is the stylized representation of the vegetation.

The same can be said about treatises on veterinary medicine. These images were often adjusted to reflect textual elements, while using models from Byzantine treatises (Lazaris 2010; Weitzmann 1951). For example, two copies of the *Book of Veterinary Medicine* (*Kitāb al-bayṭara* by Ibn al-Aḥnaf [6th/12th century]) were produced by the calligrapher ʿAlī ibn al-Ḥasan ibn Hibatallāh in Baghdad: the first was made in 605/1208–1209, and the second a year later, in 606/1210 (MSS Cairo, National Library, Ṭibb Khalīl Aghā 8f; Istanbul, Topkapı Palace Library, Ahmet III, 2115). These copies may in turn have inspired a manuscript written between 622–648/1225–1250 for a broader readership: *The Book of the Characteristics of Animals* (*Kitāb Naʿt al-ḥayawān*) by Ibn Bukhtīshūʿ ([d. after 450/1058]; MS London, BL, Or. 2784; Contadini 2012; Figure IV.7.3).

Under the rule of the Ilkhanid Mongols (r. 654–736/1256–1336), after the fall of the Abbasids in 656/1258, illustrated bestiaries continued to be produced but in a somewhat different style, as miniaturists picked up motifs and techniques from Chinese paintings. A shift from Arabic to Persian as the dominant language of culture and, to a lesser degree, of the sciences encouraged the translation of scientific texts, including bestiaries, from Arabic into Persian (Chapter II.1). An example is the translation of *The Usefulness of Animals* (*Manāfiʿ al-ḥayawān*) ordered by Ghāzān Khān (r. 694–703/1295–1304; MS New York, Pierpont Morgan Library, M. 500, Maragha, *c.* 694/1295).

Figure IV.7.3 Bestiary: The Cow, Ibn Bukhtishūʿ (d. after 450/1058), *Book of the Usefulness of Animals* (*Manāfiʿ al-ḥayawān*); MS Paris, BnF, arabe 2782, fol. 6b.

Source: © BnF, Paris

A third approach arose with the invention of lay illustrations as paratext, as may be seen in the mutilated copy of Dioscorides' *On Medicinal Substances* from 621/1224 (MS Istanbul, Süleymaniye Library, Aya Sofya 3703). In this case, the painter, for unknown reasons, displayed the making of medicines and portrayed physicians exchanging various items with one another (Collins 2000, 131–5). One of the most interesting instances of an illuminated text can be found in a book that may be an adaptation from, or a commented version of, a work by the Greek physician Galen: Pseudo-Galen's *Book of Antidotes* (*Kitāb al-Diryāq*). Probably made in northern Iraq in 595/1199 (MS Paris, BnF, Arabe 2964), its images include not only illustrations of remedies and plants, but also small painted scenes illustrating the text's anecdotes about poisoning and healing. While the images of men of science may be inspired by Byzantine or Syriac models, those of plants and, above all, the double frontispiece dedicated to the Moon are clearly innovations (Kerner 2004). Images painted in Shiʿite milieus, like those in MS Vienna, ÖNB, Cod. A.F. 9, show similarities with illustrations in collections of *adab* intended for the literati, such as al-Ḥarīrī's (446–516/1054–1122) *Assemblies* (*Maqāmāt*). Illustrated in the early 7th/13th century, one of the manuscripts of the *Assemblies* shows a physician practicing in a bazaar and provides useful information on various types of contemporary medical practices (Paris, MS Paris, BNF, arabe 5847, fol. 154v; *Maqamat al-Hariri* 2003).

IV.7.3 New books to educate princes and literati

Other illustrated scientific books in Arabic reflect a novel iconography that originated in the Islamicate world. Examples include texts on astrology, magic, medicine, cosmography and cartography.[3] These books propagated renewed forms of knowledge and were often intended to enlighten an educated readership, especially the princes to whom they were dedicated and who may have commissioned them. They offered their readers different, often hybrid perspectives on the world.

From the early 2nd/9th century through the early modern period, world maps were made not only for rulers but also for scholars of the mathematical sciences and pious educated readers (Brentjes 2009[CB]). The mathematical tradition used texts and maps by Ptolemy (*c.* 100–170) and Marinus of Tyre (*c.* 70–130) as its models. The manuscripts for other, less specialized readers can be divided into several groups. The first comprises world and regional maps from the school of Abū Zayd al-Balkhī (d. 322/ 934), to which works by al-Iṣṭakhrī (d. after 340/951) and Ibn Ḥawqal (d. 378/988) belong (Chapter I.13). They often appear in lavishly illustrated manuscripts intended for princes. Al-Sharīf al-Idrīsī's (d. after 560/1164–5) *Book of Pleasant Journeys into Faraway Lands* (*Kitāb al-nuzhat al-mushtāq fī khtirāq al-āfāq*), created for the Norman King of Sicily, Roger II (r. 1130–1154), combines the mathematical tradition with the Balkhī school. In the circular maps found in some of its manuscripts, al-Idrīsī divided the world into seven climates. This group is exemplified by the copy BnF, Arabe 2221 (Figure IV.7.4), which was made in the Maghrib in the 7th/13th century (Vernay-Nouri 2000). The text itself is arranged in a more complex taxonomy partitioning each climate in several sections, using latitudinal and longitudinal degrees, when known (Miller 1926, 53-9). The information provided for each section was visualized in a rectangular sectorial map.

Further groups are formed by the world maps of Ṭūsī Salmānī (6th/12th century), al-Qazwīnī (d. 682/1283) and al-Bīrūnī (362–d. after 444/973–d. after 1053). They targeted the

Figure IV.7.4 Cartography: *World Map*, al-Idrīsī (d. after 560/1164–5), *The Book of Pleasant Journeys into Faraway Lands* (*Kitāb al-nuzhat al-mushtāq fī khtirāq al-āfāq*); MS Paris, BnF, arabe 2221, fols 3b–4a.

Source: © BnF, Paris

broad readership of literati (Karamustafa 1992, 71–89). The first two followed the conventions of the so-called Balkhī school more or less closely, using geometrical figures to represent seas, volcanoes, golfs, and other geographical units (Chapter I.13). Al-Bīrūnī's map visualizes an argument about the distribution of oceans and seas on the globe that rejects Ptolemy's view of a closed Indian ocean.

Copies of maps made up to the 13th/19th century reflect the complex workings of tradition and innovation. Ottoman copies appropriated early modern Italian and Dutch depictions of the world, combining them with symbols of Islamic cosmology. Safavid copies introduced local innovations apparently learned from Arabic maps or Persian works on cosmography (*ʿilm al-hayʾa*) and placed in Ethiopia information about the Americas gleaned from Ottoman sources, possibly confusing the stories about Prester John with those about the New World.

The Ottoman art of topographical illustration was advanced by the work of the navigator and corsair Pīrī Reʾīs (d. 961/1554). The surviving third of his famous map of 919/1513 (Istanbul, Topkapı Palace Library, R. 1633) shows the Americas and is believed to be based partly on maps used by Columbus. It bears illustrative vignettes reminiscent of the 'Catalan' nautical charts of the late 14th and 15th centuries. The principal work of Pīrī Reʾīs was *The Book of Maritime Matters* (*Kitāb-i baḥriyye*), an atlas of the Mediterranean Sea with town views inspired by Italian island books (*isolarii*). The copy presented to Sultan Süleymān ([r. 926-974/1520-1566]; MS Istanbul, Topkapı Palace Library, R. 642) is dated 932/1525–1526 (Soucek 2013).

Owing to the high production costs of lavishly illuminated copies of this sort, only wealthy or lettered patrons could afford them. This was also the case with other manuscripts on scientific topics, such as copies of Dioscorides and the Paris manuscript of the *Book of Antidotes* – a single, particularly sophisticated copy whose recipient, an imam, was clearly well versed in botany. By the same token, veterinary and hippology treatises were sent to emirs concerned with the well-being of their horses.

One work exemplifies particularly well the process of artistic creation and dedication: al-Jazarī's (d. *c.* 602/1206) *Book of Ingenious Devices* (*Kitāb fī maʿrifat al-ḥiyal al-handasiyya*). The author, a craftsman and scholar, wrote it for a Turkish emir of the Jazira (corresponding to today's northeastern Turkey and northern Iraq), Naṣīr al-Dīn Maḥmūd (616–631/1219–1233). The autographed copy from 602/1206 (MS Istanbul, Topkapı Palace Library, Ahmet III, 3472; Chapter III.8) shows al-Jazarī's personal involvement in designing the figures, as did, in his day, al-Ṣūfī's constellations. The book, partially written in the tradition of the Banū Mūsā (3rd/9th century) and partially based on the author's practical experience, came with colored drawings whose overt purpose was twofold. First, it was meant to elicit the prince's interest in the science of automata, made to entertain princely courts. Second, it served to facilitate the construction of these same automata thanks to accurate drawings, often including a wealth of details (for a different perspective see Chapter I.9). A third function, expressing the ideology of power, can be discerned through analysis of elements such as insignia and symbols, which reflected courtly protocol (Caiozzo 2009). Prized in Syria and Mesopotamia from the 7th/13th through the 10th/16th century, the *Book of Ingenious Devices* spawned numerous painted copies in the tradition of the initial model. Surviving copies speak of a special interest in this work on the part of Mamluk notables. Copies from 715/1315 and 755/1354 were made for a Mamluk emir who was one of the dignitaries of Sultan Salāḥ al-Dīn Ṣāliḥ (r. 752–755/1351–1354; Paris, Louvre, OA 7875; al-Jazarī 1974[CB]). They are scattered today among different collections.

Encyclopaedias like al-Qazwīnī's *Wonders of Creation* (*ʿAjāʾib al-makhlūqāt fī gharāʾib al-mawjūdāt*) are true compendia of information on the celestial and terrestrial worlds combining expert knowledge, *adab,* and legends. A precious early example is MS Munich, BSB, Codex arab. 464, called after its German owner Friedrich Sarre (1865–1945) the Sarre Qazwīnī. It was produced in 679/1280. Remaining popular until the 14th/19th century (Rührdanz 2009), the various extant versions of al-Qazwīnī's work in Arabic, Persian or Ottoman Turkish are frequently illuminated. They follow different iconographic programs, three of which will be briefly described.

The first program took its inspiration from the just mentioned copy of 679/1280. Large paintings covering over a half- or a full page are its main characteristic. This feature is still recognizable in copies produced in the Ottoman world (MS Munich, BSB, Codex arab. 463), Safavid Iran and Mughal India (MS Paris, BnF, Smith-Lesouëf [oriental] 221, 11th/17th century). A second iconographic program began in the Mongol period. It is more atypical, because its miniatures depict or comment on textual passages in short narrative sketches (MSS London, BL, Or. 12120, 8th/14th century; Istanbul, Süleymaniye Library, Yeni Camı 813, beginning 7th/14th century). A third program, from the 9th/15th century, is found in Timurid and Turkmen copies. They are characterized by small-size miniatures and an iconographic register that is occasionally richer and diversified than is the case in the other programs (MSS Paris, BnF, Suppl. persan 1781 and 2051, Iran, 9th/15th century).

A century before al-Qazwīnī, Ṭūsī Salmānī had already composed a highly original cosmography. It includes angelology, the zodiac, planets, talismans, and technical inventions. Its first extant illustrated version survives from the Jalayarid dynasty (r. 735–835/1335–1432; MS Paris, BnF, Suppl. persan 332, Baghdad, 790/1388; Richard 1996, 71). In the Ottoman period it was replicated with some variations (Moor 2010).

In general, in cases of this sort, the various iconographic programs circulated through exchanges of manuscripts among scholars, the travels of scholars and students inside Islamicate countries, and gift giving among the elites (Brentjes 2018[CB]). The presents were mostly produced in aristocratic or princely workshops (like the *rabʿ-i rashīdī* at Tabriz or later the *kitāb-khāna* of Bāysunghur Mīrzā at heart), while the scholarly books were often reproduced in urban workshops.

IV.7.4 Sciences and arts in the service of ruling elites

IV.7.4.1 Illuminated works on magic and astrology

Two further iconographic traditions are paired with two other scientific registers, astrology and magic. In the case of the latter, manuscripts are scarce, but three types of sources have been identified. First, beginning in the 8th/14th century, cosmographies in Persian included representations of talismans like those found in a Timurid album belonging to Prince Iskandar Sulṭān ([r. 805–818/1403–1415]; MS Istanbul, Topkapı Palace Library, B. 411). Second, handbooks of magic such as the *Book of the Properties of Stones* (*Kitāb khawaṣṣ al-aḥjār*) (MS Paris, BnF, arabe 2775; Egypt/Syria, 9th/15th or 10th/16th century) included sketches with broad strokes and were mainly destined for people learned in magic. Third, manuscripts made for princes were illustrated with planets, planetary demons and angels, as well as lunar mansions, diagrams, and magic alphabets. An example is the 7th/13th-century compendium on magic dedicated to the Seljuq Sultan Ghiyāth al-Dīn Kay Khōsraw III (r. 663–683/1265–1284) in Konya (Figure IV.7.5).

The author-cum-copyist of this last work produced illustrations in an innovative style marked by Central Asian motifs and stylistic influences from the Byzantine world. These motifs are also found in an Ottoman copy of the *Book of Invocations* (*Daʿwat nāma*; MS Istanbul, Istanbul University Library, TY 208), in the features of the angels of the zodiac or those of the lunar mansions with many arms and in the image of the mother of demons, depicted with many faces on her body.

Figure IV.7.5 Magic*: The Subtleties of Verities* (*Daqā'iq al-ḥaqā'iq*); MS Paris, BnF, persan 174, fol. 116b.

Source: © BnF, Paris

The iconography of astrological and magical artifacts was developed under the rule of Turkish emirs in Syria and Mesopotamia during the 6th/12th and 7th/13th centuries. It spread early on to Seljuq Anatolia, as the manuscript made for Sultan Ghiyāth al-Dīn Kay Khōsraw shows. It entered the field of book arts under the Mongol Jalayarids in the 8th/14th century. Starting in the 6th/12th century, metalwork artifacts from Central Asia (mostly Khurasan) also display astrological iconography. Examples include the Bobrinski caldron at the Hermitage Museum and the Vaso Vescovali at the British Museum (Ward 1993). Imagery of this sort spread quickly and appeared in Syria and Mesopotamia a few decades later. We find it more regularly in later painted manuscripts of the Persianate world in both Persian and Arabic copies, such as the well-known *Book of Surprise (Wonders)* (*Kitāb al-bulhān*; MS Oxford, Bodleian Library, Or. 133; Carboni 1997).

The transposition of astrological models to the book arts in the Mamluk and Jalayarid periods allowed the creation of a relatively homogeneous iconographic collection, as can be seen in a copy of *The Book of Nativities* (*Kitāb al-mawālīd*) by Abū Maʿshar (171–272/787–886; Carboni 1988a). It became the norm to present each planet in its house together with its zodiac sign.[4] This tradition continued during Ottoman rule, taking inspiration from the Jalayarid and Mamluk collections. Indeed, in this period, numerous manuscripts were brought from the eastern provinces of Anatolia, Syria and Iraq to Istanbul. As Carboni has demonstrated (1988b), Sultan Murād III (r. 982–1003/1574–1595) ordered two lavishly illustrated and illuminated manuscripts in Ottoman Turkish, called *The Ascension of Propitious Stars and the Sources of Sovereignty* (*Maṭāliʿ al-saʿāda wa-manābiʿ al-siyyāda*; also named *The Book of Felicity*), as birthday gifts for his daughters Ayshe and Fatma (Figure IV.7.6). *The Ascension* was translated from and iconographically inspired by the abovementioned *Book of Surprise (Wonders)*.

Astrological images were also inscribed on coins. This occurred in particular under the Zengids (r. 521–660/1127–1262), the Artuqids (r. *c.* 494–812/*c.* 1101–1409) and the Seljuqs of Rum (r. 473–641/1081–1243; Whelan 2006). Furthermore, astrological themes were featured on ceramics and earthenware from the Kubadabad palace, the summer residence of Sultan ʿAlāʾ al-Dīn Kay Kubād (r. 616–634/1219–1236), appearing on many small objects and items of tableware (e.g., caldrons, trays, mirrors, pencil boxes, ewers and basins) produced in the 6th/12th and 7th/13th centuries. These pieces exemplify what Eva Baer calls the ruler in a cosmic setting, with the sun and the six planets embodying the political and social order enforced by the sultans and emirs from the Middle East to Yemen, Anatolia, Central Asia and India (Baer 1981).

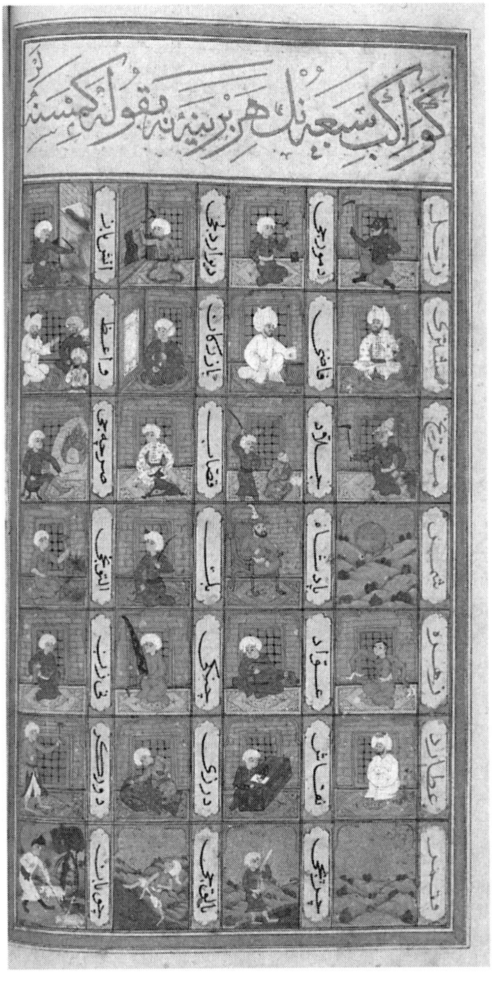

Figure IV.7.6 Astrology: Sayyid Muḥammad ibn Amīr Ḥasan al-Suʿūdī (d. 999/1591), *The Ascension of Propitious Stars and the Sources of Sovereignty* (*Maṭāliʿ al-saʿāda wa-manābiʿ al-siyyāda*; also named *The Book of Felicity*); MS Paris, BnF, Supplément turc 242, fol. 32a, made in Istanbul.

Source: © BnF, Paris

IV.7.4.2 Models and copies

Through the 7th/13th century, copyists were often also in charge of illuminating manuscripts with miniatures painted following the guidelines they found in the text, if there were any. Otherwise, they innovated. Miniatures were added after the texts had been copied so that in extant manuscript pages often display empty patches. But this changed at the turn of the century. Documents extant from the period, when the Timurids ruled from Herat (r. 771–906/1369–1500), shed light on the various modes of production of manuscripts and on the connections between the arts and the ruling elites. Under their rule a system emerged in which models circulated in albums among the princely courts of Shiraz, Herat and Samarqand. Painters would find references there not only for manuscripts of the same type but also for themes they could reproduce in other genres. The organization of the production of manuscripts in that period is known thanks to the famous *Report* (*Arẓadāsht*) by Jaʿfar Tabrīzī (d. *c.* 860/1455), the chief librarian of the *kitābkhāne* (workshop) supported by Bāysunghur Mīrzā (r. 800–837/1397–1433). In this workshop, artisans working in the book arts were divided according to skills: gilders, painters, calligraphers or specialists in particular scenes. This system was passed on to the Ottoman world.

Interest in scientific manuscripts depended on the degree of involvement of patrons with the sciences during their times. Prince Iskandar Sulṭān had a workshop in Shiraz that manufactured science and poetry anthologies as well as a remarkable horoscope (MS London, Wellcome, H. 414, Shiraz, 814/1411; Vesel 2009, 163). His interest in the celestial bodies reflected his will to govern with the consent of the heavens, in keeping with his belief in the magic power of letters and his commitment to the science of letter (*ʿilm al-ḥurūf*), also translated as lettrism (Richard 1997). The astral iconography of that period remained relatively stable through the Turkmen period of the late 8th/15th century in the manuscripts mentioned and in the folios of Timurid albums, as well as in cosmographies. Both the signs of the zodiac and the planets were typically drawn as small figures with simplified and codified but elegant shapes, each wearing its main attributes, as on Iskandar Sulṭān's horoscope. On this horoscope, Saturn appears as a black king, Mars is a warrior carrying a severed head and a sword, Venus plays the lute, Jupiter holds an astrolabe, Mercury is a scribe, and the faces of Sun and Moon are surrounded by glowing circles.

IV.7.5 Toward a hermeneutic of images in scientific manuscripts of the Islamicate world

Illustrated medieval texts served important functions for scholars and educated readers. The most beautiful manuscripts show the interest of princes in scientific matters and sometimes indicate their scholarly level, as in the case of the scientific manuscripts painted for the Timurid princes, which include astrological and astronomical material, as well as birth charts (Caiozzo 2011; Brentjes 2016). As Brentjes has indicated with respect to maps, illustrated manuscripts were used in many different ways:

> Motivations and purposes include the acquisition of knowledge through translation and adaptation, visualisation of knowledge, education, search for patronage, gift giving, propaganda, commerce, religious commitments, preparation or commemoration of military campaigns, and matters of administration.
>
> *(Brentjes 2016, 17)*

The first function of scientific images was undoubtedly didactic. They were meant to teach the appearance of curative plants or the anatomical features of men and horses (in anatomical and hippiatric texts), allowing physicians to treat men and animals, or they illustrated the appearance and use of surgical instruments. In some cases, as in al-Jazarī's treatise on automata, the depiction of individual parts of every automaton might have helped craftsmen to make these princely toys (for a different view, see Chapter I.9). The second function of scientific images was mnemonic. In particular, astronomical and geographical images helped readers to keep in mind the names and characteristics of the main stars or the location of the main places on a map. Some later anatomical images of human bodies showing bones, nerves and organs may have served a similar purpose. A third function was undoubtedly ideological. Astrological themes on metalwork objects indicated a symbolic distribution of the social levels and their functions according to the planets' patronage and rule all over the *ecumene* (Baer 1981; Caiozzo 2011).

A fourth function of scientific images was religious or spiritual, as in the case of maps that provided the *qibla* direction, cosmographical diagrams and maps for pilgrims, which showed sacred places and the sacred center of the world, that is, Mecca (King and Lorch 1992[CB]; Chapter VI.3). Because some images were linked to religious belief, reproducing them could be considered problematic. Others, such as the talismanic shapes depicted in magical works, were thought to have the power to bring about change in the human psyche or the physical world. The creation of images of this sort was sometimes preceded by the staging of a ritual to invoke spirits (angels or djinns) or other spiritual forces, in order to incorporate their powers into the magical artifact or talisman and thus render it effective. On such occasions, theurgical practices forbidden in Islam might be revealed to the reader (Caiozzo 2000).

Magical images were also considered effective agents of a kind of *apotropaia* (averting evil influences and bad luck), either according to the tradition of Solomonic magic – for instance, Solomon's seal (Milstein 1995) – or according to practices ascribed to the ancient magician Balīnūs (often identified with Apollonius of Tyana, d. *c.* 98). This was most likely the function of the zodiacal signs and planets on the coins of Turkish rulers in Anatolia and Mesopotamia (Whelan 2006) or on the medals of Mughal sultans in India in the 11th/17th century. Other examples include the metalwork artifacts decorated with astrological designs intended as propitiatory gifts and, above all, magical bowls with symbols such as a dog, a scorpion and a dragon, meant to neutralize poisons while Qurʾānic *suras* were read. In all these ways, scientific images were used to serve particular goals while remaining polysemic.

IV.7.6 Conclusion

The tradition of illustrated astronomical manuscripts was remarkably stable over a long period of time. Some of the books produced under the early modern dynasties of the Safavids, Ottomans, and Mughals were essentially copies of al-Ṣūfī's *Book on the Constellations* from the 4th/10th century. Their iconography remained stable across several centuries, save for stylistic modifications. While they remained fixed and unchanging models in a courtly context, they were less and less used by eastern astronomers due to the impact of new models brought by travelers from various European countries. The same can be said, to a certain extent, of the Arabic and Persian versions of al-Qazwīnī's cosmographies produced through the 13th/19th century, which are also marked by a lack of inventiveness.

However, the Safavid and Mughal courts also witnessed the emergence of new productions. In the realm of the occult sciences, iconographic innovations appeared in astrology, where manuscripts with the degrees of the zodiac according to the Babylonian astrologer Tankalūshā (1st century BCE) were produced (MSS Isfahan, Museum Reza-yi Abbasi, 590, 1074/1663–4; London, Wellcome, Persian 373, India, 11th/17th centuries; Vesel 2009, 166–169). Their illustrations differ from those that were created in the earlier Islamicate world. We know this, because the tradition of those older images continued in Christian Europe (e.g., MS Vatican City, BAV, Lat. Reg. 1283, 13th century). The context for the rise of the new pictorial set was the enthusiasm for soothsaying among Safavid princes, who commissioned and collected such illuminated works. Their iconography shows similarities to copies of *The History of the Prophets* (*Qiṣaṣ al-anbiyāʾ*) produced in the same period (Bağcı and Farhad 2009) and thus should be analyzed in connection with these religious books.

In the field of medical science, notable breakthroughs appeared. A distinction can be made between copies made for professional use and the luxuriously illuminated copies that were manufactured later. On the one hand, the copies made for physicians are mainly illustrated with diagrams, as in *The Book for al-Manṣūr on Medicine* (*al-Kitāb al-Manṣūrī fī l-ṭibb*) of al-Rāzī (d. c. 313/925 or 323/935) or the treatise on ophthalmology called *The Book of Ten Chapters on the Eye* (*Kitāb al-ʿashr maqālāt fī l-ʿayn*; MS Cairo, National Library, Ṭibb Taimur 100). On the other hand, more artistically made copies were painted for princely patrons beginning in the 9th/15th century, with anatomy plates as illustrations, as in the treatise *Anatomy of Manṣūr* (*Tashrīḥ i-Manṣūrī*), made by Ibn Ilyās (d. 825/1422) for Pīr Muḥammad (796–812/1394–1409), grandson of Tīmūr Lang ([771–807/1370–1405]; Tamerlane), the founder of the Timurid dynasty ([829–844/1425–1440]; MS Paris, BnF, Suppl. persan 1555; Newman 2009, 232–43).

Finally, the virtuosity of painters in Safavid Iran and Mughal India allowed the creation of superb albums of botany, astronomy, cosmography, and mechanics. While these have, to date, been studied mostly by art historians, they deserve the kind of attention historians of science have paid to lavishly illustrated books of the same kind in early modern Europe.

Notes

1 Consolidated bibliography.

2 The full title of this multi-volume work is: *The Arrangement for One Who is Not Able to Compile a Book for Himself* (*al-Taṣrīf li-man ʿajiza ʿan taʾlīf*).

3 The term cartography is used here to designate works that circulated under Arabic names that included *ṣūrat al-arḍ*, *rasm al-arḍ*, *ṣifat al-dunyā* and *ashkāl al-arḍ* (Brentjes 2016_.

4 Hellenistic astrologers had assigned one of the twelve signs of the zodiac as a house to the sun and moon, and two each to the other five planets.

Bibliography

Manuscripts

MS Leiden, University Library, Or. 289.
MS London, BL, Or. 14140.
MS Cairo, National Library, Ṭibb Khalīl Aghā 8f.
MS Cairo, National Library, Ṭibb Taimur 100.
MS Doha, Islamic Museum of Art, MI. 02.1998.80. (with the *Poem on the Stars* [*Urzūja fī kawākib*] of Ibn Ṣūfī).
MS Isfahan, Museum Reza-yi Abbasi, 590.
MS Istanbul, Istanbul University Library, TY 208.
MS Istanbul, Süleymaniye Library, Aya Sofya 3703.
MS Istanbul, Süleymaniye Library, Yeni Camı 813.
MS Istanbul, Topkapı Palace Library, Ahmet III, 2115.
MS Istanbul, Topkapı Palace Library, Ahmet III, 2127.
MS Istanbul, Topkapı Palace Library, Ahmet III, 3472.
MS Istanbul, Topkapı Palace Library, B. 411.
MS Istanbul, Topkapı Palace Library, R. 642.
MS Istanbul, Topkapı Palace Library, R. 1633.

MS London, BL, Or. 2784.
MS London, Wellcome Library for the History of Medicine, H. 414.
MS London, Wellcome Library for the History of Medicine, Persian 373.
MS Munich, BSB, Codex arab. 464.
MS New York, Pierpont Morgan Library, M. 500.
MS Oxford, Bodleian Library, Arab. d. 138.
MS Oxford, Bodleian Library, Bodl. Or. 133.
MS Paris, BnF, arabe 2221.
MS Paris, BnF, arabe 2775.
MS Paris, BnF, arabe 2782.
MS Paris, BnF, arabe 2964.
MS Paris, BnF, arabe 4947.
MS Paris, BnF, arabe 5036.
MS Paris, BnF, persan 174.
MS Paris, BnF, Smith-Lesouëf (oriental) 221.
MS Paris, BnF, Supplément persan 332.
MS Paris, BnF, Supplément persan 1555.
MS Paris, Louvre, OA 7875.
MS Vatican City, BAV, Reg. lat. 1283a.
MS Vienna, ÖNB, Cod. A.F. 10.
MS Vienna, ÖNB, Greek 1.
Paris, BnF, Supplément persan 1781 (a. f. 2374).
Paris, BnF, Supplément Turc 1126.

Research literature

Alaoui, B. 2005. *L'âge d'or des sciences arabes : exposition présentée à l'Institut du monde arabe, Paris, 25 octobre 2005–19 mars 2006.* Paris : Institut du monde arabe, Arles: Actes sud.

Baer, E. 1981. "The Ruler in Cosmic Setting: A Note on Medieval Islamic Iconography," in Daneshvari, A., ed. *Essays in Islamic Art and Architecture in Honor of Katharina Otto-Dorn.* 2 vols. Malibu, CA: Undena Publications, 1: 13–20.

Baer, E. 1999. "The Human Figure in Early Islamic Art: Some Preliminary Remarks," *Muqarnas* 16: 32–4.

Baer, E. 2004. *The Human Figure in Islamic Art: Inheritances and Islamic Transformations.* Costa Mesa: Mazda.

Bağcı, S. and Farhad, M., eds. 2009. *Falnama: The Book of Omens.* Arthur M. Sackler Gallery (Washington, DC). London: Thames & Hudson; Washington, DC: Freer Gallery of Art.

Brentjes, S. 2016. "Cartography," in *EI-3*, 2016–4: 15–29.

Burnett, C. 2007. "The 'Translation' of Diagrams and Illustrations from Arabic into Latin," in Contadini, A., ed. *Arab Painting: Text and Image in Illustrated Arabic Manuscripts.* Leiden and Boston: Brill.

Caiozzo, A. 2000. "Rituels théophaniques imagés et pratiques magiques : les anges planétaires dans le manuscrit persan 174 de Paris," *Studia Iranica* 29.1: 111–40.

Caiozzo, A. 2005. "The Horoscope of Iskandar Sultân as a Cosmological Vision in the Islamic World," in Oestmann *et al.*, eds. [CB], 115–44.

Caiozzo, A. 2009. "Iconography of the Constellations," in Vesel *et al.*, eds., 106–12.

Caiozzo, A. 2011. "Propagande dynastique et célébrations princières : mythes et images à la cour timouride," *Bulletin d'Études Orientales* 9: 177–201.

Caiozzo, A. 2013. "Rêves célestes : quelques notes relatives à la représentation des cieux du monde seldjoukide au monde ottoman," in Déroche, F., and Hitzel, F., eds. *Proceedings of the 14th International Congress of Turkish Art.* Paris: Collège de France, CETOBAC, 217–26.

Caiozzo, A. 2020. "Heavenly Camels: Some Survivals of Ancient Bedouin Beliefs and Legends in Oriental Illustrated Manuscripts," in Alexander, D., ed. *The Camels through the Ages.* 2 vols. Riyadh: Layan Cultural Foundation, 2: 67–92.

Carboni, S. 1987. "Two Fragments of a Jalayrid Astrological Treatise in the Keir Collection and in the Oriental Institute in Sarajevo," *Islamic Art* 2: 149–86.

Carboni, S. 1988a. *Il Kitāb al-Bulhān di Oxford.* Turin: Sirrenia Stampatori.

Carboni, S. 1988b. "Ricostruzione del ciclo pittorico del Kitâb al-bulhân di Oxford: le miniature delle copie ottomane mancanti nell'originale," *Annali di Ca' Foscari*, Serie orientali 27.3: 97–126.

Carboni, S. 1997. *Following the Stars, Images of the Zodiac in Islamic Art.* New York: Metropolitan Museum of Art.

Collins, M. 2000. *Medieval Herbals: The Illustrative Traditions*. London: British Library.

Contadini, A. 2007 [2010]. *Arab Painting: Text and Image in Illustrated Arabic Manuscripts*. Leiden: Brill.

Contadini, A. 2012. *A World of Beasts: A Thirteenth-century Illustrated Arabic Book on Animals (the "Kitāb Naʿt al-ḥayawān") in the Ibn Bakhtīshūʿ Tradition*. Leiden: Brill.

Gulácsi, Z. 2015. *Mani's Pictures: The Didactic Images of the Manichaeans from Sasanian Mesopotamia to Uygur Central Asia and Tang-Ming China*. Leiden: Brill.

Hoffman, E. 1992. "The Author Portrait in Thirteenth-Century Arabic Manuscripts: A New Islamic Context for a Late-Antique Tradition," *Muqarnas* 10.1: 6–20.

Karamustafa, A. 1992. "Cosmographical Diagrams," in *GTISAS*, 71–89.

Kerner, J. 2004. *Art in the Name of Science: Illustrated Manuscripts of the Kitāb al-diryāq*. PhD dissertation. New York: New York University.

Lazaris, S. 2010. *Art et science vétérinaire à Byzance : formes et fonctions de l'image hippiatrique*. Turnhout: Brepols.

Milstein, R. 1995. *King Solomon's Seal*. Jerusalem: Tower of David.

Milstein, R. 2005. *La Bible dans l'art islamique*. Paris: PUF.

Milstein, R. and Brosch, N. 1984. *Islamic Painting in the Israel Museum*. Jerusalem: Muzeon Yisrael.

Moor, B. 2010. *Popular Medicine, Divination, and Holy Geography: Sixteenth-Century Illustrations to Ṭūsī's ʿAjāʾib al-Makhlūqāt*. PhD dissertation. Jerusalem: The Hebrew University.

Newman, A. J. 2009. "Medicine: Anatomy" in Vesel, *et al.*, eds., 232–41.

Richard, F. 1997. *Splendeurs persanes : manuscrits du XIIe au XVIIe siècle: Exposition, Paris, Bibliothèque nationale de France, Galerie Mazarine, 27 novembre 1997–1er mars 1998*. Paris: BnF.

Rührdanz, K. 2009. "Cosmography," in Vesel *et al.*, eds., 177–81.

Sadek, M. M. 1983. *The Arabic Materia Medica of Dioscorides*. Quebec: Les Éditions du Sphinx.

Savage-Smith, E. 1992. "Celestial Mapping," in *GTISAS*, 12–70.

Soucek, P. P. 1992. "The Manuscripts of Iskandar Sultan: Structure and Content," in Golombek, L. and Subtelny, M., eds. *Timurid Art and Culture*. Leiden: Brill, 116–31.

Soucek, S. 2013. *Pîrî Reis & Turkish Mapmaking after Columbus*. Istanbul: Boyut Publishing.

Vernay-Nouri, A. 2000. *La géographie d'Idrīsī : un atlas du monde au XIIe siècle*. Paris: BnF.

Vesel, Ž. 2009. "Natural Sciences," in Vesel *et al.*, eds., 213–21.

Vesel, Ž. *et al.*: Vesel, Ž., Tourkin, S., Porter, Y., in collaboration with Richard, F. and Ghasemloo, F., eds. 2009. *Images of Islamic Science: Illustrated Manuscripts from the Iranian World*. Vol. 1. Teheran: Unesco, IFRI and Islamic Azad University.

Vibert-Guigue, C. and Bisheh, G. 2007. *Les peintures de Qusayr ʿAmra. Un bain omeyyade dans la bâdiya jordanienne* (Department of Antiquities of Jordan) (Bibliothèque archéologique et historique 179). Beyrouth: Institut français du Proche-Orient.

Ward, R. 1993. *Islamic Metalwork*. London: British Museum Press.

Weitzmann, K. 1951. *Greek Mythology in Byzantine Art*. Princeton: Princeton University Press.

Wellesz, E. 1965. *An Islamic Book of Constellations, The ṣuwar al-kawākib al-thābita by ʿAbd al-Rahmān al-Ṣūfī*. Oxford: Bodleian Library.

Whelan, E. J. 2006. *The Public Figure: Political Iconography in Medieval Mesopotamia*. London: Melisende.

PART V

Centers, regions, empires and the outskirts (3rd–13th/9th–19th centuries)

V.1
MATHEMATICAL KNOWLEDGE FIELDS IN THE ISLAMICATE WORLD

Similarities and differences

Ahmed Djebbar

From the end of the 2nd/8th century to the beginning of the 13th/19th century, mathematical texts produced in the Islamicate world were written, for the most part, in Arabic. But a significant number were composed in other languages, most importantly Persian, but also Turkish, Hebrew and to a lesser extent Berber and other African, Asian and European languages. In this chapter, I limit myself to works that appeared in Arabic, which have already been analyzed. These works may be divided into several categories. First are the results of research that extends Greek, Persian or Indian works translated between the end of the 2nd/8th and the middle of the 4th/10th centuries. Second are books for teaching, as well as textbooks targeting one or another category of practitioners (e.g., accountants, bailiffs, merchants, decorators, advisors on inheritance, surveyors, land dividers). The authors of these works belong to the many and varied ethnic, religious and cultural communities that existed in the Islamicate world.

There is only a limited number of comparative studies that were undertaken during the last decades on the contents of this corpus. They do not yet allow to draw definitive conclusions on questions about the existence of specific local or regional practices concerning the treated themes, the mathematical language, the orientation of research or even the birth and the development of a special part or a discipline. As will become clear later, these studies confirm the global unity of the mathematical tradition in Arabic on account of numerous similarities, despite the diverse sources that stood at the beginning of the development of this tradition. Nonetheless, the very same sources reveal certain particularities in the mathematical practices that can be observed in different periods and regions in the Muslim world.

V.1.1 The pre-Islamic heritage: The basis for mathematical activities in the Islamicate world

Although expressed in different languages alongside Arabic, the mathematics produced, taught and published in the Islamicate world developed from a common heritage nourished by different traditions, beginning in the 2nd/8th century. It was the juxtaposition or synthesis of

DOI: 10.4324/9781315170718-50

different elements from this heritage that became the basis for mathematical practices in the Islamicate world. The easily identifiable parts of this heritage circulated through translations of texts in Sanskrit and especially Greek. But there is another part with a local origin (the Fertile Crescent, Egypt) or much farther way (China). Here the ancient sources and the paths of circulation are more difficult to determine. But these sources equally nourished mathematical activities and accompanied the development of what is commonly called "scholarly knowledge".

V.1.1.1 The inheritance from Asia and Greece

The first ancient inheritance was from India. It concerned the basic concepts of trigonometry (sines, versed sines) and arithmetic (zero and the nine numbers used in place-value decimal numeration). But there are also algorithms for addition, subtraction, multiplication and division of whole numbers, the arithmetic of ordinary or sexagesimal fractions and the extraction of square roots from whole numbers or fractions (Chapter I.7). To this already rich body of work, we could add procedures for solving problems. Although these are not all explicitly associated with the Indian tradition, given their appearance in Sanskrit works from before the advent of Islam, it is possible that they originated there. But it is also possible that they originated in China and that they passed through India, as was perhaps the case for the method of false position (used to solve an equation in one unknown) and for certain classes of problems (about birds and about remainders).

The diffusion in Arabic of the heritage of Sanskrit arithmetic, and more generally the heritage of Asia, was for the first time evidenced by al-Khwārazmī's (d. *c.* 235/850) manual titled *The Book on Indian Arithmetic* (*al-Kitāb fī l-ḥisāb al-hindī*; al-Ḫwārizmī 1997, 28–107). The circulation of knowledge from India and Asia would be a factor in the emergence of specific new arithmetic genres but would not modify the nature of the goals and procedures or their respective functions. Thus, Indian numerals would be used with two different notations, that of the Islamic East and that of the West (al-Andalus and the Maghrib). For the method of false position, only the name changed from one region to the other: it was called the *method of two errors* in the East (Ibn al-Akfānī 1998, 85) and the *method of two scale pans* in writings from the Maghrib (Ibn al-Bannā' 1969, 69–71).

As for the Greek contribution, it was far more important due to its breadth and the diversity of its contents. Beginning at the close of the 2nd/8th century, it became an unavoidable reference for the mathematicians and astronomers of the scientific centers that began to appear and develop in different regions of the Islamicate world.[1] As in the case of the Indian contributions, those of ancient Greece and especially the Hellenistic period were widely diffused in the Islamicate world and began to occupy an important place in the curriculum of future mathematicians throughout the region. As an example, the *Elements* of Euclid (3rd century BCE) were studied and commentaries were written on it, starting with two Arabic versions that had the biggest circulation, those of al-Ḥajjāj ibn Yūsuf ibn Maṭar (d. after 213/828) and Isḥāq ibn Ḥunayn (d. 298/910–1), revised by Thābit ibn Qurra (d. 288/901). They would become the origin of numerous studies, including commentaries, summaries, new editions, ancillary works and new demonstrations of certain propositions. But these contributions did not introduce all the specific elements to the practice of geometry in every region. We can say the same thing for the *Conics* of Apollonius (3rd century BCE) and the *Spherics* of Menelaus (*fl.* 100), both of which benefited from new editions in the Fertile Crescent, in Asia and in al-Andalus (Hogendijk 1991, 1996).

In the theory of numbers, the Islamic West does not seem to have known the *Arithmetic* of Diophantus (3rd or 4th century). But the work of Nichomachus (d. *c.* 120) was taught, and inspired commentaries and new work, like that of Ibn Sayyid (5th/11th century) and Ibn Ṭāhir (6th/12th century) in al-Andalus, and of Ibn Munʿim (d. 626/1228) and also Ibn al-Bannāʾ (654–721/1256–1321) in the Maghrib (Djebbar 2000, 57–70). The first three authors contributed to the expansion of the contents of the *Introduction to Arithmetic.* The approach of the fourth shows a regional peculiarity, introducing combinatorial considerations that are among the unique features of the Islamic West; we will treat them in the second part of this chapter.

The Greek heritage in trigonometry, based on the concept of 'the chord of the double angle', coexisted for a time with Indian concepts that have already been mentioned and that had been integrated into the works of al-Khwārazmī and some of his colleagues, such as Ḥabash al-Ḥāsib ([d. after 255/869]; Debarnot 1997, 163–98). Beginning in the middle of the 2nd/8th century, astronomers in Baghdad introduced new objects and tools. These were the constitutive elements of a novel subject that developed up to the beginning of the 5th/11th century. Starting from this moment known scholars, such as al-Bīrūnī (362–d. after 444/973–d. after 1053), as well as anonymous scholars, like the author of the *Compendium of Astronomical Rules* (*Jāmiʿ qawānīn ʿilm al-hayʾa*), began to publish works that differed from treatises on astronomy, even though they were intended for specialists in that area (Khairetdinova 1966, 449–64). Their contents were more and more oriented to present the concepts and tools of trigonometry, thus preparing the autonomous phase of this subject, which later became a discipline in its own right.

The known sources do not show any regional variations in the development of trigonometry and its fields of application. During the 3rd/9th and 4th/10th centuries, the circulation of contributions from mathematicians and astronomers at the center of the empire first allowed scientists from the rest of the Islamicate world to benefit from the pre-Islamic heritage through the juxtaposition and reformulation of Indian and Greek sources. Second, they benefited from the extension of the domain through the definition of new concepts, the development of new tools and the demonstration of new results. This opened the way for new contributions, on the same basis as before, but this time taking place in scientific centers at the 'periphery' of the Fertile Crescent, such as Cairo (Debarnot 1997, 164, 169, 180, 190–1, 194), Córdoba (Villuendas 1979) and Maragha (Naṣīr al-Dīn al-Ṭūsī 1891; Djebbar 2004, 432–33).

V.1.1.2 The local heritage

A third part of the pre-Islamic heritage is local in origin and not associated with written sources by the first Arabic bibliographies. First there is what the 4th/10th-century mathematician al-Uqlīdisī called *Arabic arithmetic* (al-Uqlīdisī 1985, 47) and which at other times reappears elsewhere in the Islamicate world under the name *open arithmetic* (Ibn al-Akfānī 1998, 84; Chapter I.7). At the level of operations basic to arithmetic, this means a field covering only multiplication, division and ratios. Separate works were devoted to this tradition of arithmetic, for example, the *Memoir on the Foundations of Arithmetic and Inheritance Calculations* (*al-Tadhkira bi-uṣūl al-ḥisāb wa-l-farāʾiḍ*) by Ibn al-Khiḍr ([d. 460/1067]; Ibn al-Khiḍr 2001). The contents of this tradition circulated over time and certain of its details were integrated into works introducing concepts and calculation procedures from India. This is particularly true for some of the manuals published in al-Andalus and the Maghrib. The *Fertilization of Thoughts by Way of Drawing Dust Numbers* (*Talqīḥ al-afkār bi-rusūm ḥurūf al-ghubār*) by Ibn al-Yāsamīn (d. 601/1204) belongs to this category. Its order of presenting arithmetical operations on whole numbers and fractions differs from the Indian tradition. The author treats multiplication, division and ratios first, before presenting addition and subtraction (Ibn al-Yāsamīn 1993, 103–5).

The same field of arithmetic included 'arithmetic using fingers', also called 'arithmetic using joints' or 'mental arithmetic' (Sa ʿīdān 1971, 48–56, 416–20). The first Arabic manuals that treated certain aspects of this tradition have not come down to us. It is probable that manuals with the expression 'combination and separation' in their title form part of this category. The oldest among them is attributed to al-Khwārazmī and different from his book on Indian calculation methods (Djebbar 2002, 216–20). Later authors would publish their own works using the same title. But, since none of these writings has come down to us, it is impossible to tell if their contents had evolved due to local needs or practices. The only writings of which copies still exist are those containing in their titles the words *ʿuqūd* (joints) or *yad* (hand). They explain the use of fingers in counting and the operations of mental arithmetic. Although their production was not limited to a single region of the Islamicate world, it does not seem that their circulation resulted in the introduction of elements specific to one or another local practice (Ḥājjī Khalīfa 1402/1982[CB], 1: 664–5; Ibn al-Maghribī 1992).

We also need to take notice of alphabetical counting methods, inherited from the practice of Greek astronomy, which used a non-positional counting system. In Islamicate astronomy, this is expressed using the Arabic alphabet: 9 letters for the whole numbers, 9 for the tens and 9 for the hundreds. But practitioners in the Islamic West modified this scheme, changing the value of certain letters. The numbers 60, 90, 300, 800 and 900, which were designated, respectively, by the letters س (*s*), ص (*ṣ*), ش (*sh*), ض (*ḍ*), and ظ (*ẓ*) in the Islamic East, were expressed by ص (*ṣ*), ض (*ḍ*), س (*s*), ظ (*ẓ*), and غ (*gh*) in the West. The same number system appears in the Coptic tradition, expressed in symbols that seem to derive directly from letters of the Greek alphabet (Sesiano 1989; these were also called 'register numerals' or 'Fez numerals'). In 'Byzantine arithmetic' the writing and use of these numbers are no different from Coptic numerals. This number system was easily adapted to the needs of accountants, by adding a diacritical mark under each of the 27 signs designating thousands, tens of thousands, and hundreds of thousands (Djebbar and Guergour 2013, 7–52).

In the vast field of arithmetical procedures used for the solution of practical problems, or as exercises, the sources that have come down to us reveal the existence of a certain number of algorithms, the origins of which we still do not know. They seem to have been invented and used by calculators before the appearance of algebra as a discipline. The use of some procedures was limited in time and space. This is the case, for example, with the method called *bāb*, which permits the solution of a particular type of donation problems (in the legal field of inheritance calculations) with the aid of a geometrical construction (Laabid 1990, 67–79). Others diffused widely, losing their regional uniqueness. An example is the method of inversion.[2]

We may add a method that could be called 'pre-algebraic' to this category of arithmetical procedures. It consists of a series of arithmetical operations similar to those constituting the algebraic algorithm for solving an equation of the second degree but described in the manner set forth by the scribe of the cuneiform tablet BM 13901. In view of the extant mathematical sources from the Islamicate East, it would seem that this procedure did not survive the growth of algebra. It was simply abandoned by the practitioners in this region in the Islamicate world. But this did not happen in al-Andalus, since it reappears, for the first time as far as we know, in a 4th/10th-century manual, *The Epistle on Measurement* (*al-Risāla fī l-taksīr*) by Ibn ʿAbdūn ([311–after 366/923–after 976]; Box V.1.1. with Figure V.1.1.; see also Thureau-Dangin 1938, 1; Ibn ʿAbdūn 2005[CB],[3] 29). It is also present in later writings from the same region (Busard 1968, 70).

Box V.1.1 Comparison between two statements of a mathematical problem some 2,700 years apart

Epistle *of Ibn ʿAbdūn (4th/10th century)*	*BM 13901 (1750 BCE)*
Statement: If someone says to you: we have added its sides and its surface. Their sum is one hundred forty. How much is each side?	**Statement:**[4] The surface and my confrontation I have heaped: 3/4 is it.
Solution: • You add the number of sides, and the result is four. • You take its half and the result is two. • You multiply it by itself, and that is four. • You add it to one hundred forty and that is one hundred forty-four. • You take the root and that is twelve. • You subtract from what remains half of four. • This is each of its sides.	**Solution:** • 1, the projection, • you posit. The moiety of 1 you break, 1/2 and 1/2 you make hold. • 1/4 to 3/4 you join: by 1, 1 is equal. 1/2 which you have made hold • from the inside of 1 you tear out: 1/2 the confrontation.

The first developments of algebra as a discipline appeared in the 3rd/9th century and the beginning of the 4th/10th century, in the works of al-Khwārazmī (Rashed 2007[CB]), Ibn Turk ([1st half 3rd/9th century]; Sayılı 1985) and then Abū Kāmil ([*c.* 235–*c.* 317/*c.* 850–*c.* 930]; Abū Kāmil 2012). Their definitions of new concepts, their procedures for finding solutions, the types of problems they solved, and their terminology became accepted practices in different regions of the Islamicate world. But later continuations did not have the same diffusion, and certain important contributions, such as those of al-Karajī (d. *c.* 420/1029), ʿUmar al-Khayyām (439–*c.* 517/1048–*c.* 1123), al-Samawʾal (d. 570/1175) and Sharaf al-Dīn al-Ṭūsī (d. *c.* 610/1213) do not seem to have reached other scientific centers of the Islamicate empires, like those of the Maghrib and al-Andalus. Taking into account all the writings that have come down to us, they have had no effect on the development of algebra in these regions. It is possible that, starting in the 3rd–4th/9th–10th centuries, some extensions in the contents of algebra similar to those that appeared in the East may have developed independently in the West. The historian Ibn Khaldūn (732–808/1332–1406) asserts that algebraic work in al-Andalus was important before the 8th/14th century (Ibn Khaldūn 2005[CB], 5: 231–2). But the surviving bio-bibliographical and mathematical sources do not support this hypothesis. From around the middle of the 4th/10th century, it appears that essentially, new approaches in algebra were developed by mathematicians who had worked in Baghdad or in the towns of Iran and Central Asia.

Practical geometry, as opposed to the academic Greek geometry already mentioned, forms the last chapter in the local heritage. It treated properties of plane figures and simple solids, procedures for determining one part of a figure when others are known, as well as techniques of measurement and division useful in surveying and dividing land. The essential elements of this discipline appear in manuals written in different regions of the Islamicate world. Similarities can be seen in formulas for finding sought magnitudes, in parts of the terminology used and in the

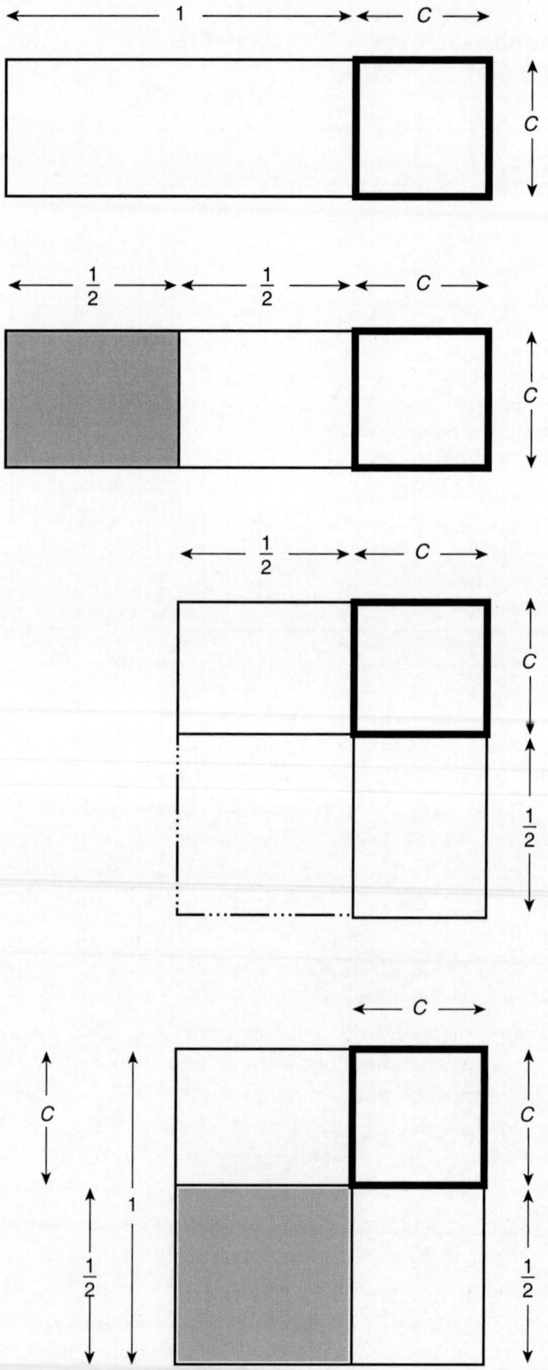

Figure V.1.1 Geometrical diagrams illustrating the cuneiform tablet BM 13901, provided by Høyrup, J. 2015. "Another Case of Stumbling Progress in the History of Algebra," *Physis*, New Series 1.1–2: 1–38.

Source: © Jens Høyrup, Copenhagen

procedures for solving certain problems. But as we will see in the second part of this chapter, specific regional contributions are also present in a certain number of manuals treating this material.

V.1.2 Specific contributions to certain episodes in mathematics from the 3rd/9th to the 9th/15th centuries

The specific contributions in certain domains of mathematics developed in the Islamicate world that have been recognized up to the present were generated or promoted by diverse motives. Some of them resulted from general cultural preoccupations in direct relation to two foundational factors: the Arabic language and the religion of Islam. Others were consequences of developments in mathematical practice responding to specific societal needs. These factors are common to all inhabitants of Islamic cities with the same cultural and religious profiles. But to these factors we must add a regional dimension, to the extent that certain centers of scientific research went further than others in developing new approaches. In what follows, I will present examples of some contributions that are characteristic of these regional traditions.

In the Islamic West, important developments in legal practices connected to inheritances seem to have made certain 6th/12th-century authors devote a longer chapter to fractions in their manuals. Examples are Ibn Munʿim's *Science of Arithmetic* (*Fiqh al-ḥisāb*; Ibn Munʿim 2005, 237–341) and al-Ḥaṣṣār's (alive in 552/1157) *Book of Demonstration and Recall on the Science of Dust Numbers* (*Kitāb al-bayān wa-l-tadhkār fī ʿilm ḥurūf al-ghubār*; MS Algiers, BN, 2712, fols 146b–162a). In the first book, fractions occupy about one third of the contents and, in the second, more than two-thirds. It is probably this process of increase in the types of fractions appearing in calculations that made it necessary to extend the arithmetic symbolism to other domains beyond counting techniques. It is not possible here to give a chronology showing when this notation was extended. But it already appears in the work by al-Ḥaṣṣār already mentioned. There we find the introduction of the bar separating numerator and denominator when writing *simple fractions* ($\frac{n}{m}$, $n < m$) and the use of other symbols for three other types: *connected fractions* $\left(\frac{n_1\, n_2}{m_1\, m_2}\left[=\frac{n_1}{m_1}+\frac{n_2}{m_2}\left(\frac{1}{m_1}\right)\right]\right)$, *differentiated fractions* $\left(\frac{n_2\, n_1}{m_2\, m_1}\left[=\frac{n_1}{m_1}+\frac{n_2}{m_2}\right]\right)$ *and partitioned fractions*

$$\left(\frac{n_2\, n_1}{m_2\, m_1}\right)$$

$$\left(\left[=\frac{n_1}{m_1}\left(\frac{n_2}{m_2}\right)\right]\right).$$

This notation was accompanied by a specific terminology for the two parts of a fraction (the numerator and denominator) and included several abbreviations explaining the arithmetical operations applied to fractions (addition, subtraction, multiplication, division and square root). We find this notation, with variations, in the book *The Fertilization of Thoughts* of al-Ḥaṣṣār's contemporary Ibn al-Yāsamīn (Ibn al-Yāsamīn 1993, 137) and among mathematicians of the 7th/13th and 8th/14th centuries. An exception is Ibn al-Bannāʾ, who takes the initiative by limiting the section on fractions to fundamental operations (Djebbar 1992).

We do not know the reasons that led - unnamed - mathematicians in al-Andalus to introduce the first algebraic symbols in chapters they devoted to this discipline. It is therefore risky to explain this new direction through cultural considerations specific to that part of the Islamicate world. We can only say that the first known examples of this 'innovation' equally date from the 6th/12th century and that they may have appeared in Seville before they diffused to the

Maghrib. The key element in this symbolism is the use of the letters ش (*sh*), م (*m*) and ك (*k*) to designate x, x^2 and x^3 and combinations of the second and third letters to express all powers higher than 3. This regional feature was extended to writing polynomials of any given degree. There were thus two ways of expressing these concepts: the Islamic East expressed them through tables where each power was associated with a column (al-Samawʾal 1979, 44–56), while in the Maghrib they were expressed by juxtaposing coefficients for individual numbers surmounted by the symbol expressing the corresponding power (Lamrabet 1981, 76–86).

Combinatorial methods are the third domain where regional differences appeared in mathematical practices. The first signs appear in the second half of the 2nd/8th century with the attempt by al-Khalīl ibn Aḥmad (d. 175/791) to list all the roots of Arabic words, represented as combinations of the 28 letters of the Arabic alphabet but without permutation or repetition of letters. After several later attempts to count all the roots, the problem was posed again in Marrakesh, at the end of the 6th/12th century, in a cultural context supportive of Arabic language studies. A complete solution was provided by Ibn Munʿim (Djebbar 1985, 18–48; Djebbar 2013, 82–107). He opened the way for the founding of a new field of mathematics, with its own definitions, its basic results (obtained by methods borrowed from number theory) and its own domain of application. In fact, this is the only area where we can see continuity in combinatorial works of mathematical practitioners leading to new results and new applications (Djebbar 1981, 41–54).

As a consequence, in the second half of the 7th/13th century, Ibn al-Bannāʾ studied the combinatorial section in Ibn Munʿim's *Science of Arithmetic* and went on to solve a new problem. He found the arithmetical expression for the combinations of n objects taken p at a time, and gave the formula expressing the result directly rather than using the "binomial triangle" (Ibn al-Bannāʾ 1988, 153–64). Then, in another work he went on to treat several problems requiring combinatorial solutions (Djebbar 2003, 39–42). Finally, beginning in the 8th/14th century, commentators on the *Epitome of the Operations of Arithmetic* (*Talkhīṣ ʿamal al-ḥisāb*) of Ibn al-Bannāʾ continued to return to this subject and to use its tools, in particular to count the multitude of fractions that they had to study (see, e.g., MS Algiers, BN, 2712, fols 72a–73a).

After a long period in which the two contributions mentioned last did not move beyond the frontiers of the Maghreb, these tools and results started to circulate, first in Cairo and then in Istanbul, beginning in the 8th/14th century. In his *Container of the Choicest Core* (*Ḥāwī l-lubāb*), Ibn al-Majdī (767–850/1366–1447) introduced one part of this symbolism. In the same work, he took up again a combinatorial application, solved in the Maghreb, and provided it with a previously unknown generalization. Then he presented a new problem about counting equations of degree n, and gave a solution for it (Djebbar 1981, 97–8, 107–12).

It is again in the Islamic West that we find two special developments in the domain of applied mathematics. The first, which is at the same time local and 'cultural,' concerns a question of law. It considers the different types of injury that an individual can suffer, establishing rates of compensation and calculating the indemnity that should be paid to a victim. Orthodox law schools had left the solution to these problems to the discretion of judges, to be decided on a case-by-case basis. As far as we know, the Ibadi school of law, which survived in the central Maghrib, in Ifrīqiya (modern Tunisia) and in Oman, was the only one to precisely determine the different elements of these problems. This led practitioners from this school to write manuals containing all the elements of law and mathematics required for the solution of problems of this type which might be presented to a judge. As an example we may cite Ismāʿīl al-Jīṭālī, a jurist of the 7th/13th and 8th/14th centuries who spent some time and was buried on the island of Djerba. In his book on the science of inheritance he devoted a chapter to the different kinds of injury, complemented by other purely mathematical chapters to teach lawyers how to solve problems in business and measurement (Djebbar 2018).

The second area where we find special developments is practical geometry, which inspired numerous authors from the 3rd/9th century onward. The rich output that has come down to us reveals some unique features at the level of the concepts studied, the methods used and the terminology. Works produced in al-Andalus, which have been studied in recent years, are a good illustration. They include the study of solid figures that are absent in manuals from the Muslim East, like the *fanīqa* (sandbag), the *ḥūt al-ṭaʿām* (a fish-shaped solid, also called *qubūrī*) and the *ʿurmat al-ṭaʿām* (heap of grain; Djebbar 2007, 113–47). There, we also find chapters devoted exclusively to dividing plane figures. This theme is already present in the pre-Islamic mathematical heritage (Moyon 2017, 79–102), and we find it again in the activities of Muslim lawyers who had to determine the portions of land for each person entitled to a part of a given inheritance. But as far as we know, only Andalusian authors devoted separate chapters to this subject (Djebbar 2016).

Notes

1 For a list of mathematical works translated from Greek, in addition to the studies they inspired and their extensions, see Sezgin 1974, 103-15; 128-35; 139-43; 154-6; 161-4; 165-6; 179.

2 It consists of starting with the last operation required in the statement of a problem and performing the inverse of the required operation successively until it leads to the starting data. As an example: If we seek a quantity c that satisfies: $2(2c-1) = 1$, we divide 1 by 2, add 1 and divide the result by 2. This produces the required result.

3 Consolidated bibliography.

4 This translation follows Høyrup (2015, 25). The illustration presented in Figure V.1.1. is taken from the same source. It is not found on the cuneiform tablet.

Bibliography

Sources

Abū Kāmil. ed., tr. and ann. Rashed, R. 2012. *Abū Kāmil, algèbre et analyse diophantienne.* Berlin: De Gruyter.

al-Ḫwārizmī. Folkerts, M., with Kunitzsch, P. 1997. *Die älteste lateinische Schrift über das indische Rechnen nach al-Ḫwārizmī.* München: Bayerischen Akademie der Wissenschaften.

al-Khwārazmī. ed., tr. and ann. Rashed, R. 2007. *Al-Khwārizmī, le commencement de l'algèbre.* Paris: Blanchard.

Ibn al-Akfānī. eds. Fakhūrī, M., Kamāl, M. and Al-Ṣaddīq, Ḥ. 1998. *Irshād al-qāṣid ilā asnā al-maqāṣid* [A Guide for One Who Aims for the Highest Goals]. Beirut: Maktabat Lubnān nāshirūn.

Ibn al-Bannā'. ed., tr. and ann. Aballagh, M. 1988. *Rafʿ al-ḥijāb d'Ibn al-Bannā'* [*The Raising of the Veil* of Ibn al-Bannā']. Thèse de Doctorat. Paris: University of Paris 1.

Ibn al-Bannā'. ed. and tr. Souissi, M. 1969. *Talkhīṣ aʿmāl al-ḥisāb* [Epitome of the Operations of Arithmetic]. Tunis: University of Tunis.

Ibn al-Khiḍr. tr. Rebstock, U. 2001. *Al-Tadhkira bi-uṣūl al-ḥisāb wa-l-farā'iḍ* [Book of Reminders on the Foundations of Arithmetic and Inheritance]. Frankfurt: Institute for the History of Arabic-Islamic Science. With facsimile.

Ibn al-Maghribī. ed. ʿAbd al-Tawwāb, A. 1992. *Lawḥ al-ḍabṭ fī ʿilm ḥisāb al-Qibṭ* [The Adjustment Board for the Coptic Science of Arithmetic]. *Revue de l'Institut des Manuscrits Arabes* 36: 119–37.

Ibn Munʿim. ed. Lamrabet, D. 2005. *Fiqh al-ḥisāb* [The Science of Arithmetic]. Rabat: Dār al-amān.

Naṣīr al-Dīn al-Ṭūsī. tr. Carathéodory, A. 1891. *Traité du quadrilatère attribué à Nassiruddin el-Toussy, d'après un manuscrit tiré de la bibliothèque de S.A. Edhem Pacha.* Constantinople: Typographie et lithographie Osmanié.

al-Samaw'al. eds. Ahmad, S. and Rashed, R. 1979. *Al-Kitāb al-bāhir fī l-jabr* [The Brilliant Book about Algebra]. Damascus: University of Damascus.

al-Uqlīdisī. ed. Saīdān, A. S. 1985. *Kitāb al-fuṣūl fī l-ḥisāb al-hindī* [Book of Sections on Indian Arithmetic]. Aleppo: I.H.A.S.

Ibn al-Yāsamīn. ed. and com. Zemouli, T. 1993. *Al-Aʿmāl al-riyāḍiyya li-Ibn al-Yāsamīn* [The Mathematical Works of Ibn al-Yāsamīn]. MA Thesis. Algiers: Ecole Normale Supérieure.

Manuscripts

MS Algiers, BN, 2712.
MS Istanbul, Süleymaniye Library, Carullah 1509.

Research literature

Busard, H. L. L. 1968. "L'algèbre au moyen âge : Le "Liber Mensurationum" d'Abû Bakr," *Journal des savants*, April–June, 65–124.

Debarnot, M.-Th. 1997. "Trigonométrie," in Rashed and Morelon, eds. [CB], eds., 2: 163–98.

Djebbar, A. 1981. *Enseignement et Recherche mathématiques dans le Maghreb des XIIIe-XIVe siècles*. Paris: Publications Mathématiques d'Orsay, no. 81–02.

Djebbar, A. 1985. *L'analyse combinatoire au Maghreb: l'exemple d'Ibn Munʿim (XIIe-XIIIe siècles)*. Paris: Publications Mathématiques d'Orsay, no. 85–01.

Djebbar, A. 1992. "Le traitement des fractions dans la tradition mathématique arabe du Maghreb," in Benoit, P., Chemla, K. and Mazard, G., eds. *Histoire de fractions, fractions d'histoire*. Berlin: Birkhäuser, 223–40.

Djebbar, A. 2000. "Figurate Numbers in the Mathematical Tradition of Andalus and the Maghrib," *Suhayl*, vol. 1.

Djebbar, A. 2002. "La circulation des mathématiques entre l'Orient et l'Occident musulmans: interrogations anciennes et éléments nouveaux," in Dold-Samplonius *et al.*, eds. [CB], 213–36.

Djebbar, A. 2003. "Mathématiques et société à travers un écrit maghrébin du XIVe siècle," in Cassinet, J., ed. *De la Chine à l'Occitanie, chemins entre arithmétique et algèbre*. Actes du colloque international. Toulouse: Editions du C.I.H.S.O., 29–54.

Djebbar, A. 2004. "La phase arabe de l'histoire de la trigonométrie," in Hebert, E., ed. *Les instruments scientifiques dans le patrimoine: quelles mathématiques?* Paris: Ellipse, 415–35.

Djebbar, A. 2007. "La géométrie du mesurage et du découpage dans les mathématiques d'Al-Andalus (Xe-XIIIe s.)," in Radelet de Grave, P., ed. *Liber Amicorum Jean Dhombres*. Turnhout: Brepols, 113–47.

Djebbar, A. 2013. "Islamic Combinatoric," in Wilson, R. and Watkins, J.-J., eds. *Combinatorics, Ancient and Modern*. Oxford: Oxford University Press, 83–108.

Djebbar, A. 2016. "Les techniques de découpage dans un ouvrage géométrique d'al-Andalus," in Bouzari, A., ed. *Actes du XIe Colloque maghrébin sur l'Histoire des mathématiques arabes*. Algiers: Dār al-Khaldūniyya, 109–38.

Djebbar, A. 2018. "Mathématiques et Droit musulman: Indemnisation des blessures et autres problèmes de la tradition juridique ibadite," in Laabid, E., ed. *Actes du 12e Colloque Maghrébin sur l'Histoire des Mathématiques Arabes*. Marrakech: Ecole Normale Supérieure, 88–108.

Djebbar, A. and Guergour, Y. 2013. "La numération rūmī dans des écrits mathématiques d'al-Andalus et du Maghreb," *Suhayl* 12: 7–52.

Hogendijk, J. P. 1991. "The Geometrical Parts of the *Istikmāl* of Yūsuf al-Mu'taman ibn Hūd (11th century). An Analytical Table of Contents," *Archives Internationales d'Histoire des Sciences* 41: 207–81.

Hogendijk, J. P. 1996. "Which Version of Menelaus' *Spherics* Was Used by al-Mu'taman ibn Hud in His Istikmal?" in Folkerts, M., ed. [CB], 17–44.

Høyrup, J. 2015. "Another Case of Stumbling Progress in the History of Algebra," *Physis*, New Series 1.1–2: 1–38.

Khairetdinova, N. G. 1966. "Trigonometričeskij traktat isfahanskogo anonima," *Istoriko- matematičeskie issledovaniya* 17: 449–64.

Laabid, E. 1990. *Arithmétique et algèbre d'héritage selon l'Islam, deux exemples: Traité d'al-Ḥubūbī (Xe-XIe s.) et pratique actuelle au Maroc*. Mémoire de Maîtrise, Montréal: Université du Québec.

Lamrabet, D. 1981. *La mathématique maghrébine au moyen-âge*. Mémoire de Post-graduation. Bruxelles: Université Libre de Bruxelles.

Moyon, M. 2017. *La géométrie de la mesure dans les traductions arabo-latines médiévales*. Turnhout: Brepols.

Proust, Ch. n.d. "Tablettes mathématiques cunéiformes: un choix de textes traduits et commentés," http://culturemath.ens.fr/materiaux/sexa/source-book/index-sourcebook.htm. (Accessed 20 November 2019).

Saʿīdān, A. S. 1971. *Taʾrīkh ʿilm al-ḥisāb al-ʿarabī*. Amman: Jamʿiyyat ʿummāl al-maṭābiʿ al-taʿāwuniyya.

Sayılı, A. 1985. *Logical necessities in mixed equations by ʿAbd al-Ḥamīd Ibn Turk and the algebra of his time*. Ankara: Türk Tarih Kurumu Basimevi.

Sesiano, J. 1989. "Koptisches Zahlensystem und (griechisch-)koptische Multiplikationstafeln nach einem arabischen Bericht," *Centaurus* 32: 53–65.
Sezgin, F. 1974. *Geschichte des arabischen Schrifttums*. Vol. V: Mathematik. Leiden and Boston: Brill.
Thureau-Dangin, F. 1938. *Textes mathématiques babyloniens*. Leiden and Boston: Brill.
Villuendas, M. V. 1979. *La trigonometria europea en el siglo XI. Estudio de la obra de Ibn Muʿāḏ, el-Kitāb maŷhūlāt* (sic). Barcelona: Instituto de Historia de la Ciencia de la Real Academia de Buenas Letras.

V.2

JEWISH MATHEMATICAL ACTIVITIES IN MEDIEVAL ISLAMICATE SOCIETIES AND BORDER ZONES

Naomi Aradi and Roy Wagner

Mathematics[1] practiced by Jews and Hebrew mathematics in medieval Islamicate societies stand at the intersection of two modern disciplines: Jewish studies and the history of mathematics. From the Jewish studies perspective, studying mathematics practiced by Jews improves our understanding of medieval Jewish conceptions of knowledge and of the interactions of Jews with the knowledge conceptions of their surrounding cultures. From the history of mathematics perspective, Jews served as cross-border carriers or disseminators of mathematical knowledge. Several Hebrew mathematical works have their roots in Arabic mathematics and were later translated into Latin (Freudenthal 1993; Lévy 2012; Lévy 1997). In fact, most of our knowledge about medieval Hebrew mathematics relates not to Islamicate societies, per se, but to their border zones.

The notion of "border zone" in this chapter is somewhat flexible. It means an area hosting a substantial community of the first generations of immigrants from Islamicate societies, where the influence of Arabic is felt strongly in Jewish communities. This means that 13th-century Provence, which does not have a physical border with any Muslim-ruled region, can still be considered a border zone, but 14th-century Provence is harder to conceive of in these terms (see section 5 in Freudenthal 1993). The border zones referred to in this chapter are the Iberian Peninsula, Provence, Sicily and southern Italy.

The general scope of Hebrew mathematics is well documented (a recent survey is available in Lévy 2012). Some important treatises and authors have been subject to detailed analysis. We also have a decent general understanding of the transmission of mathematical knowledge to and from Jewish communities in border zones, although many details remain unclear. We do not have a sufficient understanding, however, of the social aspects of learning and practicing mathematics among Jews in Islamicate societies and in border zones. This chapter tries to lean more in the latter direction, but since our knowledge is very limited, we will only be able to present suggestions and anecdotal evidence rather than a broad picture. We finish the chapter with some suggestions for future research methodologies.

DOI: 10.4324/9781315170718-51

V.2.1 Three profiles of authors of mathematical treatises in Islamicate societies and border zones

In this section, we present three "profiles" of Jewish scholars who wrote mathematical treatises. The first are the religious authorities who also wrote about mathematics, the second are the Jewish mathematical scholars who became converts, and the third are the Jewish scholars of mathematics and other sciences who did not write philosophical-theological, halachic or exegetic literature. The first category represents the bulk of known authors. The latter two categories may have been anecdotal exceptions or representatives of wider phenomena that failed to cross the threshold of manuscript production, distribution and preservation. Together, they tell us, *who* the Jewish mathematical authors were. This information will help us pose questions about their attitudes and the social context of their work.

V.2.1.1 Religious authorities who also wrote about mathematics

Most of the medieval Jewish scholars known today as authors of mathematical treatises were not known primarily for their mathematical work. They were religious authorities who wrote important treatises on theology, biblical exegesis, or Jewish law, and their interests were never restricted to science alone.

Starting with scholars who wrote in Arabic in Islamicate societies, the two most prominent figures, Saʿadia Gaʾon (268–330/882–942) and Maimonides (d. 601/1204), were also the most prominent religious authorities of their respective times. Saʿadia wrote on 'applications': the Jewish calendar and inheritance calculations, the latter being a very popular genre of the juridical and mathematical sciences in Muslim scholarship but hardly extant among the Jews (Saʿadia 1893–1899, vol. 9). Maimonides was interested in mathematics and philosophy mostly as a ladder to theology (his approach will be discussed later). He wrote on the indefinite approach of the hyperbola and its asymptote to illustrate the philosophical point that we can know a truth without being able to imagine it (assuming that we can only imagine finite ranges). This may have motivated him to write his commentary on Ibn al-Haytham's (354–*c.* 430/965–*c.* 1040) reconstruction of Apollonius' (*c.* 262–*c.* 190 BCE) Book VIII of the conics (Langermann 1984; Lévy and Rashed 2004). Mathematics was not a major concern for either Saʿadia or Maimonides. Still, their interest in mathematics testifies to its relevance to Jewish knowledge.

Moving on to major religious authorities who wrote on mathematics in Hebrew, two important figures are Abraham bar Ḥiyya (1065–*c.* 1140; Barcelona) and Elijah Mizraḥi ([d. 932/1526]; Constantinople/Istanbul). The former held the community leadership title *Nasi* and as a Barcelona court official, the title *Savasorda* (captain of the guard). The latter was head of a Yeshiva and a major community leader in Constantinople. Mathematically, the former is known for the popular *Treatise on Measurement and Area* (*Ḥibbur Hameshiḥa Vehatishboret*) and the encyclopedia *The Foundations of Understanding and Tower of Faith* (*Yesodei Hatvuna Umigdal Ha'emuna*). He also wrote and translated treatises on astronomy, astrology and moral philosophy (Katz *et al.* 2016, 296–313). The latter authored the arithmetic treatise *The Book of Number* (*Sefer Hamispar*), which includes algorithms and proofs, as well as exegetic and Jewish law literature (Katz *et al.* 2016, 244–53; Segev 2011). The former is acknowledged as the first transmitter of Arabic mathematical knowledge to the Jewish communities of the Christian ruled Catalan-Provençal area. The latter can be taken to mark the end of the medieval Hebrew mathematical tradition.

This category also includes Abraham ben Ezra (d. 562/1167). He was originally from al-Andalus, but traveled extensively in France, Italy and England. He was by far the most popular Hebrew mathematical author. His works represent more than 10% of the surviving manuscripts. He wrote the arithmetical textbook *The Book of Number* (*Sefer Hamispar*) and the arithmology *The Book of One* (*Sefer Ha'eḥad*), but he is far better known as a prolific author of astrology, astronomy, grammar, exegetic literature and theology and as a poet (Katz *et al.* 2016, 227–35, 269–73, 287–96; Sela 2003). He is exceptional in this category not only due to his popularity but also because, despite his renown and authority, he does not seem to have ever held any official public position.

Most of the known Hebrew writers of mathematics fit into this category (see Table V.2.3 in the Appendix to this chapter), although their renown is usually more modest than that of the previously discussed figures. Notably, many of the figures in this group seem to have also engaged in medicine (at least in writing). One possible explanation is the central position of medicine in the medieval system of knowledge; another is that the income and reputation stemming from medical practice may have provided the leisure and authority required for authoring and publishing mathematical work. Either explanation would require further validation.

These authors demonstrate that in (at least some) Jewish Islamicate or border zone communities, mathematics was important enough to be part of what a Jewish leader should be concerned with, but not a leader's main scholarly or scientific concern (see also Table V.2.3 in the Appendix to this chapter).

V.2.1.2 The converts

Converts testify to a very different relation between mathematics and Judaism. Instead of integrating mathematical interests into the profile of a Jewish community leader, these figures bind together non-Jewish scientific interests with an orientation away from Judaism. It is far from clear, however, whether such conjunctions are mere coincidences or express a causal relation.

Of the first figure in this category we know very little. Sanad ibn ʿAlī (3rd/9th century) was a court astrologer of Caliph al-Maʾmūn (r. 197–217/813–833) in Baghdad. He wrote on arithmetic, algebra and astronomy (Brentjes 2007, 1011; Steinschneider 1901, 49–51). The next figure, al-Samawʾal (d. 570/1175; formerly Samuel ben ʿAbbās ben Yehuda) is much better known. He was born in Baghdad, trained as a physician and moved to Maragha after having served as a doctor for some time at a court in Azerbaijan (Schmidtke 2010). His Arabic works on mathematics and astronomy have attracted substantial attention (al-Samawʾal 1972; Steinschneider 1902, §149). After his conversion, he wrote a polemic against the Jews. Despite the fact that he left an autobiographical account, the reasons for his conversion are still variously attributed to opportunism or deeper convictions (Husain 2002). It is therefore difficult to draw conclusions on whether close interaction with Muslim scholars or Arabic mathematical scholarship might have motivated conversion at the places and times in question.

Among Jews who wrote in Hebrew in border zones, we note the Castilian Abner of Burgos (d. 1347), who converted to Christianity as Alfonso de Valladolid. Abner/Alfonso authored the Hebrew geometrical treatise *Rectifier of the Curved* (*Meyyasher ʿAqov*), dealing with advanced geometric constructions and mechanical curves and attempting to shed light on the quadrature of the circle. He also wrote philosophical and anti-Jewish polemical literature which survives partly in Hebrew and partly in Castilian translation. He, too, worked as a physician. Again, there are various versions of his conversion story. It is therefore difficult to judge whether the link between immersion in non-Jewish scientific work and conversion is coincidental or causal (Katz *et al.* 2016, 345–53; Glasner and Baraness 2021).

Unlike the two converts to Islam whose Arabic work attracted further attention, Abner/Alfonso's Hebrew scientific work survives in just a single incomplete manuscript. Despite some substantial original features of his work, it does not seem to have received a lot of attention from his contemporaries. One explanation may be a wish to suppress references to converts by the Jewish community; another may be that the mathematical content was not within the range of interests of Jewish mathematical scholars.

It is difficult to estimate how many mathematical scholars converted from Judaism. Jews were not always happy to acknowledge converts. The converts themselves were not always eager to mention their Jewish background to potential readers. Moreover, conversion might make it difficult to secure a readership for one's work. Given these factors, it is difficult to estimate whether the three figures discussed earlier are anecdotal exceptions or part of a more systematic link between immersion in mathematics, or non-Jewish science in general, and conversion. We do note that the mathematical work of the converts al-Samawʾal and Abner/Alfonso was more original and 'cutting edge' than that of any of the other Jewish authors considered here (with the exception of Gersonides) and went beyond the usual reach of mathematical treatises written by Jews.

V.2.1.3 The scientific scholars

This last category includes Jewish scholars known primarily or exclusively for their scientific work who were not religious authorities. Again, we have only a few examples, but, as with the converts, it may be due to the fact that religious authorities had a much higher chance of having their work read, copied and preserved.

The most prominent candidate for this group is Jacob ben Makhir ([d. 1305], Montpellier), who translated geometrical, scientific and philosophical treatises from Arabic and wrote original astronomical treatises, including on his improved astrolabe, *The Quadrant of Israel*. According to some, he was the regent of the Montpellier medicine faculty (Schloessinger *et al.* 1904). He was involved in public intellectual debates, but we have no record of him holding a rabbinical position or writing on Jewish theology or law. Immanuel Bonfils of Tarascon in southern France (d. 1377) would be another prominent example of this profile, but he already worked in a more Latinized context (Katz *et al.* 2016, 237–9).

Other such figures are much more minor. The work of Aaron ben Isaac ([15th century]; Spain?), for example, survives in one damaged copy of a single arithmetical treatise, which mentions two other mathematical-astronomical treatises that he wrote or translated. He worked as a weaver of gold and silk and wrote the arithmetical treatise for his son, who was meant to edit it, but died prematurely (Katz *et al.* 2016, 235–7). Isaac ben Moses ʿEli (who probably fled from Spain to Constantinople at the end of the 15th century) is another author known via a single arithmetical text (Steinschneider 2008, 208).

We can also cite one example of an apparently "purely scientific" Jewish author who wrote on mathematics in Arabic, this time in the West. This is the Toledan Joseph ben Naḥmiʾas (*fl. c.* 1350) who wrote the astronomical work *Light of the World* (*Nūr al-ʿĀlam*; Joseph ibn Naḥmias 2016). The treatise includes mathematical sections and was later translated into Hebrew by the author himself. We do not know of other treatises by this author.

V.2.2 The scope of mathematics written, copied or translated by Jews

It is very difficult to evaluate the scope of mathematics written or copied by Jews in Arabic (either in Arabic or Hebrew letters), as the sources are scarce and have never been systematically researched. One exception is the Jewish Yemenite corpus in Arabic (in Hebrew letters) studied by Langermann (1987). Langermann observes a heightened interest in astrology, astronomy and mathematics among Yemenite Jews in the 7th–11th/13th–17th centuries, which reached the forefront of available knowledge, but hardly ever offered innovations. The specifically mathematical material documented by Langermann consists of scattered comments by Jewish Yemenite authors, as well as copies of Euclid's (3rd century BCE) *Elements,* parts of a commentary by Simplicius (d. after 550?), a summary of the epistle of the Ikhwān al-Ṣafā' (4th/10th century) on geometry, an arithmetic by Samuel ben ʿAbbās (al-Samaw'al) and a handful of anonymous and minor treatises.

We have a much better overview of Hebrew mathematical literature from border zones. Indeed, the need to produce Hebrew mathematics arose, with few exceptions, when Arabic mathematical knowledge was transported to communities that did not know Arabic (one possible exception is the summary of measurement formulas in the *Treatise of Measures* (*Mishnat Hamiddot*) whose time and place of authorship is unclear; see Sarfatti 1968, 58–60; Langermann 2002, 172–3). We also see a brief emergence of Hebrew mathematics in – former – Byzantium following the Ottoman conquest. This can be related to the loss of access to Greek knowledge or the influx of Jews from the West.

The most popular genre of Hebrew mathematics is the arithmetical textbook (more than 100 surviving copies witnessing a couple of dozen different books).[2] Starting with Ibn Ezra's *Book of Number* (written in Italy, 1142–1145, with over 43 surviving copies), they follow popular Arabic textbook structures, covering fundamental arithmetic principles, the basic arithmetic operations and calculation procedures (Chapter I.7). The original and well-reasoned arithmetical work of Levi ben Gershon of Avignon ([1288–1344]; Gersonides) and the presentation of decimal fractions by Immanuel Bonfils are at the very edge of the scope of this chapter, as they already belong more in a Latin rather than an Arabic context (Katz *et al.* 2016, 237–9, 253–68).[3]

Textbooks often contain word problems – narratives that describe a (usually commercial) situation and call for a mathematical solution. Over the years the collections of word problems expanded, perhaps to convince readers that arithmetic has practical applications or to provide them with recreational exercises. The result is a few independent Hebrew collections of word problems (not embedded into proper textbooks) that appear to have evolved from the textbook genre. The surviving collections were written apparently in an Italian hand around the 15th century.

Other genres of arithmetic are not very well covered in medieval Hebrew. The most prominent works are Ibn Ezra's arithmological *The Book of One* (at least 23 surviving copies), a translation of an Arabic text based on *The Introduction to Arithmetic* by Nicomachus of Gerasa ([d. *c.* 120]; eight surviving copies) and a handful of original and translated commentaries.

Geometry is represented mostly in translations. Euclid's *Elements* survive in over 30 full and partial manuscripts and at least ten copies of translated commentaries. This is more than the surviving Arabic copies, but Arabic literature had a flourishing genre of variations on Euclid, which does not seem to have been substantially developed by Hebrew writers and translators (see Brentjes 2008, for a survey). This suggests that while the *Elements* were a popular teaching resource, its study in Hebrew remained rather basic. In addition, there are about 50 manuscripts of the so-called *Middle Books*, including spherical geometry and commentaries, Archimedean works with commentaries, Euclid's *Data* and commentaries on Apollonius; note that the *Conics* itself was not translated into Hebrew.

The original production of Hebrew commentaries and treatises in this area is scarce. The two kinds of notable geometrical treatises in Hebrew include some 20 manuscripts about the hyperbola and its asymptote, which, as mentioned earlier, was brought to the attention of Jewish readers by Maimonides (see the previous discussion; also Katz *et al.* 2016, 340–4; Freudenthal

1998; Lévy 1989a, 1989b). The other kind is of a completely different nature: Abraham bar Ḥiyya's *Treatise on Measurement and Area*, which exists in around ten surviving manuscripts. This treatise, which, according to the author, depends on Arabic sources, provides a general introduction to mensuration, with a simplified exposition of some Euclidean theorems and some parts of Euclid's treatise on the divisions of figures.

Algebra did not receive substantial explicit Hebrew representation. In late 14th-century Sicily, Isaac al-Aḥdab translated and commented on the *Summary of the Operations of Calculation* (*Talkhīṣ ʿAmal al-Ḥisāb*) by Ibn al-Bannāʾ (654–721/1256–1321). But as an independent field of activity, Hebrew algebra developed further only later and outside of our context, in Northern Italy, based mostly on Italian sources.

In summary, we can say very little about Arabic mathematics written and copied by Jews beyond the fact that some Jews were familiar with the state of the art of Arabic mathematics and that mathematics was of interest in some communities, especially in the context of astronomy. When we move on to border zones, we can delineate two projects of transmission. The first is concerned with introducing elementary mathematics (mostly arithmetic) to Hebrew readers, apparently following Arabic models, emphasizing practical (or at least made-to-look practical) mathematics. The second project is the translation of the *Elements* and some of the *Middle Books* for Hebrew readers, probably in an attempt to replicate a typical Arabic course of scientific studies and in support of astrology and astronomy. The known development of original or more advanced Hebrew mathematics is very limited and occurs in border zones (Alfonso de Valladolid) and at the very margin of our scope, in Christian Provence (Levi ben Gershon and Immanuel Bonfils).

V.2.3 Mathematical education and its motivation

We do not know much about the teaching of mathematics in Jewish communities in Islamicate societies or in border zones. According to Sirat (2015) and Freudenthal (1995, 46–8), sciences were studied under masters who taught one to three students privately. In the context of mathematics, one example is Maimonides and his student Joseph ben Judah ben Simon (d. 1226), to whom the *Guide to the Perplexed* is dedicated, and whose education under Maimonides included mathematics (Sirat 1990, 160). Other examples are supplied by members of the Ibn Tibbon family (e.g., Sirat 1990, 224–5) and by Mordechai Comtino (d. after 1485), who taught Mizraḥi and the Karaite Caleb Afendopoulo (d. after 1482).

Sirat also claims that in the 15th century Jewish *Yeshivas* existed in Spain which taught philosophy and science, following a Christian university model (2015). However, it is not clear whether these institutions taught mathematics. Zonta (2006, 15, n. 61) claims that these may have been Christian institutions that admitted Jews. We also have evidence of interfaith tutoring. Sanad ibn ʿAlī is said to have joined a Muslim scholarly circle (Brentjes 2007), and Isaac al-Aḥdab testifies to having studied with Muslims in North Africa or Granada (Wartenberg 2016, 26). Al-Samawʾal, on the other hand, reports that after being trained by tutors in elementary mathematics, he could find no competent teachers and had to teach himself (Perlmann 1964, 76–7). One of his tutors, Abū l-Barakāt al-Baghdādī ([d. c. 560/1164–5]; Pines 1960), was a Jew who converted to Islam; some of his other tutors' names are more likely Muslim than Jewish. Scientific teaching of members of other denominations by Jews also emerges as an urgent and disputed issue in the times of Mizraḥi (Rozen 2002, 339–55).

This account of education fits the prevailing ambivalent attitudes to sciences among medieval Jews. Saʿadia Gaʾon (see the previous discussion) claimed that scientific knowledge cures the soul and adorns it and that the study of science is almost a specific characteristic of humanity (Saʿadia

1942, 393). Yet he criticized both those who exclusively advocate the study of science (Saʿadia 1942, 393–4) and those who argue that the study of science (specifically geometry) might lead away from faith and toward heresy (Saʿadia 1942, 26).

Saʿadia's positive assessment of geometry and arithmetic as prerequisites for knowledge of the soul and of the divine is quoted by Ibn Ezra (Ibn Ezra 1995, 22). The same position is expressed by various later authors. However, the positive assessment of mathematics was not uniform. One variety reflects Pythagorean, mystic, or Neoplatonist attitudes (as in the case of Ibn Ezra), whereas another represents a rationalist-Aristotelian position (as in the case of Maimonides). The former approach tends to correlate with an interest in number theory and arithmetic, while the latter is related to classical Greek geometry and the *Middle Books*. A third point of view concerns the position of mathematics within the system of knowledge. It may be viewed as preliminary to natural science, because of the latter's dependence on mathematics (e.g., Maimonides 1963[CB],[4] 618–9), or, because of its abstract nature, as an intermediary between natural science and theology (e.g., Mizraḥi, see Segev 2011, II-2; Judah ben Solomon HaCohen [13th century], see Sirat 1990, 250).

Even where attitudes to foreign sciences were positive, the recommended scope of mathematical education was usually fairly restricted. Both Maimonides and Abraham ben Daud ([d. 1180]; Toledo) valued the sciences and considered mathematics a necessary stage in the process to acquire the knowledge of God, but both warned against making mathematics an end in itself. Ben Daud, for example, derided an elaborate algebraic riddle about distilling alcohol. Maimonides restricted the value of "arithmetic, the study of conic sections, mechanics, the various problems of geometry, hydraulics and many others of similar nature" to training the intellect in demonstrative proof for the purpose of knowing the essence of God (Freudenthal 1995, 32, 37; Freudenthal 2016, 80–2, 95–7, 101, quoting Maimonides 1912[CB], 71). Later on, Joseph Caspi (1279–1340; Provence), when recommending the study of mathematics to his son, names only Ibn Ezra's arithmetic and Euclid's *Elements* before moving on to astronomy (Sirat 1990, 207).

Negative approaches to science usually reflect a fear of religious doubts, loss of faith and heresy. This is explicit in the 50-year ban against learning foreign sciences (literally, Greek natural and divine sciences, excluding medicine) before the age of 25, which was announced in Barcelona in the year 1305 (Freudenthal 1995, 44; Ben Sasson *et al.* 2007). The explicit mention of Greek sciences refers to a debate going back to Hellenistic times, but the extent of the ban is not clearly defined. Did the proscription include practical arithmetic and astronomy? Did it include books authored by Jews? At any rate, the ban seems not to have been enforced. In 1332, for example, Joseph Caspi publicly advised his son to study mathematics at 14 and Greek natural science at 18.

Note that similarly restrictive positions, which are usually associated with the Provençal border zone, appear at earlier times and in communities in Islamicate societies as well. In late 12th-century North Africa, for example, Joseph ben Judah ben Aknin ([d. *c.* 616/1220]; Fez) recommended to postpone the study of foreign sciences until after the age of 30 (Sirat 1990, 207). Note that Joseph lived as a clandestine Jew under Almohad rule (r. 524–668/1130–1269) and that the Barcelona boycott preceded by one year the expulsion of Jews from France. So negative attitudes to foreign science might be partly related to intolerance toward Judaism.

According to some later sources, the study of science among Jews was secretive (Leon Joseph of Carcassone's [late 14th–early 15th centuries] statement from 1394, quoted in Freudenthal 1995, 45; a short manuscript introduction to the *Book of Ṣifra* [*Sefer Ṣifra*], MS Paris, Séminaire Israélite de France (École Rabbinique) 158/1, fol. 197r, 1–6).

As noted earlier, one of the reasons of studying science was to improve the soul and prepare for studying the divine. This rationalist argument was associated with the necessary truth of deductive mathematics in the Euclidean style. Another reason was practical: everyday measurement and arithmetic. But this also had a clear religious dimension. Several authorities cite religious commandments that require some mathematical knowledge, ranging from calendrical knowledge through agricultural practices to calculating inheritance portions. Indeed, Jewish religious law penetrates many unexpected dimensions of everyday life, for example, distances between plants in gardens and architectural measurements. This is comparable to Høyrup's portrayal of the notion of Islamic "fundamentalism" (Høyrup 1987). Bar Ḥiyya, for example, complains in the introduction to his geometry that the Jews of Provence are ignorant in such matters and therefore reckon unjustly and in violation of religious law (Katz *et al.* 2016, 297–9).

A different argument for a mathematical discussion is related to the interpretation of biblical measurements. For example, the biblical measurements of Solomon's sea (a large water basin in Solomon's temple) suggest a circumference to diameter ratio of $\pi = 3$. Bar Ḥiyya brings in the thickness of the rim and the difference between the outer and the inner circumference in order to reconcile the biblical account with a circumference to diameter ratio of $\pi = 3\frac{1}{7}$. On the other hand, Simon ben Ṣemaḥ Duran (1361–1444), a leading rabbinical authority and physician, who fled from Aragon to Algiers, argued for a more pragmatic interpretation; the biblical numbers may be approximate or explained by the combination of different prevalent standards for the length of a cubit (Katz *et al.* 2016, 315–20). While for Bar Ḥiyya Solomon's sea is an occasion for subtle calculation and a pedagogical opportunity, Simon ben Ṣemaḥ prefers an approach to religious texts that would require only elementary calculations that fit the ancient sources.

One should note that stated practical motives do not necessarily conflict with an interest in reasoning, albeit not as codified and rigorous as that of Euclid. Some arithmetical textbooks are explicitly committed to high quality reasoning and explanation, without which understanding is said to be incomplete and knowledge limited to what is explicitly taught (e.g., Jacob Canpanton in *Bar Noten Ṭaʿam*, MS London, BL, Or. 1053, fol. 1v,1; fol. 3r,5–6; see Table V.2.3. in the appendix; Mizraḥi in Segev 2011, II. 5). Opposed to this maximalist approach, some authors prefer a minimalist, clear and concise presentation of rules (e.g., Isaac ben Moses ʿEli, MS Oxford, Bodleian Library, Mich. 141, fol. 17r, 1–13). Either approach covers more or less the same (mostly arithmetical) material.

The Jewish arguments for engaging in mathematics included the propaedeutic role of science for theology and the access it provides to the divine, practical applications (whether religiously sanctioned or not), and its exegetic role. Combined with resistance to non-Jewish influence, these "extrinsic" reasons kept Jewish mathematics mostly limited to nonoriginal commentarial and editorial endeavors (as argued in Freudenthal 1995). The exceptions are a handful of converts and figures at the very edge of what we call border zones, who made more original contributions.

V.2.4 Was/Were there one or several Jewish mathematical culture(s)?

Jewish and Hebrew mathematics in Islamicate societies and border zones embed neatly into their ambient cultures. Arabic mathematics written or copied by Jews does not appear to be exceptional in any way when compared to its surrounding culture, and when Hebrew enters the scene, Jewish treatises still follow Arabic standards.

Hebrew mathematics does have some unifying features. At its core, it has a unified language and a stable canon generated by its pioneer authors and translators. But this did not mature into an autonomous mathematical culture. An anecdotal example of a properly Jewish mathematical concern is an inheritance problem treated by Ibn Ezra (Katz *et al.* 2016, 232–3) according to both Muslim and Jewish law. It was adopted by several subsequent authors but did not develop into a Jewish mathematical study of inheritance in parallel to the Islamic one. Another example is the study of asymptotes following Maimonides' philosophical outlook, which, despite the attention it received, did not lead to a more elaborate Jewish or Hebrew interest in conics or asymptotes.

Another question is local variation between Hebrew mathematical cultures. For example, in Provence, the Hebrew translations of the 13th and 14th centuries show a heightened interest in higher geometry, whereas Hebrew algebra emerges mostly in a later Italian setting. But are these trends indicative of different mathematical visions?

In order to tackle this question, we provide some coarse statistics of surviving Hebrew manuscripts written no later than the 16th century. They are sorted according to themes, authors and variants of Hebrew writing. Comparing the surviving manuscripts in Byzantine, Italian and Spanish variants of Hebrew script can hint at the interests of scholars belonging to each of these regional cultures.

However, one should handle these tables with caution (Tables V.2.1 and V.2.2). First, catalogues inevitably contain mistakes and guesses. Second, it is never clear whether surviving manuscripts correctly represent the predominant mathematical engagement of their time or testify only to their preservers' preferences. Third, the variants of Hebrew writing do not necessarily correspond to the geographical areas bearing the same names. Jews migrated and took their writing variants with them. This is true especially for Spanish writing, which became widespread due to expulsions and migrations (Beit-Arié n. d.). The forms of writing therefore represent lineages of education rather than separate communities, as a person's handwriting is most likely derived from that of a teacher. Note also that all these variants of writing cover more than the Islamicate world and its border zones. Therefore, the tables are best seen as a means to generate conjectures rather than draw valid conclusions.

Based on the tables, we could suggest that Euclid, Nicomachus and Ibn Ezra were more popular in the Byzantine line of transmission than elsewhere and that Comtino and Mizrahi did not spread substantially beyond this line. This would portray the Byzantine mathematical scene as more closed and elementary. We might also suggest that asymptotes and algebra were of

Table V.2.1 Percentage of manuscripts on a given subject among surviving mathematical manuscripts in a given script family

	Euclid	*Higher geometry*	*Asymptotes*	*Other geometry*	*Nicomachus*	*Arithmology*	*Algebra*	*Other arithmetic*	*General*	*Total*
Spanish	16%	16%	3%	9%	5%	5%	2%	33%	10%	135
Byzantine	22%	5%	1%	8%	9%	8%	0%	35%	11%	76
Italian	17%	9%	9%	11%	1%	4%	12%	28%	9%	110

Table V.2.2 Percentage of manuscripts by a given author among surviving mathematical manuscripts in a given script family

	Ibn Ezra	*Bar Hiyya*	*Levi ben Gershom*	*Comtino*	*Mizraḥi*	*Total*
Spanish	12%	4%	6%	1%	1%	135
Byzantine	22%	1%	4%	8%	4%	76
Italian	11%	3%	2%	0%	1%	110

special interest in the Italian lineage of transmission, even though the discussion of asymptotes originated in Egypt and was developed by authors from the Iberian Peninsula. This indicates two opposite trends: one toward the classics and philosophy and the other toward new mathematical techniques. Finally, higher geometry seems to have been particularly popular in the Spanish line of transmission. This may suggest a stronger relation to astronomy and astrology. Verifying or rejecting these interpretations will require further research.

Variation between Hebrew mathematical sub-cultures is also attested by the nonhomogeneity of the medieval Hebrew mathematical language. As noted by Sarfatti (1968, xii–xiv), the development of the Hebrew mathematical vocabulary involved a tension between the 'purist' approach, which remained faithful to the vocabulary of ancient Hebrew sources, casting new mathematical meanings into old Hebrew terms, and the 'assimilative' approach, characterized by the acquisition of new scientific terms by means of linguistic borrowing from Arabic.

Linguistic analysis may further help trace lines of transmission within the Hebrew corpus. Several mathematical terms have various Hebrew translations by different authors. Following these terms through the corpus may help us understand which sources most inspired later authors and translators and tease out paths of dissemination. The *Peshat* online database of philosophic and scientific Hebrew terminology could be of value in this context.[5] This methodology should complement more traditional content-based comparisons. For example, Mizraḥi's *Book of Number* shares discussions and examples with the anonymous *Some Issues of the Science of Number* (*Qṣat Me'inyeney Ḥokhmat Hamispar*; Aradi 2015, 417). The language, however, is different, as the latter includes Arabic terminology absent from the former. This may indicate an unknown line of transmission from the Arabic that found its way into Byzantine Hebrew arithmetic.

V.2.5 Directions for future research

Improving the rather vague picture presented above will not be achieved by additional study of the canonical Hebrew mathematical sources alone. In order to get a better picture of Jewish mathematical education in Islamicate societies and border zones one should look into biographies and community accounts rather than just scientific and philosophical treatises. Collecting this information requires, however, scouring through large quantities of materials to collect bits and pieces, which will probably have to wait until a large quantity of material is digitized and turned into searchable text.

But even more 'internalist' questions about precise lines of transmission and the differences between the mathematical interests of different Jewish communities will require access to somewhat bigger data than that which we presently have available. The most original and canonical works may be the most interesting, but they are only peaks of knowledge and cannot be understood in context without a study of the unexplored 'minor' mathematical literature surrounding them.

In order to enable such research, we are in the process of creating *Mispar*, an open-access database of (almost) all surviving medieval Hebrew arithmetic and algebra.[6] This database will be semantically tagged according to subject matter and fully searchable. It will allow future researchers to draw larger linguistic and content-based comparisons that could answer questions of transmission and scope of knowledge. The quantity of surviving Hebrew mathematical manuscripts makes this a possible endeavor. We hope that it will prove useful to the scientific community and be extended to neighboring corpora, such as Arabic arithmetic or Hebrew geometry and astronomy.

V.2.6 Appendix

Table V.2.3 Some Jewish religious authorities who wrote about mathematics in Hebrew

Name	*Dates*	*Translations*	*Own works*	*Remarks*
Moses ben Samuel ben Tibbon	*fl.* 1240–1283, Provence	Euclid's *Elements*, Abū Bakr al-Ḥaṣṣār's arithmetic (alive in 552/1157), other Arabic scientific and philosophical treatises	exegetic treatises	worked as a physician (Schloessinger *et al.* 1904)
Qalonymos ben Qalonymos	1287–after 1329, Arles, Naples	Arabic version of Nicomachus' (d. 120) *Arithmetic* and Archimedes' (d. 212 bce) *Sphere and Cylinder*, Arabic geometric treatises, philosophical and medical works into Hebrew and Latin	the number-theoretical and arithmological *Book of Kings*, ethical and satirical treatises	held the community leadership title *Nasi* in Arles and worked at the court of Robert d'Anjou (r. 1309–1343; Katz *et al.* 2016, 337–8; Gottheil and Broydé 1904)
Isaac ben Solomon ben al-Aḥdab	b. *c.* 1350, Castile, Sicily	commentary on Ibn al-Bannā''s (654–721/1256–1321) arithmetic	astronomy, exegesis, and theology treatises; was a celebrated poet	escaped from the persecutions of the 14th century to North Africa (Katz *et al.* 2016, 362–74; Wartenberg 2016)
Jacob Canpanton	14th–15th century, Castile	medical translation from Arabic	the arithmetical treatise (with proofs) *Bar Noten Ṭaʿam*, also published on astronomy and the Bible	was a student of Hasdai Crescas and the father of the community leader Isaac Canpanton (Katz *et al.* 2016, 239–43)
Judah ibn Verga	15th century, Seville		*A Brief Treatise on Number* (*Kiṣur Hamispar*), measurement and astronomical works, a history of Jewish persecution	(Steinschneider, Mathematik, 196)
Mordechai Comtino	804–886/1402–1482, Adrianople, Constantinople		*Book on Calculation and Mensuration* (*Sefer Haḥeshbon Vehammiddot*), commentaries on Euclid and on Ibn Ezra's *Book of One*, treatises on measurement, astronomy, philosophy, and exegesis	was the teacher of Elijah Mizraḥi and of the Karaite Caleb Afendopoulo (1455–*c.* 1509), who wrote a commentary on Nicomachus' *Arithmetic* (Gottheil and Berlin 1903)

Notes

1 *Mathematics* is defined narrowly in this chapter, that is apart from astronomy, optics and music. This may be a problematic choice, as mathematics and astronomy are strongly related in medieval Islamicate societies. However, the limitations of the authors' knowledge forced us to make this choice. As a result, our comments on mathematics in astronomical contexts will be tangential and incomplete.
2 Numbers are based on the catalogue of the National Library in Jerusalem and Naomi Aradi's listing of all arithmetical and some of the geometric manuscripts. The numbers refer only to surviving manuscripts copied by the 16th century. They are not restricted to areas under Muslim rule and border zones but cover all available manuscripts.
3 Gersonides' works show that he consulted some passages of Arabic literature and reflected on the language, but it is doubtful that he could read an entire Arabic treatise. The extent of his knowledge of Latin is even less clear but probably similar (Glasner 2002). His sources were mainly Arabic treatises translated into Hebrew, but he was also well integrated into a Latin intellectual milieu (Mancha 1997). This, together with Freudenthal's periodization (1993) place him at the very edge of what should be called an Islamicate "border zones".
4 Consolidated bibliography.
5 https://peshat.gwiss.uni-hamburg.de/
6 http://mispar.ethz.ch/wiki/Main_Page

Bibliography

Sources

Ibn Ezra, Abraham ben Meir. ed. and tr. Strickman, H. N. 1995. *The Secret of the Torah: A Translation of Abraham Ibn Ezra's Sefer Yesod Mora Ve-sod Ha-Torah*. New York: Jason Aronson, Inc.

Ibn Naḥmias, Joesph. ed. and tr. Morrison, R. G. 2016. *The Light of the World*: Astronomy in al-Andalus. Berkeley and Los Angeles, CA: The University of California Press

Sa'adia ben Joseph. eds. Derenbourg, J., Derenbourg, H. and Lambert, M. 1893–99. *Œuvres complètes*. 9 vols. Vol. 9: 1897. Paris: E. Leroux.

Sa'adia ben Joseph, ed. and tr. Rosenblatt, S. 1942. *The Book of Beliefs and Opinions: Amānāt wa-l-i'tiqādāt*. New Haven: Yale University Press.

al-Samaw'al. eds. Ahmad, S. and Rashed, R. 1972. *Al-Bahir en algèbre d'as-Samaw'al*. Damascus: University Press of Damascus.

Manuscripts

MS London, BL, Or. 1053 (IMHM: f 5932; cat. Margo. 1012, 1).

MS Oxford, Bodleian Library MS Mich. 141 (IMHM: f 22111; Cat. Neub. 1297, 2).

MS Paris, Séminaire Israélite de France (École Rabbinique) 158/1 (IMHM 4102).

Research literature

Aradi, N. 2015. *The Arithmetic of the Jews in the Middle Ages: Introduction and Selected Samples*. PhD dissertation. Jerusalem: Hebrew University.

Beit-Arié, M. n.d. *Hebrew Codicology: Historical and Comparative Typology of Hebrew Medieval Codices Based on the Documentation of the Extant Dated Manuscripts in Quantitative Approach*. Preprint Internet English version 0.1. web.nli.org.il/sites/NLI/English/collections/manuscripts/hebrewcodicology/Documents/HC%20ENGLISH%20ACCUMULATED%201-5%2019.7.17%20(Autosaved).pdf (Accessed 31 May 2018).

Ben-Sasson, H. H., Jospe, R. and Schwartz, D. 2007[2]. "Maimonidean Controversy," in Skolnik, F. and Berenbaum, M., eds. *Encyclopaedia Judaica*. 22 vols. Detroit: Macmillan Reference USA, 13: 371–81.

Brentjes, S. 2007. "Sanad ibn ʿAlī: Abū al-Ṭayyib Sanad ibn ʿAlī al-Yahūdī," in Hockey *et al.*, eds. [CB], 2: 1011.

Brentjes, S. 2008. "Euclid's Elements, Courtly Patronage and Princely Education," *Iranian Studies* 41.4: 441–63.

Freudenthal, G. 1993. "Les Sciences dans les Communautés Juives Médiévales de Provence: Leur Appropriation, leur Role," *Revue des Études Juives* 93.1–2: 30–136.

Freudenthal, G. 1995. "Science in the Medieval Jewish Culture of Southern France," *History of Science* 33: 23–58.

Freudenthal, G. 1998. "Maimonides' 'Guide of the Perplexed' and the Transmission of the Mathematical Tract 'On Two Asymptotic Lines' in the Arabic, Latin and Hebrew Medieval Traditions," *Vivarium* 26.2: 113–40.

Freudenthal, G. 2016. "Abraham Ibn Daud, Avendauth, Dominicus Gundissalinus and Practical Mathematics in Mid-Twelfth Century Toledo," *Aleph* 16.1: 60–106.

Glasner, R. 2002. "On Gersonides' Knowledge of Languages," *Aleph* 2: 235–57.

Glasner, R. and Baraness, A. 2021. *Alfonso's Rectifying the Curved: A Fourteenth-Century Hebrew Geometrical-Philosophical Treatise*. Cham: Springer.

Gottheil, R. and Berlin, I. 1903 "Comtino, Mordecai ben Eliezer," in Singer, I., ed. *Jewish Encyclopedia*. 12 vols. Vol. 4. New York: Funk and Wagnalls, 203.

Gottheil, R. and Broydé, I. 1904. "Kalonymus ben Kalonymus ben Meïr," in Singer, I., ed. *Jewish Encyclopedia*. 12 vols. Vol. 10. New York: Funk and Wagnalls, 426–9.

Høyrup, J. 1987. "The Formation of 'Islamic Mathematics' Sources and Conditions," *Science in Context* 1.2: 281–329.

Husain, A. A. 2002. "Conversion to History: Negating Exile and Messianism in al-Samaw'al al-Maghribī' s Polemic Against Judaism," *Medieval Encounters* 8.1: 3–34.Katz, V. J., *et al.*: Katz, V. J., Folkerts, M., Hughes, B., Wagner, R. and J. Lennart Berggren. 2016. *Sourcebook in the Mathematics of Medieval Europe and North Africa*. Princeton: Princeton University Press.

Langermann, Y. T. 1984. "The Mathematical Writings of Maimonides," *Jewish Quarterly Review* 75.1: 57–65.

Langermann, Y. T. 1987. *Ha-Mada'im ha-meduyaḳim be-ḳerev Yehude Teman* [The Jews of Yemen and the Exact Sciences] (Hebrew). Jerusalem: Misgav Yerushalayim.

Langermann, Y. T. 2002. "On the Beginnings of Hebrew Scientific Literature and on Studying History through 'Maqbilot' (Parallels)," *Aleph* 2: 169–89.

Lévy, T. 1989a. "Le chapitre I,73 du *Guide des égarés* et la tradition mathématique hébraïque au moyen âge: Un commentaire inédit de Salomon b. Isaac," *Revue des Etudes Juives* 148.3–4: 307–36.

Lévy, T. 1989b. "L'étude des sections coniques dans la tradition médiévale hébraïque. Ses relations avec les traditions arabe et latine," *Revue d'histoire de sciences* 42.3: 198–239.

Lévy, T. 1997. "The Establishment of the Mathematical Bookshelf of the Medieval Hebrew Scholar: Translations and Translators," *Science in Context* 10.3: 431–51.

Lévy, T. 2012. "The Hebrew Mathematics Culture (Twelfth – Sixteenth Centuries)," in Freudenthal, G., ed. *Science in Medieval Jewish Cultures*. Cambridge: Cambridge University Press, 155–71.

Lévy, T. and Rashed, R., eds. 2004. *Maïmonide: philosophe et savant (1138–1204)*. Leuven: Édition Peeters.

Mancha, J. L. 1997. "Levi ben Gerson's Astronomical Work: Chronology and Christian Context," *Science in Context* 10.3: 471–93.

Perlmann, M. 1964. "Samau'al al-Maghribī *Ifḥām Al-Yahūd*: Silencing the Jews," *Proceedings of the American Academy for Jewish Research* 32: 5–104.

Pines, S. 1960. "Abu 'l-Barakāt," in *EI-2*, 1: 111–13.

Rozen, M. 2002 *A History of the Jewish Community in Istanbul: The Formative Years, 1453–1566*. Leiden: Brill.

Sarfatti, G. B. 1968. *Mathematical Terminology in Hebrew Scientific Literature of the Middle Ages*. Jerusalem: Magnes Press.

Schloessinger, M., Broydé, I. and Gottheil, R. 1904. "Ibn Tibbon," in Singer, I., ed. *Jewish Encyclopedia*. 12 vols. Vol. 6. New York: Funk and Wagnalls, 544–50.

Schmidtke, S. 2010. "Samaw'al al-Maghribī, al-," in Stillman, N. A., ed. *Encyclopedia of Jews in the Islamic World*. Leiden: Brill, 238–40.

Segev, S. 2011. *The Book of Number by Elijah Mizrahi: A Textbook from the Fifteenth Century*. PhD thesis. Jerusalem: Hebrew University.

Sela, S. 2003. *Abraham Ibn Ezra and the Rise of Medieval Hebrew Science*. Leiden: Brill.

Sirat, C. 1990. *A History of Jewish Philosophy in the Middle Ages*. Cambridge: Cambridge University Press.

Sirat, C. 2015. "*Studia* of Philosophy as Scribal Centers in Fifteenth Century Iberia," in del Barco, J., ed. *The Late Medieval Hebrew Book in the Western Mediterranean: Hebrew Manuscripts and Incunabula in Context* (Études sur le judaïsme médiéval 65). Leiden: Brill, 46–69.

Steinschneider, M. 1902. *Die arabische Literatur der Juden*. Frankfurt am Main: J. Kauffmann.

Steinschneider, M. 2008². *Die Mathematik bei den Juden*. Berlin: Mayer and Müller, 1901 (Electronic Edition: Frankfurt am Main: urn:nbn:de:hebis:30–180130610004).

Wartenberg, I. 2016. *The Epistle of the Number by Ibn al-Ahdab: The Transmission of Arabic Mathematics to Hebrew Circles in Medieval Sicily*. Piscataway, NJ: Gorgias Press.

Zonta, M. 2006. *Hebrew Scholasticism in the Fifteenth Century: A History and Source Book*. Amsterdam: Springer.

V.3
PATRONAGE AND THE PRACTICE OF ASTROLOGY IN AL-ANDALUS AND THE MAGHRIB

Julio Samsó

V.3.1 Introduction

In medieval al-Andalus and the Maghrib, astrology was evidently widely practiced with different techniques at different social levels. The simplest form of the science, which required neither a deep knowledge of astronomy nor complicated mathematical computations, was practiced in the marketplace: treatises on *ḥisbat al-sūq* (policing of the market), such as that of al-Saqaṭī (*fl. c.* 596–621/1200–1225), forbade this type of market astrology, which suggests it was common (Chalmeta 1967–1968). While we have no information about these professionals, it is likely that certain developments characteristic of Andalusī astronomy, including equatoria and perpetual almanacs designed to simplify the computation of planetary longitudes, were the result of the demands of popular astrologers who needed to sell a product (a horoscope) at a reasonable price (Samsó 2011[CB],[1] 102–10, 166–71, 199–207, 313–6). Some data do, however, survive as to the techniques used for such predictions, represented by the "system of the crosses" (*ṭarīqat aḥkām al-ṣulub*), on which see below, documented in al-Andalus since the late 2nd/the beginning of the 9th century and still used in the Maghrib in the 9th/15th (Alfonso X 1961; Samsó 1979, 1983; al-Baqqār 2018). Other sources representing this kind of popular astrology include the collection *Alchandreana*, known through Latin translations made in Catalonia in the 4th/10th century (Juste 2007), and Ramon Llull's (1232–1316) *Book of the New Astronomy* (*Liber de nova astronomia*), written in Paris in 1297, probably derived from Maghribī sources (Llull 1989, 2002).

By contrast, we are far better informed about the social history of those elite professional astrologers who moved in the circles of power, where patronage patterns followed the larger evolution of political and religious history. Most of the important astrological sources still extant are tied in some way to a patron, usually a monarch or emir, who was personally interested in the subject. With the waning of royal and elite interest in astrology, therefore, the sources, too, become scarce, and astrological practitioners more anonymous. In such periods, emphasis would sometimes be placed more on astral magic than astrology proper. As that may be, scientific patronage was mainly the prerogative of caliphs, emirs and kings, who were naturally interested in the prediction of future events, and hence in a serious kind of mathematical astrology that could guarantee a scientific result.

DOI: 10.4324/9781315170718-52

In this chapter, I begin with an analysis of the characteristics of both kinds of astrology, followed by an attempt to trace an outline of the development of this discipline on both sides of the Strait of Gibraltar between the 2nd/8th and the 9th/15th centuries, although we have much more information about such practice in al-Andalus than in the Maghrib. Nevertheless, even in the latter region the Aghlabid Emir Ibrāhīm II (261–89/875–902) founded the *House of Wisdom* (*Dār al-ḥikma*) in Raqqāda, where astronomical instruments were built by ʿUthmān al-Ṣayqāl (d. 329/941). Ismāʿīl al-Ṭallāʿ, Ibrāhīm II's court astrologer, accompanied him on his military campaigns. Information about Andalusī astrology starts in the first half of the 3rd/9th century, under the reign of ʿAbd al-Raḥmān II (r. 206–238/822–852) and its subsequent development seems to have depended on the attitude adopted toward this discipline, by the ruling emirs and caliphs. The multiplication of the number of monarchs, during the period of the *mulūk al-ṭawāʾif* (party kings, 5th/11th century) implied an increase of the number of patrons of astrology and the flowering of this discipline. This century saw the publication in Zīrid Tunis, shortly after 429/1038, of the most important Western Islamic astrological handbook, Ibn Abī l-Rijāl's *Outstanding Book on the Judgment of the Stars* (*al-Bāriʿ fī aḥkām al-nujūm*), a most surprising event given the scarcity of information about the practice of astrology in the Maghrib in this period. In the 6th/12th and 7th/13th centuries al-Andalus and the Maghrib became united under the Almoravid (*c.* 454–541/*c.* 1062–1147) and Almohad (*c.* 524–668/*c.* 1130–1269) empires, who did not favor the practice of astrology, although some evidence can be gathered about the continuity of astrological activities. This chronological survey ends with an analysis of the free development of astrology in Nasrid Granada (7th–9th/13th–15th centuries) and in Merinid Morocco (7th–9th/13th–15th centuries), exemplified by the horoscopes cast for important events such as the Battle of Faḥṣ Ṭarīf/El Salado (740/1340), the murder of Sultan Abū ʿInān (r. 748–759/1348–1358) and the astrological history of Merinid Morocco during the period 748–773/1348–1372.

V.3.2 Elite and popular western Islamic astrology in practice: two examples

Astrology, then, remained prevalent at all levels of society in al-Andalus and the Maghrib through the 9th/15th century. At the elite, courtly level, where patronage was variable and often tied to the proclivities of a given ruler, the science waxed and waned but grew steadily in sophistication after the introduction of Eastern methods. At the popular level, by contrast, the simplified "system of the crosses" seems to have been consistently used from the 3rd/9th century onward. To give a taste of how these varieties of Andalusī and Maghribī astrology were actually practiced, I give two examples in the following.

Elite patrons, of course, could afford more scientifically exact and hence more expensive astrological consultations. Here the procedure would begin with the division of the ecliptic into houses, using complicated mathematical procedures, and determination of the position of the planets using ecliptical coordinates. This was systematically done, such that two astrologers using the same set of astronomical tables would obtain the same results. When the horoscope had been cast, however, it was up to each astrologer to choose the set of variables, of which there were many, and consider the most relevant to a given case. Interpretations and predictions could thus diverge widely and did.

As an example, we can analyze what survives of the oldest known Andalusī horoscope. Cast by the poet Yaḥyā al-Ghazāl (156–249/773–864), its substance has been preserved in a short poem quoted by Ibn Ḥayyān (377–467/987–1076; 1973, 11–12), It consists of a prediction of the death of the eunuch (*fatā*) Naṣr, the powerful chamberlain (*ḥājib*; his function corresponds today to that of the prime minister) of ʿAbd al-Raḥmān II, which took place immediately after the end of Shaʿbān 236 (29 Shaʿbān 236/7 March 851; for an analysis of the poem see Samsó 2004, 482–8, 2020[CB], 156-8). The poem begins:

> Tell Eunuch Naṣr, Abū l-Fatḥ:
> Saturn is in al-Naṭḥ.
> I see it retrograde and advance,
> going backwards and forwards till it reaches *al-rumḥ*.

The star *al-naṭḥ*, α Arietis, here stands for the first lunar mansion, al-Sharaṭān, belonging to the sign of Aries, whose longitude in the first half of the 3rd/9th century was about 20°. Using al-Khwārazmī's (d. *c.* 235/850) astronomical tables, we can therefore calculate the position of Saturn for 7 March 851 as 22° 7′ 26″, which indeed corresponds with *al-naṭḥ*. In addition, Saturn, according to al-Khwārazmī's tables, reached a station (that is appeared to be motionless) on 23 Muḥarram 236/6 August 850 (longitude 23° 29′ 23″), and its retrogradation (apparent backward motion) lasted from 25 Muḥarram/8 August until 10 Jumādà al-thānī/19 December of the same year (longitude 17° 8′ 34″). The arc of retrogradation was thus 6° 20′ 50″, which seems to explain the mention of *al-rumḥ*, an angular unit used by sailors, whose values oscillate between 4;30° and something more than 14°.

The poem continues:

> I see that the malefics (*nuḥūs*) are helping it:
> So take care of yourself and follow my advice!

The *nuḥūs* are indicators of misfortune and, in the case of the planets, they are identified with Saturn and Mars. Both planets were in conjunction (at the same ecliptical longitude, Saturn 19° 53′ 45″ and Mars 19° 58′ 34″) three weeks prior (8 Shaʿbān 236/14 February 851). In addition, on 29 Shaʿbān 236/ 7 March 851 Saturn was in a position of strength, having recovered its direct (forward) motion after retrogradation and positioned to the east of the sun (*tashrīq*, orientality), the solar longitude being 347° 21′ 8″ at that date. Mars, the other malefic, whose longitude on 29 Shaʿbān/7 March was 34° 49′ 48″, was also in orientality. Furthermore, at the moment of the Saturn-Mars conjunction of 8 Shaʿbān/14 February the ascending lunar node had a longitude of 20° 59′ 18″, at a distance of only 1° from the two malefic planets. Al-Bīrūnī (362–d. after 444/973–d. after 1053) informs us that, according to the Babylonians, the ascending node increases the effect of both benefic and malefic planets. In this case, therefore, the ascending node is likewise unfortunate.

The actual prediction of Naṣr's death then appears in the following two verses:

> When you see the full moon (*badr*) in *bulaʿ*
> an end will be met in the most disgraceful way.

Here the text refers to the twenty-third lunar mansion, *saʿd bulaʿ*, in Capricorn, which in the first half of the 3rd/9thcentury was placed between 295° and 298° of longitude. The moon being full, the sun must be in opposition to it, and hence with a longitude between 115° and 118° – corresponding not to Shaʿbān-Ramaḍān 236/March 851 but to Muḥarram 237/July 851. The prediction therefore seems to correspond, approximately, to 13 Muḥarram 237/17July 851, on which the positions of the sun and the moon were 112;56,18° and 296;42,2°, respectively. This suggests that Yaḥyā al-Ghazāl must have been rather surprised when Naṣr died rather on 1 Ramaḍān 236/8 March 851, four months earlier than the date he had predicted.

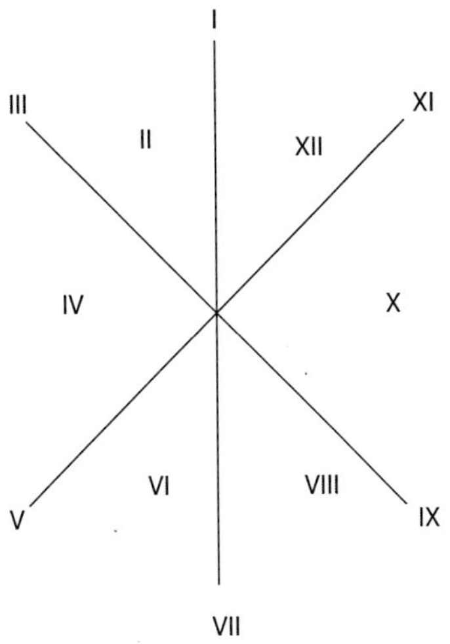

Figure V.3.1 Diagram of the horoscope of the crosses

Source: Julio Samsó. Barcelona

As for popular astrological practice as pursued in the bazaar, it is mostly invisible in the surviving sources, but we may nevertheless speculate as to the nature of the simpler techniques used for less wealthy patrons. In particular, as noted, it seems to have centered on the "system of the crosses" as described in both Arabic and Castilian sources. Here the horoscope has a format different from the standard, more complicated ones: three straight lines intersecting in the center of the figure determine twelve positions (Figure V.3.1). They correspond to the twelve "houses" that, in more sophisticated astrology, are twelve unequal divisions of the ecliptic in which the Sun, the Moon and planets are located and refer to different life sectors in which their influences are most felt. Here, however, the twelve houses are all equal length and have a one-to-one correspondence with the twelve zodiacal signs, beginning with the sign that crosses the eastern horizon at a given moment (the ascendant, marking the beginning of house I). In this way the astrologer does not need to calculate the beginning of each house with a precision of degrees and minutes. He merely needs to know which sign is ascending in the east at the moment of his casting the horoscope: if the ascending sign is Libra (house I), house II will be Scorpio, house III Sagittarius, and so on.

The predictions of the "system of the crosses" are based on many fewer variables, which can be easily listed in tables using combinatorial techniques. With this method, the astrologer needs only to establish the variables that concern a particular case and then read the corresponding prediction, already provided in the relevant list, without having to bother with the computation of planetary longitudes, the division of houses and other technicalities of standard astrology. The system of the crosses deals mainly with weather prediction, periods of drought and rains and their influence on the fluctuation of prices of agricultural products and uses mainly the positions of the superior planets (Saturn, Jupiter and Mars). As an example of a weather prediction for a full year, taken from al-Baqqār (*fl.* around 820/1418), we can read that

- if Saturn and Jupiter are in fire signs: drought and many fires due to the dryness of the land, shortage of oil, fruits and agricultural products, especially in the eastern parts of the country.

Table V.3.1 Ramon Llull's table of correspondences

Letters	*Elements*	*Zodiacal signs*	*Planets*
B	Fire (dry and ***hot***)	Aries, Leo, Sagittarius	Sun, Mars
C	Earth (***dry*** and cold)	Taurus, Virgo, Capricorn	Saturn
A	Air (***moist*** and hot)	Gemini, Libra, Aquarius	Jupiter
D	Water (moist and ***cold***)	Cancer, Scorpio, Pisces	Moon, Venus

- if Saturn and Jupiter are in earth signs: fertility both in plains and mountains, plenty of food.
- if Saturn and Jupiter are in air signs: winds, many fruits (like olives and walnuts), excessive rain and floods damaging to agriculture, higher prices, and fertility limited to the mountains.
- if Saturn and Jupiter are in water signs: fertility during most of the year, especially along the coast, abundant fruits, rain and floods, diseases in cattle and food shortages in the mountains, moderate prices (al-Baqqār 2018, 195, no. [14]).

Similar examples may be found in Ramon Llull's *Book of the New Astronomy*, which, as noted, was probably derived from Maghribī astrological sources. He rather assigns, for instance, the four elemental triplicities to four letters (A, B, C, D). Each triplicity is characterized by two qualities, one essential and one secondary (essential qualities are marked in bold and italics in Table V.3.1) and associated with the appropriate planets.

Let us analyze a hypothetical situation according to this table of correspondences (Llull 2002, 173). A conjunction of Jupiter, Mars, the Sun and Venus take place in the sign of Aries. This combines to

3 dry qualities (Aries, Mars, Sun),
4 hot qualities (Aries, Mars, Sun, Jupiter),
2 moist qualities (Venus, Jupiter), and
1 cold quality (Venus).

The problem is to determine whether the dominant planets are Venus and Jupiter (both benefic) or Mars (a malefic), the solar influence being ambiguous. At first sight, the situation seems to favor Mars, which shares with the sun a fiery nature and an association with the sign of Aries. But one should also bear in mind that Jupiter has a certain influence here as well, adding to the hot qualities, although heat is the planet's secondary quality and moistness its primary, and the two moist qualities are counterbalanced by three dry qualities. The influence of Venus is minor, as its own quality is cold (only 1 cold quality vs. 4 hot qualities), while moistness is its appropriated quality (2 moist qualities vs. 3 dry qualities). As a result, Llull concludes that "in the constellation there is more evil than good", since malefic influence dominates.

V.3.3 Astrology in al-Andalus (92–235/711–850)

As for al-Andalus, the origins of the practice of astrology are related, according to legend, to the period of the conquest: Muslim leaders like Mūsā ibn Nuṣayr (19–97/640–716) had, according to historical sources, some knowledge of astronomy and the ability to predict the future (Marín 1986; Samsó 1985–6). More detailed information survives from the end of the 2nd/8th century, when under Hishām I (r. 172–180/788–796) we find the first references to Umayyad elite interest in astrology; this emir, while known for his piety, employed al-Ḍabbī (d. *c.* 237/*c.* 852), the first known Andalusī astrologer, and even required him to calculate the length of his life and of his reign – a common but dangerous question for any astrologer (Samsó 2001).

By the time of al-Ḥakam I (r. 180–206/796–822) and especially ʿAbd al-Raḥmān II the situation is clearer. The former had his son, the future ʿAbd al-Raḥmān II, tutored in the sciences of the ancients, and both emirs had, at their courts, an important group of poet-astrologers--well attested in the sources precisely because they were poets. The publication of vol. II.1 of Ibn Ḥayyān's *The Book based on Borrowed Material* (*al-Muqtabis*, [On the History of the Men of al-Andalus], 2001, 2003; usually translated as *The Gleaner*) has brought to light new information on these individuals (al-Ḍabbī, Yaḥyā al-Ghazāl, Ibn al-Shamir/Shimr [3rd/9th century], ʿAbbās ibn Firnās [d. 274/887], Marwān ibn Ghazwān [3rd/9th century] and others), for the book contains an important chapter titled *Stories about the Astrologers and Emir ʿAbd al-Raḥmān II* (*Akhbār al-munajjimīn maʿa l-amīr ʿAbd al-Raḥmān ibn al-Ḥakam*; Forcada 2002, 2004–2005[CB]; Rius 2003). The chapter in question mainly relates an attempt by the emir to ensure the accuracy of the astrological predictions made by his astrologers; at least two of the anecdotes concerned were also attributed to eastern astrologers, such as Abū Maʿshar (171–272/787–886) and al-Bīrūnī. The new information provided by this source shows how interested ʿAbd al-Raḥman II and the members of his court were in astrology and hence astronomy: the emir himself was said to be able to calculate planetary positions, and a musician of the caliber of Ziryāb (d. 238/852?) is presented as having a thorough knowledge of both disciplines. Among the courtiers mentioned by Ibn Ḥayyān, al-Ḍabbī, Marwān ibn Ghazwān and Ibn al-ʿAdhrāʾ(2nd half 3rd–1st half 4th/2nd half 9th–1st half 10th centuries) were astrologers only, while al-Ghazāl, Ibn al-Shamir and ʿAbbās ibn Firnās were also poets. It is not yet known whether there was an official payroll (*dīwān*) of astrologers who received a royal salary similar to the payroll of poets (*dīwān al-shuʿarāʾ*) or the payroll of physicians (*dīwān al-mutaṭabbibīn)* that existed in Córdoba under Caliph al-Ḥakam II (r. 350–366/961–976). The only hint we have of such an arrangement is the fact that Ibn Saʿīd (1953, 125) reports that ʿAbd al-Raḥmān II assigned to Ibn al-Shamir a double salary (*rizq*) for his poetry and his astrology (*tanjīm*). Ibn Ḥayyān (2003, 348, 2001, 226) asserts the same about ʿAbbās ibn Firnās. It is doubtful that they were the only such cases.

Our astrologers cast horoscopes, naturally, and some of them are of momentous import: Ibn al-Shamir (correctly) predicted that ʿAbd al-Raḥmān II would attain the throne and, like al-Ḍabbī (and Mūsā ibn Nuṣayr), predicted the time of his own death. Furthermore, a poem of Ibn al-Shamir states that the line of the kings of Córdoba (that is the dynasty of the Umayyads) will end once Saturn has passed over them six times. As Saturn's revolution period is approximately 30 solar years, this is equivalent to some 180 years; assuming that the poem was written ca. 235–6/850, this astrological prognostication leads us to 850 + 180 = 1030 – corresponding to the actual end of the Umayyad Caliphate of Córdoba. These horoscopes, as well as many others, are given by Ibn Ḥayyān without technical details. As already described, however, we have a poem by Yaḥyā al-Ghazāl that predicts the death of Naṣr, the powerful chamberlain of ʿAbd al-Raḥmān II.

V.3.4 The partial eclipse of astrology between *c.* 235/850 and 422/1031

Astrology was vehemently attacked by jurists like Ibn Ḥabīb (d. 238/853); yet he accepted the existence of a legitimate tradition of knowledge about the stars, called by modern historians folk-astronomy (*anwāʾ* and *manāzil*; Ibn Ḥabīb 1994[CB], 1997[CB]). The astrological tradition does seem to have suffered partial eclipse, at least at court, under Muḥammad I (r. 237–272/852–886): most of the court astrologers of ʿAbd al-Raḥmān II seem to have continued in their positions during the reign of Muḥammad, but little is known about their activity and, according to Ibn Ḥayyān, the latter emir sentenced al-Ḍabbī to death, and imprisoned Marwān ibn Ghazwān. In both cases, however, their crime was not astrology but rather their failure to keep the secrets of the palace. This decline in courtly astrological patronage continued under the rulers of al-Andalus who followed Muḥammad, through the reign of Caliph ʿAbd al-Raḥmān III (r. 299–349/912–961). At the turn of the 4th/10th century, only one astrologer is known, Ibn al-Samīna (d. 315/927–928), who was also competent in the literary and religious sciences, as well as in medicine. In spite of this paucity of information, it is clear that Cordovan astrologers must have continued practicing their profession. An anecdote mentioned by the poet and expert in *adab* Ibn ʿAbd Rabbih (246–328/860–940), for example, shows the astrologer Ibn al-ʿAdhrāʾ and a group of colleagues attempting to predict the end of a period of drought (Ibn Marzūq 1981, 443, 1977, 365–6). Moreover, in the late 3rd and early 4th/1st half of the 10th century, a greater emphasis on astral *magic* is evident: during this period, we see the publication of the seminal grimoire *Goal of the Sage* (*Ghāyat al-ḥakīm*, its Latin translation is called *Picatrix*), authored by Maslama ibn al-Qāsim al-Qurṭubī (d. 353/964), who should not be confused with the mathematician and astronomer Maslama ibn Aḥmad al-Majrīṭī (d. before 397/1007; Fierro 1996; de Callataÿ 2013). At around the same time, Ibn al-Ḥātim (*fl.* 327/939), about whom we only know that he observed, somewhere in al-Andalus, the solar eclipse of 28 Ramaḍān 327/19 July 939, wrote a much shorter astral-magical work, known in Christian Europe through a Latin translation titled *On Celestial Images* (*De imaginibus caelestibus*; Oliveras 2009).

The rule of the chamberlain (*ḥājib*) al-Manṣūr (r. 370–392/981–1002) saw a rejection of the practice of astrology, symbolized by a selective burning of the library of al-Ḥakam II – including astrological books. According to Ibn al-Khaṭīb (713–776/1313–1374 or 1375; 1934, 89, 1956, 77), al-Manṣūr arrested, tortured and executed soothsayers (*mukahhinūn*) and astrologers (*munajjimūn*) for predicting the end of his regime. Ibn ʿIdhārī ([d. after 712/1312–3]; 1980, 2:294), for his part, states that Muḥammad ibn Abī Jumʿa (2nd half 4th/10th century) published false rumors about the existence of an astral "cutting" (*qaṭʿ*) portending the end of al-Manṣūr's regime. *Qaṭʿ* is an astrological term implying the existence of a malefic astral element, which interrupts a particular process, such as the life of the subject of a horoscope. The chamberlain (*ḥājib*) cut out Muḥammad's tongue and crucified him for this reason. The predictions that so aggravated al-Manṣūr were largely focused on the Saturn–Jupiter great conjunction in Virgo of 398/1007 (Samsó 2004, 488–96): according to al-Khwārazmī's tables, both planets had the same longitude (158° 43′) at 10 h.p.m. on 5 Rabīʿ al-awwal/19 November of that year. The phenomenon and the rumors associated with this celestial event must have produced, among the elites of Córdoba, including al-Manṣūr, a terror similar to the panic which took place in Christian countries with the change of the millennium. Four sources give information on the predictions associated with this conjunction: Ibn ʿIdhārī (1980, 3: 14–5), Ibn al-Khaṭīb (1934, 148–9, 1956, 127–8), the Alfonsine *Libro de las cruzes* (Alfonso X 1961, 9–10) and the commentary by Ibn Qunfudh (740–810/1339–1407) on Ibn Abī l-Rijāl's (d. 453/1062) astrological *urjūza* (poem written in rajaz meter versification; Ibn Qunfudh 2012, 204 [Arabic] and 356 [tr.]). All four sources predict disasters: the fall of the state, war, death and famine. One of the specialists involved in these predictions was the well-known mathematician and astronomer Maslama al-Majrīṭī.

Such evidence from the highest echelons of Andalusī society shows that, despite al-Manṣūr's distrust of astrology, the science was alive and well even in the second half of the 4th/10th century and beginning of the 5th/11th. Clearly, the chamberlain's negative attitude was motivated by concerns more political than scholarly. Nor was it total: astronomer-astrologers (here termed *muʿaddil*s or "calculators," by Ibn Bassām) yet seem to have numbered among the tutors of al-Manṣūr's son, ʿAbd al-Malik al-Muẓaffar (364–398/975–1008). The same authority, drawing on Ibn Ḥayyān's *Solid Book* (*Kitāb al-Matīn*),[2] explains that the astrologer Aḥmad ibn Fāris had studied the nativity horoscope of al-Muẓaffar and concluded that he had never seen a horoscope with happier prospects. Ibn Ḥayyān proceeds to satirize this prediction, given al-Muẓaffar's decidedly unhappy end (Ibn Bassām 1979, 7: 79).

Al-Muẓaffar's horoscope was hardly the only one to be so analyzed: there is evidence showing that the upper classes continued to patronize astrology, in spite of its persecution at court, and had horoscopes cast at the moment of their children's birth. Such was the case of the famous scholar Ibn Ḥazm (384–456/994–1064), whose father occupied important posts in al-Manṣūr's administration. The author himself reports that he was born in Córdoba on 7 November 384/994, just after the end of the dawn prayer and before sunrise, with his ascendant in Scorpio; he was thus born at about 6:30 a.m. (al-Maqqarī 1968, 2: 78–9; Ṣāʿid al-Andalusī 1985[CB], 184). That some elite opposition to astrology remained, however, is suggested by the fact that Aḥmad ibn Fāris, mentioned earlier, felt the need to defend astrology, during the reign of al-Manṣūr, by linking it to the more religiously acceptable folk-astronomy (Forcada 1996; Ibn Fāris 2000).

With the arrival of the period of civil wars (*fitna*) after the fall of the Umayyad Caliphate, however, elite interest in astrology and astronomy rebounded in al-Andalus, although not necessarily in Córdoba. Ibn al-Samḥ (d. 426/1035), for example, a disciple of Maslama al-Majrīṭī, left Córdoba for Granada, where he gained the protection of the Zīrid ruler Ḥabbūs ibn Māksan (r. 410–429/1019–1038); Ibn al-Ṣaffār (d. 426/1035), another of Maslama's disciples, fled to Denia and was patronized by Mujāhid (r. *c.* 402–436/1012–1045). While it is unclear what forced these migrations of important astronomers to Granada and Denia, it is possible that the members of Maslama's school had to leave Córdoba precisely because they had been discretely patronized and protected by the ʿĀmirid dynasty (r. 368–399/979–1009 during regency in Córdoba; 411–478/1020–1085), al-Manṣūr's lack of support notwithstanding. Here the case of Ibn al-Ṣaffār is particularly suggestive, given that Mujāhid of Denia was a client (*mawlā*) of the ʿĀmirids and his kingdom was a standard place of asylum for those who had served al-Manṣūr's family.

V.3.5 The period of the Party Kings (*mulūk al-ṭawāʾif*, *c.* 422–478/1031–1085)

Despite – or because of – political fragmentation, the next half century proved to be a golden age of Andalusī astronomy and astrology: as kings multiplied, so did patronage. This development is emphasized in Ṣāʿid's *Classes of Nations* (*Ṭabaqāt al-umam*), and contrasted with the unhappy period of al-Manṣūr's autocracy (Forcada 2015). Thus al-Muʿtamid ibn ʿAbbād of Seville (r. 461–483/1069–1091), for instance, patronized the works of Ibn al-Zarqālluh ([d. 493/1100]; al-Zarqālī), while the latter was still living in Toledo; for twenty years, he had both the Jewish astrologer Yiṣḥaq al-Baliyya (425–485/1034–1093) and the Muslim astrologer al-Khulānī (Balty-Guesdon 1992, 315, 667–8, 693) in his service.

Toledo was the great center of astronomical learning during this period. There, interest in astrology had already begun during the *fitna*: some information survives about the activities of Ibn Abī Thalla (d. 434/1043) in that city. Nor was patronage restricted to kings: the Banū Dhakwān family supported the studies of the physician and astronomer Ibn al-Ḥannāṭ (d. 436/1045; Balty-Guesdon 1992, 283, 643–4). The most important case is, however, that of the judge (*qāḍī*) Ṣāʿid of Toledo (419–462/1029–1070), author of the well-known *Classes of Nations* (*Ṭabaqāt al-umam*). Ṣāʿid, moreover, assembled the team of astronomers who compiled the *Toledan Tables*. Chief among them was Ibn al-Zarqālluh, who also, tellingly, composed a treatise on astral magic that would be summarized in Latin, titled *On the Movements of the Planets and Their Harnessing* (*Risāla fī ḥarakāt al-kawākib al-sayyāra wa-tadbīrihā*); another was al-Istijī (5th/11th century). Ṣāʿid mentions the latter twice in his *Classes* (Ṣāʿid 1985, 180, 199–200) as one of the young astronomers working in Toledo when *Classes* was completed (i.e., in 460/1068) and as an expert in astrology, who had written the excellent treatise *On Progressions and Projections of Rays, and an Explanation of Certain Principles of this Craft* (*Risāla fī l-tasyīrāt wa-maṭāriḥ al-shuʿāʿāt wa-taʿlīl baʿḍ uṣūl al-ṣināʿa*), dedicated to Ṣāʿid while he was in Cuenca (Samsó and Berrani 1999; Istijī 2005).

Ṣāʿid accordingly describes al-Istijī as "one of those who have a sound knowledge of astrology and who have read the books on this subject by ancient and modern authors alike" (Ṣāʿid 1985, 199–200). This remark is likely no mere hyberbole, as we can see by an analysis of the sources quoted in the text, most of which are also mentioned by Ṣāʿid himself. The list is particularly interesting as it provides a window onto the astrological works that were circulated in al-Andalus in the 5th/11th century. Among the ancient books are Ptolemy's (c. 100–170) *Tetrabiblos*, Hermes[3] and the anonymous *Sayings of the Persians* (*Kitāb al-amthāl li-l-Furs*); here, significantly, al-Istijī complains of the mistakes made by translators as a likely cause of the errors of practicing astrologers (al-Istijī 2005, 214–5, 240). Among authors of the Islamic period, we find Kankah al-Hindī (*c.* 158–204/775–820; Baghdad), Abū Maʿshar (*On Religious Communities and States* [*Kitāb al-milal wa-l-duwal*], *Conjunctions* [*Kitāb al-qirānāt*], the *Memoirs* [*Mudhākarāt*] and *The Great Introduction* [*al-Madkhal al-kabīr*]), al-Battānī (244–317/858–929), al-Khaṣībī (probably al-Ḥusayn/ al-Ḥasan ibn al-Khaṣīb [*fl.* around 229/844]), Aḥmad ibn Yūsuf al-Kātib (d. 329/941), al-Ḥasan ibn Aḥmad al-Hamdānī (280–334/893–945 or 946), and the 4th/10th-century *Epistles* of the Brethren of Purity (*Rasāʾil Ikhwān al-ṣafāʾ*).

Andalusī royal patronage of astrology reached its apex under the Dhū l-Nūnid ruler al-Maʾmūn of Toledo (r. 435–467/1043–1075), whose honorific title (*laqab*) is probably in mimesis of that of the famous Abbasid Caliph al-Maʾmūn, a distinguished patron of astronomy and astrology. The Toledan al-Maʾmūn's court astrologer was Ibn al-Khayyāṭ (*c.* 386–446/977–1055), a disciple of Maslama al-Majrīṭī, who had previously been court astrologer to the Umayyad Caliph Sulaymān al-Mustaʿīn (r. 399, 403–406/1009, 1013–1016; Balty-Guesdon 1992, 278–9, 643–4). According to the Emir ʿAbdallāh ibn Buluqqīn (r. 465–483/1073–1090; Ibn Buluqqīn 1986, 94), Ibn al-Khayyāṭ correctly predicted the conquest of Denia by the king of Zaragoza, al-Muqtadir ibn Hūd (r. 438–474/1046–1081) in 468/1076; al-Muqtadir's son was the famous king-mathematician al-Muʾtaman ibn Hūd (r. 474–478/1081–1085). Ibn al-Khayyāṭ dedicated to al-Maʾmūn a *Treatise on Planetary Conjunctions* (*Risāla fī l-qirānāt al-nujūmiyya*), which deals, among other things, with current events in al-Andalus and the question as to how long Muslims would be able to remain there; significantly, there is some discussion of celestial signs indicating that they would soon have to emigrate. This epistle is not known to be extant, but the Moroccan astrologer al-Baqqār has preserved 91 verses of Ibn al-Khayyāṭ's astrological *Lāmiyya* (poem rhyming in *L*) in his *Book of Rains and Prices* (*Kitāb al-amṭār wa-l-asʿār*). Al-Baqqār's text is not clear but gives enough information to conjecture that the *Lāmiyya* was written around 441/1050. It contains simple predictions based on the passage of Saturn through the four elemental triplicities and through the twelve zodiacal signs, as well as references to actual events which took place during the author's lifetime: the solar eclipse of 28 Rajab 424/29 June 1033, a conjunction of Saturn and Mars on 17 Shawwāl 424/15 September 1033, an occultation of Saturn by the Moon on 18 Jumādā al-thāniya 436/16 January 1045, an earthquake in 415/1024–1025 and famine during the period of, apparently, 439–450/1048–1058.

The end of the Taifa period – so fruitful for astrology – was marked by two somber events: the conquest of Toledo in 477/1085 by Alfonso VI of León and Castile (r. 1065–1109) and the arrival of Yūsuf ibn Tāshufīn (r. 452–500/1061–1106), leader of the Almoravids, one year later. Piquantly, al-Ma'mūn's grandson and inept successor, al-Qādir ibn Dhī l-Nūn (r. 468–477/1076–1085), the last Muslim king of Toledo, left the city holding an astrolabe with which he was attempting to establish the propitious moment of departure from Toledo. According to eye-witnesses, "he was surrounded by Christians and Muslims. The former were laughing at him, while the latter were marveling at his ignorance" (Ibn Bassām 1979, 7:167).

In 482/1090 the Almoravids conquered Granada and deposed the last king of the Zīrid dynasty, ʿAbdallāh ibn Buluqqīn, exiling him, after a trial, to Aghmāt near Marrakesh. During his exile, ʿAbdallāh wrote a fascinating book of memoirs in which he justified his alliance with Alfonso VI against the Berber invasion with reference to the influence of the stars. As one of the arguments in his self-defense, he presented his own horoscope, cast on the occasion of his fourth anniversary (Ibn Buluqqīn 1986, 174–7; Samsó 1990, 2020, 166–9). The very incomplete parameters given by this source allow us, nonetheless, to infer 5 Dhū l-Qaʿda 447/26 January 1056 as the king's birth date.

V.3.6 The Maghrib and al-Andalus during the Almoravid and Almohad Empires

As mentioned, we have very little information about the practice of astrology in the Maghrib between the 2nd/8th and the middle of the 5th/11th centuries. The lack of references in historical works and the fact that no written astrological sources seem to be extant sharply contrasts with the sudden appearance, sometime after 429/1038, of Ibn Abī l-Rijāl's *Outstanding Book on Judicial Astrology*, without doubt the most important astrological handbook of the Islamic West. It was equally influential in Christian Europe through its Castilian and Latin translations. (In this it parallels the influence of the *Picatrix* in astral magic.) Two aspects of this work should be specifically mentioned. On the one hand, it exclusively depends on a large number of Eastern sources, without mentioning a single Maghribī title; it would seem that the Maghribī astrological tradition had not yet reached the requisite level of sophistication. On the other hand, Ibn Abī l-Rijāl served, not only as an astrologer, in important functions at the court of the Zīrids Bādīs ibn al-Manṣūr (r. 385–406/996–1016) and his son al-Muʿizz ibn Bādīs (r. 406–454/1016–1062) at Qayrawan. Ibn Abī l-Rijāl's position as chief secretary to Bādīs and minister of either Bādīs or al-Muʿizz again highlights the importance of royal patronage for the development of the sciences in general and astrology in particular. The text of the *Bāriʿ* contains references to horoscopes that the author cast on Bādīs's and al-Muʿizz's orders. Ibn Abī l-Rijāl is also the author of an astrological versification (*Urjūza fī aḥkām al-nujūm*) in which he summarizes, in 467 lines, everything a competent astrologer has to master (Ibn Qunfudh 2012). Significantly in the present context, the versification emphasizes astrological issues related to political power, particularly the prediction of the course and length of a given sovereign's reign. These factors indicate that the work was written for the benefit of prominent personages at the Zīrid court in Tunis.

A lack of elite patronage is likely the main cause of the subsequent decline in public astrological practices as reported by the sources, both in al-Andalus and the Maghrib, during the Almoravid (*c.* 454–541/*c.* 1062–1147) and Almohad (*c.* 524–668/*c.* 1130–1269) periods (Forcada 2015). Nevertheless, important works on the sciences were still produced, such as the *Keys to Secrets* (*Kitāb mafātīḥ al-asrār*) of Ibn al-Kammād (active around 509/1116). Unfortunately, only the chapters describing the *animodar* (an astrological technique for calculating the beginning of a fetus's gestation) are extant, although these were well known in the Maghrib (Ibn al-Kammād 2016–7).

Despite this relative decline in elite patronage, popular belief in astrology seems, as always, to have persisted unhindered. This conclusion is supported by an anecdote dated to 580/1185:

according to al-Marrākushī (1964, 4: 210–4), in that year a document sent "by the people of Egypt" and based on an Indian source, reached Málaga. It contained the prediction of a great natural catastrophe (*ṭūfān*), involving cyclonic winds and earthquakes, which would begin on 29 Jumādā II 582/16 September 1186 and last for three days in total. The astral cause of this event is here determined to be a conjunction of the five planets in the sign of Libra. The people of India had already prepared themselves for the disaster by digging underground shelters. This prediction induced a general panic, and some undertook measures similar to those of the people of India. The jurist Abū l-Ḥajjāj ibn al-Shaykh (6th/12th century?) wrote a poem to reassure the local inhabitants, deploying religious arguments against astrology and pointing out that the predicted catastrophe had not, in fact, taken place. References to the conjunction in question are also found in Eastern sources and in Latin texts from Spain and England (de Callataÿ 2000).

V.3.7 Astrology in Nasrid Granada and in the Maghrib between the 7th/13th and 9th/15th centuries: the horoscopes of the Battle of El Salado (740/1340)

While not another golden age of astrology, the post-Almohad period saw the continued activity of professional astrologers both in al-Andalus and the Maghrib. For Nasrid Granada, we have some information derived from Ibn al-Khaṭīb's encyclopedia of Granadine history, *Comprehensive Information on the History of Granada* (*al-Iḥāṭa fī ta'rīkh Gharnāṭa* 1973–7, 1: 205–6; 2: 91). Under the reign of Ismāʿīl II (760–761/1359–1360), for example, a horoscope was cast for the moment of Muḥammad al-Fihrī's installation as minister (it predicted the unhappy nature of his office). Aḥmad ibn Muḥammad al-Anṣārī was an astrologer in the service of Muḥammad VI (r. 761–763/1360–1362); he advised the ambitious prince as to a propitious moment to rebel against Muḥammad V (754–760/1354–1359, 763–793/1362–1391) and later predicted, correctly, that Muḥammad V would recover the throne in 763/1362.[4] The natal horoscope of Muḥammad V himself is likewise recorded.

Astrology was equally openly practiced in the Maghrib, in particular under the Merinid dynasty (r. 614–869/1217–1465). Ibn Marzūq (d. 780/1379) wrote a hagiography of Sultan Abū l-Ḥasan ʿAlī (r. 731–751/1331–1351), in which we find a whole chapter on Abū l-Ḥasan's rejection of astrology – although this, it seems, did not lead to serious prosecution of astrologers and astrological beliefs (Ibn Marzūq 1981, 438–44, 1977, 361–6). The text mentions Ibn al-Bannā''s (654–721/1256–1321) fame as an astrologer, and it is evident that the celebrated mathematician of Marrakesh was, at least early in his career, interested in astrology. In fact, the short astrological texts edited by Jabbār (Djebbar) and Aballāgh (2001, 160–84) from an Escorial manuscript are rather elementary in nature, and some of them seem to be a set of student's notes copied by Ibn al-Bannā' in his youth from different sources.

The pivotal battle of El Salado (*Faḥṣ Ṭarīf*) took place in 741/1340, during Abū l-Ḥasan's reign; it pitched the Moroccan Merinid army, in coalition with the army of Yūsuf I (r. 733–754/1333–1354) of Nasrid Granada, against the armies of Alfonso XI of Castile (r. 1312–1350) and Alfonso IV of Portugal (r. 1325–1357). The Muslim armies were defeated – contrary to the prediction by the astronomer-astrologer Ibn ʿAzzūz (d. 760/1354). He identified the source of his error as the faultiness of the astronomical tables of Ibn Isḥāq (active *c.* 588–618/*c.* 1193–1222) which he had used in casting the horoscope. To prevent future failures, Ibn ʿAzzūz therefore made his own astronomical observations in Fez in around 744/1344 and on this basis composed new tables of mean planetary motions. He also cast two new horoscopes, accurately corresponding to the known results of the battle (a common procedure known as rectification).

It is not clear whether the first horoscopes (not extant) were produced in response to an official court request (not necessarily by the sultan himself), or cast by Ibn ʿAzzūz on his own initiative (Samsó 1999).

Some twenty years later, Abū l-Ḥasan al-Qusanṭīnī dedicated his own astronomical tables, with canons written in *rajaz* verse, to the Merinid Sultan al-Mustaʿīn (r. 760–767/1359–1366). As the main purpose of a *zīj* is to compute planetary positions needed for casting horoscopes, the dedication to al-Mustaʿīn is clearly significant. That astrology was of special interest to contemporary ruling elites is likewise indicated by the work of another astrologer from Constantine, Ibn Qunfudh, who dedicated his commentary on the astrological poem of Ibn Abī l-Rijāl to Abū Bakr ibn Ghāzī ibn al-Kās (deported to Majorca in 776/1375), minister of the Merinid Sultans Abū Fāris (r. 767–773/1366–1372), and Abū Zayyān (r. 773–775/1372–1374; Ibn Qunfudh 2012). The book, probably written during the latter's brief reign, contains as examples a set of eleven horoscopes. Only two of them are dated and all are unidentified, but they constitute an astrological history of Merinid Morocco during the period 748–773/1348–1372, including the dynastic crisis that ensued upon the murder of Sultan Abū ʿInān (r. 748–759/1348–1358). As the minister, Ibn Ghāzī was seriously interested in astrology; it is possible that this collection of unidentified horoscopes was presented to him as a riddle in order to see whether he could guess the identity of the people or events described by each horoscope (Samsó 2004, 2009).

Out of the eleven horoscopes in Ibn Qunfudh's commentary, five appear to deal with the prediction of the duration of the reign of five Merinid sultans. This topic also interested Ibn Masʿūd ibn Farmīja ([*fl.* 773–796/1372–1394]; possibly Firmījuh = Bermejo), who was timekeeper (*muwaqqit*) in Fez, Tunis, Jerusalem and Damascus and wrote a short, largely compilatory and unoriginal work titled *Attaining Goals and Realizing Hopes: On Those Methods Whereby the Tenures of Governors and Authorities Is Revealed* (*Ḥuṣūl al-maqāṣid wa-l-āmāl min al-ṭuruq wa-l-fawāʾid allatī tuʿlamu minhā mudud al-wulat wa-l-ʿummāl*); it contains many passages from Ibn Abī l-Rijāl, Abū Maʿshar and a certain Abū Yūsuf (al-Kindī?). Nevertheless, it notably shows a *muwaqqit* to be seriously engaged in astrology, especially political astrology, and well versed in the technique of progression (*tasyīr*; Herrera 2001).

More revealing are the theoretical works of two astrologers who worked in Fez: the *Chapters on All the Principles [of Astrology]* (*Kitāb al-Fuṣūl fī jamʿ al-uṣūl*) by the aforementioned Ibn ʿAzzūz, and the *Book of Cycles: On the Progression of the [Celestial] Lights* (*Kitāb al-Adwār fī tasyīr al-anwār*) by al-Baqqār (al-Baqqār 2001; Díaz-Fajardo 2008), as well as the latter author's *Book of Rains and Prices*. These three texts deal with the techniques of progression and the projection of rays, use Saturn–Jupiter (and Saturn–Mars) conjunction theory and discuss the different techniques (on the ecliptic or on the equator) used for the computation of progressions and the projection of rays. Such features testify to the development of a specifically Maghribī astrological tradition; while firmly rooted in the works of Eastern astrologers like Ibn Hibintā (3rd/9th century) and al-Qabīṣī (4th/10th century), from Ibn Abī l-Rijāl onward it still attained a certain level of originality and sophistication.

Al-Baqqār's *Book of Rains and Prices* is unique in this context: its treatment of astrological meteorology is entirely independent of the eastern tradition represented by al-Kindī and his followers, being based rather on the purely Andalusī "method of the crosses" (*ṭarīqat aḥkām al-ṣulub*) – a technique, seemingly of late Latin origin, that was assimilated by Andalusī astrologers toward the end of the 2nd/8th century. To this, al-Baqqār adds more sophisticated procedures for weather prediction and establishes a correlation between rainfall and the fluctuation of prices. Here he adopts the point of view of farmers and merchants: when the malefic planets are exalted or dignified, prices decrease but increase with the ascendency of the benefics.

V.3.8 Conclusion

Until the profession of timekeeper (*muwaqqit*) was established in the 8th/14th century, the practice of astrology was the only possible way by which an astronomer could earn his living *qua* astronomer. In al-Andalus and the Maghrib, professional astrologers were widely patronized by the elite and commoners alike and employed different methods and techniques depending on the fees they could command for their services. Nonelites regularly consulted astrologers in markets, and the predictions they received, based on very simple criteria, were not dissimilar to those we find today in newspapers and popular magazines based only on the zodiacal sign of the sun in the consultant's horoscope. An entirely different and far more technical kind of astrology was deployed when the customer could afford a higher price. In that case, the astrologer would predict using many variables, as well as highly complicated mathematical computations, and the process of casting and interpreting the resulting horoscope would last for many hours.

Caliphs, emirs and other rulers were heavily invested in astrological predictions, which had an obvious influence on their decision-making. To this end, they often eagerly patronized astrological and astronomical research, including astronomical observations and the preparation of new astronomical tables, hoping that this research would ensure more precise and reliable predictions. As this chapter has shown, royal patronage of astrology flourished, despite occasional lapses, on both sides of the Strait of Gibraltar.

Overall, astrology is indispensable for the writing of premodern social history since astrological texts provide a snapshot of the daily concerns of every echelon of society, from rulers to farmers. Two examples must here suffice. Books I through III of Ibn Abī l-Rijāl's *Outstanding Book* deal with interrogations (queries presented to the astrologer), while book VII is concerned with elections (choice of the propitious moment to undertake a particular activity): the casuistry presented in these four books contains a most interesting sample of the concerns of an average man or woman living in the Maghrib in the 5th/11th century. Similarly, four centuries later, we can find in al-Baqqār's *Book of Rains and Prices* long lists of –especially agricultural – products used in everyday life, and conjectures as to the fluctuation of their prices in accordance with the astrological situation at a given moment.

Notes

1 Consolidated bibliography.

2 Ibn Ḥayyān's historical work is divided into two parts: *al-Muqtabis*, dealing with events that took place before the time of Ibn Ḥayyān, for which the author needs to use earlier historical sources, and the *Matīn*, on its side, corresponds to events contemporary to the time of the historian and is the part which can be considered original. The *Matīn* is only known indirectly and Ibn Bassām is the main source.

3 In this case, the sources provide no book title.

4 Muḥammad V reigned between 754/1354 and 760/1359. In the latter year, there was a revolt and his stepbrother Ismāʿīl II accessed the throne and reigned for a few months. In 761/1360 Ismāʿīl was murdered by his cousin Muḥammad who became king as Muḥammad VI between 761/1360 and 763/1362, when he died in Tablada, near Seville. When Muḥammad V lost his throne in 760/1359, he moved to Fez, where he was protected by the Merinid Abū ʿInān. In 763/1361, he returned to the peninsula and established himself in Ronda, under the protection of Pedro I of Castile and with the help of the latter he recovered the throne of Granada in 763/1362. He reigned until 793/1391.

Bibliography

Sources

Alfonso X (Alfonso el Sabio). ed. Kasten, L. A. and Kiddle, L. B. 1961. *Libro de las Cruzes*. Madrid: Madison, Wisc.

al-Baqqār, Abū ʿAbd Allāh. ed. and com. Díaz-Fajardo, M. 2001. *La teoría de la trepidación en un astrónomo marroquí del siglo XV. Estudio y edición crítica del* Kitāb al-adwār fī tasyīr al-anwār *(parte primera) de Abū ʿAbd Allāh al-Baqqār* [The Theory of Trepidation in (the Work) of a Moroccan Astronomer of the 15th Century. Study and Critical Edition of *Book of Cycles: On the Progression of the (Celestial) Lights*]. Barcelona: Instituto "Millás Vallicrosa" de Historia de la Ciencia Árabe.

al-Baqqār, Abū ʿAbd Allāh. ed. and com. Guesmi, C. and Samsó, J. 2018.*Astrometeorología en el-Andalus y el Magrib entre los siglos VIII y XV. El* Kitāb al-amṭār wa-l-asʿār ("*Libro de las lluvias y de los precios*") *de Abū ʿAbd Allāh al-Baqqār (fl. 1411–1418)* [Book on Rains and Prices]. Turnhout: Brepols.

Ibn Bassām. ed. ʿAbbās, I. 1979. *Al-Dhakhīra fī maḥāsin ahl al-Jazīra* [The Treasure Concerning the Merits of the People of the Peninsula]. Vol. 7. Beirut: Dār al-Thaqāfa.

Ibn Buluqqīn, ʿAbd Allāh. tr. Tibi, A. 1986. *The Tibyān. Memoirs of ʿAbd Allāh b. Buluggīn Last Zīrid Amīr of Granada*. Brill: Leiden.

Ibn Fāris, Aḥmad. ed. Forcada, M. 2000. "Astrology and Folk Astronomy: The *Mukhtaṣar min al-Anwāʾ* of Aḥmad b. Fāris," *Suhayl* 1: 107–205.

Ibn Ḥayyān. ed. Makkī, M. ʿA. 1393/1973. *Al-Muqtabas min anbāʾ ahl al-Andalus* [The Gleaner on the News about the People of al-Andalus]. Beirut: Dār al-Kitāb al-ʿarabī.

Ibn Ḥayyān. ed. Makkī, M. ʿA. 1424/2003. *Al-Sifr al-thānī min Kitāb al-Muqtabis li-Ibn Ḥayyān al-Qurṭubī* [The Second Book of the *Book of the Gleaner* by Ibn Ḥayyān al-Qurṭubī]. al-Riyāḍ: Markaz al-Malik Fayṣal li-l-buḥūth wa-l-dirāsāt al-islāmiyya.

Ibn Ḥayyān. tr. Makkī, M. ʿA. and Corriente F. 2001. *Crónica de los emires Alḥakam I y ʿAbdarraḥmān II entre los años 796 y 847 [al-Muqtabis II–1]*. Zaragoza: Instituto de Estudios Islámicos y del Oriente Próximo.

Ibn ʿIdhārī. ed. Colin, G. S. and Lévi-Provençal, E. 1980 *Al-Bayān al-mughrib fī akhbār al-Andalus wa-l-Maghrib* [The Surpassing Explanation about the History of al-Andalus and the Maghrib]. Vols. 2 and 3. Beirut: Dār al-Thaqāfah.

Ibn al-Kammād. ed. Díaz-Fajardo, M. 2016–2017. "Gestation Times Correlated to Lunar Cycles. Ibn al-Kammād's Animodar of Conception Across North Africa," *Suhayl* 15: 129–229.

Ibn al-Khaṭīb. ed.ʿInān, M. A. 1973–77. *Al-Iḥāṭa fī akhbār ahl Gharnāṭa* [The Complete Source on the History of the People of Granada]. 4 vols. Cairo: Maktabat al-Khānijī.

Ibn al-Khaṭīb. ed. Levi-Provençal, E. 1934 and 1956. *Aʿmāl al-aʿlām* [The Achievements of the Eminent Men]. Rabat: Moncho, 1934, and Beirut: Dār al-Makshūf, 1956.

Ibn Marzūq. tr. Viguera, M. J. 1977. *El Musnad: hechos memorables de Abū-l-Ḥasan, sultán de los benimerines (Al-Musnad al-ṣaḥīḥ al-ḥasan fī maʾāthir wa-maḥāsin mawlānā Abī l-Ḥasan)*. Madrid: Instituto Hispano-Arabe de Cultura

Ibn Marzūq. ed. Viguera, M. J. 1981. *al-Musnad al-ṣaḥīḥ al-ḥasan fī maʾāthir wa-maḥāsin mawlānā Abī l-Ḥasan* [The Good, Authentic *Musnad*, On the Remarkable Deeds of our Lord Abū l-Ḥasan]. Algiers: al-Sharika al-waṭaniyya li-l-nashr wa-l-tawzīʿ.

Ibn Qunfudh al-Qusanṭīnī. ed. and tr. Oliveras, M. 2012. *Comentario de la Urŷūza astrológica de ʿAlī b. Abī l-Riŷāl*. Barcelona: Universitat de Barcelona, Grup Millàs Vallicrosa d'Història de la Ciència Àrab.

Ibn Saʿīd. ed. Ḍayf, S. 1953. *al-Mughrib fī ḥulā al-Maghrib* [The Extraordinary Book on the Adornments of the West]. Algiers: al-Sharika al-waṭaniyya li-l-nashr wa-l-tawzīʿ. Vol. 1. Cairo: Dār al-Maʿārif.

al-Istijī. ed. Samsó, J. and Berrani, H. 2005. "The Epistle on *Tasyīr* and the Projection of Rays by Abū Marwān al-Istijī," *Suhayl* 5: 163–242 (Reprint: Samsó 2008, XIV).

al-Kindī. ed. and tr. Bos, G. and Burnett, C. 2000. *Scientific Weather Forecasting in the Middle Ages: The Writings of al-Kindī*. London: Kegan Paul International.

Llull, Ramon. ed. Badia, L. 2002. *Començaments de medicina. Tractat d'astronomia*. Nova edició de les obres de Ramon Llull. Vol. 5. Palma de Mallorca: Patronat Ramon Llull, 121–371.

Llull, Ramon. ed. Pereira, M. 1989. *Raimundi Lulli Opera Latina*. Vol. 17. Turnhout: Brepols, 63–218.

al-Maqqarī. ed. ʿAbbās, I. 1968. *Nafḥ al-ṭīb* [Waft of Fragrance]. Vol. 2. Beirut: Dār Ṣādir.

al-Marrākushī. ed. ʿAbbās, I. 1964. *Al-Dhayl wa-l-Takmila* [Supplement and Completion]. Vol. 4. Beirut: Dār al-Thaqāfah.

Research literature

Balty-Guesdon, M. G. 1992. *Médecins et hommes de sciences en Espagne Musulmane (IIe/VIIIe- Ve/XIe s.)*. Doctoral dissertation. Paris: La Sorbonne Nouvelle (Available in microfiches from Atelier National de Reproduction des Thèses de l'Université de Lille).

Burnett, C. 2003. "Weather Forecasting in the Arab World," in Savage-Smith, E., ed. *Magic and Divination in Early Islam*. Aldershot: Ashgate-Variorum, 201–10.

Chalmeta, P. 1967–1968. "El "Kitāb fī ādāb al-ḥisba," (Libro del buen gobierno del zoco) de al-Saqaṭī [The *Kitāb fī ādāb al-ḥisba* (*The Book of the Good Governance of the Market*) of al-Saqaṭī]," *Al-Andalus* 32: 125–62, 359–97; 33: 143–95, 367–434.

de Callataÿ, G. 2000. "La grande conjonction de 1186," in Draelants, I., Tihon, A. and van den Abeele, B., eds. *Occident et Proche Orient: Contacts scientifiques au temps des Croisades*. Turnhout: Brepols, 369–84.

de Callataÿ, G. 2013. "Magia en al-Andalus: *Rasā'il Ijwān al-Ṣafā', Rutbat al-ḥakīm, Gāyat al-ḥakīm* (*Picatrix*)," *Al-Qanṭara* 34: 297–344.

Díaz-Fajardo, M. 2008. *Tasyīr y proyección de rayos en textos astrológicos magrebíes*. PhD dissertation. Barcelona: University of Barcelona. www.tdx.cat/TDX-0309109-112444.

Fierro, M. 1996. "Bāṭinism in al-Andalus, Maslama b. Qāsim al-Qurṭubī (d. 353/ 964), author of the *Rutbat al-Ḥakīm* and the *Ghāyat al-Ḥakīm* (*Picatrix*)," *Studia Islamica* 84: 87–112.

Forcada, M. 1996. "A New Andalusian Astronomical Source from the IV/Xth Century: The *Mukhtaṣar min al-Anwā'* of Aḥmad b. Fāris," in Casulleras and Samsó, eds. [CB], 769–80.

Forcada, M. 2002. "Investigating the Sources of Prosopography: The Case of the Astrologers of ʿAbd al-Raḥmān II," *Medieval Prosopography* 23: 73–100.

Forcada, M. 2015. "Astrology in al-Andalus during the Eleventh and Twelfth Centuries: Between Religion and Philosophy," in Burnett, C. and Greenbaum, D. G., eds. *From Māshā'allāh to Kepler: Theory and Practice in Medieval and Renaissance Astrology*. Ceredigion (Wales): Sophia Centre Press, 149–76.

Herrera, M. 2001. "Una aproximación al compendio astrológico de Ibn Masʿūd ibn Farmīŷa: los *Ḥuṣūl al-maqāṣid wa-l-āmāl* (ca. 1394)," *Revista del Instituto Egipcio de Estudios Islámicos en Madrid* 33: 153–64.

Jabbār (Djebbar), A. and Aballāgh, M. 2001. *Ḥayāt wa-mu'allafāt Ibn al-Bannā' al-Murrakushī* [sic] *maʿa nuṣūṣ ghayr manshūra* [The Life and Works of Ibn al-Bannā' al-Murrakushī [sic] With Unpublished Texts]. Rabat: Manshūrāt Kulliyyat al-ādāb wa-l-ʿulūm al-insāniyya bi-l-Ribāṭ.

Juste, D. 2007. *Les Alchandreana primitifs: Étude sur les plus anciens traités astrologiques latins d'origine arabe (Xe siècle)*. Leiden: Brill.

Marín, M. 1986. "ʿIlm al-nuŷūm e ʿIlm al-ḥidtān en al-Andalus," in *Actas del XII Congreso de la U.E.A.I.* (Málaga, 1984). Madrid: Union européenne d'arabisants et d'islamisants, 509–35.

Oliveras, M. 2009. "El *De imaginibus caelestibus* de Ibn al-Ḥātim," *Al-Qanṭara* 30: 171–220.

Rius, M. 2003. "La actitud de los emires cordobeses hacia los astrólogos: entre la adicción y el rechazo," in de la Puente, C., ed. *Identidades marginales. Estudios Onomástico-Biográficos de al-Andalus*. Madrid: Consejo Superior de Investigaciones Científicas, 13: 517–49.

Samsó, J. 1979. "The Early Development of Astrology in al-Andalus," *Journal for the History of Arabic Science* 3: 228–43 (Reprint: Samsó 1994, IV).

Samsó, J. 1983. "La primitiva versión árabe del Libro de las Cruces," in Vernet, J., ed. *Nuevos Estudios sobre Astronomía Española en el siglo de Alfonso X*. Barcelona: Instituto de Filología, Institución "Milá y Fontanals", Consejo Superior de Investigaciones Científicas, 149–61 (Reprint: Samsó 1994, III).

Samsó, J. 1985–1986. "Astrology, Pre-Islamic Spain and the Conquest of al-Andalus," *Revista del Instituto Egipcio de Estudios Islámicos en Madrid* 23: 79–94 (Reprint: Samsó 1994, II).

Samsó, J. 1990. "Sobre el horóscopo y la fecha de nacimiento de ʿAbd Allāh, último rey zirí de Granada," *Boletín de la Real Academia de la Historia* 187: 209–15 (Reprint: Samsó 2008[CB], XII).

Samsó, J. 1994. *Islamic Astronomy and Medieval Spain*. Aldershot: Variorum.

Samsó, J. 1999. "Horoscopes and History: Ibn ʿAzzūz and His Retrospective Horoscopes Related to the Battle of El Salado (1340)," in Nauta, L. and Vanderjagt, A., eds. *Between Demonstration and Imagination: Essays in the History of Science and Philosophy Presented to John D. North*. Leiden: Brill, 101–24 (Reprint: Samsó 2007, X).

Samsó, J. 2001. "Sobre el astrólogo ʿAbd al-Wāḥid b. Isḥāq al-Ḍabbī (fl. c. 788- c. 852)," *Anaquel de Estudios Arabes* 12: 657–69 (Reprint: Samsó 2008, X).

Samsó, J. 2004. "Cuatro horóscopos sobre muertes violentas en al-Andalus y el Magrib," in Fierro, M., ed. *De muerte violenta. Política, religión y violencia en al-Andalus. Estudios Onomástico-Biográficos de al-Andalus*. Madrid: CSIC, 14: 482–88 (Reprint: Samsó 2008, XIII).

Samsó, J. 2007. *Astronomy and Astrology in al-Andalus and the Maghrib*. Aldershot: Variorum.

Samsó, J. 2009. "La *Urŷūza* de Ibn Abī l-Riŷāl y su comentario por Ibn Qunfuḏ: Astrología e Historia en el Magrib en los siglos XI y XIV," *Al-Qanṭara* 30: 7–39, 321–60.

Samsó, J. and Berrani, H. 1999. "World Astrology in Eleventh-Century al-Andalus: The Epistle on *tasyīr* and the Projection of Rays by al-Istijjī," *Journal of Islamic Studies* 10: 293–312 (Reprint: Samsó 2007, V).

V.4

ANWĀʾ AND TIMEKEEPING (*MĪQĀT*) IN CALENDARS AND ALMANACS OF THE SOCIETIES OF AL-ANDALUS AND THE FAR MAGHRIB

Roser Puig Aguilar

In this chapter, I survey the state of current studies on *anwāʾ* and *mīqāt* (timekeeping) in general, and on the relationship of *anwāʾ* with Andalusī calendars in particular, based on Samsó (1976–2008), Muñoz (1978–1986), Varisco (1987–1991) and Forcada (1990–2000). At the same time, the science of *mīqāt* was thoroughly studied by King, who labeled the discipline "astronomy at the service of Islam" (King 2004[CB],[1] 2005[CB]).

There is no generally accepted English translation of the word *anwāʾ*, although it approximately refers to a body of knowledge, originating in pre-Islamic Arabia, about weather, in particular rainy periods and winds, the rising and setting of single stars and groups of them (asterisms) and ways to determine day, night and other temporal units. A more detailed explanation is given later. *Mīqāt*, translated into English as 'timekeeping', refers to the set of practices, usually based on astronomy, that are used to determine the beginning of each lunar month, the direction of Mecca, and the daily times of the Muslim communal prayer. Thus, although the terms *anwāʾ* and *mīqāt* are not necessarily related, they share a place in the literature regarding calendrical aspects and timekeeping that regulated the daily life of Islamicate societies from the Near East to the Western end of the Mediterranean basin.

V.4.1 *Anwāʾ* and lunar mansions (*manāzil al-qamar*)

The use of *anwāʾ* represents the beginning of astronomical activity in pre-Islamic Arabia. It reflects a compilation of knowledge that we classify as elementary astronomy, meteorology and chronology. Its immediate antecedents were traditions of empirical observation of asterisms, recognizable groups of stars usually smaller than and distinct from constellations grouped in other cultures and meteorological phenomena attested in the ancient Near East and associated with divination.

The term *anwāʾ* (plural of *nawʾ*) describes two calendrical concepts. The first is a system associated with the heliacal risings and acronical settings of certain groups or pairs of stars, permitting

DOI: 10.4324/9781315170718-53

the division of the solar year into twenty-seven periods of thirteen, plus one of fourteen days. Each of these periods contains a shorter period, from one to seven days, which was the *naw'* properly speaking. In such a shorter period, specific weather conditions and climatic effects were cyclically repeated; all of them were usually associated with the setting asterism (Forcada 1998, 306). Thus, the system served to forecast the weather throughout the year.

The second meaning of the word *naw'* is the rain that falls at the setting of certain asterisms, which was considered a seasonal marker and was often invoked by diviners. Hence, in pre-Islamic poetry and the preclassical dialects of Arabic, the word *naw'* referred to rain and rain clouds (Varisco 1989, 148, 1991, 12–3). In pre-Islamic Arabia the year was divided into a sequence of four to eight rainy periods. The terms commonly used for each specific period are related to the *anwā'* and are collected in classical Arabic dictionaries (Varisco 1987, 263–4).

After the 2nd/8th century, the *anwā'* became combined with the Indian doctrine of *nakśatras*, a lunar zodiac of 27 or 28 divisions marked out by the moon in its monthly course, which were used in weather forecasting in Indian astrology (Burnett 2004, 208). These divisions were included in the Arab tradition as stations or lunar mansions (*manāzil al-qamar*) and were merged with elements of Arab star lore (Varisco 1991, 21–2, 1987, 253; Forcada 1998, 310). The location of the moon in one or another station was considered either a good or a bad omen. This could persuade observers either to undertake or to stop certain activities.

In the early years of Islamicate society, the use of the *anwā'* system to predict rain was probably considered divination. Perhaps this is the reason of why the term *naw'* does not appear in the Qur'ān, whereas the term *manāzil* appears there (verses 10:5 and 36:39) as an indicator of time, close to the "phases of the moon", according to Ibn Kathīr's (d. 774/1373) interpretation of the two passages (Varisco 1991, 14–5; Forcada 1998, 308–9). In fact, it seems that the identification of *anwā'* as *manāzil* and their becoming inseparable concepts occurred before the coming of Islam (Forcada 1998, 308).

In Arabic astronomical works, the standard ordering of the 28 *anwā'* identified with the 28 lunar mansions is given in Table V.4.1 (Varisco 1991, 9).

The vocabulary concerning *anwā'* was first collected by lexicographers and preserved in their treatises on 'anthropology,' folklore, and other topics of social and cultural interest. The first references to *anwā'* appeared in the 2nd/8th and 3rd/9th centuries in short onomasiological dictionaries (*risāla*s). Here all available linguistic material was collected on subjects that were important in bedouin pastoral and farming life: stars, time, rain, plants, waters, agriculture and so on. References can also be found in sayings and proverbs rhymed by the *sāji'* (who writes in rhymed prose) or in short poems recited by the *shā'ir* (poet), or the *rājiz* (poet using *rajaz* meter) if the *rajaz* meter was used (Pellat 1955). In the course of the 3rd/9th century, a corpus of texts emerged that formed a new literary genre of astrometeorological treatises, called books of *anwā'* (*Kutub al-anwā'*). The oldest surviving treatise is the *Book of anwā'* (*Kitāb al-anwā'*) of

Table V.4.1 Lunar mansions

1	Sharaṭān	8	Nathra	15	Ghafr	22	Sa'd al-dhābiḥ
2	Buṭayn	9	Ṭarf	16	Zubānā	23	Sa'd bula'
3	Thurayyā	10	Jabha	17	Iklīl	24	Sa'd al-su'ūd
4	Dabarān	11	Zubra	18	Qalb	25	Sa'd al-akhbiya
5	Haq'a	12	Ṣarfa	19	Shawla	26	Fargh muqaddam
6	Han'a	13	'Awwā	20	Na'ā'im	27	Fargh mu'akhkhar
7	Dhirā'	14	Simāk	21	Balda	28	Baṭn al-ḥūt

Ibn Qutayba (d. 276/889), who was also the author of a *Book on the Winds* (*Kitāb al-riyāḥ*). More important was Ibn Qutayba's contemporary Abū Ḥanīfa al-Dīnawarī (d. after 282/895), whose writing, in al-Marzūqī's transmission (d. 421/1030), exerted great influence on later works. The administrator Ibn Khurradādhbih (d. 299/911–2) also wrote a book on *anwāʾ* (Chapter I.13).

In addition to *kutub al-anwāʾ*, 'books of the times' (*kutub al-azmina*) were compiled of terms relating to time and to *anwāʾ* asterisms that marked the seasonal periods (Ibn Fāris 2000b, 107–8). The oldest examples are the *Book of the Times* (*Kitāb al-azmina*) by Abū ʿAlī Quṭrub (d. 206/821) and the treatises of Abū Isḥāq al-Zajjāj ([d. 311/923]; Varisco 1989) and of Abū l-Qāsim al-Zajjajī (d. 337/949).

Further to be noted are the *Book of the Times and the Places* (*Kitāb al-azmina wa-l-amkina*) by the abovementioned al-Marzūqī, which combines *anwāʾ* and *azmina* genres, and the *Book of the Times* (*Kitāb al-azmina*) by Yuḥannā ibn Māsawayh (d. 243/857–8). This book adds *anwāʾ* material to an Arabic version of an almanac with useful information on each month of the year (Troupeau 1968).

By the 3rd/9th century, scholars such as Abū Maʿshar (171–272/787–886), Thābit ibn Qurra (d. 288/901) or Mūsā ibn Nawbakht (d. *c.* 328/940) became interested in *anwāʾ* and devoted themselves to locating each of the moon's mansions scientifically, recording their numbers of stars, their appearance, size and brightness and the distances between them. In his book on the *anwāʾ*, which he dedicated to Caliph al-Muʿtaḍid (r. 279–289/892–902), Sinān ibn Thābit (d. 331/943), a son of Thābit ibn Qurra, followed Ptolemy's (100–*c.* 170) *Lunar Phases*. A century later, the polymath al-Bīrūnī (362–d. after 444/973–d. after 1053) included a summary of Sinān ibn Thābit's treatise in his *Chronology of Ancient Nations* (*Kitāb al-āthār al-bāqiya ʿan al-qurūn al-khāliya*; (Samsó and Rodríguez 1976; Samsó 1976). In this work, the most important Arabic source on ancient and medieval chronology (Aguiar 2013, 17), al-Bīrūnī gathers diverse historical and astronomical information and compares calendars from different cultures.

V.4.2 Timekeeping (*mīqāt*)

The word *mīqāt*, occurring in the Qurʾān and in *ḥadīth* literature, or more precisely the expression *ʿilm al-mīqāt*, the *science of timekeeping*, coined only centuries later, designates a specialty of Islamicate astronomy that uses astronomical methods to define the times of the communal prayers (King 1996, 170). The other topics dealt with are the regulation of the lunar calendar, in which the day begins at sunset and the months begin with the first sighting of the lunar crescent (*hilāl*) after sunset, and the computation of the sacred direction of Mecca, the *qibla* (Figure V.4.1; Chapter VI.2). The *qibla* may be calculated by astronomers seeking the azimuth of the zenith of Mecca on the local horizon and determined by nonmathematical astronomy using methods involving knowledge about the sun, moon and stars and the direction of the winds. The science of timekeeping was cultivated by both astronomers and legal scholars and was documented in the first Islamic centuries in treatises on timekeeping and regulation of the times of prayer (*kutub al-mawāqīt*), as well as in treatises of sacred geography and legal texts on the *qibla* (*kutub dalāʾil al-qibla*).

Before Islam, bedouin Arabs divided the day and the night into periods that were loosely designated by the term *sāʿa* (time, hour; King 2004, 588). Several decades after the advent of Islam the division of the day was established according to the intervals determined for prayer (Aguiar 2013, 23). The definitions of the times of prayer, still in use today, were standardized in the 2nd/8th century. Each one of them must be performed within a certain time interval, whose limits are astronomically determined. The interval for the *maghrib* (sunset) prayer begin when the sun has set over the horizon. The intervals for the *ʿishāʾ* and *fajr* prayers begin at nightfall and daybreak, respectively. The *ẓuhr* (midday) prayer usually begins when the sun

Figure V.4.1 New moon at the beginning of Ramadan, 26 April 2020; London; picture taken through a 6-inch telescope, with a Sony A7ii mirrorless camera, ISO 125 0.5 sec.

Source: Photographer Liana Saif, Amsterdam

has crossed the meridian and the shadow of a gnomon is seen to increase. The interval for *ʿaṣr* (afternoon) begins when the shadow increase equals the length of the gnomon and ends when the increase is twice the length of the gnomon, or at sunset (King 1996, 170). A supplementary and nonobligatory sixth prayer can be added, the *ḍuḥā* (morning) prayer, performed in some communities at the same time before midday as the *ʿaṣr* was performed after midday (King 1996, 171–2).

The definitions of the *ḍuḥā*, *ẓuhr* and *ʿaṣr* prayer times in terms of shadow increases correspond to the third, sixth and ninth seasonal hours (*sāʿāt zamaniyya*) of daylight (King 1996, 172), where seasonal hours are one-twelfth divisions of the period between sunrise and sunset and between sunset and sunrise and, consequently, vary throughout the year. These three prayer times are also related to the way of dividing time in the ancient Christian community of Syria. It is documented in the ritual (*ʿibāda*) section of juridical texts of legal methodology (*furūʿ al-fiqh*) that in order to differentiate the Islamic ritual from others, the prayers were not to be performed at the exact time of astronomical dawn, noon and sunset (Aguiar 2013, 25).

The names of the daylight prayers appear to have been derived from the names of the seasonal hours in pre-Islamic Arabia (King 1996, 172). The information on these names comes, once again, from the extensive Arabic lexicographical corpus. Thus, among other classical philologists, al-Farrāʾ (d. 207/822) from Kufa (Iraq) devoted a chapter to the names of the times at night in his *Book of the Days and Nights* (*Kitāb al-ayyām wa-l-layālī*) and Ibn Qutayba gave a disorganized list of divisions of day and night in a chapter of his *Book of the Education of the Secretary* (*Kitāb adab al-kātib*) (King 2004, 589).

In his treatise on timekeeping *The Exhaustive Treatise on Shadows* (*Ifrād al-maqāl fī amr al-ẓilāl*), al-Bīrūnī gathered the opinions of several earlier authorities on the etymology of the names of the prayers that connect them with the times when they must be performed (al-Bīrūnī 1976). These times are the *mawāqīt*. They vary throughout the year and from one location to another, since they depend on the apparent position of the sun with respect to the local horizon.

In nonmathematical astronomy, the night prayers were regulated by observation of the lunar mansions and the daytime prayers by arithmetical schemes of shadows similar to schemes documented in earlier Hellenistic, Byzantine, Coptic and Ethiopian practices (King 1996, 172, 2004, 640). Most of the Islamic shadow schemes served mainly to regulate the times of midday (*ẓuhr*) and afternoon prayer (*ʿaṣr*). The most common bases for the gnomon length were 7 feet (*qadam*) and 12 digits (*iṣbaʿ*).

These schemes are not mentioned by the astronomers who, from the first generation of the 2nd/8th century onward, used approximate formulae for timekeeping. An example is the Indian rule found in the extant fragments of Muḥammad ibn Ibrāhīm al-Fazārī's work (d. 180 or 190/796 or 806), an astronomer from Baghdad, and discussed in al-Bīrūnī's treatise on shadows. This rule, which is also occasionally found in technical literature and more frequently in the literature of folk astronomy, is approximate and relates the increase of shadow to the time of day in seasonal hours (King 2004, 557). At the beginning of the 3rd/9th century, al-Khwārazmī (d. *c.* 235/850) prepared the first known exact tables for regulating the times of the daylight prayers in Baghdad (King 1996, 173).

Al-Fazārī, mentioned earlier, was probably responsible for the device known as the *mīzān fazārī*, a compendium of graphic scales to measure and convert shadows; Abū ʿAlī al-Ḥasan al-Marrākushī (2nd half 7th/13th century) noted it in Cairo, in his *Comprehensive Collection of the Principles and Objectives in the Science of Timekeeping* (*Jāmiʿ al-mabādiʾ wa-l-ghāyāt fī ʿilm al-mīqāt*). Other instruments for reckoning seasonal hours and prayer times mentioned by medieval sources are the portable sundial made by Abū l-Faraj ʿĪsā in Syria in 553/1159–1160 and the conical sundial (*mukḥula*) of al-Ṣiqillī (7th/13th century; King 2004, 585).

Before the 7th/13th century, prayer-time was regulated by the muezzin (*mu'adhdhin*), the person in charge of calling the faithful to prayer. The required qualifications are detailed in some *ḥisba* books (handbooks on the maintenance of public order in the market). The muezzins needed to have a good voice and to be proficient in the rudiments of astronomy, but they did not need any specific skills in astronomical tables or instruments (King 1996, 176–7, 2004, 637–9).

At the beginning of the Mamluk period (647–922/1250–1517), and over the course of the 7th/13th century, the term *mīqāt* acquired a broader meaning and encompassed a whole field including spherical astronomy, timekeeping, chronology and astronomical instruments (Charette 2003, 6–7). In 7th/13th-century Egypt, a new profession is documented, the *muwaqqit* (timekeeper), who was a trained astronomer responsible for establishing the times of prayer, computing the *qibla* and constructing sundials. Under the patronage of a mosque or a madrasa, funded mainly through religious endowments (*waqf*), he might combine his position as timekeeper with a professorship at a law school, teaching, in addition to law, astronomical timekeeping, the construction of instruments and other mathematical sciences (Brentjes 2012, 18–9).

In 8th/14th century Damascus, al-Khalīlī (d. *c.* 781/1379–80), *muwaqqit* at the Umayyad Mosque, compiled a corpus of tables for timekeeping by the sun and for regulating the times of prayer for Damascus, which remained in use until the 13th/19th century. In addition, he compiled some tables of auxiliary trigonometric functions to solve problems of spherical astronomy and a table displaying the *qibla* as a function of terrestrial longitude and latitude (King 1996, 179–81).

Al-Khalīlī was a student of al-Mizzī (d. 750/1349) and a colleague of Ibn al-Shāṭir (d. 777/1375), both distinguished astronomers and *muwaqqits*. Al-Mizzī was renowned for his astrolabes and quadrants (Figure V.4.2), and Ibn al-Shāṭir, who devised the magnificent sundial of

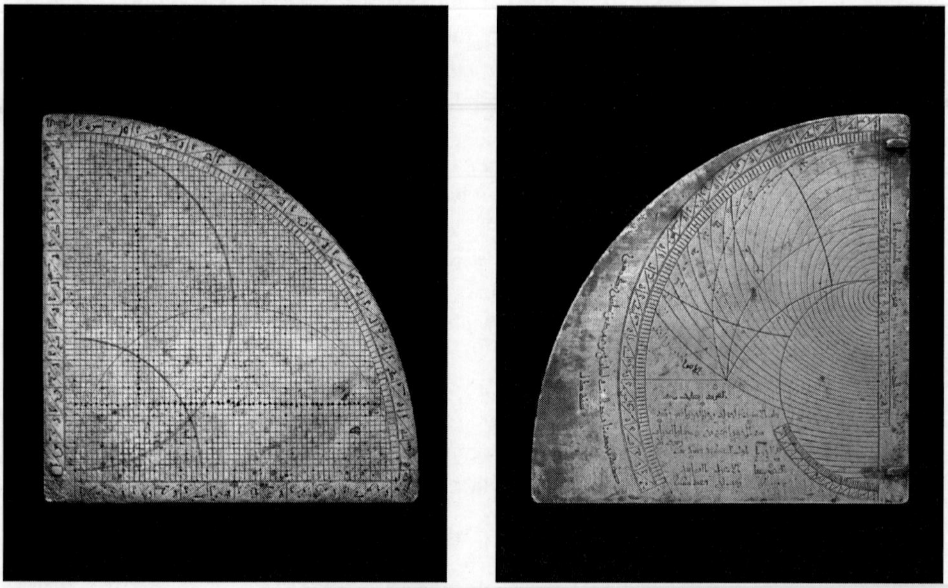

Figure V.4.2.a and Figure V.4.2.b Astrolabe quadrant made by Zayn al-Dīn (Shams al-Dīn) Muḥammad ibn Aḥmad al-Mizzī (d. 750/1349) for Sulaymān ibn Muḥammad ibn Sulaymān in Damascus, 730/1329–30; engraved brass; The David Collection, inv. no. 16/1988. Photographer Pernille Klemp.

Source: © The David Collection, Copenhagen

the Umayyad Mosque at Damascus, is more famous today for his new models of the motions of the moon and Mercury than for his invention of new astronomical instruments or his calculation of astronomical tables (Chapter II.6). His example indicates that *muwaqqits* could engage with a broad range of mathematical and astronomical themes and activities. Reflecting this state of affairs, the writers of biographical dictionaries employed numerous professional or disciplinary identifiers to describe this class of scholars and their literary, observational and instrumental practices. In addition to the term *muwaqqit* the biographers used, for instance, the term *mīqātī*. It also designates specialists in spherical astronomy and astronomical timekeeping, but those known to us were apparently not associated with any religious or educational institution.

V.4.3 *Anwā'* and timekeeping in al-Andalus and the Far Maghrib

When the Muslims arrived in the Iberian Peninsula, they brought with them elementary knowledge of timekeeping, as well as notions of astrometeorology according to the system of *anwā'* (Samsó 2011[CB], 26–7). From the 4th/10th century onward, the use of lunar mansions for magical purposes and even for making talismans is attested in Andalusī sources (Samsó 2008, 121–2), as is its use for determining the night hours and the times of prayer (Forcada 2000b; Samsó 2008). Regarding timekeeping, in the 4th/10th century the library of al-Ḥakam II (r. 350–366/961–976) apparently contained a number of books related to the science of prayer times (*'ilm awqāt al-ṣalawāt*), which escaped the burning of the library on the order of al-Manṣūr ([r. 371–392/981–1002]; Samsó 2011, 71). In the same 4th/10th century, the *Treatise of Cosmology* (*Kitāb al-hay'a*) by Qāsim ibn Muṭarrif al-Qaṭṭān (*fl. c.* 340/950), contained references to the use of a *balāṭa*, a kind of sundial, for determining the daylight hours and of a *thurayyā* or candle clock for determining the night hours (Comes 1993–4; Casulleras 1994). Al-Qaṭṭān established the rule, which became traditional, that the end of the noon prayer and the beginning of the afternoon prayer in al-Andalus and the Maghrib take place when the shadow projected by a vertical gnomon has increased over its midday minimum by one-quarter of the gnomon's length (King 1996, 170; Samsó 2011, 68).

A procedure based on the observation of successive appearances of the stars that determine the lunar mansions at night was used for determining night hours in the *anwā'* treatise of Ibn 'Āṣim ([d. 403/1013]; Forcada 1998, 322). In his treatise on the subject Ibn Khalaf al-Umawī (514–602/1120–1205 or 1206) added a simple instrument in order to facilitate the practical application of this method. The instrument consists of a semicircle, representing the sky, and a flat circle displaying the 28 lunar mansions to be rotated upon it. The instrument also displays the different phases of the moon over the course of the lunar month. In order to establish the correct time to operate it, the use of a conical sundial is recommended (Forcada 1990, 66, 1998, 318).

Except for the *faqīh* and traditionalist al-Laythī (d. 294/907), who is said to have been a *ṣāḥib* (master) in charge of the *qibla* and "facing east when making the prayer" (Samsó 2011, 60), the institution of the *muwaqqit* is not documented in al-Andalus until the middle of the Nasrid period (629–897/1232–1492) apropos of the Banū Bāṣo, a family of instrument makers. Ḥasan ibn Bāṣo (d. 716/1316) was chief of the *muwaqqits* of the mosque of Granada and his son Aḥmad ibn Bāṣo (d. 710/1310) was also a *muwaqqit*. As this and other examples indicate, the profession was often passed down from father to son (Samsó 2011, 412–3).

In an anecdote from the 9th/15th century, al-Qarābaqī (d. in the plague of 844/1440), who served as *muwaqqit* in the mosque of Baza, argued with the imam and *mufti* of Granada about the – southerly! - direction of the *qibla* in al-Andalus's mosques (Samsó 2011, 413; Rius 2008, 263). Their disagreement may reflect the variant orientations of Andalusī and Maghribī mosques between the 5th/11th and 12th/18th centuries, resulting as they did from complex architectural, astronomical and legal conditions (Rius 2000; Chapter VI.2).

Under al-Ḥakam II (r. 350–366/961–976) al-Qālī (d. 356/966) brought Abū Ḥanīfa's and Ibn Qutayba's books of *anwāʾ* with him from Baghdad, presenting them at the court of Córdoba along with the *Book of the Description of the Rain* (*Kitāb waṣf al-maṭar*) by Ibn Durayd (d. 319–20/993). Upon its arrival in al-Andalus, the *anwāʾ* tradition evolved into calendars and almanacs incorporating Visigothic practices of compiling Christian ephemerides and agricultural, medical, dietary and pharmacological information (Chapter V.2).

The oldest Andalusī document in which *anwāʾ* and astronomical materials appear together is the 4th/10th-century *Calendar of Córdoba* (Dozy and Pellat 1961; Martínez Gázquez 1981, 1991; Martínez Gázquez and Samsó 1981; Samsó and Martínez Gázquez 1981; Samsó 1983, 2011, 71–5). The astronomical matters covered by the *Calendar* include the calculation of the solstices and equinoxes, the duration of the seasons, measurements for the height of the sun at noon in Córdoba with the corresponding length of the shadow cast by a gnomon and the duration of the twilight for a given day, probably calculated with a medieval Indian formula. This document is an important source of information about different traditions of astronomical knowledge in 4th/10th-century al-Andalus, including Visigothic, Greek, Indian and Arabic (Forcada 1998, 312–5).

For centuries the *Calendar* was believed to have been composed by two authors: ʿArīb ibn Saʿīd (d. 369/980), secretary to the Umayyad monarchs of Córdoba, and Bishop Recemundus (active 2nd half 4th/10th century). Today, the *Calendar of Córdoba* is considered to be the summary of a book of *anwāʾ* by a certain al-Kātib al-Andalusī – to be identified as ʿArīb ibn Saʿīd (Forcada 1998, 328) – in combination with a Christian liturgical calendar of saints' days that corresponds to the Cordovan Mozarab ritual in the 4th/10th-century (Forcada 2000a). This liturgical calendar was derived from the calendrical part of Recemundus's *Book of the Division of Time and Benefits for Bodies* (*Kitāb Tafṣīl al-zamān wa-maṣāliḥ al-abdān*).

The *Calendar* was preceded by two sources of particular note: the *Treatise on the Stars* (*Risāla fī l-nujūm*) by Ibn Ḥabīb (d. 238/853), which, running against the genre, describes the lawful use of nonscientific astronomy and condemns astrology (Ibn Ḥabīb 1994[CB], 1997[CB]), and the *Summary of anwāʾ* (*Mukhtaṣar min al-anwāʾ*) by Aḥmad ibn Fāris (4th/10th century), of mainly astrological content (Ibn Fāris 2000).

A widely consulted book written in the 4th/10th century was the *Book of the Times and anwāʾ* (*Kitāb al-azmina wa-l-anwāʾ*) by Ibn ʿĀṣim (Ibn ʿĀṣim 1993). This book is the first faithful Andalusī reflection of the eastern books of *anwāʾ*. Ibn ʿĀṣim copies large sections from Ibn Qutayba and is clearly indebted to Abū Ḥanīfa and other classical authors of *anwāʾ*. The book begins with several rather lexicographic than technical chapters, describing the sky, the pole, the planets, the mansions and so on. They are followed by the monthly calendar describing *anwāʾ* and including, for each month, a few paragraphs on agricultural themes. These were taken from the *Nabatean Agriculture* (*al-Filāḥa al-nabatiyya*) and from the *Córdoba Calendar*. The calendar is followed by several chapters on meteorology and astronomy, including *mīqāt*, to which the author's own limited original contribution is connected. Ibn ʿĀṣim denies that the star Suhayl (Canopus), used to determine the *qibla*, is visible from Málaga, and attempts to establish an equivalence between Spanish and eastern winds (Forcada 1998, 317–9). In the 5th/11th century, the influence of Abū Ḥanīfa's treatise can be traced in the passages on *anwāʾ* in the lexicographical work *The Specialized Book* (*al-Kitāb al-mukhaṣṣaṣ*) by Ibn Sīda al-Mursī (d. 458/1066), the most important Andalusī lexicographer of the time (Forcada 1998, 319).

Politically, the 5th/11th century was, after the breakup of the Umayyad caliphate, the period of the *Taifas* ('party kings'); at the same time, it was the "golden age" of Andalusī science, particularly of astronomy. Under the Almoravid dynasty (r. 453–4541/1061–1147) territorial unity returned to al-Andalus, but scientific production declined. Many scholars emigrated to North Africa, due to advancing Christian conquests, on one hand and on the other, for more promising prospects. Professional opportunities were created by the political union of al-Andalus and the Maghrib under the Almoravids and their successors, the Almohads (*c.* 524–668/*c.* 1130–1269). The emigration of physicians to Marrakesh, such as Abū Marwān Ibn Zuhr (d. 557/1162–1163) and Ibn Ṭufayl (d. 581/1185), and of scholars from other disciplines led to the development of scientific schools in the Maghrib – heirs to the Andalusī tradition (Samsó 2011, 308–9).

In the Almohad period philosophical, medical, botanical, pharmacological and other activities recovered, but scientific astronomy did not, due to the Almohads' lack of interest in astrology (Chapter V.2). However, modern historians have suggested that nonmathematical astronomy and timekeeping (*mīqāt*) were a refuge for scientific astronomers (Samsó 2011, 520); applications for solving questions of worship gave these fields religious legitimation and led to their 'orthodox' approval (Forcada 2005, 1112).

The situation just outlined is reflected in the last important – learned! – witness to the Andalusī tradition of *anwā'* and timekeeping: the *Comprehensive, Competent and Useful Book* (*Kitāb al-Mustawʿib al-kāfī wa-l-muqniʿ al-shāfī*) by Ibn Khalaf al-Umawī (Ibn Khalaf 2017). Ibn Khalaf's work includes chapters on *anwā'*, an almanac (*taqwīm*) and chapters on astronomy based on the *Book of Times and Anwā'* by Ibn ʿĀṣim, which was written almost two hundred years earlier. A short anonymous and undated *Epistle on the Times of the Year* (*Risāla fī awqāt al-sana*) (Anonymous 1990, 27) probably followed it, combining parts of the *Córdoba Calendar* with specific details that only appear in the *Comprehensive Book.*

Ibn Khalaf al-Umawī was born in Córdoba and died in Seville. According to a manuscript of his book in the National Library of Tunis, in 546/1152 he spent Christmas in the city of Almería (Puig 2018; Figure V.4.3). He was a pious man, competent in the variant readings of the Qur'ān (*qirā'āt*) and *hadīth*, learned in the language and literature (*adab*). His knowledge of *mīqāt* and his references to two instruments, the abovementioned balance of al-Fazārī (*mīzān fazārī*) for measuring shadows and the *mukḥula* for the calculation of the hour, suggest that he may have been a *muwaqqit* somewhere. Ibn Khalaf al-Umawī wrote another, unfortunately lost book on timekeeping: the *Book of the Pearl String on Knowing Precise Times by the Stars* (*Kitāb al-Lu'lu' al-manẓūm fī maʿrifat al-awqāt bi-l-nujūm*), on determining the times of prayer (Forcada 2009, 571).

In the 7th–8th/13th–14th centuries, the *Comprehensive Book* was, together with ʿArīb's and Ibn ʿĀṣim's works, among the sources of the calendar called the *Treatise on anwā'* (*Risāla fī l-anwā'*). Its author may have been the legal scholar and consummate mathematician Ibn al-Bannā' ([654–721/1256–1321]); (Ibn al-Bannā' 1948; Forcada 1992, 1998, 322–3; Samsó 1978; al-Qadiri Boutchich and Benhamouda 2015). A hundred years later, Ibn al-Bannā'' calendar was adapted to Fez by al-Jādirī (d. 820/1416) in his *Alerting Men on What Occurs on the Days of the Year* (*Tanbīh al-anām ʿalā mā yaḥdutbu fī ayyām al-ʿām*). Ibn al-Bannā' and al-Jādirī each composed a book on arithmetical *mīqāt*, which belongs to the category of timekeeping manuals and avoids technical explanations: the *Book on the Science of the Times by Arithmetic* (*Kitāb fī ʿilm al-awqāt bi-l-ḥisāb*) and the *Plucking Lights from the Meadow of Flowers* (*Iqtiṭāf al-anwār min rawḍat al-azhār*), an abridgment of al-Jādirī's poem *The Meadow of Flowers on the Science of Night and Day* (*Rawḍat al-azhār fī ʿilm al-layl wa-l-nahār*; Calvo 2004). The Andalusī tradition of calendars was transmitted to the Merinid Maghrib and produced a Maghribī variant of the calendar genre which was of a much more practical nature. The *Comprehensive Book* is an essential part of the transmission (Table V.4.2).

بسم الله الرحمن الرحيم وصلى الله على سيدنا محمد وعلى آله وصحبه
قال الشيخ الفقيه العالم الحكيم الفاضل
ابو علي الحسن ابن علي الاموي القرطبي
رحمه الله تعالى

Figure V.4.3 Title page of Abū ʿAlī al-Ḥasan ibn Khalaf al-Umawī al-Qurṭubī's (514–602/1120–1205 or 1206) *Comprehensive, Competent and Useful Book* (*Kitāb al-Mustawʿib al-kāfī wa-l-muqniʿ al-shāfī*), MS Tunis, Bibliothèque nationale de Tunisie, A-MSS-11925, fol. 1a.

Source: © Bibliothèque nationale de Tunisie, Tunis

Table V.4.2 The Comprehensive Book (*Kitāb al-mustawʿib*). An extract of its contents on timekeeping, astrometeorology and folk astronomy

Monthly Calendar (information for each month from January to December)

- Name of the month in various languages
- *Ḥarf al-uss*. Used for determining the day of the week on which the month begins
- Number of days of the month
- *Anwāʾ* and their relationship with periods of rain; settings, risings and culmination of asterisms; lunar mansions; winds; graphic representation of the asterism that determines the daybreak
- Names of stars
- Time for the first *saḥūr*
- Shadows for the times of the day measured in feet
- Shadows for *zawāl, ẓuhr* et *ʿaṣr* times of prayer
- Meridian shadow measured in digits with the *mīzān fazārī*
- *Nawʾ* which corresponds to the crescent moon (*hilāl*)
- Seasonal hours for the first day of the month
- Nightfall
- Daybreak
- Solar meridian altitude
- Shadow projected by a person who stands up at noon
- Season of the year to which the month belong

The production of calendars and other texts on *mīqāt*, as well as instruments continued through the 13th/19th century in *madrasas* and Sufi *zāwiyas*. *Muwaqqits* continued to work in several North African cities in this later period. Close relations to Mamluk and then Ottoman Egypt and Syria can be traced in the extant manuscripts. The earlier, abovementioned texts remained important sources and were extracted by later copyists and adapted to their goals. Parallel to the continuation of this venerable heritage, new astronomical knowledge arrived in the early modern period. Its interaction with the knowledge discussed in this chapter has not yet been investigated.

Note

1 Consolidated bibliography.

Bibliography

Sources

Anonymous. ed. Navarro, M. A. 1990. *Risāla fī awqāt al-sana* [Epistle on the Times of the Year]. Granada: CSIC.

ʿArīb ibn Saʿīd al-Qurṭubī and Ibn al-Bannā' al-Marrākushī. ed. al-Qadiri Boutchich, I. and Benhamouda, S. 2015. *Risālatān fī l-anwā' li-ʿArīb b. Saʿīd al-Qurṭubī wa-Ibn al-Bannā' al-Marrākushī* [Two Epistles on the *anwā'* by ʿArīb ibn Saʿīd al-Qurṭubī and Ibn al-Bannā' al-Marrākushī]. Meknes: Kulliyyat al-ādāb wa-l-ʿulūm al-insaniyya.

al-Bīrūnī. tr. Kennedy, E. 1976. *The Exhaustive Treatise on Shadows*. Aleppo: IHAS.

Ibn ʿĀṣim. ed., tr. and ann. Forcada, M. 1993. *El* Kitāb al-anwā' wa-l-azmina – al-qawl fī l-šuhūr *de Ibn ʿĀṣim. Tratado sobre los* anwā' *y los tiempos – capítulo sobre los meses). Estudio, traducción y edición crítica.* Madrid: Instituto de Cooperación con el mundo árabe and Barcelona: Instituto Millás Vallicrosa.

Ibn al-Bannā'. ed. and tr. Rénaud, H. P. J. 1948. *Le Calendrier d'Ibn al-Bannā' de Marrakech* (Risāla fī l-anwā'). Paris: Larose.

Ibn Fāris. Aḥmad. ed., tr. and ann. Forcada, M. 2000. "Astrology and Folk Astronomy: The *Mukhtaṣar min al-anwā'* of Aḥmad ibn Fāris," *Suhayl* 1: 107–205.

Ibn Khalaf al-Umawī al-Qurṭubī. ed. Samadi, Y. 2019. *Al-Mustawʿib al-kāfī wa-l-muqniʿ al-šāfī de al-Umawī* al-Qurṭubī *(m. 1206)* [Comprehensive, Competent and Useful Book by al-Umawī al-Qurṭubī (d. 1206)]. al-Mamlaka al-maghribiyya: Wizārat al-awqāf wa-l-shu'ūn al-islamiyya.

Research literature

Aguiar, M. 2013. "Los precedentes no árabes del calendario islámico y de los momentos para la oración según el *Kitāb al-āṯār al-bāqiya ʿan al-qurūn al-jāliya* de al-Bīrūnī," in Martínez Gázquez, J. and Tolan, J. V., eds. *Ritvs Infidelivm: Miradas interconfesionales sobre las practicas religiosas en la Edad Media*. Madrid: Casa de Velázquez, 17–27.

Brentjes, S. 2012. "The Language of 'Patronage' in Islamic Societies Before 1700," *Cuadernos del CEMyR* 20: 11–22.

Burnett, C. 2004. "Weather Forecasting in the Arabic World," in Savage-Smith, E., ed. *The Formation of the Classical Islamic World. Magic and Divination in Early Islam*. Aldershot: Ashgate, Variorum, 201–10.

Calvo, E. 2004. "Two Treatises on *mīqāt* from the Maghrib (14th and 15th Centuries A.D.)," *Suhayl* 4: 159–206.

Casulleras, J. 1994. "El contenido del *Kitāb al-Hay'a* de Qāsim ibn Muṭarrif al-Qaṭṭān," in Camarasa, J. M., Mielgo, H. and Roca, A., eds. *Actes de les I Trobades d'Història de la Ciència i de la Tècnica. Trobades Científiques de la Mediterrània (Maó, 11–13 setembre 1991)*. Barcelona: Societat Catalana d'Història de la Ciència i de la Tècnica D.L., 75–94 (Translated and reedited in Fierro, M. and Samsó, J., eds. 1998. *The Formation of al-Andalus. Part. 2: Language, Religion, Culture and the Sciences*, Aldershot: Ashgate, 340–58).

Charette, F. 2003. *Mathematical Instrumentation in Fourteenth-Century Egypt and Syria. The Illustrated Treatise of Najm al-Dīn al-Miṣrī*. Leiden: Brill.

Comes, M. 1993–1994. "Un procedimiento para determinar la hora durante la noche en la Córdoba del siglo X [A Procedure to Determine the Time During Night in 10th- Century Córdoba]," *Revista del instituto egipcio de estudios islámicos en Madrid* 26: 263–72.

Dozy, R. and Pellat, Ch., eds. 1961. *Le Calendrier de Cordue*. Leiden: Brill.

Forcada, M. 1990. "*Mīqāt* en los calendarios andalusíes," *Al-Qanṭara* 11: 59–69.

Forcada, M. 1992. "Les sources andalouses du *Calendrier* d'Ibn al-Bannā' de Marrakesh," in *Historia, ciencia y sociedad. Actas del Segundo Congreso Hispano-Marroquí de Ciencias Históricas (Granada, 6–10 noviembre 1989)*. Madrid: M.A.E., Agencia Española de Cooperación Internacional, Instituto de Cooperación con el Mundo Arabe, 183–98.

Forcada, M. 1998. "The books of *anwā'* in al-Andalus," in Fierro, M. and Samsó, J., eds. *The Formation of al-Andalus. Part. 2: Language, Religion, Culture and the Sciences*. Aldershot: Ashgate, 305–28 (A reedited translation with an 'author's note' of "Los libros de *anwā'* en al-Andalus," in Vernet, J. and Samsó, J., eds. 1992. *El legado científico andalusí*. Madrid: Ministerio de Cultura, 103–15).

Forcada, M. 2000a. "The *Kitab al-anwā'* of 'Arīb b. Sa'īd and the *Calendar of Cordova*," in Folkerts, M. and Lorch, R., eds. *Sic itur ad astra. Studien zūr Geschichte der Mathematik und Naturwissenschaften. Festschrift fūr den Arabisten Paul Kunitzsch zum 70. Geburtstag*. Wiesbaden: Harrassowitz, 234–51.

Forcada, M. 2000b. "L'expression du cycle lunaire dans l'ethnoastronomie arabe," *Arabica* 47: 37–77.

Forcada, M. 2005. "Síntesis y contexto de las ciencias de los antiguos en época almohade," in Cressier, P., Fierro, M. and Molina, L., eds. *Los Almohades: problemas y perspectives*. 2 vols. Madrid: Casa de Velázquez, 2: 1091–135.

Forcada, M. 2009. "Ibn Jalaf al-Umawī, Abū 'Alī," in Lirola, J. and Puerta Vílchez, J. M., eds. *Biblioteca de al-Andalus* 3: 571–2.

King, D. 1996. "Astronomy and Islamic Society: Qibla, Gnomonics and Timekeeping," in Rashed and Morelon, eds. [CB], 1: 128–84.

Martínez Gázquez, J. 1981. "Santoral del Calendario del siglo XIII contenido en el *Liber Regius* del Museo Episcopal de Vich," *Revista Catalana de Teología* 6: 161–74.

Martínez Gázquez, J. 1991. "El texto del *Calendario de Córdoba* en el manuscrito Berlin Lat. Qu.," in Alvar, C., ed. *Studia in honorem Prof. M. de Riquer*. Barcelona: Quaderns Crema IV: 657–68.

Martínez Gázquez, J. and Samsó, J. 1981. "Una nueva traducción latina del Calendario de Córdoba (siglo XIII)," in Vernet, J., ed. *Textos y estudios sobre astronomía española en el siglo XIII*. Barcelona: CSIC, 9–78.

Muñoz, R. 1978. "Un calendario egipcio del siglo XVIII," *Awrāq* 1: 67–81.

Muñoz, R. 1984. "Un refranero árabe de contenido astronómico," in *I Congreso Hispano-Africano de las Culturas Mediterráneas "Fernando de los Ríos Urruti"*. Melilla: Escuela de Magisterio, 17–32.

Muñoz, R. 1986. "Los *kutub al-anwā'*," in *Actas del XII Congreso de la UEAI (Málaga, 1984)*. Madrid: [s. n.], 623–44.

Pellat, C. 1955. "Dictons rimés, *anwā'* et mansions lunaires chez les arabes," *Arabica* 1: 17–41.

Puig, R. 2018. "Le *Kitāb al-mustaw'ib al-kāfī wa-l-muqni' al-shāfī* d'Ibn Khalaf al-Umawī al-Qurṭubī (m. 1206) d'après le manuscrit *Aḥmadiyya* 11925/12 de la Bibliothèque Nationale de Tunis," in *Actes du XIIIème Colloque Maghrébin d'Histoire des Mathématiques*. Tunis: Graphika Imprimerie, 267–82.

Rius, M. 2000. *La alquibla en al-Andalus y al-Magrib al-Aqṣà*. Barcelona: Institut "Millás Vallicrosa" d'Història de la Ciència Àrab.

Rius, M. 2008. "Mesurar el temps al Magrib: la determinació de les hores d'oració," *Actes d'Història de la Ciència i de la Tècnica* 1.1: 261–8.

Samsó, J. 1976. "De nuevo sobre la traducción árabe de las *Pháseis* de Ptolomeo y la influencia clásica de los *kutub al-anwā'*," *Al-Andalus* 41: 471–9 (Reprint: Samsó 2008 [CB], III).

Samsó, J. 1978. "La tradición clásica en los calendarios agrícolas hispanoárabes y norteafricanos," in *Actas del Segundo Congreso internacional de estudios sobre las Culturas del Mediterráneo Occidental*. Barcelona: UAB, 177–86. (Reprint: Samsó 2008, IV).

Samsó, J. 1983. "Sobre los materiales astronómicos en el *Calendario de Córdoba* y en su versión latina del siglo XIII," in Vernet, J., ed. *Nuevos estudios sobre astronomia española en el siglo de Alfonso X*. Barcelona: CSIC, 125–38.

Samsó, J. 2008. "Lunar Mansions and Timekeeping in Western Islam," *Suhayl* 8: 121–61.

Samsó, J. and Martínez Gázquez, J. 1981. "Algunas observaciones al texto del Calendario de Córdoba," *Al-Qanṭara* 2: 319–44 (Reprint: Samsó 2008, no. VI).

Samsó, J. and Rodríguez, B. 1976. "Las *Pháseis* de Ptolomeo y el *K. al-anwā'* de Sinān b. Ṯābit," *Al-Andalus* 41: 15–48 (Reprint: Samsó 2008, no. II).
Troupeau, G. 1968. "Le livre des temps de Jean Ibn Māsawayh," *Arabica* 15: 113–42.
Varisco, D. M. 1987. "The Rain Periods in Pre-Islamic Arabia," *Arabica* 34: 251–66 (Reprint: Varisco 1997, III).
Varisco, D. M. 1989. "The Anwā' Stars According to Abū Isḥāq al-Zajjāj," *ZGAIW* 5: 145–66 (Reprint: Varisco 1997, II).
Varisco, D. M. 1991. "The Origin of Anwā' in Arab Tradition," *Studia Islamica* 74: 5–28 (Reprint: Varisco 1997, I).
Varisco, D. M. 1997. *Medieval Folk Astronomy and Agriculture in Arabia and the Yemen*. Aldershot: Ashgate, Variorum.

V.5
SCHOLARLY COMMUNITIES DEDICATED TO THE SCIENCES IN AL-ANDALUS[1]

Miquel Forcada

V.5.1 Introduction

The scientific legacy of al-Andalus was mostly created by groups of scholars and professionals who built up complex networks for the development and transmission of knowledge. The individuals who symbolize the achievements of this rich tradition, like Maslama al-Majrīṭī (d. before 397/1007), al-Zahrāwī (*fl.* late 4th/10th early 5th/11th century), Ibn al-Zarqālluh (d. 493/1100) or Ibn Rushd (520–595/1126–1198), emerged within these communities.[2] The nature of these groups varied considerably: they could be durable or ephemeral, professional or purely scholarly, religious or interreligious, court-based or non-court- based, strongly led or leaderless. Many other descriptors could be added to create a typology that would broaden our understanding of the history of science in al-Andalus. However, the following pages will focus on two preliminary yet necessary steps: first identifying the scholarly communities of al-Andalus that made significant contributions to the sciences, and second, describing them in relationship to the historical and intellectual contexts within which they developed.

V.5.2 First steps (3rd/9th century)

V.4.2.1 Christian scholarly communities and the sciences

In the intellectual history of al-Andalus, there is a phenomenon that, like dark matter, is presumed to exist but is difficult to detect: the influence of Christian institutions of learning and scholarly communities on the nascent Arabo-Islamic culture. The Iberian Peninsula of the Visigoths, the future al-Andalus, had a notable civil and religious cultural life thanks to the Roman Christian legacy.

Medicine in early al-Andalus seems to have been mostly practiced by Christians in monasteries (though we know next to nothing about the medicine practiced behind their walls) and in cities (Ibn Juljul 1985[CB],[3] 92–8; Vernet 1979). The scarce information available about urban medicine in Córdoba suggests that we think of the Christian physicians of the 3rd/9th century as members of a professional guild who provided their services to fellow residents of all religions

DOI: 10.4324/9781315170718-54

and also attended to courtly circles. There is, nevertheless, evidence that they belonged to some kind of scholarly community, in the sense of a group of learned men who cultivated a particular type of scientific culture. They followed a tradition of medical literature and practice, inherited from the very early Middle Ages, which was considerably simpler than the academic medicine of 4th/10th-century Islamicate societies. Although this medicine, of a relatively low scholarly level, left hardly any trace in the later Arabic medical literature of the Peninsula, its influence was by no means negligible at the time and lasted into the 4th/10th century (Ibn Juljul 1985, 100–1). The second area in which the Roman Christian legacy could be felt was education. Christian religious institutions of learning were still active in al-Andalus in the 3rd/9th and 4th/10th centuries. In the mid-3rd/9th century, Córdoba was a focus of Latin culture for Andalusi Christians (Herrera Roldán 1995, 35–51). Consequently, an education based on the *trivium-quadrivium* model was imparted to Christian students, not only in Córdoba but in other cities as well, and we have some evidence of the possible influence of this kind of knowledge on Muslim scholars of the mid-4th/10th century.

V.5.2.2 Muslim scholarly communities and the birth of a scientific culture in al-Andalus

Scholars and professionals devoted to scientific disciplines flocked to the Umayyad court beginning in the times of al-Ḥakam I (r. 180–206/796–822) and ʿAbd al-Raḥmān II (r. 206–238/822–852). The two emirs promoted a cultural policy that aimed to establish courtly cultural standards in imitation of Abbasid practices (Samsó 2011[CB], 45–60). The Umayyads made considerable efforts to acquire and foster scientific and other intellectual traditions originally developed in Baghdad. Two aspects of this policy proved particularly relevant. The first was its professional purpose, in so far as the Umayyads needed a more efficient administration, and the second was its religio-political implications.

The palace and its administrative circles established three professional bodies of people who transmitted mathematical knowledge: teachers, who probably taught mathematics (among other subjects) to the princes, land surveyors and astrologers (Forcada 2004–2005[CB], 18–20). The last group was by far the most important, and possibly the best paid, because it had something very like a strategic function, particularly during the time of ʿAbd al-Raḥmān II, who consulted his astrologers before making any important decision. These astrologers, a structured body that included a leader and a specific pay list (*dīwān*), reached the peak of their influence under this ruler (Samsó 2001; Rius 2003; Forcada 2004–2005, 14–8). Side by side with the professional astrologers were influential courtiers interested not only in astrology but also in other disciplines. Centered around ʿAbd al-Raḥmān II, astrologers and courtiers can be considered a scholarly community that contributed to the acculturation in al-Andalus of Baghdad's lore on technology, sciences and rational thinking.

One of the reasons the Umayyads fostered this activity seems to have been their dispute with Mālikī religious scholars regarding the hegemony over religious discourse in al-Andalus (Forcada, 217, 59–60). ʿAbd al-Raḥmān seems to have imitated, or tried to imitate, a policy implemented by Caliph al-Maʾmūn (r. 198–218/813–833) of undermining the authority of the traditionalist religious scholars by promoting scholars who held a rationalistic approach to religion. The traces of this circle of scholars disappeared during the final decades of the 3rd/9th century, probably because of a reaction on the part of the Mālikī religious scholars and the political crisis of the Umayyads.

V.5.3 Scholarly communities during the consolidation of a scientific tradition

V.5.3.1 A preliminary approach to the sciences in 4th/10th-century Córdoba

The 4th/10th century was a prosperous period for the sciences in al-Andalus, thanks to the initiative of scholars and the patronage of the dynasty. Under ʿAbd al-Raḥmān III (r. 300–350/912–961) the court consolidated its position as a center of power, and the regime became a caliphate in 316/929. The sciences were an important element not only in the promotion of the dynasty but also in the cultural life of its members and its learned elites: four preceptors of the sons of ʿAbd al-Raḥmān III had expert knowledge of mathematical and philosophical disciplines (Forcada 2017, 71). A caliph who was an intellectual in his own right, al-Ḥakam II (r. 350–366/961–976) provided renewed momentum to the scientific life of Córdoba. Out of this context emerged a complex community of scholars devoted to the sciences and philosophy and divided into several sectors.

V.5.3.2 The physicians

Physicians made up the most important group of scientific scholars in this community (Balty-Guesdon 1992, 204–15). Their main center was the court, where a regular body of physicians was installed as a *dīwān*, although we do not have a precise indication of who actually belonged to this service or how it was organized. Many physicians with connections to the court are described in the sources as qualified servants employed in high-ranking government offices or as qualified scholars with the intellectual profile of the physician-philosopher who, in tune with the model sketched out in Galen's *That the Best Physician is Also a Physician*, was profoundly interested in many disciplines. Some of these authors traveled abroad to seek knowledge from the most important centers of the time, including Baghdad, Qayrawān and Fusṭāṭ. The physicians of this epoch generated a local tradition in medicine based on Arabic translations of the Greek works that arrived in Córdoba during this period. The best examples of this tradition of scholarly medicine are, first, a tradition of pharmacological works, which emerged from the revision of the Arabic translation of Dioscorides's (1st century) *On Medical Substances* (*De materia medica*) sponsored by ʿAbd al-Raḥmān III in the middle of the 4th/10th century (Samsó 2011, 110–5); and second, the medical encyclopedia titled *Book That Makes [Medicine] Available* (*Kitāb al-Taṣrīf*), written by the enigmatic al-Zahrāwī, who was probably a court physician (Hamarneh and Sonnedecker 1963, 13–22 and 43–76). Additionally, in line with their status as multitalented experts, physicians made decisive contributions to the development of disciplines like mathematics, logic and agriculture.

V.5.3.3 Experts in mathematical disciplines

Experts in mathematics played an important role in the sciences of 4th/10th-century Córdoba. The "*faraḍīs*" were the largest group, as they practiced a legal-mathematical discipline required by society, the science of inheritance shares (*ʿilm al-farāʾiḍ*; Balty-Guerdon 1992, 191–6). A significant number of "*faraḍīs*" combined this activity with knowledge of other mathematical disciplines. The most outstanding mathematician and astronomer of the epoch, Maslama al-Majrīṭī, emerged from this sector. In fact, until Maslama, there are no recognizable circles of scholars active in the mathematical disciplines aside from the *faraḍīs* themselves. There are only a few examples of the practice of mathematical astronomy during the first part of the century, possibly because astrology was not practiced at the court at this time – probably for religious reasons. Maslama became a true master of the mathematical disciplines, particularly mathematical astronomy (Samsó 2011, 80–110, 466–73, 2020[CB], 417–428 and 688-708). Although information on his life is scarce, we know that he trained a good many disciples (Ṣāʿid al-Andalusī 1985[CB], 169–77).

Besides writing important treatises, some of which have come down to us, Maslama's pupils spread their mathematical knowledge throughout al-Andalus during the 5th/11th century. An interesting feature of Maslama's disciples is that the majority resembled the physicians described earlier: multifaceted scholars who also practiced disciplines like medicine and logic. We should probably resist the temptation to classify a particular author as a physician who knew mathematics or the reverse. Maslama and his disciples formed a solid part of the scholarly community, were connected with the main trends of intellectual life in Córdoba and with the most important people associated with it, and focused on the study and practice of arithmetic, geometry, mathematical astronomy and astrology. It is probable that the renewed interest in astrology at court during the reign of al-Ḥakam II contributed to the flourishing of astronomy in the final decades of the 4th/10th century (Forcada 2015, 152–5).

V.5.3.5 Religiously oriented communities

Jews and Christians were also important protagonists in the scientific life of 4th/10th-century Córdoba. However, while the presence of Jews in the history of science in al-Andalus steadily increased, Christians' presence was on the wane and virtually disappeared in the next century. Despite few precedents in the previous period, Jewish scholars became a force in the sciences in two cities, Córdoba and Lucena. Their first and best-known representative was the physician and diplomat Ḥasdāy ibn Shaprūṭ (d. *c.* 364/975) (Balty-Guesdon 1992, 182–9). According to contemporary sources, Ibn Shaprūṭ made the Jews of al-Andalus self-sufficient in the areas of law, history, Hebrew language and the determination of the calendar (Ṣāʿid al-Andalusī 1985, 205–7). Besides his participation in the revision of Dioscorides's *On Medical Substances*, mentioned earlier, and his concern with the Jewish calendar, he corresponded with Dūnash ibn Tamīm of Qayrawān (d. *c.* 349/960), whom he asked for a treatise on the armillary sphere (Chapter IV.2). Dunash responded to this request by dedicating a treatise on astronomy to Ḥasdāy (Mimura 2015). There is hardly any information on the scientific practices of Ibn Shaprūṭ's disciples, but notable scholars emerged from the Jewish community of Córdoba, as we will see.

During the 4th/10th century, the language and culture of the Christian community became considerably arabized. The administrative structure of the church remained functional, as did its system of education. Christians had access to the highest circles of the court, because the dynasty needed mediators for its dealings with the Christian communities inside and outside al-Andalus. At least one of these servants, the bishop Recemundus (*fl.* 2nd half 4th/10th century), is credited with a sound knowledge of mathematical disciplines, philosophy and possibly, medicine. He was the teacher of bishop Abū l-Ḥārith who, in turn, taught logic to the Muslim physician-philosopher Ibn al-Kattānī ([d. *c.* 420/1029]; Ṣāʿid al-Andalusī 1985, 193). Even though Recemundus may have been aware of the new sciences and philosophy translated into Arabic, it is reasonable to surmise that he and disciples like Abū l-Ḥārith transmitted parts of the Isidorean legacy. The oeuvre of Ibn Ḥazm (384–456/994–1064), an author who was educated in this rich interfaith context, contains interesting hints that this was the case; Ibn Ḥazm's classification of the sciences seems to follow Isidore of Seville (d. 636; Gómez Nogales 1969, 502). Some of his works show a notable influence of stoicism, a philosophy that the author could have acquired from within the Christian religious culture of Córdoba (Puig Montada 2001, 184–5).

In the 4th/10th century, a third religiously oriented group appeared, which we may call the community of the esotericists (*bāṭinīs*). One aspect of Andalusi esotericism at this time was the religious phenomenon called "Masarrism". Ibn Masarra (d. 319/931) was a religious scholar who developed a philosophizing mysticism based on several sources: mystics of the 3rd/9th century, Neoplatonism and, possibly, Ismāʿīlism, which he could have learned about while sojourning in Ifrīqiya (Ebstein 2016). His followers made up an important religious community, which was persecuted by ʿAbd al-Raḥmān III. One of the religio-political reasons for this persecution was to combat the importation of Fatimid teachings that contradicted the Sunni beliefs of the Umayyad dynasty (Fierro 2004, 135; Safran 2013, 72).

Neither the Masarrists nor Ibn Masarra himself contributed to the scientific activities of the time. However, it is difficult to understand the contributions of Maslama ibn al-Qāsim (d. 353/964) isolated from these events. He was one of the preceptors of ʿAbd al-Raḥmān III's sons and thus one of the palace teachers, who were experts in scientific and philosophical disciplines, as noted above. He is probably the author of two treatises that deal with the occult side of the sciences: *Aim of the Sage* (*Ghāyat al-ḥakīm*), on talismanic magic, and *Rank of the Sage* (*Rutbat al-ḥakīm*), on alchemy (Fierro 1996). He is also credited with having brought to al-Andalus the most important scientific encyclopedia of the time, the *Epistles of the Brethren of Purity* (*Rasāʾil Ikhwān al-Ṣafāʾ*; de Callataÿ 2013; de Callataÿ and Moureau 2017). The "Brethren" were a group of scholars, possibly Ismāʿīlis, who, influenced by Neoplatonism, believed that knowledge could lead to salvation. It is worth noting that the Andalusian historians of science Ibn Juljul (d. after 383/994) and Ṣāʿid al-Andalusī (419–462/1029–1070) do not mention these contributions of Maslama, nor do the general bibliographical sources from al-Andalus, which present him a bizarre religious scholar, who was rumored to practice magic.

Masarrist or not, Maslama ibn al-Qāsim appears in the biographical sources as a scholar concerned with mysticism and the occult and a contemporary of other scholars with the same interests (Oliveras 2009). The themes of his works, which may have provoked the anger of religious scholars, fiercely opposed to magic and alchemy, indicated the existence of a committed and discrete audience that could accept these disciplines in spite of the risk of persecution. This audience is actually alluded to by Maslama in the introduction of the *Aim of the Sage*, but we do not know its size or whether its members should be considered a scholarly community, since Maslama appears to have worked in isolation (Forcada 2017, 68–71; Fierro 2012, 135–9).

V.5.4 Scholarly communities during the flourishing of Andalusi science

V.5.4.1 A general view of the science of the 5th/11th century

The Umayyad caliphate experienced a long agony during the second decade of the 5th/11th century and disappeared from history in 422/1031. It was replaced by several regional kingdoms, known as "party" or "petty kingdoms," which governed al-Andalus during the 5th/11th century. Some of the scholars active during the last period of the caliphate emigrated to these new courts, which became new foci for the sciences. These scholars circulated fairly easily within the Iberian Peninsula and were able to move from city to city to earn their living or to find a master from whom they wished to learn. Scholars devoted to the sciences in the 5th/11th century made up a large and dispersed community kept together by the personal bonds between scholars. The new dynasties were in general receptive to the sciences, but it is rare to find a court that employed a large number of scientists. The cities of Saragossa, Seville and Toledo, however, were centers with notable concentrations of scholars.

V.5.4.2 Saragossan circles devoted to philosophy and mathematics

After the crisis of the Umayyad caliphate, Saragossa became one of the main centers of scientific activity in al-Andalus (Balty-Guesdon 1992, 325–43; Lomba 2002; Beech 2008). Several Córdobans moved to the city, then governed by the Banū Tujīb (r. 403–430/1013–1038). Three of these were physician-philosophers: Ibn al-Kattānī, whom we mentioned earlier, and the Jews Manāḥim ibn al-Fawwāl (1st half 5th/11th century) and Marwān ibn Janāḥ (d. after 430/1038). The family of Ḥasdāy ibn Shaprūt also relocated there (Stroumsa 2016). Although the master–disciple connections between the emigrants from Córdoba and the next generations of Saragossan scholars have not yet been established, they seem to have been the first to create a flourishing philosophical and scientific community centered on its Jewish members. The most important philosophers of 5th/11th-century al-Andalus/Sefarad – the latter being the Hebrew term for the region – the Jews Solomon ben Gabirol (d. *c.* 450/1058) and Ibn Paqūda (2nd half 5th/11th century) appeared in this context (Lomb 2002, 367–426).

The Banū Hūd dynasty (r. 430–503/1038–1110) enthusiastically sponsored the mathematical sciences and philosophy. The king al-Muqtadir ibn Hūd bi-Llāh (r. 438–474/1046–1081) is credited with having known astronomy, mathematics and philosophy. His son, King al-Muʾtaman ibn Hūd (r. 474–478/1081–1085), was a mathematician and philosopher in his own right. His right-hand man was Abū l-Faḍl Ḥasdāy ibn Yūsuf (2nd half 5th/11th), probably the grandson of Ḥasdāy, who was a scholar well trained in philosophy (Ṣāʿid al-Andalusī 1985, 205–6). Neoplatonism was possibly bolstered by the reintroduction of the *Epistles* of the Brethren of Purity in Saragossa by al-Kirmānī (d. 458/1066), a disciple of Maslama al-Majrīṭī. The diffusion of the ideas of the philosopher al-Fārābī ([d. 339/950–1]; Ṣāʿid al-Andalusī 1985, 137–40), gave an intellectual grounding to a new perspective on the sciences and their practitioners: mastering theoretical knowledge leads to human perfection and ultimate felicity. The seeker of this goal is a "philosopher", understood in the sense of a "metascientist", who knows all disciplines comprised in the classifications of the sciences (Chapter VI.4). One of the first products of this trend, in 5th/11th-century Saragossa, is the *Book of the Perfection* (*Kitāb al-Istikmāl*) written by King al-Muʾtaman (Hogendijk 1995; Djebbar 1997). This book can be regarded as a manual containing all the mathematical knowledge that this ideal "metascientist" needs in order to achieve his transcendental purpose (Forcada 2011, 227).

While the *Book of the Perfection* may have been a collective work written by the king and the scholars that surrounded him, there is little doubt that al-Muʾtaman's personal commitment to philosophy and the sciences was sincere. It is also possible that this dynasty promoted the theoretical disciplines in order to strengthen its legitimacy in the eyes of the Muslim and non-Muslim communities of the kingdom. Legitimacy was a severe problem to the Party Kings. By their choice, the Banū Hūd conveyed the idea that they were ruling a religiously diverse community not because they had seized the power from the previous dynasty, but because, according to rational criteria that were independent from religion, they were the most apt (Forcada 2011, 225–34).

V.5.4.2 Toledo and Seville: the circles of medicine and agricultural science

Toledo was the focus of two other important scientific trends in the 5th/11th century. The first was the continuation of the rich tradition of pharmacological works of Umayyad Córdoba. The second was the offshoot of a local tradition of agricultural works that appeared at the end of the 4th/10th century (Samsó 2011, 277–306). The key author behind both disciplines was the Toledan Ibn Wāfid (d. 466/1074), who supposedly studied with al-Zahrāwī in

Córdoba. Although this is doubtful for chronological reasons, he may have been an indirect disciple of al-Zahrāwī. Ibn Wāfid became the curator of the garden of King al-Maʾmūn ibn Dhī l-Nūn of Toledo (r. 435–467/1043–1075) and wrote important treatises on pharmacology as well as a compilation on agriculture. He trained Ibn al-Baṣṣāl (2nd half 5th/11th), who wrote a well-known agricultural treatise *Book of the Concision and Explanation* (*Kitāb al-qaṣd wa-l-bayyān*).

Ibn al-Baṣṣāl, and a disciple of his named Ibn Lunqūh (d. 498 or 499/1104–1106), moved to Seville when Toledo was conquered by the Castilians in 478/1085. The king of Seville, al-Muʿtamid, gave them shelter and appointed Ibn Baṣṣāl curator of his gardens. Seville thus became an active focus of pharmacology and agricultural science, with the presence of authors like al-Ishbīlī, Ibn al-Ḥajjāj (both 5th/11th century) and al-Ṭighnarī, who was still active during the early decades of the 6th/12th century. The interconnections between most of the authors mentioned and the quality of their works show that they formed a scholarly community devoted to the natural sciences (García Sánchez 1994). These scholars not only continued a preexisting tradition in pharmacology, but they profoundly reformed Roman-Byzantine agricultural science, beginning from a simple but fruitful principle: the earth should be studied and treated with the same principles that Hippocratic-Galenic medicine applied to the human body (García Sánchez 1992; Samsó 2011, 289–91).

V.5.4.3 Toledo and the astronomy of Ṣāʿid al-Andalusī

The Toledan judge Ṣāʿid al-Andalusī appears at the center of an intense activity in astronomy that produced, among its major achievements, the compilation of a set of astronomical tables diffused in medieval Europe under the name of *Toledan Tables*, as well as the development of universal instruments, and the project of reforming some of Ptolemy's (100–*c.* 170) planetary models (Samsó 2011, 132–244, 2020, 28–33, 440–66, 479–87, 577–610, 654–81, 719–23).

Paradoxically, the best information on the history of this circle is not given by Ṣāʿid in his work but by a later source, *Foundation of the World* (*Yesod ha-ʿolam*), written by Isaac Israeli (1st half 8th/14th century). Isaac says that Ṣāʿid sponsored a team of collaborators, some of them Jews, who assisted him in many tasks, including the observation of stars for the preparation of the *Toledan Tables* (Millás Vallicrosa 1943–1950, 6–10, 22–3). The information about Ṣāʿid's circle must be read between the lines of what he says in the *Classes of Communities* (*Ṭabaqāt al-umam*; also translated as *Categories of Nations*) and from astronomical works connected in one way or another with this circle. On this basis, we may deduce a connection between the astronomy of Ṣāʿid's circle and the tradition of Maslama, its probable sponsoring by the court and, finally, the existence of at least four Toledan scholars who left works that state that they actually had careers as astronomers or astrologers: al-Istijī, Ibn Khalaf al-Ṣaydalānī, Ibrāhīm ibn Ṣaʿīd al-Asṭurlābī (all them 2nd half 5th/11th century) and Ibn al-Zarqālluh, the most outstanding astronomer in the history of al-Andalus. Ṣāʿid mentions these men among the more promising scholars of the younger generation engaged in the "study of philosophy" (Ṣāʿid al-Andalusī 1985, 180–1). He does not speak of any master–disciple connection with them or any collaboration in a wider project like the *Toledan Tables*. We know that at least the latter connection existed in the case of al-Istijī, who mentions it in an astrological treatise addressed to Ṣāʿid (Samsó and Berrani 1999, 296–8). It is probable that similar relationships existed in other cases, particularly with Ibn al-Zarqālluh, who started his career as an instrument maker (Chapter IV.3).

Thus, it seems that during the 450s/1060s or slightly earlier, Ṣāʿid al-Andalusī created a solid scientific community, well connected with other scholars from al-Andalus, which contributed decisively to the development of virtually all areas of mathematical astronomy. Toward the end of the Toledan kingdom, or possibly earlier, some members of this group immigrated to the territories ruled by the kings of Valencia and Seville. Ibn al-Zarqālluh, who moved to Córdoba, addressed a treatise to the king of Seville, al-Muʿtamid. Seville and the cities of this kingdom became the most important focus of the sciences in al-Andalus at the end of the 5th/11th century since they attracted not only scholars from Toledo but also experts from other cities, like the Jewish astronomer and philosopher Yiṣḥaq al-Baliyya (d. 487/1094) from Granada, who served King al-Muʿtamid as court astrologer (Balty-Guesdon 1992, 315).

V.5.5 The sciences under the North African dynasties (6th/12th century)

V.5.5.1 The transition between two worlds

The "party kings" were replaced by the Almoravids (r. *c.* 454–541/1062–1147), a Maghribī dynasty that followed the teachings of ʿAbdallāh ibn Yāsīn (d. 450/1058), a legal scholar who sought the restoration of the purity of Islam as interpreted by the Mālikī legal school. For this reason, the Almoravids rejected astrology and philosophy. In addition, the number of cities that could afford to pay a learned court diminished considerably, such that patronage of the sciences decayed dramatically and only practical disciplines like medicine or agriculture were sponsored (Forcada 2011, 298–311). Men of science had to rely on their own means to survive, and some of them abandoned al-Andalus. The situation of the Jewish and Christian communities worsened, although there was no open persecution. The history of the sciences in this period is a sum of stories of individuals who studied philosophy, astrology and astronomy in isolation, if not in secrecy or in the world of the Jewish communities. It also encompasses legal scholars who performed *ʿilm al-farāʾiḍ* (Lamrabet 2014, 81–92) and physicians who practiced medicine in well-established families like the Banū Zuhr (Ibn Abī Uṣaybiʿa 1965[CB], 517–30, 2020[CB], 13.63). The latter group included some women doctors.

Within this context, the physician-philosopher Ibn Bājja (d. 533/1139) succeeded in transmitting and renewing the philosophical and scientific legacy of the philosophical circles of 5th/11th-century Saragossa. He completed his training in the late 5th/11th and early 6th/12th in Córdoba and Seville, learning mathematics, directly or indirectly, from Ibn Sayyid (d. at the end of the 5th/11th century), who lived in Valencia (Djebbar 1993, 1998). Under the influence of al-Fārābī's works, Ibn Bājja tied scientific knowledge inextricably to the ultimate felicity of humankind. In accordance with this ideal, he not only commented on the natural philosophy of Aristotle but also studied and practiced music, the mathematical disciplines, optics, astronomy and medicine. In order to establish what he considered to be the true knowledge that a philosopher must possess so as to achieve his ethically transcendent purposes, he practiced a science governed by the rules of logic and the paradigm of Aristotelian philosophy. Applying these ideas, he was the first scholar in al-Andalus to attempt to reform the works of Galen and Ptolemy, in order to eliminate their contradictions with Aristotle.

Ibn Bājja led a wandering life under the aegis of the Almoravids, whom he served in several cities in al-Andalus and the Maghrib. Nonetheless, he was able to gather around him a small but committed group of disciples, who worked with him and eventually transmitted his teachings to following generations, particularly the physician-philosopher Maimonides (d. 601/1204; Dunlop 1957, 174; Forcada 1999, 411–20). In spite of its small number and the fact that none of its members left any written work, this group is possibly the only scholarly community worthy of the name in the scientific world of the first half of the 6th/12th century in al-Andalus and the Maghrib.

V.5.5.2 The Almohad world

During the second half of the 6th/12th century and the first decades of the 7th/13th, al-Andalus was ruled by a second dynasty from the Maghrib: the Almohads, who were also religious reformists, although they differed from the Almoravids, following a religious leader, Muḥammad ibn Tūmart (d. 524/1130), who promoted a rationalist religion favoring philosophically based theology. The new regime was openly hostile to Jews, Christians and the traditional Malikism of Andalusi Islam. The Almohads created their own religious bodies, like the *ṭalaba*, a group of missionaries, who promoted their creed. The most notable philosophers of this time, Ibn Ṭufayl (d. 581/1185) and Ibn Rushd, worked for the Almohads as physicians. It is also possible that Ibn Rushd was a theologian of the Almohads (Urvoy 1998, 56–60; Brague 2010, 228), and that Ibn Rushd and Ibn Ṭufayl belonged to the *ṭalaba* (Fierro 2018), although both ascriptions are contested (Puig Montada 2010).

The Almohads not only fostered the diffusion of Ibn Tūmart's religious thinking but also promoted the writing of encyclopedic works on a wide range of subjects (Fierro 2014b, 29–30), in order to educate their elites. The presence of philosophy at the court was favored by Caliph Abū Yaʿqūb Yūsuf (r. 558–580/1163–1184), who promoted Ibn Rushd's commentaries on Aristotle.

The most important scientific organization of this period was the medical service of the Almohads, in which almost all the important physicians of the 6th/12th century served (Forcada 2005, 1118–9; Forcada 2011, 309–11, 316–8; Ibn Abī Uṣaybiʿa 1965, 530–7, 2020, 13.64). This body was led for most of its life by scholars like Ibn Ṭufayl, Ibn Rushd and Abū Jaʿfar al-Dhahabī (d. 600/1203–4), a disciple of Ibn Rushd, who was also a physician-philosopher. Among its members we find a number of scholars who possessed the multifaceted intellectual profile of the physician-philosophers. It is thus possible to conceive of this medical service as the core of a relatively large intellectual community devoted not only to the study and practice of medicine but also to the development of philosophy and the sciences along Aristotelian lines sketched out in the previous generation by Ibn Bājja under the inspiration of al-Fārābī (Forcada 2005, 1099–1103).

Ibn Rushd was at the heart of this intellectual activity, which encompassed the renewal of not only the sciences but also the religious disciplines. While literally dozens of disciples and followers of all kinds gathered around Ibn Rushd (Puig Montada 1992; Fricaud 2005, 182–5), only a few scholars actually contributed important works or ideas. Ibn Ṭufayl was the first of these major figures and appears to have been the instigator of all of their relevant works (Forcada 2005, 1106). Next was Ibn Rushd, who, on one hand, wrote *The Book of the General (Principles) of Medicine* (*Kitāb al-Kulliyyāt fī l-ṭibb*) in order to strengthen the Aristotelian basis of medicine according to al-Fārābī (Chandelier 2011; Forcada 2020, 399–404) and, on the other, criticized Ptolemy because his planetary models did not match Aristotle's astronomical and physical postulates (Samsó 2011, 337–42, 2020, 28–33, 440–66, 479–87, 577–610, 654–81, 719–23). He was followed by al-Biṭrūjī (2nd half 6th/12th century), who in his *Book on the Configuration [of the Universe]* (*Kitāb al-Hay'a*), attempted to offer alternative homocentric models for planetary movements in contrast to Ptolemy's eccentric and epicyclic models (al-Biṭrūjī 1971; Samsó 2011, 342–56; Mancha 2004; Chapter II.6). Although al-Biṭrūjī vaguely suggested that he might have been Ibn Ṭufayl's disciple, no historical source connects him with either Ibn Ṭufayl or Ibn Rushd. In his books he included a theory of planetary motion based on Neoplatonic doctrines, which were accepted by Iben Bājja but rejected by Ibn Rushd. A successor with similar astronomical ambitions, but no longer working in al-Andalus, was Joseph ben Naḥmi'as ([*fl. c.* 802/1400], Morrison 2016[CB]).

Aside from the activity centered around Ibn Ṭufayl and Ibn Rushd, there were other interesting developments. A local community of scientists appeared in Murcia in the mid-6th/12th century, at the center of which we find ʿAbd al-Raḥmān ibn Ṭāhir (d. 574/1178–9), a member of the Murcian aristocracy who was a true philosopher and who knew Ibn Ṭufayl and Ibn Rushd. He hired a disciple of Ibn Bājja, Ibn Sahl al-Ḍarīr (d. 571/1176), as the preceptor of his son, ʿAbd al-Ḥaqq ibn Ṭāhir (*fl.* late 6th/12th century), who became a competent mathematician. In addition, ʿAbd al-Raḥmān ibn Ṭāhir taught philosophy to his stepson Abū Jaʿfar ibn Ḥassān (d. 599/1202–3), who became a renowned physician of the Almohads (Forcada 2011, 300–2).

Another interesting phenomenon connected with the philosophized intellectual life of the period is the emergence of 'hidden traditions', obscure chains of transmission and scholarly communities that had been working in the background for decades. One example is that of the mathematicians. It seems that the philosophical bias imprinted on the sciences fostered the practice of pure mathematics in philosophical circles by Ibn Bājja and his disciples (Forcada 2011, 311–14). From the mid-6th/12th century onward, we find mathematicians connected with the Almohad court, including al-Ḥaṣṣār (alive in 552/1157), who quoted the disciples of Maslama, and Ibn Munʿim (d. 626/1228), who borrowed heavily from al-Muʾtaman, Ibn Sayyid and ʿAbd al-Ḥaqq ibn Ṭāhir, among others, in his *Knowledge of Arithmetic* (*Fiqh al-Ḥisāb*). These two scholars, together with others like Ibn al-Yāsamīn (d. 601/1204), who taught mathematics in Almohad Seville, made up a virtual community of mathematicians on both sides of the Straits of Gibraltar that would remain active in subsequent periods (Djebbar 2003, 105–9, Lamrabet 2014, 153–4). Another case is that of two Jewish physician-philosophers born in Córdoba in this period, Ibn Dāwūd (d. 575/1180) and Maimonides, who has been mentioned earlier. Although we know little about the formative periods of these two authors, they seem to have had some connection with the philosophical circles of the late party kings and the Almoravids. The arrival of the Almohads forced them to emigrate, Ibn Dāwūd to Christian Toledo and Maimonides to Egypt (Freudenthal 2016; Stroumsa 2009, 56–9; Chapter V.14).

V.5.6 Scholarly communities of the later periods (7th–9th/13th–15th centuries)

V.5.6.1 The Kingdom of Murcia

As we have seen, a notable group of scholars was active in Murcia, on the southeast coast of modern Spain, during the last decades of the 6th/12th century. After the Almohads had abandoned al-Andalus, Murcia was for some time an independent kingdom established by Abū ʿAbdallāh ibn Hūd al-Mutawakkil (r. 626–635/1228–1238), an adventurer who claimed to belong to the Banū Hūd dynasty. The new Banū Hūd ruled Murcia until 664/1266, and it seems that they showed the same interest in the sciences as the Banū Hūd of Saragossa (Guichard 1991, 1: 141–5). In this context, two multifaceted scholars emerged, with strong inclinations toward mysticism and known connections to the court. Both immigrated to the East: the philosopher Ibn Sabʿīn (d. 668 or 669/1270 or 1271) and Badr al-Dīn ibn Hūd (d. 700/1300), nephew of the king Ibn Hūd al-Mutawkkil (Bellver 2018). To a similar or slightly later generation belong two physician-philosophers of note, Ibn Andrās (d. 674/1275; Maḥfūẓ 1982–6, 1: 56–7) and al-Riqūṭī (last third 7th/13th century; Samsó 2011, 388–9), who, like the preceding authors, emigrated from Murcia – the first to the Maghrib (to Bejaïa) and the second to Hafsid Tunis and then to Granada. In a third or fourth generation, we find two scholars born in Murcian families, who worked in Granada: the physician-philosopher Ibn al-Raqqām (d. 715/1315) and the polymath Ibn Luyūn ([d. 750/1349]; Forcada 020–2021, 182–8).

We may conclude that Murcia had a strong scholarly community that followed the trends of the Almohad period; it produced versatile scholars, often physicians, who could study and teach a range of subjects but who were forced to emigrate by the progress of the Christian conquest. This period also saw the increasing integration of the Andalusi scholars into a western Mediterranean context of intellectual exchange (Samsó 2007, arts. IX–XII). While one of the reasons was undoubtedly the Christian conquest, it is also true that the cities of the Maghrib attracted Andalusi scholars.

V.5.6.2 The Kingdom of Granada

Ruled by the Nasrids between 634/1237 and 997/1492, Granada began in the late 7th/13th century to be an important locus for the sciences. This was probably due to imitation of the policies for promoting knowledge implemented by the Castilians and by the Marinids (r. 614–869/1217–1465) in the Maghrib. The physician-philosopher al-Riqūṭī taught at a *madrasa* built for him by Alfonso X (r. 1252–1284) in Murcia, until Muḥammad II (r. 671–701/1273–1302) invited him to teach in Granada. There he imparted medicine, mathematics and other subjects to the local elite. In 749/1349, Yūsuf I (r. 733–754/1333–1354) founded the first *madrasa* of al-Andalus,[4] where we know that at least medicine and logic were taught to the students. In the 8th/14th century, Al-Riqūṭī and other scholars, including the previously mentioned Ibn Luyūn and Ibn al-Raqqām, educated a new generation of local physicians who became experts in several scientific disciplines (Puig 1983, 1984). To this group belonged Ibn Hudhayl (d. 753/1353), Ibn al-Khaṭīb (713–776/1313–1374), Ibn Khātima (d. 770/1369) and other scholars, who were connected by well-known master–disciple relationships that included physicians of Levantine Spain (Franco Sánchez 2001, 39–43).

These relationships show the strength of a scholarly community well established in Nasrid society, which projected its influence onto the Maghrib and the Muslims who lived in the Iberian Peninsula under Christian rule. The doctrinal debates of the Almohad period were set aside. To a certain extent, these scholars tried to bring the sciences to a wider audience. For this reason, in this period we find a wealth of didactic poems on many subjects (agriculture, astronomy, medicine and so on), which were on occasion used in the *madrasas* for the teaching of future scholars. At the same time, we find for the first time *muwaqqits* (mosque officials, who regulated the times of prayer) well trained in mathematical astronomy appointed at mosques (Calvo 1992, 121–2).

During the last decades of Nasrid Granada, most of the scholars who could earn their living using their knowledge emigrated to the Maghrib or even farther afield (al-Khaṭṭābī 1988, 1: 78–82; Lamrabet 2014, 114–7). As a result, the scientific and medical activities of Granada rapidly disappeared during the years after 987/1492, when the city fell to Christian forces. In other cities of Spain, there are traces of Islamicate scientific life among the Muslims who lived under Christian rule in the 14th and 15th centuries (the so-called Mudéjares) and works in Arabic written by Jews (Samsó 2011, 388–91, 558–61). Muslims who remained in Spain after 987/1492 were forced to convert at the beginning of the 16th century and were known as "Moriscos". They preserved a certain level of popular medicine and veterinary science (García Ballester 1984). Quacks, healers, midwives and so on, they were the last scientific community of an al-Andalus that had ceased to exist. Ironically, however, a small number of Moriscos were able to study medicine in Christian institutions and therefore read Islamicate manuals that had been translated from Arabic into Latin.

V.5.7 Conclusion

Many gifted authors, like the mathematician and astronomer Ibn Muʿādh (d. 485/1093), the astronomer Jābir ibn Aflaḥ (6th/12th century) and the geoponist Ibn al-ʿAwwām (6th/12th century), have not been mentioned in the preceding pages, because they seemingly worked in isolation, although it may also be that we do not know enough about the contexts in which they lived. Even though al-Andalus is one of the most studied societies in the medieval Islamicate world, there is still much to be investigated about Andalusi scholarly communities. However, one conclusion seems clear. The development of the rational sciences in al-Andalus was neither the sum of the efforts of many individuals nor a collective work but rather a work of collectives. One of the proofs of this assertion is that the present chapter may, to some extent, serve as a summary of the history of science in al-Andalus. However trivial it may seem, this conclusion expresses an indubitable fact. In every phase of al-Andalus's history, we find a remarkable interest in science, both on the part of the individuals, who did science for the sake of science, earning their living or both things at the same time, and the rulers, who appreciated the fruits of science, considered science to be a useful political asset or both things at the same time. This "coincidence of interests" is one of the best explanations for, first, the existence of relatively high number of coherent groups of scholars devoted to the rational sciences and, second, the remarkable contribution of al-Andalus to philosophy and the sciences.

Notes

1 This chapter has been commissioned as part of the project FFI2017–88569-P, "Ciencia y sociedad en el Mediterráneo Occidental: el Calendario de Córdoba y sus tradiciones," Ministerio de Economía, Industria y Competitividad.

2 A full reference of the primary and secondary sources about the authors that are mentioned in this chapter can be found in: Fierro (2014a) and Lirola and Puerta Vílchez (2004–12). Other works that provide thorough bibliographical information are about mathematicians and astronomers, Lamrabet (2014) and Hockey (2007[CB]); about physicians, Khaṭṭābī (1988); about Jewish scholars, Stillman (2010).

3 Consolidated bibliography.

4 Al-Andalus is the part of today's Spain that was ruled by Muslim dynasties. Hence, Murcia does not count as al-Andalus after its conquest by the Christian ruler Alfonso X.

Bibliography

Sources

al-Biṭrūjī. ed. and tr. Goldstein B. R. 1971. *Al-Biṭrūjī: On the Principles of Astronomy*. 2 vols. New Haven: Yale University Press.

Research literature

al-Khaṭṭābī, M. 1988. *Al-Ṭibb wa-l-aṭibbāʾ fī l-Andalus al-islāmiyya*. 2 vols. Beirut: Dār al-gharb al-islāmī.

Balty-Guesdon, M. G. 1992. *Médecins et hommes de science en Espagne Musulmane (IIe/VIIIe–Ve/XIe s.)*. PhD dissertation. Paris: Université de la Sorbonne Nouvelle.

Beech, G. T. 2008. *The Brief Eminence and Doomed Fall of Islamic Saragossa, a Great Center of Jewish and Arabic Learning in the Iberian Peninsula during the 11th Century*. Saragossa: Instituto de Estudios Islámicos y del Oriente Próximo.

Bellver, J. 2018. "Ibn Hūd, Badr al-Dīn Ḥasan," in *EI-3*, 2017–5: 111–4. http://doi.org.sire.ub.edu/10.1163/1573-3912_ei3_COM_30824 (Accessed 4 April 2019).

Brague, R. 2010. *The Legend of the Middle Ages: Philosophical Explorations of Medieval Christianity, Judaism and Islam*. Chicago: University of Chicago Press.

Callataÿ, G. de. 2013. "Magia en al-Andalus: *Rasāʾil Ijwān al-Ṣafāʾ*, *Rutbat al-Ḥakīm* y *Gāyat al-Ḥakīm* (*Picatrix*)," *al-Qanṭara* 34: 297–344.

Callataÿ, G. de and Moureau, S. 2017. "A Milestone in the History of Andalusī Bāṭinism: Maslama b. Qāsim al-Qurṭubī's *Riḥla* in the East," *Intellectual History of the Islamicate World* 5: 86–117.

Calvo, E. 1992. "La ciencia en la Granada nazarí (ciencias exactas y tecnología)," in Vernet, J. and Samsó, J., eds. *El legado científico andalusí*. Madrid: Ministerio de Cultura, 117–26.

Chandelier, J. 2011. "Medicine and Philosophy," in Lagerlund, H., ed. *Encyclopedia of Medieval Philosophy*. 2 vols. Dordecht: Springer, 1: 745–52.

Djebbar, A. 1993. "Deux mathématiciens peu connus de l'Espagne du XIe siècle: al-Mu'taman et Ibn Sayyid," in Folkerts, M. and Hogendijk, J. P., eds. *Vestigia Mathematica: Studies in Medieval and Early Modern Mathematics in Honour of H. L. L. Busard*. Amsterdam-Atlanta: Rodopi, 79–91.

Djebbar, A. 1997. "La rédaction de l'*Istikmāl* d'al-Mu'taman (XIe siècle) par Ibn Sartaq, mathématicien de XIIIe-XIVe siècles," *Historia Mathematica* 24: 185–92.

Djebbar, A. 1998. "Abū Bakr Ibn Bājja et les mathématiques de son temps," in *Hommage à Jamal ad-Dine Alaoui. Études philosophiques et sociologiques dédiées à Jamal ad-Dine Alaoui*. Fez: Publications de la Faculté des Lettres et des Sciences Humaines Dar el-Mehraz, 5–26.

Djebbar, A. 2003. "Les activités mathématiques en al-Andalus et leur prolongement au Maghreb (IXᵉ-XVᵉ s.)," in Batlló Ortiz, J., Bernat López, P. and Puig Aguilar, R., eds. *Actes de la VII Trobada d'Història de la Ciència i de la Tècnica*. Barcelona: Societat Catalana d'Història de la Ciència i la Tècnica, 87–112.

Dunlop, D. M. 1953–57. "Remarks on the Life and Works of Ibn Bâjjah (Avempace)," in Velidi Togan, Z., ed. *Proceedings of the 22nd Congress of Orientalists*. 2 vols. Istanbul: Osman Yalcin Matbaasi, 2: 188–96.

Ebstein, M. 2016. "Ibn Masarra," in *EI-3*, 2016–5: 143–5. (Accessed 4 April 2019).

Fierro, M. 1987. *La Heterodoxia en al-Andalus durante el periodo omeya*. Madrid: Instituto Hispanoárabe de Cultura.

Fierro, M. 1996. "Bāṭinism in al-Andalus. Maslama b. Qāsim al-Qurṭubī (d. 353/964), Author of the *Rutbat al-Ḥakīm* and the *Ghāyat al-Ḥakīm* (Picatrix)," *Studia Islamica* 84: 87–112.

Fierro, M. 2004. "La política religiosa de ʿAbd al-Raḥmān III," *al-Qanṭara* 25: 119–156.

Fierro, M. 2012. "Plants, Mary the Copt, Abraham, Donkeys and Knowledge: Again on Bāṭinism During the Umayyad Caliphate in al-Andalus," in Biesterfeldt, H. and Klemm, V., eds. *Differenz und Dynamik im Islam. Festschrift für Heinz Halm zum 70. Geburtstag*. Würzburg: Ergon, 125–44.

Fierro, M. 2014a. *Historia de los Autores y Transmisores Andalusíes = History of Andalusi Authors and Transmitters* (HATA). http://kohepocu.cchs.csic.es.

Fierro, M. 2014b. "The Islamic West in the Time of Maimonides: The Almohad Revolution," in Mühlethaler, L., ed. *"Höre die Wahrheit, wer sie auch spricht": Stationen des Werks von Moses vom islamischen Spanien bis ins moderne Berlin*. Göttingen: Vandenhoeck and Ruprecht, 21–32.

Fierro, M. 2018. "Ibn Rushd's (Averroes) 'Disgrace'and his Relation with the Almohads," in Al Ghouz, A., ed. *Islamic Philosophy from the 12th to the 14th Century*. Göttingen: Bonn University Press, 73–118.

Forcada, M. 1999. "De Avempace a Averroes: la transmisión de las ciencias de los antiguos de la época taifa a la almohade," in Fierro, M. and Ávila, M. L., eds. *Estudios onomástico-biográficos de al-Andalus, IX. Bibliografías almohades I*. Madrid-Granada: CSIC, 407–24.

Forcada, M. 2005. "Síntesis y contexto de las ciencias de los antiguos en época almohade," in Cressier, P. *et al.*, eds. *Los almohades. Problemas y perspectivas*. 2 vols. Madrid: Casa Velázquez, 2: 1091–135.

Forcada, M. 2011. *Ética e ideología de la ciencia: el médico filósofo en al-Andalus (siglos X-XII)*. Almería: Fundación Ibn Ṭufayl.

Forcada, M. 2015. "Astrology in al-Andalus during the 11th and 12th Centuries: Between Religion and Philosophy," in Burnett, C. and Greenbaum, D., eds. *From Masha'allah to Kepler: The Theory and Practice of Astrology in the Middle Ages and the Renaissance*. Ceredigion (Wales): Sophia Trust and University of Wales, 149–76.

Forcada, M. 2017. "Books from Abroad: The Evolution of Science and Philosophy in Umayyad al-Andalus," *Intellectual History of the Islamicate World* 5: 55–85.

Forcada, M. 2020. "Bronze and Gold. Al-Fārābī on medicine," Oriens 48: 367–415.

Forcada, M. 2020–2021. "Didactic poems on medicine and their commentaries in medieval al-Andalus and Western Islam," Suhayl 18: 165–204.

Franco Sánchez, F. 2001 "La escuela médica *šarqī* (ss. XI-XIV): sociedad y medicina en el Levante de al-Andalus," *Dynamis* 21: 27–53.

Freudenthal, G. 2016. "Abraham Ibn Daud, Avendauth, Dominicus Gundissalinus and Practical Mathematics in Mid-Twelfth Century Toledo," *Aleph* 16: 61–106.

Fricaud, E. 2005. "Le problème de la disgrâce d'Averroès," in Bazzana, A. *et al.*, eds. *Averroès et l'averroisme (XIIe – XVe siècle). Un itinéraire historique du Haut Atlas à Paris et à Padoue. Actes du colloque international organisé à Lyon, les 4 et 5 octobre 1999 dans le cadre du temps du Maroc*. Lyon: Presses Universitaires de Lyon: 155–89.

García Ballester, L. 1984. *Los moriscos y la medicina: un capítulo de la medicina y la ciencia marginadas en la España del siglo XVI*. Barcelona: Labor.

García Sánchez, E. 1992. "Agriculture in Muslim Spain," in Jayyusi, S. K. and Marin, M., eds. *The Legac y of Muslim Spain*. Leiden: Brill, 987–99.
García Sánchez, E. 1994. "El Botánico Anónimo sevillano y su relación con la escuela agronómica andalusí," in García Sánchez, E., ed. *Ciencias de la Naturaleza en al-Andalus. Textos y Estudios III*. Granada: CSIC, 193–210.
Gómez Nogales, S. 1969. "Las artes liberales y la filosofía hispanomusulmana," in *Arts libéraux et philosophie au moyen âge: Actes du quatrième Congrès international de philosophie médiévale*. Montréal-Paris: Institut d'Etudes Médiévales-Vrin, 493–508.
Guichard, P. 1991. *Les musulmans de Valence et la Reconquête (XIe–XIIIe siècles)*. 2 vols. Damascus: Institut français de Damas.
Hamarneh, S. and Sonnedecker, G. 1963. *A Pharmaceutical View of Abulcasis al-Zahrāwī in Moorish Spain*. Leiden: Brill.
Herrera Roldán, P. P. 1995: *Cultura y lengua Latinas entre los Mozárabes Cordobeses del siglo IX*. Córdoba: Publicaciones de la Universidad de Córdoba.
Hogendijk, J. P. 1995. "Al-Mu'taman ibn Hūd, 11th Century King of Saragossa and Brilliant Mathematician," *Historia Mathematica* 22: 1–18.
Lamrabet, D. 2014. *Introduction à l'histoire des mathématiques maghrébines*. Rabat: Driss Lamrabet.
Lirola, J. and Puerta Vílchez, J. M., eds. 2004–2012. *Biblioteca de al-Andalus*. 7 vols. and appendix. Almería: Fundación Ibn Ṭufayl.
Lomba, J. 2002. *El Ebro: puente de Europa: pensamiento musulmán y judío*. Saragossa: Mira Editores.
Maḥfūẓ, M. 1982–1986. *Tarājim al-mu'allifīn al-tūnisiyyīn*. 5 vols. Beirut: Dār al-gharb al-islāmī.
Mancha, J. L. 2004. "Al-Biṭrūjī's Theory of the Motion of the Fixed Stars," *Archive for History of the Exact Sciences* 58: 143–82.
Millás Vallicrosa, J. M. 1943–1950. *Estudios sobre Azarquiel*. Madrid-Granada: CSIC.
Mimura, T. 2015. "The Arabic Original of (ps.) Māshā'allāh's *Liber de orbe*: Its Date and Authorship," *British Journal for the History of Science* 48: 321–52.
Oliveras, M. 2009. "El *De imaginibus caelestibus* de Ibn al-Ḥātim," *Al-Qanṭara* 30: 171–220.
Puig, R. 1983. "Dos notas sobre ciencia hispano-árabe a finales del siglo XIII en la *Iḥāṭa* de Ibn al-Jaṭīb," *al-Qanṭara* 4: 433–40.
Puig, R. 1984. "Ciencia y técnica en la *Iḥāṭa* de Ibn al-Jatib. Siglos XIII y XIV," *Dynamis* 4: 65–79.
Puig Montada, J. 1992. "Materials on Averroes's Circle," *Journal of Near Eastern Studies* 51: 241–60.
Puig Montada, J. 2001. "Reason and Reasoning in Ibn Ḥazm of Cordova (d. 1064)," *Studia Islamica* 92: 165–85.
Puig Montada, J. 2010. "Ibn Rušd and the Almohad Context," in Fontaine, R. *et al.*, eds. *Studies in the History of Culture and Science. A Tribute to Gad Freudenthal*. Leiden: Brill, 189–208.
Rius, M. 2003. "La actitud de los emires cordobeses hacia los astrólogos: entre la adición y el rechazo," in de la Puente, C., ed. *Estudios onomástico- biográficos de al-Andalus, 13. Identidades marginales*. Madrid: CSIC, 517–50.
Safran, J. 2013. *Defining Boundaries: Muslims, Christians, and Jews in Islamic Iberia*. Ithaca: Cornell University Press.
Samsó. J. 2001. "Sobre el astrólogo ʿAbd al-Wāḥid b. Isḥāq al-Ḍabbī (fl. c.788-c.852)", *Anaquel de estudios árabes* 12: 657-670.
Samsó, J. 2007. *Astronomy and Astrology in Al-Andalus and the Maghrib*. Aldershot: Variorum.
Samsó, J. and Berrani, H. 1999. "World Astrology in Eleventh-Century al-Andalus: The Epistle on *Tasyīr* and the Projection of Rays by al-Istijjī," *Journal of Islamic Studies* 10: 293–312.
Stillman, N. A. 2010. *Encyclopedia of Jews in the Islamic World*. 5 vols. Leiden: Brill.
Stroumsa, S. 2009. *Maimonides in His World: Portrait of a Mediterranean Thinker*. Princeton: Princeton University Press.
Stroumsa, S. 2016. "Between Acculturation and Conversion in Islamic Spain: The Case of the Banū Ḥasdāy," *Mediterranea. International Journal for the Transfer of Knowledge* 1: 9–36.
Urvoy, D. 1998. *Averroès. Les ambitions d'un intellectuel musulman*. Paris: Flammarion.
Vernet, J. 1968. "Los médicos andaluces en el *Libro de las generaciones de médicos* de Ibn Ŷulŷul," *Anuario de estudios medievales* 5: 445–62 (Reprint: Vernet, J. 1979. *Estudios sobre la historia de la ciencia medieval*. Barcelona-Bellaterra: Universidad Autónoma de Barcelona, 469–86).

V.6
POST-AVICENNAN NATURAL PHILOSOPHY

Jon McGinnis

Post-Avicennan natural philosophy (after 428/1037) is only newly charted terrain for Western scholars, and most of it remains wholly unexplored, which is particularly true for physics after 600/*c.* 1200 in the Islamicate world. Many of the texts remain unedited or poorly edited. Frequently they are also not readily accessible at all. Thus, given the incipient state of our scholarship and the relatively limited access to the material culture associated with physics in this later period, this study is more programmatic (and even impressionistic) than systematic and thus far from complete. Still, certain physical topics and figures who worked on natural philosophy during this period have received studies. In what follows, I first look at the genre in which much later physics is done – the commentary – and some key contributors to natural philosophy during this period. Second, I suggest as a hypothesis that a reorientation in the way that natural philosophy is practiced occurred in the later Islamicate world, namely, a shift from a prominently realist view concerning the aim of science to an antirealist or instrumentalist view. Third, I conclude with two specific but core physical issues addressed during this period in which this reorientation plays a role in scientific investigation, namely, conceptualizing the nature of body and providing a scientifically adequate account of motion.

V.6.1 Physical works and natural philosophers in the post-Avicennan world

While independent treatises on various aspects of natural philosophy exist during this period, as a general rule the vehicles for much post-Avicennan physics are commentaries and glosses. Between the 5th/11th and the early 8th/14th centuries, a number of short philosophical summas or 'textbooks' appeared that include lengthy sections on natural philosophy (*ṭabīʿiyyāt*), which provide the base text (*matn*) for these commentaries and physical investigations. Among the more important of these base texts is Ibn Sīnā's (d. 428/1037) *Pointers and Reminders* (*Ishārāt wa-l-tanbīhāt*), which enjoyed commentaries by Fakhr al-Dīn al-Rāzī (d. 606/1210), Sayf al-Dīn al-Āmidī (551–631/1156–1233), Naṣīr al-Dīn al-Ṭūsī (597–672/1201–1274) and Quṭb al-Dīn al-Taḥtānī (d. 766/1364), the last of whose commentary gave rise to super-commentaries from such geographically distinct individuals as al-Dawānī (830–908/1426–1502) from Shiraz (Timurid Iran) to the Ottoman Ibn Kemāl Pāshā (d. 940/1533). Al-Suhrawardī's (d. 587/1191) *Intimations of the Tablet and the Throne* (*al-Talwīḥāt al-lawḥiyya wa-l-ʿarshiyya*) also is significant and

DOI: 10.4324/9781315170718-55

commented by the Jewish philosopher Ibn Kammūna (d. 676/1277) among others. Almost equally popular among commentators as Ibn Sīnā's *Pointers* is Athīr al-Dīn al-Abharī's (d. between 660/1263 and 663/1264) *The Guide to Wisdom* (*Hidāyat al-Ḥikma*), which receives comments from al-Sharīf al-Jurjānī (740–816/1340–1413), Mīr Ḥusayn Maybudī (d. 910/1504) – whose own commentary is the subject of several super-commentaries – Ṣadr al-Dīn Dashtakī (d. 903/1497–8) and the Mullā Ṣadrā (d. 1050/1640) – whose commentary again receives several super-commentaries. Also important is *The Wisdom of the Fount* (*Ḥikmat al-ʿayn*) of Najm al-Dīn al-Kātibī (d. 675/1276 or 693/1294). Finally, from as far as the Indian subcontinent and as late as the 1840s one encounters Fażl-i Ḥaqq Khayrābādī's (d. 1861) *The Happy Gift Concerning Natural Philosophy* (*al-Hadiyya al-saʿīdiyya fī l-ḥikma al-ṭabīʿiyya*).[1]

Further important textual sources for the development and direction of natural philosophy in the later Islamicate world are Abū l-Barakāt al-Baghdādī's (d. *c.* 560/1164–5) *Reflection on Wisdom* (*al-Muʿtabar fī l-ḥikma*) and Fakhr al-Dīn al-Rāzī's *Eastern Investigations* (*al-Mabāḥith al-mashriqiyya*). One also might add Rāzī's *Compendium of Wisdom* (*al-Mulakhkhaṣ fī l-ḥikma*), which is a compendium of Avicennan philosophy. Abū l-Barakāt subjects the dominant Aristotelian–Avicennan philosophical synthesis to careful scrutiny and a thoroughgoing critique, challenging assumptions, which others frequently left unquestioned, and pressing the empirical adequacy of those theories. The response to Abū l-Barakāt is a spate of criticisms and defenses of his observations that remained an impetus for the practice of post-Avicennan natural philosophy in the East (Pines 1937, 1955; Pavlov 2017). Fakhr al-Dīn al-Rāzī's *Investigations*, which incorporates much from Abū l-Barakāt, frequently is cited among the commentators on others' works and is also extremely influential in determining how certain issues of natural philosophy are arranged, for example, discussions of impetus (*mayl*), as well as the understanding of the content of natural philosophy.

V.6.2 Scientific practice: realism versus antirealism

Let me begin with the notions of 'natural philosophy,' certainly as it pertains to the medieval Islamic world, and 'scientific practice' with reference to medieval Islamicate natural philosophy. First, medieval Islamicate physics, or natural philosophy, in its most general scope studied bodies insofar as they are subject to change or undergo motion and various processes. While there are additionally special natural sciences that consider specific kinds of bodies or processes and change, the focus here is general natural philosophy. Second, as a theoretical science, natural philosophy and its methods and practices were viewed as akin to those of mathematics and metaphysics. Thus, a natural philosopher when practicing his field would most likely spend most of his time in the proverbial armchair rather than in the lab or field collecting empirical data. Indeed, the scientific practices of the medieval Islamicate natural philosopher look very *unlike* what a contemporary physicist does and instead more like what a philosopher of science does today. In this section, I suggest that between the period up to Avicenna and the period after him there was a major theoretical shift in the way that natural philosophy was conceived, and so a major shift in the very scientific practice of natural philosophy. This section is the most theoretical in scope. Thus, let me state its main point as a hypothesis to be tested about scientific practice after the 5th/11th century rather than as a historical fact: Post-Avicennan natural philosophy experienced a shift from a robust realist approach toward scientific practice to an antirealist (or instrumentalist) approach (cf. Fazlıoğlu 2014).

By 'realism' I mean an attitude about the nature and aim of science that has two main features: one epistemic (i.e., about what one can know), one metaphysical (i.e., about the way things exist). The metaphysical feature concerns specifically the nature of truth. For realist thinkers in this period truth (*ḥaqq*) involves a reference to some mind-independent way that things are. A scientific claim, then, is true just in case the referents involved in that claim in some way correspond with the way the world actually is, and more specifically for those in the Aristotelian–Avicennan tradition, accurately correspond with the actual causal structures in the world. A common contemporary example is the claim that 'snow is white' is made true by the fact that *that stuff* (snow) is *such a color* (white); an ancient and medieval example would be the essential definition 'humans are rational, sensitive, animate, material substances' is made true just in case it uniquely identifies humans and provides what is essential to being human.

After Ibn Sīnā's synthesis of Aristotle's (384–322 BCE) philosophy, the epistemological claim of realism concerns the nature of scientific knowledge (Gk. *epistēmē*, Ar. *ʿilm*). One has scientific knowledge just in case one has true, justified belief about the very causal structure of the world. We have seen what 'true' means for scientific realists. Additionally, a belief is justified for those in this tradition, just in case it is either (1) a first principle or (2) the conclusion of a demonstration (*burhān*) whose initial premises are themselves first principles. First principles are claims that purportedly provide one with necessary essential features of the world and their causes and are known with certainty. They may include existence claims, like 'motion exists,' or essential definitions, which purportedly penetrate to the causal nature of what is being defined, like 'motion is the first actuality of potential insofar as there is potential' (Aristotle, *Physics*, 3.1, 201b4–5; Ibn Sīnā 2009, 2.1 [5]).

In contrast, antirealism is a position concerning the nature of science or scientific theories that denies or significantly weakens one or both of these two realist tenets, which again are, one, that the theoretical terms of a science must truly capture the natures of things in the world and, two, science provides a kind of necessary or certain knowledge of the way things are. It is important to note that while skepticism is a form of antirealism, not all forms of antirealism are skeptical. One might deny realism's metaphysical claim about truth while not denying that we have knowledge (perhaps acquired through revelation), or one might allow that lesser degrees of cognitive assurance than certainty, for example, high probability, count for knowledge, or one might remain agnostic with respect to the theoretical, non-observable posits of a scientific theory while accepting that the theory is true based on its predictive power or objectivity. To be sure, those in the realist camp may classify these others as skeptics, but to do so is simply to assume the realists' tenets about what counts as science or scientific knowledge.

By the time of al-Ghazālī's (d. 505/1111) *Incoherence of the Philosophers* (*Tahāfut al-falāsifa*) a fairly well-developed antirealist criticism of Aristotelian-Avicennan scientific realism is coming into place. Al-Ghazālī focuses on realism's epistemic claim, namely, that science provides necessary certainty, and argues that the philosophers have *not* demonstrated with certainty the way that certain features of the world must necessarily be. Al-Ghazālī's point is not to reject the possibility of acquiring scientific knowledge of the world but only to deny the philosophers' claim to having demonstrated the necessary causal structure of the world. In many cases, al-Ghazālī concedes that the philosophers' scientific claims are highly probable, and barring an act of God, one can be assured of their truth. Indeed, with respect to the epistemic aspect of realism, al-Ghazālī may merely be advocating a weaker form of realism rather than an outright rejection of it (Griffel 2009[CB],[2] chs. 6–7).

In contrast, al-Ghazālī's response to the metaphysical facet of Aristotelian–Avicennan realism (again that a science's theoretical terms, whether of observables or non-observables, refer to what actually exists in the world) can be classified only as antirealist. His antirealism is evident with respect to non-observables like modalities, for example, notions of necessity and possibility, and physical notions like time. For Ibn Sīnā and those following him, what is possible (*mumkin*) and what is necessary (*wājib*) are grounded in the very reality of a mind-independent world.

What makes claims like 'Necessarily, all humans are mortal' or 'Possibly, humans are black' true is human nature itself, regardless of whether such modal claims ever exist in the human mind. In contrast, for al-Ghazālī what explains the modal aspect of these claims is whether the mind recoils at them and judges the claims repugnant, as it does in the case of 'Possibly, bachelors are married,' or conversely the mind finds nothing abhorrent in joining together the various referents that make up the claim. As al-Ghazālī succinctly puts it, "possibility, necessity and impossibility reduce to propositions of the intellect" (Ghazālī 2000[CB], discussion 1, proof 4, esp. §127; see also discussion 17, §§27–37).

Some philosophers embraced and expanded on al-Ghazālī's antirealist position by actively criticizing an essentialist theory of definition. An essentialist theory of definition states that scientific knowledge requires definitions known with certainty, which specify the natures of things and are true, primitive, that is, stated in the basic notions of one's philosophy, for instance, actuality and potentiality for Aristotle or possible existents and necessary existents for Ibn Sīnā). These definitions are indemonstable as well as better known than, prior to, and explanatory of, any conclusions demonstrated from them. Abū l-Barakāt presses against the idea that first principles must always specify the natures of things, that is, explain the underlying causal structure of reality. At least in certain cases, like motion (*ḥaraka*), it is sufficient for him that the theoretical terms of one's science are merely empirically adequate (see, e.g., Abū l-Barakāt 1938–1939, 2: 28–34). In other words, for Abū l-Barakāt in certain cases, it is enough that a physical theory correctly describes the observations about the world and allows for accurate predictions even if there is no certainty concerning the theory's assumptions about the underlying unobservable causes and essential natures of things. This is not to say that Abū l-Barakāt abandons scientific realism altogether. Still, empirical observations for him are primary, while essential definitions are less important.

Taking antirealism one step further is al-Suhrawardī's trenchant attack on essential definitions. Al-Suhrawardī's argument is complex (Ziai 1990, 77–114), but briefly stated, it begins by observing that the realist theory of justification, which relies on demonstrations, presupposes that one possesses essential definitions. Essential definitions, Suhrawardī insists, purportedly provide the scientist with an account (*qawl*) of the essential and necessary features of those things that make up the physical world. Consequently, maintains al-Suhrawardī, an essential definition should provide one with a finite list of essential features that constitute the object being defined. Suhrawardī then complains that no finite list or account can ever exhaust the essential features of a natural kind; the natural necessities of a thing, that is, all the states and activities that belong to something essentially or on account of what it is, overflow any definite description, which the natural philosopher provides of that thing. In place of essential definitions, Suhrawardī posits a special intuitive and experiential type of unlimited knowledge, a "knowledge by presence."

Knowledge by presence is not unlike a Platonic intellectual vision of the Realm of the Forms or what is most True and Real. Upon acquiring such knowledge, the philosopher goes on to construct demonstrations. Now, however, a demonstrative science is more like one finite theoretical construct, among many, addressing an infinite reality that defies being reduced to any single model. Demonstrations, then, may confirm one's experience, but in contrast with the realist's epistemic tenet, demonstrations do not justify and causally explain that experience; they merely describe it in a logically rigorous fashion. Fakhr al-Dīn al-Rāzī also leveled a trenchant criticism against Aristotle and Avicenna's theory of essential definitions and even their theory of demonstration, which he sought to replace with a theory of nominal definitions (Ibrahim 2013, esp. ch. 3).

Abū l-Barakāt, al-Suhrawardī and al-Rāzī's critiques of Aristotelean–Avicennan scientific realism implicitly rely on a distinction that can be traced back to elements found in both

Aristotle's *Posterior Analytics* and Ibn Sīnā's *Book of Demonstration* (*Kitāb al-Burhān*) – their respective books on the nature of science. The historical distinction is between the *fact* (i.e., the observation) and the *reasoned fact* (the underlying essential causal explanation of the observation). For scientific realists like Aristotle and Ibn Sīnā, the aim of science is to get at the *reasoned fact*. In contrast, both Abū l-Barakāt and Suhrawardī seem satisfied with accounts and constructs that have predictive power and adequately describe the *fact*. Certainly, during the later period, this shift appears in some descriptions of the aim of a science. For example, Sayyid Sharīf al-Jurjānī, in his commentaries, observes that astronomy was concerned only with the *fact*; he then presents multiple competing physical theories that potentially explain those facts (Dhanani 2007). More generally, by the mid-1800s, Khayrābādī in effect eliminates essential definitions from much of his physics, appealing instead to immediate experience and empirically adequate descriptions as the basis of justifying scientific claims. In the next sections, I consider two concrete instances that stand as evidence in favor of the hypothesis of a shift from realism to antirealism in post-Avicennan natural philosophy.

V.6.3 Body: atomic versus continuous

Ancient and medieval physics has as its proper subject matter body insofar as it is subject to motion. Concerning the nature of body (*jism* and occasionally *jirm*), there were two competing camps in the Islamicate world: the advocates of atomism (in general theologians) and the advocates of hylomorphism, that is, the view that the most basic constituents of body are its form (*ṣura*) and matter (*hayūlā* or *mādda*), (in general philosophers). In the classical period, the atomists generally maintained that body is composed of a finite number of conceptually and physically indivisible parts (*juz' lā yatajazza'a*) and so cannot be divided beyond these atomic parts, whereas philosophers with their theory of form and matter held that body is continuous (*muttaṣil*) and so is at least potentially divisible into increasingly smaller and smaller magnitudes without end.

In *Pointers* (*namaṭ* 1.1–5), Ibn Sīnā suggests the logically possible positions that one can hold about the nature of body, which initially framed most discussions on this topic: Body either (I) has actual parts or (II) has no actual parts. If (I) the body has actual parts, then the number of those parts is either (I.A) finite or (I.B) infinite. (Of course, if (II), the body has not parts, the question of whether its parts are finite or infinite in number is mute.) Position (I.A), that of most theologians, claims that body is an aggregate of a finite number of actual parts, like a stack of cards; however, at least one theologian, namely, Ibrāhīm al-Naẓẓām (d. c. 221/845), held position (I.B) that body is composed of an infinite number of actual parts. In contrast, the philosophers held position (II) that a body need not contain any actual parts while still remaining potentially divisible, at least in thought, infinitely, in the way that imagination can divide a single card into half, half again and so on indefinitely.

In the post-Avicennan period the problem space of the constitution of body changed. In his commentary on *Pointers*, Fakhr al-Dīn al-Rāzī, probably following 'Abd al-Karīm al-Shahrastānī (d. 548/1153), now asserts: Body may be composed of either (1) a finite number of parts or (2) an infinite number of parts. Additionally, if (1), body is composed of a finite number of parts, those parts might belong to it either (1a) actually or (1b) *potentially* (the latter a position not suggested in Ibn Sīnā's initial formulation). Rāzī then continues, following Ibn Sīnā, if (2), body is composed of an infinite number of parts, those parts might belong to it either (2a) actually or (2b) potentially. Again, position (1a) and (2a), namely, that body is composed of actual parts, whether finite or infinite in number, is that of the traditional theologians, while the philosophers held position, (2b), that body has a potentially infinite number of parts. Position (1b), that body is composed of a finite number of potential parts, is new and according to the commentators, the post-Avicenna theologian al-Shahrastānī introduced it in his work *Methods and Proofs* (*Manāhij wa-l-bayyānāt*). In effect, Shahrastānī's suggestion is that when atoms are combined together they form a body that is a continuous unity, unlike the older forms of atomism corresponding with (1a) (McGinnis 2019). The continuity of the body is only destroyed when the atoms are taken in isolation and physically separated from one another so as no longer to form parts of a body.

Fakhr al-Dīn al-Rāzī integrated al-Shahrastānī's suggestion that body is composed of a finite number of potential parts into the commentary tradition and so postclassical natural philosophy.[3] In his commentary on *Pointers*, *namaṭ* 1.1–5, al-Rāzī presents Ibn Sīnā's critique of (1a) traditional atomism and of (2a) al-Naẓẓām's theory without objection and so, apparently, tentatively accepts those critiques. What al-Rāzī rejects is Ibn Sīnā's subsequent use of the purported conclusion of his critique of atomism (*Pointers*, *namaṭ* 1.6) – namely, that body is continuous – to argue for the existence of prime matter and a form of corporeality that invests matter with three-dimensionality and continuity.

Briefly, Ibn Sīnā's argument moves like this: first, from his earlier critique of atomism, he takes as proven that body *actually* possesses continuity and thus forms a unified whole, and so the body is 'one' and not just a collection of many individual atoms. He then notes that the body can also be divided, and so *potentially* possesses discontinuity. From these observations, he claims that something must underlie the actual continuity and the potential discontinuity and that this underlying thing must remain even if the initial continuity is destroyed by division. This underlying thing is prime matter (*hayūlā*). Finally, since the matter could also actually be discontinuous and, indeed, when divided actually is discontinuous, something must explain its current state of continuity, which Ibn Sīnā identifies with the form of corporeality or 'body-ness' (*ṣūra jismiyya*).

When presented with this argument, Abū l-Barakāt, followed by Sharaf al-Dīn al-Masʿūdī (d. *c.* 600/1204), in effect complain that Ibn Sīnā's argument multiplies non-observable theoretical entities beyond what an empirically adequate theory requires. Abū l-Barakāt argues, first, that continuity and discontinuity are accidental properties of body. When a body is divided, it does not cease to be a body; rather, there are now two bodies. Thus, what underlies the actual continuity and potential discontinuity, Abū l-Barakāt and al-Masʿūdī reason, is body itself, not some non-observable pure potentiality identified with matter. The form–matter view, then, unnecessarily multiplies theoretical entities.

Fakhr al-Dīn al-Rāzī when commenting on Ibn Sīnā's argument for the form–matter view mentions Abū l-Barakāt and al-Masʿūdī's objection and suggests a potential Avicennan response to it. Al-Rāzī notes that when a continuous body, B_1, is divided, the determinate corporeality (*jismiyya muʿayyana*), which makes B_1 to be the particular body that it is, is corrupted and two new determinate corporealities are generated, B_2 and B_3. Division of body then does not bring about a mere accidental change in B_1 but a substantial change: B_1 is corrupted or destroyed, and two entirely new bodies are generated, B_2 and B_3. In other words, when B_2 and B_3 determinately exist after the division of B_1, they are not strictly speaking two halves of B_1 as the right side and left side are two halves of B_1 before division. B_2 and B_3 are two new substances distinct from B_1, albeit they were two particular substances that existed potentially within B_1.

Abū l-Barakāt and al-Masʿūdī's objection relied on the loss and coming to be of accidental features of body, namely, continuity and discontinuity. Al-Rāzī recasts Ibn Sīnā's argument in terms of generation and corruption of determinate corporealities, that is, substances, rather than accidental features of substances, and so their objection no longer applies. Additionally, al-Rāzī argues that Ibn Sīnā himself is committed to a continuous body's being a composite of potential corporealities.

Having rescued Ibn Sīnā from one objection, al-Rāzī immediately launches his own critique, now against his improved version of Ibn Sīnā's argument, which effectively introduces al-Shahrastānī's position (1b) that body is composed of a finite number of potential parts. Al-Rāzī's strategy focuses on the purported underlying matter of a form-matter composite before and after a division. He lays out the logical options and argues that Ibn Sīnā must reject all the possible outcomes. For instance, the underlying matter of a composite body before division must be either one and the same (*wāḥida*) throughout that body or composed of a multiplicity of parts. Ibn Sīnā cannot concede that the matter is composed of parts, since to do so is to embrace atomism, which he rejects. Next, al-Rāzī's argument considers dividing the original form–matter composite into two new bodies. Either the original underlying matter remains one and the same with the generation of the two new bodies, and so they share that matter, or the underlying matter does not remain one and the same; likewise, either the original form of corporeality remains one and the same with the generation of the two new bodies, and so they share that form by which they are a body, or the original form of corporeality does not remain. Thus, four possible scenarios arise: (1) the two new bodies share the same matter and form of corporeality as the original body, (2) the two new bodies share the same matter as the original body but have different forms of corporeality, (3) the two new bodies have different matter from the original body but share the same form of corporeality or (4) the two new bodies have different matter and different forms of corporeality from the original body. Of these options, (1) through (3) cannot account for the generation of two new, distinct and determinate bodies after the division, which is required to avoid Abū l-Barakāt and al-Masʿūdī's criticism. As for option (4), nothing remains of the original body to underlie the generation of the two new bodies, yet Ibn Sīnā posits prime matter precisely to underlie generation. In effect, in case (4), the original body is absolutely annihilated, and the new bodies are created *de novo*. This creation of a new body and replacement of the old without an underlying matter remaining is the philosophical doctrine of occasionalism, which Ibn Sīnā also vehemently rejects.

What is noteworthy is that al-Rāzī argues only against body's being a form–matter composite, not against body's being continuous. Moreover, al-Rāzī's tacit acceptance of Ibn Sīnā's arguments against atomism need only apply to a rejection of an atomism in which body is composed of a finite number of *actual* parts (1a) not one composed of a finite number of *potential* parts (1b) as Shahrastānī envisioned body. In other words, al-Rāzī allows that a finite number of potential parts can come together to form a body that is a continuous whole and that those parts only become actual parts once they are divided and separated from the whole.

Continuity now does not mean that body is infinitely divisible but only that the body's parts are in contact at one and the same limit (cf. Aristotle, *Physics* 6, 5a1–2; Ibn Sīnā 2009, 3.2 [8]). For example, the front side of a card is continuous with its back side, since they share what is between in common, in a way that two cards, one on top of another, are not continuous, since the back side of the one card is distinct from the front side of the other. On this view, since a finite number of potential parts form one continuous whole, standard objections against traditional atomism no longer apply, for in imagination one can treat bodies and the extensions required for mathematics as continuous, even if not every imagined division into such a magnitude can be physically realized.

It is important to note that Ibn Sīnā himself recognized physical limits to the number of divisions a specific kind of body can undergo and still remain the same kind of body, for example, a watery body or a fiery body. This is just Ibn Sīnā's theory of a natural minimum (McGinnis 2015). Thus, a follower of Ibn Sīnā is committed to there being minimal bodies relative to a kind that come together to form continuous wholes, a view not entirely unlike al-Shahrastānī and al-Rāzī's. Where such a philosopher and al-Rāzī part ways is whether matter must underlie body's continuity. Ibn Sīnā clearly believes, yes. In contrast, al-Rāzī follows Abū l-Barakāt and al-Masʿūdī, making potential bodies, rather than matter, what underlies body's continuity. Thus, in the technical sense of 'matter' invoked here, as a pure potentiality that underlies forms so as to give rise to body, al-Rāzī eliminates matter as an unnecessary theoretical posit that is not required for an empirically adequate physical theory. Additionally, by denying the need for matter, al-Rāzī, now following al-Ghazālī, undercuts Ibn Sīnā's realist position concerning possibility. That is because for Ibn Sīnā, matter bears possibility and grounds it in reality. For al-Rāzī, possibility and necessity, as for al-Ghazālī and Suhrawardī, are objects of the intellect.

Al-Rāzī's word on this issue is by no means the last. Naṣīr al-Dīn al-Ṭūsī in his commentary on Ibn Sīnā's *Pointers* (Naṣīr al-Dīn al-Ṭūsī 2004–7, *namaṭ* 1.4) argues that Shahrastānī's position is still vulnerable to many of Ibn Sīnā's original criticisms against atomism, as the former posits conceptually indivisible units. He also rejects al-Rāzī's critique of the revised version of Ibn Sīnā's argument, complaining that al-Rāzī fundamentally misunderstands the passive nature of matter, for notions like being 'one and the same' or 'numerically many' belong to matter only on account of some form that it possesses not in itself. Thus, they should be understood on the 'form' side of the form/matter duality. It is perhaps enough to note that as late as the 1840s Khayrābādī still felt compelled to provide a detailed argument against atomism (Ahmed and McGinnis 2017).

V.6.4 Motion: definition versus description

Historically the subject of physics is body insofar as it undergoes motion. The realism/antirealism debate, we saw, was at the heart of discussions about body: should the theoretical entities of one's science, e.g., matter, capture real features of the world (realism) or should one's science merely provide an empirically adequate model that is consistent with observation (anti-realism). This same theme is found in discussions of motion in the later Islamicate world, and particularly in defining what motion essentially is or even if one needs an essential definition of motion or whether an empirically adequate description of motion suffices. Aristotle provided the historical basis for this discussion at *Physics* 3.1, 201b, 4–5, where he defined motion (Gk. *kinēsis*, Ar. *ḥaraka*) as the actuality of potential insofar as there is potential. How to understand Aristotle's definition, and especially the idea of actuality at its core, became an issue of concern among natural philosophers. Ibn Sīnā notes in his own *Physics* that the easiest way to understand Aristotle's definition would be to make motion the *gradual* emergence of potential into actuality (Ibn Sīnā 2009, 2.1 [2–3]). He immediately rejects this attempt because it does not provide a scientifically rigorous essential definition. Notions like *gradual*, complains Ibn Sīnā, presuppose an account of time, for what is gradual occurs over time. Time, he notes, however, is technically defined as "the number of *motion* with respect to priority and posteriority" (Ibn Sīnā 2009, 2.11 [3]). Consequently, to define motion in terms of *gradual* would render motion's definition circular and so would not provide any insight into the way the world is.

Ibn Sīnā then provides a complex analysis of motion, in which he distinguishes motion-as-traversal, which only exists in the mind, and motion-as-intermedial, which exists in the world (Ibn Sīnā 2009, 2.11 [5–6]). In more detail, when one thinks of a motion, one can think of it like a magnitude extending from *here* to *there*, like the walk from *here* (my house) to *there* (the pub): this idea of motion as something extended from starting point to ending point is motion-as-traversal. Of course, my walking to the pub is never really extended between here and there; rather, I am only ever momentarily at one or another point between here and there during the walk. This momentarily being at some point during the walk and subsequently being at another point is motion-as-intermedial. Admittedly notions like "momentarily" in this quick presentation presuppose a notion of time, but suffice it to say that Ibn Sīnā's technical explication of motion's essential definition, now in terms of priority and posteriority, arguably avoids temporal language (Hasnawi 2001; McGinnis 2006; Ahmed 2016).

Despite his labors to provide an essential definition of motion, some subsequent thinkers viewed Ibn Sīnā's efforts as superfluous at best and philosophically flawed at worse (McGinnis 2018). In his *Reflection on Wisdom* (Abū l-Barakāt 1938–1939, 28–34), Abū l-Barakāt challenged the objection that using temporal vocabulary in defining motion is viciously circular. Ibn Sīnā takes essential definitions to be first principles, and first principles must be *better known* and *prior to* any conclusion drawn from them. Since time is subsequently defined in terms of motion – namely, a magnitude *of motion* with respect to priority and posteriority – time cannot be *better known* and *prior to* motion. Abū l-Barakāt observes that both Aristotle and Ibn Sīnā use *better known* and *prior to* in two senses: better known and prior *to us* (namely, what one first sensibly observes) and better known and prior *by nature* (namely, the underlying reality that causally explains the derived phenomenon).

Drawing on this distinction, Abū l-Barakāt points out that from everyday observable experiences we are immediately aware of notions like 'gradual' and 'all at once.' These notions, *relative to us*, are better known than and prior to any essential definition of motion that is purportedly better known and prior *by nature*. Thus, Abū l-Barakāt asserts, one can use these ordinary observations to provide an empirically adequate description of motion, from which one can derive an essential definition of motion and subsequently an essential definition of time. There is no circle here because 'time' used in the definition of motion is time_u (better-known-and-prior-to-us), whereas the essential definition of time concerns time_n (better-known-and-prior-by-nature). The purported circularity involves an equivocation of the term *time*. Thus, the description of motion as the gradual emergence of potentiality into actuality, concludes Abū l-Barakāt, is a perfectly serviceable account for the natural philosopher.

While the issue of motion's proper account has its origins in Ibn Sīnā, he did not discuss motion in *Pointers*, which was the most commonly commented of his books in the post-Avicennan period. Instead, discussion of motion frequently occurred in the philosophical textbooks of other scholars and their subsequent commentaries. For instance, al-Kātibī's *The Wisdom of the Fount* begins by discussing motion in temporal terms like 'gradual' and 'not all at once' but uses this initial discussion to introduce a brief presentation of Ibn Sīnā's distinction between motion-as- traversal and motion-as-intermedial, albeit without explaining Ibn Sīnā's technical realist account of intermedial motion. In Suhrawardī's *Intimations* and al-Abharī's *Guide to Wisdom*, neither author even mentions Ibn Sīnā's technical discussion and instead describes motion as "the emergence of a thing from potentiality to actuality *not all at once*" (Suhrawardī 2012, 2: 123) and as "the *gradual* emergence of potentiality into actuality" (al-Abharī 2011, 1.9: 29).

Despite what might appear to be the success of Abū l-Barakāt's antirealist analysis of motion – or more exactly the primacy that he gives to observable features over theoretical notions – the issue of motion's proper account remained central throughout the postclassical period. Indeed, Muṣliḥ al-Dīn al-Lārī (d. 979/1572) includes it as his only discussion related to natural philosophy in his exhibition piece, *Samples of the Sciences* (*Unmūdhaj al-ʿulūm*), which Reza Pourjavady describes as intended "to ingratiate [al-Lārī] with the Ottoman court and particularly with the grand vizier, Rustam Pāshā" (Pourjavady 2014, 300). Al-Lārī wanted to demonstrate his knowledge of the important intellectual issues and trends of his day. Thus, that he chose the issue of motion's proper account gives witness that this issue remained an abiding topic of physics.

Moreover, despite a trend away from essential definitions of motion by the authors of the various philosophical encyclopedias, many of their commentators did not agree. For instance,

Mullā Ṣadrā pushed back against Abū l-Barakāt's analysis of motion in his commentary on al-Abharī's *Guide to Wisdom* (Mullā Ṣadrā 2001, 103–6). Mullā Ṣadrā's objection against Abū l-Barakāt and others in describing motion in temporal terms is precisely that they fail to appreciate Ibn Sīnā's distinction between motion-as-traversal and motion-as-intermedial. Against Abū l-Barakāt's arguement that one could use a notion of $time_u$ (better-known-and-prior-to-us) to provide an account of motion, and thereafter use that understanding of motion to provide an account of $time_n$ (better-known-and-prior-by-nature), Mullā Ṣadrā objected that Abū l-Barakāt's strategy works only if the account of motion that is explained in terms of $time_u$ is the same account of motion used to explain $time_n$, but they are not the same. Mullā Ṣadrā pointed out that $time_u$ adequately describes motion-as-traversal; however, it is motion-as-intermedial that explains $time_n$. Abū l-Barakāt's whole analysis is itself based on an equivocation of the term *motion*, or so claimed Mullā Ṣadrā.

Mullā Ṣadrā's word was far from the last on the subject. Fażl-i Ḥaqq Khayrābādī in the 19th century addressed the purported need for essential definitions in *The Happy Gift*. Thus, after discussing the realist's essential definition of motion, Khayrābādī says this:

> The truth is that the conceptualization of motion is not something that needs this definition. It is enough to say that it is the emergence from potentiality into actuality gradually, where the meanings of 'gradual,' 'little by little' and 'not all at once' are primitive conceptual notions (*al-maʿānī al-awwaliyya*), which are owing to the aid of sensation. Their conceptualization does not depend upon conceptualizing the true nature of time and the instant, even if the now and time are causes for them in existence.
>
> *(Khayrābādī 1904–1905, Fann 1, mabḥath 4, faṣl 1, 34)*

Indeed, when one turns to Khayrābādī's discussion of time, he makes no attempt to define it and says that even a child or imbecile grasps what time is. Instead, Khayrābādī focuses almost entirely on the continuity of time in what appears to be a continuation of his critique of atomism. Again, what is important to note is that Khayrābādī appears to have completely broken with a realist interpretation of natural philosophy with its insistence on essential definitions in favor of the antirealist's empirically adequate descriptions.

V.6.5 Conclusion

In the post-Avicennan period, there appears to be a shift in the practice of natural philosophy. Prior to this period, scientific practices in physics most frequently assumed a realist approach: natural philosophers aimed at true justified beliefs about the physical world with an emphasis on providing a correspondence between one's scientific theories and the necessary causal structure of reality. In contrast, in the later period certain influential figures moved away from this realist tendency toward an antirealist or instrumentalist approach to physical inquiries. For them, the emphasis is on empirical adequacy rather than on a correspondence with reality. How extensive this shift was remains a topic for future research into postclassical natural philosophy.

Notes

1 See Wisnovsky (2004) for a more complete list of textbooks and commentaries; and McGinnis (2013) for a discussion of some of these textbooks' internal structure.
2 Consolidated bibliography.
3 See Shihadeh (2014, 2016, 2017) for a detailed analysis of the following history.

Bibliography

Sources

Abharī, Athīr al-Dīn. ed. Hussayn, M. M. S. 2011. *Hidāyat al-ḥikma* [The Guide to Wisdom]. Karachi: Maktabat al-Bushra.

Abū l-Barakāt al-Baghdādī. 1938–1939. *Kitāb al-Muʿtabar fī l-ḥikma* [Book of the Reflection on Wisdom]. Hyderabad: n.p. (Reprint: Beirut, Dār wa-Maktabat Bīblīōn).

Aristotle. ed. Ross, W. D. 1936. *Aristotle's Physics*. Oxford: Clarendon Press.

Ibn Sīnā. ed. and tr. McGinnis, J. 2009. *The Physics of the Healing*. Provo, UT: Brigham Young University Press.

Kātibī, Najm al-Dīn al-Qazwīnī. ed. Ṣadrī, ʿA. 2005. *Ḥikmat al-ʿayn* [Wisdom of the Fount]. Tehran: Anjuman-i Āthār wa-mafākhir-i farhangī.

Khayrābādī, Faḍl-i Ḥaqq. ed. Barqūqī, ʿA. 1904–1905. *al-Hadiyya al-saʿīdiyya fī l-ḥikma al-ṭabīʿiyya* [The Happy Gift Concerning Natural Philosophy]. Cairo: Maṭbaʿat majallat al-manār al-islāmiyya.

Mullā Ṣadrā. ed. Fūlādkār, M. M. 2001. *Sharḥ al-Hidāyat al-Athīriyya* [Commentary on the *Guide* by al-Athīr]. Beirut: Dār Iḥyāʾ al-turāth al-ʿarabī.

Rāzī, Fakhr al-Dīn. ed. Billāh Baghdādī, M. M. 1990. *al-Mabāḥith al-mashriqiyya: fī ʿilm al-ilāhiyyāt wa-l-ṭabīʿiyyāt* [Eastern Investigations: On the Divine Science and Natural (Sciences)]. Beirut: Dār al-Kitāb al-ʿarabī.

Rāzī, Fakhr al-Dīn. ed. Najafzāda, ʿA. 2005. *Sharḥ al-Ishārāt* [Commentary on *Pointers*]. 2 vols. Tehran: Anjuman-i Āthār wa-mafākhir-i farhangī.

Suhrawardī, Shihāb al-Dīn. ed. Ḥabībī, N. 2012. *Sharḥ al-Talwīḥāt al-lawḥiyya wa-l-ʿarshiyya* [Commentary of Intimations of the Tablet and the Throne]. Mīrāṯ-i Maktūb.

Ṭūsī, Naṣīr al-Dīn. ed. Āmulī, Ḥ. Z. 2004–2007. *Sharḥ al-Ishārāt wa-l-tanbīhāt* (= *Ḥall mushkilāt al-Ishārāt*) [Commentary on *Pointers and Reminders* (= Solution of the Difficulties of *Pointers*)]. 2 vols. Qum: Būstan-i Kitāb.

Research literature

Ahmed, A. Q. 2016. "The Reception of Avicenna's Theory of Motion in the Twelfth Century," *Arabic Sciences and Philosophy* 26: 215–43.

Ahmed, A. Q. and McGinnis, J. 2017. "Faḍl-i Ḥaqq Khayrābādī (d. 1861), *al-Hadiyya al-saʿīdiyya*," in el-Rouayheb and Schmidtke, eds. [CB], 535–59.

Dhanani, A. 2007. "Jurjānī: ʿAlī ibn Muḥammad ibn ʿAlī al-Ḥusaynī al-Jurjānī," in Hockey *et al.*, eds. [CB], 1: 603–4.

Fazlıoğlu, İ. 2014. "Between Reality and Mentality: Fifteenth Century Mathematics and Natural Philosophy Reconsidered," *Nazarıyat* 1: 1–39.

Hasnawi, A. 2001. "La définition du movement dans la *Physique* du *Šifāʾ*," *Arabic Sciences and Philosophy* 11: 219–55.

Ibrahim, B. 2013. *Freeing Philosophy from Metaphysics: Fakhr al-Dīn al-Rāzī's Philosophical Approach to the Study of Natural Phenomena*. PhD thesis. Montreal: McGill University.

McGinnis, J. 2006. "A Medieval Arabic Analysis of Motion at an Instant: The Avicennan Sources to The *Forma Fluens/Fluxus Formae* Debate," *British Journal for the History of Science* 39: 189–205.

McGinnis, J. 2013. "Pointers, Guides, Founts and Gifts: The Reception of Avicennan Physics in the East," *Oriens* 41: 433–56.

McGinnis, J. 2015. "A Small Discovery: Avicenna's Theory of *Minima Naturalia*," *Journal of the History of Philosophy* 53: 1–24.

McGinnis, J. 2018. "Changing Motion: The Place (and Misplace) of Avicenna's Theory of Motion in the Post-Classical Islamic World," in Hasse, D. N. and Bertolacci, A., eds. *The Arabic, Hebrew and Latin Reception of Avicenna's Physics and Cosmology* (*Scientia Graeco-Arabica* 23). Berlin, and Boston: De Gruyter, 7–24.

McGinnis, J. 2019. "A Continuation of Atomism: Shahrastānī on the Atom and Continuity," *Journal of the History of Philosophy* 57: 595–619.

Pavlov, M. M. 2017. *Abū l-Barakāt al-Baghdādī's Scientific Philosophy: The Kitāb al-Muʿtabar*. London: Routledge.

Pines, Sh. 1937. "Études sur Awḥad al-Zamān Abu' l-Barakāt al-Baghdādī," *Revue des Études Juives* 103: 3–64 and 104: 1–33.

Pines, Sh. 1955. "Nouvelles Études sur Awḥad al-Zamān Abu' l-Barakāt al-Baghdādī," in *Mémoires de la Société des Études Juives*. Paris: Librairie Durlacher, 7–88.

Pourjavady, Reza. 2014. "Muṣliḥ al-Dīn al-Lārī and His Samples of the Sciences," *Oriens* 42: 292–322.

Shihadeh, A. 2014. "Avicenna's Corporeal Form and Proof of Prime Matter in Twelfth-Century Critical Philosophy: Abū l-Barakāt, al-Masʿūdī and al-Rāzī," *Oriens* 42: 364–96.

Shihadeh, A. 2016. *Doubts on Avicenna: A Study and Edition of Sharaf al-Dīn al-Masʿūdī's Commentary on the Ishārāt*. Leiden: Brill.

Shihadeh, A. 2017. "Al-Rāzī's (d. 1210) Commentary on Avicenna's *Pointers*: The Confluence of Exegesis and Aporetics," in el-Rouayheb and Schmidtke, eds., 296–325.

Wisnovsky, R. 2004. "The Nature and Scope of Arabic Philosophical Commentary in Post- Classical (ca. 1100–1900 AD) Islamic Intellectual History: Some Preliminary Observations," in Adamson, P., Baltussen, H. and Stone, M., eds. *Philosophy, Science & Exegesis in Greek, Arabic & Latin Commentaries*. Vol. 2. London: Institute of Classical Studies, 149–91.

Ziai, H. 1990. *Knowledge and Illumination: A Study of Suhrawardī's Ḥikmat al-Ishrāq* (Brown Judaic Studies, 97). Atlanta: Scholars Press.

V.7
COOL AND CALMING AS THE ROSE

Pharmaceutical texts as tools of regional medical practices in early modern India

Deborah J. Schlein

Medicine in the medieval and early modern Islamicate world is known for its use of Greek medical theory, its textual output and its contribution to developments in pharmacology, or the study of the uses and effects of drugs (Pormann and Savage-Smith 2007[CB],[1] 3; Chipman 2010, 1). While Greco-Islamic medical knowledge was transmitted in Arabic and Persian under the Abbasids (r. 132–656/750–1258) and vassal dynasties, subsequently in India, the Delhi sultans (r. 602–932/1206–1526) and Mughal emperors (r. 932–1273/1526–1857) sponsored the practice, study and writing down of *ṭibb* ('medicine' in Arabic and as a loan, in Persian); through their patronage, the practice of Greek-influenced medicine was diffused in South Asia as of the 7th/13th century (Alavi 2008, 28). The Greek foundations became eponymous of this medical tradition, as it was known as *ṭibb-i yūnānī* ('Greek medicine' in Arabicized Persian) during India's colonial period (under the East India Company, 1170–1274/1757–1858, and the British Crown, 1274–1366/1858–1947; Speziale 2003, 149).[2] The nomenclature of such *ṭibb* ('medicine') alluded to its basis in the humoral theories of Hippocrates (d. 370 BCE) and Galen (129–216) and to its transmission and elaboration first in Arabic and subsequently also in Persian.

V.7.1 Arabic and Persian pharmacology

Greco-Islamic medicine remained beholden to humoral physiology and pathology, positing blood, phlegm, yellow and black bile in their combinations of the four elemental qualities of hot and cold, moist and dry as primary agents. Within this frame of reference, Arabic and Persian authors recorded a range of therapies which notably transcended Greek antecedents; compared to the foundational Greek pharmacological handbook, *On Medical Substances* (*Peri hulēs iatrikēs* = *De materia medica*) by Dioscorides (d. 90), Greco-Islamic pharmacology drew on a considerably richer supply of *material medica* – vegetal, animal and mineral substances (called simples) and their compounds (Pormann and Savage-Smith 2007, 43, 120). Beginning with the 3rd/9th century, when Dioscorides became accessible in Iṣṭifān ibn Basīl's (1st half 3rd/9th century) and Ḥunayn ibn Isḥāq's (d. 260/873) Arabic translation, pharmacognosy continued to flourish, in theoretical writing as well as in therapeutic practice.[3] It stands to reason that it was enriched by and reflected the diverse climatic,

DOI: 10.4324/9781315170718-56

vegetational and cultural conditions and resources of the vast territories under Muslim rule – including, in the present case, South Asia. Its spices and 'drugs' were staples of the long-distance trade that linked the most distant parts of the Islamicate dominions (Savage-Smith 2000, 456; Amar and Lev 2017, 14, 16).

As for the Islamicate acquaintance with Indian, Ayurvedic medicine, this was based on Arabic versions from the early Abbasid period – contemporaneous with the corresponding translations from Greek. Two renowned works of Ayurvedic medicine, the compendia of Caraka (c. 3rd to 2nd century BCE) and Suśruta (revised before 500 CE) were translated into Arabic during the reign of Hārūn al-Rashīd ([r. 170–193/786–809]; Wujastyk 2003, 4, 64; Pormann and Savage-Smith 2007, 22).

In pharmacy like in the other disciplines professed by al-Bīrūnī (362– after 442 /973– after 1050), his work, *The Pharmacopeia of Medicine* (*al-Ṣaydana fī l-ṭibb*), holds outstanding rank; it also attests the diverse provenance of the pharmacist's stock. For instance, the Indian fruit myrobalan (Perso-Arabic *halīlaj*), which was prescribed against colds and coughs (al-Kindī 1966, 91; Ibn Sīnā 1988, 2: 409) was unknown to Greek and Roman authors, but at least as of the early 3rd/9th century integrated into Greco-Islamic medicine (Amar and Lev 2017, 83).[4]

Of the massive literary corpus of Greco-Islamic medicine throughout the caliphate and succeeding Muslim principalities (Chapter I.11), Ibn Sīnā's (Latin Avicenna; d. 428/1037) medical encyclopaedia in five 'books', *Canon of Medicine* (*al-Qānūn fī l-ṭibb*), is readily the most widely known and most influential work. Its impact on the study and practice of Greco-Arabic medicine lasted for centuries, and even today, the *Canon* is a cornerstone of *ṭibbī* education (Pormann and Savage-Smith 2007, 49). Its comprehensive, clear exposition of the entire system of medicine, encompassing theoretical and practical disciplines (anatomy, physiology, pathology, therapy, *materia medica*, pharmacy), won it enduring fame and inspired an immense output of abridgements, commentaries, supercommentaries, and so on, predominantly but not exclusively in Arabic and Persian.

Among the followers and commentators of Ibn Sīnā, the Central Asian Najīb al-Dīn al-Samarqandī (d. 619/1222) was one of the most influential in South Asia. His handbook *Causes and Symptoms* (*al-Asbāb wa-l-ʿalāmāt*) together with the attendant medical formulary (*al-Aqrābādhīn*) presents the gist of medical knowledge, as most persuasively set forth in the *Canon*, in an altogether more manageable format and in turn left, after some delay, a centuries-long trail of reworkings (Iskandar 1972, 464). The first of these was the – Arabic – *Commentary on* Causes and Symptoms (*Sharḥ al-Asbāb wa-l-ʿalāmāt*; MS London, British Library, Delhi Arabic 1696, fol. 2a) by Nafīs ibn ʿIwaḍ al-Kirmānī (*fl.* 813–841/1410–1437), court physician to Timur's grandson Ulugh Beg (r. 812–850/1409–1447 as governor of Samarqand, 850–853/1447–1449 as Timurid 'sultan'). After another long interval, in 1112/1700, under the Mughal emperor Awrangzīb ([r. 1068–1118/1658–1707]; Arzānī 1939, 3), Muḥammad Akbar Arzānī (d. 1134/1722) excerpted al-Kirmānī's *Commentary* and al-Samarqandī's *Causes and Symptoms* for his own Persian handbook of medicine, named after him *Akbar's Medicine* (*Ṭibb-i Akbar*). For two centuries, Arzānī's work remained the springboard for a succession of commentators and excerptors in Islamicate India. Pharmacopoeia, which inevitably had a prominent role in all of these books, reflected the multifarious, frequently South Asian, origins of the apothecary's stock. Ever since the 7th/13th century, when Greco-Islamic medicine gained a firm foothold in India, the share of Indian *materia medica* in the treatments outlined in the literature had steadily increased. Of Indian additions to the traditional dispensary, myrobalan was mentioned above; an example from among latterly introduced substances is tamarind (Amar and Lev 2008, 85; Goitein and Friedman 2008, 171, 188, 731).[5] The preference for regionally available drugs over expensive imports to be noted in Indo-Muslim sources may obviously reflect pragmatic material interest, but it also suggests a willingness to adapt the prescriptions of the authoritative works of Greco-Islamic medicine to conditions beyond the original authors' knowledge.[6]

V.7.2 Medical and reading practices

The works of al-Samarqandī and his commentators, al-Kirmānī and Arzānī, as well as the marginal annotations to be found in their Indian manuscripts are fairly representative of the evolvement of Greco-Islamic medicine (*ṭibb*) in South Asia, in particular of drug therapy.

In this chapter, headaches and their cures, as everyday phenomena regularly treated in medical writing, are examined, by way of a case study, for the changes in drug therapy from before the introduction of Greco-Islamic medicine into South Asia in the 7th/13th century until the early 12th/18th century (and beyond), when in the wake of al-Kirmānī, Arzānī and his colleagues wrote commentaries on al-Samarqandī's *Causes*. I focus on three simples that were recommended for the treatment of headache throughout the entire period – roses, opium and tamarind – and demonstrate which medicinal ingredients gained currency through trade and which ones were used as familiar *local* staples. Roses figure in al-Samarqandī' prescriptions for headaches, whereas opium and tamarind are mentioned in Arzānī's commentary and the Indian marginal annotations of the *Causes* tradition.

V.7.2.1 Headache and the constant rose

As al-Samarqandī explains in his *Causes*, common headache is caused by an excess of hot humor. At times it results from an external cause, such as hot sunshine (MS New Delhi, Jamia Hamdard, No. 1941, fol. 1b). Thus heat is the ailment's primary quality. Al-Samarqandī suggests that "the head should be cooled with fragrant (drugs), poultices and oils, and vinegar, rose water and crude oil of roses should be applied to the head" (MS New Delhi, Jamia Hamdard, No. 1941, fol. 1b). Headache, therefore, is said to respond best to a cooling regimen, following the age-old principle of 'opposites cure opposites' (Savage-Smith 2013, 92). The cooling derivates of roses serve that purpose here, as they do in al-Samarqandī's *Aqrābādhīn* (*Pharmacopeia*; al-Samarqandī 1994, 15–7).

Derivates of roses were recommended for headache treatments in the Greek sources and the standard Greco-Islamic medical works that scholars such as al-Samarqandī, al-Kirmānī and Arzānī would have consulted and possibly owned (Chapter VI.6). As we have seen, Dioscorides's *On Medicinal Substances* was a foundational source for Greco-Islamic (*ṭibbī*) pharmacology. For headaches, he prescribes a decoction of roses as well as rose oil (Diyūskūrīdus 1972, fols 39a, 54a). Galen suggests roses to help counteract the pain of 'hot' headaches (Kühn 1826, 12: 501; Everett 2012, 177). The polymath-philosopher al-Kindī (d. *c.* 256/870), to cite a trailblazing Arab-Muslim representative of Greek learning, wrote a medical formulary in which roses are generally used for 'hot' ailments of the head. A hot, flushed face, for example, can cause headache and is treated with a compound that includes pure rose oil (al-Kindī 1966, 81). Following earlier medical authors, Ibn Sīnā in the *Canon of Medicine* treats headache by rubbing the head with cold rose oil or a mixture of rose oil and vinegar (Ibn Sīnā 1988, 3: 43). Also, in the entry 'rose' in the section on 'simples', Ibn Sīnā mentions fresh roses and their decoction for headache relief (Ibn Sīnā 1988, 3: 405).

Basing himself on the standard authorities in Greco-Islamic medicine, al-Samarqandī at the same time undertook to provide his audience with a more concise, practice-oriented manual that presented the respective theory, diagnosis and therapy under a single heading. On the understanding that headache is to be treated with cooling remedies, al-Samarqandī in *Causes* suggests rose water and rose oil as part of the cure (MS New Delhi, Jamia Hamdard, No. 1941, fol. 1b). However, he sets out the details of this regimen only in his *Aqrābādhīn*, which, as explained in the introduction, he composed as a pharmaceutic guide and companion to *Causes* (al-Samarqandī 1994, 4).

Al-Kirmānī pursued a different approach in his commentary on *Causes*, combining theory, practice and formulary within one book. As a result, the *Causes'* already concise format was condensed even further; in the case of headache, the remedies from *Causes*, rose derivates, and from the *Aqrābādhīn*, such as violet and camphor are listed together (MS London, British Library, Delhi Arabic 1696, fol. 4a). By summarizing both works in his commentary, al-Kirmānī gave students of medicine an easily accessible, therapy-oriented textbook.

As mentioned above, al-Kirmānī's *Commentary* in turn was the proximate source of the first Indian work written in al-Samarqandī's succession – Arzānī's *Akbar's Medicine*. Arzānī translated al-Kirmānī's commentary into Persian, in Muslim-ruled India the language of education of both elite Muslims and Hindus. He thus added to the store of Persian medical treatises, especially formularies, available to them (Chapter V.8). In *Akbar's Medicine*, Arzānī does voice disagreement on therapy with his predecessors, but his prescription of rose derivates, camphor and violet for headache follows al-Samarqandī and al-Kirmānī (Arzānī 1939, 5). Yet he also reflects Indian practice by adding opium and tamarind: medicaments readily available in the market which were commonly used and accredited for effectiveness in medical literature.

V.7.2.2 Opium and headache

Opium is dried latex from the pods of opium poppy(*Papaver somniferum L.*). Its medicinal use dates back to remote antiquity and is, from Homer on, well attested in Greek and Roman literature, including medical writing since the Hippocratic corpus (Everett 2012, 291; Diyūskūrīdus 1972, fol. 152b). In substance and name (*afyūn* was a mere transcription from the Greek) it was transmitted to Greco-Islamic medicine and as a matter of course figures in its corpus, including, for instance, the *Book of Structures on the Essences of Medicines* (*Kitāb al-abniya ʿan ḥaqāyiq al-adviya*) by Abū Manṣūr Muwaffaq al-Harawī ([*fl.* c. 370–380/980–90]; Abū Manṣūr 1344[/1966], 52–3), the first (New) Persian *materia medica*. The author, who in search of medical knowledge also ventured into India from his home in present-day Afghanistan, is a witness to interregional cultural exchange. Of course, lively trade, both overland and sea-borne, also linked Muslim dominions and India. It stands to reason that substances such as opium, which was not native to the subcontinent, were among the goods imported there (Tibi 2006, xi).

To return to the impact of Ibn Sīnā's *Canon* on later medical writing, his prescription of the 'cold and dry' opium as analgesic in general and in particular, for chronic headache (Ibn Sīnā 1988, 2: 45) was notably omitted by al-Samarqandī as well as by al-Kirmānī, although the *Canon* was their primary source. Further, al-Kirmānī's acquaintance with the *Canon* was also mediated by Ibn al-Nafīs's (d. 687/1288) epitome *Abridgement of Medicine* (*Mūjaz fī l-ṭibb*), which he made the subject of a commentary (Iskandar 1972, 470). As concerns al-Kirmānī's motives for preferring al-Samarqandī to Ibn Sīnā in the case of opium, they remain a matter of speculation. By keeping silent about his reasons, he may have wanted to deflect attention from his effective deviation from the supreme medical authority.

During the 11th/17th and 12th/18th centuries, Ibn Sīnā's works, the *Canon* as well as his monumental re-construction of Aristotelian philosophy, *The Book of Healing* (*Kitāb al-Shifāʾ*), were assiduously studied by Indian scholars (Ahmed 2012, 212). *Akbar's Medicine* by Muḥammad Akbar Arzānī well illustrates Ibn Sīnā's continued influence on the theory and practice of medicine, including the use of opium as analgesic (be it briefly noted that it had not been known in

Indian sources before the impact of Greco-Islamic medicine). Arzānī adds opium to al-Kirmānī's prescriptions of apple and rose, or of apple, camphor and violet (Arzānī 1939, 5; MS New Delhi, Jamia Hamdard, No. 1891, fol. 4b), and in the following passage, he quotes the Buyid court-physician al-Ṭabarī (Abū l-Ḥasan Aḥmad ibn Muḥammad, [*fl.* *c.* 355–365/966–976]) on the use of opium in a compound medicine for headache (MS New Delhi, Jamia Hamdard, No. 1891, fol. 4b). Arzānī quotes the original Arabic, but in some Indian copies of *Akbar's Medicine* translations into Persian are added in the margin, including a list of ingredients according to both al-Ṭabarī and Arzānī (MSS New Delhi, Jamia Hamdard, No. 1891, fol. 4b; Aligarh, Aligarh Muslim University, Ṭibbiyya College, No. 47, fol. 5b; Alavi 2008, 30).

Marginal annotations of Arabic passages in Persian, as just seen, would seem to offer diverse evidence on medical learning and practice in Muslim-ruled India. Persian remained the language of Greco-Islamic medical study well into the 13th/19th century (Attewell 2007, 99), as did Arabic. Moreover, Arzānī's reader, listing as he did, the ingredients of the compound medicine in question, clearly had its practical use in mind. Without embarking on unfounded speculation, the tradition of Greco-Islamic medicine can be said to have shaped theory and practice alike during the Mughal and colonial periods.

V.7.2.3 Tamarind as painkiller

Tamarind (*Tamarindus indica L.*) or rather the astringent, edible pulp of its pod-like fruit, has been in medicinal use in India at least since Suśruta who prescribed it for a variety of ailments (Suśruta 1963, 1: 500) Unknown to Greek and Roman authors, including Dioscorides and Galen, tamarind became integrated into Greco-Islamic medicine during its formative period; al-Kindī refers to its use for affections of the head (al-Kindī 1966, 209). During the 4th/10th and the 5th/ 11th centuries, its knowledge spread across the entire Islamicate world, from the Iberian peninsula to Iran and present-day Afghanistan, as a cursory look at a few exemplary authors will show. The Andalusī Ibn Juljul (d. after 384/994), includes *tamr hindī* (literally 'Indian dates'; the source of the Neo-Latin botanical as well as popular modern names) in his treatise on those simples not mentioned by Dioscorides (Amar *et al.* 2014, 536). Although tamarind was by no means exclusively imported from India (al-Bīrūnī 1991, 389), the Indian provenance, registered, for instance, by Ibn Juljul and Ibn Sīnā (Ibn Sīnā 1294[1877],1: 442), determined its name. While sea-borne commerce must have facilitated its diffusion (Amar and Lev 2017, 93), it was surely overland trade which brought tamarind to al-Harawī's (al-Harawī 1344, 107) attention. Even if the authors just mentioned only gave a general outline of the curative faculties of tamarind, its cooling and anti-inflammatory effect must have recommended it for any intemperately 'hot' condition such as headaches. Yet it was only Arzānī who finally included tamarind – along with al-Samarqandī's and al-Kirmānī's violet and rose – in his prescriptions for headache (Arzānī 1939, 5; MS London, BL, Delhi Arabic 1696, fol. 4a; al-Samarqandī 1994, 13, 15–17).

Arzānī's combination of violet and rose with tamarind in treating headaches would seem to reflect an adaptation of Iranian and Central Asian textual traditions to local Indian practice, itself rooted in transmitted learning and market supply.

V.7.3 Conclusion

The makeup of Greco-Islamic pharmacy in South Asia was constituted by three major factors: (1) import of knowledge and commercial commodities from the expanding Islamicate world, (2) increasing production and supply of medical literature and (3) 'naturalization' of Greco-Islamic medicine in South Asia since the 7th/13th century. The example of headache and its treatment has illustrated the persistence of traditions and their eventual adaptation to changed environments.

Thus it was only in Arzānī's *Akbar's Medicine* that tangible change in drug therapy occurred. Opium, long a staple in Greco-Islamic medicine, reappeared in the Mughal Empire in consequence of renewed interest in Ibn Sīnā. Tamarind represented a regional South Asian contribution to the cultivation of the *ṭibbī* tradition in India. Roses, finally, were equally cherished by Greek, Arabic, Persian and Indian physicians. Belief in roses' soothing effect united them just as pain in the head itself did.

Notes

1 Consolidated bibliography.

2 Fabrizio Speziale points out that the first Indo-Persian source to refer to this medicine as *Yūnānī* was *Takmilah-i yūnānī*, written by Ahl Allāh (d. 1190/1776), the brother of the Sufi reformist Shāh Walī Allāh Dihlawī ([d. 1175/1762]; Speziale 2003, 149).

3 During the middle to late 6th/12th century, in the Artuqid emirate of South-Eastern Anatolia and Northern Mesopotamia, renewed interest in Dioscorides led to the production of two more Arabic versions of *On Medicinal Substances*; the earlier one, by Abū Sālim al-Malaṭī (*fl.* middle 6th/12th century), was probably done directly from Greek, whereas Mihrān ibn Manṣūr (d. after 553/1158) translated from Ḥunayn's Syriac version.

4 In some of the Cairo Genizah documents the vagaries of trade and fluctuations of prices, including those of medicinal substances, are recorded in lively detail. As for different qualities of myrobalan and their unstable prices in Alexandria and the Tunisian al-Mahdiyya, see Amar and Lev 2007, 528, 535.

5 In Alexandria, in the year 454/1062, a merchant wrote that there was no demand for chebulic myrobalan, yet one year later he wrote that the market was rising; it was now worth 2.5 dīnārs per *mann*, and yellow myrobalan was 10 dirhams per *qinṭār*. To put this in perspective, in 1065, the price of yellow myrobalan was 5 to 6 dīnārs while a concentrate of Chebulic myrobalan was one dirham per *mann*. 1 *qinṭār* = 100 *raṭl*s (pounds): 1 *qinṭār* = approximately 50 *mann*s and 1 *mann* = 2 *raṭl*s (pounds), "two pounds slightly lighter than the pound of Fustat."

6 To quote an example: a marginal note in a 11th/17th-century Indian copy of al-Samarqandī's *Causes* suggests, against ophthalmia (*ramad*), a decoction of three ingredients, of which two are Indian: myrobalan (*halīlaj*), plum (*ijjāṣ*) and tamarind (*al-tamar al-hindī*) (MS Hyderabad, Salar Jung Library, Arabic Ṭibb No. 4, fol. 60a).

Bibliography

Sources

Arzānī, Muḥammad Akbar. 1939. *Ṭibb-i Akbar* [Akbar's Medicine]. Lucknow: Nawal Kishor Press.

al-Bīrūnī 1991. ed. Zaryāb, ʿA. *Kitāb al-Ṣaydana fī l-ṭibb* [The Pharmacopeia of Medicine]. Tehran: Iran University Press.

Diyūskūrīdus. ed. Bābāpūr, Y. B. 1972. *al-Ḥashā'ish* [The Herbs]. Facsimile. Tehran: Intishārāt-i safīr-i ardihāl.

al-Harawī, Abū Manṣūr Muwaffaq. ed. Khānlarī, P. N. 1344 [h. sh./1966]. *Al-Abniya ʿan ḥaqāyiq al-adviya* [Structures on the Essences of Medicines] (*Silsila-yi ʿaks-i nuskhahā-yi khaṭṭī 2*). Tehran: Bunyād-i Farhang-i Īrān.

Ibn Sīnā. 1294 [AH/1877]. *al-Qānūn fī l-ṭibb* [The Canon of Medicine]. 5 bks in 3 vols. [Cairo/]Būlāq: al-Maṭbaʿa al-ʿĀmira.

Ibn Sīnā. 1988. *al-Qānūn fī l-ṭibb* [The Canon of Medicine], Books II and III. New Delhi: Jamia Hamdard.

al-Kindī. ed. and tr. Levey, M. 1966. *The Medical Formulary or Aqrābādhīn of al-Kindī*. Madison, WI: The University of Wisconsin Press.

al-Samarqandī. ed. Tomhé, G. 1963. *al-Aqrābādhīn ʿalā tartīb al-ʿilal* [Formulary According to Classified Illnesses]. Beirut: Maktabat Lubnān Nāshirūn.

al-Samarqandī. ed. Ṭuʿma, J. 1994. *al-Aqrābādhīn ʿalā tartīb al-asbāb* [Formulary According to Causes]. Beirut: Maktabat Lubnān Nāshirūn.

Suśruta. ed. and tr. Bhishagratna, K. K. 1963. *The Sushruta Samhita based on the Original Sanskrit Text*. 3 vols. Varanasi: The Chowkhamba Sanskrit Series Office.

Manuscripts

MS Aligarh, Aligarh Muslim University, No. 47.
MS Hyderabad, Salar Jung Library, Arabic Ṭibb No. 4.
MS London. BL, Delhi Arabic 1696.
MS New Delhi, Jamia Hamdard, No. 1891.
MS New Delhi, Jamia Hamdard, No. 1941.

Research literature

Ahmed, A. 2012. "The Shifā' in India I: Reflections on the Evidence of the Manuscripts," *Oriens* 40.2: 199–222.

Alavi, S. 2008. *Islam and Healing: Loss and Recovery of an Indo-Muslim Medical Tradition, 1600–1900*. Basingstoke, UK: Palgrave Macmillan.

Amar, Z. and Lev, E. 2007. "The significance of the Genizah's medical documents for the study of medieval Mediterranean trade." *Journal of the Economic and Social History of the Orient* 50.4: 524–41.

Amar, Z. and Lev, E. 2008. *Practical Materia Medica of the Medieval Eastern Mediterranean According to the Cairo Genizah*. Leiden: Brill.

Amar, Z. and Lev, E. 2017. *Arabian Drugs in Early Medieval Mediterranean Medicine*. Edinburgh: Edinburgh University Press.

Amar *et al.*: Amar, Z., Lev, E. and Serri, Y. 2014. "On Ibn Juljul and the Meaning and Importance of the List of Medicinal Substances not Mentioned by Dioscorides," *Journal of the Royal Asiatic Society* 24: 529–55.

Attewell, G. 2007. *Refiguring Unani Tibb: Plural Healing in Late Colonial India*. Hyderabad: Orient Longman.

Chipman, L. 2010. *The World of Pharmacy and Pharmacists in Mamlūk Cairo*. Leiden: Brill.

Everett, N. 2012. *The Alphabet of Galen, Pharmacy from Antiquity to the Middle Ages*. Toronto: University of Toronto Press.

Goitein, S. D. and Friedman, M. A. 2008. *India Traders of the Middle Ages: Documents from the Cairo Geniza ('India Book')*. Leiden: Brill.

Iskandar, A. Z. 1972. "A Study of al-Samarqandī's Medical Writings," *Le Muséon* 85: 451–79.

Kühn, K. G. 1826. *Claudii Galeni Opera Omnia*. Vol. 12. Leipzig: C. Knobloch.

Savage-Smith, E. 2000. "Ṭibb," in *EI-2*,10: 452–61.

Savage-Smith, E. 2013. "Were the Four Humours Fundamental to Medieval Islamic Medical Practice?" in Horden, P. and Hsu, E., eds. *The Body in Balance: Humoral Medicines in Practice*. New York: Berghahn Books, 89–106.

Speziale, F. 2003. "The Relation between Galenic Medicine and Sufism in India During the Delhi and Deccan Sultanates," *East and West* 53.1–4: 149–78.

Tibi, S. 2006. *The Medicinal Uses of Opium in Ninth-Century Baghdad*. Leiden: Brill.

Wujastyk, D. 2003. *The Roots of Ayurveda: Selections from Sanskrit Medical Writings*. London: Penguin Books.

V.8
MEDICAL PRACTICES AND CROSS-CULTURAL INTERACTIONS IN PERSIANATE SOUTH ASIA

Fabrizio Speziale

This chapter looks at how physicians' practices and experiences were recorded, gathered and interpreted in the Persianate medical culture of South Asia. Its main aim is not to describe Islamicate medical practices themselves, a subject dealt with by previous studies as well as by other chapters in this book; it rather aims to investigate the status assigned to experience by authors of Persian texts and the influence of physicians' practical experience on the formation of new features of medical knowledge and medical texts. In parallel, it looks at how the interaction with their new environment shaped Muslim physicians' efforts to adapt their therapeutic knowledge for practice in South Asia.

V.8.1 Muslim medical studies in South Asia

Specific features of the historical development, specific intellectual trends and the multicultural social environment of Persian medical studies in South Asia need to be taken into account for understanding the practices of Muslim doctors. Persian was already used to write medical texts in the Islamicate world before the foundation of Muslim-ruled kingdoms in South Asia. Persian medical texts began to be produced in South Asia after the establishment of the Sultanate of Delhi, in the early 7th/13th century (Speziale 2009a). During the Sultanate period (until the first half of 10th/16th century), Persian medical works were also written in the regional sultanates, such as those of Gujarat, Deccan and Malwa. The Mughals came into power in 932/1526 and during Akbar's reign (r. 963–1014/1556–1605) Persian became the official language of the Mughal empire. Until the early 12th/18th century, many Iranian- and Persian-speaking physicians migrated to India where they were employed at the Mughal court and at the courts of Deccan sultans such as Niẓām Shāh (r. 896–1042/1491–1633) of Ahmadnagar and Quṭb Shāh of Golkonda (r. 924–1097/1518–1687; Hameed 1986). Persian medical texts were written in the Indian Princely States that emerged with the decline of Mughal power, especially from the early 12th/18th century onward and that kept Persian as the language of administration.

DOI: 10.4324/9781315170718-57

Figure V.8.1 The *Niẓāmiyya ṣadr shifā-khāna* (1926–1939) in Hyderabad seen from the Chār Mīnār. The *Niẓāmiyya ṣadr shifā-khāna* was established by the last ruler of the Princely State of Hyderabad, Mīr ʿUthmān ʿAlī Khān (r. 1911–1948) and included a *yūnānī* hospital and medical college.

Source: © Photographer Fabrizio Speziale, Paris

Moreover, Muslim physicians in South Asia also wrote and read medical texts in Arabic language and then in Urdu during the colonial period. The practice of *yūnānī* (Greek) medicine remained widespread in South Asia during the colonial and late colonial periods when new institutions (hospitals, schools, etc.) for traditional medicines were created, such as those established in the Princely State of the Niẓāms of Hyderabad (c. 1132–1367/1720–1948), who supported both *yūnānī* and Ayurvedic institutions (Attewell 2007; Speziale 2012a; Figure V.8.1).

Persian texts written in South Asia dealt with Greco-Arabic medical knowledge and practice. Moreover, a number of them translated and incorporated a wide range of notions and practices drawn from Ayurvedic medicine. Furthermore, Muslims were not the only authors and readers of these texts: many Hindu Persian-speaking physicians wrote, copied and studied Persian texts, especially from the 11th/17th century onward. Besides texts, institutional and pedagogical settings supported the exchanges and mutual appropriation of medical practices. Hindu physicians were employed at the courts of Muslim rulers and in the hospitals established by Muslim nobles. Muslim physicians were also attached to the courts of Hindu Princely States. Until the colonial period, Muslim physicians studied with Hindu masters and vice versa, and many Hindu scholars studied at the Persian madrasas (Speziale 2018, 128–53).

V.8.2 Experience (*tajriba*) and "tested" remedies (*mujarrabāt*)

The authors of Persian medical texts often used the Arabic term *tajriba* (which also means "experiment" or "proof") to refer to the physician's "experience". The authors and readers of these texts attached a high value to *tajriba* and to direct experience of the treatments dealt with in the text. Moreover, the encounter with Ayurvedic pharmacology led Muslim physicians to engage in an empirical and textual study of Indian remedies that lasted for many generations of scholars. The function of *tajriba* is closely related to the composition and adaptation of compound (*murakkab*) drugs known as *mujarrabāt* (sing. *mujarrab*), a term that comes from the same Arabic root as *tajriba*. The *mujarrabāt* were the "tested" remedies composed and transmitted by physicians and they were often gathered into texts. In the Persianate medical culture of South Asia, to possess and transmit *mujarrabāt* was regarded as a prestigious attribute of a physician and his family. The *mujarrabāt* and other subjects such as the practice of the difficult art of pulse analysis were part of the knowledge-power kept and transmitted within families of physicians and among their students. Until the colonial period, being descended from a family of physicians, or having completed private studies with reputed masters, were among the main criteria that legitimated the practice and status of physicians.

The emphasis on *tajriba* and *mujarrabāt* led to the production of a new type of text dealing exclusively with tested remedies of physicians. Munzawī gives a long list of collections of *mujarrabāt* in his general catalogue of Persian manuscripts (Munzawī 2003, 3673–83). For instance, Akbar Arzānī, a leading medical writer of the early 12th/18th century, wrote the *Tested Remedies of Akbar* (*Mujarrabāt-i Akbarī*), a text arranged according to diseases beginning with those of the head and proceeding down through the body. It was copied several times in manuscript form, printed at least four times in lithograph versions and translated into Urdu by a Hindu scholar during the 13th/19th century (Storey 1971, 269–70, Munzawī 2003, 3675–6; on the author, see Speziale 2011).

The *Indian Healings* (*Muʿālajāt-i hindī*) further illustrates the use of *tajriba* – in this case by the commissioner of the text – as a selective criterion for the inclusion of certain knowledge into the text. This text devoted to Indian iatrochemistry (*rasaśāstra*) was written at the order of Sikandar Jāh (r. 1217–1244/1803–1829), the third Niẓām of Hyderabad, who was an expert in medicine and had a particular interest in Ayurvedic medicine (Fārūqī 1999, 158–9). Sikandar Jāh ordered that the Ayurvedic formulas he had personally tested, and that were preserved in the royal store (*tūsha-khāna*), be gathered in a text. The text was also called *Indian Pharmacopeia* (*Qarābādīn-i hindī*) and was compiled by Shaykh Ḥaydar Miṣrī, where *miṣrī* was a title given by Muslims to Ayurvedic physicians (Speziale 2018, 139; MS Hyderabad, Telangana Oriental Manuscript Library and Research Institute, Persian 339).

Hindu scholars also wrote, copied and read Persian texts on *mujarrabāt*. Bachū Lāl Tamkīn (19th century) wrote the *Tested Remedies of Bachū Lāl* (*Mujarrabāt-i Bachū Lāl*), a text arranged according to the order *az sar tā qadam* (from head to toe) of Arabic and Persian medical texts (MS Hyderabad, Telangana Oriental Manuscript Library and Research Institute, Persian 270; on Bachū Lāl see ʿAbd Allāh 1992, 249). Similarly, Raghu Nāth Singh 'Hājir', a *munshī* (secretary) of Delhi, compiled the *Treasure of Tested Remedies* (*Makhzan-i mujarrabāt*) and the *Tested Remedies of Hājir* (*Mujarrabāt-i Hājir*). In 1221/1806, Dayā Rām made a manuscript copy of Akbar Arzānī's *Tested Remedies of Akbar*, which is preserved in London (Rieu 1881, 479–80).

During the late colonial period, the *mujarrabāt* were included in the articles of biographical dictionaries of physicians written in Urdu, such as the *Secrets of Physicians* (*Rumūz al-aṭibbā'*) by Ḥakīm Muḥammad Fīrūz al-Dīn of Lahore and the *Hitting the Mark* (*Tīr ba-hadaf*) by Ẓafar al-Dīn Muḥammad Nāṣir of Hyderabad (Fīrūz al-Dīn 1913; Nāṣir 1941. See also Attewell 2007, 132–45). Another type of text used by physicians to record their remedies and personal practice is the *bayāẓ*, or "note-book", which were often anonymous (see Munzawī 2003, 3318–23). Despite being regarded as a kind of minor and secondary literature, future studies of the many *mujarrabāt* and *bayāẓ* texts which are preserved in manuscript form is needed to provide insights into medical practice in Persianate and Islamicate culture.

V.8.3 Ayurvedic medicine practiced by Muslim physicians

Various accounts show that practical concerns were the key issues prompting Indian Muslim physicians to write on Ayurvedic medicine. The *Mines of Cures of King Alexander* (*Maʿdan al-shifāʾ-i Sikandar-shāhī*) was an extensive Persian manual on Ayurveda, which gained a wide readership and was also translated into Urdu during the colonial period (Figure V.8.2). It was completed in 918/1512–1513 under the direction of Miyān Bhuwa ibn Khawāṣṣ Khān, a vizier of Sultan Sikandar Lodī (r. 894–922/1489–1517) to whom the book was dedicated (Speziale 2018, 173–8.). The preface raises two main issues faced by Muslim scholars practicing in South Asia, the issue of drug lexicon and that of the availability of drugs described in Arabic and Persian texts:

> Since the names of drugs are in the Persian and Greek languages, they are hardly identified in this country, while a number of them (*akthar-i ān*) are not present at all. Therefore, it has been necessary to study (*tatabbuʿ*) the books of Indian doctors.
>
> *(Bhuwa Khān 1877, 3)*

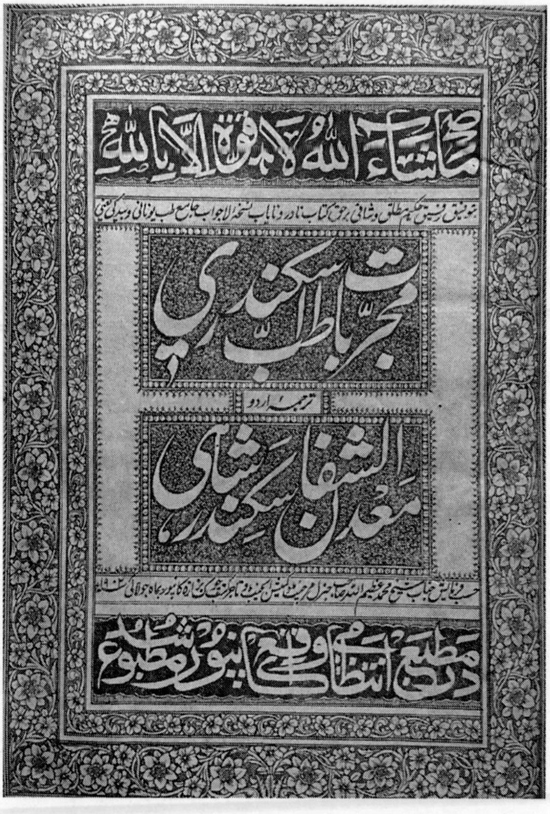

Figure V.8.2 The cover page of the Urdu translation of the *Maʿdan al-shifāʾ-i Sikandar-shāhī* compiled by Bhuwa Khān, Miyān ibn Khawāṣṣ Khān and dedicated to the Sultan of Delhi Sikandar Lodī (r. 894–922/1489–1517). The Urdu translation, titled *Mujarrabāt-i Ṭibb-i Sikandarī*, was made by Shaykh Muḥammad ʿAẓīm Allāh and published in Kanpur in 1902.

Source: © Photographer Fabrizio Speziale, Paris

One of the priorities of Muslim scholars' studies of Ayurvedic medicine was the assimilation and appropriation of the local lexicon, especially the terms most crucial for medical practice, such as those relating to drugs and diseases.

An account concerning Miyān Ṭāhā Farmulī clearly illustrates the issue of the knowledge of Indic names of drugs. Miyān Ṭāhā, a contemporary of Miyān Bhuwa, was a Muslim scholar known for his expertise in medicine and Ayurvedic treatment. The narrative recounts that the son of Ibrāhīm Khān Sarwānī, a nobleman of the time, had a wound on his abdomen that would not heal so he was taken to Miyān Ṭāhā. The latter prescribed a treatment that included powdered *gobhī* (the Hindi term for the medicinal plant *Elephantopus scaber*). The servant accompanying the child did not however understand the meaning of the word *gobhī*. Miyān Ṭāha tried to explain to him what this substance was, mentioning the different names used by villagers, yogis and the Gujaratis, but the servant was unable to understand any of these terms. Finally, Miyān Ṭāha told him the term used by the Afghans, instructing him to mention it to Ibrāhīm Khān who would understand it (Mushtāqī 2002, 176–7).

Other Muslim scholars writing on Āyurveda address similar questions in their texts. Muḥammad Qāsim Firishta (born c. 978/1570) is better known as a historian and the author of the *Rose Garden of Ibrāhīm* (*Gulshan-i Ibrāhīmī*) a Persian history of the Muslim dynasties of South Asia. He was also a physician and wrote the *Canon of Physicians* (*Dastūr al-aṭibbā'*), a Persian handbook of Ayurvedic medicine, which became quite popular. In the chapter on *rasāyana* (medical alchemy), he explained that "Indian herbalists (*'aṭṭārān*) and drug sellers (*adwiya furūshān*) know little Persian." This is why he was "compelled in this book to write the names of drugs and other [items] in the Indian language (*zabān-i hindī*)" (MS, Copenhagen, Royal Library, Pers. 22, fol. 88b). Seen from the translator's and reader's perspective, translation was not only a way to understand the other's knowledge but also a means for Muslim scholars to be understood in the local professional environment, such as with drug sellers in bazaars and with local patients. Far from being an abstract endeavor, it was an empirical strategy that allowed Indian Muslim physicians to overcome the gaps in earlier Arabic and Persian texts which did not describe in detail the Indian drugs and the Indian terms of drugs and diseases.

Moreover, evidence from practice and experience led certain Muslim scholars to criticize and revise aspects of their own thought. By the 8th/14th century, Shihāb al-Dīn Nāgawrī (*fl.* 790/1388) proposed to revise and adjust the Greco-Arabic view of humoral pathology for Muslim physicians practicing in the Indian setting, in the *Healing of Disease* (*Shifā' al-maraż*), a medical handbook in poetry. Nāgawrī's main objective was to integrate the Ayurvedic concept of wind (*vāta* in Sanskrit, *bād* in Persian) into the fourfold division of humoral pathology of Muslim physicians (Speziale 2014a). In the early 10th/16th century, the *Mines of Cures of King Alexander* openly questioned the adequacy of the Greco-Arabic humoral paradigm in the Indian environment. The text explained the limits of Greek knowledge in India and the necessity to assimilate local practices, on the basis of empirical experience (*tajārib*, pl. of *tajriba*) and by applying the same principles as the Greco-Arabic doctrine, namely, the fact that the treatment must be appropriate to the bodily temperament in a certain climate: "It is known on the basis of experience (*bi-ḥasab-i tajārib*) that Greek knowledge (*ḥikmat-i yūnān*) is not suitable (*munāsib*) for the temperaments of people in India and does not agree (*muwāfiq*) with the climate (*āb wa hawā*) of this country" (Bhuwa Khān 1294/1877, 3).

Authors of Persian texts on Āyurveda dealt with theoretical aspects and especially the features and pathology of the three *doṣa* (humors) of Indian medicine: *vāta* (wind), *kapha* (phlegm) and *pitta* (bile; Speziale 2014b). However, many of these texts focused on practical knowledge and they chiefly allowed Muslim readers to learn the therapeutic arts of their Hindu colleagues. Knowledge of Indian drugs, methods for the preparation of drugs, the medical lexicon, knowledge of pathology and treatment of diseases played key roles in this context. A number of Persian texts dealing only with Indian drugs were produced especially during the late Mughal period such as the *Compilation of Sharīf* (*Ta'līf-i Sharīfī*) by Muḥammad Sharīf Khān (d. c. 1222/1807), a scholar from Delhi (Sharīf Khān 1280/1863; on the author, see Speziale 2009b) and the *Memoir of India* (*Tadhkirat al-hind*, c. 1237/1821–2) by Riżā ʿAlī Khān, a medical officer for the Niẓām's Princely State of Hyderabad. Riżā ʿAlī Khān translated into Persian and completed a text which had been originally written in Arabic by his father, Maḥmūd ʿAlī ibn Ḥakīm Ḥażrat Allāh (Riżā ʿAlī Khān 1935, 1–3, 4–5.). These new texts classified and presented Indian drugs in the Persian alphabetical order, that is to say, in the style of the *farhang* (dictionary) typical of Islamicate literature. The *farhang*s of Indian drugs represented a major step in the process of Persianization of Ayurvedic knowledge, and at the same time, they constituted an important means to facilitate Muslim physicians' practice in the Indian environment.

A wide range of other practical notions are incorporated in Persian texts dealing with Āyurveda, such as units of weight and time, medical tools, and rules of hygiene and diet. Knowledge of the local weights of drugs and foods was a key to translating medical notions between the two cultural groups. It was indispensable not only to understand and use the formulas of the Ayurvedic texts but also to the physicians' interactions with different interlocutors in the professional setting. Hygienic and dietary practices were dealt with in the sections on *dinacaryā*, the daily hygiene rules of the Ayurvedic physicians, and *ṛtucaryā*, the seasonal regimen. Understanding the physical effects of Indian climate and seasons on the human body and its humors was indeed a major issue for Muslim physicians in a context where the incongruity of Greco Arabic thought with conditions in India was interpreted as the result of the change of bodily temperament in a different environment (see Bhuwa Khān's quotation from earlier). Practical notions were also presented in chapters dealing with *vājīkaraṇa* (virility therapy) and in Persian works on Ayurvedic veterinary medicine, especially texts on the treatment of the horse and the elephant (Speziale 2018, 207–224). Moreover, certain texts such as the *Mines of Cures of King Alexander* and the *Compilation of Sharīf* describe Indian magical beliefs and practices (Speziale 2018, 111–16).

V.8.4 Medical alchemy and rejuvenation therapy

A number of Persian texts written in South Asia included chapters on *rasaśāstra* (alchemy) and *rasāyana* (rejuvenation therapy) (Speziale 2019). Persian writings focused on the practical aspects and procedures of medical alchemy and especially those needed to learn local technical skills and to master the complex methods used to process poisonous drugs such as mercury. Persian-speaking physicians, both Muslim and Hindu, made limited efforts to incorporate in their texts Hindu philosophical and religious concepts related to alchemy. Persian medical texts dealt chiefly with notions relating to the production and administration of metals and minerals: ingredients, weights, formulas, procedures, apparatuses and the benefits and dosages of drugs. Methods to purify and calcine mercury and metals were described in detail, as all metals had to be processed before being used internally.

The 8th/14th-century *Compendium of Żiyā'* (*Majmūʿa-yi Żiyā'ī*) is emblematic in this regard. In the preface the author, Żiyā' Muḥammad Ghaznawī, states that he has based the chapter on *rasa* on the teachings (*guftār*) of Nāgārjuna, a name associated with the writing of various Sanskrit medical and alchemical texts that were probably composed by different scholars called Nāgārjuna (Meulenbeld 1999, 363–8), as well as other yogis (*jūgī*). However, he does not provide any additional explanation of the relationship between yoga and alchemy. The chapter has no introduction or conclusion and begins directly with the first compound formula (MS Hyderabad, Telangana Oriental Manuscript Library and Research Institute, Persian 344, 643).

Methods of processing mercury and other metals involved specific apparatus described in several Persian texts. Examples are the *pātanayantra* (apparatus for distillation), the *vālukāyantra* (sand apparatus) and the *dolāyantra* ('cradle-apparatus' for steaming drugs). The *Compendium of Żiyā* explained the method to prepare the *vālukāyantra* in a prescription ascribed to Nāgārjuna.

Mercury and processed sulfur in equal weights were ground and then put into a glass bottle that was wrapped several times with a cloth to which a paste was applied. The bottle was then dried and placed in a pot that had been filled with sand until the sand covered the bottle. The pot was then closed and heated on a fire for six hours. Żiyā' Muḥammad also described a method to process mercury involving the use of the *dolāyantra*. Mercury was mixed and ground with other substances, and this blend, in the form of balls, was tied in a cloth suspended over a pot containing Indian vinegar (*sirka-yi hindawī*). The piece of cloth should not touch the liquid. The pot was then covered and put over a mild fire for three days. Drawings of these apparatuses are preserved in manuscript copies of Persian medical and alchemical texts written in South Asia, such as the *Compendium of Żiyā'*, the apocryphal *Seven Friends* (*Haft ahbab*) and *Keys of Treasures* (*Maqālīd al-kunūz*) by Aḥmad ibn Arslān. Speziale (2006, 24, 28, 2019, 26, 27) reproduces the drawings of Indian alchemical apparatus included in Persian texts written in South Asia. Wujastyk (2013) describes these apparatuses in Sanskrit culture.

V.8.5 Conclusion

The interaction with the local natural and cultural environment led to several forms of adaptation in the practice of Muslim physicians in South Asia. The case of medicine was clearly different from that of other fields of knowledge concerned by the translation of Indic sources. In certain domains, the speculative interest was predominant and the practical and personal experience of the Muslim translator played a less significant role, like in the Persian texts on Hinduism that were produced during the same period in South Asia. The case of medicine was also different from that of other sciences, such as mathematics, astronomy and astrology, where applied knowledge was not radically challenged by the interaction with the new environment. The stars of the Indian sky did not differ significantly from those of Arabic- and Persian-speaking countries; the change of geographical setting did not modify the objects of study or cause gaps in the knowledge of nature as in the case of plants and drugs.

Moreover, practical experience prompted Muslim scholars to reflect on the adequacy of the Greek paradigm and how to adapt theory to evidence based on practice in India. The incorporation of Ayurvedic medical ideas and practices led to a significant renewal of the knowledge that was read and transmitted by Muslim physicians in South Asia, especially in the pharmacological field. This interaction led to the creation of new shared forms of medical treatment and pedagogy, which were transmitted by both Muslim scholars and by Persian-speaking Hindu scholars. Muslim physicians appropriated local medical practices through the translation into Persian of Indian materials, while Hindu physicians widely assimilated Islamicate knowledge by reading Persian and Arabic medical texts. These interactions and shared networks of medical knowledge remained alive during the 13th/19th century and into the late colonial period. They were erased by the new institutions for Ayurvedic and Yunani medicines established in India after independence in 1947. These institutions have supported more sectarian forms of education for Hindu and Muslim practitioners and pushed both groups toward a higher level of hybridization with Western biomedicine.

Bibliography

Sources

ʿAlī Khān, Riżā. 1935. *Tadhkirat al-hind* [Memoir of India]. 2 vols. Hyderabad: n. p.
Arzānī, Akbar.1863. *Mujarrabāt-i Akbarī* [Tested Remedies of Akbar]. Lucknow: n. p.
Bhuwa Khān, Miyān ibn Khawāṣṣ Khān. 1877. *Maʿdan al-shifāʾ-i Sikandar-shāhī* [Mines of Cures of King Alexander]. Lucknow: Nawal Kishor.
Fārūqī, Muʿīn al-Dīn Rahbar. 1999 [1937]. *Islāmī ṭibb shāhānah sar-parastiyūn men* [Islamic Medicine under the Patronage of Kings]. Hyderabad: Maktaba-yi ʿAyn al-ʿulūm.
Firishta, Muḥammad Qāsim Hindūshāh. 1874. *Gulshan-i Ibrāhīmī* (*Tārīkh-i Firishta*) [The Garden for Ibrāhīm (History of Firishta)]. 2 vols. Kanpur: Nawal Kishor.
Fīrūz al-Dīn, Hakīm Muḥammad. 1913. *Rumūz al-aṭibbā'* [Secrets of the Physicians]. 2 vols. Lahore: n. p.
Khān, Ḥakīm Muḥammad Sharīf. 1863. *Taʾlīf-i Sharīfī* [The Composition of Sharīf]. Delhi. English translation: *The Taleef Shereef or Indian Materia Medica*, Playfair, G., ed. Calcutta: The Medical and Physical Society of Calcutta, 1833.
Mushtāqī, Rizq Allāh. 1422/2002. Ṣiddīqī, I. H. and Ṣiddīqī, W. H., eds. *Wāqiʿāt-i Mushtāqī* [The Facts of Mushtāqī]. Rampur: Raza Library.
Nāṣir, Ḥakīm Ẓafar al-Dīn Muḥammad. 1941. *Tīr ba-hadaf* [Hitting the Mark]. Hyderabad: Maktaba-yi Ḥakīm-i Dakan.

Manuscripts

MS Copenhagen, Royal Library, Pers. XXII.
MS Hyderabad, Telangana Oriental Manuscript Library and Research Institute, Pers. 270.
MS Hyderabad, Telangana Oriental Manuscript Library and Research Institute, Pers. 339.
MS Hyderabad, Telangana Oriental Manuscript Library and Research Institute, Pers. 344.
MS Leiden, Universiteitsbibliotheek, Or. 22.768.
MS Rampur, Kitābkhāna-yi Rażā, Pers. 5968.

Research literature

ʿAbdallāh, S. 1992. *Adabiyāt-i fārsī men hindūon kā ḥiṣṣa*. New Delhi: Anjuman-i Taraqqī-i urdū.
Ansari, S. M. R. 2005. "Hindus' Scientific Contributions in Indo-Persian," *Indian Journal of History of Science* 40.2: 205–21.
Attewell, G. 2007. *Refiguring Unani Tibb: Plural Healing in Late Colonial India*. New Delhi: Orient Longman.
Azmi, A. A. 2004. *History of Unani Medicine in India*. New Delhi: Hamdard.
Ernst, C. W. 2003. "Muslim Studies of Hinduism? A Reconsideration of Arabic and Persian Translation from Indian Languages," *Iranian Studies* 36.2: 173–95.
Hambly, G. R. G. 1999. "Ferešta, Tārīḵ-e," in *eIr*, 9.5: 533–4.
Hameed, A. 1986. *Exchanges between India and Central Asia in the Field of Medicine*. New Delhi: Institute of History of Medicine and Medical Research.
Meulenbeld, J. G. 1999. *A History of Indian Medical Literature*. 3 vols. Groningen: Egbert Forsten. Vol. IA.
Munzawī, A. 2003. *Fihristwāra'-i kitābhā-yi fārsī*. Vol. 5. Tehran.
Rieu, C. 1879–83. *Catalogue of the Persian Manuscripts in the British Museum*. 3 vols. London: British Museum. Vol. 2. 1881.
Speziale, F. 2006. "De zeven vrienden. Een Indo-Perzische verhandeling over alchemie," in Hoftijzer, P. *et al.*, eds. *Bronnen van kennis. Wetenschap, kunst en cultuur in de collecties van de Leidse Universiteitsbibliotheek*. Leiden: Primavera Pers, 23–31.
Speziale, F. 2009a. "Indo Muslim Physicians," in *eIr*, online edition. www.iranicaonline.org/articles/india-xxxiii-indo-muslim-physicians.
Speziale, F. 2009b. "Šarif Khan, Moḥammad," in *eIr*, online edition. www.iranicaonline.org/articles/sarif-khan-mohammad-d-ca-1807-physician-at-the-court- of-the-mughal-emperor-shah-alam-ii.
Speziale, F. 2011. "Arzāni," *eIr*, online edition. www.iranicaonline.org/articles/mohammad-akbar-arzani.
Speziale, F. 2012a. "Tradition et réforme du *dār al-šifāʾ* au Deccan," in Speziale, F., ed. *Hospitals in Iran and India, 1500–1950s*. Leiden and Tehran: Brill and IFRI, 159–89.

Speziale, F. 2012b. "Hinduism and Islam: Medieval and Premodern Period," in Jacobsen, K. A., ed. *Brill's Encyclopaedia of Hinduism*. Leiden: Brill, 4: 521–9.

Speziale, F. 2014a. "A 14th Century Revision of the Avicennian and Ayurvedic Humoral Pathology: The Hybrid Model by Šihāb al-Dīn Nāgawrī," *Oriens* 42.3–4: 514–32.

Speziale, F. 2014b "The Persian Translation of the *tridoṣa*: Lexical Analogies and Conceptual Incongruities," *Asiatische Studien* 68.3: 783–96.

Speziale, F. 2018. *Culture persane et médecine ayurvédique en Asie du Sud*. Leiden: Brill.

Speziale, F. 2019. "Rasāyana and Rasaśāstra in the Persian Medical Culture of South Asia", *History of Science in South Asia* 7: 1-41. Available at: https://doi.org/10.18732/hssa.v7i0.40.

Storey, C. A. 1971. *Persian Literature: A Bio-bibliographical Survey*. Vol. 2. 2nd edition. London: Luzac.

Wujastyk, D. 2013. "Perfect Medicine. Mercury in Sanskrit Medical Literature," *Asian Medicine* 8: 15–40.

Zahuri, M. A. al-W. 1964. "On Mejmoo-a-Zia-e or Collection by Zia," *Bulletin of the Department of History of Medicine* 2.2: 81–6.

V.9
PREMODERN OTTOMAN PERSPECTIVES ON NATURAL PHENOMENA

Osman Süreyya Kocabaş

The study of natural phenomena and people's reactions to them provides us with insights into the mentalities and worldviews of the society in which they lived. This chapter surveys Ottoman knowledge of natural phenomena relying on different kinds of sources to explore how natural phenomena were understood, perceived and interpreted. Recent research has substantially enhanced our knowledge regarding people's reactions and perceptions of celestial and natural phenomena. Zachariadou's edited volume offers new insights into how Ottomans dealt with natural disasters (1997). Arslantas (2003) and Ayalon (2014) examine perceptions of earthquakes and the plague. Panzac (1985) and Varlik (2015) study the plague in various contexts. Küçük (2020) discusses the interests of early modern Ottoman scholars in natural philosophy.

This chapter focuses on the kind of texts available to the literate population in the center of the empire and the explanations of the phenomena presented in them. It does not, however, cover, the entire breadth of Ottoman conversations about such topics. To begin with, it is necessary to briefly introduce Ottoman cosmological views and ideas about the universe, as they provided the foundations for the other themes. As was the dominant view in medieval Christian and Islamic thought, Ottoman scholars often explained natural phenomena within the scientific framework of Ptolemy's geocentric world view in connection with the principles of Aristotle's (384–322 BCE) physics (Kadızade er-Rumi 2012, 277). According to a different cosmological model the Earth rests on an ox, a fish and an angel carrying them. In this view, a large ocean covers the floor of the universe and a celestial dome completely encloses it (MS Berlin, Staatsbibliothek, or. quart. 1828, fol. 161a). This tradition goes back to the 3rd/9th and 4th/10th centuries, and culminates in *The Resplendent Science of Configuration of the Universe in the Sunni [-Islamic Thought] (Al-Hay'a as-Sanīya fī al-Hay'a as-Sunnīya)* by the prominent Egyptian religious scholar al-Suyūṭī (d. 911/1505). Al-Suyūṭī rejected Aristotelian explanations of natural phenomena as interpretations of philosophers (*falāsifa*) and instead proposed an "Islamic astronomy" strictly based on mythical explanations in Qur'ān and *ḥadīth*. Various Ottoman authors referenced al-Suyūṭī's statements on natural phenomena (Heinen 1982, 7–9), but also at times simply juxtaposed them with Aristotelian concepts.

Recent studies on the history of science of the Ottomans have discussed which sources offer insights into Ottoman views about nature, and raised the question whether Ottoman thought is original or whether it is a mere repetition of medieval Islamic philosophy (Shefer-Mossensohn

DOI: 10.4324/9781315170718-58

2015; Brentjes and Morrison 2010[CB];[1] Ihsanoğlu 2002). As Shefer-Mossensohn stresses, Ottoman scholars adopted multiple scientific and cultural approaches inspired by sources from several Eurasian cultures; hence their thought was rather eclectic and multicultural (Shefer-Mossensohn 2015, 21). Despite its multicultural nature, medieval Islamic natural philosophy played the constitutive role in the emergence of Ottoman scientific thought.

Consequently, Ottoman scholars often refer to medieval Muslim philosophers such as Ibn Sīnā (d. 428/1037), who is often mentioned as *ḥaḍrat-i sheikh* or *reʾīs* in Ottoman texts, as well as his critical followers, especially Fakhr al-Dīn al-Rāzī (d. 606/1210), who adopted Aristotelian explanations in their natural philosophical perspectives. Ibn Sīnā and Fakhr al-Dīn al-Rāzī closely adhered to Aristotle's theories for explaining natural phenomena such as earthquakes, winds and comets, repeating that vapor (*bukhār*) and smoke (*dukhān*) dissolved by celestial heat are their causes (Lettinck 1999, 64). Following them and other scholars of the medieval Islamic tradition, Ottoman writers mainly based their interpretations and descriptions on the Aristotelian inhalation/exhalation theory, which posits that the interaction of the four elements (earth, water, air, fire) can cause various meteorological, terrestrial and atmospheric phenomena, such that everything in the sublunar space is in constant change, being generated and then corrupted. Ascribing dry exhalations to the element earth and moist exhalations to the element water, Aristotle taught that the moist exhalation rises from the earth as a vapor to the sphere of the element air. There, it causes the formation of rain, clouds, thunder, rainbow and other natural phenomena. Because smoke is lighter than vapor, the dry exhalation consisting of flammable materials rises as a kind of smoke and can reach the sphere of the element fire, where it causes comets, shooting stars and the Milky Way (Lettinck 1999, 34). In addition to Aristotelian theories, Ottoman perceptions of celestial phenomena included astronomical and optical theories proposed by Ptolemy (*c.* 100–170), as well as scholars from earlier Islamicate societies such as Ibn al-Haytham (354–*c.* 430/965–*c.* 1040).

The Ottomans translated Greek, Latin, Arabic and Persian scientific works during the premodern period (Chapter I.1). Thanks to those translations, the Aristotelian-Ptolemaic theories remained influential among Ottoman scholars. Many books that included descriptions of natural phenomena were dictated, glossed, quoted and partly translated into Turkish by Ottoman scholars. For this reason, major studies on natural philosophy, in which natural phenomena were discussed, have found a place in Ottoman thought, even if it represents a continuation of the medieval Islamic literature tradition.

Some epistemological and bibliographic encyclopedias, such as *The Key to Felicity and the Lamp of Sovereignty in the Subject Matters of the Sciences* (*Miftāḥ al-saʿāda wa-miṣbāḥ al-siyāda fī mawḍūʿāt al-ʿulūm*) by Ṭāshköprüzāde (d. 968/1561) and *The Removal of Doubts regarding the Names of the Books and the Arts (Kashf al-ẓunūn ʿan asamī l-kutub wa-l-funūn)* by Kātib Chelebi (d. 1067/1657), also known as Ḥajjī Khalīfa, can enlighten us about the subjects of natural phenomena and how they were positioned in the Ottoman classification of the sciences. An anonymous treatise called *The Seven Stars* (*al-Kawākib al-sabʿa*) is particularly precious in this context, since it includes a list of the fields of knowledge (*ʿulūm*), taught at Ottoman *madrasa*s in the 12th/18th century, and a description of the conditions of teachers (*müderris*) and students (*ṭalebe*). If there was interest in such topics, teachers could define and examine natural phenomena within several independent fields: (1) the science of generation and corruption (*ʿilm al-kawn wa-l-fasād*), which focused on meteorological facts such as rains, winds, clouds, thunder or lightning and their causes; (2) the science of rainbows (*ʿilm qaws quzaḥ*), which focused on the physical explanation of rainbows; and (3) the science of rain (*ʿilm nuzūl* *al-ghayth*), which focused on lightening, thunder, rain and their causes (Taşköprülüzade 2011, 303, 332; Katip Çelebi 2016, 1091, 1552; *Kevakib-i seb'a* 2015, 28–32). The science of the configuration of the universe (*'ilm al-hay'a*; planetary theory) and the science of the stars (*'ilm al-nujūm*) included descriptions of various celestial phenomena, such as lunar and solar eclipses, comets or shooting stars, even if they were not always understood in this sense. Nautical knowledge (*'ilm al-milāḥa*), geography (*'ilm al-jughrāfiyyā*), the field of special (magical) properties of regions or climes (*'ilm khawāṣṣ al-aqālīm*) and the field of meteorological prognostication (*'ilm al-malāḥim*) dealt with phenomena such as earthquakes, tides, winds, storms, comets and floods (Katip Çelebi 2016, 497, 1447; *Kevakib-i seb'a* 2015, 38–41).

V.9.1 The *Wonders of Creation*

Natural phenomena of this sort are explained neatly and briefly in *The Wonders [or: Marvels] of Creation ('Ajā'ib al-makhlūqāt)*, which has many copies in Arabic, Persian and Turkish in both medieval Islamic and early modern Ottoman literature. These works were read not only by people affiliated with the madrasa but also by statesmen and other prominent people. For this reason, the shortest way to figure out Ottoman explanations and interpretations of the natural phenomena is to look at this kind of books. The literature of *The Wonders of Creation*, and similar designations, describes many phenomena in forms that combine information from ancient Greek, Latin, Syriac, Sanskrit (and possibly other Indian languages), Middle Persian, New Persian, Arabic and Turkic literary traditions, including topics from cosmology and geography (Starkey 2007, 529–30). The *Wonders of Creation and Rarities of Existing [Beings]* (*'Ajā'ib al-makhlūqāt wa-gharā'ib al-mawjūdāt*) of Zakariyyā' al-Qazwīnī (d. 682/1283) and similar works such as *The Pearl of Wonders and the Uniqueness of Rarities* (*Khāridat al-'ajā'ib wa-farīdat al-gharā'ib*), attributed to Ibn al-Wardī (d. 861/1457), were translated into Ottoman Turkish and expounded many times (Ak 2014; Ihsanoğlu *et al.* 2000). Both were much used sources for cosmological and geographical information into the 13th/19th century (Hagen 2003[CB]; Bellino 2014, 257–97).

The Wonders of Creation provided scholars in Islamicate societies with a set of interpretations of most previously mentioned natural phenomena based partly on ancient Greek theories, partly on ancient myths and partly on Islamic religious teachings (Figure V.9.1).

Figure V.9.1 Depiction of a lunar eclipse (*khusūf*). MS Baltimore, The Walters Art Museum, W.659, fol. 10a.

Source: © The Walters Art Museum, Baltimore

When beginning to compile their own works on *The Wonders* in Turkish, Ottoman scholars were already aware of the contradictory ideas and arguments about phenomena such as the Milky Way and comets in the treatises of their predecessors. Therefore, in explaining a phenomenon, Ottoman translators added their views to the text, regardless of the original text of *The Wonders*. In other words, this literature is both the filtered version of medieval Islamic scholars' explanations and interpretations in this field and the presentation of the ideas of the authors and translators to the Ottoman readers. For instance, if we look at the explanation of earthquakes in *The Wonders*, it is stated that earthquakes occur only by the discretion of Allah. However, the author also offers a natural philosophical explanation: earthquakes occur when vapors trapped below the earth's surface cannot find a way out and thus shake the surface. In addition to this Aristotelian explanation, there is also the explanation that the angel in charge of Mount *Qaf* caused the earthquake by pulling the vein imagined to connect to the area where the earthquake occurred, by Allah's order (MS Baltimore W.659, fol. 180b).

V.9.2 Specialized treatises on individual natural phenomena

Individual works on subjects such as earthquakes and comets aimed at informing the society about these phenomena due to the actual or perceived impacts they could have on people's daily lives. Aḥmed ibn Rejeb al-İṣṭanbūlī (or al-Qusṭantīnī; [d. 1139/1727]), author of the *Epistle on Earthquakes (Risālat al-zalzala)*, one of the rare works on earthquakes in the early modern period, states that people have erroneous and superstitious ideas about earthquakes and that he wrote the treatise in simple Turkish to correct these error after an earthquake in Istanbul in 1130/1718 (MS Konya 42 Kon 4172, fols 5a–5b). In this work, the author refers to al-Suyūṭī and states that ox and fish (*ḥawāmil-i arḍ*) carry the earth, noting furthermore that the earthquake happened at the discretion of Allah through the intermediary of the angel in charge of Mount *Qaf*. (MS Konya 42 Kon 4172, fols 28b–29a).

Comets were a controversial topic among scholars in Islamicate societies. Following Aristotle, comets were defined as atmospheric phenomena or sublunar objects, and although the physical nature of a comet, called *star with a tail (kawkab al-dhanab)*, remained unclear for several centuries, comets long continued to be of great interest to Ottoman astrologers and scholars in other fields, as well as ordinary people. In scientific descriptions of comets, the Ottomans used the Aristotelian exhalation theory according to which hot vapor rises from the earth to the sphere of fire, where it is suddenly ignited. A unique treatise, treating exclusively comets and written in Arabic, is ʿAbdallāh al-Maqdisī al-Azharī's (d. after 1107/1695) *Gift of the Hearts on the Explanation of the Possessors of Tails* (*Tuḥfat al-albāb fī bayān ḥukm dhawāt al-adhnāb*). Al-Azharī compiled it after having witnessed the appearance of a comet in the night sky of 21 Ramaḍān 1078/5 March 1668 (Figure V.9.2) and dedicated it to Sultan Meḥmed IV (r. 1058–1099/1648–1687).

In this treatise, al-Azharī quoted, without naming them, some Muslim scholars who had defined comets as celestial objects, like planets. Against them he argued that comets are events caused by vapors like other sublunar phenomena. To bolster his case, he gave more space to statements by Muslim scholars who shared his views based on the exhalation theory. He sorted comets according to their shapes, discussed why their colors change, tried to explain the interaction between comets and the five planets and asked whether comets are celestial objects like stars. Additionally, he addressed the question why comets should be seen as omens (MS Istanbul, Süleymaniye Library, Hacı Beşir Ağa, nr. 674/1, fols 7a–8a; Figure V.9.3). This

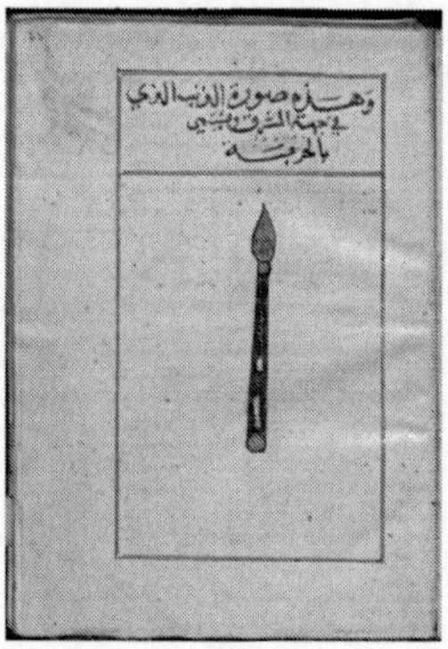

Figure V.9.2 A comet with a tail in al-Azharī's (d. after 1107/1695) autograph of his *Gift of the Hearts on the Explanation of the Possessors of Tails* finished 8 Rabīʿ al-thānī 1107/16 November 1695. MS Ann Arbor, University of Michigan, Library, Arabic 185, p. 23.

Source: © Library, University of Michigan, Ann Arbor

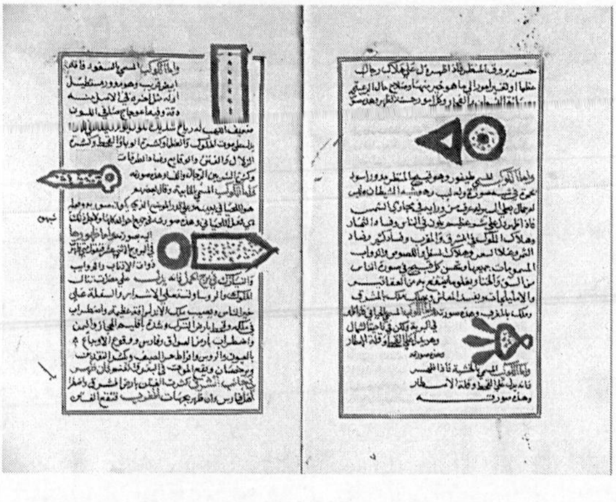

Figure V.9.3 Depictions of comets and shooting stars in al-Azharī's *Gift of the Hearts on the Explanation of the Possessors of Tails*. MS Istanbul, Süleymaniye Library, Hacı Beşir Ağa, 674/1, fols 15a–b.

Source: © Süleymaniye Library, Istanbul

treatise is an exceptional compilation, because it is devoted exclusively to comets and because the author combined a range of theoretical questions from natural philosophy with questions that belonged to other fields of knowledge as well as with observational practices. It also mentions al-Kindī's (d. *c.* 256/870) comet observations and the contents of his lost treatise on comets (Giahi Yazdi 2014, 62).

Several Ottoman scholars preferred to examine natural phenomena in different disciplinary contexts. The discipline called knowledge of the rainbow (*ʿilm qaws quzaḥ*) was a field dedicated to describing how a rainbow emerges, when it is seen, why it disappears and why it becomes colorful (Taşköprüzade 2011, 303; Katip Çelebi 2016, 1091). Treatises on this topic suggest that this discipline paved the way for examining some natural phenomena in specialized scholarly contexts. Although they do not differ greatly among themselves in terms of content, the high number of copies compared to treatises on other subjects shows that there was a regular audience in *madrasa* circles for this kind of topic. This view is supported by the observation that many of these treatises were written in Arabic rather than in Ottoman Turkish (İhsanoğlu 2017, 100).

Some such specialized treatises explain the formation of rainbows and halos by employing optical ideas (Chapter I.8), such as reflection and refraction, in a natural philosophical context. As in other cases, Ottoman discussions of rainbows and halos drew on positions taken by earlier scholars, in particular Ibn Sīnā and Ibn al-Haytham. But here, Ibn Sīnā rejected Aristotle's explanations and argued that the halo and rainbow were not material objects but imaginary things like the images in a mirror. Ibn al-Haytham tried to explain the rainbow's geometric form as the result of the reflection of the sun rays from moist air droplets in clouds that formed a spherical surface (Lettinck 1999, 277).

While it is unclear whether Ottoman authors had direct access to Ibn al-Haytham's theory on rainbows and other related topics, through the 13th/19th century one can find Ottoman scholars who owned an early version, perhaps even an autograph, of Kamāl al-Dīn al-Fārisī's (d. 718/1319) work on optics, *The Revision of Optics for Those Who Have Sight and Mind* (*Kitāb tanqīḥ al-manāẓir li-dhawī l-abṣār wa-l-baṣā'ir*); these included two well-known early modern scholars – Mīrīm Chelebi (d. 931/1525) and Taqī al-Dīn ibn Maʿrūf (d. 993/1585). But in general, Ottoman authors refer to and quote from works of Ilkhanid, Timurid and Mamluk scholars. When treating rainbows, some authors repeated and glossed al-Sharīf Jurjānī's (740–816/1340–1413) and Qāḍīzāde al-Rūmī's (d. after 844/1440) descriptions found in two commentaries mentioned repeatedly above. Mīrīm Chelebi and Nebī Efendizāde (d. 1200/1786), for instance, annotated the section on the rainbow from Qāḍīzāde's *The Commentary on the Compendium on the Science of the Configuration [of the Universe]* (*Sharḥ al-Mulakhkhaṣ fī ʿilm al-hay'a*) (İhsanoğlu *et al.* 1997[CB], 21), while Qara Veysī (d. 9th/15th) glossed a section on the rainbow from Jurjānī's *The Commentary on the Positions in the Science of kalām*. Mīrīm Chelebi, who had been educated through the works of Qāḍīzāde and his great-grandfather ʿAlī Qūshjī (d. 879/1474), chose a different approach. He compiled a treatise in Arabic on general optical topics such as light rays, vision, reflection and refraction, in addition to rainbows and halos.

Before he explained the rainbow, Mīrīm Chelebi described its colors and formation, writing that the colors change depending on the angle of the eye, the strength of the sun's rays and the position of the sun. Following him, Muṣṭafā Ḫocazāde (d. 893/1488) wrote an Arabic treatise, *Seven Premises on the Knowledge of the Rainbow* (*Muqaddimāt sabʿ fī maʿrifat qaws quzaḥ*), in which he explained that a rainbow occurs due to the reflection (*inʿikās*) of the straight rays (*khaṭṭ al-shuʿāʿ*) emanating from the radiating object (*muḍīʾ*) by the droplets (*ajzāʾ*) in the curved surface of reflection (*saṭḥ al-mir'āt*). The perceived colors depend on the angle of the eye of the person looking (*bāṣir*) at the rainbow and the distance between the droplets and the sun. While the closest droplets to the sun are seen as red, the farthest ones are purple (Figure V.9.4). Ḫocazāde wrote this treatise in the form of annotations to Athīr al-Dīn al-Abharī's (d. between 660/1263 and 663/1265) *The Guide to Philosophy* (*Hidāyat al-ḥikma*), which had a small, but stable audience, since 17 copies have been identified so far (İhsanoğlu *et al.* 1997, cxlv). Ṭoqatī, finally, was of the opinion that devout Muslims should not study natural phenomena like the rainbow, because its description belonged to the realm of philosophy (İhsanoğlu *et al.* 1997, 25).

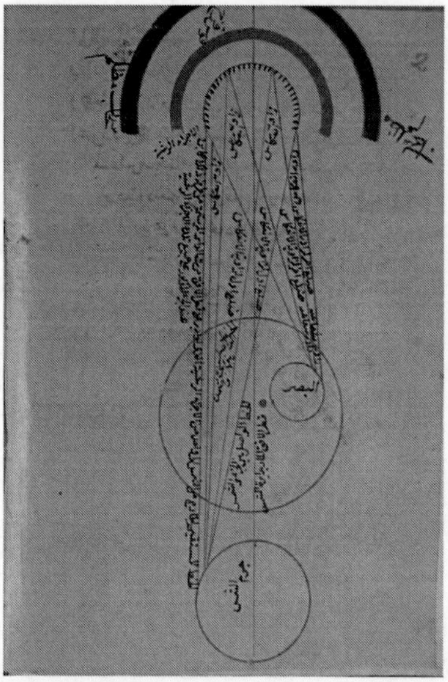

Figure V.9.4 Depiction of the rainbow by Ḫocazāde (d. 893/1488). MS Çorum, Hasan Paşa İl Halk Library, Arşiv 19, Hk 3237/2, fol. 3b.

Source: © Hasan Paşa İl Halk Library, Çorum

Although familiar from conversations with sailors and travelers or books on navigation, the tides attracted the attention of Ottoman scholars less than other phenomena. Even so, ʿAbd al-Qādir ibn Aḥmad ibn Mīmī (d. 1085/1674) dedicated a book to this topic called *The Pearl of Time on the Tides* (*Yatīmat al-ʿaṣr fī l-madd wa-l-jazr*). In this work, he claims that Aristotle's and other philosophers' explanations that the tides are caused by the sun and winds are erroneous and that their ultimate cause is the impact (*jadhb*) of the moon due to God Almighty's will (MS Istanbul Esad Efendi 551, fols 17a–17b). In other words, in conformity with standard Muslim beliefs about divine causation and nature's function according to custom, Allāh gives the Moon the ability to draw water, making it the only reasonable factor in the formation of the tides (MS Istanbul Esad Efendi 551, fol. 45a). Although Ibn Mīmī accepts an impact of the sun on the tides, he does not accept its movement as their cause, because tides also occur at night (MS Istanbul Esad Efendi 551, fol. 14b). The author believes that the moon rotates around the earth in exactly ~~is~~ 24 hours. In other words, although the moon continuously changes its position, its effects are constantly present, as it is in the sky also during daytime (MS Istanbul Esad Efendi 551, fol. 45b).

Ibn Mīmī's work reveals the complexity of beliefs about, and knowledge of, the tides among this group of Ottoman scholars (Figure V.9.5). He often mentions the cosmological views of al-Suyūṭī and Ibn al-ʿArabī ([558–638/1165–1240]; MS Istanbul, Süleymaniye Library, Esad Efendi 551, fols 27a–27b). He respectfully refers to al-Kindī and al-Bīrūnī (362–d. after 444/973–d. after 1053), suggesting that he may have been aware of their views on the tides. In numerous other parts of his work, however, he objects to the views of Aristotle or the philosophers (*falāsifa*) in general, planetary theorists (*ahl-al-hay'a*) and Islamic theologians (*mutakallimūn*).

Figure V.9.5 The giant bull al-Rayyān in a cosmological depiction of the earth in an incomplete Turkish version of Zakariyyā' ibn Muḥammad al-Qazwīnī's (d. 682/1283) *The Wonders of Creation* (*'Ajā'ib al-makhlūqāt fī gharā'ib al-mawjūdāt*), dated 960/1553. MS Washington, D.C., Library of Congress, Turkish manuscript 185, unpaginated.

Source: © Library of Congress, Washington, D.C.

V.9.3 Practical aspects of Ottoman interests in natural phenomena

Some natural phenomena were of particular interest to the Ottomans because of their practical relevance. Knowledge of winds was critical for the navy and mercantile ships. Some Ottoman sailors such as Pīrī Reʾīs (d. 961/1554) and Seydī ʿAlī Reʾīs (d. 970/1562) wrote books that contain information about seas and oceans, including knowledge based on their own experiences. Especially since the 9th/15th century, which coincided with the Ottoman Empire's imperial expansion in the Mediterranean, the Persian Gulf and the Indian Ocean, phenomena encountered at sea, such as winds or the tides, became a regular theme in maritime or geographical texts. While traveling with his uncle Kemāl Reʾīs (d. 917/1511), Pīrī Reʾīs saw the tide on the shore of Sfax. He mentions this observation in his *Book of Navigation* (*Kitāb-i baḥriye*) and claims that the occurrence of such an event depends on the position of the moon in the sky (MS Baltimore W.658, fols 6b–7a). The rising of a particular star may signal meteorological events, such as an upcoming storm (MS Baltimore W.658, fols 6b–11b). Pīrī Reʾīs emphasizes that paying attention to such natural phenomena is vital for seafarers, because they cannot be known through the use of maps and compasses. In his *Book of the Ocean* (*Kitāb al-muḥīṭ*), Seydī ʿAlī Reʾīs also acknowledges the movement of the moon as the reason for the tides. He advises sailors to anchor their ships during the flow and warns that it will be difficult to anchor during the ebb because of the strong winds (MS Istanbul, Topkapı Sarayı Müzesi Library, Revan Köşkü 1643, fols 63a–63b). Seydī ʿAlī Reʾīs connects eight winds to the compass directions and admonishes the sailors to learn which wind is blowing from which direction. If it is not possible to identify the wind accurately, he recommends consulting a compass (MS Istanbul, Topkapı Palace Library, Revan Köşkü 1643, fols 87a-89b).

As was the case in other Islamicate societies astronomy was also a science of practical, namely, religious, importance in the Ottoman world (King 2005[CB]). In Ottoman *madrasa*s, elementary planetary theory (*ʿilm al-hayʾa*) was taught using textbooks such as Qāḍīzāde's already mentioned *Commentary on the Compendium on the Science of the Configuration* [*of the Universe*]. Teaching texts also discuss celestial phenomena such as eclipses and lunar phases. This inspired some Ottoman scholars to write independent treatises on describing, predicting and calculating lunar and solar eclipses. Based on the rich Arabic and Persian astronomical literature, the Ottomans already knew how and when eclipses occur. While their dedicated treatises on the topic consist mainly of timetables, they also explain the duration of eclipses, their dates and times; methods for calculating their occurrence; and some devices for observing them. Ottoman scholars measured the phases of an eclipse (immersion, duration of totality, emersion) in terms of 1/12 of the diameter of the lunar or solar disc calling such a part 'digit' (*iṣbaʿ*) as Ptolemy had done in the *Almagest* and his various successors among Muslim astral experts after him (Ptolemy 1998[CB], 22, 302, 308; Montelle 2011, 34–5).

Because eclipses were critical events for astral experts, a number of independent works on eclipses can be found in Ottoman literature. For instance, ʿAbdallāh ibn Meḥmed Kutahyavī (*c.* 12th/18th century), wrote an *Epistle on the Knowledge of Lunar and Solar Eclipses* (*Risāla fī maʿrifat al-khusūf wa-l-kusūf*). A table prepared for eclipse predictions provides data about the size of the eclipsed part of the lunar disk, that is, the measure of the eclipse diameter, the starting time of the eclipse, the duration of its totality and the end of the eclipse (MS Çorum, Çorum Hasan Paşa İl Halk Library, 19 Hk 2989/2, fol. 191a; Figure V.9.6). The anonymous *Epistle Explaining Lunar and Solar Eclipses* (*Risāla-i Beyān al-khusūf wa-l-kusūf*) contains astrological predictions based on solar and lunar eclipses. Its author held that the magnitude of the disasters they are related to can be scientifically determined, because it is proportional to the magnitude of the occultation of the lunar or solar disc (MS Ankara, National Library of Turkey, Manuscripts Collection, 06 Mil Yz A 2053/7, fol. 85a).

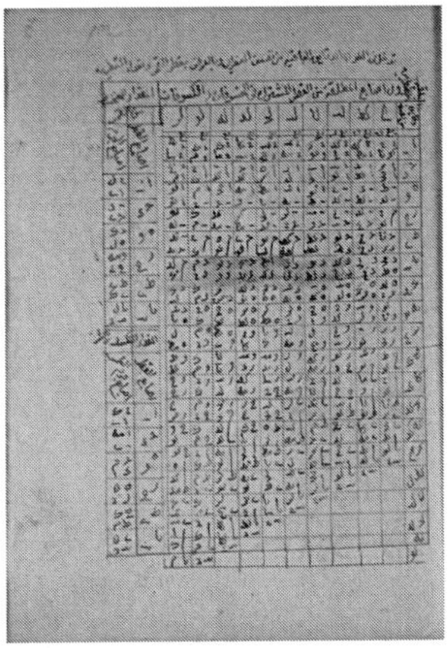

Figure V.9.6 Eclipse calculation in the treatise of al-Kütahyāvī (*c.* 12th/18th century). MS Çorum, Hasan Paşa İl Halk Library, Arşiv 19, Hk 2989/2, fol. 193b.

Source: © Hasan Paşa İl Halk Kütüphanesi, Çorum

V.9.4 Popular reactions

Ottoman popular reactions to natural phenomena are not well known. Historical chronicles sometimes mention them as causes of some societal disturbances, as divine retributions for social deviation or as signs announcing the imminent arrival of some tragic event. Lunar and solar eclipses were seen as bad omens for the state of society. Some chroniclers added astrological interpretations relating their ominous aspects to reports about eclipses. In his *History (Tārīkh)*, Süleymān ʿIzzī (d. 1168/1755) reports about a total solar eclipse on January 7, 1750, when the sun was in Capricorn, and described in detail people's reactions to the event. Astral experts had predicted the eclipse, giving rise to popular rumors that it would be almost completely dark on that day, leading some craftsmen and merchants not to open their stores and to remain at home. Because the weather was cloudy on the day of the eclipse, people remained relatively calm. Even so the sultan allegedly forbade by decree further calculations of such celestial events because they had a negative impact on the people (İzzi Süleyman Efendi 2019, 727–8). As in earlier Islamicate societies, people in the Ottoman Empire went to mosques for special prayers before or during eclipses, and some of them beat drums, tambourines, tins or cans. Some chroniclers even wrote that eclipses caused earthquakes, comets and the plague. The earthquake that shook Istanbul on September 15, 1509, wrought so much destruction that later historians remembered it as "Lesser Doomsday" (*al-qiyāma al-ṣughra*; Matrakçı Nasuh 2015, 78–9; Solakzade 1989, 1: 410–37; Sadettin Efendi 1979, 4: 3).

Comets were another natural phenomenon mentioned in historical annals. As in the case of eclipses, several chroniclers interpreted them astrologically, claiming that comets cause catastrophic incidents such as earthquakes and plagues. For example, Muṣṭafā Naʿīmā ~~Efendi~~ (1065–1128/1655–1716) described comets as catastrophic celestial signs, quoting astrologers' comments or prophecies (see also Naima Mustafa Efendi 1969, 4: 1832). The most famous comets recorded in chronicles were two seen in the west and the east during Sultan Meḥmed II's ("the Conqueror"; [r. 848–849, 855–886/1444–1446, 1451–1481]) siege of Belgrade in 860/1456; astrologers interpreted these as signaling a possible defeat, causing the sultan to withdraw his troops (Oruç Beğ 1972, 117).

V.9.5 Concluding remarks

Although Ottoman scholars named, described or predicted natural phenomena and produced information to satisfy the desire for knowledge among other scholars, state officials and sometimes the common people, not much is known about how the Ottomans examined and debated the phenomena in their scientific institutions, the courts or other public places. There is no consensus among historians of Ottoman science whether the topics of natural sciences and philosophy were included in *madrasa* classes (Shefer-Mossensohn 2015, 61; Izgi 1997, 117–27; Adıvar 1982, 176). But several topics relating to astronomy, optics and geography were undeniably taught in some Ottoman *madrasa*s (Brentjes 2018[CB], 81, 84, 91; İzgi 1997, 1: 370, 2: 249). Although natural phenomena do not seem to have been included as an independent course, it is possible that works outside the religious sciences were taught in line with the demands of students or lecturers (İhsanoğlu 2019, 200–5). The fact that more than one copy of independent treatises and studies dealing with these natural phenomena have survived, a few of which we have mentioned here, shows that the Ottoman literati had some interest in such matters.

Regrettably, it is not well documented whether natural phenomena were more than rare topics in scholarly discussions. An example is provided by Ṭāshköprüzāda, who briefly reports in *The Red Anemone on the Scholars of the Ottoman Dynasty* (*al-Shaqā'iq al-nuʿmāniyya fī ʿulamā' al-dawlat al-ʿuthmāniyya*) that ʿAlī Qūshjī and Ḫocazade discussed how the Moon affects the formation of the tides before they came into Sultan Meḥmed II's presence during the welcoming ceremony for Qūshjī (Taşköprülüzade 2019, 272). Another instance is found in Nebī Efendizāde's *Treatise on the Rainbow* (*Risāla fī qaws quzaḥ*), which is a gloss of Qāḍīzāde's description of a rainbow. There Efendizāde explains that he compiled this treatise in response to his students' insistent requests when he lectured on astronomy, probably dictating and commenting on Qāḍīzāde's *Commentary* on Jaghmīnī's *Compendium* (MS Istanbul, Süleymaniye Library, Serez 3851–003, 157–8). This suggests that discussions of rainbows, eclipses, the Milky Way or comets did in fact take place at Ottoman *madrasa*s, along with the study of books like Jaghmīnī's *Compendium* or Jurjānī's *Commentary on the Positions in the Science of kalām*. In addition to these examples, more than five copies of *The Pearl of Time on the Tides*, one of the conspicuous works compiled in the premodern era, were available in Istanbul's libraries (İhsanoğlu 2017, 314), possibly reflecting interest in the topic among literate inhabitants of the capital.

But only about a small percentage of those living in Ottoman cities may have been literate (Shefer-Mossensohn 2015,89). Whether or not these people received their education in the *madrasa*, they engaged with aspects of natural philosophy within their own means (Küçük 2020, 41–2). While we know of only a few clear records that elucidate whether and how the phenomena described above were taught or discussed in more informal meetings, given the overall working of intellectual culture in the Ottoman state, private houses and scholarly institutions, as well as the courts of sultans, viziers and other members of the elites, most likely provided occasions to do so.

Overall, conservative positions favoring existing traditions and a reluctance to adopt new perspectives may have dominated in the early modern period. But phases of change and reform also occurred. Scholars commissioned by the leading men of the state followed the inclinations of their patrons, albeit not blindly, and created their works accordingly (Shefer-Mossensohn 2015, 129). Others argued in the 12th/18th century for reforms (Menchinger 2018).

Another trend of the period consisted in the transformation of scholarly positions into bureaucratic offices, because the staff and teaching content of the Ottoman institutions were determined by the state in cooperation with the scholarly elite (Atçıl 2017, 83–119; Zilfi 1988, 81–129). As a result, scholars often found it appropriate to compile surveys of interpretations of doctrines formulated by medieval Muslim scholars, discussing and glossing them, instead of looking for new explanations outside of the intellectual canon. This attitude, as well as other behaviors of the members of the scholarly system, were repeatedly criticized by Ottoman scholars themselves (Hagen 2003, 50–76; Menchinger 2018). Beginning in the late 10th/16th century, Turkish translations of geographical, astrological, astronomical, military, mathematical, medical and natural philosophical material were made from Latin, Italian or Spanish works and later also from French and English ones (Goodrich 1990, 31–55; Hagen 2003; Shefer-Mossensohn 2015, 108–11; Chapter I.1). The engagement with early modern and modern atlases and astronomical tables and instruments was particularly intense (Ihsanoğlu *et al.* 1997, 2000; Hagen 2003; Brentjes 2005; Günergun 2011). With them, new views on natural phenomena and methods of investigation came to the attention of Ottoman scholars, whose impact can be seen in ~~numerous~~ works by Kātib Chelebi, Abū Bakr al-Dimashqī (d. 1102/1691), İbrāhīm Müteferriqa (d. d. 1160/1745), İbrāhīm Ḥaqqī Erżurūmlu (1114–1194/1703–1780) and others. How far the new knowledge about nature was integrated into the scholarly world of the Ottoman Empire deserves further study.

The focus on knowledge valuable in daily life shaped intellectual trends among Ottoman scholars. Thus the output of works on the science of timekeeping (Chapter V.4) was visibly higher than works on improving observational instruments (Ihsanoğlu *et al.* 1997). Knowledge and skills in rhetoric and logic seem to have burnished reputations better than conducting observation and experiments in the sciences about nature. Scientific explanations of natural phenomena were treated in a fragmentary fashion. The natural phenomena themselves were regarded as metaphysical or divinely inspired, rather than simply features of the natural world. This allowed Ottoman writers to understand and describe many natural phenomena not only as mathematical or philosophical objects but also, more importantly, as inauspicious facts, bad omens or signs of God's wrath incurred by people's misdeeds, as well as numinous signs foreshadowing some future event or phenomenon.

Note

1 Consolidated bibliography.

Bibliography

Sources

İzzî Süleyman Efendi. ed. Yılmazer, Z. 2019. *İzzi Tarihi* [History of Izzi]. İstanbul: Türkiye Yazma Eserler Kurumu Yayınları.

Kadızade er-Rumi. tr. Türker, Ö. 2012. *Şerhu'l-Mulahhas fi ilmi'l-hey'e* [The Commentary on the Compendium on the Science of the Configuration (of the Universe)]. Ankara: Kültür ve Turizm Bakanlığı Yayınları.

Kâtip Çelebi. ed. Balcı, R. 2016. *Keşfu'z-Zünun* [Removal of Doubts]. Ankara: Tarih Vakfi Yurt Yayınları.

Kevakib-i Seb'a Risalesi [The Epistle of the Seven Stars]. ed. Karaarslan, N. Ü. 2015. Ankara: Türk Tarih Kurumu Yayınları.

Matrakçı Nasuh (Matrakçı Nasuh). ed. Bilge, R. 2015. *Tarih-i Sultan Bayezid : Sultan II. Bayezid'in tarihi* [The History of Sultan Bayezid]. İstanbul: Arvana Yayınları.

Naima, M. E., ed. Danışman, Z. 1969. *Naima Tarihi* [History by Naima]. İstanbul: Kardeş Matbaası.

Oruç Beğ. ed. Atsız, H. N. 1972. *Oruç Beğ Tarihi* [History by Oruç Beğ]. Istanbul: Tercüman.
Peçevi, İ. E. ed. Baykal, B. S. 1981. *Peçevi Tarihi* [History by Peçevi]. Ankara: Kültür Bakanlığı Yayınları.
Sadeddin Efendi, ed. Parmaksızoğlu, İ. 1979. *Tâcü't-tevârih* [The Crown of Histories]. 4 vols. Istanbul: Millî Eğitim Basımevi.
Solakzade, Mehmet Hemdemî Çelebi. ed. Çabuk, V. 1989. *Solak-zâde tarihi* [History by Solakzāde]. [Ankara]: Kültür Bakanlığı.
Taşköprülüzade, A. E. ed. Çevik, M. 2011. *Mevzuatu'l-Ulum* [Subjects of the Sciences]. Istanbul: Üçdal Neşriyat.
Taşköprülüzade, A. E. ed. Hekimoğlu M. 2019. *Eş-Şakâ'iku'n-Nu'Mâniyye Fî Ulemâi'd-Devleti'l-Osmâniyye* [The Red Anemone on the Scholars of the Ottoman Dynasty]. Istanbul: Türkiye Yazma Eserler Kurumu Başkanlığı Yayınları.

Manuscripts

MS Ankara, National Library of Tuzrkey, Manuscripts Collection, 06 Mil Yaz A 2053/7.
MS Baltimore, The Walters Art Museum, W.658.
MS Baltimore, The Walters Art Museum, W.659.
MS Berlin, Staatsbibliothek zu Berlin, Ms. or. quart. 1828.
MS Çorum, Çorum Hasan Paşa İl Halk Library, 19 Hk 3237/2.
MS Istanbul, Süleymaniye Library, Esad Efendi, 551.
MS Istanbul, Süleymaniye Library, Hacı Beşir Ağa, 674/1.
MS Istanbul, Süleymaniye Library, Serez, 3851–003.
MS Istanbul, Topkapı Palace Library, Revan Köşkü, 1643.
MS Konya, Bölge Yazma Eserler Library, 42, Kon 3293/3.
MS Konya, Bölge Yazma Eserler Library, 42, Kon 3293/4.
MS Konya, Konya Bölge Yazma Eserler Library, 42 Kon 4172.
MS Washington, D.C., Library of Congress, Turkish 185.

Research literature

Adıvar, A. 1982. *Osmanlı Türklerinde İlim*. Istanbul: Remzi Yayınevi.
Ak, M. 1997. *Aşık Mehmed ve Menaziru'l-Avalim*. PhD thesis. Istanbul: Istanbul University.
Ak, M. 2014. "Osmanlı Coğrafya Çalışmaları – Studies on Ottoman Geography," *TALID* (*Türkiye Araştırmaları Literatür Dergisi*) 2.4: 163–212.
Arslantas, N. 2003. *İslam Dünyası'nda Depremler ve Algılanma Biçimleri*. Istanbul: İz Yayıncılık.
Atçıl, A. 2017. *Scholars and Sultans in the Early Modern Ottoman Empire*. Cambridge: Cambridge University Press.
Ayalon, Y. 2014. *Natural Disasters in the Ottoman Empire*. Cambridge: Cambridge University Press.
Bellino, F. 2014. "Sirāj al-Dīn ibn al-Wardī and the Ḫarīdat al-ʿajāʾib: Authority and Plagiarism in a Fifteenth-Century Arabic Cosmography," in Bernardini, M. and Paul, J., eds. *Scribes and Readers in Iranian, Indian and Central Asian Manuscript Traditions*. Rome: Istituto per l'Oriente Carlo Alfonso Nallino, XII, 257–97.
Brentjes, S. 2005. "Mapmaking in Ottoman Istanbul between 1650 and 1750: A Domain of Painters, Calligraphers or Cartographers?" in Imber, C., Kiyotaki, K. and Murphy, R., eds. *Frontiers of Ottoman Studies: State, Province and the West*. 2 vols. London and New York: I. B. Tauris, 2: 125–56.
Brentjes, S. 2008. "Courtly Patronage of the Ancient Sciences in Post-Classical Islamic Societies," *Al-Qanṭara* 29: 403–36.
Giahi Yazdi, H.-R. 2014. "The Fragment of Al-Kindī's Lost Treatise on Observations of Halley's Comet in A.D. 837," *Journal of History of Astronomy* 45.1: 70–2.
Goodrich, Th. 1990. *The Ottoman Turks and the New World. A Study of Tarih-i Hind-i Garbi and Sixteenth-Century Ottoman Americana*. Wiesbaden: Otto Harrassowitz.
Günergun, F. 2011. "The Ottoman Ambassador's Curiosity Coffer: Eclipse Prediction with De La Hire's 'Machine' Crafted by Bion in Paris," in Günergun, F. and Raina, Dh., eds. *Science Between Europe and Asia* (Boston Studies in the Philosophy of Science 275). New York: Springer, 103–23.

Heinen, A. M. 1982. *Islamic Cosmology, A Study of as-Suyûtî's al-Hay'a as-saniya fi l-hay'a as-Sunniya with Critical Edition, Translation, and Commentary*, Beirut: In Kommission Bei Franz Steiner Verlag Wiesbaden.
Huff, T. E. 2003. *The Rise of Early Modern Science.* Cambridge: Cambridge University Press.
İhsanoğlu, E. 2017. *Osmanlı Bilim Mirası.* Istanbul: Yapı Kredi Yayınları.
İhsanoğlu, E. 2019. *Medreseler Neydi, Ne Değildi?.* İstanbul: Kronik Yayınları.
İhsanoğlu, E. ed. 2002. *History of the Ottoman State, Society and Civilisation.* Istanbul: IRCICA.
İhsanoğlu, E. *et al.*: İhsanoğlu, E., Şeşen, R., Serdar Bekar, M. Gündüz, G. and Bulut, V., 2006. *Osmanlı Tabii ve Tatbiki Bilimler Literatürü Tarihi. History of The Literature of Natural and Applied Sciences During the Ottoman Period.* Istanbul: IRCICA.
İzgi, C. 1997. *Osmanlı Medreselerinde İlim.* İstanbul: İz Yayıncılık.
Küçük, H. 2020. *Science without Leisure. Practical Naturalism in Istanbul 1660–1732.* Pennsylvania: University of Pittsburgh Press.
Lettinck, P. 1999. *Aristotle's Meteorology and its Reception in the Arab World.* Leiden: Brill.
Menchinger, E. L. 2018. "Intellectual Creativity in a Time of Turmoil and Transition," in Salvatore A. *et al.*, eds. *The Wiley-Blackwell History of Islam.* Oxford: Wiley-Blackwell, 459–78.
Montelle, C. 2011. *Chasing Shadows – Mathematics, Astronomy, and the Early History of Eclipse Reckoning.* Baltimore: Johns Hopkins University Press.
Panzac, D. 1985. *La peste dans l'empire ottoman, 1700–1850* (Collection Turcica V). Louvain: Peeters.
Shefer-Mossensohn, M. 2015. *Science among the Ottomans.* Austin: University of Texas Press.
Starkey, P. 2007. "Al-Qazwini," in Netton, I. R., ed. *Encyclopedia of Islamic Civilization and Religion.* London and New York: Routledge, 529–30.
Stephenson, F. R. 2009. *Historical Eclipses and Earth's Rotation.* Cambridge: Cambridge University Press.
Van Brummelen, G. and Berggren, J. L. 2001. "Abu Sahl al-Kuhi on the Distance to the Shooting Stars," *Journal for the History of Astronomy* 32.2: 137–51.
Varlik, N. 2015. *Plague and Empire in the Early Modern Mediterranean World: The Ottoman Experience.* Cambridge: Cambridge University Press.
Zachariadou, E. ed. 1997. *Natural Disasters in the Ottoman Empire.* Rethymnon: Crete University Press.
Zilfi, M. C. 1988. *The Politics of Piety: The Ottoman Ulema in the Postclassical Age (1600–1800).* Minneapolis: Bibliotheca Islamica.

V.10
SCIENTIFIC PRACTICES IN SUB-SAHARAN AFRICA

Marc Moyon

V.10.1 Introduction

Over the past few decades, numerous research programs have been developed around the manuscripts of sub-Saharan Africa, especially those in Arabic. As a consequence, many initiatives have been undertaken to ensure the storage, preservation, inventory, cataloging and digitization of the manuscript heritage of the region. Given the immensity of the corpus, many of these projects are still in progress.[1] The 20th century also saw pioneering studies based on Arabic manuscripts, especially in the fields of political, cultural and religious history. These included Hunwick's works, begun in the 1960s and culminating in Brill's collection *Arabic Literature in Africa* (Hunwick and O'Fahey 1994, 1995a, 1995b, 2003; Stewart and Ould Ahmed Salim 2015. See also the first section in the Bibliography to this chapter). Today, the combination of ongoing preservation efforts with 20th-century studies allows me to ask – and begin to answer – new questions regarding scientific practices in sub-Saharan Africa. Few studies have focused specifically on Islamicate scientific practices in sub-Saharan Africa. One notable exception is research on Mauritanian mathematics (Rebstock 1990, 2007a, 2007b. See also Djebbar and Moyon 2011; Kani 1992a, 1992b; Gerdes 1994; Zaslavsky 1999). In this chapter, I examine the dissemination and appropriation of knowledge across the Sahel region from Islamic countries in the north, in order to better understand its intellectual content and its major actors as well as the factors – geographical and otherwise – that have influenced it.

The sub-Saharan Arabic manuscript corpus contains several thousand references originating from across an immense and culturally diverse territory. The consultation of these documents is extremely difficult, even impossible in some areas, and their condition is often highly degraded. The geopolitical situation of the Sahel at the beginning of the 21st century did not improve the situation (Hammer 2017).[2] Nevertheless, several public and private libraries have initiated programs to safeguard manuscripts, most often in partnership with international universities or foundations, marking their awareness of the value of these documents for the cultural history of the region.[3] Even if these programs are very unevenly distributed throughout West Africa, they are gradually contributing to a better understanding of the corpus as a whole.

DOI: 10.4324/9781315170718-59

V.10.2 General elements of the corpus

To better define the characteristics of the scientific practices of sub-Saharan scholars (especially in mathematics and astronomy) with respect to the practices of their Islamic and Arabic-speaking counterparts, I carried out a quantitative study of a corpus, excluding the *ʿajamī* manuscripts (those composed in African languages but written in Arabic letters),[4] basing my analysis on the catalogues of various libraries.[5] Quantitative analysis is necessary to understand the constitution of a corpus of texts within such a vast territory and over such a long period. Several collective phenomena such as transmission, reception and teaching can be described without precise preliminary knowledge of the players (primary or secondary), institutions and routes involved.

To adapt to or study available resources, I worked then with manuscripts from Burkina Faso, Ghana, Mali, Mauritania, Niger and Nigeria (in their modern names). I considered a total of 415 items,[6] geographically scattered and relatively heterogeneous in both form and content. Box V.10.1. provides a first representation of their geographical distribution.

Box V.10.1 Geographical distribution of the manuscript corpus

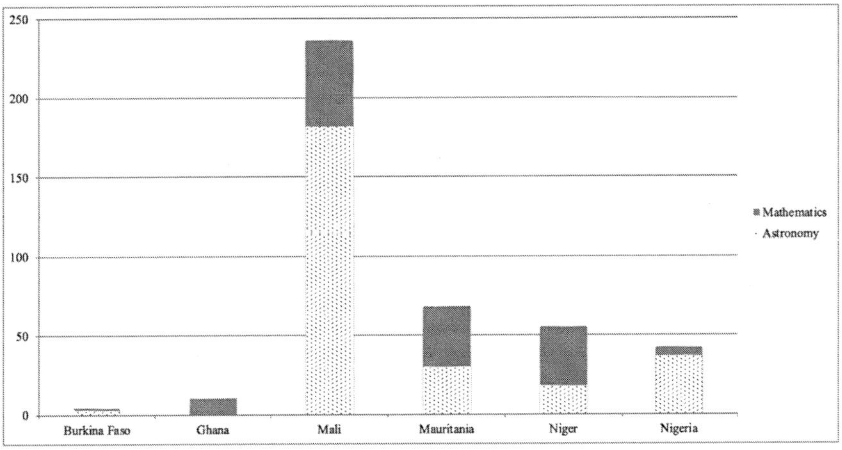

The quantitative difference between Mali and other countries can be explained by the large number of manuscripts preserved in this country and by the existence of important preservation projects supported by the two principal libraries of Timbuktu: *Mammā Ḥaydarah* and *Aḥmad Bābā*. This rich heritage is the result of the accumulation of numerous documentary resources in Timbuktu – one of the largest Sahel-Saharan cities related to the trans-Saharan trade, founded in the late 5th/11th century. Its cultural influence was immense, especially in the 10th/16th and 11th/17th centuries, when Timbuktu overtook Walāta (in present-day Mauritania) in importance (Hunwick 2003; Singleton 2004). In addition, the thematic distribution of preserved manuscripts shows the importance of works in astronomy (270 items) compared to those in mathematics (145).

Even if it is difficult to date the arrival in the region of Arabic scientific writings produced in the East, the Maghrib and al-Andalus, we know that they followed the multiple exchange routes between northern and sub-Saharan Africa (Djebbar and Moyon 2011, 93–9). These included

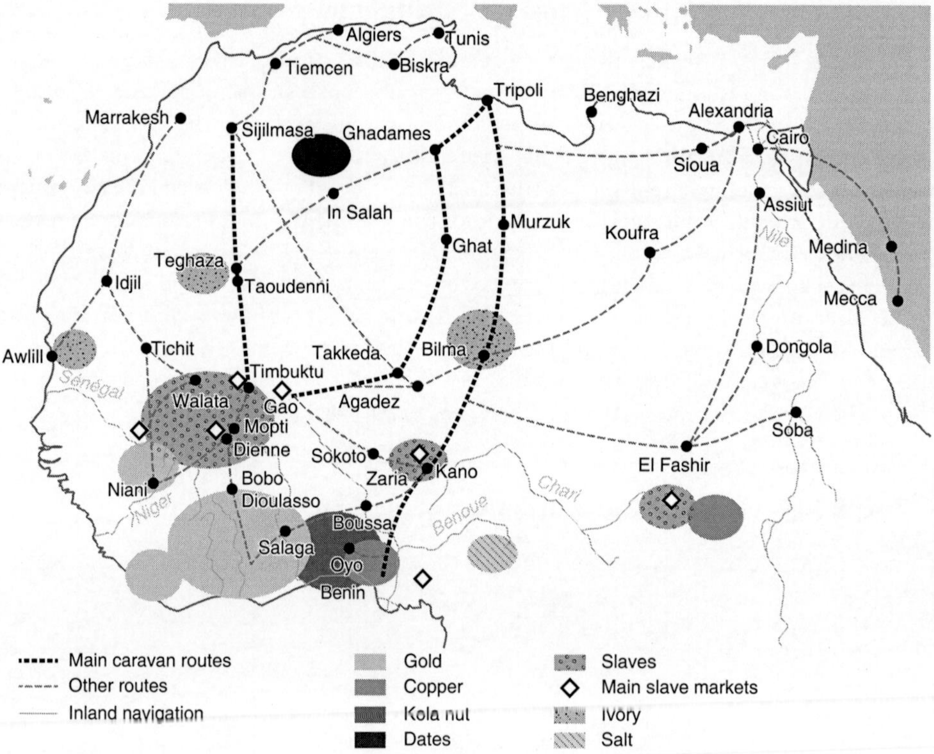

Figure V.10.1 Map of the trans-Saharan trade routes

not only commercial routes and routes of journeys in search of knowledge (*riḥla*) but also the routes of the annual pilgrimage to the holy cities of Medina and Mecca (Figure V.10.1).

V.10.3 About the authors of works in the corpus

To study the periods of activity of the authors represented in this corpus and their geographical origins, I took into account the references for which I could identify the authors (52% of the corpus) and those for which I have evidence for geographical location but not for determining exact authorship (11%). Thus, I excluded so-called anonymous texts (37%). To better highlight the dynamics of transmission, production and circulation of mathematical and astronomical knowledge over the long period represented (from the 3rd/9th to the 13th/19th centuries) and within the boundaries defined, I distinguished (especially for Box V.10.2) the total number of texts – # texts (without copy) – and the total number of manuscripts – # copies.

The most dynamic period of activity appears to run from the end of the 9th/15th century through the 12th/18th century and includes two important moments both for new scientific productions and for the number of copies made (from a limited number of texts). Historical context sheds light on this observation. Since at least the 8th/14th century, but in particular since the 10th/16th century, the region experienced a marked development of educational structures such as Qur'ānic schools. For example, following Saad's estimates from *History of the Investigator on the Stories about Localities, Armies, and High-Ranking People* (*Tārīkh al-fattāsh fī akhbār al-buldān*

Box V.10.2 Centuries of the activities of the identified authors

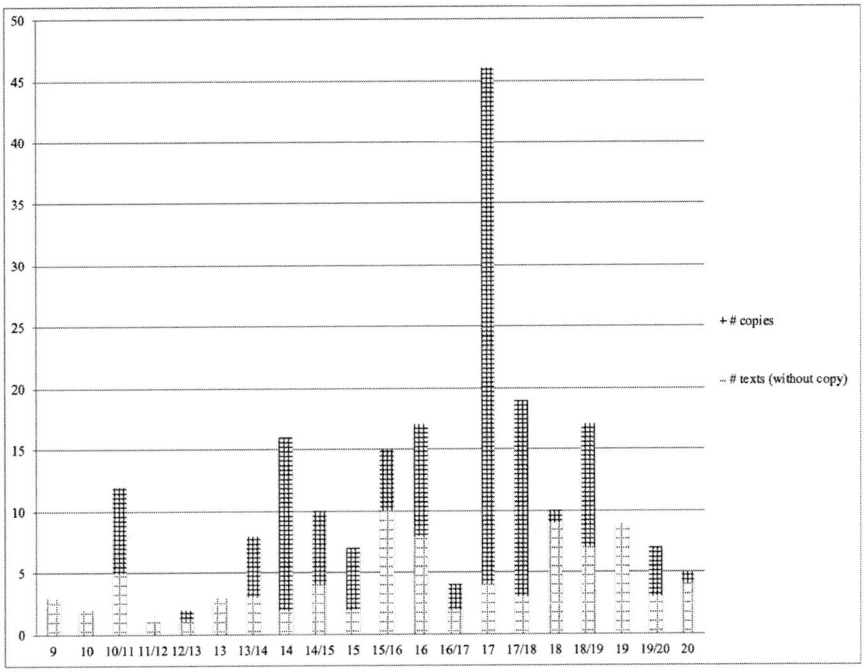

wa-l-juyūsh wa-akābir al-nās), just the city of Timbuktu would have housed 150 schools (in the 10th/16th century), each hosting an average of 50 students (Saad 1983, 90). The intense activity of the copyists made scientific texts, like others, accessible to the city's scholars (Hunwick 2003).

The geographical origins of the authors can be determined from their biographies (when available) or by their *nisba*. The *nisba* is an adjective with different meanings, often geographical, added to as a final element to a person's name. It may indicate their birthplace but not necessarily where they worked. For example, Naṣīr al-Dīn al-Ṭūsī (597–672/1201–1274) was born in Tūs but spent his working life elsewhere, notably Baghdad and Maragha. Also, because information regarding the origin of an author can be relatively old and uncertain if unconfirmed by historical sources, I worked on a corpus of 202 references (including 126 copies) for which I could make assumptions about the origin of the author. This made it possible to distinguish seven geographical areas (Box V5.10.3).

There were numerous human, cultural and economic exchanges between the north and the Sahel territories and within the sub-Saharan regions. The dissemination of manuscripts could not be independent from them (see, e.g., Dewière 2017, 54–57, on the intellectual background of the Sudanese Aḥmad ibn Furṭū [10th/16th century]). The introduction of Arabic manuscripts was undertaken by Muslim scholars from the north, who were also merchants and preachers. This was the case for Gao in the 5th/11th century or Takrūr (in present-day Ghana) in the 6th/12th century (Djebbar and Moyon 2011, 96–9). This is also what el-Nafaty (2013) clearly described with regard to relations between the Maghrib and the Hausa kingdoms (located between Niger and Lake Chad, which today corresponds to northern Nigeria).

Box V.10.3 Geographical origins of the identified authors

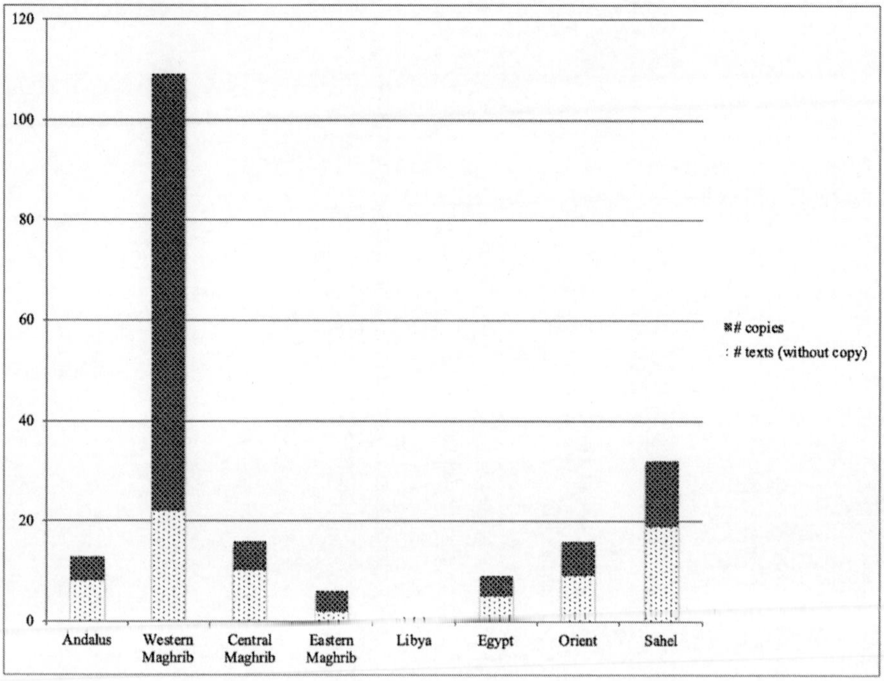

Relationships have existed between the two regions throughout the history of trans-Saharan trade. Commercial centers developed in both regions: in the North in Marrakesh, Fez, Tahert and Qayrawan, and in the South in Kumbi Saleh, Gao, Timbuktu and Jenne. Between the north and the south, there were also some 'inland ports' or 'caravan harbors' such as Wargala, Ghadames, Fezzan, Tuat, Ghat, Bilma, Taghaza, Taodani, Tadmekka and Audoghost (el-Nafaty 2013, 29).

The pilgrimage to Mecca, whose roads went almost exclusively through Egypt, also offered an opportunity to consult writers from the north, learn of new books and order and bring home copies. In this context and since at least the early 900s/1500s,[7] a significant market for books became established (Lydon 2008, 100). Through at least the 13th/19th century, special caravans were organized, in particular to the western Maghrib, to take advantage of the availability of books in Marrakesh or Fez, as evidenced by the list of 200 volumes purchased by Shaykh Sidiy-ya al-Kabīr (d. 1285/1868) in Marrakesh *c.* 1245/1830 (Stewart 1970). The western Maghrib therefore stands out significantly from other areas due to its historical and cultural ties with the major capitals of sub-Saharan West Africa.

The substantial number of copies in this corpus also reflects the fame that certain authors had acquired. For the western Maghrib, I can cite Ibn al-Bannā' (654–721/1256–1321), Abū Muqriʿ (8th/14th century) and above all al-Mirghīthī (d. 1089/1678–9) and al-Rasmūkī (d. 1133/1720–1). This was also the case for several other authors well represented here, such as al-Akhḍārī (d. 983/1575) from the central Maghreb or the Egyptian Sibṭ al-Māridīnī (d. 809/1506).

The two preceding trends appear to be cumulative: the appropriation of the texts from the western Maghrib is most important between the end of the 10th/16th century and the end of the 12th/18th century (Box V.10.4). It was not until the end of the 12th/18th century that the production of mathematical and astronomical texts in Arabic in the sub-Saharan regions would distinguish itself from copying, when the Islamization and Arabization of the local elites of the great Sahelian empires allowed the production of original writings in all major merchant cities, alongside previously imported works.

Box V.10.4 Geographical origins and periods of activity of the identified authors

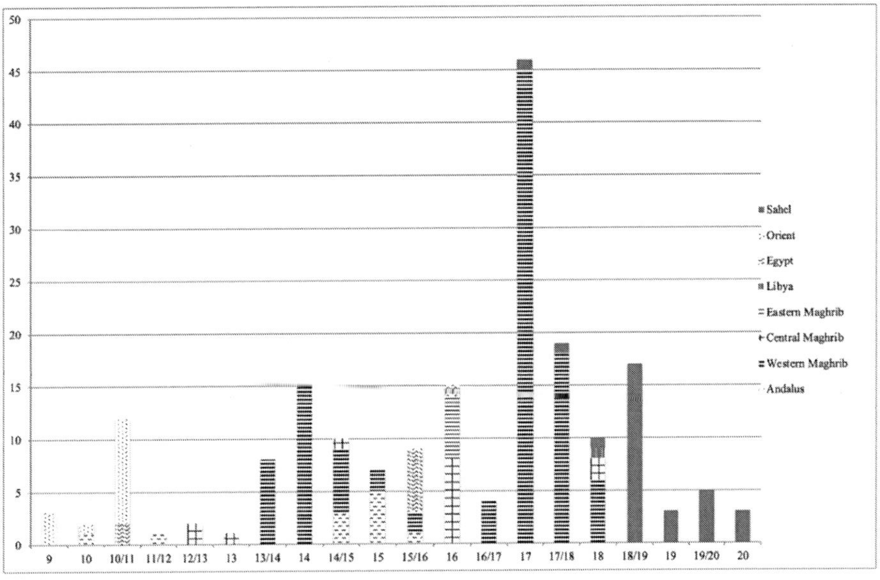

I shall now put these first results into historical perspective, situating my observations with respect to the production and the circulation of manuscripts inside the region. Last (2008) proposes a schematic historical periodization of the book market in West Africa between 800/1397 and 1300/1882. Four main phases make up this periodization which, according to Hall and Stewart (2011), may apply to a large part of the Sahel region. First, during the 9th/15th and 10th/16th centuries, copied books were imported at a high cost. Then, in the second phase, 'classic' texts and some new works available in Cairo and the Maghrib arrived in the region. Following the 10th/16th and 11th/17th centuries, copies were produced locally, especially for books in high demand, which thus became cheaper. As a result, in the third phase, an important merchandizing of copies took place in the 11th/17th and 12th/18th centuries. The fourth and last phase, in the 12th/18th and 13th/19th centuries, was marked by the development of an original, autonomous production by local scholars, who may have had difficulty accessing books from other regions. They compiled their own texts using passages or quotations taken from authors considered particularly relevant to their own teaching (Last 2008, 143). This was still the case in the 14th/20th century. A perfect example is the epistle *Elements of the Science of Reckoning* (*Nubdha fī*

ʿilm al-ḥisāb) of al-Arawānī (d. 1418/1997). This late original work was still nourished by works of the Muslim West from the 8th/14th to 10th/16th centuries, with explicit references to *The White Pearl* (*al-Durra al-bayḍā*) written by al-Akhdarī and to Ibn al-Bannā'’s *Condensed Exposition on the Operations of Arithmetic* (*Talkhīṣ aʿmāl al-ḥisāb*), among others (Djebbar and Moyon 2011, 137–55). The vast majority of books produced in the 14th/20th century was still in Arabic, even though local languages had begun to emerge as written languages. Thus, my corpus of scientific manuscripts in Arabic can be considered consistent with the general characteristics of the book market as a whole.

V.10.4 The nature of the corpus

Working from the titles indicated in the catalogues, I established a typology of the texts in the corpus. This made it possible to characterize particular kinds of scientific practice addressed by these manuscripts. I considered four distinct categories (Box V.10.5): (1) abridged texts, which corresponds to epistles (*risāla*), extracts/elements (*nubdha*), compendia/epitomes (*mukhtaṣar*) and notes (*taqyīdāt*); (2) commentaries (*sharḥ*); (3) versified texts (*qaṣīda, urjūza, manẓūma* and *naẓm*); and (4) all other texts. The last category includes the texts that the preceding categories do not take into account, in particular those with titles containing the generic terms: writing (*maktūb*), book (*kitāb*), canon (*qānūn*) or composition (*ta'līf*). Today, at this stage of the study, I lack evidence to better categorize these references; only future consultation of the manuscripts might allow me to refine the categorization.

Box V.10.5 Categorization of the corpus

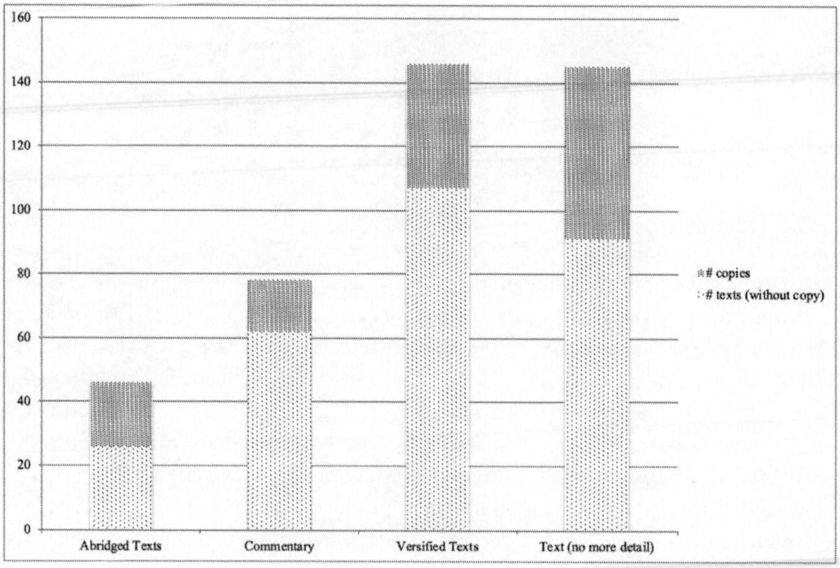

Versified texts represent the most important category (more than a third of the corpus), even without considering commentaries on poems. Versification is therefore a very important linguistic and philological dimension of the corpus.[8] It is not characteristic of sub-Saharan practices alone. Several mathematical and astronomical texts of Islamicate countries were written in verse, especially in the Muslim West (Abdeljaouad 2005; Moyon 2016).[9] Moreover, an examination of the titles in the corpus shows that many texts are commentaries (*shurūḥ*). These include notes on texts, developed *ad litteram* critical analyses and interpretations according to the sense (Gilliot 1996, 330). I should specify that some (not all) of the poems, commentaries and abridgments have a common characteristic: they were used for learning or teaching, and written either by a teacher or by a student. Hence, the majority of the corpus suggests a didactic dimension (Arazi and Ben Shammai 1995; Van Gelder 1995, 108).

V.10.5 Mathematics and astronomy

I subdivided scientific practices into two major disciplines: (1) mathematics, including geometry, the science of reckoning or arithmetic and the science of inheritances, and (2) astronomy, including folk astronomy and astrology. Each title in the corpus was attributed to one of these disciplines.[10]

First of all, in the mathematical sciences, I considered geometry in its theoretical and practical duality (*ʿilm al-handasa* and *ʿilm al-misāḥa*),[11] including optics (*ʿilm al-baṣariyyāt*). It is very poorly represented (Figure V.10.2).[12] Nevertheless, two references are remarkable. For theoretical geometry (in the Euclidean tradition), I found only one text in Mauritania: a copy of the *Book on the Method of Solving Geometrical Problems* (*Kitāb fī l-taʾattī li-istikhrāj al-aʿmāl al-handasiyya*) of Thābit ibn Qurra (d. 288/ 901). I also noted the presence of a text on measurement written by the Andalusian Ibn ʿAbdūn (311–after 366/923–after 976) – once in the Umarian library in Segou (Mali) and now preserved in Paris (Djebbar 2005[CB],[13] 2006[CB]).

Representation in the corpus is different for the science of reckoning (*ʿilm al-ḥisāb*), including Indian reckoning (*al-ḥisāb al-hindī*) and likely algebra (*al-jabr*), although it is difficult to determine it from the titles (Chapter I.7). For example, I identified copies – in Segou and Mauritania – of *The Unveiling of the Secrets Relative to the Dust Ciphers* (*Kashf al-asrār ʿan ʿilm ḥurūf al-ghubār*) written by al-Qalaṣādī (d. 891/1486). I included the science of transactions (*ʿilm al-muʿāmalāt*) in this

Figure V.10.2. *Epistle on Measurement* (*Risāla fī l-taksīr*) by Ibn ʿAbdūn (311–after 366/923–after 976), MS Paris, BnF, arabe 5311, fols 15b–16a.

Source: © BnF, Paris.

category, as advocated by Ibn Khaldūn (2002, 951), but I found very few explicit references to it. Like Sesiano (2004), I also included the science of magic squares (*wafq al-aʿdād* or *awfāq*) as an arithmetical topic, and indeed, several texts in my corpus explicitly mention this discipline. One of the best examples from sub-Saharan Africa is Muḥammad al-Fulānī al-Kishnāwī or al-Katsināwī (d. 1154/1741). Originally from Sudan, he studied in Cairo and spent some time in Bornu (northern Nigeria today; Kani 1992b, 23–4; Zaslavsky 1999, 138–51; Sesiano 2004, 18). He is the author of at least one book on magic letters and squares: *The Splendor of the Horizons and the Clarification of the Ambiguous and Hermetic in the Science of Letters and Harmonious Numbers* (*Bahjat al-āfāq wa-īḍāḥ al-labs wa-l-ighlāq fī ʿilm al-ḥurūf wa-l-awfāq*; Djebbar and Moyon 2011, 98–9).

Interestingly, several titles explicitly link harmonious numbers and astronomy, suggesting they are more about talismanic astrology than mathematics. I therefore included them in the category of astrology (discussed later).

A last category I distinguished within mathematical sciences was the science of inheritances (*ʿilm al-farāʾiḍ*). Because it was quantitatively very important, I set it off from the others so as not to increase the category "Reckoning" (Laabid 2005, 242; Chapters I.7 and V.1, Box V.10.6).[14]

Box V.10.6 Mathematical themes

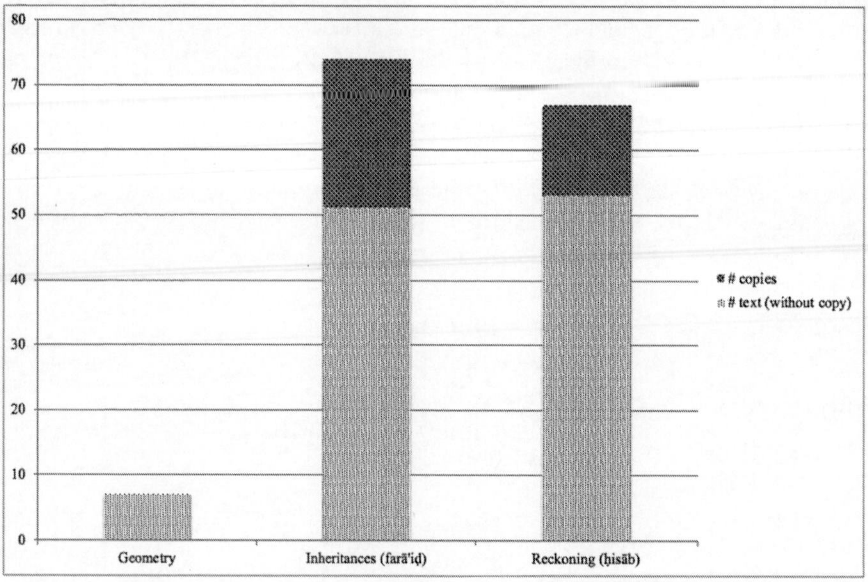

The authors of the science of inheritances practiced both mathematics and jurisprudence, like al-Rasmūkī, who wrote at least one poem, *The Jewels Hidden in the Shell of Legal Heritage* (*al-Jawāhir al-maknūna fī ṣadaf al-farāʾiḍ al-masnūna*), which he commented on himself (Figure V.10.3). He also commented on other works in the same discipline, some of which are found in sub-Saharan Africa (Djebbar and Moyon 2011, 127–8). Sibṭ al-Māridīnī (d. 912/1506) also belongs to this category. Several copies of his commentary on al-Raḥbī's *Poem on the Science of Inheritance* (*al-Urjūza al-Raḥbiyya fī ʿilm al-farāʾiḍ*) written by Ibn al-Mutaqqina al-Raḥbī (d. 577/1182–1183) are present in Mali or Mauritania (Sobieroj 2016, 268–74).

Figure V.10.3 Commentary by Aḥmad al-Jazūlī al-Rasmūkī (d. 1133/1720–1) on the *Wings of Desires on the Knowledge of Inheritance and Reckoning* (*Ajniḥat ar-righāb fī maʿrifat al-farāʾiḍ wa l-ḥisāb*), fol. 7a (?), written by Abū Sālim al-Samlālī (10th/16th century).

Source: © Library Mammā Ḥaydarah in Timbuktu. www.wdl.org/fr/item/464/#institution=mamma-haidara-commemorative-library

The works categorized under "Astronomy" are even more diverse than those in the mathematical sciences (Box V.10.7). The term *science of the celestial orbs* (*ʿilm al-falak*) is one way of describing what we call astronomy. In this context, the term designating the stars is *kawkab* (pl. *kawākib*). The designation *science of the stars* (*ʿilm al-nujūm*) is closer to astrology (also designated by the word *tanjīm*). Nevertheless, *ʿilm al-nujūm* can cover both astronomy and astrology as two different approaches to the same reality. In this case, *najm* (pl. *nujūm*) is used to designate the stars. Consequently, the *munajjim* practiced both astronomy and astrology, without distinction (Chapter I.6). Furthermore, the treatises on the *anwāʾ* deal with traditions of celestial observation. The term *nawʾ* (pl. *anwāʾ*) refers to a system of computation linked to the observation of heliacal elevations and the acronical sunsets of certain groups of stars, making it possible to divide the solar year into precise periods (Chapter V.4). After the 3rd/9th century and with the consideration of the lunar mansions (*manāzil al-qamar*), these treatises became almanacs of sorts. They gave the dates of the solar calendar for the heliacal and the acronical settings of the stars that corresponded to the lunar mansions, plus the meteorological phenomena that were traditionally linked to it (Pellat 1960; Forcada 2004–2005[CB]).

This last domain is a part of daily-life astronomy (King 1996; Forcada 2004–2005). More generally, the astrologer (*munajjim*) became interested in solving problems of the Muslim community either inherited from the Hellenistic world (as in the case of astrology) or in close connection with Islamic practices. The latter was the case for the lunar calendar, including the treatises on the *anwāʾ*, or the determination of the azimuth of the *qibla* (the orientation of the *Kaʿba* that fixes, among other things, the orientation of the prayer, for a given place) and the determination of the hours of prayer (Chapters V.4 and VI.3). More precisely, many authors in the corpus can be identified as timekeepers (*muwaqqit*), scholars affiliated with a mosque or

Box V.10.7 Astronomical themes

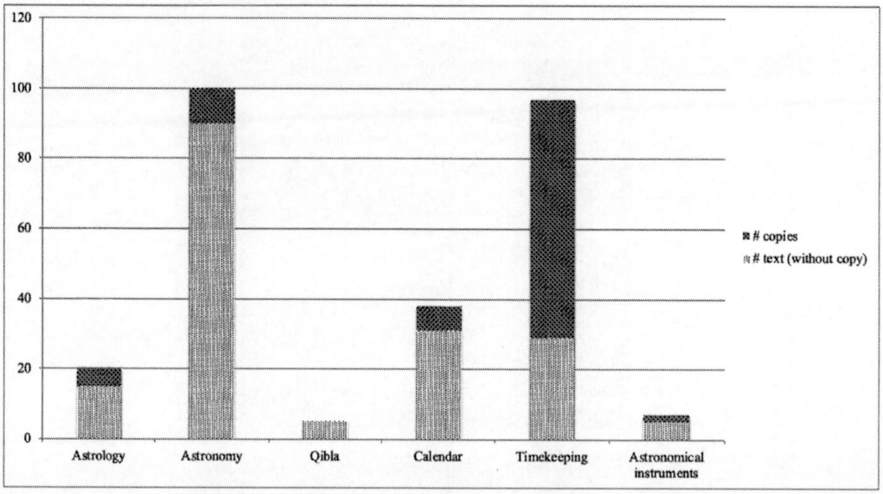

a *madrasa* in order to teach the science of timekeeping (*ʿilm al-mīqāt*) and, among other duties, maintain a sundial. They all contributed to the study and the dissemination of their sciences.

For example, numerous texts refer explicitly to Abū Muqriʿ, from the West Maghrib (Djebbar and Moyon 2011, 107–8), or to his main commentator al-Mirghīthī (Djebbar and Moyon 2011, 124–5), as well as to Sibṭ al-Māridīnī who worked in the famous al-Azhar Mosque in Cairo (Djebbar and Moyon 2011, 129–30). The knowledge (*ʿilm*) taught by Abū Muqriʿ seems to have been a very simple form of timekeeping with a few basic rules and very little technical knowledge, either astronomical or mathematical – probably one of the reasons for its success in Islamic Africa. Today, we can delimit the available corpus on timekeeping, but we do not know a lot about the training and the activities of timekeepers in sub-Saharan Africa.

Similarly, I have found texts in the astronomical literature pertaining to instruments of observation and measurement such as the astrolabe (*al-asṭurlāb*) or the sine quadrant (*al-rubʿ al-mujayyab*),[15] but we have no direct or indirect evidence of instrument makers or of instruments in sub-Saharan Africa.

V.10.6 Conclusion

My quantitative study promises to contribute to the description of scientific activities in the vast Sahel region. In particular, it allows a more precise characterization of the various mathematical and astronomical orientations among those activities and of the contexts of their productive results, based on manuscripts still available today in the sub-Saharan libraries of West Africa.

According to the history of the region and of its caravan routes, the appropriated scientific knowledge and practices came mainly from the western Maghreb. One of the main impetuses for copying and circulating books appears to have been teaching, or educational needs (Kani 1992b, 17). Copies of manuscripts – and in particular those with scientific content – were made by and for local scholars. The analysis of the present corpus provides evidence that the teaching of mathematics and astronomy in the sub-Saharan cities was largely achieved by the copying of texts produced in the Maghreb and by the process of commenting on them. My study provides new information about the content of such teachings and (to a lesser degree) about its modalities. As with the beginning of Arabic sciences in North Africa, scientific activity in the Sahel was linked to the technical needs of the Islamic community, especially for teaching. It was therefore the daily-life dimensions of astronomy that were mostly represented: the lunar calendar, determining the *qibla* and, of course, the science of timekeeping. In mathematics, the two dimensions most often privileged were also linked to Islamic education and the management of administration: the science of calculation, together with the decimal numeral system and elementary algorithms, and the science of inheritances (Laabid 2006). Thus, for both fields of scientific knowledge (astronomy and mathematics), the relationship between the scientific practices disseminated by the manuscripts and Islamic jurisprudence (*fiqh*) are undeniable.

Notes

1 Most of these programs focus on the manuscripts preserved in Timbuktu. They include "The Ancient Astronomers of Timbuktu" [www.scribesoftimbuktu.com/the-project.php], the "Tombouctou Manuscripts Project" [www.tombouctoumanuscripts.org], the "Project for the Conservation of Malian Arabic Manuscripts" (begun at the Aḥmad Bābā Center in Timbuktu) and the French project ANR VECMAS Tombouctou "Valorisation et Édition Critique des Manuscrits Arabes Subsahariens" [http://vecmas-tombouctou.ens-lyon.fr]. This is also the case for other countries or regions, as with the British Library's "Djenne Manuscript Library Collection" (part of the Endangered Archives Program) [https://eap.bl.uk/collection/EAP488-1] and the "West African Arabic Manuscript Project" [www.westafricanmanuscripts.org], which covers West Africa, not just Malian libraries.

2 In 2013, the international project "Safeguarding the Manuscripts of Timbuktu" (SAMAT) began transferring the contents of many Timbuktu libraries to secure sites in Bamako. See in particular the program of the Center for the Study of Manuscript Culture at the University of Hamburg [www.manuscript-cultures.uni-hamburg.de/timbuktu/index_e.html].

3 In particular, this is the case with the nongovernmental organization SAVAMA-DCI (Sauvegarde et Valorisation des Manuscrits-Défense de la Culture Islamique) whose aim – with the SAMAT project, for example – is to preserve and enhance the Timbuktu Arabic manuscripts. See also (Abdullahi 2010; Ayuba 2013) for Nigeria.

4 This is a development of results first presented in Moyon (2022).

5 The catalogues consulted, in addition to the database of the Northwestern Digital Library [www.westafricanmanuscripts.org], are included in the category "Sources, including catalogues" section of the bibliography here. In these references, I indicate the approximate number of manuscripts covered by each based on the estimates of Salvaing (2015).

6 The number of manuscripts dealing with mathematics or astronomy is extremely small considering the quantity of manuscripts preserved in the libraries considered (more than 20,000). The most represented themes relate to Islamic literary and religious tradition (*taṣawwuf, tafsīr, ḥadīth, fiqh, ta'rīkh*). Moreover, I have considered each handwritten piece, even fragmentary, as a separate entity even if it corresponds to the copy of a text present several times. If I no longer consider the material object as the unit but the textual reference (all the copies of the same text counting only for one), that is specified. In particular, the anonymous texts are all preserved even if there are still inevitably copies of the same text. Using these criteria, I count at most 287 references.

7 This was attested by Leo Africanus. When he visited the capital of the Empire of Mali, he wrote: "The learned are greatly revered. Also, many book manuscripts coming from Berber lands (Berbérie) are sold. More profits are realized from this sale than any other merchandise". Cited in Lydon (2004, 39–71).

8 My corpus also includes examples of the three types of poems described by Sobieroj (2016, 7) as well as of the many commentaries that accompany them.

9 Versification is no longer a process specific to the Arabic language tradition; now it is recognized as a more general movement related to the educational objectives of the texts. For example, on mathematics in ancient China, see Bréard (2014), and on Sanskrit mathematics, see Keller (2014).

10 Following the example of Triaud (2014), I am aware that the statements made in this part of my study are only indicative. Many books deal with several disciplines, in which case I chose the dominant one according to the title.

11 For a synthesis on *ʿilm al-misāḥa* in Islamic countries, see Moyon (2017, 19–77).
12 Rebstock has previously noted this lack of representation for the Mauritanian texts (Rebstock 1990).
13 Consolidated bibliography.
14 All the references recorded here do not necessarily deal with elementary arithmetic; even if I tried to be as selective as possible, they might only deal with *fiqh*.
15 For a complete description of astronomical instruments in Islamic countries, see Sezgin (2004, 78–202).

Bibliography

Sources

Bābā Yūnus, M. and al-Muḥsin Zakī, ʿA. ʿA. 2000. *Fihris makhṭūṭāt maktabāt Ghānā* [Catalogue of Manuscripts in Ghana Libraries]. London: Al-Furqān Islamic Heritage Foundation (1,969 manuscripts).

Ghali, N., Mahibou, M. and Brenner, L. 1985. *Inventaire de la bibliothèque 'Umarienne de Ségou: conservée à la Bibliothèque Nationale, Paris*. Paris: Édition du CNRS (more than 4,000 manuscripts).

Handlist of Manuscripts in the Centre de Documentation et de Recherches Historiques Aḥmad *Bābā, Timbuktu*. 1995. 5 vols. London: Al-Furqān Islamic Heritage Foundation (about 7,500 manuscripts).

Ibn Khaldūn. 2002. *Le Livre des exemples: autobiographie, Muqaddima* [The Book of Examples: Autobiography, Introduction]. Paris: Gallimard.

Mammā Ḥaydarah, ʿA. Q. and Sayyid A. F. 2000. *Fihris makhṭūṭāṭ maktabat Mammā Ḥaydarah* [Catalogue of Manuscripts in the Mammā Ḥaydarah Library]. 4 vols. London: Al-Furqān Islamic Heritage Foundation (4,004 manuscripts, and 2,000 referenced since).

Mawlāy, H. and Sayyid, A. F. 2004. *Catalogue of Islamic Manuscripts at the Institut des recherches en sciences humaines (IRSH), Niger*. 8 vols. London: Al-Furqān Islamic Heritage Foundation (about 4,000 manuscripts).

Rebstock, U. and Yahya, A. W. A. 1997. *Fihris makhṭūṭāt Shinqīṭ wa-Wādān* [Handlist of Manuscripts in Shinqīṭ and Wādān]. London: Al-Furqān Islamic Heritage Foundation (1,106 manuscripts).

Research literature

Abdeljaouad, M. 2005. "12th-Century Algebra in an Arabic Poem: Ibn al-Yāsamīn's *Urjūza fī l-jabr wa-l-muqābala*," *Llull* 28.61: 181–94.

Abdullahi, M. J. 2010. "The Role of Nigerian Universities in the Preservation of the Arabic Manuscripts," *Anyigba Journal of Arts and Humanities* 10: 12–26.

Arazi, A. and Ben Shammai, H. 1995. "Mukhtaṣar," in *EI-2*, 7: 536–40.

Ayuba, M. A. 2013. "Information and Communication Technologies in Preserving Arabic and Islamic Manuscripts," *Global Journal Al Thaqafah* 3.2: 7–14.

Bréard, A. 2014. "On the Transmission of Mathematical Knowledge in Versified Form in China," in Bernard, A. and Proust, C., eds. *Scientific Sources and Teaching Contexts Throughout History: Problems and Perspectives*. Dordrecht: Springer, 155–85.

Brigaglia, A. *Historical Context: Notes on the Arabic Literary Tradition of West Africa*. http://digital.library.northwestern.edu/arbmss/historical.html. (Accessed 2 June 2019).

Dewière, R. 2017. *Du lac Tchad à la Mecque: le sultanat du Borno et son monde (XVIe-XVIIe siècle)*. Paris: Éditions de la Sorbonne.

Djebbar, A., and Moyon, M. 2011. *Les sciences arabes en Afrique. Mathématiques et astronomie, IXe-XIXe siècles*. Brinon-sur-Sauldre: Grandvaux.

Gerdes, P. 1994. "On Mathematics in the History of Sub-Saharan Africa," *Historia Mathematica* 21.3: 345–76.

Gilliot, C. 1996. "Sharḥ," in *EI-2*, 9: 327–30.

Hall, B. S. and Stewart, C. S. 2011. "The Historic 'Core Curriculum' and the Book Market in Islamic West Africa," in Krätli, G. and Lydon, G., eds. *The Trans-Saharan Book Trade, Manuscript Culture, Arabic Literacy and Intellectual History in Muslim Africa*. Leiden and Boston: Brill, 109–74.

Hammer, J. 2017. *The Bad-Ass Librarians of Timbuktu and Their Race to Save the World's Most Precious Manuscripts*. New York: Simon & Schuster.

Hunwick, J. O. 2003. "The Timbuktu Manuscript Tradition," *Tinabantu: Journal of African National Affairs* 1.2: 1–9.

Hunwick, J. O. and O'Fahey, R. S. 1994–5. *Arabic Literature of Africa, The Writings of Eastern Sudanic Africa to c. 1900*. Vol. 1. Leiden: Brill.

Hunwick, J. O. and O'Fahey, R. S. 1995a. *Arabic Literature of Africa. The Writings of Central Sudanic Africa*. Vol. 2. Leiden: Brill.

Hunwick, J. O. and O'Fahey, R. S. 1995b. *Arabic Literature of Africa. The Writings of the Muslim Peoples of Northeastern Africa*. Vol. 3. Leiden: Brill.

Hunwick, J. O. and O'Fahey, R. S. 2003. *Arabic Literature of Africa. The Writings of Western Sudanic Africa*. Vol. 4. Leiden: Brill.

Kani, A. 1992a "Arithmetic in the Pre-Colonial Central Sudan," in Thomas-Emeagwali, G., ed. *Science and Technology in African History with Case Studies from Nigeria, Sierra Leone, Zimbabwe, and Zambia*. Lewiston, NY: Edwin Mellen Press, 33–9.

Kani, A. 1992b "Mathematics in the Central Bilâd Al-Sudân," in Thomas-Emeagwali, G., ed. *The Historical Development of Science and Technology in Nigeria*. Lewiston, NY: Edwin Mellen Press, 17–36.

Keller, A. 2014. "History of Mathematical Education in Ancient, Medieval and Pre-Modern India," in Karp, A. and Schubring, G., eds. *Handbook on the History of Mathematics Education*. New York: Springer, 70–83.

King, D. A. 1996. "Astronomy in Islamic society: Qibla, Gnomonics and Timekeeping," in Rashed and Morelon, eds. [CB], 1: 128–84.

Laabid, E. 2005. "Les mathématiques et les héritages au Maghreb des 12e-14e siècles : essai de synthèse," in *Actes du 7ème colloque maghrébin sur l'histoire des mathématiques arabes*. 2 vols. Marrakech: al-Wataniya, 1: 241–62.

Laabid, E. 2006. *Les techniques mathématiques dans la résolution des problèmes de partages successoraux dans le Maghreb médiéval à travers le* Mukhtaṣar *d'al-Hūfī (m. 1192): sources et prolongements*. PhD Thesis. University of Rabat.

Last, M. 2008. "The Book in the Sokoto Caliphate," in Jeppie, S. and Diagne, S. B., eds. *The Meanings of Timbuktu*. Cape Town: HSRC Press, 135–64.

Lydon, G. 2004. "Inkwells of the Sahara: Reflections on the Production of Islamic Knowledge in *Bilād Shinqīṭ*," in Reese, S. S., ed. *The Transmission of Learning in Islamic Africa*. Leiden: Brill, 39–71.

Lydon, G. 2008. *On Trans-Saharan Trails, Islamic law, Trade Networks, and Cross-cultural Exchange in Nineteenth-Century Western Africa*. Cambridge: Cambridge University Press.

Morelon, R. 1996. "General Survey of Arabic Astronomy," in Rashed and Morelon, eds., 1: 1–19.

Moyon, M. 2016. "Ibn Luyūn at-Tujībī (1282–1349): un nouveau témoin de la science du mesurage en occident musulman," in Bouzari, A., ed. *Actes du 11e colloque maghrébin sur l'histoire des mathématiques arabes*. Alger: Éditions Al-Khalduniya, 333–52.

Moyon, M. 2017. *La géométrie de la mesure dans les traductions arabo-latines médiévales*. Turnhout: Brepols.

Moyon, M. 2022. "Mathématiques et astronomie dans les 'manuscrits du désert' : première approche," in Kchir, K., ed. *Revisiter l'histoire des sciences, des savoirs, des techniques et des arts au Moyen Âge*. Tunis: Latrach Edition; Laboratoire du Monde Arabo-Islamique Médiéval, 159–82.

el-Nafaty, M. S. 2013. "The Intellectual and Cultural Impacts of the Maghrib on the Hausaland with Special Reference to Kano," *Dirasāt Ifriqiyya* 50: 25–52.

Pellat, C. 1960. "*Anwā'*," in *EI-2*, 1: 523–4.

Rebstock, U. 1990. "Arabic Mathematical Manuscripts in Mauretania," *Bulletin of the School of Oriental and African Studies* 53.3: 429–41.

Rebstock, U. 2007a. *Die 'Brücke Des Rechnens' von Abū Marwān. Ein Gedicht über Die Arabische Kalenderrechnung und Astrologie*. Freiburg: Universität Freiburg.

Rebstock, U. 2007b. "'Homo Ludens' at Work: Mauritanians' Skills in Determining Inheritance Shares (Farā'iḍ)," in Christmann, A., ed. *Studies in Islamic Law: A Festschrift for Colin Imber*. Oxford: Oxford University Press, 233–43.

Saad, E. N. 1983. *Social History of Timbuktu: The Role of Muslim Scholars and Notables, 1400–1900*. Cambridge: Cambridge University Press.

Salvaing, B. 2015. "À propos d'un projet en cours d'édition de manuscrits arabes de Tombouctou et d'ailleurs," *Afriques* 2015. http://journals.openedition.org/afriques/1804.

Sesiano, J. 2004. *Les carrés magiques dans les pays islamiques*. Lausanne: Presses polytechniques et universitaires romandes.

Sezgin, F. 2004. *Science et Technique en Islam: Astronomie*. Vol. 2. Algiers: Thala Éditions.

Singleton, B. D. 2004. "African Bibliophiles: Books and Libraries in Medieval Timbuktu," *Libraries and Culture* 39.1: 1–12.

Sobieroj, F. 2016. *Variance in Arabic Manuscripts: Arabic Didactic Poems from the Eleventh to the Seventeenth Centuries – Analysis of Textual Variance and Its Control in the Manuscripts*. Berlin: De Gruyter.

Stewart, C. 1970. "A New Source on the Book Market in Morocco in 1830 and Islamic Scholarship in West Africa," *Hespéris Tamuda* 11: 209–50.

Stewart, C. and Ould Ahmed Salim, S. A. 2015. *Arabic Literature of Africa. The Writings of Mauritania and the Western Sahara*. Vol. 5. Leiden. Boston: Brill.

Triaud, J. L. 2014. "Deux bibliothèques arabo-islamiques en Côte d'Ivoire au début du XXe siècle," in Müller, C. and Roiland-Rouabah, M., eds. *Les non-dits du nom. Onomastique et documents en terres d'Islam: mélanges offerts à Jacqueline Sublet*. Beyrouth: Presses de l'IFPO, 161–246.

Van Gelder, G. J. 1995. "Arabic Didactic Verse," in Drijvers, H. J. W. and MacDonald, A. A., eds. *Centres of Learning: Learning and Location in Pre-Modern Europe and the Near East*. Leiden: Brill, 103–17.

Zaslavsky, C. 1999. *Africa Counts: Number and Pattern in African Culture*. Chicago: Chicago Review Press.

URLs

http://vecmas-tombouctou.ens-lyon.fr].
https://eap.bl.uk/collection/EAP488-1.
http://journals.openedition.org/afriques/1804
http://digital.library.northwestern.edu/arbmss/historical.html
www.manuscript-cultures.uni-hamburg.de/timbuktu/index_e.html.
www.scribesoftimbuktu.com/the-project.php.
www.tombouctoumanuscripts.org.
www.westafricanmanuscripts.org.
www.wdl.org/fr/item/464/#institution=mamma-haidara-commemorative-library

V.11
MEDICAL PRACTICES IN TIBET IN INTERCULTURAL CONTEXTS

Ronit Yoeli-Tlalim

The goal of this chapter is to contribute to the study of the interactions between Muslim territories and their eastern neighbors, within the context of medical knowledge. An entry point into this topic are narratives found in Tibetan medical histories from the 16th century onward which assign a key role to Greco-Arabic medical knowledge. Following an overview and an assessment of this narrative, the chapter moves on to examine inputs that appear to be derived from Islamic cultures in Tibetan medicine, along with those coming from the Indian and Chinese cultural spheres, and discusses the discrepancy between what Tibetan medical histories tell us, and what is found in Tibetan medical texts. The chapter concludes with an overview of Tibet as the land of musk in Islamic sources.

V.11.1 Historical overview

Tibetan medicine (*Sowa Rigpa*) developed as a synthesis of Indian, Chinese and Greco-Arabic medical systems, within an overall Buddhist theoretical grounding. The multicultural character of Tibetan medicine is emphasized in Tibetan medical histories starting from the earliest extant example of this genre, the Tibetan medical history by Che rje zhang ston zhig po, dated to the early 13th century. Che rje sets medical knowledge within what is termed 'The Seven Schools' (*lugs bdun*), referring to both divine and human realms (Martin 2007, 307–25). Within the human realm, the list refers to medical systems from India, Kashmir, Urgyan (in present-day Pakistan) and Nepal (*bal po*) as well as from Arabo-Persian (*stag gzig*), Dol po (an area in present-day western Nepal), Uighur (*hor*), Tangut/Xixia (*me nyag*), Khotanese (*li*), Byzantine (*phrom*), Chinese and Tibetan sources (Martin 2011, 117–43; Yoeli-Tlalim 2010a, 195–211, 2012, 355–65).[1] Variations of this list become practically standard in subsequent Tibetan medical histories. Medical histories portray the earliest stage of Tibetan medicine at its most multicultural.

Our earliest extant sources on Tibetan medicine come from what is now known as Cave 17, or the 'library cave' in Dunhuang. This cave, which was sealed for nearly a thousand years for reasons that are still being debated by scholars, contained several tonnes of manuscripts. The Tibetan medical manuscripts from Dunhuang that date to the 9th or 10th century are largely practice-oriented but include important information for understanding some of their theoretical assumptions. The Tibetan Dunhuang medical manuscripts reflect connections with Chinese, Uighur and Khotanese medical knowledge (Yoeli-Tlalim 2019b). The main therapy to be

DOI: 10.4324/9781315170718-60

mentioned in these early manuscripts is moxibustion (*me btsa'*, the heating of particular locations on the body), but there are also references to bloodletting, fumigation, massage, horn cupping, as well as uses of *materia medica*, mostly from plants and animals.

V.11.2 The *Four Tantras*

The text known as *Four Tantras* (*Gyushi, rgyud bzhi*), thought by scholars to have been composed in the 12th century, is regarded as the key text of Tibetan medicine and is still the core of its instruction today. The *Gyushi* begins with an account of how the Buddha, manifesting as the Medicine Buddha, gave the teaching encompassed in the text. The origin and history of the *Four Tantras* have been the focus of heated debates in Tibet for centuries (Yang Ga 2010, 2014, 154–77; Yoeli-Tlalim 2007, 5: 1342–4). Traditional views ascribe the text to the Medicine Buddha and claim it was written down in Sanskrit, then hidden inside a column of the Samye Monastery, to be found when the time was ripe for its teaching.

According to this account, someone known as Yuthog the Younger later 'discovered' the text and adapted it to the local conditions of Tibet. Other views deny any Sanskrit original and ascribe the text to a Tibetan author – either the older or the younger Yuthog Yontan.

Scholars now agree that the *Four Tantras* are a native Tibetan text that has incorporated and synthesized elements from the Indian, Chinese and Greco-Arabic medical systems, combined with a Buddhist theoretical grounding. The *Gyushi* presents the same medical doctrine from four different perspectives, which is why its name is also translated as: The *Four-fold Tantra*.

The First Tantra is known as the *Root Tantra*. This part provides a general outline of the principles of Tibetan medicine, diagnosis and treatments. It discusses in brief the humors (*nyes pa*) and humoral imbalances that give rise to illness. The Second Tantra is known as the *Exegetical Tantra*. It elaborates on the theoretical basis of Tibetan medicine and discusses topics such as embryology, anatomy, channels of the body, pathology, diet and conduct, medications, external therapies, diagnosis and medical ethics. The Third Tantra is the *Instructional Tantra*. It is a very long and elaborate presentation of specific treatments for particular illnesses. The Fourth Tantra, known as the *Subsequent Tantra*, elaborates on diagnosis through pulse examination, urine analysis and observation of the tongue, the preparation of medicines, inner cleansing procedures and external therapies, such as bloodletting, moxibustion, massage and minor surgeries.

V.11.3 Greco-Islamic components of Tibetan medicine: Galen in Tibet?

An intriguing account found in a number of Tibetan medical histories from the 16th century onward tells of a famed physician from the West by the name of Galenos, who was, along with other famed doctors, from China and India, invited to Lhasa by the first king of Tibet (Yoeli-Tlalim 2012, 355–65, 2019a, 594–608; Garrett 2007, 363–87).[2] In the *Mirror of Beryl*, an account of the history of Tibetan medicine by Sangye Gyatso (1653–1705), regent of the Fifth Dalai Lama (1617–82) and author of several seminal texts on Tibetan medicine, one finds the following account:

> Once when the king was ill, the Indian doctor Bharadhaja, the Chinese doctor Hsüan Yüan Huang, and the Taksik or Phrom doctor Galenos were invited to Tibet to cure him. . . . They held many discussions and jointly composed a medical text in seven chapters called *Weapons of Fearlessness*, which they offered to the king. . . .

> Therefore, all medical science was compiled into these three main systems and propagated by them. The king gave gifts to the Indian and Chinese doctors, and they travelled back to their own lands. Galenos stayed on as royal physician. It is said that he mostly resided in Lhasa, where he composed many texts. He married and had three sons.
>
> *(Gyatso 2010, 148–9)*[3]

Sangye Gyatso's narrative, like similar narratives that followed his, locates this episode at the time of the 7th-century Tibetan king Songtsen Gampo, whose reign marks the beginning of recorded history in Tibet, and indeed, Tibetan culture at large. This captivating narrative has led to many accounts in secondary literature stating that Greek medicine was adopted in Tibet in the 8th century. One should not, however, jump to such conclusions but rather try to understand the significance of this narrative. Along with similar accounts in Tibetan medical histories, it raises many intriguing questions. The first of them concerns the mentioned Galenos's country of origin. The Tibetan denotation of Phrom or Khrom is derived from Rum, or Rome, and usually refers to Byzantium. The other option the text offers is Tazig (Taksik). Early Tibetan geographies are in agreement that Tazig is a large kingdom to the west of Tibet (Yoeli-Tlalim 2011, 1–16).[4] It could refer to Iran or more generally to the Islamic Empire (Martin 2011, 117–43) and its Arabo- or Persophone Muslim population. That either of Byzantium and 'Islam' can be suggested as country of origin of the Galenos who appeared in Lhasa is not really surprising since from a Tibetan perspective the exact demarcation between Tazig and Khrom is hard to know, and as Dan Martin has pointed out, they often occupy the same place on the map (Martin 2011).

Is the type of link described earlier between Tibetans and Muslims to the West even a possibility? Tibet is usually perceived as remote and isolated but that has not always been the case. On a map of Asia as it was more or less at the time the narrated events presumably took place, one would find three major empires abutting each other: the Abbasid Empire, founded in 132/750; the Tibetan Empire, which reached its height in the early 9th century; and Tang China (618–907). The close proximity of the Abbasid and Tibetan Empires at the time explains what people often, unduly, find surprising, continuing cultural and other connections between Tibet and the Islamic world from the 2nd/7th-8th century onward (Akasoy *et al.* 2011). Furthermore, Tibet's mediating position was significant for Asian medical knowledge during and after the time of the Tibetan Empire as well as during the Mongol era (Buell 2011, 189–208; Yoeli-Tlalim 2019b). The mentioned narratives thus present a history of Tibetan medicine as initially a synthesis of Greek, Indian and Chinese medical systems. However, the notion of a triple – Greek, Indian, Chinese – synthesis in early Tibetan medicine is not borne out by the actual sources; the two major influences to be detected in them are Indian and Chinese.

V.11.3.1 *Urine analysis*

While Tibetan medicine mostly derives from Indian and Chinese medical traditions, we can also find some Greco-Arabic influence, primarily in the diagnostic method of urine analysis, a key diagnostic method in Tibetan medicine which does not feature in Ayurvedic or Chinese medicine in any similar way.[5] Comparing the urine analysis sections in an early Tibetan medical text, the *Medical Method of the Lunar King* (*Zla ba'i rgyal po*), and in Ibn Sīnā's (d. 428/1037) *Canon of Medicine* (*al-Qānūn fī l-ṭibb*) reveals remarkable similarities not only in content, but also in the structure of the texts (Yoeli-Tlalim 2010a).

5.11.3.2 Materia medica

Already our earliest extant Tibetan medical sources, the medical manuscripts from Dunhuang, contain names of medicinal substances which seem to be derived from the West (Yoeli-Tlalim 2013). Two interesting cases are *ka phur* (camphor) and *dar ya kan* (theriac). In the ancient world, camphor, often coupled with musk, was as rare as it was costly And featured prominently in long-distance trade between Asia and Europe via the Middle-East. Its medical *and* ceremonial uses in different cultures, stem from a common body of mythical lore (Donkin 1999). Although originating primarily from Southeast Asia (the Malayan peninsula in particular), camphor trade was from ancient times dominated by Iranians (Colless 1969–70). Indeed, even in Chinese sources camphor is associated with Iran (Laufer 1919, 478–79). Camphor features frequently in pre-Islamic Iranian writings as a rare, precious and exotic substance and was therefore valued as a royal gift (A'lam 1990).

In the Dunhuang manuscripts, the Tibetan word for camphor appears in several different spellings, marking it as a loan word. Laufer already pointed out that the Tibetan forms of this word are closer to the Arabic and Persian *kāfūr* (Middle Persian: *kāpūr*) than to the Sanskrit *karpūra* (Laufer 1919, 591). Iranian dominance in the trade of camphor probably lies at the source of the form of this loan word in Tibetan.

The Tibetan word for turmeric (*gur kum*) is another loan word to be linked with Persian, *kurkum*, rather than Sanskrit, *kuṅkuma* (Laufer 1919, 321). The Persian itself is traceable to the Assyrian *karkuma*, Hebrew *karkōm* and Syriac *kurkemā*. Another foreign name in the Dunhuang *materia medica* is the Tibetan name of theriac, *dar ya kan*, a loan word that seems to be derived from the Arabic or Persian form of the potion *tiryaq*, a very popular remedy originating from Greece and very widespread in the Muslim world (Beckwith 1980; Totelin 2004; Ullmann 1970[CB],[6] 321–42; Watson 1966).

The transmission of urine analysis as well as the influence detectable in *materia medica*, however, do not quite add up to a "meta-narrative" in which Galenos, or indeed Galenic medicine, constitutes the most significant input to Tibetan medicine, as implied by the narrative quoted earlier. Indeed, in earlier examples of Tibetan medical histories, from the 13th century onward, there are no mentions of a Galenos, references to Western influences notwithstanding, nor do their accounts present Greco-Arabic teachings as in any way superior in the sense implied by Sangye Gyatso.[7] How then should one understand the primary position given to Galenos in Sangye Gyatso's account? All the preceding points suggest that the accounts of Galen coming to Tibet at the alleged time reflect more the period in which they were written, from the 17th century onward, than the period to which they supposedly refer, the 7th and 8th centuries. Galen as a representative of western medicine – here meaning Islamicate medicine – came to Tibet most probably via Mughal India, with which Tibet maintained close relations.

The autobiography of the Fifth Dalai Lama relates how in 1675 he brought a physician from India, who was known for his expertise in cataract surgery, to his court. This physician, named Manaho, is credited with a work on ophthalmology which has been preserved in the Tibetan Tanjur (*Bstan 'gyur*), the part of the Tibetan Buddhist canon composed primarily of Buddhist commentarial works. The Tibetan title of this short work on the treatment of eye diseases, translated in the Potala by Lhun grub, is *Opening the Eyes to See* (*Mig 'byed mthong ba*). According to the preface the author was a physician of the Indian emperor Shāh Jahān (r. 1037–1068/1628–1658) and came from the country of Paripura. The Fifth Dalai Lama says that Manaho taught the art of cataract removal to a local Tibetan physician, who later performed it successfully on the Fifth Dalai Lama himself (Gyatso 2015, 116–7).

More generally, this episode should be viewed within the context of the Fifth Dalai Lama's active efforts to seek out medical experts from abroad, not only for his own well-being but also with a view toward broadening Tibetan medicine's repertoire of diagnostic, therapeutic, surgical and pharmacological tools (Gyatso 2015). The person who most likely oversaw the invitation of this foreign physician and other physicians from neighboring countries was the Fifth Dalai Lama's regent, Sangye Gyatso, author of the previously mentioned Galenos narrative.

Sangye Gyatso also played a crucial role in systematizing Tibetan medicine. One should therefore consider the following aspects of his activities in conjunction: his composition of a seminal text on the history of Tibetan medicine, from which the earlier quote on Galen is taken; his systematization of Tibetan medicine; and his own connections with foreign physicians. Together they indicate that the significance Sangye Gyatso gave to Galen reflects to a great extent Galen's connection to Islamic medicine, as found at the time in Mughal India, but here projected back to the origins of Tibetan medicine.

V.11.4 Ayurvedic components of Tibetan medicine

Much of the theoretical basis of Tibetan medicine is derived from Indian sources, especially the *Aṣṭaṅgahṛdaya saṃhitā* (Yang Ga 2010). This theoretical basis defines some of the key principles of Tibetan medicine, primarily the notion of the three *nyepa*, as well as the main principles of pharmacology.

V.11.4.1 The three nyepa ('humors')

The basic view of health and illness as outlined in the *Gyushi* is based on the notion of the three *nyepa* (*nyes pa*), usually translated as 'humors,' but literally meaning 'faults' or 'troubles.' They correspond to the Indian concept of the three *doṣa*, as they are termed in Sanskrit. In a medical context the three *nyepa* refer both to the potential causes of trouble or illness and to illness itself (Gerke 2014). The three *nyepa* are wind (*rlung*), bile (*mkhris pa*) and phlegm (*bad kan*). But the translation of *nyepa* as humor is problematic, and so are the terms for each of the three – wind, bile and phlegm; they are not precise or unequivocal. Translating the *nyepa* as humor(s) not only conveys meanings but also creates meanings distinct from the original ones and posits a link with Greco-Arabic medicine, which did not necessarily exist (Gyatso 2006). Several chapters in the first and second parts of the *Four Tantras* represent the body in its healthy condition, named an 'unaltered state' in the Tibetan sources. In this state, the three *nyepa* support and maintain the functions of the body. As long as the three *nyepa*, along with the seven body constituents and the three waste products, are in a state of balance, the body will remain healthy. Any of the *nyepa* can be in a state of imbalance, which can be a state of excess, deficiency or disturbance. Factors causing the *nyepa* to be imbalanced might include inappropriate nutrition, lifestyle or behavior. When out of balance, any of the three *nyepa* can cause illness. Hence, there can be illnesses that are due to bile imbalance, phlegm imbalance, wind imbalance or more commonly a combination of two *nyepa*. Among the three *nyepa*, wind (*rlung*) is considered particularly important (Yoeli-Tlalim 2010b).

V.11.4.2 Pharmacology

Tibetan pharmacology is based on the notion that the nature of the five elements is the same, whether in the healthy body, in disease, in food and in medicinal substances. Tibetan doctors seek to understand how body and mind relate and interact with substances deriving from plants, animals and minerals. Foods and *materia medica* are classified according to the six tastes, which balance one's disturbed *nyepa*s.

V.11.5 Chinese components of Tibetan medicine

The main diagnostic method in Tibetan as in Chinese medicine is an elaborate analysis of the pulse, taken at three different locations near each wrist. In the interpretation of these diagnostic signs the age of the patient will be taken into account as well as the season, since according to the theoretical principles of Tibetan medicine, natural tendencies change according to one's age and according to season. These diagnostic methods require great skill and long practice but are able to provide much important information about the patient.

A shared therapeutic practice between Tibetan and Chinese medicine is that of moxibustion (*me btsa'*). In the Tibetan way of performing moxibustion, small cones of mugwort plant are burned on specific points of the body. Moxibustion is often prescribed for general phlegm disorders, such as digestion problems, joint pains, mild arthritis or insomnia. Moxibustion has a long history as a quick and easy type of therapy, which can be performed at the household level, from the earliest extant Tibetan moxibustion manuscripts from Dunhuang (9th or 10th century) up to present-day testimonials of its use in Tibetan households (Lo and Yoeli-Tlalim 2018).

Tibetan medicine also has various links with divination and time reckoning, which stem from Chinese notions (Yoeli-Tlalim 2014). Tibetan sources often refer to China as the land of divination. The main figures, who are predominantly mentioned as transmitters of this knowledge into Tibet, are Confucius, usually rendered in Tibetan as "Kongtse" (Kong tse, Kong tshe), and the two Chinese princesses married to Tibetan kings at the time of the Tibetan empire, Princess Wencheng and Princess Jincheng (Shen-yu 2007).

An important idea within this Sino-Tibetan context is that of a vital force circulating in the body in accordance with the lunar cycle, known in Chinese as *renshen* ('human spirit'; Harper 2003, 2005; Arrault 2010). The notion of a cyclical lunar vital force is important in Tibetan medicine and has various practical implications. This force, termed *bla*, is described as responsible for a person's vitality and well-being. For a thorough overview of the cyclical vital force see Gerke (2012, particularly Ch. 5, 137–65). According to Tibetan medical theory, sometimes, when a person suffers a great shock, the *bla* force may be lost, leading to a development of illness. The *Blue Beryl* (*Va'i du rya sngon po*), Sangye Gyatso's 17th-century commentary on the *Gyushi*, describes the movement of the *bla* in the body in locations known as the *bla gnas*, in accordance with the lunar cycle. Determining the location of the *bla* is considered important in Tibetan medical practice, since invasive therapies such as golden-needle therapy, moxibustion or bloodletting in an area where the *bla* resides at the time of treatment, are seen to be harmful for the patient. According to Tibetan medicine, these invasive therapies are also to be avoided on days of new or full moon, when the *bla* is said to pervade the entire body for a short time.

V.11.7 Tibet as the land of musk in Arabic literature

From as early as the 3rd/9th century onward, Tibet was known in Arabic literature as the 'land of musk' (Akasoy and Yoeli-Tlalim 2007; Yoeli-Tlalim 2011; King 2017). Musk, in addition to being a much desired perfume, is used both in Tibetan and Islamic medicine. Sources in Arabic, Persian, Tibetan and Hebrew discuss musk in a variety of genres – medical, geographical, zoological, religious and commercial – and reveal an active trade route which existed between Tibet and the Islamicate world from the 2nd/8th century onward. Based on a comparative study of the uses of musk as found in Tibetan and Arabic medical literature, we have reached the conclusion that alongside the lucrative trade in this substance there were also exchanges of ideas (Akasoy and Yoeli-Tlalim 2007). But while we found similarity in the uses of musk, we have also noted that the theory behind those uses, as well as the principles of *materia medica* classification, did not travel along with the substance. These were hot/cold and dry/moist in the Greco-Arabic system but six tastes and eight qualities in the Indo-Tibetan system.

V.11.8 Conclusion

Tibetan medicine developed as a synthesis of Indian, Chinese and Greco-Arabic medical systems within an overall Buddhist theoretical grounding. While Tibetan medical histories assign a central role to the Greco-Arabic input, this input is relatively small in comparison to the Indian and Chinese impact, and probably has more to do with the period in which these narratives were composed – when Islamic medicine dominated the Mughal court in India – than the formative period of Tibetan medicine which they supposedly describe. However,we can detect interesting early links, particularly by comparing the diagnostic practice of urine analysis as formulated in the early Tibetan medical text, the *Medical Method of the Lunar King* (*Zla ba'i rgyal po*), to Ibn Sīnā's *Canon of Medicine* (*al-Qānūn fī l-ṭibb*). These two texts reveal remarkable similarities not only in content but also in their structure. The process, however, by which this transmission took place is yet to be researched. Other interactions occurred in the sphere of *materia medica*: Tibetan pharmacopoeias include substances coming from (or traded in) Islamic lands, as indicated by loan words from Arabic and Persian. The study of the transmission of substances also reveals the significance of Tibetan musk within Islamic writings of various genres – medical, geographical, zoological, religious and commercial – from as early as the 3rd/9th century.

Notes

1 For a detailed analysis of the two western components of this and similar lists, that is, *stag gzig* and *phrom*, see Martin (2011; Yoeli-Tlalim (2011, 201).
2 The following section is based on Yoeli-Tlalim (2012, 2019a). On Tibetan medical histories more generally, see Garrett (2007).
3 This quotation was first discussed by Beckwith (1979, 301).
4 The various Tibetan spellings are *Ta zig, ta zhig, ta chig, stag gzig* and *stag gzigs*. On these terms, see Yoeli-Tlalim (2011, 1–16).
5 We do find a method of urine analysis in Ayurveda, but it is very different from that found in Tibetan or Greco-Arabic sources. The Ayurvedic method involves placing a drop of oil in a urine specimen. The diagnosis is then conducted according to how the oil drop changes. See Yoeli-Tlalim (2010a, 199).
6 Consolidated bibliogrphy.
7 For a study of a 13th-century exemplar of the medical history genre, see Martin (2007).

Bibliography

Sources

Gyatso, Desi Sangyé. tr. Kilty, G. 2010. *Mirror of Beryl: A Historical Introduction to Tibetan Medicine*. Boston: Wisdom.

Research literature

Akasoy *et al.* 2011: Akasoy, A., Burnett C. and Yoeli-Tlalim, R., eds. 2011. *Islam and Tibet: Interactions along the Musk Routes*. Farnham: Ashgate.
Akasoy, A. and Yoeli-Tlalim, R. 2007. "Along the Musk Routes: Exchanges between Tibet and the Islamic World," *Asian Medicine: Tradition and Modernity* 3.2: 217–40.

A'lam, H. 1990. "Camphor," in *eIr*, 4.7: 743–7.

Arrault, A. 2010. "Activités médicales et méthodes hémérologiques dans les calendriers de Dunhuang du IXe au Xe siècle: esprit humain (renshen) et esprit du jour (riyou)," in Despeux, C., ed. *Médecine, Religion et Société dans la Chine Médiévale: Étude de manuscrits chinois de Dunhuang et de Turfan*. 3 vols. Paris: Collège de France, 1: 285–332.

Beckwith, C. 1979. "The Introduction of Greek Medicine into Tibet in the Seventh and Eighth Centuries," *Journal of the American Oriental Society* 99: 297–313.

Beckwith, C. 1980. "Tibetan Treacle: A Note on Theriac in Tibet," *Tibet Society Bulletin* 15: 49–51.

Buell, P. 2011. "Tibetans, Mongols and the Fusion of Eurasian Cultures," in Akasoy, *et al.*, 189–208.

Colless, B. C. 1969–70. "The Traders of Pearl: The Mercantile and Missionary Activities of Persian and Armenian Christians in South East Asia," *Abr-Nahrain* 9: 17–38.

Donkin, R. A. 1999. *Dragon's Brain Perfume: An Historical Geography of Camphor*. Leiden: Brill.

Garrett, F. 2007. "Critical Methods in Tibetan Medical Histories," *The Journal of Asian Studies* 66.2: 363–87.

Gerke, B. 2012. *Long Lives and Untimely Deaths: Life-span Concepts and Longevity Practices among Tibetans in the Darjeeling Hills, India*. Leiden and Boston: Brill.

Gerke, B. 2014. "The Art of Tibetan Medical Practice," in Hofer, T., ed. *Bodies in Balance: The Art of Tibetan Medicine*. Seattle and London: Rubin Museum of Art and University of Washington Press, 16–31.

Gyatso, J. 2015. *Being Human in a Buddhist World: An Intellectual History of Medicine in Early Modern Tibet*. New York: Columbia University Press.

Gyatso, Y. 2006. "Nyes pa: A Brief Review of Its English Translation," *The Tibet Journal* 30–31: 109–18.

Harper, D. 2003. "Iatromancie," in Kalinowski, M., ed. *Divination et société dans la Chine médiévale: Étude des manuscrits de Dunhuang de la Bibliothèque nationale de France et de la British Library*. Paris: BnF, 471–512.

Harper, D. 2005. "Dunhuang Iatromantic Manuscripts P. 2856 R° and P. 2675 V°," in Lo, V. and Cullen, C., eds. *Medieval Chinese Medicine*. London: Routledge, 34–64.

King, A. 2017. *Scent from the Garden of Paradise: Musk and the Medieval Islamic World*. Leiden: Brill.

Laufer, B. 1919. *Sino-Iranica: Chinese Contributions to the History of Civilization in Ancient Iran*. Chicago: Field Museum of Natural History.

Lo, V. and Yoeli-Tlalim, R. 2018. "Travelling Light: Sino-Tibetan Moxa-cautery from Dunhuang," in Lo, V. and Barrett, P., eds. *Imagining Chinese Medicine*. Leiden: Brill, 271–90.

Martin, D. 2007. "An Early Tibetan History of Indian Medicine," in Scrempf, M., ed. *Soundings in Tibetan Medicine*. Leiden: Brill, 307–25.

Martin, D. 2011. "Greek and Islamic Medicines' Historical Contact with Tibet: A Reassessment in View of Recently Available But Relatively Early Sources on Tibetan Medical Eclecticism," in Akasoy *et al.*, 117–43.

Shen-yu, L. 2007. "The Tibetan Image of Confucius," *Revue d'Etudes Tibétaines* 12: 105–29.

Totelin, L. 2004. "'Mithridates' Antidote – A Pharmaceutical Ghost," *Early Science and Medicine* 9: 1–19.

Watson, G. 1966. *Theriac and Mithridaticum: A Study in Therapeutics*. London: Wellcome Historical Medical Library.

Yang, G. 2010. *The Sources for the Writing of the 'Rgyud bzhi' Tibetan Medical Classic*. Cambridge, MA: Harvard University, PhD thesis.

Yang, G. 2014. "The Origins of the *Four Tantras* and an Account of Its Author, Yuthog Yonten Gonpo," in Hofer, T., ed. *Bodies in Balance: The Art of Tibetan Medicine*. New York: Rubin Museum of Art and Washington University Press, 154–77.

Yoeli-Tlalim, R. 2007. "Yuthog Yonten (the Younger)," in Bynum, W. F. and Bynum, H., eds. *Dictionary of Medical Biography*. Westport: Greenwood Press, 5: 1342–4.

Yoeli-Tlalim, R. 2010a. "On Urine Analysis and Tibetan Medicine's Connections with the West," in Craig, S. *et al.*, eds. *Studies of Medical Pluralism in Tibetan History and Society*. Halle: International Institute for Tibetan and Buddhist Studies GmbH, 195–211.

Yoeli-Tlalim, R. 2010b. "Tibetan 'Wind' and 'Wind' Illnesses: Towards a Multicultural Approach to Health and Illness," *Studies in History and Philosophy of Biological and Biomedical Sciences* 41: 318–24.

Yoeli-Tlalim, R. 2011. "Islam and Tibet: Cultural interactions – An Introduction," in Akasoy *et al.*, 1–16.

Yoeli-Tlalim, R. 2012. "Re-visiting 'Galen in Tibet'," *Medical History* 56.3: 355–65.

Yoeli-Tlalim, R. 2013. "Central Asian Mélange: Early Tibetan Medicine from Dunhuang," in Dotson, B., Iwao, K. and Takeuchi, T., eds. *Scribes, Texts, and Rituals in Early Tibet and Dunhuang*. Wiesbaden: Reichert Verlag, 53–60.

Yoeli-Tlalim, R. 2014. "Medicine, Astrology and Divination," in Hofer, T., ed. *Bodies in Balance: The Art of Tibetan Medicine*. New York: Rubin Museum of Art and Washington University Press, 90–104.
Yoeli-Tlalim, R. 2019a. "Galen in Asia?" in Bouras-Vallianatos, P. and Zipser, B., eds. *Brill's Companion to the Reception of Galen*. Leiden: Brill, 594–608.
Yoeli-Tlalim, R. 2019b. "The Silk-Roads as a Model for Exploring Eurasian Transmissions of Medical Knowledge: Views from the Tibetan Medical Manuscripts of Dunhuang," in Smith, P. H., ed. *Entangled Itineraries of Materials, Practices, and Knowledge: Eurasian Nodes of Convergence and Transformation*. Pittsburgh: University of Pittsburgh Press, 47–62.

V.12
ISLAMICATE ASTRAL SCIENCES IN EASTERN EURASIA DURING THE MONGOL-YUAN DYNASTY (1271–1368)[1]

Yoichi Isahaya

V.12.1 Introduction

While this chapter focuses on the "Islamicate astral sciences" as practiced in eastern Eurasia[2] during the Mongol Yuan dynasty (r. 1271–1368), its timeframe is not confined to the dynastic period but begins with the rise of the Mongol empire, when the influence of Islamicate astral sciences first became notable in eastern Eurasia, and ends with the Ming dynasty (r. 1368–1644), when Jesuit missionaries began to introduce "European astronomy" to this region (Shi 2014, 42–3). According to extant sources, the earliest record of a Muslim scholar at the Chinese imperial court dates to 961, when a certain Ma Yize from Lumu/Rūm came to the court of the Song dynasty (r. 960–1279) and took part in an astronomical reform (Shi 2014, 44). While there are various ways to approach Islamicate astral sciences in eastern Eurasia, I focus here on the absence of translations, because it permits me to reconsider aspects of knowledge transmission, an important issue for historians of science, medicine, philosophy, technology and other forms of knowledge. The case I consider here, that of the Mongol Yuan dynasty, shows that a high rate of mobility of people and knowledge does not necessarily lead to the practice of translation. During the Yuan era, the exchange of scientific knowledge reached a peak, due to the political unification of large parts of Eurasia by the Mongols, yet no contemporary translations from Arabic or Persian into Chinese are known in the field of astral sciences. A couple of translation projects were launched at the beginning of the subsequent Ming period, using texts in cultural repositories left behind by the Yuan, but in a different context.

Throughout the history of eastern Eurasia, there were three major periods during which knowledge from foreign cultures became important. Of these, only the first and third have been extensively studied, relating, respectively, to their impact: "the introduction of Buddhism" (Mak 2014, 2015; Kotyk 2018) and "the arrival of Western science" (Jami 2012, 2015; Chu 2017). In contrast, despite several important works in recent decades (Weil 2018, 2022), the second impact, relating to "the transmission of Islamicate culture," has not attracted the scholarly attention it deserves. In the first and third periods, translation practices played a significant role in the process of integrating foreign sciences into the Chinese intellectual milieu. In the second case, however, translation appeared quite late, in the early Ming period. Thus, the

DOI: 10.4324/9781315170718-61

Mongol Yuan period that preceded it is distinctive in that the introduction of Islamicate astral sciences does not appear to have stimulated the practice of translation. The early phase of each impact was facilitated by cultural transmitters of scientific knowledge, namely Buddhist and Jesuit missionaries in the first and third cases, while during the Mongol period, these cultural tansmitters were described exclusively as "immigrants" (*semuren*). In Yuan Chinese nomenclature, this term referred broadly to a group of sedentary peoples that included Uighurs (mostly Buddhists), Turkestani Muslims and Middle Easterners (Muslims and occasionally Christians), as well as central Eurasian steppe peoples such as Tanguts and Önggüd (Atwood 2004, 494). These groups could be considered an "intermediate class," standing between the Mongol ruling elite and the Chinese population under Yuan rule (Atwood 2016, 304). I will discuss the communication between these immigrants and Chinese intellectuals in the sections about the Yuan dynasty.

The Mongols valued the astral sciences highly. Because they believed that their success on Earth depended on the power of Heaven (*tenggeri*), they needed experts to interpret celestial portents and rites to translate them into guidance for their earthly activities (Baumann 2013, 269–70). Under these circumstances, it was advantageous for members of conquered peoples to display or practice their astral sciences in order to promote themselves in this new political situation. The transmission of Islamicate astral sciences into eastern Eurasia was, by and large, grounded in these new sociopolitical circumstances (Yang 2019). However, this newly imported astral knowledge was not integrated into the Chinese tradition through the practice of translation, as in the earlier and later periods of impact. Extant sources document only a few instances of direct cross-cultural collaboration among astronomers/astrologers[3] with different cultural backgrounds, as in the case of the "astronomical dialogue" between Naṣīr al-Dīn al-Ṭūsī (597–672/1201–1274) and Fu Mengzhi (Isahaya 2020). In what follows I will detail the practices of the cultural transmitters and their potential recipients. First, however, I will provide a general picture of the Chinese astral tradition, especially in comparison with Islamicate astral literature.

V.12.2 The Chinese astral tradition

At the center of the Chinese astral tradition lay *li*, which, while conventionally translated as "calendar," also refers to what Sivin describes as the "art of computing the times or locations of certain future or past phenomena in the sky" (Sivin 2009, 38). In addition to describing the motions of all celestial luminaries, *li* dealt with other important phenomena such as solar and lunar eclipses, solstices, equinoxes, the exact timing of noon and midnight and planetary conjunctions. Following Nathan Sivin's usage, I call this set of components an "astronomical system" (Sivin 2009, 38–9). Several scholars specializing in the Chinese astral tradition have pointed out shared features between the Islamicate astronomical handbooks called *zīj* – a representative genre of astronomical literature in the Islamicate world – and the *li*, the Chinese astronomical system (Sivin 2009, 38; Martzloff 2009, 371–2; Chapter I.6).

This system was subject to ongoing reform throughout Chinese dynastic history, from 104 bce, when the first official astronomical system was compiled by officials of the Han dynasty (206 bce–220), to 1644, when the official *Great Concordance System* (*Datong li*) of the Ming dynasty ceased to be implemented. Despite these repeated reforms, however, the Chinese dynasties regarded astral and calendrical sciences as a part of statecraft and adhered to tradition, being loath to abandon ancient elements. This combination of ongoing reform and a tangible sense of continuity – while absorbing certain foreign influences – formed the Chinese astral tradition (Martzloff 2009, 41–53).

V.12.3 Cultural transmitters and potential recipients

For the Mongol Yuan period, as far as extant sources attest, the initial contact of East Eurasian astronomers/astrologers with the Islamicate astral sciences dates from the period of Chinggis Khan's (r. 1206–1227) western expedition, from 1219 to 1225. At that time, Yelü Chucai (1190–1243), a Sinicized Kitan and a close adviser of the khan (de Rachewiltz 1993; Biran 2012), learned that planetary motions were more exactly calculated by the "astronomical system of the western regions" (*xiyu li*) than by its Chinese counterpart. Based on this understanding, he is said to have compiled an astronomical system called the *Madaba li*. The original word and its meaning are unknown, although it was probably based on the Islamicate astral tradition, since it was regarded as "a western astronomical system (*huihu li*)," although no vestige remains to enable us to trace the contents. It was obviously put into practice to a very limited extent, if at all. Subsequently, Yelü Chucai attempted to correct the accumulated errors of the *[Revised] Great Enlightenment System* ([*Chongxiu*] *daming li*), which had been adopted as the official astronomical system in the late period of the Jin dynasty (r. 1115–1234) and was also in use in the early Mongol era. This resulted in his compilation of the *Astronomical System of the Western Expedition in the Year of Gengwu* (*Xizheng gengwu yuan li*) (*Guochao wenlei* 1965, 57: 22; Allsen 2001, 165–6). Based on the description in the *Yuan Dynastic History* (*Yuanshi*), however, Yelü Chucai's system is almost identical to the *Revised Great Enlightenment System*, except for a minor correction based on the difference in geographical longitude (*Yuanshi* 1976, 56: 1256–344; Yabuuti 1997, 12). Thus, extant sources tell us that Yelü Chucai's astral practice was fully anchored in the Chinese astral tradition.

Intermittent contact of this sort was followed by more regular exchange, first initiated by ʿĪsā, the interpreter (*kelemechi*; [*c*. 1227–1308]). ʿĪsā was the most famous Syriac-rite Christian active at the Mongol court, with great expertise in the astral and medical sciences (Kim 2020). ʿĪsā, who had come to Mongolia in the reign of Güyük (r. 1246–1248), was, to the best of our knowledge, the first immigrant astronomer/astrologer recruited from western Eurasia by the court of the Great Khan. During the period of Möngke's reign (r. 1251–1259), ʿĪsā formed a close relationship with Qubilai (r. 1260–1294). In 1263, at ʿĪsā's suggestion, Qubilai established the Office of Western Astral Sciences (*Xiyu xinglisi*) and installed ʿĪsā as its head (*Yuanshi* 1976, 8: 147 and 134: 3249; Allsen 2001, 166). But due to a paucity of sources about the office, we are unable to trace any close interaction between ʿĪsā and Chinese literati regarding the astral sciences.

The next immigrant worthy of attention is Jamāl al-Dīn (Zhamaluding; [d. *c*. 688/1289]), perhaps the most successful and best-documented immigrant astronomer/astrologer active at the Yuan court – despite probably having no command of Chinese (*Mishujian zhi* 1992, 1: 28; Yang 2017, 1232). Based on a combination of Persian and Chinese sources, it seems that Jamāl al-Dīn was already working on behalf of Möngke to establish an observatory, possibly in Karakorum (*Yuanshi* 1976, 90: 2297; *Jāmiʿ al-Tawārīkh* 1994, 2: 1024). This refutes the well-known hypothesis of Willy Hartner, who had argued that in 1267 Jamāl al-Dīn might have come from the Maragha observatory – one of the foremost intellectual centers of that period (Sayılı 1960[CB],[4] 187–223; Chapter II.6) –as part of a delegation from the Ikhanid dynasty (r. 654–736/1256–1336), bringing with him seven "western astronomical instruments" (*xiyu yixiang*; *Yuanshi* 1976, 48: 998–9; Hartner 1950, 192–3). Among these seven instruments, an armillary sphere was likely to have been designed in Alamut – the main stronghold of the Nizārī Ismāʿīlīs – given that the instrument was set for 36°, which is very close to Alamut's latitude of 36°21'.

After Alamut's fall to the Mongols in 654/1256, that is, before the construction of the Maragha observatory, the armillary sphere was brought to the Yuan court, together with some books, of which more below (Isahaya 2021). Jamāl al-Dīn was appointed as the director of the Astronomical Bureau of the Westerners (*Huihui sitiantai*) in 1271 (*Yuanshi* 1976, 7: 136). In 1273, the observatory was united with the Astronomical Bureau of the Han Chinese (*Haner sitiantai*), which was already functioning in 1260 (*Yuanshi* 1976, 90: 2297). Jamāl al-Dīn then took up a post at the Imperial Library Directorate (*Mishujian*), which controlled the united astronomical bureau, and even headed the agency for a time (*Mishujian zhi* 1992, 7: 115, 126; Allsen 2001, 167–8). Shortly after the unification of these two astronomical institutions, Liu Bingzhong – who was in Qubilai's entourage and initiated the Yuan astronomical reform – commanded that the two bureaus should report their official matters independently (*Mishujian zhi* 1992, 7: 126; Yang 2017, 1235–6). Indeed, no mutual interaction between western and Han Chinese astronomers/astrologers is evident at the united bureau. Rather, in 1288, the Han Chinese astronomers/astrologers from the united astronomical bureau waged a campaign to remove the westerners' authority over Han Chinese experts in *yinyang*, which in this context meant various practices of astral divination. The campaign was led by Yue Xuan (d. 1312), a former protégé of Liu Bingzhong (1216–1274) and a member of Qubilai's imperial guard. The Chinese astronomers/astrologers claimed that the westerners (*huihui*) who supervised them at the Palace Library did not know *yinyang*. Furthermore, they were concerned that the library's *yinyang* books would be lost (*Mishujian zhi* 1992, 1: 23–4). This campaign led to the separation of the Astronomical Bureau of the Han Chinese from the Palace Library – and thus also from the Astronomical Bureau of the Westerners – in the same year (*Mishujian zhi* 1992, 7: 128–9; Yang 2017, 1238). Thus the relationship of the two astronomical bureaus for the westerners and Han Chinese was characterized more by competition than by cooperation.

In 1273 the Astronomical Bureau of the Westerners sent to the Imperial Library Directorate a list of books at its disposal. This included Euclid's (3rd century BCE) *Elements* (*Stoicheia*), Ptolemy's (*c.* 100–170) *Almagest*, al-Ṣūfī's (290–376/903–986) *Book on the Star Constellations* (*Kitāb Ṣuwar al-kawākib al-thābita*) and Kūshyār ibn Labbān's (d. before 439/1047) *Book of Introduction to Astrology* (*Kitāb al-Madkhal fī ṣināʿat aḥkām al-nujūm*), as well as an astronomical handbook (*zīj*) and a work on theoretical astronomy (*hay'a*; *Mishujian zhi* 1992, 7: 129–31; van Dalen 2002, 340; for the list and the identification of each work, see Tasaka 1957, 99–119; van Dalen 2004, 25; Chapters I.6 and II.6). However, it is very likely that immigrant astronomers/astrologers were the only ones to make use of these texts, since no use of such books in translation was reported by Han Chinese astronomers. Furthermore, the books on the list show remarkable overlap with the referenced sources of *The Manual of Astrologers* (*Dastūr al-munajjimīn*), an Arabic astral treatise compiled by the Nizārī Ismāʿīlīs in Alamut. This suggests that the books might also have come from the Alamut library with the aforementioned armillary sphere. On the other hand, there is no evidence to prove that the works of the Maragha observatory were transmitted into the Yuan intellectual arena. The large-scale observatories in Maragha and Dadu may have had a less close relationship than one would expect (Isahaya 2021).

It has often been claimed that *The Great Astronomical Handbook for Caliph Ḥākim* (*al-Zīj al-kabīr al-ḥākimī*; *c.* 395/1005) by Ibn Yūnus (d. 399/1009) was delivered via Jamāl al-Dīn into the hands of Guo Shoujing (1231–1316), a central figure of the Yuan astronomical reform, who supposedly scrutinized the book's contents. This conclusion , based on the studies of Louis Amélie Sédillot (1808–1875), a 19th-century French Orientalist, has been challenged by more recent research, which suggests that Sédillot's conclusion depended on errors and misinterpretations of the translation of Chinese chronicles (van Dalen 2002, 341). There is no clear evidence for Guo's possession of Ibn Yūnus's tables. We have to wait until the Ming dynasty for Kūshyār's *Introduction to Astrology* to be translated into Chinese (see below). Jamāl al-Dīn, most likely together with his colleagues, prepared a new astronomical handbook for his community and the Mongol rulers.

To that end, a series of observations were undertaken, perhaps with self-made instruments, and some values were newly computed. His results were used by Chinese literati in a translated form as the *Astronomical System of the Westerners* (*Huihui li*) but not before 1383 – that is, not earlier than the beginning of the Ming dynasty (van Dalen 1999; Yano 1999).

Finally, we need to ask whether concepts, methods or tools used in Islamicate astral sciences played any role in the official astronomical reform of the Yuan dynasty. This reform resulted in the compilation of the *Season-Granting System* (*Shoushi li*), which became known as the masterpiece of traditional Chinese astronomical systems (Sivin 2009). Guo Shoujing was one of the main contributors to this reform. This is especially true for its practical phase, as he conducted systematic observations of the celestial phenomena and designed a large number of astronomical instruments in connection with it (Sivin 2009, 561–72). Guo Shoujing and his colleagues may also have been influenced by astronomical works and instruments brought from the Islamicate world, as well as by immigrant astronomers/astrologers at the one or the other Astronomical Bureau.

However, there is no evidence that Chinese astronomers took actual algorithms or tables from Islamicate *zījes* when compiling the *Season-Granting System* (van Dalen 2002, 344–5). Furthermore, despite their similar appearance, Jamāl al-Dīn's "western astronomical instruments" were in fact quite different in their usage from Guo Shoujing's instruments (Miyajima 1982; van Dalen 2002, 342).

V.12.4 The Yuan–Ming transition

Studies of the second impact of foreign knowledge on East Eurasian cultures have revealed the important role played by cultural transmitters, such as ʿĪsā, the interpreter, Jamāl al-Dīn, and other immigrants who all seem to have been active at the Astronomical Bureau of the Westerners. Their main contribution was to show off their knowledge at court and to practice their astral knowledge for the Mongol rulers as readers of divine portents. In other words, the attention of the immigrant astronomers/astrologers was not focused on the Chinese literati – their potential colleagues – but on the Mongol ruling elite. In addition, their astronomical practice – their use of instruments, their observations and their copying or compilation of astronomical handbooks – was also aimed at their own communities, for whom, for example, they compiled Muslim calendars (*Yuanshi* 1976, 94: 2404). A similar attitude is also found among Chinese literati who worked for the Mongols. As a result, it is not surprising that immigrant astronomers/astrologers seem to have exercised little influence on the Yuan astronomical reform. After all, the reform was a political project within the Chinese astral tradition: it was a Chinese astronomical reform, not a Mongol imperial one. In such a context, immigrants had little motivation to participate. In fact, the *Season-Granting Astronomical System* – the product of the reform – was implemented only within the East Asian ecumene dominated by Chinese script, despite its Mongolian translation (Ho 2006).

Outside the ecumene, the so-called Mongol imperial calendar was used instead. That calendar, while having the same calendrical structure as the Chinese one, albeit in simplified form, was represented in texts by the Turco-Mongolian twelve-year animal cycle and the Turkic months. Sometimes references were made to the seasons, such as "the beginning of autumn" (Baumann 2008, 94–7). No calendar of this sort – historically used in the Steppe nomadic empires – was ever adopted in the official documents of the Chinese dynasties (Bazin 1991). In the Yuan era, however, that calendar appeared even in Chinese writings, for example, in the form of Sino-Mongolian bilingual inscriptions. More remarkably, the imperial calendar even appear in some Persian chronicles produced during the Ilkhanid dynasty (Melville 1994; van Dalen *et al.* 1997; Isahaya 2013).

Things changed in the Ming era, after the Mongols were expelled to the northern steppes and the influx of immigrants became a thing of the past. Immediately after the foundation of the new dynasty, the first Ming emperor, Zhu Yuanzhang (r. 1368–1398), summoned Chinese and western astronomers/astrologers to the new capital, Nanjing. The Arabic and Persian books confiscated in the former Yuan capital were transferred southward to Nanjing as well. The emperor founded astronomical bureaus for both Chinese and westerners, replicating the Yuan bureaucratic structure. In fact, during the first decade of the Ming era there was little change in the practice of Islamicate astral sciences, since these were practiced at a specialized institute, mainly by Muslims and separately from their Chinese counterpart. The first emperor's last two decades, however, were a time of policy change, in which Islamicate astral knowledge began to be assimilated into the Chinese astral tradition (Weil 2018, 269–70). In 1382, the emperor embarked on a project of translating works on the Islamicate astral sciences (Lotze 2017), beginning, no later than April 1383, with a book on horoscopic astrology. The *Book on Celestial Patterns* (*Tianwen shu*) was a translation of Kūshyār's *Introduction to Astrology* (for the original and its Chinese translation, see Yano 1997). It is highly likely that the aforementioned *Astronomical System of the Westerners* was translated by the same office (Shi 2014, 50–1).

Although the translation project ended abruptly in the mid–1380s due to an inner-dynastic power struggle, and the astronomical bureau of the Muslims was abolished, Islamicate astral sciences continued to be practiced at the Department of the Muslim Astronomical System (*Huihuili ke*), founded in 1398 as a subdivision of the Chinese astronomical bureau. In this period, the focus shifted from translating to embedding Islamicate astral knowledge into the Chinese astral tradition. This phase was marked by a series of treatises compiled in Chinese by Muslims and Chinese literati working in collaboration (Weil 2018, 271–3). In the process, the *Astronomical System of the Westerners* was edited and re-edited under the scrutiny of Ming officials (Li 2016). From the 16th century onward, as the restrictions on astronomical activities the court were relaxed, this interest in assimilating aspects of Islamicate astral science extended to grassroots scholars as well as to foreigners, as in Korea, where Chinese works on Islamicate astral sciences were presented to the court. This flow continued until the beginning of the 17th century, when contemporary European science arrived in eastern Eurasia together with Jesuit missionaries (Weil 2018, 273–4).

The Mongol rulers and immigrants (*semuren*, see above) at the court had left China after the transition of dynasties. Under these circumstances, the cultural repository left by the Yuan required a new approach in order to continue being used in eastern Eurasia, and the practice of translation became indispensable for Chinese literati As a result, the second impact of foreign knowledge on eastern Eurasia now grew beyond its previously limited cultural space at the court of the Mongol cosmopolitan empire to the broader intellectual sphere of the "Chinese" Ming dynasty.

Notes

1 I wish to express my profound thanks to Qiao Yang (Max Planck Institute for the History of Science) for her thoughtful and comprehensive cooperation on this chapter. This chapter is based on a paper I read at the European Society for the History of Science Biennial Conference 2018 (UCL Institute of Education) on 15 September 2018. I benefited from the discussion at the session "Science in Translation: Looking at It from East Asia," organized by Huiyi Wu (University of Cambridge) and Mary Brazelton (University of Cambridge). Many valuable suggestions by Michal Biran (The Hebrew University of Jerusalem) proved indispensable for completing this chapter. I deeply appreciate the considerable assistance of Leigh Chipman (The Hebrew University of Jerusalem) in revising and formatting this chapter.

2 In this chapter *eastern Eurasia* is defined as the regions historically influenced by Chinese dynasties and those governed by their counter powers, for example, nomadic empires (cf. Skaff 2012, 6–8), reflecting

the vital interaction between nomadic and sedentary peoples that shaped the history of the region. Therefore, "eastern Eurasia" in our sense is not confined to the area of the so-called East Asian ecumene dominated by Chinese script (Atwood 2016, 280), the center of which was China and which historically extended to Korea, Vietnam and Japan.

3 Although "astrology" (considering celestial influences upon the terrestrial realm) is regarded as a pseudoscience in the modern world, that knowledge in fact was inseparably linked to "astronomy" (dealing with celestial phenomena) in premodern intellectual traditions. Both developed in a close relationship as "astral sciences" throughout premodern history. For the historical relationship between astronomy and astrology, see, for instance, North (2013).

4 Consolidated bibliography.

Bibliography

Sources

Guochao wenlei [Classified Literary Works from the Present Dynasty]. 1965. in *Sibucongkan Chubian*, 70 vols. Taipei: Taiwan shangwu yinshuguan.

Rashīd al-Dīn al-Hamadānī. ed. Rawshan M. and Mūsawī, M. 1994. *Jāmiʿ al-Tawārīkh* [Collected Histories]. 4 vols. Tehran: Nashr-i Alburz.

Song Lian *et al.*, eds. 1976. *Yuanshi* [Yuan Dynastic History]. 210 vols. Beijing: Zhonghua shuju.

Wang Shidian and Shang Qiweng. ed. Gao R. 1992. *Mishujian zhi* [Accounts of Imperial Library Directorate]. 11 vols. Hangzhou: Zhejiang guji chubanshe.

Research literature

Allsen, T. 2001. *Culture and Conquest in Mongol Eurasia.* New York: Cambridge University Press.

Atwood, C. 2004. *Encyclopedia of Mongolian and the Mongol Empire.* New York: Facts on File.

Atwood, C. 2016. "Buddhists as Natives: Changing Positions in the Religious Ecology of the Mongol Yuan Dynasty," in Jülch, T., ed. *The Middle Kingdom and the Dharma Wheel: Aspects of the Relationship between the Buddhist Saṃgha and the State in Chinese History*. Leiden: Brill, 278–321.

Baumann, B. 2008. *Divine Knowledge: Buddhist Mathematics According to the Anonymous Manual of Mongolian Astrology and Divination.* Leiden: Brill.

Baumann, B. 2013. "By the Power of Eternal Heaven: The Meaning of Tenggeri to the Government of the Pre-Buddhist Mongols," *Extrême-Orient Extrême-Occident* 35: 233–84.

Bazin, L. 1991. *Les systèmes chronologiques dans le monde turc ancien.* Budapest: Akadémiai Kiadó.

Biran, M. 2012. "Kitan Migrations in Eurasia: (10th–14th Centuries)," *Journal of Central Eurasian Studies* 3: 85–108.

Chu, L. 2017. "From the Jesuits' Treatises to the Imperial Compendium: The Appropriation of the Tychonic System in Seventeenth and Eighteenth-Century China," *Revue d'histoire des sciences* 70: 15–46.

Dalen, B. van. 1999. "Tables of Planetary Latitude in the *Huihui li* (I1)," in Kim, Y. and Bray, F., eds. *Current Perspectives in the History of Science in East Asia.* Seoul: Seoul National University, 316–29.

Dalen, B. van. 2002. "Islamic and Chinese Astronomy under the Mongols: A Little-Known Case of Transmission," in Dold-Samplonius *et al.*, eds. [CB], 327–56.

Dalen, B. van. 2004. "The Activities of Iranian Astronomers in Mongol China," in Pourjavady, N. and Vesel, Ž., eds. *Science, techniques et instruments dans le monde iranien (Xe-XIXe Siecle): Actes du colloque tenu a l'Universite de Teheran (7–9 juin 1998).* Tehran: Presses universitaires d'Iran and IFRI, 17–28.

Dalen, B. van, Kennedy, E. and Saiyid, M. 1997. "The Chinese-Uighur Calendar in Ṭūsī's Zīj- i Īlkhānī," *ZGAIW* 11: 111–52.

Hartner, W. 1950. "The Astronomical Instruments of Cha-ma-lu-ting, Their Identification, and Their Relations to the Instruments of the Observatory of Marāgha," *Isis* 41.2: 184–94.

Ho, K. 2006. "The Political Power and the Mongolian Translation of the Chinese Calendar during the Yuan Dynasty," *Central Asiatic Journal* 50.1: 57–69.

Isahaya, Y. 2013. "The *Tārīkh-i Qitā* in the *Zīj-i Īlkhānī*: The Chinese Calendar in Persian," *SCIAMVS* 14: 149–258.

Isahaya, Y. 2020. "Fu Mengzhi: 'The Sage of Cathay' in Mongol Iran and Astral Sciences along the Silk Roads," in Biran, M., *et al.*, eds. *Along the Mongol Silk Roads: Merchants, Generals, Religious Experts.* Berkeley: University of California Press, 238–54.

Isahaya, Y. 2021. "From Alamut to Dadu: Jamāl al-Dīn's Armillary Sphere on the Mongol Silk Roads," *Acta Orientalia* 74.1: 65–78.

Jami, C. 2012. *The Emperor's New Mathematics: Western Learning and Imperial Authority in China during the Kangxi Reign (1662–1722).* Oxford: Oxford University Press.

Jami, C. 2015. "Revisiting the Calendar Case (1664–1669): Science, Religion, and Politics in Early Qing Beijing," *Korean Journal of History of Science* 27.2: 459–77.

Kim, H. and Cho, W. tr. and ed. 2020. "Isa Kelemechi: A Translator, Official, and Envoy between Europe and Asia," in Biran, M., *et al.*, eds. *Along the Mongol Silk Roads: Merchants, Generals, Religious Experts.* Berkeley: University of California Press, 255–69.

Kotyk, J. 2018. *The Sinicization of Indo-Iranian Astrology in Medieval China.* Philadelphia: University of Pennsylvania.

Li, L. 2016. "Arabic Astronomical Tables in China: Tabular Layout and its Implications for the Transmission and Use of the *Huihui lifa*," *East Asian Science, Technology and Medicine* 44: 21–68.

Lotze, J. 2017. *Translation of Empire: Mongol Legacy, Language Policy, and the Early Ming World Order, 1368–1453.* PhD thesis. Manchester: University of Manchester.

Mak, B. 2014. "*Yusi Jing*: A Treatise of 'Western' Astral Science in Chinese and Its Versified Version *Xitian Yusi Jing*," *SCIAMVS* 15: 105–69.

Mak, B. 2015. "The Transmission of Buddhist Astral Science from India to East Asia: The Central Asian Connection," *Historia Scientiarum* 24.2: 59–75.

Martzloff, J. 2009. *Le calendrier chinois: structure et calcus (104 av. J.-C. – 1644).* Paris: Honoré Champion.

Melville, C. 1994. "The Chinese-Uighur Animal Calendar in Persian Historiography of the Mongol Period," *Iran* 32: 83–98.

Miyajima Kazuhiko (宮島一彦). 1982. "*Genshi* kisai no Isuramu temmonkiki ni tsuite 『元史』天文志記載のイスラム天文儀器について [New Identification of Islamic Astronomical Instruments Described in the *Yuan Dynastic History*]. In *Toyo no kagaku to gijutsu* 東洋の科学と技術 [Science and Skills in Asia. Festschrift for the 77th Birthday of Professor Yabuuti Kiyosi]. Kyoto: Dôhôsha, 407–27.

North, J. 2013. "Astronomy and Astrology," in Lindberg, D. and Shank, M., eds. *The Cambridge History of Science.* Vol. 2: *Medieval Science.* Cambridge: Cambridge University Press, 456–84.

Rachewiltz, I. de 1993. "Yeh-lü Ch'u-ts'ai, Yeh-lü Chu, Yeh-lü Hsi-liang," in de Rachewiltz, I., *et al.*, eds. *In the Service of the Khan: Eminent Personalities of the Early Mongol-Yüan Period.* Wiesbaden: Harrassowitz, 136–75.

~~Sayılı, A. 1960. *The Observatory in Islam and Its Place in the General History of the Observatory.* Ankara: Türk Tarih Kurumu Basımevi.~~

Shi, Y. 2014. "Islamic Astronomy in the Service of Yuan and Ming Monarchs," *Suhayl* 13: 41–61.

Sivin, N. 2009. *Granting the Seasons: the Chinese Astronomical Reform of 1280, with a Study of Its Many Dimensions and an Annotated Translation of Its Record.* New York: Springer.

Skaff, J. 2012. *Sui-Tang China and Its Turko-Mongol Neighbors: Culture, Power, and Connections, 580–800.* New York: Oxford University Press.

Tasaka, K. 1957. "An Aspect of Islam Culture Introduced into China," *Memoirs of the Research Department of Toyo Bunko* 16: 75–160.

Weil, D. 2018. "The Fourteenth-Century Transformation in China's Reception of Arabo- Persian Astronomy," in Manning, P. and Owen, A., eds. *Knowledge in Translation: Global Patterns of Scientific Exchange, 1000–1800 CE.* Pittsburgh: University of Pittsburgh Press, 262–74.

Weil, D. 2022. "Chinese-Muslims as Agents of Astral Knowledge in Late Imperial China," in Mak, B and Huntington, E., eds. *Overlapping Cosmologies in Asia: Transcultural and Interdisciplinary Approaches.* Leiden: Brill, 116–38.

Yabuuti, K. and van Dalen, B., tr. and rev. 1997. "Islamic Astronomy in China during the Yuan and Ming Dynasties," *Historia Scientiarum* 7.1: 11–43.

Yang, Q. 2017. "From the West to the East, from the Sky to the Earth: A Biography of Jamāl al-Dīn," *Asiatische Studien – Études Asiatiques* 71.4: 1231–45.

Yang, Q. 2019. "Like Stars in the Sky: Networks of Astronomers in Mongol Eurasia," *Journal of the Economic and Social History of the Orient* 62.2–3: 388–427.

Yano, M. 1997. *Kūshyār ibn Labbān's Kitāb al-Madkhal fī Ṣinā'at Aḥkām al-Nujūm (Introduction to Astrology).* Tokyo: Tokyo University of Foreign Studies.

Yano, M. 1999. "Tables of Planetary Latitude in the *Huihui li* (I)," in Kim, Y. and Bray, F. eds., *Current Perspectives in the History of Science in East Asia.* Seoul: Seoul National University, 307–15.

V.13
COLLATION AND ARTICULATION OF ARABO-PERSIAN SCIENTIFIC TEXTS IN EARLY MODERN CHINA

Dror Weil

This chapter sheds light on the textual methods that Chinese scholars applied to engage with scholarship from the Islamicate world and the transformation of these methods following changes in the Chinese sociocultural landscape between the 13th and 18th centuries. The movement of Islamicate scientific concepts, theories and data along the various routes of exchange that traversed premodern Asia into China intertwined with the development of methods of articulation, presentation, corroboration and translation. Motives ranged between the importance of preserving the original cultural setting of Islamicate scientific knowledge and a desire to naturalize and embed foreign scientific knowledge in local Chinese discourses. Local Chinese scholars used textual methods to propagate scientific ideas, engage with foreign knowledge and forge communities and identities around shared practices.

Evidence for the movement of scientific knowledge from the Islamicate world to China goes back to as early as the 10th century. An early account in Ibn al-Nadīm's (d. 380/990) *The Catalog* (*Kitāb al-Fihrist*) provides an example for such transfer of medical knowledge from the Islamicate world to China. The account reports on an unnamed Chinese disciple of the Persian polymath and physician Muḥammad ibn Zakariyyāʾ al-Rāzī ([d. 313/925 or 323/935]; known in western Europe also as Rhazes) who wished to produce a manuscript copy of *The Sixteen Books of Galen* before his return to China. Ibn al-Nadīm recounts in some detail the process of copying the lengthy manuscript and brings to light the laborious dimension of knowledge-making in a manuscript culture and the importance of teamwork in facilitating such transfers of knowledge. Similarly, later accounts report on the participation of Muslim astral experts in the calendar reform of the Song 宋 dynasty (960–1127) and their role in introducing Islamicate astronomy to the Chinese court.[1]

Scholarly communities of Arabic and Persian speakers in late medieval and early modern China[2] were instrumental in facilitating such transfers of scientific knowledge from the Islamicate world into China. Muslim, Jewish and Christian migrants from western and Central Asia, who relocated to China, established communities that maintained direct links with the various branches of Islamicate scholarship and its textual traditions in religious schools, worship halls and libraries they founded. Some rendered service to the Chinese court as translators, astral experts technical experts, physicians and pharmacists, playing a decisive role in introducing Islamicate

DOI: 10.4324/9781315170718-62

knowledge to the Chinese ruling and literati elites and maintaining the links with those foreign lands. Even Chinese students and travelers who spent time in western and central Asia and returned to China equipped with scientific knowledge and texts constituted an important path by which Islamicate scientific knowledge was brought to the attention of local Chinese audiences.

The Mongol conquests in the 13th century inflicted on the various parts of Asia a shared experience of devastation and destruction of cultural and intellectual centers, followed by extensive movement of people, ideas, objects and texts across the empire and the creation of new multicultural and polyglot societies. Among the effects of this sociocultural transformation was a new vision of universal science and its utility to the imperial cause. This vision prompted Mongol courts to recruit foreign experts. The court of the Yuan dynasty (r. 1279–1368), the Chinese branch of the Mongol Empire, employed western and Central Asian astral experts, physicians, pharmacists and technical experts, and established imperial institutions to accommodate and put into practice these branches of foreign technical and scientific knowledge.

Collecting Arabic and Persian manuscripts in designated repositories and methods of textual analysis were an integral part of science-making in the context of the Yuan but also persisted long after the fall of the dynasty. The official recognition by the Yuan and the Ming (r. 1368–1644) courts of the utility of Islamicate sciences produced reliance on foreign experts and texts and required the dynasties to react to the changing availability of such resources. The preservation of reading skills in Arabic and Persian among Chinese Muslims, their repositories of Arabic and Persian texts and their expertise in reading, writing and interpreting texts in these languages fostered collaboration in different degrees with the Yuan, Ming and Qing (r. 1644–1911) courts in fields such as the astral sciences and medicine. (Weil 2020a; Weil 2020b)

This chapter seeks to explore the ways by which Islamicate scientific knowledge was accommodated and studied in China during the 13th to 18th centuries.[3] During that period, experts in the exact sciences in China's imperial institutions and scholars commited to the propogation of Islamic theology and philosophy of nature in local communities both employed a set of scholarly practices to investigate and transmit scientific knowledge. The changing sociocultural landscape during that long period entailed certain reconfigurations of the aims of Islamicate scientific scholarship. In particular, it replaced an earlier analytical prism of investigating science through the scrutiny and collation of Arabic and Persian texts with a heuristic approach that aimed to introduce Islamicate scientific knowledge widely to local Chinese communities of readers. These differing aims prompted Chinese scholars to change their methods of approaching and exploring Islamicate scientific knowledge.

Hence, this essay divides the discussion according to the two textual forms by which Islamicate knowledge was proliferated in China – the study of Arabic and Persian manuscripts and the production of Chinese printed books. The former mirror an attempt to preserve the cultural context of Islamicate scientific knowledge by studying texts in their original languages and manuscript forms. The latter demonstrate a commitment to integrate foreign knowledge in local discourses through translation and the use of printing technologies for wider proliferation. By applying these scholarly methods and textual forms to the study of Islamicate knowledge, Chinese scholars constructed an important bridge between the scientific discourses of the Islamicate world and those of China.

V.13.1 Working with Arabic and Persian manuscripts

China's century of Mongol rule between the mid-13th and mid-14th centuries intensified its exposure to Islamicate texts and scholarly practices. Acknowledging the value of the Islamicate expertise in fields such as the astral sciences, medicine, pharmaceuticals and mechanical engineering, the Yuan dynasty recruited to its court an unprecedented number of experts, who brought with them their libraries of Arabic and Persian manuscripts, their instruments and techniques. The accommodation of Islamicate sciences at the Yuan court was characterized by the institutionalization of specific sciences, methods of storing and organizing scientific knowledge and the elaborate use of manuscripts.

The Yuan court established new institutions, such as the Directorate of Arabo-Persian Astronomy (*Huihui sitian jian* 回回司天監; Chapter V.12), the Arabo-Persian Pharmacy (*Huihui yaowu yuan* 回回藥物院) and the Arabo-Persian Medical Office (*Guanghui si* 廣惠司), which were designated to accommodate experts in selected Islamicate sciences. These institutions had libraries that housed collections of Arabic and Persian texts relevant to their work. An official document that survived in the chronicle *Records of the Palace Library* (*Mishu jianzhi* 秘書監志, known also as *Bishu jianzhi*) includes an inventory of 22 titles of Arabic and Persian texts that were once in the possession of an astronomical observatory library at the Yuan northern capital of Shangdu 上都. The document carries the date of year 1273, a few years prior to the conquest of southern China by the Mongols. The list includes works on geometry, mathematical astronomy, astronomical systems and star albums such as the *Book of Euclid*, the *Almagest*, the *Book of the Star Constellations* (*Ṣuwar al-kawākib al-thābita*) as well as works on geomancy, astrological prediction, physiology, history and even poetry (see Mishu jianzhi *juan* 7: 129; Weil 2018). Each entry on the inventory included a transliteration of the original Arabic or Persian titles in Chinese characters, a short description of the work's subject in Chinese and the total number of volumes. The list makes a distinction between works that were housed at the observatory and those that were stored at the director's residence. The former includes mainly those works that relate to mathematical astronomy, while the latter includes all other works. Another document in the same chronicle that is dated to 1275 reports that too many forbidden astrological[4] (*yinyang jinshu* 陰陽禁書) and Arabic and Persian texts (*huihui wenzi* 回回文字) were procured by the library and were placed in six new book cabinets, each including three lockable drawers (*Mishu jianzhi juan* 5:1011). While the document does not provide any information on the identity of these works, it demonstrates the ways by which Arabic and Persian scientific works were organized and stored in official repositories.

Two manuscripts from that period have survived to provide rare glimpses into the ways scientific knowledge was organized, articulated and used during the Yuan period. A manuscript copy of Sharaf al-Dīn al-Masʿūdī's (d. *c.* 600/1204) astronomical treatise, *Knowledge of the Universe* (*Jahāndānish*), is owned now by the Bibliothèque nationale de France. The manuscript, which is bound in a Chinese style, includes Chinese characters on the margins of each folio. The first two characters *tianzi* 天字 (lit. "the 'Tian' division") represent a method of data organization and the subsequent characters indicate folio number. This terse inscription suggests that the manuscript was once housed in a Chinese repository (Weil 2016). A colophon at the end of the manuscript dates the production of the copy to year 739/1338 by the hand of the scribe Aḥmad al-Jawharī from Herat. The manuscript is composed of undotted Persian written in black and red inks and accompanied by various diagrams of astral positions. The manuscript is furnished with a critical apparatus (see West 1973; Gacek 2009) in the form of interlinear markers and marginalia that indicate the textual variants in this copy from other versions. These markers suggest that multiple copies of the work were in the possession of the annotator.

An interlinear gloss in the form of Chinese words written in Arabic script appears on one of the sketches in the manuscript and sheds some light on the way this Persian astronomical treatise was read in Chinese (Weil 2016). This seems to be one of the earliest occurences of transliteration of Chinese words in Arabic script, sometimes called *xiao'er jing* 小兒經 (Feng 1982; Zavyalova 1999). The sketch displays nine positions of the moon and the sun, and includes Persian labels that indicate the duration of each phase. Glosses in *xiao'er jing* appear under four of these labels.[5] These glosses include "*chū yī*" حويي that stands for the Chinese *chuer* 初一 "the 1st day of the month; "*chū yī jih chū liyū*" حويي حه حوليو for *chuer zhi chuliu* 初一至初六 "from the 1st to the 6th day of the month"; "*chū liyū jih chū giyū*" حوليو حه حوكيو for *chuliu zhi chujiu* 初六至初九 "from the 6th to the 9th day of the month"; and "*chū giyū jih shī yī*" حو كيو حه شى يى that stands for *chujiu zhi shier* 初九至十一 "from the 9th to the 11th day of the month". Such a glossing apparatus suggests that some of the readers of the manuscript could read and write Persian but used Chinese as the main working language.

The second manuscript is an astronomical ephemeris (*zīj*) that is currently housed at the BnF in Paris. The manuscript is written in undotted Arabic in black and red inks. Chinese characters appear on the cover page and margins of each folio. Like the Chinese inscription on the previous manuscript, the first two characters here *zangzi* 藏字 (lit. "the 'zang' division") refer to a certain archival organization system and the other characters give the page number. The text is composed of lengthy explanations in Arabic in the form of questions and answers on astronomical measurements, followed by a series of ephemerides. A colophon at the beginning of the manuscript suggests that it was prepared by Abū Muḥammad ʿAṭā, son of Aḥmad, son of Khwāja Ghāzī al-Samarqandī al-Sanjufīnī in 1366 for the Mongol viceroy of Tibet (Franke 1988; van Dalen 2002). While the prose part of the text is only minimally annotated, the tables include extensive interlinear and marginal annotations. Interestingly, these annotations are written in Mongolian and Tibetan scripts. In some cases, the annotations include transliterations of Arabic terms in Mongolian and Tibetan scripts, in a similar manner to the *xiao'er jing* glosses in the previous case. These marks suggest that the text was part of a Chinese archive but was used by polyglot scholars.

As these manuscripts suggest, reading a scientific treatise required not focussing only on its contents but involved also philological concerns such as comparison of copies and glossing of selected words. Although these manuscripts were written in Arabic and Persian, oral and written translations into other languages or vernaculars was an integral part in the use of such scientific texts. Moreover, these markers and glosses bring to light the oral dimension of scientific work that accompanied the use of written texts. The various paratextual components that appear in these manuscripts mirror an additional aspect in the accommodation of Islamicate science in China – the systematized organization and storage of scientific texts in archives and libraries.

The exploration of the various aspects of the natural world was not limited to the scientific institutions or the Chinese court. The growth in China's Muslim population from the 13th century onward was accompanied by the emergence of local centers of Islamic scholarship. Chinese Muslims collected and scrutinized Arabic and Persian texts on themes such as Islamic theology, Islamic jurisprudence, Sufism, logic and grammar. The presence of topics such as cosmology, physiology, anatomy, astronomical calculation in Islamic literature captured the attention of some Chinese Muslims who meticulously sought for ways to further explore them (Weil 2022a). In a similar fashion to the Chinese court, they collected Arabic and Persian texts, set up libraries and applied various methods of collation, articulation and verification to bring out accurate knowledge from texts. In their study of aspects of the natural world in Arabic and Persian texts, these Chinese Muslims engaged with scientific exploration within their local communities, redrawing the boundaries of scientific scholarship.[6]

In the mid-16th century, Hu Dengzhou ([1522–1597]; 胡登州), a Chinese Muslim from the northern province of Shaanxi, had a vision to promote a systematic study of Arabic and Persian texts in China. The school he established in his hometown was expanded by his disciples into an empire-wide network of schools and scholars. These scholars collected Arabic and Persian texts and devised curricula for teaching them. A Chinese manuscript composed toward the end of the 17th century by an otherwise unknown Chinese author Zhao Can ([1662–1722]; 趙燦) and titled *The Genealogy of Classical Learning* (*Jingxuexi chuanpu* 經學系傳譜) charts the expansion of Hu Dengzhou's network and its various scholars and schools. It brings to light the importance that these scholars attached to manuscript collation and copying, as well as the titles of texts included in the curricula of the various schools. Despite difference in interest and focus between schools, many of these Chinese Muslim scholars were interested in the study of cosmology and the guiding principles of the natural world. They investigated theological texts, and in particular Sufi texts such as Najm al-Dīn Rāzī's (573–654/1177–1256) *The Path of God's Bondsmen* (*Mirṣad al-ʿibād*), ʿAzīz al-Dīn Nasafī's (7th/13th century) The *Furthest Goal* (*Maqṣad-i aqṣā*) and ʿAbd al-Raḥmān Jāmī's (817–898/1414–1492) *Rays of the Flashes* (*Ashiʿāt al-lamaʿāt*) and *Gleams* (*Lawāʾiḥ*) and tried to make sense of the theories and concepts that these texts introduced.

Among the various extant manuscripts produced by members of Hu Dengzhou's network is a copy of Najm al-Dīn Rāzī's *The Path of God's Bondsmen* that is in the possession of the BnF in Paris. This Sufi text includes various descriptions of the structure and operation of the natural world, including references to the positions and movements of the seven celestial spheres, embryogenesis and physiological theories such as the four humors. A Chinese inscription written in brush on the first folio gives the title of the work in Chinese *Daoxing tuiyuan jing* 道行推原經 (lit. "The Classic that Investigates the Origin of the Right Path"). The following page includes the Persian title *Mirṣad al-ʿibād*. The last folio includes an inscription written in pencil in *xiaoer jing* (Chinese written in Arabic script) that gives the full Persian and Chinese titles of the work. The work is written in black and red inks using a local Chinese round calligraphical style of Persian. The text is extensively annotated with interlinear and marginal comments.

A few types of interlinear annotations appear in this manuscript: (1) glosses of selected words in *xiaoer jing*, (2) glosses of selected words in Chinese, and (3) grammatical marks to explain the syntactical structure of the sentences, such as م for the subject of the sentence (for the Arabic term *mubtadaʾ*), خ for the predicate (for the Arabic *khabar*), ام and اه, respectively, for the possessor and possessed in a genitive construction. In other contemporary Chinese Islamic manuscripts we can find additional uses of the interlinear annotations such as morphological declension or irregular pluralization of Arabic words (sometimes marked with the letter ج), and terse explanations of the meaning of selected passages. The marginalia in this manuscript includes glosses in Chinese and *xiaoer jing* of selected words and passages. In other manuscripts, the marginal annotation is used to present intertextual references and quotes from selected commentaries, or to indicate variants across copies of the text. In a similar fashion to the reading of scientific texts, the various markers on this manuscript highlight the importance of philological examination to the understanding of contents and the use of systems of annotations to evaluate the quality and authority of the text.

With the waning of interest in, and access to, Arabic and Persian scientific texts at the Chinese court, members of Hu Denghzhou's networkwere instrumental in the accommodation of Islamicate scientific knowledge in China. They persisted in the collection of manuscripts and scrutinization of their contents.

V.13.2 Working with printed translations

One of the major transformations in the accommodation of the Islamicate sciences in China emerged following the fall of the Yuan and China's gradual withdrawal from the cross-Asian sociopolitical environment of the Mongol Empire from the 1360s onward. Acknowledging the merits of the Islamicate science, the newly established Ming dynasty transferred the rich Yuan imperial archives of Arabic and Persian texts to their new capital in Nanjing. They faced, however, a critical problem in recruiting officials that could make use of these texts and serve in court's astral and medical institutions. Following the reorganization of the Ming court, the first Ming emperor, Zhu Yuanzhang (1328–1398) launched in the 1380s a project that aimed to translate selected Arabic and Persian texts into Chinese for the benefit of the Chinese court (Weil 2018). The texts that were chosen were mainly related to astral calculation and prediction. Yet the existence of translations of works on medicine and pharmaceuticals might suggest that this project went beyond the astral sciences. In a similar fashion, a movement that encouraged the translation of works on Islamic theology, jurisprudence and Islamic practice into Chinese emerged in the mid-17th century and served as an important channel by which Islamicate scientific concepts and theories were introduced to Chinese readers outside the court.

A preface to an astrological manual titled in Chinese *Tianwen shu* 天文書 (*A Book on Astrology*) provides some clues on the method by which Arabic and Persian scientific texts were translated into Chinese at the Ming court. The work is a Chinese translation of Kūshyār ibn Labbān's (d. before 439/1047) *Introduction to the Science of the Judgement of the Stars* (*al-Madkhal fī ṣināʿat aḥkām al-nujūm*) made in the 1380s by a team of translators under the supervision of the Wu Bozong ([1334–1384]; 吳伯宗), a distinguished scholar at the first Ming emperor's court (Yano 1997). The translation involved two teams of translators – one team included foreign astral experts who previously had worked at the Yuan court and who could read Arabic and Persian, and another team of Chinese literati with high command of classical Chinese rhetoric. The first team would communicate orally the meanings of the Arabo-Persian text, while the second team would elevate the speech to the standards of Classical Chinese rhetoric. The emperor ordered the team to take extra measures to come up "with accurate articulation, not to embellish [the text] or [render it] carelessly" (*Mingyi tianwenshu*, 21: 297).

Soon after its translation, *Tianwen shu* was engraved on woodblocks and printed. The extant editions of the work display some of the changes that translation and the use of print entailed. In the type of Chinese Islamic scholarship that used sources in their original languages, reading entailed grammatical analysis and glossing of selected terms. But translated scholarship is heavily challenged by the requirement to find equivalences between the source and target languages. Technical terms, concepts and theories are especially hard to translate. In the case of *Tianwen shu*, the translators borrowed Chinese terminology to articulate Islamicate concepts. Their main objective, so it seems, was to generate commensurability between the Islamicate and Chinese scientific discourses and produce a sense of coherence at the expense of precision. This type of translation presupposes that the potential reader of the text, who might also put into practice its technical aspects, would have prior acquaintance with Islamicate sciences and theories. Many of the functionaries at the astral institutions of the Ming were indeed recruited from Chinese Muslim families and are likely to have some prior knowledge of Islamicate astral sciences. Moreover, if the study of Arabic and Persian manuscripts was characterized by an extensive use of annotations, multiplicity of hands and voices and a strong oral component, printed translations were far more unambiguous and guided. Required explanations were often embedded in the main text, producing an unequivocal reading of the text. Paratextual components, such as prefaces, appendices and diagrams substituted for interlinear and marginal annotations, and brought required information to the attention of the reader.

A comparable process of replacing the study of Arabic and Persian manuscripts with printed translations took place also in the scholarship of Chinese Muslims outside of the court. From around the 1630s, the first European missionaries arrived in China and began to proliferate their ideas through printed books. A few decades after, some members of Hu Dengzhou's network, began to publish printed books and treatises on themes such as Islamic cosmology, ritual and daily practice. They were mainly from the prosperous Jiangnan region – China's cultural hub of the time. As explicitly stated in the prefaces of some these published works, it was the aspiration to introduce Islamic concepts and theories beyond the Muslim community that prompted these scholars to published printed works in Chinese. Similar to the challenges that the court faced in its translations of scientific treatises, these Chinese Muslim scholars had to renegotiate the objectives of their scholarship and the development of new methods of knowledge making.

In their attempt to introduce Islamicate theories to the broader Chinese community of readers while lending authority to their translations and compilations in Chinese, Chinese Muslim scholars experimented with methods that could skillfully introduce foreign theories and corroborate their assertions. Such theories included descriptions of the structure and operation of the cosmos and the natural world, the movements of celestial bodies, and physiological processes among others (Weil 2022b). Wang Daiyu's (1570–1660) 王岱輿 *Great Learning of Pure and Real* (*Qingzhen daxue* 清真大學, pub. 1642), one of the earliest works on Islamic theology to be published in Chinese, for example, sought to imitate the interlinear and marginal annotation in manuscripts by including explanatory comments printed in smaller font to gloss selected terms, and placed outside the frame of the main text, or as passages in smaller font within the frame of the main text.

A generation later, a prolific Chinese Muslim scholar, Liu Zhi ([1660–1730]; 劉智), explored in his works aspects of Islamicate cosmology and ritual. He sought to rationalize Islamic theories and practices by corroborating them with passages included in Chinese scientific and philosophical literature. In the preface to one of his works, Liu provides an overview of his system of corroborating evidence that includes hermeneutical devices such as providing evidential verification, general inferences or particular inferences. In order to explain his theoretical discussions, Liu appended diagrams and sketches that would help the reader to visualize some of the intriguing structures and operations of the natural world, such the layout of the celestial bodies or human anatomy. In some cases, these diagrams are abstract and seem not to seek to provide an accurate depiction, but rather functioned as heuristic aids to facilitate reading.

Such printed translations of Islamicate scientific knowledge allowed Chinese authors to familiarize their readers with basic scientific theories and concepts without a requirement to master a foreign language or copy a manuscript. By doing so, they facilitated the spread of this knowledge beyond the walls of Islamic schools or designated institutions, and formed an important bridge between Islamicate and local Chinese scientific discourses.

V.13.3 Summary

This chapter explores the scholarly practices applied to read Arabic and Persian scientific texts in China's long early modern period between the 13th and 18th centuries. It contrasts two forms of scholarship that took place during that period – the study of Arabic and Persian manuscripts and the production of Chinese printed books on Islamicate themes. The chapter has examined methods of articulation, presentation, corroboration and translation of Islamicate scientific knowledge, juxtaposing scientific enterprises at the Chinese court with the work of local Chinese Muslims. In both cases, explorations of knowledge related to the structure and operation of the natural world came out of textual scholarship and application of shared sets of scholarly practices.

Notes

1 On the participation of Muslim astronomers in the Song court's calendar reform, see Luo (1968) and Chen (1996).

2 The labels "medieval" and "early modern" are rarely used in periodization of Chinese history. It is used in this essay in light of the cross-cultural nature of the subject matter and to be compatible with other chapters in this volume.

3 "Scientific knowledge" is broadly defined in this essay to encompass scholarly interests in the operation and structure of the natural world. It includes works on the various branches of the exact sciences and philosophy.

4 The Yuan court imposed restrictions on private circulation of books on the astral sciences, including calendars and astrological manuals (see Jiang 2021).

5 The transliteration in Arabic script, just as the Persian spelling through this manuscript, lacks diactric points. Hence the gloss حو stands for چو and حه for جه. The romanized transliteration presented here are reconstructed according to the approximate pronunciation and not according to the orthography. The accompanying text in Arabic script adheres to the manuscript's orthography.

6 On the growth of Muslim communities in China from the Mongol period onward, see Leslie and Wassel (1982); Murata (2000, 2009); Ben-Dor Benite (2005); Frankel (2011); Petersen (2018).

Bibliography

Sources

Anonymous. n.d. "Mingyi tianwenshu," in Huang X. *et al.*, eds. *Qingzhen Dadian*. Hefei shi: Huangshan shushe, 21: 295–392.

Wu, Zunqi. n.d. "Guizhen zongyi," in Huang X. *et al.*, eds. *Qingzhen Dadian*. Hefei shi: Huangshan shushe, 16: 336–492.

Research literature

Ben-Dor Benite, Z. 2005. *The Dao of Muhammad*. Cambridge, MA: Harvard University Press.

Buell, P. 2007. "How Did Persian and Other Western Medical Knowledge Move East, and Chinese West? A Look at the Role of Rashīd Al-Dīn and Others," *Asian Medicine* 3: 279–95.

Chen, J. 1996. *Huihui tianwen xue shi yanjiu*. Nanning Shi: Guangxi kexue jishu chubanshe.

Dalen, B. van. 2002. "Islamic and Chinese Astronomy Under the Mongols: A Little-Known Case of Transmission," in Dold-Samplonius *et al.*, eds. [CB],' 327–56.

Dalen, B. van and Yano, M. 1998. "Islamic Astronomy in China: Two New Sources for the *Huihui li* ('Islamic Calendar')," *Highlights of Astronomy* 11B: 697–700.

Feng, Z. 1982. "'Xiaoerjin' chutan – jieshao yizhong alabo zimu de hanyu pinyin wenzi," *Alabo shijie* 1: 37–47.

Franke, H. 1988. "Mittelmongolische Glossen in einer arabischen astronomischen Handschrift von 1366," *Oriens* 31: 95–118.

Frankel, J. 2011. *Rectifying God's Name: Liu Zhi's Confucian Translation of Monotheism and Islamic Law*. Honolulu: University of Hawai'i Press

Gacek, A. 2009. *Arabic Manuscripts: A Vademecum for Readers*. Leiden: Brill.

Jiang, X. 2021. *Chinese Astrology and Astronomy: An Outside History*. Singapore and Hackensack, NJ: World Scientific.

Leslie, D. and Wassel, M. 1982. "Arabic and Persian Sources Used by Liu Chih," *Central Asiatic Journal* 26.1–2: 78–104.

Luo, X. 1968. "Zupuzhong guanyu zhongxi jiaotong ruogan shishi zhi faxian," *Bulletin of The Institute of History and Philology, Academia Sinic* 40.1: 125–38

Murata, S. 2000. *Chinese Gleams of Sufi Light: Wang Tai-Yü's Great Learning of the Pure and Real and Liu Chih's Displaying the Concealment of the Real Realm*. Albany: State University of New York Press.

Murata, S. 2009. *The Sage Learning of Liu Zhi: Islamic Thought in Confucian Terms*. Cambridge MA: Harvard University Council on East Asian.

Murata, S. 2017. *The First Islamic Classic in Chinese: Wang Daiyu's "Real Commentary on the True Teaching"*. Albany: State University of New York.

Petersen, K. 2018. *Interpreting Islam in China: Pilgrimage, Scripture, and Language in the Han Kitab*. Oxford: Oxford University Press.

Song, X. 1999. *Huihui yaofang kaoshi*. Beijing: Zhonghua shuju.

Wang, S. and Shang, Q. 1992. *Mishujian zhi*. Hangzhou: Zhejiang guji chubanshe. Reprint.

Weil, D. 2016. "Islamicated China – China's Participation in the Islamicate Book Culture During the Seventeenth and Eighteenth Centuries," *Intellectual History of the Islamicate World* 4: 36–60.

Weil, D. 2018. "The Fourteenth-Century Transformation in China's Reception of Arabo-Persian Astronomy," in Manning, P. and Owen A., eds. *Knowledge in Translation: Global Patterns of Scientific Exchange, 1000–1800 CE.* Pittsburg: Pittsburg University Press, 262–74.

Weil, D. 2020a. "Libraries of Arabic and Persian Texts in Late Imperial China," in *EI–3*, 2020–4: 90–2.

Weil, D. 2020b. "Literacy, in Arabic and Persian, in Late imperial China," in *EI–3*, 2020–4: 92–5.

Weil, D. 2022a. "Chinese-Muslims as Agents of Astral Knowledge in Late Imperial China," in Mak, Bill M. and Huntington, Eric eds. *Overlapping Cosmologies in Asia: Transcultural and Interdisciplinary Approaches.* Leiden and Boston: Brill, 116–38.

Weil, D. 2022b. "Unveiling Nature: Liu Zhi's Translation of Arabo-Persian Physiology in Early Modern China," *Osiris* 37(1): 47–66.

West, M. 1973. *Textual Criticism and Editorial Technique Applicable to Greek and Latin Text.* Stuttgart: B.G. Teubner.

Yabuuti, K. and Dalen, B. van. 1997. "Islamic Astronomy in China During the Yuan and Ming Dynasties," *Historia Scientiarum* 7.1: 11–43.

Yang, H. and Yu, Z. 1995. *Yisilan yu zhongguo wenhua.* Yinchuan: Ningxia Renmin Chubanshe.

Yano, M. 1997. *Kūšyār Ibn Labbān's Introduction to Astrology.* Tokyo: Institute for the Study of Languages and Cultures of Asia and Africa.

Zavyalova, O. 1999. "Sino-Islamic Language Contacts Along the Great Silk Road: Chinese Texts Written in Arabic Script," *Hanxue Yanjiu (Chinese Studies)* 17.1: 285–303.

V.14

THE MULTIPLICITY OF TRANSLATING COMMUNITIES IN THE IBERIAN PENINSULA (12TH–13TH CENTURIES)

Alexander Fidora and J. L. Alexis Rivera Luque

The 12th and 13th centuries comprise the most prolific period of translating activities in the Iberian Peninsula in the European Middle Ages. This period witnessed, among other endeavors, the translation of around two hundred texts, mostly scientific, from Arabic into Latin and later into Castilian or Catalan. These translations may be attributed to single authors, or to pairs of them at most, yet these individuals almost never worked in isolation. They formed translating communities whose members connected to each other, sometimes more loosely, sometimes more strongly. It is the aim of this chapter to give an account of the diverse communities translating from Arabic into Latin that were active during this time in the Iberian Peninsula, drawing attention to the social, political and material backgrounds that provided the context for them to flourish.

The translations from this period were produced within two main areas: (1) the region of the Ebro Valley, to which Barcelona may be added, (2) and the city of Toledo, to which we can add Segovia. Chronologically, the first region to flourish was in the north of the peninsula – the Ebro Valley, Barcelona and northern Portugal – where translating activities were already occurring during the first half of the 12th century. In Toledo, translating activities started in the middle of the 12th century and continued well into the 13th century, with the translations of Hermann the German (d. 1272) and the so-called Alfonsine translations from Arabic into Castilian. The origin of this kind of activity in these places is not fortuitous but, rather, connected to their geographical location and two consecutive historical processes. Since the 11th century, or even earlier, an Arabic culture well versed in scientific and philosophical subjects had developed across both regions. Both regions were also shaped by the advance of the Christian conquest of formerly Islamic states during the last decades of the 11th century and the first of the 12th, when members of the learned Latin clergy arrived in those places.

V.14.1 The historical background

Both Toledo and the Ebro Valley provided favorable environments for contact between the Latin Christian and Arabic Islamic spheres – called by König "third places" (2015, 23, 69) – because they had formed part of the Middle and Upper Marches (respectively, *al-thaghr al-awsaṭ* and

DOI: 10.4324/9781315170718-63

al-thaghr al-aʿlā) of the Islamic territories in the Iberian Peninsula since the late 8th century. Toledo was the capital of the former until 334/946, when its capital was moved to Medinaceli.

The Ebro Valley encompassed the whole territory of the latter (Boloix 2001, 25–6; Turk 1998, 239–40). In terms of the cultural growth of these regions during the 11th century, it is crucial to consider the presence of two dynasties in the period following the disintegration of the Caliphate of Córdoba, known as the first *ṭāʾifa* period: the Banū Hūd, who ruled the *ṭāʾifa* of Zaragoza (430–503/1039–1110), which included the Ebro Valley territories, and the Banū Dhū-l-Nūn, who ruled the *ṭāʾifa* of Toledo from approximately 426/1035 until 477/1085.

The Banū Hūd, who were probably descendants of Rawḥ ibn Zinbāʿ (d. 83/703), a member of the Yemeni tribe of the Banū Judhām, took Zaragoza from the earlier ruling dynasty, the Banū Tujīb (r. 409–430/1018–1038), in 430/1038. They held power until 503/1110 when the Almoravid army conquered the city, only to be defeated by Alfonso I the Battler (r. 1104–1134), King of Aragón, in 511/1118 (Makki 1994, 54–5; Hawting 1995, 466; Bosworth 1991, 573; Dunlop 1991, 542–3). The Banū Hūd have been regarded as patrons of learning: two members of the dynasty – al-Muqtadir bi-Llāh (r. 438–474/1046–1081) and his son, al-Muʾtaman ibn Hūd (r. 474–478/1081–1085) – were prominent scholars of the mathematical sciences; al-Muʾtaman wrote a treatise called *The Perfection* (*al-Istikmāl*), which according to al-Akfānī (d. 749/1348) would have rendered obsolete previous geometrical knowledge (Hogendijk 1986, 43). It was also under the Banū Hūd that authors such as the scientist and philosopher Ibn Bājja (d. 533/1139) and the 6th/12th-century physician and botanist Ibn Baklārish would flourish (Burnett 1994, 1041–2, 1997, 802). The library collected by the Banū Hūd seems to have been "particularly rich in works on mathematics, astronomy, astrology, and magic" (Burnett 2001, 251). In all likelihood, it was moved to the stronghold of Rueda de Jalón when the Banū Hūd dominion over Zaragoza ended. Reichert (2008, 55–6) has suggested that this was done in 512/1118 by Sayf al-Dawla (d. 540/1146), the last of the Hūdids, when the city fell into Christian hands. Yet the fact that ʿImād al-Dawla (d. 524/1130) was first to retreat to Rueda after being defeated by the Almoravid army in 503/1110 suggests otherwise (Dunlop 1991, 542–3). In any case, Hugo of Cintheaux ([1st half 12th century]; formerly known as Hugo of Santalla; Santoyo 2016) informs us that Michael, Bishop of Tarazona from 1119 to 1151, obtained an Arabic manuscript for him to translate from the "Rotensi armarium" ("the library of Rueda"), which is assumed to be the library of the Banū Hūd (Haskins 1924, 70–1; Hugo of Cintheaux 1963, 95–6; Burnett 1994, 1041, 1997, 802). The library would have been moved to Toledo in 431/1140 when Sayf al-Dawla exchanged Rueda for some property in the city (Burnett 2001, 251, 2011b, 12; Dunlop 1991, 543). Some works translated by Gerard of Cremona (d. 1187) match the sources that were used by al-Muʾtaman for the composition of his book *The Perfection* (Lomba 1995, 79; Burnett 2001, 251, 2011b, 14) which seems to confirm the library's presence in the city.

The Banū Dhū l-Nūn were members of the Hawwāra Berber tribe, who established themselves in the Santaver region from approximately the second half of the 3rd/9th century and were already ruling over Toledo by 427/1035–1036 under Ismāʿīl al-Ẓāfir ibn Dhī-l-Nūn ([d. 435/1043–4]; Makki 1994, 51, 55; Dunlop 1986, 242–3). We know that Ismāʿīl al-Ẓāfir was a poet and maintained a literary environment at his court, with members who apparently arrived in Toledo with books that came from the Cordovan library collected by Caliph al-Ḥakam II (Porres 1999, 39). The reign of his son and successor, al-Maʾmūn ibn Dhī-l-Nūn (r. 434–467/1043–1075), was marked by scientific and literary prosperity. This is evident from the number of scholars working in Toledo at that time, including Ṣāʿid al-Andalusī (419–462/1029–1070), author of *The Categories of the Communities* (*Ṭabaqāt al-umam*; also translated as *The Classes of the Nations*), Ibn al-Zarqāllu (d. 493/1100), whose astronomical tables, known as the *Toledan Tables*, would be rendered into Latin before 1140 (Dunlop 1986, 242; Porres 1999, 41–2; Burnett 2001, 249–50).

The translating activities in these two regions began after they were conquered by Christians (Hasse 2006, 72; Haskins 1927, 14). While the major event in the conquest of the Ebro Valley was the capture of Zaragoza in 1118 by Alfonso I (García-Guijarro 2015, 243; Corral 1998, 58–9), there was already a growing interest among northern Spanish rulers to expand their territories, which began in the 2nd half of the 11th century, and was most prominent with the conquests of Huesca (1096) and Barbastro (1100) by Peter I of Aragón and Navarre ([r. 1094–1104]; García-Guijarro 2015, 239; Corral 1998, 53–6). More specifically, the emergence of Arabic-Latin translations can be explained through two main processes that were triggered by the Christian victories: the expansion of the Cluniac monastic order into this region, and the arrival of scholars from abroad, regular and secular alike, who would find there a suitable environment to learn Arabic and to have access to manuscripts for translation purposes.

The Benedictine abbey of Cluny would expand into the Ebro Valley, primarily between 1073 and 1179, through territorial and monetary donations by the kings of León (Reglero 2008, 73). O'Callaghan summarized these developments as follows:

> During Alfonso VI's long reign, Christian Spain experienced a constantly expanding influence from beyond the Pyrenees. The influx of pilgrims journeying to Compostela increased rapidly, while French monks introduced the Cluniac observance into many monasteries, and French knights, seeking fame and fortune, came in large numbers.
>
> *(O'Callaghan 1976, 200; see also Stalls 1995, 225–6)*

Cluny's influence on the translation activities is most visible in Peter the Venerable's (d. 1156) patronage of a translation project of Islamic texts, which is nowadays referred to as the *Corpus islamolatinum* (Martínez Gázquez 2011, 173, 2012, 40). He would use these texts to compose his refutation of Islam, the *Against the Sect or Heresy of the Saracens* (*Contra sectam sive haeresim Saracenorum*; Peter the Venerable 1985, 30–61; see D'Alverny 1947–1948; Kritzeck 1964). In fact, the abbot's journey to Iberia, which led to these translations, was a response to his invitation from Alfonso VII of León (r. 1126–1157) to come to the peninsula. Alfonso wished to make a donation to Cluny in exchange for Peter's support in Rome for the election of his candidate as Archbishop of Santiago de Compostela (Saurette 2010, 300; Bishko 1956, 164, 169, 171; Kritzeck 1964, 220–92; Peter the Venerable 1890, 313c-5b).

As for Toledo, the city was famously captured by King Alfonso VI of León and Castile (r. 1065–1109) in 1085 (Reilly 1988, 171), who, according to some accounts, styled himself as *al-Imbaraṭūr Dhū l-Millatayn* (the emperor of the two religious communities; Mackay and Benaboud 1979, 95). The city's Christian community grew because of the influx of Spaniards and Frenchmen to Toledo, while it also retained its Mozarabic community, which had burgeoned during the 11th century. This community was partly a clerical one that was, to a certain extent, interested in Latin learning (Burman 1994, 17–25). This interest of the Mozarabs is evidenced by the *Latin-Arabic Glossary* (*Glossarium Latino-Arabicum*), which would have been composed to help Mozarabs to read Latin (van Koningsveld 1977, 1; Burman 1994, 18). The clerical Mozarabic community was soon to be followed by a French one that came from the abbey of Cluny and was led by the Archbishops of Toledo Bernard of Sédirac (1086–1125) and Raymond of La Sauvetat ([1125–1152]; Burnett 2001, 250). This French community grew so rapidly that it formed its own quarter in Toledo (Burman 1994, 23).

In fact, most Arabic-Latin translators whom we can identify during the 12th and 13th centuries on the Iberian Peninsula were scholars who arrived on the peninsula from abroad and were presumably seeking knowledge of the Arabic language and science, such as Hugo of Cintheaux, Plato of Tivoli (*fl.* 1133–1147), Hermann of Carinthia (*fl.* 1138–1143), Robert of Ketton (*fl.* 1141–1157), Robert of Chester (*fl.* 1140–1150), Gerard of Cremona, Michael Scot (d. *c.* 1235), Salio of Padua (*fl. c.* 1218) and Hermann the German. Such is the rationale that Daniel of Morley (d. *c.* 1210) describes in his *On the Natures of Inferior and Superior Things* (*Philosophia* or *De naturis inferiorum et superiorum*):

> When I left England in order to study and spent some time in Paris, I realized that there were some beasts holding the chairs in the schools with great authority. . . . But since Toledo is most famous nowadays for the Arabic knowledge. . . . I sprinted there rapidly to listen to the lessons of the wisest philosophers of the world.
>
> *(ed. Maurach 1979, 212)*

V.14.2 Communities among communities

The regions of the Ebro Valley and the city of Toledo gave rise to what may be regarded as four translating communities: (1) the translators of the Ebro Valley, (2) the translators of Toledo during the 12th century, (3) the translators of Toledo during the 13th century and (4) the Alfonsine circle. These communities were far from uniform in the sense that their members showed varying degrees of interaction with each other, did not come from the same background and did not all share the same interests. Instead, these communities must be conceived of as relational spaces – spaces "made of multiple relations" (Murdoch 2006, 22–3) – composed of a mixture of individuals whose interests converged in the same environment. In that sense, each community was, so to speak, a conjunction of several scholarly initiatives. Besides, in 12th-century Iberia, there were also three independent translators working (independent in that they worked in relative geographical isolation from the main centers) – John of Seville (1st half 12th century), Plato of Tivoli and Robert of Chester – whose activity shall be examined first. John of Seville is the earliest translator among them. Probably a Mozarab (Burnett 2002, 60–3; Williams 2003, 38–40), he was already active by 1128, as he dedicated his translation of the Pseudo-Aristotelian *Secretum secretorum* (*Secret of Secrets*), a treatise on political philosophy in the manner of a mirror for princes, to Queen Teresa of Portugal (r. 1112–1128). His overall activity suggests that he was active in northern Portugal (Burnett 1995a, 227–8, 232–3, 2002, 60, 62). After Teresa's ouster by her son, Alfonso I of Portugal (r. 1128–1185), in 1128 he probably sought the patronage of Raymond of La Sauvetat, Archbishop of Toledo, to whom he had dedicated his translation of Qusṭā ibn Lūqā's (d. *c.* 299/912) *On the Difference Between Spirit and Soul* (*De differentia spiritus et animae*), a philosophical treatise on the difference between the soul and the spirit (Williams 2003, 45). It is also likely that he established contact with Maurice Bourdin, Archbishop of Braga and Antipope Gregory VIII ([1118–1121]; Williams 2003, 40–2). His connection with the three of them, says Williams, "shows that he moved in the highest circles of Christian society in the western portion of the Iberian Peninsula" (Williams 2003, 40). He was mainly interested in astrology and astronomy and translated around nine works on the subject, the most notable of which is *The Book of the Great Introduction to Judicial Astrology* (*Liber introductorii maioris ad scientiam iudiciorum astrorum*) by the Persian astrologer Abū Maʿshar (171–272/787–886), which was translated again in 1140 by Hermann of Carinthia under the title *Introduction to Astrology* (*Introductorium in astrologiam*; Burnett 1978, 126).

In chronological order, Plato of Tivoli comes next. There is no information about his background, nor does he appear to have been involved in either an ecclesiastical or courtly context. He does however seem related to French scholars, based on his dedication of a treatise on the astrolabe to John David (middle 12th century), patron of Rudolph of Bruges (*fl.* 1144), a disciple of Hermann of Carinthia (Curtze 1902, 10, 182; Rudolf of Bruges 1999, 60), and to the Jewish community of Barcelona through Abraham bar Ḥiyya (d. *c.* 1140). The latter was a Jewish polymath with whom Plato translated al-ʿImrānī's (d. 344/955–6) *Book of Choices* (*Kitāb al-ikhtiyārāt*), in Latin *De electionibus horarum* (*On Choosing the Right Time*), in Barcelona in 1133 or 1134. Abraham was the author, among others, of a book on algebra – the *Book of Areas/Surfaces* (*Liber embadorum*). Plato translated Abraham's work from Hebrew in 1145 (Millàs 1942, 328; Samsó 2004, 271–4). Abraham was called 'Savasorda' (Arabic: *ṣāḥib al-shurṭa* [chief of the guard]), which suggests that he held an official position in the administration of Barcelona (Samsó 2004, 271, 273). He was in touch with French scholars as well, for he states that he wrote his *Foundations of Understanding and Tower of Faith* (*Yesode ha-tebuna u-migdal ha-emuna*) at the behest of the Jews of Provence (Gómez Aranda 2009, 108). Plato's interests ranged from astrological-astronomical knowledge and geometry to geomancy and alchemy. He prepared the first translation of the *Emerald Tablet* (or: *Smaragdine Table*; *Tabula smaragdina*), and he is said to have translated a medical treatise called *On Pulses and Urines* (*De pulsibus et urinis*).

As for Robert of Chester, he only made one translation in Iberia, the treatise on algebra by al-Khwārazmī (d. *c.* 235/850), which he finished in Segovia in 1145 (Robert of Chester 1915, 66). He continued to pursue his career in London, where he wrote a treatise on the astrolabe in 1147 and, in either 1150 or 1170, calculated astronomical tables for the city's meridian, based on other tables calculated for the meridian of Toledo. He was probably the author of a set of political horoscopes that were made at the court of King Stephen (r. 1135–1154) in 1150 and 1151 and revised Adelard of Bath's (d. *c.* 1150) translation of the astronomical tables by al-Khwārazmī (Robert of Chester 1915, 31; Southern 1992, xlix; Burnett 2004). There is no explicit evidence of a relation to the Ebro Valley or Toledan activity, but he may have had some connection with Toledo because of his knowledge of the *Toledan Tables*. It is also possible that he could have made contact with Dominicus Gundissalinus (d. *c.* 1190), who was active in Segovia within the same span of years (Polloni 2020, 15).

V.14.2.1 The Ebro Valley community

The core of the Ebro Valley community was composed of Hermann of Carinthia, Robert of Ketton and Hugo of Cin-theaux. Hermann does not appear to have had a clerical career. He is called "scholasticus" by Peter the Venerable. Indeed, he addressed Thierry of Chartres (d. *c.* 1155) as his "most valued teacher" (*diligentissime praeceptor*) in his translation of Ptolemy's *Planispherium*. This could imply that Hermann was educated under Thierry (Hermann of Carinthia 1982, 4, 20–5, 349; Peter the Venerable 1985, 24–5). Robert of Ketton or Ketenensis, not to be confused with Robert of Chester or Cestrensis (Burnett 2004), was Archdeacon of Pamplona (1143–1152). In 1157, he appears as canon of Tudela and a close friend of King Sancho VI of Navarre ([r. 1150–1194]; Goñi 1965, 246–56; Peter the Venerable 1985, 24–5). Hugo of Cintheaux was probably a native from Normandy, as Santoyo has recently suggested, and is listed as a "magister" among the clerics of Tarazona (1145 and 1147; Santoyo 2016, 342, 348–50).

Hermann and Robert had a close relationship and shared some scientific interests (Reichert 2008; Hermann of Carinthia 1982, 237–9, 349). Robert states in his translation of al-Kindī's (d. after 256/870) forty chapters on astrology, in Latin *Iudicia* (*Judgments*), that he prepared this text at Hermann's request and adds that they were both interested in Ptolemy's *Almagest*, although it is unlikely that they translated it (Burnett 1993, 106). Moreover, Peter the Venerable informs Bernard of Clairvaux (d. 1153) that, during his sojourn in Hispania, he encountered both while they were "studying astrology in the proximities of the Ebro river" (Burnett 1993, 106). There he hired them to prepare the translations of the so-called *Corpus islamolatinum* (Peter the Venerable 1985, 24–5, 54–5), among them the Qur'ān. Peter of Toledo (d. after 1142), who was not previously related to them, joined them in this project. Peter the Venerable tells us that, although he commissioned Peter of Toledo to translate the so-called *Apology of al-Kindī* (*Apologia Alkindi*), attributed to an Arab Christian named al-Kindī, the fact that he was not as proficient in Latin as he was in Arabic prompted the abbot to provide him with the assistance of his secretary, Peter of Poitiers ([d. 1205]; Peter the Venerable 1985, 22–3). Based on this and the fact that Peter probably originated from Toledo, it is assumed that he must have been a Mozarab (González Muñoz 2011, 478). As for Hugo's relation with Hermann and Robert, Burnett purports the likelihood of there being "some kind of common study between Hermann and Hugo" (1977, 70) and suggests that the dedicatee of one of the two copies of Hugo's *Book of the Three Judges* (*Liber trium iudicum*) could be identified as being Robert (Burnett 1977, 68–70). Hugo and Hermann would most likely have had similar profiles: students of liberal arts educated in French schools, who arrived in the Upper March where they learned Arabic and were commissioned by a cleric to translate from Arabic into Latin. On top of the translations of the *Corpus islamolatinum*, the group showed a strong interest in astronomical-astrological knowledge. Robert and Hermann were interested in mathematics – both worked on Euclid's *Elements* (Hermann of Carinthia 1977; Robert of Chester 1992). Hugo was also inclined toward alchemy and divination. He translated the alchemical *On the Secrets of Nature* (*De secretis naturae*) and a treatise on geomancy and two on scapulimancy, covering divination based on soil or sand patterns and shoulder blades of animals respectively.

V.14.2.2 The Toledo community during the 12th century

The activity in Toledo during the second half of the 12th century sprang from two groups: that of Dominicus Gundissalinus and his collaborators John of Spain and Abraham ben Daud (d. 1180) (regarding the latter's identity see Freudenthal 2016, 67–8; Szilágyi 2016, 22); the other formed by Gerard of Cremona and his *socii*. Gundissalinus, Archdeacon of Cuéllar, is recorded as a member of the Toledan chapter ([1162–1178]; Polloni 2020, 14). He appears to have received his education from the so-called School of Chartres (Burnett 2001, 264; Fidora 2009, 92–6; Polloni 2020, 11–3), which would make him yet another scholar educated in French schools who came (back) to Iberia, where he produced his translations. Gundissalinus's activity consists of his original works, the translations that he produced, as well as those that he co-wrote. For the latter translations, he worked together with both Abraham ben Daud, with whom he translated parts of Ibn Sīnā's (d. 428/1037) *Book of Healing* (*Kitāb al-Shifā'*), and with the 12th-century translator John of Spain, with whom he translated for instance Solomon ben Gabirol's (d. *c.* 450/1058) book *The Fountain of Life* (*Fons vitae*). Traditionally, it has been assumed that Gundissalinus was the main proponent behind these translations. However, recent scholarship suggests that the impulse came instead from Abraham ben Daud and John of Spain (Burnett 1994, 65, 2001, 264; Bertolacci 2011, 53–4; Freudenthal 2016, 68–70). The preface to the translation of Ibn Sīnā's *Book on the Soul* (*Liber de anima*) and the *Logic* (*Logica*), prepared with Ibn Daud (Gundissalinus and Abraham b. Daud 1972, 1: 3; Ibn Sīnā 2018, 67–71), and the preface to Ibn Gabirol's *Fountain of Life* (Ibn Gabirol 2007, 680) provide some evidence for this. In these prefaces, both scholars, respectively, present themselves as the principal translators, and Gundissalinus is referred to as merely an assistant. The procedure for creating these translations – commonly and somewhat mistakenly referred to as "four-hands translation" ("a quattro mani") or also as "à deux interprètes" (D'Alverny 1989b) – is described by both Ibn Daud and John of Spain in those same prefaces. Daniel of Morley also evokes this procedure in reference to the translation of Ptolemy's

Almagest that Gerard of Cremona and his assistant Galippus prepared (Daniel of Morley 1979, 244–5), as does John of Brescia in his 1263 translation of Ibn al-Zarqāllu's treatise on the *saphaea* (an astrolabe usable at any latitude; Chapter IV.3) finished in Montpellier (Millàs 1942, 11). As can be inferred from these sources, a first translator acquainted with the Arabic language would orally translate the Arabic text into the vernacular and a second one would produce a Latin draft, which they probably polished later.

Neither Ibn Daud nor John of Spain state which intermediary language they used, but Daniel of Morley does mention that Gerard of Cremona used Mozarabic, a Romance dialect from al-Andalus, as an intermediary language, and in the 13th century, John of Brescia and his partner Jacob ben Makhir ben Tibbon (d. *c.* 1305), also called *Profatius Judaeus*, used Hebrew.

Alongside Gundissalinus's community, another one flourished in Toledo that was formed by Gerard of Cremona, the most prolific of the 12th-century translators, and his *socii* (Burnett 2001, 254–6, 273–87). Gerard was called *gloria cleri* (the glory of the clergy) by his *socii*, and there are three documents that he has signed as canon of Toledo; Gundissalinus also appears in two of these (Burnett 2001, 252, 281, 2011b, 16). It has been pointed out that even though Gerard is called "magister," this designation would be based merely on his reputation (Burnett 1995b, 224, 2001, 252–3, 2011b, 13–14; Weijers 1987, 139). Nevertheless, Gerard's *Vita*, *Commemoratio librorum* and *Eulogium* and Daniel of Morley's testimony suggest that there was a community of teaching and learning that Gerard of Cremona and his *socii* formed (Burnett 2001, 267, 275–81; Daniel of Morley 1979, 244–5). It is more than likely that his *socii* participated in making the 80 or more translations that are ascribed to him. Since Gerard's interests encompassed a wide variety of fields, his translations were organized according to subjects already recognized in the 12th century: dialectics, geometry, astrology, philosophy, physics, alchemy and geomancy (Lemay 1981, 187–8; Burnett 2001, 276–81). Regarding the choice of works of the two groups, Gundissalinus focused on Muslim philosophers, such as al-Kindī, al-Fārābī (d. 339/950–1) and Ibn Sīnā, whereas Gerard translated both philosophical and scientific texts from Muslim and ancient Greek authors, including Aristotle (Hasse 2000, 16). We know that Gundissalinus was acquainted with Gerard's translations (and those of John of Seville), and used Gerard's translation of al-Fārābī's *On the Sciences* (*De scientiis*) to produce his own rendering of the text (Burnett 2011b, 16).

V.14.2.3 The Toledo community during the 13th century

During the 13th century there were two communities in Toledo, which had two main features that set them apart from each other: one kept translating from Arabic into Latin and developed within an ecclesiastical environment; the other translated mostly into Castilian and was promoted by the Castilian court. The former may be further analyzed according to two different periods, which took place in the 1210s and in the 1240s and 1250s, approximately.

In the first phase, we find Mark of Toledo, Michael Scot and Salio of Padua. Mark of Toledo was a member of the Toledan chapter (1192–1216; Mark of Toledo 2016, xxvii–xxix). In the preface to his translation of the *On Taking the Pulse* (*De tactu pulsus*), he states that he studied medicine (D'Alverny 1989a, 39). Petrus Pons has suggested that this could have happened in Salerno or, most likely, in Montpellier (Mark of Toledo 2016, xxx–xxxi). This would make him yet another scholar educated in a French environment, who then produced his Arabic-to-Latin translations in Iberia. In this same preface, Mark claims that he was already acquainted with the Arabic language while studying medicine (Mark of Toledo 2016, xxx). This points to the fact that he could have learned Arabic from early on and may even have been of Mozarabic descent, which explains why he appears to have prepared translations on his own. The *corpus* of his translations includes medical works by Galen and Hippocrates and the translations of two Islamic texts: the second Latin translation of the Qurʾān and the translation of Ibn Tūmart's (d. 524/1130) *Profession of Faith* (*ʿAqīda*), in Latin *Libellus Habentometi de unione Dei* (Ibn Tūmart's Booklet on the Unity of God).

Michael Scot appears to be related to Toledo in three instances: a "magister Michael Scotus" accompanied the Archbishop Rodrigo Jiménez de Rada (1209–1247) to the Fourth Lateran Council in Rome in 1215, and in Toledo he translated, in 1217, al-Biṭrūjī's (*fl.* 2nd half 6th/12th century) *Book on the Configuration* [*of the Universe*] (*Kitāb fī al-hayʾa*) as *On the Motions of the Heavens* (*De motibus caelorum*), with the help of a certain "Abuteus Levita," and Aristotle's *On Animals* (*De animalibus*) in nineteen books (Scot 1992, x; D'Alverny 1982, 455–6; Haskins 1924, 277). Six or seven of Ibn Rushd's (520–595/1126–1198) commentaries on Aristotle can also be attributed to him (Hasse 2016[CB],[1] 342). There are four translations ascribed to Salio of Padua, canon of Padua, two of which – the *Book of Salcharie Albassarith* (*Liber Salcharie Albassarith*) and *Hermes's Book on the Fixed Stars* (*Liber Hermetis de stellis beibeniis*) – were written in Toledo (Burnett 2011a, 312). Another one, *On Nativities* (*De nativitatibus*), translated from Hebrew into Latin, was written in 1215 in either Barcelona or Toledo, with the help of a Jew named David (Burnett 2011a, 3110; Hasse 2016, 324).

In the second period, we encounter Hermann the German. In the translation of Ibn Rushd's commentary on Aristotle's *Rhetoric*, he states that he has translated this work on the request of John, Bishop of Burgos (d. 1246) and chancellor of the King of Castile and León Ferdinand III ([r. 1230–1252]; Pérez 1992, 276–7). Hermann dedicated his entire activity to the reception of the so-called *Corpus averroisticum* (in total seventeen Latin translations of Ibn Rushd's works were produced during the Middle Ages): he translated three *Commentaria media*, Ibn Rushd's middle commentaries on Aristotle's *Ethica nicomachea* (finished in Toledo in 1240), the *Poetica* (also finished in Toledo in 1256) and the *Rhetorica*, as well as al-Fārābī's *Didascalia in rhetoricam* (*Guide to the Rhetoric*) and the *Summa Alexandrinorum* (*Alexandrine Summary*), which epitomizes Aristotle's ethics (González Ruiz 1997, 595–600; Saccenti 2010; Hasse 2016, 330, 342).

V.14.2.4 The circle of Alfonso X

The main characteristic of the translations made under Alfonso X of Castile (r. 1252–1284) was the change of the target language from Latin to Castilian, a decision that was probably related to the fact that in 1214 it was decreed that the official language for the royal chancellery should be Castilian. Thus, it is not surprising that the impulse for translating into Romance languages came from the court. This is also evident from the fact that the Visigothic legal code *Forum judicum*, which remained valid until the 19th century, was translated from Latin into Castilian as *Fuero juzgo* under Ferdinand III in 1241 (Smith 1999, 81). This translating activity prospered under Alfonso, even before he was king, and continued throughout his reign. Both *The Book of the Stones* (*Libro de las piedras*) or *Lapidary* (*Lapidario*) and *Kalila and Dimna* (*Calila e Dimna*) were translated from Arabic into Castilian on the orders of the young Alfonso. The former was translated between 1243 and 1250 (and revised in the 1270s) by Yehudah ben Moses – who seems to have resided in Toledo and was probably the most prolific of the Alfonsine translators – with the collaboration of Garci Pérez (Vegas 1998, 83). The latter was probably written in 1251 by an unspecified translator (Yehudah ben Moses 2011, 1; Martínez 2010, 76, 124; Vicente 2002, 119; Roth 1990, 65).

The translating community that formed around Alfonso's court was the largest of this period, with approximately fifteen collaborators, of which at least four were Jews, and one, Bernardo el Arábigo, a convert from Islam (Yehudah ben Moses and Isaac ben Sid 2003, 227–9; Roth 1990;

Samsó 1981, 171). The bulk of the Alfonsine translations, which dealt mainly with astronomical or astrological subjects, was prepared by two Jewish translators, who were also the authors of the canons to the renowned Alfonsine tables (Yehudah ben Moses and Isaac ben Sid 2003, 19, 137). These translators were Yehudah ben Moses, who almost always worked in collaboration with Christian counterparts, and Isaac ben Sīd (Roth 1990, 61–8). Between 1225 and 1231, Yehudah ben Moses had already translated Ibn al-Zarqāllu's treatise on the *saphaea* into Latin with William the Englishman. It would be translated again into Castilian by Fernando of Toledo (1255–1256) and revised in 1277 by Bernardo el Arábigo and the Jewish physician Abraham ben Waqar (d. 1294) from Toledo (Yehudah ben Moses and Isaac ben Sid 2003, 227–8; Roth 1990, 70–1). Yehudah would then render into Castilian in 1254 the *Perfect Book on the Judgments of the Stars* (*Libro conplido en los iudizios de las estrellas*; Yehudah ben Moses 1954); in 1256 *The Book on the Eighth Sphere* (*Libro de la ochava sphera*), also named *Book on the Fixed Stars* (*Libro de las estrellas fixas*), with Guillén Arremón Daspa; and in 1259 the *Book of Crosses* (*Libro de las cruzes*) and the *Book of the Sphere* (*Libro de la alcora*) with Johan Daspa.

As for Isaac ben Sīd, he is considered "the most productive scientific collaborator of Alfonso" (Yehudah ben Moses and Isaac ben Sid 2003, 138). Having authored nine original treatises on the construction and use of instruments, most of which were included in the *Book on Astrological Knowledge* (*Libro del saber de astrología*; compiled between 1256 and 1280; see Alfonso X 1863–1867), he was also responsible for the translation of at least three treatises: Ibn Khalaf al-Shajjār's (5th/11th century) *Book of the Universal Plate* (*Libro de la lámina universal*; Calvo 2017), the *al-Zīj al-Ṣābi'* (*Canones*) of al-Battānī (244–317/858–929), and the *Almanac* by Ibn al-Zarqāllu (Roth 1990, 62, 68).

In addition to political, astronomical, astrological and alchemical literature, Alfonso also was interested in magic. Between 1256 and 1258, he ordered the translation of texts such as the *Picatrix* (Pingree 1981, 27). Other translations made at the Castilian court may have originated already in the time of Alfonso's father Ferdinand III, such as the *Poridat de poridades*, the Castilian name given to the *Secret of Secrets* (*Sirr al-asrār = Secretum secretorum*), which was probably translated at the end of Ferdinand's reign or at the beginning of Alfonso's (Bizzarri 2010, 19, 40). Still others were commissioned by Alfonso's brother, such as the *Sendebar*, translated in 1253 on the orders of Prince Don Fadrique ([1223–1277]; Orazi 2006, 71–2).

Extremely influential was the translation of the *Book of Ascension* (*Kitāb al-mi'rāj*), which tells the story of Muḥammad's night journey to Jerusalem and his ascension to the heavens. It was first translated in 1263–1264 by Abraham ibn Waqar into Castilian as *Muḥammad's Ladder* (*Escala de Mahoma*) and then by two unknown translators (previously identified as Bonaventura of Siena [d. after 1282]) into Latin and, from this latter rendering, into Old French (Besson and Brossard-Dandré 1991, 286–9). The translation was produced, according to its preface, "so that people may learn about Muḥammad's life and knowledge and so . . . become acquainted with the errors and unbelievable things that he recounts in this book" (Hyatte 1977, 97).

V.14.3 Final considerations

Our survey has shown how the various translation activities from Arabic into Latin and Castilian on the Iberian Peninsula, which we have analyzed, were usually the result of initiatives by one or two translators, which mostly converged within what we have called relational spaces. The geographical contexts in which these spaces emerged – mainly the Ebro Valley and Toledo – were part of the Middle and Upper Marches of the territories ruled by Muslim dynasties since the late 2nd/8th century, where a highly developed Arabic culture thrived during the first *ṭā'ifa* period (5th/11th century) and which received a decisive influx of Christian intellectuals during the

12th century. These were areas that favored from early on the transcultural contact between multiple communities from the Latin Christian, Arabic Islamic and Arabic/Hebrew Jewish spheres.

Regarding the biographical background of the translators known by name, we find the following trends. In both centuries, there was a strong influx of scholars coming from abroad: mainly from today's Italy (Plato of Tivoli, Gerard of Cremona, Salio of Padua, among whom we could include Hermann of Carinthia) and England (Robert of Ketton, Robert of Chester, Michael Scot, William the Englishman), but also from Normandy (Hugo of Cintheaux), and even from Germanic territory (Hermann the German). The ones who were native to the peninsula were either Jews (Abraham bar Ḥiyya, Abraham ben Daud, Abuteus Levita, Yehudah ben Moses, Isaac ben Sīd, Samuel ha-Levi, Abraham ben Waqar of Toledo) or Mozarabs (John of Seville, Peter of Toledo, John of Spain, Mark of Toledo). As for the translators from abroad, we have little information about the motives that drove them to the peninsula; it is likely that Hugo of Cintheaux and Hermann of Carinthia, for example, made their way there within the context of the arrival of French scholars that had occurred since the late 11th century. At the same time, it is important to remember that Daniel of Morley speaks about Toledo's fame as a place of learning even outside of Iberia, and we are told in Gerard's *Vita* that he arrived on the peninsula seeking Ptolemy's *Almagest,* which he translated. His translation was then copied twice in Toledo, first by a certain Thaddeus from Hungary (1st half 13th century) and subsequently by a Frenchman (probably Roger of Fournival [1201–1260]; Burnett 2001, 253–5).

Regarding their social and institutional background, the translators that we have presented were either related to an ecclesiastical or a courtly context (except in the case of Plato of Tivoli, who was perhaps related to Barcelona's official administration through Abraham bar Ḥiyya, and Robert of Chester, who does not show any relation to the church or a court). The patronage for their translations often came from these contexts: in the Ebro Valley group, Robert of Ketton, Hermann of Carinthia and Peter of Toledo were paid by Peter the Venerable to translate religious Islamic texts, that is the *Corpus islamolatinum*, and Hugo of Cintheaux dedicated almost all his translations to the Bishop Michael of Tarazona. The Toledan translators of the 12th and 13th centuries were related to the church – as is the case with Mark of Toledo, close collaborator of Archbishop of Toledo Rodrigo Jiménez de Rada – and the Alfonsine circle was famously sponsored by the court of Alfonso X.

Even though the translating activities may be classified broadly by the social or political contexts in which they flourished, they were by no means confined to an absolute space such as a cathedral, a court or to a relative one such as a school. Rather, the main factor that allows us to conceive of them in terms of overlapping communities of learning is that some translators were connected to each other by their interests, their knowledge about the work of the other translators or their collaborative relations but not necessarily by their specific institutional affiliations or even by contemporaneity. We know, for instance, that the works of John of Seville were read by Hermann of Carinthia, who used his translations for composing his *On Essences* (*De essentiis*; Hermann of Carinthia 1982, 370–9), and by Dominicus Gundissalinus and Abraham ben Daud, who show their acquaintance with John of Seville's *On the Difference Between the Spirit and the Soul* (*De differentia spiritus et animae*) in their translation of Ibn Sīnā's *On the Soul* (*De anima*; Hasse 2000, 16, 2002, 35). In turn, Gundissalinus was familiar with Hermann's *De essentiis* (Gundissalinus 2002, 27–9; Hermann of Carinthia 1982, 62, 243–5, 316, 318–9). In another case, we see that the translation of Ptolemy's *Almagest* reached the compilers of the 14th-century Parisian Alfonsine tables through the Castilian tradition of the Alfonsine circle (Yehudah ben Moses and Isaac ben Sid 2003, 234, 243–4, 248).

At the same time, one must highlight important differences among the various translating communities, not only concerning the disciplines that they focused on – which have been presented in detail – but also regarding the translation methods used by the authors. In this context, it is worth noting that while the Ebro Valley translators – Hermann of Carinthia, Robert of Ketton and Hugo of Cintheaux – leaned toward a translation *sensum de sensu* and even felt compelled to justify this choice, the main preference among the Toledan translators of the 12th and 13th centuries was the translation *verbum pro verbo*, which was to become the norm in the Arabic-Latin translations and had been already the method of choice of John of Seville (Burnett 1997, 63–8, 2007, 1234–7). This difference between translation methods may be seen most readily through a comparison of the largely paraphrased rendering of the Qur'ān by Robert of Ketton and the more literal one by Mark of Toledo. In light of all this, a highly differentiated picture of multiple translating communities in medieval Iberia emerges, whose continuities and discontinuities account for the overwhelming wealth of the approaches to Arabic texts and their translation.

Note

1 Consolidated bilbiography.

Bibliography

Sources

Alfonso X of Castile. ed. Rico y Sinobas, J. M. 1863–1867. *Libros del saber de astronomía del rey d. Alfonso X de Castilla.* 5 vols. Madrid: Agaudo.

Besson, G. and Brossard-Dandré, M., eds. and trs. 1991. *Le livre de l'échelle de Mahomet. Liber scale Machometi.* Paris: Librairie générale française.

Bizzarri, H. Ó., ed. 2010. *Secreto de los secretos. Poridat de las poridades. Versiones castellanas del Pseudo-Aristóteles Secretum Secretorum.* Valencia: Publicacions de la Universitat de València.

Curtze, M. 1902. "Der *Liber Embadorum* des Abraham bar Chijja Savasorda in der Übersetzung des Plato von Tivoli," in M. Curtze, M., ed. *Urkunden zur Geschichte der Mathematik im Mittelalter und der Renaissance.* 2 vols. Leipzig: B. G. Teubner, 1: 3–183.

Daniel of Morley. ed. Maurach, G. 1979. "Daniel von Morley, 'Philosophia'," *Mittellateinisches Jahrbuch* 14: 204–55.

Gundissalinus, Dominicus. ed. and tr. Laumakis, J. A. 2002. *The Procession of the World (De processione mundi)* (Medieval Philosophical Texts in Translation 39). Milwaukee: Marquette University Press.

Gundissalinus, Dominicus and Abraham ibn Daud. ed. Van Riet, S. and Verbeke, G. 1972. *Liber de anima seu Sextus de naturalibus. Partes I–III.* 2 vols. Leiden: Brill. Vol. 1: Avicenna latinus 1.

Hermann of Carinthia. ed., tr. and com. Burnett, C. 1982. *De Essentiis. A Critical Edition with Translation and Commentary.* Leiden: Brill.

Hermann of Carinthia (?). ed. Busard, H. L. L. 1968. *The Translation of the Elements of Euclid from the Arabic into Latin by Hermann of Carinthia (?), Books I–VI.* Leiden: Brill.

Hermann of Carinthia (?). ed. Busard, H. L. L. 1977. *The Translation of the Elements of Euclid from the Arabic into Latin by Hermann of Carinthia (?), Books VII–XII.* Amsterdam: Mathematisch Centrum.

Hugo of Cintheaux. ed. Millàs Vendrell, E. 1963. *El comentario de Ibn al-Muṯannā' a las Tablas Astronómicas de al-Jwārizmī. Estudio y edición crítica del texto latino en la versión de Hugo Sanctallensis.* Madrid: CSIC.

Hyatte, R., tr. 1977. *The Prophet of Islam in Old French. The Romance of Muhammad (1258) and the Book of Muhammad's Ladder (1274). English Translations, with an Introduction.* Leiden: Brill.

Ibn Gabirol. ed. Benedetto, M. 2007. *Avicebron. Fonte della vita.* Milan: Bompiani.

Ibn Sīnā. ed. Hudry, F. 2018. *Avicenne. Logica (Logique du Šifā').* Paris: Vrin.

Koningsveld, P. van, ed. 1977. *The Latin-Arabic Glossary of the Leiden University Library.* Leiden: Brill.

Mark of Toledo. ed. Petrus Pons, N. 2016. *Alchoranus Latinus quem transtulit Marcus canonicus Toletanus.* (Nueva Roma 44). Madrid: CSIC.

Orazi, V., ed. 2006. *Sendebar: Libro de los engaños de las mujeres.* Barcelona: Crítica.

Peter the Venerable. ed. Glei, R. 1985. *Petrus Venerabilis. Schriften zum Islam* (Corpus Islamo- Christianum, Series Latina 1). Altenberge: CIS-Verlag.

Peter the Venerable. ed. Migne, J. P. 1890. *Petri Venerabilis abbatis Cluniacensis noni opera omnia*. Paris: Garnier Fratres. Patrologia Latina 189.

Robert of Chester (?). eds. and trs. Busard, H. L. L. and Folkerts, M. 1992. *Robert of Chester's (?) Redaction of Euclid's Elements, the so-called Adelard II Version*. 2 vols. Basel: Birkhäuser.

Robert of Chester. ed. and trans. Karpinski, L. C. 1915. *Robert of Chester's Latin Translation of the Algebra of al-Khowarizmi*. New York: Macmillan.

Rudolf of Bruges. ed., tr. and com. Lorch, R. 1999. "The Treatise on the Astrolabe by Rudolf of Bruges," in Nauta, L. and Vanderjagt, A., eds. *Between Demonstration and Imagination. Essays in the History of Science and Philosophy Presented to John D. North*. Leiden: Brill, 55–100.

Scot, M. ed. and tr. Oppenraaij, A. M. I. 1992. *Aristotle, De animalibus. Michael Scot's Arabic- Latin Translation*. Vol. 3: *Books XV-XIX. Generation of Animals* (Aristoteles Semitico-latinus 5). Leiden: Brill.

Yehudah ben Moses. ed. Fernández Montaña, J. 2001^2 [1881]. *Lapidario del rey d. Alfonso X: códice original* [The Lapidary of King Alfonso X: the Original Codex]. Madrid: BNE.

Yehudah ben Moses. ed. Hilty, G. 1954. *El Libro conplido en los iudizios de las estrellas. Traducción hecha en la corte de Alfonso el Sabio* [The Complete Book of the Judgments of the Stars. Translation Made at the Court of Alfonso the Wise]. Madrid: Real Academia Española.

Yehudah ben Moses and Isaac ben Sid. ed. Chabás, J. and Goldstein, B. R. 2003. *The Alfonsine Tables of Toledo*. Dordrecht: Reidel.

Research literature

Bertolacci, A. 2011. "A Community of Translators: The Latin Medieval Versions of Avicenna's *Book of the Cure*," in Mews, C. J. and Crossley, J. N., eds. *Communities of Learning: Networks and the Shaping of Intellectual Identity in Europe, 1100–1500*. Turnhout: Brepols, 37–54.

Bishko, C. J. 1956. "Peter the Venerable's Journey to Spain," in Constable, G. and Kritzeck, J., eds. *Petrus Venerabilis 1156–1956* (Studia Anselmiana 40). Rome: Herder, 163–74.

Boloix Gallardo, B. 2001. "La Taifa de Toledo en el siglo XI. Aproximación a sus límites y extensión territorial," *Ṭulayṭula* 8: 23–57.

Bosworth, C. E. 1991. "Djudhām," in *EI-2*, 2: 573.

Burman, T. 1994. *Religious Polemic and the Intellectual History of the Mozarabs, c. 1050–1200*. Leiden: Brill.

Burnett, C. 1977. "A Group of Arabic-Latin Translators Working in Northern Spain in the Mid- Twelfth Century," *Journal of the Royal Asiatic Society* 109.1: 62–108.

Burnett, C. 1978. "Arabic into Latin in Twelfth-Century Spain: The Works of Hermann of Carinthia," *Mittellateinisches Jahrbuch* 13: 100–34.

Burnett, C. 1993. "Al-Kindī on Judicial Astrology: 'The Forty Chapters'," *Arabic Sciences and Philosophy* 3: 77–117.

Burnett, C. 1994. "The Translating Activity in Medieval Spain," in Jayyusi, ed., 1036–58.

Burnett, C. 1995a. "'Magister Iohannes Hispalensis et Limiensis' and Qusṭā ibn Lūqā's *De differentia spiritus et animae*: A Portuguese Contribution to the Arts Curriculum?" *Mediævalia. Textos e Estudos* 7–8: 221–67.

Burnett, C. 1995b. "The Institutional Context of Arabic-Latin Translations of the Middle Ages: A Reassessment of the 'School of Toledo'," in Weihers, O., ed. *Vocabulary of Teaching and Research between the Middle Ages and Renaissance* (CIVICIMA, Études sur le vocabulaire intellectuel du moyen âge 8). Turnhout: Brepols, 214–35.

Burnett, C. 1997. "Translating from Arabic into Latin in the Middle Ages: Theory, Practice, and Criticism," in Lofts, S. G. and Rosemann, P. W., eds. *Éditer, traduire, interpréter: essais de méthodologie philosophique*. Louvain-la-Neuve: Editions de l'Institut supérieur de philosophie; Louvain: Peeters, 55–78.

Burnett, C. 2001. "The Coherence of the Arabic-Latin Translation Program in Toledo in the Twelfth Century," *Science in Context* 14.1–2: 249–88.

Burnett, C. 2002. "John of Seville and John of Spain: *A mise au point*," *Bulletin de Philosophie Médiévale* 44: 59–78.

Burnett, C. 2003. "Translations, Scientific, Philosophical, and Literary (Arabic)," in Gerli, E. M., ed. *Medieval Iberia. An Encyclopedia*. New York: Routledge, 801–4.

Burnett, C. 2004. "Ketton, Robert of (fl. 1141–1157)," in *Oxford Dictionary of National Biography*. Oxford: Oxford University Press. https://doi.org/10.1093/ref:odnb/23723.

Burnett, C. 2007. "Translation from Arabic into Latin in the Middle Ages," in Kittel, H., *et al.*, eds. *Übersetzung. Translation. Traduction. Ein internationales Handbuch zur Übersetzungforschung. An international Encyclopedia of Translation Studies. Encyclopédie internationale de la recherche sur la traduction* (Handbücher zur Sprach- und Kommunikationswissenschaft 26.1). 3 vols. Berlin: W. de Gruyter, 2: 1231–7.

Burnett, C. 2011a. "*De meliore homine.* ʿUmar ibn al-Farrukhān al-Ṭabarī on Interrogations: A Fourth Translation by Salio of Padua?" in Sannino, A., Arfè, P. and Caiazzo Lacombe, I., eds. *Adorare caelestia, gubernare terrena. Atti del Colloquio Internazionale in onore di Paolo Lucentini (Napoli, 6–7 Novembre 2007).* Turnhout: Brepols, 295–326.

Burnett, C. 2011b. "Communities of Learning in Twelfth-Century Toledo," in Mews, C. J. and Crossley, J. N., eds. *Communities of Learning: Networks and the Shaping of Intellectual Identity in Europe, 1100–1500.* Turnhout: Brepols, 9–18.

Calvo, E. 2017. "Some Features of the Old Castilian Alfonsine Translation of ʿAlī Ibn Khalaf's Treatise on the *Lámina Universal*," *Medieval Encounters* 23: 106–23.

Corral Lafuente, J. L. 1998. "La reconquista del Valle del Ebro," *Militaria: revista de cultura militar* 12: 49–67.

D'Alverny, M.-T. 1947–1948. "Deux traductions latines du Coran au Moyen Âge" *Archives d'histoire doctrinale et littéraire du Moyen Âge* 16: 69–131.

D'Alverny, M.-T. 1982. "Translations and Translators," in Benson, R. L. and Constable, G., eds. *Renaissance and Renewal in the Twelfth Century.* Cambridge: Cambridge University Press, 421–62.

D'Alverny, M.-T. 1989a. "Marc de Tolède," in *Estudios sobre Alfonso VI y la Reconquista de Toledo. Actas del II Congreso Internacional de Estudios Mozárabes, Toledo 20–26 mayo 1985.* 4 vols. Toledo: Instituto de Estudios Visigótico Mozárabes, 3: 25–59.

D'Alverny, M.-T. 1989b. "Les traductions à deux interprètes, d'arabe en langue vernaculaire et de langue vernaculaire en latin," in Contamine, G., ed. *Traduction et traducteurs au moyen âge. Acte du colloque international du CNRS organisé à Paris, Institut de recherche et d'histoire des textes, le 26–28 mai 1986.* Paris: CNRS, 193–206.

Dunlop, D. M. 1986. "D͟hu ʾl-Nūnids," in *EI-2*, 2: 242–3.

Dunlop, D. M. 1991. "Hūdids," in *EI-2*, 3: 542–3.

Fidora, A. 2009. *Domingo Gundisalvo y la teoría de la ciencia arábigo-aristotélica.* tr. L. Langbehn. Pamplona: EUNSA.

Freudenthal, G. 2016. "Abraham Ibn Daud, Avendauth, Dominicus Gundissalinus and Practical Math ematics in Mid-Twelfth Century Toledo," *Aleph: Historical Studies in Science and Judaism* 16.1: 61–106.

García-Guijarro, L. 2015. "Reconquest and the Second Crusade in Eastern Iberia: The Christian Expansion in the Lower Ebro Valley," in Roche, J. T. and Møller Jensen, J., eds., *The Second Crusade: Holy War on the Periphery of Latin Christendom.* Turnhout: Brepols, 219–55.

Gómez Aranda, M. 2009. "Enciclopedias hebreas en la época medieval," in Alvar Ezquerra, A., ed., *Las enciclopedias en España antes de* l'Encyclopédie. Madrid: CSIC, 105–24.

Goñi Gaztambide, J. 1965. "Los obispos de Pamplona del siglo XII," *Anthologica Annua* 13: 135–358.

González Muñoz, F. 2011. "Peter of Toledo," in Thomas, D. and Mallet, A., eds. *Christian-Muslim Relations. A Bibliographical History.* 3 vols. Leiden: Brill, 3: 478–82.

González Ruiz, R. 1997. *Hombres y libros de Toledo (1086–1300).* Madrid: Fundación Ramón Areces.

Haskins, C. H. 1924. *Studies in the History of Medieval Science.* Cambridge, MA: Harvard University Press.

Haskins, C. H. 1927. *The Renaissance of the Twelfth Century.* Cambridge, MA: Harvard University Press.

Hasse, D. N. 2000. *Avicenna's De Anima in the Latin West. The Formation of a Peripatetic Philosophy of the Soul 1160–1300.* London: Warburg Institute.

Hasse, D. N. 2002. "Plato arabico-latinus: Philosophy – Wisdom Literature – Occult Sciences," in Gersh, S. and Hoenen, M. J. F. M., eds. *The Platonic Tradition in the Middle Ages.* Berlin: W. de Gruyter, 31–66.

Hasse, D. N. 2006. "The Social Conditions of the Arabic-(Hebrew-)Latin Translation Movements in Medieval Spain and in the Renaissance," in Speer, A. and Wegener, L., eds. *Wissen über Grenzen: arabisches Wissen und lateinisches Mittelalter* (Miscellanea Mediaevalia 33). Berlin: W. de Gruyter, 68–88.

Hawting, G. R. 1995. "Rawḥ b. Zinbāʿ," in *EI-2*, 8: 466.

Hogendijk, J. P. 1986. "Discovery of an 11th-Century Geometrical Compilation: The *Istikmāl* of Yūsuf al-Muʾtaman bin Hūd, King of Saragossa," *Historia Mathematica* 13.1: 43–52.

Jayyusi, S. H., ed. *The Legacy of Muslim Spain.* Leiden: Brill.

König, D. G. 2015. *Arabic-Islamic Views of the Latin West: Tracing the Emergence of Medieval Europe.* Oxford: Oxford University Press.

Kritzeck, J. 1964. *Peter the Venerable and Islam.* Princeton: Princeton University Press.

Lemay, R. 1981. "Gerard of Cremona," in Gillispie, C. C., ed. *Dictionary of Scientific Biography*. New York: Scribner, 15: 173–92.

Lomba Fuentes, J. 1995. "El Islam en el Valle del Ebro: la cultura filosófica y científica," in de la Iglesia, J. I., ed. *V Semana de estudios medievales: Nájera, 1 al 15 de agosto de 1994*. Logroño: Gobierno de La Rioja, Instituto de Estudios Riojanos, 175–90.

Mackay, A. and Benaboud, M. 1979. "Alfonso VI of León and Castile, 'al-Imbraṭūr dhū-l-Millatayn'," *Bulletin of Hispanic Studies* 56.2: 95–102.

Makki, M. 1994. "The Political History of al-Andalus (92/711–897/1492)," in Jayyusi, ed., 3–87.

Martínez, S. H. tr. Cisneros, O. 2010. *Alfonso X, the Learned: A Biography*. Leiden: Brill.

Martínez Gázquez, J. 2011. "Islamolatina: Estudios sobre el *Corpus islamolatinum* (1142–1143) y literatura de controversia islamo-judeo-cristiana," in Prieto, C. E., ed. *Arabes in patria Asturiensium*. Oviedo: Universidad de Oviedo, 171–90.

Martínez Gázquez, J. 2012. "'Islamolatina'. La percepción del Islam en la Europa cristiana. Traducciones latinas del Corán. Literatura latina de controversia," *Medievalia* 15: 39–45.

Millàs Vallicrosa, J. M. 1942. *Las traducciones orientales en los manuscritos de la Biblioteca Catedral de Toledo*. Madrid: CSIC.

Murdoch, J. 2006. *Post-Structuralist Geography. A Guide to Relational Space*. London: Sage.

O'Callaghan, J. 1976. *A History of Medieval Spain*. Ithaca: Cornell University Press.

Pérez González, M. 1992. "Herman el Alemán, traductor de la Escuela de Toledo: Estado de la cuestión," *Minerva: Revista de filología clásica* 6: 269–83.

Pingree, D. 1981. "Between the *Ghāya* and *Picatrix* I: The Spanish Version," *Journal of the Warburg and Courtauld Institutes* 44: 27–56.

Polloni, N. 2020. *The Twelfth-Century Renewal of Latin Metaphysics: Gundissalinus's Ontology of Matter and Form*, Durham: Institute of Medieval and Early Modern Studies; Toronto: Pontifical Institute of Mediaeval Studies.

Porres Martín-Cleto, J. 1999. "La dinastía de los Banu Di L-Nun de Toledo," *Tulaytula: Revista de la Asociación de Amigos del Toledo Islámico* 4. 37–47.

Reglero de la Fuente, C. 2008. *Cluny en España. Los prioratos de la provincia y sus redes sociales (1073-ca. 1270)*. León: Centro de estudios e investigación San Isidoro.

Reichert, M. 2008. "Herman of Dalmatia and Robert of Ketton: Two Twelfth-Century Translators in the Ebro Valley," in Goyens, M., De Leemans, P. and Smets, A., eds. *Science Translated. Latin and Vernacular Translations of Scientific Treatises in Medieval Europe*. Leuven: Leuven University Press, 47–57.

Reilly, B. F. 1988. *The Kingdom of León-Castilla under King Alfonso VI, 1065–1109*. Princeton: Princeton University Press.

Roth, N. 1990. "Jewish Collaborators in Alfonso's Scientific Work," in Burns, R. I., ed. *Emperor of Culture: Alfonso X the Learned of Castile and His Thirteenth-Century Renaissance*. Philadelphia, PA: University of Pennsylvania Press, 59–71.

Saccenti, R. 2010. "La *Summa Alexandrinorum*. Storia e contenuto di un'epitome dell'*Etica Nicomachea*," *Recherches de théologie et philosophie médiévales* 77.2: 201–34.

Samsó, J. 1981. "Dos colaboradores científicos musulmanes de Alfonso X," *Llull: Revista de la Sociedad Española de Historia de las Ciencias y de las Técnicas* 4.6–7: 171–9.

Samsó, J. 2004. "El procès de la transmissió científica al nord-est de la península ibèrica al segle XII: els textos," in Parés, R. and Vernet, J., eds. *La ciència en la història dels països catalans*. 3 vols. Valencia: Universitat de València, Institut d'Estudis Catalans, 1: 269–96.

Santoyo, J. C. 2016. "El normando Hugo de Cintheaux (*Hugo Sanctelliensis*), traductor en Tarazona (ca. 1145)," in Carta, C., Finci, S. and Mancheva, D., eds. *Antes se agotan la mano y la pluma que su historia. Magis deficit manus et calamus quam eius hystoria. Homenaje a Carlos Alvar*. 2 vols. San Millán de la Cogolla: Cilengua, 1: 341–57.

Saurette, M. 2010. "Peter the Venerable and Secular Friendships," in Classen, A. and Sandidge, M., eds. *Friendship in the Middle Ages and Early Modern Age*. Berlin: Springer, 281–308.

Sela, S. 2005. "Abraham bar Hiyya," in Glick, T., Livesey, S. J. and Wallis, F., eds. *Medieval Science, Technology, and Medicine. An Encyclopedia*. New York and London: Routledge, 2–4.

Smith, C. 1999. "The Vernacular," in Abulafia, D., ed. *The New Cambridge Medieval History. Volume 5: c. 1198-c. 1300*. Cambridge: Cambridge University Press, 71–83.

Southern, R. W. 1992[4]. *Robert Grosseteste: The Growth of an English Mind in Medieval Europe*. Oxford: Oxford University Press.

Stalls, C. 1995. *Possessing the Land: Aragon's Expansion into Islam's Ebro Frontier under Alfonso the Battler, 1104–1134*. Leiden: Brill.

Szilágyi, K. 2016. "A Fragment of a Book of Physics from the David Kaufmann *Genizah* Collection (Budapest) and the Identity of Ibn Daud with Avendauth," *Aleph: Historical Studies in Science and Judaism* 16.1: 11–31.

Turk, ʿA. 1998. "La marca superior como vanguardia de al-Andalus: su papel político y su espíritu de independencia," *Al-Andalus Magreb: Estudios árabes e islámicos* 6: 237–50.

Vegas González, S. 1998. *La Escuela de Traductores de Toledo en la historia del pensamiento*. Toledo: Concejalía de Cultura.

Vicente García, L. M. 2002. "La importancia del *Libro conplido en los iudizios de las estrellas* en la astrología medieval (Reflexiones sobre la selección de obras astrológicas del códice B338 del siglo XV del Archivo Catedralicio de Segovia)," *Revista de literatura medieval* 14.2: 117–34.

Weijers, O. 1987. *Terminologie des universités du XIIIe siècle* (Lessico intellettuale europeo 39). Rome: Edizioni dell'Ateneo.

Williams, S. J. 2003. *The Secret of Secrets: The Scholarly Career of a Pseudo-Aristotelian Text in the Latin Middle Ages*. Ann Arbor: University of Michigan Press.

PART VI

Encounters, conflicts, changes (4th–13th/10th–19th centuries)

VI.1
CROSS-COMMUNAL SCHOLARLY INTERACTIONS

Nathan P. Gibson[1] *and Ronny Vollandt*

The current chapter is concerned with particular fields that held a special position in cross-communal scholarly interaction: medicine, the natural sciences and the mathematical sciences. These constituted domains of knowledge that did not directly overlap with scriptural disciplines such as exegesis or law (and so are in what Goldstein [2002] calls a "neutral zone"). As a result, these were perhaps the domains where the most wholesale exchange of knowledge could take place among different communities. We attempt here to understand in concrete terms where and in which contexts cross-communal scholarly interactions took place.[2]

VI.1.1 Introduction and state of research

"Creative symbiosis" is the irenic term that was coined by Shlomo Dov Goitein (1900–1985) to describe various forms of cross-communal interaction among the members of different religious groups in the Near East.[3] His research was based on the Cairo Genizah documents, which provide an unparalleled insight into everyday life, being a particularly fertile source for examining social and economic history. The "Genizah people" or "Mediterranean people," as he would call the protagonists of his documents, were – it appears – members of a pluralistic Islamicate society and encompassed Jews, Christians and Muslims alike. Unlike in Europe, Jewish communities in the Islamicate world, as people of the book (*ahl al-kitāb* in Arabic), enjoyed the protection to exercise their faith, with similar protections extended to Christians and Zoroastrians.[4] This freedom was based on the concept of *dhimmah*, which can be translated as "contract of security," granted by Muslim rulers on the condition that protected people (*dhimmī*, collectively *ahl al-dhimma*) respected certain rules of conduct and paid a poll tax (*jizya*) and a collectively levied tax on agricultural land (*kharāj*). Furthermore, Jews and Christians were not subject to limitations or professional restrictions in the economic sphere and were given the right to self-govern. The Genizah texts give evidence that Jews, Christians and Muslims lived in very close proximity, were business partners and even owned houses jointly.[5]

It seems a somewhat abstract commonplace in modern scholarship that these everyday social relationships were echoed in scholarly exchanges (Freidenreich and Goldstein 2012; Ben-Shammai *et al.* 2013). There is a consensus that confessional boundaries seem to blur in intellectual pursuits. The permeability among the learned elite – that is, the mutual exchange among all societal components that Hodgson (1974) has termed Islamicate – naturally follows from the fact that

DOI: 10.4324/9781315170718-65

these communities shared a language. After the Islamic conquests in the 1st/7th and 2nd/8th centuries and following the practices of the chancelleries of the newly installed rulers, Arabic slowly became the common language of the entire region and the spoken tongue of most of its non-Arab inhabitants (with the exception of those living in Iran and Central Asia and regions only later coming under Muslim rule in Africa, Asia and Europe). By virtue of this paradigmatic shift, by the 3rd/9th century, a unifying Arabic literacy had come into being that encompassed both Muslim and non-Muslim writers. The educated elite and the common people, who partook much less actively in the realm of intellectual high culture, shared a similar cultural background, speaking and writing in the same language.

All texts, terminology, innovative literary models, textual practices and genres composed in Arabic could easily travel beyond communal barriers. Examples can be adduced from almost all fields of learning (which, of course, partly overlap): exegesis (Zucker 1984; Ben-Shammai 2003), philosophy (Ben-Shammai 1997), theology (*kalām*; Bertaina 2014, 2015; Griffith 1994), grammatical thought (Becker 1993, 1995; Basal 1988, 1999), legal reasoning (Freidenreich 2014; Libson 2003; Salaymeh 2015), medicine (Pormann and Savage-Smith 2007[CB])[6] and the mathematical sciences (Goldstein 2002). Hava Lazarus-Yafeh (1992, 4) described the result as "a palimpsest, layer upon layer, tradition upon tradition, intertwined to the extent that one cannot really grasp one without the other, certainly not the later without the earlier, but often also not the earlier without considering the shapes it took later." Different terms have been used to conceptualize this entangled textual commonality shared by Jews, Christians, Muslims and others inhabiting the same space. Terms such as *impact* and *influence* stress the agency of the donor community, and *acculturation* and *appropriation* stress the agency of the receptor community (Freidenreich and Goldstein 2012, 1). However, all these terms exhibit explanatory models that reduce complex forms of interaction to static binary encounters. Even the images of "cross-pollination" (Goodman 1995, 1999; Montgomery 2007) or "intertwinement" (Lazarus-Yafeh 1992), which profess reciprocity, struggle to fully capture the multiple, simultaneous dynamics of cross-communal engagement.

This chapter begins by discussing interactions that occurred primarily through texts, as scholars shared books and exchanged ideas, terminology and concepts (Section VI.1.2). But the chapter goes beyond this textual sphere, asking questions about the personal and professional networks that underlay the textual encounters and about the spheres in which scholars met. Further themes concern particular venues or occasions and social factors that encouraged cross-communal interactions. To address a much-needed area of research, we pursue these questions in a preliminary way in three sections (VI.1.3–VI.1.5).

In previous research, one very particular type of interaction has been heavily emphasized – the *majlis* (plural *majālis*). The *majālis* were public or semipublic meetings, sometimes at a caliph's or emir's court, that included disputations on a variety of religious topics. Jewish, Christian and Muslim sources provide ample details on such debates (Lazarus-Yafeh *et al.* 1999). For example, the historian al-Masʿūdī (d. 345/956) mentions in his *Book of Admonition and Revision* (*Kitāb al-Tanbīh wa-l-ishrāf*) that he had been involved in many debates with Abū Kathīr al-Kātib, the scribe, of Tiberias (d. 319/932) on the subject of the abrogation of law (*naskh*; on identifying al-Kātib, see Polliack 1997, 12 n. 39; Zucker 1984, 253 n. 266). Al-Masʿūdī also reveals that Abū Kathīr was the teacher of Saʿadia Gaʾon (268–330/882–942) and relates that the latter attended the *majlis* of the vizier Ibn al-Jarrāḥ (245–335/859–946) and his entourage (al-Masʿūdī 1894, 112–4). As fascinating as it is to trace these cross-communal *majlis* connections, they show only a sliver of the spectrum of scholarly interactions. *Majālis* appear to have been highly formalized and performative encounters among specially chosen, distinguished scholars, who followed a strict protocol and sometimes used polemic for effect.

The following discussion also wrestles with three basic difficulties underlying the topic of cross-communal scholarly interactions, which can be noted here but not fully resolved. The first is holding in tension the fact that medicine, the natural sciences and the mathematical sciences were considered faith-neutral and rational, as mentioned earlier, while grasping that they were not really "secular." On one hand, the need to reconcile Galenic and Aristotelian views with belief systems stemming from the Bible or the Qurʾān seems to have hardly hindered an active interchange and sometimes synthesis across communal boundaries. On the other hand, it was very often the case that scholars who were involved in these areas led their respective religious communities and engaged in apologetic and polemic discourse rooted in the same philosophical systems that undergirded their work in the natural sciences. With the important exception of "prophetic medicine," which was based on *ḥadīth*s (Pormann and Savage-Smith 2007, 71–5; Bürgel 2016[CB], 34–47), these sciences were neither separable from theological disciplines nor wedded to them.

The second difficulty concerns the language of borrowing, which presumes that *one* group borrows from or is influenced by the *other* (Salaymeh 2013, 412–3). Underlying such language is the notion of intellectual authenticity, that only the latter group can claim ownership over a particular scholarly practice, whereas it enters the former group as an alien influence. Thus, this practice must inevitably pass an imagined border. It is the modern interpreter who presumes to reify here a certain directional vector and the moment it passes such a border. Such language obscures, and actively ignores, the historical reality of hybridities in the Islamicate world. Sarah Stroumsa (2011) offers an alternative model, in which she suggests that the flow of ideas was never unilateral or linear, originating in one community and being transmitted to another but, rather, went back and forth. This movement created something she proposes calling a "whirlpool effect."

This brings us to the third difficulty, somewhat connected to the previous, which is the slipperiness of these scholars' religious affiliations in both primary sources and research literature (Salaymeh 2013, 413–4). Almost all medieval scholars seem to have been associated with some religious community on a social level, and some important social categories were defined in religious terms (e.g., Muslim and *dhimmī*). What this association actually meant regarding their beliefs or practices, however, could cover a very wide spectrum and should never be assumed on the basis of a label alone. For example, the scholar of mathematical sciences Ibn al-Haytham ([354–*c.* 430/965–*c.* 1040]; Alhazen in Latin; Vernet 1986b) might superficially appear to have been a Muslim – indeed, he wrote a treatise on finding the *qibla* – but he claimed to have set aside confessional disciplines deriving from scriptural revelation in order to reach epistemological certainty:

> I became engrossed in the variety of views and creeds and the kinds of religious knowledge, but I did not have the good fortune to benefit from any of them. They did not help me recognize the path of truth or follow a renewed course to certainty. So I saw that I could not get to the root of truth except by conceptions whose origins are sensory and whose forms are intelligible.
>
> *(Ibn Abī Uṣaybiʿa 1884[CB], 2: 92, 2020[CB], 14.22.4.1)*[7]

He proceeds to explain that the only approach he found adequate was an Aristotelian one, beginning with classification and logic and ending with a metaphysical account of God. Another example is Yūḥannā ibn Māsawayh (d. 243/857–8), who is called a "Christian" but who reportedly insulted the Catholicos-Patriarch with obscenities and shooed away monks from his sickbed with the remark, "A bit of rose perfume is better than the prayers of all Christians" (Ibn

Abī Uṣaybiʿa 1884, 1: 186, 2020, 8.26.7). Similarly, it is difficult to know how to concretely interpret supposed "conversions" (Stroumsa 2015). In addition to these issues regarding the spectrum of connotations for affiliation labels, one also has to question the accuracy of the sources that use them. A classic case is that of ʿAlī ibn Rabban al-Ṭabarī (3rd/9th century), who is known to have been a Christian convert to Islam but whom the Muslim biographers Ibn al-Qifṭī (568–646/1172–1248) and Ibn Abī Uṣaybiʿa (d. 668/1270; Vernet 1986a) label as originally Jewish due to confusion over his father's title, "al-Rabban" (Thomas 2000). In the following, then, it should be understood that these factors make it impossible to use confessional labels such as "Jewish," "Christian" or "Muslim" with consistent or certain meaning – the biography of each scholar must be consulted individually.

VI.1.2 Textual production and migration

At the level of texts, it is clear that there was interaction between the scholars of different communities. The presence of Christian and Muslim books on Jewish bookshelves, for example, is well attested in medieval Jewish book lists from the Cairo Genizah that document private catalogs, booksellers' lists and library inventories (Allony 2006).

The fields of natural science and medicine feature prominently in these lists. Miriam Frenkel (2017) has shown that many of the buyers and owners come from a well-defined social circle. Some were physicians and held public office in the Muslim administration; others were judges or cantors in the Jewish community. One list, recording the sale of books from the estate of Rabbi Abraham the Pious (Abraham he-Ḥasid, 7th/13th century), notes explicitly that a number of medical works were sold to Muslim colleagues (Frenkel 2017, 240; Allony 2006, n. 67) – for example, copies of Ibn Rushd's (520–595/1126–1198) *General Principles of Medicine* (*al-Kulliyyāt fī l-ṭibb*), a multiple-text manuscript with Hippocratic medicine and a separate book on ophthalmology were sold to a certain Ḥājj Bū Muḥammad. These transactions indicate that books circulated among equal-ranking members of the same profession in governmental service, irrespective of their denomination.

Specimens of such books have survived in the Cairo Genizah, where they seem to have been deposited together with manuscripts in Hebrew script that would have formed the larger part of the collections documented in the lists (Figure IV.1.1). Compositions of Christian Arabic provenance that were disseminated among Cairene Jews consist of works addressing a broader, general readership, such as medical science or philosophy (Szilágyi 2006). Among them one finds an early fragment of *Definitions of Logic* (*Kitāb ḥudūd al-manṭiq*; Ferrario and Vollandt 2010) by Ibn Bahrīz (2nd half 2nd–early 3rd century/2nd half 8th–early 9th century), *Questions on Medicine* (*Masāʾil fī l-ṭibb*) by Ḥunayn ibn Isḥāq (d. 260/873), *The Introduction to the Art of Geometry* (*al-Madkhal ilā ṣināʿat al-handasa*) of Qusṭā ibn Lūqā (d. *c.* 299/912), *The Reminder of the Oculists* (*Tadhkirat al-kaḥḥālīn*) by ʿAlī ibn ʿĪsā (d. 1st half 5th/11th century), and *The Physicians' Dinner Party* (*Daʿwat al-aṭibbāʾ*) of Ibn Buṭlān (d. 458/1066); many of these were considered standard reading for Jewish physicians. Arabic translations of the Aristotelian corpus (including commentaries thereon) can also be found (Khan 1986). A great many copies of works in Arabic by Hippocrates (particularly the *Aphorisms*) and Galen survive. Further attested are the *Diwān* of Ṭarafa ibn al-ʿAbd Abū ʿAmr al-Bakrī al-Wāʾilī (6th century) and that of al-Mutanabbī (d. 354/955), various Arabic grammars (Vidro and Kasher 2014) and books on rhetoric. Several disciplines stand out: Arabic language and literature, studied by Jews and Christians in training for governmental service, as well as science and philosophy, as preparation for the medical profession.

Although no systematic survey has been undertaken of Jewish ownership of Arabic-script manuscripts (which could be done, e.g., on the basis of owners' marks, marginalia in Hebrew

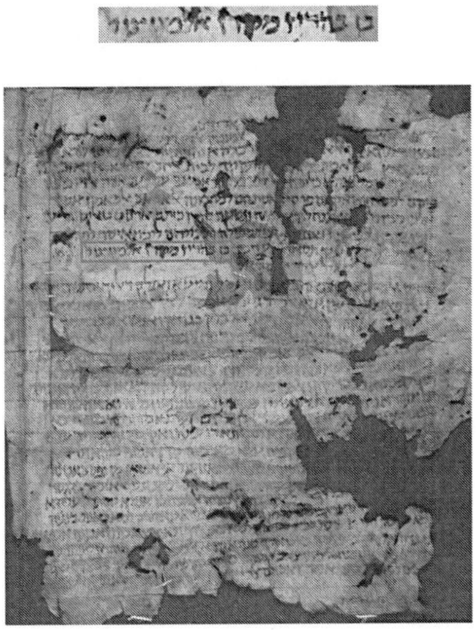

Figure VI.1.1 Detail of the name "Ibn Bahrīz Muṭrān al-Mawṣilī" (2nd half 2nd–early 3rd/late 8th–early 9th centuries) in a Genizah fragment (from Cambridge, Cambridge University Library, T-S K6.181r, 4th/10th century). The fragment is an example of Christian Arabic works circulating in a Jewish context, since it preserves in Judeo-Arabic a portion of the treatise *Definitions of Logic* (*Kitāb ḥudūd al-manṭiq*) by the Christian East Syriac metropolitan, ʿAbdīshūʿ ibn Bahrīz (Ferrario and Vollandt 2010).

letters and colophons mentioning copyists with unambiguously Jewish names), van Koningsveld (1992) has been able to identify a number of medical manuscripts from al-Andalus that were in the possession of Jewish physicians. Jews even transmitted Christian and Muslim works on grammar, medicine (with Ibn Sīnā's [d. 428/1037] *Canon of Medicine* [*al-Qānūn fī l-ṭibb*] the most popular), astronomy and astrology, philosophy, geometry and meteorology, together with various almanacs transcribed into Hebrew letters (Langermann 1996a, 1996b; Steinschneider 1893, 1897).

We also find Jewish and Muslim texts in Christian Arabic collections, with medical and scientific texts appearing to have been the most widespread. Some of the books are biblical or post-biblical in their content, such as Saʿadia Gaʾon's *Commentary* (*Tafsīr*) or an Arabic translation of *Sefer Josippon*, both originally composed in Judeo-Arabic for a Jewish readership. They were transcribed in Arabic letters in the course of transmission and disseminated among both Christians and Muslims (Vollandt 2014, 2018). As earlier, no systematic study exists, and indeed only a few collections are catalogued well enough to allow such an investigation, but a few examples will suffice to attest to the cross-communal circulation of such books. Representative of a monastic context, MS Sinai, Ar. NF Paper 11, at the Monastery of St. Catherine's, contains the *Complete Book of the Medical Art* (*Kitāb kāmil al-ṣināʿa al-ṭibbiyya*; Ullmann 1970[CB], 140–6) by ʿAlī ibn al-ʿAbbās al-Majūsī (d. *c.* 384/994), copied by the Christian scribe Khalīl ibn Hibbat Allāh ibn Abī Alūfā ([*fl.* late 7th/13th century?]; Meimaris 1985, 40/٢٤). Equally, the manuscript collection in the Coptic Orthodox Patriarchate (Simaika 1939, 2: 486–7) contains a few Muslim medical works (MSS Varia 17, 18, 20, 21, and 22) – these include the *Medical Handbook* (*Tadhkira*) of Dāwūd al-Anṭākī ([d. 1007/1599]; MS Varia 20; see Ullmann 1970, 181) and Ibn Sīnā's *Canon of Medicine* (MS Varia 21), as well as a work titled the *Splendid Book* (*Kitāb al-fākhira*;

MS Varia 17), of Jewish provenance, which is attributed to a certain ʿAbdallāh al-Isrāʾīlī, the physician.

Similarly, albeit from much later times, medical treatises in Arabic were also transmitted in Syriac script (Garshuni) among Christian communities. Examples of this include MSS Jerusalem, St. Mark's Monastery, 236 (two medical treatises on diseases and on remedies) and 238 (*Complete Book of the Medical Art* [*Kitāb kāmil al-ṣināʿa al-ṭibbiyya*] by al-Majūsī); MSS Mardin, Church of the Forty Martyrs, 556 (medical treatise) and 555/2 (treatise on medical knowledge); MS Batnaya, Chaldean Church of Batnaya, 51 (recipes); and MS Mosul, Syrian Orthodox Archdiocese of Mosul, 206 (medical treatise).[8]

VI.1.3 Teacher–student relationships and learning circles

Examples of cross-communal learning abound in medieval bibliographic and biographical works, such as Ibn al-Nadīm's (d. 380/990) *Catalogue* (*Fihrist*) and Ibn Abī Uṣaybiʿa's 7th/13th-century *History of Physicians* (*ʿUyūn al-anbāʾ fī ṭabaqāt al-aṭibbāʾ*, literally *Choice Accounts of the Classes of Physicians*). Some of these were formally acknowledged teacher–student relationships, while others were one-off consultations. Again, some examples must suffice in place of a systematic study, which is still needed.

In Baghdad and its environs, Christian scholars involved in translating classical Greek or Syriac works into Arabic would have been natural tutors for the works they translated, commented on and summarized. Biographers do not seem to have been surprised by intellectual lineages like that of the West Syriac Christian logician Yaḥyā ibn ʿAdī ([d. 363/974]; Endress 2002; Ibn al-Nadīm 1970[CB], 2: 631; Ibn Abī Uṣaybiʿa 1884, 1: 235, 2020, 10.22), who studied under the Muslim philosopher al-Fārābī (d. 339/950–1), himself a student of the East Syriac Christian philosopher Yūḥannā ibn Ḥaylān (*fl.* late 3rd–early 4th/late 9th–early 10th century; Walzer 1991; Janos 2015; Ibn Abī Uṣaybiʿa 1884, 2: 135, 2020, 15.1.2).

On issues of scriptural exegesis, scholars are known to have consulted associates from other communities, and the same type of activity is likely in other fields. (Scriptural consultations are also attested indirectly through the reception by Muslim writers of biblical material from non-Arabic languages or scripts, such as Syriac and Judeo-Arabic [Adang 1996; Griffith 2004; Gibson 2017].) One well-known example stems from the academy of Pumbedita, the Geonic academy that had moved to Baghdad at the beginning of the 4th/10th century, where Hai Gaʾon (327 or 328–429/939–1038) requested that the Sicilian Maṣliaḥ bar Eliyahu (Ibn al-Baṣaq) inquire of the East Syriac Catholicos regarding Syriac commentary traditions for an enigmatic verse, Psalm 141:5 (Dubovick 2018). When Maṣliaḥ objected, Hai Gaʾon responded, "Our pious forefathers [...] would inquire regarding languages and their explanations from members of different religions, even from shepherds and cow-hands" (Dubovick 2018, 99). A more specific example of literary exchange that involved both philosophical and medical questions is the correspondence of two Jews from Mosul with the Christian scholar and translator Yaḥyā ibn ʿAdī, mentioned above. The inquirers, Bishr ibn Samʿān and Ibn Abī Saʿīd, show that they are familiar with the work of Thābit ibn Qurra (d. 288/901), a Sabian from Harran. Yaḥyā suggests they might receive a better answer from one of Bishr's own acquaintances, the Christian physician and translator Ibn Bakkūsh (Sklare 1996, 115–6 and n. 52).

Cross-communal instruction seems to have been quite prevalent not only in Iraq, but also in Cairo throughout the Fatimid, Ayyubid and Mamluk periods. The Muslim physician Raḍī al-Dīn al-Raḥbī (534–631/1139 or 1140–1233) reportedly considered Jews and Christians unworthy to be his students (Ibn Abī Uṣaybiʿa 1884, 2: 193, 2020, 15.36.1.1). Yet the fact that a biographer would remark on his stance seems to reveal that it was rare. Moreover, al-Raḥbī made an exception for the Jew ʿImrān al-Isrāʾīlī ([560–637/1165 or 1166–1239]; Ibn Abī Uṣaybiʿa 1884, 2: 213–4, 2020, 15.42) and for the Samaritan Ibrāhīm ibn Khalaf (late 6th–early 7th/12th–13th centuries), both of whom became prominent physicians according to Ibn Abī Uṣaybiʿa. Al-Raḥbī was himself a student of the famed Egyptian Jewish scholar Ibn Jumayʿ ([d. *c.* 594/1198]; Ibn Abī Uṣaybiʿa 1884, 2: 112–5, 2020, 14.32; Nicolae 2017), personal physician to Ṣalāḥ al-Dīn (r. 564–589/1169–1193), and of the eminent Christian medical scholar Ibn al-Tilmīdh ([d. 560/1165]; Meyerhof 1986). Thus al-Raḥbī's own connections militate against any general inference that Muslim scholars were generally reticent to teach non-Muslims. Ibn Abī Uṣaybiʿa similarly comments that Saʿīd ibn Hibat Allāh ibn al-Ḥusayn ([436–495/1045–1101]; Ibn Abī Uṣaybiʿa 1884, 1: 254–5, 2020, 10.58) refused to teach Jews but made an exception for Abū l-Barakāt al-Baghdādī ([d. *c.* 560/1164–5]; Ibn Abī Uṣaybiʿa 1884, 1: 287, 2020, 10.66.1; Pines 1986).

In fact, the learning circles of Ibn Riḍwān ([388–453/998–1061 or later]; Schacht 1986b) in Cairo and later of Ibn Abī Uṣaybiʿa in Cairo and Damascus suggest the opposite. Ibn Riḍwān, a self-taught scholar and a Muslim (in some respects, anyway), became chief physician under the Fatimids and an intellectual ancestor to several Jewish physicians. He dedicated works to the doctor Yahūdā ibn Saʿāda (presumably Jewish, otherwise unknown) and taught another Jewish physician in the Fatimids' employ, Afrāʾīm ibn al-Zaffān (Schacht 1986b). The latter's Jewish student Salāma ibn Raḥmūn was well known in intellectual circles and had a son Mubārak, who, according to Ibn Abī Uṣaybiʿa (1884, 2: 106–7, 2020, 14.28), was "an eminent physician." Ibn Riḍwān's scholarship was thus disseminated throughout the Jewish community of 5th/11th-century Cairo, and manuscripts from the Genizah in fact mention his writings.

Ibn Abī Uṣaybiʿa's report of his own network is similarly diverse, showing learning in the other direction, from Jewish and Christian teachers to Muslim students.[9] On the Jewish side, he mentions that his own father, Sadīd al-Dīn al-Qāsim (575–649/1179 or 1180–1251), studied under Moses Maimonides ([d. 601/1204]; Ibn Abī Uṣaybiʿa 1884, 2: 247, 2020, 15.51.1). Later, between 631/1233 and 632/1235, Ibn Abī Uṣaybiʿa met Maimonides's son Abraham (581 or 582–635/1186–1237), a fellow physician, while working in a Cairo hospital (Ibn Abī Uṣaybiʿa 1884, 2: 118, 2020, 14.40.2). Ibn Abī Uṣaybiʿa also had the opportunity to witness the side-by-side labor of his teacher al-Dakhwār (d. 628/1230, chief physician of Egypt and Syria under the Ayyubid Sultan al-ʿĀdil I [r. 596–649/1200–1252]; Joosse 2018) with the Jewish physician ʿImrān al-Isrāʾīlī in the Nūrī Hospital of Damascus *(al-bīmāristān al-nūrī)*: "Every benefit resulted from their collaboration, and they were prepared to offer to the patients every good kind of treatment" (Ibn Abī Uṣaybiʿa 1884, 2: 214, 2020, 15.42, see also 15.50.3).

Christians and converts from Christianity were also significant in Ibn Abī Uṣaybiʿa's circles in Ayyubid Damascus and Mamluk Cairo. His intellectual lineage went back to a Christian convert to Islam, Raḍī al-Dawla (6th/12th century), son of Ibn al-Tilmīdh (Ibn Abī Uṣaybiʿa 1884, 1: 264–5, 2: 203, 2020, 10.64.16, 15.40.3). Ibn Abī Uṣaybiʿa's primary teacher, al-Dakhwār, was the "best student" of another Christian convert to Islam, Asʿad ibn al-Muṭrān (d. 587/1191) and spent much time with him (Ibn Abī Uṣaybiʿa 1884, 2: 179, see also 2: 193, 239, 2020, 15.23.4.1, see also 15.36.1, 15.50.1). The Christian Yaʿqūb ibn Siqlāb (d. *c.* 626/1229) also met frequently with al-Dakhwār, and Ibn Abī Uṣaybiʿa (1884, 2: 215, 2020, 15.43.1) describes his therapeutic skills as unrivaled. By his own account, Ibn Abī Uṣaybiʿa himself met the Samaritan vizier Yūsuf ibn Abī Saʿīd (d. 624/1227; Ibn Abī Uṣaybiʿa 1884, 2: 233–4, 2020, 15.48) and corresponded with a Samaritan convert to Islam, the vizier Amīn al-Dawla ([mid–7th/13th century]; Ibn Abī Uṣaybiʿa 1884, 2: 235–7, 2020, 15.49.6). Biographies such as those of Ibn Riḍwān and Ibn Abī Uṣaybiʿa give every indication that medical education exemplified the "whirlpool effect" of Stroumsa (2011) mentioned earlier.

VI.1.4 Patronage and clientele

No doubt a factor in the pluralistic composition of medieval scholarship in the Islamicate world was the fact that caliphs, sultans and other high-ranking personalities employed experts from all communities (Fiey 1980; Yarbrough 2012, 364, 380; Cabrol 2000; Sirry 2011). The eminent place of non-Muslim (*dhimmī*) scholars in these retinues sometimes brought down the ire of leading Muslim thinkers such as the polemicist al-Jāḥiẓ (d. 255/869), who blamed the prestige of these intellectuals for the vacillation of Muslim believers (al-Jāḥiẓ 1964–1979, 3: 315–6; Gibson 2015). He argued, moreover, that Christians were not real scholars but, rather, mere conveyors of Greek classical knowledge, which they had inherited by geographical accident. Muslim rulers who employed non-Muslims in high positions also faced a problem of public perception because of the Qurʾānic injunctions against seeking the help and friendship of nonbelievers. However, none of these arguments or principles seem to have much dampened the desire of Muslim rulers to recruit the best scholars and practitioners, wherever they might be found on the religious map. When Caliph al-Mutawakkil (r. 232–247/847–861) attempted to ban non-Muslims from positions of authority over Muslims (as other rulers both before and after him also tried to do), he exempted his personal staff from this prohibition (Yarbrough 2012, 364, 380). Indeed, the very fact that prohibitions against hiring *dhimmī*s continued to be repeated suggests that the issue remained salient for several centuries. In Iraq under the early Abbasids, it was Christian translators, secretaries and doctors who were especially visible recipients of the rulers' patronage, and it would take a few generations before Muslim or Jewish physicians would outshine the reputation of Christian ones.

Despite these perceptions, there were some prominent Jewish public intellectuals, for example in the sciences of the stars. The Jew Māshāʾallāh ([d. *c.* 199/815]; Kennedy and Pingree 1971; Pingree 1975), whose Hebrew name was Misha according to Ibn al-Nadīm (1970, 2: 650), was among the astrologers whom the second Abbasid caliph, al-Manṣūr (r. 136–158/754–775) consulted regarding the date on which to found the city of Baghdad (Samsó 1991). Other prominent Jewish astrologers in the 3rd/9th century included Sahl ibn Bishr (d. *c.* 235/850) and ʿAlī ibn Dāwūd (Goldstein 2001, 26).

In the following century, scholars would move to Qayrawan and Cairo and find support there from the Fatimid rulers, beginning several hundred years of rich, cross-communal collaboration. A picture of this emerges from comparing historiographical sources with Genizah documents, which sometimes mention the same figures and certainly depict many similar aspects of scholarship and medical practice. It is also under the Fatimids that the role of Jewish physicians would become vital in this exchange. One of the most influential of these, and a particularly notable example of cross-communal patronage, was Isḥāq ibn Sulaymān al-Isrāʾīlī ([d. after 320/932]; Ibn Abī Uṣaybiʿa 1884, 2: 36–7, 2020, 13.2; Ṣāʿid al-Andalusī 1985[CB]). He was a Neoplatonic philosopher and physician born in Egypt, who migrated to Raqqāda and Qayrawan to serve the last Aghlabid emir, Ziyādat Allāh III (r. 290–296/903–909). While there, he studied with Isḥāq ibn ʿImrān (d. around 295/908), a Muslim physician from Baghdad, who had also been recruited by Ziyādat Allāh III. Isḥāq al-Isrāʾīlī was subsequently appointed court physician by the Fatimid ruler ʿUbayd Allāh al-Mahdī (r. 297–323/910–934). His books included both medical treatises and philosophical works (Guttmann 1911; Altmann 1979; Altmann and Stern 1958; Sezgin 1970, 3: 295–7; Levin *et al.* 2018), topics about which Saʿadia Gaʾon corresponded with him while the latter was still in Egypt (Fenton 2002, 3–4, 12; see also Altmann and Stern 1958; Hirschberg 1974, 271). Isḥāq al-Isrāʾīlī's legacy in Qayrawan would live on through his pupils, Aḥmad ibn Ibrāhīm ibn Abī Khālid (Ibn al-Jazzār [d. 395/1004 or 1005]; Sezgin 1970, 304–7; Ibn Abī Uṣaybiʿa 1884, 2: 37–9, 2020, 13.3) and Dūnash ibn Tamīm ([d. *c.* 349/960];

from *dhū nās*, "master of men," translated from the Hebrew Adonim; Vajda 2002; Sezgin 1970, 295–7). The latter served as a physician to the Fatimid Caliph al-Manṣūr (r. 334–341/946–953) in Qayrawan, and both men penned influential treatises.

Somewhat later, in 405/1015, Hai Ga'on at the academy of Pumbedita (mentioned above) appointed the physician Abraham ibn Nathan ([1st half of 5th/11th century]; Ibn ʿAṭā) as "Nagid *ha-gola*" or "prince" of the diaspora, a duty which apparently overlapped with the medical services he provided to the Zirid rulers of Tunisia, Bādīs ibn al-Manṣūr ([r. 386–406/996–1016]; Idris 1986) and his son al-Muʿizz ibn Bādīs ([r. 406–454/1016–1062]; Talbi 1993; Goitein 1971, 24, 244; Ben-Sasson 1996, 1997). The title may have been, in part, a formal acknowledgment of Abraham's intercession on behalf of the Jewish community (Goitein 1971, 24). This kind of throne-room diplomacy was a role that numerous preeminent scholars, both Jewish and Christian, were called on to play as de facto heads of their communities (Goitein 1971, 243–5 and n. 12). An example is the case of Samuel ben Ḥananya (Abū Manṣūr, in office 533 or 4–553 or 4/1140–1159), who was asked to intercede regarding the tax on sugar makers (MSS Cambridge, CUL, T-S 10J15.29 + T-S 10J15.32).

The Fatimids' move to Cairo in 362/972, with physicians in attendance and a program of support for medicine, may well have been one of the catalysts that spurred on the practice of medicine in the Egyptian Jewish community. Ibn Abī Uṣaybiʿa (1884, 2: 88, 2020, 14.14.3) specifically mentions court physicians who came in the retinue of the Caliph al-Muʿizz (r. 341–365/953–975). Indeed, al-Muʿizz and his successors al-ʿAzīz (r. 365–386/975–996) and al-Ḥākim (r. 386–411/996–1021) seem to have cultivated a large cadre of Muslim, Jewish and Christian medical experts in Cairo, according to the account of Ibn Abī Uṣaybiʿa. The legacy of Christian scholarship from Baghdad was already circulating in the region, in part through Ibrāhīm ibn ʿĪsā (d. *c.* 260/873–874), who studied with the famous physician Yūḥannā ibn Māsawayh and later migrated to Fustat (near Cairo) with his employer Aḥmad ibn Ṭūlūn ([r. 254–270/868–884]; Ibn Abī Uṣaybiʿa 1884, 2: 83, 2020, 14.2). Yūḥannā ibn Māsawayh's name appears frequently in Genizah medical fragments. It was under al-Muʿizz that one of the preeminent Jewish medical families became established, that of Moses ben Eleazar (d. after 363/973), who had come from Oria in southern Italy by way of Tunisia and was another student of Isḥāq al-Isrāʾīlī (see earlier). Al-Muʿizz employed Moses together with his two sons Isḥāq (d. 363/973 or 974) and Ismāʿīl and his grandson Yaʿqūb ibn Isḥāq. A great-grandson, also named Moses, seems to have served the Fatimid court well into the next century (Ibn Abī Uṣaybiʿa 1884, 2: 86, 2020, 14.9; Goitein 1971, 2: 243 and nn. 9, 10).

Al-Muʿizz's son and heir, al-ʿAzīz, employed at least two Christian physicians: the Melkite Sahlān ibn ʿUthmān (d. 380/991) and Manṣūr ibn Sahlān ibn Muqashshir (d. before 411/1021). The latter served into the reign of al-Ḥākim and was reportedly a favorite of his. On his death, he was succeeded in al-Ḥākim's retinue by another Christian, Isḥāq ibn Ibrāhīm ibn Nasṭās ibn Jurayj, who also died during al-Ḥākim's reign and was replaced by the eminent Ibn Riḍwān (mentioned earlier), who became chief physician (Ibn Abī Uṣaybiʿa 1884, 2: 99–105, 2020, 14.25).[10]

During the reign of one of the later Fatimid caliphs, al-Āmir (r. 495–525/1101–1130), two particularly distinguished scholars migrated from Andalusia. One was Yūsuf ibn Aḥmad ibn Ḥasdāy, from the Jewish Ḥasday family of Andalusia, who attached himself to the vizier al-Maʾmūn al-Baṭāʾiḥī (held office 515–519/1121–1125). Ibn Abī Uṣaybiʿa (1884, 2: 51, 2020, 13.51.1) calls him "eminent in the medical profession" and says his reputation "became well known" during his time in Egypt. Yūsuf ibn Ḥasdāy's own religious affiliation is unclear (Stroumsa 2015, 23, 27). The other was the Muslim Abū l-Ṣalt (460–529/1067–1134), who originated from Denia and was active first in Cairo and Alexandria around the turn of the 6th/12th century, and then in Mahdiyya (Tunisia), where he died in 529/1134 (Comes 2000; Millás and Stern 1986; Ibn Abī Uṣaybiʿa 1884, 2: 52–62, 2020, 13.58). His works were particularly popular among Jewish communities in Andalusia, with some of them being translated into Hebrew (Comes 2000; Millás and Stern 1986).

Some of these scholars were actively recruited by rulers while others came of their own initiative, but it is clear that intellectuals coming to Cairo and its environs during the Fatimid period could hope to find both patronage and a lively exchange of ideas with renowned experts, regardless of their religious background. This kind of official support is equally well attested in documents of the Cairo Genizah and in Ibn Abī Uṣaybiʿa's history through the Ayyubid and into the early Mamluk periods. For some, the status of these non-Muslims as *dhimmīs* was less important than the profession they represented. It is reported that Ibn Abī l-Ḥawāfir (d. after 616/1220), chief physician under the Ayyubid al-Mālik al-ʿAzīz (r. 589–595/1193–1198), once rebuked a Jewish oculist for standing rather than sitting down to attend to the eyes of a chickpea seller: "Although you yourself may be lowly, out of regard for the profession you should sit to his side and tend his eyes rather than remain standing in the presence of a common chickpea vendor" (Ibn Abī Uṣaybiʿa 1884, 2: 119, 2020, 14.44.3).

Rulers who surrounded themselves with physicians were not only the latter's patrons but also, of course, their patients (Figure VI.1.2). However, thanks to Genizah documents we have records of many ordinary patients. The role of women in these exchanges is particularly worthy of interest, since the patients mentioned are often female. Besides women who were patients, the Genizah also mentions female oculists (Goitein 1967, 127–8, 1971, 255). Doctor–patient interactions often had a cross-communal dimension. Genizah texts reveal that it was common for physicians from any community to treat patients from other communities, and this was true of Jewish doctors as of others, as we can see in the Genizah documents from the prescriptions in a variety of scripts, with different religious formulas, and varied names (Goitein 1971, 254; Chipman and Lev 2010, 78). For example, a Muslim doctor treated a Jewish girl with dropsy (Goitein 1967, 259); a Muslim family hired a Jewish doctor for a monthly fee (MS Cambridge, CUL, T-S 13J6.16; Goitein 1971, 256); payments were due to a Jewish physician for his daily visits to apparently Muslim patients (MS Cambridge, CUL, T-S Ar.4.10, in Judeo-Arabic); and a Christian doctor is said to have treated a Jewish patient for no fee (MS Cambridge, CUL,

Figure VI.1.2 ʿAbdallāh (ʿUbaydallāh?) ibn Bukhtīshūʿ (d. 396/1006), of the renowned family of Christian physicians, converses with Emir Saʿd al-Dīn. MS London, BL, Or. 2984, fols 101b-102a, 7th/13th century.

Source: © British Library Board, London

T-S 8J20.26; Goitein 1971, 252). It is important to note that there is one indication of a religious boundary affecting medical practice, which is the absence of any documentation of Jewish patients convalescing in hospitals, even though Jewish physicians labored in them. Goitein (1971, 251–2) suggests this was to avoid transgressing dietary laws, but it is hard to make an argument from silence. In any case, the general picture is that medical services went in every direction, and that these interactions were an opportunity to exchange more than just prescriptions or medical advice.

VI.1.5 Shared workplaces

The fact that doctors from various communities worked at the court or in public hospitals provided an opportunity for them to engage with one another as colleagues, or sometimes as rivals. Schwarb cautions:

> Very few documents provide evidence for an intellectual exchange between Christians and Jews during the Fatimid and Ayyubid periods, apart from the fact that Jewish and Christian physicians worked for the same institutions, the Bīmaristān al-Nāṣirī for example, and served as officials in the various government ministries (*dawāwīn*).
>
> *(Schwarb 2014, 114; see n. 26 for references)*

Strictly speaking, it is true that most of the documentary evidence does not speak of the content of exchanges but simply puts Christians, Jews and Muslims active in the same times and places. But this in itself is quite significant as a circumstance for exchange. It may also be the case that Genizah documents indicate more about Jewish–Muslim than Jewish–Christian exchange. Nevertheless, historiographical sources can usefully complement this picture. Even though documentary texts (other than book lists) seldom inform us what colleagues discussed, they often indicate how they met. Together with literary and historiographical texts, one can synthesize from them a rich picture of cross-communal interaction in the workplace.

Scholarly rivalries were common, and a person's religious affiliation sometimes formed the vector for an attack, but it is not clear that interreligious collegial disputes were necessarily more common than intra-religious ones. Among the religiously diverse group of physician-scholars working under the Fatimids in the early 5th/11th century, the earlier dominance of Christians in the field of medicine and in the Greek classical sciences generally was being rivaled or even slowly surpassed by Jewish and Muslim expertise, according to the portrayal given by Ibn Abī Uṣaybiʿa in his *History of Physicians*. This might have been the cause for – or a symptom of – certain rivalries. For example, Ibn Riḍwān had public and vicious disputes with the Christian philosopher-physician Ibn Buṭlān of Baghdad, who visited Fustat for about three years beginning in 441/1049 (Ibn Abī Uṣaybiʿa 1884, 1: 241, 2020, 10.38.2–3). Yet their refutations of one another seem to have focused on scientific debate and some *ad hominem* attacks rather than religious wrangling (Ibn Buṭlān and ʿAlī ibn Riḍwān 1937; Schacht 1986a, 1986b), even though Ibn Riḍwān is known to have engaged in religious polemic elsewhere.[11] Ibn Abī Uṣaybiʿa does not comment on any religious dimensions of the Ibn Riḍwān-Ibn Buṭlān dispute, nor does he do so when he mentions the envy of the convert Asʿad ibn al-Muṭrān for a certain Abū l-Faraj, a Christian in the service of Ṣalāḥ al-Dīn (Ibn Abī Uṣaybiʿa 1884, 2: 176, 2020, 15.23.1.2), or the success of the Jewish physician al-Ḥaqīr al-Nāfiʿ ([late 4th–early 5th/10th–11th centuries]; Ibn Abī Uṣaybiʿa 1884, 2: 189, 2020, 14.18) in treating a leg wound that had thwarted the Christian Ibn Muqashshir.[12] Workplace tensions could involve a religious dimension, but there does not seem to be evidence that religious differences ordinarily engendered conflict in medicine or related fields. While some of these scholars wrote religious polemics,[13] this fact alone should not be taken to indicate poor relationships among these communities – polemic writing could sometimes be an exercise of expressing intellectual disagreement with others without necessarily holding personal enmity toward them.

In contrast to scholarly rivalries, physicians sometimes furthered the careers of their associates from other communities. In one of the most significant documents to shed light on cross-communal scholarly interaction, a Jewish physician, Makārim ibn Isḥāq (1st half 7th/13th century), asked the sultan for the remainder of the pay (probably a stipend or allowance) of a certain al-Asʿad (*al-bāqī min jāmakiyyat al-Asʿad al-ṭabīb*), who worked in "the Cairo hospital" (MS Cambridge, CUL, T-S Ar.40.16; Richards 1992). What exactly he means by "the remainder" has not been conclusively settled, nor has the hospital to which he refers. But the references he provides – doctors known to the sultan – are a Christian physician (Abū Ḥulayqa [591–675/1195–1277]) and one who is either Christian or Muslim (al-Rashīd al-Dimashqī [1st half 7th/13th century]). Both of these, he says, "know the excellence of the humble servant's knowledge of this art" (Richards 1992, 301). The communal ties of these experts apparently did not hinder them from putting in a good word for someone outside their community.[14] Along similar lines, an aspiring Jewish medical student in Cairo who wanted a hospital position in Alexandria was advised by his Alexandrian cousin to get letters of recommendation from various prominent people, who happened to span the confessional spectrum (MS, Cambridge, CUL, T-S 24.67; cited in Goitein 1971, 249–50).[15] Physicians and scholars were expected to know the work of their colleagues in other communities and be able to speak in their favor.

Physicians and apothecaries sometimes also physically shared workplaces with those of different confessions (Figure VI.1.3). A Genizah document records a court case in which an

Figure VI.1.3 Illustration of the preparation of a cough elixir in an Arabic manuscript of Dioscorides's *On Medicinal Substances*. MS New York, The Metropolitan Museum of Art, acc. no. 13.152.6, Rogers Fund (1913); possibly from Iraq, dated 621/1224. This pharmacopeia (handbook of medicinal drugs) was exceptionally popular among scholars of all communities and was first translated into Arabic in the 3rd/9th century by (among others) the Christian scholars Iṣṭifān ibn Basīl (1st half 3rd/9th century) and Ḥunayn ibn Isḥāq (d. 260/873). A Genizah document records a Jewish physician working with a Christian one in a medical drug shop.

Source: © The Metropolitan Museum of Art, New York Public domain.

apparently Jewish physician worked together with a Christian physician in a medical potions shop and thus had opportunity to witness the latter's affair with a Jewish woman (MS Cambridge, CUL, Or.1080 J93; Goitein 1970, 106–7, 1971, 253). Moreover, real estate dealings for the medicinal trade attested in the Genizah frequently involve relationships across communal lines (Goitein 1971, 262–4). While it may not be possible to know what arrangements were or were not typical for medicine shops, the evidence does not suggest anything unusual about sharing or transferring spaces between communities.

VI.1.6 Summary and future research directions

The differing sacred texts, observances, linguistic heritages and authority structures of Jews, Christians and Muslims in the medieval Islamicate world seem to have provided little hindrance to cross-communal scholarship. One might even go so far as to say that it was the exception rather than the norm for these distinctions to play a decisive role in scholarly exchange. Scholars from all these communities could read and transmit scientific texts in the common language of Arabic, whether in Arabic or Hebrew script, as attested by book lists and bibliographic histories. They could openly debate ideas in the *majlis*, even though they remained aware that those ideas sometimes had sensitive religious implications. Non-Muslims could usually study with leading experts and reach the pinnacle of professional success, notwithstanding their formal status as members of *dhimmī* communities (groups which were simultaneously protected and restricted). In fact, by virtue of their attainments they could often advocate on behalf of their own communities. And they could work shoulder-to-shoulder with scholars from other communities for the same patron, in the same hospital, or from the same shop.

Although this general picture of cross-communal scholarly interaction emerges from both documentary and historiographical sources throughout the medieval period, much of it has yet to be confirmed by systematic studies. Future research should focus on discovering the concrete details of both interpersonal and textual exchanges. On the interpersonal level, this would include the occasions of scholars' engagement across communities and the specific networks to which they belonged, both of which are necessary to identify the dynamics that helped or hindered collaboration. On the textual level, large-scale study of book ownership could reveal a flow of ideas in much richer detail than has previously been understood. Finally, future research must take into account the true complexity of these interactions: the slipperiness of affiliational labels and the "whirlpool effect" of multicommunal life and scholarship in Islamicate societies.

Notes

1 Work on this chapter has been funded in part by the German Federal Ministry of Education and Research as part of the project "Communities of Knowledge: Interreligious Networks of Scholars in Ibn Abi Usaybiʿa's History of the Physicians" (https://usaybia.net).

2 The lack of comprehensive studies in this area necessarily limits our scope in this chapter. We have concentrated here on Egypt, Mesopotamia and Syria prior to the 8th/14th century. For a discussion of al-Andalus, see Chapter V.2.

3 See the development of this term in Goitein's thought (1949, 1955, 1967).

4 Nineteenth-century European scholars of the Wissenschaft des Judentums constructed a vision of a "Golden Age" in the history of Judaism, a myth of an interfaith utopia, as it were, mirroring their own struggle toward cultural, legal, and political inclusion. This was eventually replaced by a countermyth stressing the inferior status and suffering of Jews under Islam (Cohen 1986, 1991).

5 For example, MS Cambridge, CUL, T-S 8.4, a Genizah fragment that is often quoted in discussions of shared ownership of properties, contains a letter about a house in Minya Zifta that was jointly owned by the Muslim judge ʿAlī ibn al-Qāsim and the son of a rabbi (Goitein 1971, 292; translated in Outhwaite *et al.* 2017, 23).

6 Consolidated bibliography.

7 Authors' translation, here and elsewhere for citations of this source. Readers may also wish to consult the parallel English translation using the cited paragraph numbers.

8 We are indebted to Adam McCollum for pointing these texts out to us.

9 It should be remembered that medieval sources sometimes used intellectual genealogies to represent scholars' pedigrees rather than their social relationships. Nevertheless, Ibn Abī Uṣaybiʿa would probably not have reason to exaggerate his own connections to Jews and Christians.

10 Schacht (1986b) thinks his promotion to chief physician could not have been during the reign of al-Ḥākim, which ended when Ibn Riḍwān was 23, but must have rather been during the reign of al-Mustanṣir (r. 427–487/1036–1094 or 5).

11 He is reported to have written a refutation against the Christian Ibn Zurʿa ([331–308/943–1008]; Abū ʿAlī ʿĪsā ibn Isḥāq ibn Zurʿa; Lewis *et al.* 1986; Monferrer Sala 2010) and the Jewish Afrāʾīm (presumably ibn al-Zaffān, Ibn Riḍwān's student, see Section VI.1.3) about the differences among religions (Ibn Abī Uṣaybiʿa 1884, 2: 104, 2020, 14.25.9).

12 Al-Ḥaqīr al-Nāfiʿ is known among biographers only by this epithet, which means something like "the contemptible one who is beneficial."

13 As well as the example of Ibn Riḍwān already discussed, a slightly earlier example is that of the Jewish philosopher ʿAbd al-Masīḥ al-Isrāʾīlī al-Raqqī (*fl.* late 4th or early 5th/late 10th or early 11th century), who became a Christian under the influence of Abū l-Fatḥ Manṣūr ibn Muqashshir (mentioned in section VI.1.4) and wrote polemical works against Judaism. He mentions Ibn Muqashshir in the inscription in his book *Dialectic* (*Kitāb al-istidlāl*; Swanson 2010, 538; Samir 1991). His works include such titles as *Refutation of the Jews* (*al-Radd ʿalā al-yahūd*) and *The Triumph of the Cross over Judaism and Paganism* (*Intiṣār al-ṣalīb ʿalā al-yahūdiyya wa-l-wathaniyya*).

14 Rashīd al-Dīn Abū Ḥulayqa seems to have been active in Cairo from 599/1202 or 1203 (Richards 1992, 302). Ibn Abī Uṣaybiʿa says that Abū Ḥulayqa's son converted to Islam (1884, 2: 130, 2020, 14.55.1), that Abū Ḥulayqa's grandfather was a Christian (1884, 2: 121, 2020, 14.49.1), and that he himself was "dedicated to the duties he undertook with much [religious] devotion" (1884, 2: 123, 2020, 14.54.1), all of which would suggest a Christian affiliation. Most of what Ibn Abī Uṣaybiʿa records about him relates to his service to al-Kāmil (r. 615–636/1218–1238), who became viceroy after coming to Egypt with his father al-ʿĀdil in 596/1200 (Ibn Abī Uṣaybiʿa 1884, 2: 123–30, 2020, 14.54; Gottschalk 1997). As suggested by Richards (1992, 303), al-Rashīd al-Dimashqī could be identified with Rashīd al-Dīn Abū Saʿīd ibn Muwaffaq al-Dīn Yaʿqūb (d. 646/1249), a Christian physician from Jerusalem who studied in Damascus, began in al-Kāmil's service in 632/1234 or 1235 and served under al-Mālik al-Ṣāliḥ Ayyūb (r. 637–647/1240–1249). Ibn Abī Uṣaybiʿa (1884, 2: 131–2, 2020, 14.56.2) speaks of his interaction with Abū Ḥulayqa while treating al-Kāmil. Al-Rashīd al-Dimashqī could alternatively be the Muslim medical scholar and mathematician Rashīd al-Dīn ʿAlī ibn Khalīfa (579–617/1183 or 4–1219), the teacher of the previously mentioned Abū Saʿīd and uncle of Ibn Abī Uṣaybiʿa. His home was Damascus, but he spent one or more periods in Cairo. From 605/1209 he was known to the Ayyubids and was in the service of some of them until his death (Ibn Abī Uṣaybiʿa 1884, 2: 246–59, 2020, 15.51; see Richards 1992, 303). For our purposes, it is not necessary to establish with certainty the identity of the al-Asʿad whose salary Makārim wants the remainder of, only to note that he could be either Muslim or Jewish (or, perhaps, neither). We cannot rule out, as Richards does, the most famous al-Asʿad during this period, the (apparently) Muslim scientist, physician, legal expert and poet Asʿad al-Dīn ʿAbd al-ʿAzīz ibn Abī l-Ḥasan ʿAlī (570–635/1174 or 5–1237 or 8), who joined al-Kāmil's service in Egypt after 626/1229 (Ibn Abī Uṣaybiʿa 1884, 2: 125–6, 132, 2020, 14.54.6, 14.57; Richards 1992, 303–4). Richards cites the disparity between the three dinars requested by Makārim and the 100 dinars per month which Ibn Abī Uṣaybiʿa says Asʿad al-Dīn ʿAbd al-ʿAzīz received in a previous post. However, if this Asʿad worked in the Cairo hospital, we do not know what his salary was there, how much of his total income it represented, or whether it may have later been redistributed (after his death?) in a way that its "remainder" would be three dinars. An alternative Jewish candidate for al-Asʿad is Asʿad al-Dīn Yaʿqūb ibn Isḥāq al-Maḥallī, who was active in Cairo late 6th/12th–early 7th/13th centuries, but it is not clear whether he was in the sultan's employ (Ibn Abī Uṣaybiʿa 1884, 2: 118, 2020, 14.42; Richards 1992, 303).

15 These are the *wālī* (chief of police), the *qāḍī* (judge), al-Muwaffaq, Ben Tammām and Ben Ṣadaqa. As suggested by Goitein, al-Muwaffaq might be the eminent Ibn Jumayʿ (Jewish physician to Ṣalāḥ al-Dīn, mentioned in Section VI.1.3), but the title al-Muwaffaq is too common to say for certain; Ibn Jumayʿ (Nicolae 2017) is associated with Alexandria in Genizah documents and wrote a treatise about the city

(Ibn Abī Uṣaybiʿa 1884, 2: 115, 2020, 14.32.5). Ben Tammām is probably Abū l-Maʿālī ibn Tammām, another Jewish physician whom Ṣalāḥ al-Dīn employed (Chipman and Lev 2006, 156). Finally, Ben Ṣadaqa could be the Samaritan Ṣadaqa ben Mīkhā ben Ṣadaqa, as Goitein apparently thinks (1971, 250), but it is difficult to see how the latter, who served the Ayyubid ruler of Damascus al-Ashraf Mūsā (d. 635/1237) and died in Ḥarrān, would be a decision-maker for an Alexandrian hospital (Ibn Abī Uṣaybiʿa 1884, 2: 118, 230–3; 2020, 15.46.3, 15.47).

Bibliography

Sources

Ibn Buṭlān and ʿAlī ibn Riḍwān. eds. Schacht, J. and Meyerhof, M. 1937. *The Medico-Philosophical Controversy between Ibn Butlan of Baghdad and Ibn Ridwan of Cairo: A Contribution to the History of Greek Learning among the Arabs*. Cairo: Egyptian University, Faculty of Arts.

al-Jāḥiẓ, Abū ʿUthmān ʿAmr ibn Baḥr al-Fuqaymī. ed. Hārūn, A. M. 1964–79. *Rasāʾil al-Jāḥiẓ* [Epistles of al-Jāḥiẓ]. 4 vols. Cairo: Maktabat al-Khājī.

al-Masʿūdī, Abū l-Ḥasan ʿAlī ibn al-Ḥusayn. ed. De Goeje, M. J. 1894. *Kitāb al-Tanbīh wa-l-ishrāf* [The Book of Admonition and Revision]. Leiden: Brill.

Research literature

Adang, C. 1996. *Muslim Writers on Judaism and the Hebrew Bible: From Ibn Rabban to Ibn Hazm*. New York: Brill.

Allony, N. ed. Frenkel, M. and Ben-Shammai, H. 2006. *The Jewish Library in the Middle Ages: Book Lists from the Cairo Genizah*. Jerusalem: Ben-Zvi Institute.

Altmann, A. 1979. "Creation and Emanation in Isaac Israeli: A Reappraisal," in Twersky, I., ed. *Studies in Medieval Jewish History and Literature*. Vol. 1. Cambridge, MA: Harvard University Press, 1–15.

Altmann, A. and Stern, S. M. 1958. *Isaac Israeli: A Neoplatonic Philosopher of the Early Tenth Century*. Oxford: Oxford University Press.

Basal, N. 1988. "Part One of *al-Kitāb al-muštamil* by Abū l-Faraj Hārūn and Its Dependence on Ibn al-Sarrāj's *Kitāb al-Uṣūl fī l-naḥw*," [in Hebrew] *Leshonenu* 61: 191–209.

Basal, N. 1999. "The Concept of *Ḥāl* in the *al-Kitāb al-muštamil* of Abū al-Faraj Hārūn in Comparison with Ibn al-Sarrāj," *Israel Oriental Studies* 19: 391–408.

Becker, D. 1993. "The Dependence of R. Yona b. Ǧanah on the Arab Grammarians," [in Hebrew] *Leshonenu* 57: 137–45.

Becker, D. 1995. "Concerning the Arabic Sources of R. Jonah ibn Janah," [in Hebrew] *Teuda* 9: 143–68.

Ben-Sasson, M. 1996. *The Emergence of the Local Jewish Community in the Muslim World (Qayrawan 800–1057)* [in Hebrew]. Jerusalem: Magnes Press.

Ben-Sasson, M. 1997. "The Emergence of the Qayrawan Jewish Community and Its Importance as a Maghrebi Community," in Golb, N., ed. *Proceedings of the Founding Conference of the Society for Judaeo-Arabic Studies*. Amsterdam: Harwood, 1–13.

Ben-Shammai, H. 1997. "Kalām in Medieval Jewish Philosophy," in Frank, D. H. and Leaman, O., eds. *History of Jewish Philosophy*. London: Routledge, 115–48.

Ben-Shammai, H. 2003. "The Tension between Literal Interpretation and Exegetical Freedom: Comparative Observations on Saadia's Method," in McAuliffe, J. D., Walfish, B. and Goering, J. W., eds. *With Reverence for the Word: Medieval Scriptural Exegesis in Judaism, Christianity, and Islam*. Oxford: Oxford University Press, 33–50.

Ben-Shammai, H., Shaked, S., and Stroumsa, S., eds. 2013. *Exchange and Transmission across Cultural Boundaries: Philosophy and Science in the Mediterranean World*. Jerusalem: The Israel Academy of Sciences and Humanities.

Bertaina, D. 2014. "*Ḥadīth* in the Christian Arabic *Kalām* of Būluṣ ibn Rajāʾ (c. 1000)," *Intellectual History of the Islamicate World* 2: 267–86.

Bertaina, D. 2015. "Christian Kalâm," in Touati, H., ed. *Encyclopedia of Mediterranean Humanism*. www.encyclopedie-humanisme.com/?Christian-Kalam.

Cabrol, C. 2000. "Une étude sur les secrétaires nestoriens sous les Abbassides 762–1258 à Bagdad," in "Actes du 5e Congrès international d'études arabes chrétiennes, Lund, août 1996," special issue, *Parole de l'Orient* 25: 407–91.

Chipman, L. and Lev, E. 2006. "Syrups from the Apothecary's Shop: A Genizah Fragment Containing One of the Earliest Manuscripts of *Minhāj al-Dukkān*," *Journal of Semitic Studies* 51.1: 137–68.

Chipman, L. and Lev, E. 2010. "Arabic Prescriptions from the Cairo Genizah," *Asian Medicine* 6.1: 75–94.

Cohen, M. R. 1986. "Islam and the Jews: Myth, Counter-Myth, History," *Jerusalem Quarterly* 38: 125–37.

Cohen, M. R. 1991. "The Neo-Lachrymose Conception of Jewish-Arab History," *Tikkun* 6: 55–60.

Comes, M. 2000. "Umayya b. ʿAbd al-ʿAzīz, Abu 'l-Ṣalt," in *EI-2*, 10: 836–7.

Dubovick, Y. M. 2018. "'Oil, Which Shall Not Quit My Head': Jewish-Christian Interaction in Eleventh-Century Baghdad," *Entangled Religions* 6: 95–123.

Endress, G. 2002. "Yaḥyā b. ʿAdī," in *EI-2*, 11: 245–6.

Fenton, P. 2002. Introduction to *Le commentaire sur le Livre de la création de Dūnaš ben Tāmīm de Kairouan (Xᵉ siècle)* by G. Vajda. Leuven: Peeters, 1–20.

Ferrario, G. and Vollandt, R. 2010. "A Judaeo-Arabic Version of Ibn Bahrīz's Treatise on Logic, T-S K6.181," *Fragment of the Month* (blog), Genizah Research Unit, Cambridge University Library, September 2010. www.lib.cam.ac.uk/collections/departments/taylor-schechter-genizah-research-unit/fragment-month/fragment-month-13-2.

Fiey, J. M. 1980. *Chrétiens syriaques sous les Abbassides, surtout à Bagdad (749–1258)* (Corpus Scriptorum Christianorum Orientalium 420). Louvain: Secrétariat du Corpus SCO.

Freidenreich, D. M. 2014. "Walking Side by Side: Engagement with Islamic Law and Theology in Rabbinic Legal Literature," *The Muslim World* 104: 413–17.

Freidenreich, D. M. and Goldstein, M., eds. 2012. *Beyond Religious Borders: Interaction and Intellectual Exchange in the Medieval Islamic World*. Philadelphia: University of Pennsylvania Press.

Frenkel, M. 2017. "Book Lists from the Cairo Genizah: A Window on the Production of Texts in the Middle Ages," *Bulletin of the School of Oriental and African Studies* 80.2: 233–52. https://doi.org/10.1017/S0041977X17000519.

Gibson, N. P. 2015. *Closest in Friendship? Al-Jāḥiẓ' Profile of Christians in Abbasid Society in 'The Refutation of Christians' (al-Radd ʿalā al-naṣārā)*. PhD thesis. Washington, DC: Catholic University of America. https://cuislandora.wrlc.org/islandora/object/cuislandora:28277.

Gibson, N. P. 2017. "A Mid-Ninth-Century Arabic Translation of Isaiah? Glimpses from al-Jāḥiẓ," in Hjälm, M. L., ed. *Senses of Scripture, Treasures of Tradition: The Bible in Arabic among Jews, Christians and Muslims* (Biblia Arabica 5). Leiden: Brill, 327–69.

Goitein, S. D. 1949. "Jewish-Arab Symbiosis," [in Hebrew] *Molad* 2: 259–66.

Goitein, S. D. 1955. *Jews and Arabs: Their Contacts Through the Ages*. New York: Schocken.

Goitein, S. D. 1967. *A Mediterranean Society*. Vol. 1: *Economic Foundations*. Berkeley: University of California Press.

Goitein, S. D. 1970. "Minority Selfrule and Government Control in Islam," *Studia Islamica* 31: 101–6. https://doi.org/10.2307/1595067.

Goitein, S. D. 1971. *A Mediterranean Society*. Vol. 2: *The Community*. Berkeley: University of California Press.

Goldstein, B. R. 2001. "Astronomy and the Jewish Community in Early Islam," *Aleph: Historical Studies in Science and Judaism* 1: 17–57.

Goldstein, B. R. 2002. "Science as a 'Neutral Zone' for Interreligious Cooperation," *Early Science and Medicine* 7.3: 290–1. https://doi.org/10.1163/157338202X00162.

Goodman, L. E. 1995. "Crosspollinations – Philosophically Fruitful Exchanges between Jewish and Islamic Thought," *Medieval Encounters* 1: 323–57.

Goodman, L. E. 1999. *Jewish and Islamic Philosophy: Crosspollinations in the Classical Age*. Edinburgh: Edinburgh University Press.

Gottschalk, H. L. 1997³. "al-Kāmil," in *EI-2*, 4: 520–1.

Griffith, S. H. 1994. "Faith and Reason in Christian *Kalām*: Theodore Abū Qurrah on Discerning the True Religion," in Samir, S. K. and Nielsen, J., eds. *Christian Arabic Apologetics during the Abbasid Period, 750–1258*. Leiden: Brill, 1–43.

Griffith, S. H. 2004. "The Gospel, the Qurʾān, and the Presentation of Jesus in al-Yaʿqūbī's *Taʾrīkh*," in Reeves, J. C., ed., *Bible and Qurʾān: Essays in Scriptural Intertextuality*. Leiden: Brill, 133–60.

Guttmann, J. 1911. *Die philosophischen Lehren des Isaak b. Salomon Israeli*. Münster: Aschendorff.

Hirschberg, H. Z. 1974². *History of the Jews in North Africa*. Vol. 1: *From Antiquity to the Sixteenth Century*, eds. Bashan, E. and Attal, R. Leiden: Brill.

Hodgson, M. G. S. 1974. *The Venture of Islam*. Vol. 1: *The Classical Age of Islam*. Chicago: University of Chicago Press.

Idris, H. R. 1986². "Bādīs," in *EI-2*, 1: 860.

Janos, D. 2015. "al-Fārābī, Philosophy," in *EI-3*, 2015–2: 108–13.

Joosse, N. P. 2018. "al-Dakhwār," in *EI-3*, 2019–1: 21–2. https://doi.org/10.1163/1573-3912_ei3_COM_32131.

Kennedy, E. S. and Pingree, D. 1971. *The Astrological History of Māshā'allah* (Harvard Monographs in the History of Science 5). Cambridge, MA: Harvard University Press.

Khan, G. 1986. "The Arabic Fragments in the Cambridge Genizah Collections," *Manuscripts of the Middle East* 1: 54–60.

Koningsveld, P. S. van. 1992. "Andalusian-Arabic Manuscripts from Christian Spain," *Israel Oriental Studies* 12: 75–110.

Langermann, Y. T. 1996a. "Arabic Writings in Hebrew Manuscripts: A Preliminary Relisting," *Arabic Sciences and Philosophy* 6: 137–60.

Langermann, Y. T. 1996b. "Transcriptions of Arabic Treatises into the Hebrew Alphabet: An Under-appreciated Mode of Transmission," in Ragep, F. J., Ragep, S. P. and Livesey, S., eds. *Tradition, Transmission, Transformation*. Leiden: Brill, 247–60.

Lazarus-Yafeh, H. 1992. *Intertwined Worlds: Medieval Islam and Bible Criticism*. Princeton: Princeton University Press.

Lazarus-Yafeh, H. *et al.*, eds. 1999. *The Majlis: Interreligious Encounters in Medieval Islam*. Wiesbaden: Harrassowitz.

Levin, L., Walker, R. D. and Sadik, S. 2018. "Isaac Israeli," in Zalta, E. N., ed. *The Stanford Encyclopedia of Philosophy (Summer 2018 Edition)*. Stanford: Metaphysics Research Lab, Stanford University. https://plato.stanford.edu/archives/sum2018/entries/israeli/.

Lewis, B., *et al.* 1986². "Ibn Zurʿa," in *EI-2*, 3: 979–80.

Libson, G. 2003. *Jewish and Islamic Law: A Comparative Study of Custom During the Geonic Period* (Harvard Series in Islamic Law). Cambridge, MA: Harvard University Press.

Meimaris, Y. 1985. *Katalogos tōn neōn aravikōn cheirographōn tēs Hieras Monēs hagías Aikaterinēs tou Orous Sina*. Athens: Ethnikon Hidryma Ereunon.

Meyerhof, M. 1986². "Ibn al-Tilmīdh," in *EI-2*, 3: 956–7.

Millás, J. M. and Stern, S. M. 1986². "Abu 'l-Ṣalt Umayya," in *EI-2*, 1: 149.

Monferrer-Sala, J. P. 2010. "Ibn Zurʿa," in Thomas, D., ed. *Christian-Muslim Relations 600–1500*. Leiden: Brill. https://doi.org/10.1163/1877-8054_cmri_COM_23344.

Montgomery, J. 2007. "Islamic Crosspollinations," in Akasoy, A., Montgomery, J. and Pormann, P., eds. *Islamic Crosspollinations: Interactions in the Medieval Middle East*. Oxford: Gibb Memorial Trust, 148–93.

Nicolae, D. 2017. "Ibn Jumayʿ," in *EI-3*, 2017–4: 138–9.

Outhwaite, B., Schmierer-Lee, M. and Burgess, C. 2017. *Discarded History: The Genizah of Medieval Cairo*. Cambridge: Cambridge University Library.

Pines, S. 1986². "Abu 'l-Barakāt," in *EI-2*, 1: 111–13.

Pingree, D. 1975. "Mashā'allāh: Some Sasanian and Syriac Sources," in Hourani, G. F., ed. *Essays on Islamic Philosophy and Science*. Albany: State University of New York Press, 5–14.

Polliack, M. 1997. *The Karaite Tradition of Arabic Bible Translation: A Linguistic and Exegetical Study of Karaite Translations of the Pentateuch from the Tenth and Eleventh Centuries C.E.* Leiden: Brill.

Richards, D. S. 1992. "A Doctor's Petition for a Salaried Post in Saladin's Hospital," *Social History of Medicine* 5.2: 297–306. https://doi.org/10.1093/shm/5.2.297.

Salaymeh, L. 2013. "Between Scholarship and Polemic in Judeo-Islamic Studies," *Islam and Christian-Muslim Relations* 24: 407–18.

Salaymeh, L. 2015. "Comparing Jewish and Islamic Legal Traditions: Between Disciplinarity and Critical Historical Jurisprudence," *Critical Analysis of Law* 2: 153–72.

Samir, Kh. 1991. "ʿAbd al-Masiḥ al-Isra'ili al-Raqqi," in Atiya, A. S., Torjesen, K. J. and Gabra, G., eds. *Claremont Coptic Encyclopedia*. Claremont: Claremont Graduate University, School of Religion, 5b–7a. http://ccdl.libraries.claremont.edu/cdm/ref/collection/cce/id/16.

Samsó, J. 1991. "Māshā' Allāh," in *EI-2*, 6: 710–12.

Schacht, J. 1986a. "Ibn Buṭlān," in *EI-2*, 3: 740–2.

Schacht, J. 1986b. "Ibn Riḍwān," in *EI-2*, 3: 907–8.

Schwarb, G. 2014. "The Reception of Maimonides in Christian-Arabic Literature," in Tobi, Y., ed. *Maimonides and His World: Proceedings of the Twelfth Conference of the Society for Judaeo-Arabic Studies* (Ben ʿEver La-ʿArav: Contacts between Arabic Literature and Jewish Literature in the Middle Ages and Modern Times 7). Haifa: Faculty of Humanities, University of Haifa, 109–75.

Sezgin, F. 1970. *Geschichte des arabischen Schrifttums*. Vol. 3. Leiden: Brill.

Simaika, M. 1939. *Catalogue of the Coptic and Arabic Manuscripts in the Coptic Museum, the Patriarchate, the Principal Churches of Cairo and Alexandria, and the Monasteries of Egypt*. 2 vols. Cairo: Government Press, Būlāq.

Sirry, M. 2011. "The Public Role of *Dhimmīs* during ʿAbbāsid Times," *Bulletin of the School of Oriental and African Studies* 74.2: 187–204. https://doi.org/10.1017/S0041977X11000024.

Sklare, D. E. 1996. *Samuel Ben Ḥofni Gaon and His Cultural World: Texts and Studies* (Études sur le judaïsme médiéval 18). Leiden: Brill.

Steinschneider, M. 1893. "Schriften der Araber in hebräischen Handschriften, ein Beitrag zur arabischen Bibliographie," *Zeitschrift der deutschen morgenländischen Gesellschaft* 47.3: 335–84.

Steinschneider, M. 1897. "An Introduction to the Arabic Literature of the Jews: I," *The Jewish Quarterly Review* 9.2: 224–39. https://doi.org/10.2307/1450587.

Stroumsa, S. 2011. "Whirlpool Effects and Religious Studies: A Response to Guy G. Stroumsa," in Krech, V. and Steinicke, M., eds. *Dynamics in the History of Religions between Asia and Europe: Encounters, Notions, and Comparative Perspectives*. Leiden: Brill, 159–62.

Stroumsa, S. 2015. "Between Acculturation and Conversion in Islamic Spain: The Case of the Banū Ḥasday," *Mediterranea: International Journal on the Transfer of Knowledge* 1: 9–36. https://doi.org/10.21071/mijtk.v0i1.5171.

Swanson, M. N. 2010. "ʿAbd al-Masīḥ al-Isrāʾīlī al-Raqqī," in Thomas, D., ed. *Christian-Muslim Relations 600–1500*. Leiden: Brill. https://doi.org/10.1163/1877-8054_cmri_COM_24989.

Szilágyi, K. 2006. "Christian Books in Jewish Libraries: Fragments of Christian Arabic Writings from the Cairo Geniza," *Ginzei Qedem* 2: 107–58.

Talbi, M. 1993. "al-Muʿizz b. Bādīs," in *EI-2*, 7: 481–4.

Thomas, D. 2000. "al-Ṭabarī," in *EI-2*, 10: 11–15.

Vajda, G. 2002. *Le commentaire sur le Livre de la création de Dūnaš ben Tāmīm de Kairouan (Xᵉ siècle)*. Leuven: Peeters.

Vernet, J. 1986a. "Ibn Abī Uṣaybiʿa," in *EI-2*, 3: 693–4.

Vernet, J. 1986b. "Ibn al-Hay<u>th</u>am," in *EI-2*, 3: 788–9.

Vidro, N. and Kasher, A. 2014. "How Medieval Jews Studied Classical Arabic Grammar: A Kufan Primer from the Cairo Genizah," *Jerusalem Studies in Arabic and Islam* 41: 173–244.

Vollandt, R. 2014. "Ancient Jewish Historiography in Arabic Garb: Sefer Josippon between South Italy and Coptic Cairo," *Zutot* 11: 70–80.

Vollandt, R. 2018. "Flawed Biblical Translations into Arabic and How to Correct Them: A Copt and a Jew Study Saadiah's Tafsīr," in Bertaina, D., *et al.*, eds. *Heirs of the Apostles: Studies on Arabic Christianity in Honor of Sidney H. Griffith*. Leiden: Brill, 56–92.

Walzer, R. 1991⁴. "al Fārābī," in *EI-2*, 2: 778–81.

Yarbrough, L. B. 2012. *Islamizing the Islamic State: The Formulation and Assertion of Religious Criteria for State Employment in the First Millennium AH*. PhD thesis. Princeton, NJ: Princeton University Press.

Zucker, M. 1984. *Saadya's Commentary on Genesis*. New York: The Jewish Theological Seminary of America.

VI.2
WHICH IS THE RIGHT *QIBLA*?

Mònica Rius-Piniés

VI.2.1 Introduction

It is a well-known fact that Muslims should face Mecca during ritual prayer. Indeed, Muslims have to know the direction of the Kaʿba not only to pray five times a day but also to carry out other daily actions such as slaughtering animals or burying the dead. They also need to know the sacred direction in order to avoid it when carrying out other common actions (such as meeting their physiological needs). In Arabic this direction is known as the *qibla*, a word found at the origin of Arabisms such as *alquibla*, in Spanish, with the article included. In English (and other languages like French), the word remains the same as in Arabic – *qibla* – but other written forms are found, like *qiblah*, *kibla* or *kiblah* (www.merriam-webster.com/dictionary/qibla?src=-search-dict-hed). In fact, the word derives from one of the meanings of the Arabic root "*qbl*" (specifically in form III of the verb) "to face". The term *qibla* indicates the direction *par excellence*, the direction of the Kaʿba. Another important point is that the term has other meanings (both in its verbal and its adjective form, *qiblī*) which may cause confusion in relation to its main meaning, since it may indicate a southern direction or due south (Wehr 1971, 740). As we will see, this second meaning comes from the position of Mecca in relation to certain other sites, above all Medina. The *qibla* is marked in mosques by the *miḥrāb* (the niche for praying) and the wall on which it is situated is called the "*qibla* wall". Today, the *qibla* is indicated in hotel rooms and even on airplanes.

But where does the need to know the *qibla* originate? The answer is both simple and complicated. Simple, because it derives from a religious prescription expressed in the Qurʾān; complicated, because the "correct" execution of this prescription can be difficult to accomplish, depending on the circumstances. For instance, some Muslim scholars used geometrical and astronomical procedures to determine the *qibla* accurately. These procedures require one to have a grasp of spherical trigonometry and to know one's geographical coordinates exactly, something that was not possible for many centuries. Other scholars deemed nonmathematical methods grounded in religious utterances and social contexts to be sufficient. The question is, then, how can a religious prescription be so difficult to comply with? In this chapter, we will discuss the extent to which the accurate determination of the *qibla* was considered truly necessary and what this meant for different members of different Muslim communities over time. Moreover, apart

DOI: 10.4324/9781315170718-66

from the religious mandate, we analyze which other elements (political, social or cultural) have influenced the orientation of Islam's most emblematic building, the mosque. This is why it is worth asking the question: Which is the right *qibla*?

VI.2.2 The *qibla* as a central element of Islamic conceptual frameworks

As the entry on the *qibla* in the second edition of the *Encyclopaedia of Islam* shows, the modern study of the history of this idea and its various practices as developed in different regions and times followed two disciplinary trajectories that remained distinct from each other: legal prescriptions and methods, and solutions offered by scholars of the mathematical sciences (Wensinck and King 1986). This bifurcated approach to *qibla* studies has come under scrutiny during the last decades. Hence, it will not be followed in this chapter. I suggest that in order to understand the historical development of this concept and its various contexts, we need to apply humanistic methods and theories alongside examining the extant scientific treatises and instruments. One option is to pick up on discussions about a typology of practices shared across religions as suggested, for instance, by Ninian Smart and surveyed succinctly by Rennie (1999). Smart proposed six to seven dimensions that he believed were common to all religions: ritual, narrative and mythological, institutional or social, material, experiential (or emotional), doctrinal and ethical (Rennie 1999, 63). As this chapter shows in the case of the *qibla*, all these dimensions are relevant elements of its histories, complementing each other and the scientific methods used for its determination across the Islamicate world. Thus, learning from anthropology and the study of religions leads us to ask in which conceptual frameworks the *qibla* and its problems were embedded. It helps gain a more nuanced and deeper understanding of the various practices that Muslims followed when debating, fighting over or simply praying according to what they considered the right *qibla*.

The conceptual frameworks of culture and religion not only help us understand some of the central axes of science (related to religion) that emerged within Islamicate societies, such as those corresponding to time and space, but are also linked to other factors that must be taken into consideration, such as economics or politics (Brentjes 2005). The fact that each culture has calculated time and space in a unique way does not imply that it has always done so independently, without any external influences. Consequently, need to take into account different types of sources, because religious, cultural, political and economic elements are all intermingled in the determination of the *qibla*.

The construction of buildings, city planning and map design all reveal how historical societies organized space. Thanks to archeoastronomy, it is known that the emblematic buildings of many civilizations have been carefully oriented (or, at least, that was the intention when they were built) and mosques are no exception (Ruggles 2015). Similarly, it can be seen that the points taken as a reference have been practically the same for thousands of years; sunrise in summer and the equinoxes or the rise of the Pleiades, to give just a few examples, were the directions toward which many emblematic temples were erected. Stonehenge is aligned in the direction of the sunrise at the summer solstice and the sunset at the winter solstice, and the pyramids of Egypt and even the Kaʿba itself are aligned astronomically (North 1997; Haack 1984; Hawkins and King 1982, 103–5). Although some have argued that it is difficult to prove scientifically that historical buildings (especially the ancient ones) were built with a precise astronomical orientation, this objection cannot be applied to Islam, since it is clearly established that mosques were constructed oriented toward the Holy Mosque (although exactly how this objective was to be achieved may be less clear).

The relationship between space and culture is also shown in the representations of the world in maps and cartography. The expression "all roads lead to Rome" is well known. Underlying the saying is the idea that this city was located at the center of the world, and this idea is illustrated, for example, in the *Tabula Peutingeriana* (possibly taken from a previous Roman model; Dilke 1987, 234–42). It is a dynamic that has been periodically repeated: the Persians used to represent the world in seven *kishwārs* (circles; regions) around a central one, which was Persia itself (Tibbetts 1992, 93–6; Chapter I.13). Jerusalem also appears as the center of the world in the magnificent world map of Heinrich Bünting (1545–1606) of 1581 (Shalev 2012, 101–2). This ethnocentrism is based not just on religious considerations but on political ones as well and shows that physical geography did not occupy the first or the only place in the work of geographers.

In the Islamicate world, sacred geography also led to the production of some magnificent depictions, such as the diagrams of Ibn al-Wardī (d. 861/1457) and ʿAlī al-Sharafī (10th/16th century), in which the world is divided into multiple sectors (in the form of mosques with their respective *miḥrābs*) around the Kaʿba at its center (King and Lorch 1992[CB], [1] 195–6; King 1999[CB], 55; Herrera Casais 2008). This geography has also been inscribed linguistically in the popular imagination; in multiple languages, Mecca is used as a synonym of centrality (e.g., when we say that Hollywood is the Mecca of cinema).

VI.2.3 Sources

Many religious rituals require astronomical calculations, such as the direction of the *qibla* and prayer times. The rituals are of incredible importantce, as visible in the textual and material record, but it should be stressed that the Qurʾān, the first source of Islamic law, does not always establish precisely how most rituals are to be carried out. Thus, the Qurʾān must be complemented with other sources, central among them being (for Sunnī Muslims) the *ḥadīth* (the deeds and sayings of the Prophet), a source that generates much debate even today. It is the *ḥadīth* rather than the Qurʾān, for example, that establishes the number of daily prayers and the times when they are to be performed (Aerts 2017).

The variety of texts thus made the narrative dimension of the Islamic religion particularly important. The range of interpretative opinions offered by the different law schools on the same issues, and even the diversity of criteria within each of them, led to the production of a great many treatises dealing specifically with religious practice throughout the Middle Ages. Muslim astronomers were obviously no strangers to this climate of interpretative variety and, with their scientific criteria, contributed to the expansion of the possibilities of decoding ritual obligations.

VI.2.4 The first *qibla*

As we have already seen, the *qibla* is first and foremost a religious precept. As such, it generated a vast body of literature. The commentators on the Qurʾān, the traditionalists and the scholars of all the law schools discussed the subject at length, and although no single solution was found to the problem, the majority agreed that the orientation should never become a reason for rupture in the heart of the Islamic community (*umma*; Rius 2000). There was no discussion, however, of the fact that there are three fundamental guidelines that anyone who wants to establish the *qibla* should study: in order of importance, these guidelines were the Qurʾān, the *ḥadīth* and the community consensus (*ijmāʿ*; al-Maṣmūdī 2000, 266–7). Following the Islamic account, Muḥammad received the precept to pray five times a day in the night of *Isrāʿ* (night journey), during which he traveled to Jerusalem and then ascended to heaven on a journey known as *miʿrāj* (Rius 2000,

71). Once the obligation to pray was established, Muḥammad faced Jerusalem when praying, but after sixteen months, he asked Gabriel for permission to change direction. The archangel transmitted the following verse to him:

> We have certainly seen the turning of your face, [O Muḥammad], toward the heaven, and We will surely turn you to a *qibla* with which you will be pleased. So, turn your face toward al-Masjid al-ḥarām. And wherever you [believers] are, turn your faces toward it [in prayer].
>
> *(Qurʾān 2:144)*

As Muḥammad was in Medina, the change was a symbolic gesture of rupture, since praying toward Mecca meant turning one's back on Jerusalem. Thus, the mosque of Medina in which he prayed became known as "the mosque of the two *qibla*s" (Watt 1974, 112–3). From that moment on, the whole community, the *umma*, turned to the Kaʿba, but that does not mean that it had received a clear explanation of how to do so – nor was there any indication of how accurately this provision should be implemented. Later, with the spread of Islam, another difficulty was added: how to determine the position of Mecca from a wide variety of locations ever farther away from the Arabian Peninsula.

VI.2.5 The *qibla* of the legal scholars

Since the 2nd/8th century, the *fuqahāʾ* – specialists in Islamic law – have written a great many educational tracts on the *qibla*. The most important law manuals always include a chapter on how to orient mosques (and those praying) toward Mecca. The issue aroused so much interest that it gave rise to a new type of legal text, the treatises on the *qibla*. This type of literature met with varying levels of success in different areas. Both in the Maghrib (West) and the Mashriq (East) the titles are numerous. Authors such as al-Mīttījī (5th/11th century), al-Maṣmūdī (8th/14th century) and al-Tājūrī al-Ṭarabūlsī (10th/16th century) made up a list that extended over time and established specific schools, with teachers and disciples, who were in charge of collecting and transmitting the opinions of the wisest scholars (MSS Paris, BnF, Arabe 5311; Rabat, Bibliothèque Ḥasaniyya, 6999). The abundance of texts in some areas was an indication that they did not achieve consensus; *ʿulamāʾ* (scholars) repeated the same arguments over the centuries in an attempt to convince other *ʿulamāʾ* and emirs and, by extension, the entire community (Rius 2009). However, and in stark contrast, there are no extant Andalusī treatises, nor have any reports with similar characteristics been found to date (Rius 2000).

In their treatises, the legal scholars analyzed a wide range of cases and offered a very detailed classification of the different types of *qibla*. The first and greatest difference is the degree of accuracy required. In this first division, Muslims are divided into two groups: those in Mecca who can see the Kaʿba directly (*fī ʿayn al-qibla*) and the rest of the Muslims (Rius 2000, 7; Rius 2009). The former are expected to observe the *samt*, or the exact orientation, while the latter can pray in a more generic direction, the *jiha*. The accuracy of the *jiha* was subject to various criteria, but a large number of legal scholars distinguished between two types: the *jiha ṣughrā*, which implies that the person praying can orient him/herself with a margin of 90° (that is to say, practically the area covered by the angle of vision) and the *jiha kubrà*, which is much less strict, since it allows the person praying within an angle of up to 180° (Rius 2009, 178).

For legal scholars, in the spirit of the Qur'anic text, the direction of prayer could not become an obstacle to worship. The categories of *qibla* received different names according to the procedure used for their establishment, from the *qibla* determined by *waḥā* (thanks to a revelation) to *qibla* established by *ijmāʿ* (by the consensus of the imams; Rius 2000, 83–4). The great distinction is made between the *qibla* obtained by *taqlīd* (tradition, imitation) and the one obtained by *ijtihād* (effort; Rius 2000, 84–5; Rius 2009, 178; Samsó 2020, 146–7). The arguments in favor of one or the other are varied but can be summarized by saying that the *qibla* determined by *taqlīd* should be used by those who, in the absence of any knowledge of their own, should imitate an already established *qibla*. As noted in the treatises, in this case the important thing is to follow a good example, the best ones being the example of Muḥammad himself, followed by the *ṣaḥāba* (companions of Muḥammad) and the *tābiʿūn* (first generation born in Islam) (al-Maṣmūdī 2000, 267, 303; Rius 2000, 72, 91–135, 143–6).

Important arguments offered by numerous legal experts, but also by scholars of the mathematical sciences, concerned the social consequences of not knowing the right *qibla* or building mosques, which contained a *qibla* considered wrong by some scholars, rulers or common people. Scholars often emphasized that the orientation toward the *qibla* was a precept that could not and should not collide with other principles, such as community consensus. According to them, prayer was a spiritual act in which belonging to the Islamic community was materialized by a physical position in the mosque; the bodies of the faithful should form regular lines, shoulder to shoulder, facing the same direction. The disparity of the criteria on the orientation of the *qibla* should thus not become a reason for rupture (*fitna*). Neither should the determination of the *qibla* place a burden on the believer. That is why scholars from very different fields of knowledge and education, including representatives of the mathematical sciences, only recommended knowledge of geometry or arithmetic if a believer was capable of grasping it. Legal scholars admitted that this type of knowledge was too difficult even for themselves and hence recommended knowledge of the rising and setting of specific stars, the path of the moon through the asterisms forming its orbit (lunar stations or mansions) or the rising and setting of the sun at the equinoxes and the solstices (see, for instance, al-Maṣmūdī 2000).

VI.2.6 The *qibla* of the scholars of the mathematical sciences

From a mathematical point of view, the *qibla* is a complex problem whose exact solution requires the use of spherical trigonometry or methods for projecting the spherical problem onto a plane and solving it there (Chapter IV.3). Based on the assumption that the Earth is a perfect sphere, the aim is to determine the angle between the North Pole, a worshipper's location and Mecca. But mathematical solutions only became available from the late 2nd/8th century onward in Baghdad and spread across Islamicate societies perhaps only a century or more later (King 1986, 85–6). By then, several of the major mosques had already been built (see for Cairo see King 1982, 304–5). In addition, for the result of the procedures to be truly correct, the precise geographical coordinates had to be known (latitudes of Mecca and the place for which the prayer direction was sought, and longitude difference between the two localities). The added disadvantage here was that the longitudes were not reliable, even at later times (Kennedy and Kennedy 1987). The earliest known approximative mathematical methods were invented by experts of the astral sciences in the 2nd/8th and early 3rd/9th centuries. The best known method became that of al-Battānī (244–317/858–929) according to which the problem is solved as if it were a problem of plane trigonometry (King 1986, 103–7; for his other works see van Dalen 2007, 101–3). As in the previous cases of non-mathematical methods, the inaccuracy of the longitudes added to the margin of error in determining *qibla*s even mathematically. But the results obtained by these means were more than acceptable, especially if one is looking for the *jiha* and not the *samt*, which, as we have seen, was considered to be the most legally orthodox.

The development of precise theoretical methods attracted the best representatives of the mathematical sciences in different Islamicate societies. Among them we find, for instance, Ḥabash al-Ḥāsib (d. after 255/869) and al-Nayrīzī (d. *c.* 310/922) in Abbasid Baghdad, al-Bīrūnī (362–d. after 444/973–d. after 1053) in Ghaznavid northern India (today Afghanistan), Ibn Yūnus (d. 400/1009) and Ibn al-Haytham (354–*c.* 430/965–*c.* 1040) in Fatimid Cairo, and Shams al-Dīn al-Khalīlī (d. *c.* 781/1379–80) in Mamluk Damascus. They used different methods, among them the theorems of Menelaus (*fl.* 100) for triangles and angles on the sphere and its derivatives or replacements, and other methods of solid geometry. They also used a series of methods learned from ancient Greek sources after their translation into Arabic that projected and folded various planes onto and into another plane, known together as *analemma*. In addition, they used algebraic interpretations and procedures, tabulations on the basis of exact trigonometric solutions and a series of ratios between auxiliary trigonometric magnitudes (Wensinck and King 1986).

In addition to theoretical solutions, scholarly experts also provided practical methods of resolution. Instructions for determining the *qibla* were included in treatises on the construction and use of instruments. Astrolabes such as the one attributed to Ḥasan (or Ḥusayn) ibn Muḥammad ibn Bāṣo (d. 716/1316) from Granada or sundials such as the one attributed to Ibn al-Ṣaffār (d. 426/1035) in Córdoba could be used to find the *qibla*, taking into account that the result was restricted to a certain locality (or localities, in the case of an astrolabe with different plates; Hernández 2018, 43–58, 211–34; Rius 2007, 566–7; Chapter IV.3). Specific universal instruments called *qibla* pointers were even designed for the determination of the *qibla* (King and Lorch 1992; King 1999, 88–124; King 2004[CB], 94–9, 2005[CB]).

One last point must be borne in mind: although any medieval scholar must have had a broad knowledge of the religious sciences, some of them also stood out in fields belonging to other disciplines and practices. An example in Marrakesh is the case of Ibn al-Bannāʾ (654–721/1256–1321, a highly skilled scholar of the mathematical sciences and at the same time also a reputed Sufi. The supposed opposition, therefore, between science and religion was far from being always a reality. The debates on the *qibla* and related religious themes, like the beginning of the month or the prayer time regulations, rather highlight the complexities of how Muslims as scholars, rulers and common people looked at and practiced religious obligations in light of their scholarly knowledge of law, exegesis or the mathematical sciences in different parts of the Islamicate world and at different times.

Today, with easy access to web pages and all kinds of applications that offer full accuracy, thanks to systems based on the use of GPS, these legal controversies have become less intensive. All one has to do is enter the geographical coordinates (or choose the location from a list) and the program immediately provides the exact direction. If one has a smartphone with geolocation, establishing the *qibla* is child's play.

VI.2.7 The *qibla* of the stars

Coming from the lands of the Arabian Peninsula, the first Muslims were experts who knew the position of the most important stars. For obvious reasons, the Sun was the most prominent; but commercial caravans often traveled at night in order to avoid the high daytime temperatures. Knowledge of pre-Islamic tradition was easily incorporated into the newly revealed religion. The Qurʾān includes verses in which this practice is permitted such as "And it is He who placed for you the stars that you may be guided by them through the darknesses of the land and sea. We have detailed the signs for a people who knows" (Qurʾān 6:97) and "And by the stars they are [also] guided" (Qurʾān 16:16). Sunrise and sunset, especially at significant times such as equinoxes and solstices, were also used to obtain directions. The Bedouins used other large, easily visible stars such as *Qalb al-ʿAqrab* (the heart of the scorpion, Antares, *α Scorpii*), *Suhayl* (Canopus, *α Carina*) and the Pole star. Orientations could also be obtained through combinations, for example, by facing *Qalb al-ʿAqrab* at the rise of *Shawla* (the stinger, *λ* Scorpii).

This method was widely used to determine the orientation of early mosques, such as those of al-Fusṭāṭ in Egypt. One could also use the human body as an instrument, as when Abū ʿAlī al-Ḥasan al-Umawī al-Qurṭubī (d. 602/1025) recommended putting the Pole on one's left shoulder: "You will attain the *qibla*, in al-Andalus, by placing the Pole on the left shoulder and then facing south. Where your eye sees, this will be the *qibla*" (Rius 2000, 181). Abū Ḥanīfa (d. 150/767), after whom one of the four main Sunni legal schools is named, quoting from one of the early books on *anwāʾ* (Chapter V.4), informed believers that the Pole is not a star but a point around which stars revolve near to a star located by different astral traditions in different constellations (Ursa minor; Pisces; Banāt Naʿsh al-ṣughrā; al-Maṣmūdī 2000, 272, 279–80; Rius 2000, 250–2; Schmidl 2015, 1927–34).

VI.2.8 The *qibla* and politics

The *qibla* has a symbolic character that transcends the merely religious and merges with the political. Different dynasties used the *qibla* for propaganda purposes when deemed useful or necessary. Ibn Simāk al-ʿAmilī's (d. 540/1146) chronicle *The Embroidered Garb* (*al-Ḥulal al-mawshiyya*; Anonymous 1936, 119) claims that the Almohad Mahdī forbade the entry to Marrakesh (Morocco) before the city was purified, which resulted in the destruction of existing mosques and the construction of new ones. The direction (toward the east approximately) used by the Almoravids (r. *c.* 454–541/1062–1147) was correct, so the fact that the Almohads (r. *c.* 525–668/1130–1269) changed the *qibla* must be interpreted as symbolic; the change of orientation was a very visible way of radically differentiating themselves from their predecessors. Furthermore, according to historical sources, during the Almohad period the *Kutubiyya* mosque in Marrakesh was erected in (542/1147) and later demolished to be rebuilt (557/1162). Archaeological studies show that the former was oriented at 154° and the latter toward 159° (all measurements are given in degrees from due North = 0°; Rius 2000, 123). The sources do not mention the reason why it was demolished and reconstructed or the criteria with which it was carried out. Beyond the economic expense this must have entailed, the technical difficulties of changing a building by only 5° are colossal. The Almohads, nonetheless, succeeded in convincing later generations of the soundness of their determinations, and thus secured their legitimacy as a political and religious power; today, however, it would be wrong to take literally something that was a clear propaganda exercise. There are also some revealing silences and absences. When reporting this episode, al-Maṣmūdī, an opponent of the Almohad regime, adds a succinct and rather skeptical comment which could be translated as "that is what they say" (Rius 2000, 150).

The treatise on mosques written by ʿAbd al-ʿAzīz al-Asfī al-Andalusī in the 12th/18th century is also of great interest, since it provides information on buildings from the Almoravid and Almohad periods that no longer exist (MS Rabat, Bibliothèque Ḥasaniyya 1110). In Safi, Morocco, as in other towns, different orientations coexisted – East and South, following the usual controversy – but according to this scholar, the best-oriented mosque in the whole region was that of the *Zāwiya Nāṣiriyya* in Tamegroute, built in the 11th/17th century. The original building had absolutely unique characteristics, as it had a hole above the *miḥrāb* through which the sun penetrated on the spring equinox and which showed that the mosque was built facing east. Al-Mirghīthī (d. 1090/1679), author of treatises on timekeeping (King 2004, 495–7; Chapter V.4), which remain famous to this day, worked as a *muwaqqit* at the *Zāwiya Nāṣiriyya*, which had a rich library. The peculiarity of this *zāwiya* and its orientation must have been well known, as its students were given the nickname "people of the *qibla*" (Rius-Piniés and Puig-Aguilar 2015, 336).

Finally, another noteworthy case is that of the double *qibla*, exemplified by the Umayyad Caliph al-Ḥakam II (r. 350–366/961–976) and the Nasrid Emir Yūsuf I (r. 733–754/1333–1354). Although the two rulers determined the *qibla* of Madīnat al-Zahrā' and of the Comares Palace at the Alhambra respectively with the utmost precision, both declined to change the "traditional" *qibla* in other major buildings of the city, even though they knew that it was inexact. In fact, before extending the Great Mosque of Cordoba, al-Ḥakam II consulted a group of legal advisors. Eventually, it was decided that the best option was to respect the criteria of the venerable ancestors and to leave the mosque with the same orientation as before (Rius 2009). The chronicles show that, on some occasions, Muslims prayed in two different directions in the same mosque. The Andalusī Abū ʿUbayda al-Laythī (d. 295/907), for instance, received the nickname of *ṣāḥib al-qibla*, because he faced eastward to pray (Samsó 2011[CB], 60; Rius 2000, 172).

VI.2.9 What the buildings say about the *qibla*

We have already talked about the close relationships that the sovereigns of all periods established with their main mosques. The importance of the orientation, underlined by the profusion of styles of *miḥrāb*s, is not restricted to the past: it continues to enjoy all its symbolic power today. One of the most magnificent *miḥrāb*s of all is the one in the mosque-cathedral in Córdoba (King 2018–2019). The primitive *miḥrāb*, built by ʿAbd al-Raḥmān I (r. 138–172/756–788), was not only replaced during the successive enlargements by ʿAbd al-Raḥmān II (r. 206–238/822–852) and al-Ḥakam II but also lost its central position after the enlargement by the chamberlain, al-Manṣūr (r. 370–392/981–1002; Chapter V.3). In all periods and in all geographical areas, the *miḥrāb* became the focal point of attention, helping the devout to pray in the right direction. Like the rest of the building, it symbolized the legacy that emirs, sultans and caliphs bequeathed to later generations. Another example is the colossal Badshāhī (Imperial) Mosque in Lahore, Pakistan, the fifth largest in the world, built between 1082/1671 and 1084/1673 under the Mughal Emperor Awrangzīb ([r. 1068–1118/1658–1707]; Koch 1992). The *miḥrāb* is richly decorated using the technique of *pietre dure* creating images with inlays of cut, polished, colored stones.

So far, we have seen that legal scholars, astronomers and rulers were all interested in how the mosques should be oriented. The question now is, How were the buildings constructed?

Let us take, for example, the plan of the so-called Great Mosque of Sidi ʿUqba in Qayrawan, Tunisia, which in its current form was built some 200 years after ʿUqba ibn Nāfiʿ (d. 63/683; Archnet.org-b; Qantara). The plan shows that the building is not correctly oriented (some 146° against 118° of the sunrise at the Winter solstice, which allegedly had been used in the 1st/7th century). In one story line, the sources record that the *qibla* was established by ʿUqba ibn Nāfiʿ, who had prayed for a revelation since no agreement could be reached among the companions and followers after observing several times the rising and setting of the Sun (Rius 2000, 54–6, 143–6). Another tradition claims that such an agreement had been reached and the building was erected on its basis (Rius 2000, 142, 146–7). How could a mishap of this kind have occurred? The justification that legal scholars presented was that, even though the exact orientation had been correctly determined by ʿUqba, there was an error in the construction of the building, which was attributable to the bricklayers or to the rulers of the Aghlabid dynasty (r.184–297/800–909), who remodeled the mosque in the 3rd/9th century (Rius 2000, 147–8). It is clear, once again, that the difficulty lies not only in determining the *qibla* but also in erecting the building in the desired direction. Another aspect of the debates on the right *qibla* that the stories about this mosque highlight consists in the multiple contradictions found in the proposed explanations and interpretations. Not only did the jurists tell different stories about the origin of this *qibla*, but they also relied in their reports on the evaluations of men of religious knowledge and scholars of the mathematical sciences. One narrator and his followers claimed that those scholars had found a substantial error and hence insisted that the mosque had to be pulled down and rebuilt (which did not take place). Another group, referring exclusively to the scholars of the mathematical sciences, minimized the error to an irrelevant small deviation, because there could be no doubt that ʿUqba had correctly observed the stars and the sun (Rius 2000, 148).

In Marrakesh, the Almohads and the Almoravids fought a famous battle in 541/1147. Although, as already mentioned, the Almohads demolished all the mosques in their "purification" of the city, on the grounds that they were not properly oriented, archaeological analyses confirm what the *qibla* treatises stated: namely, that the Almoravids had consulted astronomers and legal scholars about how to determine the correct orientation and followed their advice, which was entirely correct (Rius 2000, 151). The Great Mosque of ʿAlī ibn Yūsuf ibn Tāshufīn (r. 500–537/1106–1143) faced east, which, given its geographical location, was more than accurate. Although the city's buildings currently have divergent orientations, the only remaining Almoravid monument, the *qubba*, still stands as a vindication of the Almoravid approach (Archnet.org-a). However, as we have already seen, the all-powerful Almohad propaganda apparatus ensured that their successors did not challenge their claims. Indeed, a considerable number of mosques in Morocco have an orientation similar to that of the *Kutubiyya* mosque in Marrakesh (Rius 2000).

VI.2.10 Summary

In this chapter we have argued that the history of the *qibla* requires a multidisciplinary analysis. Being necessary to establish the direction of regular, prescribed prayers, and when undertaking other activities such as burying the dead, sacrificing animals or carrying out physiological functions, the *qibla* was omnipresent in the daily life of Muslims. Thus, the issue of determining it in an acceptable manner was addressed by a wide variety of sectors, from religious specialists to chroniclers and scholars of different disciplines. Buildings were often used as propaganda tools for a particular regime. But when it came to religious matters, the legal experts established a complex body of knowledge to guide Muslims to correctly fulfill their obligations. Scholars of the mathematical sciences were also well aware of the issue: the astronomers proposed mathematical solutions to determine the *qibla* either exactly or approximately, depending on the circumstances.

It makes no sense to examine the *qibla* from the point of view of the history of astronomy and mathematics only, without taking into account its religious, political, disputational, legal, ritual or material implications (Rius 2000). Nor should we contrast the different types of texts preserved (special treatises on the *qibla* and historical chronicles, among others) without verifying how far the theory was put into practice. The strong link between religion and politics makes it even more necessary to relate the two aspects. Emirs and caliphs were well aware of the symbolic impact of building a mosque, an action that could earn them considerable prestige; moreover, it reaffirmed their leadership of the community and ensured that their memory would last for centuries.

There is no doubt that the *qibla*, from a physical point of view, has always been in the same place in each specific locality. However, the scope for interpretation of the accuracy required in determining the *qibla* gave rise to multiple formulae. For this reason, in order to determine whether or not a mosque was correctly oriented, we must know what criteria were used. Thus, a mosque oriented toward the *jiha* – even with a margin of up to 90° from the *samt* – can be considered correct, if this had been the intention of those responsible for determining the orientation. If we add to this the technical problems involved in the construction, the nature of the terrain and the urban fabric of the city, obtaining a precisely oriented mosque was a difficult task. However, as the prescription was expressed in the Qurʾān and legal scholars were responsible for repeating it, the important thing was to comply with religion within the limitations of each particular situation.

Note

1 Consolidated bibliography.

Bibliography

Sources

Anonymous. ed. Allouche, I. S. 1936. *Kitāb al-Ḥulal al-Mawshiyya fī dhikr al-akhbār al-marrākushiyya* [Anonymous Chronical of the Almoravid and Almohad Dynasties]. Rabat: Imprimerie économique. [Now attributed to Ibn Simāk.]

al-Maṣmūdī, Abū ʿAlī Ṣāliḥ. ed. and tr. Rius, M. 2000. *Kitāb al-qibla* [The Book of the *qibla*]. Barcelona: Institut "Millás Vallicrosa" de Història de la Ciència Àrab, Universitat de Barcelona.

The Quranic Arabic Corpus. http://corpus.quran.com (Accessed 17 December 2017).

Manuscripts

MS Paris, BnF, Arabe 5311.

MS Rabat, Bibliothèque Ḥasaniyya, 1110.

MS Rabat, Bibliothèque Ḥasaniyya, 6999.

Research literature

Aerts, S. 2017. "Ascension, Descension, and Prayer Times in the Sīra and the Ḥadīth," *Der Islam* 94.2: 385–422.

Archnet.org-a. "Qubba al-Barudiyyin, Marrakech, Morocco," https://archnet.org/sites/1739 (Accessed 10 March 2019).

Archnet.org-b. "Jami' Uqba Ibn Nafi', Kairouan, Tunisia," https://archnet.org/sites/3763 (Accessed 10 March 2019).

Bashear, S. 1991. "Qibla Musharriqa and Early Muslim Prayer in Churches," *The Muslim World* 81: 267–82.

Brentjes, S. 2005. "Islamic Science," in *New Dictionary of the History of Ideas. Encyclopedia.com*. www.encyclopedia.com. (Accessed 17 December 2017).

Dalen, B. van. 2007. "Abū ʿAbd Allāh Muḥammad ibn Jābir ibn Sinān al-Battānī al-Ḥarrānī al-Ṣābiʾ," in Hockey *et al.*, eds. [CB], 101–3.

Dilke, O. A. W. 1987. "Itineraries and Geographical Maps in the Early and late Roman Empire," in Harley, J. B. and Woodward, D., eds. *History of Cartography*. 6 vols. Vol. 1: *Cartography in Prehistoric, Ancient, and Medieval Europe and the Mediterranean*. Chicago: Chicago University Press, 234–57.

Haack, S. C. 1984. "The Astronomical Orientation of the Egyptian Pyramids," *Journal for the History of Astronomy*, Archaeoastronomy Supplement 15: 119–25.

Hawkins, G. S. and King, D. A. 1982. "On the Orientation of the Kaʿbah," *Journal for the History of Astronomy* 13: 103–5.

Hernández Pérez, A. 2018. *Catálogo razonado de los astrolabios de la España medieval*. Madrid: Ediciones de la Ergastula.

Herrera Casais, M. 2008. "The Nautical Atlases of ʿAlī al-Sharafī of Sfax," *Suhayl* 8: 223–63.

Kennedy, E. S. and Kennedy, M. H. 1987. *Geographical Coordinates of Localities from Islamic Sources*. Frankfurt am Main: Institut für Geschichte der Arabisch-Islamischen Wissenschaft, Johann-Wolfgang-Goethe-Universität.

King, D. A. 1982. "Astronomical Alignments in Medieval Islamic Religious Architecture," *Annals of the New York Academy of Sciences* 385: 303–12 (Reprint: King 1993, XIII).
King, D. A. 1986. "The Earliest Islamic Mathematical Methods and Tables for Finding the Direction of Mecca," *ZGAIW* 3, with errata from 4: 82–149, 270 (Reprint: King 1993, XIV).
King, D. A. 1993. *Astronomy in the Service of Islam*. Aldershot: Variorum.
King, D. A. 2015. "Astronomy in the Service of Islam," in Ruggles, 181–96.
King, D. A. 2018–2019. "The Enigmatic Orientation of the Great Mosque of Córdoba," *Suhayl* 16–7: 33–111.
Koch, E. 1992. *Mughal Architecture*. Munich: Prestel-Verlag.
North, J. D. 1997. *Stonehenge. Ritual Origins and Astronomy*. London: Harper Collins.
Qantara. "*Mihrâb* of the Great Mosque of Kairouan," www.qantara- med.org/public/show_document.php?do_id=401.&lang=en (Accessed 10 March 2019).
"Qibla." www.merriam-webster.com (Accessed 10 June 2018).
Rennie, B. S. 1999. "The View of the Invisible World: Ninian Smart's Analysis of the Dimensions of Religion and of Religious Experience," *Bulletin/CSSR* 28.3: 63.
Rius, M. 2000. *La qibla en al-Andalus y al-Magrib al-Aqṣà*. Barcelona: Institut "Millás Vallicrosa" de Història de la Ciència Àrab, Universitat de Barcelona.
Rius, M. 2007. "Ibn al-Ṣaffār: Abū al-Qāsim Aḥmad ibn ʿAbd Allāh ibn ʿUmar al-Ghāfiqī ibn al-Ṣaffār al-Andalusī," in Hockey *et al.*, eds., 566–7.
Rius, M. 2009. "Finding the Sacred Direction: Medieval Books on the *qibla*," in Rubiño- Martín, J. A. *et al.*, eds. *Cosmology Across Cultures* (ASP Conference Series, 409). Los Angeles: Astronomical Society of the Pacific, 177–82.
Rius, M. 2015. "Qibla in the Mediterranean," in Ruggles, 1687–94.
Rius-Piniés, M. and Puig-Aguilar, R. 2015. "Al-Asfī's Description of the Zāwiya Nasiriyya: The Use of Buildings as Astronomical Tools," *Journal for the History of Astronomy* 46.3: 325–42.
Ruggles, C. L. N., ed. 2015. *Handbook of Archaeoastronomy and Ethnoastronomy*. New York: Springer
Samsó, J. 2020. "Miqat: Timekeeping and Qibla," in *On Both Sides of the Strait of Gibraltar. Studies in the History of Medieval Astronomy in the Iberian Peninsula and the Maghrib*. Leiden: Brill, 128–51.
Schmidl, P. G. 2015. "Islamic Folk Astronomy," in Ruggles, 1927–34.
Shalev, Z. 2012. *Sacred Words and Worlds. Geography, Religion, and Scholarship, 1550–1700*. Leiden: Brill.
Tibbetts, G. R. 1992. "The Beginnings of a Cartographic Tradition," in GTISAS [CB], 90–107.
Watt, W. M. 1974. *Muhammad: Prophet and Statesman*. Oxford: Oxford University Press.
Wehr, H. 1971. ed. Cowan, J. Milton. 1976³. *A Dictionary of Modern Written Arabic*. New York: Spoken Language Services, Inc.
Wensinck, A. J. and King, D. A. 1986². "Ḳibla," in *EI-2*, 5: 82–3.

VI.3
WERE PHILOSOPHERS CONSIDERED HERETICS IN ISLAM?

Frank Griffel

VI.3.1 A report by Ibn al-Qifṭī

Ibn al-Qifṭī's *Lessons for Learned Men from the History of Philosophers and Physicians* (*Ikhbār al-ʿulamāʾ fī akhbār al-ḥukamāʾ*) is a grand celebration of the Greek philosophical and scientific tradition among Arabic speaking scholars in the Middle East. Written around 616/1220 in Aleppo, the book brings together the biographies of 414 scientists and philosophers. These are mostly Arabic-speaking Muslims, Jews, Christians, and Sabians – remnants of an ancient astral religion in northern Mesopotamia – as well as many pre-Islamic Greeks and some Persians. Ibn al-Qifṭī (d. 646/1248) was a highly erudite man of letters. He grew up in Egypt and had a very successful political career in Ayyubid Aleppo. He was highly sympathetic to the practice of philosophy, the Greek tradition of medicine and many other sciences, and he praises and celebrates those who were devoted to these fields of study. Yet, when he reviews the life of Aristotle (384–322 BCE), and when he speaks about the translation of his works first into Syriac and then into Arabic, he mentions the two most important interpreters of him in Arabic and finds much fault with them. These two were al-Fārābī (d. 339/950–1) and Ibn Sīnā ([d. 428/1037]; Avicenna). According to Ibn al-Qifṭī their work of interpreting Aristotle and explaining him in Arabic was necessary because of the many distortions that had crept in through translation. Ibn al-Qifṭī goes on to say that these two thinkers restored the original meaning of Aristotle's philosophy and presented it in ways that made it interesting to Arabic readers. "The two agreed with several of Aristotle's principles (*uṣūl*) and hence they were accused of the kind of unbelief that he was accused of" (Ibn al-Qifṭī 1903[CB],[1] 51). While having the chance to reject some of Aristotle's teachings, al-Fārābī and Ibn Sīnā did not do so. Here, Ibn al-Qifṭī continues, they made a mistake, because any correct response to Aristotle's teachings should divide them into three different categories: one category that needs to be declared unbelief (*kufr*), another category that needs to be declared heterodox innovation (*bidʿa*) and a third part that need not be rejected at all and is, in fact, beneficial to adopt. According to Ibn al-Qifṭī, these three divisions apply to all six branches of the philosophical sciences that Aristotle laid the foundations of mathematics, logic, the natural sciences, metaphysics/theology (*al-ilāhiyyāt*), politics, household management and ethics.

Ibn al-Qifṭī explains that many of those six disciplines have no connection to religion, to "religious knowledge" (*ʿulūm dīniyya*) or to "religious teachings" (*maqāṣid dīniyya*), either positively or negatively. In fact, most of them are conducted in a demonstrative way and hence must

DOI: 10.4324/9781315170718-67

be accepted as true. The field of metaphysics (*al-ilāhiyyāt*), however, which includes statements about the nature of God and His relationship with His creation, is highly problematic and contains most of the mistakes of Aristotle and his followers. Although al-Fārābī and Ibn Sīnā claimed that they reached their conclusions in metaphysics through demonstrative arguments, they were, in fact, "unable to fulfill the conditions for demonstration that they had set out in their logic" (Ibn al-Qifṭī 1903, 52–3). Many of their teachings in metaphysics are wrong, Ibn al-Qifṭī says, but there are three that need to be singled out because whoever teaches them is not only wrongly guided in his or her religion, he or she is also an unbeliever (*kāfir*), "who disagrees with all the rest of the people who associate themselves with Islam" (Ibn al-Qifṭī 1903, 53). These three teachings are

(1) the view that there will be no bodily component in the afterlife and that reward and punishment are only spiritual and only afflict the soul,
(2) the view that God knows only universals and classes of beings, but not individuals and particulars, and
(3) the position that the world exists from past eternity and has no beginning in time (Ibn al-Qifṭī 1903, 53).

Earlier in his book, Ibn al-Qifṭī had already condemned two other groups of philosophers, whom he regarded as a lesser threat than al-Fārābī and Ibn Sīnā, because they existed only among ancient Greek authors. Those who believe that this world has no Creator and that things exist by themselves without depending on God's efficient causality must be called *zanādiqa* ("clandestine apostates"), Ibn al-Qifṭī says. In religious polemics and in books on religious opinions (doxographies), this group was often called "those who believe in chance or fate," yet the corresponding Arabic label "*dahriyyūn*" is probably better translated as "materialists." Finally, there was a third group of philosophical deviants, known as "those who believe in nature" (*ṭabīʿiyyūn*), who teach that there will be no afterlife and that the soul perishes together with the death of the body. These are also *zanādiqa* (Ibn al-Qifṭī 1903, 49–50).

VI.3.2 al-Ghazālī's views

Ibn al-Qifṭī's highly sophisticated views about how different practitioners of philosophy fit or rather do not fit into the community of Muslims almost all go back to al-Ghazālī (d. 505/1111), a highly influential Ashʿarite theologian and jurist (Chapters III.3, III.4). He had developed them more than a century earlier, first in his response to the philosophy of Ibn Sīnā, *The Precipitance of the Avicennans* (*Tahāfut al-falāsifa*) – on this translation see the beginning of Chapter III.4 – published in 488/1095, and later in shorter writings, most importantly his *Decisive Criterion in the Distinction Between Islam and Clandestine Apostasy* (*Fayṣal al-tafriqa bayna l-Islām wa-l-zandaqa*), where he explains the underlying theological reasoning (al-Ghazālī 1993, 25–45; English tr. 87–103). Ibn al-Qifṭī aligns his presentation most closely with al-Ghazālī's autobiography *The Deliverer from Error* (*al-Munqidh min al-ḍalāl*), where al-Ghazālī repeats the ideas from his earlier writings (al-Ghazālī 1969, 18–27, English tr. 60–70). Whole passages from that latter book appear in Ibn al-Qifṭī's text.

In *The Precipitance of the Avicennans*, al-Ghazālī tried to show that many teachings in the metaphysical part of Ibn Sīnā's philosophical system were – despite Ibn Sīnā's own claims – not proved demonstratively. Often there were flaws in the arguments or logical options that Ibn Sīnā had overlooked. If these unproven teachings violate the text of the Muslim revelation, according to al-Ghazālī, they must be rejected. Al-Ghazālī believed in a single truth that manifests itself in human reason as well as in divine revelation, and he was willing to accept demonstrative proofs wherever they could be produced successfully (Griffel 2009[CB], 98–122). Hence, he saw a tremendous value in the movement of *falsafa* – the continuation of Greek philosophy in Arabic – and he applied many of their teachings in his writings. This aspect of al-Ghazālī's attitude to *falsafa*, however, often gets lost among his condemnation. In Ibn al-Qifṭī's text it appears only where he speaks of the third category of the *falāsifa*'s teachings, "that do not need to be rejected at all" (Ibn al-Qifṭī 1903[CB], 52). This, however, is a huge domain for al-Ghazālī and it includes philosophical teachings about the universe, cosmology, the generation of human acts, the human soul and, most important, prophecy and the superior insight of Sufi masters (*awliyā*ʾ) – all fields where al-Ghazālī followed Ibn Sīnā (Treiger 2012; Griffel 2009, 215–86).

In *The Precipitance*, however, al-Ghazālī aims to show that with regard to twenty questions, Ibn Sīnā fails to produce truly demonstrative arguments and in many cases deviates from the teachings of the Qurʾān and from positions that Muslims have agreed on. At the end of that book, al-Ghazālī's asks how these twenty teachings should be judged from the perspective of Islamic law (*sharīʿa*). Himself an authority on Islamic law, al-Ghazālī issues a *fatwā* that condemns to death any scholar who teaches the three teachings that Ibn al-Qifṭī also lists. On the last page of his *Precipitance*, al-Ghazālī writes:

> If someone asks: You have explained the teachings of these people, do you then say conclusively that they are unbelievers and that the killing of someone who holds these convictions is necessary? We say: Pronouncing them unbelievers is necessary in three questions: the question of the world's eternity and their statement that all substances are pre-eternal. Second, their statement that God's knowledge does not include the particulars that come about in time among individuals. Third, their denial of the resurrection of the bodies at the Day of Judgment.
>
> *(al-Ghazālī 2000[CB], 226)*

Note that al-Ghazālī mentions the death penalty, whereas Ibn al-Qifṭī does not. Al-Ghazālī's *fatwā* is the result of a sophisticated legal argument about apostasy from Islam that he produces in his legal writings. There is nothing in Islamic law, however, that would *per se* allow the punishment of Muslims with deviant or heterodox views. In fact, Muslim lawyers have always been very reluctant to introduce anything that could be understood as a "thought-crime." Other than God, so the default argument, who truly knows what people think and what they are convinced of? This led to an attitude of tolerance toward doctrinal deviation, at least from the side of Islamic law and among its practitioners. Al-Ghazālī somewhat breaks with that tradition. Based on legal developments in the immediate generations before him, developments that were directed against the infiltration of the Sunni caliphate in Baghdad by "propagandists" (sing. *dāʿī*) of the competing Shiite Fatimid caliphate in Cairo, he found a way to employ the religious law (*sharīʿa*) as a means to persecute *falāsifa*.

Al-Ghazālī bases his legal argument for the death penalty on a ruling that applies to apostates from Islam. On the authority of a report (*ḥadīth*) that the Prophet had said "Whoever changes his religion, have his head cut off!" Muslim jurists agreed that apostasy from Islam is punishable by death (Griffel 2000, 51–5). Most of them also agreed that actions other than the explicit rejection of Islam could not constitute apostasy. Apostasy was regarded as the declared rejection of Islam and could only be sufficiently established after a person accused of apostasy had been invited three times to repent and return to Islam. The legal institution of the "invitation to repent" (*istitāba*) is mentioned neither in the Qurʾān nor in the Prophetical *ḥadīth*. It nevertheless became a necessary condition for convicting an apostate. It safeguarded that an accused apostate had a chance to return to Islam, fully avert punishment and be reinstated in all rights as a Muslim. Subsequently, only those Muslim apostates could be punished, who openly declared their breakaway from Islam and who maintained their rejection in the face of capital punishment (Griffel 2000, 67–99).

Most early jurists understood that the general application of the "invitation to repent" effectively ruled out any penalty for apostasy. They allowed persons accused of apostasy to declare their return to Islam even when it was understood to be merely nominal. Within Sunni Islam (as opposed to Shiite Islam), the four major legal schools are the Ḥanafite, Mālikite, Shāfiʿite and Ḥanbalite. The "invitation to repent" became the accepted position in the early Ḥanafite and Shāfiʿite schools of law. Only the Mālikite school ruled differently. Mālik ibn Anas (d. 179/795) ruled that "*zanādiqa*" should not be given the right to repent and could thus be killed straight away. *Zanādiqa* here means apostates, who kept their apostasy secret and who clandestinely practiced a religion other than Islam. This ruling meant that the Mālikite school of jurisprudence was, in practice, much less tolerant against heterodox Muslims than the other two major schools. It allowed Mālikite jurists to apply the death penalty against accused apostates, who had never explicitly broken with Islam. In these cases, heterodox views were regarded as evidence of clandestine apostasy (Griffel 2001).

During the 5th/11th century, the consensus of the Ḥanafite and Shāfiʿite jurists regarding the general application of the "invitation to repent" broke down. Ḥanbalite jurists had already argued that some points of religious doctrine were so central to the Muslim creed that a violation should be regarded as apostasy from Islam and punished by death. During the middle of the 5th/11th century, scholars from all schools argued that in the case of the political agents of the Fatimid Shiite counter-caliphate, no "invitation to repent" should be granted and the agents could be killed as apostates (Griffel 2000, 227–41, 282–4). This view was shared by al-Ghazālī who in his *Decisive Criterion* wrote systematically about the criteria of apostasy. Al-Ghazālī, however, did not dispense entirely with the earlier tolerant attitude toward heterodoxy in Islam; rather, he aimed at qualifying it. He stressed that heterodox groups in Islam should all be tolerated by other Muslims as long as they do not teach any of the three doctrinal offenses identified in his *fatwā*. For al-Ghazālī the Muslim community falls into two groups, the rightly guided and the heterodox ("those who commit *bidʿa*"), both of whom are legally protected from punishment as members of Islam. But there was a third group of people like Ibn Sīnā, al-Fārābī and their followers, who only pretended to be Muslims – or may even be falsely convinced that they were Muslims – yet whose convictions show them to be unbelievers (*kuffār*) and clandestine apostates from Islam (*zanādiqa*), who are liable to the death penalty for apostasy. This, however, only applies to the leaders of the movement who go out and teach any of the three teachings identified in the *fatwā* at the end of *The Precipitance*. Simple followers from among the ordinary people are so easily swayed, and they change their opinions so often that they should not be punished (Griffel 2000, 282–92).

Before al-Ghazālī, the relationship between members of the movement of *falsafa* and Muslim theologians and jurists may have been antagonistic in the sense that they rejected the other party's approach, but it was rarely characterized by violence. People may have even accused each other of "unbelief" (*kufr*), but these accusations carried no legal implications given that before al-Ghazālī's legal reinterpretation of this term, unbelief itself was not punishable. According to the majority opinion of Muslim legal scholars before the mid-5th/11th century, unbelief was a matter that God will punish in the afterlife, while in this world it would warrant no more than social sanctions for those associated with it. Consequently, accusing one's theological opponent of being an unbeliever was quite widespread (Griffel 2009, 104). Al-Ghazālī's *fatwā* on the last page of his *Precipitance* changed that.

Another change that al-Ghazālī introduced relates to the meaning of the word *falāsifa*. The Arabic words *falsafa* and *faylasūf* are calques from the Greek words *philosophía* and *philósophos*, and they refer to the discipline of philosophy, as the Arabs had inherited it from the Greeks. By the time of al-Ghazālī that discipline was so much dominated by the philosophical system of Ibn Sīnā that in his *Precipitance*, al-Ghazālī focuses only on him (*pace* Janssen 2001). This led to a change in the meaning of *falsafa* and *faylasūf* that occurred during the 6th/12th century, that is, the century after al-Ghazālī. Before the 6th/12th century, *falsafa* could refer to any school of thought within the discipline of philosophy, but after it, the word was understood as a reference to the philosophy of Ibn Sīnā. This change of meaning happened first in the Islamic East and reached the West with some delay. Arabic philosophers in the Islamic West such as Ibn Rushd (Averroes; 520–595/1126–1198) and the Jewish thinker Maimonides (d. 601/1204), who wrote his philosophical works in Arabic, understood the word *falsafa* still largely in its traditional meaning, while older contemporaries such as the philosopher Abū l-Barakāt al-Baghdādī (d. *c.* 560/1164–5) in Baghdad, who followed al-Ghazālī in much of his criticism of Ibn Sīnā, no longer did. It is important to understand that after the 6th/12th century, the Arabic plural *falāsifa* does not simply mean "philosophers" – with all the positive connotations that this word carries in English – but it merely means "followers of Ibn Sīnā" or "Avicennans." That explains, for instance, why Ibn Rushd was one of the last thinkers in Islam, who chose "*falsafa*" as a category of self-identification (Griffel 2021, 102–3). Whereas philosophy continued to be practiced widely in the Islamic world, it did so under a different label than *falsafa*. The term that becomes more widespread during the 6th/12th century is *ḥikma* (lit. "wisdom") which has Qurʾānic connotations (e.g., Q 2:129, 151, 231). Philosophers were now called *ḥukamāʾ*. The label "*falāsifa*" was now understood as a third-person identification for one particular school within philosophy that was founded by Ibn Sīnā and that, in its purest form, had no self-identifying followers after the mid-6th/12th century. During Islam's post-classical period the phrase "the *falāsifa* say . . ." referred to what Ibn Sīnā said or what a devoted follower of Ibn Sīnā would have said. As a self-description of philosophers, "*ḥukamāʾ*" replaced "*falāsifa*" (Griffel 2021, 77–107).

VI.3.3 The impact of al-Ghazālī's *fatwā*

One of the most important questions for the history of philosophy in Islam's post-classical period – meaning the period after the 6th/12th century up to the onset of Westernized modernity during the 13th/19th century – is whether al-Ghazālī's *fatwā* from the last page of *The Precipitance of the Avicennans* had any effect in curbing philosophical activity. In the past, Western scholarship naively assumed that it put an end to all philosophical activity in Islam (Goldziher 1981 [1915]). A more nuanced assumption was that after al-Ghazālī there was no philosophy that taught any of the three positions condemned in his *fatwā* (Griffel 2000, 347). Both assumptions are wrong. There was a thriving genre of philosophical compendia in post-classical Islam that began with Abū l-Barakāt al-Baghdādī and with Fakhr al-Dīn al-Rāzī (d. 606/1210), whose works teach the world's eternity implicitly and explicitly. The latter's *Eastern Investigations* (*al-Mabāḥith al-mashriqiyya*) and his highly influential *Summary on Philosophy and Logic* (*al-Mulakhkhaṣ fī l-ḥikma wa-l-manṭiq*) clearly teach – sometimes implicitly but often quite explicitly – an ongoing creation from past eternity (al-Rāzī 1990, 2:100–2, 542). They were followed by books in the same genre by Sayf al-Dīn al-Āmidī (d. 631/1233), Naṣīr al-Dīn Ṭūsī (597–672/1201–1274) or Sirāj al-Dīn al-Urmawī (d. 682/1283). The two most important books of this genre are *The Guide to Philosophy* (*Hidāyat al-ḥikma*) of Athīr al-Dīn al-Abharī (d. between 660/1263 and 663/1265) and *Wisdom of the Source* (*Ḥikmat al-ʿayn*) by Najm al-Dīn al-Kātibī (d. 675/1276 or 693/1294). These two books are short introductions to philosophy that teach the world's pre-eternity (al-Abharī 2009, 230–1, 269–70, English tr. 135–8, 177; al-Kātibī 2006, 35, 69–72, 88). They were regular textbooks in *madrasas* – institutions of higher learning in the Islamic world – during Islam's post-classical period and attracted numerous commentaries and super-commentaries (Wisnovsky 2004, 174–6; Chapter III.2).

It is important to note that many authors, like al-Rāzī, al-Āmidī or Ṭūsī, who wrote books on philosophy (*ḥikma*) where they, among other things, applied Ibn Sīnā's arguments for the pre-eternity of the world, also wrote books of *kalām* where they took a position opposite to these arguments, in favor of the world's creation in time. This brings up questions of consistency that cannot be discussed here. Post-classical philosophy in Islam developed its own methods and its own strategies of argumentation that do not respond to our modern expectations for consistency from an author in all their works. These books of *ḥikma* aimed to familiarize students with argumentative techniques that allowed them to adjudicate between a wide range of philosophical and theological positions. Teaching the right method to do philosophy was at least as important as finding the best philosophical solution – meaning the one that is most convincing given a set of premises. These books reveal a level of tolerance for ambiguity that is alien to Western philosophical writings and unknown even to those philosophical works that were produced during Islam's classical period.

Most important is that these books undermine the earlier view that al-Ghazālī's *fatwā* had a long-lasting negative effect on the history of philosophy in Islamicate societies. If many of these books teach the world's pre-eternity and God's ignorance of individuals, did al-Ghazālī's *fatwā* have any effect at all? Or, more generally, did the religious scholars of Islam aim to curb the influence of philosophy, and did they try and persecute those who were engaged in its study? Since the beginning of the Western academic study of Islam in the early half of the 13th/19th century, its scholarship has long assumed that philosophy in Islam faced a constant threat of persecution. In the first monograph study of Arabic philosophy, published in 1852, Ernest Renan (1823–1892) wrote that al-Ghazālī was "an enemy of philosophy" and spearheaded a "war" against philosophy that was waged in all countries of the Islamic world (Renan 1852, 22, 24). Earlier in 1844, Solomon Munk (1803–1867) had already said that al-Ghazālī, "struck a blow against philosophy after which it never recovered in the Orient" (Munk 1857–9, 382–3). These opinions were repeated largely unquestioned throughout the 19th and much of the 20th centuries and led Leo Strauss (1899–1973) to develop his theory that philosophical authors in the Islamic world concealed some of their most controversial ideas behind sophisticated writing techniques (Strauss 1952). Underlying Strauss's suggestion is the assumption that philosophers suffered from a constant and real threat of persecution by religious scholars.

Before we look at the extent to which philosophers were persecuted or regarded as heretics, we should remind ourselves of the methodological problem that such a question poses. Reading al-Ghazālī's *fatwā* on the last page of his *Precipitance*, one can with relative ease understand its meaning and its intention and establish the context from which it arose. The question of whether and to what degree it was applied is, however, a quantitative problem in history – and those are extremely hard to tackle. There is a certain arbitrariness about the events that are recorded in our historical sources and then even more arbitrariness about which sources have survived and which were lost. Sometimes even very important events remain unrecorded, or their records get lost. Events in certain cities such as Baghdad or Cairo, for instance, are well covered in our sources but other places, like many rural ones, left no traces. Lastly, there is the question of what to make out of the fact that sources are silent about an issue. Can silence be evidence for the absence of a particular concern or was the concern so widespread and omnipresent that it did not need to be mentioned? For all these reasons we have no full – or even adequate – picture of the extent to which al-Ghazālī's *fatwā* was heeded or dismissed, and we will likely never have that, even after all available sources in Arabic – many of those have not yet been studied – have been analyzed. These important caveats have not always been taken into account by literature on this subject (see, e.g., Levanoni 2015).

Based on the research that has been done thus far, it is prudent to distinguish between two kinds of influences that al-Ghazālī's *fatwā* had. The first is whether the death penalty against followers of Ibn Sīnā was ever applied; the second is whether al-Ghazālī's demarcation of the boundary between Islam and non-Islam was adopted. Regarding the first question we know of at least one case where a Muslim scholar was likely executed because the jurists who confirmed his death sentence were influenced by al-Ghazālī's *fatwā*. The scholar was ʿAyn al-Quḍāt al-Hamadhānī, who was crucified in his hometown Hamadan in 525/1131. We have, of course, no court records for an almost nine-hundred-year-old case, but the available evidence, particular ʿAyn al-Quḍāt's own epistle of apologetic defense, points to the fact that the jurists who sat over his case in Baghdad used criteria that come straight out of al-Ghazālī's *fatwā*. It should be noted that ʿAyn al-Quḍāt was not truly a follower of Ibn Sīnā and that his persecution may have been prompted by political rivalries. The lawyers who decided his case had a close connection to al-Ghazālī, and some may have even been his students when he taught in Baghdad forty years earlier (Griffel 2021, 127–38).

Once that close connection to al-Ghazālī's teaching activity disappears, we do not see much influence of his legal reasoning for the death penalty. Al-Suhrawardī's execution in Aleppo around 587/1191, for instance, was most probably prompted by his followers' claim that he was a prophet – a claim he did not suppress and might even have promoted (Griffel 2021, 138–52). Yet something else comes into play here: As a practicing philosopher, al-Suhrawardī was vulnerable to legal attacks, because there was a consensus among Muslim scholars about the non-Islamic character of philosophy. That consensus can, for instance, be gathered from Ibn al-Qifṭī's book, looked at earlier. Even practitioners of philosophy shared that consensus. Al-Shahrastānī (d. 548/1153) wrote a philosophical response to Ibn Sīnā, but in his famous doxography *The Book of Religions and Sects* (*Kitāb al-Milal wa-l-niḥal*) he excludes "the philosophers" (*al-falāsifa*), by whom he means the philosophical tradition from the pre-Socratics to Ibn Sīnā, from the seventy-three groups that make up the Islamic community. He rather counts them among those who err in their religious convictions, literally "the people of fanciful opinions and religions" (*ahl al-ahwāʾ wa-niḥal*; al-Shahrastānī 1951, 2: 659, 793–1216; French tr. 2: 91, 175–485; see also the analysis in the French tr. 1: 17–20). Similarly, Fakhr al-Dīn al-Rāzī, who himself authored books of philosophy, regarded "the philosophers" (*al-falāsifa*) – and here he means the Avicennans – as a religious group outside of Islam (al-Rāzī 2004, 3–4, 1938, 91–4). In his writings, he frequently juxtaposes the teachings of "the *falāsifa*" with that of "the Muslims." This is likely a result of al-Ghazālī's thorough investigation into the criteria of membership in Islam. For scholars such as al-Shahrastānī and al-Rāzī this led to a highly interesting situation: They were themselves attracted to philosophical positions that, among other things, denied the temporal creation of the world, and they even wrote books where they defended such a position. At the same time, however, they acknowledged that such a view is not truly Islamic. Unfortunately, we know of no text where any of these thinkers ever commented on this apparent conundrum. All we can say is that it did not seem to be a problem for these authors or their readers.

VI.3.4 Conclusion

On the last page of his *Precipitance of the Avicennans*, al-Ghazālī engages in an act of *takfir* – that is, declaring a Muslim an unbeliever (*kāfir*) – targeting the followers of Ibn Sīnā and their sympathizers. *Takfir*, however, can mean many different things in many different situations. Al-Ghazālī intended it as a *legal* act that persecutes the *falāsifa* as clandestine apostates from Islam. For reasons that cannot be discussed in the space of this chapter, however, the legal arguments that he gave for such a persecution were not fully convincing (Griffel 2014). Very few scholars followed al-Ghazālī in his conviction that those who engage in philosophy and argue for the world's pre-eternity, or for God's ignorance of individuals, or for a purely noncorporeal afterlife, can be persecuted and killed. Still, many scholars in Islam agreed that arguing for one of these three positions is not in line with Islam. They might have agreed that such a philosopher is a *kāfir*, but they may have disputed that this term had any legal – or even social – consequences. In the centuries before al-Ghazālī, they understood *kāfir* to mean that such a person is merely excluded from reward in the afterlife. The term *kāfir* had many different meanings for many different Muslims in a great variety of different contexts. One should, therefore, refrain from even attempting to draw general conclusions about how religious scholars thought about the "philosophers." Some hated them and tried to persecute them; others were indifferent to them and practiced toleration; again others were fascinated by their ideas, engaged with them and taught them in *madrasas*. From our current state of knowledge, it seems clear that despite a long-standing Western prejudice in favor of the opposite, the first attitude of religious persecution was – among those three – least widespread in Islamicate societies.

Were philosophers regarded as heretics in Islam? The term *heretic* (from Greek *haíresis*, meaning "one's choice") is a Christian one, and it comes with numerous connotations and associations that relate to the often violent history of Christian engagement with religious deviation. On that subject, Islam has a very different history. There is, first of all, no institution in Islam that is in any way similar to the church. "Heretic" does not match up with any of the words within the Islamic vocabulary of religious heterodoxy, which in Arabic, for instance, include "*zindīq*," "*kāfir*," "*mulḥid*," "*mubdiʿ*," "*ahl al-bidʿa*," "*ahl al-ahwāʾ*" and many more (Brentjes 2015). We do best if we acknowledge the difference between these two religions and avoid the Christian vocabulary of deviation when speaking about Islam. Ibn al-Qifṭī's book on the history of philosophy tells us that some philosophical positions, such as the view that the world has no Creator and exists by itself ("materialism") or that there is no afterlife, were regarded as *zandaqa*, which is probably one of the strongest terms of religious deviation that can be found in Islamic literature. First-person expressions of these views were indeed extremely rare in premodern Islam – if not unknown. Ibn al-Qifṭī also informs us that other philosophical views, such as the world's pre-eternity, were regarded as *kufr*. What that means, however, is hard to determine, and each and every author needs to be looked at individually. We find that the position of a pre-eternal world as an implication of Ibn Sīnā's philosophical system, for instance, was attractive to many thinkers, because it strengthened ideas about divine simplicity and immutability, and it was indeed argued for in books of philosophy even after al-Ghazālī's condemnation. Ibn al-Qifṭī himself speaks with great admiration of many philosophers, who defended that position. This only illustrates the complexity of the question of whether in Islam philosophers were regarded as religious deviants. Often they were – yet at the same time, they were cherished as great thinkers and scientists and their ideas widely discussed and taught in *madrasas*. Most often the function of branding philosophers as non-Muslims or unbelievers was not to create an atmosphere of persecution but, rather, to build boundaries of philosophical, religious or sociocultural separation (Brentjes 2015, 151–3). Such declarations were not directed outside to the philosophers who were condemned but inside to one's own community of readers, fellow Muslims and fellow philosophers.

Note

1 Consolidated bibliography.

Bibliography

Sources

al-Abharī. ed. and tr. al-Attas, S. A. T. 2009. *A Guide to Philosophy: The Hidāyat al-Ḥikmah of Athīr al-Dīn al-Mufaḍḍal ibn ʿUmar al-Abharī al-Samarqandī*. Selangor, Malaysia: Pelanduk Publications.

al-Ghazālī. ed. Bījū, M. 1993. *Fayṣal al-tafriqa bayna l-Islām wa-l-zandaqa* [The Decisive Criterion in the Distinction Between Islam and Clandestine Apostasy]. Damascus: M. Bījū.

al-Ghazālī. tr. Jackson, S. 2002. *On the Boundaries of Theological Tolerance in Islam*. Karachi: Oxford University Press.

al-Ghazālī. ed. and tr. Jabre, F. 1969. *al-Munqidh min al-ḍalāl/Erreur et délivrance*. Beirut: al-Maktaba al-sharqiyya.

al-Ghazālī. tr. McCarthy, R. J. 2000. *Deliverance From Error. An Annotated Translation of al-Munqidh min al Ḍalāl and Other Relevant Works of al-Ghazālī*. Louisville, KY: Fons Vitae.

al-Kātibī. ed. al-Shāghūl, M. ʿA. 2006. *Ḥikmat al-ʿayn* [Wisdom of the Source]. Cairo: al-Maktaba al-azhariyya li-l-turāth.

al-Rāzī. ed. al-Baghdādī, M. al-M. 1990. *al-Mabāḥith al-mashriqiyya* [Eastern Investigations]. 2 vols. Beirut: Dār al-Kitāb al-ʿarabī.

al-Rāzī, ed. Jumʿa, al-A. 2004. *al-Riyāḍ al-muʾniqa fī arāʾ ahl al-ʿilm* [The Pleasant Gardens on the Opinions of the People of Knowledge]. Tunis: Kulliyyat al-ādāb wal-l-ʿulūm al-insāniyya bi-l-Qayruwān.

al-Rāzī. ed. Nashshār, ʿA. S. 1938. *Iʿtiqādāt firaq al-muslimīn wa-l-mushriqīn* [The Beliefs of Muslim and Non-Muslim Sects]. Cairo: Maktabat al-Nahḍa al-miṣriyya.

al-Shahrastānī. ed. Badrān, M. 1951–1955. *al-Milal wa-l-niḥal* [Religions and Sects]. 2 vols. Cairo: Maṭbaʿat al-Azhar.

al-Shahrastānī. trs. Gimaret, D., Monnot, G., and Jolivet, J. 1986–1993. *Livre de religions et des sects* [Book of the Religions and Sects]. 2 vols. Leuven: Peeters/UNESCO.

Research literature

Brentjes, S. 2015. "The Vocabulary of 'Unbelief' in Three Biographical Dictionaries and Two Historical Chronicles of the 7th/13th and 8th/14th Centuries," in Adang, C. *et al.*, eds. *Accusations of Unbelief in Islam. A Diachronic Perspective on Takfīr*. Leiden: Brill, 105–54.

Goldziher, I. tr. Swartz, M. L. 1981 [1915]. "The Attitude of Orthodox Islam Toward the Ancient Sciences," in Swartz, M. L., ed. *Studies on Islam*. New York: Oxford University Press, 185–215.

Griffel, F. 2000. *Apostasie und Toleranz im Islam. Die Entwicklung zu al-Ġazālīs Urteil gegen die Philosophie und die Reaktionen der Philosophen*. Leiden and Boston: Brill.

Griffel, F. 2001. "Toleration and Exclusion: al-Shāfiʿī and al-Ghazālī on the Treatment of Apostates," *Bulletin of the School of Oriental and African Studies* 64: 339–54.

Griffel, F. 2014. "' . . . and the Killing of Someone Who Upholds These Convictions is Obligatory!' Religious Law and the Assumed Disappearance of Philosophy in Islam," in Speer, A. and Guldentops, G., eds. *Miscellanea Mediaevalia 39: Das Gesetz*. Berlin: de Gruyter, 213–26.

Griffel, F. 2021. *The Formation of Post-Classical Philosophy in Islam*. New York: Oxford University Press.

Janssen, J. 2001. "Al-Ghazzālī's *Tahāfut*: Is it Really a Rejection of Ibn Sīnā's Philosophy?" *Journal of Islamic Studies* 12: 39–66.

Levanoni, A. 2015. "*Takfīr* in Egypt and Syria During the Mamlūk Period," in Adang, C., *et al.*, eds. *Accusations of Unbelief in Islam. A Diachronic Perspective on Takfīr*. Leiden: Brill, 155–88.

Munk, S. 1857–1859. *Mélanges de philosophie juive et arabe*. Paris: A. Franck.

Renan, E. 1852. *Averroès et l'averroïsme. Essai historique*. Paris: A. Durand.

Strauss, L. 1952. *Persecution and the Art of Writing*. Glencoe, IL: The Free Press.

Treiger, A. 2012. *Inspired Knowledge in Islamic Thought. Al-Ghazālī's Theory of Mystical Cognition and Its Avicennian Foundations*. London: Routledge.

Wisnovsky, R. 2004. "The Nature and Scope of Arabic Philosophical Commentary in Post-Classical (ca. 1100–1900 AD) Islamic Intellectual History: Some Preliminary Observations," in Adamson, P., Baltussen, H. and Stone, M. W. F., eds. *Philosophy, Science and Exegesis in Greek, Arabic and Latin Commentaries*. London: Institute of Classical Studies, 149–91.

VI.4
SYSTEMS OF KNOWLEDGE
Debating organization and changing relationships

Sonja Brentjes, Nahyan Fancy and Kenan Tekin

In this chapter, we will survey briefly the vast amounts of Islamicate literature on what knowledge is, how it should be organized and how its parts should relate to each other, including which of those parts matter, which can be ignored and which are deemed forbidden. The first sections, written by Sonja Brentjes, present many different approaches to those questions from Job of Edessa (d. after 217/832) and al-Kindī (d. *c.* 256/870) up to the period of al-Ghazālī (d. 505/1111). Next, Nahyan Fancy examines specific examples of classifications from the work of the famous Mamluk physician-jurist Ibn al-Nafīs (d. 687/1288). The last part of the chapter, by Kenan Tekin, examines the rich and varied views of Ottoman scholars, taking Ṭāshköprüzāde (d. 968/1561), Meḥmed Emīn Shirvānī (d. 1036/1627) and Ibrāhīm Ḥaqqī (1114–1194/1703–1780) as examples.

VI.4.1 Introduction

Reflections on knowledge systems are documented primarily in specialized treatises and encyclopedias. But material on this topic may also be found in bibliographies, biographies, historical chronicles, encyclopedias of *adab* ("a genre of anecdotal and anthological literature which serves as a quarry of quotable materials [*muḥāḍarāt*] for the *bel esprit*"; Heinrichs 1995, 120) and occasionally in specific, scientific tracts, such as those in geography (Biesterfeldt 2000, 80). In this chapter, we focus predominantly on specialized treatises and encyclopedias of philosophy and the sciences. When modern historians have studied the organization of knowledge in Islamicate societies, they have primarily focused on editing, translating and analyzing these specialized treatises. They have also focused overwhelmingly on the relationship between three groups of sciences: first, the sciences based on transmission, also called the sciences of the moderns or the sciences based on Sharīʿa; second, the philological sciences; and third, the so-called ancient sciences, also called the sciences of the predecessors or the first (who practiced them) or the sciences based on reason (Rosenthal 1970, 1975; Endreß 1992; Jolivet 1996).

Earlier scholars saw classifications of knowledge as hierarchies subordinating disciplines of rational knowledge, specifically those that were founded on works translated from various other languages into Arabic (Chapters I.1–I.4), under disciplines based on revealed knowledge. They also claimed that the proponents of these two disciplinary divisions were often in opposition and hostile to one another. Where they encountered classification schemes that did not adopt such

DOI: 10.4324/9781315170718-68

subordinating structures, these scholars primarily understood them as trying to overcome this hostility and to integrate the three types of knowledge (Akasoy and Fidora 2015, 107).

A broader view of the aims of classification literature recognizes that these schemes do more than organize knowledge hierarchically; they also bestow order, unity and interdependence, and provide methods and tools for retrieving information (Biesterfeldt 2000, 79). Different schemes express different views about happiness and on the relationship between knowledge and actions (mental, bodily, manual, professional). They reflect competing social values, depending on whether their authors understood language as "a product of nature (i.e. a universal phenomenon) or of culture (i.e. a particular one)" (Heck 2002, 27–8). Above all, classifications of demonstrative knowledge address epistemological questions, but they can also present utopian ideals. Questions about the temporal, geographical and typological origins (divine, animal, human, or prophetic) of specific forms of knowledge are another expression of this practice of organizing and evaluating knowledge (Biesterfeldt 2000, 79; Brentjes 2013).

The scholars most often discussed by those working on this literature are al-Kindī, al-Fārābī (d. 339/950–1), Ibn Sīnā (d. 428/1037) and al-Ghazālī. Their writings on the organization of knowledge are mostly discussed as topics within philosophy and *kalām* (Rosenthal 1975; Jolivet 1996). Other encyclopedias were composed for rulers and administrators in Khurasan and Khwarazm, such as those by Abū ʿAbdallāh al-Khwārazmī (d. 387/997), Ibn Farighūn (4th/10th century) and Fakhr al-Dīn al-Rāzī (d. 606/1210). Classifications and encyclopedias of knowledge were also written by many other famous scholars, such as Ibn Ḥazm (384–456/994–1064) and Ibn Bājja (d. 533/1139) in al-Andalus, Qusṭā ibn Lūqā, a Greek Christian scholar from Baʿalbek (d. *c.* 299/912), and the group of scholars known as the Brethren of Purity (*Ikhwān al-ṣafāʾ*; 4th/10th century).

VI.4.2 Selecting and organizing knowledge

Over the millennium covered by this book, scholars selected different sets of knowledge and different principles for ordering the sciences. In the period up to al-Ghazālī, basic structural components and content elements relied on the texts and methods of Aristotle in philosophy. For the mathematical sciences, scholars relied on the sequences of Neoplatonic texts taught in Athens and other cities in late antiquity. Galen's (129–216) works were not only instrumental for the education of physicians, but his views on education supported the value of first studying the mathematical sciences and logic, where the standard text was Porphyry's (d. *c.* 305) *Introduction* (*Isagoge*).

VI.4.2.1 A physician and two philosophers: Job of Edessa, al-Kindī and Ibn Sīnā

Philosophers and physicians often ordered only parts of all possible forms of knowledge, focusing on what fell under philosophy regarding the knowledge of nature or created things. Job of Edessa, himself a physician and perhaps also a cleric, defined his goal as explaining the knowledge of created things. He thus sought to answer not only abstract, philosophical questions, such as "How we demonstrate that heat and cold are active powers, while humidity and dryness are passive ones," but also a myriad of concrete issues such as why eunuchs have no beards and thin voices, why gold does not rust or why a rainbow is formed (Job of Edessa 1935, 67, 78, 175, 206). While it is a philosophically grounded presentation and organization of knowledge, Job's *Book of Treasures* covers only a small part of the subjects found in later encyclopedias.

All of al-Kindī's treatises on how to sort and name the sciences are lost, except for his *Epistle on the Quantity of Aristotle's Books* (*Risāla fī kammiyyat kutub Arisṭūṭālīs*). There he organized knowledge according to the sequence of subjects found in Aristotle's books as presented by Andronicus of Rhodes (1st century BCE). He built his scheme through terms from Aristotle's writings on logic and metaphysics. Under the influence of translated Galenic texts, al-Kindī posited that the study of quantity and quality, terms relating to the mathematical sciences, had to come first in the education of a future philosopher. Only afterward should books on secondary substances (universals, a metaphysical category) be tackled (Akasoy and Fidora 2015, 106). This principle of organizing knowledge applied only to the human sciences, which were placed below divine knowledge. Divine knowledge was available only to prophets through revelation. Al-Kindī's students spread his teachings far to the East of the Abbasid caliphate, where they formed starting points for further reflections on how knowledge should be classified (Biesterfeldt 2000).

Ibn Sīnā designed a complex and expansive system, including knowledge fields not explicitly named in Aristotle's book list. His three philosophical encyclopedias, the *Book of Healing* (or *The Cure*; *Kitāb al-Shifā'*), the *Book of Salvation* (*Kitāb al-Najāh*) and the *Book of Knowledge* (*Dānishnāma*; in Persian) offer the most developed and novel ideas about the epistemological aspects of organizing philosophical knowledge. They all follow Aristotle's division of philosophy into three theoretical parts (metaphysics, natural philosophy, the mathematical sciences), and three practical parts (politics, economics/household management, ethics). He also wrote at least two specialized treatises on the classification of the rational sciences. At different places in his works, Ibn Sīnā presented his perspective differently. Medicine, astrology and alchemy, for instance, were absent from his encyclopedias. Copies of the *Book of Salvation* often lack the mathematical section because he left it to his student and secretary al-Jūzjānī (d. 462/1070). In the *Book on Demonstrative Proof* (*Kitāb al-burhān*), part of the *Book of Healing*, Ibn Sīnā structured philosophical knowledge through two metaphysical categories, as things "whose existence is not by our choice and action" and as things that exist due to our choice and action (Marmura 1980, 240–1). He also reflected on the interdependence of all fields of knowledge as well as their individuality. Knowledge was structured by first principles and demonstrative proofs. The higher and more general sciences contain self-evident propositions from which the scholar derives proofs for claims that are not self-evident. The subordinate and more particular sciences contain knowledge that cannot be proved within them but rely for demonstrations on principles from higher sciences and relate to each other as species to genus (Maróth 1994). Thus, medicine is subordinated to physics, since both investigate the body (genus), but medicine limits itself to the living human body (species) and regards only two aspects – health and illness. Knowledge of plants and animals is situated between physics and medicine, because it studies the living body in an absolute manner (Akasoy and Fidora 2015, 110). In his last major philosophical work, *Pointers and Reminders* (*Ishārāt wa-tanbihāt*), Ibn Sīnā even abandoned the traditional division of the disciplines according to Aristotelian as well as Neoplatonic philosophy in favor of a twofold system, which consisted only of logic and a single other unit that incorporated all other fields of knowledge, because the fundamental philosophical problems could not be formulated in a manner where each of them fit into a specific epistemic box. This modification of the macrostructure of philosophy led to changes in its microstructure and thus a substantial reorganization of philosophical knowledge as acquired through the translations made from the 2nd/8th to the later 4th/10th century. Both features are discussed in more detail in Chapter I.11.

VI.4.2.2 Knowledge schemes going beyond philosophy

Al-Fārābī, al-Khwārazmī, al-Ghazālī and Fakhr al-Dīn al-Rāzī presented sequences of knowledge fields that went beyond philosophy proper and included philological and religious disciplines. Most often those sequences have been read by modern interpreters as top-to-bottom orderings like those in philosophy, starting with the field declared from the author's specific perspective as the most important one and then branching out vertically and at times horizontally. However, this is not the only way of reading such sequences. Al-Khwārazmī, for instance, left out several religious disciplines and focused on the subject matter and its terminology in the philological and mathematical disciplines, plus philosophy, logic, medicine and alchemy, because they were less familiar to the administrators of the Samanid court for whom he wrote (Brentjes 2018[CB],[1] 206–9).

Al-Fārābī was also not interested in ordering and surveying the totality of knowledge. Rather, he wished to build a system of demonstrative knowledge grounded on six main disciplines: the science of language, logic, the mathematical sciences, natural philosophy, metaphysics and political science. But only five of these six disciplines are universal. The science of language is not general linguistics but is valid for a concrete language only. In his *Enumeration of the Sciences* (*Iḥṣā' al-ʿulūm*), each of these fields comprises subfields. Some of them such as ontology are again universal, while others such as theology, subordinated under politics, are particular (Endress 2006[CB], 107; for a concise survey on the work see Rudolph 2017, 546–7).

Recent scholars have understood the *Enumeration* in a variety of manners. Heinrichs evaluated it as a logically ordered curriculum and a partial history of the sciences (1995, 122). Akasoy and Fidora judged it an "effort to arrive at a more general and comprehensive classification of human epistemic practices" centering them on demonstrative science as defined by Aristotle in his *Posterior Analytics* (2015, 107). Biesterfeldt proposed that al-Fārābī's goal was an ideal Islamic society ruled by a perfect philosopher and prophet (Biesterfeldt 2000, 87). But the *Enumeration* was also a reply to an intense political and epistemological controversy during the 4th/10th century, defending philosophy rather than religious law as the primordial science.

Al-Ghazālī proposed seven or eight classifications of knowledge, each depending on specific arguments and goals (Treiger 2011; Rudolph 2019; Bakar 1998). According to Treiger, this intense reflection on the content, methods and relations of a broad range of religious (*sharʿiyya*) and nonreligious (*ghayr-sharʿiyya*) disciplines documents al-Ghazālī's deep involvement with philosophy, because this kind of scholarly question had originated and unfolded in philosophy (Treiger 2011, 3). In the *Intentions of the Philosophers* (*Maqāṣid al-falāsifa*), al-Ghazālī delineated a fundamentally Aristotelian division of philosophy. The three theoretical philosophical disciplines study the "state of beings" with the goal to impress on the human soul "the configuration of the universe in its hierarchical arrangement" in order to make it "virtuous in this world and entitled to felicity in the next" (Treiger 2011, 4). The three practical disciplines serve to learn about actions that enable human welfare on earth and salvation in the hereafter (Treiger 2011, 4). Although in this text al-Ghazālī followed Ibn Sīnā's Persian survey of philosophy, *Knowledge for ʿAlā' al-Dawla* (*Dānishnāma-yi ʿAlā'ī*), he also modified it through changes in terminology, descriptions and metaphors. Treiger argues that this allowed him to model the highest (religious) sciences discussed in the other classifications on the basis of the relationship between theoretical and practical philosophy outlined in the *Intentions* (Treiger 2011, 6). The elaboration of new classifications started in a work called *The Disgraces of the Batinites* (*Faḍā'iḥ al-Bāṭiniyya*) and reached its climax in Book I of the *Revival of the Religious Sciences* (*Iḥyā' ʿulūm al-dīn*). The classifications in the *Revival* had the most sustained and widespread impact on Muslim writers and readers.

As in other classifications, al-Ghazālī counsels Muslims about the right path to knowledge leading to a happy life and afterlife. He bases his advice on two closely related sets of legal concepts. First are religious obligations binding upon every Muslim (*farḍ ʿayn*), and religious obligations binding upon a Muslim community (*umma*) collectively (*farḍ kifāya*). Second are praiseworthy (*maḥmūd*), blameworthy (*madhmūm*) and permissible (*mubāḥ*) actions. To seek knowledge was, according to a *ḥadīth*, a *farḍ ʿayn*. But Muslim scholars could not agree on the kinds of knowledge that this *ḥadīth* addressed (Bakar 1998, 206–7). Al-Ghazāli tried to solve this problem by building hierarchies and offering judgments. In his system, revealed knowledge supersedes human rational knowledge. Two classes of sciences are *farḍ kifāya*: the nonreligious (*ghayr-sharʿiyya*) and the religious (*sharʿiyya*) disciplines. While all religious sciences are praiseworthy, only some members of the first class are so, among them medicine, the mathematical sciences and the philosophical science of politics. Blameworthy disciplines are magic and the sciences of talismans, trickery and deception. History, poetics or geography are permissible. Many subdivisions of the religious disciplines are discussed from different angles, but surprisingly, *kalām* is missing. Philosophy as a whole is also missing. When explaining his choice, al-Ghazāli argues that philosophy is not a single discipline but consists of four different fields of knowledge: (1) geometry and arithmetic, (2) logic, (3) metaphysics and (4) natural philosophy, with (2) and (3) subordinated under *kalām*. Geometry and arithmetic are already enumerated as part of the nonreligious sciences and thus do not need to be repeated. Physics is not included, because despite its being similar to medicine insofar as it studies material bodies, parts of it are not needed and other parts contradict religion and thus are ignorance, not knowledge (Rudolph 2019; Treiger 2011, 6–10).

A second classification, as religious (*sharʿiyya* or *dīniyya*) and rational (*aqliyya*) sciences, appears in Book XXI of the *Revival*. The rational sciences are either necessary (*ḍarūriyya*) or acquired (*muktasaba*). Necessary knowledge is axiomatic knowledge of necessary truths such as that a thing cannot be existent and nonexistent at the same time. Acquired knowledge encompasses medicine, geometry, arithmetic, astronomy and the crafts as knowledge of this world and (religious) sciences that teach rational knowledge about the hereafter (Treiger 2011, 13). The placement of the sciences about the hereafter under rational knowledge visibly contradicts their placement under the religious sciences in Book I. Treiger argues that this results from the different purposes of the two classifications. Book I discusses legal obligations of the community and the individual Muslim regarding knowledge, while Book XXI focuses on the epistemological question of how knowledge is acquired (Treiger 2011, 14).

A third classification appears in Chapter 21 of *The Balance of Action* (*Mīzān al-ʿamal*). Its main criterion is meaning, and its three divisions are (1) philology and its parts; (2) polemics, disputation, demonstration and rhetoric; and (3) the theoretical and practical disciplines that deal with meaning alone. The practical disciplines are two: Aristotelian ethics, on one hand, and economics and politics, also called law (*fiqh*), on the other. The theoretical fields combine knowledge of God, the angels, the djinns, the prophets and paradise and hell, with knowledge on the heavenly bodies and heavenly phenomena and the three terrestrial kingdoms (stones, plants, animals) as well as their orders and relationships (Treiger 2011, 16). The cosmological and terrestrial parts of these theoretical disciplines go far beyond the parts named in the earlier classifications. They encourage Muslims to acquire knowledge taught in astronomy, meteorology, physics, mineralogy, zoology and botany.

As these few, brief examples show, classifications of knowledge arose from within philosophical reflections. Classifications could cover the entire realm of philosophy, cover only parts of it, or go beyond it. These classifications could focus on epistemological perspectives, discuss concrete specific questions about nature or man, make suggestions about proper behavior and lifestyle or grapple with one of the fundamental questions of human existence: why and how do humans know. The literary formats chosen for providing the various choices, the qualifications of the individual authors and the audiences they aimed to address were equally diverse.

VI.4.3 Knowledge systems in the works of an Egyptian physician-jurist

Ibn al-Nafīs (d. 687/1288) is an example of a Sunni Muslim scholar who followed in the footsteps of al-Ghazālī and Fakhr al-Dīn al-Rāzī in classifying all kinds of knowledge – philosophical, philological and religious – under one system. Unlike al-Ghazālī, however, he included law and *ḥadīth* within his schemes, as he sought to address epistemological questions, questions of meaning and questions of practice within one overarching scheme. His classification schemes are part of the introduction of two of his technical works where they help the reader ascertain the position of the specific science under discussion in the work within the larger organization of knowledge (neither of them provides a full classification of all the sciences).

Ibn al-Nafīs begins his commentary on the first book of Ibn Sīnā's *Canon of Medicine* (*al-Qānūn fī ṭibb*) with a classification scheme that shows precisely where, he thinks, medicine fits within the larger system of knowledge. There are a number of features worth highlighting in this scheme as they reveal Ibn al-Nafīs's response to contentious debates among scholars. Heck, for example, has shown that even more primary than the division between the religious and rational sciences is a division centered on language that deeply informs the various classification schemes. He asserts that the early state secretaries privileged the Arabic linguistic sciences so much so that even philosophers like al-Fārābī proceeded to accept this division between the epistemological spaces of Arabic and Greek science (Heck 2002, 36–7). By contrast, Ibn al-Nafīs does not make such a sharp distinction. Instead, building on al-Fārābī's original analogy between grammar and logic, wherein al-Fārābī states that "the relation of logic to reason and intelligibles is like that of grammar to the language and words" (al-Fārābī 1996, 28), Ibn al-Nafīs states,

> Each science either intends its study to be for the sake of safeguarding from error in [sciences] other than it, or not like that. The first [type of science] either safeguards from error in words or their composition, regardless of whether they are rhythmically balanced or not. This science is called the science of *adab*, and it comprises grammar, prosody and others. Or, [this first type of science] protects from errors in meaning and syllogistic compositions, or definition or the like, so that the compliance with its rules safeguards against error in the sciences. Such a science is logic.
>
> *(Sharḥ al-Qānūn, fol. 1b)*

If there is a hierarchy embedded in this scheme, it is merely that one needs to know how to compose sentences correctly before one can compose definitions and syllogisms. Also, at least on the face of it, he does not separate Arabic from Greek poetry. Thus, the classification is not meant to be a separation of knowledge inherited from the Arabs versus the Greeks.

In other cases, Ibn al-Nafīs follows in the footsteps of his predecessors. For example, the Greeks had maintained a sharp distinction between art (*technē*/*ṣināʿa*) and science (*epistēmē*/*ʿilm*), often treating medicine as a mere "art" rather than a true "science" (Endress 2006, 105). In contrast, al-Fārābī and Ibn Sīnā did not adhere to such strict separations and helped ease the "traditional opposition between science and art" (Jolivet 1996, 1022). Ibn al-Nafīs, much like al-Fārābī, begins his classification of the sciences (*ʿulūm*) by first defining art (*ṣināʿa*) as "an ingrained psychological disposition (*habitus*) . . . that is necessarily comprised of knowable facts (*maʿlūmāt*)." Facts acquired through practical exercises are commonly referred to as "art," whereas those acquired through theoretical reasoning (*naẓar*) and the use of arguments are commonly referred to as "science" (*Sharḥ al-Qānūn*, fol. 1b).

In short, Ibn al-Nafīs has three organizing principles in the *Commentary on the Canon*'s classification scheme: first, whether a science protects from errors across all sciences or not; second, whether a science (or part of it) is intended for practice or not; and, third, what exactly the subject matter of the science is and how it relates to physical matter. In both the latter two cases, Ibn al-Nafīs follows Ibn Sīnā. For example, he distinguishes between purely practical sciences (such as jurisprudence and practical philosophy), purely theoretical sciences (such as metaphysics) and sciences that are partly theoretical and partly practical (such as medicine). In terms of subject matter, Ibn al-Nafīs goes through the standard order of books found in Ibn Sīnā's philosophical compendium, *The Healing*, where each book investigates a subject based on its relation to matter. For example, *Metaphysics* (*al-ʿilm al-ilāhī*) investigates bodies that do not require matter for their existence, whereas *Animals* (*Kitāb al-ḥayawān*) investigates material bodies that have temperaments due to souls that possess perception. He also defines medicine identically as a science that deals with bodies that specifically possess a soul that both perceives and possesses reason; however, medicine is technically not a part of natural science but, rather, a branch under it, exactly as Ibn Sīnā had stated in his original classification scheme (MS London, Wellcome Library, Arabic 51, fols 1b-2a; Gutas 2003, 147).

The organizing principles of the classification scheme found in his *Summary on the Principles of Ḥadīth* are fairly distinct. Here, Ibn al-Nafīs first separates the transmitted sciences (*samʿiyya*) from the rational sciences (*ʿaqliyya*). The former, he claims, use both authoritative, transmitted premises in their arguments along with rational premises, whereas the latter only employ rational premises. The transmitted sciences are then divided into those that accept their premises from authorities whose veracity (*ṣidq*) is unquestioned (e.g., God, the Prophet or consensus of the community). These are called the religious sciences (*sharʿiyya*). Where the veracity of authorities may be questioned, those fields are called the linguistic sciences (*adabiyya*). He then further subdivides the linguistic sciences based on their focus on simple or compound words, pronunciation or meaning, metered poetry or unmetered prose and so forth.

The religious sciences are divided first based on the theoretical-practical distinction. Ibn al-Nafīs further divides the theoretical religious sciences based on whether they are concerned with words or not, maintaining the hierarchy of God over the Prophet Muhammad. At the end of these divisions, he reorders all the theoretical religious sciences hierarchically, starting with *kalām*, the science that examines the nature of God and his attributes, followed by the sciences concerned with the words of God (recitation and exegesis), followed by those concerned with the words of the Prophet (sciences of *ḥadīth*), then the principles of jurisprudence and finally the science of legal disputation (Ibn al-Nafīs 1991, 95–7). In short, the organizing principles of this classification scheme are: first, whether a science is based on rational or transmitted premises; second, the truth-value of transmitted premises; third, the subject matter of the science and how is it used; and, fourth, whether a science is practical or theoretical.

Although the two classification schemes are considerably different, there are some overlapping concerns that can help us situate Ibn al-Nafīs's systematization of knowledge among his predecessors and successors. First, Ibn al-Nafīs clearly privileges purely theoretical forms of knowledge over those that are geared toward practice. In doing so, he follows a long tradition of privileging theoretical over practical knowledge in Graeco-Islamicate learning going back to Plato and Aristotle. Similarly, he privileges more abstract forms of knowledge over more concrete, whether metaphysics over physics in his *Commentary on the Canon*, or *kalām* over Qurʾānic exegesis and the sciences of *ḥadīth* in the *Summary*. He also is deeply concerned with truth across his classifications. In the medical work, this is found in his defining the linguistic sciences and logic as sciences that protect one from committing errors in all other sciences, but in the *ḥadīth* text, he privileges the religious sciences over the linguistic sciences as the former take their transmitted premises from authorities whose veracity cannot be questioned. He is also firmly committed to rational argumentation as a way to protect oneself from error in any science, rational or transmitted. Here, he parts ways with many of his predecessors and successors by clearly stating that even the transmitted sciences rely on rational premises and are steeped in rational argumentation. For example, Ibn Khaldūn (732–808/1332–1406) categorically denied any role for rational premises in the transmitted (*naqlī*) sciences (Fancy 2013[CB], 27–35). Similarly, Quṭb al-Dīn al-Shīrāzī (633–710/1236–1311) stated that only the principles of religion can be founded on both rational and transmitted premises, but all other branch sciences of religion, including the principles of jurisprudence and principles of *ḥadīth*, are based solely on transmitted premises (Bakar 1998, 257–60). Finally, and on a related note, both classification schemes are meant to truly bring all the known sciences into one, integrated, albeit hierarchical, system of knowledge. They are brought into a single system whether they are religious, rational, linguistic or even sciences such as "the science of ingenious devices" (*ʿilm al-ḥiyal*; Chapter I.9), which Ibn al-Nafīs includes as an example of practical science in his *Commentary on the Canon*.

VI.4.4 Ottoman classifications of the sciences

There are dozens of books and treatises on classifications of the sciences that were written during the long Ottoman history. These works could be divided into three groups or genres: classification or definition of sciences (*taṣnīf* or *taqsīm al-ʿulūm*), representation of sciences (*unmūzaj al-ʿulūm*) and pedagogical ordering of the sciences (*tertīb al-ʿulūm*). In this section, we examine the schemes found in each of these genres by selecting an example from each: a *taqsīm* from the encyclopedic book by Ṭāshköprüzāde, an *unmūzaj* from the work of Meḥmed Emīn Shirvānī and a *tertīb* from the didactic poem by Ibrāhīm Ḥaqqī Erżurūmlu.

There are significant differences between the orders of sciences in each one of these genres, which reflect the relative place of the author in the state and society. It is thus necessary to provide some background for these three scholars. Ṭāshköprüzāde, who was active during the reign of Süleymān the Magnificent (r. 926–974/1520–1566), represents the establishment of Ottoman epistemic self-confidence just as Süleymān's reign signifies political domination. Hence, it is not a coincidence that Ṭāshköprüzāde wrote the first biographical dictionary of Ottoman scholars, titled *The Red Anemone on the Scholars in the Ottoman State* (*al-Shaqāʾiq al-nuʿmāniyya fī ʿulamāʾ al-dawla al-ʿuthmāniyya*). Besides this work, Ṭāshköprüzāde penned a few books on the classification and order of the sciences including his magnum opus *The Key to Felicity and the Lamp to Mastery on the Subject Matters of the Sciences* (*Miftāḥ al-saʿāda wa-miṣbāḥ al-siyāda fī mawḍūʿāt al-ʿulūm*).

In the *Key of Felicity*, Ṭāshköprüzāde provided a comprehensive classification of sciences as well as a discussion of classification as a science in its own terms. In his view, the science of classifications of the sciences (*ʿilm taqāsīm al-ʿulūm*) investigated the gradation of subject matters from the most general to the most particular so that it could provide the proper place for the subject matter of the sciences in this general scheme (Ṭāşköprüzāde 1985[CB], 1: 300; Bellino 2014, 179). Ṭāshköprüzāde considered this science a branch of metaphysics since its subject matter is the most general. Accordingly, *Key to Felicity* is organized around an ontological classification of the sciences. It is his habit to enumerate books in each science and, keeping in line with that, in this entry on the science of the classifications of sciences, Ṭāshköprüzāde mentions a "nice treatise" by Ibn Sīnā (*Treatise on the Rational Sciences* [*Risāla fī l-ʿulūm al-ʿaqliyya*]) and describes his own book, which he says he is revising, as immensely useful in this regard (Ṭāşköprüzāde 1985, 1: 300). It is in fact a curious matter that Ṭāshköprüzāde does not mention other works in this science, considering that the book itself benefits from a work by the Mamluk physician and scholar of the religious sciences Ibn al-Akfānī (d. 749/1348), and acknowledges contributions of al-Fārābī in the field (especially his *Enumeration of the Sciences*) in a different context (Ṭāşköprüzāde 1985[CB], 1: 293).

Ṭāshköprüzāde states that sciences are of four species (*anwā'*) just as the things that exist (existents) are of four levels (*marātib*). These are existents in themselves (*a'yān*), existents in the mind (*adhhān*), existents in utterance (*'ibāra*) and existents in writing (*kitāba*). Since the sciences of the first level concern the state of the thing in itself, they are the real sciences. They do not change over time or across religions. This first species is further divided into two kinds: philosophical sciences (*al-'ulūm al-ḥikmiyya*), in which the investigator (*bāḥith*) looks at the conditions of reality using his own mind, and the religious sciences (*al-'ulūm al-shar'iyya*), in which the investigator looks at reality based on Islamic principles.

Ṭāshköprüzāde calls the second species of science, which concerns the second level of existence, the instrumental sciences of meaning (*al-'ulūm al-āliyya al-ma'nawiyya*). Examples are logic, dialectics and disputation. The sciences that address the remaining two levels of existence, utterance and writing, are instrumental sciences (*al-'ulūm al-āliyya*) concerning language and script. Ṭāshköprüzāde adds that the latter are overall Arabic linguistic sciences, which are also valorized by Islam. After this classification, Ṭāshköprüzāde notes that the last three types can only be acquired by investigation (*baḥth*), whereas the first type, that is sciences that concern the ultimate reality of things, can either be acquired by investigation or by purification (*taṣfiya*; Ṭāşköprüzāde 1985[CB], 1: 69–70). All in all, there are seven classes of sciences: sciences of utterance, writing, mind, philosophical sciences, which are further divided into theoretical and practical philosophy, religion and mysticism. Accordingly, the book's major chapters correspond to this classification. Ṭāshköprüzāde's endeavor in this book amounts to a synthesis of conflicting trends. He does away with a contrast between religious and rational sciences by adopting an ontological viewpoint.

Despite its huge size, Ṭāshköprüzāde's book was well received. It was translated into Turkish by Ṭāshköprüzāde's son Kemāl el-Dīn (d. 1030/1621), as well as summarized by several writers, among them Ṭāshköprüzāde himself (*City of the Sciences* [*Madīnat al-'ulūm*]). Another *Summary* (*Mukhtaṣar Miftāḥ al-sa'āda*) was composed by Ṣolaqzāde Ḫalīl ibn Meḥmed (d. 1095/1684). Another Turkish book on classification and pedagogy, namely, *The Seven Stars* (*Kevākib-i sab'a*), by an anonymous author, reproduces Ṭāshköprüzāde's classification. But not all later Ottoman scholars accepted Ṭāshköprüzāde's synthesis. For instance, Meḥmed Emīn Shirvānī rejected Ṭāshköprüzāde's looser conception of the sciences of ultimate reality. In an encyclopedic overview of the sciences, Meḥmed Emīn retained the philosophical use of the term, limiting its application to the philosophical sciences, which are differentiated from the nonphilosophical sciences. The latter are further divided into religious and nonreligious sciences.

Meḥmed Emīn Shirvānī's *Lordly Benefits of Sultan Aḥmed* (*al-Fawā'id al-khāqāniyya al-aḥmadiyya al-khāniyya*) was written in 1023/1614 in the genre of samples of sciences. The author acknowledges a few other books in this tradition including Fakhr al-Dīn al-Rāzī's *Gardens of Lights on the Realities of Secrets* (*Ḥadā'iq al-anwār fī ḥaqā'iq al-asrār*), which included sixty disciplines; Fenārī's (d. 839/1435) *Compendium of the Sciences* (*Unmūzaj al-'ulūm*), which discussed one hundred sciences; and al-Davānī's (830–908/1426–1502) *Compendium of the Sciences* (*Unmūzaj al-'ulūm*), which covered ten sciences. Shirvānī notes that he enumerated 53 sciences among the various species of the transmitted sciences and rational sciences (MS Istanbul, Süleymaniye Library, Amcazâde Hüseyin 321, fol. 3b). Most of the books written in this genre were dedicated to the rulers as evidence of the authors' wide-ranging learning. *Lordly Benefits* is no exception, and in fact, the title and the number of the sciences indicate that it was dedicated to Sultan Aḥmed I (r. 1012–1026/1603–1617).

Shirvānī organized the book in a way that was modeled on the battle divisions of the Ottoman military, as explained in the preface. Thus, the book consists of an introduction (*muqaddima*) concerning the nature of science, and four chapters devoted to classes of sciences which are grouped as those of the left flank (concerning rational sciences), those of the right flank (concerning sciences of language), those of the center (*qalb*, lit. "heart", concerning religious sciences) and those of the rearguard (*sāqa*, concerning practical philosophy). The book encompasses a total of 53 sciences, a number equal to the numerical value of the name of the dedicatee, Aḥmed I (MS Istanbul, Süleymaniye Library, Amcazâde Hüseyin 321, fol. 3b). The main feature of this genre is that, in addition to enumerating sciences, several issues from each science are discussed in order to represent their distinct subject matters.

Unlike the previous two genres of writing, the third genre, on ordering the sciences, seems to have been cherished more among scholars who were active in the Ottoman periphery. Given that provincial students lacked access to the hierarchical madrasa system of the capital and the major cities of the empire, there might have been more interest in the pedagogical ordering of the sciences (what we call today curriculum development) in the provinces where students aspired to acquire a rigorous and competitive education. Although authors of such treatises warn students against seeking worldly positions in government and rather encourage them to become perfect human beings, an old topos, the warnings themselves indicate that there was indeed such an interest. An example is a poem by İbrāhīm Ḥaqqī titled *The Order of the Sciences* (*Tertību̇ l-ʿulūm*).

Ḥaqqī composed his didactic poem *The Order of the Sciences* in 1165/1752 in the city of Erzurum, in eastern Anatolia. The poem basically provides a list of disciplines and mentions books in each discipline that should be studied in the order presented. It could be seen as reflecting the conventional curriculum in the eastern provinces of the empire or as revealing the books that Ḥaqqī had studied.

The Order of the Sciences consists of a preface and twelve sections. After the customary beginning, Ḥaqqī introduces the sciences in the following order:

> a guide to knowledge (*ʿilm*); a summary of instrumental sciences (*ijmāl-i ālet*); the Qurʾān and the art of writing (*Qurʾān u khaṭṭ*); Islamic law and lexicon (*fikh u lughat*); morphology (*tasrīf*); syntax (*nahv*); logic (*mantiq*); etiquettes of argumentation (*ādāb*); semantics (*ʿilm-i meʿānī*); philosophy (*ḥikmet*); particulars (*jüzʾiyyāt*); foundations (*uṣūl*).
> *(Fazlıoğlu 2005, 122)*

Ḥaqqī begins by enumerating instrumental sciences in general, which consist of various branches of Arabic language and literature, followed by Qurʾānic recitation, the scribal arts, pedagogy and then a further suggestion for books of grammar, logic, dialectics and semantics. Ḥaqqī's list culminates with the study of philosophical sciences and principles of religion. Most of the courses mentioned in the didactic poem of Ḥaqqī correspond to the classes mentioned in other books on the madrasa curriculum such as the anonymous *Seven Stars*. They reflect a general pattern of study in the post-classical period, particularly after the 7th/13th century, which prioritizes the study of instrumental sciences above all else. The order, thus, merely lays out a road map for studying various sciences.

The three texts by Ṭāshköprüzāde, Shirvānī and Ḥaqqī that we have briefly surveyed indicate the richness of the classifications and orders of sciences discussed by scholars in the Ottoman Empire. While Ṭāshköprüzāde provided an ontological classification of the sciences, Shirvānī combined two widespread classifications, namely the philosophical division of sciences into practical and theoretical sciences and the partition of the sciences into philological, religious and rational sciences. Ḥaqqī, on the other hand, opted for a pedagogical ordering of the sciences. All in all, we can see that the context of the writing, as well as the genre, influenced the nature of classifications.

VI.4.5 Summary

The survey of classifications of sciences by scholars from the Abbasid period to the Ottoman period shows that there was no single pattern for ordering the sciences among scholars in the Islamicate world. Their contexts as well as their intended audiences shaped the types of classification they provided. While philosophers such as Ibn Sīnā focused their classifications on the philosophical sciences as they were representing and revising the Aristotelian tradition, scholars with other intellectual commitments such as al-Ghazālī, Ibn al-Nafīs and Ṭāshköprüzāde chose different orientations and purposes, thus presenting different kinds of classifications that went beyond the philosophical disciplines.

Note

1 Consolidated bibliography.

Bibliography

SourcesIbn al-Nafīs. ed. Zaydān, Y. 1991. *Mukhtaṣar fī ʿilm uṣūl al-ḥadīth* [Epitome on the Science of the Foundations of *Ḥadīth*]. Cairo: al-Dār al-miṣriyya.

Job of Edessa. ed. and tr. Mingana, A. 1935. *Encyclopaedia of Philosophical and Natural Sciences as Taught in Baghdad about A.D. 817, or, Book of Treasures*. Cambridge: W. Heffer & Sons.

Shirvānī, Meḥmed Emin. ed. Ahmet Kamil Cihan *et al.* 2019. *El-Fevâıdü'l-Hâkâniyye: Şirvânî'nin Bilimler Tasnifi* [*The Lordly Benefits*. Shirvānī's Classification of Knowledge]. İstanbul: Yazma Eserler Kurumu Yayınları.

Manuscripts

MS Istanbul, Süleymaniye Library, Amcazâde Hüseyin 321.

MS Istanbul, Süleymaniye Library, Laleli 3757.

MS London, Wellcome Library, Arabic 51.

Research literature

Akasoy, A. A. and Fidora, A. 2015. "The Structure and Methods of the Sciences," in Taylor, R. C. and López-Farjet, L. X., eds. *The Routledge Companion to Islamic Philosophy*. London and New York: Routledge, 105–14.

Bakar, O. 1998. *Classification of Knowledge in Islam. A Study of Islamic Philosophies of Science*. Cambridge: Islamic Texts Society.

Bellino, Francesca. 2014. "The Classification of Sciences in an Ottoman Arabic Encyclopaedia: Ṭāshköprüzāda's Miftāḥ al-Saʿāda," *Quaderni di Studi Arabi* (New Series) 9: 161–80.

Biesterfeldt, H. H. 2000. "Medieval Arabic Encyclopedias of Science & Philosophy," in Harvey, S., ed. *The Medieval Hebrew Encyclopedia of Science and Philosophy*. Dordrecht: Kluwer Academic Publishers, 77–98.

Brentjes, S. 2013. "Narratives of Knowledge in Islamic Societies: What Do They Tell Us about Scholars and Their Contexts?" *Almagest* 4.1: 74–95.

Endreß, G. 1992. "Die wissenschaftliche Literatur," in Fischer, W., ed. *Grundriß der Arabischen Philologie*, Bd. III: Supplement. Wiesbaden: Reichert Verlag, 3–152.

Fazlıoğlu, Ş. 2005. "Ta'lîm ile İrşâd Arasında: Erzurumlu İbrahim Hakkı'nın Medrese Ders Müfredatı," *Divan İlmi Araştırmalar* 18: 115–73.

Gutas, D. 2003. "Medical Theory and Scientific Method in the Age of Avicenna," in Reismann, D., ed. *Before and After Avicenna: Proceedings of the First Conference of the Avicenna Study Group*. Leiden: Brill, 145–63.

Heck, P. L. 2002. "The Hierarchy of Knowledge in Islamic Civilization," *Arabica* 49.1: 27–54.

Heinrichs, W. 1995. "The Classification of the Sciences and the Consolidation of Philology in Classical Islam," in Drijvers, J. W. and MacDonald, A. A., eds. *Centres of Learning: Learning and Location in Pre-modern Europe and the Near East*. Leiden: Brill, 119–39.

Jolivet, J. 1996. "Classification of the Sciences," in Rashed and Morelon, eds. [CB], 3: 1008–25.

Marmura, M. 1980. "Avicenna on the Division of the Sciences in the *Isagoge* of His *Shifa'*," *Journal of the History of Arabic Science* 4: 239–51.

Maróth, M. 1994. *Die Araber und die antike Wissenschaftstheorie*. Leiden and New York: Brill.

Rosenthal, F. 1970. *Knowledge Triumphant: The Concept of Knowledge in Medieval Islam*. Leiden: Brill.

Rosenthal, F. 1975. *The Classical Heritage in Islam*. London: Routledge and Kegan Paul.

Rudolph, U. 2017. "Abū Naṣr al-Fārābī," in Rudolph *et al*., eds. [CB], 526–654.

Rudolph, U. 2019. "Al-Ghazālī on Philosophy and Jurisprudence," in Adamson, P., ed. *Philosophy and Jurisprudence in the Islamic World*. Berlin and Boston: De Gruyter, 67–91.

Treiger, A. 2011. "Al-Ghazālī's Classifications of the Sciences and Descriptions of the Highest Theoretical Science," *Dîvân Disciplinlerarası Çalışmalar Dergisi* 16.30: 1–32.

VI.5
EMBASSIES, TRADING POSTS, TRAVELERS AND MISSIONARIES

Simon Mills

It is well known that during the Middle Ages a substantial amount of scientific thought produced in the Jewish, Christian and Islamic societies of North Africa and Asia was familiar to Western scholars. The activities of translators from Arabic into Latin working between the 10th and the 13th centuries in centers such as Sicily and Toledo have been fairly extensively documented (Burnett 2009; Chapter V.14). Recently, however, the case has been made for the endurance into the early modern period of this westward movement of ideas (Brentjes 2010; Hasse 2016[CB])[1]. This has raised anew the question of the transmission of practical and theoretical knowledge. In the absence of a more or less comprehensive map of the translation movement, such as exists for the earlier period, historians have been led to ask how scientific ideas and practices migrated across geographical, cultural and linguistic borders.[2]

One possible line of inquiry is to uncover the channels of communication between *Dār al-Islām* and western Christendom. How exactly did early modern Jews, Christians and Muslims living on different continents come to know and experience one an other's cultural and intellectual worlds? Historians' attempts to explore these interconnections during the last 20 years or so have enriched an older narrative of the history of science. But they have brought with them new kinds of methodological problems: How to avoid effacing the individuals whose names are rarely mentioned in printed texts of the period but who were so essential to the transfer of ideas and practices? How to describe accurately the interactions of different traditions of scientific knowledge that the historical actors themselves often misunderstood and that remain the provinces of specialist historians (Brentjes 2015)? This chapter surveys some of the recent literature on the transmission of scientific ideas with the aim – if not of providing definitive answers – then at least of pointing to how such obstacles might be overcome.

The so-called capitulations (Ottoman Turkish: *ʿahdnāme*) granted by the Ottoman Sultan Selīm II (r. 974–982/1566–1574) to Charles IX of France (r. 1560–1574) in 977/1569 initiated a series of much closer diplomatic and mercantile relationships between the Ottoman Empire and western Europe. As French commerce and diplomatic influence came to supersede the long-standing links between the Ottomans and the Italian maritime states, the English (1580) and the Dutch (1612) negotiated their own capitulations with the Ottomans; they were followed during the course of the 18th century by the Habsburg Empire (1718), Russia (1774) and several other European states (Wansbrough *et al.* 2012).[3] In addition to permanent embassies in the Ottoman capital of Constantinople/Istanbul, the Ottoman-European capitulations

DOI: 10.4324/9781315170718-69

encouraged the development of settled communities of European merchants in commercial centers across the empire from Cairo to Izmir to Aleppo. In Iran, the first Safavid Shah Ismāʿīl I (r. 906–930/1501–1524) sought closer diplomatic ties with the Portuguese; during the reign of his great-grandson, ʿAbbās I (r. 995–1038/1587–1629), the new capital of Isfahan was opened up to European merchants, in particular the English and the Dutch (Matthee 2013). The French obtained comparable trading privileges after 1665. Farther east, the Portuguese controlled commerce with the Mughal Empire until they came to be challenged by English and Dutch merchants during the reign of Padishah Jahāngīr (r. 1014–1037/1605–1627). Between the 16th and the 18th centuries the number of European travelers to the Ottoman, Safavid and Mughal Empires grew steadily, a consequence of both the increased practical opportunities brought about by the unfolding of mercantile and diplomatic infrastructures, and the rising curiosity toward the peoples and cultures of faraway lands, stimulated by the spread of the printing press (Yerasimos 1991, 41–7; Touzard 2005). The Franco-Ottoman alliance, the relative religious tolerance of ʿAbbās I and Jahāngīr and the later foundation, in 1622, of the *Sacra congregatio de propaganda fide* led to the expansion of Roman Catholic missions across the extra-European world.

All of this, as far as it contributed to the movement of people, objects and ideas, would have some potential consequences for the exchange of scientific knowledge. Considering each of these developments – embassies, trading posts, travelers, and missionaries – in turn, what follows focuses, for reasons of space, on selected examples of French, Italian, English and Dutch encounters with the Ottoman Empire between the 16th and the 18th centuries.[4] Yet the map of these connections could be expanded to encompass a much broader geography, spanning the area from Spain to Russia to Persia and to India.

VI.5.1 Embassies

In the retinue sent out in 1536 with François I's (r. 1515–1547) ambassador to the Ottoman Empire, Jean de la Forêt (d. 1537) was the precocious young orientalist Guillaume Postel (1510–1581). In addition to settling some financial matters, Postel had been commissioned by the king to buy rare books for the royal library. In Istanbul, he was able to acquire a number of Arabic manuscripts on topics including medicine and mathematics (Kuntz 1981, 25). During a second trip to the East in 1549–1550, Postel joined the train of another French ambassador, Gabriel de Luetz, Baron d'Aramon (d. 1553), under whose patronage he continued his search for books (Kuntz 1981, 94). Here again, he met with some success. Among the manuscripts he brought back to Europe were several important scientific works: the *Memoir on the Science of Configuration [of the Universe]* (*Tadhkira fī ʿilm al-hayʾa*) by Naṣīr al-Dīn al-Ṭūsī (597–672/1201–1274), and the *Almanac of the Countries* (*Taqwīm al-buldān*), a geography by the Syrian scholar-prince Abū l-Fidāʾ (r. 719–732/1320–1331), the second of which would occupy European Arabists for the next three and a half centuries (Saliba 2007; Destombes 1985).

D'Aramon's embassy set a precedent for the intersection of statecraft and science that would endure until Napoleon and his *savants* reached the shores of Egypt at the end of the 18th century.[5] In addition to Postel, the ambassador's delegation contained several men with learned interests who would use their time in the East to pursue different kinds of scientific work. Among them were Pierre Gilles d'Albi (1490–1555), who would make a mark as an ichthyologist, and the naturalist Pierre Belon (1517–1564), author of a popular set of *Observations on Many Singularities* (*Observations de plusieurs singularitez*; first printed in 1553) gleaned in the course of his travels (Paviot 1987; Figure VI.5.1). Gilles's and Belon's writings are often seen as exemplary of the kind of firsthand observation characteristic of the Renaissance, and this is true to an extent. But their accounts also bear traces of their dependence on the knowledge of the peoples among

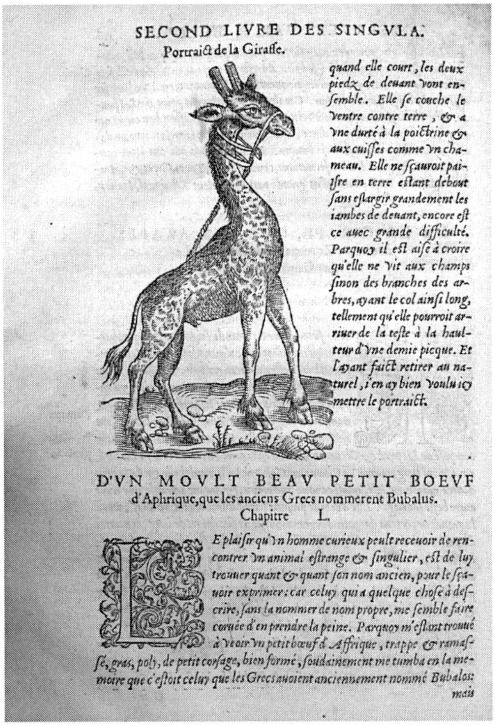
SECOND LIVRE DES SINGVLA.
Portraict de la Girafse.
quand elle court, les deux piedz de deuant vont ensemble. Elle se couche le ventre contre terre, & a vne durté à la poictrine & aux cuisses comme vn chameau. Elle ne sçauroit paisse en terre estant debout sans eslargir grandement les iambes de deuant, encore est ce auec grande difficulté. Parquoy il est aisé à croire qu'elle ne vit aux champs sinon des branches des arbres, ayant le col ainsi long, tellement qu'elle pourroit arriuer de la teste à la haulteur d'vne demie picque. Et l'ayant faict retirer au naturel, i'en ay bien voulu icy mettre le portraict.

D'VN MOVLT BEAV PETIT BOEVF
d'Aphrique, que les anciens Grecs nommerent Bubalus.
Chapitre L.

LE plaisir qu'vn homme curieux peult receuoir de rencontrer vn animal estrange & singulier, est de luy trouuer quant & quant son nom ancien, pour le sçauoir exprimer: car celuy qui a quelque chose à descrire, sans la nommer de nom propre, me semble faire coruée d'en prendre la peine. Parquoy m'estant trouué à veoir vn petit bœuf d'Affrique, trappe & ramassé, gras, poly, de petit corsage, bien formé, soudainement me tumba en la memoire que c'estoit celuy que les Grecs auoient anciennement nommé Bubalos: mais

Figure VI.5.1 Illustration of a giraffe from *Observations de plusieurs singularitez* (*Observations on Many Singularities*) by Pierre Belon (1517–1564), Paris, 1554. The Bodleian Libraries, The University of Oxford, Byw. I 2.14, fol. 118v.

Source: © The Bodleian Libraries, The University of Oxford.

whom they traveled. Belon devoted the days he spent in Aleppo in the winter of 1547 to investigating the medicinal uses of the rhubarb sold in the city's market (Belon 1554, fols 158r–v). The Greek word often used to describe the method of close inquiry, *autopsia*, is apposite to the case of Gilles. As Belon was busy with rhubarb, Gilles occupied himself in Aleppo in observing the dissection of an elephant he had brought from Iran, from which he was able to draw up a graphic description of the eyes, teeth and feet of the animal and to prove – contra Strabo – that elephants, in fact, had two joints in their legs (Gilles 1614, 12–3; Strabo 1930, 324–5 [XVI.4.10]). Yet, here too, there are traces of the foreign naturalist's reliance on local know-how; Gilles mentioned incidentally in the course of his account that the dissection had been performed by some 'Syrians, skilled in skinning camels and horses' (Gilles 1614, 11: *Syri, camelorum equorumque excoriandorum periti*). Moreover, Belon and Gilles were also interested, like Postel, in a written tradition of science. Belon could in fact learn more about rhubarb from the author known in the West as Mesue than he could from the merchants in Aleppo (De Vos 2013).[6]

VI.5.2 Trading posts

Beyond the embassies, the commercial outposts across the Ottoman Empire were a further channel for the transmission of scientific knowledge. Even before the Ottoman conquest of Syria in 922/1516, there was an important tradition of physicians serving the Venetian consulate in

Damascus who used their years abroad to familiarize themselves with medical science in Arabic. Girolamo Ramusio (d. 1486), worked on an (unfinished) improved edition of the *Canon of Medicine* (*Qānūn fī l-ṭibb*) by the physician and philosopher Ibn Sīnā (Avicenna; [d. 428/1037]), which had first been translated into Latin in the 12th century and which, by the 14th century, had become a standard text in the medical faculties of European universities (Jacquart 1989; Hasse 2016, 96, 101–6). The task begun by Ramusio was completed by his successor Andrea Alpago (d. 1522). From his arrival in Damascus around 1487, Alpago combined his roles as physician, merchant and political informant with a keen interest in philology and the medical learning of the society he encountered in Syria. His corrections to and commentary on the Latin text of the *Canon* were incorporated into subsequent printed editions, and this version (published some five years after Alpago's death) was officially adopted at the University of Padua, where Alpago was briefly an extraordinary professor after his return from the Levant. As well as his edition, Alpago also drew up indexes of technical Arabic terms reflective of his wide reading in pharmacology and (to a lesser extent) his experience of the practice of medicine in 16th-century Damascus (Veit 2006; Siraisi 1987, 13, 96, 133–4; Hasse 2016, 97, 106–107, 176).

A later example from Egypt is the case of the physician Prospero Alpini (1553–1617). Like Alpago, Alpini had been a student at Padua, whence after graduating he traveled to Cairo to serve as physician to the Venetian consul in the city. Adapting the common Renaissance belief in ancient Egypt as the source of an age-old tradition of wisdom, Alpini hoped to find among the modern Egyptians the traces of a primordial medical knowledge, passed down from antiquity through the medieval Arab physicians to contemporary practitioners. Although the account of Egyptian medicine set out in his *Four Books on the Medicine of the Egyptians* (*De medicina Aegyptiorum libri quatuor*, 1591) was critical, Alpini nevertheless adapted at least some of the practices he witnessed performed by local physicians, including a method of bloodletting, which he instituted on his return to Italy (Siraisi 2007, 223–41; Brentjes 2010, VI 233–9). Alpini also used his time to study the medicinal plants of Egypt, work that led to his being appointed on his return as lecturer in simples and afterward as prefect of the botanical garden in Padua (Siraisi 2007, 236).

In Syria, this tradition of learned physicians serving overseas continued after the Ottoman conquest, when the center of commercial life shifted from Damascus to Aleppo. Certainly by the early 1550s, there was both a doctor and an apothecary in the service of the Venetian consul in Aleppo (Ehrenberg 1576, fol. 16r). The *Travels* (*Viaggi*) of the Roman aristocrat Pietro Della Valle (1586–1652) provide a hint as to how the men who fulfilled this role might acquire a reputation for expertise in local medicaments: in Aleppo in 1616, Della Valle consulted a certain '*Medico Fiammingo*' about cinnamon; this was possibly the same '*médecin flamand*' who, as the French diplomat Julien Bordier (*fl. c.* 1609–1630) recorded, was in demand among the Ottoman élites (Della Valle 1650, 619; Bordier 1626, 853).

The activities of a later physician attached to the community of English merchants are familiar thanks to a 1667 letter written in response to several queries sent out to Syria by the members of the Royal Society of London. Thomas Harpur (*fl.* 1667–1669) thought little of the medical knowledge of the local inhabitants; his reply began by lauding the society's values of sense experience and firsthand observation. Nevertheless, like Belon and Gilles more than a century earlier, Harpur relayed shreds of medical know-how picked up from among the locals to centers of scientific inquiry back home. He reported that the infamous '*mal d'Aleppo*' (cutaneous leishmaniasis, a skin disease now understood to be transmitted by the bite of the phlebotomine sand fly) was known among the Arabs as the 'one-year disease' by virtue of its limited duration (a designation that survived until modern times); the same letter noted the purported curative properties of the thermal springs in the valley of Antioch (Hall and Hall 1966, 464–5; Abazid *et al.* 2012, 10).

The last and best known of this long tradition of scholarly physicians working among the expatriate communities of European merchants were two Scottish brothers, Alexander (1714–1768) and Patrick Russell ([1727–1805]; van den Boogert 2010; Starkey 2018). By the latter half of the 18th century, Patrick Russell had no doubts about the superiority of European science. He reproved the literati of Aleppo for what to him appeared their indifference to the latest European advances (Russell and Russell 1794, 2: 108). Nevertheless, he was not uninterested in the traditions of medical learning he encountered in Syria. He took the trouble to collate against Arabic manuscripts the recently published Arabic-Latin edition of the treatise on smallpox and measles by al-Rāzī ([d. 313 or 323/925 or 935]; Rhazes), and to acquire copies of al-Zahrāwī's ([d. 404/1013]; Albucasis) treatise and the Melkite physician Ibn al-Quff's (d. 685/1286) writings on surgery, the last of which, he noted, were held in 'considerable esteem' in Aleppo (Russell and Russell 1794, 2: (Appendix) ix, xx; van den Boogert 2005, 242, 255–6; van den Boogert 2010, 218–9; Savage-Smith 1988, 69; Dunlop 1956, 175–82).

Russell's concern with medieval medicine in Arabic leads to a broader point. One of the findings of recent inquiries into the specialized study of Arabic, which emerged in Europe in the 16th and 17th centuries, is a greater appreciation of an ongoing postmedieval engagement with the legacy of scientific writing produced in the Jewish, Christian and Muslim communities of the Middle East (Toomer 1996; Russell 1994).[7] At Leiden, one of the leading centers of the new Arabic studies, the professor Thomas Erpenius (1584–1624) exhorted his students to study the language with a promise of the riches of Arabic medicine and mathematics (Erpenius 1986, 18, 22–3). One particular work cited by Erpenius is an Arabic version of the *Conics* by the Greek geometer Apollonius of Perga (*c.* 262–*c.* 190 bce), containing three of the four books lost in the Greek version. It was brought back from Syria by another Dutchman, Jacobus Golius (1596–1667), who in the 1620s served as secretary to the Dutch consul in Aleppo (Apollonius 1990, 1: xxii–xxv).[8] The arrival of this work in Leiden was apparently an event significant enough to draw René Descartes (1596–1650) to the city. The same manuscript was later translated in early 18th-century Oxford by the Savilian professor of geometry and future astronomer royal Edmond Halley ([1656–1742]; Hattab 2009, 156; Apollonius 1990, 1: xxv-xxvi; Fried 2011).

However, it has also been claimed that the rise of experimental science in the West from the second half of the 17th century was concomitant with a decline of interest in a written tradition of science in Arabic, Persian and (to a lesser extent) Turkish (Toomer 1996, 110–1). It is certainly true that many of the figures who served in the overseas consulates from the 17th century onward were curious about the natural world and were led, consequently, to observation but not to the study of Arabic, Persian or Turkish. Such was the Venetian consul in Aleppo, Gianfrancesco Sagredo (1571–1620) who, in his connection to Galileo Galilei (1564–1642), served as a 'data-gathering scientific diplomat' (Wilding 2014, 76). Such, too, was the English consul Paul Rycaut (1629–1700), another correspondent of the London Royal Society whom Henry Oldenburg (d. 1677), the society's secretary, requested to observe a lunar eclipse from Izmir, and whom Oldenburg thought unlikely to have left England before furnishing himself with 'a good telescope' (Hall and Hall 1967, 133; Brentjes 2010, VIII 23–5).[9] However, the decline of interest in scientific texts ought not to be overstated. Some of the most sophisticated recent work on early modern European intellectual history has drawn attention to the fact that a historical interest in science – both classical and Islamic – could coincide with a commitment to experimentalism (Levitin 2015, 71–84). Exemplary in this respect is Edward Bernard (1638–1697), both Savilian professor of astronomy and a fellow of the Royal Society, who devoted much of his labor to mining Arabic and Persian mathematical and astronomical works (Apollonius 1990, I xxiv-xxv; Toomer 1996, 299–305).

The enduring quest for scientific manuscripts in these languages was pursued well into the 17th century by the chaplains appointed to minister to the English merchants in Aleppo. The English Levant Company expressed little interest in the promotion of science, beyond conferring the occasional payment to support authors of cartographical or lexicographical works (Mills 2020, 23–4). However, the chaplains in the company's employment were free to use their years in commercial centers such as Istanbul or Aleppo to pursue learned interests, typically in collaboration with a broader scholarly community. At least some of the chaplains, aided by the fruits of their study of local languages, were able to gain some insight into the learned cultures of these cities, and – often with the assistance of resident Jewish, Christian or Muslim brokers and informants – to tap into local book markets. Edward Pococke (1604–1691), who was in Aleppo in the 1630s, and would be the first incumbent of the chair in Arabic established at Oxford by William Laud (1573–1645), was interested primarily in history, poetry and biblical studies, yet among the works bought for him in Aleppo by the Sufi Aḥmad al-Gulshanī, who acted as his agent, were also books on medicine and astronomy (MS Paris, BnF, Latin 9340, fols 296v, 297v, 298v, 300v; Holt 1973, 42–5; Toomer 1996, 122–3; Kilpatrick 2010, 20–40). Among the latter, al-Ṣūfī's (291–376/903–986) *Book of the Images of the Fixed Stars* (*Kitāb ṣuwar al-kawākib al-thābita*; abbreviated English title: *Book of the Constellations*) and the astronomical tables (*zīj*) of the Timurid governor of Samarqand and later head of state Ulugh Beg (r. 812–850/1409–1447 as governor of Samarqand; 850–853/1447–1449 as head of the dynasty) were of sufficient interest in 17th-century England for new editions to be proposed (and, in the case of the second, realized; Figures VI.5.2 and VI.5.3). Pococke's student Robert Huntington (d. 1701) undertook similar work during his tenure as chaplain in Aleppo in the 1670s. Huntington's letters from Syria to Bernard depict him searching for Arabic astronomical and mathematical manuscripts, a quest in which he was to meet with some degree of success (MS Copenhagen, Kongelige Bibliotek, NKS 1675 2°: no. 37, no. 38, no. 39, no. 56).

Figure VI.5.2 Edward Pococke's (1604–1691) copy of ʿAbd al-Raḥmān al-Ṣūfī's (291–376/903–986) *Book of the Images of the Constellations* (*Kitāb ṣuwar al-kawākib al-thābita*). This manuscript was produced in Aleppo in 703/1303–4. The title page contains the notes of several former owners, among them an Aleppine physician named Abū Bakr ibn ʿUthmān. MS Oxford, Bodleian Library, Pocócke 257.

Source: © The Bodleian Libraries, The University of Oxford

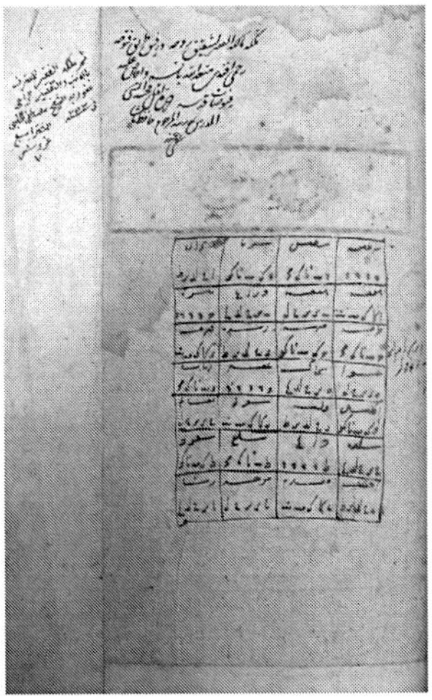

Figure VI.5.3 Edward Pococke's copy of the *New Astronomical Tables of the Sultan* (*al-Zīj al-jadīd al-sulṭānī*) by the Timurid governor of Samarqand and later head of state Ulugh Beg (r. 812–853/1409–1449). A reader's note on the title page indicates that this manuscript once belonged to a certain Shaykh Muṣṭafā al-Ḥalabī (the Aleppine) who purchased it for seven *ghurūsh* in 1043/1633–4, a decade or so before Pococke acquired it in Aleppo. MS Oxford, Bodleian Library, Pococke 226.

Source: © The Bodleian Libraries, The University of Oxford

VI.5.3 Travelers

Not all individuals who journeyed eastward in search of scientific knowledge were connected directly to embassies or trading posts. Many travelers set out independently, funded – like Della Valle – by private wealth or by patrons with learned or commercial interests. One such was the industrious Augsburg physician Leonhard Rauwolf (1535–1596). For three years in the mid-1570s, Rauwolf traveled extensively through the Middle East with the financial support of his mercantile brother-in-law. His chief interest was plants and their medicinal uses. Much like Belon and Gilles, his account combined firsthand observation, knowledge gleaned from ancient Greek and medieval Islamic authorities, and a keen interest in local culinary and pharmaceutical practices. The account of his travels published on his return to Germany in 1582 was reprinted frequently and was a valuable resource for later botanists, from John Ray (1627–1705) in the 17th century to Carl Linnaeus (1707–1778) in the 18th (Dannenfeldt 1968; Häberlein 2010). The clerical patronage conferred on oriental studies in 17th-century England supported several traveling scholars with scientific interests. John Greaves (1602–1652) was funded by William Laud, who became Archbishop of Canterbury in 1633. Greaves visited the Ottoman Empire in the 1630s, where he made astronomical observations, conversed with local scholars and acquired a substantial collection of Arabic manuscripts (Toomer 1996, 127–41; Shalev 2005). During the same period, the German Christian Ravius ([1613–1677]; Raue) traveled to the Ottoman Empire on a stipend provided by the Anglo-Irish archbishop James Ussher (1581–1656). There he assembled (not always, it seems, by wholly honorable means) another substantial collection of manuscripts, including a second copy of the Arabic *Conics*, a manuscript later put to use by the then professor of mathematics at the *Gymnasium Illustre* in Amsterdam John Pell ([1611–1685]; Apollonius 1990; 1: xxiii–xxiv; Toomer 1996; 142–5; Kilpatrick and Toomer 2016, 17–8, no.61).[10]

Such book-collecting missions were organised on a grander scale in Louis XIV's (r. 1643–1715) France. The 'extraordinarily acquisitive attitude' of the French during this period was fostered by the expanding presence of French diplomats and missionaries throughout the Levant and Iran (Wansleben 2018; 21). The magistrate and orientalist Gilbert Gaulmin (1585–1665) drew on the assistance of at least one French merchant in Egypt in assembling his library of oriental manuscripts in the 1640s (Omont 1902, 1: 11). Chancellor Pierre Séguier (1588–1672) and Cardinal Mazarin (1602–1661) both put to work missionaries and diplomats – including two French ambassadors at Istanbul, Jean de la Haye-Ventelet (1593–1665) and his son Denis de la Haye-Ventelet (1626–1722) – acquiring manuscripts for their libraries in the 1640s and 1650s.

The most successful of these expeditions were organized by Louis's minister Jean-Baptiste Colbert (1619–1683), a figure responsible, as Jacob Soll has put it, for a wholesale 'bureaucratization of scholarship' (Soll 2009, 106; Dew 2009, 16–36). Among the manuscripts shipped back from the Levant for the royal library by the men dispatched by Colbert – among them, François Pétis de La Croix (1653–1713), Johann Michael Wansleben (1635–1679) and Antoine Galland (1646–1715) – were many scientific works (Omont 1902, 1: 232, 2: 880–2; Bevilacqua 2018, 26–8). These traveling *savants* were also issued with extensive instructions for a broader project of natural history. Among the tasks assigned to Wansleben by Colbert's librarian, the mathematician Pierre de Carcavi (d. 1684), were requests to measure the height of the pole as frequently and as carefully as possible, and to amass all the data he could on the animals, minerals, waters, plants and fruits of the countries through which he traveled. Like Rycaut, Wansleben set out equipped with two telescopes, as well as two quadrants and a compass (Wansleben 2018, 22); he was even given specific advice on the best means for preserving natural specimens (the dried skins of animals should be stuffed with hay or cotton to conserve their true shape and appearance; the seeds and roots of rare plants should be retained, the dried leaves slipped between the pages of a book; Omont 1902, I 61–2; Wansleben 2018; 21). Wansleben's letters and the checklists of his acquisitions attest to his industriousness; in 1673 the skins of two crocodiles and two large lizards were loaded onto the departing ships (Omont 1902, 2: 887).[11] As with Gilles and his elephant, there are traces in Wansleben's notes of his reliance on local expertise (Omont 1902, 2: 902, listing Wansleben's accounts for 8 December 1672, including the charge 'given to two men, for flaying the crocodile and preparing the skin' [*Per scorticar il cocodrillo et per accomodar la pelle, date à due homini*]). The published reports of his expeditions relayed to European readers a wealth of information on the natural history of Egypt.

The apogee of these state-sponsored missions before Napoleon was the party of scientists and philologists sent to the Yemen in the 1760s with the financial backing of Frederik V, King of Denmark (r. 1746–1766). The party's philologist, Frederik Christian von Haven (1728–1763), collected little in the way of scientific manuscripts. More significant was the work of the botanist, the Swede and former student of Linnaeus, Pehr Forsskål (1732–1763). Forsskål's skills as a linguist meant that throughout the expedition – spotting birds on the coast at Izmir, collecting the names of plants from local peasants or quizzing the inhabitants of southern Arabia on the flowers of the tree he recognized as the Meccan balsam – he was able to draw on local knowledge of the region's natural history. All this filtered through into the results of his findings, published after his death in the Yemeni village of Yarim in the summer of 1763 (Klaver 2009, 60, 96, 101).

VI.5.4 Missionaries

One final route through which we might trace the transmission of science from Asia to Europe is via the missionaries who, from around the third decade of the 17th century, began to establish themselves in centers such as Istanbul, Aleppo and Isfahan (Heyberger 2014; Clines 2015;

Windler 2018). Like the Protestant chaplains who served the English and Dutch embassies and consulates, the Roman Catholic missionaries in the Levant had no immediate motivation for engaging in science. However, two aspects of their pastoral and missionary work meant that they were well qualified to assume an important role in the transmission of scientific knowledge. First, the missionaries were particularly diligent in studying local languages – not just in their literary variants but also the vernacular forms of Arabic, Persian, Turkish and Armenian. Some of the missionaries spent many years abroad living among the local populations and in consequence became the most linguistically adept of all expatriate Europeans. Second, some of the missionaries took up the study and practice of medicine, in the first instance as a means of gaining the trust and respect of the locals among whom they sought to proselytize.

These linguistic and medical-pharmaceutical studies could incite the missionaries, like the earlier Italian physicians, to take an interest in the medical knowledge preserved among the societies in which they worked. In Aleppo, for example, the Jesuit Jean Amieu (1587–1653) apparently used his leisure hours to compose a book of simples with the Arabic terms given in French and Latin (Poirresson 1652, 954). At least some of this work would later come to the attention of a broader European readership. Joseph de la Brosse, Ange de Saint Joseph (1636–1697), a Discalced Carmelite who put his skills as a physician to the service of his missionary work in Isfahan and Basra, was the author of an annotated translation of the *Medicine of Shifāʾī* (*Ṭibb-i shifāʾī*) by al-Shifāʾī (d. 963/1556) – published in Paris in 1681 as *Persian Pharmacopoeia* (*Pharmacopoea persica*) – a work which, Ange de Saint Joseph thought, would enrich European pharmaceutical knowledge (Labrosse 1681, 33). Moreover, many of the men who served in the missions were by no means devoid of curiosity about the natural world. The Carmelite Esprit Julien, Philippe de la Très Sainte-Trinité (1603–1671), combined his account of the missions with reflections on the topography, fauna and flora he observed in the course of his journey through Syria, Iraq, Iran and India between the 1620s and the 1640s (including another firsthand description of a genuflecting elephant; Julien 1649, 115–51, 277–308, 278).

All this led to missionaries occasionally being co-opted into the research projects of the broader European Republic of Letters. Colbert's scheme for collecting manuscripts from the Levant was delegated in Aleppo to the head of the Jesuit mission, Joseph Besson (d. 1691). By 1673, Besson had bought some 50 manuscripts for the king's library, at least some of which related to medicine and astronomy. The plan drawn up by Besson for a library of oriental books reflected his theological interests. However, it also reiterated the older arguments about the scientific value of Arabic literature: the Arabs had excelled in medicine, astronomy and mathematics, a conviction that strongly influenced the choice of books sent back to Paris (Omont 1902, 1: 223, 226).[12]

Significant in this respect, too, was the Carmelite missionary, Petrus Golius (d. 1676), known to the members of his order as Célestin de St Lidwine (Miller 2015, 125–41; Brentjes 2010, VII 43–5). During several years spent with the Carmelite mission in Syria, Célestin was able to acquire outstanding Arabic (Julien 1649, 346). He put this skill at the disposal not only of the Carmelites' missionary endeavors but also of several European correspondents with scientific interests. For his brother, the Arabist and mathematician Jacobus Golius in Leiden, Célestin directed the dervish whose acquaintance he had made in Aleppo to search for scientific works extant in Syria and Egypt, in particular, Arabic mathematical manuscripts, both original contributions and translations of lost Greek works (MS Paris BnF Dupuy 688: fols 21r-v). From Aleppo, Célestin supplied the French polymath Nicolas-Claude Fabri de Peiresc (1580–1637) with accounts of the Arabic books esteemed by local scholars: among them al-Damīrī's (d. 808/1405) zoological encyclopaedia *The Lives of Animals* (*Ḥayāt al-Ḥayawān*) and the great compendium of 51 Ismāʿīlī inclined treatises on the sciences and religious doctrines, the *Epistles of the*

Brethren of Purity (*Rasāʾil Ikhwān al-Ṣafāʾ*; MS Paris BnF Dupuy 688, fol. 19r; Aix-en-Provence Bibliothèque Méjanes MS 205 (1023), 275). Célestin was also one of several missionaries stationed across the Levant whom Peiresc encouraged to conduct astronomical observations (Miller 2015, 127, 129, 131; Brentjes 2010, VIII 23–5).

As much as any of the Europeans who resided in the Middle East between the 16th and the 18th centuries, missionaries – with their linguistic skills and with the friendships they were able to nurture during the course of many years' residence abroad – could penetrate deep into the learned culture of Muslim societies. On the other side of the Mediterranean in Aix-en-Provence, Peiresc was able to learn through his missionary contacts (in this case, the Capuchin Agathange de Vendôme [1598–1638]) of a certain '*curieux de mathematiques*' with a well-stocked library in Aleppo (Miller 2015, 130). This scholar was very likely Muḥammad al-Taqawī (d. 1061/1650–1), a noted *adīb* (man of letters), whose name has been preserved by his contemporary al-Muḥibbī (d. 1111/1699), a Muslim historian of 17th-century Syria (al-Muḥibbī [1867], 4: 304–6; Kilpatrick 2010, 35, no.24). From Célestin, Peiresc heard of the scientific curiosity of the city's dervishes (Miller 2015, 129). These reports of Muslim scholars with overlapping astronomical interests led Peiresc, as Peter Miller has shown, to envision 'a shared community of inquirers' made up of Muslim and Christian natural philosophers exchanging data through the mediation of the missionary communities (Miller 2015, 130).

This exchange was not one-sided. The books carried abroad by European travelers and diplomats, the telescopes and quadrants housed in embassies or consulates, and – most important – the knowledge of scientific theories and practices informing the actions of some of the figures surveyed above led on occasion to moments of cross-cultural intellectual collaboration. Pococke's correspondence in Oxford has captured his Sufi agent in Aleppo requesting a copy of 'the printed geography' (*al-jughrāfiyyā al-maṭbūʿa*): presumably the Medici Press edition of al-Sharīf al-Idrīsī's (d. *c.* 560/1165) *The Book of Pleasant Journeys into Faraway Lands* (*Nuzhat al-mushtāq fī 'khtirāq al-āfāq*), made from an Arabic manuscript that had come to Italy in the 16th century (MS Oxford, Bodleian Library, Pococke 142, fol. 6; Toomer 1996, 123, no. 30). The recent discovery of an inventory of Muḥammad al-Taqawī's library has revealed that at least one 17th-century Syrian bibliophile owned a printed European copy of the *Qānūn fī l-ṭibb* (Reier 2021). Some of the European doctors practicing in the Ottoman Empire passed on their medical know-how to Ottoman physicians (Günergun 2007, 200, 208). One of the most important early modern Turkish translations of a European cartographical work, *The Translation of the* Atlas Major (*Tercüme-i Atlas major*), an abridgment of Joan Blaeu's (1596–1673) atlas produced in Istanbul between around 1086/1675 and 1096/1685 by a team working under Abū Bakr al-Dimashqī (d.1102/1691), was facilitated by diplomatic relations: the Latin version had been presented to Sultan Meḥmed IV (r. 1058–1099/1648–1687) by the Dutch ambassador Justinus Colyer ([1624–1682]; Günergun 2007, 205; Brentjes 2012). The Morisco and later diplomat in the service of the Moroccan sultan, Aḥmad ibn Qāsim al-Ḥajarī (d. after 1050/1569/1640) was well versed in Western geography and astronomy, and translated into Arabic European books on these topics, as well as a Spanish work on gunnery into Arabic (al-Ḥajarī 2015).

If Peiresc's plan for a trans-Mediterranean project of astronomical inquiry was to remain, in the short term, unrealized, then his vision was something more than a polymath's dream. As historians are able to uncover more of these episodes of exchange and collaboration, a more detailed picture emerges of the mutual interchange of Eastern and Western traditions of learning well into the early modern period. To imagine Descartes with Golius in Leiden turning over the pages of a manuscript recently returned from Aleppo, to glimpse Halley translating the same work in Oxford or to picture al-Dimashqī and his team of Arabic, Turkish, Greek, Italian and Dutch (and possibly other) scholars, translators and scribes in Istanbul is to begin to piece together the small-scale histories of the voyages of manuscripts and information with some of the meta-narratives of 17th-century intellectual history. The question of the enduring influence in western Europe of the sciences produced by the Jewish, Christian, and Muslim communities of North Africa and Asia will no doubt continue to stimulate historical debate. Recovering such stories of transmission is one key to finding answers.

Notes

1 Consolidated bibliography.

2 However, see now Hasse (2016, 3–27, 317–407); Morrison (2014) revisits the well-known theory of Nicolas Copernicus's possible knowledge of the work of earlier Islamic astronomers; for the previous literature on this question, see 32–5 and the references cited there.

3 Sweden (1737), Sicily (1740), Denmark (1756), Prussia (1761) and Spain (1783); the Polish capitulations with the Ottoman Empire dated back to 1553.

4 Although beyond the scope of this chapter, piracy could also set in motion the transmission of scientific knowledge. For the dramatic instance of the capture at sea of the Moroccan Sultan Moulāy Zaydān's library, see Hershenzon (2014); Zhiri (2017).

5 For one account of another later 'scientific' embassy, see Hamilton (2005); see also, Bevilacqua (2018, 23–5), on Antoine Galland's endeavors as a collector during his employment by the French ambassador to the Ottoman Empire Charles Marie François Olier, Marquis de Nointel (1635–1685).

6 On the difficulties of identifying Mesue with either the Christian physician Ibn Māsawayh (d. 243/857–8), or Yūḥannā al-Māridīnī Māsawayh, see De Vos (2013, 683–5). De Vos makes the case that 'Mesue's works served as a conduit between Arabic and European pharmacotherapy, packaging the latest and most important elements of Arabic pharmacology in a way that was designed to address the concerns and demands of a European audience' (688); see also Hasse (2016, 123–4).

7 See also the catalogue accompanying the 2011 Royal Society exhibition 'Arabick Roots': https://royalsociety.org/~/media/exhibitions/arabick-roots/2011-06-08-arabick-roots.pdf.

8 The manuscript (the Arabic version of Apollonius's *Conics* by Thābit ibn Qurra (d. 288/901) commissioned by the Banū Mūsā) had been given to Golius in Aleppo by another Dutchman, David Leleu de Wilhem (Apollonius 1990, I: xxii, lxxxvi).

9 Rycaut's earlier work for the Royal Society is detailed in his letters to Oldenburg from the previous year.

10 Ravius's manuscript was a reworking of Thābit ibn Qurra's Arabic text by Abū l-Ḥusayn ʿAbd al-Malik ibn Muḥammad al-Shīrāzī (Apollonius 1990, 1: xxiii).

11 Wansleben reported having obtained two (then living) crocodiles in his letters to Carcavi and Colbert (Pougeois 1869, 436, 440; also, Wansleben 2018, 30, 317).

12 See also the 'Apparatus ad Bibliothecam Colbertaeam' (MS Paris, BnF, Latin 9363, fols 90r-99v), which contains descriptions of several medical and astronomical works collected by Besson.

Bibliography

Sources

Alpini, Prospero. 1591. *De medicina Aegyptiorum libri quatuor*. Venice: Franciscus de Franciscis.

Apollonius. ed., tr. and ann. Toomer, G. J. 1990. *Apollonius Conics Books V to VII: The Arabic Translation of the Lost Greek Original in the Version of the Banū Mūsā*. 2 vols. New York: Springer.

Belon, Pierre. 1554. *Les observations de plusieurs singularitez et choses mémorables, trouvées en Grece, Asie, Iudée, Egypte, Arabie, & autres pays estranges*. Paris: Gilles Corrozet.

Bordier, Julien. 1626. "Relation d'un voyage en Orient, par Julien Bordier, écuyer de Jean de Gontaut, baron de Salagnac, ambassadeur à Constantinople," in Salmon, O., ed. 2011. *Alep dans la littérature de voyage européenne pendant la période ottomane (1516–1918), Tome II: Répertoire des voyageurs européens passés à Alep XVIe, XVIIe et XVIIIe siècles*. Aleppo: Dār al-mudarris, 841–64.

Della Valle, Pietro. 1650. *Viaggi di Pietro della Valle il Pellegrino* [. . .] *Divisi in tre parti, cioè la Turchia, la Persia, e l'India*. Rome: Vitale Mascardi.

Ehrenberg, Johann von. ed. 1576. *Zwo Reise zum heiligen Grab: die Erste deß Edlen vesten Johansen von Ehrenberg, so er sampt andern vom Adel und etlichen Niderländern volbracht* [. . .] *Die ander so Daniel Ecklin von Arow gethan . . . Sampt einer kurtzen Beschreibung des gelobten Landts, und der Statt Hierusalem, wie es noch zu unserer zeit gestaltet seye: Allen guthertzigen Lesern, warhafft zum lust an tag gebracht.* Basel: Apiarius.

Erpenius, Th. tr. Jones, R. 1986. "Thomas Erpenius (1584–1624) on the Value of the Arabic Language," *Manuscripts of the Middle East* 1: 15–25.

Gilles, Pierre. 1614. *Descriptio nova elephanti.* Hamburg: Hering.

al-Ḥajarī, Aḥmad ibn Qāsim. ed., tr. and ann. van Koningsveld, P. S., al-Samarrai, Q., and Wiegers, G. A. 2015. *Kitāb Nāṣir al-Dīn ʿAlā 'l-Qawm al-Kāfirīn (The Supporter of Religion against the Infidels).* Madrid: Consejo Superior de Investigaciones Científicas.

Hall, A. R. and Hall, M. B., eds. 1966. *The Correspondence of Henry Oldenburg, Vol. 3: 1666–1667.* Madison: University of Wisconsin Press.

Hall, A. R. and Hall, M. B. eds. 1967. *The Correspondence of Henry Oldenburg, Vol. 4: 1667–1668.* Madison: University of Wisconsin Press.

Julien, Esprit (Philippe de la Très Sainte-Trinité). 1649. *Itinerarium orientale R. P. F. Philippi a SSma Trinitate carmelitae discalceati ab ipso conscriptum.* Lyon: Antoine Jullieron.

Labrosse, Joseph (Ange de Saint Joseph). 1681. *Pharmacopoea persica ex idiomate persico in latinum conversa.* Paris: Étienne Michallet.

Muḥibbī, Muḥammad Amīn ibn Faḍl Allāh. 1284/[1868]. *Tārīkh khulāṣat al-athar fī aʿyān al-qarn al-ḥādī ʿashar* [History of the Essence of the Record on the Nobles of the 11th Century]. Cairo: al-Maṭbaʿa al-wahbīya.

Poirresson, Nicolas. 1652. "Relation des missions de la Compagnie de Jésus en Syrie en l'année 1652," in Salmon, O., ed. 2011. *Alep dans la littérature de voyage européenne pendant la période ottomane (1516–1918), Tome II: Répertoire des voyageurs européens passés à Alep XVIe, XVIIe et XVIIIe siècles.* Aleppo: Dār al-mudarris, 952–7.

Russell, A. and Russell, P. 1794^2. *The Natural History of Aleppo.* 2 vols. London: G. G. & J. Robinson.

Strabo. tr. Jones, H. L. 1930. *Geography, Books 15–16.* London: William Heinemann

Wansleben, J. M. ed. and ann. Hamilton, A. 2018. *Johann Michael Wansleben's Travels in the Levant, 1671–1674: An Annotated Edition of His Italian Report.* Leiden and Boston: Brill.

Research literature

Abazid *et al.*: Abazid, N., Jones, C. and Davies, C. R. 2012. "Knowledge, Attitudes and Practices about Leishmaniasis among Cutaneous Leishmaniasis Patients in Aleppo, Syrian Arab Republic," *Eastern Mediterranean Health Journal* 18.1: 7–14.

Bevilacqua, A. 2018. *The Republic of Arabic Letters: Islam and the European Enlightenment.* Cambridge, MA: Harvard University Press.

Brentjes, S. 2010. *Travellers from Europe in the Ottoman and Safavid Empires, 16th–17th Centuries: Seeking, Transforming, Discarding Knowledge.* Farnham: Ashgate.

Brentjes, S. 2012. "On Two Manuscripts by Abū Bakr b. Bahrām al-Dimashqī (d. 1102/1691) Related to W. and J. Blaeu's *Atlas Maior*," *Osmanlı Araştırmaları. The Journal of Ottoman Studies* 40: 171–92.

Brentjes, S. 2015. "Relationships Between Early Modern Christian and Islamicate Societies in Eurasia and North Africa as Reflected in the History of Science and Medicine," *Confluence: Online Journal of World Philosophies* 3: 85–121.

Burnett, C. 2009. *Arabic into Latin in the Middle Ages: The Translators and Their Intellectual and Social Context.* Farnham: Ashgate.

Clines, R. J. 2015. "Fighting Enemies and Finding Friends: The Cosmopolitan Pragmatism of Jesuit Residences in the Ottoman Levant," *Renaissance Studies* 31.1: 66–86.

Dannenfeldt, K. H. 1968. *Leonhard Rauwolf: Sixteenth-Century Physician, Botanist, and Traveler.* Cambridge, MA: Harvard University Press.

Destombes, M. 1985. "Guillaume Postel cartographe," in *Guillaume Postel 1581–1981: Actes du colloque international d'Avranches, 5–9 septembre 1981.* Paris: Guy Trédaniel, Éditions de La Maisnie, 361–71.

De Vos, P. 2013. "The 'Prince of Medicine': Yūḥannā ibn Māsawayh and the Foundations of the Western Pharmaceutical Tradition," *Isis* 104.4: 667–712.

Dew, N. 2009. *Orientalism in Louis XIV's France.* Oxford: Oxford University Press.

Dunlop, D. M. 1956. "Arabic Medicine in England," *Journal of the History of Medicine and Allied Sciences* 11.2: 166–82.

Fried, M. N. 2011. *Edmond Halley's Reconstruction of the Lost Book of Apollonius's Conics: Translation and Commentary*. New York: Springer.

Günergun, F. 2007. "Ottoman Encounters with European Science: Sixteenth- and Seventeenth-Century Translations into Turkish," in Burke, P. and Po-chia Hsia, R., eds. *Cultural Translation in Early Modern Europe*. Cambridge: Cambridge University Press, 192–211.

Häberlein, M. 2010. "Botanisches Wissen, ökonomischer Nutzen und sozialer Aufstieg im 16. Jahrhundert: Der Augsburger Arzt und Orientreisende Leonhard Rauwolf," in Müller, G. M., ed. *Humanismus und Renaissance in Augsburg. Kulturgeschichte einer Stadt zwischen Spätmittelalter und Dreißigjährigem Krieg*. Berlin: De Gruyter, 101–16.

Hamilton, A. 2005. "'To Divest the East of all its Manuscripts and all its Rarities': The Unfortunate Embassy of Henri Gournay de Marcheville," in Hamilton, van den Boogert and Westerweel, eds., 123–50.

Hamilton, A. and van den Boogert, M. H. and Westerweel, B., eds. 2005. *The Republic of Letters and the Levant*. Leiden: Brill.

Hattab, H. 2009. *Descartes on Forms and Mechanisms*. Cambridge: Cambridge University Press.

Hershenzon, D. 2014. "Traveling Libraries: The Arabic Manuscripts of Muley Zidan and the Escorial Library," *Journal of Early Modern History* 18.6: 535–58.

Heyberger, B. 2014². *Les chrétiens du Proche-Orient au temps de la Réforme catholique (Syrie, Liban, Palestine, XVIIe – XVIIIe siècle)*. Rome: École française de Rome.

Holt, P. M. 1973. *Studies in the History of the Near East*. London: Cass.

Jacquart, D. 1989. "Arabisants du moyen age et de la renaissance: Jérôme Ramusio († 1486) correcteur de Gérard de Crémone († 1187)," *Bibliothèque de l'École des chartes* 147: 399–415.

Kilpatrick, H. 2010. "Arabic Private Correspondence from Seventeenth-Century Syria: The Letters to Edward Pococke," *Bodleian Library Record* 23.1: 20–40.

Kilpatrick, H. and Toomer, G. J. 2016. "Niqūlāwus al-Ḥalabī (c.1611 – c.1661): A Greek Orthodox Syrian Copyist and His Letters to Pococke and Golius," *Lias* 43.1: 1–159.

Klaver, J. M. I. 2009. *Scientific Expeditions to the Arab World (1761–1881)*. Oxford: Oxford University Press.

Kuntz, M. L. 1981. *Guillaume Postel, Prophet of the Restitution of All Things: His Life and Thought*. The Hague: Nijhoff.

Levitin, D. 2015. *Ancient Wisdom in the Age of the New Science: Histories of Philosophy in England, c. 1640–1700*. Cambridge: Cambridge University Press.

Matthee, R. P. 2013. "Iran's Relations with Europe in the Safavid Period: Diplomats, Missionaries, Merchants and Travel," in Langer, A., ed. *The Fascination of Persia: The Persian-European Dialogue in Seventeenth-Century Art and Contemporary Art of Tehran*. Zürich: Scheidegger and Spiess, 6–39.

Miller, P. N. 2015. *Peiresc's Mediterranean World*. Cambridge, MA: Harvard University Press.

Mills, S. 2020. *A Commerce of Knowledge: Trade, Religion, and Scholarship between England and the Ottoman Empire, c.1600–1760*. Oxford: Oxford University Press.

Morrison, R. 2014. "A Scholarly Intermediary between the Ottoman Empire and Renaissance Europe," *Isis* 105.1: 32–57.

Omont, H. 1902. *Missions archéologiques françaises en Orient aux XVIIe et XVIIIe siècles*. 2 vols. Paris: Imprimerie nationale.

Paviot, J. 1987. "Autour de l'ambassade de d'Aramon: érudits et voyageurs au Levant 1547–1553," in Céard, J. and Margolin, J.-C., eds. *Voyager à la Renaissance: actes du colloque de Tours, 30 juin – 13 juillet 1983*. Paris: Maisonneuve et Larose, 381–92.

Pougeois, A. 1869. *Vansleb savant orientaliste et voyageur: sa vie, sa disgrâce, ses* oeuvres. Paris: Didier & Ce and J. Pougeois.

Reier, B. 2021. "Bibliophilia in Ottoman Aleppo: Muḥammad al-Taqawī and His Medical Library," *Der Islam* 98.2: 473–515.

Russell, G. A. ed. 1994. *The 'Arabick' Interest of the Natural Philosophers in Seventeenth-Century England*. Leiden: Brill.

Saliba, G. 2007. "Arabic Science in Sixteenth-Century Europe: Guillaume Postel (1510–1581) and Arabic Astronomy," *Suhayl* 7: 115–64.

Savage-Smith, E. 1988. "John Channing: Eighteenth-Century Apothecary and Arabist," *Pharmacy in History* 30.2: 63–80.

Shalev, Z. 2005. "The Travel Notebooks of John Greaves," in Hamilton *et al.*, eds., 77–102.

Siraisi, N. G. 1987. *Avicenna in Renaissance Italy: The Canon and Medical Teaching in Italian Universities after 1500*. Princeton: Princeton University Press.

Siraisi, N. G. 2007. *History, Medicine, and the Traditions of Renaissance Learning*. Ann Arbor: University of Michigan Press.

Soll, J. 2009. *The Information Master: Jean-Baptiste Colbert's Secret State Intelligence System*. Ann Arbor: University of Michigan Press.

Starkey, J. 2018. *The Scottish Enlightenment Abroad: The Russells of Braidshaw in Aleppo and on the Coast of Coromandel*. Leiden: Brill.

Toomer, G. J. 1996. *Eastern Wisedome and Learning: The Study of Arabic in Seventeenth-Century England*. Oxford: Oxford University Press.

Touzard, A.-M. 2005. "Les voyageurs français en Perse de 1600 à 1730," *Eurasian Studies* 4.1: 41–74.

van den Boogert, M. H. 2005. "Patrick Russell and the Republic of Letters in Aleppo," in Hamilton *et al.*, eds., 223–64.

van den Boogert, M. H. 2010. *Aleppo Observed: Ottoman Syria through the Eyes of Two Scottish Doctors, Alexander and Patrick Russell*. Oxford: Oxford University Press.

Veit, R. 2006. "Der Arzt Andrea Alpago und sein medizinisches Umfeld im mamlukischen Syrien," in Speer, A. and Wegener, L., eds. *Wissen über Grenzen: Arabisches Wissen und lateinisches Mittelalter* (*Miscellanea Mediaevalia* 33). Berlin: De Gruyter, 305–16.

Wansbrough, J., *et al.* 2012. "Imtiyāzāt," in *EI-2*, 3: 1178–95.

Wilding, N. 2014. *Galileo's Idol: Gianfrancesco Sagredo and the Politics of Knowledge*. Chicago: University of Chicago Press.

Windler, C. 2018. *Missionare in Persien: Kulturelle Diversität und Normenkonkurrenz im globalen Katholizismus (17. – 18. Jahrhundert)*. Cologne: Böhlau.

Yerasimos, S. 1991. *Les voyageurs dans l'Empire Ottoman (XIVe–XVIe siècles): bibliographie, itinéraires et inventaire des lieux habités*. Ankara: Société turque d'histoire.

Zhiri, O. 2017. "A Captive Library between Morocco and Spain," in Keller, M. and Irigoyen-García, J., eds. *The Dialectics of Orientalism in Early Modern Europe*. Basingstoke: Palgrave Macmillan, 17–31.

VI.6
THE SCIENCES IN TWO PRIVATE LIBRARIES FROM OTTOMAN SYRIA[1]

Boris Liebrenz

If we consider the history of science – and, indeed, of knowledge in general – not only in terms of discoveries and inventions, but also in terms of transmission, preservation and access, histories of libraries and reading are a major focus. Approaches to the history of the book are not mainly concerned with ideal audiences and the intentions of authors but also with discovering the actual people with a documented interest in specific works. Who, in other words, was copying, owning, reading, teaching or endowing books in a specific time and place? A few studies engaging with these questions in the context of the Islamicate world are slowly beginning to emerge. Fewer still are those works that focus on the sciences.[2]

Of the countless private manuscript book collections, large and small, that once existed in the Islamicate world until the 13th/19th century, at present we know of only one that has survived and is accessible in the shape it was once found in its original context. The Rifāʿiyya, as it has been known since the Prussian consul Johann Gottfried Wetzstein (1815–1905) bought it in Damascus in 1853, has since remained in Leipzig, Germany (Liebrenz 2016). Made up of 488 volumes containing 865 works, its content spans the whole range of Arabic literature. A broad classification is headed by religious works (253 titles), belles lettres (155) and history (104), followed by the sciences (72) even before works on language (64) and law (51).[3] Its former possessors are not precisely known, but a connection with a family that dominated the branch of the Rifāʿiyya Sufi order in Hama while also residing in Damascus is likely.

Fortunately, many more libraries can be reconstructed, at least to some degree, through notes left by their former owners. Large parts of one such library were purchased sometime between 1803 and 1805 by the traveler Ulrich Jasper Seetzen (1767–1811),[4] a trained physician, in Aleppo.[5] They are now found in Gotha, Germany, with a few other volumes having reached other collections: Université Saint-Joseph in Beirut (Taoutel 1967, 227), ÖNB in Vienna (Loebenstein 1970, 283) and Fondation Salem in Aleppo (del Río Sánchez 2008, 21–2, 34–5). This collection of 78 volumes, containing 88 titles, belonged to several generations of Maronite physicians from Aleppo, who went by the names Shukrī and Ibn Ārūtīn. Many of the books were already in the possession of Shukr Allāh ibn Ḥannā (who dates his entries between 1112/1700–1701 and 1132/1719–1720). His son Ḥannā (whose first note is dated 1132/1719–1720 and who died in 1189/1775, according to Seetzen 1805, 102) continued to build the library and then appears to have transferred it to his sons Anṭūn and Ilyās. These men belonged to a family of many branches, some members of which had a high standing in the church hierarchy, while

DOI: 10.4324/9781315170718-70

many others practiced medicine. The brother of Ḥannā, Arsāniyūs Shukrī Ārūtīn, was ordained bishop of Aleppo in 1176/1762. Arsāniyūs was also the author of an account of his travels to many countries in Europe which his brother both possessed (orient. A 1549) and copied (Beirut, Université Saint-Joseph in 1764; Taoutel 1967, 227–8). Ḥannā wrote his own account of a journey to Istanbul in 1764 (orient. A 1550). The Shukrīs we find in the notes themselves use the religious title *shammās*, which meant 'ordained deacon', but was also "a title by which educated lay people were addressed" (Kilpatrick and Toomer 2016, 13, and the literature cited there). It is plausible to assume that those Shukrī Ārūtīns who were professional physicians were using it in the latter sense. The known content of their library breaks down as follows: the sciences make up the majority of titles, 42 out of 88. They are followed at a considerable distance by Christian religious literature (8), history (7), magic and occult sciences (5), the arts of language (5), poetry (4), travel accounts (4), *adab* (3), two Turkish guides to letter-writing and some scattered titles on miscellaneous topics.

These two outstanding book collections are chosen for study because of their relative accessibility in a developing field of collection history and because of the large extent to which they have survived. From what we know about the private and institutional libraries of the region, these two are rather typical examples in terms of size and content.[6] The study undertaken here represents only one possible approach to examining the manuscript material that has come down to us for clues on the history of science, namely, concentrating on the profiles of individual collections. Another approach would be to look at all the surviving manuscripts of scientific texts and systematically analyze all their previous readers and owners. In earlier, very preliminary attempts at this, it has already been noticed that scientific manuscripts are among those that most easily changed hands between the different religions (Liebrenz 2016, 348). Therefore, it will come as no surprise to see below some remarkable similarities between the collections of a Christian and a Muslim family.

Although they had originally been very close when seen in the context of the vast territorial, chronological and cultural expanse that we call the Islamicate world, there are still many aspects that set these two libraries apart. One was once located in 12th/18th-century Aleppo, the other bought in mid-13th/19th-century Damascus, although purportedly with roots at least in the 12th/18th. One collection was owned by Maronite Christian physicians, the other by Sunni Muslim scholars. Despite their many similarities, this results in a number of differences in topics (Christian theology and logic only in the Shukrī library), languages (more Ottoman and Persian[7] in Aleppo, Karshūnī only there) and even form. For example, the Maronites apparently shelved their books in the European manner standing up, while the Muslims followed the local tradition of putting the books on top of each other with the titles written on the edge. Thus, they represent the diversity present in the book culture of the Ottoman Arab lands, two of the many possible incarnations of institutional and private libraries.

The scientific titles from both libraries are thoroughly embedded in a much broader literary program. That program is surprisingly parallel, even in some of the more remote areas of learning. Both, for example, had a rare and curious interest in Druze literature (MSS orient. A 856; Vollers 262). They also feature shared interests in history, where both libraries hold a copy of Ibn al-Shiḥna's (d. 890/1485) *The Chosen Pearl on the History of the Principality of Aleppo* (*Durr al-muntakhab fī tārīkh mamlakat Ḥalab*; MSS Vollers 656; orient. A 1274). One can also detect overlapping tastes in poetry and *adab* (refined and entertaining prose or prosometric literature that uses techniques like cadence and rhythm usually found in poetry) with titles such as the poetry collection *Consolation for the Spectator* (*Qurrat al-nāẓir*; MSS orient. A 2159; Vollers 567 and 568) by the late Mamluk author Ibn Sūdūn (d. 868/1463–4) or Ibn Ḥijja's (d. 837/1434) *adab* anthology *Fruits of the Leaves on Disputations* (*Thamarāt al-awrāq fī l-muḥāḍarāt*; MSS Vollers 618 and 882, 7; the Rifāʿiyya even has two copies in each case).

Fundamental differences are discernible mainly where both religion and connections with traditions from outside the Islamicate world are directly concerned. Christian doctrine and Maronite authors on logic or poetry are an exclusive feature of the Shukrī library, which also held accounts of travels to the Islamic west and Europe. Islamic theology and piety, while also found in the Christian library, is much more broadly represented and indeed fundamental in the Rifāʿiyya, where the travel reports are describing Muslim holy sites in Syria and south of it.

The preceding analysis holds largely true also for the scientific content that makes up a significant part of both collections. It can be said that the possessors of both libraries aspired to a canon of literature in which the sciences had their place for both amateurs and professionals, although those topics were proportionately higher represented in the collection of the Aleppans. This is particularly true for the areas of scientific knowledge that are concerned with the health and functioning of the body.

VI.6.1 Medicine

It comes as little surprise that medicine formed the single largest section in the library of Christian physicians. The Shukrīs once possessed at least 27 volumes containing texts on different branches of medicine, more than one third of what we know of their collection. Their Damascene counterpart is scarcely less valuable a source. The Rifāʿiyya has 21 titles, significantly smaller in proportion, but still not much smaller overall than its Aleppan counterpart.

Just as the manuscripts give us an immediate impression of what literature was possibly available for interested readers, European travel reports can offer information on how outside observers perceived its application. Side by side, these two sources may inform and sometimes contradict each other. It is a fortunate coincidence that we have a very detailed account by medically trained eyewitnesses living side by side with the Maronite family in Aleppo, namely, the brothers Alexander (1714–1768) and Patrick Russell (1726–1805) and their *Natural History of Aleppo* (used here in the second enlarged edition of 1794 that added substantial commentary by Patrick to the original text of Alexander and at times changed the previous text). Both served at different times as physicians for the English Levant Company in Aleppo, but their practice extended also to the local population, and they entertained learned contacts with scholars and notables in the city. What they describe as the medical practice could very well be a realistic portrait of the Shukrī family business with which they were probably familiar. Their text, as usual not without agenda and bias but of immense and demonstrable factual value (van den Boogert 2010; Starkey 2018), will therefore accompany the following report on the manuscript holdings. Regarding medical literature, both editions say: "Medicine still being regarded as a branch of philosophy, the literati always pretend to some speculative[8] knowledge of it" (Russell and Russell 1794, 2: 116–7). As indicated, both libraries indeed have a strong focus on this art. Yet for one of them there was an aspect of practical application to it. The question of medical practice as opposed to bookish theory has occupied an important place in scholarship (Savage-Smith 2000, 307–21; Shefer-Mossensohn 2009, 45–61; Liebrenz 2014, 32–58). For Aleppo in the 12th/18th century, Patrick Russell's account tackled that question as well. And although eminent physicians, according to their testimony, are not practicing surgery (but see Liebrenz 2014), "all prepare the medicines for their own patients, and keep shops at their house, or in some more convenient situation" (Russell and Russell 1794, 2: 122). An anonymous and untitled treatise on the extraction (*qalʿ*) of essences (*ṭubūʿ*) from fruits (MS orient. A 1332) can be imagined as a useful tool for a physician acting in the way as Patrick Russell's book described. But the same goes for a collective manuscript in the Rifāʿiyya (MS Vollers 768) that comprises Muḥammad al-Khāzin's (dates unknown) *Abridgment on the Knowledge of Perfume* (*Mukhtaṣar fī maʿrifat ajnās al-ṭīb*) and Yūḥannā ibn Māsawayh's (d. 243/857–8) *Simple Essences of Perfume* (*Jawāhir al-ṭīb al-mufrada*). Both instruct in the preparation of fragrant substances and their medicinal uses. Whether either was ever used to actually make medicines is impossible to tell. But it might also be true to say that a library was simply not the only or even the preferred place for the practical training of diagnosis, treatment and use of *materia medica* to produce remedies. All this an aspiring physician presumably acquired hands on through apprenticeship and, in the case of a family business like that of the Shukrī Ārūtīns, experience in the family.

Overall, there is very little exact overlap between our two libraries in medicine. This may be surprising given the examples above where both hold the same works, but it is to be expected in collections that by the very fact of their limited size simply could not achieve comprehensive coverage in any field. The important characteristic of this sample of texts and authors appears rather to be that all that are found in one library could easily be imagined to be in the other as well. In other words, the collection of texts may not reflect a conscious choice by their collectors. Rather, the financial means and availability on the market determined what would find its way into a collection and in many instances left their owners with partial copies, such as one fragment of Ibn Sīnā's (d. 428/1037) *Canon of Medicine* in the Aleppo library (MS orient. A 1913), or none at all.

What the Rifāʿiyya does not offer its readers are examples of the early translation movement that made Greek works on medicine, often through Syriac translations, available in Arabic (Chapter I.4). These were done by scholars with a diverse set of religious backgrounds, among them Muslims, Sabians, Melkites and members of the Syriac churches, and Christian authors would continue to produce important contributions throughout the following centuries. This long tradition of Arabic writings by Christian authors on medicine is conspicuously missing in the Rifāʿiyya. The Aleppo library, on the other hand, held such important titles as Ḥunayn's *Questions on Medicine* (*Masāʾil fī l-ṭibb*; MS orient. A 2036, with two commentaries in the same library, namely, MSS orient. A 1932 and 1933), his translation of Galen's *On the Usefulness of the Parts* (*Kitāb al-a ʿḍāʾ*; orient. A 1901), Yaḥyā al-Naḥwī's (likely not John Philoponos [d. 574] but an unknown Byzantine medical writer) commentary on that work and translated by the Miaphysite Christian Ibn Zurʿa ([d. 398/1008]; MS orient. A 1906), Ibn al-Quff's (d. 685/1286) commentary *Book of the Principles on the Commentary on Hippocrates'* Aphorisms (*Kitāb al-Uṣūl fī Sharḥ al-Fuṣūl*; MS orient. A 1894), and Ibn Buṭlān's (d. 458/1066) *The Physician's Banquet* (*Daʿwat al-aṭibbāʾ*) as well as its anonymous commentary (*sharḥ*; both MS orient. A 1909). The discrepancy here is obviously not related to the confessional background of these libraries and their owners. All these texts were well received and commented on by Muslim authors, so there is no apparent reason for them not to be part of a library like the Rifāʿiyya.

The fundamental contributions by the Central Asian triumvirate Ibn Sīnā, Najīb al-Dīn al-Samarqandī (d. 619/1222), and al-Kirmānī (d. 853/1449) is as much a core element of our libraries as it had been throughout many parts of the Islamicate world (Chapter V.7). Many a work in our two Ottoman libraries is a more or less elaborate commentary or reworking of original works by one of these men. Patrick Russell noted that the Roman edition of Ibn Sīnā's *Canon* was widely used by local physicians. But whether they ever possessed one of those printed books, the Shukrīs had at least one manuscript volume (MS orient. A 1913) as well as Ibn al-Nafīs' (d. 678/1288) *Abridgment of the* Canon (*Mūjaz al-Qānūn*),[9] and a commentary on that work by al-Kāzarūnī (d. 758/1357). The Rifāʿiyya library, on the other hand, had no part of the *Canon* (their only work by Ibn Sīna was a volume containing the section on physics [*ṭabīʿiyāt*] of his philosophical compendium *The Healing*, MS Vollers 796) but participated in its commentary tradition with a unique autograph of al-Qūṣūnī's (d. after 1044/1634–5) *The Illuminating Lamp on the Small Canon on Medicine* (*al-Miṣbāḥ al-munīr ʿalā l-Qānūn al-ṣaghīr fī l-ṭibb*; MS Vollers 764).

The Damascene collection boasted two copies of Najīb al-Dīn al-Samarqandī's *The Causes and Symptons* (*al-Asbāb wa- l-ʿalāmāt*; MSS Vollers 761 and 762) as well as al-Kirmānī's commentary on it (MS Vollers 763).

Both libraries held *The Book of the New Chemical Medicine* (*Kitāb al-ṭibb al-jadīd al-kīmiyāʾī*), part adaptation of and part engagement with Paracelsian medical practices, introduced by and ascribed to their Syrian compatriot of the 11th/17th century, Ibn Sallūm (d. 1080/1669; see Bachour 2012, 37). This text circulated widely and was translated into Ottoman Turkish and Persian. Although it went by the name of Paracelsus, it was a compilation of parts translated summarily from four Latin texts: *The Medical Institutions in Five Books* (*Institutionum medicinae libri V*; 1611) and *The Book of the Agreement and Disagreement of the Chemists with the Aristotelians and Galenists* (*Liber de consensu et dissensu chymicorum cum Aristotelicis et Galenicis*; 1619) by Daniel Sennert (1572–1637), *Of the Internal Signature of Things* (*De signaturis internis rerum*; 1609) by Oswald Croll (d. 1609) and *The Special Pharmacopeia* (*Antidotarium speciale*; a register of remedies against poisons; first printed in 1561) by Johann Jacob Wecker ([1528–1586]; Savage-Smith 1987; Bachour 2012, 94–95, 121–2; Küçük 2016; Bachour 2018, 94). On the library shelves, both the works of Galenic medicine and the new iatrochemical medicine continued to exist side by side, and it is unclear in what kind of dynamic they entered for their readers. This is supported by the recent work of Natalia Bachour who, against earlier assumptions, has shown that Ibn Sallūm was not opposed to the concepts of humoral therapy. Furthermore, while one recent contribution observed that this new medicine required its practitioners to "expand their skill sets" as they now "had to master not only herbs and minerals, but also astrology, mathematics, and . . . natural philosophy" (Küçük 2016, 223), the Aleppo library, in a stark contrast to the Rifāʿiyya, was particularly light on precisely these topics. This puts into doubt whether a text like Ibn Sallūm's was always primarily seen as the foundation of an integrated philosophy. It could just as likely be used in a purely practical way as a source of recipes and diagnoses side by side with older authorities.

VI.6.2 Botany, zoology, mineralogy

Besides being yet another contribution to those aspiring for comprehensive learning, knowledge of the properties of plants, animals and stones could often be of a direct use for the practice of medicine. Several titles in both libraries deal directly with their description and taxonomy. Readers in both Aleppo and Damascus had access to a book on the properties of stones (MSS Vollers 755; orient. A 2111). The most famous of the zoological works in Arabic, al-Damīrī's (d. 808/1405) *Life of Animals* (*Ḥayāt al-ḥayawān*), was absent from both libraries, but the Rifāʿiyya held two Ottoman-era abridgments of this classic (MSS Vollers 748 and 749).

By far the most outstanding work in this group from an art-historical perspective is found in the Aleppo library (MS orient. T 122).[10] This Turkish text contains descriptions of 56 medicinal plants that are accompanied by illustrations of a naturalistic precision that I have not seen in other manuscripts of the region. Moreover, the names of the plants are usually rendered not only in Arabic and Turkish but also in Italian and Latin, the latter also in the Latin alphabet. Here we can observe one of the distinctions of the Aleppo library, namely, its occasional reception of science from Europe. Another example for this would be a short and anonymous cosmography or physical geography that, according to the cataloguer, uses European sources and was copied in 1665 (orient. A 1520). At this time, such works were more common in Istanbul, where the polymath and bibliographer Kātib Chelebi (Ḥājjī Khalīfa; [1017–1067/1609–1657]) studied Western cartography (Hagen 2003[CB]).[11] Ibn Sallūm, the Aleppan physician mentioned earlier, engaged in the capital Christian physicians to translate parts of Latin iatrochemical works and pharmacopeias, being moreover in contact with the German Dutch diplomat and scholar Levinus Warner ([1618–1665]; Vrolijk *et al.* 2012, 102–7). Although it did not host large numbers of diplomatic personnel, Aleppo, too, was likely a point of contact due to colonies of Dutch, French and especially British merchants with their retinues of physicians and scholars, combined with the presence of disproportionately large, culturally vital and linguistically skillful groups of religious minorities (Chapter VI.5). Locals with a knowledge of Italian or French were easily found for those visitors in need of a translator or language teacher. Three examples are Nicolaus Petri ([d. after 1661]; for his biography see Kilpatrick and Toomer 2016), Ḥannā Diyāb ([1688–after 1763]; see his travelogue, Diyāb 2015) and Carolus Rali Dadichi ([d. 1734]; see Seybold 1910, 591–601; Liebrenz 2008, 36–7). All three traveled from Aleppo to Europe with scholars, missionaries or diplomats, suggesting that in this atmosphere personal contacts could probably be forged more easily in Aleppo than in Damascus and facilitated the transmission of knowledge.[12]

VI.6.3 Astronomy and astrology[13]

The difference between the two libraries in Aleppo and Damascus could not be more pronounced than in the area of astronomical and astrological sciences.[14] Patrick Russell had seen many astronomical works in the libraries of Aleppo but knew of only one person who was able to predict eclipses and thus "had the reputation of a most profound Astronomer" (Russell and Russell 1794, 2: 99). A few decades later, Seetzen, too, claims to have found only one person with that capacity (Seetzen 2011, 95) but lists other local scholars famous for their bookish knowledge of the art (Seetzen 2011, 124, 249). Within a discussion of his own astronomical observations, Seetzen praises the information Ḥannā provides on distances in his travel account, a possible hint at the Maronite's active knowledge in mathematics and astronomy.[15] However, apart from one Christian calendar, conveniently bound together with a treatise ordering talismanic and other remedies according to body parts (MS orient. A 18), works from this area form no part whatsoever of the Shukrī library as we know it. Yet they are a centerpiece of the Rifāʿiyya, where they account for 48 titles, 20 of which are bound in two multitext manuscripts with 10 treatises each (MSS Vollers 814 and 820) and thereby form the largest subgroup within the sciences. Interestingly, about half the texts are anonymous. Of those with a known author, the variety among them in chronology and regional origins is remarkable, ranging from the astrological prophecies circulating under the name *Destinies of Daniel* (*Malḥamat Dāniyāl*)[16] to authors from Transoxania, Egypt and North Africa in the Ottoman period. On the technical side, the Rifāʿiyya held several volumes with tables and two astronomical handbooks, such as Ibn Abī l-Fatḥ al-Ṣūfī's (d. after 864/1460) *The Excellent Pearl, an Abridgment of the* Zīj *of Ibn al-Shāṭir* (*al-Durr al-fākhir fī khtiṣār Zīj Ibn al-Shāṭir*; d. 777/1375; MS Vollers 807) or Yaḥyā ibn Abī Manṣūr's (d. 215–217/831–833) *The Tested* Zīj *for al-Ma'mūn* (*al-Zīj al-Ma'mūnī al-mumtaḥan*; MS Vollers 821). In addition, the library contained poems to memorize constellations or houses of celestial bodies (MS Vollers 820, 2–3). Other works were prose texts that collectively covered the wide range of insights that the observation of the celestial bodies could offer. Knowing their movements and the zodiacal signs was indispensable for the casting of horoscopes. But it was also vital for the purpose of timekeeping, another major concern with many practical aspects. One was the conversion of dates in different current calendars for purposes of agriculture and taxation. Another was the definition of the hours for prayer or the advent of a month. The Rifāʿiyya also possessed several, usually smaller treatises that explain how to work with astronomical instruments such as the astrolabe.

But, yet again, we find in observations of the heavenly bodies an area of knowledge intricately linked to the medicinal arts. That the proper administration and dosage of medicines would rely on the celestial formations was a view particularly promoted by the new medicine and found in Ibn Sallūm's work which both libraries held. Casting horoscopes, too, could be part of the process of diagnosis and treatment.

VI.6.4 Mathematics

Mathematical knowledge was part of several sciences that many scholars would likely be familiar with or interested in, such as astronomy or music. The *Natural History of Aleppo* allowed for proper works on this art to "lye entombed in . . . voluminous writings", yet apart from that, "mathematical studies" was to them an all but forgotten art in 12th/18th-century Aleppo, generally not extending beyond basic arithmetic and algebra for practical purposes (Russell and Russell,1794, 2: 106–7). The manuscripts in the Shukrī library, at least, indeed reflect this situation. There are, besides what is contained in the astronomical works, six purely mathematical titles in the Rifāʿiyya, yet (so far) none in the library of the Shukrīs. Some parts of the mathematical sciences were an integral part of training for those concerned with inheritance law, and indeed two of the works in the Rifāʿiyya are specifically concerned with this aspect of computation, Ibn Ghaylān's (6th/12th century or earlier) *Impregnation of the Hearts with the Various Methods in Arithmetic* (*Talqīḥ al-albāb fī tanqīḥ ṭuruq al-abwāb fī l-ḥisāb*; MS Vollers 825) and Zayn al-ʿĀbidīn al-Durrī's (d. after 1033/1623) *Commentary on the Short Sparkles* (*Sharḥ Lumaʿ al-yasīra*; MS Vollers 826). All titles in the Rifāʿiyya, however, are concerned with arithmetic and algebra, and none purely with geometry.[17]

VI.6.5 Occult sciences

Within the Islamic sciences, those of magic squares, letters and talismans had their detractors but were widely seen as a valid part of scientific inquiry about the nature of the world. The Rifāʿiyya holds five books belonging to this area. Based on the evidence of the Maronite library with its five titles on the subject, things were no different in educated Christian circles. Within the broader framework of the sciences, this area holds one of the rare instances of actual textual overlap: both libraries include al-Būnī's *The Sun of Knowledge* (*Shams al-maʿārif*).[18] Moreover, the Rifāʿiyya holds two other works from the Būnian corpus. Another science that promised a direct path to the secrets of creation was alchemy. Again, according to the *Natural History*, there was at Aleppo "one, or more, of the medical tribe, who have acquired a sufficient smattering in alchymy to beggar themselves by the expense of laboratory" (Russell and Russell 1794, 2: 105).[19] The Maronite physicians, however, did not seem to show any interest in having this art in their library. Yet we do find two such texts, namely al-Jildakī's (d. *c.* 743/1342) *The Utmost Joy* (*Ghāyat al-surūr*) and al-Ghamrī's (d. 905/1499) *Solution of the Talisman* (*Ḥall al-ṭilasm*) in the Rifāʿiyya (MSS Vollers 836 and 877,3). The veracity of alchemy's claim to transubstantiation was repeatedly questioned, but also repeatedly affirmed by Ottoman-era Muslim writers. But more criticism was leveled at moral issues that came with the *practice* of alchemy (such as greed and overspending into bankruptcy) than at the validity of the underlying philosophy.[19]

VI.6.6 Conclusion

Even where there is little overlap in the specific works both collections held in the fields of medicine, mathematics or astronomy, they still shared fundamentally the same tradition. This tradition was based on the first translations from Greek and Syriac, such as those by Ḥunayn ibn Isḥāq, and developed by authors that were later deemed classical authorities such as Ibn Sīnā, who, in turn, were later often significantly improved, adapted or amended by their commentators. This body of knowledge was finally paired with some influences and adaptations from Western sources, such as most prominently the latest fashion of Paracelsian 'chemical' medicine. None of the modern authors seems to have pushed aside the earlier ones.

Thus, it would appear that neither library limited itself to any particular school of thought *per se*. Their holdings – if not their uses – were fundamentally all-embracing. And the kind of presentation, too, addressed all kinds of audiences, ranging from very elaborate and stylized works of professional prose and in-depth learning to very basic practical guides.

Within this broad philosophical congruence, there are remarkable nuances. The Rifāʿiyya's deep interest in astronomy and astrology was not found at all in the Shukrīs' library, nor was any sign of algebra, geometry or alchemy. At least in these cases, differences seem too profound to be considered a mere coincidence of transmission. In that sense, the Damascus collection seems to have been not only larger but also more universal in scope. Other cases are not so clear. A reader's possible wish to achieve a comprehensive overview of the sciences is reflected in Ibn al-Akfānī's (d. 749/1348) encyclopedia, *The Guidance of the Pursuer to the Most Radiant End* (*Irshād al-qāṣid ilā asnā al-maqāṣid*), which the Rifāʿiyya did possess (MS Vollers 2). Yet apart from such attempts at a systematic introduction between two covers, real comprehensiveness in the works on the sciences was very hard to reach in private and institutional libraries that rarely ever exceeded a few hundred volumes. One had to be content with what the book market had to offer, and that apparently often meant fragmentary works. This is not to say that the libraries were not indeed shaped by the real interests of their owners, interests that could be forged by individual, communal or regional tendencies. To have a clearer appreciation of those tendencies, more libraries will need to be subjected to similar analyses of their holdings.[20]

Notes

1 I am grateful to Kristina Richardson for her corrections and suggestions to a draft of this chapter. Feras Krimsti (Oxford) is working on Ḥannā Shukrī and will have more to say about his library in a monograph he is planning on the subject. After completing this chapter, I received Feras Krimsti's article "The Lives and Afterlives of the Library of the Maronite Physician Ḥannā al-Ṭabīb (c. 1702–1775) from Aleppo" as a contribution to my edited volume *The History of Books and Collections through Manuscript Notes*, which is set to appear as a special issue of the *Journal of Islamic Manuscripts* 9.2–3 (2018), before the present chapter goes to print. Here I also learned of a previous, unpublished treatment of the library in Feras Krimsti, Der Istanbul-Reisebericht des Aleppiner Arztes Ḥannā l-Ṭabīb (1764/65): Alltagsbilder und identitäre Verortungen (PhD thesis Berlin, Freie Universität 2016), 59–79, which I did not see. The manuscripts from Leipzig will be cited in the following by "Vollers", referring to the catalogue by Carl Vollers. The manuscripts from the Research Library in Gotha are cited as "orient. A" (for Arabic) and "T" (for Turkish).

2 The sciences have received focused treatment (in the form of annotated lists) within a collected volume on the 18th-century library of Cârullah Efendi (d. 1151/1738) in Istanbul, see Umut (2015) and Usluer (2015).

3 A more fine-grained analysis of the content can be found in Liebrenz (2016, 78–104).

4 Seetzen was a free-spirited enlightenment scientist and entrepreneur who was chosen to travel to Africa (via Constantinople, Aleppo, Damascus, Egypt and the Yemen, where he ultimately died), with support from the Duke of Saxony-Gotha and a commission to send back manuscripts, artifacts and natural curiosities (Seetzen 2014).

5 It is likely that the reason for Seetzen's ability to purchase so many manuscripts from this library was his first host, a lady married into the Armenian-Catholic Sceriman family of traders, whose own family is described as consisting of physicians (Seetzen 2011, 5: "Die alte war mit einem Kaufmann verheurathet, und ihre ganze Familie bestand aus Aerzten. / The old (woman) was married to a merchant, and her whole family consisted of physicians."). That this unnamed family was in fact Shukrī Ārūtīn is corroborated by a later note on Seetzen's purchase of the dictionary al-Ṣiḥāḥ, which he says he bought from

"my hostess' uncles". Four of the five volumes of that work which Seetzen bought in Aleppo (orient. A 384, 386, 387, 390, 392) definitely come from the Shukrīs (Seetzen 2011, 248: "Von den Oheimen meiner Wirthin kaufte ich unter andern das gleichfalls sehr geschätzte arabische Wörterbuch el Szeháhh (الصحاح) . . . / From the uncles of my hostess I bought among other things the highly appreciated Arab dictionary el Szeháhh (الصحاح) . . ."

6 For an overview of the development of and sources for the history of libraries in the Ottoman Arab world, see Liebrenz (forthcoming); a fuller bibliography for the history of the book in the Islamicate world is available in Liebrenz (2016, 4–34); for a broader overview of Ottoman libraries with a focus on the empire's center, see Erünsal (2008).

7 Persian texts are preserved in the Viennese codex, ÖNB Cod. Mixt. 1408, as well as orient. T 176.

8 The term *speculative* means theoretical within an Aristotelian classification scheme of the sciences, popular in the 17th- and 18th-century English Enlightenment as expressed in lists such as "speculative or theoretical; practical; and artistic" (Yeo 2003, 243).

9 See a short discussion of the enigmatic text with illustration of a double-page in Orientalische Buchkunst in Gotha (1997, 157–8).

10 It is likely that similar networks would have existed within the Christian communities of Damascus, yet in the course of the 1860 massacres the Christian quarter was burned down and many rich libraries were said to have fallen victim to the flames (Liebrenz 2016, 189).

11 Consolidated bibliography.

12 Seetzen 2011, 255: "der schätzbaren Reisebeschreibung des Arztes Hanna (the valuable travel description of the physician Hanna)", the "köstliche und einzige Werk (delicious and unique work)".

13 The role of the science of the stars (astronomy and astrology) is ill researched for the Ottoman period. For a thorough analysis of its use at the court of Bāyezīd II (r. 886–918/1481–1512) see Şen (2017); a bibliographical overview of the literature produced in this period can be obtained through İhsanoğlu *et al.* (1997[CB]).

14 Texts surviving under this title may differ considerably in terms of content and methodology; see for an overview and literature, Kohlberg (1992, 143).

15 For geometry in a Mamluk and an Ottoman biographical source, see Brentjes (2008). Note in particular her observations on the discrepancy between the Mamluk source and the manuscript record with regard to the relative absence of geometry among the latter (324–5). Note also that geometry could be treated in works on the subject of *ʿilm al-mīqāt.*

16 Gardiner (2012) identifies the different versions extant of this text; see also Witkam (2007).

17 See also Seetzen's description of the general belief in and occupation with alchemy in Seetzen (2011, 37–8).

18 For an overview of Ottoman Syrian approaches see Berger (2007, 270–1). Berger's account of al-Muḥibbī (2006, 1: 241–9), as a renunciation of alchemy is incorrect, since the author is positive about alchemy's potential but cautions against its abuse (Liebrenz 2016, 90). In Aleppo, both the Russells (1794,1: 105) and Seetzen (2011, 37–38) report general belief in the possibility of turning base metals into gold.

19 Insights of varying degrees into the holdings of libraries throughout the Islamicate world at different times are available through catalogues (e.g., Hirschler 2016), endowment deeds (e.g., Veselý 1996; al-Munajjid, 1953), estate inventories (e.g., Establet and Pascual 1999) and manuscript notes (Liebrenz 2013). A comprehensive overview using all this disparate information would require a broader study and a more systematic collection of all the aforementioned sources. This is not currently feasible in light of the many geographical and chronological gaps that impede a direct comparison.

Bibliography

Sources

Diyāb, Ḥ. tr. and ann. Fahmé-Thiéry, P., Heyberger, B. and Lentin, J. 2015. *D'Alep à Paris. Les pérégrinations d'un jeune Syrien au temps de Louis XIV*. Paris: Sindbad/Actes Sud.

al-Muḥibbī, A. ed. Ismāʿīl, M. Ḥ. 2006. *Khulāṣat al-athar fī aʿyān al-qarn al-ḥādī ʿashar* [Essence of the Report on the Nobles of the 11th Century]. 4 vols. Beirut: Dār Ṣadr.

al-Munajjid, Ṣalāḥ al-Dīn. ed. 1953. *Kitāb waqf Asʿad Bāshā al-ʿAẓm* [The Charity Book of Asʿad Pāshā al-ʿAẓm]. Damascus: al-Maʿhad al-faransī bi-Dimashq li-l-dirāsāt al-ʿarabiyya.

Russell, A. and Russell, P. 1794². *The Natural History of Aleppo*. 2 vols. London: G. G. and J. Robinson.

Seetzen, U. J. 1805. "Nachrichten von einigen Arabischen, Persischen und Türkischen Reisebeschreibungen, Topographien und andern geographischen Werken und Landkarten," *Monatliche Correspondenz zur Beförderung der Erd- und Himmelskunde* 12.August: 101–25.

Seetzen, U. J. ed. Zepter, J. 2011. *Tagebuch des Aufenthalts in Aleppo 1803–1805*. Hildesheim *et. al.*: Olms.

Research literature

Açil, B. ed. *Osmanlı Kitap Kültürü. Cârullah Efendi Kütüphanesi ve Derkenar Notları*. Istanbul and Ankara: İLEM and Nobel Yayın Dağıtım.

Bachour, N. 2012. *Oswaldus Crollius und Daniel Sennert im frühneuzeitlichen Istanbul. Studien zur Rezeption des Paracelsismus im Werk des osmanischen Arztes Ṣāliḥ b. Naṣrullāh Ibn Sallūm al-Ḥalabī*. Freiburg i. Br.: Centaurus.

Bachour, N. 2018. "Iatrochemistry and Paracelsism in the Ottoman Empire in the Sixteenth and Seventeenth Centuries," *Intellectual History of the Islamicate World* 6.1–2: 82–116.

Berger, L. 2007. *Gesellschaft und Individuum in Damaskus 1550–1791*. Würzburg: Ergon.

Boogert, M. van den. 2010. *Aleppo Observed. Ottoman Syria through* the *Eyes of Two Scottish Doctors, Alexander and Patrick Russell*. Oxford: Oxford University Press.

Brentjes, S. 2008. "The Study of Geometry According to al-Sakhāwī (Cairo, 15th c.) an al-Muḥibbī (Damascus, 17th c.)," in Dauben, J. W. *et al.*, eds. *Mathematics Celestial and Terrestrial. Festschrift for Menso Folkerts zum 65. Geburtstag*. Halle an der Saale: Deutsche Akademie der Naturforscher Leopoldina, *Acta Historica Leopoldina* 54: 323–42.

Erünsal, İ. E. 2008. *Ottoman Libraries. A Survey of the History, Development and Organization of Ottoman Foundation Libraries*. Cambridge, MA: The Department of Near Eastern Languages and Literatures, Harvard University.

Establet, C. and Pascual, J.-P. 1999. "Les livres des gens à Damas vers 1700," *Revue des Mondes Musulmans et de la Méditerranée* 87–8: 143–69.

Fancy, N. 2013. "Medical Commentaries: A Preliminary Examination of Ibn al-Nafīs's *Shurūḥ*, the *Mujaz* and Subsequent Commentaries on the *Mūjaz*," *Oriens* 41: 525–45.

Gardiner, N. 2012. "Forbidden Knowledge? Notes on the Production, Transmission, and Reception of the Major Works of Aḥmad al-Būnī, in Ghersetti, A. and Metcalf, A., eds. *The Book in Fact and Fiction in Pre-Modern Arabic Literature*," *Journal of Arabic and Islamic Studies* 12: 81–143.

Haberland, D., ed. 2014. Ulrich Jasper Seetzen (1767–1811). Jeveraner – aufgekl.rter Unternehmer – wissenschaftlicher Orientreisender. Oldenburg: Isensee.

Kilpatrick, H. and Toomer, G. J. 2016. "Niqūlāwus al-Ḥalabī (c. 1611-c. 1661): A Greek Orthodox Syrian Copyist and His Letters to Pococke and Golius," *Lias: Journal of Early Modern Intellectual Culture and Its Sources* 42: 1–159.

Kohlberg, E. 1992. *A Medieval Muslim Scholar at Work. Ibn Ṭāwūs and his Library*. Leiden: Brill.

Küçük, B. H. 2016. "New Medicine and the *Ḥikmet-i Ṭabīʿiyye* Problematic in Eighteenth- Century Istanbul," in Langermann, Y. T. and Morrison, R. G., eds. *Texts in Transit in the Medieval Mediterranean*. Philadelphia: University of Pennsylvania Press, 222–42.

Liebrenz, B. 2008. *Arabische, Persische und Türkische Handschriften in Leipzig. Geschichte ihrer Sammlung und Erschließung von den Anfängen bis zu Karl Vollers*. Leipzig: Leipziger Universitätsverlag.

Liebrenz, B. 2013. "The Library of Aḥmad al-Rabbāṭ. Books and their Audience in 12th to 13th/18th to 19th Century Syria, in Elger, R. and Pietruschka, U., eds. *Marginal Perspectives on Early Modern Ottoman Culture: Missionaries, Travelers, Booksellers*." *Orientwissenschaftliche Hefte* 32: 17–59.

Liebrenz, B. 2014. "The Social History of Surgery in Ottoman Syria: Documentary Evidence from Eighteenth-Century Hamah," *Turkish Historical Review* 5: 32–58.

Liebrenz, B. 2016. *Die Rifāʿīya aus Damaskus. Eine Privatbibliothek im osmanischen Syrien und ihr kulturelles Umfeld*. Leiden: Brill.

Liebrenz, B., forthcoming. "Libraries (in the Arab World after 1500)," in *EI-3*.

Loebenstein, H. 1970. *Katalog der arabischen Handschriften der Österreichischen Nationalbibliothek: Neuerwerbungen 1868–1968 1: Codices mixti ab Nr 744*. Wien: Hollinek.

Orientalische Buchkunst, 1997. *Orientalische Buchkunst in Gotha. Ausstellung zum 350jährigen Jubiläum der Forschungs- und Landesbibliothek Gotha*. Gotha: Forschungs- und Landesbibliothek Gotha.

Río Sánchez, F. del. 2008. *Catalogue des manuscrits de la Fondation George et Mathilde Salem*. Wiesbaden: Harrassowitz.

Savage-Smith, E. 1987. "Drug Therapy of Eye Disease in Seventeenth-Century Islamic Medicine: The Influence of the 'New Chemistry' of the Paracelsians," *Pharmacy in History* 29: 3–28.

Savage-Smith, E. 2000. "The Practice of Surgery in Islamic Lands: Myth and Reality," *Social History of Medicine* 13: 307–21.

Şen, A. T. 2017. "Reading the Stars at the Ottoman Court: Bāyezīd II (r. 886/1481–918/1512) and His Celestial Interests," *Arabica* 64: 557–608.

Seybold, C. F. 1910. "Der gelehrte Syrer Carolus Dadichi († 1734 in London), Nachfolger Salomon Negri's († 1729)," *Zeitschrift der Deutschen Morgenländischen Gesellschaft* 64: 591–601.

Shefer-Mossensohn, M. 2009. *Ottoman Medicine: Healing and Medical Institutions, 1500–1700.* Albany: State University of New York Press.

Starkey, J. 2018. *The Scottish Enlightenment Abroad. The Russells of Braidshaw in Aleppo and on the Coast of Coromandel.* Leiden: Brill.

Taoutel, F. 1967–8. "Riḥlat al-Āb Arsāniyūs Shukrī Arūtīn al-Ḥakīm [The Travel of Father Arsāniyūs Shukrī Arūtīn al-Ḥakīm]," *al-Machriq. Revue catholique orientale* 61: 227–62, 325–56, 537–90; 62: 93–120.

Umut, H. 2015. "Matematik Bilimlerine Meraklı bir Âlim: Cârullah Efendi Koleksiyonu'nun Söyledikleri," in Açil, ed. 283–95.

Usluer, F. 2015 "Cârullah Efendi'nin Cifir ve Tıp İlimlerine Dair Kitapları," in Açil, ed. 297–312.

Veselý, R. 1996. "Bibliothek eines ägyptischen Arztes aus dem 16. Jhd. A.D./10. Jhd. A.H.," in Zemánek, P., ed. *Studies in Near Eastern Languages and Literatures. Memorial Volume of Karel Petrácek.* Prague: Oriental Institute, 613–30.

Vrolijk, A., Schmidt, J., and Scheper, K. 2012. *Turcksche boucken. De oosterse verzameling van Levinus Warner, Nederlands diplomaat in zeventiende-eeuws Istanbul/Turkish Books: The Oriental Collection of Levinus Warner, Dutch Diplomat in Seventeenth-Century Istanbul.* Eindhoven: Lecturis.

Witkam, J. J. 2007. "Gazing at the Sun: Remarks on the Egyptian Magician al-Buni and His Work," in Vrolijk, A. and Hogendijk, J. P., eds. *O ye Gentlemen: Arabic Studies on Science and Literary Culture, in Honour of Remke Kruk.* Leiden: Brill, 183–99.

Yeo, R. 2003. "Classifying the Sciences," in Porter, R., ed. *The Cambridge History of Science, Vol. 4: Eighteenth Century Science.* Cambridge: Cambridge University Press, 241–66.

VI.7

13TH/19TH-CENTURY NARRATIVES AND TRANSLATIONS OF SCIENCE IN THE SOUTH ASIAN ISLAMICATE WORLD

S. Irfan Habib and Dhruv Raina

Late 12th/18th- and 13th/19th-century South Asia was witness to an accelerated integration in globalizing processes, some of which had their origins in Asia while several others had their origins in modern Europe (Frank 1998). By the end of the 12th/18th century, at least two Asian empires, central to early modern history, had entered into a phase of political decline – the Ottomans and the Mughals. But as historical scholarship of the last couple of decades has often reminded us, the decline of the Mughal Empire was not accompanied by a decline in the spheres of cultural life – music, literature, painting, poetry and other art forms (Panikkar 1995, 34–53). In fact, throughout South Asia, a form of vernacular literary expression arose and competed, although never consciously so, with the older civilizational languages, namely, Persian and Sanskrit (Pollock 2003).

In the wake of the decline of the Mughal Empire, India saw the efflorescence of regional cultural and intellectual centers in the so-called princely states. Two of these were ruled by the Nizam of Hyderabad and the Nawabs of Awadh, who were for decades deeply involved with the vernacularization of modern science. Many of these Princely States – notably Mysore, Travancore, Cochin and Baroda – effectively used the relative autonomy they enjoyed to bring modern knowledge, particularly scientific knowledge, to those who were keen to seek it in their own languages (Habib 2010a). We will return to this in some detail during our later discussion.

It is during the 18th century, and into the 19th, that literary historians locate the golden age of Urdu literature – commencing in the poetry of Mīrzā Muḥammad Rafī Saudā᾽ (1713–1781), Khwāja Mīr Dard (1720–1785) and Mīr Taqī Mīr (1723–1810) and ending with Shaykh Muḥammad Ibrāhīm Zauq (1789–1854) and Mīrzā Asad Allāh Khān Ghālib (1212–1285/1797–1869). As the Mughal Empire descended deeper and deeper into an irresolvable crisis, this was the period when colonial interests and colonial power resolutely entrenched itself – 1857 is the moment when India was formally absorbed as a colony of the British Empire (Spears 1976, 229; Chand 1974, 467).

DOI: 10.4324/9781315170718-71

But if the expansion of imperialism through colonialism was one globalizing process that entrenched itself by attempting to usher in colonial modernity, it did so at the cost of disrupting the rhythm and evolution of a South Asian modernity (Hasan 2005; Pannikar 1995). Nevertheless, the standard historical narrative suggests that colonialism was also the vector for triggering the other globalizing process of modernization, namely, the institutionalization of the scientific and technological system (Raina and Habib 2008, 735–9). In the realm of knowledge and productive practices, colonialism was the vector for the expansion of the dominion of modern science and technology (Basalla 1967, 611–22; Kumar 1995). However, inspired by the four-decade-old historical sociology of science, it could be argued that with the political embedding of colonial interests in South Asia, there was a temporary disruption of literary communication and the circulation of ideas and influences within the different areas of the Islamicate world, that may, in turn, have given rise to or resulted in the emergence of new literary genres (Tavakoli-Targhi 2001). Some of these were responses to the new cultural and political influences of European provenance (Bayly 1997; Michaels 2001). Bayly has discussed in detail how the Indian *oikumene*, by which he actually means the North Indian or, more specifically, the Indo-Persianate culture and civilization, responded to the new information order (Bayly 1997; Inden 1997; Cohn 1990).

How was the new knowledge perceived, debated, assimilated, appropriated and rejected, if it was? (Cohn 1990; Dodson 2010) What were the ensuing debates about and what is it that triggered such a variety of responses? In discussing the Bengali Bhadralok, another regional and vernacular order, which, by the 19th century, was transcultural in more than one way, David Kopf spoke of four responses, namely, revivalism, status quoism, revitalism and Westernization (Kopf 1969). In discussing the North Indian Islamicate cultural region and the response of its inhabitants to the new knowledge, it is possible over a century to identify several historical narratives as responses to a changing world order and to the new constellations of knowledge, which are inseparable from the former.

The 19th century, for our present purposes, can be partitioned into three historical moments. First, we will discuss the impressions of a South Asian traveler in Europe. Second, we focus on the narratives of mathematics educators and the so-called Delhi renaissance from the mid-century. And third, we will look at the last quarter of the 19th century, discussing how the anticolonial struggle began to revise narratives about so-called Western knowledge.

VI.7.1 South Asian Impressions of Europe

We begin with the travelogues authored by members of the Indo-Persianate elite who carefully chronicled their impressions of Europe, and of the revolutions sweeping the worlds of politics and techno-scientific knowledge at the end of the 18th and the early decades of the 19th century (Subramanyan and Alam 2007). A figure we could consider as representative of the first third of the 19th century is Mīrzā Abū Ṭālib Khān (1752–1805/6) from Lucknow and his travels to Europe (Khan 1998). When speaking of him and other travelers of the time it is important to specify that their narratives are embedded in several frames – some with premodern Islamicate elements as well as those belonging to an Indo-Persianate modernity (Tavakoli-Targhi 2003). We identify four such elements in Abū Ṭālib's writing. The first is the expression of an older literary genre widespread in the Islamicate world that celebrates the marvelous in nature, in the creations of both man and God. This aesthetic of the marvelous is the lens through which the prince (Mīrzā) describes the novelties of Europe – both the sacred and profane (read technoscience; Raina and Habib 2005). This aesthetics is referred to within the Persian and Arab literary traditions as *ʿajāʾibāt* or *ʿajāʾib* (wonders, amazing things) and prefaces the titles of many books and articles published in Persian and Urdu from the beginning of the 19th century. This aesthetics beckon a readership belonging to the regional Islamicate world of letters to a new and fascinating world (Master Ramchandra 1847 and 1848). The aesthetics run through many of the writings of the prince, as well as of those of popularizers of modern science such as Master Ramchandra (1821–1880) and Munshī Zakāʾullāh ([1832–1911]; Habib 2000, 132–46; Habib and Raina 1993, 369–75). Beyond the marvels of science, there is a fundamental reckoning with the new mechanics and astronomy practiced in Europe, which the prince feels should be assimilated in South Asia (Habib 2010b). This naturally leads us to pose the deeper question: What are the terms of assimilation? From his writings, one can cull two elements of a theory of knowledge, never expressly articulated but running below the surface. The first observation the reader trained in the philosophy of science is alerted to is the absence of any notion that the new sciences have emerged out of a rupture, an epistemological discontinuity characterizing the modern sciences, that then marks its break with the past. This may derive from – and this is the second observation – a gradualist theory of knowledge. The idea is that the world of knowledge is subject to a process of unceasing and gradual evolution, which, in turn, has produced the development of the modern sciences – and hence, its emergence does not appear as a break with the past.

There is yet another aspect, which reflects certain continuities with a historiographic frame that circulated across the Islamicate world, and this had to do with the identity of Greece in contemporary memory. Abū Ṭālib as well as others of his ken, before and after him, were overwhelmed by the inclusion of Greece in the folds of Europe's historical memory (Habib 2010b). Clearly, there were other claimants, but more important, this appropriation was considered worthy of mention and possible contestation.

VI.7.2 The Delhi renaissance

Upon turning the clock forward 30 years, we encounter the oeuvres of two polymaths from the period of the Delhi renaissance. The first half of the 19th century saw the efflorescence in science, literature and culture, which was made possible through Urdu as the local language. This renaissance also witnessed the emergence of realist writing and journalism in Urdu. As pointed out in a number of earlier papers, these figures from the late 18th and early 19th centuries are influenced by several intellectual currents. One of these is Ibn Khaldūn's (732–808/1332–1406) views on history and society presented in his book called *Introduction* or *Prolegomena* (*Muqaddima*; [779/1377]), including the rise and fall of civilizations, which in turn provides them with the explanatory devices to understand South Asia in a 19th-century comparative context (Habib 2000; Habib and Raina 1989, 51–66). But as mathematics teachers themselves schooled within a 19th-century colonial regime, they were certainly shaped by the readings of the British Indologists on the history of Indian mathematics and logic – a history premised on the dichotomy of Western geometric traditions and Eastern algebraic ones (Raina 2008, 1934–44). This operates as a theory of deficit in their pedagogical endeavors and historical writing while at the same time nourishing, possibly even inventing, historical memories of an ancient and glorious mathematical heritage (Raina 2016, 25–38). And finally, internalism – history limited to any of the sciences and their supposed development independent of the larger culture – was not a viable historiographic frame for engaging with the historical and cultural concerns that animated the times. On the contrary, the perceived decline of the Mughal Empire gave rise to discourses of lamentation and loss, of the inevitable passing of a world (Dehalvi 2017).

The early decades of the 19th century also witnessed a remarkable interest in the promotion of science through translation as well as original writing in Urdu in some of the princely states. Two prominent states as far as such activity was concerned were those of Hyderabad and Lucknow. Shams al-Umarā' II (Anglo-Indian transliteration: Shams-ul-Umra), Amīr-i Kabīr I (1781–1863) of Hyderabad occupies a unique place as an enthusiastic promoter of modern science. He was a senior official of the Niẓām of Hyderabad but was more interested in intellectual pursuits than routine administrative matters. In 1825, he established a press called Maṭbū'at-i sangī (Lithographic Press), which aimed at publishing translations of modern scientific and mathematical texts as well as original Urdu writings on science.[1] He collected useful scientific texts from England, France and other European nations and even employed some experts in these languages to help translate these books into Urdu. He was convinced that the veneration of the ancient sciences and the neglect of modern science and technology – which had led to the economic and industrial progress of Europe –would close all doors to future development (Shakeel Khan 1988, 35). We can call him one of the first Indian pioneers promoting the modern sciences in Urdu in the first half of the 19th century. The forum for the translation of European texts on science and other literature was the Dār al-tarjuma (Translation Bureau), established in 1834 in the palace called *Jahān-numā*.[2] Shams al-Umarā' II remained passionately involved in the activities of the Bureau and appointed professional translators, including some British and French intellectuals who could interpret and help translate the European scientific works into Urdu. Hashmi writes that Shams al-Umarā' IV could read and understand English and French:

> Shams ul-Umra, who was familiar with English and French among the European languages, founded a Translation Bureau where science books were translated. Poets, writers, philosophers and mathematicians, which included both Hindus as well as Muslims, and also the British and the French, all the time, surrounded him.
>
> *(Hashmi 1963, 207)*

Besides the *Dār al-tarjuma* and the Sangī Press, Fakhr al-Dīn Muḥammad founded a college and an astronomical observatory in the 1830s. The college, known as *Madarsa-yi-Faqria*, organized its curriculum around books that had been translated by the Translation Bureau. This was the first college that gave priority to imparting modern scientific knowledge in Urdu over the traditional curricula that prevailed in the madrasas all over India. The observatory was established to undertake experiments and make observations using modern instruments, which were imported from Europe (Hashmi 1963, 37). The palace had a good library where a large number of useful collections on science were available in several languages, including Urdu, Arabic, Persian, English and even French. Part of this collection is still extant in the Paigah Library at Hyderabad. This effort in translation into and teaching in Urdu, led by Shams al-Umarā' IV, extended from 1833 to 1863 and was supported by his descendants for more than 100 years.

The Princely State of Awadh (Anglo-Indian transliteration: Oudh) had a shorter but significant encounter with science writing and translation in Urdu. For this period, Awadh is known for the profligate and decadent lifestyle of its rulers and their extravagance rather than serious scholarly activity. At best, historians of the 19th century recognized the contributions of Wājid 'Alī Shāh (1822–1887) to the promotion of music and culture. However, this view is not particularly reliable (Llewellyn Jones 2014, 74). It was one widely promoted by British officials, consciously ignoring the constructive intellectual efforts of the nawabs of Awadh. The following paragraphs contain brief profiles of some of these Awadh nawabs in promoting modern science through Urdu.

In Lucknow, the short era of vernacularization of science in Urdu began with Nawāb Ghāzī al-Dīn Ḥaydar (1769–1827), but it was Naṣīr al-Dīn Ḥaydar (1803–1837), who played the most significant role in the 1830s. This era of science writing and translation in Urdu ranges between 1832 and 1853. Naṣīr al-Dīn Ḥaydar, Amjad ʿAlī Shāh (1801–1847) and Wājid ʿAlī Shāh composed several useful treatises themselves. The foundations of modern science and technology education in the region were laid during the reign of Ghāzī al-Dīn Ḥaydar following the establishment of a press called Maṭbuʿat-i sulṭānī (The Royal Press), where most of the texts were published during the following few decades (Shakeel Khan 1988, 74).

Amjad ʿAlī Shāh was himself an enthusiastic stargazer and was instrumental in founding a *Raṣad khāna* (an astronomical observatory that also published books). Colonel Richard Wilcox was appointed astronomer in 1841 (Llewellyn Jones 2014, 102–3). A large number of useful scientific texts were acquired for the *Raṣad khāna* and were translated into Urdu, in addition to which original treatises written in Urdu were commissioned. Colonel Wilcox wrote a book on the astronomical research pursued at the observatory, and the Awadh state contributed 7,000 rupees for its publication. Amjad ʿAlī Shāh also took the initiative and established a *Raṣad khāna-yi maqnatisī* (an observatory to work on magnetism and other related problems of physics). Interestingly, one Babu Rasik Mohan (19th century) from Bengal manufactured a clock for the *Raṣad khāna* (Llewellyn Jones 2014, 77).

Wājid ʿAlī Shāh took over the reins in 1847 after his father's death and immersed himself in several intellectual pursuits. The collection of books in the royal library went up to 200,000 volumes during his brief rule. The British government appointed Aloys Sprenger (1813–1893), the principal of Delhi College (1847–1850), to prepare a catalogue of all the books in the Royal Library. Sprenger spent 18 months preparing the catalogue of four volumes. Only one of the volumes was finally published, and little is known about the other three. Sprenger highly appreciated the libraries and the caretakers of these repositories of precious books (Rizvi 1977, 86; Shakeel Khan 1988).

These were the first organized efforts in India to introduce modern science through translation into Urdu. Local and regional presses were still rare and increasing control by the East India Company meant that it was hostile to such enterprises. In such a milieu, Shams al-Umarāʾ IV and the Oudh nawābs used their personal as well as state resources to ensure that science and mathematics in Urdu remained in circulation, utilizing the expertise of not only Indian experts but also Europeans who shared a sympathy with the spirit of the projects.

The post-1857 period was witness to a radical attitudinal shift in the thinking of the Muslim elite, which was led from the front by Sayyid (Anglo-Indian transliteration: Syed) Aḥmad Khān (1817–1898).[3] Prior to 1857, Sayyid Aḥmad, like most of the Muslim scholars of the period, spent his entire intellectual energy "in seeking to escape reality by having recourse to dreams of the golden age when Islamic civilization flourished in India" (Baljon Jr. 1964, 7). His two most important works, marking this stage in his career, are *The Remnants of Ancient Heroes* (*Āthār al-ṣanādīd*) (1847), an inventory and description of the old monuments of Delhi and its surroundings with an account of the famous people who once resided in Delhi.[4] He was also responsible for a recension of *The Constitution of Akbar* (*Āʾīn-ī akbarī*; Anglo-Indian transliteration: *Ain-i-Akbari*, 1856) authored by Abū l-Fażl (958–1011/1551–1602) who was an administrator in the Emperor Akbar's (r. 963–1014/1556–1605) government and leading Mughal intellectual of the 10th/16th and 11th/17th centuries.[5]

After 1857, Sayyid Aḥmad Khān consciously turned his attention toward influencing the perspective of Muslims responding to the changing times. The new emphasis was not just on acquiring a modern education but, more importantly, on inculcating a rational approach to reading and understanding the Qurʾān and the prophetic tradition. He realized early that a Muslim needed to comprehend the faith as dynamic, and not as something frozen in time.

If we look at Islam globally as well as in the Indian context today, it becomes evident that these 19th-century efforts of modernists failed to generate the response they were intended to produce. These modernists within the Islamic scholarly traditions in South Asia, including Sir Sayyid Aḥmad Khān, contested the orthodoxies of their time, raising vital questions around reason and faith.

For Sayyid Aḥmad Khān, there was no dichotomy between the 'word' and the 'work' of God in a natural religion like Islam (Syed Ahmad Khan 1963, X: xx). The proposed hermeneutic for interpreting the scripture was set out in the following steps (Troll 1978, 168–70). First, a close inquiry be made into the use, meaning and etymology of Qur'ānic language so as to yield the true meaning of the word and passage in question. Second, in order to choose between several interpretations of a passage from the scriptures, one must first ask if its truth was established by science. Such truth as is arrived at by rational proof (*dalīl-i 'aqlī*) demands firm belief. Third, if on the other hand the apparent meaning of the scriptures conflicts with demonstrable conclusions, it must be interpreted metaphorically. In this, Sir Sayyid Aḥmad Khān follows Ibn Rushd (520–595/1126–1198) in his problem of reconciling demonstrative truth (*ma'qūl*) with scriptural truth (*manqūl*). Yet he makes clear that such metaphorical and allegorical interpretation was precisely what the author of the scriptures intended.

This hermeneutic was challenged by the detractors of Sir Sayyid Aḥmad Khān, who saw this as some sort of perversion of Islam. In the 13th/19th century, efforts to bring in reason and rationalism in the reading and understanding of Islam almost failed due to the sway of *imitation* (*taqlīd*), entrenched since the 5th–6th/11th–12th centuries. This does not mean that there were no other theological viewpoints among Islamic scholars in South Asia over the centuries. Scholars like Shāh Walī Allāh (1114–1176/1703–1762) in 12th/18th-century Delhi tried to invoke *ijtihād*, and Sayyid Aḥmad was inspired by him. Jamāl al-Dīn Afghānī (1254–1314/1838–1897) and his disciple in Egypt Muḥammad Abduh (1265–1323/1849–1905) felt the need to stress the need to bring back *ijtihād*. But the dominant narrative in the second half of the 13th/19th century was that of *taqlīd*. The modernists in India and elsewhere had to confront an already well-established doctrinal system that emphasized the need for living faithfully by the dictates of tradition, with no critical engagement or *ijtihād*. Sir Sayyid Aḥmad Khān and other 13th/19th-century modernists were not promoting anything that was alien to Islam or to its history. The rationalist tradition within Islam was represented by the intellectual group called the Mu'tazilites from the 2nd/8th century who argued for free will over fatalism and quoted Qur'ānic verses showing God's displeasure with an inactive mind. According to one such verse, "the worst of creatures for Allāh are those who are (willfully) deaf and dumb, those who will not reason" (De Bellaigue 2017, xxix). This school had its eminent devotees in al-Fārābī (d. 339/950–1), Ibn Sīnā (d. 428/1037) and Ibn Rushd (Tibi 2012, 187). Within the Islamicate world the school failed to institutionalize or anchor itself within the world of learning (for different perspectives, see Chapters I.11 and III.3–III.4). In Islam instead, much before Sayyid Aḥmad and other 13th/19th-century modernists, the world of knowledge was classified broadly into two categories – praiseworthy and blameworthy knowledge. The first alluded to Islamic knowledge that encompassed the sciences, these were called the *sharī'a* sciences, while the latter referred to the so-called foreign sciences that had been derived from the processes of Hellenization and through other processes of cultural and intellectual transmission. The *'ulamā'* created a hostile distinction between the alien sciences or sciences of the ancients and the Islamic sciences. A section of the *'ulāmā'* advised Muslims to keep away from blameworthy knowledge or the alien sciences that could have kept this knowledge system dynamic and forward-looking.

VI.7.3 Institutional efforts at revitalization

When we fast forward to the end of the 19th and the early decades of the 20th century, we encounter Ḥakīm Ajmal Khān (1863–1928), the founder of the *Unānī ṭibb* (Greek medicine) and Ayurvedic College of Medicine and Surgery (Chapters V.7–V.8). The college still exists and reflects the Ḥakīm's attitude and that of several of his South Asian contemporaries toward the critical assimilation of those elements of modern medical practice that were found wanting in Unani and Ayurvedic systems of medicine. Ajmal Khān[6] and other revitalists[7] from this period were the bridge builders between different knowledge systems. The former's energies were dedicated toward revitalizing the two medical systems of South Asia in the light of modern medical developments, particularly in the domains of anatomy and surgery.

Recognizing the need for institutionally reforming the practice of traditional medicine, Ajmal Khān set in the process of revising the curriculum of Unānī medicine, a process initiated by his brother Ḥakīm ʿAbd al-Majīd Khān (d. 1901). This formalization of the curriculum moved Unani beyond the expertise of a few families of *ḥakīms*, that is Muslim physicians trained in various domains of Greco-Arabic and Greco-Persian medicine (Chapter V.9). The *Ṭibbiyya College* in Delhi was the institutional manifestation of Ajmal Khān's urge to modernize and rejuvenate the Unani as well as Ayurvedic systems. He drew up an inventory of the limitations of the two medical systems and proposed that the path to their modernization was through the incorporation of modern surgical knowledge and practice (Speziale 2009 [2011]). These attempts were condemned by his opponents as reprehensible revisions, and his family was denounced as infidel (*kāfir*) and apostate or atheist (*murtadd*). Those opposing him belonged to the same factions that wrote *fatwas* against Sayyid Ahmad Khān and mocked him by labeling him a "natury"[8] for his emphasis on faith in reason and the systematization of knowledge along modern lines. However, Ajmal Khān and the family were unfazed by the criticism and took a step further by launching a newspaper called *The Most Complete News* (*Akmal al-akhbār*), which was managed by Munshī Bihārīlāl Mushtāq (1835–1894), a Hindu disciple of the poet laureate Mīrzā Asad Allāh Khān Ghālib. The objective was to defend reform in medicine and surgery and to report on matters of general public interest. In 1902 Ajmal Khān launched yet another organ called the *Medical Journal* (*Majallat-i ṭibbiyya*), a monthly with news of the school and essays on *ṭibb* (Qarshī 1928, 22).

He too, like Ramchandra and ZakāʾAllāh, operated within the practice of the sciences framed by a theory of deficit. But Ajmal Khān as an important figure involved in the anticolonial freedom struggle, as well as the revitalization of both Unani and Ayurveda, envisioned these two distinct medical systems as composite knowledge systems of South Asia, and their route to revitalization was through the sciences (Habib and Raina 2005). It was probably in the last decades of the 13th/19th century that Jamāl al-Dīn al-Afghānī visited India and, in lectures delivered in Kolkata, alluded to this sense of a composite Asian identity or heritage – these were indeed new ideas circulating in the colonial world and beyond (Mishra 2012).

Scholarship on the Islamicate world has long since abandoned serious engagement with the frames of 'Islamic backwardness and anti-modernism'. Edward Said's (1935–2003) trenchant critique of Bernard Lewis's (1916–2018) *What Went Wrong? Western Impact and Middle Eastern Response* identifies the elements assembled together in the making of this picture. We raise this issue here in refutation of Lewis's unfounded generalization which he extends across the Muslim world that in the 19th century Muslims were primarily concerned with the art of warfare, a preoccupation arising from the reckoning that something had gone wrong (Said 2002). On the contrary, the history of the Muslim world of learning in South Asia reveals a lively and animated *oikumene* responsive to the developments in diverse fields of knowledge. In compiling this account, we do not intend to suggest that the narratives discussed here were the most popular narratives and dominant discourses among Muslims in 13th/19th-century South Asia. There were many oppositional voices, as well as a rich and diverse discourse, about modernization and modern knowledge. We have just culled out one strand that we think relevant, a strand that was woven into a tapestry of a discourse on Eastern and Western knowledge, whose participants were those involved in the dissemination of new knowledge and its practice.

Notes

1 British officials also realized that Urdu could play a crucial role in reaching out to a large number of people, and thus, John Gilchrist made some remarkable attempts at translation of European texts into Urdu. When these texts reached Hyderabad, Shams-ul-Umarāʾ rejected such efforts saying that scientific and technical books needed to be translated and not such meaningless stories, which may be of use to the British for governance but would not benefit the people.

2 But Shams-ul-Umarāʾ did not confine himself to the translation and transmission of mere Western scientific knowledge. He also got some useful Persian texts translated into Urdu.

3 Sayyid Aḥmad Khān was an educationist and social reformer who initiated a movement among the Muslims of India in the middle of the 19th century, emphasizing the need for modernization, particularly through the inculcation of a modern scientific education. He was the founder of Aligarh Muslim University at Aligarh.

4 The orientalist Joseph Héliodore Sagesse Vertu Garcin de Tassy (1794–1878) translated this archaeological work into French. The translation drew the attention of European scholars to Sayyid Aḥmad Khān's work, and he was elected Honorary Fellow of the Royal Asiatic Society, London.

5 Sayyid Aḥmad Khān approached several people, including the famous Urdu poet Mīrzā Asad Allāh Khān Ghālib, to contribute to the introduction of his edited version of Ain-i-Akbari. Ghālib saw this work as futile, wallowing in the past, and wrote to Sayyid Aḥmad Khān:

> Look at the Sahibs of England...They have gone far ahead of our oriental forbears. Wind and wave they have rendered useless. They are sailing their ships under fire and steam. They are creating music without the help of mizrab (plucker). With their magic, words fly through the air like birds...Cities are being lighted without oil lamps. This new law makes all other laws obsolete. Why must you pick up straws out of old, time-swept barns while a treasure-trove of pearls lies at your feet?
>
> (Hyder and Jafri 1970, 28)

6 Ajmal Khān belonged to an illustrious family of physicians of Delhi known as Sharīfī family after his ancestor Ḥakīm Muḥammad Sharīf Khān (d. *c.* 1222/1807) (Speziale 2009 [2011]). He was born in Delhi in 1863 and was the son of Ḥakim Maḥmūd Khān, himself a known physician. Ajmal Khān was given the best possible education and had the privilege of using his father's exhaustive library to enrich his knowledge. His mental horizon was further widened during his several trips abroad, not only to the Muslim countries of West Asia but also to Europe. He was closely involved in the freedom struggle along with Mahatma Gandhi.

7 The revitalization movement was part of a general cultural-intellectual regeneration taking place during the late 13th/19th and early 14th/20th centuries. This was not an isolated movement limited to the field of medicine or confined to a particular region of India. In medicine there is the example of Vaidyaratnam P. S. Varier (1869–1944), who was not only a contemporary of Ajmal Khān but their program of revitalization also exhibited many similarities. There is no evidence to show that they ever met or heard of each other. For P. S. Varier see (Panikkar 1992, 283–308).

8 This label refers to debates connected to Darwin's theory of evolution in India and Jamāl al-Afghānī's (1253–1314/1838–1897) denunciation of Sayyid Aḥmad Khān's teachings as materialist or naturalist (see Keddie 1968, 175–87).

Bibliography

Sources

al-Hyder, Q. and Jafri, A. S. 1970. *Ghalib and His Poetry*. Bombay: Popular Prakashan.

Qarshī, Ḥakīm M. Ḥ. 1928. *Tazkira-i Masīḥ al-Mulk* [The Memoir of Masīḥ al-Mulk]. Lahore: n. p.
Ramchandra, Master. 1847. *'Ajā'ibāt-i rūzgār* [The Wonders of the World]. Delhi: n. p.
Ramchandra, Master. 1848. *Tazkirat al-kāmilīn* [Biography of Eminent Persons]. Delhi: n. p.
Syed Ahmad Khan. ed. Shaykh Panipati, M. I.1963. *Maqālāt-i Sir Syed* [Discourses of Sir Syed]. 16 vols. Lahore: Majlis-i taraqqi-yi adab.

Research literature

Baljon Jr., J. M. S. 1964. *Syed Ahmad Khan*. Lahore: Ashraf Press.
Basalla, G. 1967. "The Spread of Western Science," *Science* 154: 611–22.
Bayly, C. A. 1997. *Empire and Information: Intelligence Gathering and Social Communication in India, 1780–1870*. Cambridge: Cambridge University Press.
Chand, T. 1974. *History of the Freedom Movement in India*. Vol. 2. New Delhi: Publications Division, Government of India.
Cohn, B. S. 1990. *Colonialism and its Forms of Knowledge: The British in India*. New Delhi: Oxford University Press.
De Bellaigue, C. 2017. *The Islamic Enlightenment: The Struggle Between Faith and Reason*. London: Bodley Head.
Dehalvi, Z. tr. Safvi, R. 2017. *Dastan-e-Ghadar: The Tale of the Mutiny*. Delhi: Penguin Books.
Dodson, M. S. 2010. *Orientalism, Empire and National Culture: India: 1770–1780*. Cambridge: Cambridge University Press.
Frank, A. G. 1998. *ReOrient: Global Economy in the Asian Age*. Oakland, CA: University of California Press.
Habib, S. I. 2000. "Munshi Zakaullah and the Vernacularisation of Science in 19th century India," in Sehgal, N., *et al.*, eds. *Uncharted Terrains: Essays on Science Popularisation in Pre-Independence India*. New Delhi: Vigyan Prasar, 132–46.
Habib, S. I. 2010a. "Vernacularization of Modern Science: An Experiment in Urdu During the Nineteenth Century in India," in Jain, A., ed. *Science and the Public*. New Delhi: Centre for Studies in Civilizations, 281–92.
Habib, S. I. 2010b. "Mirza Abu Talib and his European Sojourn: An Indian Savant's Encounter with Modernity and Science in the Early 19th Century," in Bandopadhyay, A., ed. *Science and Society in India 1750–2000*. Delhi: Manohar Publishers, 83–94.
Habib, S. I. and Raina, D. 1989. "Copernicus, Columbus, Colonialism, and the Role of Science in Nineteenth Century India," *Social Scientist* 17.3–4: 51–66.
Habib, S. I. and Raina, D. 1993. "The Discourse on Scientific Rationality: A Study of Master Ramchandra," in Niranjana, T., Sudhir, P. and Dhareshwar, V., eds. *Interrogating Modernity: Culture and Colonialism in India*. Kolkatta: Seagull Books, 369–75.
Habib, S. I. and Raina, D. 2005. "Reinventing Traditional Medicine: Method, Institutional Change, and the Manufacture of Drugs and Medication in Late Colonial India," in Alter, S. J., ed. *Asian Medicine and Globalization*. Philadelphia: University of Pennsylvania Press, 67–77.
Hashmi, N. 1963. *Daccani Culture*. Lahore: Majlis-i taraqqi-yi adab.
Inden, R. 1997. *Imagining India*. London: Basil Blackwell.
Keddie, N. 1968. *An Islamic Response to Imperialism*. Berkeley: California University Press.
Khan, G. 1998. *Indian Muslim Perceptions of the West during the Eighteenth Century*. Karachi: Oxford University Press.
Kopf, D. 1969. *British Orientalism and the Bengal Renaissance: The Dynamics of Indian Modernisation, 1773–1835*. Kolkatta: Firma KLM.
Kumar, D. 1995. *Science and the Raj*. New Delhi: Oxford University Press.
Lewis, B. 2001. *What Went Wrong? Western Impact and Middle Eastern Response*. Oxford: Oxford University Press.
Llewellyn Jones, R. 2014. *The Last King in India: Wajid Ali Shah*. New Delhi: Random House India.
Michaels, A., ed. 2001. *The Pandit: Traditional Scholarship in India*. New Delhi: Manohar Publishers.
Mishra, P. 2012. *From the Ruins of Empire: The Revolt Against the West and the Remaking of Asia*. London: Allen Lane, Penguin Books.
Mushirul Hasan. 2005. *A Moral Reckoning: Muslim Intellectuals in Nineteenth-Century Delhi*. New Delhi: Oxford University Press.
Panikkar, K. N. 1975. Presidential Address to the *36th Session of the Indian History Congress*, Section 3. December 1975 (Reprint as "Cultural Trends in Pre-colonial India," in Panikkar, K. N. 1995. *Culture, Ideology and Hegemony – Intellectual and Social Consciousness in Colonial India*. New Delhi: Tulika, 34–53).

Panikkar, K. N. 1992. "Indigenous Medicine and Cultural Hegemony: A study of the Revitalization Movement in Keralam," *Studies in History* 8.2: 283–308.

Pollock, P. 2003. "Cosmopolitan and Vernacular in History," *Public Culture* 12.3: 591–625.

Raina, D. 2008. "Science East and West," in Selin, H., ed. *Encyclopedia of Science, Technology and Medicine in Non-Western Cultures*. Heidelberg: Springer-Verlag, 1934–44.

Raina, D. 2016. "After Exceptionalism and Heritage: Thinking Through the Multiple Histories of Knowledge," in Brentjes, Edis and Richter-Bernburg [CB],[9] 25–38.

Raina, D. and Habib, S. I. 2005. "The Voyages of Abu Talib Khan and Le Gentil: A Preliminary Comparative Study of the Scientific Imagination in Eighteenth Century Indian and French Travelogues," in Saldana, J. J., ed. *Science and Cultural Diversity: Proceedings of the 21st International Congress of History of Science*. México, DF: Universidad Nacional Autónoma de México and Sociedad Mexicana de Historia de la Ciencia y la Tecnología, A.C., 2794–809.

Raina, D. and Habib, S. I. 2008. "Colonialism, Nationalism and the Institutionalisation of Science in India," in Gopal, S. and Tikhvinsky, S. L., eds. *History of Humanity: Scientific and Cultural Development*. Vol. 7: *The Twentieth Century*. Abingdon, Oxon: UNESCO/Routledge, 735–9.

Rizvi, M. H. 1977. *Wajid Ali Shah*. Lucknow: Nami Press.

Sadiq, M. 1964. *A History of Urdu Literature*. Oxford: Oxford University Press.

Said, E. 2002. "Impossible Histories: Why the Many Islams Cannot be Simplified," *Harper's*, July.

Shakeel Khan, M. 1988. *Urdu mein Scienci wa Takniki Adab*. New Delhi: Educational Publishing House.

Spears, P. 1976. *The Oxford History of Modern India, 1740–1947*. New Delhi: Oxford University Press.

Speziale, F. 2009 [2011]. "INDIA xxxiii. Indo-Muslim Physicians," www.iranicaonline.org/articles/india-xxxiii-indo-muslim-physicians (Accessed 5 June 2019).

Subramanyan, S. and Alam, M. 2007. *Indo-Persian Travels in the Age of Discoveries, 1400–1800*. Cambridge: Cambridge University Press.

Tavakoli-Targhi, M. 2001. *Refashioning Iran: Orientalism, Occidentalism and Historiography*. London: Palgrave Macmillan.

Tavakoli-Targhi, M. 2003. "Orientalist studies and its Amnesia," in Kaiwar, V. and Majumdar, S., eds. *Antinomies of Modernity-Essays on Race, Orient, Nation*. Durham: Duke University, 98–125.

Tibi, B. 2012. *Islamism and Islam*. New Haven: Yale University Press.

Troll, C. W. 1978. *Sayyid Ahmad Khan: A Reinterpretation of Muslim Theology*. New Delhi: Vikas Publishing House.

CONSOLIDATED BIBLIOGRAPHY

Sources

Abū Maʿshar al-Balkhī. 1985. *Al-Madkhal al-kabīr ilá ʿilm aḥkām al-nujūm*. (Series C, v. 21). Frankfurt am Main: Institut für Geschichte der arabisch-Islamischen Wissenschaft.

Abū Maʿshar al-Balkhī. ed. Lemay, R. 1995–1966. *Kitāb al-Madkhal al-kabīr ilā ʿilm aḥkām al-nujūm* [The Great Introduction to Astrology]. 9 vols. Naples: Instituto Universitario Orientale.

Abū Maʿshar. ed., tr. and ann. Yamamoto, K. and Burnett, C. 2019. *The Great Introduction to Astrology by Abū Maʿšar*. 2 vols. (Islamic Philosophy, Theology and Science. Texts and Studies 106). Leiden and Boston: Brill.

Banū Mūsā. tr. and com. Hill, D. R. 1979. *"The Book of Ingenious Devices" (Kitāb al-Ḥiyal) by The Banū (sons of) Mūsā bin Shākir*. Dordrecht: Reidel.

al-Battānī. ed. Nallino, C. A. 1899–1907 [1977]. *Al-Battānī sive Albatenii Opus astronomicum (al-Zīj al-Ṣābiʾ)*. 3 vols. Milan: Ulrich Hoepli. [Reprint; 1977. *Opus astronomicum: drei Bände in einem Band*. Hildesheim: Olms.]

al-Fārābī, Abū Naṣr. ed. Amīn, ʿU. 1948. *Iḥṣāʾ al-ʿulūm* [Enumeration of the Sciences]. Cairo: Dār al-fikr al-ʿarabī.

al-Fārābī. ed. ʿAlī Bū Mulḥim. 1996. *Iḥṣāʾ al-ʿulūm* [Enumeration of the Sciences]. Beirut: Dār wa-maktabat al-hilāl.

al-Ghazālī. ed. and tr. Marmura, M. E. 2000². *The Incoherence of the Philosophers/Tahāfut al-falāsifa. A Parallel English-Arabic Text*. Provo (Utah): Brigham Young University Press.

Ḥājjī Khalīfa. 1402/1982. *Kashf al-ẓunūn ʿan asāmī al-kutub wa-l-funūn* [Removal of Doubts About the Names of the Books and the Disciplines]. 6 vols. Beyrouth: Dār al-fikr.

Ḥunayn ibn Isḥāq. ed. and tr. Lamoreaux, J. C. 2016. *Ḥunayn ibn Isḥāq on His Galen Translations: A Parallel English–Arabic Text*. (Eastern Christian Texts.) Provo, Utah: Brigham Young University Press.

Ibn ʿAbdūn. ed. and tr. Djebbar, A. 2005 and 2006. "*Al-Risāla fī l-taksīr* li-Ibn ʿAbdūn, shāhid ʿalā l-mumārasāt al-sābiqa li-l-taqlīd al-jabrī al-ʿarabī [A Chapter on Measurement by Ibn ʿAbdūn: Evidence for Practices Preceding the Arabic Tradition in Algebra]," *Suhayl* 5: 7–68 and 6: 81–6.

Ibn Abī Uṣaybiʿa, Muwaffaq al-Dīn Aḥmad ibn al-Qāsim. ed. Müller, A. 1884 [1995]. *ʿUyūn al-anbāʾ fī ṭabaqāt al-aṭibbāʾ* [Sources of Information on the Classes of the Physicians]. ed. Königsberg: Müller. [Reprint of the edition Cairo 1882/1299. 1995. Frankfurt am Main: Institute for the History of Arabic-Islamic Science at the Johann Wolfgang Goethe University]

Ibn Abī Uṣaybiʿa. ed. Riḍā, N. 1965. *ʿUyūn al-anbāʾ fī ṭabaqāt al-aṭibbāʾ*. Beirut: Manshūrāt Dār Maktabat al-Ḥayāt.

Ibn Abī Uṣaybiʿa. ed. Najjār, ʿĀ. n. d. *ʿUyūn al-anbāʾ fī ṭabaqāt al-aṭibbāʾ* [*The Source Report on the Ranks of Physicians*]. 5 vols. Cairo: al-hayʾa al-miṣriyya al-ʿāmma li al-kitāb.

Ibn Abī Uṣaybiʿa, Muwaffaq al-Dīn Aḥmad ibn al-Qāsim. tr. Kopf, L. 2011. *History of Physicians*. 4 vols. Online (from an unpublished manuscript). http://www.tertullian.org/fathers/ibn_abi_usaibia_00_eintro.htm.

Ibn Abī Uṣaybiʿa. eds. and trs. Savage-Smith, E., Swain, S. and van Gelder, G. J. 2020. *A Literary History of Medicine: The ʿUyūn al-anbāʾ fī ṭabaqāt al-aṭibbāʾ of Ibn Abī Uṣaybiʿah*. Leiden: Brill.
https://dh.brill.com/scholarlyeditions/library/urn:cts:arabicLit:0668IbnAbiUsaibia/

Ibn Ḥabīb. ed. and tr. Kunitzsch, P. 1994 and 1997. "ʿAbd al-Malik ibn Ḥabībs *Book on the Stars*," *Zeitschrift für Geschichte der arabisch-islamischen Wissenschaften* 9: 161–94; 11: 179-88.

Ibn Juljul. ed. Sayyid, A. F. 1955. *Ṭabaqāt al-aṭibbāʾ wa-l-ḥukamāʾ*. [*Les générations des médecins et des sages*]. Al-Qāhira: Institut Français d'Archéologie Orientale.

Ibn Juljul. ed. Sayyid, A. F. 1985. *Ṭabaqāt al-aṭibbāʾ wa-l-ḥukamaʾ* [The Generations of the Physicians and the Sages]. Beirut: Muʾassasat al-Risāla.

Ibn Khaldūn. ed. Quatremère, M. 1858. *Al-Muqaddima* [The Introduction]. 3 vols. Paris: Didot.

Ibn Khaldūn. tr. Rosenthal, F. 1958 [1967]. *The Muqaddimah. An Introduction to History*. 3 vols. New York: Pantheon Books. [Reprint: 1967]

Ibn Khaldūn. ed. Shidādī, ʿA. 2005. *Al-Muqaddima* [Prolegomena]. Casablanca: Bayt al-funūn wa-l-ʿulūm wa-l-ādāb.

Ibn al-Nadīm, Muḥammad ibn Isḥāq. ed. and annot. by Flügel, G. in collaboration with Roediger, J. and Mueller, A. 1871-1872. *Kitāb al-Fihrist* [The Book of the Catalog]. 2 vols. Leipzig: Verlag von F. C. W. Vogel.

Ibn al-Nadīm, tr. Dodge, B. 1970. *The Fihrist of al-Nadīm. A Tenth-Century Survey of Muslim Culture*. 2 vols. New York: Columbia University Press.

Ibn al-Nadīm. ed. Tajaddud, R. 1393/1973. *Kitāb al-Fihrist* [The Book of the Catalog]. Tehran: Marwī.

Ibn al-Nadīm, Muḥammad ibn Isḥāq. Ed. Ramaḍān, I. 1415/1994. *Kitāb al-Fihrist* [The Book of the Catalog]. Beirut: Dār al-maʿrifa.

Ibn al-Nadīm, Muḥammad ibn Isḥāq. ed. Sayyid, A. F. 2014². *Kitāb al-Fihrist* [The Book of the Catalog]. London: Muʾassasat al-Furqān li-l-turāth al-islāmī.

Ibn al-Qifṭī, Abū l-Ḥasan ʿAlī ibn Yūsuf. eds. Müller, A. and Lippert, J. 1903. *Ibn al-Qifṭī's Taʾrīḫ al-ḥukamāʾ* [Ibn al-Qifṭī's History of the Sages]. Leipzig: Dieterich.

Ibn Sīnā. 1298/1880. "*Risāla fī aqsām al-ʿulūm al-ʿaqliyya*," in Ibn Sīnā. *Tisʿ rasāʾil fī l-ḥikma wa-l-ṭabīʿiyyāt wa-fī ākhirihā qiṣṣat Salāmān wa-Absāl* ["The Epistle on the Parts of the Rational Sciences," in *Ten Epistles on Metaphysics and Natural Philosophy*]. Constantinople: Maṭbaʿat al-Jawāʾib, 71–80 [Reprint: n.d. Cairo: Dār al-ʿarab, 104–8.]

Ikhwān al-Ṣafāʾ[The Brethren of Purity]. ed. Buṭrus Bustānī, B. 1957 [2008]. *Rasāʾil Ikhwān al-Ṣafāʾ* [The Epistles of the Brethren of Purity]. 4 vols. Beirut: Dār Ṣādir. [Reprint: 2008]

al-Jazarī. ed. al-Hassan, A. Y. 1979. *Kitāb fī maʿrifat al-ḥiyal al-handasiyya. "The Book of Knowledge of Ingenious Mechanical Devices" by al-Jazarī*. Aleppo: Institute for the History of Arabic Science.

al-Jazarī. tr. Hill, D. R. 1974. *The Book of Knowledge of Ingenious Mechanical Devices. Kitāb fī maʿrifat al-ḥiyal al-handasiyya by Ibn al-Razzāz al-Jazarī*. Dordrecht and Boston: D. Reidel.

Maimonides, tr. Munk, S. 1856. *Le guide des égarés: traité de théologie et de philosophie*. 3 vols. Paris: A. Franck.

Maimonides. ed. and tr. Gorfinkle, J. I. 1912. *The Eight Chapters of Maimonides on Ethics*. New York: Columbia University Press.

Maimonides. eds. Munk, S. and Joel, I. 5691/1930–1. *Dalālat al-ḥāʾirīn* (Guide of the Perplexed). Jerusalem: J. Junovitch.

Maimonides. tr. Pines, S. 1963. *The Guide of the Perplexed*. 2 vols. Chicago: University of Chicago Press.

al-Nuwayrī. 1380/1960(?). *Nihāyat al-arab fī funūn al-adab* [Fulfilled Ambition Concerning the Branches of a Complete Education]. 12 vols. Cairo: Wizārat ath-thaqāfa wa-l-irshād al-qawmī.

al-Nuwayrī, Aḥmad ibn ʿAbd al-Wahhāb. ed. Zakī, M. 1395–1410/1975–1990. *Nihāyat al-arab fī funūn al-adab* [Fulfilled Ambition Concerning the Branches of a Complete Education]. Cairo: Dār al-Kutub al-Miṣriyya.

al-Nuwayrī, Aḥmad ibn ʿAbd al-Wahhāb. ed. anonymous. 1342–1418/1923–1998. *Nihāyat al-arab fī funūn al-adab* [Fulfilled Ambition Concerning the Branches of a Complete Education]. 28 vols. in 33 pts. Cairo: Dār al-kutub al-miṣriyya *et al.* [Partial 2nd edition: 1347/1929].

Ptolemy, C. ed., tr. and ann. Toomer, G. 1998². *Ptolemy's Almagest*. Princeton: Princeton University Press.

Saʿadia Gaʾon: Saʿadia ben Joseph. ed. and tr. Rosenblatt, S. 1942. *The Book of Beliefs and Opinions: Amānāt wa-l-iʿtiqādāt*. New Haven: Yale University Press.

Ṣāʿid al-Andalusī. ed. Cheikho, L. 1912. *Ṭabaqāt al-umam*. [Book of the Nations]. Beirut: Imprimerie Catholique.

Ṣāʿid al-Andalusī. tr. Blachère, F. 1935. *Ṭabaqāt al-umam. Livre des catégories des nations*. Paris: Larose. [Reprint of Ṣāʿid 1912 and 1935. 1999. *Publications of the Institute for the History of Arabic-Islamic Science.*

Islamic Philosophy. Vol. 1: Frankfurt: Institute for the History of Arabic-Islamic Science at the Johann Wolfgang Goethe University.]

Ṣāʿid al-Andalusī. tr. Salem, S. I. and Kumar, A. 1991. *Science in the Medieval World: "Book of the Categories of Nations,"* (History of Science Series 5) Austin: University of Texas Press (to be used with caution).

Ṣāʿid al-Andalusī. ed. Ḥusayn Muʾnis. 1993. *Ṭabaqāt al-umam* [Classes of Nations]. (Dhakhāʾir al-ʿarab 74) Cairo: Dār al-Maʿārif.

Ṣāʿid of Toledo. ed. Bū ʿAlwān, Ḥ. 1985. *Ṭabaqāt al-umam* [Classes of Nations]. Beirut: Dār al-Ṭalīʿah.

Taşköprüzāde, Aḥmed Efendi. 1985. *Miftāḥ al-saʿāda wa-miṣbāḥ al-siyāda fī mawḍūʿāt al-ʿulūm* [*The Key to Happiness and the Lamp of Sovereignty in the Subject Matters of the Sciences*]. 3 vols. Beirut: Dār al-kutub al-ʿilmiyya.

Thābit ibn Qurra. ed. and tr. Morelon, R. 1987. *Oeuvres d'astronomie*. Paris: Les Belles Lettres.

Thābit ibn Qurra. ed. and tr. Lorch, R. 2008. *On the Sector-Figure and Related Texts*. Augsburg: Rauner.

Thābit ibn Qurra. ed. and tr. Sidoli, N. and Isahaya, Y. 2018. *Thābit ibn Qurra's Restoration of Euclid's Data. Text, Translation, Commentary*. Springer.

al-Yaʿqūbī, Ibn Abī Yaʿqūb ibn Wāḍiḥ. 1892. "Kitāb al-Buldān [The Book of the Countries]," in *BGA* 7.

[H1]*Research Literature*

Adamson, P. and Pormann, P. E. eds. 2017. *Philosophy and Medicine in the Formative Period of Islam*. London: Warburg Institute.

Brentjes, S. 2009. "Patronage of the mathematical sciences in Islamic societies," in Robson, E. and Stedall, J. eds. *The Oxford Handbook of the History of Mathematics*. Oxford: Oxford University Press, 301–28.

Brentjes, S. 2018. *Teaching and Learning the Sciences in Islamicate Societies (800-1700)*. Turnhout: Brepols.

Brentjes, S. with Morrison, R. G. 2010. "The Sciences in Islamic Societies," in Irwin, R., ed. *The New Cambridge History of Islam. Part IV: Learning, Arts and Culture*. Cambridge: Cambridge University Press, 564–639.

Brentjes, S., Edis, T. and Richter-Bernburg, L., eds. 1916. *1001 Distortions. How (Not) to Narrate History of Science, Medicine, and Technology in Non-Western Cultures*. Würzburg: Ergon.

Bürgel, J. C. 2016. *Ärztliches Leben im arabischen Mittelalter.* Bearbeitet von Fabian Käs. Leiden: Brill.

Casulleras, J. and Samsó, J. eds. 1996. *From Baghdad to Barcelona. Studies in the Islamic Exact Sciences in Honour of Prof. Juan Vernet*. Barcelona: Instituto Millas Vallicrosa de Historia de la Ciencia Arabe.

Charette, F. 2003. *Mathematical Instrumentation in Fourteenth-Century Egypt and Syria. The Illustrated Treatise of Najm al-Dīn al-Miṣrī*. Leiden: Brill.

Dold-Samplonius *et al.*: Dold-Samplonius, Y., Dauben, J., Folkerts, M., and van Dalen, B., eds. 2002. *From China to Paris: 2000 years transmission of mathematical ideas*. (Boethius, 46). Stuttgart: Steiner.

Endreß, G. 2001. "Philosophische Ein-Band Bibliotheken aus Isfahan," *Oriens* 36: 10–56.

Endress, G. 2006. "The Cycle of Knowledge: Intellectual Traditions and Encyclopaedies of the Rational Sciences in Arabic Islamic Hellenism," in Endress, G., ed. *Organizing Knowledge. Encyclopaedic Activities in the Pre-Eighteenth Century Islamic World*. Leiden: Brill, 103–33.

Fancy, N. 2013. *Science and Religion in Mamluk Egypt: Ibn al-Nafis, Pulmonary Transit and Bodily Resurrection*. London and New York: Routledge.

Folkerts, M., ed. 1996. *Mathematische Probleme im Mittelalter: Der Lateinische und Arabische Sprachereich*. Wiesbaden: Harrassowitz.

Forcada, M. 2004–2005. "Astronomy, Astrology and the Sciences of the Ancients in Early al-Andalus," *Zeitschrift für Geschichte der Arabisch-Islamischen Wissenschaften* 16: 1–74.

Griffel, F. 2009. *Al-Ghazālī's Philosophical Theology*. New York: Oxford University Press.

GTISAS: Harley, J. B. and Woodward, D., eds. 1992. *Cartography in the Traditional Islamic and South Asian Society*. Chicago and London: University of Chicago Press.

Gutas, D. 1998. *Greek Thought, Arabic Culture: the Graeco-Arabic Translation Movement in Baghdad and Early Abbāsid Society (2nd–4th/8th–10th centuries)*. London: Routledge.

Gutas, D. 2014[2]. *Avicenna and the Aristotelian Tradition: Introduction to Reading Avicenna's Philosophical Works*. Leiden and Boston: Brill.

Hagen, G. 2003. *Ein osmanischer Geograph bei der Arbeit. Entstehung und Gedankenwelt von Kātib Čelebis Ǧihānnümā*. Berlin: Schwarz.

al-Hassan, A. Y. and Hill, D. R. 1986. *Islamic Technology: An Illustrated History*. Cambridge: Cambridge University Press.

Hasse, D. N. 2016. *Success and Suppression: Arabic Sciences and Philosophy in the Renaissance*. Cambridge, MA: Harvard University Press.

Hirschler, K. 2016. *Medieval Damascus: plurality and diversity in an Arabic library: the Ashrafiya library catalogue*. Edinburgh: Edinburgh University Press.

Hockey, T. *et al.*, eds. 2007. *The Biographical Encyclopedia of Astronomers*. 2 vols. New York: Springer.

İhsanoğlu, E. *et al.* eds. 1997. *Osmanlı Astronomi Literatürü Tarihi. History of Astronomical Literature during the Ottoman Period*. Istanbul: IRCICA.

İhsanoğlu, E. *et al.* 2000. *Osmanlı coğrafya literatürü tarihi. History of Geographical Literature during the Ottoman Period*. İstanbul: IRCICA.

Kennedy *et al.* 1999: Kennedy, E. S., Kunitzsch, P. and Lorch, R. 1999. *The Melon-Shaped Astrolabe in Arabic Astronomy*. Stuttgart: Franz Steiner.

King, D. A. 1999. *World-Maps for Finding the Direction and Distance to Mecca. Innovation and Tradition in Islamic Science*. London: al-Furqān and Leiden: Brill.

King, D. 2000. "Too Many Cooks . . . A New Account of the Earliest Muslim Geodetic Measurements," *Suhayl* 1: 207–41.

King, D. A. 2004. *In Synchrony with the Heavens: Studies in Astronomical Timekeeping and Instrumentation in Medieval Islamic Civilization*. Vol. 1: *The Call of the Muezzin*. Leiden: Brill.

King, D. A. 2005. *In Synchrony with the Heavens: Studies in Astronomical Timekeeping and Instrumentation in Medieval Islamic Civilization*. Vol. 2: *Instruments of Mass Calculation*. Leiden: Brill.

King, D. A. and Lorch, R. P. 1992. "Qibla Charts, Qubla Maps, and Related Instruments," in GTISAS, 189–205.

King, D. A. *et al.* 2001: King, D. A. and Samsó, J. with a contribution by Goldstein, B. R. 2001. "Astronomical Handbooks and Tables from the Islamic World (750–1900): An Interim Report," *Suhayl* 2: 9–105.

Lindberg, D. C. and Shank, M. H. eds. 2013. *Medieval Science*. Cambridge: Cambridge University Press.

Morrison, R. G. 2016. *Astronomy in al-Andalus: Joseph Ibn Naḥmias' The Light of the World*. University of California Press,

Oestmann *et al.*: Oestmann, G., Rutkin, D. and Stuckrad, K. von. 2005. *Horoscopes and Public Spheres: Essays on the History of Astrology* (Religion and Society, 42.) Berlin: W. de Gruyter,

Pormann, P. E. and Savage-Smith, E, 2007. *Medieval Islamic Medicine*. Washington D.C.: Georgetown University Press.

Rashed, R. 2011. *D'al-Khwarizmi à Descartes: Études sur l'histoire des mathématiques classiques*. Paris. Hermann.

Rashed, R. 2015. "The Transmission of Greek Heritage into Arabic," in Rashed, R., ed. and Shank, M. tr. *Classical Mathematics from al-Khwārizmī to Descartes*. London and New York: Routledge, 19–56.

Rashed, R. and Morelon, R., eds. 1996. *Encyclopedia of the History of Arabic Science*. 3 vols. London and New York: Routledge. [French Version: 1997. *L'Histoire des sciences arabes*. 3 vols. Paris: Seuil.]

el-Rouayheb, K. 2015. *Islamic Intellectual History in the Seventeenth Century: Scholarly Currents in the Ottoman Empire and Maghreb*. Cambridge: Cambridge University Press.

el-Rouayheb, K. and Schmidtke, S., eds. 2016. *The Oxford Handbook of Islamic Philosophy*. Oxford: University Press.

Rudolph *et al.*: Rudolph, U., Hansberger, R. and Adamson, P., eds. 2016. *Philosophy in the Islamic World, Volume 1: 8th-10th Centuries*. (Handbook of Oriental Studies 115). Leiden and Boston: Brill.

Sabra, A. 1987. "The Appropriation and Subsequent Naturalization of Greek Science in Medieval Islam: A Preliminary Statement," *History of Science* 25: 223–43.

Saliba, G. 1994. *A History of Arabic Astronomy: Planetary Theories During the Golden Age of Islam*. New York: New York University Press.

Saliba, G. 2007. *Islamic Science and the Making of the European Renaissance*. Cambridge, MA and London: The MIT Press.

Samsó, J. 2008. *Astrometeorología y astrologia medievales*. Barcelona: Universitat de Barcelona.

Samsó, J. 2011[2]. *Las ciencias de los antiguos en al-Andalus*. with an *Addenda et Corrigenda* by Samsó, J. and Forcada, M. Roquetas (Almería): Fundación Ibn Tufayl. (1992[1]. Madrid: Mapfre).

Samsó, J. 2020. *On both sides of the Strait of Gibraltar. Studies on the history of medieval astronomy in the Iberian Peninsula and the* Maghrib. Leiden: Brill.

Savage-Smith, E. 2013. "The most authoritative copy of ʿAbd al-Raḥmān al-Ṣūfī's Guide to the Constellations," in Blair, Sh. and Bloom, J., eds., *God is Beautiful and Loves Beauty, The Object in Islamic Art and Culture*. New Haven: Yale University Press, 122–55.

Sayili, A. 1960 [1988[2]]. *The Observatory in Islam and its Place in the General History of the Observatory*. 2nd ed. Ankara: Türk Tarih Kurumu Basımevi. [Reprint: 1988]

Scheiner, J. and Janos, D. 2021. *The Place to Go: Contexts of Learning in Baghdad, 750–1000 C.E.* (SLAEI – Studies in Late Antiquity and Early Islam, Band 26). Berlin and London: Gerlach Press.

Selin, H. 1997 [2008[2]]. *Encyclopedia of Science, Technology and Medicine in Non-Western Cultures*. Heidelberg: Springer-Verlag. [Revised reprint: 2008]

Ullmann, M. 1970. *Die Medizin im Islam*. Leiden: Brill. [English translation: 1978. *Islamic Medicine*. Edinburgh: Edinburgh University Press.]

Van Bladel, K. T. 2009. *The Arabic Hermes: From Pagan Sage to Prophet of Science*. Oxford: Oxford University Press.

Van Bladel, K. T. 2014. "Eighth-Century Indian Astronomy in the Two Cities of Peace," in Sadeghi, B. *et al.*, eds. *Islamic Cultures, Islamic Contexts. Essays in Honor of Professor Patricia Crone*. Leiden: Brill, 257–94.

INDEX

Note: Page numbers in *italics* indicate figures and those in **bold** indicate tables.